VOLUME 1

CALCULUS
PREMIERE EDITION

Robert T. Smith

Millersville University of Pennsylvania

Roland B. Minton

Roanoke College

Boston   Burr Ridge, IL   Dubuque, IA   Madison, WI   New York   San Francisco   St. Louis
Bangkok   Bogotá   Caracas   Lisbon   London   Madrid
Mexico City   Milan   New Delhi   Seoul   Singapore   Sydney   Taipei   Toronto

McGraw-Hill Higher Education

A Division of The McGraw-Hill Companies

CALCULUS, VOLUME 1
PREMIERE EDITION

 This book is printed on recycled, acid-free paper containing 10% postconsumer waste.

1 2 3 4 5 6 7 8 9 0 QPD/QPD 9 0 9 8 7 6 5 4 3 2 1 0 9

ISBN 0–07–059246–2

Vice president and editorial director: *Kevin T. Kane*
Publisher: *JP Lenney*
Sponsoring editor: *Maggie Rogers*
Developmental editor: *Michelle Munn*
Marketing manager: *Mary K. Kittell*
Project manager: *Jill R. Peter*
Production supervisor: *Laura Fuller*
Design director: *Francis Owens*
Senior photo research coordinator: *Lori Hancock*
Supplement coordinator: *Stacy A. Patch*
Typeface: *10/12 Times Roman*
Printer: *Quebecor Printing Book Group/Dubuque, IA*

Cover designer: *Cristina Deh-Lee*
Cover photograph: *©PhotoDisc*

www.mhhe.com

About the Authors

Robert T. Smith is Professor of Mathematics and Chair of the Department of Mathematics at Millersville University of Pennsylvania, where he has taught since 1987. Prior to that, he was on the faculty at Virginia Tech. He earned his Ph.D. in mathematics from the University of Delaware in 1982.

Dr. Smith's mathematical interests are in the application of mathematics to problems in engineering and the physical sciences. He has published a number of research articles on the applications of partial differential equations as well as on computational problems in x-ray tomography. He is a member of the American Mathematical Society, the Mathematical Association of America and the Society for Industrial and Applied Mathematics.

Professor Smith lives in Lancaster, Pennsylvania, with his wife Pam, his daughter Katie and his son Michael. When time permits, he enjoys playing volleyball, tennis and softball. His present extracurricular goal is to learn the game of golf well enough to not come in last in his annual mathematicians / statisticians tournament.

Roland Minton is Professor of Mathematics at Roanoke College, where he has taught since 1986. Prior to that, he was on the faculty at Virginia Tech. He earned his Ph.D. from Clemson University in 1982. He is the recipient of the 1998 Roanoke College Exemplary Teaching Award.

Dr. Minton has supervised numerous student research projects in such topics as sports science, complexity theory and fractals. He has published several articles on the use of technology and sports examples in mathematics, in addition to a technical monograph on control theory. He has received grants for teacher training from the State Council for Higher Education in Virginia. He is a member of the Mathematical Association of America, American Mathematical Society and other mathematical societies.

Professor Minton lives in Salem, Virginia, with his wife Jan, daughter Kelly and son Greg. He enjoys playing golf and tennis (when health and time permit) and watching any sport on TV (even if time doesn't permit). Jan also teaches mathematics at Roanoke College and is very active in mathematics education. A busy family schedule includes band and choir performances and soccer matches. Favorite entertainers include the Marx Brothers, guitarist Danny Gatton and mystery novelist Kinky Friedman.

Professors Smith and Minton have together published three earlier books with McGraw-Hill.

To Pam, Katie and Michael
To Jan, Kelly and Greg
And our parents
Thanks for your love and inspiration.

Calculus

Volume 1

Table of Contents

Preface

The wide-ranging debate brought about by the calculus reform movement has had a significant impact on the calculus textbook market. In response to many of the questions and concerns surrounding this debate, we have written a modern calculus textbook, intended for students majoring in mathematics, physics, chemistry, engineering and related fields.

While following a relatively traditional table of contents, we have attempted to address many of the concerns of calculus reformers. We have written this book for the average student, that is, one who does not already know the subject, whose background is somewhat weak in spots and who requires significant motivation to study the subject. Our intention is that students should be able to read our book, rather than merely use it as an encyclopedia filled with the facts of calculus. We have written the text in a casual, almost conversational style, one that our students have found easy to read. Given the widespread availability of graphing calculators and powerful computer algebra systems, we assume that students using this text will have one of these systems at their disposal.

In an effort to ensure that this textbook successfully addresses our concerns about the effective teaching of calculus, as well as others' concerns, we continually asked instructors around North America for their opinions on the calculus curriculum, the strengths and weaknesses of current textbooks and our manuscript and ideas. In preparing this finished product, we had the benefit of countless insightful comments from a panel of reviewers that were carefully selected to help us with this project. Their detailed reviews of our materials and their opinions about the teaching of calculus were invaluable to us during our development of this textbook. We are deeply indebted to them for their time and effort.

PHILOSOPHY

We agree with many of the ideas that have come out of the calculus reform movement. In particular, we believe in the **Rule of Three**: that concepts should be presented **graphically**, **numerically** and **algebraically,** whenever these are appropriate. In fact, we would add **verbally** and **physically** to this list, since the communication of mathematical ideas and the modeling of physical problems are important skills that students need to develop. We also believe that, while the calculus curriculum has been in need of reform, we should not throw out those things that already work well. Our book thus represents a fresh approach to the traditional topics of calculus. We follow an essentially traditional order of presentation, while integrating technology and more thought-provoking exercises throughout.

One of the thrusts of the calculus reform movement has been to place greater emphasis on problem solving and to present students with more realistic applications, as well as open-ended problems. We have incorporated meaningful writing exercises and extended, open-ended problems into **every** problem set. You will also find a much wider range of applications than in most traditional texts. We make frequent use of applications from within students' experience to both motivate the development of new topics and to illustrate concepts we have already presented. In particular, we have included numerous examples from the physics of sports to give students a familiar context in which to think of various concepts.

We believe that a conceptual development of the calculus must drive the book. Although we have integrated technology throughout, we have not allowed the technology to drive the book. We have also not given in to the temptation to show off what technology can do, except where this has a direct bearing on learning the calculus. Our goal is to use the available technology to help students reach a conceptual understanding of the calculus as it is used today.

Perhaps the most difficult task when preparing a new calculus text is the actual *writing* of it. We have endeavored to write this text in a manner that combines an appropriate level of informality with an honest discussion regarding the difficulties that students commonly face in their study of calculus. In addition to the concepts and applications of calculus, we have also included many frank discussions about what is practical and impractical, and what is difficult and not so difficult to students of calculus. We have attempted to provide clarity of presentation in the creation of every example, application and exercise.

The book that we have written represents substantive change while still adhering to a largely traditional table of contents. By virtue of integrating technology throughout, utilizing a lively presentation style and incorporating a wider variety of problems, we believe that we meet many of the objectives of the calculus reform movement. At the same time, our relatively traditional outline retains the central strengths of mainstream calculus, allowing instructors to teach from a familiar body of material while integrating technology and modern applications.

CONTENT

We have not drastically revised the traditional table of contents. However, we have carefully reconsidered the best way in which to present each topic. Our primary objective is to keep students focused on the central concepts of calculus. To this end, we have augmented a simple algebraic presentation of certain ideas with numerical methods. For instance, when we introduce the notion of area, we emphasize the computation of area as a limit of a Riemann sum, but use regular partitions exclusively. We are careful to point out that (without the Fundamental Theorem of Calculus) the limit of Riemann sums can be computed directly only for a very small number of functions. In addition, we allow students to explore the same ideas numerically. In this case, we are not restricted to polynomials of low degree and students can observe numerical values of Riemann sums approaching a limit. With this approach, students get to see the same problem from several different viewpoints, thus improving the likelihood that they will grasp the underlying concept. Additionally, students are given a useful tool (numerical integration) that they can bring to bear on a wide variety of problems.

In our view, techniques of integration are of great importance. Our emphasis is on helping students develop the ability to carefully distinguish among similar looking integrals and identify the appropriate technique of integration to apply to each integral. The attention to detail and mathematical sophistication required by this process are invaluable skills. We do not attempt to be encyclopedic about techniques of integration, especially given the widespread use of computer algebra systems. Finally, in section 7.5, we include a discussion of integration tables and the use of computer algebra systems for performing symbolic integration.

Key Features

Use of Graphs and Tables: We have integrated more than 1000 computer-generated graphs throughout the text. Because graphs are now so easy to obtain, we emphasize that students should use them routinely to aid in solving most problems, even if only as a check on the reasonableness of an answer. Graphs and tables are used throughout the text to provide the student with a better understanding of the underlying theory presented. We have created each graph and table very carefully to ensure that the ideas presented are clear and accurate. During the development of the manuscript, we continually asked our reviewers for advice on the placement and treatment of illustrations. Thanks to the reviewer panel's insightful suggestions, we made numerous improvements to the manuscript.

Use of Technology: Considering the rate at which computer and calculator technology is advancing, we felt that it was unwise to tie the book to any one technology. Rather, we have employed a more generic use of technology, where we use those features that are shared by virtually all graphing calculators and computer algebra systems (e.g., numerical integration and common graphing features). Further, we use only those features of the technology that clearly aid in the understanding of the concepts of calculus. We do not view the technology as a gimmick to be artificially appended to the same old calculus course. Neither do we believe that a calculus course should be a course in how to use the technology. We believe that technology can and should be introduced as a natural part of a coherent development of calculus. In our presentation, technology has been seamlessly integrated throughout, but only where appropriate. Users are expected to have access to a graphing calculator or a computer algebra system and to use it routinely.

A common concern regarding substantial use of technology is that our students will become mindless button-pushers. We guard against this by making the technology secondary to understanding and by pointing out errors that can be made by an overreliance on technology. We use technology so that students can focus on the difficult and sometimes subtle connections among the different concepts of the calculus. A student who has mastered these connections will be a much more effective user of the calculus than will a student who is proficient at algebraic methods alone. By engaging students on several different levels, using different approaches, we hope to improve their understanding and empower them to try new problems on their own.

We have also chosen not to separate out "technology" exercises in the exercise sets found at the end of each section and chapter. This decision was made quite carefully, and is intended to be used as a teaching tool. Throughout the text, we provide advice and guidance on the proper use of technology and provide tools to help students determine when the use of technology is most appropriate. We feel that placing an icon next to technology-based exercises, as some texts do, detracts from teaching students to make this decision on their own. We do, however, make recommendations for technology use in the homework sets in the *Instructor's Resource Manual*, to help instructors plan their homework assignments accordingly.

Early Coverage of Trigonometric, Exponential and Logarithmic Functions: Our table of contents features an early introduction of logarithms, exponentials and the trigonometric functions. All of our students have seen these functions before they ever set foot in a calculus classroom. Why not take advantage of their familiarity? We introduce the calculus of all of these functions in Chapter 2, along with all of the other rules of differentiation. While we are not able to give complete derivations of the derivatives of the natural logarithm and exponentials at this point, we give very strong numerical evidence and nearly complete algebraic proofs. We follow this up in Chapter 6 with the definition of the natural logarithm as an integral and complete derivations of the derivative formulas. In the intervening chapters, we use these functions to provide more interesting examples in our discussion of the applications of differentiation (including graphing), integration and applications of integration. By the middle of Chapter 6, we have tied up all of the loose ends and set the calculus of the logarithmic and exponential functions on a firm theoretical footing. Thus, in the end, we have sacrificed no mathematical rigor, but students will have gained much insight from the use of a richer set of functions throughout the introductory chapters.

Emphasis on Problem Solving: This text is written with a strong problem-solving emphasis, including the introduction of many topics from graphical, numerical and algebraic points of view. In many instances, the emphasis on graphical and numerical methods for solving problems frees us to consider more realistic and complex problems than are usually presented in calculus. As a result, students should have a more complete idea of the usefulness of calculus and should be able to solve a much wider variety of problems. Further, students will see that they can approach realistic and challenging problems and obtain meaningful solutions.

Applications: We have included a wider variety of applications than is found in most traditional calculus books. For instance, in Chapter 5 (Applications of the Definite Integral), we combine in one section the traditional physics applications of work and fluid pressure and force, while adding a discussion of impulse, moments and center of mass. Our focus here is on using Riemann sums and integrals to compute the quantities of interest. We also include a section devoted to projectile motion and an optional section on probability theory.

This textbook contains numerous applications designed to engage the attention and imagination of students from a variety of backgrounds and with a diverse variety of interests. Our applications are realistic and many are unique, covering a broad spectrum of topic areas. We have included many sports-related applications that are tailored to the study of calculus. Our reviewer panel provided us with many suggestions for the inclusion of appropriate applications and commented frequently on the rich variety this text offers.

Examples: Each chapter contains a large number of worked examples ranging from the simple and concrete to more complex, abstract examples. There are examples of varying difficulty, and some examples that serve to motivate a number of different topics in a given chapter. Many topics in Chapter 9, for instance, are introduced using the amusement park ride The Scrambler as the foundation of each example.

Flexible Topic Coverage: We have included a number of optional sections that are not generally found in other calculus texts, as well as expanded coverage of selected topics. These expanded and optional sections provide instructors with the flexibility to tailor their courses to the interests and abilities of each class. In section 1.6, we explore loss-of-significance errors. In this section, we discuss how computers and calculators perform arithmetic operations and how these can cause errors, in the context of numerical approximation of limits. In section 3.7, we present a number of diverse applications of differentiation, including chemical reaction rates and heart rates. Direction fields and Euler's method for first-order ordinary differential equations are discussed in section 6.6. In Chapter 8, we follow our discussion of power series and Taylor's Theorem with a section on Fourier series.

Flexible Exercise Sets: Found at the end of each section, as well as review exercises at the end of each chapter, this text contains more than 4500 exercises. Each exercise set has been carefully designed to provide a wide variety of routine, moderate, and challenging exercises. Many of the exercises differ from the traditional set of exercises that can be found in most calculus texts. We have tried to create original and imaginative exercises that provide an appropriate review of the topics covered in each section and chapter.

In addition to the more routine exercises, each exercise set contains exploratory exercises that are optional, project-like assignments designed for students to obtain a sense of the ongoing, exciting nature of mathematics-related research. These exercises, which are clearly labeled exploratory, offer excellent opportunities for group work or discussion, depending on the structure of individual calculus courses.

Each exercise set begins with a variety of writing exercises, clearly labeled as such. As with the exploratory exercises, the writing exercises can be used as springboards for discussion and are intended to give students an opportunity to carefully consider important mathematical concepts and ideas and express these in their own words.

What's Different and Why?

Format: We introduce each chapter with a brief application related to the mathematical concepts developed in the chapter. The introductions provide motivation for students who are properly concerned with the usefulness of the calculus, and can help put each chapter into a larger problem-solving context. Topics of introductions range from aerodynamics to sports to the size of a cat's pupils.

The book has been written in a conversational style that reviewers have compared to listening to a good lecture. We explain how to use a technique and when to use it. Our sense of what works well with students has been honed by working with calculus students over the last 20 years at a variety of colleges and universities, both public and private, ranging from a small liberal arts college to large engineering schools. We hope that students will open themselves up to our approach and will thereby learn more than just the basic details of calculus manipulations.

The exercise sets have been carefully constructed to reinforce the details and concepts of calculus while encouraging individual explorations of mathematics. Each exercise set begins with several meaningful writing exercises. These are intended to help students organize their thoughts about the important ideas of the section and hopefully, restate them in their own words. These exercises can be used as discussion questions, prompts for journal entries and short essay questions on tests. At the end of each exercise set are problems we have labeled "exploratory" exercises. Although these are not the only exercises encouraging students to explore mathematics, each of these problems can be used for group work, individual project assignments or simply reading assignments to give students a better sense of the scope of calculus-based mathematics.

Content: The vast majority of the topics found in our book are part of the standard calculus curriculum that has defined the mainstream for the last 30 years or so. We believe that this curriculum still has validity in terms of both mathematical precision and student learning. Nevertheless, we have made a small number of significant changes in the table of contents. After reviewing the basic properties of exponential and trigonometric functions in Chapter 0, we make substantial use of these functions as we develop limits, derivatives and integrals. The inclusion of these functions from early in the first semester of calculus greatly increases our ability to discuss interesting applications. Our experience has been that this increase does not come at the expense of student understanding.

The treatment of differential equations varies widely among current calculus texts. We have found that many students need a sound background in integration to fully appreciate the concept of the solution of a differential equation. On the other hand, simple techniques for solving differential equations are fully accessible to second-semester calculus students and are required of many second-semester engineering students. For these reasons, we discuss separable differential equations in section 6.5, two chapters after the introduction of the integral and in conjunction with exponential growth and decay. In order to maximize the flexibility in using this text, we chose to introduce this material prior to Chapter 7 on techniques of integration.

Throughout the text, we have assumed that students have access to calculator- or computer-generated graphs. This allows us to routinely use graphs as the first step in solving a problem or as a check on the reasonableness of an answer. Being able to visualize a problem is an invaluable aid to students, and we try to take full advantage of this. One benefit of readily available graphics is the ability to solve more realistic application problems. Functions associated with realistic problems are often not mathematically simple, but we can approximate zeros or extrema graphically and numerically. Further, concepts such as the convergence of Taylor series are more meaningful when graphs are used to illustrate this convergence. This same graphical approach benefits our presentation of Fourier series, which is an important tool for understanding much of our digitally enhanced world.

Throughout the text, we have made use of the ability of calculators and computers to solve equations and quickly complete lengthy calculations. Limits, derivatives, definite integrals and infinite series are all investigated using tables of computations, where students can clearly see what convergence or divergence represents. To avoid having students blindly using technology, we include a section on loss-of-significance errors. Here, we give a highly accessible introduction to common computing errors and how to avoid them. Numerical analysis also plays a large role in our development of Euler's method for approximating solutions of differential equations. Euler's method is closely related to tangent lines and linear approximations, two places where we rely on computer graphics. Such connections are vital to what we try to accomplish using technology: not new topics added on for completeness, but new ways to visualize and understand the important concepts of calculus.

Chapter-by-Chapter Comments

Chapter 0 is largely review, where students can reacquaint themselves with some important ideas from precalculus mathematics. The chapter also focuses students' attention on those aspects of algebra and trigonometry that are going to be most useful to them as they proceed through their study of calculus. We close the chapter with a "Preview of Calculus" section, where the concept of limit is introduced in a familiar context.

Chapter 1 introduces the central concepts of limit and continuity. Some calculus books open the first chapter with applications leading directly to the derivative. While we sympathize with the desire to quickly show off some applications, we begin with limits for a number of reasons. The most important reason is that the limit is what calculus is all about and we want students to recognize this pivotal role. We hope that the limit-based applications we present here will provide the needed practical motivation. Our discussion of limits here includes limits at infinity and infinite limits. The chapter concludes with optional sections on the formal definition of limit and loss-of-significance errors in numerical computations.

In **Chapter 2**, we introduce the notion of derivative and develop the basic rules of differentiation, including the product, quotient and chain rules. We also develop the derivatives of algebraic, exponential, logarithmic and trigonometric functions. The early introduction of the calculus of exponential, logarithmic and trigonometric functions allows us to present a rich set of examples of chain rules, product rules, quotient rules and applications.

Chapter 3 presents applications of the derivative. The first section discusses Newton's method and linear approximations. We believe that it is important for students to have a basic understanding of how computers solve equations and Newton's method is an obvious starting place. Further, the idea of linear approximation is central to many different numerical methods. We utilize Newton's method throughout the chapter as we solve for zeros and extrema of functions.

Chapter 4 introduces integration. For simplicity, we limit our discussions to regular partitions of an interval. We utilize technology extensively here to compute Riemann sums and observe their convergence. We have found that such computations greatly enhance students' understanding of integration as a limit.

Chapter 5 presents applications of integration. We focus on the development of integral formulas, routinely constructing approximating sums and then passing to the limit to obtain a definite integral. Along with the standard computations of area, volume and work, we have included sections on projectile motion and probability theory. All calculus texts have a variety of projectile problems scattered throughout, but we felt that the topic was interesting and useful enough to warrant a more organized treatment. We have labeled Probability as an optional section, although the increasing use of statistical methods in modern society calls for a more prominent role for probability in students' education.

Chapter 6 provides a thorough and traditional development of the exponential and logarithmic functions. We have made use of these functions throughout the text, but with frequent warnings that our derivations of formulas have been incomplete. The details are provided here. At the same time, a discussion of exponential growth and decay leads naturally into an introduction to first-order differential equations. We also introduce the inverse trigonometric functions and the hyperbolic functions in this chapter.

Chapter 7 includes a variety of techniques of integration. We believe that students gain understanding and maturity while discriminating among different techniques, but we also recognize that in practice most users of the calculus routinely find antiderivatives using computer algebra systems. We try to strike a good balance of learning the most important techniques while leaving room in the syllabus for other topics. We include a section here on the use of integral tables and computer algebra systems for finding antiderivatives.

Chapter 8 is the longest chapter in the book, partly in response to the special difficulties many students experience when studying infinite series. We have included numerous tables of calculations and graphs to give students every chance to understand series. Since Fourier series are widely used by practitioners in engineering and the sciences, we have included a section to introduce students to this important topic and provide several interesting applications.

Chapter 9 introduces parametric equations and polar coordinates. We have included a large number of parametric graphs and applications, made more accessible by computer graphics and equation-solving capabilities.

SUPPLEMENTAL RESOURCES

For the Instructor

- **Web Site** www.mhhe.com/smithminton
  The Smith/Minton Web site is intended for both student and instructor use and will contain Web links to relevant Web resources that are directly keyed to the book, as well as instructional resources for the classroom and study aids for the student.
- **Instructor's Resource Manual**
  This manual contains the fully worked solutions to most exercises in the text, sample tests and an index of exercises that encourage the use of technology.
- **Computerized Test Bank**
  The algorithmically generated ESATest software allows instructors to efficiently select, add and organize questions to the test bank using criteria such as type of question or level of difficulty. Instructors can print tests and answer keys as well as edit the original questions or add their own. ESATest is available in both Windows and Macintosh formats.
- **Printed Test Bank**
  This print test bank contains a variety of questions and answers, including true/false, multiple-choice, short answer and free response.

For the Student

- **Student Resource Manual**
  Solutions for select odd-numbered problems found in the text.
- **Electronic Study Guide**
  This interactive, self-paced, text-specific CD-ROM allows students to practice their problem-solving skills via algorithmically generated problems and diagnostic tests.
- **Insights into Calculus**
  These exploratory technology workbooks present activities and instructions keyed to the Smith/Minton text for those who use a graphing calculator or a computer algebra system to help them learn calculus.

  Insights into Calculus is available in the following formats:

  - *Insights into Calculus Using Maple*
  - *Insights into Calculus Using DERIVE*
  - *Insights into Calculus Using Mathematica*
  - *Insights into Calculus Using the TI-83/83+*

- **Calculus Web Site** <u>www.mhhe.com/smithminton</u>
  The Smith/Minton Web site is intended for both student and instructor use and will contain Web links to relevant Web resources that are directly keyed to the book, as well as instructional resources for the classroom and study aids for the student.

ACKNOWLEDGMENTS

First and foremost, we want to express our appreciation to our sponsoring editor, Maggie Rogers, and our original editor, Jack Shira, for the encouragement and guidance to keep us going through this very long and challenging project. Jack had the confidence in us to start the project and we could never have completed it without Maggie's tireless efforts, professionalism and friendship. We could truly not think of better people with whom to work. We also wish to thank our publisher, JP Lenney, as well as our original publisher, Denise Schanck, for their strong support over the last five years.

A project of this magnitude requires the collaboration of an incredible number of talented people. Our developmental editor team, Glenn and Meg Turner of Burrston House, ran the project through their demanding schedule of reviews and meetings. Our understanding of the project and our ability to create a quality book improved tremendously through their efforts. Our production team at McGraw-Hill, particularly our project manager, Jill Peter, kept the project on schedule and helped us to produce a well-designed text. The problem solvers at Laurel Technical Services, led by Carrie Mallery, made numerous suggestions to improve the exercise sets. The TEX experts at Interactive Composition Corporation enhanced the appearance of the final manuscript.

We especially want to thank our reviewers for their many insightful comments:

David Anderson, University of Tennessee
Wilma Anderson, University of Nebraska–Omaha
Robert Beezer, University of Puget Sound
Neil Berger, University of Illinois
Mike Bonnano, Suffolk Community College
George Bradley, Duquesne University
Moody Chu, North Carolina State University
Raymond Clapsadle, University of Memphis
Joe Diestel, Kent State University
Dan Drucker, University of Puget Sound
Eugene Enneking, Portland State University
Ronald Grimmer, Southern Illinois University
Mel Hausner, New York University
Johnny Henderson, Auburn University
Robert Horvath, El Camino College
Gail Kaufmann, Tufts University
Hadi Kharaghani, University of Lethbridge (Alberta)
Masato Kimura, College of William and Mary
Robert Knott, University of Evansville
Jon Lee, University of Kentucky
John Maginnis, Kent State University
Chris McCord, University of Cincinnati
Remigijus Mikulevicius, University of Southern California
Mike Montano, Riverside Community College
Christina Pereyra, University of New Mexico
Linda Powers, Virginia Tech
Joe Rody, Arizona State University
Rod Smart, University of Wisconsin–Madison
Jerry Stonewater, Miami University of Ohio
Juan Tolosa, Richard Stockton College

Paul Weichsel, University of Illinois
Marvin Zeman, Southern Illinois University

In addition, a number of colleagues graciously gave of their time and energy to help create or improve portions of the manuscript. We would especially like to thank Bill Ergle, Ben Huddle, Jack Steehler, Deana Carideo, Jan Minton, Richard Grant and Jeff Spielman of Roanoke College for sharing their expertise in calculus and related applications; Tom Burns of General Electric for help with an industrial application; Dorothy Blum of Millersville University, who helped to field test an early version of the manuscript; Chuck Denlinger and Ron Umble of Millersville University for many conversations about the calculus and Pam Vercellone-Smith, who lent us her expertise with many of the biological applications. In addition, we would like to thank all of our students through the years who have (sometimes unknowingly) field tested innumerable ideas, some of which worked and the rest of which will not be found in this book. Finally, we gratefully acknowledge the support of the Millersville University Faculty Grants Committee and the Millersville University Faculty Professional Development Committee for the award of three released-time grants and the Millersville University administration for a sabbatical leave grant. These grants provided critical blocks of time for developing this text and the authors are most grateful.

The book was typeset in TEX using Scientific Workplace. Most of the graphs were generated using Maple V and revised using Corel Draw. Other graphs were generated using Mathematica.

Ultimately, this book is for our families. We want to thank them for their love and inspiration throughout the years that we have worked on this project. Their understanding, in both the technical and the personal sense, was essential to us. Writing a book of this magnitude could simply not have happened without our families' strong support. Our families provide us with the reason why we do all of the things we do. So, it is fitting that we especially thank our wives, Pam Vercellone-Smith and Jan Minton, our children, Katie and Michael Smith and Kelly and Greg Minton, and our parents, Anne Smith and Paul and Mary Frances Minton.

Robert T. Smith   Roland B. Minton
Lancaster, Pennsylvania   Salem, Virginia

Chapter 0
Preliminaries

In the first six sections of this chapter you will find a collection of familiar topics. You need not spend a great deal of time here, but review the material as necessary, so that you are comfortable with all of the topics discussed. We have included material that we consider **essential** for the study of calculus you are about to begin. We must emphasize that understanding is always built upon a solid foundation. While we do not intend this chapter as a comprehensive review of precalculus mathematics, we have tried to hit the highlights and provide you with some standard notation and language that we will use throughout the text.

The study of population growth is one of the oldest and most important uses of mathematics. The 16th-century mathematician Leonardo of Pisa (known as Fibonacci) analyzed a simple model of population growth that introduced the famous **Fibonacci sequence** of numbers 1, 1, 2, 3, 5, 8,..., where each number in the sequence is the sum of the preceding two numbers. The Fibonacci sequence can be seen in an astonishing variety of places, including in nature (see H.E. Huntley's *The Divine Proportion*) such as the spiral growth of the nautilus shown in the margin.

A nautilus shell.

Modern mathematical research into population growth has been vital in wildlife management, government planning, disease control, and numerous other areas. Many of the calculus techniques that you will learn in this book provide insights into population growth. For now, we can use algebra to identify significant patterns. Data for the population of the United States as measured by the Census is listed to the left.

For many people, a list of numbers is not as informative as a visual representation of the data. Graphs such as the one shown in Figure 0.1 are reviewed in sections 0.1 and 0.2.

Year	Population
1800	5,308,483
1810	7,239,881
1820	9,638,453
1830	12,866,020
1840	17,069,453
1850	23,191,876
1860	31,443,321
1870	38,558,371
1880	50,189,209
1890	62,979,766
1900	76,212,168
1910	92,228,496
1920	106,021,537
1930	123,202,624
1940	132,164,569
1950	151,325,798
1960	179,323,175
1970	203,302,031
1980	226,542,203
1990	248,709,873

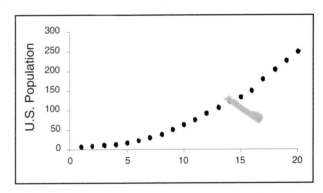

FIGURE 0.1: U.S. Population 1800 to 1990.

One obvious statement you can make based on Figure 0.1 is that the population has been increasing. One way to describe *how* it's increasing is to match the graph with the graph of a known function. For example, the first six or seven points appear to curve upward, like a parabola or exponential function. In fact, we show in section 0.6 that the data from 1800 to 1860 is closely approximated by an exponential function. In this time frame, the *percentage growth rate* of the population is almost constant; this is characteristic of exponential functions. On the other hand, the last four or five data points appear to form a straight line. Linear functions are reviewed in section 0.2 and play an important role throughout calculus.

The interplay between graphs and functions is one of the principal themes of this opening chapter. As you sharpen your algebra and trigonometry skills, pay special attention to the typical shapes of graphs of the various types of functions. By connecting the powerful equation-solving techniques of algebra with the visual images provided by graphs, you will gain significant insight into the calculus and improve your ability to make use of your mathematical skills in real-world situations.

0.1   The Real Numbers and the Cartesian Plane

Although mathematics is far more than just a study of numbers, our journey into calculus begins with the real number system. You might think that this is a fairly mundane starting place. After all, you likely mastered the properties of the real numbers a long time ago. Nevertheless, we want to give you the opportunity to brush up on those properties that are of particular interest for calculus. A solid foundation here will help you to grasp the many subtleties you will encounter in your study of the calculus.

The most familiar set of numbers is the set of **integers**. The integers consist of the whole numbers and their additive inverses: $0, \pm 1, \pm 2, \pm 3, \ldots$. Notice that if you add, subtract, or multiply two integers, the result is another integer. However, one integer divided by another is not always an integer (for example, $1 \div 2 = \frac{1}{2}$ is not an integer).

A **rational number** is any number of the form $\frac{p}{q}$, where p and q are integers and $q \neq 0$. For example, $\frac{2}{3}, -\frac{7}{3}$ and $\frac{27}{125} \ldots$ are all rational numbers. Notice that every integer n is also a rational number, since we can write it as the quotient of two integers: $n = \frac{n}{1}$.

The **irrational numbers** are all those real numbers that cannot be written in the form $\frac{p}{q}$, where p and q are integers. The irrationals include the very important numbers $\sqrt{2}, \pi$, and e. You may recall that rational numbers have a decimal expansion that either terminates or repeats. For instance, $\frac{1}{2} = 0.5$, $\frac{1}{3} = 0.333\overline{3}$, $\frac{1}{8} = 0.125$ and $\frac{1}{6} = 0.166666\overline{6}$ are all rational numbers. By contrast, irrational numbers have decimal expansions that do not repeat or terminate. For instance, we have the decimal expansions

$$\sqrt{2} = 1.41421\,35623\ldots,$$

$$\pi = 3.14159\,26535\ldots,$$

and

$$e = 2.71828\,18284\ldots$$

for the irrational numbers $\sqrt{2}, \pi$, and e.

FIGURE 0.2: The real line.

We usually picture the real numbers arranged along the number line (the **real line**) displayed in Figure 0.2, with center at 0 and extending off forever in both the positive and negative directions. We denote the set of real numbers by the symbol $\mathbb{R}$.

Recall that for real numbers a and b, where $a < b$, we define the **closed interval** $[a, b]$ to be the set of numbers between a and b, including a and b (the **endpoints**), that is,

$$[a, b] = \{x \in \mathbb{R} \,|\, a \leq x \leq b\},$$

as illustrated in Figure 0.3, where we have used the solid circles to indicate that a and b are included in $[a, b]$.

FIGURE 0.3: A closed interval.

Similarly, we define the **open interval** (a, b) to be the set of numbers between a and b, but **not** including the endpoints a and b, that is,

$$(a, b) = \{x \in \mathbb{R} \,|\, a < x < b\},$$

as illustrated in Figure 0.4, where we have used the open circles to indicate that a and b are not included in (a, b).

FIGURE 0.4: An open interval.

You should already be very familiar with the following properties of real numbers.

Theorem 1.1   If a and b are real numbers and $a < b$, then
   (i)   For any real number c, $a + c < b + c$.
   (ii)   For real numbers c and d, if $c < d$, then $a + c < b + d$.
   (iii)   For any real number $c > 0$, $a \cdot c < b \cdot c$.
   (iv)   For any real number $c < 0$, $a \cdot c > b \cdot c$.

Remark 1.1 We need the properties contained in Theorem 1.1 to solve inequalities. Notice that (i) says that you can add the same quantity to both sides of an inequality. Part (iii) says that you can multiply both sides of an inequality by a positive number. Finally, (iv) says that if you multiply both sides of an inequality by a negative number, the inequality is reversed.

We illustrate the use of Theorem 1.1 by solving a simple inequality.

Example 1.1   Solving a Linear Inequality
Solve the linear inequality $2x + 5 < 13$.

Solution   To solve for x, we need only use the properties in Theorem 1.1 to isolate the x. First, subtract 5 from both sides to obtain

$$(2x + 5) - 5 < 13 - 5$$

or

$$2x < 8.$$

Finally, divide both sides by 2 (since $2 > 0$, the inequality is not reversed) to obtain

$$x < 4.$$

We often write the solution of an inequality in interval notation. In this case, we get the interval $(-\infty, 4)$.

■

You can deal with more complicated inequalities in the same way.

Example 1.2   Solving a Two-Sided Inequality
Solve the two-sided inequality $6 < 1 - 3x \leq 10$.

Solution   Again, we need only use the properties in Theorem 1.1 to isolate the x. First, subtract 1 from all sides, to get

$$6 - 1 < (1 - 3x) - 1 \leq 10 - 1$$

or

$$5 < -3x \leq 9.$$

Now, divide by -3, but be careful. Since $-3 < 0$, the inequalities are all reversed. We have

$$\frac{5}{-3} > x \geq \frac{9}{-3}$$

or

$$-\frac{5}{3} > x \geq -3.$$

We usually write this as

$$-3 \leq x < -\frac{5}{3}$$

or in interval notation, as $\left[-3, -\frac{5}{3}\right)$.

■

You will often need to solve inequalities involving fractions. We present a typical example in the following.

Example 1.3   Solving an Inequality Involving a Fraction

Solve the quadratic inequality

$$\frac{x-1}{x+2} \geq 0.$$

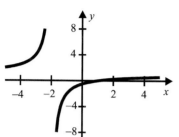

FIGURE 0.5: $y = \dfrac{x-1}{x+2}$.

Solution   In Figure 0.5, we show a graph of the function, which appears to indicate that the solution includes those $x < -2$ and $x \geq 1$. Your first thought for solving this inequality might be to multiply both sides of the inequality by $(x+2)$ to get rid of the fraction. Instead, look carefully at what the inequality means. You must find values of x which make the fraction positive or zero (i.e., nonnegative). Graphs are useful, but often fail to tell the entire story. There are only three ways to satisfy the inequality: either both numerator and denominator are positive, both are negative, or the numerator is zero. To simplify the process of finding where this is true, we draw number lines for each of the individual terms, indicating where each is positive, negative, or zero and use these to draw a third number line for the quotient, from which we can read the answer. We show these in the margin. Notice that in the third number line, we have placed an "×" above the -2 to indicate that the fraction is undefined at $x = -2$. From this last number line, you can see that the quotient is nonnegative whenever $x < -2$ **or** $x \geq 1$. We now write the solution in interval notation as $(-\infty, -2) \cup [1, \infty)$.

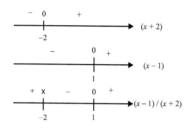

Another common type of inequality that you will encounter is one involving a polynomial of degree 2 or higher. In this case, the properties of the real numbers found in Theorem 1.1 are of little help.

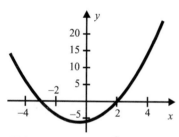

Example 1.4   Solving a Quadratic Inequality

Solve the quadratic inequality

$$x^2 + x - 6 > 0. \tag{1.1}$$

FIGURE 0.6: $y = x^2 + x - 6$.

Solution   In Figure 0.6, we show a graph of the expression on the left side of the inequality. Your first impulse for solving this may be to try to somehow isolate the x on one side of the inequality, but this will prove fruitless. (Go ahead, try it!) On the other hand, it might be worthwhile to take the time to read what this inequality says. Simply put, you want to find values of x for which the function $x^2 + x - 6$ is positive. Notice that this polynomial factors, so that (1.1) is equivalent to

$$(x+3)(x-2) > 0. \tag{1.2}$$

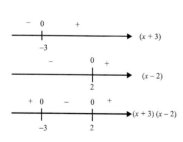

Read (1.2) carefully. It says that you want values of x for which the **product** of the two factors $(x+3)$ and $(x-2)$ is positive. This can only happen in two ways: when both factors are positive or when both factors are negative. As in example 1.3, we again draw number lines for each of the individual factors, indicating where each is positive, negative or zero and use these to draw a third number line for the product. We show these in the margin. Notice that the third number line indicates that the product is positive whenever $x < -3$ **or** $x > 2$. We write this in interval notation as $(-\infty, -3) \cup (2, \infty)$.

Remark 1.2   With any polynomial inequality, the objective is the same as in example 1.4. That is, move all the terms to one side of the inequality (so that there is a 0 on the other side). Then, factor the resulting polynomial as much as possible and determine where each individual factor is positive, negative, and zero.

No doubt, you will recall the following standard definition.

Definition 1.1   The **absolute value** of a real number x is $|x| = \begin{cases} x, & \text{if } x \geq 0 \\ -x, & \text{if } x < 0. \end{cases}$

Make certain that you read Definition 1.1 correctly. If x is negative, then $-x$ is positive. This says that $|x| \geq 0$ for all real numbers x. For instance, using the definition,

$$|-4| = -(-4) = 4.$$

Notice that for any real numbers a and b,

$$|a \cdot b| = |a| \cdot |b|.$$

However,

$$|a + b| \neq |a| + |b|.$$

(To verify this, simply take $a = 5$ and $b = -2$ and compute both quantities.)

However, it is always true that

$$|a + b| \leq |a| + |b| \tag{1.3}$$

The inequality (1.3) is referred to as the **triangle inequality**. You will find occasional uses for it as you progress through the calculus.

   NOTE: For any two real numbers a and b, $|a - b|$ gives the **distance** between a and b (see Figure 0.7).

FIGURE 0.7: The distance between a and b.

This interpretation of $|a - b|$ as distance is particularly useful for solving inequalities involving absolute values. Wherever possible, we suggest that you use this interpretation to read what the inequality means, rather than merely following a procedure to produce a solution.

Example 1.5   **Solving an Inequality Containing an Absolute Value**
   Solve the inequality

$$|x - 2| < 5. \tag{1.4}$$

Solution   Before you start trying to solve this, take a few moments to read what it **says**. Since $|x - 2|$ is the distance from x to 2, (1.4) says that the **distance** from x to 2 must be **less than 5**. So, find all numbers x whose distance from 2 is less than 5. We indicate the set of all numbers within a distance 5 of 2 in Figure 0.8. You can now read the solution directly from the figure: $-3 < x < 7$ or in interval notation: $(-3, 7)$.

FIGURE 0.8: $|x - 2| < 5$.

■

Many inequalities can be solved simply by reading the inequality correctly as in the following example.

Example 1.6   Solving an Inequality with a Sum Inside an Absolute Value
Solve the inequality

$$|x + 4| \le 7. \tag{1.5}$$

Solution   To use our distance interpretation, we must first rewrite (1.5) as

$$|x - (-4)| \le 7.$$

This now says that the distance from x to -4 is less than or equal to 7. We describe the solution graphically in Figure 0.9. You can again read the solution directly from the figure. We get $-11 \le x \le 3$ or $[-11, 3]$.

FIGURE 0.9: $|x + 4| \le 7$.

As an alternative to using the distance interpretation of $|x - a|$ for solving inequalities, we first must remind you that for any real number $r > 0$, $|x| < r$ is equivalent to the following inequality not involving absolute values:

$$-r < x < r.$$

We use this idea in the following example, where we revisit the inequality from example 1.5.

Example 1.7   An Alternative Method for Solving Inequalities
Solve the inequality

$$|x - 2| < 5.$$

Solution   This is equivalent to the inequality

$$-5 < x - 2 < 5.$$

Adding 2 to each term, we get the solution

$$-3 < x < 7$$

or in interval notation $(-3, 7)$, as before.

The Cartesian Plane

A powerful tool for exploring the relationships between numbers is the **Cartesian plane**. We take two real numbers x and y and form them into the **ordered pair** (x, y). We visualize the ordered pair $(1, 2)$ as a point in two dimensions. The Cartesian plane consists of two real number lines drawn at right angles. The horizontal line is called the x-axis and the vertical line is called the y-axis. The point where the axes cross is called the **origin**, which represents the ordered pair $(0, 0)$. To represent the ordered pair $(1, 2)$, start at the origin, move 1 unit to the right and 2 units up, and mark the point $(1, 2)$, as in Figure 0.10.

In the following example, we analyze a small set of experimental data by plotting some points in the Cartesian plane. This simple type of graph is sometimes called a **scatter plot**.

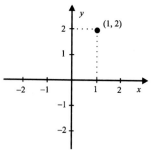

FIGURE 0.10: The Cartesian Plane.

Example 1.8   **Using a Graph Obtained from a Table of Data**

Suppose you drop an object from the top of a building and record how far the object has fallen at different times, as shown in the following table.

Time (sec)	0	0.5	1.0	1.5	2.0
Distance (ft)	0	4	16	36	64

Plot the points in the Cartesian plane and discuss any patterns you notice. In particular, use the graph to predict how far the object will have fallen in 2.5 seconds.

FIGURE 0.11: Scatter plot of data.

Solution   Taking the time to be the first coordinate (x) and distance to be the second coordinate (y), we plot the points $(0, 0)$, $(0.5, 4)$, $(1, 16)$, $(1.5, 36)$, and $(2, 64)$, as seen in Figure 0.11.

Notice that the points appear to be curving upward (like a parabola). To predict the y-value corresponding to $x = 2.5$ (i.e., the distance fallen at time 2.5 seconds), you would assume that this pattern would continue, so that the y-value would be much higher than 64. But, how much higher is reasonable? It helps now to refer back to the data. Notice that the change in height from $x = 1.5$ to $x = 2$ is $64 - 36 = 28$ feet. Since Figure 0.11 suggests that the curve is bending upward, the change in height between successive points should be getting larger and larger. You might reasonably predict that the height will change by *more than* 28. If you look carefully at the data, you might notice a pattern. Observe that the distances given at 0.5 second intervals are $0^2, 2^2, 4^2, 6^2$ and 8^2. A reasonable guess for the distance at time 2.5 seconds might then be $10^2 = 100$. Further, notice that this corresponds to a change of 36 from the distance at $x = 2.0$ seconds. At this stage, this is only an educated guess and other guesses (98 or 102, for example) might be equally reasonable.

∎

We urge that you think carefully about example 1.8. You should be comfortable with the interplay between the graph and the numerical data. This interplay will be a recurring theme in our study of calculus.

The distance between two points in the Cartesian plane is a simple consequence of the Pythagorean Theorem, as follows.

Theorem 1.2   The distance between the points (x_1, y_1) and (x_2, y_2) in the Cartesian plane is given by

$$d\{(x_1, y_1), (x_2, y_2)\} = \sqrt{(x_2 - x_1)^2 + (y_2 - y_1)^2}. \qquad (1.6)$$

Proof. We have oriented the points in Figure 0.12 so that (x_2, y_2) is above and to the right of (x_1, y_1).

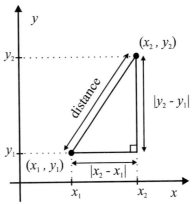

FIGURE 0.12: Distance.

Referring to the right triangle shown in Figure 0.12, notice that regardless of the orientation of the two points, the length of the horizontal side of the triangle is $|x_2 - x_1|$ and the length of the vertical side of the triangle is $|y_2 - y_1|$. The distance between the two points is the length of the hypotenuse of the triangle, given by the Pythagorean Theorem as

$$\sqrt{(x_2 - x_1)^2 + (y_2 - y_1)^2}.$$

With the distance formula (1.6) in hand, you can now do some significant analysis.

Example 1.9   Using the Distance Formula

Find the distances between each pair of points $(1, 2)$, $(3, 4)$ and $(2, 6)$. Use the distances to determine if the points form a right triangle.

Solution   The distance between $(1, 2)$ and $(3, 4)$ is

$$d\{(1, 2), (3, 4)\} = \sqrt{(3 - 1)^2 + (4 - 2)^2} = \sqrt{4 + 4} = \sqrt{8}.$$

The distance between $(1, 2)$ and $(2, 6)$ is

$$d\{(1, 2), (2, 6)\} = \sqrt{(2 - 1)^2 + (6 - 2)^2} = \sqrt{1 + 16} = \sqrt{17}.$$

Finally, the distance between $(3, 4)$ and $(2, 6)$ is

$$d\{(3, 4)(2, 6)\} = \sqrt{(2 - 3)^2 + (6 - 4)^2} = \sqrt{1 + 4} = \sqrt{5}.$$

From a plot of the points (see Figure 0.13), it is unclear whether a right angle is formed at $(3, 4)$.

However, the sides of a right triangle must satisfy the Pythagorean Theorem. This would require that

$$(\sqrt{8})^2 + (\sqrt{5})^2 = (\sqrt{17})^2.$$

Since this statement is not true, the triangle is not a right triangle.

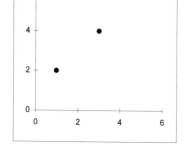

FIGURE 0.13: A right triangle?

Again, we want to draw your attention to the interplay between graphical and algebraic techniques. If you master the relationship between graphs and equations, your study of calculus will be a more rewarding and enjoyable learning experience.

Exercises 0.1

1. (writing) To understand Definition 1.1, you must believe that $|x| = -x$ for negative x's. Using $x = -3$ as an example, explain in words why multiplying x by -1 produces the same result as taking the absolute value of x.

2. (writing) A common shortcut used to write inequalities is $-4 < x < 4$ in place of "$-4 < x$ **and** $x < 4$." Unfortunately, many people mistakenly write $4 < x < -4$ in place of "$4 < x$ **or** $x < -4$." Explain why the string $4 < x < -4$ could never be true. (Hint: What does this inequality string imply about the numbers on the far left and far right? Here, you must write "$4 < x$ **or** $x < -4$.")

3. (writing) Explain the result of Theorem 1.1 (ii) in your own words, assuming that all constants involved are positive.

4. (writing) Suppose a friend has dug holes for the corner posts of a rectangular deck. Explain how to use the Pythagorean Theorem to determine whether or not the holes truly form a rectangle ($90°$ angles).

In exercises 5–32, solve the inequality.

5. $3x + 2 < 11$ **6.** $4x + 1 < 10$

7. $2x - 3 < 8$ **8.** $3x - 1 < 9$

9. $4 - 3x < 6$ **10.** $5 - 2x < 9$

11. $4 \le x + 1 < 7$ **12.** $-1 < 2 - x < 3$

13. $-2 < 2 - 2x < 3$ **14.** $0 < 3 - x < 1$

15. $x^2 + 3x - 4 > 0$

16. $x^2 + 4x + 3 < 0$

17. $x^2 - x - 6 < 0$

18. $x^2 + 1 > 0$

19. $3x^2 + 4 > 0$

20. $x^2 + 3x + 10 > 0$

21. $|x - 3| < 4$

22. $|2x + 1| < 1$

23. $|3 - x| < 1$

24. $|3 + x| > 1$

25. $|2x + 1| > 2$

26. $|3x - 1| < 4$

27. $\dfrac{x + 2}{x - 2} > 0$

28. $\dfrac{x - 4}{x + 1} < 1$

29. $\dfrac{x^2 - x - 2}{(x + 4)^2} > 0$

30. $\dfrac{3 - 2x}{(x + 1)^2} < 0$

31. $\dfrac{-8x}{(x + 1)^3} < 0$

32. $\dfrac{x - 2}{(x + 2)^3} > 0$

In exercises 33–38, find the distance between the pair of points.

33. $(2, 1), (4, 4)$

34. $(2, 1), (-1, 4)$

35. $(-1, -2), (3, -2)$

36. $(1, 2), (3, 6)$

37. $(0, 2), (-2, 6)$

38. $(4, 1), (2, 1)$

In exercises 39–42, determine if the set of points form the vertices of a right triangle.

39. $(1, 1), (3, 4), (0, 6)$

40. $(0, 2), (4, 8), (-2, 12)$

41. $(-2, 3), (2, 9), (-4, 13)$

42. $(-2, 3), (0, 6), (-3, 8)$

In exercises 43–46, the data represent populations at various times. Plot the points, discuss any patterns that are evident, and predict the population at the next step.

43. $(0, 1250), (1, 1800), (2, 2450), (3, 3200)$

44. $(0, 3160), (1, 3250), (2, 3360), (3, 3490)$

45. $(0, 4000), (1, 3990), (2, 3960), (3, 3910)$

46. $(0, 2100), (1, 2200), (2, 2100), (3, 1700)$

47. As discussed in the text, a number is rational if and only if its decimal representation terminates or repeats. Calculators and computers perform their calculations using a finite number of digits. Explain why such calculations can only produce rational numbers.

48. In example 1.8, we discussed how the tendency of the data points to "curve up" corresponds to larger increases in consecutive y-values. Explain why this is true.

49. The ancient Greeks analyzed music mathematically. They found that if pipes of length L and $\frac{L}{2}$ are struck, they make tones that blend together nicely. We say that these tones are one **octave** apart. In general, nice harmonies are produced by pipes (or strings) with rational ratios of lengths. For example, pipes of length L and $\frac{2}{3}L$ form a **fifth** (e.g., middle C and the G

above middle C). On a piano keyboard, 12 fifths are equal to 7 octaves. A glitch in piano tuning, known as the **Pythagorean comma**, results from the fact that 12 fifths with total length ratio $\left(\frac{2}{3}\right)^{12}$ do not equal 7 octaves with length ratio $\left(\frac{1}{2}\right)^7$. Show that the difference is about 1.3%.

50. For the 12 keys of a piano octave to have exactly the same length ratios (see exercise 49), the ratio of consecutive lengths should be a number x such that $x^{12} = 2$. Briefly explain why. This means that $x = \sqrt[12]{2}$. There are two problems with this **equal-tempered** tuning. First, $\sqrt[12]{2}$ is irrational. Explain why it would be difficult to get the pipe or string exactly the right length. In any case, musicians say that equal-tempered pianos sound "dull."

51. The use of squares in the Pythagorean Theorem has found a surprising use in the analysis of baseball statistics. In Bill James' *Historical Abstract*, a rule is stated that a team's winning percentage P is approximately equal to $\dfrac{R^2}{R^2 + G^2}$, where R is the number of runs scored by the team and G is the number of runs scored against the team. For example, in 1996 the Texas Rangers scored 928 runs and gave up 799 runs. The formula predicts a winning percentage of $\dfrac{928^2}{928^2 + 799^2} \approx 0.574$. In fact, Texas won 90 games and lost 72 for a winning percentage of $\dfrac{90}{162} \approx 0.556$. Fill out the following table (data from the 1996 season). What are possible explanations for teams that win more (or less) games than the formula predicts?

Team	R	G	P	wins	losses	winning %
Yankees	871	787		92	70	
Braves	773	648		96	66	
Phillies	650	790		67	95	
Dodgers	703	652		90	72	
Indians	952	769		99	62	

52. (exploratory) It can be very difficult to prove that a given number is irrational. According to legend, the following proof that $\sqrt{2}$ is irrational so upset the ancient Greek mathematicians that they drowned a mathematician who revealed the result to the general public. The proof is by **contradiction**; that is, we imagine that $\sqrt{2}$ is rational and then show that this cannot be true. If $\sqrt{2}$ were rational, we would have that $\sqrt{2} = \dfrac{p}{q}$ for some integers p and q. Assume that $\dfrac{p}{q}$ is in simplified form (i.e., any common factors have been divided out). Square the equation $\sqrt{2} = \dfrac{p}{q}$ to get $2 = \dfrac{p^2}{q^2}$. Explain why this can only be true if p is an even integer. Write $p = 2r$ and substitute to get $2 = \dfrac{4r^2}{q^2}$. Then, rearrange this expression to get $q^2 = 2r^2$. Explain why this can only be true if q is an even integer. Something has gone wrong: explain why p and q can't both be even integers. Since this can't be true, we conclude that $\sqrt{2}$ is irrational.

53. (exploratory) In the text, we stated that a number is rational if and only if its decimal representation repeats or terminates. In this exercise, we prove that the decimal representation of any rational number repeats or terminates. To start with a concrete example, use long division to show that $\frac{1}{7} = 0.\overline{142857}142857$. Note that when you get a remainder of 1, it's all over: you started with a 1 to divide into, so the sequence of digits must repeat. For a general rational number $\frac{p}{q}$, there are q possible remainders (0, 1, 2, ..., $q - 1$). Explain why when doing long division you must eventually get a remainder you have had before. Explain why the digits will then either terminate or start repeating.

54. (exploratory) The existence of irrational numbers may seem like a minor technical footnote to the rational numbers we normally use. This is far from true. Imagine that you are going to choose a number at random. Believe it or not, you would almost certainly choose an irrational number! To see why this would happen, suppose you are choosing a number between 0 and 1 by writing down its decimal representation one digit at a time. Pick the first digit, then the second digit (e.g., if you choose 4 then 6 your number starts out 0.46...) and so on. If you continue choosing the digits randomly, what is the likelihood that you will repeat a pattern forever? If you don't, you have chosen an irrational number!

0.2   Lines and Functions

Equations of Lines

By law, the federal government must conduct a nationwide census every 10 years to estimate the population. This is done for a variety of reasons, including the determination of the number of congressional seats assigned to each state. One of the more important reasons is so that the government can track the growth of various segments of the population for the purposes of planning. This is particularly important today, as programs such as Social Security, Medicare, and Medicaid consume an increasing portion of the total federal budget. Population data for the last several decades are shown in the accompanying table.

One difficulty with analyzing this data is that the numbers are so large. This problem is remedied by **transforming** the data. We can simplify the year data by defining x to be the number of years since 1960. Then, 1960 corresponds to $x = 0$, 1970 corresponds to $x = 10$, and so on. The population data can be simplified by rounding the numbers off to the nearest million. (A simpler, but less precise method is to truncate the numbers, ignoring all digits past a certain place). If we round the data, the 1960 population becomes $y = 179$ (million), the 1970 population becomes $y = 203$ (million), and so on. Notice the difference between truncating and rounding: the 1980 population rounds to 227, but truncates to 226 and the 1990 population rounds to 249, but truncates to 248. The simplified data are shown in the accompanying table and a scatter plot of these data points is shown in Figure 0.14.

Asked to describe the pattern in Figure 0.14, most people would say that the points appear to form a straight line. (Use a ruler and see if you agree.) To determine whether the points are, in fact, on the same line (such points are called **colinear**), we could use the distance formula to see if the sum of the distances between adjacent points sums to the distance between the farthest two points. (Think about how you might do this.) Again, the picture gives you a general idea of the pattern, but some analysis is required to make the conclusions precise.

As an alternative, you might consider the population growth in each of the indicated decades. From 1960 to 1970, the growth was 24 million. (That is, to move from the first point to the second you increase x by 10 and increase y by 24.) Likewise, from 1970 to 1980, the growth was 24 million. (So, to move from the second point to the third point you again increase x by 10 and y by 24.) However, from 1980 to 1990, the growth was only 22 million. (That is, to move from the third point to the fourth point, while x again increases by 10, y increases by only 22.) Since the rate of growth is not constant, this says that the data points do not fall on a line. (Notice that to stay on the same line, y would had to have increased by 24 again. Why is that?) We will return to this introductory example later in this section to see how to use this information to estimate the U.S. population in the year 2000. This second argument involves the familiar concept of **slope.**

Year	U.S. Population
1960	179, 323, 175
1970	203, 302, 031
1980	226, 542, 203
1990	248, 709, 873

x	y
0	179
10	203
20	227
30	249

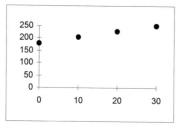

FIGURE 0.14: Population data.

Definition 2.1   For $x_1 \neq x_2$, the **slope** of the line through the points (x_1, y_1) and (x_2, y_2) is the number

$$m = \frac{y_2 - y_1}{x_2 - x_1}. \tag{2.1}$$

For $x_1 = x_2$, the line through (x_1, y_1) and (x_2, y_2) is **vertical** and has **no slope**.

We often describe slope as "the change in y divided by the change in x," written $\dfrac{\Delta y}{\Delta x}$, or more simply as $\dfrac{Rise}{Run}$ (see Figure 0.15a).

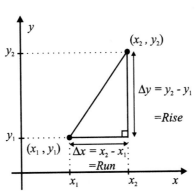

FIGURE 0.15a: Slope.   FIGURE 0.15b: Similar triangles and slope.

You should observe that the slope is the same no matter which two points on the line you select. Referring to Figure 0.15b (where the line has positive slope), notice that for any four points A, B, D, and E on the line, the two right triangles $\triangle ABC$ and $\triangle DEF$ are similar. Recall that for similar triangles, the ratios of corresponding sides must be the same. In this case, this says that

$$\frac{\Delta y}{\Delta x} = \frac{\Delta y'}{\Delta x'},$$

and so, the calculation of slope is the same no matter which two points on the line are selected. Furthermore, a line is the only geometric shape with this property (i.e., a line is the only curve with constant slope). Notice that a line is **horizontal** if and only if its slope is zero.

Example 2.1   **Finding the Slope of a Line**
Find the slope of the line through the points $(2, 5)$ and $(4, 3)$.

Solution   From (2.1), we get

$$m = \frac{y_2 - y_1}{x_2 - x_1} = \frac{5 - 3}{2 - 4} = \frac{2}{-2} = -1. \qquad\blacksquare$$

Example 2.2   **Using the Slope to Determine if Points are Colinear**
Use slopes to determine whether or not the points $(1, 2)$, $(3, 10)$, and $(4, 14)$ are colinear.

Solution   First, notice that the slope of the line joining $(1, 2)$ and $(3, 10)$ is

$$m_1 = \frac{y_2 - y_1}{x_2 - x_1} = \frac{10 - 2}{3 - 1} = \frac{8}{2} = 4.$$

Similarly, the slope through the line joining $(3, 10)$ and $(4, 14)$ is

$$m_2 = \frac{y_2 - y_1}{x_2 - x_1} = \frac{14 - 10}{4 - 3} = 4.$$

Since the slopes are the same, the points must be colinear.

$\blacksquare$

Recall that if you know that a line passes through a given point with a given slope, you have enough information to graph the line. The easiest way to graph a line is to plot two points and then draw the line through them. In this case, you only need to find a second point.

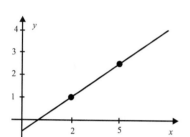

FIGURE 0.16a: Graph of straight line.

Example 2.3   Graphing a Line

If a line passes through the point $(2, 1)$ with slope $\frac{2}{3}$, find a second point on the line and then graph the line.

Solution   Since slope is given by $m = \dfrac{y_2 - y_1}{x_2 - x_1}$, we take $m = \frac{2}{3}$, $y_1 = 1$, and $x_1 = 2$, to obtain

$$\frac{2}{3} = \frac{y_2 - 1}{x_2 - 2}.$$

You are free to choose where you want the second point to be. For instance, to find the point at $x_2 = 5$, substitute this in, and solve:

$$\frac{2}{3} = \frac{y_2 - 1}{5 - 2} = \frac{y_2 - 1}{3}$$

or $y_2 = 3$. A second point is then $(5, 3)$. The graph of the line is shown in Figure 0.16a. An alternative method for finding a second point is to use the slope

$$m = \frac{2}{3} = \frac{\Delta y}{\Delta x}.$$

The slope of $\frac{2}{3}$ says that if we move three units to the right, we must move two units up, to stay on the line, as illustrated in Figure 0.16b.

$\blacksquare$

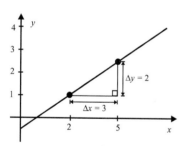

FIGURE 0.16b: Using slope to find a second point.

In example 2.3, note that the choice of $x = 5$ was entirely arbitrary; you can choose any x-value you want to find a second point. Further, since x can be any real number, you can leave x as a variable and write out an equation satisfied by any point (x, y) on the line. In the general case of the line through the point (x_0, y_0) with slope m, we have from (2.1) that

$$m = \frac{y - y_0}{x - x_0}. \tag{2.2}$$

Multiplying both sides of (2.2) by $(x - x_0)$, we get

$$y - y_0 = m(x - x_0)$$

or

$$\boxed{y = m(x - x_0) + y_0.} \tag{2.3}$$

Equation (2.3) is called the **point-slope form** of the line. With this, you can find an equation of any line.

Example 2.4   **Finding the Equation of a Line Given a Point and the Slope**
Find an equation of the line through $(3, 1)$ with slope -2 and graph the line.

Solution   It's a simple matter to use (2.3) to get the equation of the line:

$$y = -2(x - 3) + 1. \qquad (2.4)$$

To graph the line, locate the point $(3, 1)$. A second point can be obtained by plugging in whatever x-value you choose. Here, $x = 4$ is convenient and the corresponding y-value is $y = -2(4 - 3) + 1 = -1$. So, a second point is $(4, -1)$ and you can easily draw the line seen in Figure 0.17.

∎

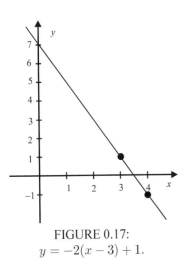

FIGURE 0.17:
$y = -2(x - 3) + 1.$

In example 2.4, you may be tempted to "simplify" the expression for y given in (2.4). As it turns out, many times, the point-slope form of the equation is the most convenient to work with. So, we will typically not ask you to rewrite this expression. At times, a form of the equation called the **slope-intercept form** is more convenient. This is an equation of the form

$$y = mx + b,$$

where m is the slope and b is the y-intercept (i.e., the place where the graph crosses the y-axis). In example 2.4, you simply multiply out (2.4) to get $y = -2x + 6 + 1$ or

$$y = -2x + 7.$$

As you can see from Figure 0.17, the graph crosses the y-axis at $y = 7$.

Example 2.5   **Finding the Equation of a Line Given Two Points**
Find an equation of the line through the points $(-2, 4)$ and $(2, 6)$.

Solution   From (2.1), the slope is

$$m = \frac{4 - 6}{-2 - 2} = \frac{-2}{-4} = \frac{1}{2}.$$

You can now use either point in (2.3) to find an equation of the line. Using $(2, 6)$, we get

$$y = \frac{1}{2}(x - 2) + 6.$$

Again, there is no particular need to simplify this.

∎

Definition 2.2   We say that two (nonvertical) lines are **parallel** if they have the same slope. Further, any two vertical lines are parallel. Two (nonvertical) lines of slope m_1 and m_2 are **perpendicular** whenever the product of their slopes is -1 (i.e., $m_1 \cdot m_2 = -1$). Also, any vertical line and any horizontal line are perpendicular.

Since we can read the slope from the equation of a line, it's a simple matter to determine when two lines are parallel or perpendicular. We illustrate this in the next several examples.

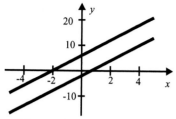

FIGURE 0.18: Parallel lines.

Example 2.6   Finding the Equation of a Parallel Line

Find an equation of the line parallel to $y = 3x - 2$ and through the point $(-1, 3)$.

Solution   It's easy to read the slope from the equation: $m = 3$. The equation of the parallel line is then

$$y = 3\left[x - (-1)\right] + 3$$

or simply $y = 3(x + 1) + 3$. We show a graph of both lines in Figure 0.18.

■

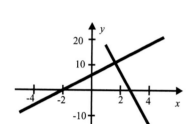

FIGURE 0.19: Perpendicular lines.

Example 2.7   Finding the Equation of a Perpendicular Line

Find an equation of the line perpendicular to $y = -2x + 4$ and intersecting the line at the point $(1, 2)$.

Solution   The slope of $y = -2x + 4$ is -2. The slope of the perpendicular line is then $-1/(-2) = \frac{1}{2}$. Since the line must pass through the point $(1, 2)$, the equation of the perpendicular line is now

$$y = \frac{1}{2}(x - 1) + 2.$$

We show a graph of the two lines in Figure 0.19.

■

We now return to the section's introductory example to use the equation of a line to estimate the population in the year 2000.

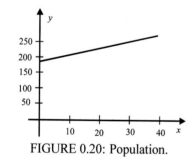

FIGURE 0.20: Population.

Example 2.8   Using a Line to Estimate Population

Given the population data for the census years 1960, 1970, 1980, and 1990, estimate the population for the year 2000.

Solution   First, recall that we began this section by showing that the points in the corresponding table were not colinear. Nonetheless, they were *nearly* colinear. So, why not use the straight line connecting the last two points $(20, 227)$ and $(30, 249)$ (recall that these correspond to the population in the years 1980 and 1990) to estimate the population in 2000? (This is a simple example of a more general procedure called **extrapolation**.) The slope of the line joining the two data points is

$$m = \frac{249 - 227}{30 - 20} = \frac{22}{10} = \frac{11}{5}.$$

The equation of the line is then

$$y = \frac{11}{5}(x - 30) + 249.$$

See Figure 0.20 for a graph of the line. If we follow this line to the point corresponding to $x = 40$ (the year 2000), we have the estimated population

$$\frac{11}{5}(40 - 30) + 249 = 271.$$

That is, the estimated population is 271 million people.

■

In this section, we have so far drawn connections between the equations and graphs of lines. This process can be extended to a variety of other curves. In fact, most of this introductory chapter is devoted to exploring such connections. Before doing so, we need some general definitions.

Functions

For any two subsets A and B of the real line, we make the following familiar definition.

Definition 2.3   A **function** f is a rule that assigns **exactly one** element y in a set B to each element x in a set A. In this case, we write

$$y = f(x).$$

We call the set A the **domain** of f. The set of all values $f(x)$ in B is called the **range** of f. That is, the range of f is $\{\,f(x)\,|\,x \in A\}$. Unless explicitly stated otherwise, the domain of a function f is the largest set of real numbers for which the function makes sense. We refer to x as the **independent variable** and to y as the **dependent variable.**

Remark 2.1   Functions are often defined by simple formulas, such as $f(x) = 3x + 2$, but in general, any correspondence meeting the requirement of matching exactly **one** y to each x defines a function.

By the **graph** of a function f, we mean the graph of the equation $y = f(x)$. That is, the graph consists of all points (x, y), where x is in the domain of f and where $y = f(x)$.

Notice that not every curve is the graph of a function, since for a function, there is only one y-value that can correspond to a given value of x. Recall that you can graphically determine which curves are the graph of a function by using the **vertical line test**: if any vertical line intersects the graph in more than one point, the curve is not the graph of a function.

Example 2.9   **Using the Vertical Line Test**
Determine which curves correspond to functions.

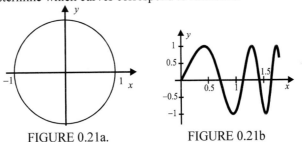

FIGURE 0.21a.   FIGURE 0.21b

Solution   Notice that the circle in Figure 0.21a is not the graph of a function, since a vertical line at $x = 0.5$ intersects the circle twice (see Figure 0.22a). The graph in Figure 0.21b is the graph of a function, even though it swings up and down repeatedly. Although horizontal lines intersect the graph repeatedly, vertical lines, such as the one at $x = 1.2$, intersect only once (see Figure 0.22b).

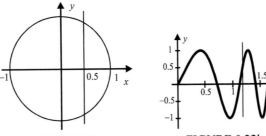

FIGURE 0.22a: FIGURE 0.22b:
Curve fails vertical line test.   Curve passes vertical line test.

You are already familiar with a number of different types of functions and we will only briefly review these here and in sections 0.5 and 0.6. The functions that you are probably most familiar with are **polynomials**. These are the simplest functions to work with because they are defined entirely in terms of arithmetic.

■

Definition 2.4   A **polynomial** is any function that can be written in the form

$$f(x) = a_n x^n + a_{n-1} x^{n-1} + \cdots + a_1 x + a_0,$$

where $a_0, a_1, a_2, \ldots, a_n$ are real numbers (the **coefficients** of the polynomial) with $a_n \neq 0$ and $n \geq 0$ is an integer (the **degree** of the polynomial).

Note that the domain of every polynomial function is the entire real line.

Example 2.10   **Sample Polynomials**

The following are all examples of polynomials:

$$f(x) = 2 \text{ (polynomial of degree 0 or } \mathbf{constant}),$$

$$f(x) = 3x + 2 \text{ (polynomial of degree 1 or } \mathbf{linear} \text{ polynomial)},$$

$$f(x) = 5x^2 - 2x + 1 \text{ (polynomial of degree 2 or } \mathbf{quadratic} \text{ polynomial)},$$

$$f(x) = x^3 - 2x + 1 \text{ (polynomial of degree 3 or } \mathbf{cubic} \text{ polynomial)},$$

$$f(x) = -6x^4 + 12x^2 - 3x + 13 \text{ (polynomial of degree 4 or } \mathbf{quartic} \text{ polynomial)},$$

and

$$f(x) = 2x^5 + 6x^4 - 8x^2 + x - 3 \text{ (polynomial of degree 5 or } \mathbf{quintic} \text{ polynomial)}.$$

We show graphs of these six functions in Figures 0.23a−0.23f

FIGURE 0.23a: $f(x) = 2.$

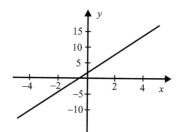
FIGURE 0.23b: $f(x) = 3x + 2.$

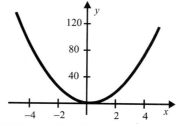
FIGURE 0.23c: $f(x) = 5x^2 - 2x + 1.$

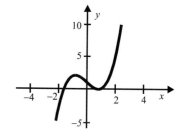
FIGURE 0.23d: $f(x) = x^3 - 2x + 1.$

FIGURE 0.23e:

$f(x) = -6x^4 + 12x^2 - 3x + 13.$

FIGURE 0.23f:

$f(x) = 2x^5 + 6x^4 - 8x^2 + x - 3.$ ∎

Definition 2.5   Any function that can be written in the form

$$f(x) = \frac{p(x)}{q(x)},$$

where p and q are polynomials is called a **rational** function.

Notice that since $p(x)$ and $q(x)$ are polynomials, they are both defined for all x, and so, the rational function $f(x) = \dfrac{p(x)}{q(x)}$ is defined for all x for which $q(x) \neq 0$.

Example 2.11   **A Sample Rational Function**
Find the domain of the function

$$f(x) = \frac{x^2 + 7x - 11}{x^2 - 4}.$$

Solution   Here, $f(x)$ is a rational function. We show a graph in Figure 0.24. Further, its domain consists of those values of x for which the denominator is nonzero. Notice that

$$x^2 - 4 = (x - 2)(x + 2),$$

and so, the denominator is zero if and only if $x = \pm 2$. This says that the domain of f is

$$\{x \in \mathbb{R} \,|\, x \neq \pm 2\} = (-\infty, -2) \cup (-2, 2) \cup (2, \infty).$$

∎

FIGURE 0.24:

$f(x) = \dfrac{x^2 + 7x - 11}{x^2 - 4}.$

The **square root** function is defined in the usual way. When we write $y = \sqrt{x}$, we mean that y is that number for which $y \geq 0$ and $y^2 = x$. In particular, $\sqrt{4} = 2$. Be careful not to write erroneous statements like $\sqrt{4} = \pm 2$. This is a common misconception. While it's true that $2^2 = 4$ and $(-2)^2 = 4$, when we write $\sqrt{4}$, we are looking for the **positive** number whose square is 4. In this way, $\sqrt{x}$ defines a function with domain $[0, \infty)$. If we allowed $\sqrt{x}$ to take on both positive and negative values, this would not define a function. In particular, be careful to write

$$\sqrt{x^2} = |x|.$$

Since $\sqrt{x^2}$ is asking for the nonnegative number whose square is x^2, we are looking for $|x|$ and not x. We can say

$$\sqrt{x^2} = x, \textbf{ only for } x \geq 0.$$

Note that the ability to interpret a graph of a function is important even when there is no known formula for the function.

Example 2.12   **Finding the Domain of a Function Involving a Square Root**

Find the domain of $f(x) = \sqrt{x^2 - 4}$.

Solution   Since square roots are defined only for nonnegative values, we need to have $x^2 - 4 \geq 0$. Notice that this is equivalent to having $x^2 \geq 4$, which occurs when $x \geq 2$ or $x \leq -2$. The domain of f is then $(-\infty, -2] \cup [2, \infty)$.

■

Example 2.13   **Interpreting a Graph**

The graph in Figure 0.25 shows the depth of snow at a ski resort for a two-week period in January. The resort has artificial snow-making machines that it uses to maintain the 55-inch base of snow necessary for good skiing. On which days did they have to run the machine?

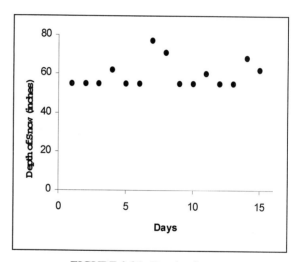

FIGURE 0.25: Depth of snow.

Solution   There is no single, "correct" answer to this type of question. Your response should be like an essay: present a plausible answer consistent with the facts. In this case, you might assume that the snow machines would be used to bring the snow up *exactly* to the 55-inch mark. Therefore, it probably snowed on days 4, 7, 11, and 14. The snow level after the big snow on day 7 dropped about 6 inches in a day. If this were generally true, there was no need to run the snow machine on day 8 or 9. By the same logic, days 5 and 15 probably did not need the machine, but day 12 needed some machine snow. That leaves days 1, 2, 3, 6, 10, and 13 for full machine use.

■

You might wonder if example 2.13 really qualifies as mathematics. It's a fair question, since mathematics courses too often seem to focus on equations and calculations. The use of hypothesis and deduction is definitely mathematics. The process of making a connection between the graph and the real-world problem is an example of **mathematical modeling**. Mathematics sometimes appears to be entirely abstract and isolated from the real world, but increasingly, mathematics (including much of the calculus) plays a vital role in our modern technological society.

Exercises 0.2

1. (writing) If the distance from point A to point B plus the distance from point B to point C equals the distance from point A to point C, explain why the points A, B, and C must be colinear.

2. (writing) If the slope between points A and B equals the slope between points B and C, explain why the points A, B, and C are colinear.

3. (writing) If a graph fails the vertical line test, it is not the graph of a function. Explain this result in terms of the definition of a function.

4. (writing) You should not automatically write the equation of a line in slope-intercept form. Compare the following forms of the same line: $y = 2.8(x - 1.8) + 0.4$ and $y = 2.4x - 3.92$. Given $x = 1.8$, which equation would you rather use to compute y? How about if you are given $x = 0$? For $x = 8$, is there any advantage to one equation over the other? Can you quickly read off the slope from either equation? Explain why neither form of the equation is "better."

In exercises 5–12, determine whether or not the points are colinear.

5. $(1, 2), (3, 6), (0, 0)$

6. $(1, 1), (3, 5), (0, -1)$

7. $(2, 1), (0, 2), (4, 0)$

8. $(3, 2), (4, 0), (1, 6)$

9. $(3, 1), (4, 4), (5, 8)$

10. $(1, 2), (2, 5), (4, 8)$

11. $(4, 1), (3, 2), (1, 3)$

12. $(3, 2), (0, 1), (6, 3)$

In exercises 13–20, find the slope of the line through the given points.

13. $(1, 2), (3, 6)$

14. $(0, 1), (2, 7)$

15. $(1, 2), (3, 3)$

16. $(0, 1), (3, 8)$

17. $(3, -6), (1, -1)$

18. $(1, -2), (-1, -3)$

19. $(0.3, -1.4), (-1.1, -0.4)$

20. $(1.2, 2.1), (3.1, 2.4)$

In exercises 21–30, find a second point on the line with slope m and point P, graph the line and find an equation of the line.

21. $m = 2, P = (1, 3)$

22. $m = 2, P = (3, 1)$

23. $m = -2, P = (1, 4)$

24. $m = -1, P = (2, 1)$

25. $m = 0, P = (-1, 1)$

26. $m = 0, P = (2, -1)$

27. $m = \frac{1}{2}, P = (2, 1)$

28. $m = -\frac{1}{4}, P = (-2, 1)$

29. $m = 1.2, P = (2.3, 1.1)$

30. $m = 0.8, P = (1.2, 0.4)$

In exercises 31–40, determine if the lines are parallel, perpendicular, or neither.

31. $y = 3(x - 1) + 2$ and $y = 3(x + 4) - 1$

32. $y = 2(x - 3) + 1$ and $y = 4(x - 3) + 1$

33. $y = -2(x + 1) - 1$ and $y = \frac{1}{2}(x - 2) + 3$

34. $y = 3(x - 1) + 1$ and $y = -\frac{1}{3}(x + 2) - 2$

35. $y = 2x - 1$ and $y = -2x + 2$

36. $y = 4x + 3$ and $y = 4x - 5$

37. $y = 4x - 2$ and $y = \frac{1}{4}x + 1$

38. $y = 3x + 1$ and $y = -\frac{1}{3}x + 2$

39. $x + 2y = 1$ and $2x + 4y = 3$

40. $x + 2y = 1$ and $2x - y = 3$

In exercises 41–46, find an equation of a line through the given point and (a) parallel to and (b) perpendicular to the given line.

41. $y = 2(x + 1) - 2$ at $(2, 1)$

42. $y = 3(x - 2) + 1$ at $(0, 3)$

43. $y = 1$ at $(0, -1)$

44. $x = 2$ at $(0, 1)$

45. $y = 2x + 1$ at $(3, 1)$

46. $y = -x + 2$ at $(1, 2)$

In exercises 47–52, find an equation of the line through the given points and compute the y-coordinate of the point on the line corresponding to $x = 4$.

47.

48.

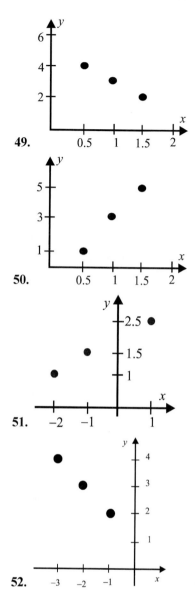

49.

50.

51.

52.

In exercises 53–58, use the vertical line test to determine whether or not the curve is the graph of a function.

53.

54.

55.

56.

57.

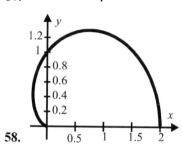

58.

In exercises 59–66, identify the given function as polynomial, rational, both or neither.

59. $f(x) = x^3 - 4x + 1$ **60.** $f(x) = -2x^4 - 3x^2 + 1$

61. $f(x) = \dfrac{x^2 + 2x - 1}{x + 1}$ **62.** $f(x) = \dfrac{x^3 + 4x - 1}{x^4 - 1}$

63. $f(x) = \sqrt{x^2 + 1}$ **64.** $f(x) = x^3 - 3\sqrt{x} + 1$

65. $f(x) = 3 - 2x + x^4$ **66.** $f(x) = 2x - x^{2/3} - 6$

In exercises 67–74, find the domain of the function.

67. $f(x) = \sqrt{x + 2}$

68. $f(x) = \sqrt{2x + 1}$

69. $f(x) = \sqrt[3]{x - 1}$

70. $f(x) = \sqrt[3]{4x - 3}$

71. $f(x) = \dfrac{4}{x^2 - 1}$

72. $f(x) = \dfrac{4x}{x^2 + 2x - 6}$

73. $f(x) = \sqrt{x^2 + 1}$

74. $f(x) = \sqrt{x^2 - 4}$

In exercises 75–78, find the indicated function values.

75. $f(x) = x^2 - x - 1$; $f(0), f(2), f(-3), f(1/2)$

76. $f(x) = \dfrac{x+1}{x-1}$; $f(0), f(2), f(-2), f(1/2)$

77. $f(x) = \sqrt{x+1}$; $f(0), f(3), f(-1), f(1/2)$

78. $f(x) = \dfrac{3}{x}$; $f(1), f(10), f(100), f(1/3)$

In exercises 79–82, a brief description is given of a physical situation. For the indicated variable, state a reasonable domain.

79. A parking deck is to be built; x = width of deck.

80. A parking deck is to be built on a 200'-by-200' lot; x = width of deck.

81. A new candy bar is to be sold; x = number of candy bars sold in first month.

82. A new candy bar is to be sold; x = cost of candy bar.

In exercises 83–86, discuss whether you think y would be a function of x.

83. y = grade you get on an exam, x = number of hours you study

84. y = probability of getting lung cancer, x = number of cigarettes smoked per day

85. y = a person's weight, x = number of minutes exercising per day

86. y = speed at which an object falls, x = weight of object

87. Figure A shows more data from the ski resort of example 2.13. Determine the days on which the snow machine was used.

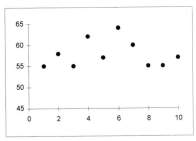

FIGURE A: Depth of snow.

88. Figure B shows the daily profit of a business as a function of time. In the time period shown, the company received two large orders and paid one large bill to a parts provider. Identify when these events occurred. If another bill is due to the parts provider, estimate the height of the graph after the bill is paid.

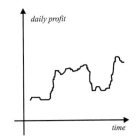

FIGURE B: Business income.

89. Figure C shows the speed of a bicyclist as a function of time. For the portions of this graph that are flat, what is happening to the bicyclist's speed? What is happening to the bicyclist's speed when the graph goes up? down? Identify the portions of the graph that correspond to the bicyclist going uphill; downhill.

FIGURE C: Bicycle speed.

90. Figure D shows the population of a small country as a function of time. During the time period shown, the country experienced two influxes of immigrants, a war and a plague. Identify these important events.

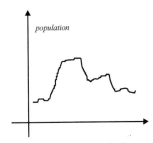

FIGURE D: Population.

In exercises 91–94, briefly describe a realistic scenario for how each quantity changes as a function of time and sketch a graph corresponding to your description (label all interesting points).

91. The electrical usage of a city (include day and night).

92. The temperature of a baked potato just placed into an oven.

93. Your level of alertness listening to the radio (include commercials, good songs, and bad songs).

94. Your speed while driving on the interstate (include speed traps and pit stops).

95. The boiling point of water (in degrees Fahrenheit) at elevation h (in thousands of feet above sea level) is given by $B(h) = -1.8h + 212$. Find h such that water boils at $98.6°$. Why would this height be dangerous to humans?

96. A **linear transformation** is defined as any function f satisfying $f(a + b) = f(a) + f(b)$ and $f(cx) = cf(x)$. Show that, according to this definition, $f(x) = 3x + 2$ is not a linear transformation but $f(x) = 3x$ is. (The function $3x + 2$ is called an **affine transformation**.)

97. The spin rate of a golf ball hit with a 9 iron has been measured at 9100 rpm for a 120-compression ball and at 10,000 rpm for a 60-compression ball. Most golfers use 90-compression balls. If the spin rate is a linear function of compression, find the spin rate for a 90-compression ball.

98. Professional golfers often use 100-compression balls. For the data in exercise 97, estimate the spin rate of a 100-compression ball.

99. (exploratory) Suppose you have a machine that will proportionally enlarge a photograph. For example, it could enlarge a 4×6 photograph to 8×12 by doubling the width and height. You could make an 8×10 picture by cropping 1 inch off each side. Explain how you would enlarge a $3\frac{1}{2} \times 5$ picture to an 8×10. A friend returns from Scotland with a $3\frac{1}{2} \times 5$ picture showing the Loch Ness monster in the outer $\frac{1}{4}$" on the right. If you use your procedure to make an 8×10 enlargement, does Nessie make the cut?

100. (exploratory) The human body's capacity to perform a variety of tasks decreases with age. William Clark's book *Sex and the Origins of Death* presents several statistics on this phenomenon. For example, average female fertility drops from 100% at age 30 to 0% at age 50. With age as x and percentage of fertility as y, show that $y = -5(x - 50)$, for $30 \le x \le 50$, assuming y is a linear function of x. Show that 50% fertility is reached at age 40. In each of the following situations, determine which age corresponds to 50% capacity.
(a) Maximum breathing capacity drops from 100% at age 20 to 40% at age 80.
(b) Maximum heart rate drops from 100% at age 20 to 68% at age 80.
(c) Maximum oxygen uptake drops from 100% at age 20 to 30% at age 80.
(d) Nerve conduction velocity drops from 100% at age 20 to 80% at age 80.

0.3   Graphing Calculators and Computer Algebra Systems

The relationships between functions and their graphs is one of the central topics in calculus. Graphing calculators and user-friendly computer software allow you to explore these relationships for a much wider variety of functions than you could with pencil and paper alone. This section presents some general framework for using technology to explore the graphs of functions.

As discussed in section 0.2, the graphs of linear functions are lines. You probably also remember that the graphs of quadratic polynomials are **parabolas**. One of the goals of this section is for you to become equally familiar with the graphs of other polynomials. The best way to become familiar is by experience, working example after example.

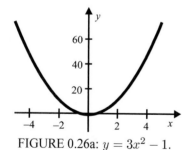

FIGURE 0.26a: $y = 3x^2 - 1$.

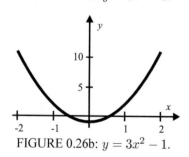

FIGURE 0.26b: $y = 3x^2 - 1$.

Example 3.1   Generating a Calculator Graph

Use your calculator or computer to sketch a graph of $f(x) = 3x^2 - 1$.

Solution   You should get an initial graph that looks something like that in Figure 0.26a. This is simply a parabola opening upward. A graph is often used to search for important points, such as x-intercepts, y-intercepts, or peaks and troughs. In this case, we could see these points better if we **zoom in**; that is, display a smaller range of x- and y-values than the technology has initially chosen for us. The graph in Figure 0.26b shows x-values from $x = -2$ to $x = 2$ and y-values from $y = -2$ to $y = 15$.

You can see more clearly in Figure 0.26b that the parabola bottoms out roughly at the point $(0, -1)$ and crosses the x-axis at approximately $x = -0.5$ and $x = 0.5$. You can make this more precise by doing some algebra. Recall that an x-intercept is a point where $y = 0$ or $f(x) = 0$. Solving $3x^2 - 1 = 0$ gives $3x^2 = 1$ or $x^2 = 1/3$ or $x = \pm\sqrt{1/3} \approx \pm 0.57735$.

■

Notice the interplay in example 3.1 between the graphics and the algebra. The graph suggested approximate values for the two x-intercepts, but you needed the algebra to find the values exactly. (It's worth noting that since $\sqrt{1/3} \approx 0.577$, our visual guess of 0.5 was not especially accurate.) Also note that by zooming in, we used the technology to show us a view of the graph that highlighted the features of the graph we wanted. But, in general, how do you know what the best view is? For example, the graphs in Figures 0.26a and 0.26b indicate a simple parabola, but how do you know that there are no interesting features of the graph lurking just off the screen? In short, you don't, without some calculus.

Before investigating other graphs, we should say a few words about what a computer- or calculator-generated graph really is. Actually, computers and calculators do not draw graphs. Yes, we usually call them graphs, but what the computer actually does is light up some tiny screen elements called **pixels**. If the pixels are small enough, the image appears to be a continuous curve or graph.

By **graphing window**, we mean the rectangle defined by the range of x- and y-values displayed. The graphing window can dramatically affect the look of the graph. For example, suppose the x's run from $x = -2$ to $x = 2$. If the computer or calculator screen is wide enough for 400 columns of pixels from left to right, then points will be displayed for $x = -2$, $x = -1.99$, $x = -1.98$, ... If there is an interesting feature of this function between $x = -1.99$ and $x = -1.98$, you will not see it unless you zoom in. In this case, zooming in would reduce the difference between adjacent x's. Similarly, suppose the y's run from $y = 0$ to $y = 3$ and there are 600 rows of pixels from top to bottom. Then, there will be pixels corresponding to $y = 0$, $y = 0.005$, $y = 0.01$, ... Now, suppose that $f(-2) = 0.0049$ and $f(-1.99) = 0.0051$. Before points are plotted, function values are rounded to the nearest y-value, in this case 0.005. You won't be able to see any difference in the y-values of these points. If the actual difference is important, you will have to zoom in further to see it.

Remark 3.1   Most calculators and computer drawing packages follow one of the following two schemes for defining the graphing window for a given function.

- **Fixed graphing window**: Most calculators follow this method. Graphs are plotted in a preselected range of x- and y-values, unless you specify otherwise. For example, the very popular Texas Instruments graphing calculators will automatically plot points in the rectangle defined by $-10 \leq x \leq 10$ and $-10 \leq y \leq 10$.
- **Automatic graphing window**: Most computer drawing packages and some calculators follow this method. Graphs are plotted for a preselected range of x values and the computer calculates the range of y values so that all of the calculated points will fit in the screen's range.

Get to know how your calculator or computer software operates and use it routinely as you progress through this course. Whether you are using a graphing calculator or a computer, you should always be able to reproduce the computer-generated graphs used in this text by adjusting your graphing window appropriately.

Graphs are drawn to provide visual displays of the significant features of a function. What qualifies as *significant* will vary from problem to problem, but often the x- and y-intercepts and points known as **extrema** are of interest. The function value $f(M)$ is a **local maximum** of the function f if $f(M) \geq f(x)$ for all x's "nearby" $x = M$ (more precisely, if there exist numbers a and b with $a < M < b$ such that $f(M) \geq f(x)$ for all x such that $a < x < b$). Similarly, the function value $f(m)$ is a **local minimum** of the function f if $f(m) \leq f(x)$ for all x's "nearby" $x = m$. A **local extremum** is a function value that is either a local maximum or local minimum. Whenever possible, you should produce graphs that show all intercepts and extrema.

Example 3.2   Sketching a Graph

Sketch a graph of $f(x) = x^3 + 4x^2 - 5x - 1$ showing all intercepts and extrema.

Solution   Depending on your calculator or computer software, you may initially get a graph that looks like one of those in Figures 0.27a or 0.27b.

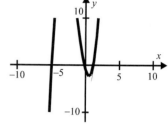

FIGURE 0.27a: $y = x^3 + 4x^2 - 5x - 1$.   FIGURE 0.27b: $y = x^3 + 4x^2 - 5x - 1$.

FIGURE 0.28:
$y = x^3 + 4x^2 - 5x - 1$.

Neither graph is completely satisfactory, although both should give you the idea of a graph that (reading left to right) rises to a local maximum near $x = -3$, drops to a local minimum near $x = 1$, and then rises again. To get a better graph, notice the scales on the x- and y-axes. The graphing window for Figure 0.27a is the rectangle defined by $-5 \leq x \leq 5$ and $-6 \leq y \leq 203$. The graphing window for Figure 0.27b is defined by the rectangle $-10 \leq x \leq 10$ and $-10 \leq y \leq 10$. From either graph, it appears that we need to show y-values bigger than 10, but not nearly as large as 203 to see the local maximum. Since all of the significant features appear to lie between $x = -6$ and $x = 6$, one choice for a better window is $-6 \leq x \leq 6$ and $-6 \leq y \leq 30$, as seen in Figure 0.28. There, you can clearly see the three x-intercepts, the local maximum, and the local minimum.

∎

Note that the graph in example 3.2 was produced by a process of trial and error with thoughtful corrections. You are unlikely to get a perfect picture on your first try. But, from this starting place, you can enlarge the graphing window (i.e., **zoom out**) if you need to see more or shrink the graphing window (i.e., **zoom in**) if the details are hard to see. You should get comfortable enough with your technology that this revision process is routine (and even fun!).

In the exercises, you will be asked to graph a variety of polynomials and discuss the shapes of the graphs of polynomials of different orders. Having some knowledge of the general shapes will help you decide whether or not you have found an acceptable graph. To get you started, we summarize the different shapes of linear, quadratic, and cubic polynomials below. Of course, the graphs of linear functions of the form $f(x) = a_1 x + a_0$ are simply straight lines of slope a_1. Two possibilities are shown in Figures 0.29a and 0.29b.

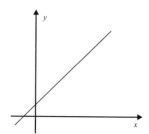

FIGURE 0.29a: Line, $a_1 < 0$.   FIGURE 0.29b: Line, $a_1 > 0$.

Whenever function values get smaller, as you move to the right, as in Figure 0.29a, we say that the function f is **decreasing**. If function values get larger, as you move to the right, as in Figure 0.29b, we say that the function is **increasing**.

You will also recall that the graphs of quadratic polynomials of the form $f(x) = a_2 x^2 + a_1 x + a_0$ are parabolas. The parabola opens upward if $a_2 > 0$ and opens downward if $a_2 < 0$. We show typical parabolas in Figures 0.30a and 0.30b.

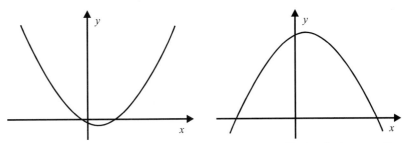

FIGURE 0.30a: Parabola, $a_2 > 0$.   FIGURE 0.30b: Parabola, $a_2 < 0$.

Notice that the parabolas indicated in Figures 0.30a and 0.30b are both increasing and decreasing on different intervals of the x-axis. One of our goals for later in the text is to learn how to determine on what intervals a given function is increasing and decreasing. The graphs of cubic functions of the form $a_3 x^3 + a_2 x^2 + a_1 x + a_0$ are somewhat S-shaped. Reading from left to right, the function begins negative and ends positive if $a_3 > 0$ and begins positive and ends negative if $a_3 < 0$, as indicated in Figures 0.31a and 0.31b.

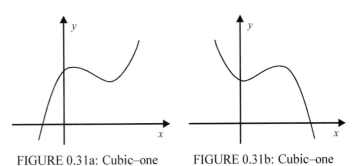

FIGURE 0.31a: Cubic–one   FIGURE 0.31b: Cubic–one
max, min, $a_3 > 0$.   max, min, $a_3 < 0$.

Cubics often have one local maximum and one local minimum, as do those in Figures 0.31a and 0.31b. Many curves (including all cubics) have what's called an **inflection point**, where the curve changes its shape (from being bent upward, to being bent downward, or vice-versa), as indicated in Figures 0.32a and 0.32b.

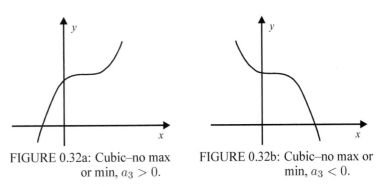

FIGURE 0.32a: Cubic–no max   FIGURE 0.32b: Cubic–no max or
or min, $a_3 > 0$.   min, $a_3 < 0$.

We will use the calculus developed over the next several chapters to see how to adjust the graphing window so that all of the significant features of a graph are shown. You can already use your knowledge of the general shapes of certain functions to see how to adjust the graphing window, as in the following example.

Example 3.3   **Sketching the Graph of a Cubic Polynomial**

Sketch a graph of the cubic polynomial $f(x) = x^3 - 20x^2 - x + 20$.

Solution   Your initial graph probably looks like one of those in Figures 0.33a or 0.33b.

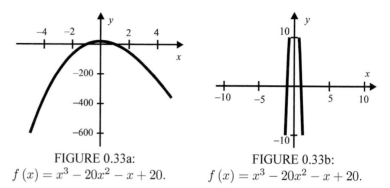

FIGURE 0.33a:
$f(x) = x^3 - 20x^2 - x + 20$.

FIGURE 0.33b:
$f(x) = x^3 - 20x^2 - x + 20$.

From the preceding discussion on the general shapes of cubics, you should recognize that neither of these graphs looks like a cubic. More than anything else, they look like a parabola. To see the S-shape behavior in the graph, we need to look on a larger range of x-values. But how much larger? Unfortunately, we can't answer that question in general, before we develop some of the concepts of calculus. For the moment, the answer is to use trial and error, until the graph resembles the shape of a cubic. You should recognize the characteristic shape of a cubic in Figure 0.33c.

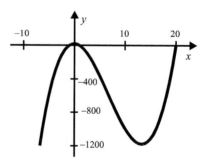

FIGURE 0.33c: $f(x) = x^3 - 20x^2 - x + 20$.

Notice from Figure 0.33c, that although we now see more of the big picture (often referred to as the **global** behavior of the function), we have lost some of the details (such as the x-intercepts), which we could clearly see in Figures 0.33a and 0.33b (often referred to as the **local** behavior of the function).

■

Rational functions have some properties not seen in polynomials, as we see in the following several examples.

Example 3.4   **Sketching the Graph of a Rational Function**

Sketch the graph of $f(x) = \dfrac{x-1}{x-2}$ and describe the behavior of the graph near $x = 2$.

Solution   Your initial graph should look like one of Figures 0.34a or 0.34b.

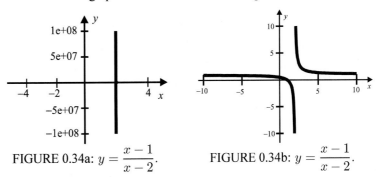

FIGURE 0.34a: $y = \dfrac{x-1}{x-2}$.   FIGURE 0.34b: $y = \dfrac{x-1}{x-2}$.

From either graph, it should be clear that something unusual is happening near $x = 2$. Zooming in closer to $x = 2$ should yield a graph like that in Figures 0.35a or 0.35b.

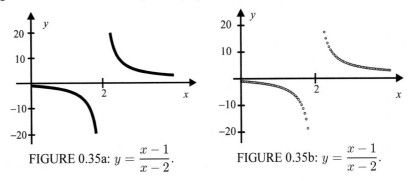

FIGURE 0.35a: $y = \dfrac{x-1}{x-2}$.   FIGURE 0.35b: $y = \dfrac{x-1}{x-2}$.

Figure 0.35a shows a *connected* (or line) plot where along with computing one point per x-value, the computer highlights extra pixels to try to smooth out the graph. The *disconnected* (or dot) plot in Figure 0.35b simply displays the computed points. In both graphs, it appears that as x increases up to 2, the function values get more and more negative, while as x decreases down to 2, the function values get more and more positive. This is confirmed in the following table of function values.

x	$f(x)$	x	$f(x)$
1.8	-4	2.2	6
1.9	-9	2.1	11
1.99	-99	2.01	101
1.999	-999	2.001	1001
1.9999	-9999	2.0001	10001

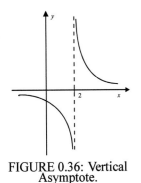

FIGURE 0.36: Vertical Asymptote.

Note that at $x = 2$, $f(x)$ is undefined. However, as x approaches 2 from the left, the graph veers down sharply. In this case, we say that $f(x)$ tends to $-\infty$. Likewise, as x approaches 2 from the right, the graph rises sharply. In this case, we say that $f(x)$ tends to ∞. In this case, we say that there is a **vertical asymptote** at $x = 2$. (We'll define this more carefully in Chapter 1.) It is common to draw a vertical dashed line at $x = 2$ to indicate this (see Figure 0.36). Since $f(2)$ is undefined, there is no point plotted at $x = 2$.   ■

Many rational functions have vertical asymptotes. Notice that there is no point plotted on the vertical asymptote since the function is undefined at such an x-value (due to the division by zero when that value of x is substituted in). Given a rational function, you can locate possible vertical asymptotes by finding where the denominator is zero. It turns out that if the numerator is not zero at that point, there is a vertical asymptote.

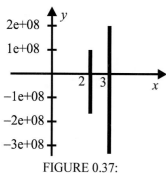

FIGURE 0.37:
$$y = \frac{x-1}{x^2 - 5x + 6}.$$

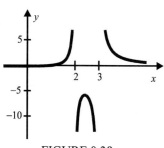

FIGURE 0.38:
$$y = \frac{x-1}{x^2 - 5x + 6}.$$

Example 3.5   A Graph with Several Vertical Asymptotes

Find all vertical asymptotes for $f(x) = \dfrac{x-1}{x^2 - 5x + 6}$.

Solution   Note that the denominator is

$$x^2 - 5x + 6 = (x-2)(x-3),$$

so that the only possible locations for vertical asymptotes are $x = 2$ and $x = 3$. Since neither x-value makes the numerator $(x-1)$ equal to zero, there are vertical asymptotes at both $x = 2$ and $x = 3$. A quick computer-generated graph gives at least some indication of how the function behaves near the asymptotes (see Figure 0.37).

We can improve the graph by zooming in in both the x- and y-directions. Figure 0.38 shows a graph of the same function using the graphing window defined by the rectangle $-1 \leq x \leq 5$ and $-13 \leq y \leq 7$. While this graph can still stand some improvement (we'll see how to use the calculus to do this in Chapter 3), the graph is adequate and clearly shows the vertical asymptotes at $x = 2$ and $x = 3$.

■

As we see in the following example, not all rational functions have vertical asymptotes.

Example 3.6   A Rational Function with No Vertical Asymptotes

Find all vertical asymptotes of $\dfrac{x-1}{x^2 + 4}$.

Solution   Solving $x^2 + 4 = 0$, you get (symbolically) $x^2 = -4$ or $x = \pm\sqrt{-4}$. There are no real solutions of this (since the square root of a negative number is not real) and so, there are no vertical asymptotes. The graph in Figure 0.39 is consistent with this observation.

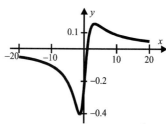

FIGURE 0.39: $y = \dfrac{x-1}{x^2 + 4}.$

■

Our final example shows why it is necessary to verify that the numerator is nonzero.

Example 3.7   A Rational Function with Only One Vertical Asymptote

Find all vertical asymptotes of $\dfrac{x^3 - 1}{x^2 - 3x + 2}$.

Solution   First, note that the denominator is

$$x^2 - 3x + 2 = (x-1)(x-2),$$

so that the only possible vertical asymptotes are located at $x = 1$ and $x = 2$. The numerator $(x^3 - 1)$ is nonzero at $x = 2$, so there is a vertical asymptote at $x = 2$. However, the numerator **is** zero at $x = 1$. As we will see in section 1.1, this means that you need to investigate the behavior near $x = 1$ more carefully. The graph shown in Figure 0.40a shows the vertical asymptote at $x = 2$ but shows no sign of an asymptote at $x = 1$. Further, the zoomed-in graph shown in Figure 0.40b gives good evidence that there is not a vertical asymptote at $x = 1$.

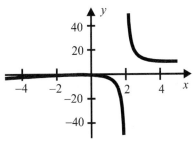

FIGURE 0.40a: $y = \dfrac{x^3 - 1}{x^2 - 3x + 2}$.

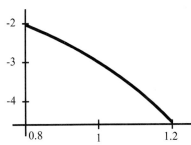

FIGURE 0.40b: $y = \dfrac{x^3 - 1}{x^2 - 3x + 2}$.

Exercises 0.3

1. (writing) Explain why there is a significant difference between Figures 0.35a and 0.35b.

2. (writing) In Figure 0.35a, the graph approaches the lower portion of the vertical asymptote from the left, whereas the graph approaches the upper portion of the vertical asymptote from the right. Use the table of function values below the figure to explain how to determine whether a graph approaches a vertical asymptote by dropping down or rising up.

3. (writing) In the text, we discussed the difference between graphing with a fixed window versus an automatic window. Discuss the advantages and disadvantages of each. (Hint: Consider the case of a first graph of a function you know nothing about and the case of hoping to see the important details of a graph for which you know the general shape).

4. (writing) Examine the graph of $y = \dfrac{x^3 + 1}{x}$ with each of the following graphing windows: (a) $-10 \leq x \leq 10$, (b) $-1000 \leq x \leq 1000$. Explain why the graph in (b) doesn't show the details that graph (a) does.

In exercises 5–44, sketch a graph of the function showing all extrema, intercepts, and asymptotes.

5. $f(x) = x^2 - 1$

6. $f(x) = 2x^2 + 1$

7. $f(x) = 3 - x^2$

8. $f(x) = 1 - x^2$

9. $f(x) = x^2 + 2x + 8$

10. $f(x) = x^2 - 3x + 6$

11. $f(x) = x^2 - 20x + 11$

12. $f(x) = x^2 + 24x - 15$

13. $f(x) = x^3 + 1$

14. $f(x) = 2x^3 - 1$

15. $f(x) = 10 - x^3$

16. $f(x) = 12 - x^3$

17. $f(x) = x^3 + 2x - 1$

18. $f(x) = x^3 - 3x + 1$

19. $f(x) = x^3 - x^2 + 1$

20. $f(x) = x^3 + 2x^2 + 1$

21. $f(x) = (x - 1)^3$

22. $f(x) = (x + 2)^3$

23. $f(x) = x^4 - 1$

24. $f(x) = 2 - x^4$

25. $f(x) = x^4 + 2x - 1$

26. $f(x) = x^4 + 6x^2 + 3$

27. $f(x) = x^5 + 2$

28. $f(x) = 12 - x^5$

29. $f(x) = x^5 + 2x^4 - x + 1$

30. $f(x) = x^5 + 5x^4 + 2x^3 + 1$

31. $f(x) = \dfrac{3}{x - 1}$

32. $f(x) = \dfrac{4}{x + 2}$

33. $f(x) = \dfrac{3x}{x - 1}$

34. $f(x) = \dfrac{4x}{x + 2}$

35. $f(x) = \dfrac{3x^2}{x - 1}$

36. $f(x) = \dfrac{4x^2}{x + 2}$

37. $f(x) = \dfrac{2}{x^2 - 4}$

38. $f(x) = \dfrac{6}{x^2 - 9}$

39. $f(x) = \dfrac{3}{x^2 + 4}$

40. $f(x) = \dfrac{6}{x^2 + 9}$

41. $f(x) = \dfrac{x + 2}{x^2 + x - 6}$

42. $f(x) = \dfrac{x - 1}{x^2 + 4x + 3}$

43. $f(x) = \dfrac{3x}{\sqrt{x^2 + 4}}$

44. $f(x) = \dfrac{2x}{\sqrt{x^2 + 1}}$

In exercises 45–54, use a graph to find all intercepts.

45. $f(x) = x^2 - 4$

46. $f(x) = x^2 - 9$

47. $f(x) = x^2 + 4$

48. $f(x) = x^2 + 9$

49. $f(x) = x^2 - 3x - 4$

50. $f(x) = x^2 + 6x + 8$

51. $f(x) = x^3 - 2x^2 - x + 2$

52. $f(x) = x^3 + 2x^2 - x - 2$

53. $f(x) = \dfrac{x+1}{x^2 - 4}$

54. $f(x) = \dfrac{x-2}{x^2 + 5x + 6}$

In exercises 55–64, find all vertical asymptotes.

55. $f(x) = \dfrac{3x}{x^2 - 4}$ **56.** $f(x) = \dfrac{x+4}{x^2 - 9}$

57. $f(x) = \dfrac{4x}{x^2 + 3x - 10}$ **58.** $f(x) = \dfrac{x+2}{x^2 - 2x - 15}$

59. $f(x) = \dfrac{4x}{x^2 + 4}$ **60.** $f(x) = \dfrac{6}{x^2 + 9}$

61. $f(x) = \dfrac{x^2 + 1}{x^3 + 3x^2 + 2x}$ **62.** $f(x) = \dfrac{3x}{x^4 - 1}$

63. $f(x) = \dfrac{3x}{\sqrt{x^2 - 9}}$ **64.** $f(x) = \dfrac{2x}{(x^3 - 1)^2}$

In exercises 65–68, a standard graphing window will not reveal all of the important details of the graph. Adjust the graphing window to find the missing details.

65. $f(x) = \frac{1}{3}x^3 - \frac{1}{400}x$

66. $f(x) = x^4 - 11x^3 + 5x - 2$

67. $f(x) = x\sqrt{144 - x^2}$

68. $f(x) = \frac{1}{5}x^5 - \frac{7}{8}x^4 + \frac{1}{3}x^3 + \frac{7}{2}x^2 - 6x$

In exercises 69–74, adjust the graphing window to identify all horizontal asymptotes.

69. $f(x) = \dfrac{3}{x - 1}$ **70.** $f(x) = \dfrac{4x}{x^2 - 1}$

71. $f(x) = \dfrac{3x^2}{x^2 - 1}$ **72.** $f(x) = \dfrac{2x}{x + 4}$

73. $f(x) = \dfrac{x^2 - 1}{\sqrt{x^4 + 1}}$ **74.** $f(x) = \dfrac{2x}{\sqrt{x^2 + 1}}$

75. Graph $y = x^2$ with the graphing window $-10 \le x \le 10$, $-10 \le y \le 10$, without drawing the x- and y-axes. Adjust the graphing window for $y = 2(x - 1)^2 + 3$ so that (without the axes showing) the graph looks identical to that of $y = x^2$.

76. Graph $y = x^2$ with the graphing window $-10 \le x \le 10$, $-10 \le y \le 10$. Separately graph $y = x^4$ with the same graphing window. Compare and contrast the graphs. Then graph the two functions on the same axes and carefully examine the differences with $-1 < x < 1$ and $x > 1$.

77. In this exercise, you will find an equation describing all points equidistant from the x-axis and the point $(0, 2)$. First, see if you can sketch a picture of what the curve ought to look like. For a point (x, y) that is on the curve, explain why $\sqrt{y^2} = \sqrt{x^2 + (y - 2)^2}$. Square both sides of this equation and solve for y. Identify the curve.

78. Find an equation describing all points equidistant from the x-axis and $(1, 4)$ (see exercise 77).

79. A falling object will approach a constant speed known as the **terminal velocity**, denoted v_T. For a sphere with weight W and diameter D, $v_T = c\sqrt{W}/D$ for some constant c. According to legend, Galileo dropped a wood ball and a lead ball of the same size from the Leaning Tower of Pisa to prove that the acceleration due to gravity does not depend on weight. (This is true, but the effect of air resistance **does** depend on weight.) If the lead ball is twice as heavy as the wood ball, how will the terminal velocities compare?

80. Fill in the blank: a ball that falls four times as fast as a wood ball is ___ times as heavy (see exercise 79).

81. (exploratory) Suppose a graphing calculator is set up with pixels corresponding to $x = 0, 0.1, 0.2, 0.3, \ldots, 2.0$ and $y = 0, 0.1, 0.2, 0.3, \ldots, 4.0$. For the function $f(x) = x^2$, compute the indicated function values and round off to give pixel coordinates [e.g., the point $(1.19, 1.4161)$ has pixel coordinates $(1.2, 1.4)$]. (a) $f(0.4)$, (b) $f(0.39)$, (c) $f(1.17)$, (d) $f(1.20)$, (e) $f(1.8)$, (f) $f(1.81)$. Repeat (c)-(d) if the graphing window is zoomed in so that $x = 1.00, 1.01, \ldots, 1.20$ and $y = 1.30, 1.31, \ldots, 1.50$. Repeat (e)-(f) if the graphing window is zoomed in so that $x = 1.800, 1.801, \ldots, 1.820$ and $y = 3.200, 3.205, \ldots, 3.300$.

82. (exploratory) Graph $y = x^2 - 1$, $y = x^2 + x - 1$, $y = x^2 + 2x - 1$, $y = x^2 - x - 1$ and $y = x^2 - 2x - 1$ and other functions of the form $y = x^2 + cx - 1$. Describe the effect(s) a change in c has on the graph.

83. (exploratory) Figures 0.31 and 0.32 provide a catalog of the possible types of graphs of cubic polynomials. In this exercise, you will compile a catalog of graphs of fourth-order polynomials (i.e., $y = ax^4 + bx^3 + cx^2 + dx + e$). Start by using your calculator or computer to sketch graphs with different values of a, b, c, d and e. Try $y = x^4$, $y = 2x^4$, $y = -2x^4$, $y = x^4 + x^3$, $y = x^4 + 2x^3$, $y = x^4 - 2x^3$, $y = x^4 + x^2$, $y = x^4 - x^2$, $y = x^4 - 2x^2$, $y = x^4 + x$, $y = x^4 - x$, and so on. Try to determine what effect each constant has.

84. (exploratory) When tracking the motion of an object such as a comet, it is often convenient to think of the object's position as a function of time. In two dimensions, the object's position would be given by functions $x(t)$ and $y(t)$. This is an example of **parametric equations.** The relatively simple polynomial graphs of this section can interact to produce complicated parametric graphs. Use a graphics calculator or computer software that graphs parametric equations to graph $x(t) = t^2 - 2$ and $y(t) = t^3 - t - 1$. Find the point (x, y) (and the corresponding values of t) where the loop begins. Try to explain in terms of the properties of $x(t)$ and $y(t)$ why it loops.

85. (exploratory) In this exercise, we look at the mathematics of random dot pictures. In the diagram, a person's eyes are located at $(e, 0)$ and $(-e, 0)$ and are focused on the point $(0, f)$. The lines of sight of the eyes intersect $y = d$ at the points $(-p, d)$ and (p, d). Show that $p = e(1 - d/f)$. The trick of the random dot pictures is to place copies of an image $2p$ units apart, analogous to the points $(\pm p, d)$: note that the pattern of dots seems to repeat as you look left to right. One way to view the image is to hold the picture to your nose and without refocusing, **slowly** pull the picture away. When the picture is the correct distance away, you will see a three-dimensional picture! Mathematically, you hold the picture d units away from your eyes while focusing your vision f units away (if the picture reflects light, focusing on your reflection accomplishes this). Your brain fuses the left-eye dot at $(-p, d)$ with the right-eye dot at (p, d) to create the illusion of a three-dimensional dot at the point $(0, f)$. Since different values of p correspond to different values of f, you perceive different depths for images which have been placed different distances apart.

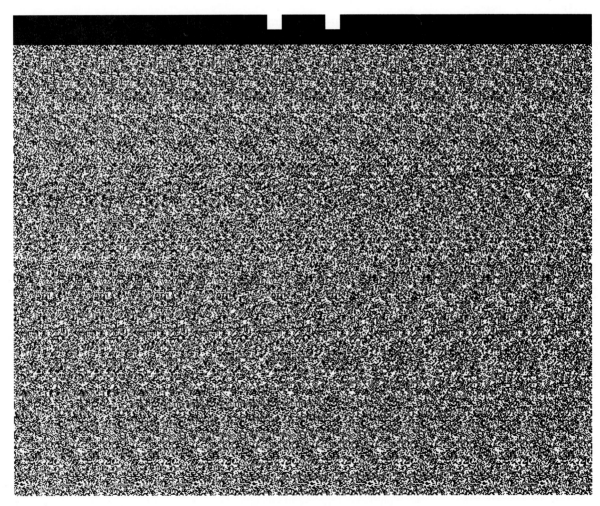

0.4   Solving Equations

There are any number of reasons why we need to solve equations. For instance, we might need to solve for the precise angle at which to launch a rocket into space, or determine the exact time at which a business should change its marketing strategy. Mathematically, we often want to find the point at which the graph of $y = f(x)$ crosses the x-axis. This is equivalent to solving the equation $f(x) = 0$. Similarly, we may want to find a point at which the graphs of $y = g(x)$ and $y = h(x)$ intersect, by solving the equation $g(x) = h(x)$. Fortunately, we don't need separate techniques to solve these two slightly different problems. Note that we can reduce the second case to the first, by setting $f(x) = g(x) - h(x) = 0$, and solving $f(x) = 0$, instead.

A **solution** of the equation $f(x) = 0$ is any number x_0 that satisfies the equation (i.e., for which $f(x_0) = 0$). In this case, x_0 is called a **zero** of the function f (since x_0 is a value which makes f zero) or a **root** of the equation $f(x) = 0$. We will briefly discuss several methods of finding zeros here.

Example 4.1   Finding Zeros by Factoring
Find the zeros of $f(x) = x^2 - 3x + 2$.

Solution   Observe that the polynomial factors and so we have

$$f(x) = x^2 - 3x + 2 = (x - 1)(x - 2) = 0.$$

You can now read-off the zeros: $x = 1$ and $x = 2$.

∎

Unfortunately, factoring is not always so easy. Of course, for quadratic equations, we have the quadratic formula available. Recall that the solution(s) of the quadratic equation

$$ax^2 + bx + c = 0$$

(for a, b and c all real numbers and $a \neq 0$) are given by the familiar **quadratic formula**:

$$x = \frac{-b \pm \sqrt{b^2 - 4ac}}{2a}.$$

Example 4.2   Finding Zeros Using the Quadratic Formula
Find the zeros of $f(x) = x^2 - 5x - 12$.

Solution   You might try to factor this one, but you probably won't have much luck. From the quadratic formula, we have

$$x = \frac{-(-5) \pm \sqrt{(-5)^2 - 4 \cdot 1 \cdot (-12)}}{2 \cdot 1} = \frac{5 \pm \sqrt{25 + 48}}{2} = \frac{5 \pm \sqrt{73}}{2}.$$

So, the two solutions are given by $x = \frac{5}{2} + \frac{\sqrt{73}}{2} \approx 6.772$ and $x = \frac{5}{2} - \frac{\sqrt{73}}{2} \approx -1.772$. (No wonder you couldn't factor the polynomial!)

∎

Finding zeros of polynomials of degree higher than 2 and other functions is usually quite a bit trickier and is sometimes impossible. At the least, you can always find an approximation of any zeros by using a graph to zoom in closer and closer to the points where the graph crosses the x-axis, as we'll illustrate shortly. A more basic question, though, is to determine **how many** zeros a given function has. In general, there is no way to answer this question without the use of calculus. For the case of polynomials, however, the following theorem (a consequence of the Fundamental Theorem of Algebra) provides a clue.

Theorem 4.1   A polynomial of degree n has **at most** n distinct zeros.

Notice that this theorem does not say how many zeros a given polynomial has, but rather, that the **maximum** number of distinct (i.e., different) zeros is the same as the degree. A polynomial may have anywhere from 0 to n distinct real zeros. (Recall that polynomials may also have complex zeros. For instance, $f(x) = x^2 + 1$ has only the complex zeros $x = \pm i$, where i is the imaginary number defined by $i = \sqrt{-1}$.) However, polynomials of odd degree must have **at least one** real zero. (Why is that?) For instance, for the case of a cubic polynomial, we have one of the three possibilities illustrated in Figures 0.41a, 0.41b, and 0.41c.

FIGURE 0.41a: One zero.

FIGURE 0.41b: Two zeros.

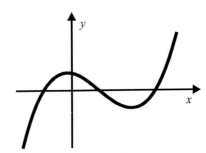

FIGURE 0.41c: Three zeros.

In these three figures, we show the graphs of cubic polynomials with 1, 2, and 3 distinct, real zeros, respectively. These are the graphs of the functions

$$f(x) = x^3 - 2x^2 + 3 = (x+1)(x^2 - 3x + 3),$$

$$g(x) = x^3 - x^2 - x + 1 = (x+1)(x-1)^2,$$

and

$$h(x) = x^3 - 3x^2 - x + 3 = (x+1)(x-1)(x-3),$$

respectively. Note that you can see from the factored form where the zeros are (and how many there are).

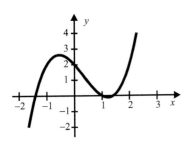

FIGURE 0.42a:
$y = x^3 - x^2 - 2x + 2$.

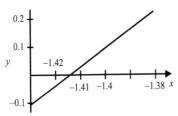

FIGURE 0.42b:
Zoomed in on zero near
$x = -1.4$.

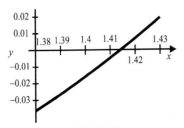

FIGURE 0.42c:
Zoomed in on zero near $x = 1.4$.

Example 4.3   Finding Zeros of a Cubic Polynomial
Find the zeros of $f(x) = x^3 - x^2 - 2x + 2$.

Solution   You might quickly recognize that one zero of this function is $x = 1$. (Just calculate $f(1)$ to verify this.) But how many other zeros are there? A graph of the function may be of some value here (see Figure 0.42a). You can see from the graph that the other two zeros of f are near $x = -1.5$ and near $x = 1.5$. You can find these zeros more precisely by using your graphing calculator or computer algebra system to zoom in on the locations of these zeros (one at a time, as in Figures 0.42b and 0.42c). From these zoomed graphs, it is clear that the two remaining zeros of f are near $x = 1.414$ and $x = -1.414$. Of course, you can make these estimates more precise by zooming in even more closely. Most graphing calculators and computer algebra systems can also find approximate zeros for you, using a built-in solve program. In Chapter 3, we present a versatile method (called Newton's method) for obtaining accurate approximations to zeros. The only way to find the exact solutions is to factor the expression (using either long division or synthetic division). Here, we have

$$f(x) = x^3 - x^2 - 2x + 2 = (x-1)\left(x^2 - 2\right),$$

from which you can see that the zeros are $x = 1$, $x = \sqrt{2}$ and $x = -\sqrt{2}$.

■

Recall that to find the points of intersection of two curves defined by $y = f(x)$ and $y = g(x)$, we set $f(x) = g(x)$ and solve for any x's satisfying the equation (the x-coordinates of the points of intersection).

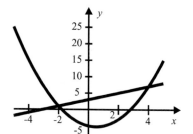

FIGURE 0.43: $y = x + 3$
and .$y = x^2 - x - 5$.

Example 4.4　　　　Finding the Intersections of a Line and a Parabola

Find the points of intersection of the parabola $y = x^2 - x - 5$ and the line $y = x + 3$.

Solution　　　First, we draw a sketch of the two curves, as seen in Figure 0.43. Notice from the graph that there are two intersections indicated, one near $x = -2$ and the other near $x = 4$. To determine these precisely, we set the two functions equal and solve for x:

$$x^2 - x - 5 = x + 3.$$

Subtracting $(x + 3)$ from both sides leaves us with

$$0 = x^2 - 2x - 8 = (x - 4)(x + 2).$$

This says that the solutions are exactly $x = -2$ and $x = 4$. From the equation of the line (or the equation of the parabola), the points of intersection are $(-2, 1)$ and $(4, 7)$. Notice that these are consistent with the intersections seen in Figure 0.43.

■

Recall that we said earlier that you can always rewrite equations in the form $f(x) = 0$. We did not say, however, that you always want to do this. It's certainly true that if you are going to solve an equation algebraically or even numerically (as we do in Chapter 3), you first want to rewrite the equation in the form $f(x) = 0$. However, if you are looking for roots of an equation graphically, this may not be your optimal choice, as we see in the following example.

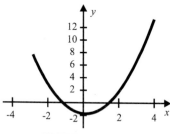

FIGURE 0.44a:
$y = x^2 - \sqrt{x + 3}$.

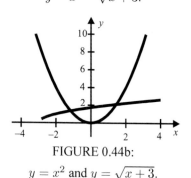

FIGURE 0.44b:
$y = x^2$ and $y = \sqrt{x + 3}$.

Example 4.5　　　　Finding Zeros Approximately

Find approximate solutions of the equation $x^2 = \sqrt{x + 3}$.

Solution　　　First, note that you can rewrite this equation as $x^2 - \sqrt{x + 3} = 0$. You can then look for zeros in the graph of $f(x) = x^2 - \sqrt{x + 3}$, seen in Figure 0.44a. Note that two zeros are clearly indicated: one near -1, the other near 1.5. However, since you know very little of the nature of the function $x^2 - \sqrt{x + 3}$, you cannot say whether or not there are any other zeros, ones that don't show up in the window seen in Figure 0.44a. On the other hand, if you graph the two functions on either side of the equation on the same set of axes, as in Figure 0.44b, you can clearly see two points where the graphs intersect (corresponding to the two zeros seen in Figure 0.44a). Further, since you know the general shapes of both of the graphs, you can infer from Figure 0.44b that there are no other intersections (i.e., there are no other zeros of f). This is important information that you cannot obtain from Figure 0.44a. Keep this useful alternative in mind when you are looking for zeros. Now that you know how many solutions there are, you need to estimate their values. As we have already illustrated, one method is to zoom in on the zeros graphically. This method is reliable but rather tedious if you want more than two or three digits of accuracy. We leave it as an exercise to verify that the zeros are approximately $x = 1.4$ and $x = -1.2$. If your calculator or computer algebra system has a solve command, you can use it to quickly obtain an accurate approximation. In this case, we get $x \approx 1.452626878$ and $x \approx -1.164035140$.

■

When using the solve command on your calculator or computer algebra system, be sure to check that the solutions make sense. If the results don't match what you've seen in your preliminary sketches and zooms, beware! Even high-tech equation solvers make mistakes occasionally.

FIGURE 0.45:
$y = 2\cos x$ and $y = 2 - x$.

Example 4.6   Finding Intersections by Calculator: An Oversight
Find all points of intersection of the graphs of $y = 2\cos x$ and $y = 2 - x$.

Solution   Notice that the intersections correspond to solutions of the equation $2\cos x = 2 - x$. Using the solve command on the TI-92 graphing calculator, we found intersections at $x \approx 3.69815$ and $x = 0$. So, what's the problem? A quick sketch of the graphs of $y = 2 - x$ and $y = 2\cos x$ (we discuss this function in the next section) clearly shows **three** intersections (see Figure 0.45).

The middle solution, $x \approx 1.10914$, was somehow passed over by our calculator's solve routine. The lesson here is to use graphical evidence to support your solutions, especially when using software and/or functions with which you are less than completely familiar.

■

We want to emphasize again that graphing should be the first step in the equation-solving process. A good graph will show you how many solutions to expect, as well as give their approximate locations. Whenever possible, you should factor or use the quadratic formula to get exact solutions. When this is impossible, approximate the solutions by zooming in on them graphically or by using your calculator's solve command. **Always** compare your results to a graph to see if there's anything you've missed.

Exercises 0.4

1. (writing) Explain in terms of graphs (see Figures 0.41a–0.41c) why a polynomial of odd degree must have at least one real zero.

2. (writing) Explain in terms of graphs why a cubic polynomial is more likely to have one zero than two zeros, but a fourth-degree polynomial is more likely to have two zeros than one zero.

3. (writing) In example 4.1 we factored the quadratic polynomial, but in example 4.2 we used the quadratic formula. Explain how you know when to factor and when to use the quadratic formula.

4. (writing) If you substitute $x = 1$ into $x^3 - 4x^2 + 2x + 1$, you get 0. Explain why this means that $x - 1$ is a factor of $x^3 - 4x^2 + 2x + 1$. Explain how to use synthetic division or long division to get the remaining (quadratic) factor.

In exercises 5–20, factor and/or use the quadratic formula to find all zeros of the given function.

5. $f(x) = x^2 - 4x + 3$ **6.** $f(x) = x^2 + x - 12$

7. $f(x) = x^2 - 2x - 15$ **8.** $f(x) = x^2 + 6x + 8$

9. $f(x) = x^2 - 4x + 2$ **10.** $f(x) = x^2 + 2x - 2$

11. $f(x) = 2x^2 + 4x - 1$ **12.** $f(x) = 2x^2 - 6x + 3$

13. $f(x) = x^3 - 3x^2 + 2x$

14. $f(x) = x^3 - 2x^2 - x + 2$

15. $f(x) = x^3 + x^2 - 4x - 4$ **16.** $f(x) = x^3 + 4x^2 + 3x$

17. $f(x) = x^4 - 1$ **18.** $f(x) = x^4 - 3x^2 + 2$

19. $f(x) = x^6 + x^3 - 2$ **20.** $f(x) = x^5 - 1$

In exercises 21–30, use a graphing calculator or computer graphing utility to estimate all zeros.

21. $f(x) = x^3 - 3x + 1$ **22.** $f(x) = x^3 - 4x^2 + 2$

23. $f(x) = x^3 - x + 1$ **24.** $f(x) = x^3 + 2x^2 - 1$

25. $f(x) = x^4 + 3x^3 - x + 1$ **26.** $f(x) = x^4 - 2x + 1$

27. $f(x) = 4x^5 + 8x - 1$ **28.** $f(x) = x^5 - 10x + 2$

29. $f(x) = x^4 - 7x^3 - 15x^2 - 10x - 1410$

30. $f(x) = x^6 - 4x^4 + 2x^3 - 8x - 2$

In exercises 31–40, determine the number of (real) solutions. Solve for the intersection points exactly if possible and estimate the points if necessary.

31. $\sqrt{x - 1} = x^2 - 1$ **32.** $\sqrt{x^2 + 4} = x^2 + 2$

33. $x^3 - 3x^2 = 1 - 3x$ **34.** $x^3 + 1 = -3x^2 - 3x$

35. $(x^2 - 1)^{2/3} = 2x + 1$ **36.** $(x + 1)^{2/3} = 2 - x$

37. $\dfrac{x^2}{x^2 + 1} = x$ **38.** $\dfrac{x^2}{x^2 + 1} = \dfrac{x}{4}$

39. $\cos x = x^2 - 1$ **40.** $\sin x = x^2 + 1$

41. Show that the quadratic formula is correct by showing that $x = \dfrac{-b + \sqrt{b^2 - 4ac}}{2a}$ and $x = \dfrac{-b - \sqrt{b^2 - 4ac}}{2a}$ are solutions of the equation $ax^2 + bx + c = 0$.

42. Find a formula for and sketch the graph of a fourth-degree polynomial with (a) no zeros; (b) one zero; (c) two zeros; (d) three zeros; (e) four zeros

43. As discussed in example 4.5, you should always compare your equation-solving results to a graph to see if you have made any mistakes. Consider the following steps in solving the equation $x - 1 = \sqrt{7 - x}$. Square both sides of the equation to get $x^2 - 2x + 1 = 7 - x$ and combine terms to get $x^2 - x - 6 = 0$. Finally, factor to get $(x - 3)(x + 2) = 0$ and conclude that the solutions are $x = 3$ and $x = -2$. Use a graph to show that $x = -2$ is not actually a solution. We say that $x = -2$ is an **extraneous solution**, produced by squaring the original equation.

44. For the equation in exercise 43, show that when you substitute $x = -2$ into both sides of the original equation, the two sides are not equal. Explain why squaring both sides of that equation makes it appear that $x = -2$ is a solution.

In exercises 45 and 46, find the true solution and an extraneous solution produced by squaring (see exercises 43 and 44).

45. $x - 2 = \sqrt{7 - 2x}$ **46.** $x + 1 = \sqrt{4x + 9}$

In exercises 47–52, use factoring and the quadratic formula to find all real and complex zeros. Sketch a graph and show that the number of real zeros corresponds to the number of x-intercepts.

47. $f(x) = x^2 + 2x + 5$ **48.** $f(x) = x^2 + x + 1$

49. $f(x) = x^2 + 2x - 5$ **50.** $f(x) = x^2 + x - 1$

51. $f(x) = x^3 - 1$ **52.** $f(x) = x^4 - 1$

In exercises 53–56, find all fixed points of the given function. (A fixed point of a function $f(x)$ is a solution of the equation $f(x) = x$. For example, to find the fixed points of $f(x) = x^2$, solve the equation $x^2 = x$. The fixed points are $x = 0$ and $x = 1$.)

53. $f(x) = x^2 - 1$ **54.** $f(x) = x^2 - 2$

55. $f(x) = x^3$ **56.** $f(x) = 4x(1 - x)$

57. Show that if $c > \frac{1}{4}$, the function $f(x) = x^2 + c$ has no real fixed points (see exercises 53 and 54). However, if $c < \frac{1}{4}$, the function $f(x) = x^2 + c$ has two fixed points. Because of the change in the number of fixed points, the value $c = \frac{1}{4}$ is called a **bifurcation point** of the function $f(x) = x^2 + c$.

58. Graph $f(x) = x^2 + \frac{1}{4}$ and $g(x) = x$ and explain why $c = \frac{1}{4}$, which we identified as a bifurcation point in exercise 57, is called a **tangent** bifurcation.

59. A baseball thrown from third base to first base follows the path of a parabola. Suppose the ball is released at the point $(0, 5)$. Based on gravity and a release angle of $17°$, an equation describing the path is $y = 5 + 0.306x - 17.5x^2/v^2$, where v is the speed (in ft/s) of the throw. Find the value of v so that the throw hits the first baseman's glove at the point $(130, 5)$. (Substitute $x = 130$ and $y = 5$ and solve for v. Convert to mph by multiplying by 3600/5280.)

60. Using the equation of exercise 59, find the speed (value of v) needed for an outfielder to hit a catcher's glove at the point $(260, 5)$. This speed is faster than any human can throw. Given this, how must the outfielder adjust the angle of the throw to get the throw to the catcher?

61. A ball tossed h feet in the air will stay in the air for $T(h) = 0.5\sqrt{h}$ seconds. For example, $T(1) = 0.5$ means that a ball tossed 1 foot in the air will come down half a second later. Solve $T(h) = 4.5$ to find how high a ball would have gone if it remained in the air for 4.5 seconds.

62. Professional jugglers sometimes claim that 10 balls is the limit for a human being to juggle. Suppose that it takes half a second to catch a ball and throw it back in the air. If you were juggling 10 balls, explain why you would want that ball to remain in the air for 4.5 seconds. Based on exercise 61, how high would you have to toss each ball? Determine how much higher you would have to toss each ball to add an 11th ball.

63. (exploratory) If you have used the built-in equation solver on a graphing calculator or computer, you may have wondered how it works. In this exercise, we introduce a simple method for finding zeros. Consider the function $f(x) = x^3 + 3x^2 - 1$. Show that $f(0) = -1$ and $f(1) = 3$. Since $f(0)$ is negative and $f(1)$ is positive, explain why you would suspect that $f(x)$ has a zero between $x = 0$ and $x = 1$. In the **method of bisections**, you next try the midpoint $x = \frac{1}{2}$. Show that $f\left(\frac{1}{2}\right)$ is negative and explain why you would now look between $x = \frac{1}{2}$ and $x = 1$ for the zero. Show that $f\left(\frac{3}{4}\right)$ is positive and explain why the zero is between $x = \frac{1}{2}$ and $x = \frac{3}{4}$. Continue in this fashion for three more steps and give your best estimate of the location of the zero. Note that while this method is tedious for you to use by hand, a computer would have no trouble in carrying this out.

64. (exploratory) In exercise 63, you explored the method of bisections. The main assumption that bisections depends on is that if $f(a)$ is negative and $f(b)$ is positive, then $f(x) = 0$ for some x between a and b. To see that this assumption is not always valid, take $f(x) = \dfrac{1}{x}$ and show that $f(-1)$ is negative and $f(1)$ is positive. Show graphically that $f(x)$ has no zeros. Briefly describe how the graph gets from the point $(-1, -1)$ to the point $(1, 1)$ without passing through the x-axis. Sketch other graphs containing the points $(-1, -1)$ and $(1, 1)$ and no x-intercepts. The property that these graphs share (we call them **discontinuous**) is precisely defined in Chapter 1.

0.5   Trigonometric Functions

Many phenomena encountered in your daily life are **waves**. For instance, music is transmitted from radio stations in the form of electromagnetic waves. Your radio receiver decodes these electromagnetic waves and causes a thin membrane inside the speakers to vibrate. This vibration, in turn, creates pressure waves in the air. When these waves reach your ears, you hear the music from your radio (see Figure 0.46).

FIGURE 0.46: Radio and sound waves.

Each of these waves is **periodic**, meaning that the basic shape of the wave is repeated over and over again. The mathematical description of such phenomena involves periodic functions, the most familiar of which are the trigonometric functions. First, we remind you of a basic definition.

Definition 5.1   A function f is **periodic** of **period** T if

$$f(x + T) = f(x)$$

for all x such that x and $x + T$ are in the domain of f.

There are several equivalent ways of defining the sine and cosine functions. Regardless of how you were originally taught these functions, we want to emphasize a simple definition from which you can easily reproduce many of the basic properties of these functions. Referring to Figure 0.47, begin by drawing the unit circle $x^2 + y^2 = 1$. Let θ be the angle measured (counterclockwise) from the positive x-axis to the point (x, y) on the circle. Here, we measure θ in **radians**, where the radian measure of the angle θ is the length of the arc indicated in the figure.

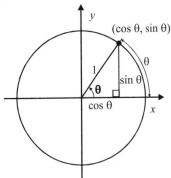

FIGURE 0.47: Definition of $\sin \theta$ and $\cos \theta$.

Again referring to Figure 0.47, we define $\sin\theta$ to be the y-coordinate of the point on the circle and $\cos\theta$ to be the x-coordinate of the point.

Remark 5.1   Unless otherwise noted, we always measure angles in radians.

Note that since the circumference of a circle ($C = 2\pi r$) of radius 1 is 2π, we have that $360°$ corresponds to 2π radians. Similarly, $180°$ corresponds to π radians, $90°$ corresponds to $\pi/2$ radians, and so on. In the table that follows, we list some common angles as measured in degrees and the corresponding radian measure.

Angle in degrees	0°	30°	45°	60°	90°	135°	180°	270°	360°
Angle in radians	0	$\dfrac{\pi}{6}$	$\dfrac{\pi}{4}$	$\dfrac{\pi}{3}$	$\dfrac{\pi}{2}$	$\dfrac{3\pi}{4}$	π	$\dfrac{3\pi}{2}$	2π

Theorem 5.1   The functions $f(\theta) = \sin\theta$ and $g(\theta) = \cos\theta$ are periodic, of period 2π.

Proof. Since a complete circle is 2π radians, adding 2π to any angle takes you all the way around the circle and back to the same point (x, y). This says that

$$\sin(\theta + 2\pi) = \sin\theta$$

and

$$\cos(\theta + 2\pi) = \cos\theta,$$

for all values of θ.

■

You are likely already familiar with the graphs of $f(x) = \sin x$ and $g(x) = \cos x$ shown in Figures 0.48a and 0.48b.

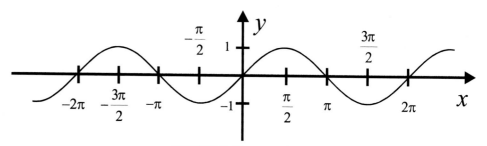

FIGURE 0.48a: $y = \sin x$.

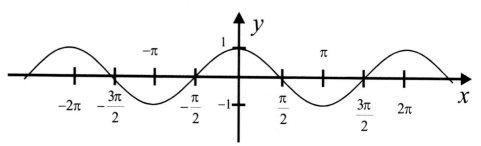

FIGURE 0.48b: $y = \cos x$.

Notice that you could slide the graph of $y = \sin x$ slightly to the left and get an exact copy of the graph of $y = \cos x$. Specifically, you might observe that

$$\sin\left(x + \frac{\pi}{2}\right) = \cos x.$$

You should also observe from Figure 0.47 that the Pythagorean Theorem gives us the familiar identity

$$\sin^2 \theta + \cos^2 \theta = 1,$$

since the hypotenuse of the indicated triangle is 1. This holds for any angle θ.

Remark 5.2   Instead of writing $(\sin \theta)^2$ or $(\cos \theta)^2$, we usually use the notation $\sin^2 \theta$ and $\cos^2 \theta$, respectively.

The accompanying table lists some common values of sine and cosine. Notice that many of these can be read directly from Figure 0.47.

We now give definitions of the remaining four trigonometric functions.

x	$\sin x$	$\cos x$
0	0	1
$\frac{\pi}{6}$	$\frac{1}{2}$	$\frac{\sqrt{3}}{2}$
$\frac{\pi}{4}$	$\frac{\sqrt{2}}{2}$	$\frac{\sqrt{2}}{2}$
$\frac{\pi}{3}$	$\frac{\sqrt{3}}{2}$	$\frac{1}{2}$
$\frac{\pi}{2}$	1	0
$\frac{2\pi}{3}$	$\frac{\sqrt{3}}{2}$	$-\frac{1}{2}$
$\frac{3\pi}{4}$	$\frac{\sqrt{2}}{2}$	$-\frac{\sqrt{2}}{2}$
$\frac{5\pi}{6}$	$\frac{1}{2}$	$-\frac{\sqrt{3}}{2}$
π	0	-1
$\frac{3\pi}{2}$	-1	0
2π	0	1

Definition 5.2   The **tangent** function is defined by $\tan x = \dfrac{\sin x}{\cos x}$.

The **cotangent** function is defined by $\cot x = \dfrac{\cos x}{\sin x}$.

The **secant** function is defined by $\sec x = \dfrac{1}{\cos x}$.

The **cosecant** function is defined by $\csc x = \dfrac{1}{\sin x}$.

We give graphs of these functions in Figures 0.49a, 0.49b, 0.49c and 0.49d.

FIGURE 0.49a: $y = \tan x$.

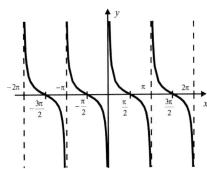

FIGURE 0.49b: $y = \cot x$.

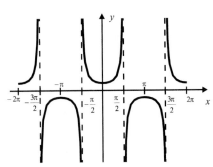

FIGURE 0.49c: $y = \sec x$.

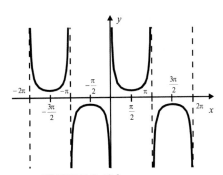

FIGURE 0.49d: $y = \csc x$.

Notice in each graph the locations of the vertical asymptotes. For the "co" functions $\cot x$ and $\csc x$, the division by $\sin x$ causes vertical asymptotes at 0, π, 2π, and so on (where

$\sin x = 0$). For $\tan x$ and $\sec x$, the division by $\cos x$ produces vertical asymptotes at $\pi/2$, $3\pi/2$, $5\pi/2$, and so on (where $\cos x = 0$). Once you have the vertical asymptotes in place, the graphs are relatively easy to draw.

Notice that $\tan x$ and $\cot x$ are periodic, of period π, while $\sec x$ and $\csc x$ are periodic, of period 2π.

Remark 5.3   Most calculators have keys for the functions $\sin x$, $\cos x$, and $\tan x$, but not for the other three trigonometric functions. This reflects the central role that $\sin x$, $\cos x$, and $\tan x$ play in applications. To calculate function values for the other three trigonometric functions, you can simply use the identities

$$\cot x = \frac{1}{\tan x}, \quad \sec x = \frac{1}{\cos x}, \quad \text{and } \csc x = \frac{1}{\sin x}.$$

It is important to learn the effect of slight modifications of these functions. We discuss transformations further in section 0.7.

FIGURE 0.50a: $y = \sin x$.

Example 5.1   **Altering Amplitude and Period**

Graph $y = 2\sin x$ and $y = \sin 2x$ and describe how each differs from the graph of $y = \sin x$ (see Figure 0.50a).

Solution   The graph of $y = 2\sin x$ is given in Figure 0.50b. Notice that this graph is similar to the graph of $y = \sin x$, except that the y-values oscillate between -2 and 2 instead of -1 and 1. Next, the graph of $y = \sin 2x$ is given in Figure 0.50c. In this case, the graph is similar to the graph of $y = \sin x$ except that the period is π instead of 2π. (The oscillations occur twice as fast).

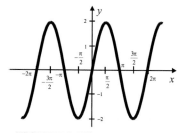

FIGURE 0.50b: $y = 2\sin x$.

The results in example 5.1 can be generalized. For $A > 0$, the graph of $y = A\sin x$ oscillates between $y = -A$ and $y = A$. In this case, we call A the **amplitude**. Notice that for any positive constant c, the period of $y = \sin cx$ is $2\pi/c$. Similarly, for the function $A\cos cx$, the amplitude is A and the period is $2\pi/c$.

The sine and cosine functions can be used to model sound waves. A pure tone (think of a single flute note) is a pressure wave described by the sinusoidal function $A\sin ct$. (Here, we are using the variable t, since the air pressure is a function of **time**.) The amplitude A determines how loud the tone is perceived to be, and the period determines the pitch of the note. In this setting, it is usually convenient to talk about the **frequency** $f = c/2\pi$ rather than the period. As you may know, the higher the frequency is, the higher the pitch of the note will be. (Frequency is measured in hertz, where 1 hertz equals 1 cycle per second.) Note that the frequency is simply the reciprocal of the period.

FIGURE 0.50c: $y = \sin(2x)$.

Example 5.2   **Finding Amplitude, Period, and Frequency**

Find the amplitude, period, and frequency of (i) $f(x) = 4\cos 3x$ and (ii) $g(x) = 2\sin(x/3)$.

Solution   (i) For $f(x)$, the amplitude is 4, the period is $2\pi/3$, and the frequency is $3/(2\pi)$. (ii) For $g(x)$, the amplitude is 2, the period is $2\pi/(1/3) = 6\pi$, and the frequency is $1/(6\pi)$.

As you will recall from your earlier study of trigonometry, there are numerous formulas, or **identities**, that are helpful in manipulating the trigonometric functions. We have already reminded you of the one absolutely essential identity: namely, that

$$\sin^2\theta + \cos^2\theta = 1$$

for any angle θ.

We list several important identities in the following theorem.

Theorem 5.2   For any real numbers α and β, the following identities hold:

$$\sin(\alpha + \beta) = \sin\alpha\cos\beta + \sin\beta\cos\alpha \tag{5.1}$$
$$\cos(\alpha + \beta) = \cos\alpha\cos\beta - \sin\alpha\sin\beta \tag{5.2}$$
$$\sin^2\alpha = \tfrac{1}{2}(1 - \cos 2\alpha) \tag{5.3}$$
$$\cos^2\alpha = \tfrac{1}{2}(1 + \cos 2\alpha). \tag{5.4}$$

From the basic identities summarized in Theorem 5.2, numerous other useful identities can be derived. We derive several of these in the following example.

Example 5.3   **Deriving New Trigonometric Identities**
Derive the identities $\sin 2\theta = 2\sin\theta\cos\theta$ and $\cos 2\theta = \cos^2\theta - \sin^2\theta$.

Solution   These can be obtained from formulas (5.1) and (5.2), respectively, by substituting $\alpha = \theta$ and $\beta = \theta$. Alternatively, the identity for $\cos 2\theta$ can be obtained by subtracting equation (5.3) from equation (5.4).   ∎

Two kinds of combinations of sine and cosine functions are especially important in applications. In the first type, a sine and cosine with the same period but different amplitudes are added.

FIGURE 0.51:
$y = 3\cos x + 4\sin x$.

Example 5.4   **Combinations of Sines and Cosines**
Graph $f(x) = 3\cos x + 4\sin x$ and describe the resulting graph.

Solution   You should get something like the graph in Figure 0.51. Notice that the graph looks very much like a sine curve with period 2π and amplitude 5, but it has been moved about 0.75 units to the left. Alternatively, you could say that it looks like a cosine curve, moved about 1 unit to the right. Using the appropriate identity, you can verify these guesses.   ∎

Example 5.5   **Writing Combinations of Sines and Cosines as a Single Sine**
Prove that $3\cos x + 4\sin x = 5\sin(x + \beta)$ for some constant β, and estimate the value of β.

Solution   Using equation (5.1), we have

$$4\sin x + 3\cos x = 5\sin(x + \beta)$$
$$= 5\sin x\cos\beta + 5\sin\beta\cos x,$$

if

$$5\cos\beta = 4 \quad\text{and}\quad 5\sin\beta = 3.$$

This will occur if we can choose a value of β so that $\cos\beta = 4/5$ **and** $\sin\beta = 3/5$. By the Pythagorean Theorem, this is possible only if $\sin^2\beta + \cos^2\beta = 1$. In this case, we have

$$\sin^2\beta + \cos^2\beta = (3/5)^2 + (4/5)^2$$
$$= 9/25 + 16/25 = 25/25 = 1.$$

For the moment, the only way to estimate β is by trial and error. Using your calculator or computer, you should find that one solution is $\beta \approx 0.6435$ radians (about 36.9 degrees).

∎

The second kind of combination of sines and cosines we explore is adding functions of different periods. (This is one of the principles behind music synthesizers.)

Example 5.6   Combinations of Sines and Cosines of Different Periods

Graph (i) $f(x) = \cos 3x + \sin 4x$ and (ii) $g(x) = \cos \pi x + \sin 4x$ and describe each graph. Determine the period if the function is periodic.

Solution   (i) We give a graph of $y = f(x)$ in Figure 0.52. This is certainly more complicated than the usual sine or cosine graph, but you should be able to identify a repetition with a period of about 6. (Notice that this is close to 2π.) To determine the actual period, note that the period of $\cos 3x$ is $2\pi/3$ and the period of $\sin 4x$ is $2\pi/4$. This says that $\cos 3x$ repeats at $2\pi/3$, $4\pi/3$, $6\pi/3$, and so on. Similarly, $\sin 4x$ repeats at $2\pi/4$, $4\pi/4$, $6\pi/4$, $8\pi/4$, and so on. Note that both $6\pi/3$ and $8\pi/4$ equal 2π. Since both terms repeat every 2π units, the function $f(x)$ has a period of 2π.

(ii) The graph of $g(x)$ is even more complicated, as you can see in Figure 0.53. In the graphing window shown in Figure 0.53 ($-10 \leq x \leq 10$ and $-2 \leq y \leq 2$), this does not appear to be a periodic function, although it's not completely different from the graph of $f(x)$. To try to find a period, you should note that $\cos \pi x$ has a period of $2\pi/\pi = 2$, and so repeats at intervals of width 2, 4, 6 and so on. On the other hand, $\sin 4x$ repeats at intervals of width $2\pi/4$, $4\pi/4$ and so on. The function is periodic if and only if there are numbers common to both lists. Since 2, 4, 6, ..., are all rational numbers and the numbers $2\pi/4$, $4\pi/4, \ldots$, are all irrational, there can't be any numbers in both lists. We conclude that the function is not periodic.

∎

In many applications, we need to calculate the length of one side of a triangle using the length of another side and an **acute** angle (i.e., an angle between 0 and $\frac{\pi}{2}$ radians). We can do this rather easily, as in the following example.

FIGURE 0.52:
$y = \cos 3x + \sin 4x$.

FIGURE 0.53:
$y = \cos \pi x + \sin 4x$.

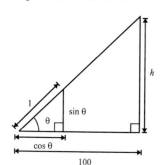

FIGURE 0.54: Height of tower.

Example 5.7   Finding the Height of a Tower

A person 100 feet from the base of a water tower measures an angle of 60° from the ground to the top of the tower (see Figure 0.54). Find the height of the tower.

Solution   First, we convert 60° to radians:

$$60° = 60\frac{\pi}{180} = \frac{\pi}{3} \text{ radians.}$$

We are given that the base of the triangle in Figure 0.54 is 100 feet. We must now compute the height of the tower h. Using the similar triangles indicated in Figure 0.54, we have that

$$\frac{\sin \theta}{\cos \theta} = \frac{h}{100},$$

so that the height of the tower is

$$h = 100\frac{\sin \theta}{\cos \theta} = 100 \tan \theta = 100 \tan \frac{\pi}{3} = 100\sqrt{3} \approx 173 \text{ feet.}$$

∎

Exercises 0.5

1. (writing) A student graphs $f(x) = \cos x$ on a graphing calculator and gets what appears to be a straight line at height $y = 1$ instead of the usual cosine curve. Upon investigation, you discover that the calculator has graphing window $-10 \le x \le 10$, $-10 \le y \le 10$ and is in degrees mode. Explain what went wrong and how to correct it.

2. (writing) Many students are comfortable using degrees to measure angles and don't understand why they must learn radian measures. As discussed in the text, radians directly measure distance along the unit circle. Distance is an important aspect of many applications. In addition, we will see later that many calculus formulas are simpler in radians form than in degrees. Aside from familiarity, discuss any and all advantages of degrees over radians. On balance, which is better?

3. (writing) In example 5.6, we saw that the sum of a certain pair of periodic functions turned out to be a periodic function, while the sum of a different pair of periodic functions was **not** periodic. Explain why the change in coefficients from 3 to π made such a big difference.

4. (writing) The trigonometric functions can be defined in terms of the unit circle (as done in the text) or in terms of right triangles for angles between 0 and $\frac{\pi}{2}$ radians. In calculus and most scientific applications, the trigonometric functions are used to model periodic phenomena (quantities that repeat). Given that we want to emphasize the periodic nature of the functions, explain why we would prefer the circular definitions to the triangular definitions.

In exercises 5 and 6, convert the given radians measure to degrees.

5. (a) $\pi/4$ (b) $\pi/3$ (c) $\pi/6$ (d) $4\pi/3$

6. (a) $3\pi/5$ (b) $\pi/7$ (c) 2 (d) 3

In exercises 7 and 8, convert the given degrees measure to radians.

7. (a) $180°$ (b) $270°$ (c) $120°$ (d) $30°$

8. (a) $40°$ (b) $80°$ (c) $450°$ (d) $390°$

In exercises 9–30, sketch a graph of the function.

9. $f(x) = \sin 2x$

10. $f(x) = \cos 3x$

11. $f(x) = \tan 2x$

12. $f(x) = \sec 3x$

13. $f(x) = 3\cos(x - \pi/2)$

14. $f(x) = 4\cos(x + \pi)$

15. $f(x) = x \sin x$

16. $f(x) = \sin x^2$

17. $f(x) = \sin 2x - 2\cos 2x$

18. $f(x) = \cos 3x - \sin 3x$

19. $f(x) = \cos 5x + 3\sin 5x$

20. $f(x) = 4\cos x - \sin x$

21. $f(x) = \sin x - 4\cos x$

22. $f(x) = 4\cos 3x + 3\sin 3x$

23. $f(x) = \sin x + \cos 2x$

24. $f(x) = \cos x + \frac{1}{2}\cos 2x$

25. $f(x) = \cos x - \frac{1}{4}\cos 8x$

26. $f(x) = \sin 2x + \sin 16x$

27. $f(x) = \sin x \sin 12x$

28. $f(x) = \sin x \cos 12x$

29. $f(x) = \cos 2x + \sin \pi x$

30. $f(x) = 2\sin 3x - \cos \pi x$

In exercises 31–38, identify the amplitude, period, and frequency.

31. $f(x) = 3\sin 2x$

32. $f(x) = 2\cos 3x$

33. $f(x) = 5\cos 3x$

34. $f(x) = 3\sin 5x$

35. $f(x) = 3\cos(2x - \pi/2)$

36. $f(x) = 4\sin(3x + \pi)$

37. $f(x) = -4\sin x$

38. $f(x) = -2\cos 3x$

In exercises 39–42, prove that the given trigonometric identity is true.

39. $\sin(\alpha - \beta) = \sin\alpha \cos\beta - \sin\beta \cos\alpha$

40. $\cos(\alpha - \beta) = \cos\alpha \cos\beta + \sin\alpha \sin\beta$

41. $\sec^2\theta = \tan^2\theta + 1$

42. $\csc^2\theta = \cot^2\theta + 1$

43. Prove that, for some constant β,

$$4\cos x - 3\sin x = 5\cos(x + \beta).$$

Then, estimate the value of β.

44. Prove that, for some constant β,

$$2\sin x + \cos x = \sqrt{5}\sin(x + \beta).$$

Then, estimate the value of β.

In exercises 45–50, determine whether or not the function is periodic. If it is periodic, find the smallest (fundamental) period.

45. $f(x) = \cos 2x + 3\sin \pi x$

46. $f(x) = \sin x - \cos \sqrt{2}x$

47. $f(x) = \sin 6x - \cos \pi x$

48. $f(x) = \cos 3x + 3\sin \sqrt{3}x$

49. $f(x) = \sin 2x - \cos 5x$

50. $f(x) = \cos 3x - \sin 7x$

51. A person sitting 2 miles from a rocket launch site measures $20°$ up to the current location of the rocket. How high up is the rocket?

52. A person who is 6 feet tall stands 4 feet from the base of a light pole and casts a 2-foot long shadow. How tall is the light pole?

53. A surveyor stands 80 feet from the base of a building and measures an angle of $50°$ to the top of the steeple on top of the building. The surveyor figures that the center of the steeple lies 20 feet inside the front of the structure. Find the height of the steeple.

54. Suppose that the surveyor of exercise 53 estimates that the center of the steeple lies between $20'$ and $21'$ inside the front of the structure. Determine how much the extra foot would change the calculation of the height of the building.

In exercises 55–60, use the range for θ to determine the indicated function value.

55. $\sin\theta = \frac{\sqrt{3}}{2}, 0 \le \theta \le \frac{\pi}{2}$, find $\cos\theta$.

56. $\cos\theta = \frac{\sqrt{2}}{2}, 0 \le \theta \le \frac{\pi}{2}$, find $\sin\theta$.

57. $\sin\theta = \frac{1}{3}, 0 \le \theta \le \frac{\pi}{2}$, find $\cos\theta$.

58. $\cos\theta = \frac{4}{5}, 0 \le \theta \le \frac{\pi}{2}$, find $\sin\theta$.

59. $\sin\theta = \frac{1}{2}, \frac{\pi}{2} \le \theta \le \pi$, find $\cos\theta$.

60. $\sin\theta = \frac{1}{2}, \frac{\pi}{2} \le \theta \le \pi$, find $\tan\theta$.

In exercises 61–68, use a graphing calculator or computer to determine the number of solutions of each equation, and numerically estimate the solutions (x is in radians).

61. $2\cos x = 2 - x$

62. $3\sin x = x$

63. $\cos x = x^2 - 2$

64. $\sin x = x^2$

65. $\sin x - x^3 + 1 = 0$

66. $x^4 - \cos x = 0$

67. $x^3 - x\sin x = 0$

68. $x^2 - x\cos x = 0$

69. In an AC circuit, the voltage is given by $v(t) = v_p \sin 2\pi ft$, where v_p is the peak voltage and f is the frequency in Hz. A voltmeter actually measures an average (called the **root-mean-square**) voltage, equal to $v_p/\sqrt{2}$. If the voltage has amplitude 170 and period $\pi/30$, find the frequency and meter voltage.

70. An old-fashioned LP record player rotated records at $33\frac{1}{3}$ rpm (revolutions per minute). What is the period (in minutes) of the rotation? What is the period for a 45 rpm single?

71. Suppose that the ticket sales of an airline (in thousands of dollars) is given by $s(t) = 110 + 2t + 15\sin(\frac{1}{6}\pi t)$, where t is measured in months. What real-world phenomenon might cause the fluctuation in ticket sales modeled by the sine term? Based on your answer, what month corresponds to $t = 0$?

Disregarding seasonal fluctuations, by what amount is the airline's sales increasing annually?

72. Piano tuners sometimes start by striking a tuning fork and then the corresponding piano key. If the tuning fork and piano note each have frequency 8, then the resulting sound is $\sin 8t + \sin 8t$. Graph this. If the piano is slightly out-of-tune at frequency 8.1, the resulting sound is $\sin 8t + \sin 8.1t$. Graph this and explain how the piano tuner can hear the small difference in frequency.

73. Many graphing calculators and computers will "graph" inequalities by shading in all points (x, y) for which the inequality is true. If you have access to this capability, graph the inequality $\sin x < \cos y$.

74. Calculator and computer graphics can be inaccurate. Using an initial graphing window of $-1 \le x \le 1$ and $-1 \le y \le 1$, graph $f(x) = \dfrac{\tan x - x}{x^3}$. Describe the behavior of the graph near $x = 0$. Zoom in closer and closer to $x = 0$, using a window with $-0.001 \le x \le 0.001$, then $-0.0001 \le x \le 0.0001$, then $-0.00001 \le x \le 0.00001$ and so on, until the behavior near $x = 0$ appears to be different. We don't want to leave you hanging: the initial graph gives you good information and the tightly zoomed graphs are inaccurate due to the computer's inability to compute $\tan x$ exactly.

75. (exploratory) Compare and contrast the graphs of $\sin \pi x + \sin 2\pi x$ and $\sin 3x + \sin 2\pi x$. Based on your results (don't draw any more graphs!) predict what the graph of $\sin \pi x + \sin 2\pi x$ will look like on the interval $[100, 102]$. Can you accurately predict the shape of $\sin 3x + \sin 2\pi x$ on the same interval?

76. (exploratory) As seen in exercise 74, computer graphics can be misleading. This exercise works best using a "disconnected" graph (individual dots, not connected). Graph $y = \sin x^2$ using a graphing window for which each pixel represents a step of 0.1 in the x- or y-direction. You should get the impression of a sine wave that oscillates more and more rapidly as you move to the left and right. Next, change the graphing window so that the middle of the original screen (probably $x = 0$) is at the far left of the new screen. You will probably see what appears to be a random jumble of dots. Continue to change the graphing window by increasing the x-values. Describe the patterns or lack of patterns that you see. You should find one pattern that looks like two rows of dots across the top and bottom of the screen; another pattern looks like the original sine wave. For each pattern that you find, pick adjacent points with x-coordinates a and b. Then change the graphing window so that $a \le x \le b$ and find the portion of the graph that is missing. Remember that, whether the points are connected or not, computer graphs always leave out part of the graph; it is part of your job to know whether the missing part is important or not.

77. (exploratory) This exercise deals with parametric equations, which most graphing calculators will graph. Try graphing $x(t) = \cos t$ and $y(t) = \sin t$. The graph will probably look like an ellipse. Use the definition of sine and cosine to explain why the graph is actually a circle. Adjust your graphing window so that the graph looks like a circle. (Hint: Adjust the window so that distances of 1 in the x- and y-directions are equivalent; that is, make the **aspect ratio** equal to 1.)

78. (exploratory) In his book and video series *The Ring of Truth*, physicist Philip Morrison performed an experiment to estimate the circumference of the earth. In Nebraska, he measured the angle to a bright star in the sky, drove 370 miles due south into Kansas, and measured the new angle to the star. Some geometry shows that the difference in angles, about $5.02°$, equals the angle from the center of the earth to the two locations in Nebraska and Kansas. If the earth is perfectly circular (it's not) and the circumference of the portion of the circle measured out by $5.02°$ is 370 miles, estimate the circumference of the earth. This experiment was based on a similar experiment by the ancient Greek scientist Eratosthenes. The ancient Greeks and the Spaniards of Columbus's day knew that the earth was round, they just disagreed about the circumference. Columbus argued for a figure about half of the actual value, since a ship couldn't survive on the water long enough to navigate the true distance.

0.6 Exponential and Logarithmic Functions

Some bacteria reproduce very quickly, as you may have discovered if you have ever had an infected cut or strep throat. Under the right circumstances, the number of bacteria in certain cultures will double in as little as an hour (or even less). In this section, we discuss some functions that can be used to model such rapid growth.

For instance, suppose that initially there are 100 bacteria at a given site, and the population doubles every hour. Call the population function $P(t)$, where t represents time (in hours) and start the clock running at time $t = 0$. The initial population of 100 means that $P(0) = 100$. After 1 hour, the population will double to 200, so that $P(1) = 200$. After another hour, the population will have doubled again to 400, making $P(2) = 400$. After a third hour, the population will have doubled again to 800, making $P(3) = 800$ and so on.

Let's compute the population after 10 hours. (Try guessing this now. Most people guess poorly on this type of problem.) You could calculate the population at 4 hours, 5 hours, and so on, or you could use the following shortcut. To find $P(1)$, you double the initial population, so that $P(1) = 2 \cdot 100$. To find $P(2)$, you double the population at time $t = 1$, so that $P(2) = 2 \cdot 2 \cdot 100 = 2^2 \cdot 100$. Similarly, $P(3) = 2^3 \cdot 100$. This pattern leads us to

$$P(10) = 2^{10} \cdot 100 = 102,400.$$

Depending on the organism, this is now a population that could cause some trouble!

The pattern discovered above suggests that the population can be modeled by the function

$$P(t) = 2^t \cdot 100.$$

We call $P(t)$ an **exponential** function because the variable t is in the exponent. There is a subtle question here: what is the domain of this function? We have so far used only integer values of t, but for what other values of t does $P(t)$ make sense? Certainly, rational powers make sense, as in $P(1/2) = 2^{1/2} \cdot 100$, where you recognize that $2^{1/2} = \sqrt{2}$. This says that the number of bacteria in the culture after a half hour is

$$P(1/2) = 2^{1/2} \cdot 100 = \sqrt{2} \cdot 100 \approx 141.$$

In this context, the function is not intended to give the exact number of bacteria at a given time, but rather, to predict an approximate number.

In any case, it's a simple matter to interpret fractional powers as roots. For instance, $x^{1/2} = \sqrt{x}$, $x^{1/3} = \sqrt[3]{x}$, $x^{1/4} = \sqrt[4]{x}$, and so on. Further, we can write

$$x^{2/3} = \sqrt[3]{x^2} = \left(\sqrt[3]{x}\right)^2,$$

$$x^{7/4} = \sqrt[4]{x^7} = \left(\sqrt[4]{x}\right)^7,$$

$$x^{3.1} = x^{31/10} = \sqrt[10]{x^{31}}$$

and so on. But, what about irrational powers? These are somewhat harder to define, but they work exactly the way you would want them to. For instance, since π is between 3.14 and 3.15, 2^π is between $2^{3.14}$ and $2^{3.15}$. This is the logic behind the definition of irrational powers. (We completely resolve this question in Chapter 6.)

If for some reason, you wanted to find the population after π hours, you can use your calculator or computer to obtain the approximate population:

$$P(\pi) = 2^\pi \cdot 100 \approx 882.$$

We rounded this answer to the nearest integer since the population describes the number of organisms present.

For your convenience, we summarize the usual rules of exponents below.

- For any integers m and n,

$$x^{m/n} = \sqrt[n]{x^m} = (\sqrt[n]{x})^m.$$

- For any real number p,

$$x^{-p} = \frac{1}{x^p}.$$

- For any real numbers p and q,

$$(x^p)^q = x^{p \cdot q}.$$

- For any real numbers p and q,

$$x^p \cdot x^q = x^{p+q}.$$

Throughout your calculus course, you will need to be able to quickly convert back and forth between exponential form and fractional or root form.

Example 6.1   Converting Expressions to Exponential Form

Convert each to exponential form: (i) $3\sqrt{x^5}$, (ii) $\dfrac{5}{\sqrt[3]{x}}$, (iii) $\dfrac{3x^2}{2\sqrt{x}}$, and (iv) $\left(2^x \cdot 2^{3+x}\right)^2$.

Solution   For (i), simply leave the 3 alone and convert the power:

$$3\sqrt{x^5} = 3x^{5/2}.$$

For (ii), use a negative exponent to write x in the numerator:

$$\frac{5}{\sqrt[3]{x}} = 5x^{-1/3}.$$

For (iii), first separate the constants from the variables and then simplify:

$$\frac{3x^2}{2\sqrt{x}} = \frac{3}{2} \cdot \frac{x^2}{x^{1/2}} = \frac{3}{2}x^{2-1/2} = \frac{3}{2}x^{3/2}.$$

For (iv), first work inside the parentheses and then square:

$$\left(2^x \cdot 2^{3+x}\right)^2 = \left(2^{x+3+x}\right)^2 = \left(2^{2x+3}\right)^2 = 2^{4x+6}.$$

■

The function in part (iv) of example 6.1 is called an exponential function with a **base** of 2.

Definition 6.1   An **exponential function** is any function of the form $f(x) = b^x$, where b is called the **base** and x is the **exponent**.

Be careful to distinguish between algebraic functions like $f(x) = x^3$ and $g(x) = x^{2/3}$ and exponential functions. Confusing these types of functions is a very common error. Notice that for exponential functions like $h(x) = 2^x$, the variable is in the exponent (hence the name), instead of in the base.

While any positive real number can be used as a base for an exponential function, for practical purposes, three bases are the most common. Base 2 arises naturally when analyzing processes that double at a regular rate (such as the bacteria at the beginning of this section). Also, most computers perform their calculations using base 2 arithmetic. Our standard counting system is base 10, so this base is commonly used. However, far and away the most useful base is the irrational number e. Like π, the number e has a surprising tendency to occur in important calculations. We define e by the following:

$$e = \lim_{n \to \infty} (1 + 1/n)^n. \tag{6.1}$$

Note that equation (6.1) has at least two serious shortcomings. First, we have not yet said what the notation $\lim_{n \to \infty}$ means. (In fact, we won't define this until Chapter 1.) Second, it's unclear why anyone would ever define a number in such a strange way. We will not be in a position to answer the second question until Chapter 6 (but the answer is worth waiting for and it will tie together some loose ends that have been hanging around for a while).

It suffices for the moment to say that equation (6.1) means that e can be approximated by calculating values of $(1 + 1/n)^n$ for large values of n and that the larger the value of n, the closer the approximation will be to the actual value of e. In particular, if you look at the sequence of numbers $(1 + 1/2)^2$, $(1 + 1/3)^3$, $(1 + 1/4)^4$ and so on, they will get progressively closer and closer to (i.e., home-in on) the irrational number e (named in honor of the famous mathematician Leonhard Euler).

To get an idea of the value of e, we compute several of these numbers:

$$(1 + 1/2)^2 = 2.25,$$
$$(1 + 1/3)^3 = 2.37037...,$$
$$(1 + 1/10)^{10} = 2.5937...,$$
$$(1 + 1/100)^{100} = 2.7048...,$$
$$(1 + 1/1000)^{1000} = 2.7169...,$$
$$(1 + 1/10,000)^{10,000} = 2.7181...$$

and so on. You should compute enough of these values to convince yourself that the first few digits of the decimal representation of e ($e \approx 2.718281828...$) are correct.

Example 6.2   **Computing Values of Exponentials**
Approximate e^4, $e^{-1/5}$, and e^0.

Solution   From a calculator, we find that

$$e^4 = e \cdot e \cdot e \cdot e \approx 54.598.$$

From the usual rules of exponents,

$$e^{-1/5} = \frac{1}{e^{1/5}} = \frac{1}{\sqrt[5]{e}} \approx 0.81873.$$

(On a calculator, it is convenient to replace $-1/5$ with -0.2.) Finally, $e^0 = 1$.

■

The graphs of the exponential functions summarize many of their important properties.

Example 6.3   Sketching Graphs of Exponentials
Sketch the graphs of the exponential functions $y = 2^x$, $y = e^x$, $y = e^{2x}$, $y = e^{x/2}$, $y = (1/2)^x$, and $y = e^{-x}$.

Solution   Using a calculator or computer, you should get graphs similar to those that follow.

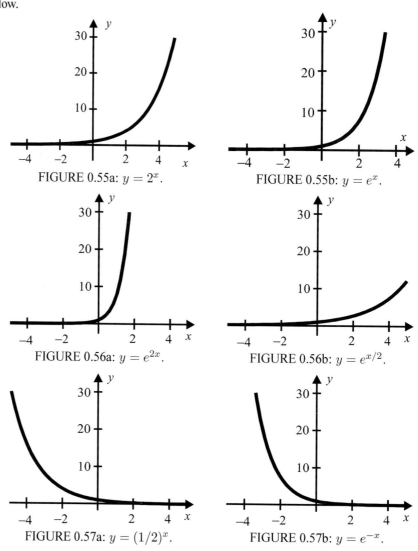

FIGURE 0.55a: $y = 2^x$.   FIGURE 0.55b: $y = e^x$.

FIGURE 0.56a: $y = e^{2x}$.   FIGURE 0.56b: $y = e^{x/2}$.

FIGURE 0.57a: $y = (1/2)^x$.   FIGURE 0.57b: $y = e^{-x}$.

Notice that each of the graphs in Figures 0.55a, 0.55b, 0.56a, and 0.56b start very near the x-axis (reading left to right), pass through the point $(0, 1)$, and then rise steeply. This is true for all exponentials with base greater than 1 and with a positive coefficient in the exponent. Note that the larger the base ($e > 2$) or the larger the coefficient in the exponent ($4 > 1 > 1/2$), the more quickly the graph rises to the right (and drops to the left). Note that the graphs in Figures 0.57a and 0.57b are the mirror images in the y-axis of

Figures 0.55a and 0.55b, respectively. The graphs rise as you move to the left and drop to the x-axis as you move to the right. It's worth noting that by the rules of exponents, $(1/2)^x = 2^{-x}$ and $(1/e)^x = e^{-x}$.

■

The exponential functions are closely related to another family of functions called the **logarithmic** functions.

Definition 6.2   For any positive number $b \neq 1$, the common **logarithm** of base b, written $\log_b x$, is defined by

$$y = \log_b x \text{ if and only if } x = b^y.$$

That is, the logarithm $\log_b x$ is the exponent to which you must raise the base b to get the given number x. For example,

$$\log_{10} 10 = 1 \text{ (since } 10^1 = 10\text{)},$$

$$\log_{10} 100 = 2 \text{ (since } 10^2 = 100\text{)},$$

$$\log_{10} 1000 = 3 \text{ (since } 10^3 = 1000\text{)}$$

and so on. The value of $\log_{10} 45$ is less clear than the preceding three values, but the idea is the same: you need to find the number y such that $10^y = 45$. The answer is between 1 and 2 (why?), but to be more precise, you will need to employ trial and error. (Of course, you can always use your calculator or computer to compute an approximate value, but you won't learn anything by doing that.) You should get $\log_{10} 45 \approx 1.6532$. Later in the course, we introduce you to a powerful method for accurately approximating the values of such functions.

As with exponential functions, the most useful bases turn out to be 2, 10, and e. We usually abbreviate $\log_{10} x$ by $\log x$. Similarly, $\log_e x$ is usually abbreviated $\ln x$ (for **natural logarithm**). It is most likely unclear what is so natural (or even interesting) about a logarithm with an irrational base (after all, why wouldn't $\log_{10} x$ be the *natural* logarithm?), but we clear this up in Chapter 6.

Example 6.4   Evaluating Logarithms
   Without using your calculator, determine $\log(1/10)$, $\log(0.001)$, $\ln e$, and $\ln e^3$.

Solution   Since $1/10 = 10^{-1}$, $\log(1/10) = -1$. Similarly, $0.001 = 10^{-3}$ implies that $\log(0.001) = -3$. Since $\ln e = \log_e e^1$, $\ln e = 1$. Similarly, $\ln e^3 = 3$.

■

Observe that one consequence of Definition 6.2 is that for any $x > 0$,

$$e^{\ln x} = x, \tag{6.2}$$

as follows. Let

$$y = \ln x = \log_e x.$$

By Definition 6.2, we have that

$$x = e^y = e^{\ln x}.$$

We can use this relationship between natural logarithms and exponentials to solve equations involving logarithms, as in the following example.

Example 6.5   Solving a Logarithmic Equation
Solve the equation $\ln (x + 5) = 3$ for x.

Solution   Taking the exponential of both sides of the equation and writing things backwards (for convenience), we have

$$e^3 = e^{\ln(x+5)} = x + 5,$$

from (6.2). Subtracting 5 from both sides gives us

$$e^3 - 5 = x.$$

■

As always, graphs provide excellent visual summaries of the important properties of functions.

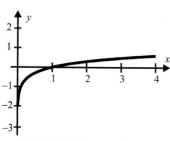

FIGURE 0.58a: $y = \log x$.

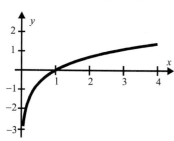

FIGURE 0.58b: $y = \ln x$.

Example 6.6   Sketching Graphs of Logarithms
Sketch graphs of $y = \log x$ and $y = \ln x$ and briefly discuss the properties of each.

Solution   From a calculator or computer, you should obtain graphs resembling those in Figures 0.58a and 0.58b. Notice that both graphs appear to have a vertical asymptote at $x = 0$ (why would that be?), cross the x-axis at $x = 1$ and very gradually increase as x increases. Neither graph has any points to the left of the y-axis. The two graphs are very similar, although not identical.

■

The properties described above graphically are summarized in the following result.

Theorem 6.1   For any positive base $b \neq 1$,
(i) $\log_b x$ is defined only for $x > 0$,
(ii) $\log_b 1 = 0$ and
(iii)   if $b > 1$, then $\log_b x < 0$ for $0 < x < 1$ and $\log_b x > 0$ for $x > 1$.

Proof. (i) Note that since $b > 0$, $b^y > 0$ for any y. So, if $\log_b x = y$, then $x = b^y > 0$.
(ii) Since $b^0 = 1$ for any number $b \neq 0$, $\log_b 1 = 0$ (i.e., the exponent to which you raise the base b to get the number 1 is 0). (iii) We leave this as an exercise.

■

Common logarithms all share a set of defining properties, as stated in the following theorem.

Theorem 6.2   For any positive base $b \neq 1$ and positive numbers x and y, we have
(i) $\log_b(xy) = \log_b x + \log_b y$
(ii) $\log_b(x/y) = \log_b x - \log_b y$
(iii) $\log_b(x^y) = y \log_b x$.

Proof. We leave the proof of these properties as an exercise.

■

As with most algebraic rules, each of these properties can dramatically simplify calculations when they apply.

Example 6.7   **Simplifying Logarithmic Expressions**

Write each as a single logarithm: (a) $\log_2 27^x - \log_2 3^x$ and (b) $\ln 8 - 3\ln(1/2)$.

Solution   First, note that there is more than one order in which to work each problem. For part (a), we have $27 = 3^3$ and so, $27^x = \left(3^3\right)^x = 3^{3x}$. This gives us

$$\log_2 27^x - \log_2 3^x = \log_2 3^{3x} - \log_2 3^x$$
$$= 3x\log_2 3 - x\log_2 3 = 2x\log_2 3 = \log_2 3^{2x}.$$

For part (b), note that $8 = 2^3$ and $1/2 = 2^{-1}$. Then,

$$\ln 8 - 3\ln(1/2) = 3\ln 2 - 3\left(-\ln 2\right)$$
$$= 3\ln 2 + 3\ln 2 = 6\ln 2 = \ln 2^6 = \ln 64.$$

■

Notice that from Theorem 6.2 and the definition of the natural logarithm, we have that for any real number x

$$\ln\left(e^x\right) = x\ln e = x. \tag{6.3}$$

We can use this relationship between exponentials and logarithms to solve equations involving exponentials, as in the following example.

Example 6.8   **Solving an Exponential Equation**

Solve the equation $e^{x+4} = 7$ for x.

Solution   Taking the natural logarithm of both sides and writing things backward (for simplicity), we have from (6.3) that

$$\ln 7 = \ln\left(e^{x+4}\right) = x + 4.$$

Subtracting 4 from both sides yields

$$\ln 7 - 4 = x.$$

■

Using the rules of exponents and logarithms, notice that we can rewrite any exponential as an exponential base e, as follows. For any base $a > 0$, we have

$$a^x = e^{\ln(a^x)} = e^{x\ln a}, \tag{6.4}$$

by Theorem 6.2 (iii), where we have used the fact that $e^{\ln y} = y$, for all $y > 0$.

Example 6.9   **Rewriting Exponentials as Exponentials with Base e**

Rewrite the exponentials 2^x, 5^x, and $(2/5)^x$ as exponentials with base e.

Solution   From (6.4), we have

$$2^x = e^{\ln(2^x)} = e^{x\ln 2},$$

$$5^x = e^{\ln(5^x)} = e^{x\ln 5},$$

and

$$\left(\frac{2}{5}\right)^x = e^{\ln[(2/5)^x]} = e^{x\ln(2/5)}.$$

■

Example 6.10   Simplifying Logarithmic Expressions
Use the rules of logarithms to simplify the expression $5 \ln x + 3 \ln y - 4 \ln z$.

Solution   From Theorem 6.2, we have

$$5 \ln x + 3 \ln y - 4 \ln z = \ln \left(x^5 \right) + \ln \left(y^3 \right) - \ln \left(z^4 \right) = \ln \left(\frac{x^5 y^3}{z^4} \right).$$

In some circumstances, it is beneficial to use the rules of logarithms to expand a given expression, as in the following example.

Example 6.11   Expanding a Logarithmic Expression
Use the rules of logarithms to expand the expression $\ln \left(\frac{x^3 y^4}{z^5} \right)$.

Solution   From Theorem 6.2, we have that

$$\ln \left(\frac{x^3 y^4}{z^5} \right) = \ln \left(x^3 y^4 \right) - \ln \left(z^5 \right) = \ln \left(x^3 \right) + \ln \left(y^4 \right) - \ln \left(z^5 \right)$$
$$= 3 \ln x + 4 \ln y - 5 \ln z.$$

Just as we were able to use the relationship between the natural logarithm and exponentials to rewrite an exponential with any positive base in terms of an exponential with base e, we can use these same properties to rewrite any logarithm in terms of natural logarithms, as follows. For any positive base b ($b \neq 1$), we will show that

$$\log_b x = \frac{\ln x}{\ln b}. \tag{6.5}$$

Let $y = \log_b x$. Then by Definition 6.2, we have that $x = b^y$. Taking the natural logarithm of both sides of this equation, by Theorem 6.2 (iii) we get that

$$\ln x = \ln \left(b^y \right) = y \ln b.$$

Dividing both sides by $\ln b$ (since $b \neq 1$, $\ln b \neq 0$) gives us

$$y = \frac{\ln x}{\ln b},$$

establishing (6.5).

One use you will find for equation (6.5) is for computing logarithms with bases other than e or 10. More than likely, your calculator has keys only for $\ln x$ and $\log x$. We illustrate this idea in the following example.

Example 6.12   Approximating the Value of Common Logarithms
Approximate the value of $\log_7 12$.

Solution   From (6.5), we have

$$\log_7 12 = \frac{\ln 12}{\ln 7} \approx 1.2769894.$$

Example 6.13   Matching Data to an Exponential Curve

Find the exponential function of the form $f(x) = ae^{bx}$ that passes through the points $(0, 5)$ and $(3, 9)$.

Solution   Notice that you can't simply solve this problem by inspection. That is, you can't just read off appropriate values for a and b. Instead, we solve for a and b, using the properties of logarithms and exponentials. First, for the graph to pass through the point $(0, 5)$, this means that

$$5 = f(0) = ae^{b \cdot 0} = a,$$

so that $a = 5$. Next, for the graph to pass through the point $(3, 9)$, we must have

$$9 = f(3) = ae^{3b} = 5e^{3b}.$$

To solve for the b in the exponent, we divide both sides of the last equation by 5 and take the natural logarithm of both sides, which yields

$$\ln\left(\frac{9}{5}\right) = \ln e^{3b} = 3b,$$

from (6.3). Finally, dividing by 3 gives us the value for b:

$$b = \frac{1}{3}\ln\left(\frac{9}{5}\right).$$

∎

Historically, logarithms were useful for handling large numbers. Given the wide availability of calculators and computers, this particular usage is now only a minor convenience (see the exercises on decibels and the Richter scale). Today, logarithms serve many other important purposes, including manipulating exponential functions.

Fitting Data to a Curve

Consider the population of the United States from 1790 to 1860, found in the accompanying table. A plot of these data points can be seen in Figure 0.59 (where the vertical scale represents the population in millions). This shows that the population was increasing, with larger and larger increases each decade. If you sketch an imaginary curve through these points, you will probably get the impression of a parabola or perhaps the right half of a cubic or exponential. And that's the question: is this data best modeled by a quadratic function, a cubic function, a fourth-order polynomial, an exponential function, or what?

Year	U.S. Population
1790	3,929,214
1800	5,308,483
1810	7,239,881
1820	9,638,453
1830	12,866,020
1840	17,069,453
1850	23,191,876
1860	31,443,321

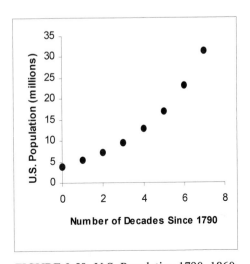

FIGURE 0.59: U.S. Population 1790–1860.

We can use the properties of logarithms from Theorem 6.2 to help determine whether a given set of data is modeled better by a polynomial or an exponential function, as follows. Suppose that the data actually comes from an exponential, say $y = ae^{bx}$ (i.e., the data points lie on the graph of this exponential). Then,

$$\ln y = \ln(ae^{bx}) = \ln a + \ln e^{bx} = \ln a + bx.$$

If you draw a new graph, where the horizontal axis shows values of x and the vertical axis corresponds to values of $\ln y$, then the graph will be the line $\ln y = bx + c$ (where the constant $c = \ln a$). On the other hand, suppose the data actually came from a polynomial. If $y = bx^n$ (for any n), then observe that

$$\ln y = \ln(bx^n) = \ln b + \ln x^n = \ln b + n \ln x.$$

In this case, a graph with horizontal and vertical axes corresponding to x and $\ln y$, respectively, will look like the graph of a logarithm, $\ln y = n \ln x + c$. Such a **semi-log graph** (i.e., graphing $\ln y$ versus x) lets us distinguish the graph of an exponential from that of a polynomial: graphs of exponentials become straight lines, while graphs of polynomials (of degree ≥ 1) become logarithmic curves. Scientists and engineers frequently use semi-log graphs to help them gain an understanding of physical phenomena represented by some collection of data.

∎

Example 6.14   Using a Semi-Log Graph to Identify a Type of Function

Determine if the population of the United States from 1790–1860 was increasing exponentially or as a polynomial.

Solution   As indicated above, the trick is to draw a semi-log graph. That is, instead of plotting $(1, 3.9)$ as the first data point, plot $(1, \ln 3.9)$ and so on. A semi-log plot of this data set is seen in Figure 0.60.

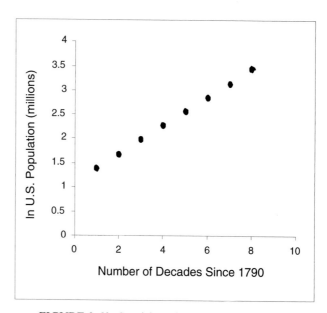

FIGURE 0.60: Semi-log plot of U.S. population.

Although the points are not exactly colinear (how would you prove this?), the plot is very close to a straight line with $\ln y$-intercept of 1 and slope 0.35. You should conclude that the population is well-modeled by an exponential function. (Keep in mind that here, as with most real problems, the data is somewhat imprecise and so, the points in the semi-log plot need **not** be perfectly colinear for you to conclude that the data is modeled quite well by an exponential.) The exponential model would be $y = ae^{bx}$, where b is the slope and $\ln a$ is the $\ln y$-intercept. Here, we get $b = 0.35$ and $\ln a = 1$ (why?), so that $a = e$. The population is then modeled by

$$P(t) = e \cdot e^{0.35t} \text{ million,}$$

where t equals the number of decades since 1780.

■

Exercises 0.6

1. (writing) Starting from a single cell, a human being is formed by 50 generations of cell division. Explain why after n divisions there are 2^n cells. Guess how many cells will be present after 50 divisions, then compute 2^{50}. Briefly discuss how rapidly exponential functions increase.

2. (writing) Explain why the graphs of $f(x) = 2^{-x}$ and $g(x) = \left(\frac{1}{2}\right)^x$ are the same.

3. (writing) Compare $f(x) = x^2$ and $g(x) = 2^x$ for $x = \frac{1}{2}$, $x = 1$, $x = 2$, $x = 3$, and $x = 4$. In general, which function is bigger for large values of x? For small values of x?

4. (writing) Compare $f(x) = 2^x$ and $g(x) = 3^x$ for $x = -2$, $x = -\frac{1}{2}$, $x = \frac{1}{2}$, and $x = 2$. In general, which function is bigger for negative values of x? For positive values of x?

In exercises 5–12, convert each exponential expression into fractional or root form.

5. 2^{-3} **6.** 4^{-2} **7.** $3^{1/2}$ **8.** $7^{1/3}$

9. $5^{2/3}$ **10.** $6^{2/5}$ **11.** $4^{-2/3}$ **12.** $6^{-1/3}$

In exercises 13–20, convert each expression into exponential form.

13. $\frac{1}{x^2}$ **14.** $\frac{1}{x^4}$ **15.** $\sqrt{x}$ **16.** $\sqrt[3]{x^2}$

17. $\frac{2}{x^3}$ **18.** $\frac{4}{x^2}$ **19.** $\frac{1}{2\sqrt{x}}$ **20.** $\frac{3}{2\sqrt{x^3}}$

In exercises 21–24, find the integer value of the given expression (no calculators!).

21. $4^{3/2}$ **22.** $8^{2/3}$ **23.** $\frac{\sqrt{8}}{2^{1/2}}$ **24.** $\frac{2}{(1/3)^2}$

In exercises 25–32, use a calculator or computer to estimate each value.

25. e^2 **26.** $3e$ **27.** $2e^{-1/2}$ **28.** $4e^{-2/3}$

29. $\frac{12}{e}$ **30.** $\frac{14}{\sqrt{e}}$ **31.** $10e^{-.01}$ **32.** $10e^{-.001}$

In exercises 33–50, sketch a graph of the given function.

33. $f(x) = e^{2x}$ **34.** $f(x) = e^{3x}$

35. $f(x) = 2e^{x/4}$ **36.** $f(x) = 3e^{x/6}$

37. $f(x) = 3e^{-2x}$ **38.** $f(x) = 10e^{-x/3}$

39. $f(x) = e^{-x^2}$ **40.** $f(x) = e^{-x^3}$

41. $f(x) = xe^{-2x}$ **42.** $f(x) = x^2e^{-2x}$

43. $f(x) = \ln 2x$ **44.** $f(x) = \ln 3x$

45. $f(x) = \ln x^2$ **46.** $f(x) = \ln x^3$

47. $f(x) = e^{2\ln x}$ **48.** $f(x) = e^{-\ln x}$

49. $f(x) = e^{-x/4}\sin x$ **50.** $f(x) = e^{-x/6}\cos x$

In exercises 51–62, solve the given equation for x.

51. $e^{2x} = 2$ **52.** $e^{4x} = 3$

53. $2e^{-2x} = 1$ **54.** $3e^{-x/2} = 2$

55. $\ln 2x = 4$ **56.** $2\ln 3x = 1$

57. $2\ln 4x - 1 = 6$ **58.** $3\ln(2-x) = 6$

59. $2e^{1-2x} = 3$ **60.** $3e^{x^2} = 4$

61. $e^{2\ln x} = 4$ **62.** $\ln(e^{2x}) = 6$

In exercises 63 and 64, use the definition of logarithm to determine the value.

63. (a) $\log_3 9$ (b) $\log_4 64$ (c) $\log_3 \frac{1}{27}$

64. (a) $\log_4 \frac{1}{16}$ (b) $\log_4 2$ (c) $\log_9 3$

In exercises 65–70, rewrite the expression as a single logarithm.

65. $\ln 3 - \ln 4$ **66.** $2\ln 4 - \ln 3$

67. $\frac{1}{2}\ln 4 - \ln 2$ **68.** $3\ln 2 - \ln\frac{1}{2}$

69. $\ln\frac{3}{4} + 4\ln 2$ **70.** $\ln 9 - 2\ln 3$

In exercises 71–76, find a function of the form $f(x) = ae^{bx}$ with the given function values.

71. $f(0) = 2$, $f(2) = 6$　　　　**72.** $f(0) = 3$, $f(3) = 4$

73. $f(0) = 4$, $f(2) = 2$　　　　**74.** $f(0) = 5$, $f(1) = 2$

75. $f(0) = -2$, $f(1) = -3$　　**76.** $f(1) = 2$, $f(2) = 4$

77. A fast-food restaurant gives every customer a game ticket. With each ticket, the customer has a 1-in-10 chance of winning a free meal. If you go 10 times, estimate your chances of winning at least one free meal. The exact probability is $1 - \left(\frac{9}{10}\right)^{10}$. Compute this number and compare it to your guess.

78. In exercise 77, if you had 20 tickets with a 1-in-20 chance of winning, would you expect your probability of winning to increase or decrease? Compute the probability $1 - \left(\frac{19}{20}\right)^{20}$ to find out.

79. In general, if you have n chances of winning with a 1-in-n chance on each try, the probability of winning at least once is $1 - \left(1 - \frac{1}{n}\right)^{n}$. As n gets larger, what number does this probability approach? (Hint: There is a very good reason that this question is in this section!)

80. If $y = a \cdot x^m$, show that $\ln y = \ln a + m \ln x$. If $v = \ln y$, $u = \ln x$, and $b = \ln a$, show that $v = mu + b$. Explain why the graph of v as a function of u would be a straight line. This graph is called the **log-log plot** of y and x.

81. For the given data, compute $v = \ln y$ and $u = \ln x$ and plot points (u, v). Find constants m and b such that $v = mu + b$ and use the results of exercise 80 to find a constant a such that $y = a \cdot x^m$.

x :	2.2	2.4	2.6	2.8	3.0	3.2
y :	14.52	17.28	20.28	23.52	27.0	30.72

82. Repeat exercise 81 for the given data.

x :	2.8	3.0	3.2	3.4	3.6	3.8
y :	9.37	10.39	11.45	12.54	13.66	14.81

83. Construct a log-log plot (see exercise 80) of the U.S. population data in example 6.14. Compared to the semi-log plot of the data in Figure 0.60, does the log-log plot look linear? Based on this, is the population data modeled better by an exponential function or a polynomial (power) function?

84. Construct a semi-log plot of the data in exercise 81. Compared to the log-log plot already constructed, does this plot look linear? Based on this, is this data better modeled by an exponential or power function?

85. The concentration $[H^+]$ of free hydrogen ions in a chemical solution determines the solution's pH, as defined by $pH = -\log[H^+]$. Find $[H^+]$ if the pH equals (a) 7, (b) 8, and (c) 9. For each increase in pH of 1, by what factor does $[H^+]$ change?

86. Gastric juice is considered an acid, with a pH of about 2.5. Blood is considered alkaline, with a pH of about 7.5. Compare the concentrations of hydrogen ions in the two substances (see exercise 85).

87. The Richter magnitude M of an earthquake is defined in terms of the energy E in joules released by the earthquake, with $\log_{10} E = 4.4 + 1.5M$. Find the energy for earthquakes with magnitudes (a) 4, (b) 5, and (c) 6. For each increase in M of one, by what factor does E change?

88. It puzzles some people who have not grown up around earthquakes that a magnitude 6 quake is considered much more severe than a magnitude 3 quake. Compare the amount of energy released in the two quakes.

89. The decibel level of a noise is defined in terms of the intensity I of the noise, with $dB = 10 \log_{10}(I/I_0)$. Here, $I_0 = 10^{-12}$ W/m^2 is the intensity of a barely audible sound. Compute the intensity levels of sounds with (a) $dB = 80$, (b) $dB = 90$, and (c) $dB = 100$. For each increase in decibels of 10, by what factor does I change?

90. At a basketball game, a courtside decibel meter shows crowd noises ranging from 60 dB to 110 dB. How much louder is the 110-dB noise compared to the 60-dB noise (see exercise 89)?

91. Graph $y = xe^{-x}$, $y = xe^{-2x}$, $y = xe^{-3x}$ and so on. Estimate the locations of the maximum for each. In general, state a rule for the location of the maximum of $y = xe^{-kx}$.

92. In golf, the task is to hit a golf ball into a small hole. If the ground near the hole is not flat, the golfer must judge how much the ball will curve. Suppose the golfer is at the point $(-13, 0)$, the hole is at the point $(0, 0)$, and the path of the ball is, for $-13 \leq x \leq 0$, $y = -1.672x + 72\ln(1 + 0.02x)$. Show that the ball goes in the hole and estimate the point on the y-axis at which the golfer aimed.

93. (exploratory) On whichever calculators and/or computers you have access to, try to evaluate $(-8)^{2/3}$. If you do not get 4, try computing $\left[(-8)^2\right]^{1/3}$. Show that, theoretically, $(-8)^{2/3} = \left[(-8)^2\right]^{1/3}$. To understand why a calculator might report a difference in the two quantities, explain why $(-8)^{2/4} \neq \left[(-8)^2\right]^{1/4}$. Investigate why some machines report that $(-8)^{2/3}$ does not exist.

94. (exploratory) Graph $y = x^2$ and $y = 2^x$ and approximate the two positive solutions of the equation $x^2 = 2^x$. Graph $y = x^3$ and $y = 3^x$ and approximate the two positive solutions of the equation $x^3 = 3^x$. Explain why $x = a$ will always be a solution of $x^a = a^x$, $a > 0$. What is different about the role of $x = 2$ as a solution of $x^2 = 2^x$ compared to the role of $x = 3$ as a solution of $x^3 = 3^x$? To determine the a-value at which the change occurs, graphically solve $x^a = a^x$ for $a = 2.1, 2.2, \ldots, 2.9$ and note that $a = 2.7$ and $a = 2.8$ behave differently. Continue to narrow down the interval of change by testing $a = 2.71, 2.72, \ldots, 2.79$. Then guess the exact value of a.

95. (exploratory) Graph $y = \ln x$ and describe the behavior near $x = 0$. Then graph $y = x \ln x$ and describe the behavior near $x = 0$. Repeat this for $y = x^2 \ln x$, $y = x^{1/2} \ln x$, and $y = x^a \ln x$ for a variety of positive constants a. Because the function "blows up" at $x = 0$, we say that $y = \ln x$ has a **singularity** at $x = 0$. The **order** of the singularity at $x = 0$ of a function $f(x)$ is the smallest value of a such that $y = x^a f(x)$ doesn't have a singularity at $x = 0$. Determine the order of the singularity at $x = 0$ for (a) $f(x) = \frac{1}{x}$, (b) $f(x) = \frac{1}{x^2}$, and (c) $f(x) = \frac{1}{x^3}$. The higher the order of the singularity, the "worse" the singularity is. Based on your work, how bad is the singularity of $y = \ln x$ at $x = 0$?

0.7   Transformations of Functions

You are now familiar with a long list of functions: polynomials, rational functions, trigonometric functions, exponentials, and logarithms. Many students at this stage expect that calculus and higher-level mathematics courses will consist of defining even more functions and solving more equations. Actually, one important goal for this course is to more fully understand the properties of the functions we have already defined. To a large extent, you will build your understanding by examining a few key properties of functions. To briefly get an idea of where we are headed, recall what the function $f(x) = x^2$ represents. To understand this function, you first had to understand how to multiply two numbers together. From practice multiplying "3 times 3" and "6 times 6" and so on, the idea of multiplying any number by itself became natural. The function x^2 is, in this sense, simply an abstraction. We have put multiplication into the broader context of functions. Now that we have functions, we need to make the manipulations of functions as natural as possible.

The first few function manipulations we define are straightforward.

Definition 7.1   Suppose that $f(x)$ and $g(x)$ are functions with domains D_1 and D_2, respectively. The functions $f + g$, $f - g$ and $f \cdot g$ are defined by

$$(f + g)(x) = f(x) + g(x),$$

$$(f - g)(x) = f(x) - g(x)$$

and

$$(f \cdot g)(x) = f(x) \cdot g(x),$$

for all x in $D_1 \cap D_2$ (i.e., $x \in D_1$ **and** $x \in D_2$). The function f/g is defined by

$$(f/g)(x) = \frac{f(x)}{g(x)},$$

for all x in $D_1 \cap D_2$ such that $g(x) \neq 0$.

In the following example, we examine various combinations of several simple functions.

Example 7.1   Combinations of Functions

If $f(x) = x - 3$ and $g(x) = \sqrt{x - 1}$, determine the functions $f + g$, $3f - g$, f/g and $g - g$, stating the domains of each.

Solution   First, note that the domain of f is the entire real line and the domain of g is the set of all $x \geq 1$. Now,

$$(f + g)(x) = x - 3 + \sqrt{x - 1}$$

and

$$(3f - g)(x) = 3(x - 3) - \sqrt{x - 1} = 3x - 9 - \sqrt{x - 1}.$$

Notice that the domain of both $(f + g)$ and $(3f - g)$ is $\{x \,|\, x \geq 1\}$. For

$$(f/g)(x) = \frac{f(x)}{g(x)} = \frac{x - 3}{\sqrt{x - 1}},$$

the domain is $\{x \,|\, x > 1\}$. (Note that we have added the restriction $x \neq 1$ to avoid dividing by 0.) Finally,

$$(g - g)(x) = \sqrt{x - 1} - \sqrt{x - 1} = 0.$$

Although the final expression is defined for all x, the domain depends on the intermediate steps. In this case, the domain of $(g - g)$ is $\{x \,|\, x \geq 1\}$. (This may seem like a picky point, but it will have important consequences later.)

■

Definition 7.1 and example 7.1 show us how to do arithmetic with functions. An operation on functions that does not directly correspond to arithmetic is the **composition** of two functions.

Definition 7.2   The **composition** of functions $f(x)$ and $g(x)$, written $f \circ g$, is defined by

$$(f \circ g)(x) = f(g(x)),$$

for all x such that x is in the domain of g and $g(x)$ is in the domain of f.

The composition of two functions is a two-step process, as in the following schematic.

$$(f \circ g)(x) = f(g(x)).$$

Be careful to notice what this definition is saying. In particular, for $f(g(x))$ to be defined, you first need $g(x)$ to be defined, so x must be in the domain of g. Next, f must be defined at the point $g(x)$, so that the number $g(x)$ will have to be in the domain of f.

Example 7.2   **Finding the Composition of Two Functions**
    For $f(x) = x^2 + 1$ and $g(x) = \sqrt{x - 2}$, find the compositions $f \circ g$ and $g \circ f$ and identify the domain of each.

Solution   First, we have

$$(f \circ g)(x) = f(g(x)) = f(\sqrt{x - 2})$$
$$= (\sqrt{x - 2})^2 + 1 = x - 2 + 1 = x - 1.$$

It's tempting to write that the domain of $f \circ g$ is the entire real line, but look more carefully. Note that for x to be in the domain of g, we must have $x \geq 2$. The domain of f is the whole real line, so no further restrictions are placed on the domain of $f \circ g$. Even though the final formula $x - 1$ is defined for all x, the domain of $(f \circ g)$ is $\{x \mid x \geq 2\}$. For the second composition,

$$(g \circ f)(x) = g(f(x)) = g(x^2 + 1)$$
$$= \sqrt{(x^2 + 1) - 2} = \sqrt{x^2 - 1}.$$

The resulting square root requires $x^2 - 1 \geq 0$ or $|x| \geq 1$. Since the "inside" function f is defined for all x, the domain of $g \circ f$ is $\{x \in \mathbb{R} \mid |x| \geq 1\}$. ∎

As you progress through calculus, you will often find yourself needing to identify the different pieces of a composition of functions. For now, it is an important skill to practice.

Example 7.3   Identifying Compositions of Functions
Identify functions f and g such that the given function can be written as $(f \circ g)(x)$ for each of (a) $\sqrt{x^2 + 1}$, (b) $(\sqrt{x} + 1)^2$, (c) $\sin x^2$, and (d) $\cos^2 x$.

Solution   (a) Notice that $x^2 + 1$ is *inside* the square root. So, one choice is to have $g(x) = x^2 + 1$ and $f(x) = \sqrt{x}$.
   (b) Here, $\sqrt{x} + 1$ is *inside* the square. So, one choice is $g(x) = \sqrt{x} + 1$ and $f(x) = x^2$.
   (c) The function can be rewritten as $\sin(x^2)$, with x^2 clearly *inside* the sine function. Then, $g(x) = x^2$ and $f(x) = \sin x$ is one choice.
   (d) The function as written is shorthand for $(\cos x)^2$. So, one choice is $g(x) = \cos x$ and $f(x) = x^2$. ∎

In general, it is quite difficult to take the graphs of $f(x)$ and $g(x)$ and produce the graph of $f(g(x))$. If one of the functions f and g is linear, however, there is a simple graphical procedure for graphing the composition. Such **linear transformations** are explored in the remainder of this section.

The first case is to take the graph of $f(x)$ and produce the graph of $f(x) + c$ for some constant c. You can deduce the general result from the following example.

Example 7.4   Vertical Translation of a Graph
Graph $y = x^2$ and $y = x^2 + 3$; compare and contrast the graphs.

Solution   You can probably sketch these by hand. You should get something like those in Figures 0.61a and 0.61b.

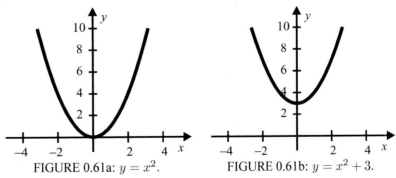

FIGURE 0.61a: $y = x^2$.   FIGURE 0.61b: $y = x^2 + 3$.

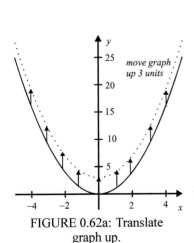

FIGURE 0.62a: Translate graph up.

Both figures show parabolas opening up. The main obvious difference is that x^2 has a y-intercept of 0 and $x^2 + 3$ has a y-intercept of 3. Some reflection should convince you that this is an important clue. For **any** given value of x, the point on the graph of $y = x^2 + 3$ will be plotted exactly 3 units higher than the corresponding point on the graph of $y = x^2$. This is shown in Figure 0.62a.

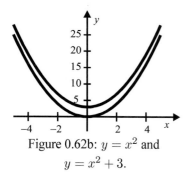

Figure 0.62b: $y = x^2$ and
$y = x^2 + 3$.

In Figure 0.62b, the two graphs are shown on the same set of axes. To many people, it does not look like the top graph is the same as the bottom graph moved up 3 units. This is an unfortunate optical illusion. Humans usually mentally judge distance between curves as the shortest distance between the curves. For these parabolas, the shortest distance is vertical at $x = 0$ but becomes increasingly horizontal as you move away from the y-axis. The distance of 3 between the parabolas is measured vertically.

■

In general, the graph of $y = f(x) + c$ is exactly the same as the graph of $f(x)$ moved up (if $c > 0$) or down (if $c < 0$) by $|c|$ units. We usually refer to $f(x) + c$ as a **translation** (up or down, by $|c|$ units).

In the following example, we explore what happens if a constant is added to x.

Example 7.5   A Horizontal Translation

Compare and contrast the graphs of $y = x^2$ and $y = (x - 1)^2$.

Solution   The graphs are shown in Figures 0.63a and 0.63b, respectively.

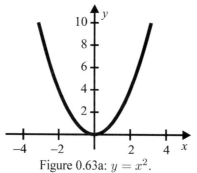

Figure 0.63a: $y = x^2$.

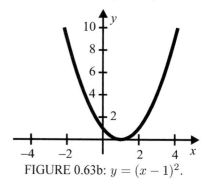

FIGURE 0.63b: $y = (x - 1)^2$.

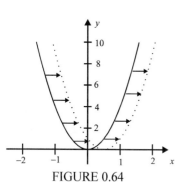

FIGURE 0.64
Move graph 1 unit to the right.

Notice that the graph of $y = (x - 1)^2$ appears to be the same as the graph of $y = x^2$, except that it is moved 1 unit to the right. This should make sense for the following reason. Pick a value of x, say $x = 13$. The value of $(x - 1)^2$ at $x = 13$ is 12^2, the same as the value of x^2 at $x = 12$, 1 unit to the left. Observe that this same pattern would hold for any x you choose. A simultaneous plot of the two functions (see Figure 0.64) shows this.

■

In general, for $c > 0$, the graph of $y = f(x - c)$ is the same as the graph of $y = f(x)$ moved c units to the right. Likewise (again, for $c > 0$), you get the graph of $f(x + c)$ by moving the graph of $y = f(x)$ to the left c units. We usually refer to $f(x - c)$ and $f(x + c)$ as translations (to the right and left, respectively, by c units). To avoid confusion on which way to translate the graph of $y = f(x)$, focus on what makes the argument (the quantity inside the parentheses) zero. For $f(x)$, this is $x = 0$, but for $f(x - c)$ you must have $x = c$ to get $f(0)$ [i.e., the same y-value as $f(x)$ when $x = 0$]. This says that the point on the graph of $y = f(x)$ at $x = 0$ corresponds to the point on the graph of $y = f(x - c)$ at $x = c$.

Example 7.6   Comparing Vertical and Horizontal Translations

Given the graph of $y = f(x)$ shown in Figure 0.65a, sketch the graphs of $y = f(x) - 2$ and $y = f(x - 2)$.

Solution   To graph $y = f(x) - 2$, simply translate the original graph down 2 units, as shown in Figure 0.65b. To graph $y = f(x - 2)$, simply translate the original graph to the right 2 units (so that the x-intercept at $x = 0$ in the original graph corresponds to an x-intercept at $x = 2$ in the translated graph), as seen in Figure 0.65c.

FIGURE 0.65a: $y = f(x)$.

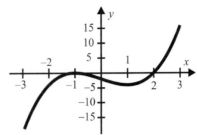

FIGURE 0.65b: $y = f(x) - 2$.

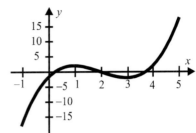

FIGURE 0.65c: $y = f(x - 2)$.

■

The following example explores the effect of multiplying or dividing x or y by a constant.

Example 7.7   Comparing Some Related Graphs

Compare and contrast the graphs of $y = x^2 - 1$, $y = 4(x^2 - 1)$ and $y = (4x)^2 - 1$.

Solution   The first two graphs are shown in Figures 0.66a and 0.66b, respectively.

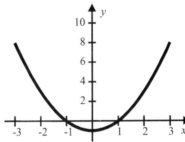

FIGURE 0.66a: $y = x^2 - 1$.

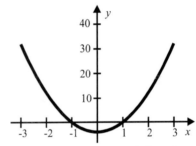

FIGURE 0.66b: $y = 4(x^2 - 1)$.

These graphs look identical until you compare the scales on the y-axes. The values in Figure 0.66b are four times as large, reflecting the multiplication of the original function by 4. The effect looks different when the functions are plotted on the same scale, as in Figure 0.66c. Here, the parabola $y = 4(x^2 - 1)$ simply looks thinner. Note that the x-intercepts remain the same. (Why would that be?)

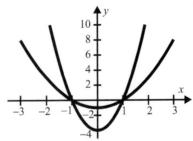

FIGURE 0.66c: $y = x^2 - 1$ and $y = 4(x^2 - 1)$.

The graphs of $y = x^2 - 1$ and $y = (4x)^2 - 1$ are shown in Figures 0.67a and 0.67b, respectively.

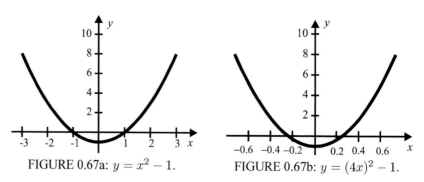

FIGURE 0.67a: $y = x^2 - 1$.   FIGURE 0.67b: $y = (4x)^2 - 1$.

Can you spot the difference here? In this case, the x-scale has now changed, by the same factor of 4 as in the function. To see this, note that plugging $x = 1/4$ into $(4x)^2 - 1$ produces $(1)^2 - 1$, exactly the same as plugging $x = 1$ into the original function. When plotted on the same set of axes (as in Figure 0.67c), the parabola $y = (4x)^2 - 1$ looks thinner. Note that here, the x-intercepts are different, but the y-intercepts are the same. (Why would that be?)

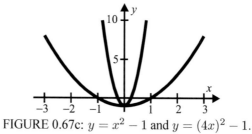

FIGURE 0.67c: $y = x^2 - 1$ and $y = (4x)^2 - 1$.

We can generalize the observations made in example 7.7. Before reading our explanation, try to state a general rule for yourself. How are the graphs of the functions $cf(x)$ and $f(cx)$ related to the graph of $y = f(x)$?

Based on example 7.7, notice that to obtain a graph of $y = cf(x)$ for some constant $c > 0$, take the graph of $y = f(x)$ and multiply the scale on the y-axis by c. To obtain a graph of $y = f(cx)$ for some constant $c > 0$, take the graph of $y = f(x)$ and multiply the scale on the x-axis by $1/c$.

These basic rules can be combined to understand more complicated graphs.

Example 7.8   A Translation and a Stretching

Describe how to get the graph of $y = 2x^2 - 3$ from the graph of $y = x^2$.

Solution   You can get from x^2 to $2x^2 - 3$ by multiplying by 2 and then subtracting 3. In terms of the graph, this has the effect of multiplying the y-scale by 2 and then lowering the graph by 3 units (see the graphs in Figures 0.68a and 0.68b).

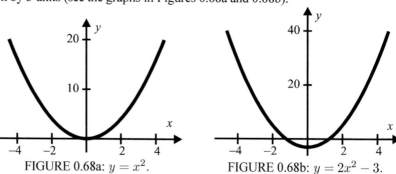

FIGURE 0.68a: $y = x^2$.   FIGURE 0.68b: $y = 2x^2 - 3$.

Example 7.9   A Translation Involving both x and y Scales

Find the graph of $y = x^2 + 4x + 3$.

Solution   We can again relate this (and the graph of **every** quadratic) to the graph of $y = x^2$. We must first **complete the square**. Recall that in this process, you take the coefficient of x (4), divide by 2 ($4/2 = 2$) and square the result ($2^2 = 4$). Add and subtract this number and then, rewrite the x terms as a perfect square. We have

$$y = x^2 + 4x + 3 = \left(x^2 + 4x + 4\right) - 4 + 3 = (x+2)^2 - 1.$$

To graph this function, take the parabola $y = x^2$ (see Figure 0.69a), translate the graph 2 units to the left and 1 unit down (see Figure 0.69b).

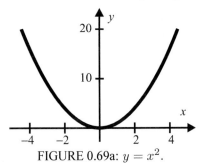

FIGURE 0.69a: $y = x^2$.

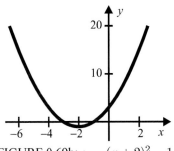

FIGURE 0.69b: $y = (x + 2)^2 - 1$.

You will explore additional transformations in the exercises.

Exercises 0.7

1. (writing) The restricted domain of example 7.2 may be puzzling. Consider the following analogy. Suppose you have an airplane flight from New York to Los Angeles with a stop for refueling in Minneapolis. If bad weather has closed the airport in Minneapolis, explain why your flight will be canceled (or at least rerouted) even if the weather is great in New York and Los Angeles.

2. (writing) Explain why the graphs of $y = 4(x^2 - 1)$ and $y = (4x)^2 - 1$ in Figures 0.66c and 0.67c appear "thinner" than the graph of $y = x^2 - 1$.

3. (writing) As illustrated in example 7.9, completing the square can be used to rewrite any quadratic function in the form $a(x-d)^2 + e$. Using the transformation rules in this section, explain why this means that all parabolas (with $a > 0$) will look essentially the same.

4. (writing) Explain why the graph of $y = f(x + 4)$ is obtained by moving the graph of $y = f(x)$ four units to the left, instead of to the right.

In exercises 5–10, find the compositions $f \circ g$ and $g \circ f$ and identify their respective domains.

5. $f(x) = x + 1$, $g(x) = \sqrt{x - 3}$

6. $f(x) = x - 2$, $g(x) = \sqrt{x + 1}$

7. $f(x) = e^x$, $g(x) = \ln x$

8. $f(x) = \sqrt{1 - x}$, $g(x) = \ln x$

9. $f(x) = x^2 + 1$, $g(x) = \sin x$

10. $f(x) = \dfrac{1}{x^2 - 1}$, $g(x) = x^2 - 2$

In exercises 11–26, identify functions $f(x)$ and $g(x)$ such that the given function equals $(f \circ g)(x)$.

11. $\sqrt{x^4 + 1}$

12. $\sqrt[3]{x + 3}$

13. $\dfrac{1}{x^2 + 1}$

14. $\dfrac{1}{x^2} + 1$

15. $(4x + 1)^2 + 3$

16. $4(x + 1)^2 + 3$

17. $\sin^3 x$

18. $\sin x^3$

19. $\cos 4x$

20. $4 \cos x$

21. $e^{x^2 + 1}$

22. $e^{4x - 2}$

23. $\sqrt{e^x + 1}$

24. $\sqrt{e^x} + 1$

25. $\ln(3x - 5)$

26. $\ln 3x - 5$

In exercises 27–34, use the graph of $y = f(x)$ given below to graph the indicated function.

27. $f(x) - 3$ **28.** $f(x + 2)$ **29.** $f(x - 3)$

30. $f(x) + 2$ **31.** $f(2x)$ **32.** $3f(x)$

33. $4f(x) - 1$ **34.** $3f(x + 2)$

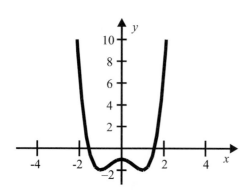

In exercises 35–42, use the graph of $y = f(x)$ given below to graph the indicated function.

35. $f(x - 4)$ **36.** $f(x + 3)$ **37.** $f(2x)$

38. $f(2x - 4)$ **39.** $f(3x + 3)$ **40.** $3f(x)$

41. $2f(x) - 4$ **42.** $3f(x) + 3$

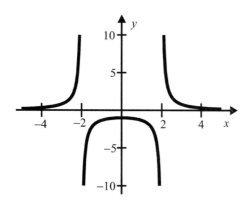

In exercises 43–48, complete the square and explain how to transform the graph of $y = x^2$ into the graph of the given function.

43. $f(x) = x^2 + 2x + 1$ **44.** $f(x) = x^2 - 4x + 4$

45. $f(x) = x^2 + 2x + 4$ **46.** $f(x) = x^2 - 4x + 2$

47. $f(x) = 2x^2 + 4x + 4$ **48.** $f(x) = 3x^2 - 6x + 2$

In exercises 49–52, graph the given function and compare to the graph of $y = x^2 - 1$.

49. $f(x) = -2(x^2 - 1)$ **50.** $f(x) = -3(x^2 - 1)$

51. $f(x) = -3(x^2 - 1) + 2$ **52.** $f(x) = -2(x^2 - 1) - 1$

In exercises 53–56, graph the given function and compare to the graph of $y = (x - 1)^2 - 1 = x^2 - 2x$.

53. $f(x) = (-x)^2 - 2(-x)$

54. $f(x) = (-2x)^2 - 2(-2x)$

55. $f(x) = (-x)^2 - 2(-x) + 1$

56. $f(x) = (-3x)^2 - 2(-3x) - 3$

57. Based on exercises 49–52, state a rule for transforming the graph of $y = f(x)$ into the graph of $y = cf(x)$ for $c < 0$.

58. Based on exercises 53–56, state a rule for transforming the graph of $y = f(x)$ into the graph of $y = f(cx)$ for $c < 0$.

59. After a car goes over a bump in the road, the car's vertical position is given by $f(t) = e^{-t/4} \cos t$. Describe what the passenger feels.

60. After a car goes over a bump in the road, the car's vertical position is given by $f(t) = e^{-4t} \cos t$. Describe what the passenger feels. Is this passenger happier than the one in exercise 59?

61. Iterations of functions are important in a variety of applications. To iterate $f(x)$, start with an initial value x_0 and compute $x_1 = f(x_0)$, $x_2 = f(x_1)$, $x_3 = f(x_2)$ and so on. For example, with $f(x) = \cos x$ and $x_0 = 1$, the **iterates** are $x_1 = \cos 1 \approx 0.54$, $x_2 = \cos x_1 \approx \cos 0.54 \approx 0.86$, $x_3 \approx \cos 0.86 \approx 0.65$ and so on. Keep computing iterates and show that they get closer and closer to 0.739085. Then pick your own x_0 (any number you like) and show that the iterates with this new x_0 also converge to 0.739085.

62. Referring to exercise 61, show that the iterates of a function can be written as $x_1 = f(x_0)$, $x_2 = f(f(x_0))$, $x_3 = f(f(f(x_0)))$, and so on. Graph $y = \cos(\cos x)$, $y = \cos(\cos(\cos x))$ and $y = \cos(\cos(\cos(\cos x)))$. The graphs should look more and more like a horizontal line. Use the result of exercise 61 to identify the limiting line.

63. Compute several iterates of $f(x) = \sin x$ (see exercise 61) with a variety of starting values. What happens to the iterates in the long run?

64. Repeat exercise 63 for $f(x) = x^2$.

65. In cases where the iterates of a function (see exercise 61) repeat a single number, that number is called a **fixed point**. Explain why any fixed point must be a solution of the equation $f(x) = x$. Find all fixed points of $f(x) = \cos x$ by solving the equation $\cos x = x$. Compare your results to that of exercise 61.

66. Find all fixed points of $f(x) = \sin x$ (see exercise 65). Compare your results to those of exercise 63.

67. (exploratory) You have explored how completing the square can transform any quadratic function into the form $y = a(x - d)^2 + e$. We concluded that all parabolas with

$a > 0$ look alike. To see that the same statement is not true of cubic polynomials, graph $y = x^3$ and $y = x^3 - 3x$. In this exercise, you will use completing the cube to determine how many different cubic graphs there are. To see what "completing the cube" would look like, first show that $(x + a)^3 = x^3 + 3ax^2 + 3a^2x + a^3$. Use this result to transform the graph of $y = x^3$ into the graphs of (a) $y = x^3 - 3x^2 + 3x - 1$ and (b) $y = x^3 - 3x^2 + 3x + 2$. Show that you can't get a simple transformation to $y = x^3 - 3x^2 + 4x - 2$. However, show that $y = x^3 - 3x^2 + 4x - 2$ can be obtained from $y = x^3 + x$ by basic transformations. Show that the following statement is true: any cubic $(y = ax^3 + bx^2 + cx + d)$ can be obtained with basic transformations from $y = ax^3 + kx$ for some constant k.

68. (exploratory) In many applications, it is important to take a section of a graph (e.g., some data) and extend it for predictions or other analysis. For example, suppose you have an electronic signal equal to $f(x) = 2x$ for $0 \leq x \leq 2$. To predict the value of the signal at $x = -1$, you would want to know whether or not the signal was periodic. If the signal is periodic, explain why $f(-1) = 2$ would be a good prediction. In some applications, you would assume that the function is **even**. That is, $f(x) = f(-x)$ for all x. In this case, you want $f(x) = 2(-x) = -2x$ for $-2 \leq x \leq 0$. Graph the **even extension** $f(x) = \begin{cases} -2x & \text{if } -2 \leq x \leq 0 \\ 2x & \text{if } 0 \leq x \leq 2 \end{cases}$. Find the even extension for (a) $f(x) = x^2 + 2x + 1$, $0 \leq x \leq 2$ and (b) $f(x) = e^{-x}$, $0 \leq x \leq 2$.

69. (exploratory) Similar to the even extension discussed in exercise 68, applications sometimes require a function to be **odd**; that is, $f(-x) = -f(x)$. For $f(x) = x^2$, $0 \leq x \leq 2$, the odd extension requires that for $-2 \leq x \leq 0$, $f(x) = -f(-x) = -(-x)^2 = -x^2$ so that $f(x) = \begin{cases} -x^2 & \text{if } -2 \leq x \leq 0 \\ x^2 & \text{if } 0 \leq x \leq 2 \end{cases}$. Graph $y = f(x)$ and discuss how to graphically rotate the right half of the graph to get the left half of the graph. Find the odd extension for (a) $f(x) = x^2 + 2x$, $0 \leq x \leq 2$ and (b) $f(x) = e^{-x} - 1$, $0 \leq x \leq 2$.

0.8   Preview of Calculus

The concepts discussed in the preceding seven sections are **not** calculus. However, these ideas are seen over and over throughout the calculus. To solve calculus problems, you will need to draw heavily on your skills in geometry, trigonometry, and especially algebra. In this section, we approach the boundary between precalculus mathematics and the calculus by investigating several important problems requiring the use of calculus.

Recall that we think of the slope of a line as the change in y divided by the change in x. This fraction is the same regardless of which two points you use to compute the slope. For example, the points $(0, 1)$, $(1, 4)$, and $(3, 10)$ all lie on the line $y = 3x + 1$. The slope of 3 can be obtained from any two of the points. For instance,

$$m = \frac{4 - 1}{1 - 0} = 3$$

or

$$m = \frac{10 - 1}{3 - 0} = 3.$$

In the calculus, we generalize this problem to find the slope of a *curve* at a point. For instance, suppose we wanted to find the slope of the curve $y = x^2 + 1$ at the point $(1, 2)$. You might think of picking a second point on the parabola, say $(2, 5)$. The slope of the line (called a **secant line**; see Figure 0.70a) through these two points is easy enough to compute. We have

$$m_{\text{sec}} = \frac{5 - 2}{2 - 1} = 3.$$

However, if you use the point $(0, 1)$ instead of $(2, 5)$, you get a different slope (see Figure 0.70b):

$$m_{\text{sec}} = \frac{2 - 1}{1 - 0} = 1.$$

Unlike with lines, the slopes of secant lines joining different points on more general curves are generally **not** the same. Refer to Figures 0.70a and 0.70b for a graphical representation of this.

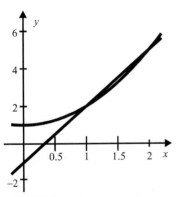

FIGURE 0.70a: Secant line, slope =3.

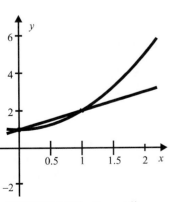

FIGURE 0.70b: Secant line, slope=1.

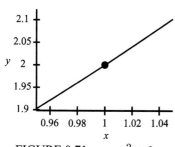

FIGURE 0.71: $y = x^2 + 1$.

If you get different slopes using different pairs of points, then what exactly does it mean for a curve to have a slope at a point? The answer can be visualized by zooming in on the specified point. Take the graph of $y = x^2 + 1$ and zoom in tight on the point $(1, 2)$. You should get a graph something like the one in Figure 0.71.

The graph looks very much like a straight line. In fact, the more you zoom in, the straighter the curve appears to be. Further, the closer the curve is to being straight, the less it matters which two points are used to compute a slope. So, here's the strategy: pick several points on the parabola, each closer to the point $(1, 2)$ than the previous one. Compute the slopes of the lines through $(1, 2)$ and each of the points. The closer the second point gets to $(1, 2)$, the closer the computed slope is to the answer you seek.

For example, the point $(1.5, 3.25)$ is on the parabola fairly close to $(1, 2)$. The slope of the line joining these points is

$$m_{\text{sec}} = \frac{3.25 - 2}{1.5 - 1} = 2.5.$$

The point $(1.1, 2.21)$ is even closer to $(1, 2)$. The slope of the secant line joining these two points is

$$m_{\text{sec}} = \frac{2.21 - 2}{1.1 - 1} = 2.1.$$

Continuing on, the point $(1.01, 2.0201)$ is closer yet to the point $(1, 2)$. The slope of the secant lines through these points is

$$m_{\text{sec}} = \frac{2.0201 - 2}{1.01 - 1} = 2.01.$$

The slopes of the secant lines $(2.5, 2.1, 2.01)$ are getting closer and closer to the slope of the parabola at the point $(1, 2)$. So what could we mean by the slope of the curve? Based on these calculations, 2 would seem like a reasonable answer.

In the differential calculus, we will develop a technique for computing such slopes exactly (and easily). For now, note what distinguishes the calculus problem from the corresponding algebra problem. The calculus problem involves a process we call a **limit**. For the moment, the slope of the curve cannot be computed exactly, but is estimated using a sequence of approximations. The limit process enables us to extend the concept of slope from simple straight lines to complicated, ever-changing curves.

Example 8.1   **Estimating the Slope of a Curve**

Estimate the slope of $y = \sin x$ at $x = 0$.

Solution   This turns out to be a very important problem, one that we will return to later. For now, choose a sequence of points near $(0, 0)$ and compute the slopes of the secant lines joining those points with $(0, 0)$. The following table shows one set of choices.

Second point	m_{sec}		Second point	m_{sec}
$(1, \sin 1)$	0.8415		$(-1, \sin(-1))$	0.8415
$(0.1, \sin 0.1)$	0.9983		$(-0.1, \sin(-0.1))$	0.9983
$(0.01, \sin 0.01)$	0.9999		$(-0.01, \sin(-0.01))$	0.9999

Note that as the second point gets closer and closer to $(0, 0)$, the slope of the secant line (m_{sec}) gets closer and closer to 1. A good estimate of the slope of the curve at the point $(0, 0)$ would then appear to be 1. (At this stage, we have no way of computing the slope exactly.)

∎

A second problem requiring the power of calculus is that of computing distance along a curved path. While this problem is of less significance than our first example (both historically and in the development of the calculus), it provides a good indication of the need for mathematics beyond simple algebra. You should pay special attention to the similarities between the development of this problem and our earlier work with slope.

Recall that the distance formula (1.6) gives you the (straight-line) distance between two points. For instance, the distance between the points $(0, 1)$ and $(3, 4)$ is

$$d\{(0,1),(3,4)\} = \sqrt{(3-0)^2 + (4-1)^2} = 3\sqrt{2} \approx 4.24264.$$

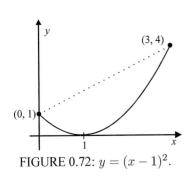

FIGURE 0.72: $y = (x-1)^2$.

However, this is not the only way we might want to compute the distance between these two points. For example, suppose that you needed to drive a car from $(0, 1)$ to $(3, 4)$ along the path defined by the curve $y = (x-1)^2$ (see Figure 0.72). In this case, you don't care about the straight-line distance connecting the two points, but only about how far you must drive along the curve (the **length** of the curve or **arc length**).

Notice that the distance along the curve must be greater than $3\sqrt{2}$ since $3\sqrt{2}$ is the straight-line distance. But how much greater? Taking a cue from the slope problem, you might want to formulate a strategy for obtaining better and better approximations. Instead of using just one line segment to get the approximation of $3\sqrt{2}$, you could use two line segments, as in Figure 0.73. Notice that the sum of the lengths of the two line segments appears to be a much better approximation to the actual length of the curve than the straight-line distance used previously. This distance is

$$d_2 = d\{(0,1),(1.5,.25)\} + d\{(1.5,.25),(3,4)\}$$
$$= \sqrt{(1.5-0)^2 + (.25-1)^2} + \sqrt{(3-1.5)^2 + (4-.25)^2} \approx 5.715\,925.$$

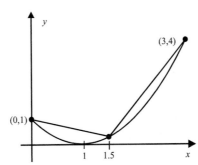

FIGURE 0.73: Two line segments.

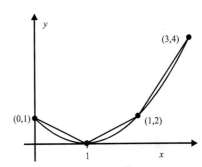

FIGURE 0.74: Three line segments.

You're probably way ahead of us by now. If approximating the length of the curve with two line segments gives an improved approximation, why not use three or four or more? Using the three line segments indicated in Figure 0.74, we get the further improved approximation

$$d_3 = d\{(0,1),(1,0)\} + d\{(1,0),(2,1)\} + d\{(2,1),(3,4)\}$$
$$= \sqrt{(1-0)^2 + (0-1)^2} + \sqrt{(2-1)^2 + (1-0)^2} + \sqrt{(3-2)^2 + (4-1)^2}$$
$$= 2\sqrt{2} + \sqrt{10} \approx 5.99070.$$

No. of Segments	Distance
1	4.24264
2	4.94026
3	5.99070
4	6.03562
5	6.06906
6	6.08713
7	6.09714

Note from Figure 0.74 that the distance along the curve is at least as large as d_3. Also, the more line segments we use, the better the approximation appears to be. While this process may be tedious, it will become much less so with the development of the definite integral in Chapter 4. For the moment, we list a number of these successively better approximations in the table found in the margin. The table suggests that the length of the curve is approximately 6.1 (quite far from the straight-line distance of 4.2). If you continued this rather tedious process using more and more line segments, the sum of their lengths would

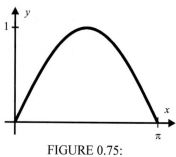

FIGURE 0.75:
$y = \sin x$ on $[0, \pi]$.

approach the actual length of the curve (about 6.126). This is the idea behind using the definite integral (developed in Chapter 4) to calculate such quantities.

Example 8.2   Estimating the Length of a Curve
Estimate the length of the curve $y = \sin x$ for $0 \leq x \leq \pi$ (see Figure 0.75).

Solution   The endpoints of the curve on this interval are $(0, 0)$ and $(\pi, 0)$. The distance between these points is $d_1 = \pi$. The point on the graph of $y = \sin x$ corresponding to the midpoint of the interval $[0, \pi]$ is $(\pi/2, 1)$. The distance from $(0, 0)$ to $(\pi/2, 1)$ plus the distance from $(\pi/2, 1)$ to $(\pi, 0)$ is

$$d_2 = \sqrt{(\pi/2)^2 + 1} + \sqrt{(\pi/2)^2 + 1} \approx 3.7242.$$

Using the five points $(0, 0)$, $(\pi/4, 1/\sqrt{2})$, $(\pi/2, 1)$, $(3\pi/4, 1/\sqrt{2})$, and $(\pi, 0)$ (i.e., four line segments), the sum of the lengths of these line segments is

$$d_4 = 2\sqrt{(\pi/4)^2 + 1/2} + 2\sqrt{(\pi/4)^2 + \left(1 - 1/\sqrt{2}\right)^2} \approx 3.7901.$$

Using eight points (i.e., seven line segments), you need a good calculator and some patience to compute the distance of 3.8125. At this stage, it would be reasonable to estimate the length of the sine curve as about 3.8.

∎

You may be relieved to know that we will develop some techniques to make such calculations much easier. For now, don't let the somewhat tedious arithmetic get in the way of your understanding the basic ideas behind the calculus. Notice that we extended the notion of distance to length of a curve by making some reasonably obvious (straight-line) approximations and then systematically improving on those approximations. As with our earlier problem on computing the slope of a curve, the essence of this is the **limit process**. In this case, the larger the number of line segments we used, the closer the approximate length got to the exact length.

It is the notion of limit that separates precalculus mathematics from the calculus. We develop this idea carefully in Chapter 1 and use limits extensively throughout the course. Another common theme is the use of algebra, graphical analysis, and numerical calculation to give us several different ways of looking at a given problem. We close this section with an indication of how this process might work.

In both example problems we discussed in this section, we used a sequence of approximations to lead us to an educated guess as to the exact solution. In the language of calculus, the exact solution is the **limit of a sequence** of approximate solutions. As a simple example, consider the sequence of numbers 1, 1/2, 1/4, 1/8, ..., where each term is half the previous term. It should be clear that the farther out you go in this sequence, the closer the terms get to 0. To make this easier to write out (and easier to recognize, for more complicated sequences), we introduce a notation for such sequences. For this sequence, write $a_0 = 1$, $a_1 = 1/2$, $a_2 = 1/4 = (1/2)^2$, $a_3 = 1/8 = (1/2)^3$ and so on. The subscript used in this notation (the **index**) is similar to the variable x for functions. In the present case, the first subscript was chosen to be 0 so that for later terms the subscript matches the exponent of $1/2$. We can write

$$a_n = (1/2)^n,$$

for $n = 0, 1, 2, \ldots$. The **limit** of the sequence is written

$$\lim_{n \to \infty} a_n = 0.$$

This notation indicates that as the subscript n gets larger and larger ($n \to \infty$), the terms of the sequence a_n get closer and closer to the limit 0.

Example 8.3   Finding the Limit of a Sequence

Write out several terms of the sequence defined by $a_n = \dfrac{2n-1}{2n+1}$, $n \geq 0$, and determine $\lim\limits_{n\to\infty} a_n$.

Solution   Substituting $n = 0$ into $a_n = \dfrac{2n-1}{2n+1}$ gives us

$$a_0 = \frac{2\cdot 0 - 1}{2 \cdot 0 + 1} = -1.$$

Similarly, we get $a_1 = \frac{1}{3}$, $a_2 = \frac{3}{5}$, $a_3 = \frac{5}{7}$, $a_4 = \frac{7}{9}$ and so on. Looking at the terms computed so far, you might guess that $\lim\limits_{n\to\infty} a_n = 1$. This can be verified algebraically by rewriting the sequence by dividing both numerator and denominator by n. This produces

$$a_n = \frac{(2n-1)\cdot(1/n)}{(2n+1)\cdot(1/n)} = \frac{2n/n - 1/n}{2n/n + 1/n} = \frac{2 - 1/n}{2 + 1/n}.$$

Notice that as n gets larger and larger, $1/n$ gets closer and closer to 0. This says that for very large n, both the numerator and denominator will be approximately equal to 2. For this reason, we write

$$\lim_{n\to\infty} a_n = \lim_{n\to\infty} \frac{2 - 1/n}{2 + 1/n} = \frac{2}{2} = 1.$$

∎

In the chapters that follow, we carefully develop the notion of limit and use limits to produce an assortment of powerful mathematical tools known collectively as the calculus.

Exercises 0.8

1. (writing) Explain why each approximation of length in example 8.2 is less than the actual length.

2. (writing) To estimate the slope of $f(x) = x^2 + 1$ at $x = 1$, you would compute the slopes of various secant lines. Note that $y = x^2 + 1$ curves up. Explain why the secant line connecting $(1, 2)$ and $(1.1, 2.21)$ will have slope greater than the slope of the tangent line. Discuss how the slope of the secant line between $(1, 2)$ and $(0.9, 1.81)$ compares to the slope of the tangent line.

In exercises 3–14, estimate the slope (as in example 8.1) of $y = f(x)$ **at** $x = a$.

3. $f(x) = x^2 + 1$, $a = 1$

4. $f(x) = x^2 + 1$, $a = 2$

5. $f(x) = \cos x$, $a = 0$

6. $f(x) = \cos x$, $a = \pi/2$

7. $f(x) = x^3 + 2$, $a = 0$

8. $f(x) = x^3 + 2$, $a = 1$

9. $f(x) = \sqrt{x+1}$, $a = 0$

10. $f(x) = \sqrt{x+1}$, $a = 3$

11. $f(x) = e^x$, $a = 0$

12. $f(x) = e^x$, $a = 1$

13. $f(x) = \ln x$, $a = 1$

14. $f(x) = \ln x$, $a = 2$

In exercises 15–22, estimate the length of the curve $y = f(x)$ **on the given interval using (a)** $n = 4$ **and (b)** $n = 8$ **line segments. (c) If you can program a calculator or computer, use larger** n**'s and conjecture the actual length of the curve.**

15. $f(x) = x^2 + 1$, $0 \leq x \leq 2$

16. $f(x) = x^3 + 2, 0 \le x \le 1$

17. $f(x) = \cos x, 0 \le x \le \pi/2$

18. $f(x) = \sin x, 0 \le x \le \pi/2$

19. $f(x) = \sqrt{x+1}, 0 \le x \le 3$

20. $f(x) = 1/x, 1 \le x \le 2$

21. $f(x) = x^2 + 1, -2 \le x \le 2$

22. $f(x) = x^3 + 2, -1 \le x \le 1$

In exercises 23–42, write out the first four terms of the sequence and determine $\lim\limits_{n \to \infty} a_n$.

23. $a_n = \dfrac{3n+1}{2n+1}, n \ge 0$

24. $a_n = \dfrac{2n+2}{n+1}, n \ge 0$

25. $a_n = \dfrac{n}{n+1}, n \ge 0$

26. $a_n = \dfrac{3n}{n+2}, n \ge 0$

27. $a_n = \dfrac{n+2}{n^2+1}, n \ge 0$

28. $a_n = \dfrac{n+3}{n^2+4}, n \ge 0$

29. $a_n = \dfrac{2n^2}{n^2+4}, n \ge 0$

30. $a_n = \dfrac{3n^2}{n^2+1}, n \ge 0$

31. $a_n = \dfrac{2}{n+1}, n \ge 0$

32. $a_n = \dfrac{3}{n^2+1}, n \ge 0$

33. $a_n = 1 - \dfrac{1}{n}, n \ge 1$

34. $a_n = 2 + \dfrac{1}{n}, n \ge 1$

35. $a_n = e^{-n}, n \ge 1$

36. $a_n = \sin(1/n), n \ge 1$

37. $a_n = \cos(1/n), n \ge 1$

38. $a_n = e^{1/n}, n \ge 0$

39. $a_n = \dfrac{n}{\sqrt{n^2+1}}, n \ge 0$

40. $a_n = \dfrac{\sqrt{n}}{n+1}, n \ge 0$

41. $a_n = 4 + \dfrac{1}{n} - \dfrac{2}{n^2}, n \ge 1$

42. $a_n = 3 - \dfrac{2}{n} + \dfrac{1}{n^2}, n \ge 1$

43. If $a_1 = 1 - 0.9$, $a_2 = 1 - 0.99$, $a_3 = 1 - 0.999$, etc., determine $\lim\limits_{n \to \infty} a_n$. Based on this limit, argue that $0.99\bar{9} = 1$.

44. As in exercise 43, argue that $0.1299\bar{9} = 0.13$.

45. (exploratory) The central concepts of calculus have been introduced in this section. An important aspect of our future development of calculus is to derive simple techniques for computing quantities such as slope and length. In this exercise, you will learn how to directly compute the slope of a curve at a point. Suppose you want the slope of $y = x^2$ at $x = 1$. You could start by computing slopes of secant lines connecting the point $(1, 1)$ with nearby points. Suppose the nearby point has x-coordinate $1 + h$, where h is a small (positive or negative) number. Explain why the corresponding y-coordinate is $(1 + h)^2$. Show that the slope of the secant line is $\dfrac{(1+h)^2 - 1}{1+h-1} = 2 + h$. As h gets closer and closer to 0, this slope better approximates the slope of the tangent line. By setting h equal to 0, show that the slope of the tangent line equals 2. In a similar way, show that the slope of $y = x^2$ at $x = 2$ is 4 and find the slope of $y = x^2$ at $x = 3$. Based on your answers, conjecture a formula for the slope of $y = x^2$ at $x = a$, for any unspecified value of a.

46. (exploratory) The **Sierpinski triangle** is an example of a **fractal**, a geometric shape with an infinite amount of detail. To construct the Sierpinski triangle, start with an equilateral triangle (Figure A). Mark the midpoints of each of the sides and connect the midpoints with line segments (Figure B). Remove the middle "upside-down" triangle. There are now three equilateral triangles. For each one, mark the midpoints, connect with line segments and remove the middle triangle (Figure C). Continue on forever in this fashion (a third stage is shown in Figure D). The eventual result is the Sierpinski triangle. For this exercise, you will track two different sequences. First, let a_0 be the area of the original triangle. If each side of the original triangle has length 1, show that $a_0 = \sqrt{3}/4$. Let a_1 be the area of the first stage of construction, as shown in Figure B. Show that $a_1 = 3\sqrt{3}/16$. Let a_2 be the area of the second stage of construction, as shown in Figure C. Find a_2. Continue to find the areas a_3 and a_4 and conjecture $\lim\limits_{n \to \infty} a_n$. Next, let p_0 be the perimeter of the original triangle. Show that $p_0 = 3$. Let p_1 be the perimeter of the first stage of construction, p_2 the perimeter of the second stage of construction, and so on. Determine $\lim\limits_{n \to \infty} p_n$. Fractals are often intricate and beautiful. They are also challenging mathematically because of apparent contradictions such as you've found: how can something have zero area and infinite length?

FIGURE A.

FIGURE B.

FIGURE C.

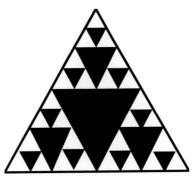

FIGURE D.

47. (exploratory) We close this section with one of the threads that runs throughout this book: there are surprising connections to be made among the different areas of mathematics. One such "coincidence" involves the Sierpinski triangle of exercise 46 and a game we'll call the **chaos game**. Start with the vertices of an equilateral triangle labeled A, B, and C. Pick one vertex at random. Again choose a vertex at random, move halfway from your starting vertex to the new vertex, and mark your location. Then repeat this sequence of steps: pick a vertex at random, move halfway from your current position to the vertex, and mark your new location. A small number of steps are shown in the figure, but yours will look different since you will choose different random vertices. What will the picture look like if you mark 1000 points? Program a graphing calculator or computer and find out! Seeing the screen fill up with dots is a great surprise to most people.

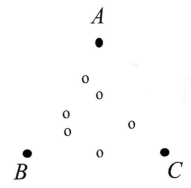

Chapter Review Exercises

In exercises 1–6, solve the inequality by isolating x.

1. $3x + 1 > 4$

2. $1 < 2x - 3 \le 5$

3. $x^2 - 2x - 8 > 0$

4. $\dfrac{x + 2}{x^2 - 4} > 0$

5. $|1 - 2x| < 3$

6. $|4 + 2x| > 2$

In exercises 7 and 8, find the distance between the pair of points.

7. $(0, 1), (2, 5)$

8. $(3, 1), (5, 2)$

In exercises 9 and 10, find the slope of the line through the given points.

9. $(2, 3), (0, 7)$

10. $(1, 4), (3, 1)$

In exercises 11 and 12, determine if the lines are parallel, perpendicular, or neither.

11. $y = 3x + 1$ and $y = 3(x - 2) + 4$

12. $y = -2(x + 1) - 1$ and $y = \frac{1}{2}x + 2$

13. Determine if the points $(1, 2)$, $(2, 4)$, and $(0, 6)$ form the vertices of a right triangle.

14. The data represents populations at various times. Plot the points, discuss any patterns and predict the population at the next time: $(0, 2100)$, $(1, 3050)$, $(2, 4100)$, and $(3, 5050)$.

15. Find an equation of the line through the given points and compute the y-coordinate corresponding to $x = 4$.

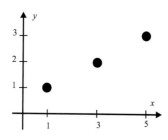

16. For $f(x) = x^2 - 3x - 4$, compute $f(0)$, $f(2)$ and $f(4)$.

In exercises 17 and 18, find an equation of the line with given slope and point.

17. $m = -\frac{1}{3}, (-1, -1)$

18. $m = \frac{1}{4}, (0, 2)$

In exercises 19 and 20, use the vertical line test to determine if the curve is the graph of a function.

19.

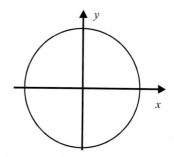

20.

In exercises 21 and 22, find the domain of the given function.

21. $f(x) = \sqrt{4 - x^2}$

22. $f(x) = \dfrac{x - 2}{x^2 - 2}$

In exercises 23–36, sketch a graph of the function showing extrema, intercepts, and asymptotes.

23. $f(x) = x^2 + 2x - 8$

24. $f(x) = x^3 - 6x + 1$

25. $f(x) = x^4 - 2x^2 + 1$

26. $f(x) = x^5 - 4x^3 + x - 1$

27. $f(x) = \dfrac{4x}{x + 2}$

28. $f(x) = \dfrac{x - 2}{x^2 - x - 2}$

29. $f(x) = \sin 3x$

30. $f(x) = \tan 4x$

31. $f(x) = \sin x + 2 \cos x$

32. $f(x) = \sec 2x$

33. $f(x) = 4e^{2x}$ **34.** $f(x) = 3e^{-4x}$

35. $f(x) = \ln 3x$ **36.** $f(x) = e^{\ln 2x}$

37. Determine all intercepts of $y = x^2 + 2x - 8$ (see exercise 23).

38. Determine all intercepts of $y = x^4 - 2x^2 + 1$ (see exercise 25).

39. Determine all asymptotes of $y = \dfrac{4x}{x + 2}$ (see exercise 27).

40. Determine all asymptotes of $y = \dfrac{x - 2}{x^2 - x - 2}$ (see exercise 28).

In exercises 41–44, find or estimate all zeros of the given function.

41. $f(x) = x^2 - 3x - 10$ **42.** $f(x) = x^3 + 4x^2 + 3x$

43. $f(x) = x^3 - 3x^2 + 2$ **44.** $f(x) = x^4 - 3x - 2$

In exercises 45 and 46, determine the number of solutions.

45. $\sin x = x^3$

46. $\sqrt{x^2 + 1} = x^2 - 1$

In exercises 47 and 48, determine whether or not the given function is periodic.

47. $f(x) = \cos 2x + 2 \sin 4x$ **48.** $f(x) = \sin 4x + 2 \cos \pi x$

49. A surveyor stands 50 feet from a telephone pole and measures an angle of $34°$ to the top. How tall is the pole?

50. Find $\sin \theta$ given that $0 < \theta < \frac{\pi}{2}$ and $\cos \theta = \frac{1}{5}$.

51. Convert to fractional or root form: (a) $5^{-1/2}$ (b) 3^{-2}.

52. Convert to exponential form: (a) $\dfrac{2}{\sqrt{x}}$ (b) $\dfrac{3}{x^2}$

53. Rewrite $\ln 8 - 2\ln 2$ as a single logarithm.

54. Solve the equation for x: $e^{\ln 4x} = 8$.

In exercises 55 and 56, solve the equation for x.

55. $3e^{2x} = 8$

56. $2\ln 3x = 5$.

In exercises 57 and 58, find $f \circ g$ and $g \circ f$ and identify their respective domains.

57. $f(x) = x^2, g(x) = \sqrt{x-1}$

58. $f(x) = x^2, g(x) = \dfrac{1}{x^2 - 1}$

In exercises 59 and 60, identify functions $f(x)$ and $g(x)$ such that $(f \circ g)(x)$ equals the given function.

59. $e^{3x^2 + 2}$

60. $\sqrt{\sin x + 2}$

In exercises 61 and 62, complete the square and explain how to transform the graph of $y = x^2$ into the graph of the given function.

61. $f(x) = x^2 - 4x + 1$ **62.** $f(x) = x^2 + 4x + 6$

In exercises 63 and 64, numerically estimate the slope of $y = f(x)$ at $x = a$.

63. $f(x) = x^2 - 2x, a = 2$ **64.** $f(x) = \sin 2x, a = 0$

In exercises 65 and 66, numerically estimate the length of the curve using (a) $n = 4$ and (b) $n = 8$ line segments.

65. $f(x) = \sin x, 0 \le x \le \frac{\pi}{4}$ **66.** $f(x) = x^2 - x, 0 \le x \le 2$

In exercises 67 and 68, write out the first four terms of the sequence and determine $\lim\limits_{n \to \infty} a_n$.

67. $a_n = \dfrac{3}{n^2 - 3}, n \ge 2$ **68.** $a_n = 2 - \dfrac{3n}{n^2 + 1}, n \ge 0$

Chapter 1
Limits and Continuity

Small pupils
(Photo courtesy of Roland Minton.)

Large pupils
(Photo courtesy of Roland Minton.)

When you enter a darkened room, your eyes adjust to the reduced level of light by increasing the size of your pupils. Enlarging the pupils allows more light to enter the eyes, which makes objects around you easier to see. By contrast, when you enter a brightly lit room, your pupils contract. Narrowing the pupils reduces the amount of light entering the eyes. This is necessary since too much light will overload your visual system. You may have experienced this kind of overload when walking outside on a sunny day after having your pupils dilated for an eye exam. The dilation enlarges your pupils, which are then unable to narrow sufficiently to make the outside light manageable.

This visual adjustment mechanism is present in many animals. Researchers study this mechanism by performing experiments and trying to find a mathematical description of the results. In this case, you might want to represent the size of the pupils as a function of the amount of light present. Two basic characteristics of such a **mathematical model** would be

1. As the amount of light (x) increases, the pupil size (y) decreases; and
2. As the amount of light (x) decreases, the pupil size (y) increases.

One function that satisfies both properties is a linear function with negative slope (for instance, $f(x) = 1 - x$).

However, there are other factors that we need to take into account. For example, the pupil size can't be negative. In fact, each animal has a minimum size beyond which the pupils can't shrink. Because of this, in extremely bright light we must squint or completely shut our eyes to block out the light. Also, there is a maximum size beyond which the pupils can't enlarge. Because of this, in extreme darkness we simply can't see well. We need to modify our modeling principles above to read

1. As the amount of light (x) increases, the pupil size (y) decreases down to a minimum value p; and
2. As the amount of light (x) decreases, the pupil size (y) increases up to a maximum value P.

Finding a function with these two properties can be a challenge. (Try it!)

In this chapter, we develop the concept of **limit**, which can be used to describe functions with specific properties like those listed above. In the pupil scenario, we want the limit of the pupil size to equal P as the amount of light decreases to zero. We also want the limit of the pupil size to equal p as the amount of light increases to infinity. The computation and visualization of these types of limits is the main goal of this chapter.

The limit is the fundamental notion of calculus. This underlying concept is the thread that binds together virtually all of the calculus you are about to study. You will find limits appearing in the definition of nearly all of the major concepts of calculus, from continuity to derivatives to the definite integral to sequences and series. In order to **understand** what calculus is all about, you will need to have more than a passing knowledge of what limits are. This is not the same as simply being able to calculate a few limits. You'll need the ability to calculate limits, but by itself, this is insufficient for understanding what follows. An investment in carefully studying limits now will have very significant payoffs throughout the remainder of your calculus experience and beyond.

1.1   The Concept of Limit

In this section, we develop the notion of limit using some common language and illustrate the idea with some simple examples. The notion turns out to be a rather subtle one, easy to think of intuitively, but a bit harder to pin down in precise terms. We leave our presentation of the precise definition of limit to the (optional) section 1.5. There, we carefully define the limit in considerable detail. The more informal notion of limit that we introduce and work with over the next several sections is adequate for most purposes. Those of you who wish to have a full exposition should work through the next several sections before tackling the challenge of section 1.5.

As a start, consider the function

$$f(x) = \frac{x^2 - 5}{x - 2}.$$

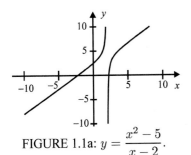

FIGURE 1.1a: $y = \dfrac{x^2 - 5}{x - 2}$.

What's the first thing that you notice about this function? You probably observed that the function is undefined at $x = 2$. So, what does this mean, beyond saying that you cannot substitute 2 for x? We can often find important clues about the behavior of a function from a calculator or computer-generated graph (see Figure 1.1a). Many calculator and computer-generated graphs will include the vertical segment indicated in Figure 1.1b; this is **not** a part of the graph. Rather, it results from your calculator or computer connecting two points, one slightly to the left of $x = 2$, the other slightly to the right of $x = 2$, where one function value is large and positive and one is very negative. The vertical line segment plotted suggests the vertical asymptote shown in Figure 1.1c. There, we correctly indicate the vertical asymptote as the dashed vertical line $x = 2$. This is a common problem in calculator or computer-generated graphs. When you see such vertical segments, you should expect that they represent vertical asymptotes. We discuss asymptotes in more detail in section 1.4.

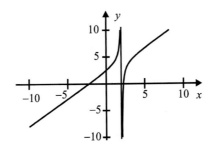

FIGURE 1.1b: A common calculator or computer-generated graph.

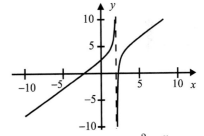

FIGURE 1.1c: $y = \dfrac{x^2 - 5}{x - 2}$.

The graphs in Figures 1.1a–c look very much like straight lines, except in the vicinity of $x = 2$, where they look quite different. This suggests that we might want to carefully examine the behavior of a function in the vicinity of a point not in its domain. Before doing so or drawing any further conclusions, we consider a closely related function:

$$g(x) = \frac{x^2 - 4}{x - 2}.$$

FIGURE 1.2a: $\dfrac{x^2 - 4}{x - 2} = 4$.

Here again, $x = 2$ is not in the domain of the function. To see what this might mean, let's look at a computer-generated graph (see Figure 1.2a). It may be surprising to see that the graph appears to be a straight line with no unusual behavior noticeable near $x = 2$. There is actually a hole in the graph at $x = 2$, but it's not visible in this graph. This is another limitation of your calculator or computer software. A computer or calculator can't show a single point removed from a graph. We indicate such a point with an open circle, as shown in Figure 1.2b. In any case, what's the big deal about $x = 2$ not being in the domain of a function? To answer this question, it will be helpful to compute some values of both functions for x close to 2, as in the following table.

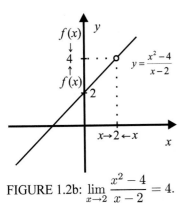

FIGURE 1.2b: $\displaystyle\lim_{x \to 2} \frac{x^2 - 4}{x - 2} = 4$.

x	$f(x) = \dfrac{x^2 - 5}{x - 2}$	$g(x) = \dfrac{x^2 - 4}{x - 2}$
1.9	13.9	3.9
1.99	103.99	3.99
1.999	1003.999	3.999
1.9999	10,003.9999	3.9999
1.99999	100,003.99999	3.99999
1.999999	10,000,004	3.999999

We also need to consider what happens to the function values for x close to 2 but larger than 2.

x	$f(x) = \dfrac{x^2 - 5}{x - 2}$	$g(x) = \dfrac{x^2 - 4}{x - 2}$
2.1	-5.9	4.1
2.01	-95.99	4.01
2.001	-995.999	4.001
2.0001	-9995.9999	4.0001
2.00001	$-99,995.99999$	4.00001
2.000001	$-999,995.999999$	4.000001

Clearly, f and g behave quite differently near $x = 2$. From both the tables and the graph, it appears that as x gets closer and closer to 2, $g(x)$ is getting closer and closer to 4. In view of this, we say that the **limit of $g(x)$ as x approaches 2 is 4**, written

$$\lim_{x \to 2} g(x) = 4$$

or

$$\lim_{x \to 2} \frac{x^2 - 4}{x - 2} = 4.$$

Notice that the behavior of $f(x)$ depends on whether $x > 2$ or $x < 2$. Once again, using both the table and the graph, it appears that as x gets closer and closer to 2 **from the left** $(x < 2)$, $f(x)$ is increasing, without bound. In this case, we say that the **one-sided limit (the limit of $f(x)$ as x approaches 2 from the left) does not exist**, written

$$\lim_{x \to 2^-} f(x) \text{ \textbf{does not exist}},$$

since there's no number that $f(x)$ is approaching. Likewise, as x gets closer and closer to 2 **from the right** $(x > 2)$, $f(x)$ is decreasing without bound. Here, we say that the one-sided limit (**the limit of $f(x)$ as x approaches 2 from the right) does not exist**, written

$$\lim_{x \to 2^+} f(x) \text{ \textbf{does not exist}}.$$

Finally, we say that

$$\lim_{x \to 2} f(x) \text{ \textbf{does not exist}}.$$

Before leaving these two examples, we should make one further observation. The notion of limit as we have described it here is intended to communicate the behavior of a function *near* some point of interest, but not actually *at* that point. Now, notice that since the expression in the numerator of $g(x)$ factors, we can write

$$\lim_{x \to 2} g(x) = \lim_{x \to 2} \frac{x^2 - 4}{x - 2}$$

$$= \lim_{x \to 2} \frac{(x - 2)(x + 2)}{x - 2}$$

$$= \lim_{x \to 2} (x + 2)$$

$$= 4,$$

where we can cancel the factors of $(x - 2)$ since in the limit as $x \to 2$, x is *close* to 2, but $x \neq 2$.

Cancel two factors of $(x - 2)$.

As x approaches 2,
$(x + 2)$ approaches 4.

FIGURE 1.3:
$$\lim_{x \to -3} \frac{3x + 9}{x^2 - 9} = -\frac{1}{2}.$$

x	$\dfrac{3x + 9}{x^2 - 9}$
-3.1	-0.491803
-3.01	-0.499168
-3.001	-0.499917
-3.0001	-0.499992
-3.00001	-0.499999

x	$\dfrac{3x + 9}{x^2 - 9}$
-2.9	-0.508475
-2.99	-0.500835
-2.999	-0.500083
-2.9999	-0.500008
-2.99999	-0.500001

Example 1.1   **A Limit Where Two Factors Cancel**

Evaluate

$$\lim_{x \to -3} \frac{3x + 9}{x^2 - 9}.$$

Solution   Here, the domain of the function includes all values of x except $x = \pm 3$. Again, we examine a graph (see Figure 1.3) and compute some function values for x near -3.

Further, note that for $x \neq \pm 3$, we have

$$\frac{3x + 9}{x^2 - 9} = \frac{3(x + 3)}{(x + 3)(x - 3)} = \frac{3}{x - 3}.$$

This suggests that

$$\lim_{x \to -3^-} \frac{3x + 9}{x^2 - 9} = \lim_{x \to -3^-} \frac{3}{x - 3} = -\frac{1}{2}$$

since $(x - 3) \to -6$ as $x \to -3$. Likewise,

$$\lim_{x \to -3^+} \frac{3x + 9}{x^2 - 9} = -\frac{1}{2}.$$

Finally, since the function approaches the **same** value as $x \to -3$ both from the right and from the left (i.e., the one-sided limits are equal), we write

$$\lim_{x \to -3} \frac{3x + 9}{x^2 - 9} = -\frac{1}{2}.$$
                                                                                      ■

In general,

$$\lim_{x \to a} f(x) = L$$

if and only if

$$\lim_{x \to a^-} f(x) = L \text{ and } \lim_{x \to a^+} f(x) = L.$$

In other words, **a limit exists if and only if both one-sided limits exist and are equal**.

Example 1.2   **A Limit That Does Not Exist**

Determine whether or not $\lim_{x \to 0} \dfrac{1}{x}$ exists.

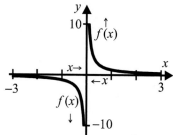

FIGURE 1.4: $\lim\limits_{x\to 0} \dfrac{1}{x}$

does not exist.

x	$1/x$
0.1	10
0.01	100
0.001	1000
0.0001	10, 000
0.00001	100, 000

x	$1/x$
−0.1	−10
−0.01	−100
−0.001	−1000
−0.0001	−10, 000
−0.00001	−100, 000

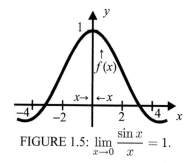

FIGURE 1.5: $\lim\limits_{x\to 0} \dfrac{\sin x}{x} = 1$.

Solution   We first graph $y = f(x)$ (see Figure 1.4) and compute some function values for x close to 0 (you can do this easily, by hand).

Notice that, as $x \to 0^+$, $f(x)$ is increasing without bound. Thus,

$$\lim_{x\to 0^+} \frac{1}{x} \text{ does not exist.}$$

Likewise, we can say that

$$\lim_{x\to 0^-} \frac{1}{x} \text{ does not exist.}$$

Since neither one-sided limit exists, we say

$$\lim_{x\to 0} \frac{1}{x} \text{ does not exist.}$$

Here, we considered both one-sided limits for the sake of completeness. Of course, you should keep in mind that if **either** one-sided limit fails to exist, then the limit does not exist.

∎

Example 1.3   Approximating the Value of a Limit

Evaluate $\lim\limits_{x\to 0} \dfrac{\sin x}{x}$.

Solution   Unlike some of the limits considered previously, there is no algebra which will simplify this expression. However, we can still draw a graph (see Figure 1.5) and compute some function values.

x	$\dfrac{\sin x}{x}$	x	$\dfrac{\sin x}{x}$
0.1	0.998334	−0.1	0.998334
0.01	0.999983	−0.01	0.999983
0.001	0.99999983	−0.001	0.99999983
0.0001	0.9999999983	−0.0001	0.9999999983
0.00001	0.999999999983	−0.00001	0.999999999983

Based on the graph and the function values in the tables, we can now make the conjectures:

$$\lim_{x\to 0^+} \frac{\sin x}{x} = 1 \text{ and } \lim_{x\to 0^-} \frac{\sin x}{x} = 1,$$

from which we conjecture that

$$\lim_{x\to 0} \frac{\sin x}{x} = 1.$$

In Chapter 2, we examine these limits with greater care (and prove that these conjectures are correct). For the moment, conjectures based on graphs and tables of values will suffice.

∎

Remark 1.1   Computer or calculator computation of limits is unreliable. We use graphs and tables of values only as (strong) evidence pointing to what a plausible answer might be. To be certain, we need to obtain careful verification of our conjectures. We see how to do this in the remaining sections of this chapter.

FIGURE 1.6: $y = (x + .01)^5$.

x	$(x + .01)^5$
0.1	1.611×10^{-5}
0.01	3.2×10^{-9}
0.001	1.611×10^{-10}
0.0001	1.051×10^{-10}

x	$(x + .01)^5$
-0.1	-5.905×10^{-6}
-0.01	0.0
-0.001	5.905×10^{-11}
-0.0001	9.51×10^{-11}

Example 1.4   **A Limit for Which Approximate Answers May be Deceptive**
Examine

$$\lim_{x \to 0} (x + 0.01)^5.$$

Solution   Let's see what happens when we examine this limit numerically and graphically. First, a graph of $y = (x + 0.01)^5$ is shown in Figure 1.6. A table of function values is also helpful and is shown in the margin.

So, what would a reasonable individual conclude from the graph and the computed function values? You might be tempted to conclude that $\lim_{x \to 0} f(x) = 0$. On closer examination, however, we can see that as $x \to 0$, $(x + 0.01) \to 0.01$. It should then follow that $(x + 0.01)^5 \to 0.01^5 = 1 \times 10^{-10}$. Add one or two more lines to these tables to convince yourself of what you may have missed. The moral: be very skeptical of machine-generated answers. Still, you might say that the answer wasn't all that far off. Unfortunately, this is not the only type of error that can occur. In section 1.6, we examine more serious and difficult to detect errors.

∎

Example 1.5   **A Limit Where One-Sided Limits are Needed**
Evaluate

$$\lim_{x \to 0} \frac{x}{|x|}.$$

Solution   The computer-generated graph shown in Figure 1.7a is revealing. As we have previously observed, your calculator or computer may draw a vertical segment connecting the two horizontal segments. Again, such a vertical segment is not actually a part of the graph, but results from the computer or calculator connecting two points on either side of the x-axis. In this case, though, there is no vertical asymptote at $x = 0$. The graph in Figure 1.7b correctly shows open circles at the intersections of the two halves of the graph with the y-axis. We also have

$$\lim_{x \to 0^+} \frac{x}{|x|} = \lim_{x \to 0^+} \frac{x}{x} \quad \text{(since } |x| = x, \text{ when } x > 0.)$$
$$= \lim_{x \to 0^+} 1$$
$$= 1$$

and

$$\lim_{x \to 0^-} \frac{x}{|x|} = \lim_{x \to 0^-} \frac{x}{-x} \quad \text{(since } |x| = -x, \text{ when } x < 0.)$$
$$= \lim_{x \to 0^-} -1$$
$$= -1.$$

It now follows that

$$\lim_{x \to 0} \frac{x}{|x|} \text{ does not exist,}$$

since the one-sided limits are not the same. You should also keep in mind that this observation is entirely consistent with what we see in the graph.

∎

FIGURE 1.7a: $y = \dfrac{x}{|x|}$.

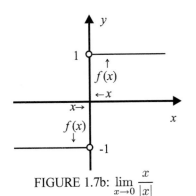

FIGURE 1.7b: $\lim_{x \to 0} \dfrac{x}{|x|}$
does not exist.

Recall that the open circles in Figure 1.7b indicate that the point is not on the curve. Alternatively, a closed (solid) circle indicates that the point is on the curve.

FIGURE 1.8:

$$y = \frac{0.68}{\omega} - \frac{1}{4\omega^2} \sin(2.72\omega).$$

Example 1.6   A Limit Describing the Movement of a Baseball Pitch

The knuckleball is one of the most exotic pitches in baseball. Batters describe the ball as unpredictably moving left, right, up and down. The motion of the knuckleball is mostly due to the raised stitches on the baseball, which act like an airplane wing to give a "lift force" to the ball. The motion is erratic because as the ball moves, the orientation of the stitches shifts and hence the lift force on the baseball changes. The left/right position p of the ball can be described by the equation

$$p = \frac{1}{\omega}t - \frac{1}{4\omega^2} \sin(4\omega t)$$

(derived from experimental data in Watts and Bahill's excellent book *Keeping Your Eye on the Ball*). Here, ω is the rotational speed of the ball, in radians per second, p is the left/right position of the ball in feet ($p = 0$ corresponds to the middle of home plate) and t is time in seconds. For a typical knuckleball speed of 60 mph, the ball takes about 0.68 second to reach the batter. Its position at that time is then

$$p = \frac{0.68}{\omega} - \frac{1}{4\omega^2} \sin(2.72\omega).$$

So, if the ball is thrown with rotational speed $\omega = 1$ radian per second, it will end up at position $\frac{0.68}{1} - \frac{1}{4} \sin(2.72) = 0.5777$ feet (about 7 inches). Likewise, if $\omega = 10$ radians per second, the ball finishes at $\frac{0.68}{10} - \frac{1}{400} \sin(27.2) = 0.0658$ feet (less than 1 inch). Folk wisdom among baseball pitchers has it that the less spin on the ball, the better the pitch. These two calculations back up this theory, but we should investigate the limit as $\omega \to 0$. As always, we look at a graph, in this case of

$$f(\omega) = \frac{0.68}{\omega} - \frac{1}{4\omega^2} \sin(2.72\omega)$$

(see Figure 1.8) and the accompanying table of function values.

ω	$\dfrac{0.68}{\omega} - \dfrac{1}{4\omega^2} \sin(2.72\omega)$
10	0.0658
1	0.5777
0.1	0.0835
0.01	0.0083
0.001	0.0008

The graphical and numerical evidence suggests that $\lim\limits_{\omega \to 0} f(\omega) = 0$.

As Watts and Bahill describe it, a knuckleball with absolutely no spin is just a very slow pitch. According to Watts and Bahill, the optimal movement appears to occur with spin rate of about $\omega = 1$ radian per second. The total spin of 0.68 radian is about one-tenth of a complete rotation of the ball on the way to the plate. Thus, a very slow rotation is best. This will be explored in the exercises.

■

Exercises 1.1

1. (writing) Suppose your professor says, "You can think of the limit of $f(x)$ as x approaches a as *what f(a) should be*." Critique this statement. What does it mean? Does it provide important insight? Is there anything misleading about it? Replace the phrase in italics with your own best description of what the limit is.

2. (writing) Your friend's professor says, "The limit is a *prediction* of what $f(a)$ will be." Compare and contrast this statement to the one in exercise 1. Does the inclusion of the word **prediction** make the limit idea seem more useful and important?

3. (writing) We have emphasized that $\lim_{x \to a} f(x)$ does not depend on the actual value of $f(a)$, or even on whether $f(a)$ exists. In principle, functions such as $f(x) = \begin{cases} x^2 & \text{if } x \neq 2 \\ 13 & \text{if } x = 2 \end{cases}$ are as "normal" as functions such as $g(x) = x^2$. With this in mind, explain why it is important that the limit concept is independent of how (or whether) $f(a)$ is defined.

4. (writing) The most common limit encountered in everyday life is the **speed limit**. Describe how this type of limit is very different from the limits discussed in this section.

5. For the function graphed below, identify each limit or state that it does not exist.

(a) $\lim_{x \to 0^-} f(x)$ (b) $\lim_{x \to 0^+} f(x)$ (c) $\lim_{x \to 0} f(x)$

(d) $\lim_{x \to 1^-} f(x)$ (e) $\lim_{x \to -1} f(x)$ (f) $\lim_{x \to 2^-} f(x)$

(g) $\lim_{x \to 2^+} f(x)$ (h) $\lim_{x \to 2} f(x)$ (i) $\lim_{x \to -2} f(x)$

(j) $\lim_{x \to 3} f(x)$

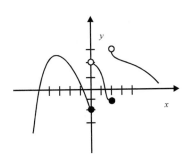

6. For the function graphed below, identify each limit or state that it does not exist.

(a) $\lim_{x \to 0^-} f(x)$ (b) $\lim_{x \to 0^+} f(x)$ (c) $\lim_{x \to 0} f(x)$

(d) $\lim_{x \to 2^-} f(x)$ (e) $\lim_{x \to -2} f(x)$ (f) $\lim_{x \to 1^-} f(x)$

(g) $\lim_{x \to 1^+} f(x)$ (h) $\lim_{x \to 1} f(x)$ (i) $\lim_{x \to -1} f(x)$

(j) $\lim_{x \to 3} f(x)$

7. Sketch the graph of $f(x) = \begin{cases} 2x & \text{if } x < 2 \\ x^2 & \text{if } x \geq 2 \end{cases}$ and identify each limit.

(a) $\lim_{x \to 2^-} f(x)$ (b) $\lim_{x \to 2^+} f(x)$ (c) $\lim_{x \to 2} f(x)$

(d) $\lim_{x \to 1} f(x)$

8. Sketch the graph of $f(x) = \begin{cases} x^3 - 1 & \text{if } x < 0 \\ 0 & \text{if } x = 0 \\ \sqrt{x+1} - 2 & \text{if } x > 0 \end{cases}$ and identify each limit.

(a) $\lim_{x \to 0^-} f(x)$ (b) $\lim_{x \to 0^+} f(x)$ (c) $\lim_{x \to 0} f(x)$

(d) $\lim_{x \to -1} f(x)$ (e) $\lim_{x \to 3} f(x)$

9. Sketch the graph of $f(x) = \begin{cases} x^2 + 1 & \text{if } x < -1 \\ 3x + 1 & \text{if } x \geq -1 \end{cases}$ and identify each limit.

(a) $\lim_{x \to -1^-} f(x)$ (b) $\lim_{x \to -1^+} f(x)$ (c) $\lim_{x \to -1} f(x)$

(d) $\lim_{x \to 0} f(x)$

10. Sketch the graph of $f(x) = \begin{cases} 2x + 1 & \text{if } x < -1 \\ 3 & \text{if } -1 \leq x < 1 \\ 2x + 1 & \text{if } x > 1 \end{cases}$ and identify each limit.

(a) $\lim_{x \to -1^-} f(x)$ (b) $\lim_{x \to -1^+} f(x)$ (c) $\lim_{x \to -1} f(x)$

(d) $\lim_{x \to 1} f(x)$ (e) $\lim_{x \to 0} f(x)$

11. Evaluate $f(1.5)$, $f(1.1)$, $f(1.01)$ and $f(1.001)$ and conjecture a value for $\lim_{x \to 1^+} f(x)$ for $f(x) = \dfrac{x-1}{\sqrt{x}-1}$. Evaluate $f(0.5)$, $f(0.9)$, $f(0.99)$ and $f(0.999)$ and conjecture a value for $\lim_{x \to 1^-} f(x)$ for $f(x) = \dfrac{x-1}{\sqrt{x}-1}$. Does $\lim_{x \to 1} f(x)$ exist?

12. Evaluate $f(-1.5)$, $f(-1.1)$, $f(-1.01)$ and $f(-1.001)$ and conjecture a value for $\lim_{x \to -1^-} f(x)$ for $f(x) = \dfrac{x+1}{x^2-1}$. Evaluate $f(-0.5)$, $f(-0.9)$, $f(-0.99)$ and $f(-0.999)$ and conjecture a value for $\lim\limits_{x \to -1^+} f(x)$ for $f(x) = \dfrac{x+1}{x^2-1}$. Does $\lim\limits_{x \to -1} f(x)$ exist?

In exercises 13–18, use numerical and graphical evidence to conjecture values for each limit.

13. $\lim\limits_{x \to 1} \dfrac{x^2-1}{x+2}$

14. $\lim\limits_{x \to -1} \dfrac{x^2+x}{x^2-x-2}$

15. $\lim\limits_{x \to 0} \dfrac{x^2+x}{\sin x}$

16. $\lim\limits_{x \to \pi} x \sin x$

17. $\lim\limits_{x \to 0} \dfrac{\tan x}{\sin x}$

18. $\lim\limits_{x \to 2} \sqrt{x^2-2x+2}$

In exercises 19–26, use numerical and graphical evidence to conjecture whether or not the limit exists. If not, describe what is happening at $x = a$ graphically.

19. $\lim\limits_{x \to 2} \dfrac{x-2}{|x-2|}$

20. $\lim\limits_{x \to -1} \dfrac{|x+1|}{x^2-1}$

21. $\lim\limits_{x \to -1} \dfrac{x^2-1}{x^2+2x-3}$

22. $\lim\limits_{x \to 1} \dfrac{x^2-1}{x-1}$

23. $\lim\limits_{x \to 1} \dfrac{x^2+1}{x-1}$

24. $\lim\limits_{x \to 1} \dfrac{x^2-1}{x^2+2x-3}$

25. $\lim\limits_{x \to 1} \dfrac{x^2-1}{x^2-2x+1}$

26. $\lim\limits_{x \to -1} \dfrac{x^2-1}{x+1}$

27. Use your results from exercises 21–26 to investigate the following. Suppose that $f(x)$ and $g(x)$ are polynomials with $g(a) = 0$ and $f(a) \neq 0$. What can you conjecture about $\lim\limits_{x \to a} \dfrac{f(x)}{g(x)}$? What can you conjecture if $g(a) = 0$ and $f(a) = 0$?

28. Use your results from exercises 21–26 to investigate the following. Suppose that $f(x)$ and $g(x)$ are polynomials with $f(a) = 0$ and $g(a) \neq 0$. What can you conjecture about $\lim\limits_{x \to a} \dfrac{f(x)}{g(x)}$?

In exercises 29–32, sketch a graph of a function with the given properties.

29. $f(-1) = 2$, $f(0) = -1$, $f(1) = 3$ and $\lim\limits_{x \to 1} f(x)$ does not exist.

30. $f(x) = 1$ for $-2 \leq x \leq 1$, $\lim\limits_{x \to 1^+} f(x) = 3$ and $\lim\limits_{x \to -2} f(x) = 1$.

31. $f(0) = 1$, $\lim\limits_{x \to 0^-} f(x) = 2$ and $\lim\limits_{x \to 0^+} f(x) = 3$.

32. $\lim\limits_{x \to 0} f(x) = -2$, $f(0) = 1$, $f(2) = 3$ and $\lim\limits_{x \to 2} f(x)$ does not exist.

33. As we see in Chapter 2, the slope of the tangent line to the curve $y = \sqrt{x}$ at $x = 1$ is given by $m = \lim\limits_{h \to 0} \dfrac{\sqrt{1+h}-1}{h}$.

Estimate the slope m. Graph $y = \sqrt{x}$ and the line with slope m through the point $(1,1)$.

34. As we see in Chapter 2, the velocity of an object that has traveled $\sqrt{x}$ miles in x hours at the $x = 1$ hour mark is given by $v = \lim\limits_{x \to 1} \dfrac{\sqrt{x}-1}{x-1}$. Estimate this limit.

35. Numerically estimate the limits $\lim\limits_{x \to 0^+} (1 + x)^{1/x}$ and $\lim\limits_{x \to 0^-} (1 + x)^{1/x}$. Note that the function values for $x > 0$ increase as x decreases, while for $x < 0$ the function values decrease as x increases. Explain why this indicates that the limit $\lim\limits_{x \to 0} (1 + x)^{1/x}$ is between function values for positive and negative x's. Approximate this limit correct to eight digits.

36. Explain what is wrong with the following logic (you may use exercise 35 to convince yourself that the answer is wrong, but discuss the logic without referring to exercise 35): as $x \to 0$, it is clear that $(1 + x) \to 1$. Since 1 raised to any power is 1, $\lim\limits_{x \to 0} (1 + x)^{1/x} = \lim\limits_{x \to 0} (1)^{1/x} = 1$.

37. Numerically estimate $\lim\limits_{x \to 0^+} x^{\sec x}$. Try to numerically estimate $\lim\limits_{x \to 0^-} x^{\sec x}$. If your computer has difficulty evaluating the function for negative x's, explain why.

38. Explain what is wrong with the following logic (note from exercise 37 that the answer is accidentally correct): since 0 to any power is 0, $\lim\limits_{x \to 0} x^{\sec x} = \lim\limits_{x \to 0} 0^{\sec x} = 0$.

39. Give an example of a function $f(x)$ such that $\lim\limits_{x \to 0} f(x)$ exists but $f(0)$ does not exist. Give an example of a function $g(x)$ such that $g(0)$ exists but $\lim\limits_{x \to 0} g(x)$ does not exist.

40. Give an example of a function $f(x)$ such that $\lim\limits_{x \to 0} f(x)$ exists and $f(0)$ exists, but $\lim\limits_{x \to 0} f(x) \neq f(0)$.

41. (exploratory) In the text, we described $\lim\limits_{x \to a} f(x) = L$ as meaning "as x gets closer and closer to a, $f(x)$ is getting closer and closer to L." As x gets closer and closer to 0, it is true that x^2 gets closer and closer to -0.01, but it is certainly not true that $\lim\limits_{x \to 0} x^2 = -0.01$. Try to modify the description of limit to make it clear that $\lim\limits_{x \to 0} x^2 \neq -0.01$. We will explore a very precise definition of limit in section 1.5.

42. (exploratory) In Figure 1.8, the final position of the knuckleball is shown as a function of the rotation rate ω. The batter must decide at time $t = 0.4$ whether or not to swing at the pitch. Graph the resulting function $h(\omega) = \dfrac{0.4}{\omega} - \dfrac{1}{4\omega^2} \sin(1.6\omega)$ and compare to Figure 1.8. Conjecture the limit of $h(\omega)$ as $\omega \to 0$. For $\omega = 0$, is there any difference in ball position between what the batter sees at $t = 0.4$ and what he hits at $t = 0.68$?

43. (exploratory) As discussed in example 1.6, the left/right motion of a knuckleball pitch in baseball can be modeled by

$P = \frac{1}{\omega}t - \frac{1}{4\omega^2}\sin 4\omega t$, where t is time measured in seconds ($0 \le t \le 0.68$) and ω is the rotation rate of the ball measured in radians per second. In example 1.6, we chose a specific t-value and evaluated the limit as $\omega \to 0$. While this gives us some information about which rotation rates produce hard-to-hit pitches, a clearer picture emerges if we look at $f(t)$ over its entire domain. Set $\omega = 10$ and graph the resulting function ($\frac{1}{10}t - \frac{1}{400}\sin 40t$) for $0 \le t \le 0.68$. Try to visualize a baseball starting at the pitcher's hand at $t = 0$ and finally reaching you, the batter, at $t = 0.68$. Repeat this with $\omega = 5$, $\omega = 1$, $\omega = 0.1$ and whatever values of ω you think would be interesting. Which values of ω produce hard-to-hit pitches?

44. (exploratory) In this exercise, the results you get will depend on the accuracy of your computer or calculator. Work this exercise with a graphing calculator, and compare your results with your classmates' results. We will investigate $\lim\limits_{x \to 0} \dfrac{\cos x - 1}{x^2}$. Start by verifying the calculations below:

x	$f(x)$
0.1	$-0.499583\ldots$
0.01	$-0.49999583\ldots$
0.001	$-0.4999999583\ldots$

Describe as precisely as possible the pattern shown here. What would you predict for $f(0.0001)$? $f(0.00001)$? Does your computer or calculator give you this answer? If you continue trying powers of 0.1 (0.000001, 0.0000001, etc.) you should eventually be given a displayed result of -0.5. Do you think this is exactly correct, or has the answer just been rounded off? Why is rounding off inescapable? It turns out that -0.5 is the exact value for the limit, so the round-off here is somewhat helpful. However, if you keep evaluating the function at smaller and smaller values of x, you will eventually see a reported function value of 0. This round-off error is not so benign; we discuss this error in section 1.6. For now, evaluate $\cos x$ at the current value of x and try to explain where the 0 came from.

1.2   Computation of Limits

Now that we have an idea of what a limit is, we need to develop some means of calculating limits of simple functions. In this section, we present some basic rules for dealing with common limit problems.

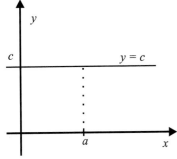

FIGURE 1.9: $\lim\limits_{x \to a} c = c$.

Theorem 2.1   For any constant c and any real number a,

$$\lim_{x \to a} c = c.$$

This certainly comes as no surprise, since the constant c does not depend on x and so, stays the same as $x \to a$ (see Figure 1.9).

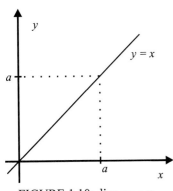

FIGURE 1.10: $\lim\limits_{x \to a} x = a$.

Theorem 2.2   For any real number a,

$$\lim_{x \to a} x = a.$$

Again, this is not a surprise, since as $x \to a$, x will approach a (see Figure 1.10). Be sure that you are comfortable enough with the limit notation to recognize how obvious the limits in Theorems 2.1 and 2.2 are. As simple as they are, we use them repeatedly in finding more complex limits. We also need the basic rules contained in the following theorem.

Theorem 2.3   Suppose that $\lim\limits_{x \to a} f(x)$ and $\lim\limits_{x \to a} g(x)$ both exist and equal L_1 and L_2, respectively and let c be any constant. The following then apply:

(i) $\lim\limits_{x \to a} [c \cdot f(x)] = c \cdot \lim\limits_{x \to a} f(x) = c \cdot L_1$,

(ii) $\lim\limits_{x \to a} [f(x) \pm g(x)] = \lim\limits_{x \to a} f(x) \pm \lim\limits_{x \to a} g(x) = L_1 \pm L_2$,

(iii) $\lim\limits_{x \to a} [f(x) \cdot g(x)] = \left[\lim\limits_{x \to a} f(x)\right] \cdot \left[\lim\limits_{x \to a} g(x)\right] = L_1 \cdot L_2$ and

(iv) $\lim\limits_{x \to a} \dfrac{f(x)}{g(x)} = \dfrac{\lim\limits_{x \to a} f(x)}{\lim\limits_{x \to a} g(x)} = \dfrac{L_1}{L_2}$ (if $L_2 \neq 0$).

The proof of the theorem is found in Appendix A and requires the formal definition of limit discussed in section 1.5. You should think of these rules as sensible results that you would certainly expect to be true, given your intuitive understanding of what a limit is.

The following result follows immediately by applying part (iii) of Theorem 2.3 with $g(x) = f(x)$. (Such a result, which follows immediately from another result, is called a **corollary**.)

Corollary 2.1   Suppose that $\lim\limits_{x \to a} f(x) = L$. Then,

$$\lim_{x \to a} [f(x)]^2 = \lim_{x \to a} [f(x) \cdot f(x)]$$
$$= \lim_{x \to a} f(x) \cdot \lim_{x \to a} f(x)$$
$$= L \cdot L = L^2.$$

Likewise, for any positive integer, n, we can apply part (iii) of Theorem 2.3 repeatedly, to yield

$$\lim_{x \to a} [f(x)]^n = \left[\lim_{x \to a} f(x)\right]^n = L^n \tag{2.1}$$

(see exercises 45 and 46).

The next result follows immediately from (2.1) and Theorem 2.2.

Corollary 2.2   For any integer $n > 0$ and any real number a,

$$\lim_{x \to a} x^n = a^n.$$

This last result says that to compute the limit of any positive power of x, you simply substitute in the value of x being approached.

Example 2.1   **Finding the Limit of a Power Function**

Evaluate $\lim\limits_{x \to -2} x^5$.

Solution   From Corollary 2.2, we have

$$\lim_{x \to -2} x^5 = (-2)^5 = -32.$$

■

Example 2.2   **Finding the Limit of a Polynomial**

Apply the rules of limits to carefully evaluate $\lim\limits_{x \to 2}(3x^2 - 5x + 4)$.

Solution   We have

By Theorem 2.3 (ii).

By Theorem 2.3 (i).

By Corollary 2.2.

$$\lim_{x \to 2}(3x^2 - 5x + 4) = \lim_{x \to 2}(3x^2) - \lim_{x \to 2}(5x) + \lim_{x \to 2}4$$
$$= 3\lim_{x \to 2}x^2 - 5\lim_{x \to 2}x + 4$$
$$= 3 \cdot (2)^2 - 5 \cdot 2 + 4 = 6.$$

∎

Example 2.3   **Finding the Limit of a Rational Function**

Apply the rules of limits to carefully evaluate $\lim\limits_{x \to 3}\dfrac{x^3 - 5x + 4}{x^2 - 2}$.

Solution   We get

By Theorem 2.3 (iv).

By Theorem 2.3 (i) and (ii).

By Corollary 2.2.

$$\lim_{x \to 3}\frac{x^3 - 5x + 4}{x^2 - 2} = \frac{\lim\limits_{x \to 3}(x^3 - 5x + 4)}{\lim\limits_{x \to 3}(x^2 - 2)}$$
$$= \frac{\lim\limits_{x \to 3}x^3 - 5\lim\limits_{x \to 3}x + \lim\limits_{x \to 3}4}{\lim\limits_{x \to 3}x^2 - \lim\limits_{x \to 3}2}$$
$$= \frac{3^3 - 5 \cdot 3 + 4}{3^2 - 2} = \frac{16}{7}.$$

∎

At this point, you may be getting a bit bored. After all, in the last two examples, we simply ended up substituting the value in for x, after taking many intermediate steps, at that. In the next example, it's not quite so simple.

Example 2.4   **Finding a Limit by Factoring**

Evaluate $\lim\limits_{x \to 1}\dfrac{x^2 - 1}{1 - x}$.

Solution   Notice that

$$\lim_{x \to 1}\frac{x^2 - 1}{1 - x} \neq \frac{\lim\limits_{x \to 1}(x^2 - 1)}{\lim\limits_{x \to 1}(1 - x)},$$

since the limit in the denominator is zero. (Recall that the limit of a quotient is the quotient of the limits **only** when both limits exist **and** the limit in the denominator is **not** zero). However, we can resolve this problem by observing that

Factoring the numerator and factoring -1 from denominator.

$$\lim_{x \to 1}\frac{x^2 - 1}{1 - x} = \lim_{x \to 1}\frac{(x - 1)(x + 1)}{-(x - 1)}$$
$$= \lim_{x \to 1}\frac{(x + 1)}{-1} = -2,$$

where the cancellation of the two factors of $(x - 1)$ is valid because in the limit as $x \to 1$, x is close to 1, but $x \neq 1$, so that $x - 1 \neq 0$.

∎

In the following theorem, we show that the limit of a polynomial is simply the value of the polynomial at that point; that is, to find the limit of a polynomial, we simply substitute the value being approached.

Theorem 2.4   For any polynomial $p(x)$ and any real number a,

$$\lim_{x \to a} p(x) = p(a).$$

Proof. Suppose that $p(x)$ is a polynomial of degree $n \geq 0$,

$$p(x) = c_n x^n + c_{n-1} x^{n-1} + \cdots + c_1 x + c_0.$$

Then, from Theorem 2.3 and Corollary 2.2,

$$\lim_{x \to a} p(x) = \lim_{x \to a} \left(c_n x^n + c_{n-1} x^{n-1} + \cdots + c_1 x + c_0 \right)$$

$$= c_n \lim_{x \to a} x^n + c_{n-1} \lim_{x \to a} x^{n-1} + \cdots + c_1 \lim_{x \to a} x + \lim_{x \to a} c_0$$

$$= c_n a^n + c_{n-1} a^{n-1} + \cdots + c_1 a + c_0 = p(a).$$

∎

Evaluating the limit of a polynomial is now easy.

Example 2.5   Evaluating the Limit of a Polynomial
Evaluate $\lim_{x \to 4} (2x^2 - 3x + 12)$.

Solution   From Theorem 2.4, we can simply substitute $x = 4$:

$$\lim_{x \to 4} (2x^2 - 3x + 12) = 2 \cdot 4^2 - 3 \cdot 4 + 12 = 32.$$

∎

The following theorem should seem reasonable enough.

Theorem 2.5   For any integer $n > 0$, and any real number a,

$$\lim_{x \to a} \sqrt[n]{x} = \sqrt[n]{a},$$

where for n even, we must assume that $a > 0$. Further, suppose that $\lim_{x \to a} f(x) = L$. Then,

$$\lim_{x \to a} \sqrt[n]{f(x)} = \sqrt[n]{\lim_{x \to a} f(x)} = \sqrt[n]{L},$$

where for n even, we assume that $L > 0$.

The proof of Theorem 2.5 must be left to Appendix A. Notice that this result says that we may (under the conditions outlined in the hypotheses) bring limits "inside" nth roots. We can then use our existing rules for computing the limit inside.

Example 2.6   Evaluating the Limit of an nth Root of a Polynomial
Evaluate $\lim_{x \to 2} \sqrt[5]{3x^2 - 2x}$.

Solution   By Theorems 2.4 and 2.5, we have

$$\lim_{x \to 2} \sqrt[5]{3x^2 - 2x} = \sqrt[5]{\lim_{x \to 2}(3x^2 - 2x)}$$
$$= \sqrt[5]{8}.$$

∎

So that we are not restricted to discussing only the algebraic functions (i.e., those that can be constructed by using addition, subtraction, multiplication, division, exponentiation and by taking nth roots), we offer the following result now, but will not prove it until later in the text.

Theorem 2.6   For any real number a, we have

(i) $\lim\limits_{x \to a} \sin x = \sin a$,

(ii) $\lim\limits_{x \to a} \cos x = \cos a$,

(iii) $\lim\limits_{x \to a} e^x = e^a$ and

(iv) $\lim\limits_{x \to a} \ln x = \ln a$, for $a > 0$.

Notice that Theorem 2.6 says that limits of the sine, cosine, exponential and natural logarithm functions are found simply by substitution.

Example 2.7   **Finding the Limit of a Quotient of Two Trigonometric Functions**

Evaluate $\lim\limits_{x \to 0} \dfrac{\sin x}{\cos x}$.

Solution   By Theorems 2.3 and 2.6, we have

$$\lim_{x \to 0} \frac{\sin x}{\cos x} = \frac{\lim\limits_{x \to 0} \sin x}{\lim\limits_{x \to 0} \cos x}$$
$$= \frac{0}{1} = 0.$$

∎

So much for limits that we can compute using elementary rules. Certainly, many of the limits that we need to find in practice are of the type discussed above. Just as many, however, cannot be determined using these elementary rules and can only be found using more careful analysis, often by an indirect approach. For instance, consider the following problem.

Example 2.8   **A Limit of a Product That Is Not the Product of the Limits**

Determine the value of

$$\lim_{x \to 0} (x \cot x).$$

Solution   Your first reaction might be to say that this is a limit of a product and so, must be the product of the limits:

This is incorrect!

$$\lim_{x \to 0} (x \cot x) = \left(\lim_{x \to 0} x \right) \left(\lim_{x \to 0} \cot x \right)$$
$$= 0 \cdot ? = 0, \tag{2.2}$$

where we've written a "?" since you probably don't know what to do with $\lim\limits_{x\to 0}\cot x$. Since the first limit is 0, do we really need to worry about the second limit? To answer this question, consider $\lim\limits_{x\to 0}\dfrac{1}{x}$. We saw in example 1.2 that this limit does not exist. So, then, where is the flaw in the following argument?

This is incorrect!

$$
\begin{aligned}
\lim_{x\to 0}\frac{1}{x} &= \lim_{x\to 0}\frac{x}{x^2} \\
&= \lim_{x\to 0}\left(x\,\frac{1}{x^2}\right) \\
&= \left(\lim_{x\to 0}x\right)\left(\lim_{x\to 0}\frac{1}{x^2}\right) \\
&= 0\cdot ? = 0.
\end{aligned}
\tag{2.3}
$$

We already know that $\lim\limits_{x\to 0}\dfrac{1}{x}$ does not exist, so it can't simultaneously equal 0. The problem here as well as in equation (2.2) is that we are attempting to apply the result of Theorem 2.3 in a case where the hypotheses are not satisfied. Specifically, Theorem 2.3 says that the limit of a product is the product of the respective limits **when all of the limits exist**. In (2.3), $\lim\limits_{x\to 0}\dfrac{1}{x^2}$ does **not** exist. (Why not?)

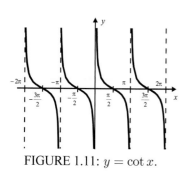

FIGURE 1.11: $y = \cot x$.

Returning to the original problem, what can be said of $\lim\limits_{x\to 0}\cot x$? Consider the graph of the function found in Figure 1.11. The graph suggests that $\lim\limits_{x\to 0}\cot x$ does not exist. You should compute some function values, as well, to convince yourself that this is in fact the case. Equation (2.2) then does not hold and we're back to square one. Since none of our rules seem to apply here, the most reasonable step is to draw a graph and compute some function values in an effort to see what is going on. The graph of $y = x\cot x$ appears in Figure 1.12. If we compute some function values, we observe that the function values appear to be approaching 1, as $x \to 0$.

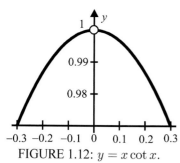

FIGURE 1.12: $y = x\cot x$.

Based on the graph and the table of function values, we make the conjecture

$$\lim_{x\to 0}(x\cot x) = 1,$$

which is definitely not 0, as you might have initially suspected. You can also think about this limit as follows:

x	$x\cot x$
± 0.1	0.9967
± 0.01	0.999967
± 0.001	0.99999967
± 0.0001	0.9999999967
± 0.00001	0.999999999967

$$
\begin{aligned}
\lim_{x\to 0}(x\cot x) &= \lim_{x\to 0}\left(x\,\frac{\cos x}{\sin x}\right) = \lim_{x\to 0}\left(\frac{x}{\sin x}\cos x\right) \\
&= \left(\lim_{x\to 0}\frac{x}{\sin x}\right)\left(\lim_{x\to 0}\cos x\right) \\
&= \frac{\lim\limits_{x\to 0}\cos x}{\lim\limits_{x\to 0}\dfrac{\sin x}{x}} = \frac{1}{1} = 1,
\end{aligned}
$$

since $\lim\limits_{x\to 0}\cos x = 1$ and where we have used the conjecture we made in example 1.3 that $\lim\limits_{x\to 0}\dfrac{\sin x}{x} = 1$. (We verify this last conjecture in section 2.5, using the Squeeze Theorem, which follows.)

■

At this point, we introduce a tool that will help us determine a number of important limits. Notice that we said *tool* and not *rule*. You should recognize that most mathematical problems are solved by careful analysis and not by simply applying a series of rules.

Theorem 2.7   (Squeeze Theorem)
    Suppose that

$$f(x) \le g(x) \le h(x)$$

for all x in some interval (c, d), except possibly at the point $a \in (c, d)$ and that

$$\lim_{x \to a} f(x) = \lim_{x \to a} h(x) = L,$$

for some number L. Then, it follows that

$$\lim_{x \to a} g(x) = L, \text{ also.}$$

The proof of the theorem is relegated to Appendix A and depends on the precise definition of limit found in section 1.5. However, if you refer to Figure 1.13, you should clearly see that if $g(x)$ lies between $f(x)$ and $h(x)$, except possibly at a itself and both $f(x)$ and $h(x)$ have the same limit as $x \to a$, then $g(x)$ gets *squeezed* between $f(x)$ and $h(x)$ and therefore should also have a limit of L. The challenge in using the Squeeze Theorem is in finding appropriate functions f and h that bound a given function g from below and above, respectively, and that have the same limit as $x \to a$.

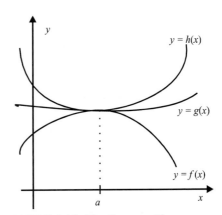

FIGURE 1.13: The Squeeze Theorem.

FIGURE 1.14: $y = \cos(1/x)$.

This is incorrect!

Example 2.9   Using the Squeeze Theorem to Verify the Value of a Limit
    Determine the value of

$$\lim_{x \to 0}[x^2 \cos(1/x)].$$

Solution   Your first reaction might be to observe that this is a limit of a product and so, might be the product of the limits:

$$\lim_{x \to 0} [x^2 \cos(1/x)] \overset{?}{=} \left(\lim_{x \to 0} x^2 \right) \left[\lim_{x \to 0} \cos(1/x) \right], \tag{2.4}$$

where we've written a "?" since you probably don't know what to do with $\lim_{x \to 0} \cos(1/x)$. Consider the graph of $y = \cos(1/x)$ found in Figure 1.14. The graph suggests that $\cos(1/x)$ oscillates back and forth between -1 and 1. Further, the closer x gets to 0, the more rapid the oscillations become. You should compute some function values, as

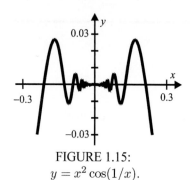

FIGURE 1.15:
$y = x^2 \cos(1/x)$.

x	$x^2 \cos(1/x)$
± 0.1	-0.008
± 0.01	8.6×10^{-5}
± 0.001	5.6×10^{-7}
± 0.0001	-9.5×10^{-9}
± 0.00001	-9.99×10^{-11}

well, to convince yourself that $\lim_{x \to 0} \cos(1/x)$ does not exist. Equation (2.4) then does not hold and we're back to square one. Since none of our rules seem to apply here, the most reasonable step is to draw a graph and compute some function values in an effort to see what is going on. The graph of $y = x^2 \cos(1/x)$ appears in Figure 1.15. If we compute some function values, we observe the same oscillation visible in the graph.

Despite the oscillation visible in the graph and in the table of function values, we make the conjecture

$$\lim_{x \to 0} \left[x^2 \cos(1/x) \right] = 0.$$

We can prove that this conjecture is correct using the Squeeze Theorem, but we will first need to find functions f and h such that

$$f(x) \le x^2 \cos(1/x) \le h(x),$$

for all $x \neq 0$ and where $\lim_{x \to 0} f(x) = \lim_{x \to 0} h(x) = 0$. How can you be expected to do that? One difficulty is that there's actually very little that you know about $\cos(1/x)$. So, ask yourself what you **do** know. Probably the only inequality you can recall concerning this function is

$$-1 \le \cos(1/x) \le 1, \tag{2.5}$$

for all $x \neq 0$. If we multiply (2.5) through by x^2 (notice that since $x^2 \ge 0$, this multiplication preserves the inequalities), we get

$$-x^2 \le x^2 \cos(1/x) \le x^2,$$

for all $x \neq 0$. Further,

$$\lim_{x \to 0} (-x^2) = 0 = \lim_{x \to 0} x^2.$$

So, from the Squeeze Theorem, it now follows that

$$\lim_{x \to 0} x^2 \cos(1/x) = 0,$$

also, as we had conjectured.

∎

Remark 2.1 To resolve the limit in example 2.9, we could not apply the rules for limits contained in Theorem 2.3. So, we resorted to an indirect method of finding the limit. This tour de force of graphics plus calculation followed by analysis is sometimes referred to as the **Rule of Three.** (The Rule of Three presents a general strategy for attacking new problems. The basic idea is to look at problems graphically, numerically and analytically.) In the case of example 2.9, the first two elements of this "rule" (the graphics in Figure 1.15 and the following table of function values) suggest a plausible conjecture, while the third element provides us with a careful mathematical verification of the conjecture.

Functions are often defined by different expressions on different intervals. Such **piecewise-defined** functions are important and we illustrate such a function in the following example.

Example 2.10   A Limit That Does Not Exist

Evaluate $\lim_{x \to 0} f(x)$, where f is defined by

$$f(x) = \begin{cases} x^2 + 2\cos x + 1, & \text{for } x < 0 \\ e^x - 4, & \text{for } x \ge 0 \end{cases}.$$

Solution   Since f is defined by different expressions for $x < 0$ and for $x \geq 0$, we must consider one-sided limits. We have

$$\lim_{x \to 0^-} f(x) = \lim_{x \to 0^-} \left(x^2 + 2\cos x + 1 \right) = 2\cos 0 + 1 = 3,$$

by Theorem 2.6. Alternatively, we have

$$\lim_{x \to 0^+} f(x) = \lim_{x \to 0^+} \left(e^x - 4 \right) = e^0 - 4 = 1 - 4 = -3.$$

Since the one-sided limits are different, we have that $\lim_{x \to 0} f(x)$ does not exist.

∎

We end the section with an example of the use of limits in computing velocity. You may recall that in section 0.7, we discussed the problem of computing the velocity of an object moving in a straight line. In section 2.1, we see that for an object whose position at time t is given by the function $f(t)$, the instantaneous velocity of that object at time $t = 1$ (i.e., the velocity at the **instant** $t = 1$, as opposed to the average velocity over some period of time) is given by the limit

$$\lim_{h \to 0} \frac{f(1+h) - f(1)}{h}.$$

Example 2.11   **Evaluating a Limit Describing Velocity**

Suppose that the position function for an object is given by

$$f(t) = t^2 + 2 \ \text{(feet)},$$

where t is measured in seconds. Find the instantaneous velocity of the object at time $t = 1$.

Solution   Given what we have just learned about limits, this is now an easy problem to solve. We have

$$\lim_{h \to 0} \frac{f(1+h) - f(1)}{h} = \lim_{h \to 0} \frac{\left[(1+h)^2 + 2 \right] - 3}{h}.$$

While we can't simply substitute $h = 0$ (Why not?), we can write

Expanding squared term.

$$\lim_{h \to 0} \frac{\left[(1+h)^2 + 2 \right] - 3}{h} = \lim_{h \to 0} \frac{(1 + 2h + h^2) - 1}{h} = \lim_{h \to 0} \frac{2h + h^2}{h}$$

$$= \lim_{h \to 0} \frac{h(2+h)}{h}$$

Canceling factors of h.

$$= \lim_{h \to 0} \frac{2+h}{1} = 2.$$

So, the instantaneous velocity of this object at time $t = 1$ is 2 feet per second.

∎

Exercises 1.2

1. (writing) Given your knowledge of the graphs of polynomials, explain why Theorems 2.1, 2.2 and 2.4 are obvious.

2. (writing) Based on your knowledge of graphs of non-polynomial functions, for which other functions can limits be

evaluated simply by substitution? Think of at least five limits which do not evaluate so easily.

3. (writing) In one or two sentences, explain the Squeeze Theorem. Use a real-world analogy (e.g., having the functions represent the locations of three people as they walk) to indicate why it is true.

4. (writing) Given the graph in Figure 1.15 and the calculations that follow, it may be unclear why we insist on using the Squeeze Theorem before concluding that $\lim_{x \to 0} \left[x^2 \cos\left(1/x\right) \right]$ is indeed 0. Review example 1.4 and section 1.1 to explain why we are being so fussy.

In exercises 5–26, evaluate the indicated limit.

5. $\lim_{x \to 0} \left(x^2 - 3x + 1 \right)$

6. $\lim_{x \to 0} \cos(x^2 + \pi)$

7. $\lim_{x \to 0} \sqrt{x^2 + 2x + 4}$

8. $\lim_{x \to 0} \sqrt[3]{2x + 1}$

9. $\lim_{x \to 0} \dfrac{x - 5}{x^2 + 4}$

10. $\lim_{x \to -2} \dfrac{x + 2}{x^2 - 4}$

11. $\lim_{x \to 3} \dfrac{x^2 - x - 6}{x - 3}$

12. $\lim_{x \to 1} \dfrac{x^2 + x - 2}{x^2 - 3x + 2}$

13. $\lim_{x \to 2} \dfrac{x^2 - x - 2}{x^2 - 4}$

14. $\lim_{x \to 0} \dfrac{x^2 + x}{x^2 - 3x}$

15. $\lim_{x \to 1} \dfrac{x^3 - 1}{x^2 + 2x - 3}$

16. $\lim_{x \to 3} \dfrac{x - 1}{x^2 + 2x - 3}$

17. $\lim_{x \to 0} \dfrac{\sin x}{\tan x}$

18. $\lim_{x \to \pi} \dfrac{\sin x}{\tan x}$

19. $\lim_{x \to 0} \dfrac{xe^{-2x+1}}{x^2 + 1}$

20. $\lim_{x \to 0} \dfrac{xe^{-2x+1}}{x^2 + x}$

21. $\lim_{x \to 0} \dfrac{\tan 2x}{x}$

22. $\lim_{x \to 0^+} x^2 \csc^2 x$

23. $\lim_{x \to 1} \dfrac{x - 1}{\sqrt{x} - 1}$

24. $\lim_{x \to 4} \dfrac{\sqrt{x} - 2}{x - 4}$

25. $\lim_{x \to 2} f(x)$ where $f(x) = \begin{cases} 3x^2 - 2x + 1 & \text{if } x < 2 \\ x^3 + 1 & \text{if } x \geq 2 \end{cases}$

26. $\lim_{x \to 0} f(x)$ where $f(x) = \begin{cases} 2x \cos x & \text{if } x < 0 \\ x^3 + 2x & \text{if } x \geq 0 \end{cases}$

27. Use numerical and graphical evidence to conjecture the value of $\lim_{x \to 0} x^2 \sin(1/x)$. Use the Squeeze Theorem to prove that you are correct: identify the functions $f(x)$ and $h(x)$, show graphically that $f(x) \leq x^2 \sin(1/x) \leq h(x)$, and justify $\lim_{x \to 0} f(x) = \lim_{x \to 0} h(x)$.

28. Why can't you use the Squeeze Theorem as in exercise 27 to prove that $\lim_{x \to 0} x^2 \sec(1/x) = 0$? Explore this limit graphically.

29. Use the Squeeze Theorem to prove that $\lim_{x \to 0^+} \left[\sqrt{x} \cos^2(1/x) \right] = 0$. Identify the functions $f(x)$ and $h(x)$, show graphically that $f(x) \leq \sqrt{x} \cos^2(1/x) \leq h(x)$, and justify $\lim_{x \to 0^+} f(x) = 0$ and $\lim_{x \to 0^+} h(x) = 0$

30. Suppose that $f(x)$ is bounded: that is, there exist constants m and n such that $m \leq f(x) \leq n$ for all x. Use the Squeeze Theorem to prove that $\lim_{x \to 0} x^2 f(x) = 0$.

In exercises 31–36, either find the limit or explain why it does not exist.

31. $\lim_{x \to 4^+} \sqrt{16 - x^2}$

32. $\lim_{x \to 4^-} \sqrt{16 - x^2}$

33. $\lim_{x \to 4} \sqrt{16 - x^2}$

34. $\lim_{x \to -2^+} \sqrt{x^2 + 3x + 2}$

35. $\lim_{x \to -2^-} \sqrt{x^2 + 3x + 2}$

36. $\lim_{x \to -2} \sqrt{x^2 + 3x + 2}$

37. Given that $\lim_{x \to 0} \dfrac{1 - \cos x}{x^2} = \dfrac{1}{2}$, quickly evaluate $\lim_{x \to 0} \dfrac{\sqrt{1 - \cos x}}{x}$.

38. Given that $\lim_{x \to 0} \dfrac{\sin x}{x} = 1$, quickly evaluate $\lim_{x \to 0} \dfrac{1 - \cos^2 x}{x^2}$.

39. Suppose $f(x) = \begin{cases} g(x) & \text{if } x < a \\ h(x) & \text{if } x > a \end{cases}$ for polynomials $g(x)$ and $h(x)$. Explain why $\lim_{x \to a-} f(x) = g(a)$ and determine $\lim_{x \to a+} f(x)$.

40. Explain how to determine $\lim_{x \to a} f(x)$ if g and h are polynomials and $f(x) = \begin{cases} g(x) & \text{if } x < a \\ c & \text{if } x = a \\ h(x) & \text{if } x > a \end{cases}$.

41. Evaluate each limit and justify each step by citing the appropriate theorem(s).

(a) $\lim_{x \to 2} \left(x^2 - 3x + 1 \right)$

(b) $\lim_{x \to 0} \dfrac{x - 2}{x^2 + 1}$

42. Evaluate each limit and justify each step by citing the appropriate theorem(s).

(a) $\lim_{x \to -1} \left[(x + 1) \sin x \right]$

(b) $\lim_{x \to 1} \dfrac{xe^x}{\tan x}$

43. As we see in Chapter 2, the slope of the tangent line to the curve $y = \sqrt{x}$ at $x = 1$ is given by $m = \lim_{h \to 0} \dfrac{\sqrt{1 + h} - 1}{h}$. Compute the slope m. Graph $y = \sqrt{x}$ and the line with slope m through the point $(1, 1)$.

44. As we see in Chapter 2, the velocity of an object which has traveled $\sqrt{x}$ miles in x hours at the $x = 1$ hour mark is given by $v = \lim_{x \to 1} \dfrac{\sqrt{x} - 1}{x - 1}$. Compute this limit.

45. Assume that $\lim_{x \to a} f(x) = L$. Use Theorem 2.3 to prove that $\lim_{x \to a} [f(x)]^3 = L^3$. Also, show that $\lim_{x \to a} [f(x)]^4 = L^4$.

46. How did you work exercise 45? You probably used Theorem 2.3 to work from $\lim_{x \to a} [f(x)]^2 = L^2$ to $\lim_{x \to a} [f(x)]^3 = L^3$, and then used $\lim_{x \to a} [f(x)]^3 = L^3$ to get $\lim_{x \to a} [f(x)]^4 = L^4$. Going one step at a time, we should be able to reach

$\lim\limits_{x \to a} [f(x)]^n = L^n$ for any positive integer n. This is the idea of **mathematical induction**. Formally, we need to show the result is true for a specific value of $n = n_0$ (Corollary 2.1 does that for $n_0 = 2$), then assume the result is true for a general $n = k \ge n_0$. If we show that we can get from the result being true for $n = k$ to the result being true for $n = k + 1$, we have proved that the result is true for any positive integer n. In one sentence, explain why this is true. Use this technique to prove that $\lim\limits_{x \to a} [f(x)]^n = L^n$ for any positive integer n.

47. Give an example of functions $f(x)$ and $g(x)$ such that $\lim\limits_{x \to 0} [f(x) + g(x)]$ exists but $\lim\limits_{x \to 0} f(x)$ and $\lim\limits_{x \to 0} g(x)$ do not exist.

48. Give an example of functions $f(x)$ and $g(x)$ such that $\lim\limits_{x \to 0} [f(x) \cdot g(x)]$ exists but at least one of $\lim\limits_{x \to 0} f(x)$ and $\lim\limits_{x \to 0} g(x)$ do not exist.

49. If $\lim\limits_{x \to a} f(x)$ exists and $\lim\limits_{x \to a} g(x)$ does not exist, is it always true that $\lim\limits_{x \to a} [f(x) + g(x)]$ does not exist? Explain.

50. Is the following true or false? If $\lim\limits_{x \to 0} f(x)$ does not exist, then $\lim\limits_{x \to 0} 1/f(x)$ does not exist. Explain.

51. Suppose a state's income tax code states the tax liability on x dollars of taxable income is given by

$$T(x) = \begin{cases} 0.14x & \text{if } x < 10,000 \\ 1500 + 0.21x & \text{if } 10,000 \le x \end{cases}.$$

Compute $\lim\limits_{x \to 0+} T(x)$; why is this good? Compute $\lim\limits_{x \to 10,000} T(x)$; why is this bad?

52. Suppose a state's income tax code states that tax liability is 12% on the first \$20,000 of taxable earnings and

16% on the remainder. Find constants a and b for the tax function $T(x) = \begin{cases} a + 0.12x & \text{if } x \le 20,000 \\ b + 0.16(x - 20,000) & \text{if } x > 20,000 \end{cases}$ such that $\lim\limits_{x \to 0+} T(x) = 0$ and $\lim\limits_{x \to 20,000} T(x)$ exists. Why is it important for these limits to exist?

53. The **greatest integer function** is denoted by $f(x) = [x]$ and equals the greatest integer which is less than or equal to x. Thus, $[2.3] = 2$, $[-1.2] = -2$ and $[3] = 3$. In spite of this last fact, show that $\lim\limits_{x \to 3} [x]$ does not exist.

54. Investigate the existence of (a) $\lim\limits_{x \to 1} [x]$, (b) $\lim\limits_{x \to 1.5} [x]$, (c) $\lim\limits_{x \to 1.5} [2x]$.

55. (exploratory) The value $x = 0$ is called a **zero of multiplicity** n $(n \ge 1)$ for the function $f(x)$ if $\lim\limits_{x \to 0} \dfrac{f(x)}{x^n}$ exists and is nonzero but $\lim\limits_{x \to 0} \dfrac{f(x)}{x^{n-1}} = 0$. Show that $x = 0$ is a zero of multiplicity 2 for x^2, $x = 0$ is a zero of multiplicity 3 for x^3 and $x = 0$ is a zero of multiplicity 4 for x^4. For polynomials, what does multiplicity describe? The reason the definition is not as straightforward as we might like is so that it can apply to nonpolynomial functions, as well. Find the multiplicity of $x = 0$ for $f(x) = \sin x$; $f(x) = x \sin x$; $f(x) = \sin x^2$. If you know that $x = 0$ is a zero of multiplicity m for $f(x)$ and multiplicity n for $g(x)$, what can you say about the multiplicity of $x = 0$ for $f(x) + g(x)$? $f(x) \cdot g(x)$? $f(g(x))$?

56. (exploratory) We have conjectured that $\lim\limits_{x \to 0} \dfrac{\sin x}{x} = 1$. Using graphical and numerical evidence, conjecture the value of $\lim\limits_{x \to 0} \dfrac{\sin 2x}{x}$, $\lim\limits_{x \to 0} \dfrac{\sin 3x}{x}$, $\lim\limits_{x \to 0} \dfrac{\sin \pi x}{x}$ and $\lim\limits_{x \to 0} \dfrac{\sin x/2}{x}$. In general, conjecture the value of $\lim\limits_{x \to 0} \dfrac{\sin cx}{x}$ for any constant c. Given that $\lim\limits_{x \to 0} \dfrac{\sin cx}{cx} = 1$ for any constant $c \ne 0$, prove that your conjecture is correct.

1.3   Continuity and Its Consequences

When you describe something as *continuous*, just what do you have in mind? For example, if you say that a machine has been in *continuous* operation for the past 60 hours, what is it that you are trying to communicate? Most of us would interpret this to mean that the machine has been in operation **all** of that time, without any interruption at all, even for a moment. Mathematicians mean very much the same thing when we say that a function is *continuous*. A function is said to be **continuous** on an interval if its graph there can be drawn without interruption, that is, without lifting your pencil from the paper.

It is helpful for us to first try to see what it is about the functions whose graphs are shown in Figures 1.16–1.19 that makes them **discontinuous** (i.e., not continuous) at the point $x = a$.

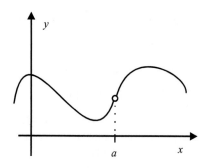

FIGURE 1.16: $f(a)$ is not defined (the graph has a hole at $x = a$).

FIGURE 1.17: $f(a)$ is defined, but $\lim\limits_{x \to a} f(x)$ does not exist (the graph has a jump at $x = a$).

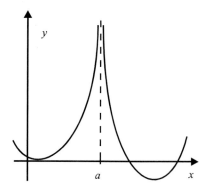

FIGURE 1.18: $\lim\limits_{x \to a} f(x)$ exists and $f(a)$ is defined, but $\lim\limits_{x \to a} f(x) \neq f(a)$ (the graph has a jump at $x = a$).

FIGURE 1.19: $\lim\limits_{x \to a} f(x)$ does not exist (the graph "blows up" at $x = a$).

This suggests the following definition of continuity at a point.

Definition 3.1   A function f is **continuous** at $x = a$ when
  (i) $f(a)$ is defined
  (ii) $\lim\limits_{x \to a} f(x)$ exists and
  (iii) $\lim\limits_{x \to a} f(x) = f(a)$.
Otherwise, f is said to be **discontinuous** at $x = a$.

NOTE: The definition of continuity all boils down to the one condition in (iii), since conditions (i) and (ii) must hold whenever (iii) is met. Further, this says that a function is continuous at a point exactly when you can compute its limit at that point by simply substituting in.

Remark 3.1   At first glance, this definition may seem a bit technical for a concept that is intuitively so clear. We'll see the need for this precise definition as we proceed. For most purposes, it is best for you to think of the intuitive notion of continuity that we've outlined above. The more technical definition should then simply follow from your intuitive understanding of the concept.

Example 3.1   **Finding Where a Rational Function is Continuous**

Determine where $f(x) = \dfrac{x^2 + 2x - 3}{x - 1}$ is continuous.

Solution   Note that

Factoring the numerator.

Cancel common factors.

$$\begin{aligned} f(x) &= \frac{x^2 + 2x - 3}{x - 1} \\ &= \frac{(x-1)(x+3)}{(x-1)} \\ &= x + 3, \text{ for } x \neq 1. \end{aligned}$$

This says that the graph of f is a straight line, but with a hole in it at $x = 1$, as indicated in Figure 1.20. So, f is discontinuous at $x = 1$, but continuous elsewhere. (Except at $x = 1$, the graph is a straight line and so, can be drawn without lifting pencil from paper.) ∎

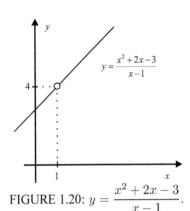

FIGURE 1.20: $y = \dfrac{x^2 + 2x - 3}{x - 1}$.

Remark 3.2   You should be careful not to confuse the continuity of a function at a point with its simply being defined there. A function can be defined at a point without being continuous there. (Look back at Figures 1.17 and 1.18.)

Example 3.2   **A Removable Discontinuity**

Make the function from example 3.1 continuous everywhere by redefining it at a single point.

Solution   In example 3.1, the function was discontinuous at $x = 1$ since it was undefined there. So, suppose we just go ahead and define it, as follows. Let

$$g(x) = \begin{cases} \dfrac{x^2 + 2x - 3}{x - 1}, & \text{if } x \neq 1 \\ 5, & \text{if } x = 1. \end{cases}$$

This function is defined for all x and equals $f(x)$ for all $x \neq 1$. So, its graph still looks like a straight line, but with a jump at $x = 1$ (instead of the hole at $x = 1$ which the graph of f has), as indicated in Figure 1.21. Here, we have

$$\begin{aligned} \lim_{x \to 1} g(x) &= \lim_{x \to 1} \frac{x^2 + 2x - 3}{x - 1} \\ &= \lim_{x \to 1} (x + 3) \\ &= 4 \neq 5 = g(1) \end{aligned}$$

and so, g is also discontinuous at $x = 1$. At this point, you should see that you can redefine g at $x = 1$ to make g continuous everywhere by defining

$$g(x) = \begin{cases} \dfrac{x^2 + 2x - 3}{x - 1}, & \text{if } x \neq 1 \\ 4, & \text{if } x = 1. \end{cases}$$

When this can be done, we call the discontinuity **removable**. There's actually a very simple way to write $g(x)$. We leave this as an exercise. ∎

FIGURE 1.21: $y = g(x)$.

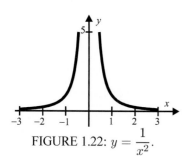

FIGURE 1.22: $y = \dfrac{1}{x^2}$.

Example 3.3   A Nonremovable Discontinuity

Find any discontinuities of $f(x) = \dfrac{1}{x^2}$.

Solution   You should observe from Figure 1.22 (also, construct a table of function values) that

$$\lim_{x \to 0} \frac{1}{x^2} \text{ does not exist.}$$

Hence, f is discontinuous at $x = 0$. Further, since $\lim_{x \to 0} f(x)$ does not exist, there is no way to redefine the function at $x = 0$ to make it continuous there. (Such a discontinuity is called **nonremovable** or **essential**.)

∎

From your experience with polynomial graphs, the following result should come as no surprise.

Theorem 3.1   All polynomials are continuous everywhere.

Proof. We have already established (in Theorem 2.4) that for any polynomial $p(x)$ and any real number a,

$$\lim_{x \to a} p(x) = p(a),$$

from which it follows that p is continuous at $x = a$.

∎

Likewise, it follows from Theorem 2.6 that $\sin x$, $\cos x$ and e^x are continuous everywhere and $\ln x$ is continuous for $x > 0$. From these very basic continuous functions, we can build a large variety of continuous functions, using the following result.

Theorem 3.2   Suppose that f and g are continuous at $x = a$. Then all of the following are true.
    (i) $(f \pm g)$ is continuous at $x = a$,
    (ii) $(f \cdot g)$ is continuous at $x = a$, and
    (iii) (f/g) is continuous at $x = a$ if $g(a) \neq 0$ and discontinuous at $x = a$ if $g(a) = 0$.

From Theorem 2.3.

Since f and g are continuous at a.

Proof. (i) If f and g are continuous at $x = a$, then

$$\lim_{x \to a} [f(x) \pm g(x)] = \lim_{x \to a} f(x) \pm \lim_{x \to a} g(x)$$
$$= f(a) \pm g(a)$$
$$= (f \pm g)(a),$$

by the usual rules of limits. Thus, $(f \pm g)$ is also continuous at $x = a$.

Parts (ii) and (iii) are proved in a similar way and are left as exercises.

∎

FIGURE 1.23:
$$y = \frac{x^4 - 3x^2 + 2}{x^2 - 3x - 4}.$$

Example 3.4   Continuity for a Rational Function

Determine where f is continuous, for

$$f(x) = \frac{x^4 - 3x^2 + 2}{x^2 - 3x - 4}.$$

Solution   Here, f is a quotient of two polynomial (hence continuous) functions. The graph of the function indicated in Figure 1.23 suggests a vertical asymptote at around $x = 4$, but doesn't indicate any other discontinuity. From Theorem 3.2, f will be continuous at all x where the denominator

$$x^2 - 3x - 4 = (x + 1)(x - 4) \neq 0.$$

Thus, f is continuous for $x \neq -1, 4$. (Think about why you didn't see anything peculiar about the graph at $x = -1$.) ∎

With the addition of the following result, we will have the basic tools needed to determine the continuity or lack of continuity of most elementary functions.

Theorem 3.3   Suppose $\lim_{x \to a} g(x) = L$ and f is continuous at L. Then,

$$\lim_{x \to a} f(g(x)) = f(\lim_{x \to a} g(x)) = f(L).$$

Notice that this says that if f is continuous then we can bring the limit "inside."

Corollary 3.1   Suppose that g is continuous at a and f is continuous at $g(a)$. Then, the composition, $f \circ g$ is continuous at a.

Proof. From Theorem 3.3, we have

$$\lim_{x \to a} (f \circ g)(x) = \lim_{x \to a} f(g(x))$$
$$= f(\lim_{x \to a} g(x))$$
$$= f(g(a))$$
$$= (f \circ g)(a).$$

Since g is continuous at a.

∎

Example 3.5   Continuity for a Composite Function

Determine where $h(x) = \cos(x^2 - 5x + 2)$ is continuous.

Solution   Note that

$$h(x) = f(g(x)),$$

where $g(x) = x^2 - 5x + 2$ and $f(x) = \cos x$. Since both f and g are continuous for all x, h is continuous for all x, by Corollary 3.1. ∎

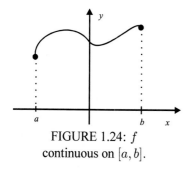

FIGURE 1.24: f
continuous on $[a, b]$.

Definition 3.2 If f is continuous at every point on an open interval (a, b), we say that f is **continuous on** (a, b). Following Figure 1.24, we say that f is **continuous on the closed interval** $[a, b]$, if f is continuous on the open interval (a, b) and

$$\lim_{x \to a^+} f(x) = f(a) \text{ and } \lim_{x \to b^-} f(x) = f(b).$$

Finally, if f is continuous on all of $(-\infty, \infty)$, we simply say that f is **continuous** (i.e., when we don't specify an interval, we mean continuous everywhere).

For many functions, it's a simple matter to determine the intervals on which the function is continuous. We illustrate this in the following example.

Example 3.6   **Intervals Where a Function is Continuous**
          Determine the interval(s) where f is continuous, for

$$f(x) = \frac{\sin(3x - 7)}{e^x - 1}.$$

Solution   Note that f is the quotient of two continuous functions and hence, is continuous for all x for which the denominator is nonzero. We show a graph of the function in Figure 1.25. Since

$$e^x - 1 = 0$$

only for $x = 0$, we have that f is continuous for all $x \neq 0$. Thus, f is continuous on both of the intervals $(-\infty, 0)$ and $(0, \infty)$.

∎

FIGURE 1.25:
$y = \dfrac{\sin(3x - 7)}{e^x - 1}.$

Example 3.7   **Interval of Continuity for a Logarithm**
          Determine the interval(s) where $f(x) = \ln(x - 3)$ is continuous.

Solution   Here, f is continuous for $(x - 3) > 0$ or $x > 3$. Thus, f is continuous on the interval $(3, \infty)$.

∎

The Internal Revenue Service presides over some of the most despised functions in existence. Look up the current Tax Rate Schedules. In 1998, the first few lines (for single taxpayers) looked like:

For taxable amount over	but not over	your tax liability is	of amount over
$0	$25,350	15%	$0
$25,350	$61,400	$3802.50 + 28%	$25,350
$61,400	$128,100	$13,896.50 + 31%	$61,400

Where do the numbers $3802.50 and $13,896.50 come from? If we write the tax liability $T(x)$ as a function of the taxable amount x (assuming that x can be any real value and not just a whole dollar amount), we have

$$T(x) = \begin{cases} 0.15x & \text{if } 0 < x \leq 25{,}350 \\ 3802.5 + 0.28(x - 25{,}350) & \text{if } 25{,}350 < x \leq 61{,}400 \\ 13{,}896.5 + 0.31(x - 61{,}400) & \text{if } 61{,}400 < x \leq 128{,}100. \end{cases}$$

Be sure you understand our translation so far. Note that it is important that this be a continuous function: think of the fairness issues that would arise if it were not!

Example 3.8   Continuity of Federal Tax Tables
    Verify that the federal tax rate function $T(x)$ is continuous at the "joints" $x = 25,350$ and $x = 61,400$ and find a and b to complete the table:

For taxable amount over	but not over	your tax liability is	of amount over
$128,100	$278,450	$(a) + 36\%$	$128,100
$278,450	——	$(b) + 39.6\%$	$278,450

Solution   For $T(x)$ to be continuous at $x = 25,350$, we must have

$$\lim_{x \to 25,350^-} T(x) = \lim_{x \to 25,350^+} T(x).$$

Since both functions $0.15x$ and $3802.5 + 0.28(x - 25,350)$ are continuous, we can compute the one-sided limits by substituting $x = 25,350$. Thus,

$$\lim_{x \to 25,350^-} T(x) = 0.15(25,350) = 3802.5$$

and

$$\lim_{x \to 25,350^+} T(x) = 3802.5 + 0 = 3802.5.$$

Since the one-sided limits agree and equal the value of the function at that point, $T(x)$ is continuous at $x = 25,350$. Similarly, we compute the one-sided limits at $x = 61,400$ by substituting $x = 61,400$ into the second and third formulas given. We get

$$\lim_{x \to 61,400^-} T(x) = 3802.5 + 0.28(61,400 - 25,350) = 13,896.5$$

and

$$\lim_{x \to 61,400^+} T(x) = 13,896.5 + 0.31(0) = 13,896.5.$$

Again, since the one-sided limits agree and equal the value of the function at that point, the function is continuous at $x = 61,400$. (It's worth noting that the function could be written with equal signs on all of the inequalities; this would be incorrect if the function was discontinuous). To finish the table, we choose a and b to get the remaining one-sided limits to match. First, we need

$$\lim_{x \to 128,100^-} T(x) = 13,896.5 + 0.31(128,100 - 61,400) = a + 0.36(0) = \lim_{x \to 128,100^+} T(x)$$

or

$$34,573.5 = a.$$

Second, we need the one-sided limits to match at $x = 278,450$. So,

$$\lim_{x \to 278,450^-} T(x) = 34,573.5 + 0.36(27,8450 - 128,100) = b + 0.396(0) = \lim_{x \to 27,8450^+} T(x)$$

or

$$88,699.5 = b.$$

∎

The following theorem should seem an obvious consequence of our intuitive definition of continuity.

Theorem 3.4   (Intermediate Value Theorem)

Suppose that f is continuous on the closed interval $[a, b]$ and W is any number between $f(a)$ and $f(b)$. Then, there is a number $c \in [a, b]$ for which $f(c) = W$.

Remark 3.3   Theorem 3.4 says that if f is continuous on $[a, b]$, then f must take on **every** value between $f(a)$ and $f(b)$ at least once. That is, a continuous function cannot skip over any numbers between its values at the two endpoints. To do so, the graph would need to leap across the horizontal line $y = W$, something that continuous functions cannot do (see Figure 1.26a). Of course, a given function may take on a given value W more than once (see Figure 1.26b). We must point out that, although these graphs make this result seem reasonable, like any other result, Theorem 3.4 requires proof. As it turns out, the proof is more complicated than you might imagine and we must refer you to a more advanced text.

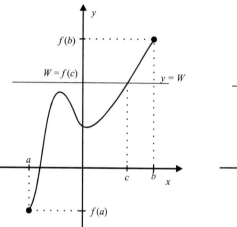

FIGURE 1.26a: An illustration of   FIGURE 1.26b: More than
the Intermediate Value Theorem.   one value of c.

In the corollary that follows, we see an immediate and useful application of the Intermediate Value Theorem.

Corollary 3.2   Suppose that f is continuous on $[a, b]$ and $f(a)$ and $f(b)$ have opposite signs (i.e., $f(a) \cdot f(b) < 0$). Then, there is at least one number $c \in (a, b)$ for which $f(c) = 0$. (In this event, c is called a **zero** of f.)

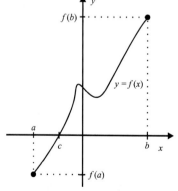

FIGURE 1.27: Intermediate
Value Theorem where c is
a zero of f.

The corollary is simply the special case of the Intermediate Value Theorem where $W = 0$ and follows immediately (see Figure 1.27).

In the following example, we see how Corollary 3.2 can help us locate the zeros of a function.

Example 3.9   Finding Zeros by the Method of Bisections

Find the zeros of

$$f(x) = x^5 + 4x^2 - 9x + 3.$$

Solution   If f were a quadratic or even a cubic polynomial, you could certainly find all of its zeros. But, what can we do here? You certainly don't have any formulas for finding zeros of polynomials of degree 5. The only alternative then, is to try to somehow approximate the zeros. A good starting place would be to use your graphing utility to draw a graph of $y = f(x)$ like the one in Figure 1.28. There are three zeros visible on the graph. (Can you convince yourself that f has **only** three zeros?) Note that since f is a polynomial, it is continuous everywhere. Hence, Corollary 3.2 says that there must be a zero on any interval on which the function changes sign. From the graph, you can see that there must be zeros between -3 and -2, between 0 and 1 and between 1 and 2. Without the graph, we could observe this by computing say, $f(0) = 3$ and $f(1) = -1$. Although we've now found intervals that contain zeros, the question remains as to how we can **find** the zeros themselves.

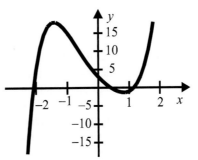

Figure 1.28: $y = x^5 + 4x^2 - 9x + 3$.

You could certainly use a root-finding utility to obtain an accurate approximation, but that's not the point. The issue here is not so much to get an answer as it is to understand the behavior of functions. We suggest the following primitive yet effective root-finding method, called the **method of bisections**.

To find the zero between 0 and 1, we make a *guess* as to its location. A reasonable guess might be the midpoint, 0.5. Since $f(0.5) = -0.469 < 0$ and $f(0) = 3 > 0$, there's a sign change between 0 and 0.5 and so, the zero is there. Next, look at the midpoint of $[0, 0.5]$. Since $f(0.25) = 1.00098 > 0$, the zero is on the interval $(0.25, 0.5)$ and so on. We continue to narrow the interval on which there's a zero until the interval is sufficiently small so that any point in the interval can serve as an adequate approximation to the actual zero. We do this in the following table.

a	b	$f(a)$	$f(b)$	*midpoint*	$f(midpoint)$
0	1	3	-1	0.5	-0.469
0	0.5	3	-0.469	0.25	1.00098
0.25	0.5	1.00098	-0.469	0.375	0.195
0.375	0.5	0.195	-0.469	0.4375	-0.156
0.375	0.4375	0.195	-0.156	0.40625	0.015
0.40625	0.4375	0.015	-0.156	0.421875	-0.072
0.40625	0.421875	0.015	-0.072	0.4140625	-0.029
0.40625	0.4140625	0.015	-0.029	0.41015625	-0.007
0.40625	0.4140625	0.015	-0.007	0.408203125	0.004

If you were to continue this process through many more steps, you would ultimately arrive at the approximate zero $x = 0.40892288$, which is accurate to at least nine decimal places.

∎

Remark 3.4 This method of bisections certainly is a tedious process, at least if you're working it with pencil and paper. Even though this is not a particularly efficient method of approximating zeros, it is interesting because it's reliable and it's likely the first general method that you have seen. We all have access to a computer or calculator root-finding utility that is much more accurate and efficient. Such tools are very useful, but their inner workings remain unseen by the user. Our purpose in this calculus text is to provide you with some tools for understanding, as well as with the ability to calculate some answers. Among other things, this understanding should leave you more capable of appreciating programs such as root-finding utilities. We discuss some more powerful methods for root-finding in Chapter 3.

Exercises 1.3

1. (writing) Think about the following "real-life" functions, each of which is a function of the independent variable time: the height of a falling object, the velocity of an object, the amount of money in a bank account, the cholesterol level of a person, the heart rate of a person, the amount of a certain chemical present in a test tube and a machine's most recent measurement of the cholesterol level of a person. Which of these are continuous functions? For each function you identify as discontinuous, what is the real-life meaning of the discontinuities?

2. (writing) Whether a process is continuous or not is not always clear-cut. When you watch television or a movie, the action seems to be continuous. This is an optical illusion, since both movies and television consist of individual "snapshots" that are played back at many frames per second. Where does the illusion of continuous motion come from? Given that the average person blinks several times per minute, is our perception of the world actually continuous? [A famous experiment in cognitive psychology showed that the human brain first decides whether a stimulus is important enough to merit conscious consideration. If so, the brain "pre-dates" the stimulus so that the person correctly identifies when the stimulus actually occurred (hence, the different reaction times we have to stimuli)].

3. (writing) When you sketch the graph of the parabola $y = x^2$ with pencil or pen, is your sketch (at the molecular level) actually the graph of a continuous function? Is your calculator or computer's graph actually the graph of a continuous function? On many calculators, you have the option of a connected or disconnected graph. At the pixel level, does a connected graph show the graph of a function? Does a disconnected graph show the graph of a continuous function? Do we ever have problems correctly interpreting a graph due to these limitations? In the exercises in section 1.6 we examine one case where our perception of a computer graph depends on which choice is made.

4. (writing) For each of the graphs in Figures 1.16–1.19, describe (with an example) what the formula for $f(x)$ might look like to produce the given discontinuity.

In exercises 5–10, use the given graph to identify all discontinuities of the functions.

5.

6.

7.

8.

9.

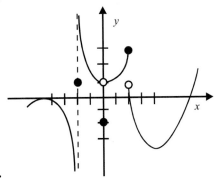

10.

In exercises 11–16, explain why each function is discontinuous at the given point.

11. $f(x) = \dfrac{x}{x-1}$ at $x = 1$

12. $f(x) = \sin\dfrac{1}{x}$ at $x = 0$

13. $f(x) = \dfrac{x^2 - 1}{x - 1}$ at $x = 1$

14. $f(x) = \begin{cases} x^2 & \text{if } x < 2 \\ 3x - 1 & \text{if } x \geq 2 \end{cases}$ at $x = 2$

15. $f(x) = \begin{cases} x^2 & \text{if } x < 2 \\ 3 & \text{if } x = 2 \\ 3x - 2 & \text{if } x > 2 \end{cases}$ at $x = 2$

16. $f(x) = \begin{cases} x^2 & \text{if } x < 2 \\ 3x - 2 & \text{if } x > 2 \end{cases}$ at $x = 2$

In exercises 17–28, find all discontinuities of $f(x)$ and determine which are removable.

17. $f(x) = \dfrac{x-1}{x^2 - 1}$

18. $f(x) = \dfrac{4x}{x^2 + x - 2}$

19. $f(x) = \dfrac{4x}{x^2 + 4}$

20. $f(x) = \dfrac{3x}{x^2 - 2x - 4}$

21. $f(x) = x^2 \tan x$

22. $f(x) = \csc x$

23. $f(x) = x \cot x$

24. $f(x) = \dfrac{\sec x}{x^2}$

25. $f(x) = \begin{cases} 2x & \text{if } x < 1 \\ x^2 & \text{if } x \geq 1 \end{cases}$

26. $f(x) = \begin{cases} \dfrac{\sin x}{x} & \text{if } x \neq 0 \\ 1 & \text{if } x = 0 \end{cases}$

27. $f(x) = \begin{cases} 3x - 1 & \text{if } x \leq -1 \\ x^2 + 5x & \text{if } -1 < x < 1 \\ 3x^3 & \text{if } x \geq 1 \end{cases}$

28. $f(x) = \begin{cases} 2x & \text{if } x < 0 \\ \sin x & \text{if } 0 < x \leq \pi \\ x - \pi & \text{if } x > \pi \end{cases}$

In exercises 29–36, determine the intervals on which $f(x)$ is continuous.

29. $f(x) = \sqrt{x + 3}$

30. $f(x) = \sqrt{x^2 - 4}$

31. $f(x) = \sqrt[3]{x + 2}$

32. $f(x) = (x - 1)^{3/2}$

33. $f(x) = \sin(x^2 + 2)$

34. $f(x) = \cos\left(\dfrac{1}{x}\right)$

35. $f(x) = \ln(x + 1)$

36. $f(x) = \ln(4 - x^2)$

37. Suppose that a state's income tax code states that the tax liability on x dollars of taxable income is given by

$$T(x) = \begin{cases} 0 & \text{if } x \leq 0 \\ 0.14x & \text{if } 0 < x < 10,000 \\ c + 0.21x & \text{if } 10,000 \leq x \end{cases}.$$

Determine the constant c that makes this function continuous at all x. Give a rationale why such a function should be continuous.

38. Suppose a state's income tax code states that tax liability is 12% on the first \$20,000 of taxable earnings and 16% on the remainder. Find constants a and b for the tax function

$$T(x) = \begin{cases} 0 & \text{if } x \leq 0 \\ a + 0.12x & \text{if } 0 < x \leq 20,000 \\ b + 0.16(x - 20,000) & \text{if } x > 20,000 \end{cases}$$

such that $T(x)$ is continuous at all x.

39. In example 3.8, find a formula for $T(x)$ for $128,100 < x \leq 278,450$.

40. In example 3.8, find a formula for $T(x)$ for $x > 278,450$.

In exercises 41–46, use the Intermediate Value Theorem to verify that $f(x)$ has a zero in the given interval. Then use the method of bisections to find an interval of length 1/32 which contains the zero.

41. $f(x) = x^2 - 7$, $[2, 3]$

42. $f(x) = x^3 - 4x - 2,\ [2,3]$

43. $f(x) = x^3 - 4x - 2,\ [-1,0]$

44. $f(x) = x^3 - 4x - 2,\ [-2,-1]$

45. $f(x) = \cos x - x,\ [0,1]$

46. $f(x) = e^x + x,\ [-1,0]$

A function is continuous from the right at $x = a$ if $\lim_{x \to a^+} f(x) = f(a)$. **In exercises 47–50, determine if $f(x)$ is continuous from the right at $x = 2$.**

47. $f(x) = \begin{cases} x^2 & \text{if } x < 2 \\ 3x - 1 & \text{if } x \geq 2 \end{cases}$

48. $f(x) = \begin{cases} x^2 & \text{if } x < 2 \\ 3 & \text{if } x = 2 \\ 3x - 3 & \text{if } x > 2 \end{cases}$

49. $f(x) = \begin{cases} x^2 & \text{if } x \leq 2 \\ 3x - 3 & \text{if } x > 2 \end{cases}$

50. $f(x) = \begin{cases} x^2 & \text{if } x < 2 \\ 3x - 2 & \text{if } x > 2 \end{cases}$

51. Define what it means for a function to be **continuous from the left** at $x = a$ and determine which of the functions in exercises 47–50 are continuous from the left at $x = 2$.

52. Suppose that $f(x) = \dfrac{g(x)}{h(x)}$ and $h(a) = 0$. Determine whether each of the following statements is definitely true, definitely false, or maybe true/maybe false. Explain. (a) $\lim_{x \to a} f(x)$ does not exist. (b) $f(x)$ is discontinuous at $x = a$.

53. The sex of newborn Mississippi alligators is determined by the temperature of the eggs in the nest. The eggs fail to develop unless the temperature is between $26°C$ and $36°C$. All eggs between $26°C$ and $30°C$ develop into females and eggs between $34°C$ and $36°C$ develop into males. The percentage of females and males changes proportionally between $30°C$ and $34°C$. If $f(T)$ is the percentage of females developing from an egg at $T°C$, explain why $f(T) = \begin{cases} 100 & \text{if } 26 \leq T \leq 30 \\ g(T) & \text{if } 30 < T < 34 \\ 0 & \text{if } 34 \leq T \leq 36 \end{cases}$ for some function $g(T)$. Explain why it is reasonable that $f(T)$ be continuous. Determine a function $g(T)$ such that $0 \leq g(T) \leq 100$ for $30 \leq T \leq 34$ and the resulting function $f(T)$ is continuous. (Hint: It may help to draw a graph first and make $g(T)$ linear.)

54. Give an example of functions $f(x)$ and $g(x)$ such that
$$\lim_{x \to a} f(g(x)) \neq f(\lim_{x \to a} g(x)).$$

55. If you push on a large box resting on the ground, at first nothing will happen because of the static friction force that opposes motion. If you push hard enough, the box will start sliding, although there is again a friction force that opposes

the motion. Suppose you are given the following description of the friction force. Up to 100 pounds, friction matches the force you apply to the box. Over 100 pounds, the box will move and the friction force will equal 80 pounds. Sketch a graph of friction as a function of your applied force based on this description. Where is this graph discontinuous? What is significant physically about this point? Do you think the friction force actually ought to be continuous? Modify the graph to make it continuous while still retaining most of the characteristics described.

56. For $f(x) = 2x - \dfrac{400}{x}$, we have $f(-1) > 0$ and $f(2) < 0$. Does the Intermediate Value Theorem guarantee a zero of $f(x)$ between $x = -1$ and $x = 2$? What happens if you try the method of bisections?

57. On Monday morning, a saleswoman leaves on a business trip at 7:13 A.M. and arrives at her destination at 2:03 P.M.. The following morning, she leaves for home at 7:17 A.M. and arrives at 1:59 P.M.. The woman notices that at a particular stoplight along the way, a nearby bank clock changes from 10:32 A.M. to 10:33 A.M. on both days. Therefore, she must have been at the same location at the same time on both days. Her boss doesn't believe that such an unlikely coincidence could occur. Use the Intermediate Value Theorem to argue that it **must** be true that at some point on the trip, the saleswoman was at exactly the same place at the same time on both Monday and Tuesday.

58. Suppose you ease your car up to a stop sign at the top of a hill. Your car rolls back a couple of feet and then you drive through the intersection. A police officer pulls you over for not coming to a complete stop. Use the Intermediate Value Theorem to argue that there was an instant in time when your car was stopped (in fact, there were at least two). What is the difference between this stopping and the stopping that the police officer wanted to see?

59. Suppose a worker's salary starts at \$40,000 with \$2000 raises every 3 months. Graph the salary function $s(t)$; why is it discontinuous? How does the function $f(t) = 40,000 + \frac{2000}{3}t$ (t in months) compare? Why might it be easier to do calculations with $f(t)$ than $s(t)$?

60. Prove the final two parts of Theorem 3.2.

61. Suppose that $f(x)$ is a continuous function with consecutive zeros at $x = a$ and $x = b$; that is, $f(a) = f(b) = 0$ and $f(x) \neq 0$ for $a < x < b$. Further, suppose that $f(c) > 0$ for some number c between a and b. Use the Intermediate Value Theorem to argue that $f(x) > 0$ for all $a < x < b$.

62. (exploratory) Look up the current federal Tax Rate Schedule (choose single or married as you prefer) and write out the tax function as we did in example 3.8.

63. (exploratory) In the text, we discussed the use of the method of bisections to find an approximate solution of equations such as $f(x) = x^3 + 5x - 1 = 0$. We can start by noticing that $f(0) = -1$ and $f(1) = 5$. Since $f(x)$ is continuous, the Intermediate Value Theorem tells us that there is a solution between $x = 0$ and $x = 1$. For the method of bisections, we guess the midpoint, $x = 0.5$. Is there any reason to suspect that the solution is actually closer to $x = 0$ than to $x = 1$? Using the function values $f(0) = -1$ and $f(1) = 5$, devise your own method of guessing the location of the solution. Generalize your method to using $f(a)$ and $f(b)$, where one function value is positive and one is negative. Compare your method to the method of bisections on the problem $x^3 + 5x - 1 = 0$; for both methods, stop when you are within 0.001 of the solution, $x \approx 0.198437$. Which method performed better? Before you get overconfident in your method, compare the two methods again on $x^3 + 5x^2 - 1 = 0$. Does your method get close on the first try? See if you can determine graphically why your method works better on the first problem.

64. (exploratory) You have probably seen the turntables on which luggage rotates at the airport. Suppose that such a turntable has two long straight parts with semicircles on the end (see the figure). We will attempt to model the motion of someone's lost luggage (we realize this is not a terribly important problem, but it is a nice modeling exercise and may give you some amusement the next time your luggage is lost). We focus on the left/right movement of the luggage. Suppose the straight part is 40 feet long, extending from $x = -20$ to $x = 20$. Assume that our luggage starts at time $t = 0$ at location $x = -20$, and that it takes 60 seconds for the luggage to reach $x = 20$. Suppose the radius of the circular motion is 5 feet, and it takes the luggage 30 seconds to complete the half-circle. We model the straight-line motion with a linear function $x(t) = at + b$. Find constants a and b so that $x(0) = -20$ and $x(60) = 20$. For the circular motion, we use a cosine (Why is this a good choice?) $x(t) = 20 + d \cdot \cos(et + f)$ for constants d, e and f. The requirements are $x(60) = 20$ (since the motion is continuous), $x(75) = 25$ and $x(90) = 20$. Find values of d, e and f to make this work. Find equations for the position of the luggage along the backstretch and the other semicircle. What would the motion be from then on?

Luggage Carousel

1.4   Limits Involving Infinity

In this section, we revisit some old limit problems in order to give more informative answers and examine some related questions.

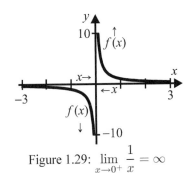

Figure 1.29: $\displaystyle\lim_{x \to 0^+} \frac{1}{x} = \infty$

and $\displaystyle\lim_{x \to 0^-} \frac{1}{x} = -\infty$.

x	$1/x$
0.1	10
0.01	100
0.001	1000
0.0001	10,000
0.00001	100,000

x	$1/x$
-0.1	-10
-0.01	-100
-0.001	-1000
-0.0001	$-10,000$
-0.00001	$-100,000$

Example 4.1   **A Simple Limit Revisited**

Examine

$$\lim_{x \to 0} \frac{1}{x}.$$

Solution   Of course, we can draw a graph (see Figure 1.29) and compute a table of function values easily, by hand (see the tables in the margin).

While we would say that the limits $\displaystyle\lim_{x \to 0^+} \frac{1}{x}$ and $\displaystyle\lim_{x \to 0^-} \frac{1}{x}$ do not exist, the behavior of the function is clearly quite different for $x > 0$ than for $x < 0$. Specifically, as $x \to 0^+$, $1/x$ increases without bound, while as $x \to 0^-$, $1/x$ decreases without bound. Although the two one-sided limits both fail to exist, they do so for different reasons. To communicate as much as we can about the behavior of the function near $x = 0$, we write

$$\lim_{x \to 0^+} \frac{1}{x} = \infty \tag{4.1}$$

and

$$\lim_{x \to 0^-} \frac{1}{x} = -\infty. \tag{4.2}$$

This says that the graph of $y = 1/x$ approaches the vertical line $x = 0$, as $x \to 0$, as seen in Figure 1.29. When this occurs, we say that the line $x = 0$ is a **vertical asymptote**. It is important to note that the limits (4.1) and (4.2) **do not exist**, but we say that they "equal" ∞ and $-\infty$, respectively, only to be specific as to **why** they do not exist. Finally, in view of the one-sided limits (4.1) and (4.2), we say that

$$\lim_{x \to 0} \frac{1}{x} \text{ does not exist.}$$

■

Remark 4.1   It may seem contradictory to first say that $\displaystyle\lim_{x \to 0^+} \frac{1}{x}$ does not exist and then to write $\displaystyle\lim_{x \to 0^+} \frac{1}{x} = \infty$. Note that since ∞ is **not** a real number, there is no contradiction here. (When we say that a limit "does not exist," we are saying that there is no real number L which the function values are approaching.) We say that $\displaystyle\lim_{x \to 0^+} \frac{1}{x} = \infty$ to indicate that as $x \to 0^+$, the function values are increasing without bound.

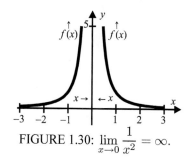

FIGURE 1.30: $\displaystyle\lim_{x \to 0} \frac{1}{x^2} = \infty$.

x	$1/x^2$
0.1	100
0.01	10000
0.001	1×10^6
0.0001	1×10^8
0.00001	1×10^{10}

x	$1/x^2$
-0.1	100
-0.01	10000
-0.001	1×10^6
-0.0001	1×10^8
-0.00001	1×10^{10}

Example 4.2   A Function Whose One-Sided Limits Both Tend to Infinity
Examine the behavior of

$$f(x) = \frac{1}{x^2}$$

near $x = 0$.

Solution   The graph (in Figure 1.30) seems to indicate a vertical asymptote at $x = 0$. A table of values is easily constructed by hand (see the tables in the margin).
     From this, we can see that

$$\lim_{x \to 0^+} \frac{1}{x^2} = \infty$$

and

$$\lim_{x \to 0^-} \frac{1}{x^2} = \infty.$$

Since both one-sided limits agree (i.e., both tend to $+\infty$), even though neither one exists, we say that

$$\lim_{x \to 0} \frac{1}{x^2} = \infty.$$

This one concise statement says that the limit does not exist, but also that $f(x)$ has a vertical asymptote at $x = 0$ where $f(x) \to +\infty$ as $x \to 0$ from either side.

■

Remark 4.2   Note that saying $\displaystyle\lim_{x \to 0} \frac{1}{x^2} = \infty$ is more specific than saying $\displaystyle\lim_{x \to 0} \frac{1}{x^2}$ does not exist, since the first statement not only says that the limit does not exist, but also indicates **why** the limit does not exist. Thus, whenever both statements apply, we prefer to use the more specific of the two. This is analogous to your manner of answering someone who asks you where you were born. You might reasonably answer "Philadelphia" or you could say "North America." They might both be correct, but the former is more informative than the latter. It is this way with limits, too: if you have a more specific answer, give it!

FIGURE 1.31:

$$\lim_{x \to 5^+} \frac{1}{(x-5)^3} = \infty$$

$$\text{and } \lim_{x \to 5^-} \frac{1}{(x-5)^3} = -\infty.$$

Example 4.3   A Case Where the One-Sided Limits Disagree

Examine the behavior of

$$f(x) = \frac{1}{(x-5)^3}$$

as x approaches 5.

Solution   In Figure 1.31, we show a graph of the function. From the graph, you should get a pretty clear idea that there's a vertical asymptote at $x = 5$ and just how the function is blowing up there (to $+\infty$ from the right side and to $-\infty$ from the left). You can verify this behavior algebraically, by noticing that as $x \to 5$, the denominator approaches 0, while the numerator approaches 1. This says that the fraction grows large in absolute value, without bound as $x \to 5$. Specifically,

$$\text{as } x \to 5^+, (x-5)^3 \to 0 \text{ and } (x-5)^3 > 0.$$

We indicate the sign of each factor by printing a small "+" or "–" sign above or below each one. This allows you to see the signs of the various terms at a glance. In this case, we have

Since $(x-5)^3 > 0$, for $x > 5$.

$$\lim_{x \to 5^+} \frac{\overset{+}{1}}{\underset{+}{(x-5)^3}} = +\infty.$$

Likewise,

$$\text{as } x \to 5^-, (x-5)^3 \to 0 \text{ and } (x-5)^3 < 0.$$

In this case, we have

Since $(x-5)^3 < 0$, for $x < 5$.

$$\lim_{x \to 5^-} \frac{\overset{+}{1}}{\underset{-}{(x-5)^3}} = -\infty.$$

Finally, we say that

$$\lim_{x \to 5} \frac{1}{(x-5)^3} \text{ does not exist.}$$

Notice that since the one-sided limits are different, this is the most specific statement we can make here.

■

Learning from the lessons of the last several examples, you should recognize that if the denominator tends to 0 and the numerator does not, then the limit in question does not exist. In this event, we can determine whether the limit tends to $+\infty$ or $-\infty$ by carefully examining the signs of the various factors.

Example 4.4   Another Case Where the One-Sided Limits Disagree

Evaluate $\lim_{x \to -2} \dfrac{x+1}{(x-3)(x+2)}$.

Solution   First, notice from the graph of the function shown in Figure 1.32, there appears to be a vertical asymptote at $x = -2$.

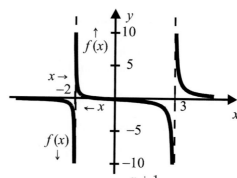

FIGURE 1.32: $\lim\limits_{x \to -2^+} \dfrac{x+1}{(x-3)(x+2)}$ does not exist.

Further, the function appears to tend to $+\infty$, as $x \to -2^+$ and to $-\infty$, as $x \to -2^-$. You can verify this behavior, by observing that

Since $(x+1) < 0$, $(x-3) < 0$ and $(x+2) > 0$, for $-2 < x < -1$.

$$\lim_{x \to -2^+} \frac{\overset{-}{x+1}}{\underset{-\quad\;\; +}{(x-3)(x+2)}} = +\infty$$

and

Since $(x+1) < 0$, $(x-3) < 0$ and $(x+2) < 0$, for $x < -2$.

$$\lim_{x \to -2^-} \frac{\overset{-}{x+1}}{\underset{-\quad\;\; -}{(x-3)(x+2)}} = -\infty.$$

So, we can see that $x = -2$ is indeed a vertical asymptote and that

$$\lim_{x \to -2} \frac{x+1}{(x-3)(x+2)} \text{ does not exist.}$$

∎

Limits at Infinity

We are also interested in examining the limiting behavior of functions as x increases without bound (written $x \to +\infty$) and as x decreases without bound (written $x \to -\infty$). Returning to $f(x) = 1/x$, we can see that as $x \to +\infty$, $1/x \to 0$. In view of this, we write

$$\lim_{x \to +\infty} \frac{1}{x} = 0.$$

Similarly,

$$\lim_{x \to -\infty} \frac{1}{x} = 0.$$

Notice that in Figure 1.33, the graph appears to approach the horizontal line $y = 0$, as $x \to +\infty$ or $x \to -\infty$. In this case, we call $y = 0$ a **horizontal asymptote**.

FIGURE 1.33: $\lim\limits_{x \to +\infty} \dfrac{1}{x} = 0$

and $\lim\limits_{x \to -\infty} \dfrac{1}{x} = 0.$

Example 4.5   Finding Horizontal Asymptotes

Look for any horizontal asymptotes of

$$f(x) = 2 - \frac{1}{x}.$$

Solution   We show the graph in Figure 1.34. Since as $x \to \pm\infty$, $1/x \to 0$, we get that

$$\lim_{x \to +\infty} \left(2 - \frac{1}{x} \right) = 2$$

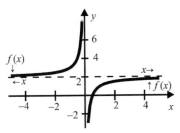

FIGURE 1.34:

$$\lim_{x \to \infty} \left(2 - \frac{1}{x} \right) = 2$$

and $\lim\limits_{x \to -\infty} \left(2 - \frac{1}{x} \right) = 2.$

and

$$\lim_{x \to -\infty} \left(2 - \frac{1}{x} \right) = 2.$$

Thus, the line $y = 2$ is a horizontal asymptote.

∎

As you can see in the following theorem, the behavior of $\frac{1}{x^t}$, for any positive rational power t, as $x \to \pm\infty$, is largely the same as we observed for $f(x) = \frac{1}{x}$.

Theorem 4.1   For any rational number $t > 0$,

$$\lim_{x \to \pm\infty} \frac{1}{x^t} = 0,$$

where for the case where $x \to -\infty$, we assume that $t = \frac{p}{q}$ where q is odd.

Remark 4.3   All of the usual rules for limits stated in Theorem 2.3 also hold for limits as $x \to \pm\infty$.

In the following theorem, we see that the behavior of a polynomial at infinity is easy to determine.

Theorem 4.2   For a polynomial of degree $n > 0$, $p_n(x) = a_n x^n + a_{n-1} x^{n-1} + \cdots + a_0$, we have

$$\lim_{x \to \infty} p_n(x) = \begin{cases} +\infty, \text{ if } a_n > 0 \\ -\infty, \text{ if } a_n < 0 \end{cases}$$

and

$$\lim_{x \to -\infty} p_n(x) = \begin{cases} -\infty, \text{ if } a_n > 0 \\ +\infty, \text{ if } a_n < 0 \end{cases}.$$

Proof. We have

$$\lim_{x \to \infty} p_n(x) = \lim_{x \to \infty} \left(a_n x^n + a_{n-1} x^{n-1} + \cdots + a_0 \right)$$

$$= \lim_{x \to \infty} \left[x^n \left(a_n + \frac{a_{n-1}}{x} + \cdots + \frac{a_0}{x^n} \right) \right]$$

$$= +\infty,$$

if $a_n > 0$, since

$$\lim_{x \to \infty} \left(a_n + \frac{a_{n-1}}{x} + \cdots + \frac{a_0}{x^n} \right) = a_n$$

and $\lim\limits_{x \to \infty} x^n = +\infty$. The remaining three parts are proved similarly and are left as exercises.

∎

In the following example, we again see the need for caution when applying our basic rules for limits (Theorem 2.3).

Example 4.6   **A Limit of a Quotient That Is Not the Quotient of the Limits**
Evaluate

$$\lim_{x \to \infty} \frac{5x - 7}{4x + 3}.$$

Solution   You might be tempted to write

$$\lim_{x \to \infty} \frac{5x - 7}{4x + 3} = \frac{\lim\limits_{x \to \infty} (5x - 7)}{\lim\limits_{x \to \infty} (4x + 3)}$$

$$= \frac{\infty}{\infty} = 1. \tag{4.3}$$

As a check on this conjecture, we draw a graph (see Figure 1.35) and compute some function values. At first glance, the graph appears to be consistent with the conjecture. The computed values, however (see the accompanying table), appear to suggest that the conjecture is incorrect. Can you spot the two errors in (4.3)? The first error was saying that the limit of the quotient was the quotient of the limits. Recall that this rule only applies when **both** limits exist (and the limit in the denominator is nonzero). Since both the limit in the denominator and that in the numerator tend to $+\infty$, this means that the limits **do not exist**. We further confounded the situation when we wrote

$$\frac{\infty}{\infty} = 1. \quad \textbf{(This is incorrect!)}$$

Since ∞ is not a real number, this operation is meaningless. In fact, when a limit looks like $\frac{\infty}{\infty}$, the actual value of the limit can be anything at all. For this reason, we call $\frac{\infty}{\infty}$ an **indeterminate form,** meaning that the value of the expression cannot be determined solely by noticing that both numerator and denominator tend to ∞. (In section 7.6, we carefully examine $\frac{\infty}{\infty}$ and other indeterminate forms such as $\frac{0}{0}$, $\infty - \infty$ and $0 \cdot \infty$.)

Rule of Thumb: When faced with the indeterminate form $\frac{\infty}{\infty}$ in calculating the limit of a rational function, divide numerator and denominator by the highest power of x appearing in the **denominator**.
    Here, we have

$$\lim_{x \to \infty} \frac{5x - 7}{4x + 3} = \lim_{x \to \infty} \left[\frac{5x - 7}{4x + 3} \cdot \frac{(1/x)}{(1/x)} \right]$$

$$= \lim_{x \to \infty} \frac{5 - 7/x}{4 + 3/x}$$

$$= \frac{\lim\limits_{x \to \infty} (5 - 7/x)}{\lim\limits_{x \to \infty} (4 + 3/x)}$$

$$= \frac{5}{4} = 1.25,$$

which is consistent with what we observed both graphically and numerically earlier.

∎

    In the following two examples, we apply our rule of thumb to several common limit problems.

This is an incorrect use of Theorem 2.3 (iv), since the limits in the numerator and the denominator do not exist.

FIGURE 1.35:
$$\lim_{x \to \infty} \frac{5x - 7}{4x + 3} = \frac{5}{4}.$$

x	$\dfrac{5x - 7}{4x + 3}$
10	1
100	1.223325
1000	1.247315
10,000	1.249731
100,000	1.249973

Multiply numerator and denominator by $\frac{1}{x}$.

Multiply through by $\frac{1}{x}$.

By Theorem 2.3 (iv).

Example 4.7   **Finding Horizontal Asymptotes**
Evaluate

$$\lim_{x \to -\infty} \frac{6x^3 + 3x^2 - 8}{7x^4 + 16x^2 + 2}.$$

Solution   Before we do any computations, we look at a graph. From Figure 1.36, it appears as if the function tends to 0, as $x \to \pm\infty$.

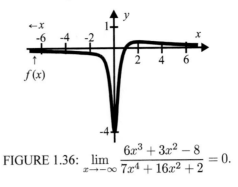

FIGURE 1.36: $\displaystyle\lim_{x \to -\infty} \frac{6x^3 + 3x^2 - 8}{7x^4 + 16x^2 + 2} = 0.$

Using our rule of thumb and Theorem 4.1, we find

Multiply numerator and denominator by $\frac{1}{x^4}$.

Multiply through by $\frac{1}{x^4}$.

By Theorem 2.3 (iv).

$$\lim_{x \to -\infty} \frac{6x^3 + 3x^2 - 8}{7x^4 + 16x^2 + 2} = \lim_{x \to -\infty} \left[\frac{6x^3 + 3x^2 - 8}{7x^4 + 16x^2 + 2} \cdot \frac{(1/x^4)}{(1/x^4)} \right]$$

$$= \lim_{x \to -\infty} \frac{6/x + 3/x^2 - 8/x^4}{7 + 16/x^2 + 2/x^4}$$

$$= \frac{\displaystyle\lim_{x \to -\infty} (6/x + 3/x^2 - 8/x^4)}{\displaystyle\lim_{x \to -\infty} (7 + 16/x^2 + 2/x^4)}$$

$$= \frac{0}{7} = 0.$$

In view of this, the line $y = 0$ is a horizontal asymptote. Look again at Figure 1.36. This does not say that the graph cannot cross the line $y = 0$ (as this one clearly does), but simply that the graph approaches the line $y = 0$, as $x \to \infty$.

∎

Example 4.8   **Finding Slant Asymptotes**

Evaluate

$$\lim_{x \to \infty} \frac{4x^3 + 5}{-6x^2 - 7x}.$$

Solution   As usual, we first examine a graph (see Figure 1.37a). Note that here, the graph appears to tend to $-\infty$ as $x \to \infty$. Further, you might notice that outside of the interval $[-2, 2]$, the graph looks very much like a straight line. If we look at the graph in a somewhat larger window, this linearity is even more apparent (see Figure1.37b).

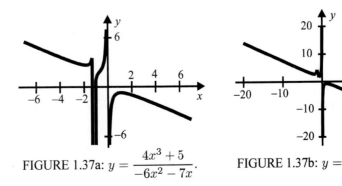

FIGURE 1.37a: $y = \dfrac{4x^3 + 5}{-6x^2 - 7x}.$ FIGURE 1.37b: $y = \dfrac{4x^3 + 5}{-6x^2 - 7x}.$

First, let's compute the limit and then, we will try to sort out what it means for this graph to look like a straight line for x large.

Using our rule of thumb, we have

Multiply numerator and denominator by $\frac{1}{x^2}$.

$$\lim_{x\to\infty}\frac{4x^3+5}{-6x^2-7x}=\lim_{x\to\infty}\left[\frac{4x^3+5}{-6x^2-7x}\cdot\frac{(1/x^2)}{(1/x^2)}\right]$$

Multiply through by $\frac{1}{x^2}$.

$$=\lim_{x\to\infty}\frac{4x+5/x^2}{-6-7/x}$$

$$=-\infty,$$

since as $x\to\infty$, the numerator tends to ∞ and the denominator tends to -6.

To further explain the behavior seen in Figure 1.37b, we perform a long division. We have

$$\frac{4x^3+5}{-6x^2-7x}=-\frac{2}{3}x+\frac{7}{9}+\frac{5+\frac{49}{9}x}{-6x^2-7x}.$$

Notice that the third term in this expansion tends to 0 as $x\to\infty$. (You should verify this as an exercise.) Thus, the function values approach those of the linear function

$$-\frac{2}{3}x+\frac{7}{9},$$

as $x\to\infty$. When this occurs, we say that the function has a **slant** (or oblique) asymptote. That is, instead of approaching a vertical or horizontal line, as happens with vertical or horizontal asymptotes, the graph is approaching the slanted straight line $y=-\frac{2}{3}x+\frac{7}{9}$. (This is the behavior we're seeing in Figure 1.37b.)

■

Exercises 1.4

1. (writing) It may seem odd that we use ∞ in describing limits but do not count ∞ as a real number. Discuss the existence of ∞: is it a number or a concept?

2. (writing) In example 4.6, we dealt with the "indeterminate form" $\frac{\infty}{\infty}$. Thinking of a limit of ∞ as meaning "getting very large" and a limit of 0 as meaning "getting very close to 0" explain why the following are indeterminate forms: $\frac{\infty}{\infty}$, $\frac{0}{0}$, $\infty-\infty$, and $\infty\cdot 0$. Determine what the following noninde-terminate forms represent: $\infty+\infty$, $-\infty-\infty$, $\infty+0$ and $0/\infty$.

3. (writing) On your computer or calculator, graph $y=1/(x-2)$ and look for the horizontal asymptote $y=0$ and the vertical asymptote $x=2$. Most computers will draw a vertical line at $x=2$ and will show the graph completely flattening out at $y=0$ for large x's. Is this accurate? misleading? Most computers will compute the locations of points for adjacent x's and try to connect the points with a line. Why might this result in a vertical line at the location of a vertical asymptote?

4. (writing) Many students learn that asymptotes are lines which the graph gets closer and closer to without ever reaching. This is true for many asymptotes, but not all. Explain why vertical asymptotes are never reached or crossed. Explain why horizontal or slant asymptotes may, in fact, be crossed any number of times; draw one example.

In exercises 5–40, determine the following limits (answer as appropriate, with a number, ∞, $-\infty$ or does not exist).

5. $\displaystyle\lim_{x\to 1^-}\frac{1-2x}{x^2-1}$

6. $\displaystyle\lim_{x\to 1^+}\frac{1-2x}{x^2-1}$

7. $\displaystyle\lim_{x\to 1}\frac{1-2x}{x^2-1}$

8. $\displaystyle\lim_{x\to -1^-}\frac{1-2x}{x^2-1}$

9. $\displaystyle\lim_{x\to -1^+}\frac{1-2x}{x^2-1}$

10. $\displaystyle\lim_{x\to -1}\frac{1-2x}{x^2-1}$

11. $\displaystyle\lim_{x\to 2^-}\frac{4-x}{(x-2)^2}$

12. $\displaystyle\lim_{x\to 2^+}\frac{4-x}{(x-2)^2}$

13. $\displaystyle\lim_{x\to 2}\frac{4-x}{(x-2)^2}$

14. $\displaystyle\lim_{x\to -2^+}\frac{-x}{\sqrt{4-x^2}}$

15. $\displaystyle\lim_{x\to 2^-}\frac{-x}{\sqrt{4-x^2}}$

16. $\displaystyle\lim_{x\to 3^+}(x^2-2x-3)^{-2/3}$

17. $\displaystyle\lim_{x\to -1^-}(x^2-2x-3)^{-2/3}$

18. $\displaystyle\lim_{x\to\infty} \frac{-x}{\sqrt{4+x^2}}$

19. $\displaystyle\lim_{x\to-\infty} \frac{-x}{\sqrt{4+x^2}}$

20. $\displaystyle\lim_{x\to\infty} \frac{2x^2-x+1}{4x^2-3x-1}$

21. $\displaystyle\lim_{x\to-\infty} \frac{x^3-2x+1}{3x^3+4x^2-1}$

22. $\displaystyle\lim_{x\to\infty} \frac{2x^2-1}{4x^3-5x-1}$

23. $\displaystyle\lim_{x\to\infty} \frac{x^3-2x+5}{3x^2+4x-1}$

24. $\displaystyle\lim_{x\to-\infty} \frac{x^2-\sin x}{x-5}$

25. $\displaystyle\lim_{x\to\infty} \frac{x^2-\sin x}{x^2+4x-1}$

26. $\displaystyle\lim_{x\to-\infty} \frac{x+\cos x}{3x+2}$

27. $\displaystyle\lim_{x\to\infty} \frac{3x^3-x+5}{4x^3+4x^2-1}$

28. $\displaystyle\lim_{x\to\infty} \frac{x^2-x+2}{\sqrt{x^4+2x+3}}$

29. $\displaystyle\lim_{x\to\infty} \left(\sqrt{x^2+3}-x\right)$

30. $\displaystyle\lim_{x\to\infty} \left(\sqrt{4x^2-2x+1}-2x\right)$

31. $\displaystyle\lim_{x\to\infty} e^{2x}$

32. $\displaystyle\lim_{x\to-\infty} e^{2x}$

33. $\displaystyle\lim_{x\to\infty} e^{-4x}$

34. $\displaystyle\lim_{x\to-\infty} e^{-4x}$

35. $\displaystyle\lim_{x\to\infty} \left(e^{-3x}\cos 2x\right)$

36. $\displaystyle\lim_{x\to-\infty} \left(e^{4x}\sin x\right)$

37. $\displaystyle\lim_{x\to\infty} \ln 2x$

38. $\displaystyle\lim_{x\to0^+} \ln 2x$

39. $\displaystyle\lim_{x\to0^+} \left(x\ln 2x\right)$

40. $\displaystyle\lim_{x\to\infty} \left(x/\ln 2x\right)$

In exercises 41–50, determine all horizontal, slant and vertical asymptotes.

41. $f(x)=\dfrac{x}{\sqrt{4+x^2}}$

42. $f(x)=\dfrac{x}{\sqrt{4-x^2}}$

43. $f(x)=\dfrac{x}{4-x^2}$

44. $f(x)=\dfrac{x^2}{4-x^2}$

45. $f(x)=\dfrac{x^3}{4-x^2}$

46. $f(x)=\dfrac{x^4}{4-x^2}$

47. $f(x)=\dfrac{x^3}{4+x^2}$

48. $f(x)=\dfrac{\sqrt{x^4+2}}{4-x^2}$

49. $f(x)=\dfrac{2x^3}{x^2-x+1}$

50. $f(x)=\dfrac{x^2+2}{x^2-2x+1}$

51. Sketch the graph of $f(x)=e^{-x}\cos x$. Identify the horizontal asymptote. Is this asymptote approached in both directions (as $x\to\infty$ and $x\to-\infty$)? How many times does the graph cross the horizontal asymptote?

52. Explain why it is reasonable that $\displaystyle\lim_{x\to\infty} f(x)=\lim_{x\to0^+} f(1/x)$ and $\displaystyle\lim_{x\to-\infty} f(x)=\lim_{x\to0^-} f(1/x)$.

53. In the exercises of section 1.1, we found that $\displaystyle\lim_{x\to0^+}(1+x)^{1/x}=\lim_{x\to0^-}(1+x)^{1/x}$. (It turns out that both limits equal the irrational number e.) Use this result and exercise 52 to argue that $\displaystyle\lim_{x\to\infty}(1+1/x)^x=\lim_{x\to-\infty}(1+1/x)^x$.

54. One of the reasons for saying that infinite limits do not exist is that we would otherwise invalidate Theorem 2.3 in section 1.2. Find examples of functions with infinite limits such that parts (ii) and (iv) of Theorem 2.3 are not correct.

55. Suppose that the length of a small animal t days after birth is $h(t)=\dfrac{300}{1+9(0.8)^t}$ mm. What is the length of the animal at birth? What is the eventual length of the animal (i.e., the length as $t\to\infty$)?

56. Suppose that the length of a small animal t days after birth is $h(t)=\dfrac{100}{2+3(0.4)^t}$ mm. What is the length of the animal at birth? What is the eventual length of the animal (i.e., the length as $t\to\infty$)?

57. Suppose that the size of the pupil of a certain animal is given by $f(x)$ (mm), where x is the intensity of the light on the pupil. If $f(x)=\dfrac{160x^{-0.4}+90}{4x^{-0.4}+15}$, find the size of the pupil with no light and the size of the pupil with an infinite amount of light.

58. After an injection, the concentration of drug in a muscle varies according to a function of time $f(t)$. Suppose that t is measured in hours and $f(t)=e^{-0.02t}-e^{-0.42t}$. Find the limit of $f(t)$ both as $t\to0$ and $t\to\infty$ and interpret both limits in terms of the concentration of drug.

59. You've heard the sports cliche to "keep your eye on the ball." In the diagram, x is the distance from the ball to home plate and θ is an angle indicating the direction of the player's gaze. We denote the speed of the ball by x' and the rate of change of the player's gaze by θ'. For a 90-mph baseball pitch, $x'=-132$ ft/s. It can be shown that $\theta'=\dfrac{264}{4+x^2}$ radians/second. From this formula, explain why θ' increases as x decreases. Explain the same result from physical principles. Finally, compute $\displaystyle\lim_{x\to0}\theta'$, the maximum rate of change of the player's gaze. This is not an infinite limit. However, given that human beings cannot maintain focus at a rate of more than $\dfrac{\pi}{2}$ radians/second, how big is the maximum θ'? Is it possible for a baseball player to keep his or her eyes on the ball?

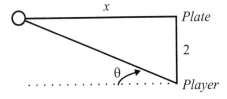

60. According to Einstein's theory of relativity, the mass of an object traveling at speed v is given by $m=m_0/\sqrt{1-v^2/c^2}$, where c is the speed of light (about 9.8×10^8 ft/s). Compute $\displaystyle\lim_{v\to0} m$ and explain why m_0 is called the "rest mass." Compute $\displaystyle\lim_{v\to c^-} m$ and discuss the implications. (What would

happen if you were traveling in a spaceship approaching the speed of light?) How much does the mass of a 192-pound man ($m_0 = 6$) increase at the speed of 9000 ft/s (about 4 times the speed of sound)?

61. Suppose an object with initial velocity $v_0 = 0$ ft/s and (constant) mass m slugs is accelerated by a constant force F pounds for t seconds. According to Newton's laws of motion, the object's speed will be $v_N = Ft/m$. According to Einstein's theory of relativity, the object's speed will be $v_E = Fct/\sqrt{m^2c^2 + F^2t^2}$ where c is the speed of light. Compute $\lim\limits_{t\to\infty} v_N$ and $\lim\limits_{t\to\infty} v_E$.

62. The speed of a 128-lb sky-diver is given by (approximately) $v = \frac{128}{k}(1 - e^{-kt/4})$ ft/s, where k is a constant that depends on the orientation of the sky-diver's body. Compute $\lim\limits_{t\to\infty} v$. Explain why this is called "terminal velocity." Compare the terminal velocity of sky-divers in the spread-eagle position ($k = 3$) and the tuck position ($k = 1$). How are sky-divers able to control their speed?

63. Explain why $\lim\limits_{t\to\infty} (e^{-at} \sin t) = 0$ for any positive constant a. Although this is theoretically true, it is not necessarily useful in practice. The function $e^{-at} \sin t$ is a simple model for a spring-mass system, such as the shock absorber on a car. Suppose t is measured in seconds and the car passengers cannot feel any vibrations less than 0.01 (inches). If shock absorber A has the vibration function $e^{-t} \sin t$ and shock absorber B has the vibration function $e^{-t/4} \sin t$, determine graphically how long it will take before the vibrations damp out, that is, $|f(t)| < 0.01$. Is the result $\lim\limits_{t\to\infty} (e^{-at} \sin t) = 0$ much consolation to the owner of shock absorber B?

64. It is very difficult to find simple statements in calculus that are always true; this is one reason that a careful development of theory is so important. You may have heard the simple rule: to find the vertical asymptotes of $f(x) = \dfrac{g(x)}{h(x)}$, simply set the denominator equal to 0 (i.e., solve $h(x) = 0$). Give an example where $h(a) = 0$ but there is *not* a vertical asymptote at $x = a$.

65. In exercise 64, you needed to find an example indicating that the following statement is not (necessarily) true: if $h(a) = 0$, then $f(x) = \dfrac{g(x)}{h(x)}$ has a vertical asymptote at $x = a$. This is not true, but perhaps its converse is true: if $f(x) = \dfrac{g(x)}{h(x)}$ has a vertical asymptote at $x = a$, then $h(a) = 0$. Is this statement true?

66. (exploratory) Suppose you are shooting a basketball from a (horizontal) distance of L feet, releasing the ball from a location h feet below the basket. To get a perfect swish, it is necessary that the initial velocity v_0 and initial release angle θ_0 satisfy the equation $v_0 = \sqrt{gL}/ \sqrt{2\cos^2 \theta_0(\tan \theta_0 - h/L)}$. For a free throw, take $L = 15$, $h = 2$ and $g = 32$ and graph v_0 as a function of θ_0. What is the significance of the two vertical asymptotes? Explain in physical terms what type of shot corresponds to each vertical asymptote. Assume that the basket is 2 feet in diameter and the ball is 1 foot in diameter. For a free throw, $L = 15$ feet is perfect. What is the maximum horizontal distance the ball could travel and still go in the basket (don't count bouncing off the backboard)? What is the minimum horizontal distance? Call these numbers $L_{\max}$ and $L_{\min}$. Start with an angle of $\theta_0 = 50°$. Find the velocity v_1 corresponding to θ_0 and $L_{\max}$; find the velocity v_2 corresponding to θ_0 and $L_{\min}$. At an initial angle of $50°$, an initial velocity anywhere between v_1 and v_2 will produce a good shot. By trial and error, try to find the angle that produces the maximum margin of error, that is the maximum difference $|v_1 - v_2|$.

67. (exploratory) In applications, it is common to compute $\lim\limits_{x\to\infty} f(x)$ to determine the "stability" of the function $f(x)$. Consider the function $f(x) = xe^{-x}$. As $x \to \infty$, the first factor in $f(x)$ goes to ∞, but the second factor goes to 0. What does the product do when one term is getting smaller and the other term is getting larger? It depends on which one is changing faster. What we want to know is which function "dominates." Use graphical and numerical evidence to conjecture the value of $\lim\limits_{x\to\infty} (xe^{-x})$. Which function dominates? In the limit $\lim\limits_{x\to\infty} (x^2 e^{-x})$, which function dominates? Also, try $\lim\limits_{x\to\infty} (x^5 e^{-x})$. Based on your investigation, is it always true that exponentials dominate polynomials? Are you positive? Try to determine which type of function, polynomials or logarithms, dominates.

68. (exploratory) A different type of limit at infinity that will be very important to us is the limit of a sequence. Investigating the area under a parabola in Chapter 4, we will compute the following approximations: $\frac{2(3)}{6(1)} = 1$, $\frac{3(5)}{6(4)} = 0.625$, $\frac{4(7)}{6(9)} \approx 0.518$, $\frac{5(9)}{6(16)} \approx 0.469$ and so on. Do you see a pattern? If we name our approximations a_1, a_2, a_3 and a_4, verify that $a_n = \dfrac{(n+1)(2n+1)}{6n^2}$. The area under the parabola is the limit of these approximations as n gets larger and larger. Find the area.

1.5   Formal Definition of the Limit (Optional)

We have now spent many pages discussing various aspects of the computation of limits. This may seem a bit odd, in that we have never actually *defined* what a limit is. Oh, sure, we have an *idea* of what a limit is, but that's about all. Once again, the intuitive notion of a limit goes something like this. We say that

$$\lim_{x \to a} f(x) = L$$

if $f(x)$ gets closer and closer to L as x gets closer and closer to a.

So far, we have been quite happy with this somewhat vague, although intuitive, description. In fact, for many purposes, this notion is quite adequate. In this section, however, we will make this more precise. In doing so, you will begin to see how **mathematical analysis** (that branch of mathematics of which the calculus is the most elementary study) works.

Studying more advanced mathematics without an understanding of the precise definition of limit is somewhat akin to studying brain surgery without bothering with all that background work in chemistry and biology. A brain surgeon certainly doesn't seem to need these things to perform his or her job, but neither of the authors would consent to surgery by one who did not have a **thorough** understanding of these areas. The reasons are obvious. In medicine, it has only been through a careful examination of the microscopic world that a deeper understanding of our own macroscopic world has developed, and good surgeons need to understand what they are doing **and why** they are doing it. Mere technical proficiency is simply not enough. Likewise, in mathematical analysis, it is through an understanding of the microscopic behavior of functions (here, the precise definition of limit) that a deeper understanding of the mathematics will come about.

We begin with the careful examination of an elementary example. At this point, it should not be difficult to believe that

$$\lim_{x \to 2}(3x + 4) = 10.$$

This is the limit of a polynomial and we've already seen that such limits are found by simply substituting the value that x is approaching. Suppose that you were asked to explain the meaning of this particular limit to a younger brother or sister. You would probably repeat the intuitive explanation we have used so far: that as x gets closer and closer to 2, $(3x + 4)$ gets arbitrarily close to 10. But, exactly what do we mean by *close*? Close is a relative term, imprecise by its nature. For example, if one of your classmates asks you where you grew up and you say "close to Mars," they might think that you are just a bit eccentric. On the other hand, while visiting the next galaxy, if you were to give the very same response to an alien (who happens to know where Mars is), your statement would make perfect sense. So, when we say *close*, in the context of limit, what are we trying to communicate? One answer is that if $\lim_{x \to 2}(3x + 4) = 10$, we should be able to make $(3x + 4)$ as close as we like to 10, just by making x sufficiently close to 2. But can we actually do this? For instance, can we force $(3x + 4)$ to be within distance 1 of 10? (We might as well start conservatively.) To see what values of x will guarantee this, we write an inequality which says that $(3x + 4)$ is within 1 unit of 10:

$$|(3x + 4) - 10| < 1.$$

Eliminating the absolute values, we see that this is equivalent to

$$-1 < (3x + 4) - 10 < 1$$

or

$$-1 < 3x - 6 < 1.$$

Keep in mind that we want to isolate $x - 2$, instead of x. So, dividing by 3, we get

$$-\frac{1}{3} < x - 2 < \frac{1}{3}$$

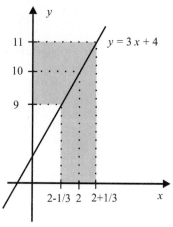

FIGURE 1.38:
$2 - \frac{1}{3} < x < 2 + \frac{1}{3}$
guarantees that
$|(3x + 4) - 10| < 1$.

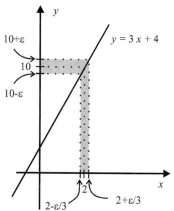

FIGURE 1.39: The range
of x-values that keep
$|(3x + 4) - 10| < \varepsilon$.

or

$$|x - 2| < \frac{1}{3}. \qquad (5.1)$$

Reading the inequality (5.1), we see that if x is within distance $\frac{1}{3}$ of 2, then $(3x + 4)$ will be within the specified distance (1) of 10. (See Figure 1.38 for a graphical interpretation of this.) So, does this convince you that we can make $(3x + 4)$ as close as you want to 10? Probably not, but if we used a smaller distance, perhaps you'd be more convinced. Can we make $(3x + 4)$ within distance $\frac{1}{100}$ of 10? This is equivalent to

$$|(3x + 4) - 10| < \frac{1}{100}.$$

Eliminating the absolute values, we get

$$-\frac{1}{100} < (3x + 4) - 10 < \frac{1}{100}$$

or

$$-\frac{1}{100} < 3x - 6 < \frac{1}{100}.$$

Dividing by 3 yields

$$-\frac{1}{300} < x - 2 < \frac{1}{300},$$

which is equivalent to

$$|x - 2| < \frac{1}{300}.$$

So, are you now convinced that we can make $(3x + 4)$ as close as desired to 10? While you should be more convinced than a moment ago, you should still retain some skepticism. All we've been able to show is that we can make $(3x + 4)$ pretty close to 10. So, how close do we need to be able to make it? The answer is: arbitrarily close, as close as anyone would ever demand. We can show that this is possible by repeating the preceding argument, this time for an unspecified distance, call it ε (*epsilon*, where $\varepsilon > 0$). That is, we verify that we can make $(3x + 4)$ within any distance ε of 10 (no matter how small ε is), just by making x sufficiently close to 2. The objective, then, is to determine the range of x-values which will guarantee that $(3x + 4)$ stays within ε of 10 (see Figure 1.39 for a sketch of this range). We have

$$|(3x + 4) - 10| < \varepsilon.$$

This is equivalent to

$$-\varepsilon < (3x + 4) - 10 < \varepsilon$$

or

$$-\varepsilon < 3x - 6 < \varepsilon.$$

Dividing by 3, we get

$$-\frac{\varepsilon}{3} < x - 2 < \frac{\varepsilon}{3}$$

or

$$|x - 2| < \frac{\varepsilon}{3}.$$

Notice that each of the preceding steps is reversible, so that $|x - 2| < \frac{\varepsilon}{3}$ also implies that $|(3x + 4) - 10| < \varepsilon$. This says that as long as x is within distance $\frac{\varepsilon}{3}$ of 2, $(3x + 4)$ will be within the required distance, ε, of 10. That is,

$$|(3x + 4) - 10| < \varepsilon \text{ whenever } |x - 2| < \frac{\varepsilon}{3}.$$

Take a moment or two to recognize what we've done. By using an **unspecified** distance, ε, we have verified that we can indeed make $(3x+4)$ as close to 10 as might be demanded (i.e., arbitrarily close; just name whatever $\varepsilon > 0$ you would like), just by making x sufficiently close to 2. Further, we have explicitly spelled out what "sufficiently close to 2" means in the context of the present problem. Thus, no matter how close we are asked to make $(3x + 4)$ to 10, we can accomplish this simply by taking x to be in the specified interval.

Next, we examine this more precise notion of limit in the case of a function which is not defined at the point in question.

Example 5.1   Proving That a Limit is Correct

Prove that

$$\lim_{x \to 1} \frac{2x^2 + 2x - 4}{x - 1} = 6.$$

Solution   It is easy to use the usual rules of limits to establish this result. (We leave this as an exercise.) It is yet another matter to verify that this is correct using our new and more precise notion of limit. In this case, we want to know how close x must be to 1 to ensure that

$$f(x) = \frac{2x^2 + 2x - 4}{x - 1}$$

is within an unspecified distance $\varepsilon > 0$ of 6.

First, notice that f is undefined **at** $x = 1$. Thus, we seek a distance δ (*delta*, $\delta > 0$), such that if x is within distance δ of 1 (but $x \neq 1$), then $|f(x) - 6| < \varepsilon$. This is equivalent to saying that

$$0 < |x - 1| < \delta \text{ guarantees that } |f(x) - 6| < \varepsilon.$$

Notice that we have specified that $0 < |x - 1|$ to ensure that $x \neq 1$. Further, $|f(x) - 6| < \varepsilon$ is equivalent to

$$-\varepsilon < \frac{2x^2 + 2x - 4}{x - 1} - 6 < \varepsilon$$

or adding 6 to all three terms,

$$6 - \varepsilon < \frac{2x^2 + 2x - 4}{x - 1} < 6 + \varepsilon.$$

Since the numerator factors, we get

$$6 - \varepsilon < \frac{2(x - 1)(x + 2)}{x - 1} < 6 + \varepsilon.$$

Since $x \neq 1$, we can cancel the two factors of $(x - 1)$ to yield

$$6 - \varepsilon < 2(x + 2) < 6 + \varepsilon.$$

Dividing by 2, we get

$$3 - \frac{\varepsilon}{2} < x + 2 < 3 + \frac{\varepsilon}{2}.$$

Recall that the strategy is to isolate $x - 1$, here. So, subtracting 3 from each term, we get

$$-\frac{\varepsilon}{2} < x - 1 < \frac{\varepsilon}{2}.$$

So, taking $\delta = \frac{\varepsilon}{2}$ and requiring that x satisfies

$$0 < |x - 1| < \delta$$

guarantees that

$$\left| \frac{2x^2 + 2x - 4}{x - 1} - 6 \right| < \varepsilon.$$

We illustrate this graphically in Figure 1.40.

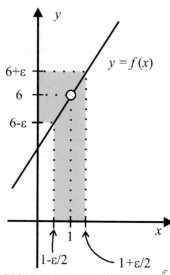

FIGURE 1.40: $0 < |x - 1| < \dfrac{\varepsilon}{2}$
guarantees that
$$6 - \varepsilon < \frac{2x^2 + 2x - 4}{x - 1} < 6 + \varepsilon.$$

What we have seen so far motivates us to make the following general definition, illustrated in Figure 1.41.

Definition 5.1   (Precise Definition of Limit) For a function f defined in some open interval containing a (but not necessarily **at** a itself), we say

$$\lim_{x \to a} f(x) = L$$

if given any number $\varepsilon > 0$, there is another number $\delta > 0$, such that $|f(x) - L| < \varepsilon$ whenever $0 < |x - a| < \delta$.

Remark 5.1   We want to emphasize that this formal definition of limit is not a new idea. Rather, it is a more precise mathematical statement of the same intuitive notion of limit that we have been using all along. Also, we must in all honesty point out that it is extremely difficult to explicitly find δ as a function of ε, for all but a few simple examples. Despite this, learning how to work through the definition, even for a small number of problems, will shed considerable light on a deep concept.

The following example, although only slightly more complex than the last several problems, provides an unexpected challenge.

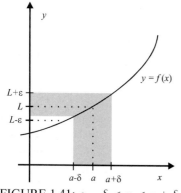

FIGURE 1.41: $a - \delta < x < a + \delta$
guarantees that
$$L - \varepsilon < f(x) < L + \varepsilon.$$

Example 5.2   Using the Precise Definition of Limit

Use Definition 5.1 to prove that

$$\lim_{x \to 2} (x^2 + 1) = 5.$$

Solution   Again, this is the comparatively simple case of a limit of a polynomial. If this is correct, then given any $\varepsilon > 0$, there must be a $\delta > 0$ for which $0 < |x - 2| < \delta$ guarantees that

$$\left| (x^2 + 1) - 5 \right| < \varepsilon.$$

Notice that

Factor the difference of two squares.

$$\begin{aligned}\left| (x^2 + 1) - 5 \right| &= \left| x^2 - 4 \right| \\ &= |x + 2|\,|x - 2|.\end{aligned} \tag{5.2}$$

Our strategy is to isolate $|x - 2|$ and so, we'll need to do something with the term $|x + 2|$. Since we're only interested in what happens near $x = 2$, anyway, we only consider x's within a distance of 1 from 2, that is, x's that lie in the interval $[1, 3]$. This means that we want $|x - 2| < 1$. Notice that this will be true if we require $\delta \leq 1$ and $|x - 2| < \delta$. In this case, notice that

$$|x + 2| \leq 5, \text{ (since } x \in [1, 3])$$

and so, from (5.2) we have

$$\left|(x^2+1)-5\right| = |x+2|\,|x-2|$$
$$\leq 5\,|x-2|. \tag{5.3}$$

Finally, if we require that

$$5\,|x-2| < \varepsilon, \tag{5.4}$$

then we will also have from (5.3) that

$$\left|(x^2+1)-5\right| \leq 5\,|x-2| < \varepsilon.$$

Of course, (5.4) is equivalent to

$$|x-2| < \frac{\varepsilon}{5}.$$

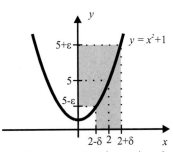

FIGURE 1.42: $0 < |x-2| < \delta$
guarantees that
$\left|(x^2+1)-5\right| < \varepsilon.$

So, it now follows from (5.3) that if we choose $\delta = \min\left\{1, \frac{\varepsilon}{5}\right\}$ (i.e., the **minimum** of 1 and $\frac{\varepsilon}{5}$), then having

$$0 < |x-2| < \delta \leq \frac{\varepsilon}{5}$$

will guarantee that

$$\left|(x^2+1)-5\right| < \varepsilon,$$

as desired. We illustrate this in Figure 1.42.

Exploring the Definition of Limit Graphically

As you can see from example 5.2, this business of finding δ's for a given ε is not easily accomplished. There, we found that even for the comparatively simple case of a quadratic polynomial, the job can be quite a challenge. Unfortunately, there is no procedure that will work for all problems. However, we can explore the definition graphically in the case of more complex functions. First, we reexamine example 5.2 graphically.

Example 5.3   Exploring the Precise Definition of Limit Graphically
Explore the precise definition of limit graphically, for

$$\lim_{x\to 2}(x^2+1) = 5.$$

Solution   In example 5.2, we discovered that

$$0 < |x-5| < \delta \leq \frac{\varepsilon}{5} \text{ implies that } \left|(x^2+1)-5\right| < \varepsilon.$$

FIGURE 1.43: $y = x^2 + 1$.

This says that if we draw a graph of $y = x^2 + 1$ and restrict the x-values to lie in the interval $\left(2 - \frac{\varepsilon}{5}, 2 + \frac{\varepsilon}{5}\right)$, then the y-values will lie in the interval $(5 - \varepsilon, 5 + \varepsilon)$. Take $\varepsilon = \frac{1}{2}$, for instance. If we draw the graph in the window defined by $2 - \frac{1}{10} \leq x \leq 2 + \frac{1}{10}$ and $4.5 \leq y \leq 5.5$, then the graph will stay in that window (i.e., the graph will stay on the screen; see Figure 1.43). Of course, we can draw virtually the same picture for any given value of ε, since we have an explicit formula for finding δ given ε. For most limit problems, we are not so fortunate.

Example 5.4   Exploring the Definition of Limit for a Trigonometric Function

Graphically find a $\delta > 0$ corresponding to (a) $\varepsilon = \frac{1}{2}$ and (b) $\varepsilon = 0.1$ for

$$\lim_{x \to 2} \sin \frac{\pi x}{2} = 0.$$

Solution   This limit certainly seems plausible enough. After all, $\sin \dfrac{2\pi}{2} = 0$ and we've already observed that $f(x) = \sin x$ is a continuous function. However, we are trying to verify this very carefully. Given any $\varepsilon > 0$, we would like to be able to find a $\delta > 0$, for which

$$0 < |x - 2| < \delta \text{ guarantees that } \left| \sin \frac{\pi x}{2} - 0 \right| < \varepsilon.$$

Note that since we have no algebra for simplifying $\sin \dfrac{\pi x}{2}$, we have little hope of accomplishing this symbolically. Instead, we'll try to graphically find δ's corresponding to the specific ε's given. First, for $\varepsilon = \frac{1}{2}$, we would like to find a $\delta > 0$ for which if $0 < |x - 2| < \delta$, then

$$-\frac{1}{2} < \sin \frac{\pi x}{2} - 0 < \frac{1}{2}.$$

Drawing the graph of $y = \sin \dfrac{\pi x}{2}$ on the interval $[1, 3]$, we get Figure 1.44a.

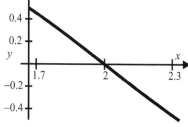

FIGURE 1.44a: $y = \sin \frac{\pi x}{2}$.   FIGURE 1.44b: $y = \sin \frac{\pi x}{2}$.

FIGURE 1.45: $y = \sin \frac{\pi x}{2}$.

FIGURE 1.46: $y = \sin \frac{\pi x}{2}$.

If you trace along the curve, you will notice that the graph stays on the screen (i.e., the y-values stay in the interval $[-0.5, 0.5]$) for $x \in [1.666667, 2.333333]$. Thus, we have determined experimentally that for $\varepsilon = \frac{1}{2}$,

$$\delta = 2.333333 - 2 = 2 - 1.666667 = 0.333333$$

will work. (Of course, any value of δ smaller than 0.333333 will also work.) This becomes even more clear if we redraw the last graph, but restrict x to lie in the interval $[1.666667, 2.333333]$ (see Figure 1.44b). In this case, the graph stays in the window over the entire range of displayed x-values. Taking $\varepsilon = 0.1$, we wish to determine those values of x near $x = 2$ that will guarantee that $\sin \dfrac{\pi x}{2}$ stays between -0.1 and 0.1. We first redraw the graph from Figure 1.44a, with the y-range restricted to the interval $[-0.1, 0.1]$ (see Figure 1.45). Again, tracing along the graph tells us that the y-values will stay in the desired range for $x \in [1.936508, 2.063492]$. Thus, we have experimentally determined that

$$\delta = 2.063492 - 2 = 2 - 1.936508 = 0.063492$$

will work here. Again, this becomes even more clear if we redraw the graph in Figure 1.45 using the new range of x-values (see Figure 1.46), since the graph remains in the window for all values of x in the indicated interval.

It is important to recognize that we are not **proving** that the above limit is correct. To prove this requires us to symbolically find a δ for **every** $\varepsilon > 0$. The idea here is to use these graphical illustrations to become more familiar with the definition and with what δ and ε represent.

Example 5.5   Exploring the Definition of Limit Where the Limit Does Not Exist

Determine whether or not

$$\lim_{x \to 0} \frac{x^2 + 2x}{\sqrt{x^3 + 4x^2}} = 1.$$

x	$\dfrac{x^2 + 2x}{\sqrt{x^3 + 4x^2}}$
0.1	1.03711608
0.01	1.0037461
0.001	1.00037496
0.0001	1.0000375

FIGURE 1.47: $y = \dfrac{x^2 + 2x}{\sqrt{x^3 + 4x^2}}$.

FIGURE 1.48: $y = \dfrac{x^2 + 2x}{\sqrt{x^3 + 4x^2}}$.

Solution   We first construct a table of function values. From the table alone, we might conjecture that the limit is 1. But we would be making a **huge** error. We have not considered negative values of x. Further, we haven't bothered to draw a graph. This kind of carelessness is dangerous. You should always consider values of x on both sides of the value being approached and draw a graph (see Figure 1.47 for the default graph drawn by our computer algebra system). In this graph, the function values do not quite look like they are approaching 1 as $x \to 0$ (at least as $x \to 0^-$). We now investigate the limit graphically for $\varepsilon = \frac{1}{2}$. We need to find a $\delta > 0$ for which $0 < |x| < \delta$ guarantees that

$$1 - \frac{1}{2} < \frac{x^2 + 2x}{\sqrt{x^3 + 4x^2}} < 1 + \frac{1}{2}$$

or

$$\frac{1}{2} < \frac{x^2 + 2x}{\sqrt{x^3 + 4x^2}} < \frac{3}{2}.$$

We try $\delta = 0.1$ to see if this is sufficiently small. So, we set the x-range to the interval $[-0.1, 0.1]$ and the y-range to the interval $[0.5, 1.5]$ and redraw the graph in this window (see Figure 1.48). Notice that no points are plotted in the window for any $x < 0$. According to the definition, the y-values must lie in the interval $(0.5, 1.5)$ for **all** x in the interval $(-\delta, \delta)$. Further, you can see that $\delta = 0.1$ clearly does not work since $x = -0.05$ lies in the interval $(-\delta, \delta)$, but $f(-0.05) \approx -0.981$ is not in the interval $(0.5, 1.5)$. Finally, you should convince yourself that no matter how small we make δ, there is an x in the interval $(-\delta, \delta)$ such that $f(x) \notin (0.5, 1.5)$. (Notice that for all x's in the interval $(-1, 0)$, $f(x) < 0$.) That is, there is no choice of δ which makes the defining inequality true for $\varepsilon = \frac{1}{2}$. Thus, the conjectured limit of 1 is incorrect.

There are several observations to be made here. First, if we had constructed a table of function values that included negative values of x, or had zoomed in our initial graph somewhat, we would not have made the erroneous conjecture in the first place. You should also note here that, while we've only shown that the limit is not 1, it's somewhat more complicated to show that the limit does not exist. This is left as an exercise.

■

Limits Involving Infinity

Recall that we had observed that $\lim_{x \to 0} \dfrac{1}{x^2}$ does not exist, but to be more descriptive, we had written

$$\lim_{x \to 0} \frac{1}{x^2} = \infty.$$

By this statement, we mean that the limit does not exist specifically because the function increases without bound as $x \to 0$. Just as with our initial intuitive notion of $\lim_{x \to a} f(x) = L$, this description is imprecise and needs to be more carefully defined. When we say that $\frac{1}{x^2}$ increases without bound as $x \to 0$, we mean that we can make $\frac{1}{x^2}$ as large as we like, simply by making x sufficiently close to 0. So, just how large must $\frac{1}{x^2}$ get? Well, given any large positive number, M, we must be able to make $\frac{1}{x^2} > M$, for x sufficiently close to 0. We measure closeness here the same way as we did before. If you think that this sounds like it has the makings of a definition, you're right. We have the following.

Definition 5.2   For a function f defined in some open interval containing a (but not necessarily at a itself), we say

$$\lim_{x \to a} f(x) = \infty$$

if given any number $M > 0$, there is another number $\delta > 0$, such that $f(x) > M$ whenever $0 < |x - a| < \delta$.

Similarly, we had said that if f decreases without bound as $x \to a$, then $\lim_{x \to a} f(x) = -\infty$. Think of how you would make this more precise and then consider the following definition.

Definition 5.3   For a function f defined in some open interval containing a (but not necessarily at a itself), we say

$$\lim_{x \to a} f(x) = -\infty$$

if given any number $N < 0$, there is another number $\delta > 0$, such that $f(x) < N$ whenever $0 < |x - a| < \delta$.

It's easy to keep these definitions straight if you think of their meaning. Don't simply try to memorize them. We can work with these definitions in much the same way that we worked with Definition 5.1.

Example 5.6   **Using the Definition of Limit Where the Limit is Infinite**
Prove that

$$\lim_{x \to 0} \frac{1}{x^2} = \infty.$$

Solution   Given any (large) number $M > 0$, we need to determine if there is a distance $\delta > 0$ such that if x is within δ of 0 (but not equal to 0) then

$$\frac{1}{x^2} > M. \tag{5.5}$$

Since both M and x^2 are positive, (5.5) is equivalent to

$$x^2 < \frac{1}{M}.$$

Taking the square root of both sides and recalling that $\sqrt{x^2} = |x|$, we get

$$|x| < \sqrt{\frac{1}{M}}.$$

So, for any $M > 0$, if we take $\delta = \sqrt{\frac{1}{M}}$, we have that $0 < |x - 0| < \delta$ guarantees that

$$\frac{1}{x^2} > M,$$

as desired. (Keep in mind that all of the preceding steps are reversible.) Note that this says, for instance, that for $M = 100$, $\frac{1}{x^2} > 100$, whenever $0 < |x| < \sqrt{\frac{1}{100}} = \frac{1}{10}$. (Verify that this works, as an exercise.)

■

There are two remaining limits that we have yet to place on a careful footing. Before reading on, try to figure out for yourself what appropriate definitions would look like.

First, if we write $\lim_{x \to \infty} f(x) = L$, we mean that as x grows larger and larger, without bound, $f(x)$ gets closer and closer to L. That is, we can make $f(x)$ as close to L as we like, simply by choosing x sufficiently large. More precisely, we have the following definition.

Definition 5.4   For a function f defined on an interval (a, ∞), for some $a > 0$, we say

$$\lim_{x \to \infty} f(x) = L$$

if given any $\varepsilon > 0$, there is a number $M > 0$ such that $x > M$ guarantees that

$$|f(x) - L| < \varepsilon.$$

Similarly, we have said that $\lim_{x \to -\infty} f(x) = L$ means that as x decreases without bound, $f(x)$ gets closer and closer to L. So, in this case, we should be able to make $f(x)$ as close to L as desired, just by making x sufficiently negative. We have the following definition.

Definition 5.5   For a function f defined on an interval $(-\infty, a)$, for some $a < 0$, we say

$$\lim_{x \to -\infty} f(x) = L$$

if given any $\varepsilon > 0$, there is a number $N < 0$ such that $x < N$ guarantees that

$$|f(x) - L| < \varepsilon.$$

We have now given definitions of five different limits. You will be able to keep these straight if you think about what they mean. Again, don't try to get away with memorizing a bunch of meaningless phrases. We use Definitions 5.4 and 5.5 in much the same way as we do Definitions 5.1–5.3, as we see in the following example.

Example 5.7   **Using the Definition of Limit Where x Is Becoming Infinite**
We observed earlier that

$$\lim_{x \to -\infty} \frac{1}{x} = 0.$$

Solution   To prove this using the definition of limit, we must show that given any $\varepsilon > 0$, we can make $\frac{1}{x}$ within ε of 0, simply by making x sufficiently large in absolute value and negative. So, we need to determine for which x

$$\left| \frac{1}{x} - 0 \right| < \varepsilon$$

or

$$\left|\frac{1}{x}\right| < \varepsilon. \tag{5.6}$$

Since $x < 0$, $|x| = -x$, and so (5.6) becomes

$$\frac{1}{-x} < \varepsilon.$$

Dividing both sides by ε and multiplying by x (remember that $x < 0$ and $\varepsilon > 0$, so that this will change the direction of the inequality), we get

$$-\frac{1}{\varepsilon} > x.$$

So, if we take $N = -\frac{1}{\varepsilon}$, we have satisfied the definition and thereby proved that the limit is correct.

■

Remark 5.2   You should take care to note the commonality among the definitions of the five limits we have given. They are all strikingly similar. The reason for this is that all five deal with a precise description of what it means to be "close." (If you continue in your study of mathematics, you will pay particularly careful attention to questions of this sort in a course in advanced calculus or elementary real analysis.) It is of considerable bene-fit to work through these definitions until you can provide your own words for each. Don't bother to memorize the formal definitions as stated here. Rather, work toward understand-ing what they mean and come to appreciate the exacting language mathematicians use.

We want to point out that these limit definitions are not here so that we can use them to prove each and every limit which comes along. Actually, we only use them to prove a very few basic limits and to prove the limit theorems that we've been using for some time without proof. Once the limit theorems have been carefully justified, further use of these theorems then provides solid justification of each new limit. As an illustration, we prove the rule for a limit of a sum.

Theorem 5.1   Suppose that for a real number a, $\lim_{x \to a} f(x) = L_1$ and that $\lim_{x \to a} g(x) = L_2$. Then,

$$\lim_{x \to a}[f(x) + g(x)] = \lim_{x \to a} f(x) + \lim_{x \to a} g(x) = L_1 + L_2.$$

Proof. Since $\lim_{x \to a} f(x) = L_1$, we know that given any number $\varepsilon_1 > 0$, there is a number $\delta_1 > 0$ for which

$$|f(x) - L_1| < \varepsilon_1 \text{ whenever } 0 < |x - a| < \delta_1. \tag{5.7}$$

Likewise, since $\lim_{x \to a} g(x) = L_2$, we know that given any number $\varepsilon_2 > 0$, there is a number $\delta_2 > 0$ for which

$$|g(x) - L_2| < \varepsilon_2 \text{ whenever } 0 < |x - a| < \delta_2. \tag{5.8}$$

Now, in order to get

$$\lim_{x \to a}[f(x) + g(x)] = (L_1 + L_2),$$

we must show that, given any number $\varepsilon > 0$, there is a number $\delta > 0$ such that

$$|[f(x) + g(x)] - (L_1 + L_2)| < \varepsilon, \text{ whenever } 0 < |x - a| < \delta.$$

Notice that

$$|[f(x) + g(x)] - (L_1 + L_2)| = |[f(x) - L_1] + [g(x) - L_2]|$$
$$\leq |f(x) - L_1| + |g(x) - L_2|, \qquad (5.9)$$

by the triangle inequality. Of course, both terms on the right-hand side of (5.9) can be made arbitrarily small, from (5.7) and (5.8). In particular, if we take $\varepsilon_1 = \varepsilon_2 = \dfrac{\varepsilon}{2}$, then as long as

$$0 < |x - a| < \delta_1 \text{ and } 0 < |x - a| < \delta_2,$$

we get from (5.7), (5.8) and (5.9) that

$$|[f(x) + g(x)] - (L_1 + L_2)| \leq |f(x) - L_1| + |g(x) - L_2|$$

$$< \frac{\varepsilon}{2} + \frac{\varepsilon}{2} = \varepsilon,$$

as desired. Of course, this will happen if we take

$$0 < |x - a| < \delta = minimum\{\delta_1, \delta_2\}.$$

■

The other rules for limits are proven similarly, using the definition of limit. These are found in Appendix A.

Exercises 1.5

1. (writing) In his 1726 masterpiece *Mathematical Principles of Natural Philosophy*, which introduces many of the fundamentals of calculus, Isaac Newton described the important limit $\lim\limits_{h \to 0} \dfrac{f(a + h) - f(a)}{h}$ (which we study at length in Chapter 2) as "the limit to which the ratios of quantities decreasing without limit do always converge, and to which they approach nearer than by any given difference, but never go beyond, nor ever reach until the quantities vanish." If you ever get weary of all the notation that we use in calculus, think of what it would look like in words! Critique Newton's definition of limit, addressing the following questions in the process. What restrictions do the phrases "never go beyond" and "never reach" put on the limit process? Give an example of a simple limit, not necessarily of the form $\lim\limits_{h \to 0} \dfrac{f(a + h) - f(a)}{h}$, that violates these restrictions. Give your own (English language) description of the limit, avoiding restrictions such as Newton's. Why do mathematicians consider the $\varepsilon - \delta$ definition simple and elegant?

2. (writing) You have computed numerous limits before seeing the definition of limit. Explain how this definition changes and/or improves your understanding of the limit process.

3. (writing) Each word in the $\varepsilon - \delta$ definition is carefully chosen and precisely placed. Describe what is wrong with each of the following slightly incorrect "definitions" (use examples!):
(a) There exists $\varepsilon > 0$ such that there exists a $\delta > 0$ such that if $0 < |x - a| < \delta$, then $|f(x) - L| < \varepsilon$.
(b) For all $\varepsilon > 0$ and for all $\delta > 0$, if $0 < |x - a| < \delta$, then $|f(x) - L| < \varepsilon$.
(c) For all $\varepsilon > 0$ there exists $\delta > 0$ such that $0 < |x - a| < \delta$ and $|f(x) - L| < \varepsilon$.

4. (writing) In order for the limit to exist, given every $\varepsilon > 0$, we must be able to find a $\delta > 0$ such that the if/then inequalities are true. To prove that the limit does not exist, we must find a particular $\varepsilon > 0$ such that the if/then inequalities are not true for any choice of $\delta > 0$. To understand the logic behind the swapping of the "for every" and "there exists" roles, draw an analogy with the following situation. Suppose the statement, "Everybody loves somebody" is true. If you wanted to verify the statement, why would you have to talk to every person on earth? But, suppose that the statement is not true. What would you have to do to disprove it?

In exercises 5–12, numerically and graphically determine a δ corresponding to the given ε. Graph the function in the $\varepsilon - \delta$ window [x-range is $(a - \delta, a + \delta)$ and y-range is $(L - \varepsilon, L + \varepsilon)$] to verify that your choice works.

5. $\lim\limits_{x \to 0} (x^2 + 1) = 1; \varepsilon = 0.1$

6. $\lim\limits_{x \to 0} (x^2 + 1) = 1; \varepsilon = 0.05$

7. $\lim\limits_{x \to 0} \cos x = 1; \varepsilon = 0.1$

8. $\lim\limits_{x \to 0} \cos x = 1; \varepsilon = 0.05$

9. $\lim\limits_{x \to 1} \sqrt{x+3} = 2;\ \varepsilon = 0.1$

10. $\lim\limits_{x \to 1} \sqrt{x+3} = 2;\ \varepsilon = 0.05$

11. $\lim\limits_{x \to 1} \dfrac{x+2}{x^2} = 3;\ \varepsilon = 0.1$

12. $\lim\limits_{x \to 1} \dfrac{x+2}{x^2} = 2;\ \varepsilon = 0.05$

In exercises 13–24, symbolically find δ in terms of ε.

13. $\lim\limits_{x \to 0} 3x = 0$

14. $\lim\limits_{x \to 1} 3x = 3$

15. $\lim\limits_{x \to 2} (3x+2) = 8$

16. $\lim\limits_{x \to 1} (3x+2) = 5$

17. $\lim\limits_{x \to 1} (3-4x) = -1$

18. $\lim\limits_{x \to -1} (3-4x) = 7$

19. $\lim\limits_{x \to -1} (1 + x/2) = 1/2$

20. $\lim\limits_{x \to 1} (1 + x/2) = 3/2$

21. $\lim\limits_{x \to 1} (x^2 - 1) = 0$

22. $\lim\limits_{x \to 1} (x^2 - x + 1) = 1$

23. $\lim\limits_{x \to 2} (x^2 - 1) = 3$

24. $\lim\limits_{x \to 0} (x^3 + 1) = 1$

25. Determine a formula for δ in terms of ε for $\lim\limits_{x \to a} (mx + b)$ (Hint: Use exercises 13–20). Does the formula depend on the value of a? Try to explain this answer graphically.

26. Based on exercises 21 and 23, does the value of δ depend on the value of a for $\lim\limits_{x \to a} (x^2 + b)$? Try to explain this graphically.

27. Modify the $\varepsilon - \delta$ definition to define the one-sided limits $\lim\limits_{x \to a^-} f(x)$ and $\lim\limits_{x \to a^+} f(x)$.

28. Symbolically find the largest δ corresponding to $\varepsilon = 0.1$ in the definition of $\lim\limits_{x \to 1^-} 1/x = 1$. Symbolically find the largest δ corresponding to $\varepsilon = 0.1$ in the definition of $\lim\limits_{x \to 1^+} 1/x = 1$. Which δ could be used in the definition of $\lim\limits_{x \to 1} 1/x = 1$? Briefly explain. Then prove that $\lim\limits_{x \to 1} 1/x = 1$.

In exercises 29–34, find a δ corresponding to $M = 100$ or $N = -100$ (as appropriate) for each limit.

29. $\lim\limits_{x \to 1^+} \dfrac{2}{x-1} = \infty$

30. $\lim\limits_{x \to 1^-} \dfrac{2}{x-1} = -\infty$

31. $\lim\limits_{x \to 0^+} \cot x = \infty$

32. $\lim\limits_{x \to \pi^-} \cot x = -\infty$

33. $\lim\limits_{x \to 2^-} \dfrac{2}{\sqrt{4-x^2}} = \infty$

34. $\lim\limits_{x \to 0^+} \ln x = -\infty$

In exercises 35–40, find an M corresponding to $\varepsilon = 0.1$ for each limit at infinity.

35. $\lim\limits_{x \to \infty} \dfrac{x^2 - 2}{x^2 + x + 1} = 1$

36. $\lim\limits_{x \to \infty} \dfrac{x - 2}{x^2 + x + 1} = 0$

37. $\lim\limits_{x \to -\infty} \dfrac{x^2 + 3}{4x^2 - 4} = 0.25$

38. $\lim\limits_{x \to -\infty} \dfrac{3x^2 - 2}{x^2 + 1} = 3$

39. $\lim\limits_{x \to \infty} e^{-2x} = 0$

40. $\lim\limits_{x \to \infty} \dfrac{e^x + x}{e^x - x^2} = 1$

In exercises 41–48, prove that the limit is correct using the appropriate definition (assume that k is an integer).

41. $\lim\limits_{x \to \infty} \dfrac{1}{x^3} = 0$

42. $\lim\limits_{x \to -\infty} \dfrac{1}{x^3} = 0$

43. $\lim\limits_{x \to \infty} \dfrac{1}{x^k} = 0$, for $k > 0$

44. $\lim\limits_{x \to -\infty} \dfrac{1}{x^k} = 0$, for $k > 0$

45. $\lim\limits_{x \to -\infty} \left(\dfrac{1}{x^2 + 2} - 3 \right) = -3$

46. $\lim\limits_{x \to \infty} \dfrac{1}{(x-7)^2} = 0$

47. $\lim\limits_{x \to -3} \dfrac{-1}{(x+3)^4} = -\infty$

48. $\lim\limits_{x \to 7} \dfrac{1}{(x-7)^2} = \infty$

49. $\lim\limits_{x \to 5} \dfrac{1}{(x-5)^2} = \infty$

50. $\lim\limits_{x \to -4} \dfrac{-1}{(x+4)^6} = -\infty$

In exercises 51–54, identify a specific $\varepsilon > 0$ for which no $\delta > 0$ exists to satisfy the definition of limit.

51. $f(x) = \begin{cases} 2x & \text{if } x < 1 \\ x^2 + 3 & \text{if } x > 1 \end{cases}$; $\lim\limits_{x \to 1} f(x) \neq 2$

52. $f(x) = \begin{cases} x^2 - 1 & \text{if } x < 0 \\ -x - 2 & \text{if } x > 0 \end{cases}$; $\lim\limits_{x \to 0} f(x) \neq -2$

53. $f(x) = \begin{cases} 2x & \text{if } x < 1 \\ 5 - x^2 & \text{if } x > 1 \end{cases}$; $\lim\limits_{x \to 1} f(x) \neq 2$

54. $f(x) = \begin{cases} x^2 - 1 & \text{if } x < 0 \\ x - 2 & \text{if } x > 0 \end{cases}$; $\lim\limits_{x \to 0} f(x) \neq -2$

55. A metal washer of (outer) radius r inches weighs $2r^2$ ounces. A company manufactures 2-inch washers for different customers who have different error tolerances. If the customer demands a washer of weight $8 \pm \varepsilon$ ounces, what is the error tolerance for the radius? That is, find δ such that a radius of r within the interval $(2 - \delta, 2 + \delta)$ guarantees a weight within $(8 - \varepsilon, 8 + \varepsilon)$.

56. A fiberglass company ships its glass as spherical marbles. If the volume of each marble has to be within ε of $\pi/6$, how close does the radius have to be to $1/2$?

57. (exploratory) We hope that working through this section has provided you with extra insight into the limit process. However, we have not yet solved any problems we could not already solve in previous sections. We do so now, while investigating an unusual function. Recall that rational numbers can be written as fractions p/q, where p and q are integers. We will assume that p/q has been simplified by dividing out common factors (e.g., $1/2$ and not $2/4$). Define

$$f(x) = \begin{cases} 0 & \text{if } x \text{ is irrational} \\ 1/q & \text{if } x = \frac{p}{q} \text{ is rational} \end{cases}.$$ We will try to show that $\lim_{x \to 2/3} f(x)$ exists. Without graphics, we need a good definition to answer this question. We know that $f(2/3) = 1/3$, but recall that the limit is independent of the actual function value. We need to think about x's close to $2/3$. If such an x is irrational, $f(x) = 0$. A simple hypothesis would then be $\lim_{x \to 2/3} f(x) = 0$. We'll try this out for $\varepsilon = 1/6$. We would like to guarantee that $|f(x)| < 1/6$ whenever $0 < |x - 2/3| < \delta$.

Well, how many x's have a function value greater than 1/6? The only possible function values are 1/5, 1/4, 1/3, 1/2 and 1. The x's with function value 1/5 are 1/5, 2/5, 3/5, 4/5 and so on. The closest of these x's to 2/3 is 3/5. Find the closest x (not counting $x = 2/3$) to 2/3 with function value 1/4. Repeat for $f(x) = 1/3$, $f(x) = 1/2$ and $f(x) = 1$. Out of all these closest x's, how close is the absolute closest? Choose δ to be smaller than this number, and argue that if $0 < |x - 2/3| < \delta$ we are guaranteed that $|f(x)| < 1/6$. Argue that a similar process can find a δ for any ε.

1.6   Limits and Loss-of-Significance Errors (Optional)

"Pay no attention to that man behind the curtain . . ." (from *The Wizard of Oz*)

Things are not always what they appear to be. In fact, a great deal of our lives are spent learning to distinguish reality from mere appearances. We make a great many mistakes along the way and hopefully, develop some level of skepticism. As children (and as parents), we learn to look past the apparent kindness of a stranger offering candy to see an ominous sign of danger. It's not that most strangers would offer candy with anything but the best of intentions, but that the prospect of danger, however remote, cannot be overlooked when it comes to the safety of our children. Some amount of skepticism is thus very healthy. You may have already come to realize that mathematicians are a skeptical lot. This is of necessity, for you simply can't accept things at face value.

One of the dangers of modern technology often cited in the popular media is that we might develop an overreliance on it. There is some truth to this, as we all have a tendency to accept a computer's answer as a fact not subject to debate. In this section, we want to briefly examine one of the shortcomings of using a computer to evaluate a function.

When we are using a computer (or calculator) to generate function values for us, we must always keep in mind that these devices carry out calculations only approximately. Most of the time, this will cause us no difficulty whatsoever. Modern computational devices generally carry out calculations to a very high degree of accuracy. Occasionally, however, the results of round-off errors in a string of calculations are disastrous. In this section, we briefly investigate these errors and learn how to recognize and avoid some of them.

We first consider a relatively tame-looking example.

Example 6.1   **A Limit with Unusual Graphical and Numerical Behavior**

Evaluate

$$\lim_{x \to \infty} \frac{(x^3 + 4)^2 - x^6}{x^3}.$$

Solution   At first glance, the numerator looks like $\infty - \infty$, which is indeterminate, while the denominator tends to ∞. Algebraically, the only reasonable step to take is to square-out the first term in the numerator. Before we do that, let's draw a graph and compute some function values. [Here, all calculations were performed on a TI-85 graphics calculator. The graphs were produced by the computer algebra system (CAS) Maple running on a Pentium II-powered IBM-compatible computer. A TI-85 graph of the same function shows even greater oscillation than that in Figure 1.49.

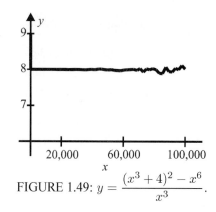

FIGURE 1.49: $y = \dfrac{(x^3 + 4)^2 - x^6}{x^3}$.

We point this out because different machines (even different software packages running on the same machine) will produce somewhat different errors.] In Figure 1.49, the function appears nearly constant, until it begins oscillating around $x = 40,000$. Notice that the accompanying table of function values is inconsistent with Figure 1.49.

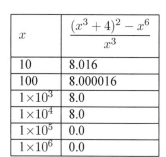

x	$\dfrac{(x^3 + 4)^2 - x^6}{x^3}$
10	8.016
100	8.000016
1×10^3	8.0
1×10^4	8.0
1×10^5	0.0
1×10^6	0.0

The last two values in the table may have surprised you. Up until that point, the function values seemed to be settling down to 8.0 very nicely. So, what happened here and what is the correct value of the limit? Obviously, something unusual has occurred between $x = 1 \times 10^4$ and $x = 1 \times 10^5$. We should look carefully at function values in that interval. A more detailed table is shown in the margin.

FIGURE 1.50: $y = \dfrac{(x^3 + 4)^2 - x^6}{x^3}$.

x	$\dfrac{(x^3 + 4)^2 - x^6}{x^3}$
2×10^4	8.0
3×10^4	8.14815
4×10^4	7.8125
5×10^4	0

In Figure 1.50, we have blown up the graph to enhance the oscillation observed between $x = 1 \times 10^4$ and $x = 1 \times 10^5$. The picture that is emerging is even more confusing. The deeper we look into this limit, the more erratically the function appears to behave. We use the word **appears** because all of the oscillatory behavior we are seeing is an illusion, created by the finite precision of the computer used to perform the calculations.

■

Computer Representation of Real Numbers

The reason for the unusual behavior seen in example 6.1 boils down to the way in which computers represent real numbers. Without getting into all of the intricacies of computer arithmetic, it suffices to think of computers and calculators as storing real numbers internally in scientific notation. For example, the number 1,234,567 would be stored as 1.234567×10^6. The number preceding the power of 10 is called the **mantissa** and the power is called the **exponent**. Thus, the mantissa here is 1.234567 and the exponent is 6.

All computing devices have finite memory and consequently have limitations on the size mantissa and exponent that they can store (this is called **finite precision**). In the case of the TI-85, it carries a 14-digit mantissa and a 3-digit exponent. So, on a 14-digit computer, this would suggest that the computer would only retain the first 14 digits in the decimal expansion of any given number.

Example 6.2   Computer Representation of a Rational Number

Determine how $\frac{1}{3}$ is stored internally on a 10-digit computer and how $\frac{2}{3}$ is stored internally on a 14-digit computer.

Solution   On a 10-digit computer, $\frac{1}{3}$ is stored internally as $\underbrace{3.333333333}_{\text{10 digits}} \times 10^{-1}$.

On a 14-digit computer (such as the TI-85), $\frac{2}{3}$ is stored internally as $\underbrace{6.6666666666667}_{\text{14 digits}}$

$\times 10^{-1}$.

■

For most purposes, this finite precision presents no problem. However, we do occasionally come across a disastrous error caused by finite precision. In the next example, we subtract two relatively close numbers and examine the resulting catastrophic error.

Example 6.3   A Computer Subtraction of Two "Close" Numbers

Compare the exact value of

$$1.\underbrace{0000000000000}_{13 \text{ zeros}}4 \times 10^{18} - 1.\underbrace{0000000000000}_{13 \text{ zeros}}1 \times 10^{18}$$

with the result obtained from a calculator or computer with a 14-digit (or smaller) mantissa.

Solution   Notice that

$$1.\underbrace{0000000000000}_{13 \text{ zeros}}4 \times 10^{18} - 1.\underbrace{0000000000000}_{13 \text{ zeros}}1 \times 10^{18} = 0.\underbrace{0000000000000}_{13 \text{ zeros}}3 \times 10^{18}$$

$$= 30,000. \qquad (6.1)$$

However, if this calculation is carried out on a computer or calculator with a 14-digit (or smaller) mantissa, both numbers on the left-hand side of (6.1) are stored by the computer as 1×10^{18} and hence, the result is calculated as 0. Try this calculation for yourself now.

■

Example 6.4   Another Subtraction of Two "Close" Numbers

Compare the exact value of

$$1.\underbrace{0000000000000}_{13 \text{ zeros}}6 \times 10^{20} - 1.\underbrace{0000000000000}_{13 \text{ zeros}}4 \times 10^{20}$$

with the result obtained from a calculator or computer with a 14-digit (or smaller) mantissa.

Solution   Notice that

$$1.\underbrace{0000000000000}_{13 \text{ zeros}}6 \times 10^{20} - 1.\underbrace{0000000000000}_{13 \text{ zeros}}4 \times 10^{20}$$

$$= 0.\underbrace{0000000000000}_{13 \text{ zeros}}2 \times 10^{20} = 2,000,000.$$

However, if this calculation is carried out on a machine with a 14-digit mantissa, the first number is represented as $1.0000000000001 \times 10^{20}$, while the second number is represented by 1.0×10^{20}, due to the finite precision and rounding. The difference between the two values is then computed as $0.0000000000001 \times 10^{20}$ or $10,000,000$. Again, this is a very serious error.

◼

In examples 6.3 and 6.4, we witnessed a gross error caused by the subtraction of two numbers whose significant digits are very close to one another. This type of error is called a **loss-of-significant-digits error** or simply a **loss-of-significance error**. These are subtle, often disastrous computational errors. Returning now to example 6.1, we will see that it was this type of error that caused the unusual behavior noted.

Example 6.5   A Loss-of-Significance Error

In example 6.1, we considered the function

$$f(x) = \frac{(x^3 + 4)^2 - x^6}{x^3}.$$

Follow the calculation of $f(5 \times 10^4)$ one step at a time, as a 14-digit computer would do it.

Solution   We have

$$f(5 \times 10^4) = \frac{\left[\left(5 \times 10^4\right)^3 + 4\right]^2 - \left(5 \times 10^4\right)^6}{\left(5 \times 10^4\right)^3}$$

$$= \frac{\left(1.25 \times 10^{14} + 4\right)^2 - 1.5625 \times 10^{28}}{1.25 \times 10^{14}}$$

$$= \frac{(125,000,000,000,000 + 4)^2 - 1.5625 \times 10^{28}}{1.25 \times 10^{14}}$$

$$= \frac{(1.25 \times 10^{14})^2 - 1.5625 \times 10^{28}}{1.25 \times 10^{14}} = 0,$$

since $125,000,000,000,004$ is rounded off to $125,000,000,000,000$.

Note that the real culprit here was not the rounding of $125,000,000,000,004$, but the fact that this was followed by a subtraction of a nearly equal value. Further, note that this is not a problem unique to the numerical computation of limits, but one that occurs in numerical computation, in general.

◼

Remark 6.1   If at all possible, avoid subtractions of nearly equal values. Sometimes, this can be accomplished by some algebraic manipulation of the function.

In the case of the function from example 6.5, we can avoid the subtraction and hence, the loss of significance error by rewriting the function as follows:

$$f(x) = \frac{(x^3 + 4)^2 - x^6}{x^3}$$

$$= \frac{(x^6 + 8x^3 + 16) - x^6}{x^3}$$

$$= \frac{8x^3 + 16}{x^3},$$

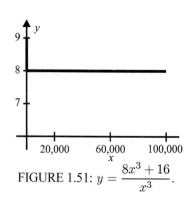

FIGURE 1.51: $y = \dfrac{8x^3 + 16}{x^3}$.

where we have eliminated the subtraction. Using this new (and equivalent) expression for the function, we can compute a table of function values reliably. Notice, too, that if we redraw the graph in Figure 1.49 using the new expression (see Figure 1.51), we no longer see the oscillation present in Figures 1.49 and 1.50.

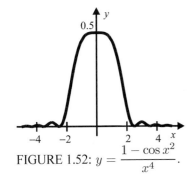

x	$\dfrac{8x^3 + 16}{x^3}$
10	8.016
100	8.000016
1×10^3	8.000000016
1×10^4	8.00000000002
1×10^5	8.0
1×10^6	8.0
1×10^7	8.0

FIGURE 1.52: $y = \dfrac{1 - \cos x^2}{x^4}$.

So, by avoiding the indicated subtraction, we have found an expression that we can use to accurately compute values of the function for a much larger range of values (see the accompanying table). Finally, we can use the graph and corrected table of function values to conjecture that

$$\lim_{x \to \infty} \frac{(x^3 + 4)^2 - x^6}{x^3} = 8.$$

We leave it as an exercise to demonstrate that this conjecture is correct.

In the following example, we examine a loss-of-significance error that occurs for x close to 0.

Example 6.6   **Loss-of-Significance Involving a Trigonometric Function**

Evaluate

$$\lim_{x \to 0} \frac{1 - \cos x^2}{x^4}.$$

Solution   As usual, we look at a graph (see Figure 1.52) and some function values.

x	$\dfrac{1 - \cos x^2}{x^4}$	x	$\dfrac{1 - \cos x^2}{x^4}$
0.1	0.499996	−0.1	0.499996
0.01	0.5	−0.01	0.5
0.001	0.5	−0.001	0.5
0.0001	0.0	−0.0001	-1×10^{-4}
0.00001	0.0	−0.00001	-1×10^{-5}

As in example 6.1, note that the function values seem to be approaching 0.5, but then suddenly take a jump down to 0.0 or thereabouts. Once again, we are seeing a loss-of-significance error. In this particular case, this occurs because we are subtracting nearly equal values ($\cos x^2$ and 1). We can again avoid the error by eliminating the subtraction. Note that

Multiply numerator and denominator by $(1 + \cos x^2)$.

$1 - \cos^2(x^2) = \sin^2(x^2)$.

x	$\dfrac{\sin^2(x^2)}{x^4(1 + \cos x^2)}$
±0.1	0.499996
±0.01	0.4999999996
±0.001	0.5
±0.0001	0.5
±0.00001	0.5

$$\begin{aligned}
\frac{1 - \cos x^2}{x^4} &= \frac{1 - \cos x^2}{x^4} \cdot \frac{1 + \cos x^2}{1 + \cos x^2} \\
&= \frac{1 - \cos^2(x^2)}{x^4(1 + \cos x^2)} \\
&= \frac{\sin^2(x^2)}{x^4(1 + \cos x^2)}.
\end{aligned}$$

Since this last (equivalent) expression has no subtraction indicated, we should be able to use it to reliably generate values without the worry of loss-of-significance error. Using this to compute function values, we get the accompanying table.

Using the graph and the new table, we conjecture that

$$\lim_{x \to 0} \frac{1 - \cos x^2}{x^4} = \frac{1}{2}.$$

■

We offer one final example where a loss-of-significance error occurs, even though no subtraction is explicitly indicated.

Example 6.7   **A Loss-of-Significance Error Involving a Sum**

Evaluate

$$\lim_{x \to -\infty} x \left[\left(x^2 + 4 \right)^{1/2} + x \right].$$

Solution   Initially, you might think that since there is no subtraction (explicitly) indicated, there will be no loss of significance error. We first draw a graph (see Figure 1.53) and compute a table of values.

FIGURE 1.53:

$$y = x \left[\left(x^2 + 4 \right)^{1/2} + x \right].$$

x	$x \left[\left(x^2 + 4 \right)^{1/2} + x \right]$
-10	-1.9804
-100	-1.9998
-1×10^3	-1.999998
-1×10^4	-2.0
-1×10^5	-2.0
-1×10^6	-2.0
-1×10^7	0.0
-1×10^8	0.0

You should quickly notice the sudden jump in values in the table and the wild oscillation visible in the graph. Although a subtraction is not explicitly indicated, there is indeed a subtraction here, since $x < 0$ and $(x^2 + 4)^{1/2} > 0$. We can again remedy this with some algebraic manipulation, as follows.

Multiply numerator and denominator by the conjugate.

$$x \left[\left(x^2 + 4 \right)^{1/2} + x \right] = x \left[\left(x^2 + 4 \right)^{1/2} + x \right] \cdot \frac{\left[\left(x^2 + 4 \right)^{1/2} - x \right]}{\left[\left(x^2 + 4 \right)^{1/2} - x \right]}$$

Simplify the numerator.

$$= x \cdot \frac{\left[(x^2 + 4) - x^2 \right]}{\left[\left(x^2 + 4 \right)^{1/2} - x \right]}$$

$$= \frac{4x}{\left[\left(x^2 + 4 \right)^{1/2} - x \right]}.$$

We use this last expression to generate a graph in the same window as that used for Figure 1.53 and to generate the accompanying table of values. In Figure 1.54, we can see none of the wild oscillation observed in Figure 1.53 and the graph appears to be a horizontal line.

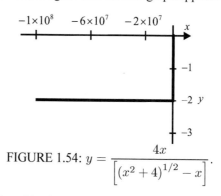

FIGURE 1.54: $y = \dfrac{4x}{\left[\left(x^2 + 4 \right)^{1/2} - x \right]}.$

x	$\dfrac{4x}{\left[(x^2+4)^{1/2} - x \right]}$
-10	-1.9804
-100	-1.9998
-1×10^3	-1.999998
-1×10^4	-1.99999998
-1×10^5	-2.0
-1×10^6	-2.0
-1×10^7	-2.0
-1×10^8	-2.0

Further, the values displayed in the table no longer show the sudden jump indicative of a loss-of-significance error. We can now confidently conjecture that

$$\lim_{x \to -\infty} x \left[\left(x^2 + 4 \right)^{1/2} + x \right] = -2.$$

In each of the preceding examples, we demonstrated a calculation that suffered from a catastrophic loss-of-significance error. In each case, we showed how we could rewrite the expression to avoid this error. We have by no means exhibited a general procedure for recognizing and repairing such errors. Rather, we hope that by seeing a few of these subtle errors, and by seeing how to fix even a limited number of them, you will become a more skeptical and intelligent user of technology.

Exercises 1.6

1. (writing) It is probably clear that caution is important in using technology. Equally important is redundancy. This property is sometimes thought to be a negative (wasteful, unnecessary), but its positive role is one of the lessons of this section. By redundancy, we mean investigating a problem using graphical, numerical and symbolic tools. Why is it important to use multiple methods? Answer this from a practical perspective (think of the problems in this section) and a theoretical perspective (if you have learned multiple techniques, do you understand the mathematics better?).

2. (writing) The drawback of caution and redundancy is that it they take extra time. In computing limits, when should you stop and take extra time to make sure an answer is correct, and when is it safe to go on to the next problem? Should you always look at a graph? compute function values? do symbolic work? an $\varepsilon - \delta$ proof? Prioritize the techniques in this chapter.

3. The limit $\lim_{h \to 0} \dfrac{f(a+h) - f(a)}{h}$ will be very important in the next chapter. For a specific function and specific a, we could compute a table of values of the fraction for smaller and smaller values of h. Why should we be wary of loss-of-significance errors?

4. Notice that we rationalized the numerator in example 6.7. The old rule of rationalizing the denominator is another example of rewriting an expression to try to minimize computational errors. Before computers, square roots were very difficult to compute. To see one reason why you might want the square root in the numerator, suppose that you can only get one decimal place of accuracy, so that $\sqrt{3} \approx 1.7$. Compare $\frac{6}{1.7}$ to $\frac{6}{\sqrt{3}}$ and then compare $2(1.7)$ to $\frac{6}{\sqrt{3}}$. Which of the approximations could you do in your head?

In exercises 5–12, use graphics and numerics to conjecture a value of the limit. Find a value of x for which a loss-of-significance error occurs. Rewrite the function to avoid loss-of-significance error.

5. $\lim_{x \to \infty} x(\sqrt{4x^2 + 1} - 2x)$ **6.** $\lim_{x \to -\infty} x(\sqrt{4x^2 + 1} + 2x)$

7. $\lim_{x \to \infty} \sqrt{x}(\sqrt{x + 4} - \sqrt{x + 2})$

8. $\lim_{x \to \infty} x^2(\sqrt{x^4 + 8} - x^2)$

9. $\lim_{x \to \infty} x(\sqrt{x^2 + 4} - \sqrt{x^2 + 2})$

10. $\lim_{x \to \infty} x(\sqrt{x^3 + 8} - x^{3/2})$

11. $\lim_{x \to 0} \dfrac{1 - \cos 2x}{12x^2}$ **12.** $\lim_{x \to 0} \dfrac{1 - \cos x^3}{x^6}$

In exercises 13 and 14, compare the limits to show that small errors can have disastrous effects.

13. $\lim_{x \to 1} \dfrac{x^2 + x - 2}{x - 1}$ and $\lim_{x \to 1} \dfrac{x^2 + x - 2.01}{x - 1}$

14. $\lim_{x \to 2} \dfrac{x - 2}{x^2 - 4}$ and $\lim_{x \to 2} \dfrac{x - 2}{x^2 - 4.01}$

15. Compare $f(x) = \sin \pi x$ and $g(x) = \sin 3.14x$ for $x = 1$ (radian), $x = 10$, $x = 100$ and $x = 1000$.

16. For exercise 5 above, follow the calculation of the function for $x = 10^5$ as it would proceed for a machine computing with a 14-digit mantissa. Identify where the round-off error occurs.

In exercises 17 and 18, compare the exact answer to one obtained by a computer with a six-digit mantissa.

17. $(1.000003 - 1.000001) \times 10^7$

18. $(1.000006 - 1.000001) \times 10^7$

19. (exploratory) In this exercise, we look at one aspect of the mathematical field of chaos. First, iterate the function $f(x) = x(3.57 - x)$ starting at $x_0 = 0.5$. That is, compute $x_1 = f(0.5)$, then $x_2 = f(x_1)$, then $x_3 = f(x_2)$ and so on. Although the sequence of numbers stays bounded, the numbers never repeat (except by the accident of round-off errors). An impressive property of chaotic functions is the **butterfly effect** (more properly referred to as *sensitive dependence on initial conditions*). The butterfly effect applies to the chaotic nature of weather, and states that the amount of air stirred by a butterfly flapping its wings in Brazil can create or disperse a tornado in Texas a few days later. Therefore, long-range weather prediction is impossible. To illustrate the butterfly effect, iterate $f(x)$ starting at $x_0 = 0.5$ and $x_0 = 0.51$. How many iterations does it take before the iterations are more than 0.1 apart? Try this again with $x_0 = 0.5$ and $x_0 = 0.501$. Repeat this exercise for the function $g(x) = x(3.55 - x)$. Even though the functions are almost identical, $g(x)$ is not chaotic and behaves very differently. This represents an important idea in modern medical research called dynamical diseases: a small change in one of the constants in a function (e.g., the rate of an electrical signal within the human heart) can produce a drastic change in the behavior of the system (e.g., the pumping of blood from the ventricles).

20. (exploratory) Just as we are subject to round-off error in using calculations from a computer, so are we subject to errors in a computer-generated graph. After all, the computer has to compute function values before it can decide where to plot points. On your computer or calculator, graph $y = \sin x^2$ (a disconnected graph, —or point plot—, is preferable). You should see the oscillations you expect from the sine function, but with the oscillations getting faster as x gets larger. Shift

your graphing window to the right several times. At some point, the plot will become very messy and almost unreadable. Depending on your technology, you may see patterns in the plot. Are these patterns real or an illusion? To explain what is going on, recall that a computer graph is a finite set of pixels, with each pixel representing one x and one y. Suppose the computer is plotting points at $x = 0$, $x = 0.1$, $x = 0.2$ and so on. The y-values would then be $\sin 0^2$, $\sin 0.1^2$, $\sin 0.2^2$ and so on. Investigate what will happen between $x = 15$ and $x = 16$. Compute all the points $(15, \sin 15^2)$, $(15.1, \sin 15.1^2)$ and so on. If you were to graph these points, what pattern would emerge? To explain this pattern, argue that there is approximately half a period of the sine curve missing between each point plotted. Also, investigate what happens between $x = 31$ and $x = 32$.

Chapter Review Exercises

In exercises 1 and 2, use numerical and graphical evidence to conjecture the value of the limit.

1. $\lim\limits_{x \to -2} \dfrac{x+2}{|x+2|}$

2. $\lim\limits_{x \to 0}(1+2x)^{1/x}$

In exercises 3 and 4, identify the limits from the graph of $f(x)$.

3. (a) $\lim\limits_{x\to-1^-} f(x)$ (b) $\lim\limits_{x\to-1^+} f(x)$
(c) $\lim\limits_{x\to-1} f(x)$ (d) $\lim\limits_{x\to0} f(x)$

4. (a) $\lim\limits_{x\to1^-} f(x)$ (b) $\lim\limits_{x\to1^+} f(x)$
(c) $\lim\limits_{x\to1} f(x)$ (d) $\lim\limits_{x\to2} f(x)$

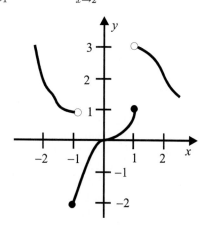

5. Identify the discontinuities in the function graphed above.

6. Sketch a graph of a function with $f(-1)=0$, $f(0)=0$, $\lim\limits_{x\to1^-} f(x) = 1$ and $\lim\limits_{x\to1^+} f(x) = -1$.

In exercises 7–26, evaluate the limit. Answer with a number, ∞, $-\infty$ or does not exist.

7. $\lim\limits_{x\to2} \dfrac{x-2}{x^2-4}$

8. $\lim\limits_{x\to1} \dfrac{x^2-1}{x^2+x-2}$

9. $\lim\limits_{x\to-2} \dfrac{x+2}{\sqrt{x^2-4}}$

10. $\lim\limits_{x\to0} \dfrac{e^{2x}}{x-2}$

11. $\lim\limits_{x\to0} \dfrac{\tan x}{\sec x}$

12. $\lim\limits_{x\to0} \dfrac{\sin x^2}{x^2}$

13. $\lim\limits_{x\to1} f(x)$, where $f(x) = \begin{cases} 2x+1 & \text{if } x<1 \\ x^2+1 & \text{if } x\geq1 \end{cases}$

14. $\lim\limits_{x\to2} f(x)$, where $f(x) = \begin{cases} 3x-1 & \text{if } x<2 \\ x^2+1 & \text{if } x\geq2 \end{cases}$

15. $\lim\limits_{x\to1^+} \sqrt{x^2+1}$

16. $\lim\limits_{x\to2^+} \sqrt{x^2-4}$

17. $\lim\limits_{x\to1^+} \dfrac{x+1}{x^2-1}$

18. $\lim\limits_{x\to2^-} \dfrac{1-x}{x^2-4x+4}$

19. $\lim\limits_{x\to\infty} \dfrac{x^2-4}{3x^2+x+1}$

20. $\lim\limits_{x\to\infty} \dfrac{2x}{\sqrt{x^2+4}}$

21. $\lim\limits_{x\to\infty} e^{-3x}$

22. $\lim\limits_{x\to-\infty} e^{-x^2}$

23. $\lim\limits_{x\to\infty} \ln 2x$

24. $\lim\limits_{x\to0^+} \ln 3x$

25. $\lim\limits_{x\to-\infty} \dfrac{2x}{x^2+3x-5}$

26. $\lim\limits_{x\to-2} \dfrac{2x}{x^2+3x+2}$

27. Use the Squeeze Theorem and the fact that $0 \leq \dfrac{x^2}{x^2+1} < 1$ to prove that $\lim\limits_{x\to0} \dfrac{2x^3}{x^2+1} = 0$.

28. Use the Intermediate Value Theorem to verify that $f(x) = x^3 - x - 1$ has a zero in the interval $[1, 2]$. Use the method of bisections to find an interval of length $1/32$ which contains a zero.

In exercises 29–32, find all discontinuities and determine which are removable.

29. $f(x) = \dfrac{x-1}{x^2+2x-3}$

30. $f(x) = \dfrac{x+1}{x^2-4}$

31. $f(x) = \begin{cases} \sin x & \text{if } x < 0 \\ x^2 & \text{if } 0 \le x \le 2 \\ 4x-3 & \text{if } x > 2 \end{cases}$

32. $f(x) = x \cot x$

In exercises 33–36, find all intervals of continuity.

33. $f(x) = \dfrac{x+2}{x^2-x-6}$

34. $f(x) = \ln(3x-4)$

35. $f(x) = \sin(1+e^x)$

36. $f(x) = \sqrt{x^2-4}$

In exercises 37–40, determine all vertical, horizontal and slant asymptotes.

37. $f(x) = \dfrac{x+1}{x^2-3x+2}$

38. $f(x) = \dfrac{x+2}{x^2-2x-8}$

39. $f(x) = \dfrac{x^2}{x^2-1}$

40. $f(x) = \dfrac{x^3}{x^2-x-2}$

41. Numerically and graphically determine a δ corresponding to $\epsilon = 0.1$ for $\lim_{x \to \pi/2} \sin x = 1$.

42. Numerically and graphically determine a δ corresponding to $\epsilon = 0.01$ for $\lim_{x \to \pi/2} \sin x = 1$.

43. Symbolically find δ in terms of ϵ for $\lim_{x \to 0}(2x+1) = 1$.

44. Symbolically find δ in terms of ϵ for $\lim_{x \to 1}(x^2+1) = 2$.

45. Find M corresponding to $\epsilon = 0.1$ for $\lim_{x \to \infty} \dfrac{x^2}{x^2+1} = 1$.

46. Find M corresponding to $\epsilon = 0.1$ for $\lim_{x \to \infty} \dfrac{2}{x^2} = 0$.

47. (exploratory) As discussed in this chapter, the limit of a function is the single number, if one exists, that the function approaches as x approaches its limiting value. The limit concept can be generalized to limiting behavior more complicated than a single number. The field of **chaos** (more properly called **nonlinear dynamics**) makes use of this extended concept. We will explore chaos theory at various times during our calculus journey. For now, we look at some examples of different limiting behaviors. We first *iterate* the function $f_2(x) = x(2-x)$. That means we start at some initial x-value, say $x_0 = 0.5$, and compute $x_1 = f_2(x_0)$, then compute $x_2 = f_2(x_1)$, then $x_3 = f_2(x_2)$, and so on. Use a calculator or computer to verify that $x_1 = 0.5(2-0.5) = 0.75$, $x_2 = 0.9375$, $x_3 = 0.99609375$, $x_4 = 0.9999847412$ and so on. You should conclude that the limit of this sequence of calculations is 1. Now, try iterating $f_{3.2}(x) = x(3.2-x)$ starting at $x_0 = 0.5$. Verify that $x_1 = 1.35$, $x_2 = 2.4975$, $x_3 = 1.75449375$, $x_4 = 2.536131681$ and so on. If you continue this process, you will see a different type of limiting behavior: alternation between the values of (approximately) 1.641742431 and 2.558257569. In what way might you compare this limiting behavior to a periodic function? To find other periodic limits, try the functions $f_{3.48}(x) = x(3.48-x)$, $f_{3.555}(x)$, $f_{3.565}(x)$ and $f_{3.569}(x)$. What is the pattern of the size of the periods? Note that the parameter (subscript) of this family of functions is being changed less and less to produce the higher periods. The limit of these subscripts is also of interest. We explore this in exercise 49.

48. (exploratory) In exercise 47, we looked at some examples from the family of functions $f_c(x) = x(c-x)$ for various values of the parameter c. In particular, as we gradually increased c from $c = 2$ to $c = 3.57$, we saw the limiting behavior (called the **attractor**) change from convergence to a single number (called a **one-cycle**) to alternation between two numbers (a **two-cycle**) to alternation among four numbers (a **four-cycle**) to, eventually, chaos (bounded but aperiodic). The transitions from one type of limiting behavior to another occur at special values of c called **bifurcation points**. By trial and error, find the first bifurcation point; that is, find the number b such that $f_c(x)$ has an attracting one-cycle if $c < b$ and an attracting two-cycle if $c > b$.

49. (exploratory) In this exercise, we look at another aspect of the mathematical field of chaos. In the language of exercises 47 and 48, we start by iterating the function $f(x) = x(3.57-x)$ starting at $x_0 = 0.5$. That is, compute $x_1 = f(0.5)$, then $x_2 = f(x_1)$, then $x_3 = f(x_2)$ and so on. Although the sequence of numbers stays bounded, the numbers never repeat (except by the accident of round-off errors). This is called **mathematical chaos**: bounded but not periodic. The weather is one example of a natural process which is thought to be chaotic. In what sense is the weather (take, for example, the local temperature) bounded but not periodic? Explain why it is inherently impossible to have accurate long-range weather forecasts.

Chapter 2
Differentiation

How fast can a human being run? This question has fascinated people for literally thousands of years, as countless cultures dating back to prehistory have held foot races. In modern times, the winner of the 100-meter dash at the Olympic Games has been given the title of "World's Fastest Human." The choice of 100 meters as the appropriate distance for this title is not universally accepted, though. For instance, the 1996 100-meter champion, Donovan Bailey and the 200- and 400-meter champion, Michael Johnson, held a much publicized race of 150 meters to determine 1997's World's Fastest Human. (Bailey won.)

The debate about the appropriate distance leads us to a significant mathematical question: exactly what do we mean by *fastest*? Many of us would imagine speedometers somehow attached to runners, where whoever records the highest speed is declared the fastest. You might reasonably argue that it's also important to maintain a high speed for a certain period of time, but let's focus on the **instantaneous speed** that a speedometer measures.

Without some kind of speedometer, how can we determine a runner's peak speed? At present, we have only the elementary formula

$$distance = rate \times time.$$

Solving this for the rate (speed), we get that

$$speed = \frac{distance}{time}.$$

However, this formula actually gives us an **average speed** over whatever span of time and distance is used. Suppose that a runner covers 100 meters in 10 seconds. The average speed is then $\frac{100 \, m}{10 \, s} = 10$ m/s. But, is this the runner's peak speed? A little thought should convince you that it's not. Sprinters are not allowed running starts, so they use the first several strides of a race to build up speed. Further, by the end of the race, a sprinter may be getting tired or losing rhythm. The entire 100 meters will then include some periods where the runner is not at their peak running speed.

Notice that if we were to average over shorter intervals of time and distance, we should have a more accurate approximation of peak speed. For instance, if our runner covers 10 meters in 0.91 seconds, the average speed is $\frac{10 \, m}{0.91 \, s} \approx 11$ m/s. The objections we listed for averaging over the entire 100 meters don't seem as important over a 10-meter distance, particularly if this is measured in the middle of the race. However, it's still possible for a runner's speed to change in five or six strides, so an even shorter distance would be better. Using shorter intervals, it has been estimated that Carl Lewis reached a peak speed of 28 mph in winning a gold medal at the 1992 Olympic Games.

In fact, the smaller the interval we use, the more accurate the calculation is likely to be. In section 2.1, we define velocity to be the limit of average velocities as the time interval shrinks to zero. As it turns out, this same limit arises in a vast number of contexts, so many so that in section 2.2, we give this limit a name, the **derivative**. In sections 2.3 through 2.8, we develop a large number of rules concerning derivatives, enabling us to compute the derivative of nearly any function that we can write down.

Along with computing velocity, the derivative enables us to calculate the rate at which a variety of quantities change. Several applications of the derivative are explored at length in Chapter 3. An understanding of the concept of rate of change is one of the primary focuses of elementary calculus.

2.1   Tangent Lines and Velocity

The hammer throw has been an Olympic event since 1900. To most of us today, it looks a little strange. The "hammer" thrown consists of a 15-pound ball on the end of a 3-foot chain with a handle. Athletes build up speed by completing three full rotations, then release the entire hammer apparatus. The world record is nearly 300 feet

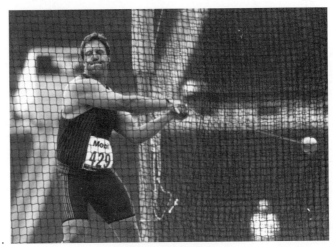

A hammer thrower.
(Photo courtesy of Erik Palmer.)

FIGURE 2.1: Hammer throw.

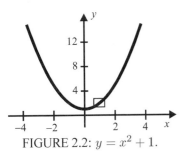

FIGURE 2.2: $y = x^2 + 1$.

FIGURE 2.3: $y = x^2 + 1$ and lines intersecting the curve only at the origin.

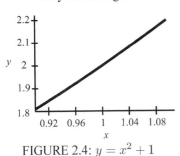

FIGURE 2.4: $y = x^2 + 1$

A simple two-dimensional illustration of the hammer throw is shown in Figure 2.1, as we look at the event from above. We assume that the three preliminary rotations describe a single circle, and we ignore the weight of the chain and handle (which actually weigh about a pound). In Figure 2.1, we see a ball traveling in a circle that is released at the point P.

Many people mistakenly expect that the ball will curve to the left after its release, but Newton's first law of motion tells us that the path is straight. As it turns out, the hammer follows a path along the tangent line to the circle at the point P.

If we wanted to determine the path followed by the hammer, we could do so, as tangent lines to circles are relatively easy to find. (Recall from elementary geometry that a tangent line to a circle is a line that intersects a circle in exactly one point.) Our aim in this section is to extend the notion of tangent line to more general curves. Among other things, we will see that such tangent lines can give us important information about the motion of objects in general.

To make our discussion more concrete, suppose that we want to find the tangent line to the curve $y = x^2 + 1$ at the point $(1, 2)$ (see Figure 2.2). Unfortunately, the notion that a tangent line is a line that intersects the curve at only one point does not work here. For instance, in Figure 2.3, we show a number of straight lines all of which intersect the curve $y = x^2 + 1$ only at the point $(0, 1)$. Are these all tangent lines? Certainly not; only the horizontal line $y = 1$ acts like the tangent line to a circle and hugs the curve near the point of tangency. What this has in common with the tangent line to a circle is that it has the same direction as the curve at the point of tangency. That is, if you were standing on the curve at the point of tangency and had to take a small step and try to stay on the curve, you would step in the direction of the tangent line. Another way to think of this is to observe that this curve appears to be *locally linear*. That is, if we zoom in sufficiently far, the graph appears to approximate that of a straight line. In Figure 2.4, we show the graph of $y = x^2 + 1$ zoomed in on the small rectangular box indicated in Figure 2.2. (Be aware that the "axes" indicated in Figure 2.4 do not intersect at the origin. We provide them only as a guide as to the scale used to produce the figure.) We now choose two points from the curve: for example, $(1, 2)$ and $(1.05, 2.1025)$, and compute the slope, m_{sec} of the secant line joining these two points:

$$m_{\text{sec}} = \frac{2.1025 - 2}{1.05 - 1} = 2.05.$$

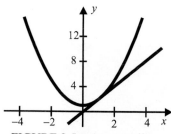

FIGURE 2.5a: A secant line.

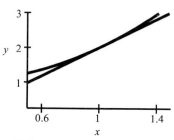

FIGURE 2.5b: Closeup of secant line.

An equation of the secant line is then determined by

$$\frac{y-2}{x-1} = 2.05,$$

so that

$$y = 2.05(x-1)+2.$$

As can be seen in Figure 2.5a, the secant line looks very much like a tangent line, even when zoomed in quite far, as in Figure 2.5b.

Notice that we can improve this approximation, simply by zooming in farther before choosing points for the secant line. For instance, instead of choosing our second point at $x = 1.05$, we could choose it at $x = 1.01$. The slope of the secant line through $(1, 2)$ and $(1.01, 2.0201)$ is

$$m_{\text{sec}} = \frac{2.0201-2}{1.01-1} = 2.01.$$

An equation of the secant line is then determined by

$$\frac{y-2}{x-1} = 2.01,$$

so that

$$y = 2.01(x-1)+2.$$

The graph of this secant line is virtually indistinguishable from the secant line seen in Figure 2.5a. Rather than continue this process by choosing more points close to $(1, 2)$, we compute the slope of the secant line joining $(1, 2)$ and the unspecified point $(1+h, f(1+h))$, for some value of h close to 0. The slope of this secant line is

Multiply-out and cancel.

Factor-out common h and cancel.

$$m_{\text{sec}} = \frac{f(1+h)-2}{(1+h)-1} = \frac{\left[(1+h)^2+1\right]-2}{h}$$

$$= \frac{(1+2h+h^2)-1}{h} = \frac{2h+h^2}{h}$$

$$= \frac{h\,(2+h)}{h} = 2+h.$$

Notice that as h approaches 0, the slope of the secant line approaches 2, which we define to be the slope of the tangent line.

Remark 1.1   There's one more observation we'd like to make before moving on to the general case of tangent lines. That is, unlike the case for a circle, tangent lines may intersect a curve at more than one point, as indicated in Figure 2.6.

The General Case

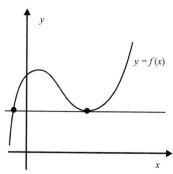

FIGURE 2.6: Tangent line intersecting a curve at more than one point.

We have now developed a graphical process for getting better and better approximations of the slope of a tangent line. The general problem is to find an equation of the tangent line to $y = f(x)$ at $x = a$. The first step is to zoom in on the graph and pick two points on the curve. It is convenient to choose the point of tangency, $(a, f(a))$ as one of the points. The second point depends on how far we have zoomed in. We will name the x-coordinate of the second point $x = a + h$, for some small number h. The corresponding y-coordinate is then $f(a + h)$. It is natural to think of h as being positive, as shown in Figure 2.7a, although h can also be negative, as shown in Figure 2.7b.

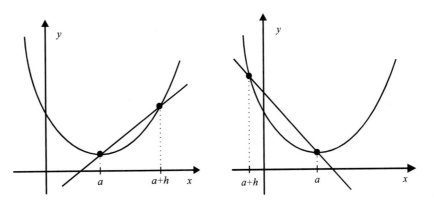

FIGURE 2.7a: Secant line ($h > 0$).   FIGURE 2.7b: Secant line ($h < 0$).

The slope of the secant line through the points $(a, f(a))$ and $(a + h, f(a + h))$ is given by

$$m_{\text{sec}} = \frac{f(a + h) - f(a)}{(a + h) - a} = \frac{f(a + h) - f(a)}{h}. \qquad (1.1)$$

Notice that the expression in (1.1) (called a **difference quotient**) gives the slope of the secant line for any second point we might choose (i.e., for any $h \neq 0$). Recall that in order to obtain better and better approximations to the tangent line, we zoom in farther and farther toward the point of tangency. This makes the two points closer together, which in turn makes h closer to 0. Just how far should we zoom in? The farther, the better; this means that we want h to approach 0. This process is illustrated in Figures 2.8a–2.8c, where we have plotted secant lines for three values of $h > 0$ as h gets successively smaller. Notice that as the point Q approaches the point P, the secant lines are approaching the tangent line at P.

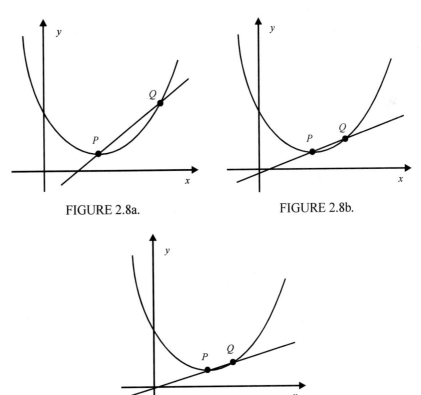

FIGURE 2.8a.   FIGURE 2.8b.

FIGURE 2.8c.

We define the slope of the tangent line to be the limit of the slopes of the secant lines as h tends to 0, whenever this limit exists. From (1.1), we obtain the following definition.

Definition 1.1   The **slope** $m_{\tan}$ of the tangent line to $y = f(x)$ at $x = a$ is given by

$$m_{\tan} = \lim_{h \to 0} \frac{f(a+h) - f(a)}{h}, \tag{1.2}$$

provided the limit exists.

The tangent line is then the line passing through the point $(a, f(a))$ with slope $m_{\tan}$ and so, is determined from

$$m_{\tan} = \frac{y - f(a)}{x - a},$$

so that the equation of the tangent line is

$$y = m_{\tan}(x - a) + f(a).$$

Example 1.1   **Finding the Equation of a Tangent Line**

Find an equation of the tangent line to $y = x^2 + 1$ at $x = 1$.

Solution   We compute the slope using (1.2):

Multiply-out and cancel.

$$
\begin{aligned}
m_{\tan} &= \lim_{h \to 0} \frac{f(1+h) - f(1)}{h} \\
&= \lim_{h \to 0} \frac{[(1+h)^2 + 1] - [1+1]}{h} \\
&= \lim_{h \to 0} \frac{1 + 2h + h^2 + 1 - 2}{h} \\
&= \lim_{h \to 0} \frac{2h + h^2}{h} \\
&= \lim_{h \to 0} \frac{h(2+h)}{h} = \lim_{h \to 0} (2+h) = 2.
\end{aligned}
$$

Factor-out common h and cancel.

Notice that the point corresponding to $x = 1$ is $(1, 2)$ and the line with slope 2 through the point $(1, 2)$ has equation

$$y = 2(x - 1) + 2 \text{ or } y = 2x.$$

Note how closely this corresponds to the two secant lines computed earlier. We show a graph of the function and this tangent line in Figure 2.9.

∎

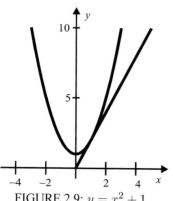

FIGURE 2.9: $y = x^2 + 1$ and the tangent line at $x = 1$.

Example 1.2   **Tangent Line to the Graph of a Rational Function**

Find an equation of the tangent line to $y = \dfrac{2}{x}$ at $x = 2$.

Solution   From (1.2), we have

$$m_{\tan} = \lim_{h \to 0} \frac{f(2+h) - f(2)}{h} = \lim_{h \to 0} \frac{\frac{2}{2+h} - \frac{2}{2}}{h}$$

Since $f(2+h) = \frac{2}{2+h}$.

$$= \lim_{h \to 0} \frac{\left[\dfrac{2(2) - (2+h)2}{(2+h)2}\right]}{h}$$

Add fractions, multiply-out and cancel.

$$= \lim_{h \to 0} \frac{\left[\dfrac{4 - 4 - 2h}{(2+h)2}\right]}{h}$$

$$= \lim_{h \to 0} \frac{-2h}{(2+h)2h}$$

Cancel h's and 2s.

$$= \lim_{h \to 0} \frac{-1}{2+h}$$

$$= \frac{-1}{2+0} = -\frac{1}{2}.$$

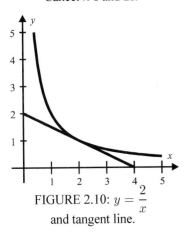

FIGURE 2.10: $y = \dfrac{2}{x}$
and tangent line.

The point corresponding to $x = 2$ is $(2, 1)$. An equation of the tangent line can then be determined from

$$\frac{y-1}{x-2} = -\frac{1}{2},$$

so that

$$y = -\frac{1}{2}(x - 2) + 1.$$

We show a graph of the function and this tangent line in Figure 2.10.

∎

Velocity

Finding the slope of a tangent line was an interesting adventure, but you might ask why anyone would care to do that. It turns out that the slopes of tangent lines have many applications. As you may have gathered from our brief discussion of the hammer throw, one of these is in computing velocity. The term **velocity** is certainly familiar to you, but can you say precisely what it is? We often describe velocity as a quantity determining the speed **and direction** of an object. Alright, so what exactly is speed? If your car did not have a speedometer, how would you determine your speed? You would probably use the familiar formula

$$distance = rate \ \times time. \tag{1.3}$$

Using (1.3), if you want to know the rate (speed), you simply divide the distance by the time. Unfortunately, this is a little too simple. The rate in (1.3) refers to the **average velocity**, not the instantaneous velocity we are interested in. The following story should indicate the difference.

Suppose that you are stopped on the highway by a police officer. The officer steps up to your car and asks that dreaded question, "Do you know how fast you were going?" An overzealous student might answer, "Absolutely. I've been driving for 3 years, 2 months, 7 days, 5 hours and 45 minutes. During that time I've driven exactly $45,259.7$ miles. Therefore, my speed was

$$rate = \frac{distance}{time} = \frac{45,259.7 \text{ miles}}{27,917.75 \text{ hours}} \approx 1.62118 \text{ mph.}$$

Of course, most police officers would not be impressed with your analysis. To be sure, it's ridiculous. But, *why* is it wrong? Certainly there's nothing wrong with formula (1.3) or the arithmetic. However, the police officer could reasonably argue that you were not even in your car during most of that this 3-year period and, hence, the results of your computation are invalid.

Suppose that you substitute the following argument instead, "I left my house at precisely 6:17 P.M. and by the time you pulled me over at 6:43 P.M., I had driven exactly 17 miles. Therefore, my speed was

$$rate = \frac{17 \text{ miles}}{26 \text{ minutes}} \cdot \frac{60 \text{ minutes}}{1 \text{ hour}} = 39.23077 \text{ mph},$$

well under the posted 45 mph speed limit."

Certainly, this is a much better estimate of your velocity than the 1.6 mph computed previously, but we are still computing an average velocity using too long of a time period. Since cars can speed up and slow down very quickly, we must compute the velocity using a much smaller time interval. But, how small is small enough?

As we did with finding slopes of tangent lines, it is helpful to rework our arguments symbolically. Suppose that the function $f(t)$ gives the distance covered from time 0 until time t. Then, for two times a and b (where $a < b$), $f(b) - f(a)$ gives the distance covered between times $t = a$ and $t = b$. The **average velocity** v_{avg} is then given by

$$\boxed{v_{avg} = \frac{distance}{time} = \frac{f(b) - f(a)}{b - a}.} \tag{1.4}$$

Example 1.3   Finding Average Velocity

Suppose that you are driving in a straight line and that the distance that you've traveled in the first t minutes is given by the function

$$s(t) = \frac{1}{2}t^2 - \frac{1}{12}t^3, 0 \leq t \leq 4.$$

Approximate the velocity at time $t = 2$.

Solution   As a start, we might find the average velocity over the first 2 minutes (i.e., from $t = 0$ to $t = 2$). From (1.4), we have

$$v_{avg} = \frac{s(2) - s(0)}{2 - 0} = \frac{1.333333333 - 0}{2}$$
$$= 0.666666667 \text{ miles/minute}$$
$$= 40 \text{ mph.}$$

Of course, the average velocity of a car over a 2-minute-long interval is not a particularly good approximation to its velocity at an instant, since cars can speed up and slow down a great deal in 2 minutes. An improved approximation is found by averaging over the second minute. Again, from (1.4), we have

$$v_{avg} = \frac{s(2) - s(1)}{2 - 1} = \frac{1.333333333 - 0.416666667}{1}$$
$$= 0.9166666663 \text{ miles/minute}$$
$$= 54.99999998 \text{ mph.}$$

Of course, this latest estimate is considerably better than the first one, since the time interval is so much shorter, but is it accurate enough? As we make the time interval shorter and shorter, the average velocity should be getting closer and closer to the velocity at the instant $t = 2$. It stands to reason that for $h > 0$, if we compute the average velocity over the time interval $[2 - h, 2]$, and then let $h \to 0$, the resulting average velocities should be getting closer and closer to the velocity at the instant $t = 2$. On the interval $[2 - h, 2]$,

h	$\dfrac{s(2) - s(2-h)}{h}$
1.0	0.9166666667
0.1	0.9991666667
0.01	0.9999916666
0.001	0.999999916
0.0001	1.0
0.00001	1.0

h	$\dfrac{s(2+h) - s(2)}{h}$
1.0	0.9166666667
0.1	0.9991666667
0.01	0.9999916667
0.001	0.999999917
0.0001	1.0
0.00001	1.0

we have

$$v_{\text{avg}} = \frac{s(2) - s(2-h)}{2 - (2-h)} = \frac{s(2) - s(2-h)}{h}.$$

In the accompanying table, we compute a sequence of these values for successively smaller values of $h > 0$.

Of course, it also makes sense to average the velocity over intervals beginning at $t = 2$ and going slightly to the right, that is, over intervals of the form $[2, 2+h]$, for small values of $h > 0$. We get

$$v_{\text{avg}} = \frac{s(2+h) - s(2)}{(2+h) - 2} = \frac{s(2+h) - s(2)}{h}.$$

A sequence of these average velocities are displayed in the accompanying table.

Notice that the first formula for v_{avg} is equivalent to the second one if we allow h to be negative. From both sets of results, it appears as if the average velocity is approaching 1 mile/minute (60 mph), as $h \to 0$. This limiting value is what we refer to as **instantaneous velocity**.

∎

This leads us to make the following definition.

Definition 1.2 If $f(t)$ represents the distance of an object from some fixed location at time t, then the **instantaneous velocity** at time $t = a$ is given by

$$v(a) = \lim_{h \to 0} \frac{f(a+h) - f(a)}{(a+h) - a} = \lim_{h \to 0} \frac{f(a+h) - f(a)}{h}, \tag{1.5}$$

provided the limit exists.

Notice that if (for example) t is measured in seconds and $f(t)$ is measured in feet, velocity (average or instantaneous) is measured in feet per second (ft/s). The term **velocity** is always used to refer to instantaneous velocity.

Example 1.4   Finding Average and Instantaneous Velocity

Suppose that the height of a falling object t seconds after being dropped from a height of 64 feet is given by $f(t) = 64 - 16t^2$ feet. Find the average velocity between times $t = 1$ and $t = 2$; the average velocity between times $t = 1.5$ and $t = 2$; the average velocity between times $t = 1.9$ and $t = 2$ and the instantaneous velocity at time $t = 2$.

Solution   The average velocity between times $t = 1$ and $t = 2$ is

$$v_{\text{avg}} = \frac{f(2) - f(1)}{2 - 1} = \frac{64 - 16(2)^2 - [64 - 16(1)^2]}{1} = -48 \text{ ft/s}.$$

The average velocity between times $t = 1.5$ and $t = 2$ is

$$v_{\text{avg}} = \frac{f(2) - f(1.5)}{2 - 1.5} = \frac{64 - 16(2)^2 - [64 - 16(1.5)^2]}{0.5} = -56 \text{ ft/s}.$$

The average velocity between times $t = 1.9$ and $t = 2$ is

$$v_{\text{avg}} = \frac{f(2) - f(1.9)}{2 - 1.9} = \frac{64 - 16(2)^2 - [64 - 16(1.9)^2]}{0.1} = -62.4 \text{ ft/s}.$$

The instantaneous velocity is the limit of such average velocities. From (1.5), we have

$$v(2) = \lim_{h \to 0} \frac{f(2+h) - f(2)}{(2+h) - 2}$$

$$= \lim_{h \to 0} \frac{[64 - 16(2+h)^2] - [64 - 16(2)^2]}{h}$$

Multiply out and cancel.

$$= \lim_{h \to 0} \frac{[64 - 16(4 + 4h + h^2)] - [64 - 16(2)^2]}{h}$$

$$= \lim_{h \to 0} \frac{-64h - 16h^2}{h}$$

Factor-out common h and cancel.

$$= \lim_{h \to 0} \frac{-16h(h + 4)}{h}$$

$$= \lim_{h \to 0} [-16(h + 4)] = -64 \text{ ft/s}.$$

Notice that the average velocities in this example appear to be approaching the instantaneous velocity as the time interval gets shorter and shorter. But, what does the negative sign on the velocity indicate? We noted earlier that velocity indicates both speed and direction. In this problem, $f(t)$ measures the height above the ground. A negative velocity thus indicates that the object is moving in the negative (or downward) direction. The **speed** of the object at the 2-second mark is then 64 ft/s. (Speed is simply the absolute value of velocity.)

■

You should observe that the formulas for instantaneous velocity (1.5) and for the slope of a tangent line (1.2) are identical. We want to make this connection as strong as possible. Let's first illustrate example 1.4 graphically. We graph the distance function $f(t) = 64 - 16t^2$ for $0 \le t \le 2$. The average velocity between $t = 1$ and $t = 2$ is the slope of the secant line between the points at $t = 1$ and $t = 2$ (see Figure 2.11a). Similarly, the average velocity between $t = 1.5$ and $t = 2$ is the slope of the corresponding secant line (see Figure 2.11b). Finally, the instantaneous velocity at time $t = 2$ is the slope of the tangent line at $t = 2$ (see Figure 2.11c).

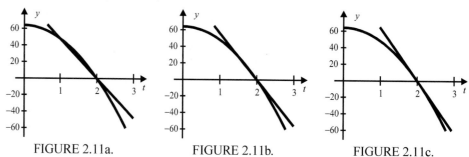

FIGURE 2.11a.   FIGURE 2.11b.   FIGURE 2.11c.

We often think of velocity as a *rate* (in the present case, the instantaneous rate of change of height with respect to time). Further, for any quantity, we can generalize this notion of instantaneous rate of change. In general, we have that the **average rate of change** of a given function $f(x)$ between $x = a$ and $x = b$ $(a \ne b)$ is given by

$$\frac{f(b) - f(a)}{b - a}.$$

The **instantaneous rate of change** of $f(x)$ at $x = a$ is given by

$$\lim_{h \to 0} \frac{f(a+h) - f(a)}{h},$$

provided the limit exists.

Example 1.5   **Finding Average and Instantaneous Rates of Change**

Suppose that the population of a city is estimated to be $f(t) = \sqrt{100 + 8t}$ million people t years from now. Find the average rate of change of the population for the next 2 years and the instantaneous rate of change of the population 2 years from now.

Solution   The average rate of change is

$$\frac{f(2) - f(0)}{2 - 0} = \frac{\sqrt{100 + 16} - \sqrt{100}}{2} \approx 0.3852$$

million people per year. The instantaneous rate of change is then

$$\lim_{h \to 0} \frac{f(2 + h) - f(2)}{h} = \lim_{h \to 0} \frac{\sqrt{100 + 8(2 + h)} - \sqrt{116}}{h}$$

Multiply numerator and denominator by the conjugate: $\sqrt{116 + 8h} + \sqrt{116}$.

$$= \lim_{h \to 0} \frac{(\sqrt{116 + 8h} - \sqrt{116})}{h} \frac{(\sqrt{116 + 8h} + \sqrt{116})}{(\sqrt{116 + 8h} + \sqrt{116})}$$

$$= \lim_{h \to 0} \frac{(116 + 8h) - 116}{h(\sqrt{116 + 8h} + \sqrt{116})}$$

Cancel common h's.

$$= \lim_{h \to 0} \frac{8h}{h(\sqrt{116 + 8h} + \sqrt{116})}$$

$$= \lim_{h \to 0} \frac{8}{\sqrt{116 + 8h} + \sqrt{116}}$$

$$= \frac{8}{\sqrt{116} + \sqrt{116}}$$

$$= \frac{4}{\sqrt{116}} \approx 0.3714 \text{ million people per year.}$$

Since the instantaneous rate of change at $t = 2$ is less than the average rate of change from $t = 0$ to $t = 2$, we might conjecture that the rate of population growth is slowing down. A graph of $f(t)$ will confirm this. (We leave this as an exercise.) Finally, when discussing population change, we typically report the rate of change as a percentage of the current population. Since the population at time $t = 2$ is $\sqrt{116}$ million people, the percentage change at $t = 2$ is $\dfrac{4/\sqrt{116}}{\sqrt{116}} \approx 0.034$ or about 3.4% per year.

■

Additional applications of the slope of a tangent line are too numerous too list. These include the rate of a chemical reaction, the inflation rate in economics and learning growth rates in psychology. In short, rates of change in almost any discipline you can name can be thought of as slopes of tangent lines. We explore many of these applications as we progress through the book.

You may have noticed that we tacked the phrase "provided the limit exists" onto the end of the definitions of the slope of a tangent line, the instantaneous velocity and the instantaneous rate of change. You might have already wondered whether these defining limits don't always exist. They do not, as the following example illustrates.

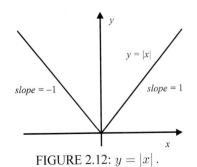

FIGURE 2.12: $y = |x|$.

Example 1.6   **A Graph with No Tangent Line at a Point**

Determine if there is a tangent line to $y = |x|$ at $x = 0$.

Solution   We can look at this problem graphically, numerically and symbolically. The graph is shown in Figure 2.12. Our graphical technique is to zoom in on the point of tangency until the graph appears straight. But, no matter how far we zoom in on $(0, 0)$, the graph continues to look like Figure 2.12. Try this for yourself. (This is one reason why we left the scale off.) From this evidence alone, we would conjecture that the tangent line does not exist. Recall that the tangent line can be thought of as the limiting position of a sequence of secant lines, as the second point approaches the point of tangency. Numerically, we observe that the secant line through $(0, 0)$ and $(1, 1)$ has slope 1, as does the secant line through $(0, 0)$ and $(0.1, 0.1)$. In fact, if h is any positive number, the slope of the secant line through $(0, 0)$ and $(h, |h|)$ is 1. However, the secant line through $(0, 0)$ and $(-1, 1)$ has slope -1, as does the secant line through $(0, 0)$ and $(h, |h|)$ for any negative number h. We therefore conjecture that the one-sided limits are different and hence, the limit (and also the tangent line) does not exist. To prove this conjecture, we must work symbolically. Taking our cue from the numerical work, we look at the one-sided limits: if $h > 0$, then $|h| = h$, so that

$$\lim_{h \to 0^+} \frac{f(0 + h) - f(0)}{h} = \lim_{h \to 0^+} \frac{|h| - 0}{h} = \lim_{h \to 0^+} \frac{h}{h} = 1.$$

On the other hand, if $h < 0$, then $|h| = -h$ (remember that if $h < 0$, $-h > 0$), so that

$$\lim_{h \to 0^-} \frac{f(0 + h) - f(0)}{h} = \lim_{h \to 0^-} \frac{|h| - 0}{h} = \lim_{h \to 0^-} \frac{-h}{h} = -1.$$

Since the one-sided limits are different, we conclude that

$$\lim_{h \to 0} \frac{f(0 + h) - f(0)}{h} \quad \text{does not exist}$$

and hence, the tangent line does not exist.

∎

Exercises 2.1

1. (writing) What does the phrase "off on a tangent" mean? Relate the common meaning of the phrase to the image of a tangent to a circle (use the hammer throw example, if that helps). In what way does the zoom image of the tangent promote the opposite view of the relationship between a curve and its tangent?

2. (writing) In general, the instantaneous velocity of an object cannot be computed directly; the limit process is the only way to compute velocity *at an instant*. Given this, how does a car's speedometer compute velocity? (Hint: Look this up in a reference book or on the Internet. An important aspect of the car's ability to do this seemingly difficult task is that it performs **analog** calculations. For example, the pitch of a fly's buzz gives us an analog device for computing the speed of a fly's wings, since pitch is proportional to speed.)

3. (writing) Look in the news media (TV, newspaper, Internet) and find references to at least five different *rates*. We have defined a rate of change as the limit of the difference quotient of a function. For your five examples, state as precisely as possible what the original function is. Is the rate given quantitatively or qualitatively? If it is given quantitatively, is the rate given as a percentage or a number? In calculus, we usually compute rates (quantitatively) as numbers; is this in line with the standard usage?

4. (writing) Sketch the graph of a function that is discontinuous at $x = 1$. Explain why there is no tangent line at $x = 1$.

In exercises 5–8, sketch in a plausible tangent line at the given point (Hint: Mentally zoom in on the point and use the zoom image of the tangent).

5. 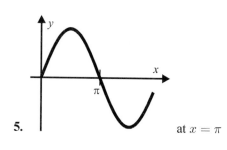 at $x = \pi$

6. at $x = 1$

7. at $x = 0$

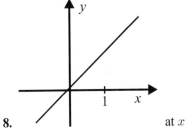

8.   at $x = 1$

In exercises 9 and 10, estimate the slope of the tangent line to the curve at $x = 1$.

9.

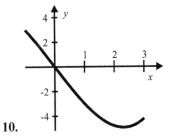

10.

In exercises 11 and 12, list the points A, B, C and D in order of increasing slope of tangent line.

11.

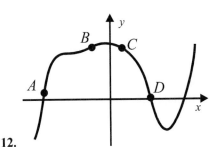

12.

In exercises 13–16, compute the slope of the secant line between the points at (a) $x = 1$ and $x = 2$, (b) $x = 2$ and $x = 3$, (c) $x = 1.5$ and $x = 2$, (d) $x = 2$ and $x = 2.5$, (e) $x = 1.9$ and $x = 2$ and (f) $x = 2$ and $x = 2.1$ and (g) estimate the slope of the tangent line at $x = 2$.

13. $f(x) = x^3 - x$ **14.** $f(x) = \sqrt{x^2 + 1}$

15. $f(x) = \cos x^2$ **16.** $f(x) = \tan 2x$

In exercises 17–20, find the slope of the tangent line to $y = f(x)$ at $x = a$ using both (equivalent) definitions:
$$\lim_{h \to 0} \frac{f(a+h) - f(a)}{h} \text{ and } \lim_{x \to a} \frac{f(x) - f(a)}{x - a}.$$

17. $f(x) = x^2 - 1, a = 0$

18. $f(x) = x^2 - 2, a = 3$

19. $f(x) = x^3 - 3x, a = 2$

20. $f(x) = x^4 + x, a = -1$

In exercises 21–28, find the equation of the tangent line to $y = f(x)$ at $x = a$. Graph $y = f(x)$ and the tangent line to verify that you have the correct equation.

21. $f(x) = x^2 - 2, a = 1$

22. $f(x) = x^2 - 2, a = 0$

23. $f(x) = x^2 - 3x, a = -2$

24. $f(x) = x^3 + x, a = 1$

25. $f(x) = \dfrac{2}{x + 1}, a = 1$

26. $f(x) = \dfrac{x}{x - 1}, a = 0$

27. $f(x) = \sqrt{x + 3}, a = -2$

28. $f(x) = \sqrt{x^2 + 1}, a = 1$

In exercises 29–34, use graphical and numerical evidence to determine whether or not the tangent line to $y = f(x)$ exists at $x = a$. If it does, estimate the slope of the tangent; if not, explain why not.

29. $f(x) = |x - 1|$ at $a = 1$

30. $f(x) = \dfrac{4x}{x - 1}$ at $a = 1$

31. $f(x) = \begin{cases} -2x^2 & \text{if } x < 0 \\ x^3 & \text{if } x \geq 0 \end{cases}$ at $a = 0$

32. $f(x) = \begin{cases} -2x & \text{if } x < 1 \\ x - 3 & \text{if } x \geq 1 \end{cases}$ at $a = 1$

33. $f(x) = \begin{cases} x^2 - 1 & \text{if } x < 0 \\ x^3 + 1 & \text{if } x \geq 0 \end{cases}$ at $a = 0$

34. $f(x) = \begin{cases} -2x & \text{if } x < 0 \\ x^2 - 2x & \text{if } x > 0 \end{cases}$ at $a = 0$

In exercises 35–38, the function represents the distance (in feet) traveled by an object in t seconds. Find the average velocity between (a) $t = 0$ and $t = 2$, (b) $t = 1$ and $t = 2$, (c) $t = 1.9$ and $t = 2$ and (d) $t = 1.99$ and $t = 2$ and (e) estimate the instantaneous velocity at $t = 2$.

35. $f(t) = 16t^2 + 10$ **36.** $f(t) = 3t^3 + t$

37. $f(t) = \sqrt{t^2 + 8t}$ **38.** $f(t) = 100\sin(t/4)$

In exercises 39–42, use the distance function $f(t)$ to find the velocity at time $t = a$.

39. $f(t) = -16t^2 + 5$, $a = 1$

40. $f(t) = -16t^2 + 5$, $a = 2$

41. $f(t) = \sqrt{t + 16}$, $a = 0$

42. $f(t) = 2\sqrt{t + 14}$, $a = 2$

43. The graph below shows the elevation of a person on a hike up a mountain as a function of time. When did the hiker reach the top? When was the hiker going the fastest on the way up? When was the hiker going the fastest on the way down? What do you think occurred at places where the graph is level?

44. The graph below shows the amount of water in a city water tank as a function of time. When was the tank the fullest? the emptiest? When was the tank filling up at the fastest rate? When was the tank emptying at the fastest rate? What time of day do you think the level portion represents?

45. Suppose a hot cup of coffee is left in a room for 2 hours. Sketch a reasonable graph of what the temperature would look like as a function of time. Then sketch a graph of what the rate of change of the temperature would look like.

46. Sketch a graph representing the height of a bungee-jumper. Sketch the graph of the person's velocity (use + for upward velocity and − for downward velocity).

47. In the hammer throw, a different type of velocity is important during the three preliminary rotations. **Angular velocity** is defined by $\lim\limits_{h \to 0} \dfrac{\theta(a + h) - \theta(a)}{h}$ where $\theta(t)$ is the angle of rotation at time t. If the angle of a hammer thrower is $\theta(t) = 0.4t^2$, what is the angular velocity at the time of release? (Hint: Which value of t (radians) corresponds to three rotations?)

48. Find the angular velocity of the hammer thrower in exercise 47 after two rotations. Explain why the third rotation is helpful.

49. You may have noticed in exercise 35 that there is a wrong way to get the right answer. For quadratic functions (but definitely *not* most other functions), the average velocity between $t = r$ and $t = s$ equals the average of the velocities at $t = r$ and $t = s$. To show this, assume that $f(t) = at^2 + bt + c$ is the distance function. Show that the average velocity between $t = r$ and $t = s$ equals $a(s + r) + b$. Show that the velocity at $t = r$ is $2ar + b$ and the velocity at $t = s$ is $2as + b$. Finally, show that $a(s + r) + b = \dfrac{(2ar + b) + (2as + b)}{2}$.

50. Find a cubic function (try $f(t) = t^3 + \cdots$) and numbers r and s such that the average velocity between $t = r$ and $t = s$ is different from the average of the velocities at $t = r$ and $t = s$.

51. In example 1.5, we analyzed a population $f(t) = \sqrt{100 + 8t}$. Graph $y = f(t)$. Argue that the rate of change of the population (the slope of the tangent line) gets smaller as t gets larger. To illustrate this, compute the rate of change at $t = 3$. Compare your answer to the rate of change at $t = 2$, found in example 1.5.

52. In exercise 51, compute the percentage change at $t = 3$ and compare to example 1.5.

53. Show that $\lim\limits_{h \to 0} \dfrac{f(a + h) - f(a)}{h} = \lim\limits_{x \to a} \dfrac{f(x) - f(a)}{x - a}$. (Hint: Let $h = x - a$.)

54. (exploratory) Many optical illusions are caused by our brain's (unconscious) use of the tangent line in determining the positions of objects. Suppose you are in the desert 100 feet from a palm tree. You see a particular spot 10 feet up on the palm tree due to light reflecting from that spot to your eyes. Normally, it is a good approximation to say that the light follows a straight line (top path in the figure).

Two paths of light from tree to person.

However, when there is a large temperature difference in the air, light may follow nonlinear paths. If, as in the desert, the air near the ground is much hotter than the air higher up, light will bend as indicated by the bottom path in the figure. Our brains always interpret light coming in straight paths, so you would think the spot on the tree is at $y = 10$ because of the top path and also at some other y because of the bottom path. If the bottom curve is $y = 0.002x^2 - 0.24x + 10$, find an equation of the tangent line at $x = 100$ and show that it crosses the y-axis at $y = -10$. That is, you would "see" the spot at $y = 10$ and also at $y = -10$, a perfect reflection.

Two perceived locations of tree.

How do reflections normally occur in nature? From water! You would perceive a tree and its reflection in a pool of water. This is the desert mirage!

55. (exploratory) You can use a VCR to estimate speed. Most VCRs play at 30 frames per second, so with a frame-by-frame advance you can estimate time as the number of frames divided by 30. If you know the distance covered, you can compute the average velocity by dividing distance by time. Try this to estimate how fast you can throw a ball, run 50 yards, hit a tennis ball, or whatever speed you find interesting. Some of the possible inaccuracies are explored in the following exercise.

56. (exploratory) What is the peak speed for a human being? It has been estimated that Carl Lewis reached a peak speed of 28 mph while winning a gold medal in the 1992 Olympics. Suppose that we have the following data for a sprinter. We want to estimate peak speed. We could start by computing $\frac{\text{distance}}{\text{time}} = \frac{100 \text{ m}}{10 \text{ s}} = 10$ m/s, but this is the average speed over the entire race, not the peak speed. Argue that we want to compute average speeds only using adjacent measurements (e.g., 40 and 50 meters, or 50 and 56 meters). Do this for all 11 adjacent pairs and find the largest speed (if you want to convert to mph, divide by 0.447). We will explore how accurate this estimate might be below.

Meters	Seconds	Meters	Seconds
30	3.2	62	6.26666
40	4.2	64	6.46666
50	5.16666	70	7.06666
56	5.76666	80	8.0
58	5.93333	90	9.0
60	6.1	100	10.0

Notice that all times are essentially multiples of 1/30 since the data was obtained using the VCR technique in exercise 55. Given this, why is it suspicious that all the distances are whole numbers? To get an idea of how much this might affect your calculations, change some of the distances. For instance, if you change 60 (meters) to 59.8, how much do your average velocity calculations change? One possible way to identify where mistakes have been made is to look at the pattern of average velocities: does it seem reasonable? Would a sprinter speed up and slow down in such a pattern? In places where the pattern seems suspicious, try adjusting the distances and see if you can produce a more realistic pattern. Taking all this into account, try to quantify your error analysis: what is the highest (lowest) the peak speed could be?

2.2   The Derivative

In section 2.1, we investigated two seemingly unrelated concepts: tangent lines and velocity. What we discovered was that both can be expressed in terms of the **same** limit. This is an indication of the power of mathematics, that otherwise unrelated notions are described by the **same** mathematical expression. This particular limit turns out to be so useful that we give it a special name and pause now to devote more study to it.

Definition 2.1   The **derivative** of the function $f(x)$ at $x = a$ is defined as

$$f'(a) = \lim_{h \to 0} \frac{f(a+h) - f(a)}{h}, \qquad (2.1)$$

provided the limit exists. If the limit exists, we say that f is **differentiable** at $x = a$.

Remark 2.1   An alternative form of (2.1) is

$$f'(a) = \lim_{b \to a} \frac{f(b) - f(a)}{b - a}. \tag{2.2}$$

You may recall that we had left it to you to establish the proof of the equivalence of (2.1) and (2.2) in exercise 53 in section 2.1.

Example 2.1   **Finding the Derivative at a Point**

Compute the derivative of $f(x) = 3x^3 + 2x - 1$ at $x = 1$.

Solution   From (2.1), we have

$$f'(1) = \lim_{h \to 0} \frac{f(1 + h) - f(1)}{h}$$

Multiply out and cancel.

$$= \lim_{h \to 0} \frac{3(1 + h)^3 + 2(1 + h) - 1 - (3 + 2 - 1)}{h}$$

$$= \lim_{h \to 0} \frac{3(1 + 3h + 3h^2 + h^3) + (2 + 2h) - 1 - 6}{h}$$

$$= \lim_{h \to 0} \frac{3 + 9h + 9h^2 + 3h^3 + 2 + 2h - 7}{h}$$

Factor-out common h and cancel.

$$= \lim_{h \to 0} \frac{11h + 9h^2 + 3h^3}{h} = \lim_{h \to 0}(11 + 9h + 3h^2) = 11.$$

∎

Suppose that in example 2.1 we also needed to find $f'(2)$ and $f'(3)$. We've just gone through a long limit calculation to find $f'(1)$. Must we now repeat largely the same steps to find each of $f'(2)$ and $f'(3)$? The solution is to compute the derivative without specifying a value for x. This may make the limit slightly more difficult to calculate, but we will be left with a function from which we can calculate $f'(a)$ for any a, simply by substituting a for x.

Example 2.2   **Finding the Derivative at an Unspecified Point**

Find the derivative of $f(x) = 3x^3 + 2x - 1$ at an unspecified value of x, then evaluate the derivative at $x = 1$, $x = 2$ and $x = 3$.

Solution   Replacing a with x in the definition of the derivative, we have

$$f'(x) = \lim_{h \to 0} \frac{f(x + h) - f(x)}{h}$$

$$= \lim_{h \to 0} \frac{3(x + h)^3 + 2(x + h) - 1 - (3x^3 + 2x - 1)}{h}$$

Multiply out and cancel.

$$= \lim_{h \to 0} \frac{3(x^3 + 3x^2h + 3xh^2 + h^3) + (2x + 2h) - 1 - 3x^3 - 2x + 1}{h}$$

$$= \lim_{h \to 0} \frac{3x^3 + 9x^2h + 9xh^2 + 3h^3 + 2x + 2h - 3x^3 - 2x}{h}$$

$$= \lim_{h \to 0} \frac{9x^2h + 9xh^2 + 3h^3 + 2h}{h}$$

Factor-out common h and cancel.

$$= \lim_{h \to 0}(9x^2 + 9xh + 3h^2 + 2) = 9x^2 + 0 + 0 + 2 = 9x^2 + 2.$$

Notice that in this case, we have derived a new **function**, $f'(x) = 9x^2 + 2$. Simply substituting in for x, we get $f'(1) = 9 + 2 = 11$ (the same as we got in example 2.1!), $f'(2) = 9(4) + 2 = 38$ and $f'(3) = 9(9) + 2 = 83$.

∎

This leads us to the following definition.

Definition 2.2   The **derivative** of $f(x)$ is the function $f'(x)$ given by

$$f'(x) = \lim_{h \to 0} \frac{f(x+h) - f(x)}{h}, \tag{2.3}$$

provided the limit exists. The process of computing a derivative is called **differentiation**.

We compute derivative functions in the next several examples.

Example 2.3   **Finding the Derivative of a Linear Polynomial**
        If $f(x) = 3x - 5$, find $f'(x)$.

Solution   We have from (2.3) that

Multiply out and cancel.

$$\begin{aligned}
f'(x) &= \lim_{h \to 0} \frac{f(x+h) - f(x)}{h} \\
&= \lim_{h \to 0} \frac{[3(x+h) - 5] - (3x - 5)}{h} \\
&= \lim_{h \to 0} \frac{3x + 3h - 5 - 3x + 5}{h} \\
&= \lim_{h \to 0} \frac{3h}{h} = \lim_{h \to 0} 3 = 3.
\end{aligned}$$

Cancel common h's.

Notice that this agrees with the slope of the line, 3.

■

Example 2.4   **Finding the Derivative of a Quadratic Polynomial**
        If $f(x) = 2x^2 - 5x + 7$, find $f'(x)$.

Solution   From (2.3) we have that

$$\begin{aligned}
f'(x) &= \lim_{h \to 0} \frac{f(x+h) - f(x)}{h} \\
&= \lim_{h \to 0} \frac{[2(x+h)^2 - 5(x+h) + 7] - (2x^2 - 5x + 7)}{h} \\
&= \lim_{h \to 0} \frac{2(x^2 + 2xh + h^2) - 5x - 5h + 7 - 2x^2 + 5x - 7}{h} \\
&= \lim_{h \to 0} \frac{4xh + 2h^2 - 5h}{h} \\
&= \lim_{h \to 0} \frac{h(4x + 2h - 5)}{h} \\
&= \lim_{h \to 0} (4x + 2h - 5) = 4x - 5.
\end{aligned}$$

Multiply out and cancel.

Factor out common h's and cancel.

■

Even after only two examples, you should observe that the name of the game here seems to be to write down the defining limit and then to find some way of evaluating that limit (all of which initially have the indeterminate form $\frac{0}{0}$).

Example 2.5   Finding the Derivative of a Simple Rational Function

If $f(x) = \dfrac{1}{x}$ ($x \neq 0$), find $f'(x)$.

Solution   We have

Since $f(x+h) = \frac{1}{x+h}$.

$$f'(x) = \lim_{h \to 0} \frac{f(x+h) - f(x)}{h}$$

$$= \lim_{h \to 0} \frac{\left(\dfrac{1}{x+h} - \dfrac{1}{x}\right)}{h}$$

Add fractions and cancel.

$$= \lim_{h \to 0} \frac{\left[\dfrac{x - (x+h)}{x(x+h)}\right]}{h}$$

Cancel h's.

$$= \lim_{h \to 0} \frac{-h}{hx(x+h)}$$

$$= \lim_{h \to 0} \frac{-1}{x(x+h)} = -\frac{1}{x^2}$$

or $f'(x) = -x^{-2}$.

$\blacksquare$

Example 2.6   The Derivative of the Square Root Function

If $f(x) = \sqrt{x}$ (for $x \geq 0$), find $f'(x)$.

Solution   We have

$$f'(x) = \lim_{h \to 0} \frac{f(x+h) - f(x)}{h}$$

$$= \lim_{h \to 0} \frac{\sqrt{x+h} - \sqrt{x}}{h}$$

Multiply numerator and denominator by the conjugate: $\sqrt{x+h} + \sqrt{x}$.

$$= \lim_{h \to 0} \frac{\sqrt{x+h} - \sqrt{x}}{h} \frac{\sqrt{x+h} + \sqrt{x}}{\sqrt{x+h} + \sqrt{x}}$$

Multiply out and cancel.

$$= \lim_{h \to 0} \frac{(x+h) - x}{h\left[\sqrt{x+h} + \sqrt{x}\right]}$$

Cancel common h's.

$$= \lim_{h \to 0} \frac{h}{h\left[\sqrt{x+h} + \sqrt{x}\right]}$$

$$= \lim_{h \to 0} \frac{1}{\sqrt{x+h} + \sqrt{x}}$$

$$= \frac{1}{2\sqrt{x}} = \frac{1}{2}x^{-1/2}.$$

Notice that $f'(x)$ is defined only for $x > 0$.

$\blacksquare$

The benefits of having a derivative **function** go well beyond simplifying the computation of a derivative at multiple points. As we'll see, the derivative function tells us a great deal about the original function.

Keep in mind that the value of a derivative at a point is the slope of the tangent line at that point. To get an idea of how this translates graphically, consider Figures 2.13a – 2.13c. There, we have graphed a function along with its tangent lines at three different points. Note that the slope of the tangent line in Figure 2.13a is negative, corresponding to the function dropping (**decreasing**) as we move from left to right. The slope of the tangent line in Figure 2.13c is positive, corresponding to the function rising (**increasing**) as we move from left to right. The slope of the tangent line in Figure 2.13b is zero, corresponding to the function reaching its minimum value as it changes from decreasing to increasing.

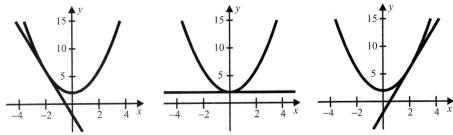

FIGURE 2.13a: $m_{\tan} < 0$.   FIGURE 2.13b: $m_{\tan} = 0$.   FIGURE 2.13c: $m_{\tan} > 0$.

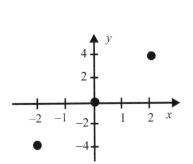

FIGURE 2.14: $y = f'(x)$ (three points).

Based on these three tangent lines, we could sketch in three points on the graph of the derivative function (see Figure 2.14), by estimating the value of $f'(x)$ at the three points. Thus, as x changes, the slope of the tangent line changes and hence $f'(x)$ changes. The derivative function keeps track of these changes. With some effort (thought), you will be able to determine the behavior of a function from information about its derivative (and vice versa).

Example 2.7   Sketching the Graph of $f'(x)$ Given the Graph of $f(x)$
Given the graph of $f(x)$ in Figure 2.15, sketch a plausible graph of $f'(x)$.

FIGURE 2.15: $y = f(x)$.

Solution   We will not worry about exact values of the slope, but only wish to get the general shape right. As we did in Figures 2.13 and 2.14, we will pick a few important points to analyze carefully. Which points are important? In general, you should pay attention to any discontinuities and any places where the graph of f turns around. For the moment, we **do not** need to know the x- or y-intercepts of f.

The graph levels out at approximately $x = -2$ and $x = 2$. At these points, the derivative is 0. As we move left to right, the graph rises for $x < -2$, drops for $-2 < x < 2$ and rises again for $x > 2$. This means that $f'(x) > 0$ for $x < -2$, $f'(x) < 0$ for $-2 < x < 2$ and finally $f'(x) > 0$ for $x > 2$. We can say even more. As x approaches -2 from the left, the tangent lines get less steep. (Draw a few of these to convince yourself of this.) Therefore, $f'(x)$ becomes less positive as x approaches -2 from the left. Moving to the right from $x = -2$, the graph gets steeper until about $x = 0$, then gets less steep until it levels out at $x = 2$. Thus, $f'(x)$ gets more negative until $x = 0$, then less negative until $x = 2$. Finally, the graph gets steeper as we move to the right from $x = 2$. Putting this all together, we have the possible graph of $f'(x)$ shown in Figure 2.16. ■

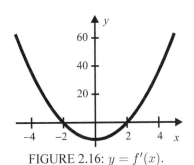

FIGURE 2.16: $y = f'(x)$.

The opposite question to that asked in example 2.7 is even more interesting. That is, given the graph of a derivative, what might the graph of the original function look like. We explore this in the following example.

Example 2.8   Sketching the Graph of $f(x)$ Given the Graph of $f'(x)$
Given the graph of $f'(x)$ in Figure 2.17, sketch a plausible graph of $f(x)$.

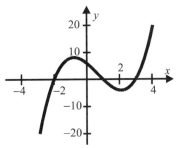

FIGURE 2.17: $y = f'(x)$.

Solution   We are not concerned here about getting exact values of the function, but rather are concerned only with getting the general shape of the graph. (In fact, we see later that it's not possible to determine function values given values of the derivative.) It's most significant to notice from the graph of $y = f'(x)$ that $f'(x) < 0$ for $x < -2$. This

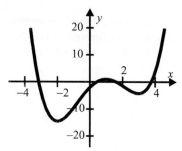

FIGURE 2.18: A plausible
graph of $y = f(x)$.

says that on this interval, the slopes of the tangent lines to $y = f(x)$ are negative, so that
the function is decreasing. Then, on the interval $(-2, 1)$, $f'(x) > 0$, indicating that the
tangent lines to the graph of $y = f(x)$ are positive and the function is increasing. Further,
this says that the graph turns around (i.e., goes from decreasing to increasing) at $x = -2$.
We have drawn a graph exhibiting this behavior in Figure 2.18. Further, $f'(x) < 0$ on
the interval $(1, 3)$, so that the tangent lines again have negative slope (and the function
decreases) here. Finally, for $x > 3$, we have that $f'(x) > 0$, so that the tangent lines have
positive slope and the function is increasing here. We show a graph exhibiting all of this
behavior in Figure 2.18. You're probably wondering why we drew the graph of f so that
the small "valley" on the right side of the y-axis was not as deep as the one on the left
side of the y-axis. There is a reason. Look carefully at the graph of $f'(x)$ and notice that
$|f'(x)|$ gets much larger on $(-2, 1)$ than on $(1, 3)$. This says that the tangent lines and
hence, the graph will be much steeper on the interval $(1, 3)$ than on $(-2, 1)$.

Alternative Notations

We have denoted the derivative function by $f'(x)$. There are other commonly used notations, each with advantages and disadvantages. One of the coinventors of the calculus,
Gottfried Leibniz, used the notation $\dfrac{df}{dx}$ (Leibniz notation) for the derivative. If we write
$y = f(x)$, the following are all alternatives for denoting the derivative:

$$f'(x) = y' = \frac{dy}{dx} = \frac{df}{dx} = \frac{d}{dx}f(x).$$

The expression $\dfrac{d}{dx}$ is called a **differential operator** and tells you to take the derivative of
whatever expression follows.

   In section 2.1, we observed that although $f(x) = |x|$ is continuous everywhere, it does
not have a tangent line at $x = 0$ (i.e., it is not differentiable at $x = 0$). Thus, there are continuous functions which are not differentiable. You might have already wondered whether
the reverse is true. That is, are there differentiable functions that are not continuous? The
answer is provided by the following theorem.

Theorem 2.1   If $f(x)$ is differentiable at $x = a$, then $f(x)$ is continuous at $x = a$.

Proof. For f to be continuous at $x = a$, we need only show that $\lim\limits_{x \to a} f(x) = f(a)$. We
consider

Multiply and divide by $(x - a)$.

$$\lim_{x \to a}[f(x) - f(a)] = \lim_{x \to a}\left[\frac{f(x) - f(a)}{x - a}(x - a)\right]$$

By Theorem 2.3 (iii)
from section 1.2.

$$= \lim_{x \to a}\left[\frac{f(x) - f(a)}{x - a}\right]\lim_{x \to a}(x - a)$$

$$= f'(a)\,(0) = 0,$$

where we have used the alternative definition of derivative (2.2) discussed earlier.
■

   Note that Theorem 2.1 says that if a function is **not** continuous at a point then it **cannot**
have a derivative at that point.

   It also turns out that functions are not differentiable at any point where their graph has
a "sharp" corner.

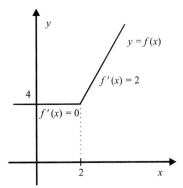

FIGURE 2.19: A sharp corner.

Multiply-out and cancel.

Cancel common h's.

Example 2.9   **Showing That a Function is Not Differentiable at a Point**

Show that $f(x) = \begin{cases} 4 & \text{if } x < 2 \\ 2x & \text{if } x \geq 2 \end{cases}$ is not differentiable at $x = 2$.

Solution   Looking at the graph (see Figure 2.19), we see a "bend" at $x = 2$. This qualifies as a sharp corner, so you might expect that the derivative does not exist. To verify this, we investigate the derivative by evaluating one-sided limits. For $h > 0$, note that $(2 + h) > 2$ and so, $f(2 + h) = 2(2 + h)$. This gives us

$$\lim_{h \to 0^+} \frac{f(2 + h) - f(2)}{h} = \lim_{h \to 0^+} \frac{2(2 + h) - 4}{h}$$
$$= \lim_{h \to 0^+} \frac{4 + 2h - 4}{h}$$
$$= \lim_{h \to 0^+} \frac{2h}{h} = 2.$$

Likewise, if $h < 0$, $(2 + h) < 2$ and so, $f(2 + h) = 4$. Thus, we have

$$\lim_{h \to 0^-} \frac{f(2 + h) - f(2)}{h} = \lim_{h \to 0^-} \frac{4 - 4}{h} = 0.$$

Since the one-sided limits do not agree ($0 \neq 2$), $f'(2)$ does not exist (i.e., f is not differentiable at $x = 2$).

∎

The following graphs show a variety of functions for which $f'(0)$ does not exist. In Figure 2.20a, we have a jump discontinuity, in Figure 2.20b, a vertical asymptote, in Figure 2.20c, a cusp and, in figure 2.20d, a vertical tangent.

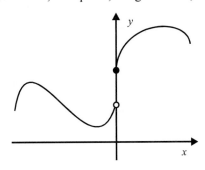

FIGURE 2.20a: A jump discontinuity.

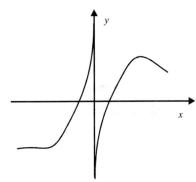

FIGURE 2.20b: A vertical asymptote.

FIGURE 2.20c: A cusp.

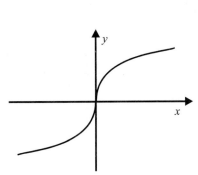

FIGURE 2.20d: A vertical tangent line.

Numerical Differentiation

In the next few sections, we develop rules to make it easy to compute the derivative of nearly any function you can write down. However, there are many times in applications when it is not possible or practical to compute derivatives symbolically. One reason for this might be a function for which no symbolic representation is available. This is frequently the case in applications where we have only some data (i.e., a table of values) representing an **unknown** function. You will need an understanding of the limit definition to compute reasonable estimates of the derivative.

Example 2.10   Approximating a Derivative Numerically

Numerically estimate the derivative of $f(x) = x^2 \sqrt{x^3 + 2}$ at $x = 1$.

Solution   Although we will see how to compute this derivative symbolically later, neither of the authors is anxious to struggle with the limit definition for this function. The definition tells us, however, that the derivative at $x = 1$ is the limit of slopes of secant lines. We compute some of these below:

h	$\dfrac{f(1+h) - f(1)}{h}$	h	$\dfrac{f(1+h) - f(1)}{h}$
0.1	4.7632	-0.1	3.9395
0.01	4.3714	-0.01	4.2892
0.001	4.3342	-0.001	4.3260

Notice that the slopes seem to be converging to approximately 4.33 as h approaches 0. Thus, we make the approximation $f'(1) \approx 4.33$.

■

Example 2.11   Estimating Velocity Numerically

Suppose that a sprinter reaches the following distances in the given times. Estimate the velocity of the sprinter at the 6-second mark.

t (s)	5.0	5.5	5.8	5.9	6.0	6.1	6.2	6.5	7.0
$f(t)$ (ft)	123.7	141.01	151.41	154.90	158.40	161.92	165.42	175.85	193.1

Time Interval	Average Velocity
$(5.9, 6.0)$	35.0 ft/s
$(6.0, 6.1)$	35.2 ft/s

Time Interval	Average Velocity
$(5.5, 6.0)$	34.78 ft/s
$(5.8, 6.0)$	34.95 ft/s
$(5.9, 6.0)$	35.00 ft/s
$(6.0, 6.1)$	35.20 ft/s
$(6.0, 6.2)$	35.10 ft/s
$(6.0, 6.5)$	34.90 ft/s

Solution   The instantaneous velocity is the limit of the average velocity as the time interval shrinks. We might first compute the average velocities over the shortest intervals given, from 5.9 to 6.0 and from 6.0 to 6.1.

Since these are the best individual estimates available from the data, we could just split the difference and estimate a velocity of 35.1 ft/s. However, there is useful information in the rest of the data. Let's look at more average velocities.

Based on the accompanying table, we can conjecture that the sprinter was reaching a peak speed at about the 6-second mark. Thus, we might accept the higher estimate of 35.2 ft/s. We should emphasize that there is not a single correct answer to this question, since the data is incomplete (i.e., we only know the distance at discrete times, rather than over a continuum of times).

■

Exercises 2.2

1. (writing) The derivative is important because of its many different uses and interpretations. Describe four aspects of the derivative: graphical (think of tangent lines), symbolic (the derivative function), numerical (approximations) and applications (velocity and others).

2. (writing) Mathematicians often use the word "smooth" to describe functions with certain (desirable) properties. Graphically, how are differentiable functions smoother than functions that are continuous but not differentiable, or functions which are not continuous?

3. (writing) Briefly describe what the derivative tells you about the original function. In particular, if the derivative is positive at a point, what do you know about the trend of the function at that point? What is different if the derivative is negative at the point?

4. (writing) In example 2.3, we found that the derivative of $f(x) = 3x - 5$ is $f'(x) = 3$. Explain in terms of slope why this is true.

In exercises 5–10, compute $f'(a)$.

5. $f(x) = 3x + 1$, $a = 1$ **6.** $f(x) = 3x + 1$, $a = 3$

7. $f(x) = 3x^2 + 1$, $a = 1$ **8.** $f(x) = 3x^2 + 1$, $a = 3$

9. $f(x) = \sqrt{3x + 1}$, $a = 1$

10. $f(x) = 3/(x + 1)$, $a = 2$

In exercises 11–20, compute the derivative function $f'(x)$.

11. $f(x) = 3x + 1$ **12.** $f(x) = -2x + 3$

13. $f(x) = 3x^2 + 1$ **14.** $f(x) = x^2 - 2x + 1$

15. $f(x) = \dfrac{3}{x + 1}$ **16.** $f(x) = \dfrac{2}{2x - 1}$

17. $f(x) = \sqrt{3x + 1}$ **18.** $f(x) = \sqrt{2x + 3}$

19. $f(x) = x^3 + 2x - 1$ **20.** $f(x) = x^4 - 2x^2 + 1$

In exercises 21–26, match the graphs of the functions on the left with the graphs of their derivatives on the right.

22.   (b)

23.   (c)

24.   (d)

25.   (e)

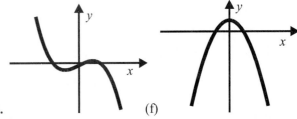

26.   (f)

In exercises 27–30, use the given graph of $f(x)$ to sketch a graph of $f'(x)$.

21.   (a)

27.   **28.**

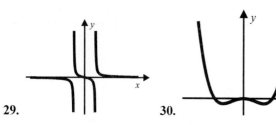

29.   **30.**

In exercises 31 and 32, use the given graph of $f'(x)$ to sketch a plausible graph of $f(x)$.

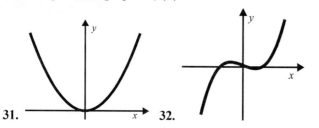

31.   **32.**

In exercises 33 and 34, identify all points at which the function is not differentiable.

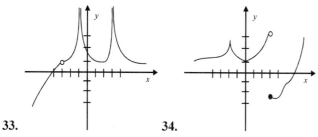

33.   **34.**

In exercises 35 and 36, estimate the slope of the tangent line at $x = 1$.

35.

x	0.2	0.4	0.6	0.8	1.0	1.2	1.4	1.6	1.8
$f(x)$	2.4	3.1	3.9	4.8	5.8	6.8	7.7	8.5	9.2

36.

x	0.2	0.4	0.6	0.8	1.0	1.2	1.4	1.6	1.8
$f(x)$	4.0	4.6	5.3	6.1	7.0	7.8	8.6	9.3	9.9

In exercises 37 and 38, use the distances $f(t)$ to estimate the velocity at $t = 2$.

37.

t	1.6	1.7	1.8	1.9	2.0	2.1	2.2	2.3	2.4
$f(t)$	2.4	3.1	3.9	4.8	5.8	6.8	7.7	8.5	9.2

38.

t	1.6	1.7	1.8	1.9	2.0	2.1	2.2	2.3	2.4
$f(t)$	4.0	4.6	5.3	6.1	7.0	7.8	8.6	9.3	9.9

39. If $f(x) = x^{2/3}$, show graphically and numerically that $f(x)$ is continuous at $x = 0$ but $f'(0)$ does not exist.

40. If $f(x) = \begin{cases} 0 & \text{if } x < 0 \\ 2x & \text{if } x \geq 0 \end{cases}$, show graphically and numerically that $f(x)$ is continuous at $x = 0$ but $f'(0)$ does not exist.

In exercises 41 and 42, compute the right-hand derivative
$$D_+ f(0) = \lim_{h \to 0^+} \frac{f(h) - f(0)}{h} \text{ and the left-hand derivative}$$
$$D_- f(0) = \lim_{h \to 0^-} \frac{f(h) - f(0)}{h}.$$

41. $f(x) = \begin{cases} 2x + 1 & \text{if } x < 0 \\ 3x + 1 & \text{if } x \geq 0 \end{cases}$

42. $f(x) = \begin{cases} x^2 & \text{if } x < 0 \\ x^3 & \text{if } x \geq 0 \end{cases}$

43. Assume that $f(x) = \begin{cases} g(x) & \text{if } x < 0 \\ k(x) & \text{if } x \geq 0 \end{cases}$. If $f(x)$ is continuous at $x = 0$ and $g(x)$ and $k(x)$ are differentiable at $x = 0$, prove that $D_+ f(0) = k'(0)$ and $D_- f(0) = g'(0)$. Which statement is not true if $f(x)$ has a jump discontinuity at $x = 0$?

44. Explain why the derivative $f'(0)$ exists if and only if the one-sided derivatives exist and are equal.

45. If $f'(x) > 0$ for all x, use the tangent line interpretation to argue that $f(x)$ is an **increasing function**; that is, if $a < b$, then $f(a) < f(b)$.

46. If $f'(x) < 0$ for all x, use the tangent line interpretation to argue that $f(x)$ is a **decreasing function**; that is, if $a < b$, then $f(a) > f(b)$.

47. One model for the spread of a disease is the following: at first the disease spreads very slowly, gradually the infection rate increases to a maximum, then the infection rate decreases back to zero, marking the end of the epidemic. If $I(t)$ represents the number of people infected at time t, sketch a graph of both $I(t)$ and $I'(t)$, assuming that those who get infected do not recover.

48. One model for urban population growth is as follows: at first, the population is growing very rapidly, then the growth rate decreases until the population starts decreasing. If $P(t)$ is the population at time t, sketch a graph of both $P(t)$ and $P'(t)$.

49. Give an example showing that the following is not always true: if $f(x) \leq x$, then $f'(x) \leq 1$.

50. Determine if the following is always true: if $f(0) = 0$, $f'(x)$ exists for all x and $f(x) \leq x$, then $f'(x) \leq 1$.

51. Psychologists study the way humans think by analyzing the results of experiments like the following. If you take the shape $\lceil$ and rotate it, does it rotate into the shape $\lfloor$ or the shape $\rfloor$? To which shape does $\neg$ rotate? Research indicates that the length of time to answer such questions is proportional to the angle the given figure needs to rotate to reach a target figure. The conclusion is that we actually mentally rotate the given figure. (Did you do this consciously?) Pinker measures the speed of rotation as $11°$ per microsecond (0.001 second). Given this rate, how long should it take to correctly identify $\lceil$? $\neg$? Did you find $\neg$ easier to identify?

52. (exploratory) Compute the derivative function for x^2, x^3 and x^4. Based on your results, identify the pattern and conjecture a general formula for the derivative of x^n. Test your conjecture on the functions $\sqrt{x} = x^{1/2}$ and $1/x = x^{-1}$.

53. (exploratory) One of the important uses of the derivative is to obtain **linear approximations**. That is, the tangent line is a line that approximates the function near the point of tangency. As an example of its usefulness, how would you compute $\sqrt{1.24}$ without using a calculator? The answer is obviously slightly larger than 1, but how much larger? Find an equation of the tangent line to $y = \sqrt{x}$ at $x = 1$ and then compute the y-value of the tangent line at $x = 1.24$. For numbers close to 1, describe a method of approximating the square root without using a calculator. (We see more of this method in section 3.1.)

54. (exploratory) Suppose there is a function $F(x)$ such that $F(1) = 1$ and $F(0) = f_0$, where $0 < f_0 < 1$. If $F'(1) > 1$, show graphically that the equation $F(x) = x$ has a solution q where $0 < q < 1$. (Hint: Graph $y = x$ and a plausible $F(x)$

and look for intersections.) Sketch a graph where $F'(1) < 1$ and there are no solutions to the equation $F(x) = x$ between 0 and 1 (although $x = 1$ is a solution). Solutions have a connection with the probability of the extinction of animals or family names. Suppose you and your descendants have children according to the following probabilities: $f_0 = 0.2$ is the probability of having no children, $f_1 = 0.3$ is the probability of having exactly one child, and $f_2 = 0.5$ is the probability of having two children. Define $F(x) = 0.2 + 0.3x + 0.5x^2$ and show that $F'(1) > 1$. Find the solution of $F(x) = x$ between $x = 0$ and $x = 1$; this number is the probability that your "line" will go extinct some time into the future. Find nonzero values of f_0, f_1 and f_2 such that the corresponding $F(x)$ satisfies $F'(1) < 1$ and hence the probability of your line going extinct is 1.

2.3   Computation of Derivatives: The Power Rule

You have now computed numerous derivatives using the limit definition. In fact, you may have computed enough that you have started taking some shortcuts. In exercise 52 in section 2.2, we asked you to compile a list of derivatives of basic functions and to generalize. We continue that process in this section.

The Power Rule

We first revisit the limit definition of derivative to compute two very simple derivatives.

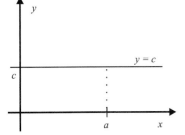

FIGURE 2.21: A horizontal line.

Theorem 3.1   For any constant c, $\dfrac{d}{dx}c = 0$.

Note that the theorem says that for any constant c, the horizontal line $y = c$ has a tangent line with zero slope. That is, the tangent line to a horizontal line is the same horizontal line (see Figure 2.21).

Proof. Let $f(x) = c$, for all x. From the definition (2.3), we have

$$\frac{d}{dx}c = f'(x) = \lim_{h \to 0} \frac{f(x+h) - f(x)}{h}$$

$$= \lim_{h \to 0} \frac{c - c}{h} = \lim_{h \to 0} 0 = 0.$$

■

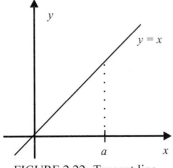

FIGURE 2.22: Tangent line to $y = x$.

Theorem 3.2 $\dfrac{d}{dx}x = 1$.

Notice that the theorem says that the tangent line to the line $y = x$ is a line of slope one (i.e., $y = x$; see Figure 2.22). This is unsurprising, since intuitively, it should be clear that the tangent line to any line is that same line.

Proof. Let $f(x) = x$. From the definition (2.3), we have

$$\frac{d}{dx}x = f'(x) = \lim_{h \to 0} \frac{f(x+h) - f(x)}{h}$$

$$= \lim_{h \to 0} \frac{(x+h) - x}{h}$$

$$= \lim_{h \to 0} \frac{h}{h} = \lim_{h \to 0} 1 = 1.$$

$f(x)$	$f'(x)$
1	0
x	1
x^2	$2x$
x^3	$3x^2$
x^4	$4x^3$

To the left is a short list of derivatives calculated previously either as examples or in the exercises using the limit definition. Can you identify any pattern to the derivatives shown in the table? There are two features to note. First, the power of x in the derivative is always one less than the power of x in the original function. Second, the coefficient of x in the derivative is equal to the power of x in the original function. Putting these ideas into symbolic form, we make and then prove a reasonable conjecture.

Theorem 3.3   (Power Rule) For any integer $n > 0$,

$$\frac{d}{dx}x^n = nx^{n-1}.$$

Proof. From the limit definition (2.3), if $f(x) = x^n$, then

$$\frac{d}{dx}x^n = f'(x) = \lim_{h \to 0} \frac{f(x+h) - f(x)}{h} = \lim_{h \to 0} \frac{(x+h)^n - x^n}{h}. \qquad (3.1)$$

To evaluate the limit, we will need to simplify the expression in the numerator. If n is a positive integer, we can multiply out $(x+h)^n$. Recall that $(x+h)^2 = x^2 + 2xh + h^2$ and $(x+h)^3 = x^3 + 3x^2h + 3xh^2 + h^3$. More generally, you may recall from the binomial theorem that

$$(x+h)^n = x^n + nx^{n-1}h + \frac{n(n-1)}{2}x^{n-2}h^2 + \cdots + nxh^{n-1} + h^n. \qquad (3.2)$$

Substituting (3.2) into (3.1), we get

$$f'(x) = \lim_{h \to 0} \frac{x^n + nx^{n-1}h + \frac{n(n-1)}{2}x^{n-2}h^2 + \cdots + nxh^{n-1} + h^n - x^n}{h}$$

Cancel two x^n terms.

$$= \lim_{h \to 0} \frac{nx^{n-1}h + \frac{n(n-1)}{2}x^{n-2}h^2 + \cdots + nxh^{n-1} + h^n}{h}$$

Factor out common h and cancel.

$$= \lim_{h \to 0} \frac{h\left[nx^{n-1} + \frac{n(n-1)}{2}x^{n-2}h^1 + \cdots + nxh^{n-2} + h^{n-1}\right]}{h}$$

Since every term but the first has a factor of h.

$$= \lim_{h \to 0} \left[nx^{n-1} + \frac{n(n-1)}{2}x^{n-2}h^1 + \cdots + nxh^{n-2} + h^{n-1}\right] = nx^{n-1}.$$

The power rule is very easy to apply, as in the following example.

Example 3.1   Using the Power Rule

Find the derivative of x^8 and t^{107}.

Solution   We have

$$\frac{d}{dx}x^8 = 8x^{8-1} = 8x^7.$$

Similarly,

$$\frac{d}{dt}t^{107} = 107\,t^{107-1} = 107\,t^{106}.$$

■

Recall that in section 2.2, we showed that

$$\frac{d}{dx}\left(\frac{1}{x}\right) = -\frac{1}{x^2}. \tag{3.3}$$

Notice that we could rewrite (3.3) as

$$\frac{d}{dx}x^{-1} = (-1)\,x^{-2}.$$

That is, the derivative of x^{-1} follows the same pattern as the power rule which we just stated and proved for **positive** integer exponents.

Likewise, in section 2.2, we used the limit definition to show that

$$\frac{d}{dx}\sqrt{x} = \frac{1}{2\sqrt{x}}. \tag{3.4}$$

We can also rewrite (3.4) as

$$\frac{d}{dx}x^{1/2} = \frac{1}{2}x^{-1/2}.$$

Here, notice that the derivative of a rational power of x also follows the same pattern as the power rule that we proved for positive **integer** exponents.

Remark 3.1   As we will see, the power rule is true for **any** power of x. Although we will not be able to prove this fact for some time now (the proof of Theorem 3.3 does **not** generalize because the binomial theorem only holds for positive integer exponents!), we will use the rule freely for any power of x. We have

---

Theorem 3.4   (General Power Rule) For any real number r,

$$\frac{d}{dx}x^r = rx^{r-1}. \tag{3.5}$$

The power rule is the simplest of derivative rules to apply, as we see in the following example.

Example 3.2   Using the General Power Rule

Find the derivative of $\dfrac{1}{x^{19}}$, $\sqrt[3]{x^2}$ and x^{π}.

Solution   From (3.5), we have

$$\frac{d}{dx}\left(\frac{1}{x^{19}}\right) = \frac{d}{dx}x^{-19} = -19\,x^{-19-1} = -19\,x^{-20}.$$

Be careful here to avoid a common error:

$$\frac{d}{dx}x^{-19} \neq -19x^{-18}.$$

The power rule says to *subtract* 1 from the exponent (even if the exponent is negative).

If we rewrite $\sqrt[3]{x^2}$ as a fractional power of x, we can use (3.5) to compute the derivative, as follows.

$$\frac{d}{dx}\sqrt[3]{x^2} = \frac{d}{dx}x^{2/3} = \frac{2}{3}x^{2/3-1} = \frac{2}{3}x^{-1/3}.$$

Finally, we have

$$\frac{d}{dx}x^{\pi} = \pi x^{\pi-1}.$$

Notice that here there is the additional conceptual problem (which we resolve in Chapter 6) of deciding what x^{π} means. Since the exponent isn't rational, what exactly do we mean when we raise a number to the irrational power π?

■

General Derivative Rules

The power rule has given us a large class of functions whose derivatives we can quickly compute without using the limit definition. The following rules for combining derivatives further expand the number of derivatives we can compute without resorting to the definition.

Theorem 3.5   If $f(x)$ and $g(x)$ are differentiable at x and c is any constant, then

(i) $\dfrac{d}{dx}[f(x) + g(x)] = f'(x) + g'(x)$,

(ii) $\dfrac{d}{dx}[f(x) - g(x)] = f'(x) - g'(x)$ and

(iii) $\dfrac{d}{dx}[c\,f(x)] = c\,f'(x)$.

Proof. We prove only part (i). The proofs of parts (ii) and (iii) are left as exercises. Let $k(x) = f(x) + g(x)$. Then, from the limit definition of the derivative (2.3), we get

By definition of $k(x)$.

Grouping the f terms together and the g terms together.

Recognizing the derivative of f and the derivative of g.

$$\frac{d}{dx}[f(x) + g(x)] = k'(x) = \lim_{h \to 0}\frac{k(x+h) - k(x)}{h}$$

$$= \lim_{h \to 0}\frac{[f(x+h) + g(x+h)] - [f(x) + g(x)]}{h}$$

$$= \lim_{h \to 0}\frac{[f(x+h) - f(x)] + [g(x+h) - g(x)]}{h}$$

$$= \lim_{h \to 0}\frac{f(x+h) - f(x)}{h} + \lim_{h \to 0}\frac{g(x+h) - g(x)}{h}$$

$$= f'(x) + g'(x).$$

■

We illustrate Theorem 3.5 by working through the calculation of a derivative step by step, showing all of the details.

Example 3.3   **Finding the Derivative of a Sum**
Find the derivative of $f(x) = 2x^6 + 3\sqrt{x}$.

Solution   We have

By Theorem 3.5 (i).

By Theorem 3.5 (iii).

By the power rule.

Multiply the constants.

$$f'(x) = \frac{d}{dx}(2x^6) + \frac{d}{dx}(3\sqrt{x})$$

$$= 2\frac{d}{dx}(x^6) + 3\frac{d}{dx}(x^{1/2})$$

$$= 2(6x^5) + 3\left(\frac{1}{2}x^{-1/2}\right)$$

$$= 12x^5 + \left(\frac{3}{2\sqrt{x}}\right).$$

∎

Example 3.4   **Rewriting a Function before Computing the Derivative**

Find the derivative of $f(x) = \dfrac{4x^2 - 3x + 2\sqrt{x}}{x}$.

Solution   Note that we don't yet have any rule for computing the derivative of a quotient. So, we must first rewrite $f(x)$ by dividing out the x in the denominator. We have

$$f(x) = \frac{4x^2}{x} - \frac{3x}{x} + \frac{2\sqrt{x}}{x} = 4x - 3 + 2x^{-1/2}.$$

From Theorem 3.5 and the power rule (3.5), we get

$$f'(x) = 4\frac{d}{dx}(x) - 3\frac{d}{dx}(1) + 2\frac{d}{dx}(x^{-1/2}) = 4 - 0 + 2(-\frac{1}{2}x^{-3/2}) = 4 - x^{-3/2}.$$

∎

Example 3.5   **Finding the Equation of the Tangent Line**

Find an equation of the tangent line to $f(x) = 3 - 4x + \dfrac{2}{x}$ at $x = 1$.

Solution   First, notice that $f(x) = 3 - 4x + 2x^{-1}$. From Theorem 3.5 and the power rule (3.5), we have

$$f'(x) = 0 - 4 - 2x^{-2}.$$

At $x = 1$, the slope of the tangent line is then $f'(1) = -4 - 2 = -6$. The line with slope -6 through the point $(1, 1)$ has equation

$$y - 1 = -6(x - 1).$$

∎

Higher Order Derivatives

Notice that one consequence of having the derivative function is that we can compute the derivative of the derivative. It turns out that such **higher order** derivatives have important applications.

Suppose we start with a function $f(x)$ and compute its derivative $f'(x)$. We can then compute the derivative of $f'(x)$, called the **second derivative** of f and written $f''(x)$. We can then compute the derivative of $f''(x)$, called the **third derivative** of f, written $f'''(x)$. We can continue to take derivatives indefinitely. Below, we show common notations for the first five derivatives of f [where we assume that $y = f(x)$].

Order	Prime notation	Leibniz notation
1	$y' = f'(x)$	$\dfrac{df}{dx}(x)$
2	$y'' = f''(x)$	$\dfrac{d^2 f}{dx^2}(x)$
3	$y''' = f'''(x)$	$\dfrac{d^3 f}{dx^3}(x)$
4	$y^{(4)} = f^{(4)}(x)$	$\dfrac{d^4 f}{dx^4}(x)$
5	$y^{(5)} = f^{(5)}(x)$	$\dfrac{d^5 f}{dx^5}(x)$

Computing higher order derivatives is no different from computing a first derivative, as we see in the following example.

Example 3.6   **Computing Higher Order Derivatives**

If $f(x) = 3x^4 - 2x^2 + 1$, compute as many derivatives as possible.

Solution   We have

$$f'(x) = \frac{df}{dx} = \frac{d}{dx}(3x^4 - 2x^2 + 1) = 12x^3 - 4x.$$

Then,

$$f''(x) = \frac{d^2 f}{dx^2} = \frac{d}{dx}(12x^3 - 4x) = 36x^2,$$

$$f'''(x) = \frac{d^3 f}{dx^3}(x) = \frac{d}{dx}(36x^2) = 72x,$$

$$f^{(4)}(x) = \frac{d^4 f}{dx^4} = \frac{d}{dx}(72x) = 72,$$

$$f^{(5)}(x) = \frac{d^5 f}{dx^5} = \frac{d}{dx}(72) = 0,$$

and so on. We now have

$$f^{(n)}(x) = \frac{d^n f}{dx^n} = 0, \text{ for } n \geq 5.$$

∎

Acceleration

What information does the second derivative of a function give us? Graphically, we get a property called **concavity,** which we develop in Chapter 3. One important application of the second derivative is acceleration, which we briefly discuss now.

Acceleration is probably a word that is familiar to you. Unfortunately, common experience with your car's accelerator pedal and sports jargon may give a somewhat distorted idea of what acceleration is. As a calculus student, you can define it simply: **acceleration is the derivative of velocity**. If the velocity of an object at time t is given by $v(t)$, then the acceleration is

$$a(t) = v'(t) = \frac{dv}{dt}.$$

Since derivatives are always rates of change, a longer definition is: acceleration is the instantaneous rate of change of velocity.

Example 3.7   **Computing the Acceleration of a Skydiver**

Suppose that the height of a skydiver t seconds after jumping from an airplane is given by $f(t) = 640 - 20t - 16t^2$ feet. Find the person's acceleration at time t.

Solution   Since acceleration is the derivative of velocity, we first compute velocity:

$$v(t) = f'(t) = 0 - 20 - 32t = -20 - 32t \text{ ft/s}.$$

Computing the derivative of this function gives us

$$a(t) = v'(t) = -32.$$

To finish the problem, we need to determine the units of acceleration. We have measured velocity here in terms of feet per second and the time variable is measured in seconds, so the units of the derivative are feet per second per second. This could be written ft/s/s, but it is more common to use the shorthand ft/s^2 (feet per second squared). Our answer says that the velocity changes by -32 ft/s every second. In this case, the speed in the downward (negative) direction increases by 32 ft/s every second due to gravity.

■

Exercises 2.3

1. (writing) Explain to a non-calculus-speaking friend how to (mechanically) use the power rule. Decide if it is better to give separate explanations for positive and negative exponents; integer and noninteger exponents; other special cases.

2. (writing) In the 1700s, mathematical "proofs" were, by modern standards, a bit fuzzy and lacked rigor. In 1734, the Irish metaphysician Bishop Berkeley wrote *The Analyst* to an "infidel mathematician" (thought to be Edmund Halley of Halley's comet fame). The accepted proof at the time of the power rule may be described as follows.

If x is incremented to $x + h$, then x^n is incremented to $(x + h)^n$. It follows that $\dfrac{(x+h)^n - x^n}{(x+h) - x} = nx^{n-1} + \dfrac{n^2 - n}{2}hx^{n-2} + \cdots$. Now, let the increment h vanish, and the derivative is nx^{n-1}.

Bishop Berkeley objected to this argument.

"But it should seem that the reasoning is not fair or conclusive. For when it is said, let the increments vanish, the former supposition that the increments were something, or that there were increments, is destroyed, and yet a consequence of that supposition is retained. Which . . . is a false way of reasoning. Certainly when we suppose the increments to vanish, we must suppose . . . everything derived from the supposition of their existence to vanish with them."

Do you think Berkeley's objection is fair? Is it logically acceptable to assume something exists to draw one deduction, and then assume the something does not exist to avoid having to accept other consequences? For example, can you use the principle of "always tell the truth" to get a friend to tell you something, but then get mad when part of the truth is unflattering to you? Mathematically speaking, how does the limit avoid Berkeley's objection of the increment h both existing and not existing?

3. (writing) The historical episode in exercise 2 is just one part of an ongoing conflict between people who blindly use mathematical techniques without proof and those who insist on a full proof before permitting anyone to use the technique. Which side are you sympathetic to? Defend your position in an essay. Try to anticipate and rebut the other side's arguments.

4. (writing) Explain the first two terms in the expansion $(x + h)^n = x^n + nhx^{n-1} + \cdots$ where n is a positive integer. Think of multiplying out $(x + h)(x + h)(x + h) \cdots (x + h)$; how many terms would include x^n? x^{n-1}?

In exercises 5–24, find the derivative of each function.

5. $f(x) = x^3 - 2x + 1$ **6.** $f(x) = 2x^4 - 3x^2 + x$

7. $f(x) = 3x^2 - 4$ **8.** $f(x) = x^9 - 3x^5 + 4x^2 + x$

9. $g(x) = 4$ **10.** $g(x) = -4x$

11. $f(t) = 3t^3 - 2\sqrt{t}$

12. $f(s) = 5\sqrt{s} - 4s^2 + 3$

13. $f(x) = \dfrac{3}{x} - 8x + 1$

14. $f(x) = \dfrac{2}{x^4} - x^3 + 2$

15. $h(x) = \dfrac{10}{\sqrt{x}} - 2x$

16. $h(x) = 12x - x^2 - \dfrac{3}{\sqrt{x}}$

17. $f(s) = 2s^{3/2} - 3s^{-1/3}$

18. $f(t) = 3t^\pi - 2t^{1.3}$

19. $f(x) = 2\sqrt[3]{x} + 3$

20. $f(x) = 4x - 3\sqrt[3]{x^2}$

21. $f(x) = x(3x^2 - \sqrt{x})$

22. $f(x) = (x + 1)(3x^2 - 4)$

23. $f(x) = \dfrac{3x^2 - 3x + 1}{2x}$

24. $f(x) = \dfrac{4x^2 - x + 3}{\sqrt{x}}$

In exercises 25–34, compute the indicated derivative.

25. $f''(x)$ for $f(x) = x^4 + 3x^2 - 2$

26. $f''(x)$ for $f(x) = x^3 - 6x + 2/x$

27. $\dfrac{d^2 f}{dx^2}$ for $f(x) = x^6 - \sqrt{x}$

28. $\dfrac{d^2 f}{dx^2}$ for $f(x) = 2x^4 - 3/\sqrt{x}$

29. $f'''(t)$ for $f(t) = 4t^2 - 12 + 4/t^2$

30. $f'''(t)$ for $f(t) = t^6 - 3t^3 + 2$

31. $f^{(4)}(x)$ for $f(x) = x^4 + 3x^2 - 2$

32. $f^{(5)}(x)$ for $f(x) = x^{10} - 3x^4 + 2x - 1$

33. $f'''(x)$ for $f(x) = \dfrac{x^2 - x + 1}{\sqrt{x}}$

34. $f^{(4)}(t)$ for $f(t) = (t^2 - 1)(\sqrt{t} + t)$

In exercises 35–38, use the given position function to find the velocity and acceleration functions.

35. $s(t) = -16t^2 + 40t + 10$

36. $s(t) = 12t^3 - 6t - 1$

37. $s(t) = \sqrt{t} + 2t$

38. $s(t) = 10 - \dfrac{10}{t}$

In exercises 39–42, the given function represents the height of an object. Compute the velocity and acceleration at time $t = a$. Is the object going up or down? Is the object speeding up or slowing down?

39. $h(t) = -16t^2 + 40t + 5$, $a = 1$

40. $h(t) = -16t^2 + 40t + 5$, $a = 2$

41. $h(t) = 10t^2 - 24t$, $a = 2$

42. $h(t) = 10t^2 - 24t$, $a = 1$

In exercises 43–46, find an equation of the tangent line to $y = f(x)$ at $x = a$.

43. $f(x) = 4\sqrt{x} - 2x$, $a = 1$

44. $f(x) = x^2 - 2x + 1$, $a = 2$

45. $f(x) = x^2 - 2$, $a = 0$

46. $f(x) = 3x + 4$, $a = 2$

In exercises 47 and 48, determine the value(s) of x for which the tangent line to $y = f(x)$ is horizontal. Graph the function and determine the graphical significance of each point.

47. $f(x) = x^3 - 3x + 1$

48. $f(x) = x^4 - 2x^2 + 2$

In exercises 49 and 50, determine the value(s) of x for which the slope of the tangent line to $y = f(x)$ does not exist. Graph the function and determine the graphical significance of each point.

49. $f(x) = x^{2/3}$

50. $f(x) = x^{1/3}$

In exercises 51 and 52, one curve represents a function $f(x)$ and the other two represent $f'(x)$ and $f''(x)$. Determine which is which.

51. (a)   **(b)**

(c)

52. (a)   **(b)**

(c)

In exercises 53 and 54, find a general formula for the nth derivative $f^{(n)}(x)$.

53. $f(x) = \sqrt{x}$ **54.** $f(x) = 2/x$

55. Find a second-degree polynomial (of the form $ax^2 + bx + c$) such that $f(0) = -2$, $f'(0) = 2$ and $f''(0) = 3$.

56. Find a second-degree polynomial (of the form $ax^2 + bx + c$) such that $f(0) = 0$, $f'(0) = 5$ and $f''(0) = 1$.

57. Show that the formula $\dfrac{d}{dx}(1) = 0$ is a special case of the power rule. Explain the more general formula $\dfrac{d}{dx}(c) = 0$ graphically.

58. Show that the formula $\dfrac{d}{dx}(x) = 1$ is a special case of the power rule. Explain the more general formula $\dfrac{d}{dx}(cx) = c$ graphically.

59. A public official solemnly proclaims, "We have achieved a reduction in the rate at which the national debt is increasing." If $d(t)$ represents the national debt, which derivative of $d(t)$ is being reduced? What can you conclude about the size of $d(t)$ itself?

60. A rod made of an inhomogeneous material extends from $x = 0$ to $x = 4$. The mass of the portion of the rod from $x = 0$ to $x = t$ is given by $m(t) = 3t^2$ kg. Compute $m'(t)$ and explain why it represents the density of the rod.

61. For most land animals, the relationship between leg width w and body length b follows an equation of the form $w = c \times b^{3/2}$ for some constant c. Show that if b is large enough, $w'(b) > 1$. Conclude that for larger animals, leg width (necessary for support) increases faster than body length. Why does this put a limitation on the size of land animals?

62. Suppose the function $v(d)$ represents the average speed in m/s of the world record running time for d meters. For example, if the fastest 200-meter time ever is 19.32 s, then $v(200) = 200/19.32 \approx 10.35$. Compare the function $f(d) = 26.7d^{-0.177}$ to the values of $v(d)$, which you will have to research and compute, for distances ranging from $d = 400$ to $d = 2000$. Explain what $v'(d)$ would represent.

In exercises 63–70, find a function with the given derivative.

63. $f'(x) = 4x^3$ **64.** $f'(x) = 5x^4$

65. $f'(x) = x^4$ **66.** $f'(x) = x^5$

67. $f'(x) = 3x - 1$ **68.** $f'(x) = 4 - x^2$

69. $f'(x) = \sqrt{x}$ **70.** $f'(x) = \dfrac{1}{x^2}$

71. (exploratory) A plane is cruising at an altitude of 2 miles at a distance of 10 miles from an airport. Choosing the airport to be at the point $(0, 0)$, the plane starts at $(10, 2)$. The plane starts its descent and lands at the airport. Sketch a picture of a reasonable flight path $y = f(x)$. (Think about it as you draw!) Explain what the derivative $f'(x)$ represents. (Hint: It's not velocity.) Explain why it is important and/or necessary to have $f(0) = 0$, $f(10) = 2$, $f'(0) = 0$ and $f'(10) = 0$. The simplest polynomial that can meet these requirements is a cubic polynomial $f(x) = ax^3 + bx^2 + cx + d$ (note: four requirements, four constants). Find values of the constants a, b, c and d to fit the flight path. (Hint: Start by setting $f(0) = 0$ and then set $f'(0) = 0$.) Graph the resulting function; does it look right? Suppose that airline regulations prohibit a derivative of $\frac{2}{10}$ or larger. Why might such a regulation exist? Show that the flight path you found is illegal. Argue that in fact all flight paths meeting the four requirements are illegal. Therefore, the descent needs to start further away than 10 miles. Find a flight path with descent starting 20 miles that meets all requirements.

72. (exploratory) We discuss a graphical interpretation of the second derivative in Chapter 3. You can discover the most important aspects of that here. For $f(x) = x^4 - 2x^2 - 1$, solve the equations $f'(x) = 0$ and $f''(x) = 0$. What do the solutions of the equation $f'(x) = 0$ represent graphically? The solutions of the equation $f''(x) = 0$ are a little harder to interpret. Looking at the graph of $f(x)$ near $x = 0$, would you say that the graph is curving up or curving down? Compute $f''(0)$. Looking at the graph near $x = 2$ and $x = -2$, is the graph curving up or down? Compute $f''(2)$ and $f''(-2)$. Where does the graph change from curving up to curving down and vice versa? Hypothesize a relationship between $f''(x)$ and the curving of the graph of $y = f(x)$. Test your hypothesis on a variety of functions. (Try $y = x^4 - 4x^3$.)

73. (exploratory) In the enjoyable book *Surely You're Joking Mr. Feynman*, physicist Richard Feynman tells the story of a contest he had pitting his brain against the technology of the day (an abacus). The contest was to compute the cube root of 1729.03. Feynman came up with 12.002 before the abacus expert gave up. Feynman admits to some luck in the choice of the number 1729.03: he knew that a cubic foot contains 1728 cubic inches. Explain why this told Feynman that the answer is slightly greater than 12. How did he get three digits of accuracy? "I had learned in calculus that for small fractions, the cube root's excess is one-third of the number's excess. The excess, 1.03, is only one part in nearly 2000. So all I had to do is find the fraction 1/1728, divide by 3 and multiply by 12." To see what he did, find an equation of the tangent line to $y = x^{1/3}$ at $x = 1728$ and find the y-coordinate of the tangent line at $x = 1729.03$.

74. (exploratory) Suppose that you want to find solutions of the equation $x^3 - 4x^2 + 2 = 0$. Show graphically that there is a solution between $x = 0$ and $x = 1$. We will approximate this solution in stages. First, find an equation of the tangent line to $y = x^3 - 4x^2 + 2$ at $x = 1$. Then, determine where this tangent line crosses the x-axis. Show graphically that the x-intercept is considerably closer to the solution than is $x = 1$. Now, repeat the process: for the new x-value, find the equation of the tangent line, determine where it crosses the x-axis and show that this is closer still to the desired solution. This process of using tangent lines to produce continually improved approximations is referred to as **Newton's method**. We discuss this in some detail in section 3.1.

2.4   The Product and Quotient Rules

We have now seen how to quickly compute derivatives of a variety of functions. Given the general formulas we developed for the derivative of a sum or difference of two functions, you might expect that there would be corresponding formulas for the derivative of a product or quotient. In particular, you might wonder whether the derivative of a product of two functions is the same as the product of the derivatives. We test this conjecture with a simple example.

Product Rule

Consider $\frac{d}{dx}\left[(x^2)(x^5)\right]$. We already know how to compute this derivative by first combining the two terms:

$$\frac{d}{dx}\left[(x^2)(x^5)\right] = \frac{d}{dx}x^7 = 7x^6.$$

Is this derivative the same as the product of the two individual derivatives? Notice that

$$\left(\frac{d}{dx}x^2\right)\left(\frac{d}{dx}x^5\right) = (2x)(5x^4)$$

$$= 10x^5 \neq 7x^6 = \frac{d}{dx}\left[(x^2)(x^5)\right]. \tag{4.1}$$

You can now plainly see from (4.1) that the derivative of a product is **not** the product of the respective derivatives. As it turns out, there is a rule for computing the derivative of a product. We certainly don't need such a rule for computing derivatives like the one examined above, but we'll need this for computing a derivative like

$$\frac{d}{dx}\left(x^2 \sin x\right).$$

The only other way to do this is by using the limit definition, which would certainly strain our abilities. (In section 2.5, we work quite hard to find the derivative of $\sin x$ alone.)

The following theorem states a general rule for computing the derivative of a product of two differentiable functions.

Theorem 4.1   (Product Rule) Suppose that f and g are differentiable at x. Then

$$\frac{d}{dx}[f(x)g(x)] = f'(x)g(x) + f(x)g'(x). \tag{4.2}$$

Proof. First, realize that since we are trying to prove a general rule, we have only the limit definition of derivative to work with. For $p(x) = f(x)g(x)$, we have

$$\frac{d}{dx}[f(x)g(x)] = p'(x) = \lim_{h\to 0}\frac{p(x+h)-p(x)}{h}$$

$$= \lim_{h\to 0}\frac{f(x+h)g(x+h)-f(x)g(x)}{h}. \tag{4.3}$$

Notice that the elements of the derivatives of f and g are present (limit, $f(x+h)$, $f(x)$, etc.), but we need to get them into the right form. The trick is to add and subtract $f(x)g(x+h)$ in the numerator.

From (4.3), we have

Subtract and add $f(x)g(x+h)$.

$$p'(x) = \lim_{h \to 0} \frac{f(x+h)\,g(x+h) - f(x)\,g(x+h) + f(x)\,g(x+h) - f(x)\,g(x)}{h}$$

Break into two pieces.

$$= \lim_{h \to 0} \frac{f(x+h)\,g(x+h) - f(x)\,g(x+h)}{h} + \lim_{h \to 0} \frac{f(x)\,g(x+h) - f(x)\,g(x)}{h}$$

$$= \lim_{h \to 0} \left[\frac{f(x+h) - f(x)}{h}\,g(x+h) \right] + \lim_{h \to 0} \left[f(x)\,\frac{g(x+h) - g(x)}{h} \right]$$

Recognize the derivative of f and the derivative of g.

$$= \left[\lim_{h \to 0} \frac{f(x+h) - f(x)}{h} \right] \left[\lim_{h \to 0} g(x+h) \right] + f(x) \lim_{h \to 0} \frac{g(x+h) - g(x)}{h}$$

$$= f'(x)\,g(x) + f(x)\,g'(x).$$

In this last step, we have used the assumption that f and g are differentiable to identify the limits of the difference quotients as derivatives. There is also a subtle technical detail in the last step. Since g is differentiable at x, recall that it is also continuous at x. From this, it follows that $g(x+h) \to g(x)$ as $h \to 0$. ∎

Example 4.1   A Simple Use of the Product Rule

We pause here briefly to return to our introductory example, to illustrate the product rule. From (4.2), we have

$$\frac{d}{dx}\left[\left(x^2\right)\left(x^5\right)\right] = \left(\frac{d}{dx}x^2\right)x^5 + x^2\left(\frac{d}{dx}x^5\right)$$

$$= 2x\left(x^5\right) + x^2\left(5x^4\right)$$

$$= 2x^6 + 5x^6 = 7x^6,$$

which is exactly what we get when we combine terms and differentiate:

$$\frac{d}{dx}\left[\left(x^2\right)\left(x^5\right)\right] = \frac{d}{dx}\left(x^7\right) = 7x^6.$$

∎

Example 4.2   Using the Product Rule

Find $f'(x)$ if $f(x) = (2x^4 - 3x + 5)(x^2 - \sqrt{x} + 2/x)$.

Solution   We do have the option of multiplying the expression out or using the product rule. This should be an easy choice. Using the product rule, we get

$$f'(x) = \frac{d}{dx}(2x^4 - 3x + 5)\,(x^2 - \sqrt{x} + 2/x) + (2x^4 - 3x + 5)\,\frac{d}{dx}\left(x^2 - \sqrt{x} + 2/x\right)$$

$$= (8x^3 - 3)\,(x^2 - \sqrt{x} + 2/x) + (2x^4 - 3x + 5)\left(2x - \frac{1}{2\sqrt{x}} - \frac{2}{x^2}\right).$$

This form of the answer may bother you. The product rule usually leaves derivatives in such an "unsimplified" form. So, what is the proper way to simplify? A more basic question is: "What is the need to simplify?" The answer to this depends on what you are going to do with the derivative. For the moment, unless you have some particular reason to simplify (or some particularly obvious simplification to make), just leave the results of the product rule alone.

∎

Example 4.3   **Finding the Equation of the Tangent Line**
Find an equation of the tangent line to

$$y = (x^4 - 3x^2 + 2x)(x^3 - 2x + 3)$$

at $x = 0$.

Solution   From the product rule, we have

$$y' = (4x^3 - 6x + 2)(x^3 - 2x + 3) + (x^4 - 3x^2 + 2x)(3x^2 - 2).$$

Evaluating at $x = 0$, we have $y'(0) = (2)(3) + (0)(-2) = 6$. The line with slope 6 and passing through the point $(0,0)$ [Why $(0,0)$?] has equation $y = 6x$.

■

Quotient Rule

Given our experience with the product rule, you probably have no expectation that the derivative of a quotient will turn out to be the quotient of the derivatives. But, just to be sure, let's try a simple experiment. Note that

$$\frac{d}{dx}\left(\frac{x^5}{x^2}\right) = \frac{d}{dx}(x^3) = 3x^2,$$

while

$$\frac{\frac{d}{dx}(x^5)}{\frac{d}{dx}(x^2)} = \frac{5x^4}{2x^1} = \frac{5}{2}x^3 \neq 3x^2 = \frac{d}{dx}\left(\frac{x^5}{x^2}\right).$$

Since these are obviously not the same, we know that the derivative of a quotient is clearly not the quotient of the respective derivatives.

The following theorem provides us with a general rule for computing the derivative of a quotient of two differentiable functions.

Theorem 4.2   **(Quotient Rule)** Suppose that f and g are differentiable at x and $g(x) \neq 0$. Then

$$\frac{d}{dx}\left[\frac{f(x)}{g(x)}\right] = \frac{f'(x)\,g(x) - f(x)\,g'(x)}{[g(x)]^2}. \tag{4.4}$$

Proof. You should be able to guess the structure of the proof. For $Q(x) = \dfrac{f(x)}{g(x)}$, we have from the limit definition of derivative that

By the definition of derivative.

$$\frac{d}{dx}\left[\frac{f(x)}{g(x)}\right] = Q'(x) = \lim_{h \to 0}\frac{Q(x+h) - Q(x)}{h}$$

$$= \lim_{h \to 0}\frac{\frac{f(x+h)}{g(x+h)} - \frac{f(x)}{g(x)}}{h}$$

Adding the fractions.

$$= \lim_{h \to 0}\frac{\left[\frac{f(x+h)\,g(x) - f(x)\,g(x+h)}{g(x+h)\,g(x)}\right]}{h}$$

Simplifying.

$$= \lim_{h \to 0}\left[\frac{f(x+h)\,g(x) - f(x)\,g(x+h)}{h\,g(x+h)\,g(x)}\right]$$

As in the proof of the product rule, we look for the right term to add and subtract in the numerator, so that we can isolate the limit definitions of $f'(x)$ and $g'(x)$.

Here, we add and subtract $f(x)\,g(x)$, to get

$$Q'(x) = \lim_{h\to 0}\left[\frac{f(x+h)\,g(x) - f(x)\,g(x+h)}{h\,g(x+h)\,g(x)}\right]$$

Subtracting and adding $f(x)g(x)$.

$$= \lim_{h\to 0}\frac{f(x+h)\,g(x) - f(x)\,g(x) + f(x)\,g(x) - f(x)\,g(x+h)}{h\,g(x+h)\,g(x)}$$

Grouping first two and last two terms together and factoring-out common terms.

$$= \lim_{h\to 0}\frac{\dfrac{f(x+h) - f(x)}{h}\,g(x) - f(x)\,\dfrac{g(x+h) - g(x)}{h}}{g(x+h)\,g(x)}$$

$$= \frac{\displaystyle\lim_{h\to 0}\left[\frac{f(x+h) - f(x)}{h}\right]g(x) - f(x)\lim_{h\to 0}\left[\frac{g(x+h) - g(x)}{h}\right]}{\displaystyle\lim_{h\to 0}g(x+h)\,g(x)}$$

$$= \frac{f'(x)\,g(x) - f(x)\,g'(x)}{[g(x)]^2},$$

where we have recognized the derivatives of f and g and used the fact that g is differentiable to imply that g is continuous, so that

$$\lim_{h\to 0}g(x+h) = g(x).$$

∎

Notice that the numerator in the quotient rule looks very much like the product rule, but with a minus sign between the two terms. For this reason, you need to be very careful with the order.

Example 4.4   A Simple Use of the Quotient Rule

We first return to our introductory example to illustrate the quotient rule. We have from (4.4) that

$$\frac{d}{dx}\left(\frac{x^5}{x^2}\right) = \frac{\dfrac{d}{dx}(x^5)\,x^2 - x^5\,\dfrac{d}{dx}(x^2)}{(x^2)^2}$$

$$= \frac{5x^4\,(x^2) - x^5\,(2x^1)}{x^4}$$

$$= \frac{5x^6 - 2x^6}{x^4}$$

$$= \frac{3x^6}{x^4} = 3x^2,$$

as we had determined previously.

∎

Of course, we would never use the quotient rule to compute the derivative of such a simple function, but there are times when there is no escaping the quotient rule, as in the following example.

Example 4.5   Using the Quotient Rule

Compute the derivative of $f(x) = \dfrac{x^2 - 2}{x^2 + 1}$.

Solution   Using the quotient rule, we have

$$f'(x) = \frac{\frac{d}{dx}(x^2-2)(x^2+1) - (x^2-2)\frac{d}{dx}(x^2+1)}{(x^2+1)^2}$$

$$= \frac{2x(x^2+1) - (x^2-2)(2x)}{(x^2+1)^2}$$

$$= \frac{6x}{(x^2+1)^2}.$$

In this case, we rewrote the numerator because it simplified significantly. This often occurs with the quotient rule.

■

Now that we have the quotient rule, we can justify the use of the power rule for negative integer exponents. (Recall that we have been using this rule without proof since section 2.3.)

Theorem 4.3   (Power Rule) For any integer exponent n,

$$\frac{d}{dx}x^n = nx^{n-1}.$$

Proof. We have already proven this for **positive** integer exponents. So, suppose that $n < 0$ and let $M = -n > 0$. Then, using the quotient rule, we get

Since $x^{-M} = 1/x^M$.

$$\frac{d}{dx}x^n = \frac{d}{dx}x^{-M} = \frac{d}{dx}\left(\frac{1}{x^M}\right)$$

By the quotient rule.

$$= \frac{\frac{d}{dx}(1)\,x^M - (1)\,\frac{d}{dx}(x^M)}{(x^M)^2}$$

By the power rule, since $M > 0$.

$$= \frac{(0)\,x^M - (1)\,Mx^{M-1}}{x^{2M}}$$

By the usual rules of exponents.

$$= \frac{-Mx^{M-1}}{x^{2M}} = -Mx^{M-1-2M}$$

Since $n = -M$.

$$= (-M)x^{-M-1} = nx^{n-1},$$

where we have used the fact that $\frac{d}{dx}x^M = Mx^{M-1}$, since $M > 0$.

■

Example 4.6   A Case Where the Product and Quotient Rules Are Not Needed
Compute the derivative of $f(x) = x\sqrt{x} + \dfrac{2}{x^2}$.

Solution   Although it may be tempting to use the product rule for the first term and the quotient rule for the second term, notice that if we take a moment to rewrite the function, we can considerably simplify our work. (It often pays to first think about the best way to approach a problem, before blasting away with derivative rules.) In the present case, we can combine the two powers of x in the first term and since the second term is a fraction with a constant numerator, we can more simply write it using a negative exponent. We have

$$f(x) = x\sqrt{x} + \frac{2}{x^2} = x^{3/2} + 2x^{-2}.$$

Using the power rule, we have simply

$$f'(x) = \frac{3}{2}x^{1/2} - 4x^{-3}.$$

■

Applications

We hope that you will not be especially surprised to learn that numerous applications require the use of the product and quotient rules. That is, after all, why we have taken the trouble to derive and practice these rules. You will see important uses of these rules throughout your mathematical and scientific careers. We start you off with a couple of simple applications now.

Example 4.7   Investigating the Rate of Change of Revenue

Suppose that a product currently sells for \$25, with the price increasing at the rate of \$2 per year. At this price, consumers will buy 150 thousand items, but the number being bought is decreasing at the rate of 8 thousand per year. At what rate is the total revenue changing? Is the total revenue increasing or decreasing?

Solution   To answer these questions, we need the basic relationship

$$revenue = quantity \times price$$

(e.g., if you sell 10 items at \$4 each, you earn \$40). Since we think of these quantities as changing in time, we write $R(t) = Q(t) P(t)$, where $R(t)$ is revenue, $Q(t)$ is quantity sold and $P(t)$ is the price, all at time t. We don't have formulas for any of these functions, but we do have enough information to use the product rule. The product rule gives us

$$R'(t) = Q'(t) P(t) + Q(t) P'(t).$$

We have information about each of these terms: the initial price, $P(0)$ is 25 (dollars), the rate of change of the price is $P'(t) = +2$ (dollars per year), the initial quantity, $Q(0)$ is 150 (thousand items), and the rate of change of quantity is $Q'(t) = -8$ (thousand items per year). Note that the negative sign of $Q'(t)$ denotes a decrease in Q (negative rates of change indicate a decrease). Thus,

$$R'(0) = (-8)(25) + (150)(2) = +100 \text{ dollars per year.}$$

Since the rate of change is positive, the revenue is increasing at present. This may be a surprise since one of the two factors in the equation is decreasing, and the rate of decrease of the quantity is more than the rate of increase in the price.

■

Example 4.8   Using the Derivative to Solve a Baseball Problem

The geometry of baseball is one of the pleasures of the game: a batter hits a long, towering fly ball that soars to the outfield along a graceful arc, where the outfielder calmly reaches up a glove and catches it. How does the outfielder know where to stand? This is a computational puzzle to which several physicists have proposed a variety of solutions. We can study one aspect of the problem now. Suppose the flight of the ball is along

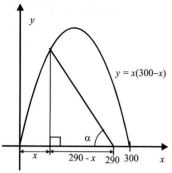

FIGURE 2.23: A fly ball.

the parabola $y = x(300 - x)$. This fly ball is hit at the origin $(0,0)$ and lands at the 300-foot mark $(300,0)$. Suppose the outfielder stands at $x = 290$ (not far enough back). We name the vertical angle at which the outfielder views the ball α (see Figure 2.23). Using elementary trigonometry, we have

$$\tan \alpha = \frac{y}{290 - x} = \frac{x(300 - x)}{290 - x} = \frac{300x - x^2}{290 - x}.$$

Using the quotient rule to compute the derivative of $\tan \alpha$, we get

$$\frac{d}{dx}(\tan \alpha) = \frac{(300 - 2x)(290 - x) - (300x - x^2)(-1)}{(290 - x)^2} = \frac{x^2 - 580x + 87000}{(290 - x)^2}.$$

Thus, the angle changes as the ball moves along the path. If the player stands at the correct location for catching the ball ($x = 300$), we have

$$\tan \alpha = \frac{y}{300 - x} = \frac{x(300 - x)}{300 - x} = x.$$

In this case, the derivative is

$$\frac{d}{dx}(\tan \alpha) = \frac{d}{dx}(x) = 1.$$

Thus, the tangent of the angle at which the player sees the ball changes at a constant rate (with respect to the distance x) if the player is standing at the right place, but at a non-constant rate if the player is in the wrong place. Isn't this remarkable? Unfortunately, in spite of the mathematical beauty of this theory, the available evidence suggests that this is not the primary method ballplayers use to correctly judge fly balls.

■

Exercises 2.4

1. (writing) The product and quotient rules give you the ability to symbolically calculate the derivative of a wide range of functions. However, many calculators and almost every computer algebra system (CAS) can do this work for you. Discuss why you should learn these basic rules anyway. (Keep example 4.6 in mind.)

2. (writing) Gottfried Leibniz is recognized (along with Isaac Newton) as a coinventor of calculus. Many of the fundamental methods and (perhaps more importantly) much of the notation of calculus are due to Leibniz. The product rule was worked out by Leibniz in 1675, in the form $d(xy) = (dx)y + x(dy)$. His "proof," as given in a letter written in 1699, follows. "If we are to differentiate xy we write:

$$(x + dx)(y + dy) - xy = xdy + ydx + dxdy.$$

But here $dxdy$ is to be rejected as incomparably less than $xdy + ydx$. Thus in any particular case the error is less than any finite quantity." Answer Leibniz' letter with one describing your own "discovery" of the product rule for $d(xyz)$. Use Leibniz' notation.

3. (writing) In example 4.2, we cautioned you against multiplying out the terms of the derivative. To see one reason for this warning, suppose that you want to find solutions of the equation $f'(x) = 0$. (In fact, we do this routinely in Chapter 3.) Explain why having a factored form of $f'(x)$ is very helpful. Discuss the extent to which the product rule gives you a factored form.

4. (writing) Many students prefer the product rule to the quotient rule. Many computer algebra systems actually use the product rule to compute the derivative of $f(x)[g(x)]^{-1}$ instead of using the quotient rule on $\frac{f(x)}{g(x)}$ (see exercise 26 below). Given the simplifications in problems like example 4.5, explain why the quotient rule can be preferable.

In exercises 5–24, find the derivative of each function.

5. $f(x) = (x^2 + 3)(x^3 - 3x + 1)$

6. $f(x) = (x^3 - 2x^2 + 5)(x^4 - 3x^2 + 2)$

7. $f(x) = (3x + 4)(x^3 - 2x^2 + x)$

8. $f(x) = (x^2 - 4)(x^4 + 2x + 1)$

9. $f(x) = (\sqrt{x} + 3x)(5x^2 - 3/x)$

10. $f(x) = (x^{3/2} - 4x)(x^4 - 3/x^2 + 2)$

11. $f(x) = \dfrac{3x - 2}{5x + 1}$　　**12.** $f(x) = \dfrac{2x + 5}{5x - 1}$

13. $f(x) = \dfrac{x - 2}{x^2 + x + 1}$　　**14.** $f(x) = \dfrac{x^2 + 2x + 5}{x^2 - 5x + 1}$

15. $f(x) = \dfrac{3x - 6\sqrt{x}}{5x^2 - 2}$　　**16.** $f(x) = \dfrac{6x - 2/x}{x^2 + \sqrt{x}}$

17. $f(x) = \dfrac{(x + 1)(x - 2)}{x^2 - 5x + 1}$　　**18.** $f(x) = \dfrac{x^2 - 2x}{x^2 + 5x}$

19. $f(x) = \dfrac{x^2 + 3x - 2}{\sqrt{x}}$　　**20.** $f(x) = \dfrac{2x}{x^2 + 1}$

21. $f(x) = x(\sqrt[3]{x} + 3)$　　**22.** $f(x) = \dfrac{x^2}{3} + \dfrac{5}{x^2}$

23. $f(x) = (x^2 - 1)\dfrac{x^3 + 3x^2}{x^2 + 2}$　　**24.** $f(x) = (x + 2)\dfrac{x^2 - 1}{x^2 + x}$

25. Write out the product rule for the function $f(x)\, g(x)\, h(x)$. (Hint: Group the first two terms together.) Describe the general product rule: for n functions, what is the derivative of the product $f_1(x)\, f_2(x)\, f_3(x) \cdots f_n(x)$? How many terms are there? What does each term look like?

26. Use the quotient rule to show that the derivative of $(g(x))^{-1}$ is $-g'(x)\,(g(x))^{-2}$. Then use the product rule to compute the derivative of $f(x)\,(g(x))^{-1}$. How does your result compare to the form in which your CAS reports the derivative of a quotient? Write out the "quotient rule" used by your CAS as completely as possible.

In exercises 27–30, find the derivative of each function using the general product rule developed in exercise 25.

27. $f(x) = x^{2/3}(x^2 - 2)(x^3 - x + 1)$

28. $f(x) = (x + 4)(x^3 - 2x^2 + 1)(3 - 2/x)$

29. $f(x) = (x + 1)(x^3 + 4x)(x^5 - 3x^2 + 1)$

30. $f(x) = \sqrt{x}(x - 2)(x + 1)(x + 5)$

31. Suppose that for some toy the quantity sold, $Q(t)$, decreases at a rate of 4%; explain why this translates to $Q'(t) = -0.04Q(t)$. Suppose also that the price increases at a rate of 3%; write out a similar equation for $P'(t)$ in terms of $P(t)$. The revenue for the toy is $R(t) = Q(t)\, P(t)$. Substituting the expressions for $Q'(t)$ and $P'(t)$ into the product rule $R'(t) = Q'(t)\, P(t) + Q(t)\, P'(t)$, show that the revenue decreases at a rate of 1%. Explain why this is "obvious."

32. As in exercise 31, suppose that the quantity sold decreases at a rate of 4%. By what rate must the price be increased to keep a constant revenue?

33. Suppose the price of an object is $20 and $20,000$ units are sold. If the price increases at a rate of $1.25 per year and the quantity sold increases at a rate of 2000 per year, at what rate will revenue increase?

34. Suppose the price of an object is $14 and $12,000$ units are sold. The company wants to increase the quantity sold by

1200 units per year, while increasing the revenue by $20,000$ per year. At what rate would the price have to be increased to reach these goals?

35. In example 4.8, interpret the derivative $\dfrac{d}{dx}(\tan \alpha)$. If you were trying to catch a baseball, would you be able to determine if this derivative stayed constant?

36. In example 4.8, show that $\dfrac{d}{dx}(\tan \alpha) > 1$ except at $x = 300$. So, a baseball player could locate the correct place to stand by minimizing the derivative.

37. Suppose that $F(x) = f(x)\, g(x)$ for infinitely differentiable functions $f(x)$ and $g(x)$ (that is, $f'(x)$, $f''(x)$, etc. exist for all x). Show that $F''(x) = f''(x)\, g(x) + 2f'(x)\, g'(x) + f(x)\, g''(x)$. Compute $F'''(x)$. Compare $F''(x)$ to the binomial formula for $(a + b)^2$ and compare $F'''(x)$ to the formula for $(a + b)^3$.

38. Given that $(a + b)^4 = a^4 + 4a^3b + 6a^2b^2 + 4ab^3 + b^4$, write out a formula for $F^{(4)}(x)$. (See exercise 37.)

39. Use the product rule to show that if $g(x) = [f(x)]^2$ and $f(x)$ is differentiable, then $g'(x) = 2f(x)\, f'(x)$. This is an example of the **chain rule**, to be discussed in section 2.7.

40. Use the result from exercise 39 and the product rule to show that if $g(x) = [f(x)]^3$ and $f(x)$ is differentiable, then $g'(x) = 3[f(x)]^2\, f'(x)$. Hypothesize the correct chain rule for the derivative of $[f(x)]^n$.

41. The relationship among the pressure P, volume V and temperature T of a gas or liquid is given by **van der Waal's equation** $\left(P + \dfrac{n^2 a}{V^2}\right)(V - nb) = nRT$ for positive constants a, b, n and R. Solve the equation for P. Treating T as a constant and V as the variable, find the critical point (T_c, P_c, V_c) such that $P'(V) = P''(V) = 0$. For temperatures above T_c, the substance can only exist in its gaseous form; below T_c, the substance is a gas or liquid (depending on the pressure and volume). For water, take $R = 0.08206$ l-atm/mo-K, $a = 5.464$ l²-atm/mo² and $b = 0.03049$ l/mo. Find the highest temperature at which $n = 1$ mo of water may exist as a liquid. (Note: your answer will be in degrees Kelvin; add 273.15 to get degrees Celsius.)

42. (exploratory) In the text, we explored an unusual rule for catching a baseball. Our analysis used a parabola for the flight of the ball. This path is accurate only in a vacuum; air resistance causes the path to be asymmetric. Replace the parabolic path of example 7 with the cubic path $y = x(x - 300)(x - 600)$, $0 \le x \le 300$. Is it still true that the derivative of $\tan \alpha$ is constant if the fielder is standing in the right place?

43. (exploratory) In many sports, the collision between a ball and a striking implement is central to the game. Suppose the ball has weight w and velocity v before the collision, and the striker (bat, tennis racket, golf club, etc.) has weight

W and velocity $-V$ before the collision (the negative indicates the striker is moving in the opposite direction from the ball). The velocity of the ball after the collision will be $u = \dfrac{WV(1+c) + v(cW - w)}{W + w}$ where the parameter c, called the **coefficient of restitution**, represents the "bounciness" of the ball in the collision. Treating W as the independent variable (like x) and the other parameters as constants, compute the derivative and verify that $\dfrac{du}{dW} = \dfrac{V(1+c)w + cvw + vw}{(W+w)^2} \geq 0$ since all parameters are non-negative. Explain why this implies that if the athlete uses a bigger striker (bigger W) with all other things equal, the speed of the ball increases. Does this match your intuition? What is doubtful about the assumption of all other things being equal? Similarly compute and interpret $\dfrac{du}{dw}, \dfrac{du}{dv}, \dfrac{du}{dV}$ and $\dfrac{du}{dc}$. (Hint: c is between 0 and 1 with 0 representing a dead ball and 1 the liveliest ball possible.)

2.5   Derivatives of Trigonometric Functions

Springs are fundamental to many mechanical systems, including shock absorbing systems for cars, stereo equipment and other sensitive devices. Imagine a weight hanging from a spring suspended from the ceiling (see Figure 2.24). If we set the weight in motion (e.g., by tapping it down), the weight will bounce up and down in ever-shortening strokes until it eventually is again at rest (equilibrium). To mathematically describe such spring motion, we need functions other than the algebraic functions we have focused on so far in this chapter. It turns out that, for short periods of time, the motion is nearly periodic. To describe this mathematically, we first need to identify what we are measuring. Suppose we keep track of the vertical distance of the weight from its natural resting position (equilibrium position), with up being positive and down negative (see Figure 2.24).

FIGURE 2.24: Spring-mass system.

When we pull the weight down, its vertical displacement is negative. The weight then swings up to where the displacement is positive, swings down to a negative displacement, and so on. What kind of function might describe this sort of oscillation? The only functions we've experienced that exhibit this kind of behavior are the sine and cosine functions. For situations like this, either function can be used (recall that the cosine graph is identical to the sine graph except for a horizontal shift). The sine and cosine functions are used in almost any model where there are oscillations. We calculate the derivatives of these and the other trigonometric functions in this section.

We can get a good idea of what the derivatives of $\sin x$ and $\cos x$ are from their graphs. Let's analyze the graph of $y = \sin x$ in Figure 2.25a. The graph has horizontal tangents at $x = -3\pi/2, -\pi/2, \pi/2$ and $3\pi/2$. Thus, at these x-values, the derivative must equal 0. The slope is positive for $-2\pi < x < -3\pi/2$, negative for $-3\pi/2 < x < -\pi/2$ and so on. For each interval on which the derivative is positive (or negative), the graph appears to be steepest in the middle of the interval: for example, from $x = -\pi/2$, the graph gets steeper until about $x = 0$ and then gets less steep until leveling out at $x = \pi/2$. A sketch of the derivative graph should then look like the graph in Figure 2.25b.

Notice how much this looks like the graph of $y = \cos x$. We show below that this conjecture is in fact, correct. In the exercises, you are asked to perform a similar graphical analysis to conjecture that the derivative of $f(x) = \cos x$ equals $-\sin x$.

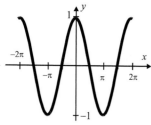

FIGURE 2.25a: $y = \sin x$.

FIGURE 2.25b: The derivative of $f(x) = \sin x$.

Before we move to the calculation of the derivatives of the six trigonometric functions, we first consider a few limits involving trig functions. (We refer to these results as *Lemmas* — minor theorems that lead up to some more significant result.) You will see shortly why we must consider these first.

Lemma 5.1 $\displaystyle\lim_{\theta \to 0} \sin\theta = 0$.

This result certainly seems reasonable, especially when we consider the graph of $y = \sin x$. In fact, we have been using this for some time now, but will now prove the result.

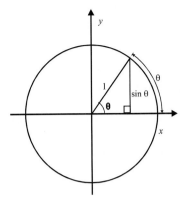

FIGURE 2.26: Definition of $\sin\theta$.

Proof. For $0 < \theta < \dfrac{\pi}{2}$, consider Figure 2.26. From the figure, observe that

$$0 \le \sin\theta \le \theta. \tag{5.1}$$

It is a simple matter to see that

$$\lim_{\theta\to 0^+} 0 = 0 = \lim_{\theta\to 0^+} \theta. \tag{5.2}$$

From the Squeeze Theorem (see section 1.2), it now follows from (5.1) and (5.2) that

$$\lim_{\theta\to 0^+} \sin\theta = 0,$$

also. Similarly, you can show that

$$\lim_{\theta\to 0^-} \sin\theta = 0.$$

This is left as an exercise. Since both one-sided limits are the same, it follows that

$$\lim_{\theta\to 0} \sin\theta = 0.$$

∎

Lemma 5.2 $\lim\limits_{\theta\to 0} \cos\theta = 1.$

The proof of this result follows from Lemma 5.1 and the Pythagorean Theorem. We leave this as an exercise.

The following result was conjectured to be true (based on a graph and some computations) when we first examined limits in section 1.1. We can now prove the result.

Lemma 5.3 $\lim\limits_{\theta\to 0} \dfrac{\sin\theta}{\theta} = 1.$

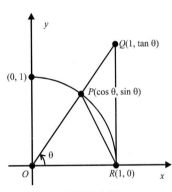

FIGURE 2.27

Proof. Assume $0 < \theta < \dfrac{\pi}{2}$.

Referring to Figure 2.27, observe that

$$0 < Area\ \triangle OPR < Area\ \text{sector}\ OPR < Area\ \triangle OQR. \tag{5.3}$$

You can see from Figure 2.28 that

$$Area\ \text{sector}\ OPR = \pi(radius)^2\ (\text{fraction of circle included})$$

$$= \pi\left(1^2\right)\frac{\theta}{2\pi} = \frac{\theta}{2}. \tag{5.4}$$

Also,

$$Area\ \triangle OPR = \frac{1}{2}(base)(height)$$

$$= \frac{1}{2}\ (1)\ \sin\theta \tag{5.5}$$

and

$$Area\ \triangle OQR = \frac{1}{2}\ (1)\ \tan\theta. \tag{5.6}$$

Thus, from (5.3), (5.4), (5.5) and (5.6), we have

$$0 < \frac{1}{2}\sin\theta < \frac{\theta}{2} < \frac{1}{2}\tan\theta. \tag{5.7}$$

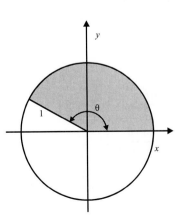

FIGURE 2.28: A circular sector.

If we divide (5.7) by $\frac{1}{2}\sin\theta$ (recall that this is positive, so that the inequalities are not affected), we get

$$1 < \frac{\theta}{\sin\theta} < \frac{\tan\theta}{\sin\theta} = \frac{1}{\cos\theta}.$$

Taking reciprocals (again, recall that everything here is positive), we find

$$1 > \frac{\sin\theta}{\theta} > \cos\theta. \tag{5.8}$$

The inequality (5.8) also holds if $-\frac{\pi}{2} < \theta < 0$. (This is left as an exercise.) Finally, note that

$$\lim_{\theta\to 0}\cos\theta = 1 = \lim_{\theta\to 0}1.$$

Thus, it follows from (5.8) and the Squeeze Theorem that

$$\lim_{\theta\to 0}\frac{\sin\theta}{\theta} = 1$$

also.

∎

We need one additional limit result before we tackle the derivatives of the trig functions.

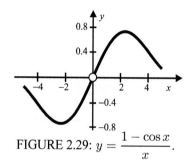

FIGURE 2.29: $y = \dfrac{1-\cos x}{x}$.

Lemma 5.4 $\displaystyle\lim_{\theta\to 0}\frac{1-\cos\theta}{\theta} = 0.$

Before we prove this, we want to make this conjecture seem reasonable. We draw a graph of $y = \dfrac{1-\cos x}{x}$ in Figure 2.29. The tables of function values that follow should prove equally convincing.

x	$\dfrac{1-\cos x}{x}$
0.1	0.04996
0.01	0.00499996
0.001	0.0005
0.0001	0.00005

x	$\dfrac{1-\cos x}{x}$
-0.1	-0.04996
-0.01	-0.00499996
-0.001	-0.0005
-0.0001	-0.00005

Now that we have strong evidence for the conjecture, we prove the Lemma.

Multiply numerator and denominator by $1 + \cos\theta$.

Multiply out numerator and denominator.

Since $\sin^2\theta + \cos^2\theta = 1$.

Split up terms.

Proof.

$$\lim_{\theta\to 0}\frac{1-\cos\theta}{\theta} = \lim_{\theta\to 0}\left(\frac{1-\cos\theta}{\theta}\right)\left(\frac{1+\cos\theta}{1+\cos\theta}\right)$$

$$= \lim_{\theta\to 0}\frac{1-\cos^2\theta}{\theta(1+\cos\theta)}$$

$$= \lim_{\theta\to 0}\frac{\sin^2\theta}{\theta(1+\cos\theta)}$$

$$= \lim_{\theta\to 0}\left[\left(\frac{\sin\theta}{\theta}\right)\left(\frac{\sin\theta}{1+\cos\theta}\right)\right]$$

$$= \left(\lim_{\theta\to 0}\frac{\sin\theta}{\theta}\right)\left(\lim_{\theta\to 0}\frac{\sin\theta}{1+\cos\theta}\right)$$

$$= (1)\left(\frac{0}{1+1}\right) = 0,$$

as conjectured.

∎

We are finally in a position to compute the derivatives of the six trig functions.

Theorem 5.1 $\dfrac{d}{dx}\sin x = \cos x.$

Proof. From the limit definition of derivative, for $f(x) = \sin x$, we have

$$\frac{d}{dx}\sin x = f'(x) = \lim_{h\to 0}\frac{\sin(x+h) - \sin(x)}{h}$$

$$= \lim_{h\to 0}\frac{\sin x\,\cos h + \sin h\,\cos x - \sin x}{h}$$

$$= \lim_{h\to 0}\frac{\sin x\,\cos h - \sin x}{h} + \lim_{h\to 0}\frac{\sin h\,\cos x}{h}$$

$$= \sin x \lim_{h\to 0}\frac{\cos h - 1}{h} + \cos x \lim_{h\to 0}\frac{\sin h}{h}$$

$$= \sin x\,(0) + \cos x\,(1) = \cos x,$$

from Lemmas 5.3 and 5.4.

By the definition of derivative.

Trig identity: $\sin(\alpha + \beta) = \sin\alpha\cos\beta + \sin\beta\cos\alpha$.

Grouping terms with $\sin x$ and terms with $\sin h$ separately.

Factoring $\sin x$ from the first term and $\cos x$ from the second term.

∎

Theorem 5.2 $\dfrac{d}{dx}\cos x = -\sin x.$

The proof of Theorem 5.2 is left as an exercise.

For the remaining four trig functions, we can use the quotient rule in conjunction with the derivatives of $\sin x$ and $\cos x$.

Theorem 5.3 $\dfrac{d}{dx}\tan x = \sec^2 x.$

Proof. For $f(x) = \tan x = \dfrac{\sin x}{\cos x}$, we have from the quotient rule that

$$\frac{d}{dx}\tan x = f'(x) = \frac{d}{dx}\left[\frac{\sin x}{\cos x}\right]$$

$$= \frac{\dfrac{d}{dx}(\sin x)\,(\cos x) - (\sin x)\dfrac{d}{dx}(\cos x)}{(\cos x)^2}$$

$$= \frac{\cos x\,(\cos x) - \sin x\,(-\sin x)}{(\cos x)^2}$$

$$= \frac{\cos^2 x + \sin^2 x}{(\cos x)^2} = \frac{1}{(\cos x)^2} = \sec^2 x,$$

where we have used the quotient rule and the preceding results on the derivatives of $\sin x$ and $\cos x$ (Theorems 5.1 and 5.2).

∎

The derivatives of the remaining trigonometric functions are left as exercises. The derivatives of all six trigonometric functions are summarized below.

$$\frac{d}{dx}\left(\sin x\right) = \cos x \qquad \frac{d}{dx}(\cos x) = -\sin x$$

$$\frac{d}{dx}(\tan x) = \sec^2 x \qquad \frac{d}{dx}(\cot x) = -\csc^2 x$$

$$\frac{d}{dx}(\sec x) = \sec x \tan x \qquad \frac{d}{dx}\left(\csc x\right) = -\csc x \cot x$$

The following example shows where the product rule is necessary.

Example 5.1   A Derivative That Requires the Product Rule

Find the derivative of $f(x) = x^5 \cos x$.

Solution   From the product rule, we have

$$\frac{d}{dx}\left(x^5 \cos x\right) = \frac{d}{dx}\left(x^5\right)\cos x + x^5 \frac{d}{dx}\left(\cos x\right)$$
$$= 5x^4 \cos x - x^5 \sin x.$$

∎

Example 5.2   Computing Some Routine Derivatives

Compute the derivatives of (a) $f(x) = \sin^2 x$ and (b) $g(x) = 4\tan x - 5\csc x$.

Solution   For (a), we first rewrite the function as $f(x) = (\sin x)(\sin x)$ and use the product rule. We have

$$f'(x) = \cos x \sin x + \sin x \cos x = 2 \sin x \cos x.$$

(In section 2.7, we present the chain rule, which gives us a simpler way to compute this derivative.) For (b), we have

$$g'(x) = 4\sec^2 x + 5\csc x \cot x.$$

∎

Applications

The trigonometric functions arise quite naturally in the solution of numerous physical problems of interest. For this reason, they are indispensable.

It can be shown that the vertical displacement of a weight suspended from a spring, in the absence of damping (i.e., when resistance, such as air resistance, to the motion is negligible) is given by

$$u(t) = a\cos(\omega t) + b\sin(\omega t),$$

where ω is the frequency, t is time and a and b are constants (see Figure 2.30 for a depiction of such a spring-mass system).

displacement = u(t)

equilibrium position

FIGURE 2.30: Spring-mass system.

Example 5.3   Analysis of a Spring-Mass System

Suppose that $u(t)$ measures the displacement (measured in inches) of a weight suspended from a spring t seconds after it is released and that

$$u(t) = 4\cos t.$$

Find the velocity at any time t, and determine the maximum velocity.

Solution   Since $u(t)$ represents distance, the velocity is given by $u'(t)$. We have

$$u'(t) = 4(-\sin t) = -4\sin t,$$

where $u'(t)$ is measured in inches per second. Of course, $\sin t$ oscillates between -1 and 1, and hence, the largest that $u'(t)$ can be is $-4(-1) = 4$ inches per second. This occurs when $\sin t = -1$, that is, at $t = 3\pi/2$, $t = 7\pi/2$ and so on. Notice that at these times $u(t) = 0$, so that the weight is moving fastest when it is passing through its resting position.

∎

1 farad

1 henry

$2t^2$ volts

FIGURE 2.31: A simple circuit.

Example 5.4   Analysis of a Simple Electrical Circuit
    The diagram in Figure 2.31 shows a simple electrical circuit. If the capacitance is 1 (farad), the inductance is 1 (henry) and the impressed voltage is $2t^2$ (volts) at time t, then a model for the total charge $Q(t)$ in the circuit at time t is

$$Q(t) = 2\sin t + 2t^2 \text{ (coulombs).}$$

The current is defined to be the rate of change of the charge with respect to time and so is given by

$$I(t) = \frac{dQ}{dt} \text{ (amperes).}$$

Compare the current at times $t = 0$ and $t = 1$.

Solution   In general, the current is given by

$$I(t) = \frac{dQ}{dt} = 2\cos t + 4t \text{ (amperes).}$$

Notice that at time $t = 0$, $I(0) = 2$ (amperes). At time $t = 1$,

$$I(1) = 2\cos 1 + 4 \approx 5.08 \text{ (amperes).}$$

This represents an increase of

$$\frac{3.08}{2}(100) = 154\%.$$

∎

Example 5.5   Finding the Equation of the Tangent Line
    Find an equation of the tangent line to

$$y = 3\tan x - 2\csc x$$

at $x = \pi/3$.

Solution   The derivative is

$$y' = 3\sec^2 x - 2(-\csc x \cot x) = 3\sec^2 x + 2\csc x \cot x.$$

At $x = \pi/3$, we have

$$y'(\pi/3) = 3(2)^2 + 2(2/\sqrt{3})(1/\sqrt{3}) = 12 + 4/3 = 40/3 \approx 13.33333.$$

The tangent line with slope 40/3 and point of tangency $(\pi/3, 3\sqrt{3} - 4/\sqrt{3})$, has equation

$$y = \frac{40}{3}(x - \pi/3) + 3\sqrt{3} - \frac{4}{\sqrt{3}}.$$

∎

Exercises 2.5

1. (writing) Most people draw sine curves that are very steep and rounded. Given the results of this section, discuss the actual shape of the sine curve. Starting at $(0,0)$, how steep should the graph be drawn? What is the steepest the graph should be drawn anywhere? In which regions is the graph almost straight and where does it curve a lot?

2. (writing) In many common physics and engineering applications, the term $\sin x$ makes calculations difficult. A common simplification is to replace $\sin x$ with x, accompanied by the justification "$\sin x$ approximately equals x for small angles." Discuss this approximation in terms of the tangent line to $y = \sin x$ at $x = 0$. How small is the "small angle" for which the approximation is good? The tangent line to $y = \cos x$ at $x = 0$ is simply $y = 1$, but the simplification "$\cos x$ approximately equals 1 for small angles" is almost never used. Why would this approximation be less useful than $\sin x \approx x$?

3. Use a graphical analysis as in the text to argue that the derivative of $\cos x$ is $-\sin x$.

4. In the proof of $\dfrac{d}{dx}(\sin x) = \cos x$, where do we use the assumption that x is in radians? If you have access to a CAS, find out what the derivative of $\sin x°$ is; explain where the $\pi/180$ came from.

In exercises 5–24, find the derivative of each function.

5. $f(x) = 4\sin x - x$ **6.** $f(x) = x^2 + 2\cos x$

7. $f(x) = \tan x - \csc x$ **8.** $f(x) = 4\sec x$

9. $f(x) = x\cos x$ **10.** $f(x) = 4x^2 - 3\tan x$

11. $f(x) = 4\sqrt{x} - 2\sin x$ **12.** $f(x) = 3\sec x \tan x$

13. $f(x) = \dfrac{\sin x}{x}$ **14.** $f(x) = \dfrac{x^2}{\csc x}$

15. $f(t) = 3\csc t - 8\cot t$ **16.** $f(t) = 3\sin t + 4\cos t$

17. $f(x) = \dfrac{\cos x - 1}{x^2}$ **18.** $f(x) = \dfrac{\tan x}{x^2}$

19. $f(x) = 2\sin x \cos x$ **20.** $f(x) = \cos^2 x - \sin^2 x$

21. $f(x) = 4x^2 \tan x$ **22.** $f(x) = \dfrac{\tan x}{\sin x}$

23. $f(x) = 4\sin^2 x$ **24.** $f(x) = 2\tan^2 x$

In exercises 25–28, find an equation of the tangent line to $y = f(x)$ **at** $x = a$.

25. $f(x) = \sin x, a = \frac{\pi}{2}$ **26.** $f(x) = \tan x, a = 0$

27. $f(x) = \cos x, a = \frac{\pi}{2}$ **28.** $f(x) = x\sin x, a = \frac{\pi}{2}$

In exercises 29–32, use the position function to find the velocity at time $t = a$.

29. $s(t) = t^2 - \sin t, a = 0$

30. $s(t) = t\cos t, a = 0$

31. $s(t) = \dfrac{\cos t}{t}, a = \pi$

32. $s(t) = 4 + 3\sin t, a = \pi$

33. Suppose an object is traveling in a circular path counterclockwise at constant speed. If the circle has radius 1 and it takes 2π seconds to make one revolution, explain why the x-coordinate of the object can be modeled by $x(t) = \cos t$. What is the y-coordinate? What changes if the radius is 3? What changes if it takes π seconds for a revolution? What if it takes 2 seconds? What if the object travels clockwise? What changes if at time $t = 0$ the object is at the top of the circle?

34. Based on your answers to exercise 33, explain the meanings of the parameters a, b and c in the general model $x(t) = a\cos(bt + c)$, $y(t) = a\sin(bt + c)$.

35. A spring hanging from the ceiling vibrates up and down. Its vertical position is given by $f(t) = 4\sin t$. Find the velocity of the spring at time t.

36. In exercise 35, for what time values is the velocity 0? What is the location of the spring when its velocity is 0?

In exercises 37 and 38, refer back to example 5.4.

37. If the total charge in an electrical circuit is given by $Q(t) = 3\sin t + t + 4$ coulombs, compare the current at times $t = 0$ and $t = 1$.

38. If the total charge in an electrical circuit is given by $Q(t) = 4\cos t - 3t + 1$ coulombs, compare the current at times $t = 0$ and $t = 1$.

39. Using the identities $\sin 2x = 2\sin x \cos x$ and $\cos 2x = \cos^2 x - \sin^2 x$, prove that if $f(x) = \sin 2x$, then $f'(x) = 2\cos 2x$.

40. Using the identities $\sin 2x = 2\sin x \cos x$ and $\cos 2x = \cos^2 x - \sin^2 x$, prove that if $f(x) = \cos 2x$, then $f'(x) = -2\sin 2x$.

41. For $f(x) = \sin x$, find $f^{(75)}(x)$ and $f^{(150)}(x)$.

42. For $f(x) = \cos x$, find $f^{(77)}(x)$ and $f^{(120)}(x)$.

43. Use the quotient rule to derive the formulas for the derivatives of $\cot x$, $\sec x$ and $\csc x$.

44. Knowing that the derivative of $\sin x$ is $\cos x$, argue that the equation $\sin x = x$ has only one solution.

45. Use the basic limits $\lim\limits_{x\to 0}\dfrac{\sin x}{x}=1$ and $\lim\limits_{x\to 0}\dfrac{\cos x-1}{x}=$ 0 to find the following limits:

(a) $\lim\limits_{x\to 0}\dfrac{\sin 3x}{x}$ (b) $\lim\limits_{t\to 0}\dfrac{\sin t}{4t}$ (c) $\lim\limits_{x\to 0}\dfrac{\cos x-1}{5x}$

(d) $\lim\limits_{x\to 0}\dfrac{\sin x^2}{x^2}$ (e) $\lim\limits_{t\to 0}\dfrac{2t}{\sin t}$ (f) $\lim\limits_{x\to 0}\dfrac{\sin x}{5x+1}$

46. Prove Lemma 5.2. (Hint: First show that for $-\pi/2<\theta<\pi/2$, $\cos\theta=\sqrt{1-\sin^2\theta}$ and then use Lemma 5.1.)

47. (exploratory) The sex of newborn Mississippi alligators is determined by the temperature of the developing egg. If the temperature is not between $26°$ and $36°$, the egg does not develop at all. Between $26°$ and $30°$, the eggs always develop into females. Between $34°$ and $36°$, the eggs always develop into males. Between $30°$ and $34°$, the probability of getting a female or male increases proportionally. Thus, the probability of getting a female as a function of temperature is

$$f(t)=\begin{cases} 1 & \text{if } 26\le t\le 30 \\ g(t) & \text{if } 30\le t\le 34 \\ 0 & \text{if } 34\le t\le 36 \end{cases}$$

Find a function of the form $g(t)=a\cos(bt+c)$ to make $f(t)$ differentiable at $t=30$ and $t=34$. At this point, there is

not enough information to compute the probability of a Mississippi alligator egg developing into a female. What information is missing?

48. (exploratory) We have seen that the approximation $\sin x\approx x$ for small x derives from the tangent line to $y=\sin x$ at $x=0$. We can also think of it as arising from the result $\lim\limits_{x\to 0}\dfrac{\sin x}{x}=1$. Explain why the limit implies $\sin x\approx x$ for small x. How can we get a better approximation? Instead of using the tangent line, we can try a quadratic function. To see what to multiply x^2 by, compute $\lim\limits_{x\to 0}\dfrac{\sin x-x}{x^2}=a$. Then $\sin x-x\approx ax^2$ or $\sin x\approx x+ax^2$ for small x. How about a cubic approximation? Compute $\lim\limits_{x\to 0}\dfrac{\sin x-(x+ax^2)}{x^3}=b$. Then $\sin x-(x+ax^2)\approx bx^3$ or $\sin x\approx x+ax^2+bx^3$ for small x. Starting with $\lim\limits_{x\to 0}\dfrac{\cos x-1}{x}=0$, find a cubic approximation of the cosine function. If you have a CAS, take the approximations for sine and cosine out to a seventh-order polynomial and identify the pattern. (Hint: Write the coefficients — a, b, etc. — in the form $1/n!$ for some integer n.) We will take a longer look at these approximations, called **Taylor polynomials**, in Chapter 8.

2.6   Derivatives of Exponential and Logarithmic Functions

Exponential and logarithmic functions are among the most common functions encountered in applications. We begin with a simple application in business.

Suppose that you have an investment that doubles in value every year. If you start with \$100, then the value of the investment after 1 year is \$100 (2) or \$200. After 2 years, its value is \$100 (2) (2) = \$400 and after 3 years, its value is \$100 (2^3) = \$800. In general, after t years, the value of your investment is \$100 (2^t). How would you describe the *rate* of return on this investment? Since the value doubles every year, you might answer 100%. This is usually called the **effective interest rate** or APR (**annual percentage rate**). To a calculus student, the term *rate* should suggest the derivative.

Recall from our discussion in Chapter 0, that for any rational exponent, $r=\frac{p}{q}$ (for p and q integers), we define

$$a^r=a^{p/q}=\sqrt[q]{a^p}.$$

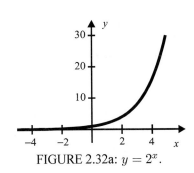

FIGURE 2.32a: $y=2^x$.

For irrational exponents, x, we defined a^x so that the gaps in the graph of $y=a^x$ were "filled-in." At this point, you might recognize that we did this so that the resulting function would be continuous on the entire real line. We make the definition of a^x for x irrational more precise in Chapter 6.

We first consider $f(x)=a^x$ for some constant $a>1$ (a is called the **base**). Recall from our discussion in Chapter 0, that the graph will look something like the graph of $f(x)=2^x$, shown in Figure 2.32a.

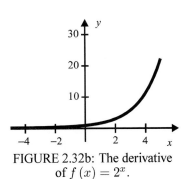

FIGURE 2.32b: The derivative of $f(x)=2^x$.

Observe that as you look from left to right, the graph always rises. Thus, the slopes of the tangent lines and so, too the values of the derivative, are always positive. Also, note that the farther to the right you look, the steeper the graph is, and so, the more positive the derivative is. Further, to the left of the origin, the tangent lines are nearly horizontal and hence, the derivative is positive but close to zero. The sketch of $y=f'(x)$ shown in Figure 2.32b is consistent with all of the above information. (Use your computer algebra system to generate a series of graphs of $y = a^x$ for various values of $a > 0$, along with the graphs of the corresponding derivatives and you may detect a pattern.)

Look closely at Figure 2.32b. The sketch of the derivative closely resembles the graph of the function itself. This should reinforce our earlier suggestion that the description of the function ("doubles every year") also describes its rate of change.

Derivatives of the Exponential Functions

Let's start by using the usual limit definition to try to compute the derivative of $f(x) = a^x$, for $a > 1$. We have

$$f'(x) = \lim_{h \to 0} \frac{f(x+h) - f(x)}{h} = \lim_{h \to 0} \frac{a^{x+h} - a^x}{h}$$

From the usual rules of exponents.

$$= \lim_{h \to 0} \frac{a^x a^h - a^x}{h}$$

Factor out the common term of a^x.

$$= a^x \lim_{h \to 0} \frac{a^h - 1}{h}, \tag{6.1}$$

where we have used the usual rules of exponents. Unfortunately, we have, at present, no means of computing the limit in (6.1). Nonetheless, **assuming** the limit exists, (6.1) says that

$$\frac{d}{dx} a^x = (\text{constant}) \quad a^x. \tag{6.2}$$

Further, (6.2) is consistent with what we observed graphically in Figures 2.32a and 2.32b. The question that we now face is does

$$\lim_{h \to 0} \frac{a^h - 1}{h}$$

exist for all (or any) $a > 1$? We explore this limit numerically in the following table for $a = 2$.

h	$\dfrac{2^h - 1}{h}$		h	$\dfrac{2^h - 1}{h}$
0.1	0.7177346		-0.1	0.6696701
0.01	0.6955550		-0.01	0.6907505
0.001	0.6933875		-0.001	0.6929070
0.0001	0.6931712		-0.0001	0.6931232
0.00001	0.6931496		-0.00001	0.6931448
0.000001	0.6931474		-0.000001	0.6931469
0.0000001	0.6931474		-0.0000001	0.6931472

The numerical evidence suggests that the limit in question exists and that

$$\lim_{h \to 0} \frac{2^h - 1}{h} \approx 0.693147.$$

Similarly for $a = 3$, we have the following table.

h	$\dfrac{3^h - 1}{h}$	h	$\dfrac{3^h - 1}{h}$
0.1	1.1612317	-0.1	1.0404154
0.01	1.1046692	-0.01	1.0925996
0.001	1.0992160	-0.001	1.0980090
0.0001	1.0986726	-0.0001	1.0985519
0.00001	1.0986183	-0.00001	1.0986063
0.000001	1.0986129	-0.000001	1.0986117
0.0000001	1.0986120	-0.0000001	1.0986122

Once again, the numerical evidence suggests that the limit in question exists and that

$$\lim_{h \to 0} \frac{3^h - 1}{h} \approx 1.098612.$$

The approximate values of these limits are largely unremarkable until you observe that

$$\ln 2 \approx .693147 \text{ and } \ln 3 \approx 1.098612.$$

Next, consider the exponential function $f(x) = a^x$ for $0 < a < 1$, for instance, $f(x) = \left(\frac{1}{2}\right)^x$. Notice that we may write such a function as $f(x) = b^{-x}$, where $b = \frac{1}{a} > 1$. For example, for $b = \frac{1}{2}$, we have

$$f(x) = \left(\frac{1}{2}\right)^x = 2^{-x}.$$

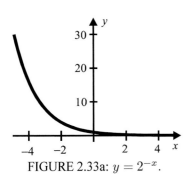

FIGURE 2.33a: $y = 2^{-x}$.

The graph of this type of exponential looks something like the graph of $y = 2^{-x}$ shown in Figure 2.33a.

In this case, the slopes of the tangent lines and hence too, the values of the derivative are always negative, but get less negative and tend toward zero as you look from left to right. This leads us to the sketch of $f'(x)$ shown in Figure 2.33b. This time, the graph of the derivative looks like the reflection of the original graph through the x-axis. Look back at how we derived equation (6.1) and notice that the argument does not depend on having $a > 1$. So, even if $0 < a \leq 1$, we have

$$f'(x) = a^x \lim_{h \to 0} \frac{a^h - 1}{h}.$$

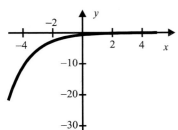

FIGURE 2.33b: The derivative of $f(x) = 2^{-x}$.

Exploring this limit numerically for $a = \frac{1}{2}$ we get the following table.

h	$\dfrac{.5^h - 1}{h}$	h	$\dfrac{.5^h - 1}{h}$
0.1	-0.6696701	-0.1	-0.7177346
0.01	-0.6907505	-0.01	-0.6955550
0.001	-0.6929070	-0.001	-0.6933875
0.0001	-0.6931232	-0.0001	-0.6931712
0.00001	-0.6931448	-0.00001	-0.6931496
0.000001	-0.6931469	-0.000001	-0.6931474
0.0000001	-0.6931472	-0.0000001	-0.6931470

Once again, the numerical evidence suggests that the limit in question exists and that

$$\lim_{h \to 0} \frac{0.5^h - 1}{h} \approx -0.693147.$$

We also observe that

$$\ln 0.5 \approx -0.693147.$$

With all of this numerical evidence, we can now conjecture the following result.

Theorem 6.1   For any constant $a > 0$,

$$\frac{d}{dx}a^x = a^x \ln a. \qquad (6.3)$$

While the proof of this result is complete except for the evaluation of the limit in question, we will not be in a position to complete this work until Chapter 6. For the moment, you should be content with the strong numerical, graphical and (nearly complete) algebraic arguments supporting this conjecture.

You should notice that (6.3) is consistent with what we observed in Figures 2.32a, 2.32b, 2.33a and 2.33b. Recall that for $0 < a < 1$, $\ln a < 0$. This accounts for why the graph of the derivative looks like the reflection of the graph of the original function through the x-axis.

Example 6.1   Finding the Rate of Change of an Investment

If the value of an investment doubles every year, its value after t years is given by $v(t) = 100 \, 2^t$. Find the instantaneous percentage rate of change of the worth.

Solution   The instantaneous rate of change is the derivative

$$v'(t) = 100 \, 2^t \ln 2.$$

The relative rate of change is then

$$\frac{v'(t)}{v(t)} = \frac{100 \, 2^t \ln 2}{100 \, 2^t} = \ln 2 \approx 0.693.$$

The percentage change is about 69.3% per year. This is surprising to most people. A percentage rate of 69.3% will double your investment each year if it is compounded "continuously." We explore the notion of continuous compounding of interest further in the exercises.

■

The most commonly used base (by far) is the (naturally occurring) irrational number e. You should immediately notice the significance of this choice. Since $\ln e = 1$, the derivative of $f(x) = e^x$ is simply e^x. We have the following results.

Corollary 6.1 $\dfrac{d}{dx}e^x = e^x$ and $\dfrac{d}{dx}e^{-x} = -e^{-x}.$

Proof. The proof of the first of these follows immediately from Theorem 6.1. The second result follows from the first, using the quotient rule. This is left as an exercise.

■

You will probably agree that these are the two easiest derivative formulas to remember. In section 2.5, we looked at a simple model of the oscillations of a weight hanging from a spring. Using the cosine function, we had a model that produced the oscillations of the spring, but did not reproduce the physical reality of the spring eventually coming to rest. Coupling the trigonometric functions with the exponential functions produces a more realistic model.

FIGURE 2.34: Spring-mass system.

FIGURE 2.35: $u\,(t) = e^{-t}\cos t$.

FIGURE 2.36a: $y = \ln x$.

FIGURE 2.36b: The derivative of $f\,(x) = \ln x$.

Example 6.2   Finding the Velocity of a Hanging Weight

If we build damping (i.e., resistance to the motion due to friction, for example) into our model spring-mass system (see Figure 2.34), the vertical displacement of a weight hanging from a spring is sometimes described by

$$u(t) = Ae^{\alpha t}\cos(\omega t) + Be^{\alpha t}\sin(\omega t),$$

where A, B, α and ω are constants. Suppose that

$$u(t) = e^{-t}\cos t.$$

Sketch a graph of the motion of the weight and find its velocity at any time t.

Solution   Figure 2.35 displays a graph of the motion of the weight. Notice that it seems to briefly oscillate and then quickly come to rest at $u = 0$. You should observe that this is precisely the kind of behavior you expect from your car's suspension system (the spring-mass system most familiar to you) as you hit bumps in the road.

The velocity of the weight is given by the derivative. From the product rule, we get

$$u'(t) = \frac{d}{dt}(e^{-t})\cos t + e^{-t}\frac{d}{dt}(\cos t)$$

$$= -e^{-t}\cos t - e^{-t}\sin t$$

$$= -e^{-t}(\cos t + \sin t).$$

■

Derivative of the Natural Logarithm

The natural logarithm function $\ln x$ is closely connected with exponential functions. We've already seen it arise as a part of the general exponential derivative formula (6.3). Recall from Chapter 0 that the graph of the natural logarithm looks like the one shown in Figure 2.36a.

The function is defined only for $x > 0$ and as you look to the right, the graph always rises. Thus, the slopes of the tangent lines and hence, also the values of the derivative are always positive. Further, as $x \to \infty$, the slopes of the tangent lines become less positive and seem to approach 0. On the other hand, as x approaches 0 from the right, the graph gets steeper and steeper and hence, the derivative gets larger and larger, without bound. The graph shown in Figure 2.36b is consistent with all of the preceding observations. Do you recognize it?

Using the definition of derivative, we get the following for $f(x) = \ln x$.

$$f'(x) = \lim_{h\to 0}\frac{f(x+h) - f(x)}{h}$$

$$= \lim_{h\to 0}\frac{\ln(x+h) - \ln x}{h}.$$

Unfortunately, we don't yet have the mathematical machinery necessary to evaluate this limit or even to determine if it exists. (We'll develop all of this in Chapter 6.) Our only option is then to investigate this limit numerically. In the following table, we compute some values for $x = 2$.

h	$\dfrac{\ln(2+h)-\ln 2}{h}$	h	$\dfrac{\ln(2+h)-\ln 2}{h}$
0.1	0.4879016	−0.1	0.5129329
0.01	0.4987542	−0.01	0.5012542
0.001	0.4998750	−0.001	0.5001250
0.0001	0.4999875	−0.0001	0.5000125
0.00001	0.4999986	−0.00001	0.5000013
0.000001	0.4999999	−0.000001	0.5000001
0.0000001	0.5000002	−0.0000001	0.5000000

This leads us to the conjecture

$$f'(2) = \lim_{h \to 0} \frac{\ln(2+h)-\ln 2}{h} \approx \frac{1}{2}.$$

Similarly, we compute some values for $x = 3$.

h	$\dfrac{\ln(3+h)-\ln 3}{h}$	h	$\dfrac{\ln(3+h)-\ln 3}{h}$
0.1	0.3278982	−0.1	0.3390155
0.01	0.3327790	−0.01	0.3338901
0.001	0.3332778	−0.001	0.3333889
0.0001	0.3333278	−0.0001	0.3333389
0.00001	0.3333328	−0.00001	0.3333339
0.000001	0.3333333	−0.000001	0.3333334

This leads us to the conjecture

$$f'(3) = \lim_{h \to 0} \frac{\ln(3+h)-\ln 3}{h} \approx \frac{1}{3}.$$

You should compute approximations to a few more values, for instance, $f'(4)$, $f'(5)$ and so on. Look back one more time at Figure 2.36b. Do you recognize the function, now? You should easily be convinced to conjecture the following result.

Theorem 6.2 For $x > 0$,

$$\frac{d}{dx}(\ln x) = \frac{1}{x}. \tag{6.4}$$

The proof of this result will have to wait until Chapter 6. However, given the significant graphical and numerical evidence, the result should be believable for the present.

Example 6.3   **Derivatives of Logarithms**

Find the derivative of $f(x) = x \ln x$ and $g(x) = \ln x^2$.

Solution   Using the product rule, we get

$$f'(x) = (1) \ln x + x \left(\frac{1}{x}\right) = \ln x + 1.$$

We do not yet have a derivative rule for the composition of functions making up $g(x)$. We introduce such a rule, called the **chain rule**, in the next section. However, using the properties of logarithms, recall that we can rewrite $g(x) = \ln x^2 = 2\ln x$, and using (6.4), we get

$$g'(x) = 2\frac{d}{dx}(\ln x) = 2\left(\frac{1}{x}\right) = \frac{2}{x}.$$

∎

Exercises 2.6

1. (writing) The graph of $f(x) = e^x$ curves upward in the interval from $x = -1$ to $x = 1$. Interpreting $f'(x) = e^x$ as the slopes of tangent lines and noting that the larger x is, the larger e^x is, explain why the graph curves upward. For larger values of x, the graph of $f(x) = e^x$ appears to shoot straight up with no curve. Using the tangent line, determine if this is correct or just an optical illusion.

2. (writing) The graph of $f(x) = \ln x$ appears to get flatter as x gets larger. Interpret the derivative $f'(x) = \frac{1}{x}$ as the slopes of tangent lines to determine if this is correct or just an optical illusion.

3. (writing) Graphically compare and contrast the functions x^2, x^3, x^4 and e^x for $x > 0$. Sketch the graphs for large x (and *very* large y's) and compare the relative growth rates of the functions. In general, how does the exponential function compare to polynomials?

4. (writing) Graphically compare and contrast the functions $x^{1/2}$, $x^{1/3}$, $x^{1/4}$ and $\ln x$ for $x > 1$. Sketch the graphs for large x and compare the relative growth rates of the functions. In general, how does $\ln x$ compare to $\sqrt[n]{x}$?

In exercises 5–28, find the derivative of the function.

5. $f(x) = 4e^x - x$ **6.** $f(x) = \sin x - 3e^x$

7. $f(x) = xe^x$ **8.** $f(x) = e^x \cos x$

9. $f(x) = x + 2^x$ **10.** $f(x) = x4^x$

11. $f(x) = 2e^{x+1}$ **12.** $f(x) = (1/e)^x$

13. $f(x) = (1/3)^x$ **14.** $f(x) = 4^{-x}$

15. $f(x) = 4^{-x+1}$ **16.** $f(x) = (1/2)^{1-x}$

17. $f(x) = \dfrac{e^x}{x}$ **18.** $f(x) = \dfrac{x}{e^x}$

19. $f(x) = \ln 2x$ **20.** $f(x) = \ln 8x$

21. $f(x) = \ln x^2$ **22.** $f(x) = x\ln x$

23. $f(x) = e^{2x} = e^x e^x$ **24.** $f(x) = e^{3x}$

25. $f(x) = x^2 e^{-x}$ **26.** $f(x) = e^x \ln x$

27. $f(x) = \dfrac{\ln x}{x}$ **28.** $f(x) = \dfrac{e^x}{2^x}$

In exercises 29–36, find an equation of the tangent line to $y = f(x)$ at $x = 1$.

29. $f(x) = 3e^x$ **30.** $f(x) = 2e^{x-1}$

31. $f(x) = 3^x$ **32.** $f(x) = 2^x$

33. $f(x) = xe^x$ **34.** $f(x) = \dfrac{x}{e^x}$

35. $f(x) = x^2 \ln x$ **36.** $f(x) = 2\ln x^3$

In exercises 37–40, the value of an investment is given by $v(t)$. Find the instantaneous percentage rate of change.

37. $v(t) = 100\, 3^t$ (an investment of \$100 triples every year)

38. $v(t) = 100\, 4^t$ (an investment of \$100 quadruples every year)

39. $v(t) = 100\, e^t$ **40.** $v(t) = 100\, e^{-t}$

41. An investment of A dollars receiving r percent (per year) interest compounded continuously will be worth $f(t) = Ae^{rt}$ dollars after t years. APR can be defined as $f(1)/A$, the relative increase of worth in one year. Find the APR for the following interest rates:

(a) 5%   (b) 10%   (c) 20%   (d) $\ln 2\%$
(e) 100%

42. Determine the interest rate needed to obtain an APR of (a) 100% and (b) 10%.

43. The motion of a spring is described by $f(t) = e^{-t}\cos t$. Compute the velocity at time t. Graph the velocity function. When is the velocity zero? What is the position of the spring when the velocity is zero? When does the velocity reach a maximum or minimum value? What is the position of the spring when the velocity reaches a maximum or minimum? In a paragraph, describe the motion of the spring.

44. The motion of a spring is given by $f(t) = e^{-t}\sin t$. Compute the velocity at time t. Graph the velocity function. When is the velocity zero? What is the position of the spring when the velocity is zero? When does the velocity reach a maximum or minimum value? What is the position of the spring when the velocity reaches a maximum or minimum? In a paragraph, describe the motion of the spring.

45. The **Padé approximation** of e^x is the function of the form $f(x) = \dfrac{a + bx}{1 + cx}$ for which the values of $f(0)$, $f'(0)$ and $f''(0)$

match the corresponding values of e^x. Show that these values all equal 1 and find the values of a, b and c that make $f(0) = 1$, $f'(0) = 1$ and $f''(0) = 1$. Compare the graphs of $f(x)$ and e^x.

46. In a *World Almanac* or Internet resource, look up the population of the United States by decade for as many years as are available. Make sure there is a column indicating growth by decade both numerically and by percentage. The United States has had periods of both linear and exponential growth. Explain why linear growth corresponds to a constant numerical increase; during which decades has the numerical growth been (approximately) constant? Explain why exponential growth corresponds to a constant percentage growth. During which decades has the percentage growth been (approximately) constant?

47. (exploratory) In Chapter 0, we defined e as the limit $e = \lim\limits_{n \to 0}(1 + 1/n)^n$. There are a variety of other interesting properties of this important number. We discover one here. Graph the functions x^2 and 2^x. For $x > 0$, how many times do they intersect? Graph the functions x^3 and 3^x. For $x > 0$, how many times do they intersect? Try the pair $x^{2.5}$ and 2.5^x and the pair x^4 and 4^x. What would be a reasonable conjecture for the number of intersections ($x > 0$) of the functions x^a and a^x? By trial and error, verify graphically that e is the value of a at which the number of intersections changes from 2 to 0. How many intersections are there for $a = e$?

48. (exploratory) Using the properties of exponential and logarithmic functions, we can discover derivative rules that preview the chain rule of the next section. Since $e^{2x} = (e^x)^2$, show that $\frac{d}{dx}(e^{2x}) = 2e^{2x}$. Since $e^{3x} = e^x e^{2x}$, show that $\frac{d}{dx}(e^{3x}) = 3e^{3x}$. Use mathematical induction to prove that $\frac{d}{dx}(e^{nx}) = ne^{nx}$ for any positive integer n. As we will see, the formula holds for any real number n. Using $\ln x^a = a \ln x$, derive a formula for the derivative of $\ln x^a$. Using $\ln ax = \ln a + \ln x$, find the derivative of $\ln ax$.

2.7   The Chain Rule

Over the last several sections, we have been steadily refining a model for the motion of a weight hanging from a spring. In section 2.5, we found that the cosine function could be used to describe the oscillations of the weight. In section 2.6, we found that if we combined this with an exponential function, we could describe an oscillatory motion that gradually dies out. However, the model $u(t) = e^{-t}\cos t$ represents only one possible motion (specifically, one where the motion dies out quite rapidly). If we wanted to change the rate at which the motion dies out, we could change the exponential function used. If we wanted to change the period of the oscillations, we could change the trigonometric function used. For example, $u(t) = e^{-t/6}\cos 4t$ describes a somewhat different motion (see Figure 2.37).

FIGURE 2.37: $y = e^{-t/6}\cos 4t$.

This looks much more like the gradual dying out of a spring's motion that you might expect, compared to the motion shown in Figure 2.35. Unfortunately, we do not presently have any rules for computing the derivative of $e^{-t/6}$ or $\cos(4t)$. Rather than resorting to the definition of derivative, we will derive a rule for computing the derivative of the composition of two functions. Using this, we can, for example, combine our knowledge of the derivatives of $-t/6$ and e^t to obtain the derivative of the composition function $e^{-t/6}$.

Recall that in example 5.2, we used the product rule to show that

$$\frac{d}{dx}\left(\sin^2 x\right) = 2\sin x \cos x.$$

Using this result and the product rule, notice that

$$\frac{d}{dx}\left(\sin^3 x\right) = \frac{d}{dx}\left(\sin x \sin^2 x\right)$$

$$= \cos x \sin^2 x + \sin x \left(2\sin x \cos x\right)$$

$$= 3\sin^2 x \cos x.$$

We leave it as a straightforward exercise to extend this result to

$$\frac{d}{dx}\left(\sin^4 x\right) = 4\sin^3 x \cos x.$$

You should notice that in each case, we have brought the exponent down, lowered the power by one and then multiplied by the derivative of $\sin x$. Notice that we can write $\sin^4 x$ as the composite function

$$\sin^4 x = (\sin x)^4 = f\left(g\left(x\right)\right),$$

where $g\left(x\right) = \sin x$ and $f\left(x\right) = x^4$. Further, observe that the derivative of the composite function is

$$\frac{d}{dx}\left[f\left(g\left(x\right)\right)\right] = \frac{d}{dx}\left(\sin^4 x\right) = 4\sin^3 x \cos x = f'\left(g\left(x\right)\right) g'\left(x\right).$$

This is an example of the **chain rule**, which has the following general form.

Theorem 7.1   (Chain Rule) If g is differentiable at x and f is differentiable at $g(x)$, then

$$\frac{d}{dx}\left[f(g(x))\right] = f'(g(x))\, g'(x).$$

Proof. The theorem is true even if $g'(x) = 0$, but our proof requires $g'(x) \neq 0$. Let $F(x) = f(g(x))$. Then,

Since $F(x) = f(g(x))$.

$$\frac{d}{dx}\left[f(g(x))\right] = F'(x) = \lim_{h \to 0} \frac{F(x+h) - F(x)}{h}$$

$$= \lim_{h \to 0} \frac{f(g(x+h)) - f(g(x))}{h}$$

Multiply numerator and denominator by $g(x+h) - g(x)$.

$$= \lim_{h \to 0} \frac{f(g(x+h)) - f(g(x))}{h} \frac{g(x+h) - g(x)}{g(x+h) - g(x)}$$

Regroup terms.

$$= \lim_{h \to 0} \frac{f(g(x+h)) - f(g(x))}{g(x+h) - g(x)} \lim_{h \to 0} \frac{g(x+h) - g(x)}{h}$$

Since as $x \to 0$, $g(x+h) \to g(x)$.

$$= \lim_{g(x+h) \to g(x)} \frac{f(g(x+h)) - f(g(x))}{g(x+h) - g(x)} \lim_{h \to 0} \frac{g(x+h) - g(x)}{h}$$

$$= f'(g(x))\, g'(x).$$

You will be asked in the exercises to fill in some of the gaps in this argument. In particular, you should identify why we need $g'(x) \neq 0$ in this proof. ∎

It is often helpful to think of the chain rule in Leibniz notation. If $y = f(u)$ and $u = g(x)$, then $y = f(g(x))$ and the chain rule says that

$$\frac{dy}{dx} = \frac{dy}{du}\frac{du}{dx}. \tag{7.1}$$

Example 7.1   Using the Chain Rule
Differentiate $y = (x^3 + x - 1)^5$.

Solution   For $u = x^3 + x - 1$, note that $y = u^5$. From (7.1), we have

$$\frac{dy}{dx} = \frac{dy}{du}\frac{du}{dx}$$

Since $y = u^5$.

$$= \frac{d}{du}\left(u^5\right) \frac{du}{dx}$$

$$= 5u^4 \frac{d}{dx}(x^3 + x - 1)$$

$$= 5(x^3 + x - 1)^4\,(3x^2 + 1).$$

The last term is the derivative of the expression inside the parentheses (often referred to as the *derivative of the inside*).

∎

Example 7.2   Using the Chain Rule for an Exponential Function

Find $\dfrac{d}{dt}\left(e^{-t/6}\right)$.

Solution   Let $u = -t/6$. Then, from (7.1),

$$\frac{d}{dt}\left(e^{-t/6}\right) = \frac{d}{dt}\left(e^{u}\right) = \frac{d}{du}\left(e^{u}\right)\frac{du}{dt}$$

$$= e^{u}\frac{d}{dt}(-t/6) = -\frac{1}{6}e^{-t/6}.$$

Notice that here, the derivative of the *inside* is the derivative of the exponent.

∎

Example 7.3   Finding the Velocity of a Hanging Weight

Suppose the position of a weight hanging from a spring is given by $u(t) = e^{-t/6}\cos 4t$. Find the velocity of the weight at any time t.

Solution   Using the product rule and the chain rule, we have

By the product rule.

$$v(t) = u'(t) = \frac{d}{dt}(e^{-t/6})(\cos 4t) + e^{-t/6}\frac{d}{dt}(\cos 4t)$$

By the chain rule.

$$= e^{-t/6}\underbrace{\frac{d}{dt}\left(-\frac{t}{6}\right)}_{\text{derivative of the inside}}(\cos 4t) + e^{-t/6}(-\sin 4t)\underbrace{\frac{d}{dt}(4t)}_{\text{derivative of the inside}}$$

$$= e^{-t/6}\left(-\frac{1}{6}\right)\cos 4t + e^{-t/6}(-\sin 4t)(4)$$

$$= -\frac{1}{6}e^{-t/6}\cos 4t - 4e^{-t/6}\sin 4t.$$

∎

You are now in a position to calculate the derivative of virtually every function you can write down, by using the rules we have at our disposal. In many cases, the chain rule must be combined with other differentiation rules.

Example 7.4   The Derivatives of Some Similar Trigonometric Functions

Compute the derivative of $f(x) = \cos x^3$, $g(x) = \cos^3 x$, and $h(x) = \cos 3x$.

Solution   Note the differences in these three functions. Using the implied parentheses we normally do not bother to include, we have $f(x) = \cos(x^3)$, $g(x) = (\cos x)^3$, and $h(x) = \cos(3x)$. For the first function, we have

$$f'(x) = \frac{d}{dx}\cos(x^3) = -\sin(x^3)\underbrace{\frac{d}{dx}x^3}_{\text{derivative of the inside}} = -\sin(x^3)\left(3x^2\right) = -3x^2\sin x^3.$$

Next, we have

$$g'\left(x\right) = \frac{d}{dx}(\cos x)^3 = 3(\cos x)^2 \underbrace{\frac{d}{dx}\cos x}_{\text{derivative of the inside}}$$

$$= 3(\cos x)^2 \left(-\sin x\right)$$

$$= -3\sin x \cos^2 x.$$

Finally, we have

$$h'\left(x\right) = \frac{d}{dx}\cos(3x) = -\sin(3x) \underbrace{\frac{d}{dx}(3x)}_{\text{derivative of the inside}} = -\sin(3x)\,(3) = -3\sin 3x.$$

■

Example 7.5   **A Derivative Involving a Chain Rule and a Quotient Rule**

Find the derivative of $f\left(x\right) = \sin\left(\dfrac{2x}{x+1}\right)$.

Solution   We have

By the chain rule.

$$f'\left(x\right) = \cos\left(\frac{2x}{x+1}\right) \underbrace{\frac{d}{dx}\left(\frac{2x}{x+1}\right)}_{\text{derivative of the inside}}$$

By the quotient rule.

$$= \cos\left(\frac{2x}{x+1}\right)\frac{2\left(x+1\right) - 2x\left(1\right)}{(x+1)^2}$$

$$= \cos\left(\frac{2x}{x+1}\right)\frac{2}{(x+1)^2}.$$

■

Example 7.6   **A Derivative Involving a Quotient Rule and Several Chain Rules**

Find the derivative of $f\left(x\right) = \dfrac{\tan 2x^2}{(2x+1)^2}$.

Solution   We have

By the quotient rule.

$$f'\left(x\right) = \frac{\frac{d}{dx}\left(\tan 2x^2\right)(2x+1)^2 - \tan 2x^2\frac{d}{dx}(2x+1)^2}{[(2x+1)^2]^2}$$

By the chain rule.

$$= \frac{\sec^2 2x^2\frac{d}{dx}(2x^2)(2x+1)^2 - \tan 2x^2\,(2)\,(2x+1)^1\frac{d}{dx}(2x+1)}{(2x+1)^4}$$

$$= \frac{\sec^2 2x^2\,(4x)\,(2x+1)^2 - \tan 2x^2\,(2)\,(2x+1)\,(2)}{(2x+1)^4}$$

$$= \frac{(2x+1)\left[4x(2x+1)\sec^2 2x^2 - 4\tan 2x^2\right]}{(2x+1)^4}$$

$$= \frac{4x(2x+1)\sec^2 2x^2 - 4\tan 2x^2}{(2x+1)^3}.$$

■

Next, we consider a simple application.

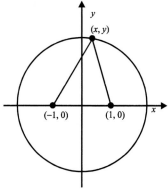

FIGURE 2.38: The path of light.

Since $y^2 = 4 - x^2$.

Example 7.7   The Shortest Path for Reflecting Light

Imagine there is a light source at the point $(1, 0)$ and we want to get light to the point $(-1, 0)$ by reflecting off a circular mirror described by $x^2 + y^2 = 4$ (see Figure 2.38). We want to find the point (x, y) that minimizes the total distance traveled. (It is a remarkable fact that light follows the fastest path available.) Write out the functions for the total distance traveled by the light and use the derivative to determine the shortest (fastest) path.

Solution   First, note that the distance from $(1, 0)$ to any point (x, y) on the circle is found from the usual distance formula:

$$
\begin{aligned}
d\{(x, y), (1, 0)\} &= \sqrt{(x-1)^2 + (y-0)^2} \\
&= \sqrt{(x-1)^2 + y^2} \\
&= \sqrt{(x-1)^2 + 4 - x^2} \\
&= \sqrt{x^2 - 2x + 1 + 4 - x^2} \\
&= \sqrt{5 - 2x}, \tag{7.2}
\end{aligned}
$$

where we have substituted $y^2 = 4 - x^2$, since (x, y) is a point on the circle. (Note that $-2 \le x \le 2$ for such points.) Similarly, the distance from (x, y) to $(-1, 0)$ is

$$
\begin{aligned}
d\{(x, y), (-1, 0)\} &= \sqrt{(x+1)^2 + y^2} \\
&= \sqrt{(x^2 + 2x + 1) + (4 - x^2)} \\
&= \sqrt{5 + 2x}. \tag{7.3}
\end{aligned}
$$

The total distance traveled by the light is then found by adding the distances in (7.2) and (7.3):

$$
d(x) = \sqrt{5 - 2x} + \sqrt{5 + 2x}
$$

and the derivative of the distance function is

$$
\begin{aligned}
d'(x) &= \frac{1}{2\sqrt{5 - 2x}} (-2) + \frac{1}{2\sqrt{5 + 2x}} (2) \\
&= \frac{-1}{\sqrt{5 - 2x}} + \frac{1}{\sqrt{5 + 2x}}.
\end{aligned}
$$

If $x < 0$, notice that

$$
5 - 2x > 5 + 2x.
$$

It then follows that

$$
\frac{1}{\sqrt{5 - 2x}} < \frac{1}{\sqrt{5 + 2x}}
$$

and so, $d'(x) > 0$ on the interval $[-2, 0)$. Similarly, if $x > 0$, we have that $d'(x) < 0$. This says that on the interval $[-2, 0)$, the slopes of the tangent lines are positive and hence, the graph of the function rises as you look from left to right. Consequently, the distance is increasing as x increases on the interval $[-2, 0)$. Likewise, since $d'(x) < 0$ on the interval $(0, 2]$, the slopes of the tangent lines are negative and hence, the graph of the function falls as you look from left to right on the interval $(0, 2]$. This says that $d(x)$ (the total distance) is smallest (i.e., the distance is **minimized**) when $x = 2$ or $x = -2$ (the distance is the same at both points). A sketch of the graph of $y = d(x)$ (see Figure 2.39) indicates that $x = 2$ and $x = -2$ do indeed give the minimum distance.

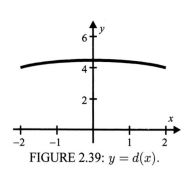

FIGURE 2.39: $y = d(x)$.

■

Exercises 2.7

1. (writing) If Fred can run 10 mph and Greg can run twice as fast as Fred, how fast can Greg run? The answer is obvious for most people. Formulate this simple problem as a chain rule calculation and conclude that the chain rule (in this context) is obvious.

2. (writing) The biggest challenge in computing the derivatives of $e^{x \sin x}$, $e^x \sin x$ and $xe^{\sin x}$ is knowing which rule (product, chain, etc.) to use when. Discuss how you know which rule to use when. (Hint: Think of the order in which you would perform operations to compute the value of each function for a specific choice of x.)

3. (writing) One simple implication of the chain rule is: if $g(x) = f(x-a)$, then $g'(x) = f'(x-a)$. Explain this derivative graphically: how does $g(x)$ compare to $f(x)$ graphically and why do the slopes of the tangent lines relate as the formula indicates?

4. (writing) Another simple implication of the chain rule is: if $h(x) = f(2x)$, then $h'(x) = 2f'(2x)$. Explain this derivative graphically: how does $h(x)$ compare to $f(x)$ graphically and why do the slopes of the tangent lines relate as the formula indicates?

In exercises 5–8, find the derivative with and without using the chain rule.

5. $f(x) = (x^3 - 1)^2$　　　**6.** $f(x) = (x^2 + 2x + 1)^2$

7. $f(x) = e^{-2x}$　　　**8.** $f(x) = \sin^2 x$

In exercises 9–40, find the derivative of each function.

9. $f(x) = \sqrt{x^2 + 4}$　　　**10.** $f(x) = x^3 + 4x$

11. $f(x) = (x^3 + x - 1)^3$　　**12.** $f(x) = (2x + 3)^{2/3}$

13. $f(x) = \sin(2x^2 + 3)$　　**14.** $f(x) = \sqrt[3]{\cos x}$

15. $f(x) = \cos^3 x$　　　**16.** $f(x) = \cos x^3$

17. $f(x) = \tan^2 x$　　　**18.** $f(x) = \tan x^2$

19. $f(x) = e^{x^2}$　　　**20.** $f(x) = (e^x)^2$

21. $f(x) = e^{1/x}$　　　**22.** $f(x) = (e^{2x} + 1)^{1/3}$

23. $f(x) = \ln(x^2 + 1)$　　**24.** $f(x) = \ln(\cos x)$

25. $f(x) = x^2 \sin 4x$　　**26.** $f(x) = xe^{x^2}$

27. $f(x) = \sec^3 4x$　　**28.** $f(x) = e^{\sin 2x}$

29. $f(x) = \dfrac{\sin x^2}{x^2}$　　**30.** $f(x) = \dfrac{3}{\sqrt{x^2 + 4}}$

31. $f(x) = \ln(x^3 + 3x)$　　**32.** $f(x) = \sin(\ln x)$

33. $f(x) = \sqrt{\sin x^2}$　　**34.** $f(x) = \cos\sqrt{x^2 + 1}$

35. $f(x) = (\ln x)^8$　　　**36.** $f(x) = e^{\ln x^2}$

37. $f(x) = \dfrac{\sin x^2}{\tan x^2}$　　**38.** $f(x) = x^2 \tan x$

39. $f(x) = e^{-3x} \sin 2x$　　**40.** $f(x) = e^{-2x} \cos 4x$

In exercises 41 and 42, find an equation of the tangent line to $y = f(x)$ at $x = a$.

41. $f(x) = \sqrt{x^2 + 16}$, $a = 3$

42. $f(x) = \sin x^2$, $a = \sqrt{\pi}$

In exercises 43 and 44, use the position function to find the velocity at time $t = 2$.

43. $s(t) = e^{-4t} \sin 3t$　　**44.** $s(t) = e^{-5t} \cos 2t$

In exercises 45–48, compute $f''(x)$, $f'''(x)$ and $f^{(4)}(x)$ and identify a pattern for the nth derivative $f^{(n)}(x)$.

45. $f(x) = \sqrt{2x + 1}$　　**46.** $f(x) = \sin 3x$

47. $f(x) = e^{2x}$　　　**48.** $f(x) = \dfrac{2}{x + 1}$

In exercises 49–52, find a function $g(x)$ such that $g'(x) = f(x)$.

49. $f(x) = (x^2 + 3)^2 (2x)$　　**50.** $f(x) = 2xe^{x^2}$

51. $f(x) = 3 \sin 2x$　　**52.** $f(x) = \cos 3x$

53. In the figure, light reflects from the point $(L, 0)$ to the point (x, y) on the circle $x^2 + y^2 = r^2$ to the point $(-L, 0)$. The distance traveled by the light is then $D = \sqrt{(x - L)^2 + y^2} + \sqrt{(x + L)^2 + y^2}$. Graphically find x to maximize the distance traveled (replace y with $\sqrt{r^2 - x^2}$). Show that the value of the derivative of D at that point is 0. Are there other points at which the derivative is 0? Physically what do these points represent?

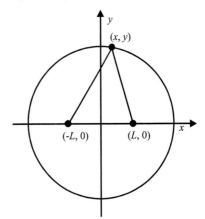

54. The motion of an object traveling in a circular path at constant speed can be described by the equations $x(t) = a \cos bt$

and $y(t) = a \sin bt$ for constants a and b. Compute the speed $\sqrt{(x'(t))^2 + (y'(t))^2}$. To model an object with nonconstant speed, the argument of the trig functions needs to be adjusted. What is the speed if $x(t) = \cos t^2$ and $y(t) = \sin t^2$? Find a function $g(t)$ such that $x(t) = \cos g(t)$ and $y(t) = \sin g(t)$ has speed equal to $3t^2$.

55. Given that $a^x = e^{x \ln a}$ for $a > 0$, use the chain rule to find the derivative of $f(x) = a^x$.

56. Which steps in our outline of the proof of the chain rule are not well documented? Where do we use the assumption that $g'(x) \neq 0$?

57. (exploratory) A guitar string of length L, density T and tension p will vibrate at the frequency $f = \frac{1}{2L}\sqrt{\frac{T}{p}}$. Compute the derivative $\frac{df}{dp}$, where we think of p as the independent variable and treat T and L as constants. Interpret this derivative in terms of a guitarist tightening or loosening the string to "tune" it. Compute the derivative $\frac{df}{dL}$ and interpret it in terms of a guitarist playing notes by pressing the string against a fret.

58. (exploratory) Newton's second law of motion is $F = ma$, where m is the mass of the object that undergoes an acceleration a due to an applied force F. This law is accurate at low speeds. At high speeds, we use the corresponding formula from Einstein's theory of relativity, $F = m\dfrac{d}{dt}\left(\dfrac{v(t)}{\sqrt{1 - v^2(t)/c^2}} \right)$, where $v(t)$ is the velocity function and c is the speed of light. Compute $\dfrac{d}{dt}\left(\dfrac{v(t)}{\sqrt{1 - v^2(t)/c^2}} \right)$. What has to be "ignored" to simplify this expression to the acceleration $a = v'(t)$ in Newton's second law?

2.8   Implicit Differentiation and Related Rates

Compare the following two equations describing familiar curves:

$$y = x^2 + 3 \text{ (parabola)}$$

and

$$x^2 + y^2 = 4 \text{ (circle)}.$$

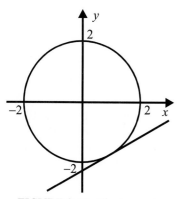

FIGURE 2.40: The tangent line at the point $\left(1, -\sqrt{3}\right)$.

The first equation defines y as a function of x **explicitly**, since for each x, the equation gives an explicit formula $y = f(x)$ for finding the corresponding value of y. On the other hand, the second equation defines y as a function of x **implicitly**, since there is no explicit formula for y in terms of x. In this case, for each x, the equation must be solved to obtain the corresponding value(s) of y.

Suppose we want to find the slope of the tangent line to the circle $x^2 + y^2 = 4$ at the point $(1, -\sqrt{3})$ (see Figure 2.40). Our derivative formulas are not directly applicable because we do not have y written explicitly as a function of x. We can, of course, think of the circle as the graph of two semicircles, $y = \sqrt{4 - x^2}$ and $y = -\sqrt{4 - x^2}$. Since we are interested in the point $(1, -\sqrt{3})$, we use the equation describing the bottom semicircle, $y = -\sqrt{4 - x^2}$. This is now an explicit formula for y in terms of x and we can use this to compute the derivative

$$y'(x) = -\frac{1}{2\sqrt{4 - x^2}}(-2x) = \frac{x}{\sqrt{4 - x^2}}.$$

So, the slope of the tangent line at the point $(1, -\sqrt{3})$ is then $y'(1) = \dfrac{1}{\sqrt{3}}$.

This calculation was not that bad, although we will soon see an easier way to do it. However, it's not always possible to explicitly solve for a function defined implicitly by a given equation. For example, van der Waal's equation relating the pressure P, volume V and temperature T of a gas has the form

$$\left(P + \frac{an^2}{V^2} \right)(V - nb) = nRT, \tag{8.1}$$

where a, n, b and R are constants. Notice the difficulty in solving this for V as a function of P. If we want the derivative $\dfrac{dV}{dP}$, we will need a method for computing the derivative directly from the implicit representation given in (8.1).

Consider each of the following calculations:

$$\frac{d}{dx}(x^3 + 4)^4 = 4(x^3 + 4)^3 (3x^2),$$

$$\frac{d}{dx}(\sin x - 3x)^4 = 4(\sin x - 3x)^3 (\cos x - 3)$$

and

$$\frac{d}{dx}(\tan x + 2)^4 = 4(\tan x + 2)^3 \sec^2 x.$$

Notice that each of these calculations has the form

$$\frac{d}{dx}[y(x)]^4 = 4[y(x)]^3 y'(x),$$

for some choice of the function $y(x)$. This last equation is simply an expression of the chain rule. We can use this notion to find the derivatives of functions defined implicitly by equations.

We first return to the simple case of the circle $x^2 + y^2 = 4$. Assuming this equation defines y as a function of x: $y = y(x)$, the equation is

$$x^2 + [y(x)]^2 = 4. \tag{8.2}$$

Differentiating both sides of equation (8.2) with respect to x, to we obtain

$$\frac{d}{dx}\left\{x^2 + [y(x)]^2\right\} = \frac{d}{dx}(4),$$

and hence,

$$2x + 2y(x)\,y'(x) = 0. \tag{8.3}$$

We can solve equation (8.3) for the derivative $y'(x)$, as follows. Subtracting $2x$ from both sides gives us

$$2y(x)\,y'(x) = -2x$$

and dividing by $2y(x)$ (assuming this is not zero), we get

$$y'(x) = \frac{-2x}{2y(x)} = \frac{-x}{y(x)}.$$

Notice that here $y'(x)$ is expressed in terms of both x and y. To get the slope at the point $(1, -\sqrt{3})$, substitute $x = 1$ and $y = -\sqrt{3}$. We have

$$y'(1) = \frac{-x}{y(x)}\bigg|_{x=1} = \frac{-1}{-\sqrt{3}} = \frac{1}{\sqrt{3}}.$$

(Notice that this is the same as we had found earlier by first solving for y explicitly and then differentiating.) This process of differentiating both sides of an equation with respect to x and then solving for $y'(x)$ is called **implicit differentiation**.

When faced with an equation implicitly defining y as a function of x, differentiate both sides with respect to x, taking care to use the chain rule wherever applicable. In particular, be careful to recognize that differentiating any function of y will require the chain rule:

$$\frac{d}{dx}g(y) = g'(y)\,y'(x).$$

Then, gather any terms with a factor of $y'(x)$ on one side of the equation, with the remaining terms on the other side of the equation and solve for $y'(x)$. We illustrate this process in the next several examples.

Example 8.1   Finding Slope Implicitly

Find $y'(x)$ for $x^2 + y^3 - 2y = 3$. Then, find the slope of the tangent line at the point $(2,1)$.

Solution   Since we can't (easily) solve for y in terms of x explicitly, we compute the derivative implicitly. Differentiating both sides with respect to x, we get

$$\frac{d}{dx}[x^2 + y^3 - 2y] = \frac{d}{dx}(3)$$

and so,

$$2x + 3y^2\,y'(x) - 2y'(x) = 0.$$

To solve for $y'(x)$, simply write all terms involving $y'(x)$ on one side of the equation and all other terms on the other side. We have

Subtracting $2x$ from both sides.

$$3y^2\,y'(x) - 2y'(x) = -2x$$

and hence, after factoring, we have

Factoring $y'(x)$ from both terms on the left side.

$$(3y^2 - 2)\,y'(x) = -2x.$$

Solving for $y'(x)$, we get

Dividing by $(3y^2 - 2)$.

$$y'(x) = \frac{-2x}{3y^2 - 2}.$$

Substituting $x = 2$ and $y = 1$, we find that the slope of the tangent line at the point $(2,1)$ is

$$y'(2) = \frac{-4}{3-2} = -4.$$

The equation of the tangent line is then found from

$$-4 = \frac{y-1}{x-2}$$

or

$$y - 1 = -4\,(x - 2).$$

We have plotted a graph of the equation and the tangent line in Figure 2.41 using the implicit plot mode in our computer algebra system.   ∎

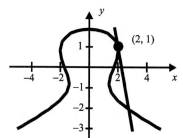

FIGURE 2.41: Tangent line at $(2,1)$.

Example 8.2   A Tangent Line for an Implicit Function

Find $y'(x)$ for $x^2y^2 - 2x = 4 - 4y$. Then, find an equation of the tangent line at the point $(2, -2)$.

Solution   Differentiating both sides with respect to x, we get

$$\frac{d}{dx}\left(x^2y^2 - 2x\right) = \frac{d}{dx}\left(4 - 4y\right).$$

Since the first term is the product of x^2 and y^2, we must use the product rule. We get

$$2x\,y^2 + x^2\,(2y)\,y'(x) - 2 = 0 - 4\,y'(x).$$

Grouping the terms with $y'(x)$ on one side and terms without $y'(x)$ on the other, we get

$$(2x^2y + 4)\, y'(x) = 2 - 2xy^2,$$

and we can solve for $y'(x)$:

$$y'(x) = \frac{2 - 2xy^2}{2x^2y + 4}.$$

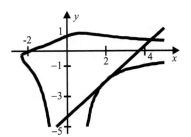

Substituting $x = 2$ and $y = -2$, we get the slope of the tangent line,

$$y'(2) = \frac{2 - 16}{-16 + 4} = \frac{7}{6}.$$

Finally, an equation of the tangent line is given by

$$y + 2 = \frac{7}{6}(x - 2).$$

FIGURE 2.42: Tangent line at $(2, -2)$.

We have plotted the curve and the tangent line at $(2, -2)$ in Figure 2.42 using the implicit plot mode in our computer algebra system.

■

You can use implicit differentiation to find a needed derivative from virtually any equation you can write down. We illustrate this next for an application.

Example 8.3   Rate of Change of Volume with Respect to Pressure
Suppose that van der Waal's equation for a specific gas is

$$\left(P + \frac{5}{V^2}\right)(V - 0.03) = 9.7. \tag{8.4}$$

Thinking of the volume V as a function of pressure P, use implicit differentiation to find the derivative $\dfrac{dV}{dP}$ at the point $(5, 1)$.

Solution   Differentiating both sides of (8.4) with respect to P, we have

$$\frac{d}{dP}\left[\left(P + 5V^{-2}\right)(V - 0.03)\right] = \frac{d}{dP}9.7.$$

From the product rule and the chain rule, we get

$$\left(1 - 10V^{-3}\frac{dV}{dP}\right)(V - 0.03) + \left(P + 5V^{-2}\right)\frac{dV}{dP} = 0.$$

Separating the terms that contain a factor of $\dfrac{dV}{dP}$ from those that do not, we get

$$\left[-10V^{-3}(V - 0.03) + P + 5V^{-2}\right]\frac{dV}{dP} = 0.03 - V.$$

Solving for $\dfrac{dV}{dP}$, we get

$$\frac{dV}{dP} = \frac{0.03 - V}{-10V^{-3}(V - 0.03) + P + 5V^{-2}}$$

FIGURE 2.43: Graph of implicitly defined function $V = f(P)$ and the tangent line at the point $(5, 1)$.

and so,

$$V'(5) = \frac{0.03 - 1}{-10(1)(0.97) + 5 + 5(1)} = \frac{-0.97}{0.03} = -\frac{97}{3}.$$

(The units are in terms of volume per unit pressure.) We show a graph of V as a function of P along with the tangent line to the graph at the point $(5, 1)$ in Figure 2.43.

■

Recall that, up to this point, we have proved the power rule

$$\frac{d}{dx}x^r = rx^{r-1}$$

only for **integer** exponents (see Theorems 3.3 and 4.3), although we have been freely using this result for any real exponent, r. Now that we have developed implicit differentiation, however, we finally have the tools we need to prove the power rule for the case of any **rational** exponent.

Theorem 8.1   For any rational exponent, r, $\dfrac{d}{dx}x^r = rx^{r-1}$.

Proof. Suppose that r is any rational number. Then $r = \dfrac{p}{q}$, for some integers p and q. Let

$$y = x^r = x^{p/q}. \tag{8.5}$$

Then, raising both sides of equation (8.5) to the qth power, we get

$$y^q = x^p. \tag{8.6}$$

If we differentiate both sides of equation (8.6) with respect to x, we get

$$\frac{d}{dx}(y^q) = \frac{d}{dx}(x^p).$$

From the chain rule, we have

$$q\,y^{q-1}\frac{dy}{dx} = p\,x^{p-1}.$$

It's now a simple matter to solve for $\dfrac{dy}{dx}$. We have

$$\frac{dy}{dx} = \frac{px^{p-1}}{qy^{q-1}} = \frac{px^{p-1}}{q\left(x^{p/q}\right)^{q-1}}$$

$$= \frac{px^{p-1}}{qx^{p-\frac{p}{q}}} = \frac{p}{q}x^{p-1-p+p/q}$$

$$= \frac{p}{q}x^{p/q-1} = rx^{r-1},$$

Since $y = x^{p/q}$.

Using the usual rules of exponents.

Since $\dfrac{p}{q} = r$.

as desired.

■

Related Rates

As we have mentioned, you can use implicit differentiation to find a derivative from nearly any equation. In many situations, each quantity in the equation changes with time (or some other variable). In this case, any derivative we find is called a **related rate**, since we can see from the differentiated equation how the derivatives (rates) are related. We illustrate this idea in the next several examples.

Example 8.4   A Related Rates Problem

An oil tanker has an accident and oil pours out at the rate of 150 gallons per minute. Suppose that the oil spreads onto the water in a circle at a thickness of $\frac{1}{10}''$.

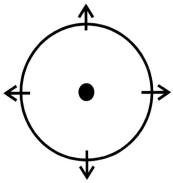

FIGURE 2.44: Oil spill.

(see Figure 2.44). Given that 1 ft³ equals 7.5 gallons, determine the rate at which the radius of the spill is increasing when the radius reaches 50 feet.

Solution   Since the area of a circle of radius r is πr^2, the volume of oil is given by

$$V = (depth)\ (area) = \frac{1}{120}\pi r^2,$$

since the depth is $\frac{1}{10}'' = \frac{1}{120}$ ft. Both volume and radius are functions of time, so

$$V(t) = \frac{\pi}{120}[r(t)]^2.$$

Differentiating both sides of the equation, we get

$$V'(t) = \frac{\pi}{120}2[r(t)]\,r'(t).$$

We are given a radius of 50 feet. The volume increases at a rate of 150 gallons per minute, or $\frac{150}{7.5} = 20$ ft³/min. Substituting in $V'(t) = 20$ and $r = 50$, we have

$$20 = \frac{\pi}{120}2(50)r'(t).$$

Solving for $r'(t)$, we find that the radius is increasing at the rate of $\frac{24}{\pi} = 7.6427$ feet per minute.

∎

Although the details change from problem to problem, the general pattern of solutions is the same for all related rates problems. Looking back, you should be able to identify each of the following steps in example 8.4.

1. Set up an equation relating all of the relevant quantities.
2. Differentiate both sides of the equation.
3. Substitute in values for all but one of the quantities and derivatives.
4. Solve for the remaining rate.

Example 8.5   Another Related Rates Problem
   A car is traveling at 50 mph due south at a point $\frac{1}{2}$ mile north of an intersection. A police car is traveling at 40 mph due west at a point $\frac{1}{4}$ mile due east of the same intersection. At that instant, the radar in the police car measures the rate at which the distance between the two cars is changing. What does the radar gun register?

Solution   First, we sketch a picture and name the vertical distance of the first car y and the horizontal distance of the police car x (see Figure 2.45). Notice that this says that $\dfrac{dx}{dt} = -40$, since the police car is moving in the direction of the negative x-axis and $\dfrac{dy}{dt} = -50$, since the other car is moving in the direction of the negative y-axis. From the Pythagorean Theorem, the distance between the two cars is $d = \sqrt{x^2 + y^2}$. Since all quantities are changing with time, our equation is

$$d(t) = \sqrt{[x(t)]^2 + [y(t)]^2} = \left([x(t)]^2 + [y(t)]^2\right)^{1/2}.$$

Differentiating both sides with respect to t, we have by the chain rule that

$$d'(t) = \frac{1}{2}\left([x(t)]^2 + [y(t)]^2\right)^{-1/2}2\,[x\,(t)\,x'\,(t) + y(t)y'(t)]$$
$$= \frac{x(t)\,x'(t) + y(t)\,y'(t)}{\sqrt{[x(t)]^2 + [y(t)]^2}}.$$

FIGURE 2.45: Cars approaching an intersection.

Substituting in $x(t) = \frac{1}{4}$, $x'(t) = -40$, $y(t) = \frac{1}{2}$ and $y'(t) = -50$, we have

$$d'(t) = \frac{-\frac{1}{4}40 - \frac{1}{2}50}{\sqrt{\frac{1}{4} + \frac{1}{16}}} = \frac{-140}{\sqrt{5}} \approx -62.6,$$

so that the radar gun registers 62.6 mph.

$\blacksquare$

Exercises 2.8

1. (writing) For implicit differentiation, we assume that y is a function of x: we write $y(x)$ to remind ourselves of this. However, for the circle $x^2 + y^2 = 1$, it is not true that y is a function of x. Since $y = \pm\sqrt{1 - x^2}$, there are actually two functions of x implicitly defined. Explain why this is not really a contradiction; that is, explain exactly what we are assuming when we do implicit differentiation.

2. (writing) To perform implicit differentiation on an equation such as $x^2 y^2 + 3 = x$, we start by differentiating all terms. We get $2xy^2 + x^2 2yy' = 1$. Many students learn the rules this way: take "regular" derivatives of all terms, and tack on a y' every time you take a y-derivative. Explain why this works, and rephrase the rule in a more accurate and understandable form.

3. (writing) In implicit differentiation, the derivative is typically a function of both x and y; for example, on the circle $x^2 + y^2 = r^2$, we have $y' = -x/y$. If we take the derivative $-x/y$ and plug in any x and y, will it always be the slope of a tangent line? That is, are there any requirements on which x's and y's we can plug in?

4. (writing) In each example in this section, after we differentiated the given equation we were able to rewrite the resulting equation in the form $f(x, y) y'(x) = g(x, y)$ for some expressions $f(x, y)$ and $g(x, y)$. Explain why this can always be done; that is, why doesn't the chain rule ever produce a term like $[y'(x)]^2$ or $\frac{1}{y'(x)}$?

In exercises 5–8, compute the slope of the tangent line at the given point both explicitly (first solve for y as a function of x) and implicitly.

5. $x^2 + 4y^2 = 8$ at $(2, 1)$

6. $x^3 y - 4\sqrt{x} = x^2 y$ at $(2, \sqrt{2})$

7. $y - 3x^2 y = \cos x$ at $(0, 1)$

8. $y^2 + 2xy + 4 = 0$ at $(-2, 2)$.

In exercises 9–20, find the derivative $y'(x)$ implicitly.

9. $x^2 y^2 + 3y = 4x$

10. $3xy^3 - 4x = 10y^2$

11. $\sqrt{xy} - 4y^2 = 12$

12. $\sin xy = x^2 - 3$

13. $\dfrac{x + 3}{y} = 4x + y^2$

14. $3x + y^3 - 4y = 10x^2$

15. $e^{x^2 y} - e^y = x$

16. $xe^y - 3y \sin x = 1$

17. $\sqrt{x + y} - 4x^2 = y$

18. $\cos y - y^2 = 8$

19. $e^{4y} - \ln y = 2x$

20. $e^{x^2} y - 3y = x^2 + 1$

In exercises 21–28, find an equation of the tangent line at the given point. If you have a CAS that will graph implicit curves, sketch the curve and the tangent line.

21. $x^2 - 4y^2 = 0$ at $(2, 1)$

22. $x^3 - 4y^2 = 4$ at $(2, 1)$

23. $x^2 - 4y^3 = 0$ at $(2, 1)$

24. $x^2 y^2 = 4x$ at $(1, 2)$

25. $x^2 y^2 = 4y$ at $(2, 1)$

26. $x^3 y^2 = -3xy$ at $(-1, -3)$

27. $x^3 y^3 = 9y$ at $(1, 3)$

28. $x^2 y^3 = 9y$ at $(1, -3)$

29. A baseball player stands 2 feet from home plate and watches a pitch fly by. In the diagram, x is the distance of the ball from home plate and θ is the angle indicating the direction of the player's gaze. Find the rate θ' at which his eyes must move to watch a fastball with speed $x' = -130$ ft/s as it crosses home plate at $x = 0$.

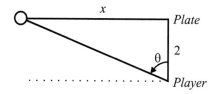

30. In the situation of exercise 29, humans can maintain focus only when $\theta' \leq \frac{\pi}{2}$ (see Watts and Bahill's book *Keep Your Eye on the Ball*). Find the fastest pitch that you could actually watch cross home plate.

31. A camera tracks the launch of a spacecraft. The camera is located 2 miles from the launchpad. If the spacecraft is 3 miles up and traveling at 0.2 miles per second, at what rate is the camera angle changing?

32. Repeat exercise 31 for the spacecraft at 1 mile up (assume the same velocity). Which rate is higher? Explain in commonsense terms why it is larger.

33. Assume that the infected area of an injury is circular. If the radius of the infected area is 3 mm and growing at a rate of 1 mm/hr, at what rate is the infected area increasing?

34. For the injury of exercise 33, find the rate of increase of the infected area when the radius reaches 6 mm. Explain in commonsense terms why this rate is larger than that of exercise 33.

35. A plane is $x = 40$ miles away from an airport at a height of h miles. A radar at the airport detects that the distance $s(t)$ between the plane and airport is changing at the rate of $s'(t) = -240$ mph. What is the speed $x'(t)$ of the airplane if the height is $h = 4$?

36. Repeat exercise 35 with a height of 6 miles. Based on your answers, how important is it to know the actual height of the airplane?

37. Rework example 8.5 if the police car is not moving. Does this make the radar gun's measurement more accurate?

38. Show that the radar gun of example 8.5 gives the correct speed if the police car is at the origin.

39. Show that the radar gun of example 8.5 gives the correct speed if the police car is at $x = \frac{1}{2}$ moving at a speed of $(\sqrt{2} - 1)50$ mph.

40. Find a position and speed for which the radar gun of example 8.5 has a slower reading than the actual speed.

41. Suppose that the average yearly cost per item for producing x items of a business product is $\bar{C}(x) = 10 + \frac{100}{x}$. If the current production is $x = 10$ and production is increasing at a rate of 2 items per year, find the rate of change of the average cost.

42. Suppose that the average yearly cost per item for producing x items of a business product is $\bar{C}(x) = 12 + \frac{94}{x}$. The three most recent yearly production figures are given in the table.

Year	0	1	2
Prod. (x)	8.2	8.8	9.4

Estimate the value of x' and the current (year 2) rate of change of the average cost.

43. For a small company spending $\$x$ thousand per year in advertising, suppose that annual sales in thousands of dollars equal $s(x) = 60 - 40e^{-0.5x}$. The three most recent yearly advertising figures are given in the table.

Year	0	1	2
Adver.	16,000	18,000	20,000

Estimate the value of x' and the current (year 2) rate of change of sales.

44. For a small company spending $\$x$ thousand per year in advertising, suppose that annual sales in thousands of dollars equal $s(x) = 80 - 20e^{-0.1x}$. If the current advertising budget is $x = 40$ and the budget is increasing at a rate of $\$1500$ per year, find the rate of change of sales.

45. Suppose a 6-ft tall person is 12 ft away from a 18-ft tall lamppost (see the figure). If the person is moving away from the lamppost at a rate of 2 ft/s, at what rate is the length of the shadow changing? (Hint: Show that $\frac{x+s}{18} = \frac{s}{6}$.)

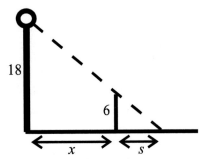

Figure for exercise 45.

46. Rework exercise 45 if the person is 6 ft away from the lamppost and is walking toward the lamppost at a rate of 3 ft/s.

47. Suppose that a raindrop evaporates in such a way that it maintains a spherical shape. Given that the volume of a sphere of radius r is $V = \frac{4}{3}\pi r^3$ and its surface area is $A = 4\pi r^2$, if the radius changes in time, show that $V' = Ar'$. If the rate of evaporation (V') is proportional to the surface area, show that the radius changes at a constant rate.

48. Suppose a forest fire spreads in a circle with radius changing at a rate of 5 feet per minute. When the radius reaches 200 feet, at what rate is the area of the burning region increasing?

In exercises 49 and 50, find the locations of all horizontal and vertical tangents.

49. $x^2 + y^3 - 3y = 4$ **50.** $xy^2 - 2y = 2$

In exercises 51 and 52, find the second derivative $y''(x)$.

51. $x^2 + y^2 = 4$ **52.** $x^2 + y^3 - 2y = 3$

53. In example 8.1, it is easy to find a y-value for $x = 2$, but other y-values are not so easy to find. Try solving for y if $x = 1.9$. Use the tangent line found in example 8.1 to estimate a y-value. Repeat for $x = 2.1$.

54. Use the tangent line found in example 8.2 to estimate a y-value corresponding to $x = 0.9$; $x = 1.1$.

55. Use implicit differentiation to find $y'(x)$ for $x^2y - 2y = 4$. Based on this equation, why would you expect to find vertical tangents at $x = \pm\sqrt{2}$ and horizontal tangents at $y = 0$? Show that there are no points for these values. To see what's going on, solve the original equation for y and sketch the graph. Describe what's happening at $x = \pm\sqrt{2}$ and $y = 0$.

56. (exploratory) Suppose a hammer thrower (see section 2.1) rotates counterclockwise along the circle $x^2 + y^2 = 9$ and releases the hammer at the point $(2.9, 0.77)$. If the throw is 300 feet long, where will it land? [Hint: find the tangent line at $(2.9, 0.77)$ and find the point (x, y) on that line such that the distance is $\sqrt{(x - 2.9)^2 + (y - 0.77)^2} = 300$.]

57. (exploratory) A landowner's property line runs along the path $y = 6 - x$. The landowner wants to run an irrigation ditch from a reservoir bounded by the ellipse $4x^2 + 9y^2 = 36$. The landowner wants to build the shortest ditch possible from the reservoir to the closest point on the property line. We explore how to find the best path. Sketch the line and ellipse, and draw in a tangent line to the ellipse that is parallel to the property line. Argue that the ditch should start at the point of tangency and run perpendicular to the two lines. We start by identifying the point on the right side of the ellipse with tangent line parallel to $y = 6 - x$. Find the slope of the tangent line to the ellipse at (x, y) and set it equal to -1. Solve for x and substitute into the equation of the ellipse. Solve for y and you have the point on the ellipse at which to start the ditch. Find an equation of the (normal) line through this point perpendicular to $y = 6 - x$, and find the intersection of the normal line and $y = 6 - x$. This point is where the ditch ends.

2.9   The Mean Value Theorem

In this section, we present a result called the Mean Value Theorem, which is at the heart of the calculus. While the conclusion is easy to understand graphically, we will also carefully prove the theorem. The Mean Value Theorem is so significant that we will be deriving new ideas from it for many chapters to come. Before considering the main result, we look at a special case.

Theorem 9.1   (Rolle's Theorem) Suppose that $f(x)$ is continuous on the interval $[a, b]$, differentiable on the interval (a, b) and $f(a) = f(b)$. Then there is a number $c \in (a, b)$ such that $f'(c) = 0$.

Sketch of Proof: We present the main ideas of the proof from a graphical perspective. First, note that if $f(x)$ is constant on $[a, b]$, then $f'(x) = 0$ for *all* x's between a and b. If $f(x)$ is not constant, then, starting at $x = a$ and moving to the right, the graph either rises or falls (see Figure 2.46a and 2.46b). In either case, the graph must return to the level it started at, since $f(b) = f(a)$.

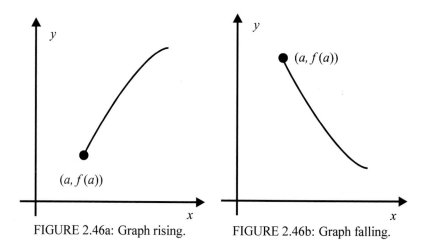

FIGURE 2.46a: Graph rising.   FIGURE 2.46b: Graph falling.

Therefore, there is at least one point where the graph turns around (see Figure 2.47a and 2.47b). We name this point $x = c$. We want to argue that $f'(c) = 0$. Since we know that $f'(c)$ exists, we have either $f'(c) > 0$, $f'(c) < 0$ or $f'(c) = 0$. The graphs clearly suggest that $f'(c) = 0$. To establish this, it is easier to show that it is *not* true that $f'(c) > 0$ or $f'(c) < 0$. If it were true that $f'(c) > 0$, then

$$\lim_{h \to 0} \frac{f(c + h) - f(c)}{h} > 0.$$

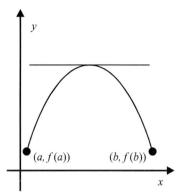

FIGURE 2.47a: Graph rises and turns around to fall back to where it started.

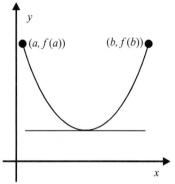

FIGURE 2.47b: Graph falls and then turns around to rise back to where it started.

If $|h|$ is sufficiently small, then

$$\frac{f(c+h) - f(c)}{h} > 0. \tag{9.1}$$

If $h < 0$, (9.1) implies that $f(c + h) - f(c) < 0$ and so, for $s = c + h$, we have $f(c + h) = f(s) < f(c)$. However, if $h > 0$, then $f(c + h) - f(c) > 0$, so that $f(c + h) > f(c)$. Setting $t = c + h$ (for $h > 0$), we have that

$$f(c) < f(c+h) = f(t).$$

Putting this together, we have

$$f(s) < f(c) < f(t).$$

This can't happen if $x = c$ is a point where the graph "turns around." We conclude that it is not true that $f'(c) > 0$. Similarly, we can show that it is not true that $f'(c) < 0$. Therefore, $f'(c) = 0$, as desired.   ∎

We now give an illustration of the conclusion of Rolle's Theorem.

Example 9.1   An Illustration of Rolle's Theorem
Find a value of c satisfying the conclusion of Rolle's Theorem for

$$f(x) = x^3 - 3x^2 + 2x + 2$$

on the interval $[0, 1]$.

Solution   First, we verify that the hypotheses of the theorem are satisfied: f is differentiable and continuous for all x (since $f(x)$ is a polynomial and all polynomials are continuous and differentiable everywhere). Also, $f(0) = f(1) = 2$. We have

$$f'(x) = 3x^2 - 6x + 2.$$

We now look for values of c such that

$$f'(c) = 3c^2 - 6c + 2 = 0.$$

By the quadratic formula, we get $c = 1 + \frac{1}{3}\sqrt{3} \approx 1.5774$ [not in the interval $(0, 1)$] and $c = 1 - \frac{1}{3}\sqrt{3} \approx 0.42265 \in (0, 1)$.

∎

Remark 9.1   We want to emphasize that example 9.1 is merely an *illustration* of Rolle's Theorem. Finding the number(s) c satisfying the conclusion of Rolle's Theorem is **not** the point of our discussion. Rather, Rolle's Theorem is of interest to us primarily because we can use it to prove one of the fundamental results of elementary calculus, the Mean Value Theorem.

Although Rolle's Theorem is a simple result, we can use it to derive numerous properties of functions. For example, we are often interested in finding the zeros of a function f (that is, solutions of the equation $f(x) = 0$). In practice, it is often difficult to determine how many zeros a given function has. Rolle's Theorem can be of help here.

Theorem 9.2   If $f(x)$ is continuous on the interval $[a, b]$, differentiable on the interval (a, b) and $f(x) = 0$ has two solutions in $[a, b]$, then $f'(x) = 0$ has at least one solution in (a, b).

Proof. This is just a special case of Rolle's Theorem. Call the two zeros of $f(x)$, say, $x = s$ and $x = t$, where $s < t$. Since $f(s) = f(t)$, Rolle's Theorem guarantees that there is a number c such that $s < c < t$ (and hence $a < c < b$) where $f'(c) = 0$.   ∎

We can easily generalize the result of Theorem 9.2, as in the following theorem.

Theorem 9.3   For any integer $n > 0$, if $f(x)$ is continuous on the interval $[a, b]$ and differentiable on the interval (a, b) and $f(x) = 0$ has n solutions in $[a, b]$, then $f'(x) = 0$ has at least $(n - 1)$ solutions in (a, b).

Proof. From Theorem 9.2, between every pair of solutions of $f(x) = 0$ is at least one solution of $f'(x) = 0$. In this case, there are $(n - 1)$ consecutive pairs of solutions of $f(x) = 0$ and so, the result follows.   ∎

We can use Theorems 9.2 and 9.3 to investigate the number of zeros a given function has.

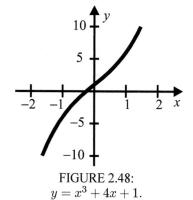

FIGURE 2.48:
$y = x^3 + 4x + 1$.

Example 9.2   Determining the Number of Zeros of a Function
Prove that $x^3 + 4x + 1 = 0$ has exactly one solution.

Solution   The graph shown in Figure 2.48 makes the result seem reasonable: we can see one solution, but how can we be sure there are no others outside of the displayed window? Notice that if $f(x) = x^3 + 4x + 1$, then

$$f'(x) = 3x^2 + 4 > 0$$

for all x. By Theorem 9.2, if $f(x) = 0$ had two solutions, $f'(x) = 0$ would have at least one solution. But, since $f'(x) \neq 0$ for all x, it can't be true that $f(x) = 0$ has two (or more) solutions. Therefore, $f(x) = 0$ has exactly one solution.   ∎

We now generalize Rolle's Theorem to one of the most significant results of elementary calculus.

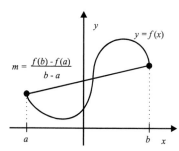

FIGURE 2.49: Secant line.

Theorem 9.4   (Mean Value Theorem) Suppose that f is continuous on the interval $[a, b]$ and differentiable on the interval (a, b). Then there exists a number $c \in (a, b)$ such that

$$f'(c) = \frac{f(b) - f(a)}{b - a}. \tag{9.2}$$

Proof. Note that the hypotheses are identical to those of Rolle's Theorem, except that there is no assumption about the values of f at the endpoints. You should recognize the expression $\dfrac{f(b) - f(a)}{b - a}$ as the slope of the secant line connecting the endpoints, $(a, f(a))$ and $(b, f(b))$. [Note that in the special case where $f(a) = f(b)$, (9.2) simplifies to the conclusion of Rolle's Theorem, that $f'(c) = 0$.] The theorem states that there is a line tangent to this section of the curve which has the same slope as (and hence, is parallel to) the secant line (see Figures 2.49 and 2.50). If you tilt your head so that the line segment looks horizontal, Figure 2.50 will look like a figure for Rolle's Theorem (Figures 2.47a

FIGURE 2.50: Mean
Value Theorem.

and 2.47b). The proof is to in effect "tilt" the function and then apply Rolle's Theorem. First, note that the secant line through the endpoints has slope

$$m = \frac{f(b) - f(a)}{b - a}.$$

The equation of the secant line is then

$$y - f(a) = m(x - a).$$

Define the "tilted" function g to be the difference between f and the function whose graph is the secant line

$$g(x) = f(x) - [m(x - a) + f(a)]. \qquad (9.3)$$

Then, g is continuous on $[a, b]$ and differentiable on (a, b), since f is. Further,

$$g(a) = f(a) - [0 + f(a)] = 0$$

and

Since $m = \frac{f(b)-f(a)}{b-a}$.

$$\begin{aligned} g(b) &= f(b) - [m(b - a) + f(a)] \\ &= f(b) - [f(b) - f(a) + f(a)] = 0. \end{aligned}$$

Since $g(a) = g(b)$, we have by Rolle's Theorem that there exists a number c in the interval (a, b) such that $g'(c) = 0$. But, differentiating (9.3), we get

$$g'(c) = f'(c) - m. \qquad (9.4)$$

Finally, solving (9.4) for $f'(c)$ gives us

$$f'(c) = m = \frac{f(b) - f(a)}{b - a},$$

as desired.   ■

Before we demonstrate some of the power of the Mean Value Theorem, we first briefly illustrate its conclusion.

Example 9.3   An Illustration of the Mean Value Theorem
Find a value of c satisfying the conclusion of the Mean Value Theorem for

$$f(x) = x^3 - x^2 - x + 1$$

on the interval $[0, 2]$.

Solution   Notice that f is continuous on $[0, 2]$ and differentiable on $(0, 2)$ (since f is a polynomial and all polynomials are continuous and differentiable everywhere). Here, we have $f(0) = 1$ and $f(2) = 3$. The Mean Value Theorem then says that there is a number c between 0 and 2 for which

$$f'(c) = \frac{3 - 1}{2 - 0} = 1.$$

To find this number c, we set

$$f'(c) = 3c^2 - 2c - 1 = 1$$

or

$$3c^2 - 2c - 2 = 0.$$

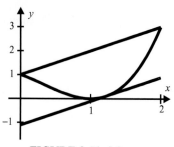

FIGURE 2.51: Mean
Value Theorem.

From the quadratic formula, we get $c = \dfrac{1 \pm \sqrt{7}}{3}$. In this case, only one of these, $c = \dfrac{1 + \sqrt{7}}{3}$ is in the interval $(0, 2)$. In Figure 2.51, we show the graphs of $y = f(x)$, the secant line joining the endpoints of the portion of the curve on the interval $[0, 2]$ and the tangent line at $x = \dfrac{1 + \sqrt{7}}{3}$.

■

Remark 9.2   Be careful not to confuse the significance of the Mean Value Theorem with the illustration in example 9.3 where we found the number c whose existence is guaranteed by the theorem. In practice, these c's remain unknown. The point of the Mean Value Theorem is in relating a difference of function values to the difference of the corresponding x-values, as in (9.5). The ability to relate these two quantities is a crucial result.

Note that if we take the conclusion of the Mean Value Theorem (9.2) and multiply both sides by the quantity $(b - a)$, we get

$$f(b) - f(a) = f'(c)(b - a). \tag{9.5}$$

As it turns out, many of the most important results in the calculus (including one known as the **Fundamental Theorem of Calculus**) follow from the Mean Value Theorem. For now, we derive a result essential to our work in Chapter 4. The question concerns how many functions share the same derivative.

Recall that for any constant c,

$$\frac{d}{dx}(c) = 0.$$

This is of course, the easiest derivative around. Here's a question that you probably haven't thought to ask, though: Are there any other functions whose derivative is zero? The answer is no, as we see in the following theorem.

Theorem 9.5   Suppose that $f'(x) = 0$ for all x in some open interval I. Then, $f(x)$ is constant on I.

Proof. Pick any two numbers, say a and b in I, with $a < b$. Since f is differentiable in I and $(a, b) \subset I$, f is differentiable on (a, b) and continuous on $[a, b]$. By the Mean Value Theorem, we know that

$$\frac{f(b) - f(a)}{b - a} = f'(c), \tag{9.6}$$

for some number $c \in (a, b) \subset I$. But, $f'(x) = 0$ for all $x \in I$. So, $f'(c) = 0$ and it follows from (9.6) that

$$f(b) - f(a) = 0 \text{ or } f(b) = f(a).$$

Since a and b were arbitrary points in I, this says that f is constant on I, as desired.

■

A question closely related to Theorem 9.5 is the following. We know, for example, that

$$\frac{d}{dx}(x^2 + 2) = 2x.$$

But are there any other functions with the same derivative? You should quickly come up with several. For instance, $x^2 + 3$ and $x^2 - 4$ also have the derivative $2x$. In fact,

$$\frac{d}{dx}(x^2 + c) = 2x,$$

for any constant c. Are there any other functions, though, with the derivative $2x$? Is there perhaps some complicated function whose derivative involves the chain rule and quotient rule, but that magically simplifies to $2x$? The following result says that there are no such functions.

Corollary 9.1   Suppose that $g'(x) = f'(x)$ for all x in some open interval I. Then, for some constant c,

$$g(x) = f(x) + c, \text{ for all } x \in I.$$

Proof. Define the function $h(x) = g(x) - f(x)$. Then

$$h'(x) = g'(x) - f'(x) = 0$$

for all x in I. From Theorem 9.5, $h(x) = c$, for some constant c. The result then follows immediately from the definition of $h(x)$.
∎

We see in Chapter 4 that Corollary 9.1 has significant implications when we try to reverse the process of differentiation (called **antidifferentiation**). We take a look ahead to this in the following simple example.

Example 9.4   Finding Every Function with a Given Derivative
    Find all functions that have a derivative equal to $3x^2 + 1$.

Solution   We first write down (from our experience with derivatives) one function with the correct derivative: $x^3 + x$. Then, Corollary 9.1 tells us that any other function with the same derivative differs by at most a constant. So, **every** function whose derivative equals $3x^2 + 1$ has the form $x^3 + x + c$, for some constant c.
∎

As our final example, we demonstrate how the Mean Value Theorem can be used to establish a useful inequality.

Example 9.5   Proving an Inequality for $\sin x$
    Prove that

$$|\sin a| \le |a| \text{ for all } a.$$

Solution   First, note that $f(x) = \sin x$ is continuous and differentiable on any interval and that

$$|\sin a| = |\sin a - \sin 0|,$$

since $\sin 0 = 0$. From the Mean Value Theorem, we have that (for $a \ne 0$)

$$\frac{\sin a - \sin 0}{a - 0} = f'(c) = \cos c, \tag{9.7}$$

for some number c between a and 0. Notice that if we multiply both sides of (9.7) by a and take absolute values, we get

$$|\sin a| = |\sin a - \sin 0| = |\cos c| \, |a - 0| = |\cos c| \, |a|. \tag{9.8}$$

But, $|\cos c| \leq 1$, for all real numbers c and so, from (9.8), we have

$$|\sin a| = |\cos c|\,|a| \leq (1)\,|a| = |a|,$$

as desired.

∎

Exercises 2.9

1. (writing) Notice that for both Rolle's Theorem and the Mean Value Theorem, we have assumed that the function is continuous on the closed interval $[a, b]$ and differentiable on the open interval (a, b). Recall that if a function is differentiable at $x = a$, then it is also continuous at $x = a$. Therefore, if we had assumed that the function was differentiable on the closed interval, we would not have had to mention continuity. We do not do so because the assumption of being differentiable at the endpoints is not necessary. The "ethics" of the statement of a theorem is to only include assumptions that are absolutely necessary. Discuss the virtues of this tradition. Is this common practice in our social dealings, such as financial obligations or personal gossip?

2. (writing) One of the results in this section is that if $f'(x) = g'(x)$, then $g(x) = f(x) + c$ for some constant c. Explain this result graphically.

3. (writing) Explain the result of Corollary 9.1 in terms of position and velocity functions. That is, if two objects have the same velocity functions, what can you say about the relative positions of the two objects?

4. (writing) As we mentioned, you can derive Rolle's Theorem from the Mean Value Theorem simply by setting $f(a) = f(b)$. Given this, it may seem odd that Rolle's Theorem rates its own name and portion of the book. To explain why we do this, discuss ways in which Rolle's Theorem is easier to understand than the Mean Value Theorem.

In exercises 5–8, explain why it is not valid to use the Mean Value Theorem. When the hypotheses are not true, the theorem does not tell you anything about the truth of the conclusion. In three of the four cases, show that there is no value of c which makes the conclusion of the theorem true. In the fourth case, find the value of c.

5. $f(x) = \dfrac{1}{x}$, $[-1, 1]$

6. $f(x) = \dfrac{1}{x^2}$, $[-1, 2]$

7. $f(x) = \tan x$, $[0, \pi]$

8. $f(x) = x^{1/3}$, $[-1, 1]$

In exercises 9–14, check the hypotheses of Rolle's Theorem and the Mean Value Theorem and find a value of c that makes the appropriate conclusion true. Illustrate the conclusion with a graph.

9. $f(x) = x^2 + 1$, $[-2, 2]$

10. $f(x) = x^2 + 1$, $[0, 2]$

11. $f(x) = x^3 + x^2$, $[0, 1]$

12. $f(x) = x^3 + x^2$, $[-1, 1]$

13. $f(x) = \sin x$, $[0, \pi/2]$

14. $f(x) = \sin x$, $[-\pi, 0]$

15. If $f'(x) > 0$ for all x, prove that $f(x)$ is an **increasing** function: that is, if $a < b$, then $f(a) < f(b)$.

16. If $f'(x) < 0$ for all x, prove that $f(x)$ is a **decreasing** function: that is, if $a < b$, then $f(a) > f(b)$.

In exercises 17—24, determine if the function is increasing, decreasing or neither.

17. $f(x) = x^3 + 5x + 1$

18. $f(x) = x^5 + 3x^3 - 1$

19. $f(x) = -x^3 - 3x + 1$

20. $f(x) = x^4 + 2x^2 + 1$

21. $f(x) = e^x$

22. $f(x) = e^{-x}$

23. $f(x) = \ln x$

24. $f(x) = \ln x^2$

25. Prove that $x^3 + 5x + 1 = 0$ has exactly one solution.

26. Prove that $x^3 + 4x - 3 = 0$ has exactly one solution.

27. Prove that $x^4 + 3x^2 - 2 = 0$ has exactly two solutions.

28. Prove that $x^4 + 6x^2 - 1 = 0$ has exactly two solutions.

29. Prove that $x^3 + ax + b = 0$ has exactly one solution for $a > 0$.

30. Prove that $x^4 + ax^2 - b = 0$ $(a > 0, b > 0)$ has exactly two solutions.

31. Prove that $x^5 + ax^3 + bx + c = 0$ has exactly one solution for $a > 0, b > 0$.

32. Prove that a third-degree (cubic) polynomial has at most three zeros (you may use the quadratic formula).

33. Prove that a fourth-degree polynomial has at most four zeros.

34. Prove that an nth degree polynomial has at most n zeros.

In exercises 35–42, find all functions $g(x)$ such that $g'(x) = f(x)$.

35. $f(x) = x^2$

36. $f(x) = x^3$

37. $f(x) = x^4$

38. $f(x) = 9x^4$

39. $f(x) = 1/x^2$

40. $f(x) = \sqrt{x}$

41. $f(x) = \sin x$ **42.** $f(x) = \cos x$

43. Assume that $f(x)$ is a differentiable function such that $f(0) = f'(0) = 0$ and $f''(0) > 0$. Argue that there exists a positive constant $a > 0$ such that $f(x) > 0$ for all x in the interval $(0, a)$. Can anything be concluded about $f(x)$ for negative x's?

44. Show that for any real numbers u and v, $|\cos u - \cos v| \le |u - v|$. (Hint: Use the Mean Value Theorem.)

45. For $f(x) = \begin{cases} 2x & \text{if } x \le 0 \\ 2x - 4 & \text{if } x > 0 \end{cases}$ show that $f(x)$ is continuous on the interval $(0, 2)$, differentiable on the interval $(0, 2)$ and has $f(0) = f(2)$. Show that there *does not* exist a value of c such that $f'(c) = 0$. Which hypothesis of Rolle's Theorem is not satisfied?

46. Assume that $f(x)$ is a differentiable function such that $f(0) = f'(0) = 0$. Show by example that it is not necessarily true that $f(x) = 0$ for all x. Find the flaw in the following bogus "proof." Using the Mean Value Theorem with $a = x$ and $b = 0$, we have $f'(c) = \dfrac{f(x) - f(0)}{x - 0}$. Since $f(0) = 0$ and $f'(c) = 0$, we have $0 = \dfrac{f(x)}{x}$ so $f(x) = 0$.

47. (exploratory) In section 2.1, we gave an example of computing the velocity of a moving car. The point of the story was that computing instantaneous velocity requires us to compute a limit. However, we left an interesting question unanswered. If you have an average velocity of 60 mph over 1 hour and the speed limit is 65 mph, you are unable to prove that you never exceeded the speed limit. What is the longest time interval over which you can average 60 mph and still guarantee

no speeding? We can use the Mean Value Theorem to answer the question after clearing up a couple of preliminary questions. First, argue that we need to know the maximum acceleration of a car, and the maximum positive acceleration may differ from the maximum negative acceleration. Based on your experience, what is the fastest your car could accelerate (speed up)? What is the fastest your car could decelerate (slow down)? Back up your estimates with some real data (e.g., my car goes from 0 to 60 in 15 minutes). Call the larger number A (use units of mph per second). Next, argue that if acceleration (the derivative of velocity) is constant, then the velocity function is linear. Therefore, if the velocity varies from 55 mph to 65 mph at constant acceleration, the average velocity will be 60 mph. Now, apply the Mean Value Theorem to the velocity function $v(t)$ on a time interval $[0, T]$ where the velocity changes from 55 mph to 65 mph at constant acceleration A: $v'(c) = \dfrac{v(T) - v(0)}{T - 0}$ becomes $A = \dfrac{65 - 55}{T - 0}$ and so $T = 10/A$. How long is the guarantee good for?

48. (exploratory) Suppose that a pollutant is dumped into a lake at the rate of $p'(t) = t^2 - t + 4$ tons per month. The amount of pollutant dumped into the lake in the first two months is $A = p(2) - p(0)$. Using $c = 1$ (the midpoint of the interval), estimate A by applying the Mean Value Theorem to $p(t)$ on the interval $[0, 2]$. To get a better estimate, apply the Mean Value Theorem to the intervals $[0, 1/2]$, $[1/2, 1]$, $[1, 3/2]$ and $[3/2, 2]$. (Hint: $A = p(1/2) - p(0) + p(1) - p(1/2) + p(3/2) - p(1) + p(2) - p(3/2)$.) If you have access to a CAS, get better estimates by dividing the interval $[0, 2]$ into more and more pieces and try to conjecture the limit of the estimates. You will be well on your way to understanding Chapter 4 on integration.

Chapter Review Exercises

1. Estimate the value of $f'(1)$ from the given data.

x	0	0.5	1	1.5	2
$f(x)$	2.0	2.6	3.0	3.4	4.0

2. List the points A, B, C and D in order of increasing slope of the tangent line.

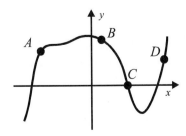

In exercises 3–8, use the limit definition to find the indicated derivative.

3. $f'(2)$ for $f(x) = x^2 - 2x$ **4.** $f'(1)$ for $f(x) = 1 + \dfrac{1}{x}$

5. $f'(1)$ for $f(x) = \sqrt{x}$ **6.** $f'(0)$ for $f(x) = x^3 - 2x$

7. $f'(x)$ for $f(x) = x^3 + x$ **8.** $f'(x)$ for $f(x) = \dfrac{3}{x}$

In exercises 9–14, find an equation of the tangent line.

9. $y = x^4 - 2x + 1$ at $x = 1$

10. $y = \sin 2x$ at $x = 0$

11. $y = 3e^{2x}$ at $x = 0$

12. $y = \sqrt{x^2 + 1}$ at $x = 0$

13. $y - x^2 y^2 = x - 1$ at $(1, 1)$

14. $y^2 + xe^y = 4 - x$ at $(2, 0)$

In exercises 15–18, use the given distance function to find velocity and acceleration.

15. $s(t) = -16t^2 + 40t + 10$

16. $s(t) = -9.8t^2 - 22t + 6$

17. $s(t) = 10e^{-2t} \sin 4t$

18. $s(t) = \sqrt{4t + 16} - 4$

19. In exercise 15, $s(t)$ gives the height of a ball. Find the ball's velocity at $t = 1$; is the ball going up or down? Find the ball's velocity at $t = 2$; is the ball going up or down?

20. In exercise 17, $s(t)$ gives the position of a spring. Compare the velocities at $t = 0$ and $t = \pi$. Is the spring moving in the same direction or opposite directions? At which time is the spring moving faster?

In exercises 21 and 22, compute the slopes of the secant lines between (a) $x = 1$ and $x = 2$, (b) $x = 1$ and $x = 1.5$, and (c) $x = 1$ and $x = 1.1$ and estimate the slope of the tangent line at $x = 1$.

21. $f(x) = \sqrt{x + 1}$ **22.** $f(x) = e^{2x}$

In exercises 23–46, find the derivative of the given function.

23. $f(x) = x^4 - 3x^3 + 2x - 1$

24. $f(x) = x^{2/3} - 4x^2 + 5$

25. $f(x) = \dfrac{3}{\sqrt{x}} + \dfrac{5}{x^2}$ **26.** $f(x) = \dfrac{2 - 3x + x^2}{\sqrt{x}}$

27. $f(t) = t^2(t + 2)^3$

28. $f(t) = (t^2 + 1)(t^3 - 3t + 2)$

29. $g(x) = \dfrac{x}{3x^2 - 1}$ **30.** $g(x) = \dfrac{3x^2 - 1}{x}$

31. $f(x) = x^2 \sin x$ **32.** $f(x) = \sin x^2$

33. $f(x) = \tan \sqrt{x}$ **34.** $f(x) = \sqrt{\tan x}$

35. $f(t) = t \csc t$ **36.** $f(t) = \sin 3t \cos 4t$

37. $u(x) = 2e^{-x^2}$ **38.** $u(x) = (2e^{-x})^2$

39. $f(x) = x \ln x^2$ **40.** $f(x) = \sqrt{\ln x + 1}$

41. $f(x) = \sqrt{\sin 4x}$ **42.** $f(x) = \cos^2 3x$

43. $f(x) = \left(\dfrac{x + 1}{x - 1}\right)^2$ **44.** $f(x) = e^{\sqrt{3x}}$

45. $f(t) = te^{4t}$ **46.** $f(x) = \dfrac{6x}{(x - 1)^2}$

In exercises 47 and 48, use the graph of $f(x)$ to sketch the graph of $f'(x)$.

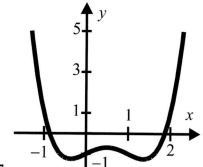

47.

48.

In exercises 49–56, find the indicated derivative.

49. $f''(x)$ for $f(x) = x^4 - 3x^3 + 2x^2 - x - 1$

50. $f'''(x)$ for $f(x) = \sqrt{x + 1}$

51. $f'''(x)$ for $f(x) = xe^{2x}$

52. $f''(x)$ for $f(x) = \dfrac{4}{x + 1}$

53. $f''(x)$ for $f(x) = \tan x$

54. $f^{(4)}(x)$ for $f(x) = x^6 - 3x^4 + 2x^3 - 7x + 1$

55. $f^{(26)}(x)$ for $f(x) = \sin 3x$

56. $f^{(31)}(x)$ for $f(x) = e^{-2x}$

57. Revenue equals price times quantity. Suppose that the current price is \$2.40 and 12,000 items are sold at that price. If the price is increasing at the rate of 10 cents per year and the quantity sold decreases at the rate of 1500 items per year, at what rate is the revenue changing?

58. The value (in dollars) of an investment as a function of time (years) is given by $v(t) = 200 \left(\frac{3}{2}\right)^t$. Find the instantaneous percentage rate of change of the value of the investment.

59. The position of a spring in motion vertically is given by $f(t) = 4 \cos 2t$. Find the position of the spring when it has (a) zero velocity, (b) maximum velocity and (c) minimum velocity.

60. The position of a spring in motion vertically is given by $f(t) = e^{-2t} \sin 3t$. Find the velocity of the spring at any time t.

In exercises 61–64, find the derivative $y'(x)$.

61. $x^2y - 3y^3 = x^2 + 1$

62. $\sin(xy) + x^2 = x - y$

63. $\dfrac{y}{x+1} - 3y = \tan x$

64. $x - 2y^2 = 3e^{x/y}$

65. A car is traveling at 50 mph due north and a police car is traveling 60 mph due east. A radar gun measures the rate at which the distance between the cars is changing. If the radar gun is activated when the car is $\frac{1}{4}$-mile north of an intersection and the police car is $\frac{1}{2}$-mile east of the same intersection, what does the gun register?

66. An oil spill is circular in shape with a depth of $\frac{1}{50}$ ft. If the oil is spilling at the rate of 40 ft^3 per second, at what rate is the radius of the spill increasing when the radius equals 50 feet?

In exercises 67–70, find all points at which the tangent line to the curve is (a) horizontal (b) vertical.

67. $y = x^3 - 6x^2 + 1$ **68.** $y = x^{2/3}$

69. $x^2y - 4y = x^2$ **70.** $y = x^4 - 2x^2 + 3$

71. Prove that the equation $x^3 + 7x - 1 = 0$ has exactly one solution.

72. Prove that the equation $x^4 + 2x^2 - 3 = 0$ has exactly two solutions.

In exercises 73 and 74, find a value of c as guaranteed by the Mean Value Theorem.

73. $f(x) = x^2 - 2x$ on the interval $[0, 2]$

74. $f(x) = x^3 - x$ on the interval $[0, 2]$

In exercises 75 and 76, find all functions $g(x)$ such that $g'(x) = f(x)$.

75. $f(x) = 3x^2 - \cos x$ **76.** $f(x) = x^3 - e^{2x}$

77. (exploratory) Knowing where to aim a ball is an important skill in many sports. If the ball doesn't follow a straight path (because of gravity or other factors), aiming can be a difficult task. When throwing a baseball, for example, the player must take gravity into account and aim higher than the target. Ignoring air resistance and any lateral movement, the motion of a thrown ball may be approximated by $y = -\dfrac{16}{v^2 \cos^2 \theta}x^2 + (\tan \theta)x$. Here, the ball is thrown from the position $(0,0)$ with initial speed v ft/s at angle θ from the horizontal.

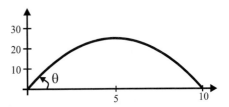

Given such a curve, we can compute the slope of the tangent line at $x = 0$, but how can we compute the proper angle θ? Show that if m is the slope of the tangent line at $x = 0$, then $\tan \theta = m$. (Hint: Draw a triangle using the tangent line and x-axis, and recall that slope is rise over run.) Tangent is a good name, isn't it? Now, for some baseball problems. We will look at how high players need to aim to make throws that are easy to catch. Throwing height is also a good catching height. If L (ft) is the length of the throw and we want the ball to arrive at the same height as it is released (as shown in the figure), the parabola can be solved for the following relationship between angle and velocity: $\sin 2\theta = 32L/v^2$. A third baseman throwing at 130 ft/s (about 90 mph) must throw 120 ft to reach first base. Find the angle of release (substitute L and v and, by trial and error, find a value of θ that works), the slope of the tangent line and the height at which the third baseman must aim (that is, the height at which the ball would arrive if there were no gravity). How much does this change for a soft throw at 100 ft/s? How about for an outfielder's throw of 300 feet at 130 ft/s? Most baseball players would deny that they are aiming this high; what in their experience would make it difficult for them to believe the calculations?

78. (exploratory) We continue our explorations in the theory of mathematical chaos here. (Related exploratory exercises can be found in the review exercises of Chapter 1.) We have seen how to iterate functions, and that such iterations may result in **attractors** consisting of a single number, two numbers alternating (a two-cycle), a four-cycle, other size cycles or chaos (bounded but not repeating). We now have the tools to determine which cycles attract and which do not. First, try iterating $\cos x$: start at any x_0 and compute $x_1 = \cos x_0$, $x_2 = \cos x_1$ and so on. Without knowing what your choice of x_0 is, we can predict that the iterations will converge to the mysterious-looking number 0.739085133215. From our previous work, you should be able to explain where this number comes from: it is the only intersection of $y = \cos x$ and $y = x$. Similarly, iterates of $\sin x$ will converge to 0, the only solution of the equation $\sin x = x$. You might hypothesize that iterates of $f(x)$ always converge to a solution of the equation $f(x) = x$. However, iterates for most starting points of $f(x) = x(3.2 - x)$ converge to a two-cycle (approximately 2.5582 and 1.6417) and not to either solution of the equation $x(3.2 - x) = x$. Those **fixed points**, as they're called, are $x = 0$ and $x = 2.2$. They are called **repelling fixed points**: pick x_0 very close to one of them and the iterates will get "repelled" away from the fixed points. For example, if $x_0 = 2.22$, then $x_1 = 2.1756$, $x_2 = 2.22868$, $x_3 = 2.1647$, $x_4 = 2.24105$, and so on with succeeding iterates getting farther and farther from 2.2 (and closer to the two-cycle 2.5582 and 1.6417). A simple rule tells whether a fixed point will be attracting or repelling. If a is a **fixed point** of $f(x)$ (that is, a is a solution of the equation $f(x) = x$) and $|f'(a)| < 1$, then a is attracting. If $|f'(a)| > 1$, then a is repelling. Verify that 0.739085133215 is an attracting fixed point of $\cos x$, that 0 is an attracting fixed point of $\sin x$, that 0 is a repelling fixed point of $x(3.2 - x)$ and that 2.2 is a repelling fixed point of $x(3.2 - x)$.

79. (exploratory) Based on exercise 78, we can find a **bifurcation point** for the family of functions $f_c(x) = x(c - x)$. Show that $x = 0$ and $x = c - 1$ are the fixed points of $f_c(x)$. If $-1 < c < 1$, show that 0 attracts and $c - 1$ repels. Illustrate this by computing iterates of $f_0(x)$ and $f_{1/2}(x)$ starting at $x_0 = 0.2$. If $1 < c < 3$, show that 0 repels and $c - 1$ attracts. Illustrate this by computing iterates of $f_2(x)$ and $f_{2.8}(x)$ starting at $x_0 = 0.5$. If $c > 3$, show that both 0 and $c - 1$ repel. Illustrate this by computing iterates of $f_{3.1}(x)$ and $f_{3.2}(x)$ starting at $x_0 = 0.5$. We call $c = 3$ a bifurcation point because of the abrupt change of attractor from a one-cycle for $c < 3$ to a two-cycle for $c > 3$. Many diseases, such as some forms of heart disease, epilepsy and Parkinson's disease, are now viewed by some researchers as bifurcations from healthy behavior to unhealthy behavior.

80. (exploratory) In exercise 78, we learned a simple rule for determining whether a fixed point is attracting or repelling. If there are no attracting fixed points, the iterates may converge to a two-cycle (two numbers alternating) or some other pattern. We can do analysis on the two-cycles similar to our work on the fixed points. Numbers a and b are a two-cycle if $f(a) = b$ and $f(b) = a$; both numbers are solutions of the equation $f(f(x)) = x$. For $f(x) = x^2 - 1$, set up and find all four solutions of the equation $f(f(x)) = x$. Show that 0 and -1 form a two-cycle; what do the other two solutions represent? Compute iterates starting at $x_0 = 0.2$. Does it appear that the two-cycle attracts or repels? The test for attracting or repelling two-cycles is similar to the test for fixed points, except applied to the function $g(x) = f(f(x))$. If a is part of a two-cycle and $|g'(a)| < 1$, then the two-cycle is attracting; if $|g'(a)| > 1$, the two-cycle is repelling. Show that if a and b are a two-cycle, then $g'(a) = f'(f(a)) f'(a) = f'(b) f'(a)$. Using this rule, show that 0 and -1 form an attracting two-cycle. Explore the two-cycle for $x^2 - 3$.

Chapter 3
Applications of Differentiation

One relatively new test available to physicians for diagnosing injuries and disease is the MRI. Magnetic resonance imaging (MRI) is used to visualize internal structures, such as torn cartilage in a knee. The ability to *see* the physical status of a knee or an internal organ without surgery is an invaluable aid to physicians and their patients. However, it still takes an experienced physician to distinguish the important features of an MRI from insignificant ones. If you have ever looked at an MRI or even a conventional x-ray, you have probably been amazed at the details that your physician could quickly identify. In the MRI to the left, can you identify any damage to the knee? Of course, it always helps to know what you are looking for.

MRI of a knee.
(FONAR Corporation)

The ability to accurately read graphs is one of the primary goals of this chapter. By the end of section 3.5, you should have a good idea of what the significant features of a graph are. Although we will only be looking at two-dimensional graphs of functions, the language and skills that you acquire here will transfer to plots of seismic readings, sonar mappings of the ocean floor and other graphical displays of information that you may encounter.

Most people do not recognize the vast amount of mathematical computation required to produce a viewable image from an MRI. In an MRI, magnetic fields and pulses of radio waves are used to determine the distribution of hydrogen atoms in the body (see *Visualization* by R. Friedhoff and W. Benzon for more details). The presence of hydrogen atoms, in turn, is deduced from the release of energy during the magnetization process. (This is a long way from a standard x-ray image!) By solving countless equations and performing lengthy calculations, a computer transforms the energy data into an accurate image of the interior of a human body.

Likewise, it may surprise you how many calculations we must perform to draw an accurate graph of a function. At each stage of the graphing process, we must solve equations to identify significant features of the graph. Because of the central role that equation solving plays in this chapter, we devote a portion of the first section to a discussion of a powerful method that you can use to approximate solutions of difficult equations.

But, you may ask, if a computer or calculator can do all of the calculations, why do you need to know what it's doing? One answer is that most computer algorithms are imperfect approximations that can occasionally result in significant errors on certain types of problems. By understanding how such algorithms work, you can anticipate and identify when a computer is in error. For example, the picture on the right below is a computer enhancement of the out-of-focus picture on the left.

(Images courtesy of Vicom Systems.)

Notice that the picture on the right shows a faint halo around the airplane. This is not a real ghost or aura or even sound waves, but instead a by-product of the computer algorithm used to sharpen the picture. If you understand the algorithm, you will not misinterpret the ghost. A ghost on an airplane is not serious, but the mathematics used to sharpen the picture is also used to produce MRIs, where a misinterpreted ghost could have serious consequences. (How would you feel if an MRI appeared to show a tumor that was not actually there?)

3.1   Linear Approximations and Newton's Method

For what purpose do you use a scientific calculator? If you think about it, you'll discover that there are two distinctly different jobs that calculators do for you. First, they perform arithmetic operations (addition, subtraction, multiplication and division) much faster than any of us could hope to do them. It's not that you don't know how to multiply 1024 times 1673, but rather that it is time-consuming and even annoying to carry out this (albeit well-understood) calculation with pencil and paper. For such problems, calculators are a tremendous convenience, which none of us would like to live without. Perhaps more significantly, we also use our calculators to compute values of transcendental functions such as sine, cosine, tangent, exponentials and logarithms. In the case of these function evaluations, the calculator is much more than a mere convenience.

If asked to calculate $\sin(1.2345678)$ without a calculator, you would understandably draw a blank. Don't worry, there's nothing wrong with your background. (Also, don't worry that anyone will ever ask you to do this without a calculator.) The problem is that the sine function is not *algebraic*. That is, there is no formula for $\sin x$ involving only the arithmetic operations. So, how does your calculator "know" that $\sin(1.2345678) = 0.9440056953$? In short, it doesn't *know* this at all. Rather, the calculator has a built-in program that generates **approximate** values of the sine and other transcendental functions.

In this section, we take a small step into the (very large) world of approximation, by developing two simple approximation methods. The first of these is somewhat crude, but points the way for more sophisticated approximation techniques to follow later in the text. Our primary intent is to give you a taste of how you might approach the problem of approximation.

Linear Approximations

Suppose we wanted to find an approximation for $f(x_1)$, where $f(x_1)$ is unknown, but where $f(x_0)$ is known for some x_0 "close" to x_1. For instance, the value of $\cos(1)$ is unknown, but we do know that $\cos \frac{\pi}{3} = \frac{1}{2}$ exactly and $\frac{\pi}{3}$ is "close" to 1. We could always use $\frac{1}{2}$ as an approximation to $\cos(1)$, but we can do better.

Recall that the tangent line to the curve $y = f(x)$ at $x = x_0$ stays close to the curve near the point of tangency. Referring to Figure 3.1, notice that if x_1 is "close" to x_0 and we follow the tangent line at $x = x_0$ to the point corresponding to $x = x_1$, then the y-coordinate of that point should be "close" to the y-coordinate of the point on the curve $y = f(x)$ (i.e., $f(x_1)$).

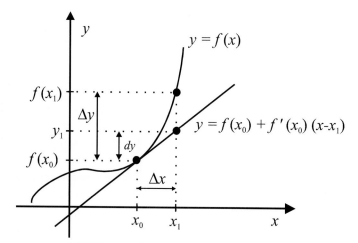

FIGURE 3.1: Linear approximation of $f(x_1)$.

Since the slope of the tangent line to $y = f(x)$ at $x = x_0$ is $f'(x_0)$, the equation of the tangent line to $y = f(x)$ at $x = x_0$ is found from

$$m_{\text{tan}} = f'(x_0) = \frac{y - f(x_0)}{x - x_0}. \tag{1.1}$$

Solving equation (1.1) for y gives us

$$y = f(x_0) + f'(x_0)(x - x_0). \tag{1.2}$$

Notice that (1.2) is the equation of the straight line tangent to the graph of $y = f(x)$ at $x = x_0$. We give the linear function defined by this equation a name, as follows.

Definition 1.1   The **linear** (or **tangent line**) **approximation** of $f(x)$ at $x = x_0$ is the function $L(x) = f(x_0) + f'(x_0)(x - x_0)$.

Notice that the y-coordinate of the point on the tangent line corresponding to $x = x_1$ is simply found by substituting $x = x_1$ in equation (1.2). This gives us

$$y_1 = f(x_0) + f'(x_0)(x_1 - x_0). \tag{1.3}$$

We define the **increments** Δx and Δy by

$$\Delta x = x_1 - x_0$$

and

$$\Delta y = f(x_1) - f(x_0).$$

Using this notation, we now have from equation (1.3) the approximation

$$\boxed{f(x_1) \approx y_1 = f(x_0) + f'(x_0)\,\Delta x.} \tag{1.4}$$

We sometimes rewrite (1.4) by subtracting $f(x_0)$ from both sides, to yield

$$\Delta y = f(x_1) - f(x_0) \approx f'(x_0)\,\Delta x = dy, \tag{1.5}$$

where $dy = f'(x_0)\,\Delta x$ is called the **differential** of y.

We can use linear approximations to produce approximate values of transcendental functions, as in the following example.

Example 1.1   Finding a Linear Approximation

Find the linear approximation to $f(x) = \cos x$ at $x_0 = \frac{\pi}{3}$ and use it to approximate $\cos(1)$.

Solution   From Definition 1.1, the linear approximation is defined as $L(x) = f(x_0) + f'(x_0)(x - x_0)$. Here, $x_0 = \frac{\pi}{3}$, $f(x) = \cos x$ and $f'(x) = -\sin x$ and so, we have

$$L(x) = \cos\left(\tfrac{\pi}{3}\right) - \sin\left(\tfrac{\pi}{3}\right)\left(x - \tfrac{\pi}{3}\right) = \tfrac{1}{2} - \tfrac{\sqrt{3}}{2}\left(x - \tfrac{\pi}{3}\right).$$

In Figure 3.2, we show a graph of $y = \cos x$ and the linear approximation to $\cos x$ for $x_0 = \frac{\pi}{3}$. Notice that the linear approximation (i.e., the tangent line at $x_0 = \frac{\pi}{3}$) stays close to the graph of $y = \cos x$ only for x close to $\frac{\pi}{3}$. In fact, for $x < 0$ or $x > \pi$, the linear approximation is obviously quite poor. This is typical of linear approximations (tangent lines). They generally stay close to the curve only nearby the point of tangency.

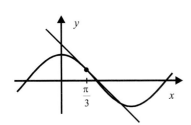

FIGURE 3.2: $y = \cos x$ and the linear approximation at $x_0 = \frac{\pi}{3}$.

Although we gave you the value of x_0 here, observe that if you had been given the choice, one reason you might choose $x_0 = \frac{\pi}{3}$ is that $\frac{\pi}{3}$ is the value closest to 1 at which we know the value of the cosine exactly. Finally, an estimate of $\cos(1)$ is

$$L(1) = \tfrac{1}{2} - \tfrac{\sqrt{3}}{2}\left(1 - \tfrac{\pi}{3}\right) \approx 0.5409.$$

Notice that on your calculator you would find $\cos(1) = 0.5403$, so that we have found a fairly good approximation to the desired value.

■

In the following example, you can see that the farther x_1 is from x_0, the less accurate the approximation is. Take one more look at Figure 3.1 to convince yourself that this is generally true. We also illustrate the use of the increments Δx and Δy here.

Example 1.2   Linear Approximation to Some Cube Roots
Approximate $\sqrt[3]{8.02}$, $\sqrt[3]{8.07}$ and $\sqrt[3]{8.15}$.

Solution   Here we are approximating values of the function $f(x) = \sqrt[3]{x} = x^{1/3}$. So, $f'(x) = \frac{1}{3}x^{-2/3}$. Of course, the closest number to any of 8.02, 8.07 or 8.15 whose cube root we know exactly is 8. Now,

Add and subtract $f(8)$.

$$
\begin{aligned}
f(8.02) &= f(8) + [f(8.02) - f(8)] \\
&= f(8) + \Delta y.
\end{aligned}
\tag{1.6}
$$

From (1.5), we have

Since $\Delta x = 8.02 - 8$.

$$
\begin{aligned}
\Delta y \approx dy &= f'(8)\,\Delta x \\
&= \left(\frac{1}{3}\right) 8^{-2/3}(8.02 - 8) = \frac{1}{600}.
\end{aligned}
\tag{1.7}
$$

Using (1.6) and (1.7), we get

$$f(8.02) \approx f(8) + dy = 2 + \frac{1}{600} \approx 2.0016667,$$

while your calculator accurately returns $\sqrt[3]{8.02} \approx 2.0016653$. Similarly, we get

$$f(8.07) \approx f(8) + \frac{1}{3}8^{-2/3}(8.07 - 8) \approx 2.0058333,$$

while your calculator returns $\sqrt[3]{8.07} \approx 2.005816$. Finally, we have

$$f(8.15) \approx f(8) + \frac{1}{3}8^{-2/3}(8.15 - 8) \approx 2.0125,$$

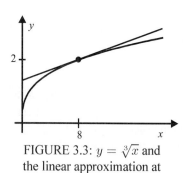

FIGURE 3.3: $y = \sqrt[3]{x}$ and the linear approximation at $x_0 = 8$.

while your calculator returns $\sqrt[3]{8.15} \approx 2.012423$. It is important to recognize here that the farther the value of x gets from the point of tangency, the worse the approximation gets. You can see this clearly in Figure 3.3, where we plot $y = \sqrt[3]{x}$, along with the linear approximation at $x = 8$.

■

Once again, we want to emphasize that linear approximations do not typically produce highly accurate approximations. The primary objective in studying them is to take one small step toward learning how to make reasonable approximations.

In practice, we often use linear approximations in situations in which we are unable to check the answer. Our first two examples were intended to familiarize you with the technique and to give you a feel for how good (or bad) linear approximations tend to be. In the following example, there is no exact answer to compare to. Our use of the linear approximation here is referred to as **linear interpolation**.

Example 1.3   Using a Linear Approximation to Perform Linear Interpolation

The price of an item affects consumer demand for that item. Suppose that based on market research, a company estimates that $f(x)$ thousand small cameras can be sold at the price of $\$x$, as given in the accompanying table. Estimate the number of cameras that can be sold at $\$7$.

x	6	10	14
$f(x)$	84	60	32

Solution   The closest x-value to $x = 7$ in the table is $x = 6$. (In other words, this is the closest value of x at which we know the value of $f(x)$.) The linear approximation of $f(x)$ at $x = 6$ would look like

$$L(x) = f(6) + f'(6)(x - 6).$$

From the table, we know that $f(6) = 84$, but we do not know $f'(6)$. Further, we can't compute $f'(x)$, since we don't have a formula for $f(x)$. The best we can do with the given data is approximate the derivative by

$$f'(6) \approx \frac{f(10) - f(6)}{10 - 6} = \frac{60 - 84}{4} = -6.$$

The linear approximation is then

$$L(x) = 84 - 6(x - 6).$$

Using this, we estimate that the number of cameras sold at $x = 7$ would be $L(7) = 84 - 6 = 78$. That is, we would expect to sell approximately 78 thousand cameras at a price of $\$7$. We show a graphical interpretation of this in Figure 3.4, where the straight line is the linear approximation (in this case, the secant line joining the first two data points).

∎

FIGURE 3.4: Linear interpolation.

Newton's Method

We now return to the question of finding zeros of a function. In section 1.3, we introduced the method of bisections as a tedious, yet reliable, method of finding zeros of continuous functions. In this section, we explore a method which is usually much more efficient than bisections. We are again looking for values of x such that $f(x) = 0$. These values are called **roots** of the equation $f(x) = 0$ or **zeros** of the function f. If

$$f(x) = ax^2 + bx + c,$$

there is no challenge to doing this, since we have an explicit formula for the solution(s) (the quadratic formula). But, what if we want to find zeros of

$$f(x) = \tan x - x \ ?$$

This function is not algebraic and there are no formulas available for finding the zeros. Even so, we can clearly see zeros in Figure 3.5 (in fact, there are infinitely many of them). The question is, how are we to **find** them?

FIGURE 3.5: $y = \tan x - x$.

In general, if we wish to find approximate solutions to $f(x) = 0$, we first make a reasonable guess as to the location of a solution. We will call this an **initial guess**, denoted x_0. Once again, since the tangent line to $y = f(x)$ at $x = x_0$ tends to hug the curve, we can follow the tangent line to where it intersects the x-axis (see Figure 3.6).

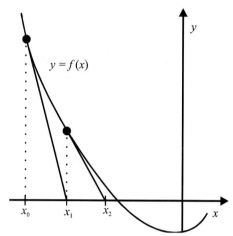

FIGURE 3.6: Newton's method.

Notice that this appears to provide an improved approximation to the zero. The equation of the tangent line to $y = f(x)$ at $x = x_0$ is found from

$$m_{\text{tan}} = f'(x_0) = \frac{y - f(x_0)}{x - x_0}.$$

Hence, the equation of the tangent line is

$$y - f(x_0) = f'(x_0)(x - x_0). \tag{1.8}$$

We denote the x-intercept of the tangent line by x_1 (found by setting $y = 0$ in (1.8)). We then have

$$0 - f(x_0) = f'(x_0)(x_1 - x_0)$$

and so,

$$x_1 = x_0 - \frac{f(x_0)}{f'(x_0)}.$$

If we repeat this process, using x_1 as our new guess, we should produce an improved approximation,

$$x_2 = x_1 - \frac{f(x_1)}{f'(x_1)}$$

and so on (see Figure 3.6). In this way, we generate a sequence of **successive approximations** determined from

$$\boxed{x_{n+1} = x_n - \frac{f(x_n)}{f'(x_n)}, \text{ for } n = 0, 1, 2, 3, \dots} \tag{1.9}$$

This procedure is called the **Newton-Raphson method** or simply **Newton's method**. If Figure 3.6 is any indication, x_n should get closer and closer to a zero as n increases.

Newton's method is generally a very fast, accurate method for approximating the zeros of a function, as we illustrate with the following example.

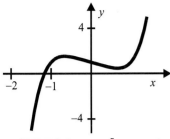

FIGURE 3.7: $y = x^5 - x + 1$.

Example 1.4   Using Newton's Method to Approximate a Zero

Find a zero of $f(x) = x^5 - x + 1$.

Solution   From Figure 3.7, it appears as if the only zero of f is located between $x = -2$ and $x = -1$. We further observe that $f(-1) = 1 > 0$ and $f(-2) = -29 < 0$. Since f is continuous (all polynomials are continuous!), the Intermediate Value Theorem (Theorem 3.4 in section 1.3) says that f must have a zero on the interval $(-2, -1)$. Further, because the zero appears to be closer to $x = -1$, we make the initial guess $x_0 = -1$. Finally, $f'(x) = 5x^4 - 1$ and so, Newton's method gives us

$$x_{n+1} = x_n - \frac{f(x_n)}{f'(x_n)}$$

$$= x_n - \frac{x_n^5 - x_n + 1}{5x_n^4 - 1}, \ n = 0, 1, 2, \ldots.$$

Using the initial guess $x_0 = -1$, we get

$$x_1 = -1 - \frac{(-1)^5 - (-1) + 1}{5(-1)^4 - 1}$$

$$= -1 - \frac{1}{4} = -\frac{5}{4}.$$

Likewise, from $x_1 = -\frac{5}{4}$, we get the improved approximation

$$x_2 = -\frac{5}{4} - \frac{\left(-\frac{5}{4}\right)^5 - \left(-\frac{5}{4}\right) + 1}{5\left(-\frac{5}{4}\right)^4 - 1}$$

$$= -1.178459394$$

and so on, we find that

$$x_3 = -1.167537384$$

$$x_4 = -1.167304083$$

and

$$x_5 = -1.167303978 = x_6.$$

Since $x_5 = x_6$, we will make no further progress by calculating additional steps. As a final check on the accuracy of our approximation, we compute

$$f(x_6) \approx 3 \times 10^{-12}.$$

Since this is very close to zero, we say that $x_6 = -1.167303978$ is an **approximate zero** of f.

∎

You can bring Newton's method to bear on a variety of approximation problems. As we illustrate in the following example, you may first need to rephrase the problem as a rootfinding problem.

Example 1.5   Using Newton's Method to Approximate a Cube Root

Use Newton's method to approximate $\sqrt[3]{7}$.

Solution   This may seem like a problem better suited to solution by a tangent line approximation. However, Newton's method will quickly provide us with an accurate approximation. Recall that Newton's method is used to solve equations of the form $f(x) = 0$. We

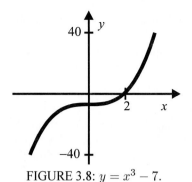

FIGURE 3.8: $y = x^3 - 7$.

can rewrite the current problem in this form, as follows. Suppose $x = \sqrt[3]{7}$. Then, $x^3 = 7$, which can be rewritten as

$$f(x) = x^3 - 7 = 0.$$

Here, $f'(x) = 3x^2$ and we obtain an initial guess from a graph of $y = f(x)$ (see Figure 3.8). Notice that there is a zero near $x = 2$ and so we take $x_0 = 2$. Newton's method then yields

$$x_1 = 2 - \frac{2^3 - 7}{3(2^2)} = \frac{23}{12} \approx 1.916666667.$$

Continuing this process, we have

$$x_2 \approx 1.912938458$$

and

$$x_3 \approx 1.912931183 \approx x_4.$$

Further,

$$f(x_4) \approx -5 \times 10^{-12}$$

and so, x_4 is an approximate zero of f. This also says that

$$\sqrt[3]{7} \approx 1.912931183.$$

Compare this with the value of $\sqrt[3]{7}$ produced by your calculator, to see how very accurate Newton's method was here.

Remark 1.1   Although it seemed to be very efficient in the last two examples, Newton's method does not always work. We urge you to make sure that the values coming from the method are getting progressively closer and closer together (zeroing-in, we hope, on the desired solution). Don't stop until you've reached the limits of accuracy of your computing device. Also, be sure to compute the value of the function at the suspected approximate zero. If the function value is not close to zero, do not accept the value as an approximate zero.

As we illustrate in the following example, Newton's method requires a good initial guess in order to find an accurate approximation.

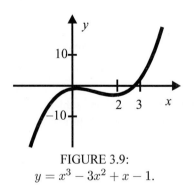

FIGURE 3.9:
$y = x^3 - 3x^2 + x - 1$.

Example 1.6   The Effect of a Bad Guess on Newton's Method

Use Newton's method to find an approximate zero of $f(x) = x^3 - 3x^2 + x - 1$.

Solution   From the graph in Figure 3.9, there appears to be a zero on the interval $(2, 3)$. If you were to use the (not particularly good) initial guess $x_0 = 1$, you would get $x_1 = 0$, $x_2 = 1$, $x_3 = 0$ and so on. Try this for yourself. Newton's method is sensitive to the initial guess and you just made a bad initial guess. If you had instead started with the improved initial guess $x_0 = 2$, Newton's method would have quickly converged to the approximate zero 2.769292354. (Again, try this for yourself.)

As we see in the following example, a poor initial guess may have disastrous consequences. However, simply picking a good initial guess will not guarantee the rapid convergence of Newton's method. For some functions, the convergence will be slow no matter how good your initial guess is. By slow convergence, we mean that it takes many iterations to see significant improvement in the approximation.

n	x_n
1	-9.5
2	-65.9
3	-2302
4	-2654301
5	-3.5×10^{12}
6	-3.1×10^{24}

Newton's method iterations for $x_0 = -2$.

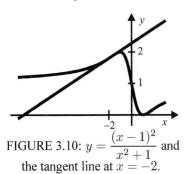

FIGURE 3.10: $y = \dfrac{(x-1)^2}{x^2+1}$ and the tangent line at $x = -2$.

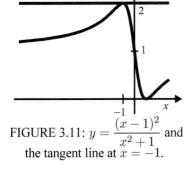

FIGURE 3.11: $y = \dfrac{(x-1)^2}{x^2+1}$ and the tangent line at $x = -1$.

Example 1.7   Unusually Slow Convergence for Newton's Method

Use Newton's method with (a) $x_0 = -2$, (b) $x_0 = -1$ and (c) $x_0 = 0$ to try to locate the zero of $f(x) = \dfrac{(x-1)^2}{x^2+1}$.

Solution   Of course, there's no mystery here: f has only one zero, located at $x = 1$. But, watch what happens if we try to use Newton's method with the specified guesses.

(a) If we take $x_0 = -2$ and apply Newton's method, we calculate the values in the accompanying table.

Obviously, Newton's method is blowing up for the given initial guess. To see why, look at Figure 3.10, which shows the graphs of both $y = f(x)$ and the tangent line at $x = -2$. Notice that if you follow the tangent line to where it intersects the x-axis, you will be going **away** from the zero (far away). Since all of the tangent lines for $x \le -2$ have positive slope (compute $f'(x)$ to see why this is true), each subsequent step takes you farther from the zero. (Draw your own graph showing several tangent lines to see why this is true.)

(b) If we use the improved initial guess $x_0 = -1$, note that we cannot even compute x_1. In this case, $f'(x_0) = 0$ and so, Newton's method fails. Notice that graphically, this means that the tangent line to $y = f(x)$ at $x = -1$ is horizontal (see Figure 3.11), so that the tangent line never intersects the x-axis.

(c) With the even better initial guess $x_0 = 0$, we obtain the successive approximations in the following table.

n	x_n		n	x_n
1	0.5		7	0.9881719
2	0.70833		8	0.9940512
3	0.85653		9	0.9970168
4	0.912179		10	0.9985062
5	0.95425		11	0.9992525
6	0.976614		12	0.9996261

Newton's method iterations for $x_0 = 0$.

Finally, we happened upon an initial guess for which Newton's method converged to the root $x = 1$. What is unusual here is that the successive approximations shown in the table are converging to 1 much more slowly than in previous examples. By comparison, note that in example 1.4, the iterations stop changing at x_5. Here, x_5 is not particularly close to the desired zero of $f(x)$. In fact, in this example, x_{12} is not as close to the zero as x_5 was in example 1.4. We will look further into this type of behavior in the exercises.

∎

Despite the small problems experienced in examples 1.6 and 1.7, you should view Newton's method as a generally reliable and efficient method of locating zeros approximately. Just use a bit of caution and common sense. If the successive approximations are converging to some value that does not appear consistent with the graph, then you need to scrutinize your results more carefully and perhaps try some other initial guesses.

Exercises 3.1

1. (writing) We constructed a variety of linear approximations in this section. Approximations can be "good" approximations or "bad" approximations. Explain why it can be said that $y = x$ is a good approximation to $y = \sin x$ near $x = 0$ but $y = 1$ is not a good approximation to $y = \cos x$ near $x = 0$. (Hint: look at the graphs of $y = \sin x$ and $y = x$ on the same axes, then do the same with $y = \cos x$ and $y = 1$.)

2. (writing) In example 1.5, we mentioned that you might think of using a linear approximation instead of Newton's method. Explain the relationship between a linear approximation to $\sqrt[3]{7}$ and a Newton's method approximation to $\sqrt[3]{7}$. (Hint: Compare the first step of Newton's method to a linear approximation.)

3. (writing) Briefly explain in terms of tangent lines why the approximation in example 1.2 gets worse as x gets farther from 8.

4. (writing) Explain why Newton's method fails computationally if $f'(x_0) = 0$. In terms of tangent lines intersecting the x-axis, explain why having $f'(x_0) = 0$ is a problem.

In exercises 5–12, find the linear approximation to $f(x)$ at $x = x_0$.

5. $f(x) = \sqrt{x}$, $x_0 = 1$

6. $f(x) = (x+1)^{1/3}$, $x_0 = 0$

7. $f(x) = \sqrt{2x+9}$, $x_0 = 0$ **8.** $f(x) = \frac{2}{x}$, $x_0 = 1$

9. $f(x) = \sin 3x$, $x_0 = 0$ **10.** $f(x) = \sin x$, $x_0 = \pi$

11. $f(x) = e^{2x}$, $x_0 = 0$ **12.** $f(x) = e^{x^2}$, $x_0 = 0$

In exercises 13–16, find the linear approximation at $x = 0$ to show that the following commonly used approximations are valid for "small" x. Compare the approximate and exact values for $x = 0.01$, $x = 0.1$ and $x = 1$.

13. $\tan x \approx x$ **14.** $\sqrt{1+x} \approx 1 + \frac{1}{2}x$

15. $\sqrt{4+x} \approx 2 + \frac{1}{4}x$ **16.** $e^x \approx 1 + x$

In exercises 17–22, use linear approximations to estimate the quantity.

17. $\sin 1$ **18.** $\sin 9/4$

19. $\sqrt[4]{16.04}$ **20.** $\sqrt[4]{16.08}$

21. $\sqrt[4]{16.16}$ **22.** $\ln 2.8$ (Hint: $\ln e = 1$)

23. For exercises 19-21, compute the error (the absolute value of the difference between the exact value and the linear approximation).

24. Thinking of exercises 19-21 as numbers of the form $\sqrt[4]{16 + \Delta x}$, denote the errors as $e(\Delta x)$ (where $\Delta x = 0.04$, $\Delta x = 0.08$ and $\Delta x = 0.16$). Based on these three computations, determine a constant c such that $e(\Delta x) \approx c(\Delta x)^2$.

25. Given the graph of $y = f(x)$, draw in the tangent lines used in Newton's method to determine x_1 and x_2 after starting at $x_0 = 2$. Which of the zeros will Newton's method converge to?

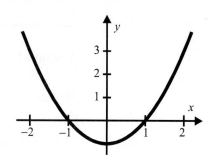

26. Repeat exercise 25 with $x_0 = -2$ and $x_0 = 0.4$.

In exercises 27–30, use Newton's method with the given x_0 to (a) compute x_1 and x_2 by hand and (b) use a computer or calculator to find the root to at least 5-digit accuracy.

27. $x^3 + 3x^2 - 1 = 0$, $x_0 = 1$

28. $x^3 + 4x^2 - x - 1 = 0$, $x_0 = -1$

29. $x^4 - 3x^2 + 1 = 0$, $x_0 = 1$

30. $x^4 - 3x^2 + 1 = 0$, $x_0 = -1$

In exercises 31–40, use Newton's method to find an approximate root (at least 6-digit accuracy). Sketch the graph and explain how you determined your initial guess.

31. $x^3 + 4x^2 - 3x + 1 = 0$ **32.** $x^4 - 4x^3 + x^2 - 1 = 0$

33. $x^5 + 3x^3 + x - 1 = 0$ **34.** $x^3 + 2x + 1 = 0$

35. $\cos x - x = 0$ **36.** $x - e^{-x} = 0$

37. $\sin x = x^2 - 1$ **38.** $\cos x^2 = x$

39. $e^x = -x$ **40.** $e^{-x} = \sqrt{x}$

In exercises 41–46, Newton's method fails. Explain why the method fails and, if possible, find a root by correcting the problem.

41. $4x^3 - 7x^2 + 1 = 0$, $x_0 = 0$

42. $4x^3 - 7x^2 + 1 = 0$, $x_0 = 1$

43. $x^2 + 1 = 0$, $x_0 = 0$

44. $x^2 + 1 = 0$, $x_0 = 1$

45. $\dfrac{4x^2 - 8x + 1}{4x^2 - 3x - 7} = 0$, $x_0 = -1$

46. $\left(\dfrac{x+1}{x-2}\right)^{1/3} = 0$, $x_0 = 0.5$

47. Show that Newton's method applied to $x^2 - a = 0$ produces the iterative scheme $x_{n+1} = \frac{1}{2}(x_n + a/x_n)$ for approximating $\sqrt{a}$. This scheme was known to the ancient Babylonians. To understand how they approached the problem, suppose that your initial guess at $\sqrt{a}$ is a little too small. How would a/x_0 compare to $\sqrt{a}$? Explain why the average of x_0 and a/x_0 would give a better approximation to $\sqrt{a}$.

48. By applying Newton's method to the appropriate equation, explain why $x_{n+1} = 2x_n - ax_n^2$ should converge to $1/a$.

49. Newton's theory of gravitation states that the weight of a person at elevation x feet above sea level is $W(x) = PR^2/(R+x)^2$, where P is the person's weight at sea level and R is the radius of the earth (approximately $20,900,000$ feet). Find the linear approximation of $W(x)$ at $x = 0$. Use the linear approximation to estimate the elevation required to reduce the weight of a 120-pound person by one percent.

50. One important aspect of Einstein's theory of relativity is that mass is not constant. For a person with mass m_0 at rest, the mass will equal $m = m_0/\sqrt{1 - v^2/c^2}$ at velocity v (here, c is the speed of light). Thinking of m as a function of v, find the linear approximation of $m(v)$ at $v = 0$. Use the linear approximation to show that mass is essentially constant for small velocities.

51. In the figure, a beam is clamped at its left end and simply supported at its right end. A force P (called an **axial load**) is applied at the right end.

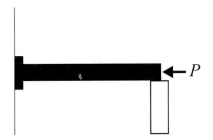

If enough force is applied, the beam will buckle. Obviously, it is important for engineers to be able to compute this **buckling load**. For certain types of beams, the buckling load is the smallest positive solution of the equation $\tan \sqrt{x} = \sqrt{x}$. The shape of the buckled beam is given by $y = \sqrt{L} - \sqrt{L}x - \sqrt{L}\cos\sqrt{L}x + \sin\sqrt{L}x$, where L is the buckling load. Find L and the shape of the buckled beam.

52. (exploratory) If a function has more than one zero, which zero will Newton's method converge to (assuming it converges)? The answer is far more interesting than you might guess. In this exercise, we will examine zeros of the function $f(x) = x^3 - 3x^2 + 2x$. First, show that the zeros of $f(x)$ are $x = 0$, $x = 1$ and $x = 2$. Then run Newton's method with starting values $x_0 = 0.1$, $x_0 = 1.1$ and $x_0 = 2.1$. A reasonable hypothesis at this stage would be that Newton's method converges to the zero closest to the guess. To see that this isn't true, try $x_0 = 0.54$, $x_0 = 0.55$ and $x_0 = 0.56$. Are you surprised by the results? Try to explain graphically why Newton's method is "confused" in this region of x's. We are interested in finding the **basin of attraction** for each zero. That is, for which starting values x_0 does Newton's method converge to 0? For which starting values x_0 does Newton's method converge to 1 and to 2? To answer these questions, we can start by determining to which zero Newton's method converges with $x_0 = 0, 0.1, 0.2, \ldots, 0.99, 1.0$. You will want

to write a program to do this! Most of the basin boundaries are well defined, but (as we have already seen) there is some confusion between $x = 0.5$ and $x = 0.6$. So, let's magnify the interval $[0.5, 0.6]$. That is, run Newton's method with $x_0 = 0.5, 0.501, 0.502, \ldots, 0.599, 0.6$. Again, some clear boundaries emerge, but the picture is not sharp between 0.55 and 0.56. If you magnify $[0.55, 0.56]$, you will find erratic behavior in the interval $[0.552, 0.553]$. Continue this process, and you will continue to see confusion in 1 out of 10 subintervals. (This is a fractal process!) Computer pictures of basins of attraction using **complex numbers** are often beautiful. The following fractals show the basin of attraction (in black) for $x = 1$. In the first figure, we display the fractal for $-1.5 \leq x \leq 3.5$. In the second figure, we have zoomed in to the portion for $0.26 \leq x \leq 0.56$. The third shows an even tighter zoom: $0.5 \leq x \leq 0.555$.

53. (exploratory) Another important question involving Newton's method is how fast it converges to a given zero. Intuitively, we can distinguish between the rate of convergence for $f(x) = x^2 - 1$ (with $x_0 = 1.1$) and that for $g(x) = x^2 - 2x + 1$ (with $x_0 = 1.1$). But how can we measure this? One method is to take successive approximations x_{n-1} and x_n and compute the difference $\Delta_n = x_n - x_{n-1}$. To discover the importance of this quantity, run Newton's method with $x_0 = 1.5$ and then compute the ratios Δ_3/Δ_2, Δ_4/Δ_3, Δ_5/Δ_4 and so on, for each of the following functions

$$F_1(x) = (x-1)(x+2)^3 = x^4 + 5x^3 + 6x^2 - 4x - 8,$$
$$F_2(x) = (x-1)^2(x+2)^2 = x^4 + 2x^3 - 3x^2 - 4x + 4,$$
$$F_3(x) = (x-1)^3(x+2) = x^4 - x^3 - 3x^2 + 5x - 2,$$
$$F_4(x) = (x-1)^4 = x^4 - 4x^3 + 6x^2 - 4x + 1.$$

In each case, conjecture a value for the limit $r = \lim\limits_{n\to\infty} \dfrac{\Delta_{n+1}}{\Delta_n}$.

If the limit exists and is nonzero, we say that Newton's method **converges linearly**. How does r relate to your intuitive sense of how fast the method converges? For $f(x) = (x - 1)^4$, we say that the zero $x = 1$ has **multiplicity** 4. For $f(x) = (x-1)^3(x+2)$, $x = 1$ has multiplicity 3 and so on. How does r relate to the multiplicity of the zero? Based on this analysis, why did Newton's method converge faster for $f(x) = x^2 - 1$ than for $g(x) = x^2 - 2x + 1$? Finally, use Newton's method to compute the rate r and hypothesize the multiplicity of the zero $x = 0$ for $f(x) = x \sin x$ and $g(x) = x \sin x^2$.

3.2   Maximum and Minimum Values

It seems that no matter where we turn today, we hear about the need to maximize this or minimize that. In order to remain competitive in a global economy, businesses need to minimize waste and maximize the return on their investment. Managers for massively complex projects like the International Space Station must constantly readjust their programs to squeeze the most out of dwindling resources. In the extremely competitive personal computer industry, companies must continually evaluate how low they can afford to set their prices and still earn a profit adequate to survive. Computer buyers want to maximize their computing speed and other features while minimizing their cost. With this backdrop, it should be apparent that one of the main thrusts of our increasingly mathematical society is to use mathematical methods to maximize and minimize various quantities of interest. In this section, we investigate the notion of maximum and minimum from a purely mathematical standpoint. In section 3.6, we study how to apply these notions to problems of an applied nature.

First, we give careful mathematical definitions of some familiar terms.

Definition 2.1   For a function f defined on a set S of real numbers and a number $c \in S$,
(i) $f(c)$ is the **absolute maximum** of f on S if $f(c) \geq f(x)$ for all $x \in S$ and
(ii) $f(c)$ is the **absolute minimum** of f on S if $f(c) \leq f(x)$ for all $x \in S$.
An absolute maximum or an absolute minimum is referred to as an **absolute extremum**. If a function has more than one extremum, we refer to these as **extrema** (the plural form of extremum).

The first question you might ask is whether every function has an absolute maximum and an absolute minimum. The answer is no, as we can see from Figures 3.12a and 3.12b.

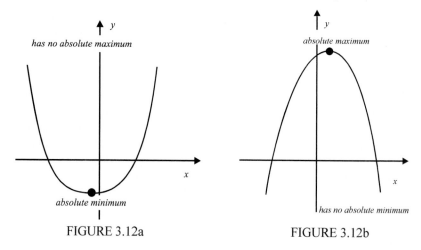

FIGURE 3.12a

FIGURE 3.12b

Example 2.1   Absolute Maximum and Minimum Values
(a) Locate any absolute extrema of $f(x) = x^2 - 9$ on the interval $(-\infty, \infty)$. (b) Locate any absolute extrema of $f(x) = x^2 - 9$ on the interval $(-3, 3)$. (c) Locate any absolute extrema of $f(x) = x^2 - 9$ on the interval $[-3, 3]$.

Solution   (a) From the graph in Figure 3.13, notice that f has an absolute minimum value of $f(0) = -9$, but has no absolute maximum value.
(b) From the graph in Figure 3.14, we see that f has an absolute minimum value of $f(0) = -9$, but still has no absolute maximum value. Your initial reaction might be to say that f has an absolute maximum of 0, but $f(x) \neq 0$ for any $x \in (-3, 3)$, since this is an open interval and hence, does not include the endpoints -3 and 3.

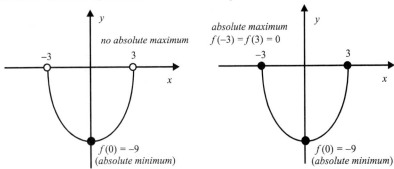

FIGURE 3.14: $y = x^2 - 9 \ (-3, 3)$.   FIGURE 3.15: $y = x^2 - 9 \ [-3, 3]$.

(c) In this case, the endpoints 3 and -3 are **in** the interval $[-3, 3]$. Here, f assumes its absolute maximum at two points: $f(3) = f(-3) = 0$ (see Figure 3.15). ∎

From example 2.1, we see that even nice, continuous functions may fail to have absolute extrema, depending on the interval on which we're looking. In example 2.1, the function failed to have an absolute maximum, except on the closed, bounded interval, $[-3, 3]$. This provides some clues, but the question remains as to when a function is **guaranteed** to have an absolute maximum and an absolute minimum on a given interval. The following example provides one more piece of the puzzle.

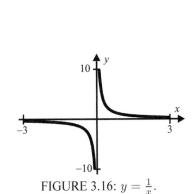

FIGURE 3.16: $y = \frac{1}{x}$.

Example 2.2   A Function with No Absolute Maximum or Minimum
Locate any absolute extrema of $f(x) = \dfrac{1}{x}$, on the interval $[-3, 3]$.

Solution   From the graph in Figure 3.16, f clearly fails to have either an absolute maximum or an absolute minimum on $[-3, 3]$. The following table of values for $f(x)$ for x close to 0 suggests the same conclusion.

x	$1/x$	x	$1/x$
1	1	-1	-1
0.1	10	-0.1	-10
0.01	100	-0.01	-100
0.001	1,000	-0.001	$-1,000$
0.0001	10,000	-0.0001	$-10,000$
0.00001	100,000	-0.00001	$-100,000$
0.000001	1,000,000	-0.000001	$-1,000,000$

∎

The most obvious difference between the functions in examples 2.1 and 2.2 is that $f(x) = \dfrac{1}{x}$ is discontinuous at a point in the interval $[-3, 3]$. We offer the following theorem without proof.

FIGURE 3.13:
$y = x^2 - 9$ on $(-\infty, \infty)$.

228   Chapter 3   Applications of Differentiation

Theorem 2.1   (Extreme Value Theorem) A continuous function f defined on a **closed** interval $[a,b]$ assumes both an absolute maximum and an absolute minimum on that interval.

As we have already seen, you do not need to have a continuous function or a closed interval to have an absolute extremum. However, the theorem says that continuous functions are **guaranteed** to have an absolute maximum and an absolute minimum on a closed, bounded interval.

In the following example, we return to the function from example 2.2, but look on a different interval.

Example 2.3   Finding Absolute Extrema of a Continuous Function
Find the absolute extrema of $f(x) = \dfrac{1}{x}$ on the interval $[1,3]$.

Solution   Notice that on the interval $[1,3]$, f is continuous. Consequently, the Extreme Value Theorem guarantees that f has both an absolute maximum and an absolute minimum on $[1,3]$. Judging from the graph in Figure 3.17, it appears that $f(x)$ reaches its maximum at $x = 1$ and its minimum at $x = 3$.

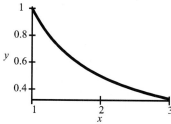

FIGURE 3.17: $y = 1/x$ on $[1,3]$.

■

Our objective is to determine how to locate the absolute extrema of a given function. Before we do this, we need to consider one additional type of extremum.

Definition 2.2   (i) $f(c)$ is a **local maximum** of f if $f(c) \geq f(x)$ for all x in some **open** interval containing c.
   (ii) $f(c)$ is a **local minimum** of f if $f(c) \leq f(x)$ for all x in some **open** interval containing c.
   In either case, we call $f(c)$ a **local extremum** of f.

Local maxima and minima (the plural forms of maximum and minimum, respectively) are sometimes referred to as **relative** maxima and minima, respectively. In Figure 3.18, we see a function with several local extrema.

FIGURE 3.18: Local extrema.

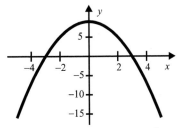

FIGURE 3.19a: $y = 9 - x^2$.

You should notice from Figure 3.18 that each local extremum seems to occur at a point where the tangent line is horizontal (i.e., $f'(x) = 0$), at a point where the tangent line is vertical (where $f'(x)$ is undefined) or at a corner (again, where $f'(x)$ is undefined). We can see this behavior quite clearly in the following two examples.

Example 2.4   A Function with a Zero Derivative at a Local Maximum
    Locate any local extrema for $f(x) = 9 - x^2$ and describe the behavior of the derivative at the local extremum.

Solution   We can see from Figure 3.19a that there is a local maximum at $x = 0$. Further, note that $f'(x) = -2x$ and so, $f'(0) = 0$. In Figure 3.19b, we show the graph of $y = f(x)$ together with its tangent line at $x = 0$.

■

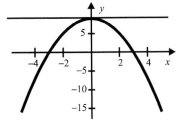

FIGURE 3.19b: $y = 9 - x^2$ and the tangent line at $x = 0$.

Example 2.5   A Function with an Undefined Derivative at a Local Minimum
    Locate any local extrema for $f(x) = |x|$ and describe the behavior of the derivative at the local extremum.

Solution   We can see from Figure 3.20 that there is a local minimum at $x = 0$. As we have noted in Chapter 2, the graph has a corner at $x = 0$ and hence, $f'(0)$ is undefined. (Recall that all the tangent lines to the graph for $x < 0$ have slope -1, while all the tangent lines for $x > 0$ have slope 1. Since the limit of the slopes of secant lines from the right and from the left are not the same, we know that $f'(0)$ does not exist.)

■

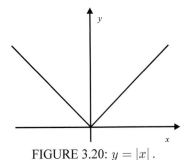

FIGURE 3.20: $y = |x|$.

    The graphs shown in Figures 3.18–3.20 are not unusual. Here is a small challenge: spend a little time now drawing graphs of functions with local extrema. It should not take long to convince yourself that local extrema occur only at points where the derivative is either zero or undefined. Because of this, we give these points a name.

Definition 2.3   A number c in the domain of a function f is called a **critical number** of f if $f'(c) = 0$ or $f'(c)$ is undefined.

It turns out that our earlier observation regarding the location of extrema is correct. That is, local extrema occur **only** at points where the derivative is zero or undefined. We state this formally in the following theorem.

Theorem 2.2   (Fermat's Theorem) Suppose that $f(c)$ is a local extremum (local maximum or local minimum). Then c must be a critical number of f.

Proof. Suppose that f is differentiable at $x = c$. (If not, c is a critical number of f and we are done.) Suppose further that $f'(c) \neq 0$. Then, either $f'(c) > 0$ or $f'(c) < 0$.
    If $f'(c) > 0$, we have by the definition of derivative that

$$f'(c) = \lim_{h \to 0} \frac{f(c+h) - f(c)}{h} > 0.$$

So, for all h sufficiently small,

$$\frac{f(c+h) - f(c)}{h} > 0. \tag{2.1}$$

For $h > 0$, (2.1) says that

$$f(c+h) - f(c) > 0$$

and so,

$$f(c+h) > f(c).$$

Thus, $f(c)$ is not a local maximum.
    For $h < 0$, (2.1) says that

$$f(c+h) - f(c) < 0$$

and so,

$$f(c+h) < f(c).$$

Thus, $f(c)$ is not a local minimum.
    Since we had assumed that $f(c)$ was a local extremum, this is a contradiction. Consequently, $f'(c) \leq 0$.
    Similarly, if $f'(c) < 0$, we obtain the same contradiction. This is left as an exercise. The only remaining possibility is to have $f'(c) = 0$ and this proves the theorem.

■

Remark 2.1   Fermat's Theorem says that local extrema can only occur at critical numbers. This does not say that there is a local extremum at **every** critical number. In fact, this is false, as the following two examples illustrate.

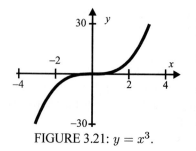

FIGURE 3.21: $y = x^3$.

Example 2.6   A Horizontal Tangent at a Point That Is Not a Local Extremum
    Consider $f(x) = x^3$. It should be clear from Figure 3.21 that f has no local extrema. However, $f'(x) = 3x^2 = 0$ for $x = 0$ (the only critical number of f). In this case, f has

a horizontal tangent line at $x = 0$, but does not have a local extremum there (or anywhere else).   ∎

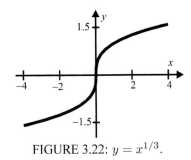

FIGURE 3.22: $y = x^{1/3}$.

Example 2.7   A Vertical Tangent at a Point That Is Not a Local Extremum
Consider $f(x) = x^{1/3}$. As in example 2.6, f has no local extrema (see Figure 3.22). Here, $f'(x) = \frac{1}{3} x^{-2/3}$ and so, f has a critical number at $x = 0$ (in this case the derivative is undefined at $x = 0$). However, f does not have a local extremum at $x = 0$ (or anywhere else).   ∎

We can use Fermat's Theorem and calculator- or computer-generated graphs to find local extrema, as in the following two examples.

Example 2.8   Finding Local Extrema of a Polynomial
Find the critical numbers and local extrema of $f(x) = 2x^3 - 3x^2 - 12x + 5$.

Solution   Here,

$$f'(x) = 6x^2 - 6x - 12$$
$$= 6(x^2 - x - 2)$$
$$= 6(x - 2)(x + 1).$$

Thus, f has two critical numbers, $x = -1$ and $x = 2$. Notice from the graph in Figure 3.23 that these correspond to the locations of a local maximum and a local minimum, respectively.   ∎

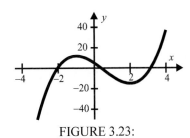

FIGURE 3.23:
$y = 2x^3 - 3x^2 - 12x + 5$.

Example 2.9   An Extremum at a Point where the Derivative is Undefined
Find the critical numbers and local extrema of $f(x) = (3x + 1)^{2/3}$.

Solution   Here, we have

$$f'(x) = \frac{2}{3}(3x + 1)^{-1/3} (3)$$
$$= \frac{2}{(3x + 1)^{1/3}}.$$

Of course, $f'(x) \neq 0$ for all x, but $f'(x)$ is undefined at $x = -\frac{1}{3}$. Be sure to note that $-\frac{1}{3}$ is in the domain of f. Thus, $x = -\frac{1}{3}$ is the only critical number of f. From the graph in Figure 3.24, we see that this corresponds to the location of a local minimum (also the absolute minimum). You should try to use your graphing utility to produce a graph of $y = f(x)$. If you get only half of the graph displayed in Figure 3.24, you have discovered one of the dangers in relying too heavily on technology. The algorithms used by most calculators and many computers will return a complex number (or an error) when asked to compute certain fractional powers of negative numbers. This annoying shortcoming presents only occasional difficulties. We mention this here only to make you aware that technology has limitations.   ∎

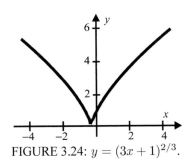

FIGURE 3.24: $y = (3x + 1)^{2/3}$.

You must be certain to check that a given value is in the domain of the function before declaring it a critical number, as in the following example.

Example 2.10   Finding Critical Numbers of a Rational Function

Find all the critical numbers of $f(x) = \dfrac{2x^2}{x+2}$.

Solution   You should note that the domain of f consists of all real numbers other than $x = -2$. Here, we have

From the quotient rule.

$$f'(x) = \frac{4x(x+2) - 2x^2\,(1)}{(x+2)^2}$$
$$= \frac{2x\,(x+4)}{(x+2)^2}.$$

Notice that $f'(x) = 0$ for $x = 0, -4$ and $f'(x)$ is undefined for $x = -2$. However, -2 is not in the domain of f and consequently, the only critical numbers are $x = 0$ and $x = -4$.   ∎

Remark 2.2   When we use the terms maximum, minimum or extremum without specifying absolute or local, we will **always** be referring to absolute extrema.

So far in this section, we have been dancing all around the question of how to locate extrema. We have said that local extrema occur only at critical numbers and that continuous functions must have an absolute maximum and an absolute minimum on a closed, bounded interval. But, so far, we haven't really been able to say how to find these extrema. The following theorem is particularly useful.

Theorem 2.3   Suppose that f is continuous on the closed interval $[a, b]$. Then, the absolute extrema of f must occur at an endpoint (a or b) or at a critical number.

Proof.   First, recall that by the Extreme Value Theorem, f will assume its maximum and minimum values on $[a, b]$, since f is continuous. Let $f(c)$ be an absolute extremum. If c is **not** an endpoint (i.e., $c \neq a$ and $c \neq b$), then c must be in the open interval (a, b). Thus, $f(c)$ must be a local extremum, also. Finally, by Fermat's Theorem, c must be a critical number, since local extrema occur only at critical numbers.   ∎

Remark 2.3   Notice that Theorem 2.3 says that in order to find the absolute extrema of a continuous function on a closed, bounded interval, we need only compare the values of the function at the endpoints and at the critical numbers. The largest of these will be the absolute maximum and the smallest will be the absolute minimum.

We illustrate Theorem 2.3 for the case of a polynomial function in the following example.

FIGURE 3.25:
$y = 2x^3 - 3x^2 - 12x + 5$.

Example 2.11   Finding Absolute Extrema on a Closed Interval

Find the absolute extrema of $f(x) = 2x^3 - 3x^2 - 12x + 5$ on the interval $[-2, 4]$.

Solution   From the graph in Figure 3.25, the maximum appears to be at the endpoint $x = 4$, while the minimum appears to be at a local minimum near $x = 2$. In example 2.8, we found that the critical numbers of f are $x = -1$ and $x = 2$. Further, both of these are in the interval $[-2, 4]$. So, we compare the values at the endpoints:

$$f(-2) = 1 \quad \text{and} \quad f(4) = 37$$

and the values at the critical numbers:

$$f(-1) = 12 \quad \text{and} \quad f(2) = -15.$$

Theorem 2.3 says that the absolute extrema must be among these four values. Thus, $f(4) = 37$ is the absolute maximum and $f(2) = -15$ is the absolute minimum. You should take care to note that these values are consistent with what we saw in the graph in Figure 3.25.

∎

Of course, most real problems of interest are unlikely to result in derivatives that are quadratic polynomials with integer zeros. Consider the following somewhat less user-friendly example.

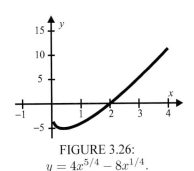

FIGURE 3.26:
$y = 4x^{5/4} - 8x^{1/4}$.

Example 2.12   Finding Extrema for a Function with Fractional Exponents
   Find the absolute extrema of $f(x) = 4x^{5/4} - 8x^{1/4}$ on the interval $[0, 4]$.

Solution   First, we draw a graph of the function to get an idea of where the extrema are located (see Figure 3.26). From the graph, it appears that the maximum occurs at the endpoint $x = 4$ and the minimum near $x = \frac{1}{2}$. Next, observe that

$$f'(x) = 5x^{1/4} - 2x^{-3/4}$$
$$= \frac{5x - 2}{x^{3/4}}.$$

Thus, the critical numbers are $x = \frac{2}{5}$ (since $f'\left(\frac{2}{5}\right) = 0$) and $x = 0$ (since f' is undefined at $x = 0$ and 0 is in the domain of f.) We now need only compare

$$f(0) = 0, \quad f(4) \approx 11.3137 \quad \text{and} \quad f\left(\frac{2}{5}\right) \approx -5.0897.$$

So, the absolute maximum is $f(4) \approx 11.3137$ and the absolute minimum is $f\left(\frac{2}{5}\right) \approx -5.0897$, which is consistent with what we expected from Figure 3.26.

∎

In practice, the critical numbers are not always as easy to find as they were in examples 2.11 and 2.12. In the following example, it is not even known *how many* critical numbers there are. We can only estimate the number and locations of these from a careful analysis of computer-generated graphs.

FIGURE 3.27:
$y = f(x) = x^3 - 5x + 3\sin x^2$.

Example 2.13   Finding Absolute Extrema Approximately
   Find the absolute extrema of $f(x) = x^3 - 5x + 3\sin x^2$ on the interval $[-2, 2.5]$.

Solution   Once again, we first draw a graph to get an idea of where the extrema will be located (see Figure 3.27). From the graph, we can see that the maximum seems to occur near $x = -1$, while the minimum seems to occur near $x = 2$. Next, we compute

$$f'(x) = 3x^2 - 5 + 6x\cos x^2.$$

Unlike examples 2.11 and 2.12, there is no algebra we can use to find the zeros of f'. Our only alternative is to find the zeros approximately. You can do this by using Newton's method to solve $f'(x) = 0$. (You can also use any other rootfinding method built into your calculator or computer.) First, we'll need adequate initial guesses. We obtain these from the graph of $y = f'(x)$ found in Figure 3.28. From the graph, it appears that there are four zeros of $f'(x)$ on the interval in question, located near $x = -1.3, 0.7, 1.2$ and 2.0. Further, referring back to Figure 3.27, these four zeros correspond with the four local extrema seen in the graph of $y = f(x)$. We now apply Newton's method to solve $f'(x) = 0$, using

FIGURE 3.28:
$y = f'(x) = 3x^2 - 5 + 6x\cos x^2$.

the preceding four values as our initial guesses. This leads us to four approximate critical numbers of f on the interval $[-2, 2.5]$. We have

$$a \approx -1.26410884789$$

$$b \approx 0.674471354085$$

$$c \approx 1.2266828947$$

and

$$d \approx 2.01830371473.$$

While not exact, these are very accurate approximations of the critical numbers. We now need only compare the values of f at the endpoints and the approximate critical numbers:

$$f(a) \approx 7.3, \quad f(b) \approx -1.7, \quad f(c) \approx -1.3$$

$$f(d) \approx -4.3, \quad f(-2) \approx -0.3 \quad \text{and} \quad f(2.5) \approx 3.0.$$

Thus, the absolute maximum is approximately

$$f(-1.26410884789) \approx 7.3$$

and the absolute minimum is approximately

$$f(2.01830371473) \approx -4.3.$$

It is important (especially in light of how much of our work here was approximate and graphical) to verify that the approximate extrema correspond with what we expect from the graph of $y = f(x)$. Since these correspond closely, we have great confidence in their accuracy.

■

We have now seen how to locate the absolute extrema of a continuous function on a closed interval. In section 3.3, we see how to find local extrema.

Exercises 3.2

1. (writing) Using a graph(s), explain why the Extreme Value Theorem is true. Is the conclusion true if we drop the hypothesis that $f(x)$ is a continuous function? Is the conclusion true if we drop the hypothesis that the interval is closed?

2. (writing) Using a graph(s), argue that Fermat's Theorem is true. Discuss how Fermat's Theorem is used. Restate the theorem in your own words to make its use more clear.

3. (writing) Suppose that $f(t)$ represents your elevation after t hours on a mountain hike. If you stop to rest, explain why $f'(t) = 0$. Discuss the circumstances under which you would be at a local maximum, local minimum or neither.

4. (writing) Mathematically, an if/then statement is usually strictly one-directional. When we say "If A then B" it is generally **not** the case that "If B then A" is also true: when both are true, we say "A if and only if B" which is abbreviated to "A iff B." Unfortunately, common English usage is not always this precise. This occasionally causes a problem

interpreting a mathematical theorem. To get this straight, consider the statement, "If you wrote a best-selling book, **then** you made a lot of money." Is this true? How does this differ from its **converse**, "If you made a lot of money, **then** you wrote a best-selling book." Is the converse always true? sometimes true? Apply this logic to both the Extreme Value Theorem and Fermat's Theorem: state the converse and decide if it is sometimes true or always true.

In exercises 5–32, find all critical numbers and determine whether each represents a local maximum, local minimum or neither.

5. $f(x) = x^2 + 5x - 1$ **6.** $f(x) = -x^2 + 4x + 2$

7. $f(x) = x^3 - 3x + 1$ **8.** $f(x) = x^3 + 3x + 1$

9. $f(x) = x^3 - 3x^2 + 3x$ **10.** $f(x) = x^3 - 3x^2 + 6x$

11. $f(x) = x^4 - 3x^3 + 2$

12. $f(x) = x^4 + 8x^3 + 2$

13. $f(x) = x^4 - 2x^2 + 3$

14. $f(x) = x^4 + 6x^2 - 2$

15. $f(x) = x^3 - 2x^2 - 4x$

16. $f(x) = x^5 - 20x^2 + 1$

17. $f(x) = \sin x \cos x, [0, 2\pi]$

18. $f(x) = x \sin x, [0, 2\pi]$

19. $f(x) = \dfrac{x+1}{x-1}$

20. $f(x) = \dfrac{x-3}{x+4}$

21. $f(x) = \dfrac{x}{x^2+1}$

22. $f(x) = \dfrac{3x}{x^2-1}$

23. $f(x) = \frac{1}{2}(e^x + e^{-x})$

24. $f(x) = \frac{1}{2}(e^x - e^{-x})$

25. $f(x) = 4x^{1/3}$

26. $f(x) = (x+4)^{2/3}$

27. $f(x) = 2x\sqrt{x+1}$

28. $f(x) = x/\sqrt{x^2+1}$

29. $f(x) = e^{-x^2}$

30. $f(x) = xe^{-x}$

31. $f(x) = \sin x^2, [0, \pi]$

32. $f(x) = \sin^2 x, [0, 2\pi]$

In exercises 33–44, find the absolute extrema of the given function on the given interval.

33. $f(x) = x^3 - 3x + 1, [0, 2]$

34. $f(x) = x^3 - 3x + 1, [-3, 2]$

35. $f(x) = x^4 - 8x^2 + 2, [-3, 1]$

36. $f(x) = x^4 - 8x^2 + 2, [-1, 3]$

37. $f(x) = x^{2/3}, [-4, -2]$

38. $f(x) = x^{2/3}, [-1, 3]$

39. $f(x) = \dfrac{x-1}{x+2}, [-4, 0]$

40. $f(x) = \dfrac{x-1}{x+2}, [-1, 2]$

41. $f(x) = \sin x + \cos x, [0, 2\pi]$

42. $f(x) = \sin x + \cos x, [\pi/2, \pi]$

43. $f(x) = x \sin x + 3, [0, 2\pi]$

44. $f(x) = x^2 + e^x, [-2, 2]$

45. Repeat exercises 33-38, except instead of finding extrema on the closed interval, find the extrema on the open interval, if they exist.

46. Briefly outline a procedure for finding extrema on an open interval (a, b), a procedure for the half-open interval $(a, b]$ and a procedure for the half-open interval $[a, b)$.

47. Sketch a graph of a function $f(x)$ such that the absolute maximum of $f(x)$ on the interval $[-2, 2]$ equals 3 and the absolute minimum does not exist.

48. Sketch a graph of a continuous function $f(x)$ such that the absolute maximum of $f(x)$ on the interval $(-2, 2)$ does not exist and the absolute minimum equals 2.

49. Sketch a graph of a continuous function $f(x)$ such that the absolute maximum of $f(x)$ on the interval $(-2, 2)$ equals 4 and the absolute minimum equals 2.

50. Sketch a graph of a function $f(x)$ such that the absolute maximum of $f(x)$ on the interval $[-2, 2]$ does not exist and the absolute minimum does not exist.

51. Sketch a graph of $f(x) = \dfrac{x^2}{x^2+1}$ for $x > 0$ and determine where the graph is steepest (that is, find where the slope is a maximum).

52. Sketch a graph of $f(x) = e^{-x^2}$ and determine where the graph is steepest. (Note: This is an important problem in probability theory.)

53. Sketch a graph showing that $y = f(x) = x^2 + 1$ and $y = g(x) = \ln x$ do not intersect. Find x to minimize $f(x) - g(x)$. At this value of x, show that the tangent lines to $y = f(x)$ and $y = g(x)$ are parallel.

54. Explain graphically why it makes sense that the tangent lines in exercise 53 are parallel, given that at this point the graphs come closest to each other.

55. Give an example showing that the following statement is false (not always true): between any two local minima of $f(x)$ there is a local maximum.

56. The statement in exercise 55 is true if $f(x)$ is "smooth" enough. What is the least that you can assume about the properties of $f(x)$ and $f'(x)$ to make the statement true?

57. A section of roller coaster is in the shape of $y = x^5 - 4x^3 - x + 10$, where x is between -2 and 2. Find all relative extrema and explain what portions of the roller coaster they represent. Find the location of the steepest part of the roller coaster.

58. Suppose a large computer file is sent over the Internet. If the probability that it reaches its destination without any errors is x, then the probability that an error is made is $1 - x$. The field of information theory studies such situations. An important quantity is **entropy** (a measure of unpredictability), defined by $H = -x \ln x - (1 - x) \ln(1 - x)$, for $0 < x < 1$. Find the value of x that maximizes this quantity. Explain why this value makes sense as the probability that maximizes entropy.

59. Researchers in a number of fields (including population biology, economics and the study of animal tumors) make use of the Gompertz growth curve, $W(t) = a\,e^{-b\,e^{-t}}$. As $t \to \infty$, show that $W(t) \to a$ and $W'(t) \to 0$. Find the maximum growth rate.

60. (exploratory) In this exercise, we will explore the family of functions $f(x) = x^3 + cx + 1$, where c is a constant. How many and what types of relative extrema are there? (Your answer will depend on the value of c.) Assuming that this family is indicative of all cubic functions, list all types of cubic functions. Without looking at specific examples, try to list all types of fourth-order polynomials, sketching a graph of each.

61. (exploratory) Explore the graphs of e^{-x}, xe^{-x}, x^2e^{-x} and x^3e^{-x}. Find all local extrema and determine the behavior as $x \to \infty$. You can think of the graph of $x^n e^{-x}$ as showing the results of a tug-of-war: $x^n \to \infty$ as $x \to \infty$ but $e^{-x} \to 0$ as $x \to \infty$. Describe the graph of $x^n e^{-x}$ in terms of this tug-of-war.

62. (exploratory) Johannes Kepler (1571-1630) is best known as an astronomer, especially for his three laws of planetary motion. However, his discoveries were primarily due to his brilliance as a mathematician. While serving in Austrian Emperor Matthew I's court, Kepler observed the ability of Austrian vintners to quickly and mysteriously compute the capacities of a variety of wine casks. Each cask (barrel) had a hole in the middle of its side (see Figure a). The vintner would insert a rod in the hole until it hit the far corner and then announce the volume. Kepler first analyzed the problem for a cylindrical barrel (see Figure b). The volume of a cylinder is $V = \pi r^2 h$. In Figure b, $r = y$ and $h = 2x$ so $V = 2\pi y^2 x$. Call the rod measurement z. By the Pythagorean Theorem, $x^2 + (2y)^2 = z^2$. Kepler's mystery was how to compute V given only z. The key observation made by Kepler was that Austrian wine casks were made with the same height-to-diameter ratio (for us, x/y). Let $t = x/y$ and show that $z^2/y^2 = t^2 + 4$. Use this to replace y^2 in the volume formula.

Then replace x with $\sqrt{z^2 - 4y^2}$. Show that $V = \dfrac{2\pi z^3 t}{(4+t^2)^{3/2}}$. In this formula, t is a constant so the vintner could measure z and quickly estimate the volume. We haven't told you yet what t equals. Kepler assumed that the vintners would have made a smart choice for this ratio. Find the value of t which maximizes the volume for a given z. This is, in fact, the ratio used in the construction of Austrian wine casks!

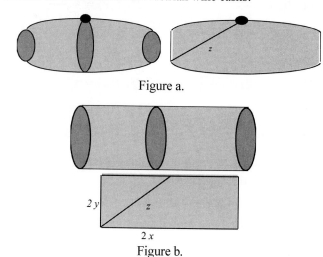

Figure a.

Figure b.

3.3   Increasing and Decreasing Functions

One of the questions left unanswered from the preceding section is how to determine where a function has a local maximum or local minimum. We were able to determine that local extrema occur only at critical numbers. Unfortunately, not all critical numbers correspond to local extrema. In this section, we develop a means of deciding which critical numbers correspond to local extrema. At the same time, we'll learn more about the connection between the derivative and graphing. We begin with a very simple notion.

We are all familiar with the terms **increasing** and **decreasing**. If your employer informs you that your salary will be increasing steadily over the term of your employment, you have in mind that as time goes on, your salary will rise. If you plotted your salary against time, the graph might look something like Figure 3.29. If you take out a loan to purchase a car or a home or to pay for your college education, once you start paying back the loan, your indebtedness will decrease over time. If you plotted your debt against time, the graph might look something like Figure 3.30.

We now carefully define these notions. Notice that the following definition is merely a formal statement of something you already understand.

FIGURE 3.29: Increasing salary.

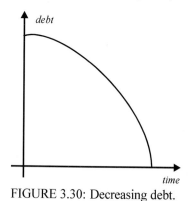

FIGURE 3.30: Decreasing debt.

Definition 3.1   A function f is **(strictly) increasing** on the interval (a, b) if for every $x_1, x_2 \in (a, b)$ with $x_1 < x_2$, $f(x_1) < f(x_2)$ (i.e., $f(x)$ gets larger as x gets larger).

A function f is **(strictly) decreasing** on the interval (a, b) if for every $x_1, x_2 \in (a, b)$ with $x_1 < x_2$, $f(x_1) > f(x_2)$ (i.e., $f(x)$ gets smaller as x gets larger).

Why do we bother with such an obvious definition? Of course, anyone can look at a graph of a function and immediately see where that function is increasing and decreasing.

The real challenge is to determine where a function is increasing and decreasing, given only a mathematical formula for the function. For example, can you determine where $f(x) = x^2 \sin x$ is increasing and decreasing, **without** looking at a graph? Even with a graph, can you determine where this happens precisely? Look carefully at Figure 3.31 to see if you can notice what happens at every point at which the function is increasing or decreasing.

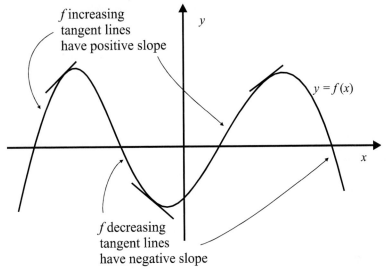

FIGURE 3.31: Increasing and decreasing.

You should observe that on intervals where f is increasing, the tangent lines all have positive slope, while on intervals where f is decreasing, the tangent lines all have negative slope. Of course, we already know that the slope of the tangent line at a point is given by the value of the derivative at that point. So, whether a function is increasing or decreasing on an interval seems to be connected to the sign of its derivative on that interval. This conjecture, although it's based only on a single picture, sounds like a theorem and it is.

Theorem 3.1   Suppose that f is differentiable on the interval (a, b).
  (i) If $f'(x) > 0$ for all $x \in (a, b)$, then f is increasing on (a, b).
  (ii) If $f'(x) < 0$ for all $x \in (a, b)$, then f is decreasing on (a, b).

Proof. (i) Pick any two points $x_1, x_2 \in (a, b)$, with $x_1 < x_2$. Applying the Mean Value Theorem (Theorem 9.4 in section 2.9) to f on the interval (x_1, x_2), we get

$$\frac{f(x_2) - f(x_1)}{x_2 - x_1} = f'(c), \qquad (3.1)$$

for some number $c \in (x_1, x_2)$. (Why can we apply the Mean Value Theorem here?) By hypothesis, $f'(c) > 0$ and since $x_1 < x_2$ (so that $x_2 - x_1 > 0$), we have from (3.1) that

$$f(x_2) - f(x_1) > 0$$

or

$$f(x_2) > f(x_1). \qquad (3.2)$$

Since (3.2) holds for all $x_1 < x_2$, f is increasing on (a, b).
    The proof of (ii) is nearly identical and is left as an exercise.

■

What You See May Not Be What You Get

One aim in the next few sections is to learn how to draw fairly representative graphs of functions (i.e., graphs that display all of the significant features of a function: where it is increasing or decreasing, any extrema, asymptotes and two features we'll introduce in section 3.4: concavity and inflection points). You might think that there is little to be concerned about, given the ease with which you can draw graphs by machine, but there is a significant issue here. When we draw a graph, we are drawing in a particular viewing window (i.e., a particular range of x- and y-values). When we use a computer or calculator to draw graphs, the window is often chosen by the machine. So, when is the window properly adjusted to show enough of the graph to be representative of the behavior of the function? We need to know when significant features are hidden outside of a given window and how to determine the precise locations of features that we can see in a given window. As we'll see, the only way we can resolve these questions is with a healthy dose of calculus.

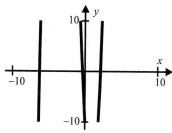

FIGURE 3.32:
$y = 2x^3 + 9x^2 - 24x - 10.$

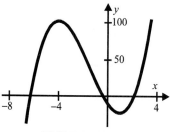

FIGURE 3.33a:
$y = 2x^3 + 9x^2 - 24x - 10.$

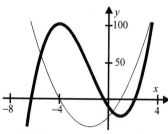

FIGURE 3.33b: $y = f(x)$
and $y = f'(x)$.

Example 3.1   Drawing a Graph

Draw a graph of $f(x) = 2x^3 + 9x^2 - 24x - 10$ showing all local extrema.

Solution   The most popular graphing calculators use the window defined by $-10 \leq x \leq 10$ and $-10 \leq y \leq 10$ as their default. Using this window, the graph of $y = f(x)$ looks like that displayed in Figure 3.32. The three segments seen in Figure 3.32 are not particularly revealing. Instead of blindly manipulating the window in the hope that a reasonable graph will magically appear, we stop briefly to determine where the function is increasing and decreasing. First, we need

$$f'(x) = 6x^2 + 18x - 24$$
$$= 6(x^2 + 3x - 4)$$
$$= 6(x - 1)(x + 4).$$

Note that the critical numbers are 1 and -4, so these are the only candidates for local extrema. We can see where the two factors and consequently the derivative are positive and negative from the number lines in the margin. From this, note that

$$f'(x) > 0 \text{ on } (-\infty, -4) \cup (1, \infty)$$

and

$$f'(x) < 0 \text{ on } (-4, 1).$$

Accordingly, f is increasing on $(-\infty, -4) \cup (1, \infty)$ and decreasing on $(-4, 1)$. So that you can conveniently see this information at a glance, we have placed arrows indicating where the function is increasing and decreasing beneath the last number line. In particular, notice that this suggests that we are not seeing enough of the graph in the window in Figure 3.32. In Figure 3.33a, we redraw the graph in the window defined by $-8 \leq x \leq 4$ and $-25 \leq y \leq 105$. Here, we have selected the y-range so that the critical points $(-4, 102)$ and $(1, -23)$ (and every point in between) are displayed. Since we know that f is increasing on all of $(-\infty, -4)$, we know that the function is still increasing to the left of the portion displayed in Figure 3.33a. Likewise, since we know that f is increasing on all of $(1, \infty)$, we know that the function continues to increase to the right of the displayed portion. The graph in Figure 3.33a conveys all of this significant behavior of the function. In Figure 3.33b, we have plotted both $y = f(x)$ (thick line) and $y = f'(x)$ (thin line). Notice the connection between the two graphs. When $f'(x) > 0$, f is increasing; when $f'(x) < 0$, f is decreasing and also notice what happens to $f'(x)$ at the local extrema of f. (We'll say more about this shortly.)

You may be tempted to think that you can dispense with using any calculus to draw a graph. After all, it's easy to get lured into thinking that your graphing calculator or computer can draw a much better graph than you can in only a fraction of the time and with virtually no effort. To some extent, this is true. You can draw graphs by machine and with a little fiddling with the graphing window, get a reasonable looking graph. Unfortunately, this frequently doesn't work. The calculus tells you what you should look for in a graph. Without it, you're simply fooling around and hoping you get something reasonable. (You often will.) We illustrate this with the following example, where the default graph looks reasonable, but is incorrect.

Example 3.2   Uncovering Hidden Behavior in a Graph

Graph $f(x) = 3x^4 + 40x^3 - 0.06x^2 - 1.2x$ showing all local extrema.

Solution   We first show the default graph drawn by our computer algebra system (see Figure 3.34a). We show a common default graphing calculator graph in Figure 3.34b. You can certainly make Figure 3.34b look more like Figure 3.34a by fooling around with the window some. But it takes some calculus to know what you need to be looking for.

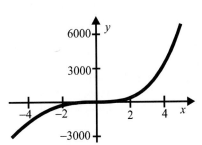

FIGURE 3.34a: Default CAS graph of $y = 3x^4 + 40x^3 - 0.06x^2 - 1.2x$.

FIGURE 3.34b: Default calculator graph of $y = 3x^4 + 40x^3 - 0.06x^2 - 1.2x$.

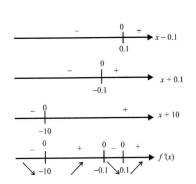

First, notice that

$$f'(x) = 12x^3 + 120x^2 - 0.12x - 1.2$$
$$= 12\left(x^2 - 0.01\right)(x + 10)$$
$$= 12(x - 0.1)(x + 0.1)(x + 10).$$

We show number lines for the three factors in the margin. Observe that

$$f'(x) \begin{cases} > 0 & \text{on } (-10, -0.1) \cup (0.1, \infty) \\ < 0 & \text{on } (-\infty, -10) \cup (-0.1, 0.1) \end{cases}.$$

So, f is increasing on $(-10, -0.1) \cup (0.1, \infty)$ and decreasing on $(-\infty, -10) \cup (-0.1, 0.1)$. This says that neither of the machine-generated graphs seen in Figures 3.34a or 3.34b are adequate, as the behavior on $(-\infty, -10) \cup (-0.1, 0.1)$ cannot be seen in either graph. As it turns out, no single graph captures all of the behavior of this function. However, by increasing the range of x-values to the interval $[-15, 5]$, we get the graph seen in Figure 3.35a. This shows the *big picture*, what we refer to as the **global** behavior of the function. Here, you can see the local minimum at $x = -10$, which was missing in our earlier graphs, but the behavior for values of x close to zero is not clear. To see this, we need a separate graph, restricted to a smaller range of x-values, as seen in Figure 3.35b.

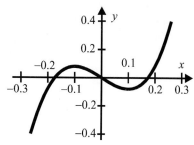

FIGURE 3.35a: The global behavior of $f(x) = 3x^4 + 40x^3 - 0.06x^2 - 1.2x$.

FIGURE 3.35b: Local behavior of $f(x) = 3x^4 + 40x^3 - 0.06x^2 - 1.2x$.

FIGURE 3.36a: $y = f(x)$ and $y = f'(x)$ (global behavior).

Notice that here, we can see the behavior of the function for x close to zero quite clearly. In particular, the local maximum at $x = -0.1$ and the local minimum at $x = 0.1$ are clearly visible. We often say that a graph such as Figure 3.35b shows the **local** behavior of the function. In Figures 3.36a and 3.36b, we show graphs indicating the global and local behavior of $f(x)$ (thick line) and $f'(x)$ (thin line) on the same set of axes. Pay particular attention to the behavior of $f'(x)$ in the vicinity of local extrema of $f(x)$.   ∎

You may have already noticed a connection between local extrema and the intervals on which a function is increasing and decreasing. We state this in the following theorem.

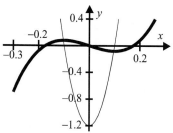

FIGURE 3.36b: $y = f(x)$ and $y = f'(x)$ (local behavior).

Theorem 3.2   (First Derivative Test) Suppose that f is continuous on the interval $[a, b]$ and $c \in (a, b)$ is a critical number.

(i) If $f'(x) > 0$ for all $x \in (a, c)$ and $f'(x) < 0$ for all $x \in (c, b)$ (i.e., f changes from increasing to decreasing at c), then $f(c)$ is a local maximum.

(ii) If $f'(x) < 0$ for all $x \in (a, c)$ and $f'(x) > 0$ for all $x \in (c, b)$ (i.e., f changes from decreasing to increasing at c), then $f(c)$ is a local minimum.

(iii) If $f'(x)$ has the **same** sign on (a, c) and (c, b), then $f(c)$ is **not** a local extremum.

It's easiest to think of this result graphically. If f is increasing to the left of a critical number and decreasing to the right, then there must be a local maximum at the critical number (see Figure 3.37a). Likewise, if f is decreasing to the left of a critical number and increasing to the right, then there must be a local minimum at the critical number (see Figure 3.37b). By the way, the preceding argument suggests a proof of the theorem. The job of writing out all of the details is left as an exercise.

FIGURE 3.37a: Local maximum.

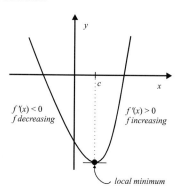

FIGURE 3.37b: Local minimum.

Example 3.3   **Finding Local Extrema Using the First Derivative Test**
Find the local extrema of the function from example 3.1, $f(x) = 2x^3 + 9x^2 - 24x - 10$.

Solution   We had found in example 3.1 that

$$f'(x) > 0 \text{ on } (-\infty, -4) \cup (1, \infty)$$

and

$$f'(x) < 0 \text{ on } (-4, 1).$$

It now follows from the first derivative test that f has a local maximum located at $x = -4$ and a local minimum located at $x = 1$.

∎

Theorem 3.2 works equally well for a function with critical points where the derivative is undefined.

Example 3.4   **Finding Local Extrema of a Function with Fractional Exponents**
Find the local extrema of $f(x) = x^{5/3} - 3x^{2/3}$.

Solution   We have

$$f'(x) = \frac{5}{3}x^{2/3} - 3\left(\frac{2}{3}\right)x^{-1/3}$$

$$= \frac{5x - 6}{3x^{1/3}},$$

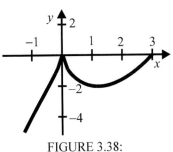

so that the critical numbers are $\frac{6}{5}$ ($f'(\frac{6}{5}) = 0$) and 0 ($f'(0)$ is undefined). Again drawing number lines for the factors, we determine where f is increasing and decreasing. Here, we have placed an ⊠ above the 0 on the number line for $f'(x)$ to indicate that $f'(x)$ is not defined at $x = 0$. From this, we can see at a glance where f' is positive and negative:

$$f'(x) \begin{cases} > 0, \text{ on } (-\infty, 0) \cup (\frac{6}{5}, \infty) \\ < 0, \text{ on } (0, \frac{6}{5}). \end{cases}$$

So, f is increasing on $(-\infty, 0) \cup (\frac{6}{5}, \infty)$ and decreasing on $(0, \frac{6}{5})$. Consequently, f has a local maximum at $x = 0$ and a local minimum at $x = \frac{6}{5}$. These local extrema are both clearly visible in the graph in Figure 3.38.

∎

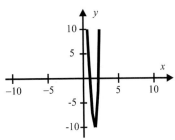

FIGURE 3.38:
$y = x^{5/3} - 3x^{2/3}$.

Example 3.5   **Finding Local Extrema Approximately**
Find the local extrema of $f(x) = x^4 + 4x^3 - 5x^2 - 31x + 29$ and draw a graph.

Solution   A graph of $y = f(x)$ using the most common graphing calculator default window appears in Figure 3.39. Without further analysis, we do not know if this graph shows all of the significant behavior of the function. [Note that some fourth-degree polynomials (e.g., $f(x) = x^4$) have graphs that look very much like the one in Figure 3.39.] First, we compute

$$f'(x) = 4x^3 + 12x^2 - 10x - 31.$$

FIGURE 3.39: $f(x) = x^4 + 4x^3 - 5x^2 - 31x + 29$.

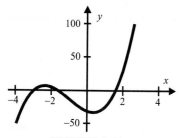

FIGURE 3.40:
$f'(x) = 4x^3 + 12x^2 - 10x - 31$.

Unlike the last several examples, this derivative does not easily factor. A graph of $y = f'(x)$ (see Figure 3.40) reveals three zeros, one near each of $x = -3, -1.5$, and 1.5. Since a cubic polynomial has at most three zeros, there are no others. Using Newton's method or some other root-finding method (applied to $f'(x)$), we can find approximations to the three zeros of f'. We get

$$a = -2.96007700922,$$

$$b = -1.63816222103$$

and

$$c = 1.59823923025.$$

From Figure 3.40, we can see that $f'(x)$ is positive for $a < x < b$ and for $x > c$ and is negative elsewhere. That is,

$$f'(x) > 0 \text{ on } (-2.96\cdots, -1.638\cdots) \cup (1.598\cdots, \infty)$$

and

$$f'(x) < 0 \text{ on } (-\infty, -2.96\cdots) \cup (-1.638\cdots, 1.598\cdots).$$

You can quickly read off the local extrema: a local minimum at $a = -2.96007700922$, a local maximum at $b = -1.63816222103$ and a local minimum at $c = 1.59823923025$. Since only the local minimum at $x = c$ is visible in the graph in Figure 3.39, this graph is clearly **not** representative of the behavior of the function. By narrowing the range of displayed x-values and widening the range of displayed y-values, we obtain the far more useful graph seen in Figure 3.41. You should convince yourself, using the preceding analysis, that the local minimum at $x = c = 1.59823923025$ is also the **absolute** minimum.

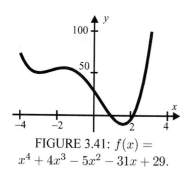

FIGURE 3.41: $f(x) = x^4 + 4x^3 - 5x^2 - 31x + 29$.

∎

Exercises 3.3

1. (writing) Suppose that $f(0) = 2$ and $f(x)$ is an increasing function. To sketch the graph of $y = f(x)$, you could start by plotting the point $(0, 2)$. Filling in the graph to the left, would you move your pencil up or down? How does this fit with the definition of increasing?

2. (writing) Suppose you travel east on an east-west interstate highway. You reach your destination, stay a while and then return home. Explain the First Derivative Test in terms of your velocities (positive and negative) on this trip.

3. (writing) Suppose that you have a differentiable function $f(x)$ with two critical numbers. Your computer has shown you a graph that looks like the one below.

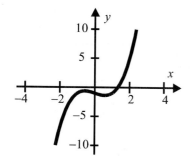

Discuss the possibility that this is a representative graph: that is, is it possible that there are any important points not shown in this window?

4. (writing) Suppose that the function in exercise 3 has three critical numbers. Explain why the graph is not a representative graph. Explain how you would change the graphing window to show the rest of the graph.

In exercises 5–14, find (by hand) the intervals where the function is increasing and decreasing. Verify your answers with a graph.

5. $y = x^3 - 3x + 2$

6. $y = x^3 + 2x^2 + 1$

7. $y = x^3 - 6x^2 + 9x - 1$

8. $y = x^3 - 3x^2 - 9x + 1$

9. $y = x^4 - 8x^2 + 1$

10. $y = (x + 1)^{2/3}$

11. $y = \sin 3x$

12. $y = \sin^2 x$

13. $y = e^{x^2 - 1}$

14. $y = \ln(x^2 - 1)$

In exercises 15–36, find the x-coordinates of all extrema and sketch a graph.

15. $y = x^3 + 2x^2 - x - 1$

16. $y = x^3 + 4x - 2$

17. $y = x^4 + 2x^2 - x + 2$

18. $y = x^5 + 2x^4 - x^2 + 1$

19. $y = x\sqrt{x^2 + 1}$

20. $y = x/\sqrt{x^2 + 1}$

21. $y = xe^{-2x}$

22. $y = x^2 e^{-x}$

23. $y = x^3 - 13x^2 - 10x + 1$

24. $y = x^3 + 15x^2 - 70x + 2$

25. $y = \ln x^2$

26. $y = e^{-x^2}$

27. $y = \dfrac{x}{x^2 - 1}$

28. $y = \dfrac{x^2}{x^2 - 1}$

29. $y = \dfrac{x^3}{x^2 - 1}$

30. $y = \dfrac{x^2}{x^2 + 1}$

31. $y = \sin x + \cos x$

32. $y = \cos x - x$

33. $y = \sqrt{x^3 + 3x^2}$

34. $y = 2x^{1/2} - 4x^{-1/2}$

35. $y = x^{2/3} - 2x^{-1/3}$

36. $y = x^{4/3} + 4x^{1/3}$

In exercises 37–40, sketch a graph of a function with the given properties.

37. $f(0) = 1$, $f(2) = 5$, $f'(x) < 0$ for $x < 0$ and $x > 2$, $f'(x) > 0$ for $0 < x < 2$.

38. $f(-1) = 1$, $f(2) = 5$, $f'(x) < 0$ for $x < -1$ and $x > 2$, $f'(x) > 0$ for $-1 < x < 2$, $f'(-1) = 0$, $f'(2)$ does not exist.

39. $f(3) = 0$, $f'(x) < 0$ for $x < 0$ and $x > 3$, $f'(x) > 0$ for $0 < x < 3$, $f'(3) = 0$, $f(0)$ and $f'(0)$ do not exist.

40. $f(1) = 0$, $\lim\limits_{x \to \infty} f(x) = 2$, $f'(x) < 0$ for $x < 1$, $f'(x) > 0$ for $x > 1$, $f'(1) = 0$.

41. Suppose an object has position function $s(t)$, velocity function $v(t) = s'(t)$ and acceleration function $a(t) = v'(t)$. If $a(t) > 0$, then the velocity $v(t)$ is increasing. Sketch two possible velocity functions, one with velocity getting less negative and one with velocity getting more positive. In both cases, sketch possible position graphs. The position graph should be curved (nonlinear). Does it look more like part of an upward-curving parabola or a downward-curving parabola? We look more closely at curving in section 3.4.

42. Repeat exercise 41 for the case where $a(t) < 0$.

43. Prove Theorem 3.2 (The First Derivative Test).

44. Give a graphical argument that if $f(a) = g(a)$ and $f'(x) > g'(x)$ for all $x > a$, then $f(x) > g(x)$ for all $x > a$. Use the Mean Value Theorem to prove it.

In exercises 45–48, use the result of exercise 44 to verify the inequality.

45. $2\sqrt{x} > 3 - 1/x$ for $x > 1$

46. $x > \sin x$ for $x > 0$

47. $e^x > x + 1$ for $x > 0$

48. $x - 1 > \ln x$ for $x > 1$

49. Give an example showing that the following statement is false. If $f(0) = 4$ and $f(x)$ is a decreasing function, then the equation $f(x) = 0$ has exactly one solution.

50. Determine if the following statement is true or false: If $f(0) = 4$ and $f(x)$ is an increasing function, then the equation $f(x) = 0$ has no solutions.

51. Suppose that the total sales of a product after t months is given by $s(t) = \sqrt{t + 4}$ thousand dollars. Compute and interpret $s'(t)$.

52. In exercise 51, show that $s'(t) > 0$ for all $t > 0$. Explain in business terms why it is impossible to have $s'(t) < 0$.

53. (exploratory) In this exercise, we look at the ability of fireflies to synchronize their flashes. (To see a remarkable demonstration of this ability, see David Attenborough's video series *Trials of Life*.) Let the function $f(t)$ represent an individual firefly's rhythm, so that the firefly flashes whenever $f(t)$ equals an integer. Let $e(t)$ represent the rhythm of a neighboring firefly, where again $e(t) = n$, for some integer n, whenever the neighbor flashes. One model of the interaction between fireflies is $f'(t) = \omega + A \sin[e(t) - f(t)]$ for constants ω and A. If the fireflies are synchronized ($e(t) = f(t)$), then $f'(t) = \omega$, so the fireflies flash every $1/\omega$ time units. Assume that the difference between $e(t)$ and $f(t)$ is less than π. Show that if $f(t) < e(t)$, then $f'(t) > \omega$. Explain why this means that the individual firefly is speeding up its flash to match its neighbor. Similarly, discuss what happens if $f(t) > e(t)$.

54. (exploratory) The HIV virus attacks specialized T cells that trigger the human immune system response to a foreign substance. If $T(t)$ is the population of uninfected T cells at time t (days) and $V(t)$ is the population of infectious HIV in the bloodstream, a model which has been used to study AIDS is given by the following **differential equation** which describes the rate at which the population of T cells changes. $T'(t) = 10\left[1 + \dfrac{1}{1 + V(t)}\right] - 0.02T(t) + 0.01\dfrac{T(t)V(t)}{100 + V(t)} - 0.000024T(t)V(t)$. If there is no HIV present (that is, $V(t) = 0$) and $T(t) = 1000$, show that $T'(t) = 0$. Explain why this means that the T-cell count will remain constant at 1000 (cells per cubic mm). Now, suppose that $V(t) = 100$. Show that if $T(t)$ is small enough, then $T'(t) > 0$ and the T-cell population will increase. On the other hand, if $T(t)$ is large enough, then $T'(t) < 0$ and the T-cell population will decrease. For what value of $T(t)$ is $T'(t) = 0$? Even though this population would remain stable, explain why this would be bad news for the infected human.

55. (exploratory) In this exercise, you will play the role of professor and construct a tricky graphing exercise. The first goal is to find a function with local extrema so close together that they're difficult to see. For instance, suppose you want local extrema at $x = -0.1$ and $x = 0.1$. Explain why you could start with $f'(x) = (x-0.1)(x+0.1) = x^2 - 0.01$. Look for a function whose derivative is as given. Graph your function to see if the extrema are "hidden." Next, construct a polynomial of degree 4 with two extrema very near $x = 1$ and another near $x = 0$.

3.4   Concavity

In section 3.3, we saw how to determine where a function is increasing and decreasing and how this relates to drawing the graph of a function. At this point, we need to see how to refine the shape of a graph. First, you must realize that simply knowing where a function increases and decreases is not sufficient to draw good graphs. In Figures 3.42a and 3.42b, we show two very different shapes of increasing functions joining the same two points.

FIGURE 3.42a: Increasing function.

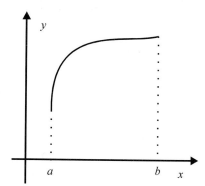
FIGURE 3.42b: Increasing function.

So, given that a graph joins two particular points and is increasing, which of the two shapes shown do we draw? Without further information, there's no way to tell. Realize that this is an important distinction to make. For example, suppose that Figure 3.42a or 3.42b depicts the federal budget deficit. Both indicate a deficit that is growing. Politicians (from both ends of the political spectrum) have occasionally bragged that through their efforts, the rate of increase in the budget deficit has been slowed. Notice that they do not say that they have reduced or eliminated the deficit (and balanced the federal budget). Rather, the rate of growth has been slowed. This describes the growth in Figure 3.42b, while the rate of growth depicted in Figure 3.42a is increasing. Which would you want to have describe the federal budget deficit? Why? Figures 3.43a and 3.43b are the same as Figures 3.42a and 3.42b, respectively, but with a few tangent lines drawn in.

FIGURE 3.43a: Concave upward.

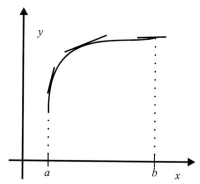
FIGURE 3.43b: Concave downward.

Notice that although all of the tangent lines have positive slope (since $f'(x) > 0$), the slopes of the tangent lines in Figure 3.43a are increasing, while those in Figure 3.43b are decreasing. We call the graph in Figure 3.43a **concave upward** and the graph in Figure 3.43b **concave downward**. The situation is similar for decreasing functions. In Figures 3.44a and 3.44b, we show two different shapes of decreasing function. Again, although both functions are decreasing, the one shown in Figure 3.44a is concave upward (slopes of tangent lines increasing) and the one shown in Figure 3.44b is concave downward (slopes of tangent lines decreasing). We have the following definition.

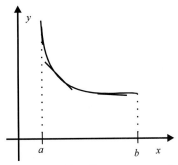

FIGURE 3.44a: Concave upward.

Definition 4.1   For a function f that is differentiable on the interval (a, b), the graph of f is

(i) **concave upward** on (a, b) if f' is increasing on (a, b) or

(ii) **concave downward** on (a, b) if f' is decreasing on (a, b).

How can you tell when f' is increasing or decreasing? The derivative of f' (i.e., f'') yields that information. The following theorem connects this definition with what we already know about increasing and decreasing functions. The proof of the theorem is a straightforward application of Theorem 3.1 to Definition 4.1.

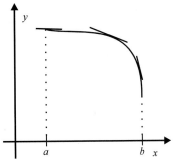

FIGURE 3.44b: Concave downward.

Theorem 4.1   Suppose that f'' exists on the interval (a, b).
(i)   If $f''(x) > 0$ on (a, b), then the graph of f is concave upward on (a, b).
(ii) If $f''(x) < 0$ on (a, b), then the graph of f is concave downward on (a, b).

Remark 4.1   We usually abbreviate the terms concave upward and concave downward to simply **concave up** and **concave down**, respectively.

Example 4.1   **Determining Concavity**
Determine where the graph of $f(x) = 2x^3 + 9x^2 - 24x - 10$ is concave up and concave down and draw a graph showing all significant behavior of the function.

Solution   Here, we have

$$f'(x) = 6x^2 + 18x - 24$$

and

$$f''(x) = 12x + 18 \begin{cases} > 0, \text{ for } x > -\frac{3}{2} \\ < 0, \text{ for } x < -\frac{3}{2}. \end{cases}$$

Thus, the graph is concave up for $x > -\frac{3}{2}$ and concave down for $x < -\frac{3}{2}$. We saw in example 3.3 that

$$f'(x) \begin{cases} > 0 \text{ on } (-\infty, -4) \cup (1, \infty) \\ < 0 \text{ on } (-4, 1). \end{cases}$$

So, f is increasing on $(-\infty, -4) \cup (1, \infty)$ and decreasing on $(-4, 1)$. Using all of this information, we are able to draw the graph shown in Figure 3.45. Notice that at the point $\left(-\frac{3}{2}, f\left(-\frac{3}{2}\right)\right)$, the graph changes from concave down to concave up. Such points are called **inflection points**. More precisely, we have the following definition.

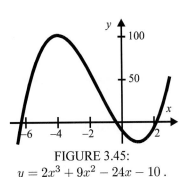

FIGURE 3.45:
$y = 2x^3 + 9x^2 - 24x - 10$.

Definition 4.2 Suppose that f is continuous on the interval (a, b) and that the graph changes concavity at a point $c \in (a, b)$ (i.e., the graph is concave down on one side of c and concave up on the other). Then, the point $(c, f(c))$ is called an **inflection point** of f.

NOTE: If $(c, f(c))$ is an inflection point, then either $f''(c) = 0$ or $f''(c)$ is undefined. So, finding all points where $f''(x)$ is zero or is undefined gives you all possible candidates for inflection points. But beware: not all points where $f''(x)$ is zero or undefined correspond to inflection points.

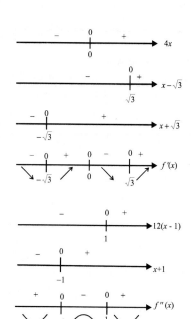

Example 4.2   Determining Concavity and Locating Inflection Points

Determine where the graph of $f(x) = x^4 - 6x^2 + 1$ is concave up and concave down, find any inflection points and draw a graph showing all significant features.

Solution   Here, we have

$$f'(x) = 4x^3 - 12x = 4x \left(x^2 - 3\right)$$
$$= 4x \left(x - \sqrt{3}\right) \left(x + \sqrt{3}\right).$$

We have drawn number lines for the factors of $f'(x)$ in the margin. From this, notice that

$$f'(x) \begin{cases} > 0 \text{ on } (-\sqrt{3}, 0) \cup (\sqrt{3}, \infty) \\ < 0 \text{ on } (-\infty, -\sqrt{3}) \cup (0, \sqrt{3}). \end{cases}$$

So, f is increasing on $(-\sqrt{3}, 0) \cup (\sqrt{3}, \infty)$ and decreasing on $(-\infty, -\sqrt{3}) \cup (0, \sqrt{3})$. Next, we have

$$f''(x) = 12x^2 - 12 = 12(x - 1)(x + 1).$$

We have drawn number lines for the two factors in the margin. From this, we can see that

$$f''(x) \begin{cases} > 0, \text{ on } (-\infty, -1) \cup (1, \infty) \\ < 0, \text{ on } (-1, 1). \end{cases}$$

Thus, the graph is concave up on $(-\infty, -1) \cup (1, \infty)$ and concave down on $(-1, 1)$. So that you can see this information at a glance, we have indicated the concavity below the bottom number line, with small concave up and concave down segments. Finally, observe that since the graph changes concavity at $x = -1$ and $x = 1$, there are inflection points located at $(-1, 4)$ and $(1, 4)$. Using all of this information, we are able to draw the graph shown in Figure 3.46. For your convenience, we have reproduced the number lines for $f'(x)$ and $f''(x)$ together above the figure.

As we see in the following example, having $f''(x) = 0$ does not imply the existence of an inflection point.

FIGURE 3.46: $y = x^4 - 6x^2 + 1$.

Example 4.3   A Graph with No Inflection Points

Determine the concavity of $f(x) = x^4$ and locate any inflection points.

Solution   There's nothing tricky about this function. We have $f'(x) = 4x^3$ and $f''(x) = 12x^2$. Since $f'(x) > 0$ for $x > 0$ and $f'(x) < 0$ for $x < 0$, we know that f is increasing for $x > 0$ and decreasing for $x < 0$. Further, $f''(x) > 0$ for all $x \neq 0$,

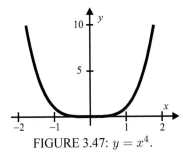

FIGURE 3.47: $y = x^4$.

while $f''(x) = 0$. So, the graph is concave up for $x \neq 0$. Further, even though $f''(x) = 0$, there is **no** inflection point at $x = 0$. We show a graph of the function in Figure 3.47.

∎

There is also a connection between second derivatives and extrema. Suppose that $f'(c) = 0$ and that the graph of f is concave down in some open interval containing c. Then, nearby $x = c$, the graph looks like that in Figure 3.48 and hence, $f(c)$ is a local maximum. Likewise, if $f'(c) = 0$ and the graph of f is concave up in some open interval containing c, then nearby $x = c$, the graph looks like that in Figure 3.49 and hence, $f(c)$ is a local minimum.

FIGURE 3.48: Local maximum.

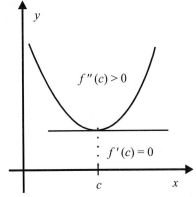

FIGURE 3.49: Local minimum.

More precisely, we have the following theorem.

Theorem 4.2   (Second Derivative Test) Suppose that f is continuous on the interval (a, b) and $f'(c) = 0$, for some number $c \in (a, b)$.
  (i)   If $f''(c) < 0$, then $f(c)$ is a local maximum and
  (ii) if $f''(c) > 0$, then $f(c)$ is a local minimum.

We leave a formal proof of this theorem as an exercise. When applying the theorem, simply think about Figures 3.48 and 3.49.

Example 4.4   Using the Second Derivative Test to Find Extrema
Use the second derivative test to find the local extrema of $f(x) = x^4 - 8x^2 + 10$.

Solution   Here,

$$f'(x) = 4x^3 - 16x$$
$$= 4x(x^2 - 4)$$
$$= 4x(x - 2)(x + 2).$$

Thus, the critical numbers are $x = 0, 2$ and -2. We can test these using the second derivative test as follows. We have

$$f''(x) = 12x^2 - 16$$

and so,

$$f''(0) = -16 < 0,$$

$$f''(-2) = 32 > 0$$

FIGURE 3.50:
$y = x^4 - 8x^2 + 10$.

and
$$f''(2) = 32 > 0.$$

Thus, $f(0)$ is a local maximum and $f(-2)$ and $f(2)$ are local minima. We show a graph of $y = f(x)$ in Figure 3.50.

∎

Remark 4.2   If $f''(c) = 0$ or $f''(c)$ is undefined, the second derivative test yields no conclusion. That is, $f(c)$ may be a local maximum, a local minimum or neither. In this event, we must rely solely on first derivative information (i.e., the first derivative test) to determine if $f(c)$ is a local extremum or not. We illustrate this with the following example.

Example 4.5   **Functions for Which the Second Derivative Test is Inconclusive**

Use the second derivative test to try to classify any local extrema for (a) $f(x) = x^3$, (b) $g(x) = x^4$ and (c) $h(x) = -x^4$.

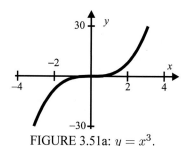

FIGURE 3.51a: $y = x^3$.

Solution   (a) Note that $f'(x) = 3x^2$ and $f''(x) = 6x$. So, the only critical number is $x = 0$ and $f''(0) = 0$, also. We leave it as an exercise to show that the point $(0, 0)$ is not a local extremum (see Figure 3.51a).

(b) We have $g'(x) = 4x^3$ and $g''(x) = 12x^2$. Again, the only critical number is $x = 0$ and $g''(0) = 0$. In this case, though, $g'(x) < 0$ for $x < 0$ and $g'(x) > 0$ for $x > 0$. So, by the first derivative test, $(0, 0)$ is a local minimum (see Figure 3.51b).

(c) Finally, we have $h'(x) = -4x^3$ and $h''(x) = -12x^2$. Once again, the only critical number is $x = 0$, $h''(0) = 0$ and we leave it as an exercise to show that $(0, 0)$ is a local maximum for h (see Figure 3.51c).

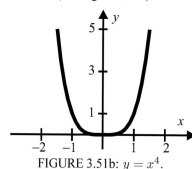

FIGURE 3.51b: $y = x^4$.   FIGURE 3.51c: $y = -x^4$.

We can use first and second derivative information to help produce a meaningful graph of a function, as in the following example.

∎

Example 4.6   **Drawing a Graph of a Rational Function**

Draw a graph of $f(x) = x + \dfrac{25}{x}$, showing all local extrema, inflection points and vertical and horizontal asymptotes.

Add the fractions.

Solution   The first thing that you should notice here is that the domain of f consists of all real numbers other than $x = 0$. Next, we have

$$f'(x) = 1 - \frac{25}{x^2}$$
$$= \frac{x^2 - 25}{x^2}$$
$$= \frac{(x-5)(x+5)}{x^2}.$$

So, the only two critical numbers are $x = -5, 5$. (Why is $x = 0$ **not** a critical number?)

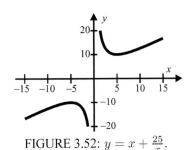

Looking at the three factors in $f'(x)$, we get the number lines shown in the margin. Thus,

$$f'(x) \begin{cases} > 0, \text{ on } (-\infty, -5) \cup (5, \infty) \\ < 0, \text{ on } (-5, 0) \cup (0, 5) \end{cases}$$

and hence, f is increasing on $(-\infty, -5) \cup (5, \infty)$ and f is decreasing on $(-5, 0) \cup (0, 5)$. Further,

$$f''(x) = \frac{50}{x^3} \begin{cases} > 0, \text{ on } (0, \infty) \\ < 0, \text{ on } (-\infty, 0). \end{cases}$$

This says that the graph is concave up on $(0, \infty)$ and concave down on $(-\infty, 0)$. Be careful here. There is **no** inflection point on the graph, even though the graph is concave up on one side of $x = 0$ and concave down on the other. (Why not?) We can now use either the first derivative test or the second derivative test to determine the local extrema. By the second derivative test,

$$f''(5) = \frac{50}{125} > 0$$

and

$$f''(-5) = -\frac{50}{125} < 0,$$

so that there is a local minimum at $x = 5$ and a local maximum at $x = -5$. Finally, before we can draw a representative graph, we need to know what happens to the graph near $x = 0$, since 0 is not in the domain of f. We have

$$\lim_{x \to 0^+} f(x) = \lim_{x \to 0^+} \left(x + \frac{25}{x} \right) = \infty$$

and

$$\lim_{x \to 0^-} f(x) = \lim_{x \to 0^-} \left(x + \frac{25}{x} \right) = -\infty.$$

So, there is a vertical asymptote at $x = 0$. Putting together all of this information, we get the graph shown in Figure 3.52. The only feature of the graph which we have not examined is the behavior as $x \to \pm\infty$. This is left as an exercise. (Hint: Look carefully at Figure 3.52.)

FIGURE 3.52: $y = x + \frac{25}{x}$.

■

In example 4.6, we computed $\lim_{x \to 0^+} f(x)$ and $\lim_{x \to 0^-} f(x)$ to uncover the behavior of the function near $x = 0$. We needed to consider this since $x = 0$ was not in the domain of f. In the following example, we'll see that since $x = -2$ is not in the domain of f' (although it **is** in the domain of f), we must compute $\lim_{x \to -2^+} f'(x)$ and $\lim_{x \to -2^-} f'(x)$. This will tell us about the behavior of the tangent lines near $x = -2$.

Example 4.7   A Function with a Vertical Tangent Line at an Inflection Point
Draw a graph of $f(x) = (x + 2)^{1/5} + 4$, showing all significant features.

Solution   First, notice that the domain of f is the entire real line. We also have

$$f'(x) = \frac{1}{5}(x + 2)^{-4/5}$$
$$> 0, \text{ for } x \neq -2.$$

So, f is increasing everywhere, except at $x = -2$ (the only critical number, where $f'(-2)$ is undefined). This also says that f has no local extrema. Further,

$$f''(x) = -\frac{4}{25}(x+2)^{-9/5} \begin{cases} > 0, \text{ on } (-\infty, -2) \\ < 0, \text{ on } (-2, \infty). \end{cases}$$

Thus, the graph is concave up on $(-\infty, -2)$ and concave down on $(-2, \infty)$, so that there is an inflection point at $x = -2$. In this case, $f'(x)$ is undefined at $x = -2$. Since -2 is in the domain of f, but not in the domain of f', we consider

$$\lim_{x \to -2^-} f'(x) = \lim_{x \to -2^-} \frac{1}{5}(x+2)^{-4/5} = \infty$$

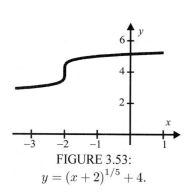

FIGURE 3.53:
$y = (x+2)^{1/5} + 4.$

and

$$\lim_{x \to -2^+} f'(x) = \lim_{x \to -2^+} \frac{1}{5}(x+2)^{-4/5} = \infty.$$

This says that the graph has a vertical tangent line at $x = -2$. We obtain the graph in Figure 3.53.

∎

Exercises 3.4

1. (writing) It is often said that a graph is concave up if it "holds water." This is certainly true for parabolas like $y = x^2$, but is it true for graphs like $y = 1/x^2$? It is always helpful to put a difficult concept into everyday language, but the danger is in oversimplification. Do you think that "holds water" is helpful or can it be confusing? Give your own description of concave up, using everyday language. (Hint: One popular image involves smiles and frowns.)

2. (writing) Find a reference book with the population of the United States since 1800. From 1800 to 1900, the numerical increase by decade increased. Argue that this means that the population curve is concave up. From 1960 to 1990, the numerical increase by decade has been approximately constant. Argue that this means that the population curve is near a point of zero concavity. Why does this not necessarily mean that we are at an inflection point? Argue that we should hope, in order to avoid overpopulation, that it is indeed an inflection point.

3. (writing) The goal of investing in the stock market is to buy low and sell high. But, how can you tell whether a price has peaked or not? Once a stock price goes down, you can see that it *was* at a peak but then it's too late to do anything about it! Concavity can help. Suppose a stock price is increasing and the price curve is concave up. Why would you suspect that it will continue to rise? Is this a good time to buy? Now, suppose the price is increasing but the curve is concave down. Why should you be preparing to sell? Finally, suppose the price is decreasing. If the curve is concave up, should you buy or sell? What if the curve is concave down?

4. (writing) Suppose that $f(t)$ is the amount of money in your bank account at time t. Explain in terms of spending and saving what would cause $f(t)$ to be decreasing and concave down; increasing and concave up; decreasing and concave up.

In exercises 5–8, estimate the intervals where the function is concave up and concave down. (Hint: Estimate where the slope is increasing and decreasing.)

5.

6.

7.

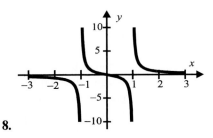

8.

In exercises 9–40, find the intervals of increase and decrease, all local extrema, the intervals of concavity, all inflection points and sketch a graph.

9. $f(x) = x^3 - 3x^2 + 4$

10. $f(x) = x^3 + 3x^2 - 6x$ **11.** $f(x) = x^4 - 2x^2 + 1$

12. $f(x) = x^4 + 4x - 2$ **13.** $f(x) = x + 1/x$

14. $f(x) = x^2 - 16/x$ **15.** $f(x) = x^3 - 6x + 1$

16. $f(x) = x^3 + 3x - 1$ **17.** $f(x) = x^4 + 4x^3 - 1$

18. $f(x) = x^4 + 4x^2 + 1$ **19.** $f(x) = xe^{-x}$

20. $f(x) = e^{-x^2}$ **21.** $f(x) = x^2\sqrt{x^2 - 9}$

22. $f(x) = x\sqrt{x^2 - 9}$ **23.** $f(x) = (x^2 + 1)^{2/3}$

24. $f(x) = x\ln x$ **25.** $f(x) = x^2/(x^2 - 9)$

26. $f(x) = x/(x + 2)$ **27.** $f(x) = \sin x + \cos x$

28. $f(x) = x + \cos x$ **29.** $f(x) = e^x + e^{-x}$

30. $f(x) = e^x - e^{-x}$ **31.** $f(x) = x^{1/3} - 1$

32. $f(x) = x^{2/3} + 1$ **33.** $f(x) = \sqrt[3]{2x + 1}$

34. $f(x) = \sqrt[4]{x - 1}$ **35.** $f(x) = x^4 - 26x^3 + x$

36. $f(x) = 2x^4 - 11x^3 + 17x^2$

37. $f(x) = \dfrac{x^2 - 5x + 4}{x}$

38. $f(x) = \dfrac{x^2 - 1}{x}$

39. $f(x) = x^4 - 16x^3 + 42x^2 - 39.6x + 14$

40. $f(x) = x^4 + 32x^3 - 0.02x^2 - 0.8x$

In exercises 41–46, sketch a graph with the given properties.

41. $f(0) = 2$, $f'(x) > 0$ for all x, $f'(0) = 1$, $f''(x) > 0$ for $x > 0$, $f''(x) < 0$ for $x < 0$, $f''(0) = 0$.

42. $f(0) = 1$, $f'(x) \geq 0$ for all x, $f'(0) = 0$, $f''(x) > 0$ for $x > 0$, $f''(x) < 0$ for $x < 0$, $f''(0) = 0$.

43. $f(0) = 0$, $f'(x) > 0$ for $x < -1$ and $-1 < x < 1$, $f'(x) < 0$ for $x > 1$, $f''(x) > 0$ for $x < -1$, $0 < x < 1$ and $x > 1$, $f''(x) < 0$ for $-1 < x < 0$.

44. $f(0) = 2$, $f'(x) > 0$ for all x, $f'(0) = 1$, $f''(x) > 0$ for $x < 0$, $f''(x) < 0$ for $x > 0$.

45. $f(0) = 0$, $f(-1) = -1$, $f(1) = 1$, $f'(x) > 0$ for $x < -1$ and $0 < x < 1$, $f'(x) < 0$ for $-1 < x < 0$ and $x > 1$, $f''(x) < 0$ for $x < 0$ and $x > 0$.

46. $f(1) = 0$, $f'(x) < 0$ for $x < 1$, $f'(x) > 0$ for $x > 1$, $f''(x) < 0$ for $x < 1$ and $x > 1$.

In exercises 47 and 48, estimate the intervals of increase and decrease, the locations of local extrema, intervals of concavity and locations of inflection points.

47.

48.

49. Repeat exercise 47 if the given graph is of $f'(x)$ instead of $f(x)$.

50. Repeat exercise 48 if the given graph is of $f'(x)$ instead of $f(x)$.

51. Suppose that $w(t)$ is the depth of water in a city's water reservoir. Which would be better news at time $t = 0$, $w''(0) = 0.05$ or $w''(0) = -0.05$ or would you need to know the value of $w'(0)$ to determine which is better?

52. Suppose that $T(t)$ is a sick person's temperature at time t. Which would be better news at time t, $T''(0) = 2$ or $T''(0) = -2$ or would you need to know the value of $T'(0)$ and $T(0)$ to determine which is better?

53. Suppose that a company which spends $\$x$ thousand on advertising sells $\$s(x)$ of merchandise, where $s(x) = -3x^3 + 270x^2 - 3600x + 18{,}000$. Find the value of x that maximizes the rate of change of sales. (Hint: Read the question carefully!)

54. For the sales function in exercise 53, find the inflection point and explain why in advertising terms this is the "point of diminishing returns."

55. Suppose that it costs a company $C(x) = 0.01x^2 + 40x + 3600$ dollars to manufacture x units of a product. For this **cost function**, the **average cost function** is $\bar{C}(x) = \frac{C(x)}{x}$. Find the value of x which minimizes the average cost.

56. In exercise 55, the cost function can be related to the efficiency of the production process. Explain why a cost function that is concave down indicates better efficiency than a cost function that is concave up.

57. Show that there is an inflection point at $(0,0)$ for any function of the form $f(x) = x^4 + cx^3$, where c is a nonzero constant. What role(s) does c play in the graph of $y = f(x)$?

58. The following examples show that there is not a perfect match between inflection points and places where $f''(x) = 0$. First, for $f(x) = x^6$, show that $f''(0) = 0$, but there is no inflection point at $x = 0$. Then, for $g(x) = x|x|$, show that there is an inflection point at $x = 0$, but that $f''(0)$ does not exist.

59. Give an example of a function showing that the following statement is false. If the graph of $y = f(x)$ is concave down for all x, the equation $f(x) = 0$ has at least one solution.

60. Determine if the following statement is true or false. If $f(0) = 1$, $f''(x)$ exists for all x and the graph of $y = f(x)$ is concave down for all x, the equation $f(x) = 0$ has at least one solution.

61. One basic principle of physics is that light follows the path of minimum time. Assuming that the speed of light in the earth's atmosphere decreases as altitude decreases, argue that the path that light follows is concave down. Explain why this means that the setting sun appears higher in the sky than it really is.

62. Prove Theorem 4.2 (the Second Derivative Test). (Hint:

Think about what the definition of $f''(c)$ says when $f''(c) > 0$ or $f''(c) < 0$.)

63. (exploratory) The linear approximation that we defined in section 3.1 is the line having the same location and the same slope as the function being approximated. Since two points determine a line, two requirements (point, slope) are all that a linear function can satisfy. However, a quadratic function can satisfy three requirements since three points determine a parabola (and there are three constants in a general quadratic function $ax^2 + bx + c$). Suppose we want to define a **quadratic approximation** to $f(x)$ at $x = a$. Building on the linear approximation, the general form is $g(x) = f(a) + f'(a)(x - a) + c(x - a)^2$ for some constant c to be determined. In this way, show that $g(a) = f(a)$ and $g'(a) = f'(a)$. That is, $g(x)$ has the right position and slope at $x = a$. The third requirement is that $g(x)$ have the right concavity at $x = a$, so that $g''(a) = f''(a)$. Find the constant c that makes this true. Then, find such a quadratic approximation for each of the functions $\sin x$, $\cos x$ and e^x at $x = 0$. In each case, graph the original function, linear approximation and quadratic approximation and describe how close the approximations are to the original functions.

64. (exploratory) In this exercise, we explore a basic problem in genetics. Suppose that a species reproduces according to the following probabilities: p_0 is the probability of having no children, p_1 is the probability of having one child, p_2 is the probability of having two children,..., p_n is the probability of having n children and n is the largest number of children possible. Explain why for each i, we have $0 \leq p_i \leq 1$ and $p_0 + p_1 + p_2 + \cdots + p_n = 1$. We define the function $F(x) = p_0 + p_1 x + p_2 x^2 + \cdots + p_n x^n$. Solutions of the equation $F(x) = x$ for $0 \leq x \leq 1$ represent the probability that the species becomes extinct. Show graphically that if $p_0 > 0$ and $F'(1) > 1$, then there is a solution of $F(x) = x$ with $0 < x < 1$. Thus, there is a positive probability of survival. However, if $p_0 > 0$ and $F'(1) < 1$, show that there are no solutions of $F(x) = x$ with $0 < x < 1$. (Hint: First show that F is increasing and concave up.)

3.5   Overview of Curve Sketching

You might ask yourself why we need to spend any more time on curve sketching. We have already drawn numerous graphs over the last three sections. Besides, with a graphing calculator or computer algebra system at your disposal, why must you even consider drawing graphs by hand?

Of course, graphing calculators or computer algebra systems are indispensable tools today in the study or application of mathematics. As the authors of this text, we admit it. We have made extensive use of a computer algebra system (Maple) in preparing the manuscript for this text and even in designing our problems. So then, why should we condemn you to drawing graphs by hand? We're certain that late night talk show hosts would have a list of the top 10 reasons why sadistic mathematics professors would want to inflict such pain on their students, but there's really only one reason. For better or worse, graphing calculators and computer algebra systems do not actually draw graphs. What they do is plot points (albeit lots of them) and then connect the points with a smooth curve. Of course, this works exceptionally well for a simple polynomial, but leaves something to be desired for more complicated functions. The problem boils down to the window in which you draw the graph and how many points you plot in that window. The only way to know how to choose the window or how many points to plot in that window is to use the calculus to determine the properties of the graph that you are interested in seeing. We have already made this point a number of times. The calculus tells us the properties that a representative graph should exhibit. We can then adjust the window and the number of points plotted in order to produce such a graph.

We begin this section by summarizing the various tests that you should perform on a function when trying to draw a graph of $y = f(x)$.

- **Domain**: You should always determine the domain of f **first**.
- **Vertical Asymptotes**: For any isolated point not in the domain of f, check the limiting value of the function as x approaches that point, to see if there is a vertical asymptote or a jump or removable discontinuity at that point.
- **First Derivative Information**: Determine where f is increasing and decreasing and find any local extrema.
- **Vertical Tangent Lines**: At any isolated point not in the domain of f', **but in the domain of** f, check the limiting values of $f'(x)$, to determine if there is a vertical tangent line at that point.
- **Second Derivative Information**: Determine where the graph is concave up and concave down and locate any inflection points.
- **Horizontal Asymptotes**: Check the limit of $f(x)$ as $x \to \infty$ and as $x \to -\infty$.
- **Intercepts**: Locate x- and y-intercepts, if any. If this can't be done exactly, then do so approximately (e.g., using Newton's method).

We start with a very simple example.

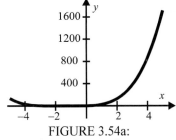

FIGURE 3.54a:
$y = x^4 + 6x^3 + 12x^2 + 8x + 1$
(one view).

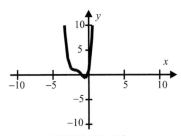

FIGURE 3.54b:
$y = x^4 + 6x^3 + 12x^2 + 8x + 1$
(standard calculator view).

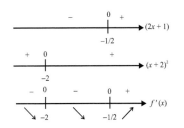

Example 5.1   Drawing a Graph of a Polynomial

Draw a graph of $f(x) = x^4 + 6x^3 + 12x^2 + 8x + 1$ showing all significant features.

Solution   Why shouldn't we simply ask our computer algebra system for a graph? Computer algebra systems and graphing calculators usually do one of two things to determine the window in which they will display a graph. (Some give the user an option as to how to set the default method). One method is to compute a set number of function values over a given standard range of x-values. The y-range is then chosen so that all of the calculated points can be displayed. This might result in a graph that looks like the one in Figure 3.54a, produced using the computer algebra system Maple. Another method is to draw a graph in a fixed, default window. For instance, most graphing calculators use the default window defined by

$$-10 \le x \le 10 \text{ and } -10 \le y \le 10.$$

Using this window, we get the graph shown in Figure 3.54b. Of course, these two graphs are very different. Without the calculus, it's difficult to tell which, if either of these is truly representative of the behavior of f. Some analysis will help to clear up the situation. First, note that the domain of f is the entire real line. Further, since $f(x)$ is a polynomial, it doesn't have any vertical or horizontal asymptotes. (Think about this!) Next, note that

$$f'(x) = 4x^3 + 18x^2 + 24x + 8 = 2(2x+1)(x+2)^2.$$

Drawing number lines for the individual factors in $f'(x)$, we have that

$$f'(x) \begin{cases} > 0, \text{ on } \left(-\frac{1}{2}, \infty\right) \\ < 0, \text{ on } (-\infty, -2) \cup \left(-2, -\frac{1}{2}\right) \end{cases}. \tag{5.1}$$

FIGURE 3.55:
$y = x^4 + 6x^3 + 12x^2 + 8x + 1$.

FIGURE 3.56: $y = \frac{x^2-3}{x^3}$.

FIGURE 3.57: $y = \frac{x^2-3}{x^3}$.

From (5.1), f is increasing on $\left(-\frac{1}{2}, \infty\right)$ and decreasing on $(-\infty, -2) \cup \left(-2, -\frac{1}{2}\right)$. This also tells us that there is a local minimum at $x = -\frac{1}{2}$ and that there are no local maxima. Next, we have

$$f''(x) = 12x^2 + 36x + 24 = 12 \, (x + 2) \, (x + 1).$$

Drawing number lines for the factors of $f''(x)$, we have

$$f''(x) \begin{cases} > 0, \text{ on } (-\infty, -2) \cup (-1, \infty) \\ < 0, \text{ on } (-2, -1) \end{cases}. \tag{5.2}$$

From (5.2), the graph is concave up on $(-\infty, -2) \cup (-1, \infty)$ and concave down on $(-2, -1)$. Further, there are inflection points at $x = -2$ and at $x = -1$. Finally, to find the x-intercepts, we need to solve $f(x) = 0$ approximately. Doing this (we leave the details as an exercise: use Newton's method or your calculator's solver), we find that there are two x-intercepts: $x = -1$ (exactly) and $x \approx -0.160713244786$. Notice that the significant x-values that we have identified are $x = -2$, $x = -1$ and $x = -\frac{1}{2}$. Computing the corresponding y-values from $y = f(x)$, we get the points $(-2, 1)$, $(-1, 0)$ and $\left(-\frac{1}{2}, -\frac{11}{16}\right)$. We summarize the first and second derivative information in the number lines in the margin. In Figure 3.55, we include all of these important points by setting the x-range to be $-3 \leq x \leq 1$ and the y-range to be $-2 \leq y \leq 8$.

∎

In the following example, we examine a function that has local extrema, inflection points and both vertical and horizontal asymptotes.

Example 5.2   Drawing a Graph of a Rational Function

Draw a graph of $f(x) = \dfrac{x^2 - 3}{x^3}$ showing all significant features.

Solution   The default graph drawn by our computer algebra system (Maple) appears in Figure 3.56. Notice that this doesn't seem to be a particularly useful graph, since very little is visible (or at least distinguishable from the axes). The graph drawn using the most common graphing calculator default window (with a range of $-10 \leq x \leq 10$ and $-10 \leq y \leq 10$) is seen in Figure 3.57. This is arguably an improvement over Figure 3.56, but does this graph convey all that it could about the function (e.g., about local extrema, inflection points, etc.)? We can answer this question only after we do some calculus. We follow the outline given at the beginning of the section.

First, observe that the domain of f includes all real numbers $x \neq 0$. Since $x = 0$ is an isolated point not in the domain of f, we scrutinize the limiting behavior of f as x approaches 0. We have

$$\lim_{x \to 0^+} f(x) = \lim_{x \to 0^+} \frac{\overset{-}{x^2 - 3}}{\underset{+}{x^3}} = -\infty \tag{5.3}$$

and

$$\lim_{x \to 0^-} f(x) = \lim_{x \to 0^-} \frac{\overset{-}{x^2 - 3}}{\underset{-}{x^3}} = \infty. \tag{5.4}$$

From (5.3) and (5.4), we see that the graph has a vertical asymptote at $x = 0$.

Next, we look for whatever information the first derivative will yield. We have

Quotient rule.

$$f'(x) = \frac{2x\left(x^3\right) - \left(x^2 - 3\right)\left(3x^2\right)}{\left(x^3\right)^2}$$

Factor out an x^2.

$$= \frac{x^2\left[2x^2 - 3\left(x^2 - 3\right)\right]}{x^6}$$

Combine terms.

$$= \frac{9 - x^2}{x^4}$$

Factor difference of two squares.

$$= \frac{(3-x)(3+x)}{x^4}.$$

Looking at the individual factors in $f'(x)$, we have the number lines shown below.

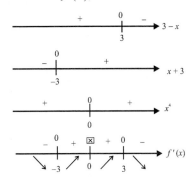

Thus,

$$f'(x) \begin{cases} > 0, \text{ on } (-3,0) \cup (0,3) \\ < 0, \text{ on } (-\infty, -3) \cup (3, \infty) \end{cases} \tag{5.5}$$

and so, f is increasing on $(-3,0) \cup (0,3)$ and decreasing on $(-\infty, -3) \cup (3, \infty)$. Note that this says that f has a local minimum at $x = -3$ and a local maximum at $x = 3$.

Next, we look at

Quotient rule.

$$f''(x) = \frac{-2x\left(x^4\right) - \left(9 - x^2\right)\left(4x^3\right)}{\left(x^4\right)^2}$$

Factor out $-2x^3$.

$$= \frac{-2x^3\left[x^2 + \left(9 - x^2\right)(2)\right]}{x^8}$$

Combine terms.

$$= \frac{-2\left(18 - x^2\right)}{x^5}$$

Factor difference of two squares.

$$= \frac{2\left(x - \sqrt{18}\right)\left(x + \sqrt{18}\right)}{x^5}.$$

Looking at the individual factors in $f''(x)$, we obtain the number lines found in the margin. Thus, we have

$$f''(x) \begin{cases} > 0, \text{ on } (-3,0) \cup (3, \infty) \\ < 0, \text{ on } (-\infty, -3) \cup (0,3) \end{cases} \tag{5.6}$$

and so, the graph is concave up on $(-3,0) \cup (3, \infty)$ and concave down on $(-\infty, -3) \cup (0,3)$. Further, there are inflection points at $x = \pm\sqrt{18}$. (Why is there no inflection point at $x = 0$?)

To determine the limiting behavior as $x \to \pm\infty$, we consider

$$\lim_{x \to \infty} f(x) = \lim_{x \to \infty} \frac{x^2 - 3}{x^3}$$

$$= \lim_{x \to \infty}\left(\frac{1}{x} - \frac{3}{x^3}\right) = 0. \tag{5.7}$$

Likewise, we have

$$\lim_{x \to -\infty} f(x) = 0. \tag{5.8}$$

So, the line $y = 0$ is a horizontal asymptote both as $x \to \infty$ and as $x \to -\infty$. Finally, the x-intercepts are where

$$0 = f(x) = \frac{x^2 - 3}{x^3},$$

that is, at $x = \pm\sqrt{3}$. Notice that there are no y-intercepts, since $x = 0$ is not in the domain of the function. We now have all of the information that we need to draw a representative graph. With some experimentation, you can set the x- and y-ranges so that most of the significant features of the graph (i.e., vertical and horizontal asymptotes, local extrema, inflection points, etc.) are displayed, as in Figure 3.58. Notice that the graph in Figure 3.58 is consistent with all of the information which we accumulated on the function in (5.3) – (5.8). Although the existence of the inflection points is clearly indicated by the change in concavity, their precise location is as yet a bit fuzzy in this graph. Notice, however, that both vertical and horizontal asymptotes and the local extrema are clearly indicated, something which cannot be said about either of Figures 3.56 or 3.57.

■

In the following example, there are multiple vertical asymptotes, only one extremum and no inflection points.

FIGURE 3.58: $y = \frac{x^2-3}{x^3}$.

FIGURE 3.59: $y = \frac{x^2}{x^2-4}$.

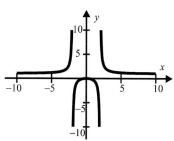

FIGURE 3.60: $y = \frac{x^2}{x^2-4}$.

Example 5.3   A Graph with Two Vertical Asymptotes

Draw a graph of $f(x) = \dfrac{x^2}{x^2 - 4}$ showing all significant features.

Solution   The default graph produced by our computer algebra system is seen in Figure 3.59, while the default graph drawn by most graphing calculators looks like the graph seen in Figure 3.60. Notice that the domain of f includes all x except $x = \pm 2$ (since the denominator is zero at $x = \pm 2$). Figure 3.60 suggests that there are vertical asymptotes at $x = \pm 2$, but let's establish this carefully. We have

$$\lim_{x \to 2^+} \frac{x^2}{x^2 - 4} = \lim_{x \to 2^+} \frac{\overset{+}{x^2}}{\underset{+}{(x - 2)} \underset{+}{(x + 2)}} = \infty. \tag{5.9}$$

Likewise, we get

$$\lim_{x \to 2^-} \frac{x^2}{x^2 - 4} = -\infty, \quad \lim_{x \to -2^+} \frac{x^2}{x^2 - 4} = -\infty \tag{5.10}$$

and

$$\lim_{x \to -2^+} \frac{x^2}{x^2 - 4} = \infty. \tag{5.11}$$

These are left as exercises. Thus, there are vertical asymptotes at $x = \pm 2$. Next, we have

$$f'(x) = \frac{2x(x^2 - 4) - x^2(2x)}{(x^2 - 4)^2}$$

$$= \frac{-8x}{(x^2 - 4)^2}.$$

Since the denominator is positive, it is a simple matter to see that

$$f'(x) \begin{cases} > 0, \text{ on } (-\infty, -2) \cup (-2, 0) \\ < 0, \text{ on } (0, 2) \cup (2, \infty) \end{cases}. \tag{5.12}$$

From (5.12), we have that f is increasing on $(-\infty, -2)\cup(-2, 0)$ and decreasing on $(0, 2)\cup (2, \infty)$. In particular, notice that the only critical number is $x = 0$ (since $x = -2, 2$ are not in the domain of f). Thus, the only local extremum is the local maximum located at $x = 0$. Next, we have

Quotient rule.

Factor out $8(x^2 - 4)$.

Combine terms.

Factor difference of two squares.

$$f''(x) = \frac{-8(x^2 - 4)^2 + (8x)\ 2(x^2 - 4)^1\ (2x)}{(x^2 - 4)^4}$$

$$= \frac{8(x^2 - 4)\left[-(x^2 - 4) + 4x^2\right]}{(x^2 - 4)^4}$$

$$= \frac{8(3x^2 + 4)}{(x^2 - 4)^3}$$

$$= \frac{8(3x^2 + 4)}{(x - 2)^3(x + 2)^3}.$$

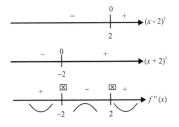

Since the numerator is positive for all x, we need only consider the terms in the denominator, as seen in the margin. We then have

$$f''(x)\begin{cases} > 0, \text{ on } (-\infty, -2) \cup (2, \infty) \\ < 0, \text{ on } (-2, 2) \end{cases}. \qquad (5.13)$$

From (5.13), we have that the graph is concave up on $(-\infty, -2) \cup (2, \infty)$ and concave down on $(-2, 2)$. However, since $x = 2, -2$ are not in the domain of f, there are no inflection points. It is an easy exercise to verify that

$$\lim_{x \to \infty} \frac{x^2}{x^2 - 4} = 1 \qquad (5.14)$$

and

$$\lim_{x \to -\infty} \frac{x^2}{x^2 - 4} = 1. \qquad (5.15)$$

From (5.14) and (5.15), we have that $y = 1$ is a horizontal asymptote, both as $x \to \infty$ and as $x \to -\infty$. Finally, we observe that the only x-intercept is at $x = 0$. We can now summarize this information in (5.9) – (5.15) in the graph seen in Figure 3.61.

■

In the following example, we need to use computer-generated graphs, as well as a rootfinding method to determine the behavior of the function.

FIGURE 3.61: $y = \frac{x^2}{x^2 - 4}$.

Example 5.4   Graphing Where the Domain and Extrema Must be Approximated

Draw a graph of $f(x) = \dfrac{1}{x^3 + 3x^2 + 3x + 3}$ showing all significant features.

Solution   Our computer algebra system draws the default graph shown in Figure 3.62.

FIGURE 3.62: $y = \dfrac{1}{x^3 + 3x^2 + 3x + 3}$.

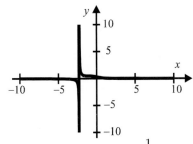

FIGURE 3.63: $y = \dfrac{1}{x^3 + 3x^2 + 3x + 3}$.

The default graph drawn by most graphing calculators looks something like the one shown in Figure 3.63. Both of these seem to reveal more about the function than did their counterparts in example 5.3, but we can only determine this through doing some calculus.

Since f is a rational function, it is defined for all x, except for where the denominator is zero, that is, where

$$x^3 + 3x^2 + 3x + 3 = 0.$$

If you don't see how to factor the expression to find the zeros exactly, you must rely on approximate methods. First, to get an idea of where the zero(s) might be, draw a graph of the cubic (see Figure 3.64). The graph does not need to be elaborate, merely detailed enough to get an idea of where and how many zeros there are. In the present case, we see that there is only one zero, around $x = -2$. We can verify that this is the only zero, since

$$\frac{d}{dx}\left(x^3 + 3x^2 + 3x + 3\right) = 3x^2 + 6x + 3 = 3\left(x+1\right)^2 \ge 0.$$

Since the derivative is never negative, the function cannot decrease to cross the x-axis a second time. We can obtain an approximation of this zero, as accurate as needed, using Newton's method or your calculator's solver. Here, we get the approximate zero $x = a \approx -2.25992104989$. Even though we don't know this zero exactly, we can use the graph in Figure 3.64 to help us compute the limits

FIGURE 3.64:
$y = x^3 + 3x^2 + 3x + 3$.

and

$$\lim_{x \to a^+} f(x) = \lim_{x \to a^+} \frac{\overset{+}{1}}{\underset{+}{x^3 + 3x^2 + 3x + 3}} = \infty \qquad (5.16)$$

$$\lim_{x \to a^-} f(x) = \lim_{x \to a^-} \frac{\overset{+}{1}}{\underset{-}{x^3 + 3x^2 + 3x + 3}} = -\infty. \qquad (5.17)$$

From (5.16) and (5.17), f has a vertical asymptote at $x = a$. Turning to the derivative information, we have

$$f'(x) = -(x^3 + 3x^2 + 3x + 3)^{-2}\,(3x^2 + 6x + 3)$$

$$= -3\left[\frac{(x+1)^2}{(x^3 + 3x^2 + 3x + 3)^2}\right]$$

$$= -3\left(\frac{x+1}{x^3 + 3x^2 + 3x + 3}\right)^2$$

$$< 0, \text{ for } x \ne a. \qquad (5.18)$$

Thus, f is decreasing for all $x \ne a$. Also, notice that the only critical number is $x = -1$, but since f is decreasing everywhere except at $x = a$, there are no local extrema. Turning to the second derivative, we get

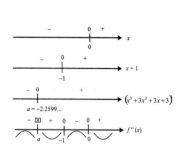

$$f''(x) = -6\left(\frac{x+1}{x^3 + 3x^2 + 3x + 3}\right)\frac{1\left(x^3 + 3x^2 + 3x + 3\right) - \left(x+1\right)\left(3x^2 + 6x + 3\right)}{(x^3 + 3x^2 + 3x + 3)^2}$$

$$= \frac{-6\left(x+1\right)}{(x^3 + 3x^2 + 3x + 3)^3}\left(-2x^3 - 6x^2 - 6x\right)$$

$$= \frac{12\,x\left(x+1\right)\left(x^2 + 3x + 3\right)}{(x^3 + 3x^2 + 3x + 3)^3}.$$

Since $(x^2 + 3x + 3) > 0$ for all x (Why is that?), we need not consider this factor. Considering the remaining factors, we have the number lines shown in the margin. Thus, we have that

$$f''(x)\begin{cases} > 0, \text{ on } (a, -1) \cup (0, \infty) \\ < 0, \text{ on } (-\infty, a) \cup (-1, 0) \end{cases} \qquad (5.19)$$

and hence, the graph is concave up on $(a, -1) \cup (0, \infty)$ and concave down on $(-\infty, a) \cup (-1, 0)$. It also follows that there are inflection points at $x = 0$ and at $x = -1$. Notice that in Figures 3.62 and 3.63, the concavity information is not very clear and the inflection points are difficult to discern.

We note the obvious fact that the function is never zero and hence, there are no x-intercepts. Finally, we consider the limits

$$\lim_{x \to \infty} \frac{1}{x^3 + 3x^2 + 3x + 3} = 0 \tag{5.20}$$

and

$$\lim_{x \to -\infty} \frac{1}{x^3 + 3x^2 + 3x + 3} = 0. \tag{5.21}$$

Using all of the information in (5.16)–(5.21), we draw the graph seen in Figure 3.65. Here, we can clearly see the vertical and horizontal asymptotes, the changes in concavity and the fact that the function is decreasing across its entire domain. Since these are all of the features of the graph which we discovered through the calculus, this graph is representative of the major features in the behavior of the function.

∎

In the following example, we consider the graph of a transcendental function with a vertical asymptote.

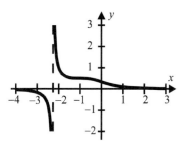

FIGURE 3.65: $y = \frac{1}{x^3 + 3x^2 + 3x + 3}$.

FIGURE 3.66: $y = e^{1/x}$.

FIGURE 3.67: $y = e^{1/x}$.

Example 5.5   Graphing Where Some Features Are Difficult to See

Draw a graph of $f(x) = e^{1/x}$ showing all significant features.

Solution   The default graph produced by our computer algebra system is not particularly helpful (see Figure 3.66). The default graph produced by most graphing calculators (see Figure 3.67) is certainly better, but we can't be sure if this is adequate without further analysis. First, notice that the domain of f is $(-\infty, 0) \cup (0, \infty)$. Thus, we consider

$$\lim_{x \to 0^+} e^{1/x} = \infty, \tag{5.22}$$

since $\frac{1}{x} \to \infty$ as $x \to 0^+$. Also, since $\frac{1}{x} \to -\infty$ as $x \to 0^-$ (and $e^t \to 0$, as $t \to -\infty$),

$$\lim_{x \to 0^-} e^{1/x} = 0. \tag{5.23}$$

From (5.22) and (5.23), there is a vertical asymptote at $x = 0$, but an unusual one, in that $f(x) \to \infty$ on one side of 0 and $f(x) \to 0$ on the other side. Next,

$$f'(x) = e^{1/x} \frac{d}{dx} \left(\frac{1}{x} \right)$$
$$= e^{1/x} \left(\frac{-1}{x^2} \right)$$
$$< 0, \text{ for all } x \neq 0, \tag{5.24}$$

since $e^{1/x} > 0$, for all x. From (5.24), we have that f is decreasing for all $x \neq 0$. We also have

$$f''(x) = e^{1/x} \left(\frac{-1}{x^2} \right) \left(\frac{-1}{x^2} \right) + e^{1/x} \left(\frac{2}{x^3} \right)$$
$$= e^{1/x} \left(\frac{1}{x^4} + \frac{2}{x^3} \right)$$
$$= e^{1/x} \left(\frac{1 + 2x}{x^4} \right) \begin{cases} < 0, \text{ on } \left(-\infty, -\frac{1}{2} \right) \\ > 0, \text{ on } \left(-\frac{1}{2}, 0 \right) \cup (0, \infty) \end{cases}. \tag{5.25}$$

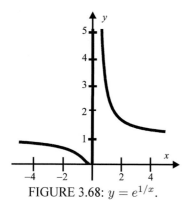

FIGURE 3.68: $y = e^{1/x}$.

From (5.25), the graph is concave down on $(-\infty, -\frac{1}{2})$ and concave up on $(-\frac{1}{2}, 0) \cup (0, \infty)$. Since $x = 0$ is not in the domain of f, the only inflection point is at $x = -\frac{1}{2}$. Next, note that

$$\lim_{x \to \infty} e^{1/x} = 1, \tag{5.26}$$

since $\frac{1}{x} \to 0$ as $x \to \infty$ and $e^t \to 1$ as $t \to 0$. Likewise,

$$\lim_{x \to -\infty} e^{1/x} = 1. \tag{5.27}$$

From (5.26) and (5.27), $y = 1$ is a horizontal asymptote, both as $x \to \infty$ and as $x \to -\infty$. Finally, since

$$e^{1/x} > 0,$$

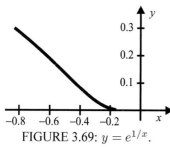

FIGURE 3.69: $y = e^{1/x}$.

for all $x \neq 0$, there are no x-intercepts. It is worthwhile noting that all of these features of the graph were discernible from Figure 3.67, except for the inflection point at $x = -\frac{1}{2}$. Note that in almost any graph you draw, it is difficult to see all of the features of the function. This happens because the inflection point $(-\frac{1}{2}, e^{-2})$ or $(-0.5, 0.135335\ldots)$ is so close to the x-axis. Since the horizontal asymptote is the line $y = 1$, it is difficult to see both of these features on the same graph (without drawing the graph on a very large piece of paper). We settle for the graph seen in Figure 3.68, which shows all of the features except the inflection point and the concavity on the interval $(-\frac{1}{2}, 0)$. To clearly see the behavior near the inflection point, we draw a graph which is zoomed-in on the area of the inflection point (see Figure 3.69). Here, while we have resolved the problem of the concavity near $x = 0$ and the inflection point, we have lost the details of the "big picture." ∎

In our final example, we consider the graph of a function that is the sum of a trigonometric function and a polynomial.

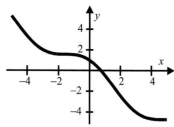

FIGURE 3.70: $y = \cos x - x$.

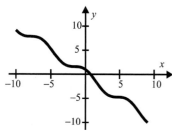

FIGURE 3.71: $y = \cos x - x$.

Example 5.6   Graphing the Sum of a Polynomial and a Trigonometric Function
Draw a graph of $f(x) = \cos x - x$, showing all significant features.

Solution   The default graph provided by our computer algebra system can be seen in Figure 3.70. The graph produced by most graphing calculators looks like that in Figure 3.71. As always, we will use the calculus to determine the behavior of the function more precisely. First, notice that the domain of f is the entire real line. Consequently, there are no vertical asymptotes. Next, we have

$$f'(x) = -\sin x - 1 \leq 0, \text{ for all } x. \tag{5.28}$$

Further, $f'(x) = 0$ if and only if $\sin x = -1$. So, there are critical numbers (here, these are all locations of horizontal tangent lines), but since $f'(x)$ does not change sign, there are **no** local extrema. Even so, it is still of interest to find the locations of the horizontal tangent lines. Recall that

$$\sin x = -1 \text{ for } x = \frac{3\pi}{2}$$

or for

$$x = \frac{3\pi}{2} + 2n\pi,$$

for any integer n. Next, we see that

$$f''(x) = -\cos x$$

and on the interval $[0, 2\pi]$, we have

$$\cos x \begin{cases} > 0, \text{ on } [0, \frac{\pi}{2}) \cup (\frac{3\pi}{2}, 2\pi] \\ < 0, \text{ on } (\frac{\pi}{2}, \frac{3\pi}{2}) \end{cases}.$$

So,

$$f''(x) = -\cos x \begin{cases} < 0, \text{ on } [0, \frac{\pi}{2}) \cup (\frac{3\pi}{2}, 2\pi] \\ > 0, \text{ on } (\frac{\pi}{2}, \frac{3\pi}{2}) \end{cases}. \tag{5.29}$$

Outside of $[0, 2\pi]$, $f''(x)$ simply repeats this pattern. In particular, this says that the graph has infinitely many inflection points, located at odd multiples of $\frac{\pi}{2}$.

To determine the behavior as $x \to \pm\infty$, we examine the limits

$$\lim_{x \to \infty} (\cos x - x) = -\infty \tag{5.30}$$

and

$$\lim_{x \to -\infty} (\cos x - x) = \infty. \tag{5.31}$$

Finally, to determine the x-intercept(s), we need to solve

$$f(x) = \cos x - x = 0.$$

This can't be solved exactly, however. Since $f'(x) \leq 0$ for all x and Figures 3.70 and 3.71 show a zero around $x = 1$, there is only one zero and we must approximate this (use Newton's method or your calculator's solver). We get $x \approx 0.739085\ldots$ is an approximation to the only x-intercept. Assembling all of the information in (5.28) - (5.31), we can draw the graph seen in Figure 3.72. Notice that Figure 3.71 shows the behavior just as clearly as Figure 3.72, but for a smaller range of x-values. Which of these is more "representative" is open to discussion.

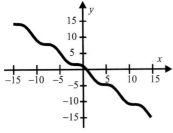

FIGURE 3.72: $y = \cos x - x$.

∎

Exercises 3.5

1. (writing) We have talked about sketching representative graphs, but it is often impossible to draw a graph correctly to scale that shows all of the properties we might be interested in. For example, try to generate a computer or calculator graph that shows all three local extrema of $x^4 - 25x^3 - 2x^2 + 80x - 3$. When two extrema have y-coordinates of approximately -60 and 50, it takes a very large graph to also show a point with $y = -40,000$! If an accurate graph cannot show all the points of interest, perhaps a "caricature graph" like the one shown below is needed.

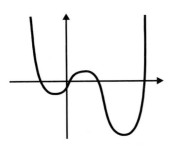

There is no scale shown on the graph because we have distorted different portions of the graph in an attempt to show all of the interesting points. Discuss the relative merits of an "honest" graph with a consistent scale but not showing all the points of interest versus a caricature graph which distorts the scale but does show all the points of interest.

2. (writing) While studying for a test, a friend of yours says that a graph is not allowed to touch an asymptote. While it is often the case that graphs don't touch asymptotes, there is definitely not any rule against it. Explain why graphs can intersect a horizontal asymptote any number of times (Hint: Look at the graph of $e^{-x} \sin x$) but can't pass through a vertical asymptote.

3. (writing) Explain why polynomials never have vertical or horizontal asymptotes.

4. (writing) Explain how the graph of $f(x) = \cos x - x$ in example 5.6 relates to the graphs of $y = \cos x$ and $y = -x$. Based on this discussion, explain how to sketch the graph of $y = x + \sin x$.

In exercises 5–44, graph the function and completely discuss the graph as in example 5.2.

5. $f(x) = x^3 - 3x^2 + 3x$ **6.** $f(x) = x^3 - 9x + 1$

7. $f(x) = x^4 - 3x^2 + 2x$ **8.** $f(x) = x^4 + 8x - 2$

9. $f(x) = x^5 - 2x^3 + 1$

10. $f(x) = x^5 - 2.5x^4 - 5x^3$

11. $f(x) = x\sqrt{x^2 - 4}$ **12.** $f(x) = x/\sqrt{x^2 - 4}$

13. $f(x) = x + 4/x$ **14.** $f(x) = (x^2 - 1)/x$

15. $f(x) = \sin x - \cos x$ **16.** $f(x) = \cos^3 x$

17. $f(x) = e^{-x^2/4}$ **18.** $f(x) = xe^{-4x}$

19. $f(x) = x \ln x$ **20.** $f(x) = x \ln x^2$

21. $f(x) = \sqrt{x^2 + 1}$ **22.** $f(x) = \sqrt{2x - 1}$

23. $f(x) = \dfrac{2x^2}{x^2 + 2}$ **24.** $f(x) = \dfrac{x^2 + 1}{3x^2 - 1}$

25. $f(x) = \dfrac{2x^2}{x^3 + 1}$ **26.** $f(x) = \dfrac{x^2}{x^4 + 1}$

27. $f(x) = \dfrac{4x}{x^2 - x + 1}$ **28.** $f(x) = \dfrac{x - 1}{x^2 - x - 2}$

29. $f(x) = \dfrac{4x^3}{x^2 - x + 1}$ **30.** $f(x) = \dfrac{x^3}{x^2 - 2x - 3}$

31. $f(x) = e^{-x} \cos 2x$ **32.** $f(x) = e^x \sin 2x$

33. $f(x) = x^5 - 5x$ **34.** $f(x) = x^3 - \frac{3}{400}x$

35. $f(x) = \dfrac{2x}{\sqrt{x^2 + 2}}$ **36.** $f(x) = \sqrt{3x^4 + x^3}$

37. $f(x) = x^{1/5}(x + 1)$ **38.** $f(x) = \sqrt[3]{x^3 + 1}$

39. $f(x) = \sin x - \frac{1}{2}\sin 2x$ **40.** $f(x) = \sin x + \frac{1}{3}\sin 3x$

41. $f(x) = e^{-2/x}$ **42.** $f(x) = e^{1/x^2}$

43. $f(x) = xe^{-x^2/2}$ **44.** $f(x) = x^{10}e^{-x}$

In exercises 45–50, the "family of functions" contains a parameter c. The value of c affects the properties of the functions. Determine what differences, if any, there are for c being zero, positive or negative. Then determine what the graph would look like for very large positive c's and for very negative c's.

45. $f(x) = x^4 + cx^2$ **46.** $f(x) = x^4 + cx^2 + x$

47. $f(x) = \dfrac{x^2}{x^2 + c^2}$ **48.** $f(x) = e^{-x^2/c}$

49. $f(x) = \sin(cx)$ **50.** $f(x) = x^2\sqrt{c^2 - x^2}$

51. The "FM" in FM radio stands for **frequency modulation,** a method of transmitting information encoded in a radio wave by modulating (or varying) the frequency. A basic example of such a modulated wave is $f(x) = \sin(x + \cos x)$. Use computer-generated graphs of $f(x)$, $f'(x)$ and $f''(x)$ to try to locate all relative extrema of $f(x)$.

52. A rational function is a function of the form $\dfrac{p(x)}{q(x)}$ where $p(x)$ and $q(x)$ are polynomials. Is it true that all rational functions have vertical asymptotes? Is it true that all rational functions have horizontal asymptotes?

53. In a variety of applications, researchers model a phenomenon whose graph starts at the origin, rises to a single maximum and then drops off to a horizontal asymptote of $y = 0$. For example, the probability density function of events such as the time from conception to birth of an animal and the amount of time surviving after contracting a fatal disease might have these properties. Show that the family of functions xe^{-bx} has these properties for all positive constants b. What effect does b have on the location of the maximum? In the case of the time since conception, what would b represent? In the case of survival time, what would b represent?

54. It can be useful to identify asymptotes other than vertical and horizontal. For example, the parabola x^2 is an asymptote of $f(x)$ if $\lim\limits_{x \to \infty} [f(x) - x^2] = 0$ and/or $\lim\limits_{x \to -\infty} [f(x) - x^2] = 0$. Show that x^2 is an asymptote of $f(x) = \dfrac{x^4 - x^2 + 1}{x^2 - 1}$. Graph $y = f(x)$ and zoom out until the graph looks like a parabola. (Note: the effect of zooming out is to emphasize large values of x.)

A function $f(x)$ has a slant asymptote $y = mx + b$ $(m \neq 0)$ if $\lim\limits_{x \to \infty} [f(x) - (mx + b)] = 0$ and/or $\lim\limits_{x \to -\infty} [f(x) - (mx + b)] = 0$. In exercises 55–60, find the slant asymptote (use long division to rewrite the function). Then, graph the function and its asymptote on the same axes.

55. $f(x) = \dfrac{3x^2 - 1}{x}$ **56.** $f(x) = \dfrac{3x^2 - 1}{x - 1}$

57. $f(x) = \dfrac{x^3 - 2x^2 + 1}{x^2}$ **58.** $f(x) = \dfrac{x^3 - 1}{x^2 - 1}$

59. $f(x) = \dfrac{x^4}{x^3 + 1}$ **60.** $f(x) = \dfrac{x^4 - 1}{x^3 + x}$

In exercises 61–64, find a function whose graph has the given asymptotes.

61. $x = 1$, $x = 2$ and $y = 3$

62. $x = -1$, $x = 1$ and $y = 0$

63. $x = -1$, $x = 1$, $y = -2$ and $y = 2$

64. $x = 1$, $y = 2$ and $x = 3$

65. (exploratory). For each function, find a polynomial $p(x)$ such that $\lim\limits_{x \to \infty} [f(x) - p(x)] = 0$.

(a) $\dfrac{x^4}{x + 1}$ (b) $\dfrac{x^5 - 1}{x + 1}$ (c) $\dfrac{x^6 - 2}{x + 1}$

Show by zooming out that $f(x)$ and $p(x)$ look similar for large x. The *first term* of a polynomial is the term with the highest power (e.g., x^3 is the first term of $x^3 - 3x + 1$). Can you zoom out enough to make the graph of $f(x)$ look like the first term of its polynomial asymptote? State a very quick rule enabling you to look at a rational function and determine the first term of its polynomial asymptote (if one exists).

66. (exploratory) One of the natural enemies of the balsam fir tree is the spruce budworm, which attacks the leaves of the fir tree in devastating outbreaks. Define $N(t)$ to be the number of worms on a particular tree at time t. A mathematical model of the population dynamics of the worm must include a term to indicate the worm's death rate due to its predators (e.g., birds). The form of this term is often taken to be $\dfrac{B[N(t)]^2}{A^2 + [N(t)]^2}$ for positive constants A and B. Graph the functions $\dfrac{x^2}{4 + x^2}$, $\dfrac{2x^2}{1 + x^2}$, $\dfrac{x^2}{9 + x^2}$ and $\dfrac{3x^2}{1 + x^2}$ for $x > 0$. Based on these graphs, discuss why $\dfrac{B[N(t)]^2}{A^2 + [N(t)]^2}$ is a plau-

sible model for the death rate by predation. What role do the constants A and B play? The possible stable population levels for the spruce budworms are determined by intersections of the graphs $y = r(1 - x/k)$ and $y = \dfrac{x}{1 + x^2}$. Here, $x = N/A$, r is proportional to the birthrate of the budworms and k is determined by the amount of food available to the budworms. Note that $y = r(1 - x/k)$ is a line with y-intercept r and x-intercept k. How many solutions are there to the equation $r(1 - x/k) = \dfrac{x}{1 + x^2}$? (Hint: The answer depends on the values of r and k.) One current theory is that outbreaks are caused in situations where there are three solutions and the population of budworms jumps from a small population to a large population.

3.6   Optimization

One of the most frequently heard questions in industry today is "Is this design optimal?" We are finally in a position to bring the power of the calculus to bear on questions of this sort. The tools that we've been developing over the last five sections are exactly what we need. Everywhere we look in business and industry today, we see people struggling to minimize waste and maximize productivity. Certainly some of this is done by using commonsense methods to reduce unnecessary expenditures, but much of this is accomplished by using a careful mathematical analysis of the problem. You might also wonder whether you can solve such problems simply by reading the answer from a calculator- or computer-generated graph. The answer to this is that you can do this sometimes. Remember that functions (even reasonably familiar ones) sometimes have behavior that is hidden in the standard graph you get from a computer or calculator. Further, the calculus can be used to solve more general questions, such as the one following example 6.1, which you cannot solve by using a machine alone.

This section contains a number of illustrations of how to apply the methods of calculus to problems requiring you to find a maximum or a minimum. Pay close attention to how we have solved the problems. We start by giving a few general guidelines. (Notice that we said *guidelines* and not *rules*.) Do not memorize them, but rather keep these in mind as you work through the section.

- If there's a picture to draw, draw it! Don't try to envision how things look in your head. Put a picture down on paper and label it.
- Determine what the variables are and how they are related.
- Decide what quantity needs to be maximized or minimized.
- Write an expression for the quantity to be maximized or minimized in terms of only **one** variable. To do this, you may need to solve for any other variables in terms of this one variable.
- Determine the minimum and maximum allowable values (if any) of the variable you're using.
- Solve the problem. (Be sure to answer the question that is asked!)

We begin with a simple example where the goal is to accomplish what businesses face every day: getting the most from limited resources.

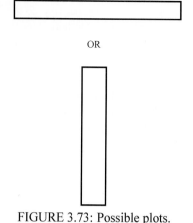

OR

FIGURE 3.73: Possible plots.

Example 6.1   Constructing a Rectangular Garden of Maximum Area

You have 40 (linear) feet of fencing with which to enclose a rectangular space for a garden. What is the **largest** area that can be enclosed with this much fencing?

Solution   First, note that there are lots of possibilities. We could enclose a plot that is very long but narrow, or one that is very wide, but not very long (see Figure 3.73). How are we to decide which configuration is optimal? We first draw a picture and label it appropriately (see Figure 3.74).

FIGURE 3.74: Rectangular plot.

We want to find the dimensions of the largest possible area, that is, maximize

$$A = x\,y.$$

You might immediately notice that this function has two variables and so, cannot be dealt with via the means we have available. However, there is another piece of information that we can use here. If we want the maximum area, then **all** of the fencing must be used. This says that the perimeter of the resulting fence must be $40'$ and hence,

$$40 = Perimeter = 2\,x + 2\,y. \tag{6.1}$$

Notice that we can use (6.1) to solve for one variable (either one) in terms of the other. We have

$$2\,y = 40 - 2\,x$$

and hence,

$$y = 20 - x.$$

Substituting for y, we get that

$$A = x\,y = x\,(20 - x).$$

So, our job is to find the maximum value of the function

$$A(x) = x\,(20 - x).$$

Hold it! We know what you're thinking. You want to multiply out this expression. While it doesn't matter very much here (this was to be a very simple problem, after all), this is, in practice, a very bad idea. In short, don't do algebra just for the sake of doing it. Unless you have a clear picture of how it will affect your later work, leave the function alone!

Before we attempt to maximize $A(x)$, we need to determine if there is an interval in which x is constrained to lie. First, notice that since x is a distance, we must have $0 \le x$. Further, since the perimeter is $40'$, we must have $x \le 20$. (Why don't we have $x \le 40$?) So, we want to find the maximum value of $A(x)$ on the closed interval $[0, 20]$. This is now a simple problem. As a check on what a reasonable answer should be, we draw a graph of $y = A(x)$ (see Figure 3.75). Here, we have only drawn that portion of the graph on the interval $[0, 20]$ and simply adjusted the y-range to display all of the plotted values. The

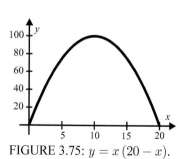

FIGURE 3.75: $y = x\,(20 - x)$.

maximum value appears to occur around $x = 10$. Now, let's analyze the problem carefully. We have

$$A'(x) = 1(20 - x) + x(-1)$$
$$= 20 - 2x$$
$$= 2(10 - x).$$

So, the only critical number is $x = 10$ and this is in the interval under consideration. Recall that the maximum and minimum value of a continuous function on a closed and bounded interval must occur at either the endpoints or a critical number. This says that we need only compare the function values

$$A(0) = 0, \quad A(20) = 0 \quad \text{and} \quad A(10) = 100.$$

Thus, the maximum area that can be enclosed with $40'$ of fencing is 100 ft^2. We also want the dimensions of the plot. (This result is only of theoretical value, if we don't know how to construct the rectangle with the maximum area.) We have that $x = 10$ and

$$y = 20 - x = 10.$$

That is, the rectangle of perimeter $40'$ with maximum area is a **square** $10'$ on a side.

∎

A more general problem that you can now solve is to show that (given a fixed perimeter) the rectangle of maximum area is a square. This is virtually identical to example 6.1 and is left as an exercise. It's worth noting here that this more general problem is one that cannot be solved by simply using a calculator to draw a graph. You'll need to use some calculus here.

Manufacturing companies routinely make countless decisions that affect the efficiency of their production processes. One decision that is surprisingly important is how to economically package their products for shipping. The following example is a simple version of this problem.

Example 6.2   Constructing a Box of Maximum Volume

A square sheet of cardboard $18''$ on a side is made into an open box (i.e., there's no top), by cutting squares of equal size out of each corner (see Figure 3.76a) and folding up the sides along the dotted lines (see Figure 3.76b). Find the dimensions of the box with the maximum volume.

FIGURE 3.76a: A sheet of cardboard.

FIGURE 3.76b: Rectangular box.

Solution   Recall that the volume of a rectangular parallelepiped (a box) is given by

$$V = l \times w \times h.$$

From Figure 3.76b, we can see that the height is $h = x$, while the length and width are $l = w = 18 - 2x$. Thus, we can write the volume in terms of the one variable x as

$$V = V(x) = (18 - 2x)^2 \, (x).$$

Once again, don't multiply this out, just out of habit. Notice that since x is a distance, we have $x \geq 0$. Further, we have $x \leq 9$, since cutting squares of side 9 out of each corner will cut up the entire sheet of cardboard. Thus, we are faced with finding the absolute maximum of the continuous function

$$V = V(x) = (18 - 2x)^2 \, (x)$$

on the closed interval

$$0 \leq x \leq 9.$$

This should be a simple matter. The graph of $y = V(x)$ on the interval $[0, 9]$ is seen in Figure 3.77. From the graph, the maximum volume seems to be somewhat over 400 and seems to occur around $x = 3$. Now, we solve the problem precisely. We have

Product rule.

Factor out $(18 - 2x)$.

$$\begin{aligned} V'(x) &= 2\,(18 - 2x)\,(-2)\,x + (18 - 2x)^2\,(1) \\ &= (18 - 2x)\,[-4x + (18 - 2x)] \\ &= (18 - 2x)\,(18 - 6x). \end{aligned}$$

So, V has two critical numbers: 3 and 9 and these are both in the interval under consideration. We now need only compare the value of the function at the endpoints and the critical numbers. We have

$$V(0) = 0, \quad V(9) = 0 \quad \text{and} \quad V(3) = 432.$$

Obviously, the maximum possible volume is 432 cubic inches. We can achieve this volume if we cut squares of side $3''$ out of each corner. You should note that this corresponds with what we expected from the graph of $y = V(x)$ in Figure 3.77. Finally, observe that the dimensions of this optimal box are $12''$ long by $12''$ wide by $3''$ deep.

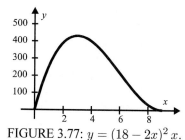

FIGURE 3.77: $y = (18 - 2x)^2 \, x$.

When a new building is built, it must be connected to existing telephone and power cables, water and sewer lines and paved roads. If the cables, water or sewer lines or road are curved, then it may not be obvious how to make the shortest (i.e., least expensive) connection possible. The next two examples consider the general problem of finding the shortest distance from a point to a curve.

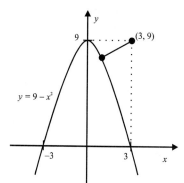

FIGURE 3.78: $y = 9 - x^2$.

Example 6.3   **Finding the Closest Point to a Parabola**

Find the point on the parabola $y = 9 - x^2$ closest to the point $(3, 9)$ (see Figure 3.78).

Solution   Using the usual distance formula, we find that the distance between the point $(3, 9)$ and any point (x, y) is

$$d = \sqrt{(x - 3)^2 + (y - 9)^2}.$$

If the point (x, y) is on the parabola, note that its coordinates satisfy the equation $y = 9 - x^2$ and so, we can write the distance in terms of the single variable x as follows

$$\begin{aligned} d(x) &= \sqrt{(x - 3)^2 + [(9 - x^2) - 9]^2} \\ &= \sqrt{(x - 3)^2 + (-x^2)^2} \\ &= \sqrt{(x - 3)^2 + x^4}. \end{aligned}$$

Although we can certainly solve the problem in its present form, we can simplify our work by observing that $d(x)$ is minimized if and only if the quantity under the square root is minimized. (We leave it as an exercise to show why this is true.) So, instead of minimizing $d(x)$ directly, we minimize the **square** of $d(x)$:

$$f(x) = [d(x)]^2 = (x-3)^2 + x^4$$

instead. Notice from Figure 3.78 that any point on the parabola to the left of the y-axis is farther away from $(3,9)$ than is the point $(0,9)$. Likewise, any point on the parabola below the x-axis is farther from $(3,9)$ than is the point $(3,0)$. So, it suffices to look for the closest point with

$$0 \le x \le 3.$$

This is the now familiar problem of minimizing a continuous function over a closed and bounded interval. See Figure 3.79 for a graph of $y = f(x)$ over the interval of interest. The minimum value of f (the square of the distance) seems to be around 10 and seems to occur near $x = 1$. We have

$$f'(x) = 2(x-3)^1 + 4x^3$$
$$= 4x^3 + 2x - 6.$$

You might notice that $f'(x)$ factors. (One way to see this is to recognize that $x = 1$ is a zero of $f'(x)$, which makes $(x-1)$ a factor. You can then get the factorization by dividing $f'(x)$ by $(x-1)$.) We have

$$f'(x) = 2(x-1)(2x^2 + 2x + 3).$$

So, $x = 1$ is a critical number. In fact it's the only critical number, since $(2x^2 + 2x + 3)$ has no zeros. (Why not?) We now need only compare the value of f at the endpoints and the critical number. We have

$$f(0) = 9, \quad f(3) = 81 \quad \text{and} \quad f(1) = 5.$$

Thus, the minimum value of $f(x)$ is 5. This says that the minimum distance from the point $(3,9)$ to the parabola is $\sqrt{5}$ and the closest point on the parabola is $(1,8)$. Again, notice that this corresponds with what we expected from the graph of $y = f(x)$.

∎

The following example is very similar to example 6.3, except that we need to use approximate methods to find the critical number.

Example 6.4   Finding Minimum Distance Approximately

Find the point on the parabola $y = 9 - x^2$ closest to the point $(5,11)$ (see Figure 3.80).

Solution   As in example 6.3, we want to minimize the distance from a fixed point (in this case, the point $(5,11)$) to a point (x,y) on the parabola. Using the distance formula, the distance from any point (x,y) on the parabola to the point $(5,11)$ is

$$d = \sqrt{(x-5)^2 + (y-11)^2}$$
$$= \sqrt{(x-5)^2 + [(9-x^2)-11]^2}$$
$$= \sqrt{(x-5)^2 + (x^2+2)^2}.$$

Again, it is equivalent (and simpler) to minimize the quantity under the square root:

$$f(x) = [d(x)]^2 = (x-5)^2 + (x^2+2)^2.$$

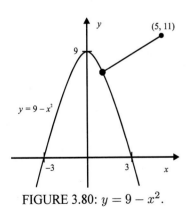

FIGURE 3.79:
$y = (x-3)^2 + x^4$.

$y = 9 - x^2$

FIGURE 3.80: $y = 9 - x^2$.

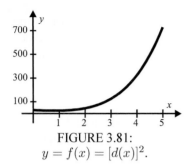

FIGURE 3.81:
$y = f(x) = [d(x)]^2$.

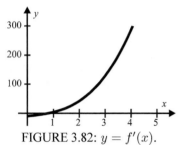

FIGURE 3.82: $y = f'(x)$.

As in example 6.3, we can see from Figure 3.80 that any point on the parabola to the left of the y-axis is farther from $(5, 11)$ than is $(0, 9)$. Likewise, any point on the parabola to the right of $x = 5$ is farther from $(5, 11)$ than is $(5, -16)$. Thus, we minimize $f(x)$ for $0 \leq x \leq 5$. In Figure 3.81 we see a graph of $y = f(x)$ on the interval of interest. The minimum value of f seems to be about 25 and seems to occur around $x = 1$. We can make this more precise as follows:

$$f'(x) = 2(x - 5) + 2(x^2 + 2)(2x)$$
$$= 4x^3 + 10x - 10.$$

Unlike in example 6.3, the expression for $f'(x)$ has no obvious factorization. Our only choice then is to find zeros of $f'(x)$ approximately. First, we draw a graph of $y = f'(x)$ on the interval of interest (see Figure 3.82). The only zero appears to be slightly less than 1. Using $x_0 = 1$ as an initial guess in Newton's method (applied to $f'(x) = 0$) or using your calculator's solver, you should get the approximate root $x_c = 0.79728\ldots$. We now compare function values:

$$f(0) = 29, \quad f(5) = 729 \quad \text{and} \quad f(0.79728\ldots) \approx 24.6\ldots.$$

Thus, the minimum distance from $(5, 11)$ to the parabola is approximately $\sqrt{24.6\ldots} \approx 4.9607998$ and the closest point on the parabola is located at approximately $(0.79728\ldots, 8.364\ldots)$. Notice that this example was essentially the same as example 6.3, except that we needed to use approximate methods, as there was no algebra available to us.

■

Notice that in both Figures 3.78 and 3.80 the shortest path appears to be perpendicular to the tangent line to the curve at the point where the path intersects the curve. We leave it as an exercise to prove that this is always the case. This observation is an important geometric principle which you can apply to many problems of this type.

Remark 6.1   At this point you might be tempted to forgo the comparison of function values at the endpoints and at the critical numbers. After all, in all of the examples we have seen so far, the desired maximizer or minimizer (i.e., the point at which the maximum or minimum occurred) was the only critical number in the interval under consideration. You might suspect that if there is only one critical number it will correspond to the maximizer or minimizer you are searching for. Unfortunately, this is not always the case. In 1945, two prominent aeronautical engineers derived a function to model the range of an aircraft. Their intention was to use this function to discover how to maximize the range. They found two critical numbers of this function and reasoned that one of them (the critical number corresponding to distributing virtually all of the plane's weight in the wings), gave the **maximum** range. The result was the famous "Flying Wing" aircraft. Some years later, it was argued that they had in fact stumbled onto a critical number corresponding to a local **minimum** of the range function. In the engineers' defense, they did not have easy, accurate computational power at their fingertips, as we do today. Remarkably, this design strongly resembles the modern B-2 Stealth bomber. This story came out as controversy brewed over the production of the B-2 (see *Science*, **244**, pp. 650–651, May 12, 1989, also see the *Monthly* of the Mathematical Association of America, October, 1993, pp. 737–738). The moral should be crystal clear: check the function values at the critical numbers and at the endpoints. Do not simply *assume* (even by virtue of having only one critical number) that a given critical number corresponds to the extremum you are seeking.

Next, we consider an optimization problem that cannot be restricted to a closed interval. We will use the fact that a single local extremum must be an absolute extremum. (Think about why this is true.)

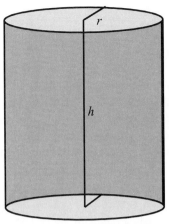

FIGURE 3.83: Soda can.

Example 6.5   Designing a Soda Can That Uses a Minimum Amount of Material

A soda can is to hold 12 fluid ounces. Find the dimensions that will minimize the amount of material used in its construction, assuming that the thickness of the material is uniform (i.e., the thickness of the aluminum is the same everywhere in the can).

Solution   First, we draw and label a picture of a typical soda can (see Figure 3.83). Here we have drawn a right circular cylinder of height h and radius r. Assuming uniform thickness of the aluminum, notice that we can minimize the amount of material by minimizing the surface area of the can. We have

$$Area = Area\ of\ top + bottom + curved\ surface\ area$$
$$= 2\pi r^2 + 2\pi rh. \qquad (6.2)$$

We can eliminate one of the variables by using the fact that the volume (using 1 fluid ounce ≈ 1.80469 in^3) must be

$$12\ fluid\ ounces = 12\ \text{fl oz} \times 1.80469\ \frac{\text{in}^3}{\text{fl oz}}$$
$$= 21.65628\ \text{in}^3.$$

Further, the volume of a right circular cylinder is

$$Vol = \pi r^2 h,$$

and so,

$$h = \frac{Vol}{\pi r^2} = \frac{21.65628}{\pi r^2}. \qquad (6.3)$$

Thus, from (6.2) and (6.3), the surface area is

$$A(r) = 2\pi r^2 + 2\pi r\,\frac{21.65628}{\pi r^2}$$
$$= 2\pi \left(r^2 + \frac{21.65628}{\pi r} \right).$$

So, our job is to minimize $A(r)$, but here, there is no closed and bounded interval of allowable values. In fact, all we can say is that $r > 0$. We can have r as large or small as you can imagine, simply by taking h to be correspondingly small or large, respectively. That is, we must find the absolute minimum of $A(r)$ on the open and unbounded interval $(0, \infty)$. To get an idea of what a plausible answer might be, we graph $y = A(r)$ (see Figure 3.84). There appears to be a local (and possibly an absolute) minimum between $r = 1$ and $r = 2$. This minimum value seems to be slightly less than 50. Next, we compute

$$A'(r) = \frac{d}{dr}\left[2\pi \left(r^2 + \frac{21.65628}{\pi r} \right) \right]$$
$$= 2\pi \left(2r - \frac{21.65628}{\pi r^2} \right)$$
$$= 2\pi \left(\frac{2\pi r^3 - 21.65628}{\pi r^2} \right).$$

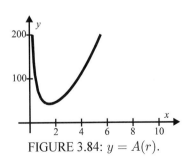

FIGURE 3.84: $y = A(r)$.

Notice that the only critical numbers are those for which the numerator of the fraction is zero:

$$0 = 2\pi r^3 - 21.65628.$$

This occurs if and only if

$$r^3 = \frac{21.65628}{2\pi}$$

and hence, the only critical number is

$$r = r_c = \sqrt[3]{\frac{21.65628}{2\pi}} \approx 1.51054846.$$

Further, notice that for $0 < r < r_c$, $A'(r) < 0$ and for $r_c < r$, $A'(r) > 0$. That is, $A(r)$ is decreasing on the interval $(0, r_c)$ and increasing on the interval (r_c, ∞). Thus, $A(r)$ has not only a local minimum, but also an absolute minimum at $r = r_c$. Notice, too that this corresponds with what we expected from the graph of $y = A(r)$ in Figure 3.84. This says that the can that uses a minimum of material has radius $r_c \approx 1.51054846$ and height $h = \dfrac{21.65628}{\pi r_c^2} \approx 3.0211.$

We should observe that a standard 12-ounce soda can has a radius of about $1.156''$. (How would you explain the discrepancy?) We study the problem of designing a more realistic soda can further in the exercises.

In our final example, we consider a problem where most of the work must be done numerically and graphically.

Example 6.6   Minimizing the Cost of Highway Construction

The state wants to build a new stretch of highway to link an existing bridge with a turnpike interchange, located 8 miles to the east and 8 miles to the south of the bridge. There is a 5-mile-wide stretch of marsh land adjacent to the bridge which must be crossed (see Figure 3.85). Given that the highway costs $10 million per mile to build over the marsh and only $7 million to build over dry land, how far to the east of the bridge should the highway be when it crosses out of the marsh?

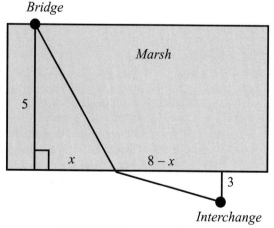

FIGURE 3.85: A new highway.

Solution   You might guess that the highway should cut directly across the marsh, so as to minimize the amount built over marsh land. We will use the calculus to decide this question. We let x represent the distance in question (see Figure 3.85). Then, the interchange lies $8 - x$ miles to the east of the point where the highway leaves the marsh. Thus, the total cost (in millions of dollars) is

$$Cost = 10\,(\text{distance across marsh}) + 7\,(\text{distance across dry land}).$$

Using the Pythagorean theorem on the two right triangles seen in Figure 3.85, we get the cost function

$$C(x) = 10\sqrt{x^2 + 25} + 7\sqrt{(8-x)^2 + 9}.$$

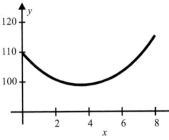

FIGURE 3.86: $y = C(x)$.

FIGURE 3.87: $y = C'(x)$.

From Figure 3.85, you can see that $0 \leq x \leq 8$. (Making $x < 0$ means going **away** from the interchange and making $x > 8$ means the highway would be beyond the interchange at the point at which it exits the marsh.) So, we have the routine problem of minimizing a continuous function $C(x)$ over the closed and bounded interval $[0, 8]$. Or is it really that routine? First, we draw a graph of $y = C(x)$ on the interval in question to get an idea of a plausible answer (see Figure 3.86). From the graph, the minimum appears to be slightly less than 100 and occurs around $x = 4$. To solve the problem precisely, we need to compute the derivative

$$C'(x) = \frac{d}{dx}\left(10\sqrt{x^2 + 25} + 7\sqrt{(8-x)^2 + 9}\right)$$

$$= 5\left(x^2 + 25\right)^{-1/2}(2x) + \frac{7}{2}\left[(8-x)^2 + 9\right]^{-1/2}(2)(8-x)^1(-1)$$

$$= \frac{10x}{\sqrt{x^2 + 25}} - \frac{7(8-x)}{\sqrt{(8-x)^2 + 9}}.$$

First, note that the only critical numbers are where $C'(x) = 0$. (Why?) The only way to find these is to approximate them. A graph of $y = C'(x)$ is seen in Figure 3.87. The only zero of $C'(x)$ on the interval $[0, 8]$ appears to be between $x = 3$ and $x = 4$. We approximate this zero numerically (e.g., with bisections or your calculator's solver), to obtain the approximate critical number

$$x_c \approx 3.5600515203.$$

Now, we need only compare the value of $C(x)$ at the endpoints and at this one critical number:

$$C(0) \approx \$109.808023 \text{ million}$$

$$C(8) \approx \$115.339811 \text{ million}$$

and

$$C(x_c) \approx \$98.888375 \text{ million}.$$

So, by using a little calculus, we can save the taxpayers more than \$10 million over cutting directly across the marsh and more than \$16 million over cutting diagonally across the marsh (not a bad reward for a few minutes work).

■

The examples that we've presented in this section together with the exercises should give you the basis for solving a wide collection of applied optimization problems. Be careful when solving these problems to draw good pictures, as well as graphs of the functions involved. Make sure that the answer you obtain computationally is consistent with what you expect from the graphs. If not, further analysis is required to see what you have missed. Also, make sure that the solution makes physical sense, when appropriate. All of these multiple checks on your work will reduce the likelihood of error.

Exercises 3.6

1. (writing) Suppose some friends complain to you that they can't work any of the problems in this section. When you ask to see their work, they say that they couldn't even get started. In the text, we have emphasized sketching a picture and defining variables. Part of the benefit of this is to help you get started writing something (anything) down. Do you think this advice helps? What do you think is the most difficult aspect of these problems? Give your friends the best advice you can.

2. (writing) We have neglected one important aspect of optimization problems, an aspect that might be called "common sense." For example, suppose you are finding the optimal dimensions for a fence and the mathematical solution is to build a square fence of length $10\sqrt{5}$ feet on each side. At the meeting with the carpenter who is going to build the fence, what length fence do you order? Why is $10\sqrt{5}$ probably not the best way to express the length? We can approximate $10\sqrt{5} \approx 22.36$. What would you tell the carpenter? Suppose the carpenter accepts measurements down to the tenth of a foot. Assuming that the building constraint was that the perimeter of the fence could not exceed a certain figure, why should you truncate to 22.3 feet instead of rounding up to 22.4 feet?

3. (writing) In example 6.3, we stated that $d(x) = \sqrt{f(x)}$ is minimized by exactly the same x-value(s) as $f(x)$. Use the fact that $\sqrt{x}$ is an increasing function to explain why this is true.

4. (writing) Suppose that $f(x)$ is a continuous function with a single critical number, and $f(x)$ has a local minimum at that critical number. Explain why $f(x)$ also has an absolute minimum at the critical number.

5. A three-sided fence is to be built next to a straight section of river, which forms the fourth side of a rectangular region. The enclosed area is to equal 1800 ft^2. What dimensions of the fence will minimize the perimeter?

6. A three-sided fence is to be built next to a straight section of river, which forms the fourth side of a rectangular region. There is 96 feet of fencing available. What dimensions of the fence will maximize the enclosed area?

7. A two-pen corral is to be built. The outline of the corral forms two identical adjoining rectangles. If there is 120 ft of fencing available, what dimensions of the corral will maximize the enclosed area?

8. A showroom for a department store is to be rectangular with walls on three sides, 6-foot door openings on the two facing sides and a 10-foot door opening on the remaining wall. The showroom is to have 800 ft^2 of floor space. What dimensions will minimize the length of wall used?

9. Show that the rectangle of maximum area for a given perimeter P is always a square.

10. Show that the rectangle of minimum perimeter for a given area A is always a square.

11. Find the point on the curve $y = x^2$ closest to the point $(0, 1)$.

12. Find the point on the curve $y = x^2$ closest to the point $(3, 4)$.

13. Find the point on the curve $y = \cos x$ closest to the point $(0, 0)$.

14. Find the point on the curve $y = \cos x$ closest to the point $(1, 1)$.

15. In exercises 11 and 12, find the slope of the line through the given point and the closest point on the given curve. Show that, in each case, this line is perpendicular to the tangent line to the curve at the given point.

16. Sketch the graph of some function $y = f(x)$ and mark a point not on the curve. Explain why the result of exercise 15 is true. (Hint: Pick a point for which the joining line is *not* perpendicular and explain why you can get closer.)

17. A box with no top is to be built by taking a 6″-by-10″ sheet of cardboard and cutting x-inch squares out of each corner and folding up the sides. Find the value of x that maximizes the volume of the box.

18. A box with no top is to be built by taking a 12″-by-16″ sheet of cardboard and cutting x-inch squares out of each corner and folding up the sides. Find the value of x that maximizes the volume of the box.

19. A water line runs east-west. A town wants to connect two new housing developments to the line by running lines from a single point on the existing line to the two developments. One development is 3 miles south of the existing line, the other development is 4 miles south of the existing line and 5 miles east of the first development. Find the place on the existing line to make the connection to minimize the total length of new line.

20. A company needs to run an oil pipeline from an oil rig 25 miles out to sea to a storage tank that is 5 miles inland. The shoreline runs east-west and the tank is 8 miles east of the rig. Assume it costs \$50 thousand dollars per mile to construct the pipeline underwater and \$20 thousand dollars per mile to construct the pipeline on land. The pipeline will be built in a straight line from the rig to a selected point on the shoreline, then in a straight line to the storage tank. What point on the shoreline should be selected to minimize the total cost of the pipeline?

21. A city wants to build a new section of highway to link an existing bridge with an existing highway interchange, which lies 8 miles to the east and 10 miles to the south of the bridge. The first 4 miles south of the bridge is marsh land. Assume that the highway costs \$5 million dollars per mile over marsh and \$2 million dollars per mile over dry land. The highway will be built in a straight line from the bridge to the edge of the marsh, then in a straight line to the existing interchange. At what point should the highway emerge from the marsh in order to minimize the total cost of the new highway? How much is saved over building the new highway in a straight line from the bridge to the interchange? (Hint: Use similar triangles to find the point on the boundary corresponding to a straight path and evaluate your cost function at that point.)

22. After construction has begun on the highway in exercise 21, the cost per mile over marsh land is reestimated at \$6 million dollars. Find the point on the marsh/dry land boundary that would minimize the total cost of the highway with the new cost function. If the construction is too far along to change paths, how much extra cost is there in using the path from exercise 21?

23. After construction has begun on the highway in exercise 21, the cost per mile over dry land is reestimated at \$3 million dollars. Find the point on the marsh/dry land boundary that would minimize the total cost of the highway with the new cost function. If the construction is too far along to change paths, how much extra cost is there in using the path from exercise 21?

24. In an endurance contest, contestants 2 miles at sea need to reach a location 2 miles inland and 3 miles east (the shoreline runs east-west). Assume a contestant can swim 4 mph and run 10 mph. To what point on the shoreline should the person swim to minimize the total time? Compare the amount of time spent in the water and the amount of time spent on land.

25. Suppose that light travels from point A to point B as shown in the figure. (Recall that light always follows the path that minimizes time.) Assume that the velocity of light above the boundary line is v_1 and the velocity of light below the boundary is v_2. Show that the total time to get from point A to point B is $T(x) = \dfrac{\sqrt{1+x^2}}{v_1} + \dfrac{\sqrt{1+(2-x)^2}}{v_2}$. Write out the equation $T'(x) = 0$ and argue (by graphing $T(x)$ for different choices of v_1 and v_2) that this critical point represents the minimum time. Replace the square roots using the sines of the angles in the figure and show that light obeys Snell's Law $\dfrac{\sin\theta_1}{\sin\theta_2} = \dfrac{v_1}{v_2}$.

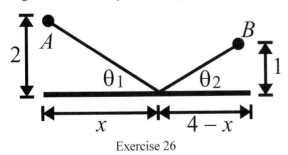

Exercise 25

26. Suppose that light reflects off a mirror to get from point A to point B as indicated in the figure. Assuming a constant velocity of light, we can minimize time by minimizing the distance traveled. Find the point on the mirror that minimizes the distance traveled. Show that the angles in the figure are equal (the angle of incidence equals the angle of reflection).

Exercise 26

27. A soda can is to hold 12 fluid ounces. Suppose that the bottom and top are twice as thick as the sides. Find the dimensions of the can which minimize the amount of material used. (Hint: Instead of minimizing surface area, minimize the cost,

which is proportional to the product of the thickness and the area.)

28. Following example 6.5, we mentioned that real soda cans have a radius of about $1.156''$. Show that this radius minimizes the cost if the top and bottom are 2.23 times as thick as the sides.

29. The human cough is intended to increase the flow of air to the lungs, by dislodging any particles blocking the windpipe and changing the radius of the pipe. Suppose a windpipe under no pressure has radius r_0. The velocity of air through the windpipe at radius r is approximately $V(r) = cr^2(r_0 - r)$ for some constant c. Find the radius that maximizes the velocity of air through the windpipe. Does this mean the windpipe expands or contracts?

30. To supply blood to all parts of the body, the human artery system must branch repeatedly. Suppose an artery of radius r branches off from an artery of radius R ($R > r$) at an angle θ. The energy lost due to friction is approximately $E(\theta) = \dfrac{\csc\theta}{r^4} + \dfrac{1 - \cot\theta}{R^4}$. Find the value of θ that minimizes the energy loss.

31. In an electronic device, individual circuits may serve many purposes. In some cases, the flow of electricity must be controlled by reducing the power instead of amplifying it. In the circuit shown below, a voltage V volts and resistance R ohms are given. We want to determine the size of the remaining resistor (x ohms). The power absorbed by the resistance x is $p(x) = \dfrac{V^2 x}{(R+x)^2}$. Find the value of x that maximizes the power absorbed.

32. In an AC circuit with voltage $V(t) = v\sin 2\pi f t$, a voltmeter actually shows the average (root-mean-square) voltage of $v/\sqrt{2}$. If the frequency is $f = 60$ (Hz) and the meter registers 115 volts, find the maximum voltage reached. (Hint: this is "obvious" if you determine v and think about the graph of $V(t)$.)

33. A Norman window has the outline of a semicircle on top of a rectangle, as shown below. Suppose there is $8 + \pi$ feet of wood trim available. Discuss why a window designer might want to maximize the area of the window. Find the dimensions of the rectangle (and, hence, the semicircle) that will maximize the area of the window.

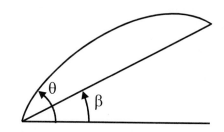

34. Suppose a wire 2 feet long is to be cut into two pieces, each of which will be formed into a square. Find the size of each piece to maximize the total area of the two squares.

35. An advertisement consists of a rectangular printed region plus 1-inch margins on the sides and 2-inch margins at top and bottom. If the area of the printed region is to be 92 in.^{2}, what dimensions should the advertisement be to minimize the total area?

36. An advertisement consists of a rectangular printed region plus 1-inch margins on the sides and 1.5-inch margins at top and bottom. If the total area of the advertisement is to be 120 in^2, what dimensions should the advertisement be to maximize the area of the printed region?

37. A farmer relishes growing cucumbers. A government subsidy is available to growers of more than 2 acres of cucumbers. The farmer has 10 acres available to plant. The net income for the farmer planting A acres is $2A^3 - 33A^2 + 108A - 310$ dollars. Determine the best strategy for the farmer.

38. The owners of the Big Belly Deli run a special on bologna sandwiches, with a limit of six per customer. The amount of heartburn obtained from n sandwiches is $h(n) = n^3 + 12n - 11$ gasu's (gastronomical units). How many sandwiches gives the wurst case of heartburn?

39. In sports where balls are thrown or hit, the ball often finishes at a different height than it starts. Examples include a downhill golf shot and a basketball shot. In the diagram, a ball is released at an angle θ and finishes at an angle β above the horizontal (for downhill trajectories, β would be negative). Neglecting air resistance and spin, the horizontal range is given by

$$R = \frac{2v^2 \cos^2 \theta}{g}(\tan \theta - \tan \beta)$$

if the initial velocity is v and g is the gravitational constant. In the following cases, find θ to maximize R (treat v and g as constants): (a) $\beta = 10°$ (b) $\beta = 0°$ (c) $\beta = -10°$. Verify that $\theta = 45° + \beta°/2$ maximizes the range.

40. For your favorite sport in which it is important to throw or hit a ball a long way, explain in one sentence the result of exercise 39 in the language of your sport.

41. A ball is thrown from $s = b$ to $s = a$ (where $a < b$) with initial speed v_0. Assuming that air resistance is proportional to speed, the time it takes the ball to reach $s = a$ is

$$T = -\frac{1}{c} \ln \left(1 - c \frac{b - a}{v_0} \right)$$

where c is a constant of proportionality. A baseball player is 300 feet from home plate and throws a ball directly toward home plate with an initial speed of 125 ft/s. Suppose that $c = 0.1$. How long does it take the ball to reach home plate? Another player standing x feet from home plate has the option of catching the ball and then, after a delay of 0.1 second, relaying the ball toward home plate with an initial speed of 125 ft/s. Find x to minimize the total time for the ball to reach home plate. Is the straight throw or the relay faster? What, if anything, changes if the delay is 0.2 second instead of 0.1 second?

42. For the situation in exercise 41, for what length delay is it equally fast to have a relay and not have a relay? Do you think that you could catch and throw a ball in such a short time? Why do you think it is considered important to have a relay option in baseball?

43. Repeat exercises 41 and 42 if the second player throws the ball with initial speed 100 ft/s.

44. For a delay of 0.1 second in exercise 41, find the value of the initial speed of the second player's throw for which it is equally fast to have a relay and not have a relay.

45. (exploratory) In exercise 62 in section 3.2, you did a preliminary investigation of Kepler's wine cask problem. You showed that a height-to-diameter ratio (x/y) of $\sqrt{2}$ for a cylindrical barrel will maximize the volume (see Figure a). However, real wine casks are bowed out (like beer kegs). Kepler continued his investigation of wine cask construction by approximating a cask with the straight-sided barrel in Figure b. It can be shown (we told you Kepler was good!) that the volume of this barrel is $V = \frac{2}{3}\pi[y^2 + (w-y)^2 + y(w-y)]\sqrt{z^2 - w^2}$. Treating w and z as constants, show that $V'(y) = 0$ if $y = w/2$. Recall that such a critical point can correspond to a maximum or minimum of $V(y)$, but it also could correspond to something else (e.g., inflection point). To discover which one we have here, redraw Figure b to scale (show the correct relationship between $2y$ and w). In physical terms (think about increasing and decreasing y), argue that this critical point is neither a maximum nor minimum. Interestingly enough, such a non-extreme critical point would have a definite advantage to the Austrian vintners. Recall that their goal was to convert the measurement z into an estimate of the volume. The vintners would hope that small imperfections in the dimensions of the cask would have little effect on the volume. Explain why $V'(y) = 0$ means that small variations in y would convert to small errors in the volume V.

Figure b

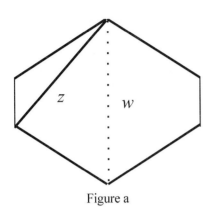

Figure a

46. (exploratory) The following problem is fictitious, but involves the kind of ambiguity that can make technical jobs challenging. The Band Candy Company decides to liquidate one of its candies. The company has 600,000 bags in inventory that it wants to sell. The candy had cost 35 cents per bag to manufacture and originally sold for 90 cents per bag. A marketing study indicates that if the candy is priced at p cents per bag, approximately $Q(p) = -p^2 + 40p + 250$ thousand bags will be sold. Your task as consultant is to recommend the best selling price for the candy. As such, you should do the following: (a) find p to maximize $Q(p)$; (b) find p to maximize $10pQ(p)$, which is the actual revenue brought in by selling the candy. Then form your opinion, based on your evaluation of the relative importance of getting rid of as much candy as possible and making the most money possible.

3.7   Rates of Change in Applications (Optional)

In this section, we round out our exposition of the derivative by presenting a collection of applications selected from a variety of fields. It has often been said that mathematics is the language of nature. Today, the concepts of calculus are being applied in virtually every field of human endeavor. The applications in this section represent but a small sampling of some elementary uses of the derivative. These are not all of the uses of the derivative nor are they necessarily the most important uses. Our intent is simply to present some interesting applications in a variety of settings.

Recall that the derivative of a function gives the instantaneous rate of change of that function. So, when you see the word *rate,* you should be thinking *derivative.* You can hardly pick up a newspaper without finding reference to some rates (e.g., inflation rate, interest rate, unemployment rate, etc.). These can all be thought of as derivatives! There are also many quantities with which you are familiar, but that you might not recognize as rates of change. Our first example is of this type.

Suppose that $Q(t)$ represents the electrical charge in a wire at time t. Then, the derivative $Q'(t)$ gives the **current** flowing through the wire. To see this, consider the cross section of a wire as shown in Figure 3.88.

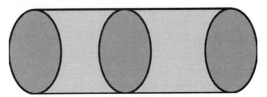

FIGURE 3.88: An electrical wire.

Between times t_1 and t_2, the net charge passing through such a cross section is $Q(t_2) - Q(t_1)$. The **average current** (charge per unit time) over this time interval is then defined as

$$\frac{Q(t_2) - Q(t_1)}{t_2 - t_1}.$$

The **instantaneous current** $I(t)$ at any time t can then be found by computing the limit

$$I(t) = \lim_{t_1 \to t} \frac{Q(t) - Q(t_1)}{t - t_1}. \tag{7.1}$$

Notice that (7.1) is simply the alternative definition of derivative discussed in section 2.2. Thus, we have that $I(t) = Q'(t)$.

Example 7.1   Modeling Electrical Current in a Wire

The electrical circuit shown in Figure 3.89 includes a 5-ohm resistor, a 12-henry inductor, a 10-farad capacitor and a battery supplying 8 volts of AC current modeled by the oscillating function $8 \sin 2t$, where t is measured in seconds. Find the current in the circuit at any time t.

Solution   It can be shown (using the elementary laws of electricity) that the charge in this circuit is given by

$$Q(t) = 12 \sin(4t - \pi/3) + 4 \sin 2t \text{ coulombs.}$$

The current is then

$$Q'(t) = 48 \cos(4t - \pi/3) + 8 \cos 2t \text{ amps (coulombs per second).}$$

■

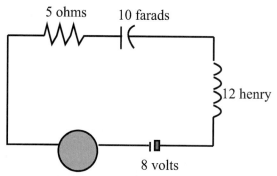

FIGURE 3.89: A simple electrical circuit.

The next example we offer comes from chemistry. It is very important for chemists to have a handle on the rate at which a given reaction proceeds. Reaction rates give chemists information about the nature of the chemical bonds being formed and broken, as well as information about the type and quantity of product to expect. A simple situation is depicted in the schematic

$$A + B \longrightarrow C$$

which indicates that chemicals A and B (the *reactants*) combine to form chemical C (the *product*). Let $[C](t)$ denote the concentration (in moles per liter) of the product. The average reaction rate between times t_1 and t_2 is

$$\frac{[C](t_2) - [C](t_1)}{t_2 - t_1}.$$

The instantaneous reaction rate at any given time t_1 is then given by

$$\lim_{t \to t_1} \frac{[C](t) - [C](t_1)}{t - t_1} = \frac{d[C]}{dt}(t_1).$$

Depending on the details of the reaction, it is often possible to write down an equation relating the reaction rate $\frac{d[C]}{dt}$ to the concentrations of the reactants, $[A]$ and $[B]$.

Example 7.2   Modeling the Rate of a Chemical Reaction

In an **autocatalytic** chemical reaction, the reactant and the product are the same. The reaction continues until some saturation level is reached. From experimental evidence, chemists know that the reaction rate is jointly proportional to the amount of the product present and the difference between the saturation level and the amount of the product. If the initial concentration of the chemical is 0 and the saturation level is 1 (corresponding to 100%), then the concentration $x(t)$ of the chemical satisfies the equation

$$x'(t) = r\, x(t)\, [1 - x(t)].$$

Find the concentration of chemical for which the reaction rate is a maximum.

Solution   To clarify the problem, we write the reaction rate as

$$f(x) = r\, x\, (1 - x).$$

Our aim is then to find $x \geq 0$ that maximizes $f(x)$. We have

$$f'(x) = r\,(1)\,(1 - x) + r\, x\,(-1)$$
$$= r\,(1 - 2x)$$

and so, the only critical number is $x = \frac{1}{2}$. Notice that the graph of $y = f(x)$ is a parabola opening downward and hence, the critical number must correspond to the absolute maximum. Although the mathematical problem here was easy to solve, the result gives an experimenter some precise information. At the time the reaction rate reaches a maximum, the concentration of chemical equals exactly half of the saturation level.

■

Calculus and elementary physics are quite closely connected historically. It should come as no surprise, then, that physics provides us with such a large number of important applications of the calculus. We have already explored the concepts of velocity and acceleration. Another important application in physics where the derivative plays a role involves density. There are many different kinds of densities that we could consider. For example, we could study population density (number of people per unit area) or color density (depth of color per unit area) used in the study of radiographs. However, the most common type of density discussed is **mass density** (mass per unit volume). You probably already have some idea of what we mean by mass density, but how would you define it? If the object of interest is made of some homogeneous material (i.e., the mass of any portion of the object of a given volume is the same), then the mass density is simply

$$mass\ density = \frac{mass}{volume}$$

and this quantity is constant throughout the object. However, if the mass of a given volume varies in different parts of the object, then this formula only represents the *average density* of the object. In the next example we find a means of computing the mass density at a specific point in a nonhomogeneous object.

Suppose that the function $f(x)$ gives us the mass (in kilograms) of the first x meters of a thin rod (see Figure 3.90).

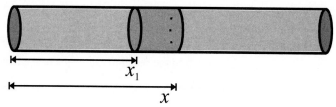

FIGURE 3.90: A thin rod.

The total mass between marks x and x_1 ($x > x_1$) is given by $f(x) - f(x_1)$ kg. The **average linear density** (i.e., mass per unit length) between x and x_1 is then defined as $\dfrac{f(x) - f(x_1)}{x - x_1}$. Finally, the **linear density at** x is defined as

$$\rho(x) = \lim_{x_1 \to x} \frac{f(x) - f(x_1)}{x - x_1} = f'(x). \tag{7.2}$$

Example 7.3   Density of a Thin Rod

Suppose that the mass in a thin rod is given by $f(x) = \sqrt{2x}$. Compute the linear density at $x = 2$ and at $x = 8$.

Solution   From (7.2), we have

$$\rho(x) = f'(x) = \frac{1}{2\sqrt{2x}}\,(2) = \frac{1}{\sqrt{2x}}.$$

Thus, $\rho(2) = 1/\sqrt{4} = 1/2$ and $\rho(8) = 1/\sqrt{16} = 1/4$. Notice that this says that the rod is *inhomogeneous* (i.e., the mass density in the rod is not constant). Specifically, we have that the rod is less dense at $x = 8$ than at $x = 2$.

∎

Our next example comes from medicine (cardiology, to be precise). The rate of change in this application is not actually a derivative. However, heart rate is one of the most familiar and important rates of change in our lives. The language and concepts of calculus enable us to understand some of the exciting research currently being conducted.

Cardiologists have long used graphs to help them identify heart problems. You are probably familiar with the term *electrocardiogram* (ECG or more commonly, EKG), a graph depicting some of the electrical activity of the heart. A small section of a normal EKG is shown in Figure 3.91.

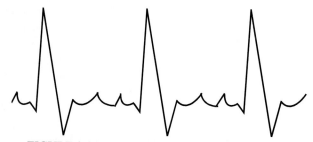

FIGURE 3.91: A normal electrocardiogram (EKG).

A well-trained cardiologist can determine an amazing amount of information from an EKG strip. Each section of the EKG (containing one peak) corresponds to a primary phase of heart activity, and the particular shape of the curve can indicate various problems with the heart. You should notice that the plot appears to be periodic (that is, the same shape is repeated over and over). It is a simple matter to measure the length of the period from peak to peak (each peak corresponds to a contraction of the ventricles of the heart, and the distance between peaks gives the heart rate). Specifically, if successive peaks occur at times t_1 and t_2 seconds, the **heart rate** equals

$$\frac{1 \text{ beat}}{t_2 - t_1 \text{ seconds}} = \frac{60}{t_2 - t_1} \text{ beats per minute.}$$

Note that heart rate is **not** a derivative. To get a derivative, we would need to have a function representing heart beats as a function of time and would then take the limit as $t_2 \to t_1$. From an EKG, all we have is a number of observations at specific times and therefore cannot compute a limit. Having defined heart rate, what can we learn about the human body?

Example 7.4   **Heart Rates and Electrocardiograms**

The following graphs depict heart **rate** over time. Which one would you say corresponds to a healthy individual at rest?

Constant rate.

Smooth change.

Erratic rate.

Solution   Before answering, we should clarify a couple of points. First, since heart rate is only measured at discrete time intervals (the intervals separating the peaks of the EKG), a graph of heart rate over time is like a computer graph: a finite set of points. If the points are connected and there are many points, the plot may appear to be smooth like the idealized plots we show. Second, be sure to read the graphs carefully! Each graph shows a person resting with a heart rate of about 60 beats per minute (bpm). The difference is in the variation from this average. The first person has a constant heart rate of exactly 60 bpm. The graph depicts a steady one beat per second, and is not the "flat-line" of a person whose heart has stopped beating. The second person shows a smooth rise and fall in heart rate, and the third person has a somewhat erratic heart rate.

Which heart rate do you think is healthy? Research by cardiologists in the 1980s and 1990s indicate that healthy hearts generally have a very erratic heart rate, even at rest (You're right if you picked plot 3!). Highly periodic heart rates like those in plot 2 have been observed in patients who were resting comfortably prior to experiencing a heart attack. The constant heart rate in plot 1 is typical of a person experiencing a heart attack. We should point out that the variations in the healthy third plot are too small to discern without sophisticated equipment. These observations have led to the design of a new type of heart monitor that can be worn by at-risk patients. Current research is investigating the possibility of designing "smart" heart pacemakers that can identify and avoid an impending crisis.

The following example comes from economics. Much as with heart rate, the rate of change discussed here is not precisely a derivative, but the derivative has proved to be a useful tool in economic modeling. In economics, the term **marginal** is used to indicate a rate. Thus, **marginal cost** is the derivative of the cost function, **marginal profit** is the derivative of the profit function, and so on.

Example 7.5   **Analyzing the Marginal Cost of Producing a Commercial Product**
Suppose that

$$C(x) = 0.02x^2 + 2x + 4000$$

is the total cost (in dollars) for a company to produce x units of a certain product. Compute the marginal cost at $x = 100$ and compare this to the actual cost of producing the 100th unit.

Solution   You might think that it is an unfair assumption to start with a function that purports to represent cost. After all, cost is determined by accountants after a product is produced. That's true, but in order to project what your cost would be for quantities you haven't actually produced, it is helpful to develop a mathematical model of cost. In practice, this means that you make some observations of the cost of producing a number of different quantities and then try to fit that data to the graph of a known function, which you can then analyze using the tools of calculus. (This is one way in which the calculus is brought to bear on real-world problems.) The marginal cost function is

$$C'(x) = 0.04x + 2$$

and so, the marginal cost at $x = 100$ is $C'(100) = 4 + 2 = 6$ dollars per unit. On the other hand, the actual cost of producing item number 100 would be $C(100) - C(99)$. (Why?) We have

$$C(100) - C(99) = 200 + 200 + 4000 - (196.02 + 198 + 4000)$$
$$= 4400 - 4394.02 = 5.98 \text{ dollars.}$$

Note that this is very close to the marginal cost of $6 and is arguably harder to compute.

■

Our final example comes from psychology. You have probably heard references to the "learning curve" for a piece of computer software or other technical product. The phrase comes from research attempting to quantify and understand the dynamics of learning something new.

Example 7.6   **Analyzing a Learning Curve**
Suppose that the percentage of problems a person works correctly on a test is approximately

$$f(t) = \frac{80}{1 + 3e^{-0.4t}}$$

after t hours of training. Graph the learning curve $y = f(t)$ and compute and interpret $f'(2)$ and $f'(10)$.

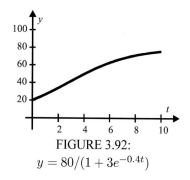

FIGURE 3.92:
$y = 80/(1 + 3e^{-0.4t})$

Solution   The graph looks like the one shown in Figure 3.92.
Note that with no training the person gets 20% correct and after some steady improvement seems to have trouble getting beyond the 80% mark. To compute the derivative, note that we do not need to use the quotient rule, since the numerator is simply a constant. We rewrite the function as

$$f(t) = 80(1 + 3e^{-0.4t})^{-1}$$

and obtain

$$f'(t) = -80(1 + 3e^{-0.4t})^{-2}(-1.2e^{-0.4t})$$
$$= 96e^{-0.4t}(1 + 3e^{-0.4t})^{-2}.$$

After 2 hours, we then have

$$f'(2) = 96e^{-0.8}(1 + 3e^{-0.8})^{-2} \approx 7.8243 \text{ percent per hour.}$$

That is, after 2 hours, a person can expect to add about 7.8 points to the test score by training for an extra hour. At the 10-hour mark,

$$f'(10) = 96e^{-4}(1 + 3e^{-4})^{-2} \approx 1.5799 \text{ percent per hour.}$$

Thus, the next hour of training only increases the test score by about 1.6 points.

∎

We have now discussed examples of six rates of change drawn from engineering and the sciences. Add these to the examples we have seen in previous sections (velocity, population growth, etc.) and we have an impressive list of applications of the derivative. Even so, we have barely begun to scratch the surface. In any field where it is possible to quantify and analyze the properties of a function, calculus and the derivative are powerful tools. This list includes at least some portion of every major field of study. The continued study of calculus will give you the ability to read (and understand) technical studies in a wide variety of fields and to see (as we have in this section) the underlying unity that mathematics brings to human endeavors.

Exercises 3.7

1. (writing) A variable that is defined only at isolated points is called a **discrete** variable. For example, the size of a calculus class is discrete, since it must be an integer. A variable that can assume an interval of values is called **continuous**. In the six applications discussed in the text, in which cases is time a discrete variable and in which cases is it continuous? Which functions are discrete and which are continuous? Among the six examples given in the text, in which cases were the rates of change truly derivatives? Explain this answer in terms of discrete and continuous variables.

2. (writing) An important model of population growth is the so-called **logistics equation** $x'(t) = x(t)[1-x(t)]$. Here, $x(t)$ represents not the actual population size but the proportion of sustainable capacity: for instance, $x(t) = 0.5$ means that the population is half of the total number of organisms that the environment can support and $x(t) = 1.1$ means that there are 10% more organisms than the available resources can support. Note that the differential equation here is the same as was used to describe an autocatalytic chemical reaction. The equation has two competing contributions to the rate of change $x'(t)$. The term $x(t)$ by itself would mean that the larger $x(t)$ is, the faster the population (or concentration of chemical) grows. This is balanced by the term $1 - x(t)$, which indicates that the closer $x(t)$ gets to 1, the slower the population growth is. With these two terms together, the model has the property that for small $x(t)$, slightly larger $x(t)$ means greater growth, but as $x(t)$ approaches 1, the growth tails off. Explain in terms of population growth and the concentration of chemical why the model is reasonable.

3. (writing) The federal budget deficit and national debt are constantly in the news, but the terms are often confused with each other. To take an example, suppose a company finishes a fiscal year owing $5,000. That is their **debt**. Suppose that in the following year the company has revenues of $106,000 and expenses of $109,000. The company's **deficit** for the year is $3,000 and the company's debt has increased to $8,000. Briefly explain why deficit is the derivative of debt.

4. (writing) Many people find the healthy heart example (example 7.4) surprising. To make it more believable, explain how your level of activity affects your heart rate. Also, explain how breathing rate and emotional status can affect heart rate. Given these and many other factors, discuss whether you should expect your heart rate to be exactly constant.

5. Suppose that the charge in an electrical circuit is $Q(t) = e^{-2t}(\cos 3t - 2\sin 3t)$ coulombs. Find the current.

6. Suppose that the charge in an electrical circuit is $Q(t) = e^t(3\cos 2t + \sin 2t)$ coulombs. Find the current.

7. Suppose that the charge at a particular location in an electrical circuit is $Q(t) = e^{-3t}\cos 2t + 4\sin 3t$ coulombs. What happens to this function as $t \to \infty$? Explain why the term $e^{-3t}\cos 2t$ is called a **transient** term and $4\sin 3t$ is known as the **steady-state** or **asymptotic** value of the charge function. Find the transient and steady-state values of the current function.

8. As in exercise 7, find the steady-state and transient values of the current function if the charge function is given by $Q(t) = e^{-2t}(\cos t - 2\sin t) + te^{-3t} + 2\cos 4t$.

9. If the concentration of a chemical changes according to the equation $x'(t) = 2x(t)[4 - x(t)]$, find the concentration $x(t)$ for which the reaction rate is a maximum.

10. If the concentration of a chemical changes according to the equation $x'(t) = 0.5x(t)[5 - x(t)]$, find the concentration $x(t)$ for which the reaction rate is a maximum.

11. Show that in exercise 9, the maximum concentration is 4. Find the maximum concentration in exercise 10.

12. Find the equation for an autocatalytic reaction in which the maximum concentration is $x(t) = 16$ and the reaction rate equals 12 when $x(t) = 8$.

13. Mathematicians often study equations of the form $x'(t) = r\,x(t)[1 - x(t)]$ instead of the more complicated $x'(t) = c\,x(t)[K - x(t)]$, justifying the simplification with the statement that the second equation "reduces to" the first equation. Starting with $y'(t) = c\,y(t)[K - y(t)]$, substitute $y(t) = Kx(t)$ and show that the equation reduces to the form $x'(t) = r\,x(t)[1 - x(t)]$. How does the constant r relate to the constants c and K?

14. Suppose a chemical reaction follows the equation $x'(t) = c\,x(t)[K - x(t)]$. Suppose that at time $t = 4$ the concentration is $x(4) = 2$ and the reaction rate is $x'(4) = 3$. At time $t = 6$, suppose that the concentration is $x(6) = 4$ and the reaction rate is $x'(6) = 4$. Find the values of c and K for this chemical reaction.

15. In a general **second-order chemical reaction**, chemicals A and B (the **reactants**) combine to form chemical C (the **product**). If the initial concentrations of the reactants A and B are a and b, respectively, then the concentration $x(t)$ of the product satisfies the equation $x'(t) = [a - x(t)][b - x(t)]$. What is the rate of change of the product when $x(t) = a$? At this value, is the concentration of product increasing, decreasing or staying the same? Assuming that $a < b$ and there is no product present when the reaction starts, explain why the maximum concentration of product is $x(t) = a$.

16. For the second-order reaction defined in exercise 15, find the (mathematical) value of $x(t)$ that minimizes the reaction rate. Show that the reaction rate for this value of $x(t)$ is negative. Explain why the concentration $x(t)$ would never get this large, so that this mathematical solution is not physically relevant. Explain why $x(t)$ must be between 0 and a, and find the maximum and minimum reaction rates on this closed interval.

17. It can be shown that a solution of the equation $x'(t) = [a - x(t)][b - x(t)]$ is given by $x(t) = \dfrac{a[1 - e^{-(b-a)t}]}{1 - (a/b)e^{-(b-a)t}}$. Find $x(0)$, the initial concentration of chemical, and $\lim_{t \to \infty} x(t)$, the limiting concentration of chemical (assume $a < b$). Graph $x(t)$ on the interval $[0, \infty)$ and describe in words how the concentration of chemical changes over time.

18. For the solution in exercise 17, find and graph $x'(t)$. Compute $\lim_{t \to \infty} x'(t)$ and describe in words how the reaction rate changes over time.

In exercises 19–22, the mass of the first x meters of a thin rod is given by the function $m(x)$ on the indicated interval. Find the linear density function for the rod. Based on what you find, briefly describe the composition of the rod.

19. $m(x) = 4x - \sin x$ dynes for $0 \le x \le 6$

20. $m(x) = (x - 1)^3 + 6x$ dynes for $0 \le x \le 2$

21. $m(x) = 4x$ dynes for $0 \le x \le 2$

22. $m(x) = 4x^2$ dynes for $0 \le x \le 2$

In exercises 23–26, you are given times (in seconds) at which peaks in a heart patient's EKG occurred. Analyze the heart rate and diagnose the patient as resting comfortably or in danger of a heart attack.

23. 0.0, 0.99, 2.12, 3.19, 4.12, 5.17, 6.08, 6.95, 8.16, 9.26, 10.24, 11.28, 12.22, 13.14, 14.21, 15.18

24. 0.0, 0.99, 2.02, 3.01, 4.02, 5.03, 6.01, 6.99, 8.01, 9.02, 10.02, 11.02, 12.02, 13.01, 14.01, 15.02

25. 0.0, 0.98, 1.96, 2.90, 3.80, 4.74, 5.72, 6.70, 7.70, 8.72, 9.74, 10.80, 11.90, 13.06, 14.08, 15.10

26. 0.0, 0.98, 1.96, 3.02, 3.98, 5.04, 6.22, 7.17, 8.15, 9.17, 10.13, 11.08, 12.11, 13.06, 14.08, 15.10

27. If the cost of manufacturing x items is $C(x) = x^3 + 20x^2 + 90x + 15$, find the marginal cost function and compare the marginal cost at $x = 50$ and the actual cost of manufacturing the 50th item.

28. If the cost of manufacturing x items is $C(x) = x^4 + 14x^2 + 60x + 35$, find the marginal cost function and compare the marginal cost at $x = 50$ and the actual cost of manufacturing the 50th item.

29. If the cost of manufacturing x items is $C(x) = x^3 + 21x^2 + 110x + 20$, find the marginal cost function and compare the marginal cost at $x = 100$ and the actual cost of manufacturing the 100th item.

30. If the cost of manufacturing x items is $C(x) = x^3 + 11x^2 + 40x + 10$, find the marginal cost function and compare the marginal cost at $x = 100$ and the actual cost of manufacturing the 100th item.

31. Suppose the cost of manufacturing x items is $C(x) = x^3 - 30x^2 + 300x + 100$ dollars. Find the inflection point and discuss the significance of this value in terms of the cost of manufacturing.

32. A baseball team owner has determined that if tickets are priced at $10, the average attendance at a game will be $27,000$ and if tickets are priced at $8, the average attendance will be $33,000$. Using a linear model, we would then estimate that tickets priced $9 would produce an average attendance of $30,000$. Discuss whether or not you think the use of a linear model here is reasonable. Then, using the linear model, determine the price at which the revenue is maximized.

33. The learning curve $f(t) = \dfrac{80}{1 + 3e^{-0.4t}}$ was discussed in example 7.6. Show that $\lim\limits_{t \to \infty} f(t) = 80$ and show that $f(t)$ is an increasing function for $t > 0$. Discuss whether you think both results are reasonable properties for a learning curve.

34. Suppose the maximum score that a person will ever score on a test is 90. Further suppose that the person scores 60 after 2 hours of studying. Find a learning curve of the form $f(t) = \dfrac{a}{1 + 3e^{-bt}}$ for this person on this test.

35. Suppose that a person scores $f(t) = \dfrac{180}{2 + 4e^{-0.2t}}$ after t hours of studying. What is the person's score after 3 hours of studying? Find $f'(3)$ and estimate how many additional points the person would earn by studying a fourth hour. Find $f'(10)$ and estimate how many additional points the person would earn by studying an eleventh hour.

36. Suppose that a person scores $f(t) = \dfrac{10}{1 + 4e^{-0.5t}}$ after t hours of studying. What is the person's score after 4 hours of studying? Find $f'(4)$ and estimate how many additional points the person would earn by studying a fifth hour. Find $f'(10)$ and estimate how many additional points the person would earn by studying an eleventh hour.

37. The function $f(t) = \dfrac{a}{1 + 3e^{-bt}}$ has also been used to model the spread of a rumor. Suppose that $a = 70$ and $b = 0.2$. Compute $f(2)$, the percentage of the population that has heard the rumor after 2 hours. Compute $f'(2)$ and describe what it represents. Compute $\lim\limits_{t \to \infty} f(t)$ and describe what it represents.

38. After an injection, the concentration of drug in a muscle varies according to a function of time, $f(t)$. Suppose that t is measured in hours and $f(t) = e^{-0.02t} - e^{-0.42t}$. Determine the time when the maximum concentration of drug occurs.

39. Suppose that the size of the pupil of an animal is given by $f(x)$ (mm), where x is the intensity of the light on the pupil. If $f(x) = \dfrac{160x^{-0.4} + 90}{4x^{-0.4} + 15}$, show that $f(x)$ is a decreasing function. Interpret this result in terms of the response of the pupil to light.

40. Suppose that the body temperature 1 hour after receiving x mg of a drug is given by $T(x) = 102 - \frac{1}{6}x^2(1 - x/9)$ for $0 \le x \le 6$. The absolute value of the derivative, $|T'(x)|$, is defined as the **sensitivity** of the body to the drug dosage. Find the dosage which maximizes sensitivity.

41. Let $C(x)$ be the cost of manufacturing x items and define $\bar{C}(x) = \dfrac{C(x)}{x}$ as the average cost function. Suppose that $C(x) = 0.01x^2 + 40x + 3600$. Show that $C'(100) < \bar{C}(100)$ and show that increasing the production (x) by 1 will decrease the average cost.

42. For the cost function in exercise 41, show that $C'(1000) > \bar{C}(1000)$ and show that increasing the production (x) by 1 will increase the average cost.

43. For the cost function in exercise 41, prove that average cost is minimized at the x-value where $C'(x) = \bar{C}(x)$.

44. If the cost function is linear, $C(x) = a + bx$ with a and b positive, show that there is no minimum average cost and that $C'(x) \ne \bar{C}(x)$ for all x.

45. Let $R(x)$ be the revenue and $C(x)$ be the cost from manufacturing x items. **Profit** is defined as $P(x) = R(x) - C(x)$. Show that at the value of x which maximizes profit, marginal revenue equals marginal cost.

46. Find the maximum profit if $R(x) = 10x - 0.001x^2$ and $C(x) = 2x + 5000$.

47. In the titration of a weak acid and strong base, the pH is given by $c + \ln \frac{f}{1-f}$ where c is a constant (closely related to the **acid dissociation constant**) and f is the fraction ($0 < f < 1$) of converted acid (see Harris' *Quantitative Chemical Analysis* for more details). Find the value of f at which the rate of change of pH is the smallest. What happens as f approaches 1?

48. In exercise 47, you found the significance of one inflection point of a titration curve. A second inflection point, called the **equivalence point**, corresponds to $f = 1$. In the generalized titration curve below, identify on the graph both inflection points and briefly explain why chemists prefer to measure the equivalence point and not the inflection point of exercise 47. (Note: the horizontal axis of a titration curve indicates the amount of base added to the mixture. This is directly proportional to the amount of converted acid in the region where $0 < f < 1$.)

49. (exploratory) Epidemiology is the study of the spread of infectious diseases. A simple model for the spread of fatal diseases such as AIDS divides people into the categories of susceptible (but not exposed), exposed (but not infected) and infected. The proportions of people in each category at time t are denoted $S(t)$, $E(t)$ and $I(t)$, respectively. The general equations for this model are

$$S'(t) = mI(t) - bS(t)I(t),$$
$$E'(t) = bS(t)I(t) - aE(t),$$
$$I'(t) = aE(t) - mI(t),$$

where m, b and a are positive constants. Notice that each equation gives the rate of change of one of the categories.

Each rate of change has both a positive and negative term. Explain why the positive term represents people who are entering the category and the negative term represents people who are leaving the category. In the first equation, the term $mI(t)$ represents people who have died from the disease (the constant m is the reciprocal of the life expectancy of someone with the disease). This term is slightly artificial: the assumption is that the population is constant, so that when one person dies, a baby is born who is not exposed or infected. The dynamics of the disease is that susceptible (healthy) people get infected by contact with infected people. Explain why the number of contacts between susceptible people and infected people is proportional to $S(t)$ and $I(t)$. The term $bS(t)I(t)$, then, represents susceptible people who have been exposed by contact with infected people. Explain why this same term shows up as a positive in the second equation. Explain the rest of the remaining two equations in this fashion. (Hint: The constant a represents the reciprocal of the average latency period. In the case of AIDS, this would be how long it takes an HIV-positive person to actually develop AIDS.)

50. (exploratory) Without knowing how to solve differential equations (we hope you will go far enough in your study of mathematics to learn to do so!), we can nonetheless deduce some important properties of the solutions of differential equations. For example, consider the equation for an autocatalytic reaction $x'(t) = x(t)[1 - x(t)]$. Suppose $x(0)$ lies between 0 and 1. Show that $x'(0)$ is positive by determining the possible values of $x(0)[1 - x(0)]$. Explain why this indicates that the value of $x(t)$ will increase from $x(0)$, and will continue to increase as long as $0 < x(t) < 1$. Explain why if $x(0) < 1$ and $x(t) > 1$ for some $t > 0$, then it must be true that $x(t) = 1$ for some $t > 0$. However, if $x(t) = 1$, then $x'(t) = 0$ and the solution $x(t)$ stays constant (equal to 1). Therefore, we can conjecture that $\lim_{t \to \infty} x(t) = 1$. Similarly, show that if $x(0) > 1$, then $x(t)$ decreases and we could again conjecture that $\lim_{t \to \infty} x(t) = 1$. Changing equations, suppose that $x'(t) = -0.05x(t) + 2$. This is a model of an experiment in which a radioactive substance is decaying at the rate of 5% but the substance is being replenished at the constant rate of 2. Find the value of $x(t)$ for which $x'(t) = 0$. Pick various starting values of $x(0)$ less than and greater than the constant solution and determine if the solution $x(t)$ will increase or decrease. Based on these conclusions, conjecture the value of $\lim_{t \to \infty} x(t)$, the limiting amount of radioactive substance in the experiment.

Chapter Review Exercises

In exercises 1 and 2, find the linear approximation to $f(x)$ at x_0.

1. $f(x) = e^{3x}, x_0 = 0$ **2.** $f(x) = \sqrt{x^2 + 3}, x_0 = 1$

In exercises 3 and 4, use a linear approximation to estimate the quantity.

3. $\sqrt[3]{7.96}$ **4.** $\sin 3$

In exercises 5 and 6, use Newton's method to find an approximate root.

5. $x^3 + 5x - 1 = 0$ **6.** $x^3 = e^{-x}$

7. Explain why Newton's method fails on $x^3 - 3x + 2 = 0$ with $x_0 = 1$.

8. Show that the approximation $\dfrac{1}{1-x} \approx 1 + x$ is valid for "small" x.

In exercises 9–18, do the following by hand. (a) Find all critical numbers, (b) identify all intervals of increase and decrease, (c) determine whether each critical number represents a local maximum, local minimum or neither, (d) determine all intervals of concavity and (e) find all inflection points.

9. $f(x) = x^3 + 3x^2 - 9x$ **10.** $f(x) = x^4 - 4x + 1$

11. $f(x) = x^4 - 4x^3 + 2$ **12.** $f(x) = x^3 - 3x^2 - 24x$

13. $f(x) = xe^{-4x}$ **14.** $f(x) = x^2 \ln x$

15. $f(x) = x\sqrt{x^2 - 4}$ **16.** $f(x) = (x^2 - 1)^{2/3}$

17. $f(x) = \dfrac{x}{x^2 + 4}$ **18.** $f(x) = \dfrac{x}{\sqrt{x^2 + 2}}$

In exercises 19–22, find the absolute extrema of the function on the interval.

19. $f(x) = x^3 + 3x^2 - 9x$ on $[0, 4]$

20. $f(x) = x^3 + 3x^2 - 9x$ on $[-4, 0]$

21. $f(x) = x^{4/5}$ on $[-2, 3]$

22. $f(x) = x^2 e^{-x}$ on $[-1, 4]$

In exercises 23–26, find the x-coordinates of all local extrema.

23. $f(x) = x^3 + 4x^2 + 2x$ **24.** $f(x) = x^4 - 3x^2 + 2x$

25. $f(x) = x^5 - 2x^2 + x$ **26.** $f(x) = x^5 + 4x^2 - 4x$

27. Sketch a graph of a function with $f(-1) = 2$, $f(1) = -2$, $f'(x) < 0$ for $-2 < x < 2$, $f'(x) > 0$ for $x < -2$ and $x > 2$.

28. Sketch a graph of a function with $f'(x) > 0$ for $x \neq 0$, $f'(0)$ undefined, $f''(x) > 0$ for $x < 0$ and $f''(x) < 0$ for $x > 0$.

In exercises 29–38, sketch a graph of the function and completely discuss the graph.

29. $f(x) = x^4 + 4x^3$

30. $f(x) = x^4 + 4x^2$

31. $f(x) = x^4 + 4x$

32. $f(x) = x^4 - 4x^2$

33. $f(x) = \dfrac{x}{x^2 + 1}$

34. $f(x) = \dfrac{x}{x^2 - 1}$

35. $f(x) = \dfrac{x^2}{x^2 + 1}$

36. $f(x) = \dfrac{x^2}{x^2 - 1}$

37. $f(x) = \dfrac{x^3}{x^2 - 1}$

38. $f(x) = \dfrac{4}{x^2 - 1}$

39. Find the point on the graph of $y = 2x^2$ that is closest to $(2, 1)$.

40. Show that the line through the two points of exercise 39 is perpendicular to the tangent line to $y = 2x^2$ at $(2, 1)$.

41. A city is building a highway from point A to point B, which is 4 miles east and 6 miles south of point A. The first 4 miles south of point A is swamp land, where the cost of building the highway is \$6 million per mile. On dry land, the cost is \$2 million per mile. Find the point on the boundary of swamp land and dry land to which the highway should be built to minimize the total cost.

42. Repeat exercise 41 with a cost of \$16 million per mile on swamp land. Explain why the optimal point is this exercise is west of the optimal point found in exercise 41.

43. A soda can in the shape of a cylinder is to hold 16 ounces. Find the dimensions of the can that minimizes the surface area of the can.

44. Suppose that $C(x) = 0.02x^2 + 4x + 1200$ is the cost of manufacturing x items. Show that $C'(x) > 0$ and explain in business terms why this has to be true. Show that $C''(x) > 0$ and explain why this indicates that the manufacturing process is not very efficient.

45. The charge in an electrical circuit at time t is given by $Q(t) = e^{-3t} \sin 2t$ coulombs. Find the current.

46. If the concentration $x(t)$ of a chemical in a reaction changes according to the equation $x'(t) = 0.3x(t)[4 - x(t)]$, find the concentration at which the reaction rate is a maximum.

47. Suppose that the mass of the first x meters of a thin rod is given by $m(x) = 20 + x^2$ for $0 \leq x \leq 4$. Find the density of the rod and briefly describe the composition of the rod.

48. A person scores $f(t) = \dfrac{90}{1 + 4e^{-0.4t}}$ points on a test after t hours of studying. What does the person score without studying at all? Compute $f'(0)$ and estimate how many points one hour of studying will add to the score.

49. The cost of manufacturing x items is given by $C(x) = 0.02x^2 + 20x + 1800$. Find the marginal cost function. Compare the marginal cost at $x = 20$ to the actual cost of producing the 20th item.

50. For the cost function in exercise 49, find the value of x which minimizes the average cost $\bar{C}(x) = C(x)/x$.

51. (exploratory) Let $n(t)$ be the number of photons in a laser field. One model of the laser action is $n'(t) = an(t) - b[n(t)]^2$, where a and b are positive constants. If $n(0) = a/b$, what is $n'(0)$? Based on this calculation, would $n(t)$ increase, decrease or neither? If $n(0) > a/b$, is $n'(0)$ positive or negative? Based on this calculation, would $n(t)$ increase, decrease or neither? If $n(0) < a/b$, is $n'(0)$ positive or negative? Based on this calculation, would $n(t)$ increase, decrease or neither? Putting this information together, conjecture the limit of $n(t)$ as $t \to \infty$. Repeat this analysis under the assumption that $a < 0$. (Hint: Because of its definition, $n(t)$ is positive, so ignore any negative values of $n(t)$.)

52. (exploratory) One way of numerically approximating a derivative is by computing the slope of a secant line. For example, $f'(a) \approx \dfrac{f(b) - f(a)}{b - a}$, if b is close enough to a. In this exercise, we will develop an analogous approximation to the second derivative. Graphically, we can think of the secant line as an approximation of the tangent line. Similarly, we can match the second derivative behavior (concavity) with a parabola. Instead of finding the secant line through two points on the curve, we find the parabola through three points on the curve. The second derivative of this approximating parabola will serve as an approximation of the second derivative of the curve. The first step is messy, so we recommend using a CAS if one is available. Find a function of the form $g(x) = ax^2 + bx + c$ such that $g(x_1) = y_1$, $g(x_2) = y_2$ and $g(x_3) = y_3$. Since $g''(x) = 2a$, you actually only need to find the constant a. The so-called **second difference** approximation to $f''(x)$ is the value of $g''(x) = 2a$ using the three points $x_1 = x - \Delta x$ [$y_1 = f(x_1)$], $x_2 = x$ [$y_2 = f(x_2)$] and $x_3 = x + \Delta x$ [$y_3 = f(x_3)$]. Find the second difference for $f(x) = \sqrt{x + 4}$ at $x = 0$ with $\Delta x = 0.5$, $\Delta x = 0.1$ and $\Delta x = 0.01$. Compare to the exact value of the second derivative, $f''(0)$.

Chapter 4
Integration

HIV virus.

The outbreak of AIDS in the 1980s has provided a dramatic public health challenge. While infectious diseases caused by bacteria had been largely controlled by the use of antibiotics, viruses (even those associated with the common cold) were largely beyond the reach of modern medicine. So, the appearance of a deadly virus like HIV, which causes AIDS (shown to the left), has rated as a full-blown public health emergency. A significant percentage of the funds for AIDS research have been spent on basic medical studies, in an effort to understand the mechanisms by which HIV infects humans and to devise effective treatments.

An equally important, but very different research role was played by epidemiologists, who worked frantically to learn how HIV was transmitted. Epidemiologists study the spread of diseases, often using sophisticated mathematical models to predict the severity of an outbreak. Many of the basic mathematical quantities used by epidemiologists will be developed in this chapter.

When analyzing the spread of a disease, one vital statistic is the number of new cases reported in a given time period. A careful analysis of this data indicates the **growth rate**, which may be used to give accurate predictions of the number of cases at some point in the future. Mathematically speaking, the idea is to use the rate of change (i.e., the derivative) of an unknown function to determine properties of the function itself. One aspect of this problem, called **antidifferentiation**, is introduced in section 4.1.

The study of AIDS is complicated by the fact that once a person is infected with HIV (i.e., becomes "HIV positive"), the amount of time that passes before the symptoms of AIDS emerge is highly variable. So, a realistic model for the spread of AIDS must start with the number of people infected and take into account this variable "incubation period." A simple example should help to illustrate the difficulty.

Suppose that for the sake of argument, the infection rates for HIV in a 4-month period are as shown in the following table.

Month	1	2	3	4
Number of new infections	20	40	40	30

That is, in the first month, 20 new infections per month are reported, the rate rises to 40 new infections per month for the next 2 months and then drops to 30 new infections per month. Further, suppose that the incubation period is as indicated in the following table.

Number of months since infection	1	2	3	4
Percentage developing AIDS	20	50	20	10

(Note: These figures are **not** realistic.) Let's predict the number of new AIDS cases in the fourth and fifth months.

Note that the people who develop AIDS in the fourth month fall into three different categories: those infected in the first month who have a 3-month incubation period, those infected in the second month who have a 2-month incubation period, and those infected in the third month with a 1-month incubation period. The numbers of new AIDS cases in each category are then 20% of 20, 50% of 40 and 20% of 40. The total number of new AIDS cases in the fourth month is then

$$0.2(20) + 0.5(40) + 0.2(40) = 4 + 20 + 8 = 32.$$

A similar calculation gives us the following prediction for the number of new AIDS cases in fifth month:

$$0.1(20) + 0.2(40) + 0.5(40) + 0.2(30) = 36.$$

Notice that even though the HIV infection rate drops in the fourth month, the number of new cases of AIDS increases in the fifth month.

Sums like those used above are the focus of section 4.2. Our initial aim is to use such sums to tackle the geometric problem of computing area under a curve in section 4.3 and then generalize that technique to define the **definite integral** in section 4.4. At that point, we are ready to present one of the greatest results of all of mathematics: the Fundamental Theorem of Calculus, which reveals the connection between derivatives and integrals. This connection is central to the ability of epidemiologists, engineers and scientists to use calculus to make fundamental discoveries.

4.1   Antiderivatives

Calculus provides us with a powerful set of tools for understanding the world around us. When engineers originally designed the space shuttle for NASA, it was equipped with aircraft engines to power its flight through the atmosphere after reentry. In order to cut costs, the aircraft engines were scrapped and the space shuttle became a huge glider. One consequence of this decision is that once the shuttle has begun its reentry, there is only **one** choice of landing site. What's more, shuttle astronauts must land their spacecraft on their first approach. Without aircraft engines, there are **no** second chances. Obviously, the shuttle's flight path must be chosen (and controlled) with the utmost precision. NASA engineers use the calculus to provide precise answers to these immensely important problems. While we are not in a position to deal with the vast complexities of a space shuttle flight, we can consider an idealized model. (We should note that for a realistic model of a system as complex as a space shuttle, we must consider much more than the simple concepts discussed here. For a very interesting presentation of this problem, see the article by Long and Weiss in the February, 1999, issue of *The American Mathematical Monthly*.)

Of course, no one looks at the flight of a space shuttle and sees an equation. In order to solve real-world problems, we begin with a physical principle(s) and use this to produce a *mathematical model* of the physical system. We then solve the mathematical problem and interpret the solution in terms of the physical problem.

Space Shuttle Endeavor.

Let's start by considering a space shuttle hurtling toward the earth. We can identify the forces affecting the shuttle's motion: gravity pulling downward and air resistance pushing in the direction opposite the motion. The physical principle describing such a problem is Newton's second law of motion:

$$Force = mass \times acceleration \ \text{ or } \ F = ma.$$

That is, the sum of all forces acting on an object is directly proportional to acceleration. Recall that acceleration is the derivative of velocity and the second derivative of the position function.

Our task might sound simple enough: identify the forces, use Newton's second law to write down an equation and then solve it. Unfortunately, each of these steps can be quite involved. For instance, how should we write down the force of air resistance? From riding in a car at different speeds, you know that the faster you move, the greater the air resistance is. Aeronautical engineers have determined experimentally that the force due to air resistance (the drag), F_d is proportional to the square of the speed of the object and acts in the direction opposite the motion. That is, when the velocity $v > 0$, the force is negative and when $v < 0$, the force is positive. In either case, we have

$$F_d = -kv^2,$$

for some constant $k > 0$.

Next, gravity is well understood and easy to describe. The force due to gravity is simply the weight of the object, which would be essentially constant in this case (at least at low altitudes). Weight is related to mass by $W = -mg$, where the gravitational constant g is approximately 32 ft/s^2. (The minus sign indicates that the force of gravity acts downward.) From Newton's second law of motion, we now have

$$F = ma = -mg - kv^2.$$

Recognizing that $a = v'(t)$, we have the **differential equation**

$$mv'(t) = -mg - kv^2(t). \tag{1.1}$$

At this time, we don't have the mathematical machinery to solve this equation for $v(t)$. To get our feet wet, we simplify the problem (simplification is always one of the first steps in mathematical modeling). Assuming that gravity is the only force acting on the shuttle (i.e., neglecting air resistance), we can take $k = 0$ in (1.1), which gives us

$$mv'(t) = -mg \text{ or } v'(t) = -g.$$

Now, let $y(t)$ be the position function (the elevation of the shuttle in feet t seconds after reentry). Since velocity is the derivative of position, we have

$$y''(t) = -32.$$

You are quite experienced at computing derivatives, but the task before you now is analogous to the TV quiz show *Jeopardy*: you know that the second derivative of $y(t)$ is -32, but what is the original function? More generally, we need to find a way to *undo* differentiation. That is, given a function, $f(x)$, we'd like to find another function $F(x)$ such that $F'(x) = f(x)$. We call such a function F an **antiderivative** of f.

Example 1.1   Finding Several Antiderivatives of a Given Function
Find an antiderivative of $f(x) = x^2$.

Solution   Notice that $F(x) = \frac{1}{3}x^3$ is an antiderivative of $f(x)$, since

$$\frac{d}{dx}\left(\frac{1}{3}x^3\right) = x^2.$$

Further,

$$\frac{d}{dx}\left(\frac{1}{3}x^3 + 5\right) = x^2,$$

so that $G(x) = \frac{1}{3}x^3 + 5$ is also an antiderivative of f. In fact, for any constant c, we have

$$\frac{d}{dx}\left(\frac{1}{3}x^3 + c\right) = x^2.$$

Thus, $H(x) = \frac{1}{3}x^3 + c$ is also an antiderivative of $f(x)$, for any choice of the constant c.

■

In general, if F is any antiderivative of f and c is any constant, then

$$\frac{d}{dx}[F(x) + c] = F'(x) + 0 = f(x).$$

Thus, $F(x) + c$ is also an antiderivative of $f(x)$, for any constant c. At this point, you might ask if there are any other antiderivatives of $f(x)$ besides $F(x) + c$. The answer, as provided in the following theorem, is no.

Theorem 1.1 Suppose that F and G are both antiderivatives of f on an interval $[a, b]$. Then,

$$G(x) = F(x) + c,$$

for some constant c.

Proof. Since F and G are both antiderivatives for f, we have that $G'(x) = F'(x)$. It now follows from Corollary 9.1 in section 2.9, that $G(x) = F(x) + c$, for some constant, c, as desired.

■

NOTE: This says that given **any** antiderivative F of f, **every** possible antiderivative of f can be written in the form $F(x) + c$, for some constant, c. We call this most general antiderivative the **indefinite integral** of $f(x)$ (with respect to x), written

$$\int f(x)\,dx = F(x) + c.$$

The process of computing an integral is called **integration**. Here, $f(x)$ is called the **integrand** and c is an arbitrary constant (the **constant of integration**). The term dx identifies x as the **variable of integration**.

Example 1.2   **An Indefinite Integral**
 Evaluate $\int 3x^2 dx$.

Solution   You should recognize $3x^2$ as the derivative of x^3 and so,

$$\int 3x^2 dx = x^3 + c.$$

■

Example 1.3   **Determining the Coefficient in an Indefinite Integral**
 Evaluate $\int t^5 dt$.

Solution   We know that $\dfrac{d}{dt}t^6 = 6t^5$ and so, $\dfrac{d}{dt}\left(\frac{1}{6}t^6\right) = t^5$. Therefore,

$$\int t^5 dt = \frac{1}{6}t^6 + c.$$

■

We should point out that every differentiation rule gives rise to a corresponding integration rule. For instance, recall that for every rational power, r, $\dfrac{d}{dx}x^r = r\,x^{r-1}$. Likewise, we have

$$\frac{d}{dx}x^{r+1} = (r+1)\,x^r.$$

This proves the following result.

Theorem 1.2   (Power Rule) For any rational power, $r \neq -1$,

$$\int x^r dx = \frac{x^{r+1}}{r+1} + c. \qquad (1.2)$$

Remark 1.1   Theorem 1.2 says that to integrate a power of x (other than x^{-1}), you simply raise the power by 1 and divide by the new power. Notice that this rule obviously doesn't work for $r = -1$, since this would produce a division by 0. Later in this section, we develop a rule to cover this case.

Example 1.4   Using the Power Rule
Evaluate $\int x^{17}dx$.

Solution   From the power rule, we have

$$\int x^{17}dx = \frac{x^{17+1}}{17+1} + c = \frac{x^{18}}{18} + c.$$

■

Example 1.5   The Power Rule with a Negative Exponent
Evaluate $\int \dfrac{1}{x^3}\,dx$.

Solution   We can use the power rule if we first rewrite the integrand. We have

$$\int \frac{1}{x^3}dx = \int x^{-3}dx = \frac{x^{-3+1}}{-3+1} + c = -\frac{1}{2}x^{-2} + c.$$

■

Example 1.6   The Power Rule with a Fractional Exponent
Evaluate $\int \sqrt{x}dx$.

Solution   As in example 1.5, we first rewrite the integrand and then apply the power rule. We have

$$\int \sqrt{x}dx = \int x^{1/2}dx = \frac{x^{1/2+1}}{\frac{1}{2}+1} + c = \frac{x^{3/2}}{3/2} + c = \frac{2}{3}x^{3/2} + c.$$

Notice that the fraction $\frac{2}{3}$ in the last expression is exactly what it takes to cancel the new exponent $3/2$. (This is what happens if you differentiate.)

■

Notice that since $\dfrac{d}{dx}(\sin x) = \cos x$, we have

$$\int \cos x \, dx = \sin x + c.$$

Again, by reversing any derivative formula, we get a corresponding integration formula. A table containing a number of important formulas follows.

$$
\begin{array}{rcl}
\int \sin x \, dx & = & -\cos x + c \\[1.5em]
\int \cos x \, dx & = & \sin x + c \\[1.5em]
\int \sec^2 x \, dx & = & \tan x + c \\[1.5em]
\int \csc^2 x \, dx & = & -\cot x + c \\[1.5em]
\int \sec x \tan x \, dx & = & \sec x + c \\[1.5em]
\int \csc x \cot x \, dx & = & -\csc x + c \\[1.5em]
\int e^x \, dx & = & e^x + c \\[1.5em]
\int e^{-x} \, dx & = & -e^{-x} + c
\end{array}
$$

The proofs of these are left as straightforward, yet important, exercises. Notice that we do not yet have integration formulas for several familiar functions: $\frac{1}{x}$, $\ln x$, $\tan x$, $\cot x$ and others. At this point, we are simply reversing the most basic derivative rules we know. We will develop more sophisticated techniques later. For now, we need a general rule to allow us to combine our basic integration formulas.

Theorem 1.3   Suppose that $f(x)$ and $g(x)$ have antiderivatives. Then, for any constants, a and b,

$$\int [a\,f(x) + b\,g(x)]\,dx = a \int f(x)\,dx + b \int g(x)\,dx. \qquad (1.3)$$

Proof. We have that $\dfrac{d}{dx}\int f(x)\,dx = f(x)$ and $\dfrac{d}{dx}\int g(x)\,dx = g(x)$. It then follows that

$$\frac{d}{dx}\left[a \int f(x)\,dx + b \int g(x)\,dx\right] = a\,f(x) + b\,g(x),$$

as desired.

■

Note that this theorem says that we can easily compute integrals of sums, differences and constant multiples of functions. However, it turns out that we have no product or quotient rule for integrals.

Example 1.7   An Indefinite Integral of a Sum

Evaluate $\int \left(3 \cos x + 4x^8 \right) \, dx$.

Solution

From (1.3).

$$\int \left(3 \cos x + 4x^8 \right) \, dx = 3 \int \cos x \, dx + 4 \int x^8 \, dx$$

$$= 3 \sin x + 4 \frac{1}{9} x^9 + c$$

$$= 3 \sin x + \frac{4}{9} x^9 + c.$$

■

Example 1.8   An Indefinite Integral of a Difference

Evaluate $\int \left(3e^x - 2 \sec^2 x \right) \, dx$.

Solution

$$\int \left(3e^x - 2 \sec^2 x \right) \, dx = 3 \int e^x \, dx - 2 \int \sec^2 x \, dx = 3e^x - 2 \tan x + c.$$

■

We can easily evaluate the integral of any function on our list if x is replaced by ax for any constant a. To see how to do this, first recall that

$$\frac{d}{dx} \sin 2x = 2 \cos 2x,$$

where the factor of 2 comes from the chain rule. As with any other differentiation rule, this leads to a corresponding integration rule. You should observe that

$$\int \cos 2x \, dx = \frac{1}{2} \sin 2x + c,$$

since

$$\frac{d}{dx} \left(\frac{1}{2} \sin 2x + c \right) = \cos 2x.$$

The general rule follows.

Theorem 1.4   If $\int f(x) \, dx = F(x) + c$, then for any constant $a \neq 0$,

$$\int f(ax) \, dx = \frac{1}{a} F(ax) + c.$$

Proof. Since $\int f(x) \, dx = F(x) + c$, it follows that $F'(x) = f(x)$. By the chain rule, we have

$$\frac{d}{dx} \left[\frac{1}{a} F(ax) \right] = \frac{1}{a} \frac{d}{dx} [F(ax)] = \frac{1}{a} a F'(ax) = F'(ax) = f(ax).$$

Thus, $\frac{1}{a} F(ax)$ is an antiderivative of $f(ax)$, as desired.

■

With practice, working problems like those in the following example will become automatic.

Example 1.9 **Indefinite Integrals of Functions of the Form** $f(ax)$

Evaluate (a) $\int \sin 3x\, dx$, (b) $\int 5e^{4x}\, dx$ and (c) $\int 8\sec^2 5x\, dx$.

Solution

(a) Since $\int \sin x\, dx = -\cos x + c$,

$$\int \sin 3x\, dx = -\frac{1}{3}\cos 3x + c.$$

(b) Since $\int e^x\, dx = e^x + c$,

$$\int 5e^{4x}\, dx = 5\int e^{4x}\, dx = \frac{5}{4}e^{4x} + c.$$

(c) Since $\int \sec^2 x\, dx = \tan x + c$,

$$\int 8\sec^2 5x\, dx = 8\int \sec^2 5x\, dx = \frac{8}{5}\tan 5x + c.$$

■

From the power rule, we know how to evaluate $\int x^r\, dx$ for any rational exponent **except** $r = -1$. We can deal with this exceptional case if we make the following observation. Note that for $x \neq 0$, we can define $\ln |x|$. For $x > 0$, we have $\ln |x| = \ln x$ and hence,

$$\frac{d}{dx}\ln |x| = \frac{d}{dx}\ln x = \frac{1}{x}.$$

Similarly, for $x < 0$, $\ln |x| = \ln(-x)$, and hence,

$$\frac{d}{dx}\ln |x| = \frac{d}{dx}\ln(-x)$$
$$= \frac{1}{-x}\frac{d}{dx}(-x)$$
$$= \frac{1}{-x}(-1) = \frac{1}{x}.$$

Of course, we got the same derivative in either case. This proves the following theorem.

Theorem 1.5 For $x \neq 0$, $\dfrac{d}{dx}\ln |x| = \dfrac{1}{x}$.

Example 1.10 **The Derivative of the Log of an Absolute Value**

For any x for which $\tan x \neq 0$, evaluate $\dfrac{d}{dx}\ln |\tan x|$.

Solution From Theorem 1.5 and the chain rule, we have

$$\frac{d}{dx}\ln |\tan x| = \frac{1}{\tan x}\frac{d}{dx}\tan x$$
$$= \frac{1}{\tan x}\sec^2 x.$$

■

Of course, with the new differentiation rule in theorem 1.5, we get a new integration rule:

$$\int \frac{1}{x}\, dx = \ln|x| + c, \text{ for } x \neq 0.$$

More generally, notice that if $f(x) \neq 0$ and f is differentiable, we have by the chain rule that

$$\frac{d}{dx} \ln|f(x)| = \frac{1}{f(x)} f'(x) = \frac{f'(x)}{f(x)}.$$

This gives us the integration rule

$$\int \frac{f'(x)}{f(x)}\, dx = \ln|f(x)| + c. \tag{1.4}$$

Example 1.11   **The Indefinite Integral of a Fraction of the Form $\dfrac{f'(x)}{f(x)}$**

Evaluate $\displaystyle\int \frac{\sec^2 x}{\tan x}\, dx$.

Solution   Notice that the numerator ($\sec^2 x$) is the derivative of the denominator ($\tan x$). From (1.4) we get

$$\int \frac{\sec^2 x}{\tan x}\, dx = \ln|\tan x| + c.$$

∎

Before concluding the section by returning to the space shuttle, we should emphasize that we have developed only a small number of integration rules. Further, unlike with derivatives, we will never have rules to cover all of the functions we are familiar with. Thus, it is important to recognize when you *cannot* find an antiderivative.

Example 1.12   **Identifying Integrals That We Cannot Yet Evaluate**

Which of the following integrals can you evaluate given the rules developed in this section? (a) $\displaystyle\int \frac{1}{\sqrt[3]{x}}\, dx$ (b) $\displaystyle\int \sec x\, dx$ (c) $\displaystyle\int \frac{2x}{x^2+1}\, dx$
(d) $\displaystyle\int \frac{x^3+1}{x}\, dx$ (e) $\displaystyle\int (x+1)(x-1)\, dx$ and (f) $\displaystyle\int x \sin 2x\, dx$.

Solution   First, notice that each of these integrands are at least slightly different from the antiderivatives we have so far identified. (This is usually the case.) However, we can rewrite problems (a), (c), (d) and (e) into forms where we can recognize an antiderivative, as follows.

$$\int \frac{1}{\sqrt[3]{x}}\, dx = \int x^{-1/3}\, dx = \frac{x^{-1/3+1}}{-\frac{1}{3}+1} + c = \frac{3}{2} x^{2/3} + c.$$

In part (c), notice that $\frac{d}{dx}(x^2+1) = 2x$ (the numerator). From (1.4), we then have

$$\int \frac{2x}{x^2+1}\, dx = \ln|x^2+1| + c.$$

In part (d), if we divide out the integrand, we find

$$\int \frac{x^3+1}{x}\, dx = \int (x^2 + x^{-1})\, dx = \frac{1}{3} x^3 + \ln|x| + c.$$

Finally, in part (e), if we multiply out the integrand, we get

$$\int (x+1)(x-1)\,dx = \int \left(x^2 - 1\right)\,dx = \frac{1}{3}x^3 - x + c.$$

There are no product or quotient rules for integrals, and so we cannot yet evaluate integrals (b) or (f).

∎

Now that we know how to find antiderivatives for a number of functions, we return to the problem of the plummeting space shuttle that opened the section.

Example 1.13   Finding the Position of a Falling Object Given Its Acceleration

If a space shuttle's downward acceleration is given by $y''(t) = -32$ ft/s^2, find the position function $y(t)$. Assume that the shuttle's initial velocity is $y'(0) = -100$ ft/s, and its initial position is $y(0) = 100,000$ feet.

Solution   We have to undo two derivatives, so we compute two antiderivatives. First, we have

$$y'(t) = \int y''(t)\,dt = \int (-32)\,dt = -32t + c.$$

Recall that $y'(t)$ is the velocity of the shuttle, given in units of feet per second. We can evaluate the constant c using the given initial velocity. Since

$$v(t) = y'(t) = -32t + c$$

and $v(0) = -100$, we must have

$$-100 = v(0) = -32(0) + c = c,$$

so that $c = -100$. Thus, the velocity is $y'(t) = -32t - 100$. Next, we have

$$y(t) = \int y'(t)\,dt = \int (-32t - 100)\,dt = -32\left(\frac{1}{2}\right)t^2 - 100t + c = -16t^2 - 100t + c.$$

Recall that $y(t)$ gives the height of the shuttle, measured in feet. Using the initial position, we have

$$100,000 = y(0) = -16(0) - 100(0) + c = c.$$

Thus, $c = 100,000$ and

$$y(t) = -16t^2 - 100t + 100,000.$$

Keep in mind that this models the space shuttle's altitude assuming that the only force acting on the shuttle is gravity (i.e., there is no air resistance). In this case, we have a disastrous (and hopefully unrealistic) outcome. (Compute the velocity of the shuttle on landing, to see why.)

∎

Exercises 4.1

1. (writing) In the text, we emphasized that the indefinite integral represents *all* antiderivatives of a given function. To understand why this is important, consider a situation where you know the net force, $F(t)$, acting on an object. By Newton's second law, $F = ma$. For the position function $s(t)$, this translates to $a(t) = s''(t) = F(t)/m$. To compute $s(t)$, you need to compute two antiderivatives of the force function $F(t)/m$. But, suppose you were unable to find *all* antiderivatives. How would you know whether you had computed the antiderivative that corresponded to the position function? For this setting, explain why it is reasonable to expect that there is only one antiderivative (give or take a constant that is determined by the initial conditions).

2. (writing) In the text, we presented a one-dimensional model of a space shuttle flight. We ignored some of the forces on the shuttle so that the resulting mathematical equation would be one that we could solve. You may wonder what the benefit of doing this is. Weigh the relative worth of having an unsolvable but realistic model versus having a solution of a model that is only partially accurate. Keep in mind that when you toss trash into a wastebasket you do not take the curvature of the earth into account.

3. (writing) Verify that $\int xe^{x^2} dx = \frac{1}{2}e^{x^2} + c$ and $\int xe^x dx = xe^x - e^x + c$ by computing derivatives of the proposed antiderivatives. Which derivative rules did you use? Why does this make it unlikely that we will find a general product (antiderivative) rule for $\int f(x)g(x)\, dx$?

4. (writing) We stated in the text that we do not yet have a formula for the antiderivative of several elementary functions, including $\ln x$, $\sec x$ and $\csc x$. Given a function $f(x)$, explain what determines whether or not we have a simple formula for $\int f(x)\, dx$. For example, why is there a simple formula for $\int \sec x \tan x\, dx$ but not $\int \sec x\, dx$?

In exercises 5–40 find the general antiderivative.

5. $\int 3x^4\, dx$

6. $\int 5x^2\, dx$

7. $\int (3x^4 - 3x)\, dx$

8. $\int (x^3 - 2)\, dx$

9. $\int 3\sqrt{x}\, dx$

10. $\int (4x - 2\sqrt{x})\, dx$

11. $\int \left(3 - \dfrac{1}{x^4}\right) dx$

12. $\int \left(2x^{-2} + \dfrac{1}{\sqrt{x}}\right) dx$

13. $\int \dfrac{x - 3}{x^3}\, dx$

14. $\int \dfrac{x + 2}{\sqrt{x}}\, dx$

15. $\int (2\sin x + \cos x)\, dx$

16. $\int (3\cos x - \sin x)\, dx$

17. $\int 2\sec x \tan x\, dx$

18. $\int 4\csc^2 x\, dx$

19. $\int 5\sec^2 x\, dx$

20. $\int 4\csc x \cot x\, dx$

21. $\int (3e^x - 2)\, dx$

22. $\int (4x - 2e^x)\, dx$

23. $\int (3\cos x - 1/x)\, dx$

24. $\int (2x^{-1} + \sin x)\, dx$

25. $\int \dfrac{4x}{x^2 + 4}\, dx$

26. $\int \dfrac{2x^3}{x^4 + 1}\, dx$

27. $\int \left(5x - \dfrac{3}{e^x}\right) dx$

28. $\int (2\cos x - \sqrt{e^x})\, dx$

29. $\int 5\sin 2x\, dx$

30. $\int 4\cos 5x\, dx$

31. $\int (e^{3x} - x)\, dx$

32. $\int (4\sin 3x - 1)\, dx$

33. $\int 3\sec 2x \tan 2x\, dx$

34. $\int 3\sec^2 3x\, dx$

35. $\int \dfrac{e^x}{e^x + 3}\, dx$

36. $\int \dfrac{\cos x}{\sin x}\, dx$

37. $\int \dfrac{e^x + 3}{e^x}\, dx$

38. $\int \dfrac{(e^x)^2 - 2}{e^{2x}}\, dx$

39. $\int (x + 1)(x - 4)\, dx$

40. $\int (x^2 + 2)(x - 3)\, dx$

In exercises 41–52, 6 of the 12 antiderivatives can be determined using basic algebra and the derivative formulas we have presented. Find the antiderivatives of those 6 and label the others "N/A."

41. $\int \sqrt{x^3 + 4}\, dx$

42. $\int (\sqrt{x^3} + 4)\, dx$

43. $\int \dfrac{3x^2 - 4}{x^2}\, dx$

44. $\int \dfrac{x^2}{3x^2 - 4}\, dx$

45. $\int 2\sec x\, dx$

46. $\int \sec^2 x\, dx$

47. $\int 2\sin 4x\, dx$

48. $\int 2\sin^4 x\, dx$

49. $\int e^{x^2}\, dx$

50. $\int (e^x)^2\, dx$

51. $\int \left(\dfrac{1}{x^2} - 1\right) dx$

52. $\int \dfrac{1}{x^2 - 1}\, dx$

In exercises 53–58, find a function $f(x)$ satisfying the given conditions.

53. $f'(x) = 4x^2 - 1$, $f(0) = 2$

54. $f'(x) = 4\cos x$, $f(0) = 3$

55. $f'(x) = 3e^x + x$, $f(0) = 4$

56. $f'(x) = 3\sin 2x$, $f(0) = 1$

57. $f''(x) = 12$, $f'(0) = 2$, $f(0) = 3$

58. $f''(x) = 2x$, $f'(0) = -3$, $f(0) = 2$

In exercises 59–62, find all functions satisfying the given conditions.

59. $f''(x) = 3\sin x + 4x^2$

60. $f''(x) = \sqrt{x} - 2\cos x$

61. $f'''(x) = 4 - 2/x^3$

62. $f'''(x) = \sin 2x - e^x$

63. Determine the position function if the velocity function is $v(t) = 3 - 12t$ and the initial position is $s(0) = 3$.

64. Determine the position function if the velocity function is $v(t) = 3e^{-t} - 2$ and the initial position is $s(0) = 0$.

65. Determine the position function if the acceleration function is $a(t) = 3\sin t + 1$, the initial velocity is $v(0) = 0$ and the initial position is $s(0) = 4$.

66. Determine the position function if the acceleration function is $a(t) = t^2 + 1$, the initial velocity is $v(0) = 4$ and the initial position is $s(0) = 0$.

67. Suppose that a car can accelerate from 30 mph to 50 mph in 4 seconds. Assuming a constant acceleration, find the acceleration (in miles per second squared) of the car and find the distance traveled by the car during the 4 seconds.

68. Suppose that a car can come to rest from 60 mph in 3 seconds. Assuming a constant (negative) acceleration, find the acceleration (in miles per second squared) of the car and find the distance traveled by the car during the 3 seconds (i.e., the stopping distance).

In exercises 69 and 70, sketch the graph of a function $f(x)$ corresponding to the given graph of $y = f'(x)$.

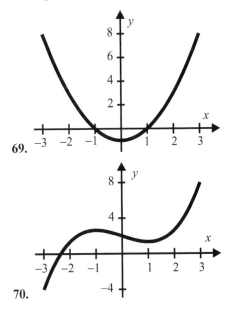

69.

70.

71. For the shuttle in example 1.13, find the time when it reaches the ground and its velocity at that time. Why did we say that this would be a disastrous outcome?

72. (exploratory) Compute the derivatives of $e^{\sin x}$ and e^{x^2}. Given these derivatives, evaluate the antiderivatives $\displaystyle\int \cos x \, e^{\sin x}\, dx$ and $\displaystyle\int 2x\, e^{x^2}\, dx$. Next, evaluate $\displaystyle\int x\, e^{x^2}\, dx$. (Hint: $\displaystyle\int x\, e^{x^2}\, dx = \tfrac{1}{2}\int 2x\, e^{x^2}\, dx$.) Similarly, evaluate $\displaystyle\int x^2\, e^{x^3}\, dx$. In general, evaluate

$$\int f'(x)\, e^{f(x)}\, dx.$$

Next, evaluate $\displaystyle\int e^x \cos(e^x)\, dx$, $\displaystyle\int 2x\cos(x^2)\, dx$ and the more general

$$\int f'(x)\cos(f(x))\, dx.$$

As we have stated, there is not a general rule for the antiderivative of a product, $\displaystyle\int f(x)g(x)\, dx$. Instead, there are many special cases that you evaluate case by case.

73. (exploratory) A **differential equation** is an equation involving an unknown function and one or more of its derivatives, for instance, $y'(x) = 2x + 3$. To solve this differential equation, you simply find the antiderivative $y(x) = \int (2x+3)\, dx = x^2 + 3x + c$. Notice that solutions of a differential equation are functions. In general, differential equations can be challenging to solve. For example, to solve $y'(x) = \sqrt{x^2 + [y(x)]^2}$, we would need to find a function whose derivative equals the square root of x^2 plus the square of the function. It is difficult to find a function whose derivative is written in terms of $[y(x)]^2$ when $y(x)$ is precisely what is unknown. We can nonetheless construct a graphical representation of the solution using what is called a **direction field**. Suppose we want to construct a solution passing through the point $(0, 1)$. With $x = 0$ and $y = 1$, we know that the slope of the solution is $y' = \sqrt{0^2 + 1^2} = 1$. Starting at $(0, 1)$, sketch in a short line segment with slope 1. Such a line segment would connect to the point $(1, 2)$ if we extended it that far (but make yours much shorter). At $x = 1$ and $y = 2$, the slope of the solution is $y' = \sqrt{1^2 + 2^2} = \sqrt{5}$. Sketch in a short line segment with slope $\sqrt{5} \approx 2.2$ starting at the point $(1, 2)$. This line segment points to $(2, 4.2)$. At this point, $y' = \sqrt{2^2 + 4.2^2} \approx 4.6$. Sketch in a short line segment with slope 4.6 at $(2, 4.2)$. Do you see a graphical solution starting to emerge? Is the solution increasing or decreasing? concave up or concave down? If your CAS has a direction field capability, sketch the direction field and try to visualize the solution starting at points $(0, 1)$, $(2, 0)$, $(0, -2)$ and $(-1, -1)$.

4.2   Sums and Sigma Notation

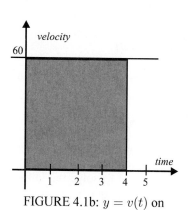

FIGURE 4.1a: $y = v(t)$ on $[0, 2]$.

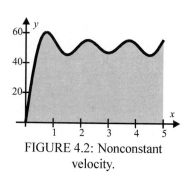

FIGURE 4.1b: $y = v(t)$ on $[0, 4]$.

In section 4.1, we discussed how to calculate backward from a function describing the acceleration of an object to arrive at the function giving the position of the object at any time, t. We would now like to investigate the same process graphically. In this section, we develop an important skill necessary for this new interpretation. First, we pause to indicate the connection to our ongoing problem of determining the position of an object from its acceleration.

Suppose that you are cruising on a highway at 60 mph. In 2 hours, you will have traveled 120 miles; in 4 hours, you will have traveled 240 miles. There's no surprise here, but notice that you can see this graphically by looking at some graphs of the (constant) velocity function $v(t) = 60$. In Figure 4.1a, we have shaded in the area under the graph from $t = 0$ to $t = 2$. Notice that the area of this region equals 120, the distance covered in the time span from $t = 0$ to $t = 2$. In Figure 4.1b, the shaded region from $t = 0$ to $t = 4$ has area equal to the distance of 240 miles.

So, it appears that the distance traveled over a particular time interval equals the area of the region bounded by $y = v(t)$ and the t-axis on the indicated time interval. For the case of constant velocity, this is no surprise, as we have that

$$d = r \times t = velocity \times time.$$

This works well when the velocity is constant, but how could we compute the area under the curve (equal to the distance traveled) for a nonconstant velocity function, such as the one shown in Figure 4.2 for the time interval $[0, 5]$?

Our work in this section provides the first step toward a powerful technique for computing such areas. To indicate the direction we will take, suppose that the velocity curve in Figure 4.2 is replaced by the approximation in Figure 4.3, where the velocity is assumed to be constant over each of five 1-hour time intervals.

The area on the interval from $t = 0$ to $t = 5$ is then approximately the sum of the areas of the five rectangles:

$$A \approx 60 + 45 + 50 + 55 + 50 = 260 \text{ miles.}$$

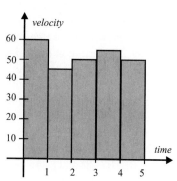

FIGURE 4.2: Nonconstant velocity.

Of course, this is a fairly crude estimate of the area in Figure 4.2, but you should observe that we could get a better estimate by approximating the area using more (and smaller) rectangles. Certainly, we had no problem adding up the areas of five rectangles, but for 5000 rectangles, you will want some means for simplifying and automating the process. Dealing with such sums is the topic of the remainder of this section.

We start the simplification process by introducing some notation. Suppose that you want to sum the squares of the first 20 positive integers. Notice that

$$1 + 4 + 9 + \cdots + 400 = 1^2 + 2^2 + 3^2 + \cdots + 20^2.$$

Of course, the pattern is obvious; each term in the sum has the form i^2, for $i = 1, 2, 3, \ldots,$ 20. To reduce the amount of writing, we use the Greek capital letter sigma, $\sum$, as a symbol for *sum* and write the sum in **summation notation** as

$$\sum_{i=1}^{20} i^2 = 1^2 + 2^2 + 3^2 + \cdots + 20^2,$$

to indicate that we add together terms of the form i^2, starting with $i = 1$ and ending with $i = 20$. The variable i is called the **index of summation**.

In general, for any real numbers $a_1, a_2, \ldots, a_n$, we have

$$\sum_{i=1}^{n} a_i = a_1 + a_2 + \cdots + a_n.$$

FIGURE 4.3: Approximate area.

Example 2.1   Using Summation Notation
Write in summation notation: $\sqrt{1}+\sqrt{2}+\sqrt{3}+\cdots+\sqrt{10}$ and $3^3+4^3+5^3+\cdots+45^3$.

Solution   We have

$$\sqrt{1}+\sqrt{2}+\sqrt{3}+\cdots+\sqrt{10}=\sum_{i=1}^{10}\sqrt{i}$$

and

$$3^3+4^3+5^3+\cdots+45^3=\sum_{i=3}^{45}i^3.$$

∎

Example 2.2   Summation Notation for a Sum Involving Odd Integers
Write in summation notation: the sum of the first 200 odd integers.

Solution   First, notice that $(2i)$ is even for every integer i and hence, both $(2i-1)$ and $(2i+1)$ are odd. So, we have

$$1+3+5+\cdots+399=\sum_{i=1}^{200}(2i-1).$$

Alternatively, we can write this as the equivalent expression $\sum_{i=0}^{199}(2i+1)$.

∎

NOTE: The index of summation is a **dummy variable**, since it is only used as a counter, to keep track of terms and the value of the summation does not depend on the letter used as the index. For this reason, you may use any letter you like as an index. By tradition, we most frequently use i, j, k, m and n, but any index will do. For instance,

$$\sum_{i=1}^{n}a_i=\sum_{j=1}^{n}a_j=\sum_{k=1}^{n}a_k.$$

Example 2.3   Computing Sums Given in Summation Notation
Write out all terms and compute the sums $\sum_{i=1}^{8}(2i+1)$, $\sum_{i=2}^{6}\sin(2\pi i)$ and $\sum_{i=4}^{10}5$.

Solution   We have

$$\sum_{i=1}^{8}(2i+1)=3+5+7+9+11+13+15+17=80$$

and

$$\sum_{i=2}^{6}\sin(2\pi i)=\sin 4\pi+\sin 6\pi+\sin 8\pi+\sin 10\pi+\sin 12\pi=0.$$

(Note that the sum started at $i = 2$.) Finally,

$$\sum_{i=4}^{10} 5 = 5 + 5 + 5 + 5 + 5 + 5 + 5 = 35.$$

■

As example 2.3 suggests, there are sometimes shortcuts for computing sums. For instance, the easy way of evaluating the third sum above is to notice that 5 appears 7 times, and 7 times 5 is 35. We now state this and two other useful formulas.

Theorem 2.1   If n is any positive integer and c is any constant, then

(i) $\displaystyle\sum_{i=1}^{n} c = c\,n$ **(sum of constants)**,

(ii) $\displaystyle\sum_{i=1}^{n} i = \frac{n(n+1)}{2}$ **(sum of the first n positive integers)** and

(iii) $\displaystyle\sum_{i=1}^{n} i^2 = \frac{n(n+1)(2n+1)}{6}$ **(sum of the squares of the first n positive integers)**.

Proof. (i) $\displaystyle\sum_{i=1}^{n} c$ indicates to add the same constant to itself n times and hence, the sum is simply c times n.

(ii) The following clever proof has been credited to 10-year-old Karl Friedrich Gauss (1777–1855), one of the greatest mathematicians of all time. First notice that

$$\sum_{i=1}^{n} i = \underbrace{1 + 2 + 3 + \cdots + (n-2) + (n-1) + n}_{n \text{ terms}}. \qquad (2.1)$$

Since the order in which we add the terms does not matter, we add the terms in (2.1) in reverse order, to get

$$\sum_{i=1}^{n} i = \underbrace{n + (n-1) + (n-2) + \cdots + 3 + 2 + 1}_{\text{same } n \text{ terms (backwards)}}. \qquad (2.2)$$

Adding equations 2.1 and 2.2 term by term, we get

Adding each term in parentheses.

$$2\sum_{i=1}^{n} i = (1+n) + (2+n-1) + (3+n-2) + \cdots + (n-1+2) + (n+1)$$

$$= \underbrace{(n+1) + (n+1) + (n+1) + \cdots + (n+1) + (n+1) + (n+1)}_{n \text{ terms}}$$

$$= n(n+1),$$

since $(n+1)$ appears n times in the sum. Dividing both sides by 2 gives us

$$\sum_{i=1}^{n} i = \frac{n(n+1)}{2},$$

as desired. The proof of (iii) requires a more sophisticated proof using mathematical induction and we defer it to the end of this section.

■

We also have the following general rule for expanding sums. The proof is straightforward and is left as an exercise.

Theorem 2.2   For any constants c and d,

$$\sum_{i=1}^{n}(c\,a_i + d\,b_i) = c\sum_{i=1}^{n}a_i + d\sum_{i=1}^{n}b_i.$$

Using Theorems 2.1 and 2.2, we can now compute several simple sums with ease. Note that we have no more difficulty summing 800 terms than we do summing 8.

Example 2.4   Computing Sums Using Theorems 2.1 and 2.2

Compute $\displaystyle\sum_{i=1}^{8}(2i+1)$ and $\displaystyle\sum_{i=1}^{800}(2i+1)$.

Solution   From Theorems 2.1 and 2.2, we have

$$\sum_{i=1}^{8}(2i+1) = 2\sum_{i=1}^{8}i + \sum_{i=1}^{8}1 = 2\,\frac{8(9)}{2} + (1)\,(8) = 72 + 8 = 80.$$

Similarly,

$$\sum_{i=1}^{800}(2i+1) = 2\sum_{i=1}^{800}i + \sum_{i=1}^{800}1 = 2\,\frac{800(801)}{2} + (1)\,(800)$$
$$= 640,800 + 800 = 641,600.$$

∎

Example 2.5   Computing Sums Using Theorems 2.1 and 2.2

Compute $\displaystyle\sum_{i=1}^{20}i^2$ and $\displaystyle\sum_{i=1}^{20}(i/20)^2$.

Solution   From Theorems 2.1 and 2.2, we have

$$\sum_{i=1}^{20}i^2 = \frac{20(21)(41)}{6} = 2870$$

and

$$\sum_{i=1}^{20}\left(\frac{i}{20}\right)^2 = \frac{1}{20^2}\sum_{i=1}^{20}i^2 = \frac{1}{400}\,2870 = 7.175.$$

∎

We return to the study of general sums in Chapter 8. Recall that our initial motivation for studying sums was to calculate distance from velocity. In the beginning of this section, we approximated distance by summing several values of the velocity function. In section 4.3, we will further develop these sums to allow us to compute areas exactly.

Example 2.6   Computing a Sum of Function Values
Sum the values of $f(x) = x^2 + 3$ evaluated at $x = 0.1$, $x = 0.2$, $\ldots$, $x = 1.0$.

Solution   We first formulate this in summation notation, so that we can use the rules we have developed in this section. The terms to be summed are $a_1 = f(0.1) = 0.1^2 + 3$, $a_2 = f(0.2) = 0.2^2 + 3$ and so on. Note that since each of the x-values is a multiple of 0.1, we can write the x's in the form $0.1i$, for $i = 1, 2, \ldots, 10$. In general, we have

$$a_i = f(0.1\,i) = (0.1\,i)^2 + 3, \text{ for } i = 1, 2, \ldots, 10.$$

We thus want the sum

From Theorem 2.1 (i) and (iii).

$$\sum_{i=1}^{10} a_i = \sum_{i=1}^{10} f(0.1\,i) = \sum_{i=1}^{10} [(0.1\,i)^2 + 3] = 0.1^2 \sum_{i=1}^{10} i^2 + \sum_{i=1}^{10} 3$$

$$= 0.01 \frac{10(11)(21)}{6} + (3)(10) = 3.85 + 30 = 33.85.$$

∎

Example 2.7   A Sum of Function Values at Equally Spaced x's
Sum the values of $f(x) = 3x^2 - 4x + 2$ evaluated at $x = 1.05$, $x = 1.15$, $x = 1.25, \ldots, x = 2.95$.

Solution   You will need to think carefully about the x's. The distance between successive x-values is 0.1 and there are 20 such values. (Be sure to count these for yourself.) Notice that we can write the x's in the form $0.95 + 0.1\,i$, for $i = 1, 2, ..., 20$. (As always, there are other ways to represent the x's. For instance, you could write the x's in the form $1.05 + 0.1i$, for $i = 0, 1, \ldots, 19$.)
We now have

$$\sum_{i=1}^{20} f(0.95 + 0.1\,i) = \sum_{i=1}^{20} \left[3(0.95 + 0.1\,i)^2 - 4(0.95 + 0.1\,i) + 2 \right]$$

$$= \sum_{i=1}^{20} \left(0.03i^2 + 0.17i + 0.9075 \right)$$

From Theorem 2.2.

$$= 0.03 \sum_{i=1}^{20} i^2 + 0.17 \sum_{i=1}^{20} i + \sum_{i=1}^{20} 0.9075$$

From Theorem 2.1 (i), (ii) and (iii).

$$= 0.03 \frac{20(21)(41)}{6} + 0.17 \frac{20(21)}{2} + 0.9075\,(20) = 139.95.$$

∎

Over the next several sections, we will see how sums such as those found in examples 2.6 and 2.7 play a very significant role. We end this section by looking at a powerful mathematical principle.

Principle of Mathematical Induction (Optional)

For any proposition that depends on a positive integer, n, we first show that the result is true for a specific value $n = n_0$. We then **assume** that the result is true for an **unspecified** $n = k \geq n_0$. (This is called the **induction assumption**.) If we can show that it follows that the proposition is true for $n = k + 1$, we have proved that the result is true for any positive integer $n \geq n_0$. [Explain why this must be true. (Hint: If P_1 is true and you know that P_k true implies P_{k+1} is true, then P_1 true implies P_2 true, which in turn implies P_3 true and so on.)]

We can now use mathematical induction to prove the last part of Theorem 2.1, which states that for any positive integer n, $\displaystyle\sum_{i=1}^{n} i^2 = \frac{n(n+1)(2n+1)}{6}$.

Proof of Theorem 2.1 (iii). For $n = 1$, we have

$$1 = \sum_{i=1}^{1} i^2 = \frac{1(2)(3)}{6},$$

as desired. So, the proposition is true for $n = 1$. Next, **assume** that

$$\sum_{i=1}^{k} i^2 = \frac{k(k+1)(2k+1)}{6}, \quad \text{(induction assumption)} \qquad (2.3)$$

for some integer $k \geq 1$.

In this case, we have by the induction assumption that for $n = k + 1$,

Splitting off the last term.	$\displaystyle\sum_{i=1}^{n} i^2 = \sum_{i=1}^{k+1} i^2 = \sum_{i=1}^{k} i^2 + \sum_{i=k+1}^{k+1} i^2$
From (2.3).	$= \dfrac{k(k+1)(2k+1)}{6} + (k+1)^2$
Adding the fractions.	$= \dfrac{k(k+1)(2k+1) + 6(k+1)^2}{6}$
Factor out $(k+1)$.	$= \dfrac{(k+1)[k(2k+1) + 6(k+1)]}{6}$
Combine remaining terms.	$= \dfrac{(k+1)[2k^2 + 7k + 6]}{6}$
Factor the quadratic.	$= \dfrac{(k+1)(k+2)(2k+3)}{6}$
Rewrite the terms.	$= \dfrac{(k+1)[(k+1)+1][2(k+1)+1]}{6}$
Since $n = k + 1$.	$= \dfrac{n(n+1)(2n+1)}{6},$

as desired.   ■

Exercises 4.2

1. (writing) In the text, we mentioned that one of the benefits of using the summation notation is the simplification of calculations. To help understand this, write out in words what is meant by $\displaystyle\sum_{i=1}^{40}(2i^2 - 4i + 11)$.

2. (writing) Following up on exercise 1, calculate the sum $\displaystyle\sum_{i=1}^{40}(2i^2 - 4i + 11)$ and then describe in words how you did so. Be sure to describe any formulas and your use of them in words.

In exercises 3–6, a calculation is described in words. Translate each into summation notation and then compute the sum.

3. The sum of the squares of the first 50 positive integers.

4. The square of the sum of the first 50 positive integers.

5. The sum of the square roots of the first 10 positive integers.

6. The square root of the sum of the first 10 positive integers.

In exercises 7–14, write out all terms and compute the sums.

7. $\displaystyle\sum_{i=1}^{6} 3i^2$ **8.** $\displaystyle\sum_{i=1}^{5} (2i-1)$

9. $\displaystyle\sum_{i=3}^{7} (i^2 + i)$ **10.** $\displaystyle\sum_{i=6}^{10} (4i + 2)$

11. $\displaystyle\sum_{i=0}^{7} (3i - 1)$ **12.** $\displaystyle\sum_{i=-1}^{5} 2i^3$

13. $\displaystyle\sum_{i=2}^{5} (\sqrt{i} + i)$ **14.** $\displaystyle\sum_{i=6}^{6} (i^2 + 2)$

In exercises 15–26, use summation rules to compute the sum.

15. $\displaystyle\sum_{i=1}^{70} (3i - 1)$ **16.** $\displaystyle\sum_{i=1}^{45} (3i - 4)$

17. $\displaystyle\sum_{i=1}^{40} (4 - i^2)$ **18.** $\displaystyle\sum_{i=1}^{50} (8 - i)$

19. $\displaystyle\sum_{i=1}^{100} (i^2 - 3i + 2)$ **20.** $\displaystyle\sum_{i=1}^{140} (i^2 + 2i - 4)$

21. $\displaystyle\sum_{i=1}^{200} (4 - 3i - i^2)$ **22.** $\displaystyle\sum_{i=1}^{250} (i^2 + 8)$

23. $\displaystyle\sum_{i=1}^{n} (i^2 - 3)$ **24.** $\displaystyle\sum_{i=1}^{n} (i^2 + 5)$

25. $\displaystyle\sum_{i=1}^{n} (4i^2 - i)$ **26.** $\displaystyle\sum_{i=1}^{n} (i^2 + 4i)$

In exercises 27–30, compute the sum and the limit of the sum as $n \to \infty$.

27. $\displaystyle\sum_{i=1}^{n} \frac{1}{n}\left[\left(\frac{i}{n}\right)^2 + 2\left(\frac{i}{n}\right)\right]$

28. $\displaystyle\sum_{i=1}^{n} \frac{1}{n}\left[\left(\frac{i}{n}\right)^2 - 5\left(\frac{i}{n}\right)\right]$

29. $\displaystyle\sum_{i=1}^{n} \frac{1}{n}\left[4\left(\frac{2i}{n}\right)^2 - \left(\frac{2i}{n}\right)\right]$

30. $\displaystyle\sum_{i=1}^{n} \frac{1}{n}\left[\left(\frac{2i}{n}\right)^2 + 4\left(\frac{i}{n}\right)\right]$

31. Suppose that a car has velocity 50 mph for 2 hours, velocity 60 mph for 1 hour, velocity 70 mph for 30 minutes and velocity 60 mph for 3 hours. Find the distance traveled.

32. Suppose that a car has velocity 50 mph for 1 hour, velocity 40 mph for 1 hour, velocity 60 mph for 30 minutes and velocity 55 mph for 3 hours. Find the distance traveled.

33. Suppose that a runner has velocity 15 mph for 20 minutes, velocity 18 mph for 30 minutes, velocity 16 mph for 10 minutes and velocity 12 mph for 40 minutes. Find the distance run.

34. Suppose that a runner has velocity 12 mph for 20 minutes, velocity 14 mph for 30 minutes, velocity 18 mph for 10 minutes and velocity 15 mph for 40 minutes. Find the distance run.

In exercises 35–38, use the formula $\displaystyle\sum_{i=1}^{n} i^3 = \frac{n^2(n+1)^2}{4}$ to compute the sums.

35. $\displaystyle\sum_{i=1}^{10} (i^3 - 3i + 1)$ **36.** $\displaystyle\sum_{i=1}^{20} (i^3 + 2i)$

37. $\displaystyle\sum_{i=1}^{100} (i^3 - 2i^2)$ **38.** $\displaystyle\sum_{i=1}^{100} (2i^3 + 2i + 1)$

39. Prove Theorem 2.2.

40. (exploratory) Suppose that the velocity of a car is given by $v(t) = 3\sqrt{t} + 30$ mph at time t hours ($0 \le t \le 4$). We will try to determine the distance traveled in the 4 hours. To start, we can note that the velocity at $t = 0$ is $v(0) = 3\sqrt{0} + 30 = 30$ mph and the velocity at time $t = 1$ is $v(1) = 3\sqrt{1} + 30 = 33$ mph. The average of these velocities is 31.5 mph, so we could estimate that the car traveled 31.5 miles in the first hour. Carefully explain why this is not necessarily correct. Even so, it will serve as a first approximation. Since $v(1) = 33$ mph and $v(2) = 3\sqrt{2} + 30 \approx 34$ mph, we can estimate that the car traveled 33.5 mph in the second hour. Using $v(3) \approx 35$ mph and $v(4) = 36$ mph, find similar estimates for the distance traveled in the third and fourth hours, and then estimate the total distance. To improve this estimate, we can find an estimate for the distance covered each half-hour. The first estimate would take $v(0) = 30$ mph and $v(0.5) \approx 32.1$ mph and estimate an average velocity of 31.05 mph and a distance of 15.525 miles. Estimate the average velocity and then the distance for the remaining 7 half-hours and estimate the total distance. We can improve this estimate, too. By estimating the average velocity every quarter-hour, find a third estimate of the total distance. Based on these three estimates, conjecture the limit of these approximations as the time interval considered goes to zero.

41. (exploratory) In this exercise, we investigate a generalization of sums called an **infinite series**. Suppose a bouncing ball has **coefficient of restitution** equal to 0.6. This means that if the ball hits the ground with velocity v ft/s, it rebounds with velocity $0.6v$. Ignoring air resistance, a ball launched with velocity v ft/s will stay in the air $v/16$ seconds before hitting the ground. Suppose a ball with coefficient of restitution 0.6 is launched with initial velocity 60 ft/s. Explain why the total time in the air is given by

$60/16 + (0.6)(60)/16 + (0.6)(0.6)(60)/16 + \cdots$. It might seem like the ball would continue to bounce forever. To see otherwise, compute sums using more and more terms follow-ing this pattern. Find the limit that these sums approach. The limit is the number of seconds that the ball continues to bounce.

4.3   Area

One basic problem we have been exploring in this chapter is how to compute the distance traveled from a given velocity function. We examined this in terms of antiderivatives in section 4.1 and reworked this as an area problem in section 4.2. In this section, we will develop the general problem of calculation of areas in some detail.

You are all familiar with the formulas for computing the area of a rectangle, a circle and a triangle. From endless use of these formulas over the years, you probably have good intuition about what area is: one measure of the size of a two-dimensional region. But, how can we compute the area of a region that's not a rectangle, circle or triangle? For instance, how would you find the area of the region bounded by the x-axis and the curve $y = \sin x$ on the interval $[0, \pi]$ shown in Figure 4.4?

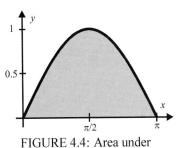

FIGURE 4.4: Area under $y = \sin x$.

Notice that we've used the word *compute*. For most regions, areas are not measured directly, but rather are computed using some one-dimensional measurements and a formula.

The problem of computing areas is more important than you might think. There are fairly obvious examples such as the area of your front yard (how much grass seed is needed?), the area of the living room floor (how much carpet is needed?) and the area of a bedroom's walls (how much paint is needed?). We saw in section 4.2 that areas are also computed to determine the distance traveled. We see later how to interpret area to give the probability of a random event or the volume of oil flowing through a pipeline.

What we need, then, is a more general description of area, one that can be used to find the area of almost any two-dimensional region imaginable. In this section, we develop a general process for computing area. It turns out that this process (which we generalize to the notion of the *definite integral* in section 4.4) has significance far beyond the calculation of area. In fact, this powerful and flexible tool is one of the central ideas of calculus, with applications in a wide variety of fields.

Let's start our investigation of area by looking at a fairly simple example: we'll try to compute the area bounded between the graph of $y = 2x - 2x^2$ and the x-axis. Since $2x - 2x^2 = 0$ for $x = 0$ and $x = 1$, the region we are interested in extends from $x = 0$ to $x = 1$ (see Figure 4.5).

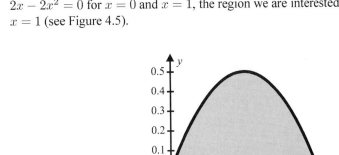

FIGURE 4.5: Area under $2x - 2x^2$.

The region is clearly not a rectangle, circle or triangle, so we have no handy formula available for computing the area. Since we don't know how to compute the area exactly, we first try to approximate it, using what we already know about area. We'll then suggest a means for systematically improving the approximation.

Although we often think of graphics calculators and computers as drawing curves, what they actually do is plot points, usually by coloring in small squares on the screen called *pixels*. If you look closely enough at a calculator display of the graph of $y = 2x - 2x^2$, you will probably see a picture similar to Figure 4.6. (Computer screens look the same way, but their resolution is so fine as to make individual pixels indistinguishable to the naked eye.)

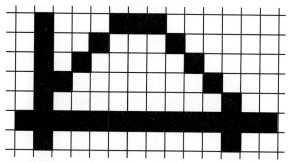

FIGURE 4.6: Calculator display of $y = 2x - 2x^2$.

In the case of Figure 4.6, each pixel represents a square of side 0.1. To estimate the area, then, we simply count the number of pixels in the region of interest and multiply the total by 0.01, the area of 1 pixel. Try this now: counting the highlighted pixels forming the curve, but not the pixels forming the x-axis. You should find 33 pixels, yielding an approximate area 0.33. We can simplify the counting and gain some valuable insight by rethinking the problem somewhat. In Figure 4.7, we have organized the pixels we want to count into columns: the first column is 2 pixels high, the second column is 3 pixels high and so on.

FIGURE 4.7: Counting pixels to approximate area.

Notice that the area indicated in Figure 4.5 can be approximated by the sum of the areas of the rectangles indicated in Figure 4.7. Each rectangle has width 0.1 (1 pixel wide) but the heights differ. What determines the height of each rectangle? Since the top pixel is supposed to represent the graph of $y = f(x)$, the height is determined by the function values. For example, we have $f(0.1) = 0.18$ and 0.18 rounds off to 0.2 or two pixels, so that the first column is 2 pixels high. Similarly, $f(0.2) = 0.32$ and so, the second column is 3 pixels high. Ignoring the rounding off, each rectangle has width 0.1 and height $f(x)$, for some x. The total area is then

$$A = 0.1 \, f(0.1) + 0.1 \, f(0.2) + 0.1 \, f(0.3) + \cdots + 0.1 \, f(0.9) + 0.1 \, f(1.0).$$

After factoring out the common width of 0.1 and evaluating the function, we get

$$A = (0.18 + 0.32 + 0.42 + 0.48 + 0.5 + 0.48 + 0.42 + 0.32 + 0.18 + 0) \, (0.1) = 0.33.$$

We will see later that the exact area in Figure 4.5 is 1/3, so we have found a fairly good approximation.

How can we improve on this approximation? It should be clear graphically that the smaller the pixels are, the higher the resolution of the picture will be. The same is true computationally. In general, the more rectangles we use, the better the area approximation. Since more rectangles means a better approximation, it should make sense that the actual area is the limit of the approximations as the number of rectangles tends to infinity. The following definitions make this idea more precise.

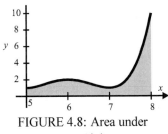

FIGURE 4.8: Area under
$y = f(x)$.

The general problem is to estimate the area below the graph of $y = f(x)$ and above the x-axis for $a \le x \le b$. For now, we assume that $f(x) \ge 0$ and f is continuous on the interval $[a, b]$, as in Figure 4.8.

We start by dividing the interval $[a, b]$ into n equal pieces. (This is called a **regular partition** and is equivalent to marking off the boundaries of the pixels.) The width of each subinterval in the partition is then $\dfrac{b - a}{n}$, which we denote by Δx (meaning a small change in x). The points in the partition are denoted by $x_0 = a$, $x_1 = x_0 + \Delta x$, $x_2 = x_1 + \Delta x$, and so on. In general,

$$x_i = x_0 + i\, \Delta x, \ i = 1, 2, \ldots, n$$

(think about this some). See Figure 4.9 for an illustration of a regular partition for the case where $n = 6$.

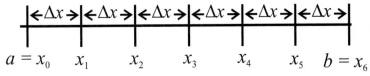

FIGURE 4.9: Regular partition of $[a, b]$.

On each subinterval, $[x_{i-1}, x_i]$ $(i = 1, 2, \ldots, n)$, construct a rectangle of height $f(x_i)$ (the value of the function at the right endpoint of the subinterval), as illustrated in Figure 4.10, for the case where $n = 4$. It should be clear from Figure 4.10 that the area under the curve, A, is approximately the same as the sum of the areas of the four rectangles,

$$A \approx f(x_1)\, \Delta x + f(x_2)\, \Delta x + f(x_3)\, \Delta x + f(x_4)\, \Delta x = A_4.$$

FIGURE 4.10: $A \approx A_4$.

In particular, notice that although two of these rectangles enclose more area than that under the curve and two enclose less area, on the whole the sum of the areas of the four rectangles provides an approximation to the total area under the curve. More generally, if we construct n rectangles of equal width on the interval $[a, b]$, we have

$$A \approx f(x_1)\, \Delta x + f(x_2)\, \Delta x + \cdots + f(x_n)\, \Delta x$$

$$= \sum_{i=1}^{n} f(x_i)\, \Delta x = A_n. \qquad (3.1)$$

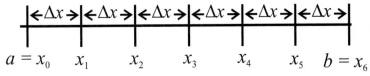

FIGURE 4.11: $A \approx A_{10}$.

Example 3.1   Approximating an Area with Rectangles

Approximate the area under the curve $y = f(x) = 2x - 2x^2$ on the interval $[0, 1]$, using 10 rectangles.

Solution   The partition divides the interval into 10 subintervals, each of length $\Delta x = 0.1$, namely $[0, 0.1], [0.1, 0.2], \ldots, [0.9, 1.0]$. In Figure 4.11, we have drawn in rectangles of height $f(x_i)$ on each subinterval $[x_{i-1}, x_i]$, $i = 1, 2, \ldots, n$. Notice that the sum of the areas of the 10 rectangles indicated provides an approximation to the area under the curve. That is,

$$A \approx A_{10} = \sum_{i=1}^{10} f(x_i)\, \Delta x$$

$$= [f(0.1) + f(0.2) + \cdots + f(1.0)] \,(0.1)$$

$$= (0.18 + 0.32 + 0.42 + 0.48 + 0.5 + 0.48 + 0.42 + 0.32 + 0.18 + 0)\,(0.1)$$

$$= 0.33$$

This should look familiar to you as the same estimate of the area in Figure 4.5 that we had arrived at earlier.

■

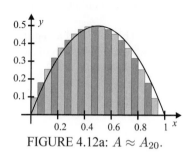

FIGURE 4.12a: $A \approx A_{20}$.

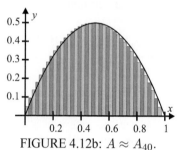

FIGURE 4.12b: $A \approx A_{40}$.

n	A_n
10	0.33
20	0.3325
30	0.332963
40	0.333125
50	0.3332
60	0.333241
70	0.333265
80	0.333281
90	0.333292
100	0.3333

Example 3.2   A Better Approximation Using More Rectangles

Repeat example 3.1, with $n = 20$.

Solution   Here, we partition the interval $[0, 1]$ into 20 subintervals, each of width

$$\Delta x = \frac{1 - 0}{20} = \frac{1}{20} = 0.05.$$

We then have $x_0 = 0$, $x_1 = 0 + \Delta x = 0.05$, $x_2 = x_1 + \Delta x = 2\,(0.05)$ and so on, so that $x_i = (0.05)i$, for $i = 0, 1, 2, \ldots, n$. So, from (3.1), the area is approximately

$$A \approx \sum_{i=1}^{20} f(x_i)\,\Delta x = \sum_{i=1}^{20} (2x_i - 2x_i^2)\,\Delta x$$

$$= \sum_{i=1}^{20} 2\left[0.05\,i - (0.05\,i)^2\right](0.05) = 0.3325.$$

Figure 4.12a shows an approximation using 20 rectangles and in Figure 4.12b, we see 40 rectangles.

Based on Figures 4.11 4.12a and 4.12b, you should expect that the larger we make n, the better A_n will approximate the actual area, A. The obvious drawback to this idea is the length of time it would take to compute A_n, for n large. However, your CAS or programmable calculator can compute these sums for you, with ease. The table shown in the margin indicates approximate values of A_n for various values of n.

Notice that as n gets larger and larger, A_n seems to be approaching $\frac{1}{3}$.

■

Examples 3.1 and 3.2 give strong evidence that the larger the number of rectangles we use, the better our approximation of the area is. Thinking this through, we arrive at the following definition of the area under a curve.

Definition 3.1   For a function f defined on the interval $[a, b]$, if f is continuous on $[a, b]$ and $f(x) \geq 0$ on $[a, b]$, the **area A under the curve** $y = f(x)$ on $[a, b]$ is given by

$$A = \lim_{n \to \infty} A_n = \lim_{n \to \infty} \sum_{i=1}^{n} f(x_i)\,\Delta x. \qquad (3.2)$$

In the following example, we use the limit defined in (3.2) to find the exact area under the curve from examples 3.1 and 3.2.

Example 3.3   Computing the Area Exactly

Find the area under the curve $y = f(x) = 2x - 2x^2$ on the interval $[0, 1]$.

Solution   Here, we have

$$\Delta x = \frac{1 - 0}{n} = \frac{1}{n}$$

and so, $x_0 = 0$, $x_1 = \frac{1}{n}$, $x_2 = x_1 + \Delta x = \frac{2}{n}$ and so on, so that $x_i = \frac{i}{n}$, for

$i = 0, 1, 2, \ldots, n$. So, from (3.1), the area is approximately

$$A \approx A_n = \sum_{i=1}^{n} f\left(\frac{i}{n}\right)\left(\frac{1}{n}\right) = \sum_{i=1}^{n}\left[2\frac{i}{n} - 2\left(\frac{i}{n}\right)^2\right]\left(\frac{1}{n}\right)$$

$$= \sum_{i=1}^{n}\left[2\left(\frac{i}{n}\right)\left(\frac{1}{n}\right)\right] - \sum_{i=1}^{n}\left[2\left(\frac{i^2}{n^2}\right)\left(\frac{1}{n}\right)\right]$$

$$= \frac{2}{n^2}\sum_{i=1}^{n} i - \frac{2}{n^3}\sum_{i=1}^{n} i^2$$

From Theorem 2.1 (ii) and (iii).

$$= \frac{2}{n^2}\frac{n(n+1)}{2} - \frac{2}{n^3}\frac{n(n+1)(2n+1)}{6}$$

$$= \frac{n+1}{n} - \frac{(n+1)(2n+1)}{3n^2}$$

$$= \frac{(n+1)(n-1)}{3n^2}.$$

(You should provide the algebraic details for the last step.) Note that since we have a formula for A_n, for any n, we can compute various values with ease. We have

$$A_{200} = \frac{(201)(199)}{3(40000)} = 0.333325,$$

$$A_{500} = \frac{(501)(499)}{3(250000)} = 0.333332$$

and so on. Finally, we can compute the limiting value of A_n explicitly. We have

$$\lim_{n\to\infty} A_n = \lim_{n\to\infty} \frac{n^2 - 1}{3n^2} = \lim_{n\to\infty} \frac{1 - 1/n^2}{3} = \frac{1}{3}.$$

Therefore, the exact area in Figure 4.5 is 1/3.

■

Example 3.4   Estimating the Area Under a Curve
    Find the area under the curve $y = f(x) = \sqrt{x+1}$ on the interval $[1, 3]$.

Solution   Here, we have

$$\Delta x = \frac{3-1}{n} = \frac{2}{n}$$

and $x_0 = 1$, so that

$$x_1 = x_0 + \Delta x = 1 + \frac{2}{n},$$

$$x_2 = 1 + 2\left(\frac{2}{n}\right)$$

and so on, so that

$$x_i = 1 + \frac{2i}{n}, \text{ for } i = 0, 1, 2, \ldots, n.$$

Thus, we have from (3.1) that

$$A \approx A_n = \sum_{i=1}^{n} f(x_i)\,\Delta x = \sum_{i=1}^{n} \sqrt{x_i + 1}\,\Delta x$$

$$= \sum_{i=1}^{n} \sqrt{\left(1 + \frac{2i}{n}\right) + 1}\,\left(\frac{2}{n}\right)$$

$$= \frac{2}{n} \sum_{i=1}^{n} \sqrt{2 + \frac{2i}{n}}.$$

n	A_n
10	3.50595
50	3.45942
100	3.45357
500	3.44888
1000	3.44830
5000	3.44783

You should quickly recognize that we have no formulas like those in Theorem 2.1 for simplifying this last sum (unlike the sum in example 3.3). Our only choice, then, is to compute A_n for a number of values of n. Of course, this is easy to do using a CAS or programmable calculator. The accompanying table lists values of A_n displayed to six digits. Although we can't compute the area exactly (as yet), you should get the sense that the area is approximately 3.4478.

∎

We pause now to define some of the mathematical objects we have been examining.

Definition 3.2   Let $\{x_0, x_1, \ldots, x_n\}$ be a regular partition of the interval $[a, b]$, with $x_i - x_{i-1} = \Delta x = \dfrac{b - a}{n}$, for all i. Pick points $c_1, c_2, \ldots, c_n$, where c_i is any point in the subinterval $[x_{i-1}, x_i]$, for $i = 1, 2, \ldots, n$. (These are called **evaluation points**.) The **Riemann sum** for this partition and set of evaluation points is

$$\sum_{i=1}^{n} f(c_i)\,\Delta x.$$

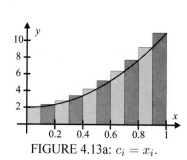

FIGURE 4.13a: $c_i = x_i$.

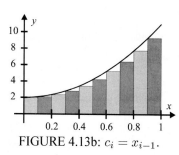

FIGURE 4.13b: $c_i = x_{i-1}$.

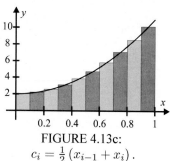

FIGURE 4.13c:
$c_i = \frac{1}{2}\left(x_{i-1} + x_i\right)$.

NOTE: So far in our development, we have shown that for a continuous, nonnegative function f, the area under the curve $y = f(x)$ is the limit of the Riemann sums:

$$A = \lim_{n \to \infty} \sum_{i=1}^{n} f(c_i)\,\Delta x, \tag{3.3}$$

where $c_i = x_i$, for $i = 1, 2, \ldots, n$. Surprisingly, for any continuous function f, the limit in (3.3) is the same for **any choice** of the evaluation points $c_i \in [x_{i-1}, x_i]$ (although the proof is beyond the level of this course). In examples 3.3 and 3.4, we used the evaluation points $c_i = x_i$, for each i (the right endpoint of each subinterval). This is usually the most convenient choice when working by hand, but usually does **not** produce particularly accurate approximations).

Remark 3.1   Most often, we **cannot** compute the limit of Riemann sums exactly (at least not directly). However, we can **always** obtain an approximation to the area by calculating Riemann sums for some large values of n. The most common (and obvious) choices for the evaluation points c_i are x_i (the right endpoint), x_{i-1} (the left endpoint) and $\frac{1}{2}(x_{i-1} + x_i)$ (the midpoint). As it turns out, the midpoint usually provides the best approximation, for a given n. See Figures 4.13a, 4.13b and 4.13c for the right endpoint, left endpoint and midpoint approximations, respectively, for $f(x) = 9x^2 + 2$, on the interval $[0, 1]$, using $n = 10$. You should note that in this case (as with any increasing function), the rectangles corresponding to the right endpoint evaluation (Figure 4.13a) give too much area on each subinterval, while the rectangles corresponding to left endpoint evaluation (Figure 4.13b) give too little area. We leave it to you to observe that the reverse is true for a decreasing function.

Example 3.5 **Computing Riemann Sums with Different Evaluation Points**

Compute Riemann sums for $f(x) = \sqrt{x+1}$ on the interval $[1, 3]$, for $n = 10, 50, 100, 500, 1000$ and 5000, using the left endpoint, right endpoint and midpoint of each subinterval as the evaluation points.

Solution The numbers given in the following table are from a program written for a programmable calculator. We suggest that you test your own program or one built into your CAS against these values (rounded off to six digits).

n	left endpoint	midpoint	right endpoint
10	3.38879	3.44789	3.50595
50	3.43598	3.44772	3.45942
100	3.44185	3.44772	3.45357
500	3.44654	3.44772	3.44888
1000	3.44713	3.44772	3.44830
5000	3.44760	3.44772	3.44783

There are several conclusions to be drawn from these numbers. First, there is good evidence that all three sets of numbers are converging to a common limit of approximately 3.4477. You should notice that the limit is independent of the particular evaluation point used. Second, even though the limits are the same, the different rules approach the limit at different rates. You should try computing left and right endpoint sums for larger values of n, to see that these eventually approach 3.44772, also.

The Midpoint Rule usually approaches the limit far faster than left or right endpoint rules. If you think about the rectangles being drawn, you may be able to explain why. Finally, the left and right endpoint sums in example 3.5 approach the limit from opposite directions and at about the same rate. We take advantage of this in an approximation technique called the Trapezoid Rule, to be discussed in section 4.7. If your CAS or graphics calculator does not have a command for calculating Riemann sums, we suggest that you write a program for computing them yourself.

Exercises 4.3

1. (writing) It turns out that for many functions, the limit of the Riemann sums is independent of the choice of evaluation points. Discuss why this is a somewhat surprising result. To make the result more believable, consider a continuous function $f(x)$. As the number of partition points gets larger, the distance between the endpoints gets smaller. For the continuous function $f(x)$, explain why the difference between the function values at any two points in a given subinterval will have to get smaller.

2. (writing) Rectangles are not the only basic geometric shapes for which we have an area formula. Discuss how you might approximate the area under a parabola using circles or triangles. Which geometric shape do you think is the easiest to use?

In exercises 3–10, list the evaluation points corresponding to the midpoint of each subinterval, sketch the function and approximating rectangles and evaluate the Riemann sum.

3. $f(x) = x^2 + 1$, $[0, 1]$, $n = 4$

4. $f(x) = x^2 + 1$, $[0, 2]$, $n = 4$

5. $f(x) = x^3 - 1$, $[1, 2]$, $n = 4$

6. $f(x) = x^3 - 1$, $[1, 3]$, $n = 4$

7. $f(x) = \sin x$, $[0, \pi]$, $n = 4$

8. $f(x) = \sin x$, $[0, \pi]$, $n = 8$

9. $f(x) = 4 - x^2$, $[-1, 1]$, $n = 4$

10. $f(x) = 4 - x^2$, $[-3, -1]$, $n = 4$

312

Chapter 4   Integration

In exercises 11–26, approximate the area under the curve on the given interval using n rectangles and the indicated evaluation rule.

11. $y = x^2$ on $[0,1]$, $n = 8$, midpoint evaluation

12. $y = x^2$ on $[0,1]$, $n = 8$, right-endpoint evaluation

13. $y = x^2$ on $[-1,1]$, $n = 8$, left-endpoint evaluation

14. $y = x^2$ on $[-1,1]$, $n = 8$, midpoint evaluation

15. $y = \sqrt{x+2}$ on $[1,4]$, $n = 8$, midpoint evaluation

16. $y = \sqrt{x+2}$ on $[1,4]$, $n = 8$, right-endpoint evaluation

17. $y = e^{-2x}$ on $[-1,1]$, $n = 16$, left-endpoint evaluation

18. $y = e^{-2x}$ on $[-1,1]$, $n = 16$, midpoint evaluation

19. $y = \cos x$ on $[0, \pi/2]$, $n = 50$, midpoint evaluation

20. $y = \cos x$ on $[0, \pi/2]$, $n = 100$, right-endpoint evaluation

21. $y = 3x - 2$ on $[1,4]$, $n = 4$, midpoint evaluation

22. $y = 3x - 2$ on $[1,4]$, $n = 40$, midpoint evaluation

23. $y = x^3 - 1$ on $[1,3]$, $n = 100$, midpoint evaluation

24. $y = x^3 - 1$ on $[1,3]$, $n = 100$, right-endpoint evaluation

25. $y = x^3 - 1$ on $[-1,1]$, $n = 100$, left-endpoint evaluation

26. $y = x^3 - 1$ on $[-1,1]$, $n = 100$, right-endpoint evaluation

In exercises 27–30, construct a table of Riemann sums as in Example 3.5 to show that sums with right-endpoint, midpoint and left-endpoint evaluation all converge to the same value as $n \to \infty$.

27. $f(x) = 4 - x^2$, $[-2,2]$ **28.** $f(x) = \sin x$, $[0, \pi/2]$

29. $f(x) = x^3 - 1$, $[1,3]$ **30.** $f(x) = x^3 - 1$, $[-1,1]$

In exercises 31–36, use the given function values to estimate the area under the curve using left-endpoint and right-endpoint evaluation.

31.
x	0.0	0.1	0.2	0.3	0.4	0.5	0.6	0.7	0.8
$f(x)$	2.0	2.4	2.6	2.7	2.6	2.4	2.0	1.4	0.6

32.
x	0.0	0.1	0.2	0.3	0.4	0.5	0.6	0.7	0.8
$f(x)$	3.0	2.2	1.6	0.7	0.6	0.4	-0.2	0.4	0.6

33.
x	0.0	0.2	0.4	0.6	0.8	1.0	1.2	1.4	1.6
$f(x)$	1.0	1.4	2.1	2.7	2.6	2.8	3.0	3.4	3.6

34.
x	0.0	0.2	0.4	0.6	0.8	1.0	1.2	1.4	1.6
$f(x)$	2.0	2.2	1.6	1.4	1.6	2.0	2.2	2.4	2.0

35.
x	1.0	1.1	1.2	1.3	1.4	1.5	1.6	1.7	1.8
$f(x)$	1.8	1.4	1.1	0.7	1.2	1.4	1.8	2.4	2.6

36.
x	1.0	1.2	1.4	1.6	1.8	2.0	2.2	2.4	2.6
$f(x)$	0.0	0.4	0.6	0.8	1.2	1.4	1.2	1.4	1.0

37. Give an example of a function (its graph and/or formula) and interval where the Riemann sum with $n = 2$ would be closer to the actual area using right-endpoint evaluation than using midpoint evaluation.

38. For the function $f(x) = x^2$ on the interval $[0,1]$, discuss geometrically why the midpoint evaluation produces a Riemann sum smaller than the actual area (of $1/3$). By trial and error, find evaluation points such that the Riemann sum equals exactly $1/3$.

39. Show that for right-endpoint evaluation on the interval $[a,b]$ with each subinterval of length $\Delta x = (b-a)/n$, the evaluation points are $c_i = a + i\,\Delta x$, for $i = 1, 2, \ldots, n$.

40. Show that for left-endpoint evaluation on the interval $[a,b]$ with each subinterval of length $\Delta x = (b-a)/n$, the evaluation points are $c_i = a + (i-1)\,\Delta x$, for $i = 1, 2, \ldots, n$.

41. As in exercises 39 and 40, find a formula for the evaluation points for midpoint evaluation.

42. As in exercises 39 and 40, find a formula for evaluation points which are one-third of the way from the left-endpoint to the right-endpoint.

43. (exploratory) Riemann sums can also be defined on **irregular partitions** for which subintervals are not of equal size. An example of an irregular partition of the interval $[0,1]$ is $x_0 = 0$, $x_1 = 0.2$, $x_2 = 0.6$, $x_3 = 0.9$, $x_4 = 1$. Explain why the corresponding Riemann sum would be

$$f(c_1)\,(0.2) + f(c_2)\,(0.4) + f(c_3)\,(0.3) + f(c_4)\,(0.1),$$

for evaluation points c_1, c_2, c_3 and c_4. Identify the interval from which each c_i must be chosen and give examples of evaluation points. To see why irregular partitions might be useful, consider the function $f(x) = \begin{cases} 2x & \text{if } x < 1 \\ x^2 + 1 & \text{if } x \geq 1 \end{cases}$ on the interval $[0,2]$. One way to approximate the area under the graph of this function is to compute Riemann sums using midpoint evaluation for $n = 10$, $n = 50$, $n = 100$ and so on. Show graphically and numerically that with midpoint evaluation, the Riemann sum with $n = 2$ gives the correct area on the subinterval $[0,1]$. Then explain why it would be wasteful to compute Riemann sums on this subinterval for larger and larger values of n. A more efficient strategy would be to compute the areas on $[0,1]$ and $[1,2]$ separately and add them together. The area on $[0,1]$ can be computed exactly using a small value of n, while the area on $[1,2]$ must be approximated using larger and larger values of n. Use this technique to estimate the area for $f(x)$ on the interval $[0,2]$. Try to determine the area to within an error of 0.01 (discuss why you believe your answer is this accurate).

44. (exploratory) Graph the function $f(x) = e^{-x^2}$. You may recognize this curve as the so-called "bell curve," which is of fundamental importance in statistics. We define the **area function** $g(t)$ to be the area between this graph and the x-axis between $x = 0$ and $x = t$ (for now, assume that $t > 0$). Sketch the area that defines $g(1)$ and $g(2)$ and argue that $g(2) > g(1)$. Explain why the function $g(x)$ is increasing and hence $g'(x) > 0$ for $x > 0$. Further, argue that $g'(2) < g'(1)$. Explain why $g'(x)$ is a decreasing function. Thus, $g'(x)$ has the same general properties (positive, decreasing) that $f(x)$ does. In fact, we will discover in section 4.5 that $g'(x) = f(x)$. To collect some evidence for this result, use Riemann sums to estimate $g(2)$, $g(1.1)$, $g(1.01)$ and $g(1)$. Use these values to estimate $g'(1)$ and compare to $f(1)$.

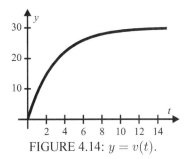

FIGURE 4.14: $y = v(t)$.

4.4   The Definite Integral

Throughout this chapter, we have been considering the problem of computing distance from a given velocity function. For instance, suppose that a sky diver starts her jump by stepping out of an airplane (so that she starts with zero downward velocity). The jumper gradually picks up speed until a maximum speed is reached. Such a *terminal velocity* is the speed at which the force due to air resistance cancels out the force due to gravity. A function that has these properties is $f(x) = 30(1 - e^{-x/3})$ (see Figure 4.14), which we can think of as modeling the (downward) velocity x seconds into the jump.

As we discussed in section 4.2, the area under this curve on the interval $0 \le x \le t$ corresponds to the distance fallen in the first t seconds. On a given interval (i.e., for a given value of t), we can approximate this area by first partitioning the interval into n subintervals $[x_{i-1}, x_i]$, $i = 1, 2, \ldots, n$, of equal width, Δx. On each subinterval, we construct a rectangle of height $f(c_i)$, for any choice of $c_i \in [x_{i-1}, x_i]$. Finally, summing the areas of the rectangles gives us an approximation to the area,

$$A \approx \sum_{i=1}^{n} f(c_i) \Delta x.$$

Further, we have observed that the exact area is the limit of these Riemann sums,

$$A = \lim_{n \to \infty} \sum_{i=1}^{n} f(c_i) \Delta x. \qquad (4.1)$$

If we consider a plummeting bungee jumper instead of a sky diver, notice that the velocity should initially follow a similar pattern. However, at some point in the bungee jump, the bungee cord (hopefully) exerts a force to stop the downward velocity and bounce the jumper back upward. From our current perspective, this would result in a negative velocity. We would like to be able to compute and interpret the limit of sums in (4.1) even when the velocity is negative.

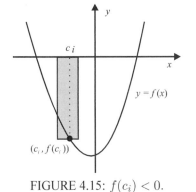

FIGURE 4.15: $f(c_i) < 0$.

Think carefully about our definition of area under the curve given by the limit in (4.1). How could we interpret such a limit in the case where f is both positive and negative on the interval $[a, b]$? Notice that if $f(c_i) < 0$, for some i, then the height of the rectangle shown in Figure 4.15 is $-f(c_i)$ and so,

$$f(c_i) \Delta x = -\text{Area of the } i\text{th rectangle.}$$

To see the effect this has on the sum, consider the following example.

Example 4.1   A Riemann Sum for a Function with Positive and Negative Values

For $f(x) = \sin x$ on $[0, 2\pi]$, give an area interpretation of $\lim_{n \to \infty} \sum_{i=1}^{n} f(c_i) \Delta x$.

FIGURE 4.16a: 10 rectangles.

FIGURE 4.16b: 20 rectangles.

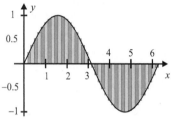

FIGURE 4.16c: 40 rectangles.

Solution   For this illustration, we take c_i to be the midpoint of $[x_{i-1}, x_i]$, for $i = 1, 2, \ldots, n$. In Figure 4.16a, we see 10 rectangles constructed between the x-axis and the curve $y = f(x)$.

The first five rectangles (where $f(c_i) > 0$) lie above the x-axis and have height $f(c_i)$. The remaining five rectangles (where $f(c_i) < 0$) lie below the x-axis and have height $-f(c_i)$. Thus,

$$\sum_{i=1}^{10} f(c_i)\, \Delta x$$

$$= (\text{Area of rectangles above the } x\text{-axis}) - (\text{Area of rectangles below the } x\text{-axis}).$$

In Figures 4.16b and 4.16c, we show 20 and 40 rectangles, respectively, constructed in the same way. From this, observe that

$$\lim_{n \to \infty} \sum_{i=1}^{n} f(c_i)\, \Delta x = (\text{Area above the } x\text{-axis}) - (\text{Area below the } x\text{-axis}).$$

For the general case (f positive or negative), we make the following definition.

Definition 4.1   The **definite integral** of f from a to b is

$$\int_a^b f(x)\, dx = \lim_{n \to \infty} \sum_{i=1}^{n} f(c_i)\, \Delta x,$$

for any function f defined on $[a, b]$ for which the limit exists and is the same for any choice of evaluation points, $c_1, c_2, \ldots, c_n$. When the limit exists, we say that f is **integrable** on $[a, b]$.

Remark 4.1   Definition 4.1 is adequate for any function that you might encounter in this text (i.e., functions that are continuous except for at most a finite number of discontinuities). For more general functions, we must broaden the definition to include partitions with subintervals of different lengths. You will find a suitably generalized definition in any advanced calculus text.

The notation should be easier to understand if we make the meaning behind it more clear. In the Riemann sum, the Greek letter $\sum$ indicates a sum; so does the elongated "S", $\int$ used as the integral sign. (You may have seen S's written this way in old formal documents.) The **lower** and **upper limits of integration**, a and b, respectively, indicate the interval over which you are integrating; this is only implicitly indicated in the Riemann sum. The dx in the integral corresponds to the increment Δx in the Riemann sum, and also indicates the variable of integration (a dummy variable). Here, $f(x)$ is called the **integrand**.

NOTE: If f is continuous and $f(x) \geq 0$ on $[a, b]$, then

$$\int_a^b f(x)\, dx = \text{The area under the curve on } [a, b].$$

Returning to the sky diver example used to introduce the section, notice that the definite integral $\int_0^{15} 30(1 - e^{-t/3})\, dt$ represents the distance fallen in the first 15 seconds. How can we calculate this distance? To date, we have two options: if the function is simple enough (a polynomial of degree 2 or less) we can symbolically compute the Riemann sum and compute the limit as $n \to \infty$. Otherwise, we can numerically compute several specific Riemann sums and approximate the value of the limit. Typically, we would use the **Midpoint Rule**, which uses the midpoints as the evaluation points for the Riemann sum.

Example 4.2   A Midpoint Rule Approximation of a Definite Integral
Use the Midpoint Rule to estimate $\int_0^{15} 30(1 - e^{-t/3})\, dt$.

Solution   The Midpoint Rule is a Riemann sum using the midpoint of each subinterval as evaluation point:

$$\int_0^{15} 30(1 - e^{-t/3})\, dt \approx \sum_{i=1}^{n} f(c_i)\, \Delta x = 30 \sum_{i=1}^{n} \left(1 - e^{-c_i/3}\right) \left(\frac{15 - 0}{n}\right),$$

n	R_n
10	361.5
20	360.8
50	360.6
100	360.6

where $c_i = \dfrac{x_i + x_{i-1}}{2}$. Using a CAS or a calculator program, you can get the sequence of approximations found in the accompanying table.

Among the issues in this problem are how many digits to keep and when to stop increasing n. If the point of the calculation is to compute the distance fallen by a sky diver (in feet), one digit after the decimal is adequate and may even be excessive. In this case, we continued to increase n until it seemed clear that 361 feet was a reasonable approximation.

■

Example 4.3   Using Riemann Sums to Compute a Definite Integral
Compute $\int_0^2 \left(x^2 - 2x \right) dx$ exactly.

Solution   The definite integral is the limit of a sequence of Riemann sums, where we can choose any evaluation points we wish. It is usually easiest to write out the formula using right endpoints, as we do here. In this case,

$$\Delta x = \frac{2 - 0}{n} = \frac{2}{n}.$$

We then have $x_0 = 0$, $x_1 = x_0 + \Delta x = \dfrac{2}{n}$,

$$x_2 = x_1 + \Delta x = \frac{2}{n} + \frac{2}{n} = \frac{2\,(2)}{n}$$

and so on. We then have $c_i = x_i = \dfrac{2i}{n}$. The Riemann sum is then

$$R_n = \sum_{i=1}^{n} f\,(x_i)\, \Delta x = \sum_{i=1}^{n} \left(x_i^2 - 2x_i\right) \Delta x$$

From Theorem 2.1 (ii) and (iii).

$$= \sum_{i=1}^{n} \left[\left(\frac{2i}{n}\right)^2 - 2\left(\frac{2i}{n}\right)\right]\left(\frac{2}{n}\right) = \sum_{i=1}^{n} \left(\frac{4i^2}{n^2} - \frac{4i}{n}\right)\left(\frac{2}{n}\right)$$

$$= \frac{8}{n^3} \sum_{i=1}^{n} i^2 - \frac{8}{n^2} \sum_{i=1}^{n} i = \left(\frac{8}{n^3}\right) \frac{n(n + 1)(2n + 1)}{6} - \left(\frac{8}{n^2}\right) \frac{n(n + 1)}{2}$$

$$= \frac{4(n + 1)(2n + 1)}{3n^2} - \frac{4(n + 1)}{n} = \frac{8n^2 + 12n + 4}{3n^2} - \frac{4n + 4}{n}.$$

Taking the limit of R_n as $n \to \infty$ gives us the exact value of the integral:

$$\int_0^2 \left(x^2 - 2x \right) dx = \lim_{n \to \infty} \left(\frac{8n^2 + 12n + 4}{3n^2} - \frac{4n + 4}{n}\right) = \frac{8}{3} - 4 = -\frac{4}{3}.$$

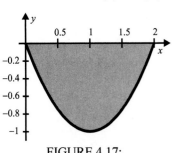

FIGURE 4.17:
$y = x^2 - 2x$ on $[0, 2]$.

Don't be disturbed by the negative answer. Definite integrals can be positive or negative. To interpret this answer in terms of area, look at a graph of $y = x^2 - 2x$ (see Figure 4.17).

On the interval we're interested in, $[0, 2]$, the function is always negative. Notice that the *absolute value* of the integral, $\frac{4}{3}$, corresponds to the area between the curve $y = x^2 - 2x$ and the x-axis. The negative value of the integral indicates that the area lies *below* the x-axis. We refer to this as **signed area**, which we define below.

■

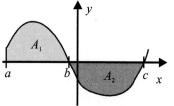

FIGURE 4.18: Signed area.

Definition 4.2   Suppose that $f(x) \geq 0$ on the interval $[a, b]$ and A_1 is the area bounded between the curve $y = f(x)$ and the x-axis for $a \leq x \leq b$. Further, suppose that $f(x) \leq 0$ on the interval $[b, c]$ and A_2 is the area bounded between the curve $y = f(x)$ and the x-axis for $b \leq x \leq c$. The **signed area** between $y = f(x)$ and the x-axis for $a \leq x \leq c$ is $A_1 - A_2$ and the **total area** between $y = f(x)$ and the x-axis for $a \leq x \leq c$ is $A_1 + A_2$ (see Figure 4.18).

The following example examines the general case where the integrand may be both positive and negative on the interval of integration.

Example 4.4   Relating Definite Integrals to Signed Area
Compute three related integrals: $\int_0^2 \left(x^2 - 2x\right) dx$, $\int_2^3 \left(x^2 - 2x\right) dx$ and $\int_0^3 \left(x^2 - 2x\right) dx$ and interpret each in terms of area.

Solution   From example 4.3, we already know that $\int_0^2 \left(x^2 - 2x\right) dx = -\frac{4}{3}$. (See Figure 4.17 to interpret this result graphically.)
On the interval $[2, 3]$, we have $\Delta x = \frac{1}{n}$ and $x_0 = 2$, $x_1 = x_0 + \Delta x = 2 + \frac{1}{n}$,

$$x_2 = x_1 + \Delta x = \left(2 + \frac{1}{n}\right) + \frac{1}{n} = 2 + \frac{2}{n}$$

and so on. Using right endpoint evaluation, we have $c_i = x_i = 2 + \frac{i}{n}$. This gives us the Riemann sum

$$R_n = \sum_{i=1}^{n} f(x_i) \Delta x = \sum_{i=1}^{n} \left(x_i^2 - 2x_i\right) \Delta x$$

$$= \sum_{i=1}^{n} \left[\left(2 + \frac{i}{n}\right)^2 - 2\left(2 + \frac{i}{n}\right)\right]\left(\frac{1}{n}\right)$$

$$= \sum_{i=1}^{n} \left(4 + 4\frac{i}{n} + \frac{i^2}{n^2} - 4 - \frac{2i}{n}\right)\left(\frac{1}{n}\right)$$

From Theorem 2.1 (ii) and (iii).

$$= \frac{1}{n^3} \sum_{i=1}^{n} i^2 + \frac{2}{n^2} \sum_{i=1}^{n} i$$

$$= \left(\frac{1}{n^3}\right)\frac{n(n+1)(2n+1)}{6} + \left(\frac{2}{n^2}\right)\frac{n(n+1)}{2}$$

$$= \frac{(n+1)(2n+1)}{6n^2} + \frac{(n+1)}{n}.$$

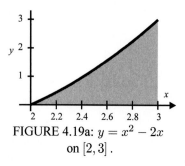

FIGURE 4.19a: $y = x^2 - 2x$ on $[2, 3]$.

The integral is then the limit of this Riemann sum as $n \to \infty$:

$$\int_2^3 \left(x^2 - 2x\right) dx = \lim_{n \to \infty}\left[\frac{(n+1)(2n+1)}{6n^2} + \frac{(n+1)}{n}\right] = \frac{2}{6} + 1 = \frac{4}{3}.$$

A graph of $y = x^2 - 2x$ on the interval $[2, 3]$ is shown in Figure 4.19a.

Notice that since the function is always positive on the interval $[2, 3]$, the integral corresponds to the area under the curve.

Finally, on the interval $[0, 3]$, we have $\Delta x = \frac{3}{n}$ and $x_0 = 0$, $x_1 = x_0 + \Delta x = \frac{3}{n}$,

$$x_2 = x_1 + \Delta x = \frac{3}{n} + \frac{3}{n} = \frac{3(2)}{n}$$

and so on. Using right endpoint evaluation, we have $c_i = x_i = \frac{3i}{n}$. This gives us the Riemann sum

$$R_n = \sum_{i=1}^{n} \left[\left(\frac{3i}{n}\right)^2 - 2\left(\frac{3i}{n}\right)\right]\left(\frac{3}{n}\right) = \sum_{i=1}^{n} \left(\frac{9i^2}{n^2} - \frac{6i}{n}\right)\left(\frac{3}{n}\right)$$

$$= \frac{27}{n^3}\sum_{i=1}^{n} i^2 - \frac{18}{n^2}\sum_{i=1}^{n} i$$

From Theorem 2.1 (ii) and (iii).

$$= \left(\frac{27}{n^3}\right)\frac{n(n+1)(2n+1)}{6} - \left(\frac{18}{n^2}\right)\frac{n(n+1)}{2}$$

$$= \frac{9(n+1)(2n+1)}{2n^2} - \frac{9(n+1)}{n}.$$

Thus,

$$\int_0^3 (x^2 - 2x)\, dx = \lim_{n\to\infty}\left[\frac{9(n+1)(2n+1)}{2n^2} - \frac{9(n+1)}{n}\right] = \frac{18}{2} - 9 = 0.$$

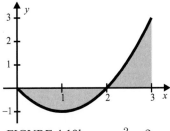

FIGURE 4.19b: $y = x^2 - 2x$ on $[0, 3]$.

On the interval $[0, 2]$, notice that the curve $y = x^2 - 2x$ lies below the x-axis and the area bounded between the curve and the x-axis is $\frac{4}{3}$. On the interval $[2, 3]$, the curve lies above the x-axis and the area bounded between the curve and the x-axis is also $\frac{4}{3}$. Notice that the integral of 0 on the interval $[0, 3]$ indicates that the signed areas have canceled out one another. (See Figure 4.19b for a graph of $y = x^2 - 2x$ on the interval $[0, 3]$.) You should also observe that the total area A bounded between $y = x^2 - 2x$ and the x-axis is the sum of the two areas indicated above, $A = \frac{4}{3} + \frac{4}{3} = \frac{8}{3}$.

We can also interpret signed area in terms of velocity and distance. Suppose that $v(t)$ is the velocity function for an object moving back and forth in a straight line. Notice that the velocity may be both positive and negative. When the velocity is negative, it indicates that the object is moving in the negative direction. If the velocity is positive on the interval $[t_1, t_2]$, then $\int_{t_1}^{t_2} v(t)\, dt$ equals the distance traveled in the positive direction. If the velocity is negative on the interval $[t_3, t_4]$, then $\int_{t_3}^{t_4} v(t)\, dt$ equals the distance traveled in the negative direction. If the object starts moving at time 0 and stops at time T, then $\int_0^T v(t)\, dt$ equals the distance traveled in the positive direction minus the distance traveled in the negative direction. That is, $\int_0^T v(t)\, dt$ corresponds to the *overall change* in distance from start to finish.

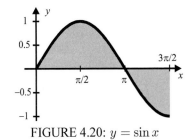

FIGURE 4.20: $y = \sin x$ on $[0, \frac{3\pi}{2}]$.

Example 4.5   Estimating Areas Above and Below the x-axis

Estimate the total area bounded between $y = \sin x$ and the x-axis on the interval $[0, \frac{3\pi}{2}]$.

Solution   From the graph (see Figure 4.20), notice that $\sin x \geq 0$ for $0 \leq x \leq \pi$ and $\sin x \leq 0$ for $\pi \leq x \leq 3\pi/2$. Thus, the total area will be

$$A = \int_0^\pi \sin x\, dx - \int_\pi^{3\pi/2} \sin x\, dx.$$

You can use the Midpoint Rule to get the following Riemann sums:

n	$R_n \approx \int_0^\pi \sin x\, dx$	n	$R_n \approx \int_\pi^{3\pi/2} \sin x\, dx$
10	2.0082	10	-1.0010
20	2.0020	20	-1.0003
50	2.0003	50	-1.0000
100	2.0001	100	-1.0000

Observe that the sums are converging to 2 and -1, respectively. There is always reason to be suspicious of such "nice" answers (were you expecting to get integer values here?), but we will soon be able to show that these are indeed the correct values. Thus, the total area bounded between $y = \sin x$ and the x-axis on $[0, \frac{3\pi}{2}]$ is

$$\int_0^\pi \sin x\, dx - \int_\pi^{3\pi/2} \sin x\, dx = 2 + 1 = 3.$$

∎

We have defined the definite integral of a function in terms of a limit, but we have not yet discussed the circumstances under which the limit actually exists. The following theorem indicates that many of the functions with which you are familiar are indeed integrable.

Theorem 4.1   If $f(x)$ is continuous on the closed interval $[a, b]$, then f is integrable on $[a, b]$.

The proof of Theorem 4.1 is too technical to include here. However, if you think about the area interpretation of the definite integral, the result should seem plausible.

∎

Next, we give some general rules for integrals.

Theorem 4.2   If f and g are integrable on $[a, b]$ and c is any constant, then the following are true.
(i) $\int_a^b [f(x) + g(x)]\, dx = \int_a^b f(x)\, dx + \int_a^b g(x)\, dx$,
(ii) $\int_a^b [f(x) - g(x)]\, dx = \int_a^b f(x)\, dx - \int_a^b g(x)\, dx$,
(iii) $\int_a^b c\, f(x)\, dx = c \int_a^b f(x)\, dx$ and
(iv) $\int_a^b f(x)\, dx = \int_a^c f(x)\, dx + \int_c^b f(x)\, dx$, for any c in $[a, b]$.

Proof. We can prove parts (i), (ii) and (iii) all at once by showing that for any constants c and d,

$$\int_a^b [c\, f(x) + d\, g(x)]\, dx = c \int_a^b f(x)\, dx + d \int_a^b g(x)\, dx.$$

By definition, we have

$$\int_a^b [c\, f(x) + d\, g(x)]\, dx = \lim_{n \to \infty} \sum_{i=1}^n [c\, f(c_i) + d\, g(c_i)]\, \Delta x$$

$$= \lim_{n \to \infty} \left[c \sum_{i=1}^n f(c_i)\, \Delta x + d \sum_{i=1}^n g(c_i)\, \Delta x \right]$$

From Theorem 2.2.

$$= c \lim_{n \to \infty} \sum_{i=1}^{n} f(c_i) \, \Delta x + d \lim_{n \to \infty} \sum_{i=1}^{n} g(c_i) \, \Delta x$$

$$= c \int_{a}^{b} f(x) \, dx + d \int_{a}^{b} g(x) \, dx,$$

where we have used our usual rules for summations plus the fact that f and g are integrable. We leave the proof of part (iv) to the exercises, but note that we have already illustrated the idea in Example 4.5. ∎

We now make a pair of reasonable definitions. First, for any integrable function f, if $a < b$, we define

$$\int_{b}^{a} f(x) \, dx = - \int_{a}^{b} f(x) \, dx.$$

This should appear reasonable in that if we integrate "backward" along an interval, the width of the rectangles corresponding to a Riemann sum (Δx) would seem to be negative. Secondly, if $f(a)$ is defined, we define

$$\int_{a}^{a} f(x) \, dx = 0.$$

If you think of the definite integral as area, this says that the area from a up to a is zero.

It turns out that a function is integrable even when it has a finite number of jump discontinuities, but is otherwise continuous. (Such a function is called **piecewise continuous**; see Figure 4.21 for the graph of such a function.)

In the following example, we give an integral of a discontinuous function.

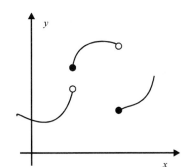

FIGURE 4.21: Piecewise continuous.

Example 4.6   An Integral with a Discontinuous Integrand

Evaluate $\int_{0}^{3} f(x) \, dx$, where $f(x)$ is defined by

$$f(x) = \begin{cases} 2x, & \text{if } x \leq 2 \\ 1, & \text{if } x > 2 \end{cases}.$$

Solution   We start by looking at a graph of $y = f(x)$ in Figure 4.22a. Notice that although f is discontinuous at $x = 2$, it has only a single jump discontinuity and so, is piecewise continuous on $[0, 3]$. By Theorem 4.2 (iv), we have that

$$\int_{0}^{3} f(x) \, dx = \int_{0}^{2} f(x) \, dx + \int_{2}^{3} f(x) \, dx.$$

Referring to Figure 4.22b, observe that $\int_{0}^{2} f(x) \, dx$ corresponds to the area of the triangle of base 2 and altitude 4 shaded in the figure, so that

$$\int_{0}^{2} f(x) \, dx = \frac{1}{2} \, (base) \, (height) = \frac{1}{2} \, (2) \, (4) = 4.$$

Next, also notice from Figure 4.22b that $\int_{2}^{3} f(x) \, dx$ corresponds to the area of the square of side 1, so that

$$\int_{2}^{3} f(x) \, dx = 1.$$

We now have

$$\int_{0}^{3} f(x) \, dx = \int_{0}^{2} f(x) \, dx + \int_{2}^{3} f(x) \, dx = 4 + 1 = 5.$$

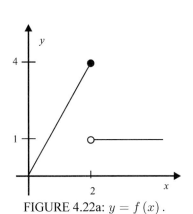

FIGURE 4.22a: $y = f(x)$.

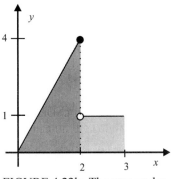

FIGURE 4.22b: The area under the curve $y = f(x)$ on $[0, 3]$.

Notice that in this case, the areas corresponding to the two integrals could be computed using simple geometric formulas and so, there was no need to compute Riemann sums here.

∎

Average Value of a Function

We all have a good idea of what is meant by an *average*. For instance, to compute the average age of students in your calculus class, you need only add up each student's age and divide the total by the number of students in your class. By contrast, suppose that you wanted to find the average depth of a cross section of a lake. In this case, there are an infinite number of points on the surface of the lake from which you might measure the depth. Even if it were somehow possible to measure the depth at these infinitely many points, what would you do: add together infinitely many numbers and then divide by infinity? A reasonable approach is to sample the depth of the lake at a number of points spread out along the length of the lake and then average these depths, as indicated in Figure 4.23.

FIGURE 4.23: Average Depth of a cross section of a lake.

More generally, we often want to calculate the average value of a function, f on some interval $[a, b]$. To do this, we first pick a number of points at which we will sample the value of the function and form the partition of $[a, b]$:

$$a = x_0 < x_1 < \cdots < x_n = b,$$

where the difference between successive points is $\Delta x = \dfrac{b - a}{n}$. The **average value**, f_{ave} is then given approximately by the average of the function values at $x_1, x_2, \ldots, x_n$:

$$f_{ave} \approx \frac{1}{n} \left[f(x_1) + f(x_2) + \cdots + f(x_n) \right]$$

$$= \frac{1}{n} \sum_{i=1}^{n} f(x_i)$$

Multiply and divide by $(b - a)$.

$$= \frac{1}{b - a} \sum_{i=1}^{n} f(x_i) \left(\frac{b - a}{n} \right)$$

Since $\Delta x = \frac{b-a}{n}$.

$$= \frac{1}{b - a} \sum_{i=1}^{n} f(x_i) \, \Delta x.$$

Notice that the last summation is a Riemann sum. Observe that the more points we sample, the better our approximation should be. So, letting $n \to \infty$, we arrive at an integral representing average value:

$$\boxed{f_{ave} = \lim_{n \to \infty} \left[\frac{1}{b - a} \sum_{i=1}^{n} f(x_i) \, \Delta x \right] = \frac{1}{b - a} \int_a^b f(x) \, dx.} \qquad (4.2)$$

FIGURE 4.24 $y = \sin x$ and its average.

Example 4.7   Computing the Average Value of a Function
    Compute the average value of $f(x) = \sin x$ on the interval $[0, \pi]$.

Solution   From (4.2), we have

$$f_{ave} = \frac{1}{\pi - 0} \int_0^\pi \sin x \, dx.$$

We can approximate the value of this integral by calculating some Riemann sums, to obtain the approximate average, $f_{ave} \approx 0.6366198$. (This is left as an exercise.) In Figure 4.24, we show a graph of $y = \sin x$ and its average value on the interval $[0, \pi]$. You should note that the area under both curves is the same.

∎

FIGURE 4.25: A lake and its average depth.

Returning to the problem of finding the average depth of a lake, imagine the dirt at the bottom of the lake settling out to form a flat bottom. The depth of the lake would then be constant and equal to the average value of the depth of the original lake. In the settling out process, the depth of one (and possibly more) points would not change (see Figure 4.25). That is, the average depth of the lake exactly equals the depth of the lake at one or more points. A precise statement of this result is given below as the Integral Mean Value Theorem.

Notice that for any constant, c,

$$\int_a^b c\,dx = \lim_{n\to\infty} \sum_{i=1}^n c\,\Delta x = c\sum_{i=1}^n \Delta x = c\,(b-a),$$

since $\sum_{i=1}^n \Delta x$ is simply the sum of the lengths of the subintervals in the partition.

Let f be any continuous function defined on $[a,b]$. Recall that by the Extreme Value Theorem, since f is continuous, it has a minimum, m and a maximum, M on $[a,b]$. It follows that

$$m \le f(x) \le M, \text{ for all } x \in [a,b]$$

and consequently,

$$\int_a^b m\,dx \le \int_a^b f(x)\,dx \le \int_a^b M\,dx.$$

Since m and M are constants, we get

$$m\,(b-a) \le \int_a^b f(x)\,dx \le M\,(b-a).$$

Finally, dividing by $(b-a) > 0$, we obtain

$$m \le \frac{1}{b-a}\int_a^b f(x)\,dx \le M.$$

That is, $\frac{1}{b-a}\int_a^b f(x)\,dx$ (the average value of f on $[a,b]$) is a number between the minimum and the maximum values of f on $[a,b]$. Since f is a continuous function, we have by the Intermediate Value Theorem that there must be some $c \in (a,b)$ for which

$$f(c) = \frac{1}{b-a}\int_a^b f(x)\,dx.$$

We have just proved a theorem:

Theorem 4.3   (Integral Mean Value Theorem) If f is continuous on $[a,b]$, there is a number $c \in (a,b)$ for which

$$f(c) = \frac{1}{b-a}\int_a^b f(x)\,dx.$$

The Integral Mean Value Theorem is a fairly simple idea (that a continuous function will take on its average value at some point), but it has some significant applications. The first of these will be found in the following section, in the proof of one of the most significant results in the calculus, the Fundamental Theorem of Calculus.

Exercises 4.4

1. (writing) You will be asked to prove in exercise 49 below that if $f(x) \geq g(x)$ for all x in the interval (a, b), then $\int_a^b f(x)\,dx \geq \int_a^b g(x)\,dx$, assuming that both $f(x)$ and $g(x)$ are integrable on the interval (a, b). Sketch functions for which $f(x) \geq g(x) \geq 0$ and explain in graphical terms why $\int_a^b f(x)\,dx \geq \int_a^b g(x)\,dx$.

2. (writing) To get a physical interpretation of the result in exercise 1, suppose that $f(x)$ and $g(x)$ are velocity functions for two different objects starting at the same position. If $f(x) \geq g(x) \geq 0$, explain why it follows that $\int_a^b f(x)\,dx \geq \int_a^b g(x)\,dx$.

3. (writing) The Integral Mean Value Theorem says that if $f(x)$ is continuous on the interval $[a, b]$, then there exists a number c between a and b such that $f(c)\,(b - a) = \int_a^b f(x)\,dx$. By thinking of the left-hand side of this equation as the area of a rectangle, sketch a picture that illustrates this result, and explain why the result follows.

4. (writing) Write out the Integral Mean Value Theorem as applied to the derivative $f'(x)$. Then write out the Mean Value Theorem for derivatives (see section 2.9). If the c-values identified by each theorem are the same, what does $\int_a^b f'(x)\,dx$ have to equal? Explain why, at this point, we don't know whether or not the c-values are the same.

In exercises 5–10, use Riemann sums to estimate the value of the integral (obtain two digits of accuracy).

5. $\displaystyle\int_0^3 (x^3 + x)\,dx$ **6.** $\displaystyle\int_1^2 (x^2 - 1)\,dx$

7. $\displaystyle\int_2^4 \frac{1}{x^2}\,dx$ **8.** $\displaystyle\int_0^3 \sqrt{x^2 + 1}\,dx$

9. $\displaystyle\int_0^\pi \sin x^2\,dx$ **10.** $\displaystyle\int_{-2}^2 e^{-x^2}\,dx$

In exercises 11–16, evaluate the integral by computing the limit of Riemann sums.

11. $\displaystyle\int_0^1 2x\,dx$ **12.** $\displaystyle\int_1^2 2x\,dx$

13. $\displaystyle\int_0^2 x^2\,dx$ **14.** $\displaystyle\int_0^3 (x^2 + 1)\,dx$

15. $\displaystyle\int_1^3 (2x - 1)\,dx$ **16.** $\displaystyle\int_{-2}^2 (x^2 - 1)\,dx$

In exercises 17–26, write the given (total) area as an integral or sum of integrals.

17. The area above the x-axis and below $y = 4 - x^2$

18. The area above the x-axis and below $y = 4x - x^2$

19. The area below the x-axis and above $y = x^2 - 4$

20. The area below the x-axis and above $y = x^2 - 4x$

21. The area of the region bounded by $y = x^2$, $x = 2$ and the x-axis

22. The area of the region bounded by $y = x^3$, $x = 3$ and the x-axis

23. The area between $y = \sin x$ and the x-axis for $0 \leq x \leq \pi$

24. The area between $y = \sin x$ and the x-axis for $-\pi/2 \leq x \leq \pi/4$

25. The area between $y = x^3 - 3x^2 + 2x$ and the x-axis for $0 \leq x \leq 2$

26. The area between $y = x^3 - 4x$ and the x-axis for $-2 \leq x \leq 3$

In exercises 27–30, use the velocity function to estimate the distance traveled in the given time interval.

27. $v(t) = 60 - 16t$, $[0, 2]$

28. $v(t) = 20 + 10t$, $[1, 3]$

29. $v(t) = 40(1 - e^{-2t})$, $[0, 4]$

30. $v(t) = 30e^{-t/4}$, $[0, 4]$

In exercises 31–34, use Theorem 4.2 to write the expression as a single integral.

31. $\displaystyle\int_0^2 f(x)\,dx + \int_2^3 f(x)\,dx$

32. $\displaystyle\int_0^3 f(x)\,dx - \int_2^3 f(x)\,dx$

33. $\displaystyle\int_0^2 f(x)\,dx + \int_2^1 f(x)\,dx$

34. $\displaystyle\int_{-1}^2 f(x)\,dx + \int_2^3 f(x)\,dx$

35. Prove part (iv) of Theorem 4.2.

36. Show that parts (i)–(iii) of Theorem 4.2 must follow if it is true that

$$\int_a^b [c\,f(x) + d\,g(x)]\,dx = c\int_a^b f(x)\,dx + d\int_a^b g(x)\,dx.$$

(Hint: Make clever choices of c and d, like $d = 0$ or $c = d = 1$.)

In exercises 37–40, use the graph to determine whether $\int_0^2 f(x)\,dx$ is positive or negative.

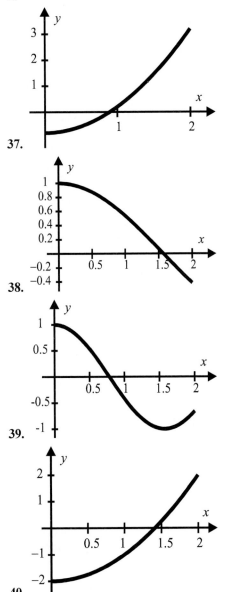

37.

38.

39.

40.

41. For the functions $f(x) = \begin{cases} 2x & \text{if } x < 1 \\ x^2 + 2 & \text{if } x \geq 1 \end{cases}$ and

$g(x) = \begin{cases} 2x & \text{if } x \leq 1 \\ x^2 + 2 & \text{if } x > 1 \end{cases}$ assume that $\int_0^2 f(x)\,dx$ and $\int_0^2 g(x)\,dx$ exist. Explain why the approximating Riemann sums with midpoint evaluations are equal for any even value of n. Argue that this result implies that the two integrals are equal.

42. For the functions defined in exercise 41, explain why the integrals are both equal to the sum $\int_0^1 2x\,dx + \int_1^2 (x^2 + 2)\,dx$.

In exercises 43–46, estimate $\int_0^4 f(x)\,dx$.

43. $f(x) = \begin{cases} 2x & \text{if } x < 1 \\ x^2 + 2 & \text{if } x \geq 1 \end{cases}$

44. $f(x) = \begin{cases} x^2 & \text{if } x < 2 \\ 3x & \text{if } x \geq 2 \end{cases}$

45. $f(x) = \begin{cases} 2x^2 & \text{if } x < 3 \\ x + 1 & \text{if } x \geq 3 \end{cases}$

46. $f(x) = \begin{cases} x - 1 & \text{if } x < 1 \\ x^2 & \text{if } x \geq 1 \end{cases}$

47. Suppose that, for a particular population of organisms, the birth rate is given by $b(t) = 410 - 0.3t$ organisms per month and the death rate is given by $a(t) = 390 + 0.2t$ organisms per month. Explain why $\int_0^{12}[b(t) - a(t)]\,dt$ represents the net change in population in the first 12 months. Determine for which values of t it is true that $b(t) > a(t)$. At which times is the population increasing? decreasing? Determine the time at which the population reaches a maximum.

48. Suppose that, for a particular population of organisms, the birth rate is given by $b(t) = 400 - 3\sin t$ organisms per month and the death rate is given by $a(t) = 390 + 0.2t$ organisms per month. Explain why $\int_0^{12}[b(t) - a(t)]\,dt$ represents the net change in population in the first 12 months. Graphically determine for which values of t it is true that $b(t) > a(t)$. At which times is the population increasing? decreasing? Estimate the time at which the population reaches a maximum.

49. Prove that if $f(x) \geq g(x)$ for all x in the interval $[a, b]$ and both f and g are integrable on $[a, b]$, then

$$\int_a^b f(x)\,dx \geq \int_a^b g(x)\,dx.$$

50. Find a counterexample for the following statement. (That is, find an example that shows that the statement is not always true.) As n increases, the Riemann sum $R_n(f)$ gets closer to $\int_a^b f(x)\,dx$.

51. For a particular ideal gas at constant temperature, pressure P and volume V are related by $PV = 10$. The work required to increase the volume from $V = 2$ to $V = 4$ is given by the integral $\int_2^4 P(V)\,dV$. Estimate the value of this integral.

52. Suppose that the temperature t months into the year is given by $T(t) = 64 - 24\cos\frac{\pi}{6}t$ (degrees Fahrenheit). Estimate the average temperature over an entire year. Explain why this answer is obvious from the graph of $T(t)$.

In exercises 53–60, estimate the average value of the function on the given interval.

53. $f(x) = 2x + 1$, $[0, 4]$ **54.** $f(x) = 3x - 1$, $[1, 3]$

55. $f(x) = x^2 - 1$, $[1, 3]$ **56.** $f(x) = x^2 + 2x$, $[0, 1]$

57. $f(x) = 2x - 2x^2$, $[0, 1]$

58. $f(x) = x^3 - 3x^2 + 2x$, $[1, 2]$

59. $f(x) = \cos x$, $[0, \pi/2]$ **60.** $f(x) = \sin x$, $[0, \pi/2]$

61. The *impulse-momentum equation* states the relationship between a force $F(t)$ applied to an object of mass m and the resulting change in velocity Δv of the object. The equation is $m \, \Delta v = \int_a^b F(t) \, dt$, where $\Delta v = v(b) - v(a)$. Suppose that the force of a baseball bat on a ball is approximately $F(t) = 9 - 10^8 (t - 0.0003)^2$ thousand pounds for t between 0 and 0.0006 seconds. What is the maximum force on the ball? Using $m = 0.01$ for the mass of a baseball, estimate the change in velocity Δv (in ft/s).

62. Measurements taken of the feet of badminton players lunging for a shot indicate a vertical force of approximately $F(t) = 1000 - 25,000(t - 0.2)^2$ Newtons for t between 0 and 0.4 seconds (see *The Science of Racquet Sports*). For a player of mass $m = 5$, use the impulse-momentum equation in exercise 61 to estimate the change in vertical velocity of the player.

63. Use the Integral Mean Value Theorem to prove the following fact for a continuous function. If the evaluation point is chosen properly, the Riemann sum approximation of $\int_a^b f(x) \, dx$ with $n = 1$ can be made to be exact.

64. Use the Integral Mean Value Theorem to prove the following fact for a continuous function. For any positive integer n, there exists a set of evaluation points for which the Riemann sum approximation of $\int_a^b f(x) \, dx$ is exact.

65. (exploratory) Many of the basic quantities used by epidemiologists to study the spread of disease are described by integrals. In the case of AIDS, a person becomes infected with the HIV virus and, after an incubation period, develops AIDS. Our goal is to derive a formula for the number of AIDS cases given the HIV infection rate $g(t)$ and the incubation distribution $F(t)$. To take a simple case, suppose that the infection rate the first month is 20 people per month, the infection rate the second month is 30 people per month and the infection rate the third month is 25 people per month. Then $g(1) = 20$, $g(2) = 30$ and $g(3) = 25$. Also, suppose that 20% of those infected develop AIDS after 1 month, 50% develop AIDS after 2 months and 30% develop AIDS after 3 months (fortunately, these figures are not at all realistic). Then $F(1) = 0.2$, $F(2) = 0.5$ and $F(3) = 0.3$. Explain why the number of people developing AIDS in the fourth month would be $g(1) \, F(3) + g(2) \, F(2) + g(3) \, F(1)$. Compute this number. Next, suppose that $g(0.5) = 16$, $g(1) = 20$, $g(1.5) = 26$, $g(2) = 30$, $g(2.5) = 28$, $g(3) = 25$ and $g(3.5) = 22$. Further, suppose that $F(0.5) = 0.1$, $F(1) = 0.1$, $F(1.5) = 0.2$, $F(2) = 0.3$, $F(2.5) = 0.1$, $F(3) = 0.1$ and $F(3.5) = 0.1$. Compute the number of people developing AIDS in the fourth month. If we have $g(t)$ and $F(t)$ defined at all real numbers t, explain why the number of people developing AIDS in the fourth month equals $\int_0^4 g(t) \, F(4 - t) \, dt$.

4.5   The Fundamental Theorem of Calculus

In this section, we present a pair of results known collectively as the Fundamental Theorem of Calculus. What could these results be to make them **fundamental** to calculus? On a practical level, the Fundamental Theorem provides us with a much-needed shortcut for computing definite integrals. Just as the power rule and other basic differentiation formulas relieved us of the burden of using the limit definition of the derivative, so does the Fundamental Theorem give us a powerful tool for computing integrals symbolically without struggling to find limits of Riemann sums. Given that we can presently compute such limits exactly only for a very small number of functions, this theorem takes on even greater significance.

On a conceptual level, the Fundamental Theorem unifies the seemingly disconnected studies of differentiation and integration. In the rush of learning the rules for computing derivatives and integrals, there is barely time to wonder why these concepts are in the same course. The Fundamental Theorem shows us that differentiation and integration are in fact, inverse processes. In this sense, the theorem is truly **fundamental** to calculus as a coherent discipline.

We have dropped a few hints as to the nature of the first part of the Fundamental Theorem. First, recall that we used suspiciously similar notation for indefinite and definite integrals. We have also used both antidifferentiation and area calculations to compute distance from velocity. However, the Fundamental Theorem makes much stronger statements about the relationship between differentiation and integration.

Theorem 5.1   (Fundamental Theorem of Calculus, Part I) If f is continuous on $[a, b]$ and $F(x)$ is any antiderivative of f, then

$$\int_a^b f(x)\, dx = F(b) - F(a). \tag{5.1}$$

NOTE: The Fundamental Theorem, Part I says that to compute a definite integral, we need only find an antiderivative and then evaluate it at the two limits of integration. Observe that this is a vast improvement over computing limits of Riemann sums, which we could compute exactly for only a few simple cases.

Proof. First, we partition $[a, b]$:

$$a = x_0 < x_1 < x_2 < \cdots < x_n = b,$$

where $x_i - x_{i-1} = \Delta x = \frac{b-a}{n}$, for $i = 1, 2, \ldots, n$. Working backward, note that by virtue of all the cancellations, we can write

$$F(b) - F(a) = [F(x_1) - F(x_0)] + [F(x_2) - F(x_1)] + \cdots + [F(x_n) - F(x_{n-1})]$$
$$= \sum_{i=1}^{n} [F(x_i) - F(x_{i-1})]. \tag{5.2}$$

Since F is an antiderivative of f, F is differentiable on (a, b) and continuous on $[a, b]$. By the Mean Value Theorem, we have for each $i = 1, 2, \ldots, n$, that

$$F(x_i) - F(x_{i-1}) = F'(c_i)\,(x_i - x_{i-1}) = f(c_i)\,\Delta x, \tag{5.3}$$

for some $c_i \in (x_{i-1}, x_i)$. Thus, owing to (5.2) and (5.3), we have

$$F(b) - F(a) = \sum_{i=1}^{n} [F(x_i) - F(x_{i-1})] = \sum_{i=1}^{n} f(c_i)\,\Delta x. \tag{5.4}$$

You should recognize this last expression as a Riemann sum for f on $[a, b]$. Taking the limit of both sides of (5.4), as $n \to \infty$, we find that

$$\int_a^b f(x)\, dx = \lim_{n \to \infty} \sum_{i=1}^{n} f(c_i)\,\Delta x = \lim_{n \to \infty} [F(b) - F(a)]$$
$$= F(b) - F(a),$$

as desired, since this last quantity is a constant.   ∎

Remark 5.1   We will often use the notation

$$F(x)\Big|_a^b = F(b) - F(a).$$

This allows us to write down the antiderivative before evaluating it at the endpoints.

Example 5.1   Using the Fundamental Theorem

Compute $\int_0^2 \left(x^2 - 2x \right) dx.$

Solution   Notice that $f(x) = x^2 - 2x$ is continuous on the interval $[0, 2]$ and so, we can apply the Fundamental Theorem. We find an antiderivative from the power rule and simply evaluate:

$$\int_0^2 (x^2 - 2x)\, dx = \left(\frac{1}{3}x^3 - x^2\right)\Big|_0^2 = \left(\frac{8}{3} - 4\right) - (0) = -\frac{4}{3}.$$

■

Recall that we had already evaluated the integral in example 5.1 by computing the limit of Riemann sums (see example 4.3). Given a choice, which method would you prefer?

While you had a choice in example 5.1, the integrals in examples 5.2, 5.3, 5.4 and 5.5 **cannot** be evaluated by computing the limit of a Riemann sum directly, as we have no formulas for the summations involved.

Example 5.2   Computing a Definite Integral Exactly

Compute $\int_1^4 \left(\sqrt{x} - \frac{1}{x^2}\right) dx.$

Solution   Observe that since $f(x) = x^{1/2} - x^{-2}$ is continuous on $[1, 4]$, we can apply the Fundamental Theorem. An antiderivative is $F(x) = \frac{2}{3}x^{3/2} + x^{-1}$, so

$$\int_1^4 \left(\sqrt{x} - \frac{1}{x^2}\right) dx = \frac{2}{3}x^{3/2} + x^{-1}\Big|_1^4 = \left[\frac{2}{3}(4)^{3/2} + 4^{-1}\right] - \left(\frac{2}{3} + 1\right) = \frac{47}{12}.$$

■

Example 5.3   Using the Fundamental Theorem to Compute Areas

Find the area under the curve $f(x) = \sin x$ on the interval $[0, \pi]$.

Solution   Since $\sin x \geq 0$ and $\sin x$ is continuous on $[0, \pi]$, we have that

$$Area = \int_0^\pi \sin x\, dx.$$

Notice that an antiderivative of $\sin x$ is $F(x) = -\cos x$. By the Fundamental Theorem, then, we have

$$\int_0^\pi \sin x\, dx = F(\pi) - F(0) = (-\cos \pi) - (-\cos 0) = -(-1) - (-1) = 2.$$

■

Example 5.4   A Definite Integral Involving an Exponential Function

Compute $\int_0^4 e^{-2x}\, dx.$

Solution   Since $f(x) = e^{-2x}$ is continuous, we can apply the Fundamental Theorem. Notice that an antiderivative of e^{-2x} is $-\frac{1}{2}e^{-2x}$ so that

$$\int_0^4 e^{-2x}\, dx = -\frac{1}{2}e^{-2x}\Big|_0^4 = -\frac{1}{2}e^{-8} - \left(-\frac{1}{2}e^0\right) \approx 0.49983.$$

Note that it is not always true that an antiderivative evaluated at $x = 0$ will be zero. Also, note that the exact integral is $-\frac{1}{2}e^{-8} + \frac{1}{2}$, while 0.49983 is simply a decimal approximation.

■

Example 5.5   A Definite Integral Involving a Logarithm

Evaluate $\displaystyle\int_{-3}^{-1} \frac{2}{x}\, dx$.

Solution   Since $f(x) = \dfrac{2}{x}$ is continuous on $[-3, -1]$, we can apply the Fundamental Theorem. First, recall that an antiderivative for $f(x)$ is $2\ln|x|$. (It's a common error to leave off the absolute values. In this case, the error is fatal! Look carefully at the following to see why.)

$$\int_{-3}^{-1} \frac{2}{x}\, dx = 2\ln|x|\Big|_{-3}^{-1}$$

$$= 2\left[\ln|-1| - \ln|-3|\right] = -2\ln 3.$$ ∎

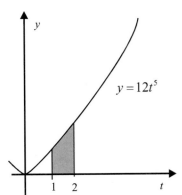

FIGURE 4.26a: Area from $t = 1$ to $t = 2$.

Example 5.6   A Definite Integral with a Variable Upper Limit

Evaluate $\displaystyle\int_{1}^{x} 12t^5\, dt$.

Solution   Even though the upper limit of integration is a variable, we can use the Fundamental Theorem to evaluate this, since $f(t) = 12t^5$ is continuous on any interval. We have

$$\int_{1}^{x} 12t^5\, dt = 12\frac{t^6}{6}\Big|_{1}^{x} = 2\left(x^6 - 1\right).$$

It's not surprising that the definite integral will be a function of x, since the one limit of integration involves x. The following observation may be surprising, though. Note that

$$\frac{d}{dx}\left[2\left(x^6 - 1\right)\right] = 12\,x^5.$$

This is exactly the same as the original integrand, except that the (dummy) variable of integration, t has been replaced by the variable in the upper limit of integration, x.   ∎

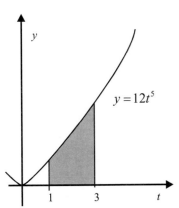

FIGURE 4.26b: Area from $t = 1$ to $t = 3$.

The seemingly odd coincidence observed in example 5.6 is in fact, not an isolated occurrence, as we see in the next theorem. First, you need to be clear about what a function like $F(x) = \int_{1}^{x} 12t^5\, dt$ means. Notice that the function value at $x = 2$ is found by replacing x by 2:

$$F(2) = \int_{1}^{2} 12t^5\, dt,$$

which corresponds to the area under the curve $y = 12t^5$ from $t = 1$ to $t = 2$ (see Figure 4.26a). Similarly, the function value at $x = 3$ is

$$F(3) = \int_{1}^{3} 12t^5\, dt,$$

which corresponds to the area under the curve $y = 12t^5$ from $t = 1$ to $t = 3$ (see Figure 4.26b). More generally, for any $x > 1$, $F(x)$ gives the area under the curve $y = f(t)$ from $t = 1$ up to $t = x$ (see Figure 4.26c). For this reason, the function F is sometimes called an **area function**. Notice that for $x > 1$, as x increases, $F(x)$ gives more and more of the area under the curve to the right of $t = 1$.

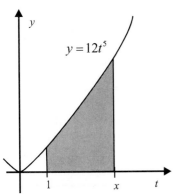

FIGURE 4.26c: Area from $t = 1$ to $t = x$.

Theorem 5.2   (Fundamental Theorem of Calculus, Part II) If f is continuous on $[a, b]$ and $F(x) = \displaystyle\int_{a}^{x} f(t)\, dt$, then $F'(x) = f(x)$, on $[a, b]$.

Proof. Using the definition of derivative, we have

$$F'(x) = \lim_{h \to 0} \frac{F(x+h) - F(x)}{h} = \lim_{h \to 0} \frac{1}{h} \left[\int_a^{x+h} f(t)\, dt - \int_a^x f(t)\, dt \right]$$

$$= \lim_{h \to 0} \frac{1}{h} \left[\int_a^{x+h} f(t)\, dt + \int_x^a f(t)\, dt \right] = \lim_{h \to 0} \frac{1}{h} \int_x^{x+h} f(t)\, dt. \quad (5.5)$$

Look very carefully at the last term in (5.5). You may recognize it as the limit of the average value of $f(x)$ on the interval $[x, x+h]$ (if $h > 0$). By the Integral Mean Value Theorem (Theorem 4.3), we have

$$\frac{1}{h} \int_x^{x+h} f(t)\, dt = f(c), \quad (5.6)$$

for some number c between x and $x + h$. Finally, since c is between x and $x + h$, we have that $c \to x$, as $h \to 0$. Since f is continuous, it follows from (5.5) and (5.6) that

$$F'(x) = \lim_{h \to 0} \frac{1}{h} \int_x^{x+h} f(t)\, dt = \lim_{h \to 0} f(c) = f(x),$$

as desired.

∎

NOTE: Part II of the Fundamental Theorem says that **every** continuous function, f has an antiderivative, namely, $\int_a^x f(t)\, dt$.

Example 5.7   Using the Fundamental Theorem, Part II

For $F(x) = \displaystyle\int_1^x \left(t^2 - 2t + 3 \right) dt$, compute $F'(x)$.

Solution   Here, the integrand is $f(t) = t^2 - 2t + 3$. By Theorem 5.2, the derivative is

$$F'(x) = f(x) = x^2 - 2x + 3,$$

that is, $F'(x)$ is the function in the integrand with t replaced by x. It really is that easy!

∎

Before moving on to more complicated examples, let's look at example 5.7 in more detail, just to get more comfortable with the meaning of part II of the Fundamental Theorem. First, we can use Part I of the Fundamental Theorem to find

$$F(x) = \int_1^x (t^2 - 2t + 3)\, dt = \frac{1}{3} t^3 - t^2 + 3t \Big|_1^x = \left(\frac{1}{3} x^3 - x^2 + 3x \right) - \left(\frac{1}{3} - 1 + 3 \right).$$

It's easy to differentiate this directly, to get

$$F'(x) = \frac{1}{3} 3x^2 - 2x + 3 - 0 = x^2 - 2x + 3.$$

Notice that the lower limit of integration (in this case, 1) has no effect on the value of $F'(x)$. In the definition of $F(x)$, the lower limit of integration merely determines the value of the constant that is subtracted at the end of the calculation of $F(x)$. Since the derivative of a constant is 0, this value does not affect $F'(x)$.

Example 5.8   Using the Chain Rule and the Fundamental Theorem, Part II

If $F(x) = \displaystyle\int_2^{x^2} \cos t\, dt$, compute $F'(x)$.

Solution   Let $u = x^2$, so that

$$F(x) = \int_2^u \cos t \, dt.$$

From the chain rule,

$$F'(x) = \cos u \, \frac{du}{dx} = \cos u \, (2x) = 2x \cos x^2.$$

■

Remark 5.2   The general form of the chain rule used in example 5.8 is:

if $F(x) = \int_a^{h(x)} f(t) \, dt$, then $F'(x) = f(h(x)) \, h'(x)$ or

$$\frac{d}{dx} \int_a^{h(x)} f(t) \, dt = f(h(x)) \, h'(x).$$

Example 5.9   **An Integral with Variable Upper and Lower Limits**

If $F(x) = \int_{2x}^{x^2} \sqrt{t^2 + 1} \, dt$, compute $F'(x)$.

Solution   The Fundamental Theorem only applies to definite integrals with variables in the upper limit, so we will first rewrite the integral as

$$F(x) = \int_{2x}^0 \sqrt{t^2 + 1} \, dt + \int_0^{x^2} \sqrt{t^2 + 1} \, dt = -\int_0^{2x} \sqrt{t^2 + 1} \, dt + \int_0^{x^2} \sqrt{t^2 + 1} \, dt.$$

Using the chain rule as in example 5.8, we get

$$F'(x) = -\sqrt{(2x)^2 + 1} \, \frac{d}{dx}(2x) + \sqrt{(x^2)^2 + 1} \, \frac{d}{dx}(x^2)$$

$$= 2x\sqrt{x^4 + 1} - 2\sqrt{4x^2 + 1}.$$

■

Before discussing the theoretical significance of the two parts of the Fundamental Theorem, we present two examples that remind you of why you would want to compute integrals and derivatives.

Example 5.10   **Computing the Distance Fallen by an Object**

Suppose the (downward) velocity of a sky diver is given by $v(t) = 30(1 - e^{-t})$ ft/s for the first 5 seconds of a jump. Compute the distance fallen.

Solution   Recall that the distance d is given by the definite integral (area)

$$d = \int_0^5 \left(30 - 30e^{-t}\right) dt = 30t + 30e^{-t} \Big|_0^5$$

$$= (150 + 30e^{-5}) - (0 + 30e^0) = 120 + 30e^{-5} \approx 120.2 \text{ feet.}$$

■

Recall that velocity is the *instantaneous rate of change* of the distance function with respect to time. We see in example 5.10 that the definite integral of velocity gives the *total change* of the distance function over the given time interval. A similar interpretation of derivative and the definite integral holds for many quantities of interest. In the following example, we look at the rate of change and total change of water in a tank.

Example 5.11   Rate of Change and Volume of a Water Tank

Suppose that water can flow in and out of a storage tank. The net rate of change (that is, the rate in minus the rate out) of water is $f(t) = 20(t^2 - 1)$ gallons per minute. (a) For $0 \leq t \leq 3$, determine when the water level is increasing and when the water level is decreasing. (b) If the tank has 200 gallons of water at time $t = 0$, determine how many gallons are in the tank at $t = 3$.

Solution   Let $w(t)$ be the number of gallons in the tank at time t. (a) Notice that the water level decreases if $w'(t) = f(t) < 0$. We have

$$f(t) = 20(t^2 - 1) < 0, \ \text{ if } 0 \leq t < 1.$$

Alternatively, the water level increases if $w'(t) = f(t) > 0$. In this case, we have

$$f(t) = 20(t^2 - 1) > 0, \ \text{ if } 1 < t \leq 3.$$

(b) We start with $w'(t) = 20(t^2 - 1)$. Integrating from $t = 0$ to $t = 3$, we have

$$\int_0^3 w'(t)\, dt = \int_0^3 20(t^2 - 1)\, dt$$

Evaluating the integrals on both sides yields

$$w(3) - w(0) = 20 \left(\frac{t^3}{3} - t \right)_{t=0}^{t=3}.$$

Since $w(0) = 200$, we have

$$w(3) - 200 = 20(9 - 3)$$

and hence,

$$w(3) = 200 + 120 = 320.$$

so that the tank will have 320 gallons at time 3.

∎

In the following example, we use Part II of the Fundamental Theorem to determine information about a seemingly very complicated function. Notice that although we don't know how to evaluate the integral, we can use the Fundamental Theorem to obtain some important information about the function.

Example 5.12   Finding a Tangent Line for a Function Defined as an Integral

For the function $F(x) = \int_4^{x^2} \ln(t^3 + 4)\, dt$, find an equation of the tangent line at $x = 2$.

Solution   Notice that there are almost no function values that we can compute exactly, yet we can easily find an equation of a tangent line! From Part II of the Fundamental Theorem and the chain rule, we get the derivative

$$F'(x) = \ln\left[(x^2)^3 + 4\right] \frac{d}{dx}(x^2) = \ln\left[(x^2)^3 + 4\right](2x).$$

So, the slope at $x = 2$ is $F'(2) = \ln(68)(4) \approx 16.878$. The tangent passes through the point $x = 2$ and $y = F(2) = \int_4^4 \ln(t^3 + 4)\, dt = 0$ (since the upper limit equals the lower limit). An equation of the tangent line is then

$$y = 4 \ln 68 \, (x - 2).$$

∎

At the beginning of this section, we briefly discussed the theoretical importance of the Fundamental Theorem in unifying the differential and integral calculus. Both parts of the theorem involve both derivatives and integrals, and this connection is certainly important. Now that we've discussed the computational significance of the results, we want to close the section by observing that the two parts of the Fundamental Theorem are different sides of the same theoretical coin. In part I, we observe that if we can find an antiderivative for the function in the integrand, then we can easily evaluate the definite integral. In Part II, we observe that every continuous function has an antiderivative, in the form of a definite integral with a variable limit of integration.

Recall the conclusions of Parts I and II of the Fundamental Theorem:

$$\int_a^b F'(x) \, dx = F(b) - F(a)$$

and

$$\frac{d}{dx} \int_a^x f(t) \, dt = f(x).$$

In both cases, we are saying that differentiation and integration are in some sense *inverse operations*: their effects in a certain sense (with appropriate hypotheses) cancel each other out.

Exercises 4.5

1. (writing) To explore part I of the Fundamental Theorem graphically, first suppose that $F(x)$ is increasing on the interval $[a, b]$. Explain why both of the expressions $F(b) - F(a)$ and $\int_a^b F'(x) \, dx$ will be positive. Further, explain why the faster $F(x)$ increases, the larger each expression will be. Similarly, explain why if $F(x)$ is decreasing both expressions will be negative.

2. (writing) You can think of part I of the Fundamental Theorem in terms of distance $s(t)$ and velocity $v(t) = s'(t)$. Start by assuming that $v(t) \geq 0$. Explain why $\int_a^b v(t) \, dt$ gives the total distance traveled, and explain why this equals $s(b) - s(a)$. Discuss what changes if $v(t) < 0$.

3. (writing) To explore part II of the Fundamental Theorem graphically, consider the function $g(x) = \int_a^x f(t) \, dt$. If $f(t)$ is positive on the interval $[a, b]$, explain why $g'(x)$ will also be positive. Further, the larger $f(t)$ is, the larger $g'(x)$ will be. Similarly, explain why if $f(t)$ is negative then $g'(x)$ will also be negative.

4. (writing) Suppose you are helping a friend review for a test on integration. As part of this test, a number of definite integrals $\int_a^b f(x) \, dx$ will be given that can be computed using the Fundamental Theorem. Other integrals will be given that must be approximated using Riemann sums or other techniques. Based on your knowledge so far, make a short list of

guidelines to help your friend identify when to use the Fundamental Theorem and when to use Riemann sums.

In exercises 5–30, use part I of the Fundamental Theorem to compute each integral exactly.

5. $\displaystyle\int_0^2 (2x - 3) \, dx$

6. $\displaystyle\int_0^3 (x^2 - 2) \, dx$

7. $\displaystyle\int_{-1}^1 (x^3 + 2x) \, dx$

8. $\displaystyle\int_0^2 (x^3 + 3x - 1) \, dx$

9. $\displaystyle\int_0^4 (\sqrt{x} + 3x) \, dx$

10. $\displaystyle\int_1^2 (4x - 2/x^2) \, dx$

11. $\displaystyle\int_0^1 (x\sqrt{x} + x^{-1/2}) \, dx$

12. $\displaystyle\int_0^8 (\sqrt[3]{x} - x^{2/3}) \, dx$

13. $\displaystyle\int_0^{\pi/2} 2 \sin x \, dx$

14. $\displaystyle\int_{-\pi}^0 4 \cos x \, dx$

15. $\displaystyle\int_0^{\pi/4} \sec x \tan x \, dx$

16. $\displaystyle\int_0^{\pi/4} \sec^2 x \, dx$

17. $\displaystyle\int_{\pi/2}^{\pi} (2 \sin x - \cos x) \, dx$

18. $\displaystyle\int_0^{\pi} 3 \sin 2x \, dx$

19. $\displaystyle\int_0^1 (e^x - e^{-x}) \, dx$

20. $\displaystyle\int_{-1}^1 (e^x + e^{-x}) \, dx$

21. $\int_0^3 \left(3e^{2x} - x^2\right) dx$ **22.** $\int_0^2 \left(3x - e^{-3x}\right) dx$

23. $\int_{-1}^1 \left(e^x + e^{-x}\right)^2 dx$ **24.** $\int_0^1 \dfrac{e^{2x} - 1}{e^x} dx$

25. $\int_1^4 \dfrac{x-3}{x} dx$ **26.** $\int_1^2 \dfrac{x^2 - 3x + 4}{x^2} dx$

27. $\int_0^4 x(x-2) dx$ **28.** $\int_0^{\pi/3} \dfrac{3}{\cos^2 x} dx$

29. $\int_0^3 \left(x^3 - \sin x\right) dx$ **30.** $\int_{-2}^0 \left(2x + e^x\right) dx$

In exercises 31–40, use the Fundamental Theorem if possible or estimate the integral using Riemann sums. (Hint: Six problems can be worked using antiderivative formulas we have covered so far.)

31. $\int_0^2 \sqrt{x^2 + 1}\, dx$ **32.** $\int_0^2 (\sqrt{x} + 1)^2 dx$

33. $\int_0^1 (e^x + 1)^2 dx$ **34.** $\int_0^1 e^{x^2 + 1} dx$

35. $\int_1^4 \dfrac{x^2}{x^2 + 4} dx$ **36.** $\int_1^4 \dfrac{x^2 + 4}{x^2} dx$

37. $\int_0^\pi \sin x^2\, dx$ **38.** $\int_0^\pi \sin^2 x\, dx$

39. $\int_0^{\pi/4} \dfrac{\sin x}{\cos^2 x} dx$ **40.** $\int_0^{\pi/4} \dfrac{\tan x}{\sec x} dx$

In exercises 41–46, find the derivative $f'(x)$.

41. $f(x) = \int_0^x (t^2 - 3t + 2)\, dt$

42. $f(x) = \int_2^x (t^2 - 3t - 4)\, dt$

43. $f(x) = \int_0^{x^2} (e^{-t^2} + 1)\, dt$

44. $f(x) = \int_2^{x^2} (e^{t^2} - 1)\, dt$

45. $f(x) = \int_x^{-1} \ln(t^2 + 1)\, dt$

46. $f(x) = \int_x^2 \sec t\, dt$

In exercises 47–50, find an equation of the tangent line at the given value of x.

47. $y = \int_0^x \sin \sqrt{t^2 + \pi^2}\, dt, \ x = 0$

48. $y = \int_{-1}^x \ln(t^2 + 2t + 2)\, dt, \ x = -1$

49. $y = \int_2^x \cos(\pi t^3)\, dt, \ x = 2$

50. $y = \int_0^x e^{-t^2 + 1}\, dt, \ x = 0$

51. Use the derivative in exercise 41 to locate and identify all relative extrema of $f(x) = \int_0^x (t^2 - 3t + 2)\, dt$.

52. Katie drives a car at speed $f(t) = 55 + 10\cos \frac{\pi}{2} t$ mph and Michael drives a car at speed $g(t) = 50 + 2t$ mph at time t minutes. Suppose that Katie and Michael are at the same location at time $t = 0$. Compute $\int_0^x [f(t) - g(t)]\, dt$ and interpret the integral in terms of a race between Katie and Michael.

In exercises 53 and 54, (a) explain how you know the proposed integral value is wrong and (b) find all mistakes.

53. $\int_{-1}^1 \dfrac{1}{x^2} dx = -\dfrac{1}{x} \Big|_{x=-1}^{x=1} = -1 - (1) = -2$

54. $\int_0^\pi \sec^2 x\, dx = \tan x \Big|_{x=0}^{x=\pi} = \tan \pi - \tan 0 = 0$

In exercises 55–62, find the position function $s(t)$ from the given velocity or acceleration function and initial value(s). Assume that units are feet and seconds.

55. $v(t) = 40 - \sin t, \ s(0) = 2$

56. $v(t) = 30 + 4\cos 3t, \ s(0) = 0$

57. $v(t) = 25(1 - e^{-2t}), \ s(0) = 0$

58. $v(t) = 10e^{-t}, \ s(0) = 2$

59. $a(t) = 4 - t, \ v(0) = 8, \ s(0) = 0$

60. $a(t) = 16 - t^2, \ v(0) = 0, \ s(0) = 30$

61. $a(t) = 24 + e^{-t}, \ v(0) = 0, \ s(0) = 0$

62. $a(t) = 3e^{-2t}, \ v(0) = -4, \ s(0) = 0$

63. If $\theta(t)$ is the angle between the path of a moving object and a fixed ray (see the figure that follows), the **angular velocity** of the object is $\omega(t) = \theta'(t)$ and the **angular acceleration** of the object is $\alpha(t) = \omega'(t)$.

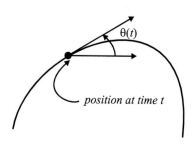

$\theta(t)$

position at time t

Suppose a baseball batter swings with a constant angular acceleration of $\alpha(t) = 10$ rad/s^2. If the batter hits the ball 0.8 s later, what is the angular velocity? The (linear) speed of the part of the bat located 3 feet from the pivot point (the batter's body) is $v = 3\omega$. How fast is this part of the bat moving at the moment of contact? Through what angle was the bat rotated during the swing?

64. Suppose a golfer rotates a golf club through an angle of $3\pi/2$ with a constant angular acceleration of α rad/s^2. If the clubhead is located 4 feet from the pivot point (the golfer's body), the (linear) speed of the clubhead is $v = 4\omega$. Find the value of α that will produce a clubhead speed of 100 mph at impact.

65. From 1970 to 1974, the function $16.1e^{0.07t}$ closely approximated the number of billions of barrels of oil consumed in the United States, where $t = 0$ corresponds to 1970. When prices rose in 1974, the consumption was approximated by the function $21.3e^{0.04(t-4)}$. Show that both functions give (approximately) the same consumption value for 1974. Compute $\int_4^{10} 16.1e^{0.07t}\ dt$ and $\int_4^{10} 21.3e^{0.04(t-4)}\ dt$ and compute the amount of oil saved by the reduced consumption.

66. The amount of work done by a force $F(x)$ moving an object from $x = a$ to $x = b$ is given by $\int_a^b F(x)\ dx$. As the object continues to move, the endpoint changes in time. Then $b = b(t)$ and the work done is a function of time. The derivative of this function defines **power**. Show that the power $\frac{d}{dt}\int_a^{b(t)} F(x)\ dx$ equals $F(b(t))\ b'(t)$.

In exercises 67 and 68, use the result of exercise 66 to compute the horsepower in the following situations. (Note: 1 hp $= 550$ ft-lb/s.)

67. $F(x) = 1000$ lb, $b'(t) = 130$ ft/s.

68. $F(x) = 1000e^{-x}$ lb, $b'(t) = 100 - t$ ft/s.

In exercises 69–74, find the average value of the function on the given interval.

69. $f(x) = x^2 - 1$, $[1, 3]$

70. $f(x) = x^2 + 2x$, $[0, 1]$

71. $f(x) = 2x - 2x^2$, $[0, 1]$

72. $f(x) = x^3 - 3x^2 + 2x$, $[1, 2]$

73. $f(x) = \cos x$, $[0, \pi/2]$

74. $f(x) = \sin x$, $[0, \pi/2]$

In exercises 75 and 76, use the graph to list $\int_0^1 f(x)\ dx$, $\int_0^2 f(x)\ dx$ and $\int_0^3 f(x)\ dx$ in order, from smallest to largest.

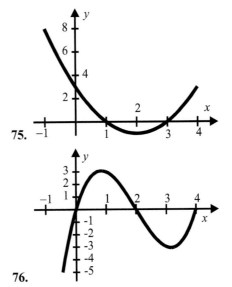

75.

76.

77. (exploratory) A boat is rowed across a stream of width L ft. At x ft from the starting shoreline, the boat is steered at an angle $\theta(x)$ radians from the shoreline. With a constant current and rowing speed v ft/s, the transit time is

$$T = \int_0^L \frac{1}{v\cos\theta(x)}\ dx \text{ seconds.}$$ Use "common sense" to find the steering angle function $\theta(x)$ that will minimize the transit time. Explain your solution mathematically and physically. Is your result true even if the rowing speed is allowed to be a function $v(x)$?

78. (exploratory) When solving differential equations of the form $\frac{dy}{dt} = f(y)$ for the unknown function $y(t)$, it is often convenient to make use of a **potential function** $V(y)$. This is a function such that $-\frac{dV}{dy} = f(y)$. For the function $f(y) = y - y^3$, find a potential function $V(y)$. Find the locations of the local minima of $V(y)$ and use a graph of $V(y)$ to explain why this is called a "double-well" potential. Explain each step in the calculation

$$\frac{dV}{dt} = \frac{dV}{dy}\frac{dy}{dt} = -f(y)f(y) \leq 0.$$

Since $\frac{dV}{dt} \leq 0$, does the function V increase or decrease as time goes on? Use your graph of V to predict the possible values of $\lim_{t\to\infty} y(t)$. Thus, you can predict the limiting value of the solution of the differential equation without ever solving the equation itself. Use this technique to predict $\lim_{t\to\infty} y(t)$ if $y' = 2 - 2y$.

4.6   Integration by Substitution

The Fundamental Theorem of Calculus gives us a powerful tool for evaluating definite integrals. The only problem is that to use it, we first need to find an antiderivative. Our goal in this section is to expand our ability to compute antiderivatives, through a useful technique called **integration by substitution**. As you have already seen, you need the chain rule to compute the derivatives of many functions. Integration by substitution gives us a process for helping to recognize when an integrand is the result of a chain rule derivative.

Example 6.1   Finding an Antiderivative by Trial and Error

Evaluate $\int 2xe^{x^2}\,dx$.

Solution   The idea is to find a function $F(x)$ for which $F'(x) = 2xe^{x^2}$. You might be tempted to guess that since x^2 is an antiderivative of $2x$, that

$$F(x) = x^2 e^{x^2}$$

is an antiderivative of $2xe^{x^2}$. To see that this is incorrect, simply compute the derivative of $x^2 e^{x^2}$. From the product rule, we have that

$$\frac{d}{dx}\left(x^2 e^{x^2}\right) = 2xe^{x^2} + x^2 e^{x^2}(2x) \neq 2xe^{x^2}.$$

So much for our guess. Before making another guess, look more closely at what went wrong with the first guess. Notice that the product rule produces two terms and our target derivative $2xe^{x^2}$ has only one term. So, we should avoid guesses that are products. A very important clue is that x^2 is an antiderivative of $2x$ and x^2 already appears, as the exponent of e^{x^2}. To see the significance of this, you need only compute the derivative of $F(x) = e^{x^2}$. Notice that

$$F'(x) = e^{x^2}(2x) = 2x\,e^{x^2},$$

which is exactly our integrand. Note that the factor of $2x$ in the derivative calculation comes from the chain rule, not the product rule. To finish this example, recall that we need to add an arbitrary constant, to get

$$\int 2xe^{x^2}\,dx = e^{x^2} + c.$$

■

We want to develop the lessons learned in example 6.1 into a process that we can apply to a wide range of integrals. Notice that the factor $2x$ in the integrand $2xe^{x^2}$ is the derivative of the exponent x^2. More generally, you should recognize that when one part of an integrand is the derivative of another part of the integrand, you may be looking at a chain rule derivative. We will develop a simple procedure for testing this idea below. First, we should take a careful look at the chain rule.

Note that, in general, if F is any antiderivative of f, then from the chain rule, we have

$$\frac{d}{dx}[F(u)] = F'(u)\frac{du}{dx} = f(u)\frac{du}{dx}.$$

From this, we have that

$$\int f(u)\frac{du}{dx}\,dx = \int \frac{d}{dx}[F(u)]\,dx = F(u) + c = \int f(u)\,du, \qquad (6.1)$$

since F is an antiderivative of f. If you read the expressions on the far left and the far right sides of (6.1), this suggests that

$$du = \frac{du}{dx}\, dx.$$

So, if we have an integral, $\int h(x)\,dx$, which we cannot compute directly, we often look for a new variable, u and function $f(u)$ for which

$$\int h(x)\,dx = \int f(u(x))\frac{du}{dx}\,dx = \int f(u)\,du.$$

In deciding how to choose a new variable, there are several things to look for:

- terms that are derivatives of other terms (or pieces thereof) and
- terms that are particularly troublesome. (You can often substitute your troubles away.)

Example 6.2   Using Substitution to Evaluate an Integral
    Evaluate $\int (x^3 + 5)^{100}\,(3x^2)\,dx.$

Solution   You probably cannot evaluate this as it stands. (Don't feel bad about this; this is usually the case.) However, you might observe that

$$\frac{d}{dx}\left(x^3 + 5\right) = 3x^2,$$

which is part of the integrand. We make the substitution $u = x^3 + 5$, so that $du = 3x^2\,dx$. This gives us

$$\int \underbrace{(x^3 + 5)^{100}}_{u^{100}}\,\underbrace{(3x^2)\,dx}_{du} = \int u^{100}\,du = \frac{u^{101}}{101} + c.$$

We are not quite done yet. Since we invented the new variable u, we need to convert back to the original variable, x. We get

$$\int (x^3 + 5)^{100}\,(3x^2)\,dx = \frac{u^{101}}{101} + c = \frac{(x^3 + 5)^{101}}{101} + c.$$

It's always a good idea to perform a quick check on the antiderivative. (Remember that integration and differentiation are inverse processes!) Here, we compute

$$\frac{d}{dx}\left[\frac{(x^3 + 5)^{101}}{101}\right] = \frac{101\,(x^3 + 5)^{100}\,(3x^2)}{101} = (x^3 + 5)^{100}\,(3x^2),$$

which is the original integrand. This confirms that we have indeed found an antiderivative.

■

   The technique of integration by substitution illustrated in example 6.2 consists of the following general steps.

- **Choose an expression for** u: a common choice is the innermost expression or "inside" term of a composition of functions. (In example 6.2, note that $x^3 + 5$ is the inside term of $(x^3 + 5)^{100}$.)
- **Compute** $du = \dfrac{du}{dx}\,dx.$
- **Replace all terms** in the original integrand with expressions involving u and du.
- **Evaluate** the resulting (u) integral. If you still can't evaluate the integral, you may need to try a different choice of u.
- **Replace all the** u **terms** in the antiderivative with the corresponding expression in x.

Always keep in mind that finding antiderivatives is the reverse process of finding derivatives. There aren't any new *rules* to learn for finding antiderivatives. This will help you to find antiderivatives and to check your answers. In the next example, we are not so fortunate as to have the exact derivative we want in the integrand.

Example 6.3   Using Substitution: A Power Function Inside a Cosine

Evaluate $\int x \cos x^2 dx$.

Solution   Notice that

$$\frac{d}{dx}x^2 = 2x.$$

Unfortunately, we don't quite have this in the integrand. This should get you thinking, though. Once again, you can push constants back and forth past an integral sign. Notice that we can rewrite the integral as

$$\int x \cos x^2\, dx = \frac{1}{2}\int 2x \cos x^2\, dx.$$

We now substitute $u = x^2$, so that $du = 2x\, dx$ and we have

$$\int x \cos x^2\, dx = \frac{1}{2}\int \underbrace{\cos x^2}_{\cos u} \underbrace{(2x)\, dx}_{du}$$

$$= \frac{1}{2}\int \cos u\, du = \frac{1}{2}\sin u + c = \frac{1}{2}\sin x^2 + c.$$

Again, as a check, observe that

$$\frac{d}{dx}\left(\frac{1}{2}\sin x^2\right) = \frac{1}{2}\cos x^2\,(2x) = x \cos x^2,$$

which is the original integrand.

■

Example 6.4   Using Substitution: A Trigonometric Function Inside a Power

Evaluate $\int (3\sin x + 4)^5 \cos x\, dx$.

Solution   Like most integrals, you probably can't evaluate this one as it stands. So, what do you notice about the integrand? Keep in mind that, among other things, you're looking for terms that are derivatives of other terms. There's a $\sin x$ term and a $\cos x$ term and $\frac{d}{dx}(\sin x) = \cos x$. So, we let $u = 3\sin x + 4$, so that $du = 3\cos x\, dx$. We then have

$$\int (3\sin x + 4)^5 \cos x\, dx = \frac{1}{3}\int \underbrace{(3\sin x + 4)^5}_{u^5} \underbrace{(3\cos x)\, dx}_{du}$$

$$= \frac{1}{3}\int u^5\, du = \left(\frac{1}{3}\right)\frac{u^6}{6} + c$$

$$= \frac{1}{18}(3\sin x + 4)^6 + c.$$

■

Sometimes you will need to look a bit deeper into an integral to see terms that are derivatives of other terms, as in the following example.

Example 6.5   Using Substitution: A Root Function Inside a Sine

Find $\int \dfrac{\sin \sqrt{x}}{\sqrt{x}}\,dx$.

Solution   This integral is anything but obvious. It never hurts to try something, though. If you had to substitute for something, what would you choose? You might notice that $\sin \sqrt{x} = \sin x^{1/2}$ and letting $u = \sqrt{x} = x^{1/2}$ (the "inside"), we get $du = \frac{1}{2}\,x^{-1/2}\,dx = \frac{1}{2\sqrt{x}}\,dx$. Since there is a $\frac{1}{\sqrt{x}}\,dx$ in the integrand, we can proceed. We have

$$\int \frac{\sin \sqrt{x}}{\sqrt{x}}\,dx = 2 \int \underbrace{\sin \sqrt{x}}_{\sin u} \underbrace{\left(\frac{1}{2\sqrt{x}}\right)\,dx}_{du}$$

$$= 2 \int \sin u\,du = -2\cos u + c = -2\cos \sqrt{x} + c.$$

■

Example 6.6   Where the Numerator Is the Derivative of the Denominator

Evaluate $\int \dfrac{x^2}{x^3 + 5}\,dx$.

Solution   Since $\frac{d}{dx}(x^3 + 5) = 3x^2$, you might notice that we should let $u = x^3 + 5$, so that $du = 3x^2\,dx$ and so,

$$\int \frac{x^2}{x^3 + 5}\,dx = \frac{1}{3} \int \underbrace{\frac{1}{x^3 + 5}}_{u} \underbrace{\left(3x^2\right)\,dx}_{du} = \frac{1}{3} \int \frac{1}{u}\,du$$

$$= \frac{1}{3} \ln|u| + c = \ln\left|x^3 + 5\right| + c.$$

■

This last problem is an illustration of a very common type of integral, one where the numerator is the derivative of the denominator. More generally, we have the following theorem.

Theorem 6.1   For any continuous and nonzero function, f

$$\int \frac{f'(x)}{f(x)}\,dx = \ln|f(x)| + c.$$

Proof.   Let $u = f(x)$. Then $du = f'(x)\,dx$ and

$$\int \frac{f'(x)}{f(x)}\,dx = \int \underbrace{\frac{1}{f(x)}}_{u} \underbrace{f'(x)\,dx}_{du}$$

$$= \int \frac{1}{u}\,du = \ln|u| + c = \ln|f(x)| + c,$$

as desired. As an alternative to this proof, you might simply compute $\frac{d}{dx}\ln|f(x)|$ directly, to obtain the integrand.

■

You should recall that we already stated this result in section 4.1. It is important enough to repeat here in the context of substitution.

Example 6.7   An Antiderivative for the Tangent Function

Evaluate $\int \tan x \, dx$.

Solution   Note that this is **not** one of our basic integration formulas. However, you might notice that

$$\int \tan x \, dx = \int \frac{\sin x}{\cos x} \, dx = - \int \underbrace{\frac{1}{\cos x}}_{u} \underbrace{(-\sin x) \, dx}_{du}$$

$$= - \int \frac{1}{u} \, du = -\ln|u| + c = -\ln|\cos x| + c,$$

where we have used the fact that $\frac{d}{dx}(\cos x) = -\sin x$.

∎

So far, every one of our examples has been solved by spotting a term in the integrand that was a derivative of another term. We present an integral now where this is not the case, but where a substitution is made to deal with a particularly troublesome term in the integrand.

Example 6.8   A Substitution That Lets You Expand the Integrand

Evaluate $\int x\sqrt{2-x} \, dx$.

Solution   You certainly cannot evaluate this as it stands. If you look for terms that are derivatives of other terms, you will come up empty-handed. The real problem here is that there is a square root of a sum (or difference) in the integrand. A reasonable step would be to substitute for the expression under the square root. We let $u = 2 - x$, so that $du = -dx$. That doesn't seem so bad, but what are we to do with the extra x in the integrand? Well, since $u = 2 - x$, it follows that $x = 2 - u$. Making these substitutions in the integral, we get

$$\int x\sqrt{2-x} \, dx = (-1) \int \underbrace{x}_{2-u} \underbrace{\sqrt{2-x}}_{\sqrt{u}} \underbrace{(-1) \, dx}_{du}$$

$$= -\int (2-u)\sqrt{u} \, du.$$

You might wonder if this integral is any better than the one we started with. While we can't evaluate it directly, if we multiply out the terms, we get

$$\int x\sqrt{2-x} \, dx = -\int (2-u)\sqrt{u} \, du$$

$$= -\int \left(2u^{1/2} - u^{3/2} \right) du$$

$$= -2 \frac{u^{3/2}}{\left(\frac{3}{2}\right)} + \frac{u^{5/2}}{\left(\frac{5}{2}\right)} + c$$

$$= -\frac{4}{3} u^{3/2} + \frac{2}{5} u^{5/2} + c$$

$$= -\frac{4}{3} (2-x)^{3/2} + \frac{2}{5} (2-x)^{5/2} + c.$$

You should check the validity of this antiderivative via differentiation.

∎

Substitution in Definite Integrals

There is only one slight difference in using substitution for evaluating a definite integral: If you change variables, you must also change the limits of integration to correspond to the new variable. The procedure here is then precisely the same as that used for examples 6.2 through 6.8, with one additional step. When you introduce the new variable u, you must also change the limits of integration from $x = a$ and $x = b$ to the corresponding limits for $u : u = u(a)$ and $u = u(b)$. We have

$$\int_a^b f(u(x))\, u'(x)\, dx = \int_{u(a)}^{u(b)} f(u)\, du.$$

Example 6.9   **Using Substitution in a Definite Integral**

Evaluate $\int_1^2 x^3 \sqrt{x^4 + 5}\, dx$.

Solution   Of course, you probably can't evaluate this as it stands. However, notice that $\frac{d}{dx}\left(x^4 + 5\right) = 4x^3$. This suggests the substitution $u = x^4 + 5$, so that $du = 4x^3\, dx$. For the limits of integration, note that when $x = 1$,

$$u = x^4 + 5 = 1^4 + 5 = 6$$

and when $x = 2$,

$$u = x^4 + 5 = 2^4 + 5 = 21.$$

We now have

CAUTION: You must change the limits of integration as soon as you change variables!

$$\int_1^2 x^3 \sqrt{x^4 + 5}\, dx = \frac{1}{4}\int_1^2 \underbrace{\sqrt{x^4 + 5}}_{\sqrt{u}}\underbrace{\left(4x^3\right)\, dx}_{du} = \frac{1}{4}\int_6^{21} \sqrt{u}\, du$$

$$= \frac{1}{4}\left.\frac{u^{3/2}}{\left(\frac{3}{2}\right)}\right|_6^{21} = \left(\frac{1}{4}\right)\left(\frac{2}{3}\right)\left[21^{3/2} - 6^{3/2}\right].$$

Notice that because we changed the limits of integration to match the new variable, we did **not** need to convert back to the original variable at the conclusion of the problem, as we do when we make a substitution in an indefinite integral. (Indeed, if we had switched the variables back, we would also have needed to switch the limits of integration back to their original values before evaluating!)

■

There is another way of dealing with definite integrals that may have occurred to you. You could always use u-substitution to find an antiderivative and then return to the original variable to do the evaluation. Although this method will work for most of the problems you will encounter in this text, we recommend that you avoid it, for several reasons. First, changing the limits of integration is not very difficult and results in a much more readable mathematical expression. Second, in many theoretical applications of u-substitution, you will **need** to change the limits of integration.

Example 6.10   **Substitution in a Definite Integral Involving an Exponential**

Compute $\int_0^{15} te^{-t^2/2}\, dt$.

Solution   As always, we look for terms that are derivatives of other terms. Here, you should notice that $\frac{d}{dt}\left(\frac{-t^2}{2}\right) = -t\, dt$. So, we set $u = -\frac{t^2}{2}$ and compute $du = -t\, dt$. For the upper limit of integration, we have that $t = 15$ corresponds to $u = -\frac{(15)^2}{2} = -\frac{225}{2}$. For the lower limit, we have that $t = 0$ corresponds to $u = 0$. This gives us

$$\int_0^{15} te^{-\frac{t^2}{2}}\, dt = -\int_0^{15} \underbrace{e^{-\frac{t^2}{2}}}_{e^u}\underbrace{(-t)\, dt}_{du}$$

$$= -\int_0^{-225/2} e^u\, du = -e^u\Big|_0^{-112.5} = -e^{-112.5} + 1.$$

■

Exercises 4.6

1. (writing) It is never *wrong* to make a substitution in an integral, but sometimes it is not very helpful. For example, using the substitution $u = x^2$, you can correctly conclude that

$$\int x^3\sqrt{x^2 + 1}\, dx = \int \frac{1}{2}u\sqrt{u+1}\, du,$$

but the new integral is no easier than the original integral. In this case, a better substitution makes this workable. (Can you find it?) However, the general problem remains of how you can tell whether or not to give up on a substitution. Give some guidelines for answering this question, using the integrals $\int x\sin x^2\, dx$ and $\int x\sin x^3\, dx$ as illustrative examples.

2. (writing) It is not uncommon for students learning substitution to use incorrect notation in the intermediate steps. Be aware of this – it can be harmful to your grade! Carefully examine the following string of equalities and find each mistake. Using $u = x^2$,

$$\int_0^2 x\sin x^2\, dx = \int_0^2 (\sin u)x\, dx = \int_0^2 (\sin u)\frac{1}{2}\, du$$

$$= -\frac{1}{2}\cos u\Big|_0^2 = -\frac{1}{2}\cos x^2\Big|_0^2$$

$$= -\frac{1}{2}\cos 4 + 1.$$

The final answer is correct, but is this work worthy of full credit? By analogy, how much credit would you give the writer of the following paragraph, who "merely" got confused on a gender substitution: "He needed an A on this test, but the professor's tests are hard. When he got the test back, he found that he had given him a B. When they saw each other that afternoon in the library, he started shouting at her so profanely that the security guard had to escort him out." Even assuming that the final line is correct, how much has been communicated and how confident are you that the writer knows what is going on?

3. (writing) Suppose that an integrand has a term of the form $e^{f(x)}$. For example, suppose you are trying to evaluate $\int x^2 e^{x^3}\, dx$. Discuss why you should immediately try the substitution $u = f(x)$. If this substitution does not work, what could you try next? (Hint: Think about $\int x^2 e^{\ln x}\, dx$.)

4. (writing) Suppose that an integrand has a composite function of the form $f(g(x))$. Explain why you should look to see if the integrand also has the term $g'(x)$. Discuss possible substitutions.

In exercises 5–8, use the given substitution to evaluate the indicated integral.

5. $\int x^2\sqrt{x^3 + 2}\, dx$, $u = x^3 + 2$

6. $\int x^3(x^4 + 1)^{-2/3}\, dx$, $u = x^4 + 1$

7. $\int \frac{(\sqrt{x} + 2)^3}{\sqrt{x}}\, dx$, $u = \sqrt{x}$

8. $\int \sin x\cos x\, dx$, $u = \sin x$

In exercises 9–46, evaluate the indicated integral.

9. $\int x(x^2 - 3)^4\, dx$

10. $\int x^3\sqrt{x^4 + 3}\, dx$

11. $\int (2x + 1)(x^2 + x)^3\, dx$

12. $\int (x^2 + 2x)(x^3 + 3x^2)^2\, dx$

13. $\int \cos x\sqrt{\sin x + 1}\, dx$

14. $\int \sec^2 x\sqrt{\tan x}\, dx$

15. $\int \frac{\sin x}{\sqrt{\cos x}}\, dx$

16. $\int \sin^3 x\cos x\, dx$

17. $\int x^2\cos x^3\, dx$

18. $\int \sec^2 x\cos(\tan x)\, dx$

19. $\int \sin x(\cos x + 3)^{3/4}\, dx$

20. $\int x\csc x^2\cot x^2\, dx$

21. $\int xe^{x^2+1}\, dx$

22. $\int e^x\sqrt{e^x + 4}\, dx$

23. $\displaystyle\int \frac{e^{\sqrt{x}}}{\sqrt{x}}\, dx$

24. $\displaystyle\int \frac{e^{2x}}{(e^{2x}+1)^3}\, dx$

25. $\displaystyle\int \frac{x^2}{\sqrt{x^3-2}}\, dx$

26. $\displaystyle\int \frac{x+1}{(x^2+2x-1)^2}\, dx$

27. $\displaystyle\int \frac{(\ln x + 2)^2}{x}\, dx$

28. $\displaystyle\int \frac{\sqrt{\ln x}}{x}\, dx$

29. $\displaystyle\int \frac{2x+1}{x^2+x-1}\, dx$

30. $\displaystyle\int \frac{\cos x}{\sin x + 2}\, dx$

31. $\displaystyle\int \frac{1}{\sqrt{x}(\sqrt{x}+1)^2}\, dx$

32. $\displaystyle\int \frac{x}{x^2+4}\, dx$

33. $\displaystyle\int \cos x\, e^{\sin x}\, dx$

34. $\displaystyle\int \tan 2x\, dx$

35. $\displaystyle\int \sin x(\cos x - 1)^3\, dx$

36. $\displaystyle\int x^2 \sec^2 x^3\, dx$

37. $\displaystyle\int \frac{4}{x(\ln x + 1)^2}\, dx$

38. $\displaystyle\int \frac{\sin\sqrt{x}}{\sqrt{x}}\, dx$

39. $\displaystyle\int \frac{e^x - e^{-x}}{e^x + e^{-x}}\, dx$

40. $\displaystyle\int \frac{6x}{(x^2-3)^4}\, dx$

41. $\displaystyle\int \frac{1}{x \ln\sqrt{x}}\, dx$

42. $\displaystyle\int \frac{\cos x}{\sin^2 x}\, dx$

43. $\displaystyle\int \frac{2x+3}{x+7}\, dx$

44. $\displaystyle\int \frac{3x+4}{2x+7}\, dx$

45. $\displaystyle\int \frac{x^2}{\sqrt[3]{x+3}}\, dx$

46. $\displaystyle\int \frac{x^2+2}{\sqrt{x-5}}\, dx$

In exercises 47–56, evaluate the definite integral.

47. $\displaystyle\int_0^2 x\sqrt{x^2+1}\, dx$

48. $\displaystyle\int_1^3 x\sin(\pi x^2)\, dx$

49. $\displaystyle\int_{-1}^1 \frac{x}{(x^2+1)^2}\, dx$

50. $\displaystyle\int_0^2 x^2 e^{x^3}\, dx$

51. $\displaystyle\int_{\pi/2}^\pi \frac{4\cos x}{(\sin x + 1)^2}\, dx$

52. $\displaystyle\int_0^{\pi^2} \frac{\cos\sqrt{x}}{\sqrt{x}}\, dx$

53. $\displaystyle\int_{\pi/4}^{\pi/2} \cot x\, dx$

54. $\displaystyle\int_1^e \frac{\ln x}{x}\, dx$

55. $\displaystyle\int_1^4 \frac{x-1}{\sqrt{x}}\, dx$

56. $\displaystyle\int_0^1 \frac{x}{\sqrt{x^2+1}}\, dx$

In exercises 57–66, evaluate the integral exactly, if possible. Otherwise, estimate it numerically.

57. $\displaystyle\int_0^\pi \sin x^2\, dx$

58. $\displaystyle\int_0^\pi x\sin x^2\, dx$

59. $\displaystyle\int_{-1}^1 xe^{-x^2}\, dx$

60. $\displaystyle\int_{-1}^1 e^{-x^2}\, dx$

61. $\displaystyle\int_0^2 \frac{4}{(x^2+1)^2}\, dx$

62. $\displaystyle\int_0^2 \frac{4x}{(x^2+1)^2}\, dx$

63. $\displaystyle\int_0^2 \frac{4x^2}{(x^2+1)^2}\, dx$

64. $\displaystyle\int_0^2 \frac{4x^3}{(x^2+1)^2}\, dx$

65. $\displaystyle\int_0^{\pi/4} \sec x\, dx$

66. $\displaystyle\int_0^{\pi/4} \sec^2 x\, dx$

67. A function $f(x)$ is said to be **even** if $f(-x) = f(x)$ for all x. $f(x) = x^2$ and $f(x) = x^4$ are even, since $(-x)^2 = x^2$ and $(-x)^4 = x^4$. A function $f(x)$ is said to be **odd** if $f(-x) = -f(x)$. Show that $\cos x$ and $x\sin x$ are even, but $\sin x$ and $x\cos x$ are odd.

68. Suppose $f(x)$ is continuous for all x. For any positive constant a, $\int_{-a}^a f(x)\, dx = \int_{-a}^0 f(x)\, dx + \int_0^a f(x)\, dx$. Using the substitution $u = -x$ in $\int_{-a}^0 f(x)\, dx$, show that if $f(x)$ is even, then $\int_{-a}^a f(x)\, dx = 2\int_0^a f(x)\, dx$. Also, if $f(x)$ is odd, show that $\int_{-a}^a f(x)\, dx = 0$.

In exercises 69–76, determine if the integrand is even or odd (or neither), rewrite the integral accordingly and compute (or estimate) the integral.

69. $\displaystyle\int_{-1}^1 x\cos x\, dx$

70. $\displaystyle\int_{-1}^1 x^2 \sin x\, dx$

71. $\displaystyle\int_{-1}^1 (x^4 - 2x^2 + 1)\, dx$

72. $\displaystyle\int_{-1}^1 xe^{-x^2}\, dx$

73. $\displaystyle\int_{-1}^1 x^2 \cos x\, dx$

74. $\displaystyle\int_{-1}^1 x^2 e^{-x}\, dx$

75. $\displaystyle\int_{-1}^1 (x^2 + \sin x)\, dx$

76. $\displaystyle\int_{-1}^1 (x^3 - 2x)\, dx$

77. The location $(\bar{x}, \bar{y})$ of the center of gravity (balance point) of a flat plate bounded by $y = f(x) > 0$, $a \le x \le b$ and the x-axis is given by $\bar{x} = \dfrac{\int_a^b xf(x)\, dx}{\int_a^b f(x)\, dx}$ and $\bar{y} = \dfrac{\int_a^b [f(x)]^2\, dx}{\int_a^b f(x)\, dx}$. For the semi-circle $y = f(x) = \sqrt{4-x^2}$, use symmetry as in exercises 69–76 to argue that $\bar{x} = 0$ and $\bar{y} = \dfrac{1}{4\pi}\int_0^2 (4-x^2)\, dx$. Compute $\bar{y}$.

78. Suppose that the population density of a group of animals can be described by $f(x) = xe^{-x^2}$ for $0 \le x \le 2$, where x is the distance from a river. Graph $y = f(x)$ and briefly describe where these animals are likely to be found. Find the total population $\int_0^2 f(x)\, dx$.

79. The voltage in an AC (alternating current) circuit is given by $V(t) = V_p \sin(2\pi f t)$, where f is the frequency. A voltmeter does not indicate the amplitude V_p. Instead, the voltmeter reads the **root-mean-square** (rms), the square root of the average value of the square of the voltage over one cycle. That is, rms $= \sqrt{f \int_0^{1/f} V^2(t)\, dt}$. Use the trigonometric identity $\sin^2 x = \frac{1}{2} - \frac{1}{2}\cos 2x$ to show that rms $= V_p/\sqrt{2}$.

80. Graph $y = f(t)$ and find the root-mean-square of

$$f(t) = \begin{cases} -1 & \text{if } -2 \leq t < -1 \\ t & \text{if } -1 \leq t \leq 1 \\ 1 & \text{if } 1 < t \leq 2 \end{cases}, \text{ where rms} = \sqrt{\frac{1}{4} \int_{-2}^{2} f^2(t)\, dt}.$$

81. (exploratory) A **predator-prey system** is a set of differential equations modeling the change in population of interacting species of organisms. A simple model of this type is

$$\begin{cases} x'(t) = x(t)\,[a - by(t)] \\ y'(t) = y(t)\,[dx(t) - c] \end{cases}$$

for positive constants a, b, c and d. Each equation includes a term of the form $x(t)y(t)$, which is intended to represent the result of confrontations between the species. Noting that the contribution of this term is negative to $x'(t)$ but positive to $y'(t)$, explain why it must be that $x(t)$ represents the population of the prey and $y(t)$ the population of the predator. If $x(t) = y(t) = 0$, compute $x'(t)$ and $y'(t)$. In this case, will x and y increase or decrease or stay constant? Explain why this makes sense physically. Determine $x'(t)$ and $y'(t)$ and the subsequent change in x and y at the so-called **equilibrium point** $x = c/d$, $y = a/b$. If the population is periodic, we can show that the equilibrium point gives the average population (even if the population does not remain constant). To do so, note that $\dfrac{x'(t)}{x(t)} = a - by(t)$. Integrating both sides of the equation from $t = 0$ to $t = T$ (the period of $x(t)$ and $y(t)$), we get $\int_{0}^{T} \dfrac{x'(t)}{x(t)}\, dt = \int_{0}^{T} a\, dt - \int_{0}^{T} by(t)\, dt$. Evaluate each integral to show that $\ln x(T) - \ln x(0) = aT - \int_{0}^{T} by(t)\, dt$. Assuming that $x(t)$ has period T, we have $x(T) = x(0)$ and so, $0 = aT - \int_{0}^{T} by(t)\, dt$. Finally, rearrange terms to show that $1/T \int_{0}^{T} y(t)\, dt = a/b$; that is, the average value of the population $y(t)$ is the equilibrium value $y = a/b$. Similarly, show that the average value of the population $x(t)$ is the equilibrium value $x = c/d$.

4.7   Numerical Integration

Thus far, our development of the integral has paralleled our development of the derivative. In both cases, we began with a limit definition that was difficult to use for calculation and then, proceeded to develop simplified rules for calculation. At this point, you should be able to find the derivative of nearly any function you can write down. You might expect that with a few more rules you will be able to do the same for integrals. Unfortunately, this is not the case. There are many functions for which **no** elementary antiderivative is available. (By elementary antiderivative, we mean an antiderivative expressible in terms of the elementary functions with which you are familiar: the algebraic, trigonometric, exponential and logarithmic functions.) For instance,

$$\int_{0}^{2} \cos\left(x^2\right) dx$$

cannot be calculated exactly, since $\cos\left(x^2\right)$ does not have an elementary antiderivative. (Try to find one, but don't spend much time, as mathematicians have proven that it can't be done.)

In fact, most definite integrals cannot be calculated exactly. When we can't compute the value of an integral exactly, we do the next best thing: we approximate its value numerically. In this section, we develop three methods of approximating definite integrals. None will replace the built-in integration routine on your calculator or computer. However, by exploring these methods, you will be exposed to some of the basic ideas used to develop more sophisticated numerical integration routines.

You should recognize that you already have a number of approximation methods at your disposal. Since a definite integral is the limit of a sequence of Riemann sums, any Riemann sum serves as an approximation of the integral,

$$\int_{a}^{b} f(x)\, dx \approx \sum_{i=1}^{n} f(c_i)\, \Delta x,$$

where c_i is any point chosen from the subinterval $[x_{i-1}, x_i]$, for $i = 1, 2, \ldots, n$. From the definition of definite integral, observe that the larger n is, the better the approximation tends to be. The reason we say that Riemann sums provide us with numerous approximation schemes is that we are free to choose the evaluation points, c_i, $i = 1, 2, \ldots, n$. The most common choice leads to a method called the **Midpoint Rule**:

$$\int_a^b f(x)\,dx \approx \sum_{i=1}^n f(c_i)\,\Delta x,$$

where c_i is the midpoint of the subinterval $[x_{i-1}, x_i]$,

$$c_i = \frac{1}{2}(x_{i-1} + x_i), \quad \text{for } i = 1, 2, \ldots, n.$$

Example 7.1   Using the Midpoint Rule

Write out the Midpoint Rule approximation of $\int_0^1 3x^2\,dx$ with $n = 4$.

Solution   For $n = 4$, the regular partition of the interval $[0, 1]$ is $x_0 = 0$, $x_1 = 1/4$, $x_2 = 1/2$, $x_3 = 3/4$ and $x_4 = 1$. The midpoints are then $c_1 = 1/8$, $c_2 = 3/8$, $c_3 = 5/8$ and $c_4 = 7/8$. With $\Delta x = 1/4$, the Riemann sum is then

$$\left[f\left(\frac{1}{8}\right) + f\left(\frac{3}{8}\right) + f\left(\frac{5}{8}\right) + f\left(\frac{7}{8}\right) \right]\left(\frac{1}{4}\right) = \left(\frac{3}{64} + \frac{27}{64} + \frac{75}{64} + \frac{147}{64}\right)\left(\frac{1}{4}\right)$$

$$= \frac{252}{256} = 0.984375.$$

Of course, from the Fundamental Theorem, the exact value of the integral in example 7.1 is

$$\int_0^1 3x^2\,dx = \frac{3x^3}{3}\bigg|_0^1 = 1.$$

So, our approximation in example 7.1 is not especially accurate. To obtain greater accuracy, notice that you could always compute an approximation using more rectangles. One way to simplify this process is to write a short program for your calculator or computer to implement the Midpoint Rule. A suggested outline for such a program follows.

1. Store $f(x)$, a, b and n.
2. Compute $\Delta x = (b - a)/n$.
3. Compute $c_1 = a + \Delta x/2$ and start the sum with $f(c_1)$.
4. Compute the next $c_i = c_{i-1} + \Delta x$ and add $f(c_i)$ to the sum
5. Repeat step 4 until $i = n$ (i.e., perform step 4 a total of $(n-1)$ times).
6. Multiply the sum by Δx.

Example 7.2   Using a Program for the Midpoint Rule

Repeat example 7.1 using a program to compute the Midpoint Rule approximations for $n = 8, 16, 32, 64$ and 128.

Solution   You should confirm the values in the following table. We include a column displaying the error in the approximation at each step (i.e., the difference between the exact value of 1 and the approximate values).

n	**Midpoint Rule**	**Error**
4	0.984375	0.015625
8	0.99609375	0.00390625
16	0.99902344	0.00097656
32	0.99975586	0.00024414
64	0.99993896	0.00006106
128	0.99998474	0.00001526

You should note that each time the number of steps is doubled, the error is reduced approximately by a factor of 4. Although this precise reduction in error will not occur with all integrals, this rate of improvement in the accuracy of the approximation is typical of the Midpoint Rule.

■

Notice that in general, we won't know the error in a Midpoint Rule approximation, except where we know the value of the integral exactly. In such a case, of course, there's no point to approximating the value of something you already know exactly. We started with a simple integral, whose value we knew exactly, so that you could get a sense of how accurate the Midpoint Rule approximation is. As it turns out, it is possible to obtain bounds on the errors in such approximations, although the derivation of such error bounds is beyond the level of this course. For this, we refer the interested reader to a text on numerical analysis.

Note that in the following example we can't compute an exact value of the integral, since we do not know an antiderivative for the integrand.

Example 7.3   Finding an Approximation with a Given Accuracy

Use the Midpoint Rule to approximate $\int_0^1 \sqrt{x^2 + 1}\, dx$ accurate to three decimal places.

Solution   Given the instructions, how do we know when to stop? We want to make sure that three decimal places are correct, so continue increasing n until it is unlikely the third decimal could change: the size of n will vary substantially from integral to integral, so this is more a matter of common sense than a precise rule. You should test your program against the numbers in the accompanying table.

From the table, we can make the reasonable approximation

$$\int_0^1 \sqrt{x^2 + 1}\, dx \approx 1.148.$$

n	Midpoint Rule
10	1.14749
20	1.14772
30	1.14776
40	1.14778

■

Remark 7.1   Computer and calculator programs that estimate the value of integrals face the same challenge we did in example 7.3; that is, knowing when a given approximation is good enough. As it turns out, such software generally includes sophisticated algorithms for estimating the accuracy of its own approximations. You can find an introduction to such algorithms in most texts on numerical analysis.

Another important reason for pursuing numerical methods is for the case where we don't know the function that we're trying to integrate. That's right: we often only know some *values* of a function at a collection of points, while an algebraic representation of a function is unavailable. This is often the case in the physical and biological sciences and engineering, where the only information available about a function comes from measurements made at a finite number of points.

Example 7.4   Estimating an Integral from a Table of Function Values

Estimate $\int_0^1 f(x)\, dx$, where we have measurements of the unknown function $f(x)$ as given in the table shown in the margin.

Solution   Approaching the problem graphically, we have five data points (see Figure 4.27). How can we estimate the area under the curve from five points? Conceptually, we have two tasks. First, we need a reasonable way to connect the given points. Secondly, we need to compute the area of the resulting region. So, how should we connect the dots? The most obvious way is the same way the authors' children would: connect the dots with straight-line segments as in Figure 4.28.

x	$f(x)$
0	1
0.25	0.8
0.5	1.3
0.75	1.1
1	1.6

FIGURE 4.27:
Data from an unknown function.

Notice that the region bounded by the graph and the x-axis on the interval $[0, 1]$ consists of four trapezoids (see Figure 4.29).

FIGURE 4.28: Connecting the dots.

FIGURE 4.29: Four trapezoids.

Recall that the area of a trapezoid with sides h_1 and h_2 and base b is given by $\left(\dfrac{h_1 + h_2}{2}\right) b$. (It's an easy exercise to verify this.) This can be thought of as the average of the areas of the rectangle whose height is the value of the function at the left endpoint and the rectangle whose height is the value of the function at the right endpoint.

The total area is then

$$\frac{f(0) + f(0.25)}{2}0.25 + \frac{f(0.25) + f(0.5)}{2}0.25 + \frac{f(0.5) + f(0.75)}{2}0.25$$
$$+ \frac{f(0.75) + f(1)}{2}0.25$$
$$= [f(0) + 2f(0.25) + 2f(0.5) + 2f(0.75) + f(1)]\frac{0.25}{2} = 1.125.$$

∎

More generally, for any continuous function f defined on the interval $[a, b]$, we partition $[a, b]$ as follows:

$$a = x_0 < x_1 < x_2 < \cdots < x_n = b,$$

where the points in the partition are equally spaced, with spacing $\Delta x = \dfrac{b - a}{n}$. On each subinterval, approximate the area under the curve by the area of the trapezoid whose sides have length $f(x_{i-1})$ and $f(x_i)$, as indicated in Figure 4.30. The area under the curve on the interval $[x_{i-1}, x_i]$ is then approximately

$$A_i \approx \frac{1}{2}[f(x_{i-1}) - f(x_i)]\,\Delta x,$$

FIGURE 4.30: Trapezoidal Rule.

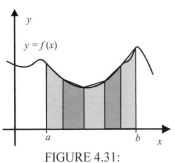

FIGURE 4.31: The $(n + 1)$-point Trapezoid Rule.

for each $i = 1, 2, \ldots, n$. Adding together the approximations for the area under the curve on each subinterval, we get that

$$\int_a^b f(x)\,dx \approx \left[\frac{f(x_0) + f(x_1)}{2} + \frac{f(x_1) + f(x_2)}{2} + \cdots + \frac{f(x_{n-1}) + f(x_n)}{2}\right]\Delta x$$
$$= \frac{b - a}{2n}[f(x_0) + 2\,f(x_1) + 2\,f(x_2) + \cdots + 2\,f(x_{n-1}) + f(x_n)] = T_n(f).$$

This is called the $(n + 1)$-point **Trapezoid Rule** and is illustrated in Figure 4.31. Notice that each of the middle terms are multiplied by 2, since each one is used in two trapezoids, once as the height of the trapezoid at the right endpoint and once as the height of the trapezoid at the left endpoint.

There are two ways to write a program for the Trapezoid Rule. One way is to add together $[f(x_{i-1}) + f(x_i)]$ for $i = 1, 2, \ldots, n$ and then multiply by $\Delta x/2$. As discussed in the exercises, an alternative is to add together the Riemann sums using left and right endpoint evaluations, and then divide by 2.

Example 7.5   **Using the Trapezoid Rule**

Compute the Trapezoid Rule approximations with $n = 4$ (by hand) and $n = 8, 16,$ 32, 64 and 128 (use a program) for $\int_0^1 3x^2 \, dx$.

Solution   As we saw in examples 7.1 and 7.2, the exact value of this integral is 1. For the Trapezoid Rule with $n = 4$, we have

$$T_4(f) = \frac{1-0}{(2)(4)} \left[f(0) + 2f\left(\frac{1}{4}\right) + 2f\left(\frac{1}{2}\right) + 2f\left(\frac{3}{4}\right) + f(1) \right]$$

$$= \frac{1}{8}\left(0 + \frac{3}{8} + \frac{12}{8} + \frac{27}{8} + 3\right) = \frac{66}{64} = 1.03125.$$

In Figure 4.29, we showed the four trapezoids comprising the approximation $T_4(f)$. In this case, notice how well the trapezoids fit the area under the curve. From a program, you can easily get the values in the following table, where we have computed $T_n(f)$ for the values of n indicated.

n	$T_n(f)$	error
4	1.03125	0.03125
8	1.0078125	0.0078125
16	1.00195313	0.00195313
32	1.00048828	0.00048828
64	1.00012207	0.00012207
128	1.00003052	0.00003052

We have included a column showing the error (the absolute value of the difference between the exact value of 1 and the approximate value) for your convenience. Notice that (much as with the Midpoint Rule) as the number of steps doubles, the error is reduced by approximately a factor of 4. You should also observe that the results here, both for the size of the error for a given value of n and the rate at which the error decreases as n increases are very comparable to those same features for the Midpoint Rule, discussed previously.

■

NOTE: Since the Trapezoid Rule formula is an average of two Riemann sums, we have

$$\int_a^b f(x) \, dx = \lim_{n \to \infty} T_n(f).$$

Simpson's Rule

Recall that in the Trapezoid Rule, we take a regular partition of the interval $[a, b]$ and for each subinterval, $[x_{i-1}, x_i]$, connect the points $(x_{i-1}, f(x_{i-1}))$ and $(x_i, f(x_i))$ with a straight line segment. We then approximate the area under the curve by the area under the line segment (i.e., the area of a trapezoid). We suggest the following alternative to the Trapezoid Rule. Again, construct a regular partition of the interval $[a, b]$:

$$a = x_0 < x_1 < x_2 < \cdots < x_n = b,$$

where

$$x_i - x_{i-1} = \frac{b-a}{n} = \Delta x,$$

for each $i = 1, 2, ..., n$ and where n is an **even** number. Instead of connecting each pair of points with a straight line segment, we connect each set of three consecutive points, $(x_{i-2}, f(x_{i-2}))$, $(x_{i-1}, f(x_{i-1}))$ and $(x_i, f(x_i))$ for $i = 2, 4, ..., n$, with a parabola (see

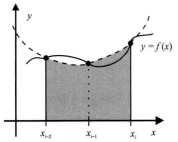

FIGURE 4.32: Simpson's Rule.

Figure 4.32). That is, we look for the quadratic function $p(x)$ whose graph passes through these three points, so that

$$p(x_{i-2}) = f(x_{i-2}), \ p(x_{i-1}) = f(x_{i-1}) \text{ and } p(x_i) = f(x_i).$$

We can use this to approximate the value of the integral of f on the interval $[x_{i-2}, x_i]$. We have

$$\int_{x_{i-2}}^{x_i} f(x)\,dx \approx \int_{x_{i-2}}^{x_i} p(x)\,dx.$$

Notice why we want to approximate f by a polynomial: polynomials are easy to integrate. A straightforward though tedious computation (try this; your CAS may help) gives

$$\int_{x_{i-2}}^{x_i} f(x)\,dx \approx \int_{x_{i-2}}^{x_i} p(x)\,dx$$
$$= \frac{x_i - x_{i-2}}{6} \left[f(x_{i-2}) + 4f(x_{i-1}) + f(x_i) \right]$$
$$= \frac{b-a}{3n} \left[f(x_{i-2}) + 4f(x_{i-1}) + f(x_i) \right].$$

Adding together the integrals over each subinterval $[x_{i-2}, x_i]$, for $i = 2, 4, 6, \ldots, n$, we get

$$\int_a^b f(x)\,dx$$
$$\approx \frac{b-a}{3n} \left[f(x_0) + 4f(x_1) + f(x_2) \right] + \frac{b-a}{3n} \left[f(x_2) + 4f(x_3) + f(x_4) \right] + \cdots$$
$$+ \frac{b-a}{3n} \left[f(x_{n-2}) + 4f(x_{n-1}) + f(x_n) \right]$$
$$= \frac{b-a}{3n} \left[f(x_0) + 4f(x_1) + 2f(x_2) + 4f(x_3) + 2f(x_4) + \cdots + 4f(x_{n-1}) + f(x_n) \right].$$

This is called the $(n+1)$-point **Simpson's Rule**, $S_n(f)$. Be sure to notice the pattern that the coefficients follow.

Next, we compare the accuracy of Simpson's Rule against the accuracy of the Midpoint and Trapezoid Rules.

Example 7.6   Using Simpson's Rule

Approximate the value of $\int_0^1 3x^2\,dx$ using Simpson's Rule with $n = 4$.

Solution   We have

$$S_4(f) = \frac{1-0}{(3)(4)} \left[f(0) + 4f\left(\frac{1}{4}\right) + 2f\left(\frac{1}{2}\right) + 4f\left(\frac{3}{4}\right) + f(1) \right] = 1,$$

which is in fact, the **exact** value. Notice that this is far more accurate than the Midpoint or Trapezoid Rules and yet, requires no more effort.

◼

Recall that Simpson's Rule computes the area beneath approximating parabolas. Given this, it shouldn't surprise you that Simpson's Rule gives the exact area in example 7.6. As you will discover in the exercises, Simpson's Rule gives exact values of integrals for *any* polynomial of degree 3 or less.

The following example illustrates an integral that you do not know how to compute exactly.

Example 7.7   Using a Program for Simpson's Rule

Compute Simpson's Rule approximations with $n = 4$ (by hand), $n = 8, 16, 32, 64$ and 128 (use a program) for $\int_0^2 \sqrt{x^2 + 1}\, dx$.

Solution   For $n = 4$, we have

$$S_4(f) = \frac{2 - 0}{(3)(4)} \left[f(0) + 4f\left(\frac{1}{2}\right) + 2f(1) + 4f\left(\frac{3}{2}\right) + f(2) \right]$$

$$= \left(\frac{1}{6}\right)\left[1 + 4\sqrt{\frac{5}{4}} + 2\sqrt{2} + 4\sqrt{\frac{13}{4}} + \sqrt{5} \right] \approx 2.95795560.$$

n	$S_n(f)$
4	2.9579556
8	2.9578835
16	2.95788557
32	2.95788571
64	2.95788571
128	2.95788572

From a program, you can easily get the values in the accompanying table.

Based on these calculations, we would expect 2.9578857 to be a very good approximation of $\int_0^2 \sqrt{x^2 + 1}\, dx$. ∎

Since most graphs curve somewhat, you might expect the parabolas of Simpson's Rule to better track the curve than the line segments of the Trapezoid Rule. As the following example shows, Simpson's Rule can be much more accurate than either the Midpoint Rule or the Trapezoid Rule.

Example 7.8   Comparing the Midpoint, Trapezoid and Simpson's Rules

Compute the Midpoint, Trapezoid and Simpson's Rule approximations of

$$\int_0^1 \frac{4}{x^2 + 1}\, dx$$ with $n = 10$, $n = 20$, $n = 50$ and $n = 100$. Compare to the exact value of π. (You'll see how to get this exact value in Chapter 6.)

Solution

n	Midpoint Rule	Trapezoid Rule	Simpson's Rule
10	3.142425985	3.139925988	3.141592613
20	3.141800986	3.141175986	3.141592652
50	3.141625986	3.141525986	3.141592653
100	3.141600986	3.141575986	3.141592653

Compare these values to the exact value of $\pi \approx 3.141592653$. Note that the Midpoint Rule tends to be slightly closer to π than the Trapezoid Rule, but neither is as close with $n = 100$ as Simpson's Rule is with $n = 10$. ∎

Remark 7.2   Notice that for a given value of n, the number of computations (and hence the effort) required to produce the Midpoint, Trapezoid and Simpson's Rule approximations are all nearly the same. So, example 7.8 gives an indication of how much more efficient Simpson's Rule is than the other two methods. This becomes a significant issue when the function $f(x)$ is difficult to evaluate. For instance, in the case of experimental data, each function value $f(x)$ could be the result of an expensive and time-consuming experiment.

In our final example, we revise our estimate of the area in Figure 4.27, first examined in example 7.4.

Example 7.9   Using Simpson's Rule With Data

Use Simpson's Rule to estimate $\int_0^1 f(x)\, dx$, where the only information known about f is given in the table of values shown in the margin.

Solution   From Simpson's Rule with $n = 4$, we have

$$\int_0^1 f(x)\,dx \approx \frac{1-0}{(3)(4)}\left[f(0) + 4f(0.25) + 2f(0.5) + 4f(0.75) + f(1)\right]$$

$$= \left(\frac{1}{12}\right)\left[1 + 4(0.8) + 2(1.3) + 4(1.1) + 1.6\right] \approx 1.066667.$$

x	$f(x)$
0	1.0
0.25	0.8
0.5	1.3
0.75	1.1
1.0	1.6

Since Simpson's Rule is generally much more accurate than the Trapezoid Rule (for a given number of points), we expect that this approximation is more accurate than the approximation of 1.125 arrived at in example 7.4 via the Trapezoid Rule. ■

Remark 7.3   Most graphing calculators and computer algebra systems have built-in programs for numerical approximation of definite integrals. Generally, these programs are very fast and accurate. Some ask you to specify a tolerance and will then calculate a value accurate to within that tolerance. One thing to note, though, is that these methods all return a single value. Recall that with all three methods developed here, we were careful to always report a sequence of values, so that you would have an idea of the accuracy of the result. Most of the routines built into calculators and CASs use *adaptive quadrature*. That is, they automatically calculate how many points are needed to obtain the desired accuracy. You should feel comfortable using these programs. However, if the integral you are approximating is a critical part of an important project, it's always a good idea to check your result. One way to do this is to use Simpson's Rule, $S_n(f)$, for a sequence of values of n. Of course, if all you know about a function is its value at a fixed number of points, most calculator and CAS programs will not help you, but the three methods discussed here will, as we saw in examples 7.4 and 7.9. We will pursue this idea further in the exercises.

Exercises 4.7

1. (writing) Ideally, approximation techniques should be both simple and accurate. How do the numerical integration methods presented in this section compare in terms of simplicity and accuracy? Which criterion would be more important if you were working entirely by hand? Which method would you use? Which criterion would be more important if you were using a very fast computer? Which method would you use?

2. (writing) Suppose you were going to construct your own rule for approximate integration. (Name it after yourself!) In the text, new methods were obtained both by choosing evaluation points for Riemann sums (Midpoint Rule) and by geometric construction (Trapezoid Rule and Simpson's Rule). Without working out the details, explain how you would develop a very accurate but simple rule.

3. (writing) Many calculators and computers are equipped with powerful numerical integration routines. Discuss the type of information you need to provide to your calculator; in particular, what determines how accurate the approximation is? Does this correspond directly to the value of n? Conjecture how your numerical integration routine might work.

4.   (writing) Test your calculator or computer on $\int_0^1 \sin(1/x)\,dx$. Discuss what your options are when your technology does not immediately return an accurate approximation. Based on a quick sketch of $y = \sin(1/x)$, describe why a numerical integration routine would have difficulty with this integral.

In exercises 5–8, compute Midpoint, Trapezoid and Simpson's Rule approximations by hand for $n = 4$.

5. $\displaystyle\int_0^1 (x^2 + 1)\,dx$ **6.** $\displaystyle\int_0^2 (x^2 + 1)\,dx$

7. $\displaystyle\int_1^3 \frac{1}{x}\,dx$ **8.** $\displaystyle\int_{-1}^1 (2x - x^2)\,dx$

In exercises 9 and 10, use the graph to estimate (a) Riemann sum with left-endpoint evaluation, (b) Midpoint Rule and (c) Trapezoid Rule approximations with $n = 4$ of $\int_0^2 f(x)\,dx$.

9.

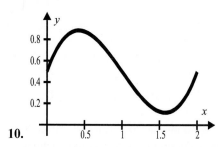

10.

that the value of C_m depends on the function and the interval of integration.) Then use this value to estimate how large a value of n will reduce the error in the Midpoint Rule below 10^{-10}.

In exercises 11–16, use a computer or calculator to compute the Midpoint, Trapezoid and Simpson's Rule approximations with $n = 10$, $n = 20$ and $n = 50$. Compare these values to the approximation given by your calculator or computer.

11. $\int_0^\pi \cos x^2 \, dx$ **12.** $\int_0^{\pi/4} \sin \pi x^2 \, dx$

13. $\int_0^2 e^{-x^2} \, dx$ **14.** $\int_0^3 e^{-x^2} \, dx$

15. $\int_0^\pi e^{\cos x} \, dx$ **16.** $\int_0^1 \sqrt[3]{x^2 + 1} \, dx$

In exercises 17–20, compute the exact value and compute the error (the difference between the approximation and the exact value) in each of the Midpoint, Trapezoid and Simpson's Rule approximations using $n = 10$, $n = 20$, $n = 40$ and $n = 80$.

17. $\int_0^1 5x^4 \, dx$ **18.** $\int_1^2 \frac{1}{x} \, dx$

19. $\int_0^\pi \cos x \, dx$ **20.** $\int_0^{\pi/4} \cos x \, dx$

21. Based on the evidence from exercises 17–20, which power of 2 (2, 4, 8, etc.) best finishes the following statements. If you double n, the error in the Midpoint Rule is divided by _____. If you double n, the error in the Trapezoid Rule is divided by _____. If you double n, the error in Simpson's Rule is divided by _____.

22. Use your answers to exercise 21 to explain why the following inequalities are true. Denote the errors in the Midpoint, Trapezoid and Simpson's rules by E_m, E_t and E_s, respectively. There are constants C_m, C_t and C_s such that

$$|E_m| \le \frac{C_m}{n^2}, \quad |E_t| \le \frac{C_t}{n^2} \quad \text{and} \quad |E_s| \le \frac{C_s}{n^4}.$$

Based on the evidence from exercises 17–20, conjecture the relative sizes of C_m and C_t.

23. It turns out that the constants C_m and C_t in exercise 22 both depend on the size of $f''(x)$, the second derivative of the function being integrated. Use graphs and concavity to explain why the larger $f''(x)$ is, the larger the error in the Midpoint and Trapezoid rules will be.

24. Use your results from exercise 17 to estimate the value of C_m for the integral in exercise 17. (It is important to realize

25. Use your results from exercise 17 to estimate the value of C_t for the integral in exercise 17. (It is important to realize that the value of C_t depends on the function and the interval of integration.) Then use this value to estimate how large a value of n will reduce the error in the Trapezoid Rule below 10^{-10}.

26. Use your results from exercise 17 to estimate the value of C_s for the integral in exercise 17. (It is important to realize that the value of C_s depends on the function and the interval of integration.) Then use this value to estimate how large n needs to be to reduce the error in Simpson's Rule below 10^{-10}.

In exercises 27–30, use (a) Trapezoid Rule and (b) Simpson's Rule to estimate $\int_0^2 f(x) \, dx$ from the given data.

27.

x	0.0	0.25	0.5	0.75	1.0
$f(x)$	4.0	4.6	5.2	4.8	5.0

x	1.25	1.5	1.75	2.0
$f(x)$	4.6	4.4	3.8	4.0

28.

x	0.0	0.25	0.5	0.75	1.0
$f(x)$	1.0	0.6	0.2	-0.2	-0.4

x	1.25	1.5	1.75	2.0
$f(x)$	0.4	0.8	1.2	2.0

29.

x	0.0	0.2	0.4	0.6	0.8	1.0
$f(x)$	2.4	2.6	2.9	3.2	3.4	3.6

x	1.2	1.4	1.6	1.8	2.0
$f(x)$	3.8	3.9	4.0	4.1	4.2

30.

x	0.0	0.2	0.4	0.6	0.8	1.0
$f(x)$	1.2	0.8	0.4	0.2	-0.4	-0.6

x	1.2	1.4	1.6	1.8	2.0
$f(x)$	-0.4	-0.2	0.0	0.4	1.2

In exercises 31 and 32, the table gives the measurements (in feet) of the width of a plot of land at 10-foot intervals. Estimate the area of the plot.

31.

x	0	10	20	30	40	50	60
$f(x)$	56	54	58	62	58	58	62

x	70	80	90	100	110	120
$f(x)$	56	52	48	40	32	22

32.

x	0	10	20	30	40	50	60
$f(x)$	26	30	28	22	28	32	30

x	70	80	90	100	110	120
$f(x)$	33	31	28	30	32	22

In exercises 33 and 34, the velocity of an object at various times is given. Use the data to estimate the distance traveled.

33.

t (s)	0	1	2	3	4	5	6
$v(t)$ (ft/s)	40	42	40	44	48	50	46

t (s)	7	8	9	10	11	12
$v(t)$ (ft/s)	46	42	44	40	42	42

34.

t (s)	0	2	4	6	8	10	12
$v(t)$ (ft/s)	26	30	28	30	28	32	30

t (s)	14	16	18	20	22	24
$v(t)$ (ft/s)	33	31	28	30	32	32

In exercises 35 and 36, the data comes from a pneumotachograph which measures air flow through the throat (in liters per second). The integral of the air flow equals the volume of air exhaled. Estimate this volume.

35.

t (s)	0	0.2	0.4	0.6	0.8	1.0	1.2
$f(t)$ (l/s)	0	0.2	0.4	1.0	1.6	2.0	2.2

t (s)	1.4	1.6	1.8	2.0	2.2	2.4
$f(t)$ (l/s)	2.0	1.6	1.2	0.6	0.2	0

36.

t (s)	0	0.2	0.4	0.6	0.8	1.0	1.2
$f(t)$ (l/s)	0	0.1	0.4	0.8	1.4	1.8	2.0

t (s)	1.4	1.6	1.8	2.0	2.2	2.4
$f(t)$ (l/s)	2.0	1.6	1.0	0.6	0.2	0

In exercises 37–42, use the given the information about $f(x)$ and its derivatives to determine whether the (a) Midpoint Rule would be exact, underestimate or overestimate the integral (or there's not enough information to tell). Repeat for (b) Trapezoid Rule and (c) Simpson's Rule.

37. $f''(x) > 0, f'(x) > 0$

38. $f''(x) > 0, f'(x) < 0$

39. $f''(x) < 0, f'(x) > 0$

40. $f''(x) < 0, f'(x) < 0$

41. $f''(x) = 4, f'(x) > 0$

42. $f''(x) = 0, f'(x) > 0$

43. Sketch a trapezoid with base L and heights h_1 and h_2. Show that the area of the trapezoid is given by $L(h_1 + h_2)/2$.

44. Suppose that R_L and R_R are the Riemann sum approximations of $\int_a^b f(x)\,dx$ using left- and right-endpoint evaluation rules, respectively, for some $n > 0$. Show that the trapezoid approximation T_n is equal to $(R_L + R_R)/2$.

45. For the data in Figure 4.27, sketch in the two approximating parabolas for Simpson's Rule. Compare the Simpson's Rule approximation to the Trapezoid Rule approximation. Explain graphically why the Simpson's Rule approximation is smaller.

46. For the data in Figure 4.27, sketch a smooth curve that passes through the five data points. Which more closely approximates your smooth curve, the line segments in Figure 4.28 or the parabolas you drew in exercise 45? Would your answer change if you had twice as many data points?

47. In most of the calculations that you have done, it is true that the Trapezoid Rule and Midpoint Rule are on opposite sides of the exact integral (i.e., one is too large, the other too small). Also, you may have noticed that the Trapezoid Rule tends to be about twice as far from the exact value as the Midpoint Rule. Given this, explain why the linear combination $\frac{1}{3}T_n + \frac{2}{3}M_n$ should give a good estimate of the integral. (Here, T_n represents the Trapezoid Rule approximation using n partitions and M_n the corresponding Midpoint Rule approximation.)

48. Show that the approximation rule $\frac{1}{3}T_n + \frac{2}{3}M_n$ in exercise 47 is identical to Simpson's Rule.

49. (exploratory) Compute the Trapezoid Rule approximations T_4, T_8 and T_{16} of $\int_0^1 3x^2\,dx$ and compute the error (the difference between the approximation and the exact value of 1). Verify that when the step size is cut in half, the error is divided by four. When such patterns emerge, they can be taken advantage of using **extrapolation**. The idea is simple: if the approximations continually get smaller, then the value of the integral is smaller and we should be able to predict (extrapolate) how much smaller the integral is. Given that $(T_4 - I) = 4(T_8 - I)$ where $I = 1$ is the exact integral, show that $I = T_8 + \dfrac{T_8 - T_4}{3}$. Also, show that $I = T_{16} + \dfrac{T_{16} - T_8}{3}$. In general, we have the *approximations* $(T_4 - I) \approx 4(T_8 - I)$ and $I \approx T_8 + \dfrac{T_8 - T_4}{3}$. Then the extrapolation $E_{2n} = T_{2n} + \dfrac{T_{2n} - T_n}{3}$ is closer to the exact integral than either of the individual Trapezoid Rule approximations T_{2n} and T_n. Show that, in fact, E_{2n} equals the Simpson's Rule approximation for $2n$.

50. (exploratory) The geometric construction of Simpson's Rule makes it clear that Simpson's Rule will compute integrals such as $\int_0^1 3x^2\,dx$ exactly. Briefly explain why. Now, compute Simpson's Rule with $n = 2$ for $\int_0^1 4x^3\,dx$. It turns out that Simpson's Rule also computes integrals of cubics exactly! In this exercise, we want to understand why a method that uses parabolas can compute integrals of cubics exactly. But first, sketch out the Midpoint Rule approximation of $\int_0^1 2x\,dx$ with $n = 1$. On part of the interval, the midpoint rectangle is above the straight line and on part of the interval, the midpoint rectangle is below the line. Explain why the Midpoint Rule computes the area exactly. Now, back to Simpson's Rule. To see how Simpson's Rule works on $\int_0^1 4x^3\,dx$, we need to determine the actual parabola being used. The parabola must pass through the points $(0, 0)$, $(1/2, 1/2)$ and $(1, 4)$. Find the quadratic function $y = ax^2 + bx + c$ which

accomplishes this. (Hint: Explain why $0 = 0 + 0 + c$, $1/2 = a/4 + b/2 + c$ and $1 = a + b + c$ and then solve for a, b and c.) Graph this parabola and $y = 4x^3$ on the same axes, carefully choosing the graphing window so that you can see what is happening on the interval $[0, 1]$. Where is the vertex of the parabola? How do the integrals of the parabola and cubic compare on the subinterval $[0, 1/2]$? $[1/2, 1]$? Why does Simpson's Rule compute the integral exactly?

Chapter Review Exercises

In exercises 1–20, find the antiderivative.

1. $\int (4x^2 - 3)\, dx$

2. $\int (x - 3x^5)\, dx$

3. $\int \dfrac{4}{x}\, dx$

4. $\int \dfrac{4}{x^2}\, dx$

5. $\int 2 \sin 4x\, dx$

6. $\int 3 \sec^2 x\, dx$

7. $\int (x - e^{4x})\, dx$

8. $\int 3\sqrt{x}\, dx$

9. $\int \dfrac{x^2 + 4}{x}\, dx$

10. $\int \dfrac{x}{x^2 + 4}\, dx$

11. $\int e^x (1 - e^{-x})\, dx$

12. $\int e^x (1 + e^x)^2\, dx$

13. $\int x\sqrt{x^2 + 4}\, dx$

14. $\int x(x^2 + 4)\, dx$

15. $\int 6x^2 \cos x^3\, dx$

16. $\int 4x \sec x^2 \tan x^2\, dx$

17. $\int \dfrac{e^{1/x}}{x^2}\, dx$

18. $\int \dfrac{\ln x}{x}\, dx$

19. $\int \tan x\, dx$

20. $\int \sqrt{3x + 1}\, dx$

21. Find a function $f(x)$ satisfying $f'(x) = 3x^2 + 1$ and $f(0) = 2$.

22. Find a function $f(x)$ satisfying $f'(x) = e^{-2x}$, $f(0) = 3$.

23. Determine the position function if the velocity is $v(t) = -32t + 10$ and the initial position is $s(0) = 2$.

24. Determine the position function if the acceleration is $a(t) = 6$ with initial velocity $v(0) = 10$ and initial position $s(0) = 0$.

25. Write out all terms and compute $\sum\limits_{i=1}^{6} (i^2 + 3i)$

26. Translate into summation notation and compute: the sum of the squares of the first 12 integers.

In exercises 27 and 28, use summation rules to compute the sum.

27. $\sum\limits_{i=1}^{100} (i^2 - 1)$

28. $\sum\limits_{i=1}^{100} (i^2 + 2i)$

29. Compute the sum $\dfrac{1}{n^3} \sum\limits_{i=1}^{n} (i^2 - i)$ and the limit of the sum as n approaches ∞.

30. For $f(x) = x^2 - 2x$ on the interval $[0, 2]$, list the evaluation points for the Midpoint Rule with $n = 4$, sketch the function and approximating rectangles and evaluate the Riemann sum.

In exercises 31–34, approximate the area under the curve using n rectangles and the given evaluation rule.

31. $y = x^2$ on $[0, 2]$, $n = 8$, midpoint evaluation

32. $y = x^2$ on $[-1, 1]$, $n = 8$, right-endpoint evaluation

33. $y = \sqrt{x + 1}$ on $[0, 3]$, $n = 8$, midpoint evaluation

34. $y = e^{-x}$ on $[0, 1]$, $n = 8$, left-endpoint evaluation

In exercises 35 and 36, use the given function values to estimate the area under the curve using (a) left-endpoint evaluation, (b) right-endpoint evaluation, (c) Trapezoid Rule and (d) Simpson's Rule.

35.

x	0.0	0.2	0.4	0.6	0.8	1.0	1.2	1.4	1.6
$f(x)$	1.0	1.4	1.6	2.0	2.2	2.4	2.0	1.6	1.4

36.

x	1.0	1.4	1.8	2.2	2.6	3.0	3.4	3.8	4.2
$f(x)$	4.0	3.4	3.6	3.0	2.6	2.4	3.0	3.6	3.4

37. In exercises 35 and 36, which of the four area estimates would you expect to be the most accurate? Briefly explain.

38. If $f(x)$ is positive and concave up, will the Midpoint Rule give an overestimate or underestimate of the actual area? Will the Trapezoid Rule give an overestimate or underestimate of the actual area?

In exercises 39 and 40, evaluate the integral by computing the limit of Riemann sums.

39. $\int_0^1 2x^2\, dx$

40. $\int_0^2 (x^2 + 1)\, dx$

In exercises 41 and 42, write the total area as an integral or sum of integrals, and then evaluate it.

41. The area above the x-axis and below $y = 3x - x^2$.

42. The area between the x-axis and $y = x^3 - 3x^2 + 2x$, $0 \le x \le 2$.

In exercises 43 and 44, use the velocity function to compute the distance traveled in the given time interval.

43. $v(t) = 40 - 10t$, $[1, 2]$

44. $v(t) = 20e^{-t/2}$, $[0, 2]$

In exercises 45 and 46, compute the average value of the function on the interval.

45. $f(x) = e^x$, $[0, 2]$ **46.** $f(x) = 4x - x^2$, $[0, 4]$

In exercises 47–58, evaluate the integral.

47. $\displaystyle\int_0^2 (x^2 - 2)\,dx$ **48.** $\displaystyle\int_{-1}^1 (x^3 - 2x)\,dx$

49. $\displaystyle\int_0^{\pi/2} \sin 2x\,dx$ **50.** $\displaystyle\int_0^{\pi/4} \sec^2 x\,dx$

51. $\displaystyle\int_0^{10} (1 - e^{-t/4})\,dt$ **52.** $\displaystyle\int_0^1 te^{-t^2}\,dt$

53. $\displaystyle\int_0^2 \frac{x}{x^2 + 1}\,dx$ **54.** $\displaystyle\int_1^2 \frac{\ln x}{x}\,dx$

55. $\displaystyle\int_0^2 x\sqrt{x^2 + 4}\,dx$ **56.** $\displaystyle\int_0^2 x(x^2 + 1)\,dx$

57. $\displaystyle\int_0^1 (e^x - 2)^2\,dx$ **58.** $\displaystyle\int_{-\pi}^{\pi} \cos(x/2)\,dx$

In exercises 59 and 60, find the derivative.

59. $f(x) = \displaystyle\int_2^x (\sin t^2 - 2)\,dt$

60. $f(x) = \displaystyle\int_0^{x^2} \sqrt{t^2 + 1}\,dt$

In exercises 61 and 62, compute the (a) Midpoint Rule, (b) Trapezoid Rule and (c) Simpson's Rule approximations with $n = 4$ by hand.

61. $\displaystyle\int_0^1 \sqrt{x^2 + 4}\,dx$ **62.** $\displaystyle\int_0^2 e^{-x^2/4}\,dx$

63. Repeat exercise 61 using a computer or calculator and $n = 20$; $n = 40$.

64. Repeat exercise 62 using a computer or calculator and $n = 20$; $n = 40$.

65. (exploratory) Suppose that $f(t)$ is the rate of occurrence of some event (e.g., the birth of an animal or the lighting of a firefly). Then the average rate of occurrence R over a time interval $[0, T]$ is $R = \frac{1}{T}\int_0^T f(t)\,dt$. We will assume that the function $f(t)$ is periodic with period T. (That is, $f(t + T) = f(t)$ for all t.) **Perfect asynchrony** means that the event is equally likely to occur at all times. Argue that this corresponds to a constant rate function $f(t) = c$ and find the value of c (in terms of R and T). **Perfect synchrony** means that the event only occurs once every period (e.g., the fireflies all light at the same time, or all babies are born simultaneously). We will see what the rate function $f(t)$ looks like in this case. First, define the **degree of synchrony** to be $\dfrac{\text{area under } f \text{ and above } R}{RT}$. Show that if $f(t)$ is constant then the degree of synchrony is 0. Then graph and find the degree of synchrony for the following functions (assuming $T > 2$):

$$f_1(t) = \begin{cases} (RT)(t - \frac{T}{2}) + RT & \text{if } \frac{T}{2} - 1 \le t \le \frac{T}{2} \\ (-RT)(t - \frac{T}{2}) + RT & \text{if } \frac{T}{2} \le t \le \frac{T}{2} + 1 \\ 0 & \text{otherwise} \end{cases}$$

$$f_2(t) = \begin{cases} (4RT)(t - \frac{T}{2}) + 2RT & \text{if } \frac{T}{2} - \frac{1}{2} \le t \le \frac{T}{2} \\ (-4RT)(t - \frac{T}{2}) + 2RT & \text{if } \frac{T}{2} \le t \le \frac{T}{2} + \frac{1}{2} \\ 0 & \text{otherwise} \end{cases}$$

$$f_3(t) = \begin{cases} (9RT)(t - \frac{T}{2}) + 3RT & \text{if } \frac{T}{2} - \frac{1}{3} \le t \le \frac{T}{2} \\ (-9RT)(t - \frac{T}{2}) + 3RT & \text{if } \frac{T}{2} \le t \le \frac{T}{2} + \frac{1}{3} \\ 0 & \text{otherwise} \end{cases}$$

What would you conjecture as the limit of the degrees of synchrony of $f_n(t)$ as $n \to \infty$? The "function" that $f_n(t)$ approaches as $n \to \infty$ is called an **impulse function** of strength RT. Discuss the appropriateness of this name.

Chapter 5
Applications of the Definite Integral

force

stretch

recoil

compression

When you watch or play a sport, you probably don't think much about engineering or science. However, the design of sports equipment has evolved into a highly technical, mathematically intensive enterprise. Even a superficial comparison of the wooden tennis racket of 1980 and the Kevlar tennis racket of 1990 illustrates the dramatic impact that technology has had on sports. (See the photograph in the margin.)

One of the principal goals in designing a tennis racket is to create the largest possible "sweet spot." The sweet spot refers to the area of a racket that produces an ideal shot. To some players, ideal means maximum power, while for other players, ideal means minimal vibration of the racket. Both of these criteria can be precisely translated into engineering problems that can be analyzed using definite integrals.

The design of a tennis racket begins with understanding of the physics of the collision between a ball and a racket. At high speeds, a tennis ball flattens almost completely and the strings of the tennis racket are stretched. (A remarkable photograph of this is found in Lee Torrey's *Stretching the Limits*, where you can find more details about the sweet spot.) The manner in which the ball and strings recover their shape determines the characteristics of the resulting flight of the ball. Detailed measurements can provide us with graphs of the forces and displacements of the strings, both as they are stretched and as they return to their natural positions. The figure to the left shows a typical graph. It turns out that the amount of energy lost to friction is proportional to the area between the two curves. The amount of twisting of a racket depends on its center of mass and moment of inertia. We mention these here, because the area between two curves, the center of mass and the moment of inertia of the racket are all calculated using definite integrals. We discuss the area between two curves in section 5.1, while center of mass and moments are found in section 5.6.

The collision between a ball and a racket determines the velocity and spin with which the ball leaves the racket. These characteristics, in turn, determine the flight path of the ball (with any luck, a graceful curve to a corner of the court). In section 5.4, we discuss how to compute the length of such a curve and the motion of a projectile (such as a tennis ball) is the subject of section 5.5. The velocity of the ball depends on a quantity called impulse, which we introduce in section 5.6.

In this chapter, we will demonstrate what a versatile tool the definite integral is. While we developed the integral as a means of finding the area under a curve, we'll see that it can be used to solve a much wider range of problems. The applications included in this chapter represent only a sampling of elementary problems encountered in mathematics, statistics, engineering and physics. We have already considered the definite integral from a variety of perspectives: graphical (signed areas), numerical (Riemann sum approximations) and symbolic (the Fundamental Theorem of Calculus). You will need to keep all three of these perspectives in mind as we now consider the applications of integration to other problems. As you study each new application in this chapter, pay particular attention to the role that each of these plays in connecting the new problem to integration. By doing so, you will see that the common thread running through so many diverse applications is the definite integral.

5.1   Area between Curves

Recall that we initially developed the definite integral (in Chapter 4) to compute the area under a curve. In particular, for any function f, where f is continuous on $[a, b]$ and $f(x) \geq 0$ on $[a, b]$, we wanted to find the area under the curve $y = f(x)$ on the interval $[a, b]$. We started by dividing (**partitioning**) the interval $[a, b]$ into n subintervals of equal size, $\Delta x = \dfrac{b-a}{n}$. The points in the partition are denoted $x_0 = a$, $x_1 = x_0 + \Delta x$, $x_2 = x_1 + \Delta x$,

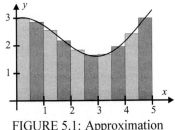

FIGURE 5.1: Approximation
of area.

and so on. That is,

$$x_i = x_0 + i\,\Delta x, \text{ for } i = 1, 2, \ldots, n.$$

On each subinterval $[x_{i-1}, x_i]$, we then constructed a rectangle, of height $f(c_i)$, for some $c_i \in [x_{i-1}, x_i]$, as in Figure 5.1. We took the sum of the areas of the n rectangles as an approximation of the area, A under the curve:

$$A \approx \sum_{i=1}^{n} f(c_i)\,\Delta x.$$

We then observed that as we took more and more rectangles, the sums of the areas of the rectangles approached the exact area under the curve and we ultimately defined this limit as the definite integral,

$$A = \lim_{n \to \infty} \sum_{i=1}^{n} f(c_i)\,\Delta x = \int_a^b f(x)\,dx.$$

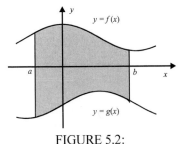

FIGURE 5.2:
Area between two curves.

In this section, we want to extend this notion ever so slightly. First, suppose that we have two continuous functions, f and g and $f(x) \geq g(x)$ for all x on $[a, b]$. We would like to find the area bounded between the two curves $y = f(x)$ and $y = g(x)$ on the interval $[a, b]$ (see Figure 5.2). Since we don't yet know how to find this area exactly, we first use rectangles to approximate the area. As before, we partition the interval $[a, b]$ into n subintervals of equal size, $\Delta x = \dfrac{b - a}{n}$, so that the points in the partition are

$$x_i = a + i\,\Delta x, \text{ for } i = 1, 2, \ldots, n.$$

Now, on each subinterval $[x_{i-1}, x_i]$, we construct a rectangle, stretching from the lower curve $y = g(x)$ to the upper curve, $y = f(x)$, as shown in Figure 5.3a. Referring to Figure 5.3b, these rectangles have height $h_i = f(c_i) - g(c_i)$, for some $c_i \in [x_{i-1}, x_i]$.

So, the area of the ith rectangle is

$$Area = length \times width = h_i\,\Delta x = [f(c_i) - g(c_i)]\,\Delta x.$$

The total area is then approximately equal to the sum of the areas of the n indicated rectangles,

$$A \approx \sum_{i=1}^{n} [f(c_i) - g(c_i)]\,\Delta x.$$

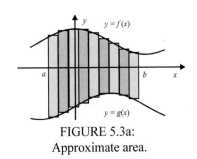

FIGURE 5.3a:
Approximate area.

Finally, you should observe that in the limit as $n \to \infty$ exists, we will get the exact area, which we recognize as a definite integral:

$$\boxed{A = \lim_{n \to \infty} \sum_{i=1}^{n} [f(c_i) - g(c_i)]\,\Delta x = \int_a^b [f(x) - g(x)]\,dx.} \qquad (1.1)$$

FIGURE 5.3b: Area of
ith rectangle.

Remark 1.1   Formula (1.1) is valid only when $f(x) \geq g(x)$ on the interval $[a, b]$. In general, the area between $y = f(x)$ and $y = g(x)$ for $a \leq x \leq b$ is given by $\int_a^b |f(x) - g(x)|\,dx$ Notice that to evaluate this integral, you must evaluate $\int_c^d [f(x) - g(x)]\,dx$ on all subintervals where $f(x) \geq g(x)$, then evaluate $\int_c^d [g(x) - f(x)]\,dx$ on all subintervals where $g(x) \geq f(x)$ and finally, add the integrals together.

Example 1.1   **Finding the Area between Two Curves**
Find the area bounded by the graphs of $y = 3 - x$ and $y = x^2 - 9$ (see Figure 5.4).

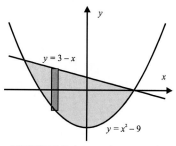

FIGURE 5.4: $y = 3 - x$ and $y = x^2 - 9$.

Solution   Notice that the region in Figure 5.4 is determined by the intersection of the two curves. The limits of integration will then correspond to the x-coordinates of the points of intersection. To find the limits, we set the two functions equal and solve for x. We have

$$3 - x = x^2 - 9 \quad \text{or} \quad 0 = x^2 + x - 12 = (x - 3)(x + 4).$$

Thus, the curves intersect at $x = -4$ and $x = 3$. You must be careful to notice from the graph which curve forms the upper boundary of the region and which one forms the lower boundary. In this case, the upper boundary is formed by $y = 3 - x$. So, for a fixed value of x, the height of a rectangle (such as the one indicated in Figure 5.4) is

$$h(x) = (3 - x) - (x^2 - 9).$$

Don't worry that $(x^2 - 9) < 0$ on part of the interval. The height of a given rectangle is still given by $h(x)$. From (1.1), the area is then

$$A = \int_{-4}^{3} \left[(3 - x) - (x^2 - 9) \right] \, dx$$

$$= \int_{-4}^{3} (-x^2 - x + 12) = \left[-\frac{x^3}{3} - \frac{x^2}{2} + 12x \right]_{-4}^{3}$$

$$= \left[-\frac{3^3}{3} - \frac{3^2}{2} + 12(3) \right] - \left[-\frac{(-4)^3}{3} - \frac{(-4)^2}{2} + 12(-4) \right] = \frac{343}{6}.$$

∎

Unfortunately, a given upper or lower boundary may not be defined by a single function, as in the case of intersecting graphs.

Example 1.2   Finding the Area between Two Curves Which Cross

Find the area bounded by the graphs of $y = x^2$ and $y = 2 - x^2$ for $0 \le x \le 2$.

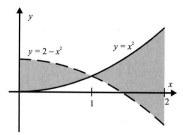

FIGURE 5.5: $y = x^2$ and $y = 2 - x^2$.

Solution   As always, a graph helps us devise the proper strategy (see Figure 5.5). Notice that since the two curves intersect in the middle of the interval, we will need to compute two integrals, one on the interval where $2 - x^2 \ge x^2$ and a second integral on the interval where $x^2 \ge 2 - x^2$. It appears that the graphs cross at about $x = 1$, with $2 - x^2 \ge x^2$ for $0 \le x \le 1$ and $x^2 \ge 2 - x^2$ for $1 \le x \le 2$. To find the point of intersection exactly, we solve $x^2 = 2 - x^2$, so that $2x^2 = 2$ or $x^2 = 1$ or $x = \pm 1$. Since $x = -1$ is outside the interval of interest, the only intersection of note is at $x = 1$. From (1.1), the area is

$$A = \int_{0}^{1} \left[(2 - x^2) - x^2 \right] \, dx + \int_{1}^{2} \left[x^2 - (2 - x^2) \right] \, dx$$

$$= \int_{0}^{1} (2 - 2x^2) \, dx + \int_{1}^{2} (2x^2 - 2) \, dx = \left[2x - \frac{2x^3}{3} \right]_{0}^{1} + \left[\frac{2x^3}{3} - 2x \right]_{1}^{2}$$

$$= \left(2 - \frac{2}{3} \right) - (0 - 0) + \left(\frac{16}{3} - 4 \right) - \left(\frac{2}{3} - 2 \right) = \frac{4}{3} + \frac{4}{3} + \frac{4}{3} = 4.$$

∎

In example 1.2, note that even though $2 - x^2 < 0$ on a portion of the interval, the calculation of area depends only on whether $2 - x^2 \ge x^2$ or $x^2 \ge 2 - x^2$. Also, you should realize that this example was designed so that the intersection point was easy to find. In the following example, the intersection points must be found using a numerical approximation method.

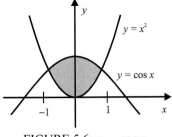

FIGURE 5.6: $y = \cos x$ and $y = x^2$.

Example 1.3   A Case Where the Intersection Points Are Known Only Approximately
Find the area bounded by the graphs of $y = \cos x$ and $y = x^2$.

Solution   Again, we must first determine the points of intersection of the curves. The graph in Figure 5.6 indicates intersections at about $x = -1$ and $x = 1$, so the x-coordinates of these points will be the limits of integration. The intersections occur where $\cos x = x^2$. You should recognize that this equation cannot be solved algebraically. So instead, we use a rootfinding method to find the approximate solutions $x = \pm 0.80107$. (For instance, you can use Newton's method to find values of x for which $f(x) = \cos x - x^2 = 0$.) From the graph, we can see that between these two x-values, $\cos x \geq x^2$ and so, the desired area is given by

$$A \approx \int_{-0.80107}^{0.80107} \left(\cos x - x^2 \right) dx = \left[\sin x - \frac{1}{3} x^3 \right]_{-0.80107}^{0.80107}$$

$$= \sin 0.80107 - \frac{1}{3}(0.80107)^3 - \left[\sin(-0.80107) - \frac{1}{3}(-0.80107)^3 \right]$$

$$\approx 1.0935.$$

Note that we have approximated both the limits of integration and the final calculations.

∎

Finding the area of some regions may require breaking the region up into several pieces, each having different upper and/or lower boundaries.

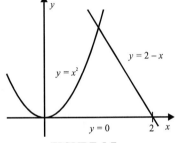

FIGURE 5.7a: $y = x^2$ and $y = 2 - x$.

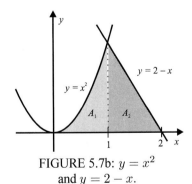

FIGURE 5.7b: $y = x^2$ and $y = 2 - x$.

Example 1.4   The Area of a Region Determined by Three Curves
Find the area bounded by the graphs of $y = x^2$, $y = 2 - x$ and $y = 0$.

Solution   A sketch of the three defining curves is shown in Figure 5.7a. Notice that the top boundary of the region is the curve $y = x^2$ on the first portion of the interval and the line $y = 2 - x$ on the second portion. To determine this precisely, you need to solve for the point of intersection. We have

$$2 - x = x^2 \text{ or } 0 = x^2 + x - 2 = (x + 2)(x - 1).$$

Since $x = -2$ is to the left of the y-axis, the intersection we seek occurs at $x = 1$. This corresponds to breaking the region into two pieces, as in Figure 5.7b and finding the area of each separately. The area of the region is then

$$A = A_1 + A_2 = \int_0^1 \left(x^2 - 0 \right) dx + \int_1^2 [(2 - x) - 0] \, dx$$

$$= \left. \frac{x^3}{3} \right|_0^1 + \left[2x - \frac{x^2}{2} \right]_1^2 = \frac{5}{6}.$$

∎

Although it was certainly not difficult to break up the region in example 1.4 into several pieces, we want to suggest an alternative that will prove to be surprisingly useful. Notice that if you turn the page sideways, Figure 5.7a will look like a region with a single curve determining each of the upper and lower boundaries. Of course, by turning the page sideways, you are essentially reversing the roles of x and y, which is the key to computing the area of this type of region: treat the left and right boundaries of the region as functions of y.

More generally, for two continuous functions, f and g of y, where $f(y) \geq g(y)$ for all y on the interval $c \leq y \leq d$, we can extend our notion of area to find the area bounded

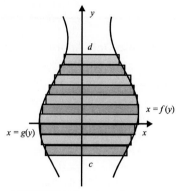

FIGURE 5.8a: Area between $x = g(y)$ and $x = f(y)$.

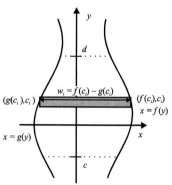

FIGURE 5.8b: Area of ith rectangle.

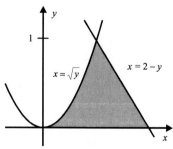

FIGURE 5.9: $y = x^2$ and $y = 2 - x$.

between the two curves $x = f(y)$ and $x = g(y)$. In this case, we partition the interval, $[c, d]$ into n equal subintervals, each of width $\Delta y = \dfrac{d - c}{n}$ (see Figure 5.8a). We denote the points in the partition by $y_0 = c$, $y_1 = y_0 + \Delta y$, $y_2 = y_1 + \Delta y$ and so on. That is,

$$y_i = c + i\,\Delta y, \text{ for } i = 1, 2, \ldots, n.$$

On each subinterval $[y_{i-1}, y_i]$ (for $i = 1, 2, \ldots, n$), we then construct a rectangle, of width $w_i = [f(c_i) - g(c_i)]$, for some $c_i \in [y_{i-1}, y_i]$, as seen in Figure 5.8b. The area of the ith rectangle is then given by

$$Area = length \times width = [f(c_i) - g(c_i)]\,\Delta y.$$

The total area between the two curves is then given approximately by

$$A \approx \sum_{i=1}^{n} [f(c_i) - g(c_i)]\,\Delta y.$$

As we have done several times now, we get the exact area by taking the limit as $n \to \infty$ and recognizing the limit as a definite integral. We get

$$A = \lim_{n \to \infty} \sum_{i=1}^{n} [f(c_i) - g(c_i)]\,\Delta y = \int_{c}^{d} [f(y) - g(y)]\,dy. \qquad (1.2)$$

Example 1.5   An Area Computed by Integrating with Respect to y

Repeat example 1.4, but integrate with respect to y instead.

Solution   The area bounded by the graphs of $y = x^2$, $y = 2 - x$ and $y = 0$ is shown in Figure 5.9. Notice that the left-hand boundary of the region is formed by the graph of $y = x^2$ and the right-hand boundary of the region is formed by the line $y = 2 - x$ and so, a single integral with respect to y will suffice. However, we must first write these left- and right-hand boundaries as functions of y. To do this, simply solve the equation $y = x^2$ for x. We get $x = \sqrt{y}$ (note that only the right half of the parabola forms the left boundary). Likewise, $y = 2 - x$ is equivalent to $x = 2 - y$. Finally, these curves intersect where $\sqrt{y} = 2 - y$. Squaring both sides gives us

$$y = (2 - y)^2 = 4 - 4y + y^2$$

or

$$0 = y^2 - 5y + 4 = (y - 1)(y - 4).$$

So, the curves intersect at $y = 1$ and $y = 4$. From Figure 5.9, it is clear that $y = 1$ is the solution we need. (What does the solution $y = 4$ correspond to?) From (1.2), the area is given by

$$A = \int_{0}^{1} [(2 - y) - \sqrt{y}]\,dy = \left[2y - \frac{1}{2}y^2 - \frac{2}{3}y^{3/2} \right]_{0}^{1} = 2 - \frac{1}{2} - \frac{2}{3} = \frac{5}{6}.$$

■

Example 1.6   The Area of a Region Bounded by Functions of y

Find the area bounded by the graphs of $x = y^2$ and $x = 2 - y^2$.

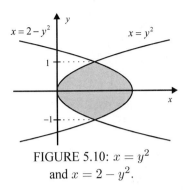

FIGURE 5.10: $x = y^2$
and $x = 2 - y^2$.

Solution   Of course, the graphs are parabolas opening to the right and left, respectively, as indicated in Figure 5.10. Notice that it's easiest to compute this area by integrating with respect to y, since integrating with respect to x would require us to break the region into two pieces and then find separate equations for the top and bottom boundaries of each of the two regions. To integrate with respect to y, we must first find the two intersections of the curves. These occur where $y^2 = 2 - y^2$, or $y^2 = 1$, so that $y = \pm 1$. On the interval $[-1, 1]$, notice that $2 - y^2 \geq y^2$ (since the curve $x = 2 - y^2$ stays to the right of the curve $x = y^2$). So, from (1.2), the area is given by

$$A = \int_{-1}^{1} [(2 - y^2) - y^2]\, dy = \int_{-1}^{1} (2 - 2y^2)\, dy$$

$$= \left[2y - \frac{2}{3}y^3 \right]_{-1}^{1} = \left(2 - \frac{2}{3} \right) - \left(-2 + \frac{2}{3} \right) = \frac{8}{3}.$$

∎

Exercises 5.1

1. (writing) Suppose the functions f and g satisfy $f(x) \geq g(x) \geq 0$ for all x in the interval $[a, b]$. Explain in terms of the areas $A_1 = \int_a^b f(x)\, dx$ and $A_2 = \int_a^b g(x)\, dx$ why the area between the curves $y = f(x)$ and $y = g(x)$ is given by $\int_a^b |f(x) - g(x)|\, dx$.

2. (writing) Suppose the functions f and g satisfy $f(x) \leq g(x) \leq 0$ for all x in the interval $[a, b]$. Explain in terms of the areas $A_1 = \int_a^b f(x)\, dx$ and $A_2 = \int_a^b g(x)\, dx$ why the area between the curves $y = f(x)$ and $y = g(x)$ is given by $\int_a^b |f(x) - g(x)|\, dx$.

3. (writing) Suppose that the speeds of racing cars A and B are $v_A(t)$ and $v_B(t)$ mph, respectively. If $v_A(t) \geq v_B(t)$ for all t, $v_A(0) = v_B(0)$ and the race lasts from $t = 0$ to $t = 2$ hours, explain why car A will win the race by $\int_0^2 [v_A(t) - v_B(t)]\, dt$ miles.

4. (writing) Suppose that the speeds of racing cars A and B are $v_A(t)$ and $v_B(t)$ mph, respectively. If $v_A(t) \geq v_B(t)$ for $0 \leq t \leq 0.5$ and $1.1 \leq t \leq 1.6$ and $v_B(t) \geq v_A(t)$ for $0.5 \leq t \leq 1.1$ and $1.6 \leq t \leq 2$, describe the difference between $\int_0^2 |V_A(t) - V_B(t)|\, dt$ and $\int_0^2 [V_A(t) - V_B(t)]\, dt$. Which integral will tell you which car wins the race?

In exercises 5–12, find the area between the curves on the given interval.

5. $y = x^3$, $y = x^2 - 1$, $1 \leq x \leq 3$

6. $y = \cos x$, $y = x^2 + 2$, $0 \leq x \leq 2$

7. $y = e^x$, $y = x - 1$, $-2 \leq x \leq 0$

8. $y = e^{-x}$, $y = x^2$, $1 \leq x \leq 4$

9. $y = x^2 - 1$, $y = 1 - x$, $0 \leq x \leq 2$

10. $y = x^2 - 3$, $y = x - 1$, $0 \leq x \leq 3$

11. $y = x^3 - 1$, $y = 1 - x$, $-2 \leq x \leq 2$

12. $y = x^4 + x - 2$, $y = x - 1$, $-2 \leq x \leq 2$

In exercises 13–20, find the area of the region determined by the intersections of the curves.

13. $y = x^2 - 1$, $y = 7 - x^2$

14. $y = x^2 - 1$, $y = \frac{1}{2}x^2$

15. $y = x^2 + 1$, $y = 3x - 1$

16. $y = x^2 - x - 4$, $y = x + 4$

17. $y = x^3$, $y = 3x + 2$

18. $y = x^3 - 2x^2$, $y = x^2$

19. $y = x^3$, $y = x^2$

20. $y = \sqrt{x}$, $y = x^2$

In exercises 21–26, estimate the area determined by the intersections of the curves.

21. $y = e^x$, $y = 1 - x^2$

22. $y = x^4$, $y = 1 - x$

23. $y = \sin x$, $y = x^2$

24. $y = \cos x$, $y = x^4$

25. $y = x^4$, $y = 2 + x$

26. $y = \ln x$, $y = x^2 - 2$

In exercises 27–34, find the area of the region bounded by the given curves. Choose the variable of integration so that the area is written as a single integral.

27. $y = x$, $y = 2 - x$, $y = 0$

28. $y = 2x \, (x > 0)$, $y = 3 - x^2$, $x = 0$

29. $x = 3y$, $x = 2 + y^2$

30. $x = y^2$, $x = 1$

31. $x = y$, $x = -y$, $x = 1$

32. $y = x$, $y = -x$, $y = 2$

33. $y = x$, $y = 2$, $y = 6 - x$, $y = 0$

34. $x = y^2$, $x = 4$

35. The average value of a function $f(x)$ on the interval $[a, b]$ is $A = \dfrac{1}{b-a} \displaystyle\int_a^b f(x)\,dx$. Compute the average value of $f(x) = x^2$ on $[0, 3]$ and show that the area above $y = A$ and below $y = f(x)$ equals the area below $y = A$ and above the x-axis.

36. Prove that the result of exercise 35 is always true by showing that $\displaystyle\int_a^b [f(x) - A]\,dx = 0$.

37. The United States oil consumption for the years 1970–1974 was approximately equal to $f(t) = 16.1e^{0.07t}$ million barrels per day, where $t = 0$ corresponds to 1970. Following an oil shortage in 1974, the country's consumption changed and was better modeled by $g(t) = 21.3e^{0.04(t-4)}$ million barrels per day, $t \geq 4$. Show that $f(4) = g(4)$ and explain what this number represents. Compute the area between $f(t)$ and $g(t)$ for $4 \leq t \leq 10$. Use this number to estimate the number of barrels of oil saved by Americans' reduced oil consumption from 1974 to 1980.

38. Suppose that a nation's fuelwood consumption is given by $76e^{0.03t}$ m³/yr and new tree growth is $50 - 6e^{0.09t}$ m³/yr. Compute and interpret the area between the curves for $0 \leq t \leq 10$.

39. Suppose that the birth rate for a certain population is $b(t) = 2e^{0.04t}$ million people per year, and the death rate for the same population is $d(t) = 2e^{0.02t}$ million people per year. Show that $b(t) \geq d(t)$ for $t \geq 0$, and explain why the area between the curves represents the increase in population. Compute the increase in population for $0 \leq t \leq 10$.

40. Suppose that the birth rate for a population is $b(t) = 2e^{0.04t}$ million people per year, and the death rate for the same population is $d(t) = 3e^{0.02t}$ million people per year. Find the intersection T of the curves ($T > 0$). Interpret the area between the curves for $0 \leq t \leq T$ and the area between the curves for $T \leq t \leq 30$. Compute the net change in population for $0 \leq t \leq 30$.

41. In collisions between a ball and a striking object (e.g., baseball bat or tennis racket), the ball changes shape, first compressing and then expanding. If x represents the change in size of the ball (e.g., in inches) for $0 \leq x \leq m$ and $f(x)$ represents the force between ball and striking object (e.g., in pounds), the area under the curve $y = f(x)$ is proportional to the energy transferred. Suppose that $f_c(x)$

is the force during compression and $f_e(x)$ is the force during expansion. Explain why $\int_0^m [f_c(x) - f_e(x)]\,dx$ is proportional to the energy lost by the ball (due to friction) and thus $\int_0^m [f_c(x) - f_e(x)]\,dx / \int_0^m f_c(x)\,dx$ is the proportion of energy lost in the collision. For a baseball and bat, reasonable values are shown (see Adair's book *The Physics of Baseball*):

x (in):	0	0.1	0.2	0.3	0.4
$f_c(x)$ (lb):	0	250	600	1200	1750
$f_e(x)$ (lb):	0	10	100	270	1750

Use Simpson's rule to estimate the proportion of energy retained by the baseball.

42. Using the same notation as in exercise 41, values for the force $f_c(x)$ during compression and force $f_e(x)$ during expansion of a golf ball are given by

x (in):	0	0.045	0.09	0.135	0.18
$f_c(x)$ (lb):	0	200	500	1000	1800
$f_e(x)$ (lb):	0	125	350	700	1800

Use Simpson's rule to estimate the proportion of energy retained by the golf ball.

43. Much like the compression and expansion of a ball discussed in exercises 41 and 42, the force exerted by a tendon as a function of its extension determines the loss of energy. Suppose that x is the extension of the tendon, $f_s(x)$ is the force during stretching of the tendon and $f_r(x)$ is the force during recoil of the tendon. The data given is for a hind leg tendon of a wallaby (see Alexander's book *Exploring Biomechanics*):

x (mm):	0	0.75	1.5	2.25	3.0
$f_s(x)$ (N):	0	110	250	450	700
$f_r(x)$ (N):	0	100	230	410	700

Use Simpson's rule to estimate the proportion of energy returned by the tendon.

44. The arch of a human foot acts like a spring during walking and jumping, storing energy as the foot stretches (i.e., the arch flattens) and returning energy as the foot recoils. In the data, x is the vertical displacement of the arch, $f_s(x)$ is the force on the foot during stretching and $f_r(x)$ is the force during recoil (see Alexander's book *Exploring Biomechanics*):

x (mm):	0	2.0	4.0	6.0	8.0
$f_s(x)$ (N):	0	300	1000	1800	3500
$f_r(x)$ (N):	0	150	700	1300	3500

Use Simpson's rule to estimate the proportion of energy returned by the arch.

45. (exploratory) At this stage, you can compute the area of any "simple" planar region. For a general figure bounded on the left by a function $x = l(y)$, on the right by a function $x = r(y)$, on top by a function $y = t(x)$ and on the bottom by a function $y = b(x)$, write the area as a sum of integrals. (Hint: Divide the region into subregions which can be written as the integral of $r(y) - l(y)$ or $t(x) - b(x)$.)

46. (exploratory) This exercise follows up on exercise 41. Suppose that $f_c(x) = 10,000x^2$ for $0 \le x \le 0.4$. Suppose that $f_e(x)$ is also a quadratic function, $f_e(x) = ax^2$ for some constant a. Find the value of a such that exactly 50% of the energy is lost.

5.2   Volume

As we shall see throughout this chapter, the integral is an amazingly versatile tool. In section 5.1, we saw how to use definite integrals to compute area, a principal measure of the size of a two-dimensional region. In this section, we'll see how to use integrals to compute volume, a principal measure of the size of a three-dimensional solid. We begin with a simple problem.

Suppose that a swimming pool has a rectangular top surface, of length 50 feet, oriented along the x-axis and width 15 feet. Further, suppose that the top surface of the pool is described by $y = 0$ and the bottom of the pool is described by $y = \dfrac{1}{165,000}(x^4 - 67x^3) - 3$.

(Note that this says that the depth at any given point is given by $-y$.) Find the volume of water needed to fill the pool up to the level $y = 0$. In Figure 5.11, we show a three-dimensional representation of the top and bottom surfaces of the pool.

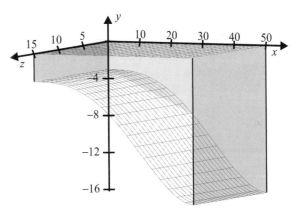

FIGURE 5.11: Side view of a swimming pool.

Before we try to find the volume of the pool, we should ask you a basic question. For what kinds of solids do you presently know how to compute the volume? You can probably only think of the volume of a box ($V = length \times width \times height$), a sphere ($V = \frac{4}{3}\pi r^3$) and a right circular cylinder ($V = \pi r^2 h$). In a way, the first and the last solids are the same: they're both *cylinders*. That's right; a cylinder is *any* solid whose cross sections (perpendicular to some axis running through the solid) are all the same. So, for instance, a rectangular box is a cylinder with rectangular cross sections. Now, notice some connection between the volume formulas for these two cylinders. The volume of a right circular cylinder is

$$V = \underbrace{\left(\pi r^2\right)}_{\text{cross-sectional area}} \times \underbrace{h}_{\text{height}}.$$

Likewise, in the case of a box, we have

$$V = \underbrace{(length \times width)}_{cross\text{-}sectional\ area} \times height.$$

In general, the volume of any cylinder is found by

$$V = (cross\text{-}sectional\ area) \times (height).$$

Looking at the pool in Figure 5.11, observe that it is a cylinder, since every vertical cross section parallel to the x-axis is the same. This says that the volume of the pool is the product of its cross-sectional area and the constant width of 15 feet. Notice that the cross-sectional area A is the area bounded between the curves $y = 0$ and $y = \dfrac{1}{165,000}(x^4 - 67x^3) - 3$. From (1.1), we have

$$A = \int_0^{50} \left\{ 0 - \left[\frac{1}{165,000}(x^4 - 67x^3) - 3 \right] \right\} dx$$

and so, the volume is given by

$$V = Area \times 15 = -15 \int_0^{50} \left[\frac{1}{165,000}(x^4 - 67x^3) - 3 \right] dx$$

$$= \left[\frac{-1}{2200}x^5 + \frac{65}{2750}x^4 + 45x \right]_0^{50} = 12,477 \text{ ft}^3.$$

Volumes by Slicing

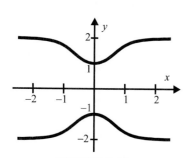

FIGURE 5.12:
Swimming pool (top view).

If either the cross-sectional area or width of a solid is not constant, we will need to modify our approach somewhat. For instance, when viewed from above, the pool shown in Figure 5.12 has the shape of an hourglass. Unless the depth of the pool is constant, this solid is not a cylinder and we cannot use the method developed for our introductory example. Since we don't know how to find the volume, we take the approach we've taken a number of times now: first approximate the volume and then improve the approximation.

For any solid that extends from $x = a$ to $x = b$, we start by partitioning the interval $[a, b]$ on the x-axis into n subintervals, each of width $\Delta x = \dfrac{b - a}{n}$. As usual, we denote $x_0 = a$, $x_1 = a + \Delta x$ and so on, so that

$$x_i = a + i\,\Delta x, \quad \text{for } i = 0, 1, 2, \ldots, n.$$

We then slice the solid perpendicular to the x-axis at each of the $(n-1)$ points $x_1, x_2, \ldots, x_{n-1}$ (see Figure 5.13a). We next need to find a reasonable way to approximate the volume of each of the n slices. First, notice that if n is large, then each slice of the solid will be thin. Again, for how many solids do you know how to find the volume? Notice that since each slice is thin, the cross-sectional area is nearly constant throughout any given slice. Suppose that the area of the cross section corresponding to any particular value of x is given by $A(x)$. If we consider the slice between $x = x_{i-1}$ and $x = x_i$, observe that the slice is nearly a cylinder (see Figure 5.13b). So, for any point c_i in the interval $[x_{i-1}, x_i]$, the area of the cross sections on that interval are all approximately $A(c_i)$. We can then approximate the volume V_i of the ith slice by the volume of the cylinder lying along the interval $[x_{i-1}, x_i]$, with constant cross-sectional area $A(c_i)$ (see Figure 5.13c).

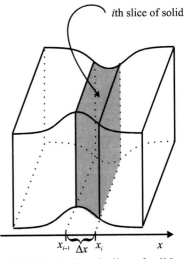

FIGURE 5.13a: Sliced solid.   FIGURE 5.13b: ith slice of solid.

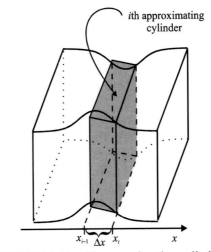

FIGURE 5.13c: ith approximating cylinder.

We have

$$V_i \approx \underbrace{A(c_i)}_{\text{cross-sectional area}} \underbrace{\Delta x}_{\text{width}},$$

where Δx is the width of the slice.

   Repeating this process for each of the n slices, we can approximate the total volume of the solid by

$$V \approx \sum_{i=1}^{n} A(c_i)\, \Delta x.$$

Notice that as the number of slices increases, the volume approximation should improve, and we get the exact volume by computing

$$V = \lim_{n \to \infty} \sum_{i=1}^{n} A(c_i)\, \Delta x,$$

assuming the limit exists. You should recognize the sum as a Riemann sum, so that the limit defines an integral. The exact volume is then given by

$$V = \int_a^b A(x)\,dx.$$

(2.1)

Remark 2.1   We should emphasize that we will use the same process followed above to derive many important formulas. We compute area, volume, arc length and surface area using essentially the same process: we divide an object into n smaller pieces, approximate the quantity of interest for each of the small pieces, add up the approximations and then take a limit, only to recognize that we have derived a definite integral. For this reason, it is essential that you understand the concept behind formula (2.1). Memorizing the formula will not do this for you. However, if you understand how the various pieces of this puzzle work and how they fit together (with the common theme of partitioning, approximation, summing the approximations and finally, taking a limit to arrive at a definite integral), all of the rest of this chapter should fall into place for you nicely.

Example 2.1   Computing Volume from Cross-Sectional Areas

Suppose that the hourglass shape of the swimming pool pictured in Figure 5.12 is defined by the curves $y(x) = \pm\left(\dfrac{x^2}{\sqrt{x^4+1}}+1\right)$ feet, for $-3 \le x \le 3$ and the depth of the pool is given by $d(x) = 6 + x$ feet. Find its volume.

Solution   Note that for any x, the cross section of the pool perpendicular to the x-axis is a rectangle with dimensions $2\left(\dfrac{x^2}{\sqrt{x^4+1}}+1\right)$ by $6 + x$. From (2.1), the volume is given by

$$V = \int_{-3}^{3} \underbrace{\left(2\frac{x^2}{\sqrt{x^4+1}}+1\right)(6+x)}_{\text{cross-sectional area}}\,dx \approx 123.81 \text{ ft}^3,$$

where we have calculated the integral numerically. (You can do this using Simpson's Rule or using the built-in integration procedure on your calculator or computer.)

■

In example 2.1, we knew how to compute the cross-sectional area exactly. In many important applications, the cross-sectional area is not known exactly, but must be approximated using measurements. In such cases, we can still find the volume (approximately), but we'll need to use numerical integration.

Example 2.2   Estimating Volume from Cross-Sectional Data

In medical imaging, such as CT (computerized tomography) and MRI (magnetic resonance imaging) processes, numerous measurements are taken and processed by a computer to produce a three-dimensional image that resembles the tissue the physician wishes to study. The process is similar to the slicing process we have used to find the volume of a solid. In this case, however, mathematical representations of various slices of the tissue are combined to produce a three-dimensional image that can be displayed and sliced back apart for a physician to determine the health of the tissue. Suppose that an MRI scan indicates that the cross-sectional areas of adjacent slices of a tumor are given by the values in the following table.

x (cm) :	0	0.1	0.2	0.3	0.4	0.5	0.6	0.7	0.8	0.9	1.0
$A(x)$ (cm^2) :	0.0	0.1	0.4	0.3	0.6	0.9	1.2	0.8	0.6	0.2	0.1

Estimate the volume of the tumor.

Solution   To find the volume of the tumor, we would compute (following (2.1))

$$V = \int_0^1 A(x)\, dx,$$

except that we don't know $A(x)$ (except approximately at a finite number of points). Does this ring any bells? Notice that we can use Simpson's rule (see section 4.7) with $\Delta x = 0.1$ to estimate the value of this integral:

$$V = \int_0^1 A(x)\, dx$$
$$\approx \frac{b-a}{3n}\left[\begin{array}{c} A(0) + 4A(0.1) + 2A(0.2) + 4A(0.3) + 2A(0.4) + 4A(0.5) \\ +2A(0.6) + 4A(0.7) + 2A(0.8) + 4A(0.9) + A(1) \end{array}\right]$$
$$= \frac{0.1}{3}\left(0 + 0.4 + 0.8 + 1.2 + 1.2 + 3.6 + 2.4 + 3.2 + 1.2 + 0.8 + 0.1\right)$$
$$= 0.49667 \text{ cm}^3.$$

■

In example 2.1, the cross-sectional areas of the swimming pool were easy to compute because each cross section was a rectangle. We will expend considerable effort analyzing solids in which the cross sections are circles formed by revolving a curve about a line. Such solids, called **solids of revolution**, are common in nature and are frequently encountered in industrial processes.

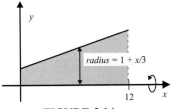

FIGURE 5.14a:
Revolve about the x-axis.

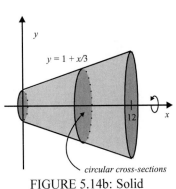

FIGURE 5.14b: Solid
of revolution.

Example 2.3   The Volume of a Solid of Revolution
Suppose that the line segment $y = 1 + x/3$, $0 \le x \le 12$, (see Figure 5.14a) is revolved about the x-axis. The resulting solid looks like a megaphone (see Figure 5.14b). Compute the volume of this solid.

Solution   First, notice that for any given values of x, the cross section perpendicular to the x-axis is a circle whose radius is given by $r = 1 + x/3$. (Think about this some!) The area of each cross section then has the form

$$A(x) = \pi r^2 = \pi\left(1 + x/3\right)^2.$$

From (2.1), the volume of the solid is then

$$V = \int_0^{12} \underbrace{\pi(1 + x/3)^2}_{\text{cross-sectional area}=\pi\,r^2}\, dx = \int_0^{12} \pi(1 + 2x/3 + x^2/9)\, dx$$
$$= \pi\left[x + \frac{x^2}{3} + \frac{x^3}{27}\right]_0^{12} = 2(12 + 48 + 64) = 124\pi.$$

■

Example 2.3 illustrates the **method of disks** used to compute the volume of a solid formed by revolving a curve about a vertical or horizontal line. We consider this general method next.

The Method of Disks

Suppose that $f(x) \geq 0$ and f is continuous on the interval $[a, b]$. Note that revolving the curve $y = f(x)$, for $a \leq x \leq b$, about the x-axis generates a solid (see Figures 5.15a and 5.15b).

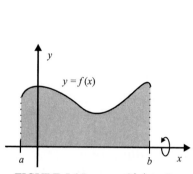

FIGURE 5.15a: $y = f(x) \geq 0$.

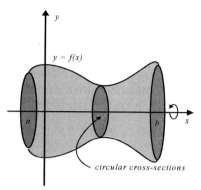

FIGURE 5.15b: Solid of revolution.

Notice that we can find the volume of this solid by slicing it perpendicular to the x-axis and recognizing that each cross section is a circle of radius $r = f(x)$ (see Figure 5.15b). From (2.1), we then have that the volume of the solid is

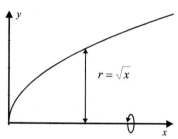

FIGURE 5.16a: $y = \sqrt{x}$.

$$V = \int_a^b \underbrace{\pi \left[f(x) \right]^2}_{\text{cross-sectional area} \,=\, \pi\, r^2} dx. \qquad (2.2)$$

Since the cross sections of such a solid of revolution are all disks, we refer to this method of finding volume as the **method of disks**.

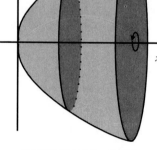

FIGURE 5.16b: Solid of revolution.

Example 2.4   Using the Method of Disks to Compute Volume

Revolve the region under the curve $y = \sqrt{x}$ on the interval $[0, 4]$ about the x-axis and find the volume of the resulting solid of revolution.

Solution   We should emphasize that it's important to draw a picture of the region and the solid of revolution, so that you get a clear idea of the radii of the circular cross sections. You can see from Figures 5.16a and 5.16b that the radius of each cross-section is given by $r = \sqrt{x}$. From (2.2), we then get the volume:

$$V = \int_0^4 \underbrace{\pi \left[\sqrt{x} \right]^2}_{\text{cross-sectional area} \,=\, \pi\, r^2} dx = \pi \int_0^4 x \, dx = \pi \left. \frac{x^2}{2} \right|_0^4 = 8\pi.$$

$\blacksquare$

In a similar way, suppose that $g(y) \geq 0$ and g is continuous on the interval $[c, d]$. Then, revolving the curve $x = g(y)$, for $c \leq y \leq d$, about the y-axis generates a solid (see Figures 5.17a and 5.17b).

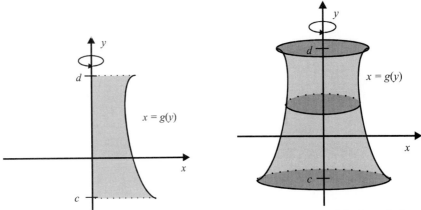

FIGURE 5.17a: Revolve about the y-axis.   FIGURE 5.17b: Solid of revolution.

Once again, notice from Figure 5.17b that the cross sections of the resulting solid of revolution are circles of radius $r = g(y)$. You should notice that all that has changed here is that we have interchanged the roles of the variables x and y. The volume of the solid is then given by

$$V = \int_c^d \underbrace{\pi[g(y)]^2}_{\text{cross-sectional area} = \pi\, r^2} dy. \qquad (2.3)$$

Remark 2.2   When using the method of disks, the question of whether to integrate with respect to x or y is **not** dependent on the geometry of the region (as it was when we were finding the area between two curves in section 5.1). Rather, the variable of integration depends solely on the axis about which you revolve the two-dimensional region: revolution about the x-axis requires integration with respect to x, while revolution about the y-axis requires integration with respect to y. This question is easily decided by looking at a sketch of the solid. Don't make the mistake of simply looking for what you can plug in where. This is a recipe for disaster, for the rest of this chapter will require you to make similar choices, each based on distinctive requirements of the problem at hand.

Example 2.5   **Using the Method of Disks with y as the Independent Variable**
        Find the volume of the solid resulting from revolving the portion of the curve $y = 2 - x^2/2$ from $x = 0$ to $x = 2$ about the y-axis.

Solution   You will find a graph of the curve in Figure 5.18a and of the solid in Figure 5.18b.

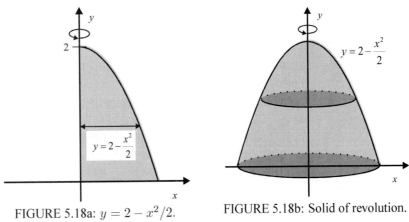

FIGURE 5.18a: $y = 2 - x^2/2$.   FIGURE 5.18b: Solid of revolution.

FIGURE 5.18c: Wireframe
of solid.

A wireframe representation of the solid is shown in Figure 5.18c. Notice from Figures 5.18a and 5.18b that the radius of any of the circular cross sections is given by x. So, we must solve the equation $y = 2 - x^2/2$, for x, to get $x = \sqrt{4 - 2y}$. Since the surface extends from $y = 0$ to $y = 2$, the volume is given by (2.3) to be

$$V = \int_0^2 \underbrace{\pi \left(\sqrt{4 - 2y} \right)^2}_{\pi r^2} dy = \int_0^2 \pi \left(4 - 2y \right) dy$$

$$= \pi \left[4y - 2 \frac{y^2}{2} \right]_0^2 = \pi \left(8 - 2^2 \right) = 4\pi.$$

∎

The Method of Washers

There are two complications that can be found in the types of volume calculations we have been studying. The first of these is that you may need to compute the volume of a solid that has a cavity or "hole" in it. The second of these occurs when a region is revolved about a line other than the x-axis or the y-axis. In either case, if you look carefully at the figures, these will not present you with any substantial difficulties. We illustrate these ideas in the examples that follow.

Example 2.6   Computing Volumes of Solids with and without Cavities
Let R be the region bounded by the graphs of $y = x^2$, $x = 0$ and $y = 1$. Compute the volume of the solid formed by revolving R about (a) the y-axis, (b) the x-axis and (c) the line $y = 2$.

Solution   (a) The region R is shown in Figure 5.19a and the solid formed by revolving it about the y-axis is shown in Figure 5.19b. A wireframe representation of the solid is found in Figure 5.19c. Notice that this part of the problem is similar to example 2.5.

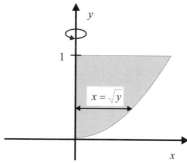

FIGURE 5.19a: $x = \sqrt{y}$.

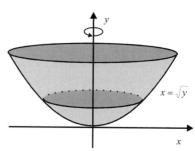

FIGURE 5.19b: Solid of revolution.

FIGURE 5.19c: Wireframe
of solid.

From (2.3), the volume is given by

$$V = \int_0^1 \pi (\sqrt{y})^2 \, dy = \pi \left. \frac{1}{2} y^2 \right|_0^1 = \frac{\pi}{2}.$$

Part (b) is slightly different from what we have already seen. Revolving the region about the x-axis produces a cavity in the middle of the solid. See Figure 5.20a for a graph of the region R and Figure 5.20b for a picture of the solid. A wireframe representation of the solid is shown in Figure 5.20c, where we have rotated the solid slightly to get a better view of the interior cavity.

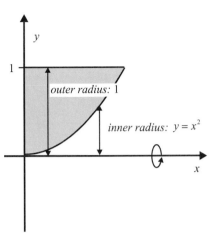

FIGURE 5.20a: $x = \sqrt{y}$.

FIGURE 5.20b: Solid with cavity.

washer-shaped
cross sections

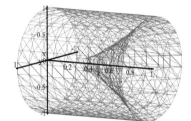

FIGURE 5.20c: Wireframe
of solid.

Our strategy for computing the volume of such a solid is to compute the volume of the outside of the object (as if it were filled in) and then subtract the volume of the cavity. Before diving into a computation, be sure to visualize the geometry behind this. For the present example, the outside surface of the solid is formed by revolving the line $y = 1$ about the x-axis. The cavity is formed by revolving the curve $y = x^2$ about the x-axis. Look carefully at Figures 5.20a and 5.20b and make certain that you see this. Observe that the outer radius, r_O is the distance from the x-axis to the line $y = 1$, or $r_O = 1$. The inner radius, r_I is the distance from the x-axis to the curve $y = x^2$, or $r_I = x^2$. Applying (2.2) twice, we see that the volume is given by

$$V = \int_0^1 \underbrace{\pi(1)^2}_{\pi\,(outer\ radius)^2}\ dx - \int_0^1 \underbrace{\pi(x^2)^2}_{\pi\,(inner\ radius)^2}\ dx$$

$$= \pi \int_0^1 \left(1 - x^4\right)\ dx = \pi \left(x - \frac{1}{5}\,x^5 \right)\Big|_0^1 = \pi \left(1 - \frac{1}{5} \right) = \frac{4}{5}\pi.$$

The solid in part (c) is formed by revolving the region R about the line $y = 2$. This time, the rotation produces a washer-like solid with a cylindrical hole in the middle. The region R is shown in Figure 5.21a and the solid is shown in Figure 5.21b. A wireframe representation of the solid is shown in Figure 5.21c.

FIGURE 5.21c: Wireframe
of solid.

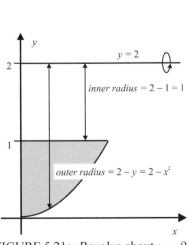

FIGURE 5.21a: Revolve about $y = 2$.

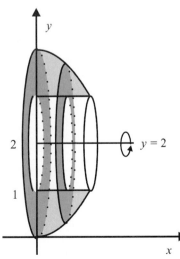

FIGURE 5.21b: Solid of revolution.

The volume is computed in the same way as in part (b), by subtracting the volume of the cavity from the volume of the outside solid. Look carefully at Figures 5.21a and 5.21b. Notice that the radius of the outer surface is the distance from the line $y = 2$ to the curve $y = x^2$. This outer radius is then $r_O = 2 - x^2$. The radius of the inner hole is the distance from the line $y = 2$ to the line $y = 1$. This inner radius is then $r_I = 2 - 1 = 1$. From (2.2), the volume is given by

$$V = \int_0^1 \underbrace{\pi(2 - x^2)^2}_{\pi\,(outer\ radius)^2}\ dx - \int_0^1 \underbrace{\pi(2 - 1)^2}_{\pi\,(inner\ radius)^2}\ dx$$

$$= \pi \int_0^1 \left[(4 - 4x^2 - x^4) - 1\right]\ dx = \pi \left[3x - \frac{4}{3}x^3 + \frac{1}{5}x^5\right]_0^1$$

$$= \pi \left(3 - \frac{4}{3} + \frac{1}{5}\right) = \frac{28}{15}\,\pi.$$

In parts (b) and (c) of example 2.6, the volume was computed by subtracting an inner volume from an outer volume in order to compensate for a cavity inside the solid. This technique is a slight generalization of the method of disks and is referred to as the **method of washers**, since the cross sections of the solids look like washers.

Example 2.7   Revolving a Region about Different Lines

Let R be the region bounded by $y = 4 - x^2$ and $y = 0$. Find the volume of the solids obtained by revolving R about each of the following: (a) the y-axis, (b) the line $y = -3$, (c) the line $y = 7$ and (d) the line $x = 3$.

Solution   For part (a), we draw the region R in Figure 5.22a and the solid of revolution in Figure 5.22b.

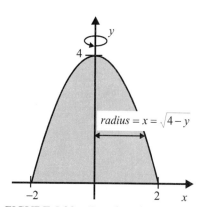

FIGURE 5.22a: Revolve about y-axis.

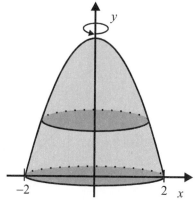

FIGURE 5.22b: Solid of revolution.

From Figures 5.22a and 5.22b, you should notice that the cross sections of the solid are simply circles, each of whose radius is the distance from the y-axis to the graph of $y = 4 - x^2$, namely x. Solving for x, we get $x^2 = 4 - y$, so that $x = \sqrt{4 - y}$, where we have selected x to be positive, since in this context, x represents a distance. From (2.3), the volume of the solid of revolution is given by

$$V = \int_0^4 \underbrace{\pi \left(\sqrt{4 - y}\right)^2}_{\pi\,(radius)^2}\ dy = \pi \int_0^4 (4 - y)\ dy$$

$$= \pi \left[4y - \frac{y^2}{2}\right]_0^4 = 8\pi.$$

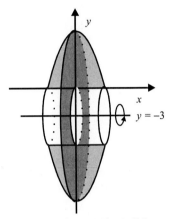

FIGURE 5.23a: Revolve
about $y = -3$.

There is at least one easy mistake to make here, namely to use $2\sqrt{4-y}$ as the radius. (What volume does this give you?)

For part (b), we have sketched the region R in Figure 5.23a and the solid of revolution in Figure 5.23b. Notice from Figures 5.23a and 5.23b that the cross sections of the solid are shaped like washers. Further, you can see that the outer radius r_O is the distance from the axis of revolution $y = -3$ to the curve $y = 4 - x^2$. That is,

$$r_O = y - (-3) = (4 - x^2) - (-3) = 7 - x^2,$$

while the inner radius is the distance from the bottom edge of the region R (i.e., the x-axis) to the line $y = -3$. That is,

$$r_I = 0 - (-3) = 3.$$

From (2.2), this gives us the volume

$$V = \int_{-2}^{2} \underbrace{\pi(7 - x^2)^2}_{\pi\,(outer\ radius)^2}\ dx - \int_{-2}^{2} \underbrace{\pi\,(3)^2}_{\pi\,(inner\ radius)^2}\ dx = \frac{1472}{15}\,\pi,$$

where we have left the details of the computation as an exercise.

Part (c) (revolving about the line $y = 7$) is very similar to part (b). You can see the region in Figure 5.24a and the solid in Figure 5.24b.

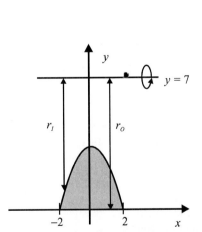

FIGURE 5.23b: Solid
of revolution.

FIGURE 5.24a: Revolve about $y = 7$.   FIGURE 5.24b: Solid of revolution.

In this case, you should note from Figures 5.24a and 5.24b that the cross sections of the solid are again shaped like washers. However, this time, the outer radius is the distance from the line $y = 7$ to the x-axis, that is, $r_O = 7$. The inner radius is the distance from the line $y = 7$ to the curve $y = 4 - x^2$, that is,

$$r_I = 7 - (4 - x^2) = 3 + x^2.$$

From (2.2), the volume of the solid is then

$$V = \int_{-2}^{2} \underbrace{\pi\,(7)^2}_{\pi\,(outer\ radius)^2}\ dx - \int_{-2}^{2} \underbrace{\pi\,(3 + x^2)^2}_{\pi\,(inner\ radius)^2}\ dx = \frac{576}{5}\,\pi,$$

where we again leave the details of the calculation as an exercise.

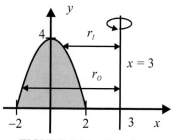

FIGURE 5.25a: Revolve
about $x = 2$.

FIGURE 5.25b: Solid
of revolution.

Finally, for part (d) (revolving about the line $x = 3$), we show the region R in Figure 5.25a and the solid of revolution in Figure 5.25b. In this case, the cross sections of the solid are washers, but the inner and outer radii are a bit trickier to determine than in the previous parts. Look carefully at Figures 5.25a and 5.25b and observe that the outer radius is the distance between the line $x = 3$ and the **left** half of the parabola, while the inner radius is the distance between the line $x = 3$ and the **right** half of the parabola. The parabola is given by $y = 4 - x^2$ and solving for x, we get $x^2 = 4 - y$, so that

$$x = \pm\sqrt{4 - y}.$$

Notice that $x = \sqrt{4 - y}$ corresponds to the right half of the parabola, while $x = -\sqrt{4 - y}$ describes the left half of the parabola. This gives us

$$r_I = 3 - \sqrt{4 - y} \quad \text{and} \quad r_O = 3 - \left(-\sqrt{4 - y}\right) = 3 + \sqrt{4 - y}.$$

Consequently, we get the volume

$$V = \int_0^4 \underbrace{\pi \left(3 + \sqrt{4 - y}\right)^2}_{\pi \, (outer \; radius)^2} dy - \int_0^4 \underbrace{\pi \left(3 - \sqrt{4 - y}\right)^2}_{\pi \, (inner \; radius)^2} dy = 64\pi,$$

where we leave the details of this rather messy calculation to you. (We recommend using your computer algebra system for this one.) In section 5.3, we present an alternative method for finding the volume of a solid of revolution, which for the present problem will produce much simpler integrals.

■

Remark 2.3 You will be most successful in finding volumes of solids of revolution if you spend the time it takes to draw reasonable figures and to label them carefully. Don't simply look for what to plug in where. You only need to keep in mind how to find the area of a cross section of the solid. Integration does the rest. Be careful!

Exercises 5.2

1. (writing) Discuss the relationships (e.g., perpendicular or parallel) to the x-axis and y-axis of the disks in examples 2.4 and 2.5. Explain how this relationship enables you to correctly determine the variable of integration.

2. (writing) The methods of disks and washers were developed separately in the text, but each is a special case of the general volume formula. Discuss the advantages of learning separate formulas versus deriving each example separately from the general formula. For example, would you prefer to learn the extra formulas or have to work each problem from basic principles? How many formulas is too many to learn?

3. (writing) To find the area of a triangle of the form △ in section 5.1, explain why you would use y-integration. In this section, would it be easier to compute the volume of this triangle revolved about the x-axis or y-axis? Explain your preference.

4. (writing) In part (a) of example 2.7, Figure 5.22a extends from $x = -\sqrt{4 - y}$ to $x = \sqrt{4 - y}$ but we used $\sqrt{4 - y}$ as the radius. Explain why this is the correct radius and not $2\sqrt{4 - y}$.

In exercises 5–8, find the volume of the solid with cross-sectional area $A(x)$.

5. $A(x) = x + 2, -1 \le x \le 3$

6. $A(x) = 10e^{0.01x}, 0 \le x \le 10$

7. $A(x) = \pi(4 - x)^2, 0 \le x \le 2$

8. $A(x) = 2(x + 1)^2, 1 \le x \le 4$

In exercises 9–16, set up an integral and compute the volume.

9. A swimming pool has outline $y = \pm\left(\dfrac{x^2}{\sqrt{x^4 + 1}} + 1\right)$ for $-3 \le x \le 3$ (assume units of feet for all dimensions). The depth of the pool is $6 + 3\sin\left(\frac{\pi}{6}x\right)$. Sketch a picture of the pool and compute its volume.

10. A swimming pool with the same outline as in exercise 9 has a depth of 3 feet for $-3 \le x \le -1$, a depth of 9 feet for $1 \le x \le 3$ and a depth that increases linearly from 3 to 9 for

$-1 \leq x \leq 1$. Sketch a picture of the pool and compute its volume.

11. The great pyramid at Gizeh is 500 feet high rising from a square base of side 750 feet. Compute its volume.

12. Suppose that instead of completing a pyramid, the builders at Gizeh had stopped at height 250 feet (with a square plateau top of side 375 feet). Compute the volume of this structure. Explain why the volume is greater than half the volume of the pyramid in exercise 11.

13. A church steeple is 30 feet tall with square cross sections. The square at the base has side 3 feet, the square at the top has side 6 inches, and the side varies linearly in between. Compute the volume.

14. A house attic has rectangular cross sections parallel to the ground and triangular cross sections perpendicular to the ground. The rectangle is 30 feet by 60 feet at the bottom of the attic and the triangles have base 30 feet and height 10 feet. Compute the volume of the attic.

15. A pottery jar has circular cross sections of radius $4 + \sin \frac{x}{2}$ inches for $0 \leq x \leq 2\pi$. Sketch a picture of the jar and compute its volume.

16. A pottery jar has circular cross sections of radius $4 - \sin \frac{x}{2}$ inches for $0 \leq x \leq 2\pi$. Sketch a picture of the jar and compute its volume.

17. Suppose an MRI scan indicates that cross-sectional areas of adjacent slices of a tumor are as given in the table. Use Simpson's rule to estimate the volume.

x (cm):	0.0	0.1	0.2	0.3	0.4	0.5
$A(x)$ (cm^2):	0.0	0.1	0.2	0.4	0.6	0.4
x (cm):	0.6	0.7	0.8	0.9	1.0	
$A(x)$ (cm^2):	0.3	0.2	0.2	0.1	0.0	

18. Suppose an MRI scan indicates that cross-sectional areas of adjacent slices of a tumor are as given in the table. Use Simpson's rule to estimate the volume.

x (cm):	0.0	0.2	0.4	0.6	0.8	1.0	1.2
$A(x)$ (cm^2):	0.0	0.2	0.3	0.2	0.4	0.2	0.0

19. Estimate the volume from the cross-sectional areas.

x (ft):	0.0	0.5	1.0	1.5	2.0
$A(x)$ (ft^2):	1.0	1.2	1.4	1.3	1.2

20. Estimate the volume from the cross-sectional areas.

x (m):	0.0	0.1	0.2	0.3	0.4
$A(x)$ (m^2):	2.0	1.8	1.7	1.6	1.8
x (m):	0.5	0.6	0.7	0.8	
$A(x)$ (m^2):	2.0	2.1	2.2	2.4	

In exercises 21–36, compute the volume of the solid formed by revolving the given region about the given line.

21. Region bounded by $y = 2 - x$, $y = 0$ and $x = 0$ about the x-axis

22. Region bounded by $y = x^2$, $y = 0$ and $x = 2$ about the x-axis

23. Region bounded by $y = 2 - x$, $y = 0$ and $x = 0$ about $y = 3$

24. Region bounded by $y = x^2$, $y = 0$ and $x = 2$ about $y = 4$

25. Region bounded by $y = \sqrt{x}$, $y = 2$ and $x = 0$ about the y-axis

26. Region bounded by $y = 2x$, $y = 2$ and $x = 0$ about the y-axis

27. Region bounded by $y = \sqrt{x}$, $y = 2$ and $x = 0$ about $x = 4$

28. Region bounded by $y = 2x$, $y = 2$ and $x = 0$ about $x = 1$

29. Region bounded by $y = e^x$, $x = 0$, $x = 2$ and $y = 0$ about the y-axis

30. Region bounded by $y = e^x$, $x = 0$, $x = 2$ and $y = 0$ about $y = -2$

31. Region bounded by $y = \cos x$, $x = 0$ and $y = 0$ about $y = 1$

32. Region bounded by $y = \cos x$, $x = 0$ and $y = 0$ about $y = -1$

33. Region bounded by $y = x^3$, $y = 0$ and $x = 1$ about the y-axis

34. Region bounded by $y = x^3$, $y = 0$ and $x = -1$ about the x-axis

35. Region bounded by $y = x^3$, $y = 0$ and $x = 1$ about $x = 1$

36. Region bounded by $y = x^3$, $y = 0$ and $x = -1$ about $y = -1$

37. Let R be the region bounded by $y = 3 - x$, the x-axis and the y-axis. Compute the volume of the solid formed by revolving R about the given line.
(a) y-axis   (b) x-axis   (c) $y = 3$ (d) $y = -3$
(e) $x = 3$ (f) $x = -3$

38. Let R be the region bounded by $y = x^2$ and $y = 4$. Compute the volume of the solid formed by revolving R about the given line.
(a) $y = 4$ (b) y-axis   (c) $y = 6$ (d) $y = -2$

39. Let R be the region bounded by $y = x^2$, $y = 0$ and $x = 1$. Compute the volume of the solid formed by revolving R about the given line.

(a) y-axis   (b) x-axis   (c) $x = 1$ (d) $y = 1$
(e) $x = -1$ (f) $y = -1$

40. Let R be the region bounded by $y = x$, $y = -x$ and $x = 1$. Compute the volume of the solid formed by revolving R about the given line.
(a) x-axis   (b) y-axis   (c) $y = 1$ (d) $y = -1$

41. Let R be the region bounded by $y = ax^2$, $y = h$ and the y-axis (where a and h are positive constants). Compute the volume of the solid formed by revolving this region about the y-axis. Show that your answer equals half the volume of a cylinder of height h and radius $\sqrt{h/a}$. Sketch a picture to illustrate this.

42. Use the result of exercise 41 to immediately write down the volume of the solid formed by revolving the region bounded by $y = ax^2$, $x = \sqrt{h/a}$ and the x-axis about the y-axis.

43. Suppose that the square with $-1 \le x \le 1$ and $-1 \le y \le 1$ is revolved about the y-axis. Show that the volume of the resulting solid is 2π.

44. Suppose that the circle $x^2 + y^2 = 1$ is revolved about the y-axis. Show that the volume of the resulting solid is $\frac{4}{3}\pi$.

45. Suppose that the triangle with vertices $(-1, -1)$, $(0, 1)$ and $(1, -1)$ is revolved about the y-axis. Show that the volume of the resulting solid is $\frac{2}{3}\pi$.

46. Sketch the square, circle and triangle of exercises 43–45 on the same axes. Show that the relative volumes of the revolved regions (cylinder, sphere and cone, respectively) are 3:2:1.

47. Verify the formula for the volume of a sphere by revolving the circle $x^2 + y^2 = r^2$ about the y-axis.

48. (exploratory) Generalize the result of exercise 46 to any rectangle. That is, sketch the rectangle with $-a \le x \le a$ and $-b \le y \le b$, the ellipse $\dfrac{x^2}{a^2} + \dfrac{y^2}{b^2} = 1$ and the triangle with vertices $(-a, -b)$, $(0, b)$ and $(a, -b)$. Show that the relative volumes of the solid formed by revolving these regions about the y-axis are 3:2:1.

49. (exploratory) Take the circle $(x - 2)^2 + y^2 = 1$ and revolve it about the y-axis. The donut-shaped solid is called a **torus**. Compute its volume. Show that the volume equals the area of the circle times the distance travelled by the center of the circle. This is an example of **Pappus' Theorem**, dating from the fourth century B.C. Verify that the result also holds for the triangle in exercise 40, parts (c) and (d)

5.3   Volumes by Cylindrical Shells

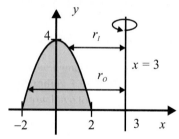

FIGURE 5.26a: Revolve about $x = 2$.

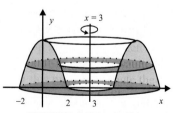

FIGURE 5.26b: Solid of revolution.

In this section, we present an alternative to the method of washers discussed in section 5.2. This method will help with solving some problems such as example 2.7, part (d), where the method of washers led to a rather awkward integral. Beyond learning the specifics of the method, it is most important that you learn how the geometry of a given problem dictates the method of solution.

As an introduction, we return to part (d) of example 2.7. There, we let R be the region bounded by the graphs of $y = 4 - x^2$ and $y = 0$ (see Figure 5.26a). As shown in the graph, R extends from $x = -2$ to $x = 2$. If R is revolved about the line $x = 3$, as indicated in Figure 5.26a, how would you compute the volume of the resulting solid (shown in Figure 5.26b)?

The geometry of the region R made it awkward to integrate with respect to y, since the left-hand and right-hand boundaries of R are the left and right halves of the parabola, respectively. On the other hand, since R is nicely defined on top by $y = 4 - x^2$ and on bottom by $y = 0$, it might be much easier to integrate with respect to x. Unfortunately, the method of washers requires the y-integration, in this case. The solution lies in an alternative method of computing volumes that uses the opposite variable of integration.

Before returning to this example, we consider the general case for a region revolved about the y-axis. Let R denote the region bounded by the graph of $y = f(x)$ and the x-axis on the interval $[a, b]$, where $0 < a < b$ and $f(x) \ge 0$ on $[a, b]$ (see Figure 5.27a). If we revolve this region about the y-axis, we get the solid shown in Figure 5.27b.

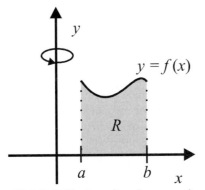

FIGURE 5.27a: Revolve about y-axis.

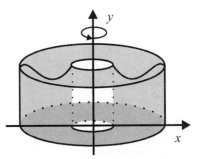

FIGURE 5.27b: Solid of revolution.

Look carefully at Figures 5.27a and 5.27b and observe that finding this volume by the method of washers is quite challenging, requiring the region to be split into several pieces. Instead, we develop a new method, as follows. As usual, we first partition the interval $[a, b]$ into n subintervals of equal width $\Delta x = \dfrac{b-a}{n}$. On each subinterval $[x_{i-1}, x_i]$, pick a point c_i and construct the rectangle of height $f(c_i)$ as indicated in Figure 5.28a. Revolving this rectangle about the y-axis forms a thin cylindrical shell (i.e., a hollow cylinder, like a soda can), as in Figure 5.28b.

FIGURE 5.28a: ith rectangle.   FIGURE 5.28b: Cylindrical shell.

To find the volume of this thin cylindrical shell, imagine cutting the cylinder from top to bottom and then flattening the shell out. After doing this, you should have essentially a thin rectangular sheet, as seen in Figure 5.28c.

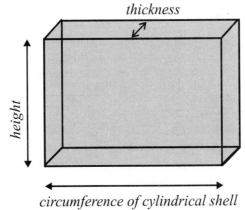

FIGURE 5.28c: Flattened cylindrical shell.

Notice that the length of such a thin sheet corresponds to the circumference of the cylindrical shell, which is $2\pi \cdot radius = 2\pi c_i$. So, the volume V_i of the ith cylindrical

shell is approximately

$$V_i \approx length \times width \times height$$
$$= (2\pi \times radius) \times thickness \times height$$
$$= (2\pi c_i)\, \Delta x\, f(c_i).$$

The total volume V of the solid can then be approximated by the sum of the volumes of the n cylindrical shells:

$$V \approx \sum_{i=1}^{n} 2\pi \underbrace{c_i}_{radius} \underbrace{f(c_i)}_{height} \underbrace{\Delta x}_{thickness} .$$

As we have done many times now, we can get the exact volume of the solid by taking the limit as $n \to \infty$ and recognizing the resulting definite integral. We have

$$V = \lim_{n \to \infty} \sum_{i=1}^{n} 2\pi c_i\, f(c_i)\, \Delta x = \int_a^b 2\pi \underbrace{x}_{radius} \underbrace{f(x)}_{height} \underbrace{dx}_{thickness} . \qquad (3.1)$$

WARNING: Do not rely on simply memorizing this formula. Once again, you must strive to understand the meaning of the components. It's simple to do if you just think of how they correspond to the volume of a cylindrical shell: $2\pi\,(radius)\,(height)\,(thickness)$. If you think of volumes in this way, you will be able to solve any problem you encounter.

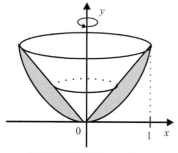

FIGURE 5.29a: Sample rectangle generating a cylindrical shell.

FIGURE 5.29b: Solid of revolution.

Example 3.1   Using the Method of Cylindrical Shells

Revolve the region bounded by the graphs of $y = x$ and $y = x^2$ in the first quadrant about the y-axis.

Solution   From Figure 5.29a, notice that the region has an upper boundary of $y = x$ and a lower boundary of $y = x^2$ and runs from $x = 0$ to $x = 1$.

In Figure 5.29a, we have drawn a sample rectangle that generates a cylindrical shell. The resulting solid of revolution can be seen in Figure 5.29b. We can write down an integral for the volume by analyzing the various components of the solid in Figures 5.29a and 5.29b. From (3.1), we have

$$V = \int_0^1 2\pi \underbrace{x}_{radius} \underbrace{(x - x^2)}_{height} \underbrace{dx}_{thickness}$$

$$= 2\pi \int_0^1 (x^2 - x^3)\, dx = 2\pi \left(\frac{x^3}{3} - \frac{x^4}{4} \right) \Bigg|_0^1 = \frac{\pi}{6}.$$

We can now generalize this method to solve the introductory example.

Example 3.2   A Volume Where Shells are Simpler Than Washers

Find the volume of the solid formed by revolving the region bounded by the graph of $y = 4 - x^2$ and the x-axis about the line $x = 3$.

Solution   Look carefully at Figure 5.30a, where we have drawn a sample rectangle that generates a cylindrical shell and at the solid shown in Figure 5.30b. Notice that the radius of a cylindrical shell is the distance from the line $x = 3$ to the shell:

$$r = 3 - x.$$

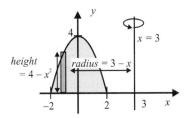

FIGURE 5.30a:
Typical cylindrical shell.

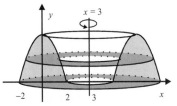

FIGURE 5.30b: Solid
of revolution.

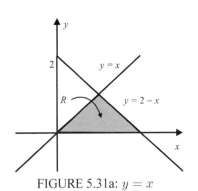

FIGURE 5.31a: $y = x$
and $y = 2 - x$.

This gives us the volume

$$V = \lim_{n \to \infty} \sum_{i=1}^{n} 2\pi \underbrace{(3 - c_i)}_{radius} \underbrace{\left(4 - c_i^2\right)}_{height} \underbrace{\Delta x}_{thickness}$$

$$= \int_{-2}^{2} 2\pi (3 - x) \left(4 - x^2\right) \, dx$$

$$= 2\pi \int_{-2}^{2} \left(x^3 - 3x^2 - 4x + 12\right) \, dx = 64\pi,$$

where the routine details of the calculation of the integral are left as an exercise. ∎

With the method of cylindrical shells to augment the method of washers, your first step in a volume calculation is to analyze the geometry of the solid to decide if it's easier to integrate with respect to x or y. Note that for a given solid, the variable of integration in the method of shells is exactly **opposite** that of the method of washers. You should choose the method that best fits the geometry of a given problem.

Example 3.3   Computing Volumes Using Shells and Washers

Let R be the region bounded by the graphs of $y = x$, $y = 2 - x$ and $y = 0$. Compute the volume of the solid formed by revolving R about the lines (a) $y = 2$, (b) $y = -1$ and (c) $x = 3$.

Solution   The region is shown in Figure 5.31a. Look carefully at the differences among the following three volumes.

(a) Revolving about the line $y = 2$, observe that the radius of a cylindrical shell is the distance from the line $y = 2$ to the shell: $2 - y$, for $0 \le y \le 1$ (see Figure 5.31b). The height is the difference in the x-values on the two curves: solving for x, we have $x = y$ and $x = 2 - y$. Following (3.1), we get the volume

$$V = \int_0^1 2\pi \underbrace{(2 - y)}_{radius} \underbrace{[(2 - y) - y]}_{height} \underbrace{dy}_{thickness} = \frac{10}{3}\pi,$$

where we leave the routine details of the calculation to you.

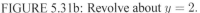

FIGURE 5.31b: Revolve about $y = 2$.

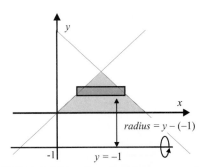

FIGURE 5.31c: Revolve about $y = -1$.

(b) Revolving about the line $y = -1$, notice that the height of the cylindrical shells is the same as in part (a), but the radius r is the distance from the line $y = -1$ to the shell: $r = y - (-1) = y + 1$ (see Figure 5.31c). This gives us the volume

$$V = \int_0^1 2\pi \underbrace{[y - (-1)]}_{radius} \underbrace{[(2 - y) - y]}_{height} \underbrace{dy}_{thickness} = \frac{8}{3}\pi.$$

FIGURE 5.31d: Revolve about $x = 3$.

(c) Finally, revolving about the line $x = 3$, notice that we would need to break the calculation into several pieces since the height of the cylindrical shells would be different for $x \in [0, 1]$ than for $x \in [1, 2]$. (Think about this some.) On the other hand, this is done easily by the method of washers. Observe that the outer radius is the distance from the line $x = 3$ to the line $x = y : r_O = 3 - y$, while the inner radius is the distance from the line $x = 3$ to the line $x = 2 - y : r_I = 3 - (2 - y)$ (see Figure 5.31d). This gives us the volume

$$V = \int_0^1 \pi \left[\underbrace{(3 - y)^2}_{outer\ radius^2} - \underbrace{(3 - (2 - y))^2}_{inner\ radius^2} \right] dy = 4\pi.$$

∎

Note the importance of sketching a picture and visualizing the solid. The calculation itself is of secondary importance. (You could always use your CAS.) What is most significant here is to see how to set up the problem. Based on the geometry, determine which variable you want to use in the integration (i.e., determine whether you will use washers or shells). Then carefully look at your picture to determine the components of the appropriate integral [i.e., the radius (radii) and possibly the height].

FIGURE 5.32: $y = \cos x$, $y = x^2$.

Example 3.4   Computing Volumes Using Shells and Washers

Let R be the region bounded by the graphs of $y = \cos x$ and $y = x^2$. Compute the volume of the solid formed by revolving R about the lines (a) $x = 2$ and (b) $y = 2$.

Solution   First, we sketch the region R (see Figure 5.32). We next look for the points of intersection of the two curves, by solving the equation $\cos x = x^2$. Since we can't solve this exactly, we must use an approximate method (e.g., Newton's method) to obtain the approximate intersections at $x = \pm 0.824132$.

(a) If we revolve the region about the line $x = 2$, we should use cylindrical shells (see Figure 5.33a). In this case, observe that the radius r of a cylindrical shell is the distance from the line $x = 2$ to the shell: $r = 2 - x$, while the height of a shell is $\cos x - x^2$. We get the volume

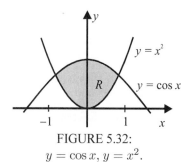

FIGURE 5.33a: Revolve about $x = 2$.

$$V \approx \int_{-0.824132}^{0.824132} 2\pi \underbrace{(2 - x)}_{radius} \underbrace{(\cos x - x^2)}_{height} dx \approx 13.757,$$

where we have approximated the value of the integral numerically. (Notice that at present, you do not know an antiderivative for this integrand, but we'll see how to find one in Chapter 7.)

(b) If we revolve the region about the line $y = 2$ (see Figure 5.33b), we use the method of washers. In this case, observe that the outer radius of a washer is the distance from the line $y = 2$ to the curve $y = x^2 : r_O = 2 - x^2$, while the inner radius is the distance from the line $y = 2$ to the curve $y = \cos x : r_I = 2 - \cos x$ (again, see Figure 5.33b). This gives us the volume

FIGURE 5.33b: Revolve about $y = 2$.

$$V \approx \int_{-0.824132}^{0.824132} \pi \left[\underbrace{(2 - x^2)^2}_{outer\ radius^2} - \underbrace{(2 - \cos x)^2}_{inner\ radius^2} \right] dx \approx 10.08,$$

where we have approximated the value of the integral numerically.

∎

We close this section with a summary of strategies for computing volumes of solids of revolution.

- Sketch the region to be revolved.
- Determine the variable of integration (x if the region has a well-defined top and bottom, y if the region has well-defined left and right boundaries).
- Based on the axis of revolution and the variable of integration, determine the method (disks or washers for x-integration about a horizontal axis or y-integration about a vertical axis, shells for x-integration about a vertical axis or y-integration about a horizontal axis).
- Set up the integral(s) and evaluate.

Exercises 5.3

1. (writing) Explain why the method of cylindrical shells produces an integral with x as the variable of integration when revolving about a vertical axis. (Describe where the shells are and which direction to move in to go from shell to shell.)

2. (writing) Explain why the method of cylindrical shells has the same form whether or not the solid has a hole or cavity. That is, there is no need for separate methods analogous to disks and washers.

3. (writing) Suppose that the region bounded by $y = x^2 - 4$ and $y = 4 - x^2$ is revolved about the line $x = 2$. Carefully explain which method (disks, washers or shells) would be easiest to use to compute the volume.

4. (writing) Suppose that the region bounded by $y = x^3 - 3x - 1$ and $y = -4$, $-2 \le x \le 2$, is revolved about $x = 3$. Explain what would be necessary to compute the volume using the method of washers, and what would be necessary to use the method of cylindrical shells. Which method would you prefer and why?

In exercises 5–12, sketch the region, draw in a typical shell, identify the radius and height of each shell and compute the volume.

5. The region bounded by $y = x^2$ and the x-axis, $-1 \le x \le 1$, revolved about $x = 2$

6. The region bounded by $y = x^2$ and the x-axis, $-1 \le x \le 1$, revolved about $x = -2$

7. The region bounded by $y = x$, $y = -x$ and $x = 1$ revolved about the y-axis

8. The region bounded by $y = x$, $y = -x$ and $x = 1$ revolved about $x = 1$

9. The region bounded by $y = x$, $y = -x$ and $y = 1$ revolved about $y = 2$

10. The region bounded by $y = x$, $y = -x$ and $y = 1$ revolved about $y = -2$

11. The right half of $x^2 + (y - 1)^2 = 1$ revolved about the x-axis

12. The right half of $x^2 + (y - 1)^2 = 1$ revolved about $y = 2$

In exercises 13–20, use cylindrical shells to compute the volume.

13. The region bounded by $y = x^2$ and $y = 2 - x^2$, revolved about $x = -2$

14. The region bounded by $y = x^2$ and $y = 2 - x^2$, revolved about $x = 2$

15. The region bounded by $x = y^2$ and $x = 1$ revolved about $y = -2$

16. The region bounded by $x = y^2$ and $x = 1$ revolved about $y = 2$

17. The region bounded by $y = x$ and $y = x^2 - 2$ revolved about $x = 2$

18. The region bounded by $y = x$ and $y = x^2 - 2$ revolved about $x = 3$

19. The region bounded by $x = (y - 1)^2$ and $x = 1$ revolved about the x-axis

20. The region bounded by $x = (y - 1)^2$ and $x = 1$ revolved about $y = 2$

In exercises 21–30, use the best method available to find each volume.

21. The region bounded by $y = 4 - x$, $y = 4$ and $y = x$ revolved about
(a) x-axis   (b) y-axis   (c) $x = 4$ (d) $y = 4$

22. The region bounded by $y = x + 2$, $y = -x - 2$ and $x = 0$ revolved about
(a) $y = -2$ (b) $x = -2$ (c) y-axis   (d) x-axis

23. The region bounded by $y = x$ and $y = x^2 - 6$ revolved about
(a) $x = 3$ (b) $y = 3$ (c) $x = -3$ (d) $y = -6$

24. The region bounded by $x = y^2$ and $x = 2 + y$ revolved about
(a) $x = -1$ (b) $y = -1$ (c) $x = -2$ (d) $y = -2$

25. The region bounded by $y = \cos x$ and $y = x^4$ revolved about

(a) $x = 2$ (b) $y = 2$ (c) x-axis   (d) y-axis

26. The region bounded by $y = \sin x$ and $y = x^2$ revolved about

(a) $y = 1$ (b) $x = 1$ (c) y-axis   (d) x-axis

27. The region bounded by $y = x^2$, $y = 2 - x$ and $x = 0$ revolved about

(a) x-axis   (b) y-axis   (c) $x = 1$ (d) $y = 2$

28. The region bounded by $y = 2 - x^2$, $y = x$ $(x > 0)$ and the y-axis revolved about

(a) x-axis   (b) y-axis   (c) $x = -1$ (d) $y = -1$

29. The region bounded by $y = 2 - x$, $y = x - 2$ and $x = y^2$ revolved about

(a) x-axis   (b) y-axis

30. The region bounded by $y = e^x - 1$, $y = 2 - x$ and the x-axis revolved about

(a) x-axis   (b) y-axis

31. A jewelry bead is formed by drilling a $\frac{1}{2}$-cm hole from the center of a 1-cm sphere. Explain why the volume is given by $\int_{1/2}^{1} 4\pi x \sqrt{1 - x^2} \, dx$. Evaluate this integral or compute the volume in some easier way.

32. Find the size of the hole in exercise 31 such that exactly half the volume is removed.

33. An anthill is in the shape formed by revolving the region bounded by $y = 1 - x^2$ and the x-axis about the y-axis. A researcher removes a cylindrical core from the center of the hill. What should the radius be to give the researcher 10% of the dirt?

34. (exploratory) From a sphere of radius R, a hole of radius r is drilled out of the center. Compute the volume removed in terms of R and r. Compute the length L of the hole in terms of R and r. Rewrite the volume in terms of L. Is it reasonable to say that the volume removed depends on L and not on R?

5.4   Arc Length and Surface Area

In this section, we use the definite integral to investigate two more measures of geometric size. Length and area are quantities you already understand intuitively. But, as you have learned with area, the calculation of these quantities can be surprisingly challenging for many geometric shapes. The calculations that we explore here are complicated by an increase in the dimension. Specifically, we will compute the length (a one-dimensional measure) of a curve in two dimensions and we will compute the area (a two-dimensional measure) of a surface in three dimensions. As always, pay particular attention to the derivations. As we have done a number of times now, we start with an approximation and then proceed to the exact solution, using the notion of limit.

Arc Length

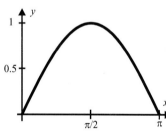

FIGURE 5.34: $y = \sin x$.

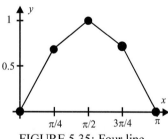

FIGURE 5.35: Four line segments approximating $y = \sin x$.

What could we mean by the *length* of the portion of the sine curve shown in Figure 5.34? (We call the length of a curve its **arc length**.) If the curve represented a road, you could measure the length on your car's odometer by driving along that section of road. f the curve were actually a piece of string, you could straighten out the string and then measure its length with a ruler. Both of these ideas are very helpful intuitively. They both involve turning the problem of measuring length in two dimensions into the (much easier) problem of measuring the length in one dimension.

To accomplish this mathematically, we first approximate the curve with several line segments joined together. In Figure 5.35, the line segments connect the points $(0, 0)$, $(\pi/4, 1/\sqrt{2})$, $(\pi/2, 1)$, $(3\pi/4, 1/\sqrt{2})$ and $(\pi, 0)$ on the curve $y = \sin x$. An approximation of the arc length s of the curve is given by the sum of the lengths of the line segments:

$$s \approx \sqrt{(\pi/4)^2 + (1/\sqrt{2})^2} + \sqrt{(\pi/4)^2 + (1 - 1/\sqrt{2})^2}$$
$$+ \sqrt{(\pi/4)^2 + (1/\sqrt{2} - 1)^2} + \sqrt{(\pi/4)^2 + (1/\sqrt{2})^2} \approx 3.79.$$

You might notice that this estimate is too small. (Why is that?) We could improve our approximation by using more than four line segments. In the accompanying table, we

n	length
8	3.8125
16	3.8183
32	3.8197
64	3.8201
128	3.8201

show estimates of the length of the curve using n line segments for larger values of n. You should expect that as we increase the number of line segments, the approximation of length will get closer to the actual length of the curve. This general idea should sound familiar.

We develop this notion further now for the more general problem of finding the arc length of the curve $y = f(x)$ on the interval $[a, b]$. Here, we'll assume that f is continuous on $[a, b]$ and differentiable on (a, b). (Where have you seen hypotheses like these before?) As usual, we begin by partitioning the interval $[a, b]$ into n equal pieces: $a = x_0 < x_1 < \cdots < x_n = b$, where

$$x_i - x_{i-1} = \Delta x = \frac{b - a}{n},$$

for each $i = 1, 2, \ldots, n$.

Between each pair of adjacent points on the curve, $(x_{i-1}, f(x_{i-1}))$ and $(x_i, f(x_i))$, we approximate the arc length s_i by the straight-line distance between the two points (from the usual distance formula). We have

$$s_i \approx d\left\{(x_{i-1}, f(x_{i-1})), (x_i, f(x_i))\right\} = \sqrt{(x_i - x_{i-1})^2 + [f(x_i) - f(x_{i-1})]^2}.$$

Since f is continuous on all of $[a, b]$ and differentiable on (a, b), f is also continuous on the subinterval $[x_{i-1}, x_i]$ and is differentiable on (x_{i-1}, x_i). Recall that by the Mean Value Theorem, we have

$$f(x_i) - f(x_{i-1}) = f'(c_i)\,(x_i - x_{i-1}),$$

for some number $c_i \in (x_{i-1}, x_i)$. This gives us the approximation

$$
\begin{aligned}
s_i &\approx \sqrt{(x_i - x_{i-1})^2 + [f(x_i) - f(x_{i-1})]^2} \\
&= \sqrt{(x_i - x_{i-1})^2 + [f'(c_i)\,(x_i - x_{i-1})]^2} \\
&= \sqrt{1 + [f'(c_i)]^2}\ \underbrace{(x_i - x_{i-1})}_{\Delta x} = \sqrt{1 + [f'(c_i)]^2}\,\Delta x.
\end{aligned}
$$

Adding together the lengths of these n line segments, we get an approximation of the total arc length,

$$s \approx \sum_{i=1}^{n} \sqrt{1 + [f'(c_i)]^2}\,\Delta x.$$

Notice that as n gets larger, this approximation should approach the exact arc length, that is,

$$s = \lim_{n \to \infty} \sum_{i=1}^{n} \sqrt{1 + [f'(c_i)]^2}\,\Delta x.$$

You should recognize this as the limit of a Riemann sum for $\sqrt{1 + [f'(x)]^2}$, so that the arc length is given exactly by the definite integral:

Arc length of $y = f(x)$ on the interval $[a, b]$

$$\boxed{s = \int_a^b \sqrt{1 + [f'(x)]^2}\,dx,} \tag{4.1}$$

whenever the limit exists.

Remark 4.1   The formula for arc length is very simple. Unfortunately, very few functions produce arc length integrals that can be evaluated exactly. You should expect to use a numerical integration method on your calculator or computer to find most arc lengths.

Example 4.1   Using the Arc Length Formula

Find the arc length of $y = \sin x$ for $0 \le x \le \pi$. (Recall that we estimated this as 3.79 in our introductory example.)

Solution   From (4.1), the arc length is

$$s = \int_0^{\pi} \sqrt{1 + (\cos x)^2}\, dx.$$

Try to find an antiderivative of $\sqrt{1 + \cos^2 x}$, but don't try for too long. Using a numerical integration method, the arc length is

$$s = \int_0^{\pi} \sqrt{1 + (\cos x)^2}\, dx \approx 3.8202.$$

Even for very simple curves, evaluating the arc length integral exactly can prove quite challenging.

Example 4.2   Estimating an Arc Length

Find the arc length of the curve $y = x^2$ for $0 \le x \le 1$.

Solution   Using the arc length formula (4.1), we get

$$s = \int_0^1 \sqrt{1 + (2x)^2}\, dx = \int_0^1 \sqrt{1 + 4x^2}\, dx \approx 1.4789,$$

where we have again evaluated the integral numerically. (In this case, you can find an antiderivative using an integral table or a CAS, but the decimal approximation is just as meaningful for our present purposes.)

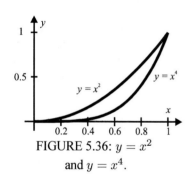

FIGURE 5.36: $y = x^2$
and $y = x^4$.

The graphs of $y = x^2$ and $y = x^4$ look surprisingly similar on the interval $[0, 1]$ (see Figure 5.36). They both connect the points $(0, 0)$ and $(1, 1)$, are increasing and are concave up. If you graph them simultaneously, you will note that $y = x^4$ starts out flatter and then becomes steeper from about $x = 3/4$ on. (Try **proving** that this is true!) Arc length gives us one way to quantify the difference between the two graphs.

Example 4.3   A Comparison of Arc Lengths of Power Functions

Find the arc length of $y = x^4$ for $0 \le x \le 1$ and compare to the arc length of $y = x^2$.

Solution   From (4.1), the arc length for $y = x^4$ is given by

$$\int_0^1 \sqrt{1 + (4x^3)^2}\, dx = \int_0^1 \sqrt{1 + 16x^6}\, dx \approx 1.6002.$$

Notice that this arc length is about 8% larger than that of $y = x^2$.

In the exercises, you will be asked to explore the trend in the lengths of the curves $y = x^6$, $y = x^8$ and so on, on the interval $[0, 1]$. Can you guess now what happens to the arc length of $y = x^n$, on the interval $[0, 1]$, as $n \to \infty$?

As with many words, everyday usage of the word *length* can be ambiguous and misleading. For instance, when somebody says that the *length* of a frisbee throw was 100 yards, the length refers to the horizontal distance covered, not to the arc length of the frisbee's flight path. In this case, reporting the horizontal distance makes more sense (and is easier to measure) than arc length. In many other cases, arc length is the quantity of interest. For example, suppose you are going to hang a banner between two poles that are 20 feet apart. If you only have 20 feet of rope to work with, you are going to be in trouble. The length of rope required is determined by the arc length and not the horizontal distance.

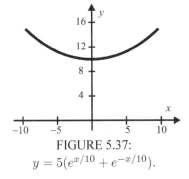

FIGURE 5.37:
$y = 5(e^{x/10} + e^{-x/10})$.

Example 4.4   Estimating the Length of Cable Hanging between Two Poles

A cable is to be hung between two poles of equal height that are 20 feet apart. It can be shown that such a hanging cable assumes the shape of a **catenary**. In this case, suppose that the cable takes the shape of $y = 5(e^{x/10} + e^{-x/10})$, $-10 \leq x \leq 10$, as seen in Figure 5.37. How long is the cable?

Solution   From (4.1), the arc length of the curve is given by

$$s = \int_{-10}^{10} \sqrt{1 + \left(\frac{e^{x/10}}{2} - \frac{e^{-x/10}}{2} \right)^2} \, dx$$

$$= \int_{-10}^{10} \sqrt{1 + \frac{1}{4} \left(e^{x/5} - 2 + e^{-x/5} \right)} \, dx$$

$$\approx 23.504 \text{ feet},$$

or 20 feet plus about $3\frac{1}{2}$ feet of slack.

∎

Surface Area

In sections 5.2 and 5.3, we saw how to compute the volume of a solid formed by revolving a two-dimensional region about a fixed axis. In addition to computing the volume of the solid, we often want to determine the area of the *surface* that is generated by the revolution. For instance, when revolving the line $y = x + 1$, for $0 \leq x \leq 1$ about the x-axis, the surface generated looks like a megaphone (ends not included), as in Figure 5.38.

Notice that this is the bottom portion of a right circular cone that has had its top cut off by plane parallel to its base. Before we go any further, we should find the curved surface area of a right circular cone. In Figure 5.39a, we show a right circular cone of base radius r and **slant height** l. (As you'll see later, it is more convenient in this context to specify the slant height than the altitude.)

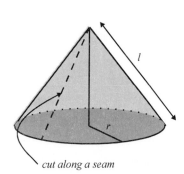

FIGURE 5.38:
Surface of revolution.

FIGURE 5.39a: Right circular cone.   FIGURE 5.39b: Flattened cone.

Suppose that we cut the cone along a seam and flatten it out. Then, we get the circular sector shown in Figure 5.39b. Notice that the curved surface area of the cone is the same as the area A of the circular sector. This is the area of a circle of radius l multiplied by the fraction of the circle included: θ out of a possible 2π radians, or

$$A = \pi \ (radius)^2 \ \frac{\theta}{2\pi} = \pi l^2 \frac{\theta}{2\pi} = \frac{\theta}{2} \, l^2. \tag{4.2}$$

The only problem with this is that we don't know θ. However, notice that by the way we constructed the sector (i.e., by flattening the cone), the circumference of the sector is the same as the circumference of the base of the cone. That is,

$$2\pi r = 2\pi l \, \frac{\theta}{2\pi} = l \, \theta.$$

Dividing by l gives us

$$\theta = \frac{2\pi r}{l}.$$

FIGURE 5.40: Frustum of a cone.

From (4.2), the curved surface area of the cone is now

$$A = \frac{\theta}{2} \, l^2 = \frac{\pi r}{l} \, l^2 = \pi r l.$$

Recall that we were originally interested in finding the surface area of only a portion of a right circular cone. For the **frustum** of a cone shown in Figure 5.40, the curved surface area is given by

$$A = \pi \ (r_1 + r_2) \ L.$$

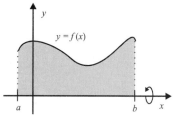

FIGURE 5.41a: Revolve about x-axis.

You can verify this by subtracting the curved surface area of two cones, where you must use similar triangles to find the height of the larger cone from which the frustum is cut. We leave the details as an exercise.

Returning to the original problem of revolving the line $y = x + 1$ on the interval $[0, 1]$ about the x-axis (seen in Figure 5.38), we have $r_1 = 1$, $r_2 = 2$ and $L = \sqrt{2}$ (from the Pythagorean Theorem). The curved surface area is then

$$A = \pi \ (1 + 2) \ \sqrt{2} = 3\pi\sqrt{2} \approx 13.\,329.$$

For the general problem of finding the curved surface area of a surface of revolution, consider the case where $f(x) \geq 0$ and where f is continuous on the interval $[a, b]$ and differentiable on (a, b). If we revolve the graph of $y = f(x)$ about the x-axis on the interval $[a, b]$ (see Figure 5.41a), we get the surface of revolution seen in Figure 5.41b.

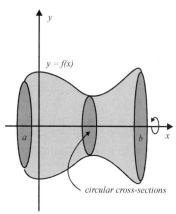

FIGURE 5.41b: Surface of revolution.

As we have done many times now, we first partition the interval $[a, b]$ into n pieces of equal size: $a = x_0 < x_1 < \cdots < x_n = b$, where $x_i - x_{i-1} = \Delta x = \dfrac{b - a}{n}$, for each $i = 1, 2, \ldots, n$. On each subinterval $[x_{i-1}, x_i]$, we can approximate the curve by the straight line segment joining the points $(x_{i-1}, f(x_{i-1}))$ and $(x_i, f(x_i))$, as in Figure 5.42. Notice that if we revolve this line segment around the x-axis, we get the frustum of a cone. The surface area of this frustum will give us an approximation to the actual surface area on the interval $[x_{i-1}, x_i]$. First, observe that the slant height of this frustum is

$$L_i = d\{(x_{i-1}, f(x_{i-1})), (x_i, f(x_i))\} = \sqrt{(x_i - x_{i-1})^2 + [f(x_i) - f(x_{i-1})]^2},$$

from the usual distance formula. Because of our assumptions on f, we can apply the Mean Value Theorem, to obtain

$$f(x_i) - f(x_{i-1}) = f'(c_i) \ (x_i - x_{i-1}),$$

for some number $c_i \in (x_{i-1}, x_i)$. This gives us

$$L_i = \sqrt{(x_i - x_{i-1})^2 + [f(x_i) - f(x_{i-1})]^2} = \sqrt{1 + [f'(c_i)]^2} \ \underbrace{(x_i - x_{i-1})}_{\Delta x}.$$

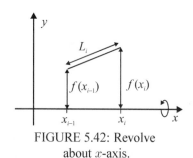

FIGURE 5.42: Revolve about x-axis.

The surface area S_i of that portion of the surface on the interval $[x_{i-1}, x_i]$ is approximately the surface area of the frustum of the cone,

$$S_i \approx \pi \left[f(x_i) + f(x_{i-1}) \right] \sqrt{1 + [f'(c_i)]^2} \, \Delta x$$

$$\approx 2\pi f(c_i) \sqrt{1 + [f'(c_i)]^2} \, \Delta x,$$

since if Δx is small,

$$f(x_i) + f(x_{i-1}) \approx 2 f(c_i).$$

Repeating this argument for each subinterval $[x_{i-1}, x_i]$, $i = 1, 2, \ldots, n$ gives us an approximation to the total surface area S,

$$S \approx \sum_{i=1}^{n} 2\pi f(c_i) \sqrt{1 + [f'(c_i)]^2} \, \Delta x.$$

As we have observed many times now, as n gets larger, this approximation approaches the actual surface area. Taking the limit as $n \to \infty$ gives us

$$S = \lim_{n \to \infty} \sum_{i=1}^{n} 2\pi f(c_i) \sqrt{1 + [f'(c_i)]^2} \, \Delta x.$$

You should recognize this limit as the limit of a Riemann sum, which gives us the integral

Surface area

$$\boxed{S = \int_a^b 2\pi f(x) \sqrt{1 + [f'(x)]^2} \, dx,} \tag{4.3}$$

whenever the integral exists.

You should notice that the factor of $\sqrt{1 + [f'(x)]^2} \, dx$ in the integrand in (4.3) corresponds to the arc length of a small section of the curve $y = f(x)$, while the factor $2\pi f(x)$ corresponds to the circumference of the solid of revolution. This should make sense to you, as follows. For any small segment of the curve, if we approximate the surface area by revolving a small segment of the curve, of radius $f(x)$ around the x-axis, the surface area generated is simply the surface area of a cylinder,

$$S = 2\pi r h = 2\pi f(x) \sqrt{1 + [f'(x)]^2} \, dx,$$

since the radius of such a small cylindrical segment is $f(x)$ and the height of the cylinder is $h = \sqrt{1 + [f'(x)]^2} \, dx$. It is far better to think about the surface area formula in this way than to simply memorize the formula.

Remark 4.2   There are exceptionally few functions f for which the integral in (4.3) can be computed exactly. Don't worry; we have numerical integration for just such occasions.

Example 4.5   Using the Surface Area Formula

Find the surface area of the surface generated by revolving $y = x^4$, $0 \le x \le 1$, about the x-axis.

Solution   Using the surface area formula (4.3), we have

$$S = \int_0^1 2\pi x^4 \sqrt{1 + (4x^3)^2} \, dx = \int_0^1 2\pi x^4 \sqrt{1 + 16x^6} \, dx \approx 3.4365,$$

where, as usual, we have used a numerical method to approximate the value of the integral.

■

Exercises 5.4

1. (writing) Explain in words how the arc length integral is derived from the lengths of the approximating secant line segments.

2. (writing) Explain why any Riemann sum approximation of arc length must be too small. Why couldn't one be too large?

3. (writing) Discuss whether the arc length integral is more accurately called a formula or a definition (i.e., can you precisely define the length of a curve without using the integral?).

4. (writing) Suppose you graph the trapezoid bounded by $y = x + 1$, $y = -x - 1$, $x = 0$ and $x = 1$, cut it out and roll it up. Explain why you would not get Figure 5.38. (Hint: compare areas and carefully consider Figures 5.39a and 5.39b.)

In exercises 5–8, approximate the length of the curve using n secant lines for $n = 2$; $n = 4$.

5. $y = x^2$, $0 \le x \le 1$

6. $y = x^4$, $0 \le x \le 1$

7. $y = \cos x$, $0 \le x \le \pi$

8. $y = \ln x$, $1 \le x \le 3$

In exercises 9–18, set up the integral for arc length, then approximate the integral with a numerical method.

9. $y = x^3$, $-1 \le x \le 1$

10. $y = x^3$, $-2 \le x \le 2$

11. $y = 2x - x^2$, $0 \le x \le 2$

12. $y = \tan x$, $0 \le x \le \pi/4$

13. $y = x^3 + x$, $0 \le x \le 3$

14. $y = x^{1/3}$, $1 \le x \le 2$

15. $y = \cos x$, $0 \le x \le \pi$

16. $y = \ln x$, $1 \le x \le 3$

17. $y = x \sin x$, $0 \le x \le \pi$

18. $y = e^{-x} \sin x$, $0 \le x \le \pi$

19. A rope is to be hung between two poles 20 feet apart. If the rope assumes the shape of the catenary $y = 4(e^{x/10} - e^{-x/10})$, $-10 \le x \le 10$, compute the length of the rope.

20. A rope is to be hung between two poles 20 feet apart. If the rope assumes the shape of the catenary $y = 4(e^{x/30} - e^{-x/30})$, $-30 \le x \le 30$, compute the length of the rope.

21. In example 4.4, compute the "sag" in the cable; that is, the difference between the y-values in the middle ($x = 0$) and at the poles ($x = 10$). Given this, is the arc length calculation surprising?

22. Compute the sag in exercises 19 and 20.

23. A football punt follows the path $y = \frac{1}{15}x(60 - x)$ yards. Sketch a graph. How far did the punt go horizontally? How high did it go? Compute the arc length. If the ball was in the air for 4 seconds, what was the ball's average velocity?

24. A baseball outfielder's throw follows the path $y = \frac{1}{300}x(100 - x)$ yards. Sketch a graph. How far did the ball go horizontally? How high did it go? Compute the arc length. Explain why the baseball player would want a small arc length, while the football player in exercise 23 would want a large arc length.

In exercises 25–32, set up the integral for the surface area of the surface of revolution, and approximate the integral with a numerical procedure.

25. $y = x^2$, $0 \le x \le 1$, revolved about the x-axis

26. $y = \sin x$, $0 \le x \le \pi$, revolved about the x-axis

27. $y = 2x - x^2$, $0 \le x \le 2$, revolved about the x-axis

28. $y = x^3 - 4x$, $-2 \le x \le 0$, revolved about the x-axis

29. $y = e^x$, $0 \le x \le 1$, revolved about the x-axis

30. $y = \ln x$, $1 \le x \le 2$, revolved about the x-axis

31. $y = \cos x$, $0 \le x \le \pi/2$, revolved about the x-axis

32. $y = \sqrt{x}$, $1 \le x \le 2$, revolved about the x-axis

Arc length can provide information about how close a given curve is to a straight line. In exercises 33–38, compute the arc length L_1 of the curve and the length L_2 of the secant line connecting the endpoints of the curve. Compute the ratio L_2/L_1; the closer this number is to 1, the straighter the curve is.

33. $y = \sin x$, $-\pi/6 \le x \le \pi/6$

34. $y = \cos x$, $-\pi/6 \le x \le \pi/6$

35. $y = \sin x$, $\pi/6 \le x \le \pi/2$

36. $y = \cos x$, $\pi/3 \le x \le 2\pi/3$

37. $y = e^x$, $3 \le x \le 5$

38. $y = e^x$, $-5 \le x \le -3$

39. For $y = x^6$, $y = x^8$ and $y = x^{10}$, compute the arc length for $0 \le x \le 1$. Using results from examples 4.2 and 4.3, identify the pattern for the length of $y = x^n$, $0 \le x \le 1$, as n increases. Conjecture the limit as $n \to \infty$.

40. To help understand the result of exercise 39, determine $\lim_{n \to \infty} x^n$ for each x such that $0 \le x < 1$. Compute the length of this limiting curve. Connecting this curve to the endpoint $(1, 1)$, what is the total length?

41. Prove that $y = x^4$ is flatter than $y = x^2$ for $0 < x < \sqrt{1/2}$ and steeper for $x > \sqrt{1/2}$.

42. Compare the flatness and steepness of $y = x^6$ and $y = x^4$.

43. Suppose that the square with $-1 \leq x \leq 1$ and $-1 \leq y \leq 1$ is revolved about the y-axis. Compute the surface area.

44. Suppose that the circle $x^2 + y^2 = 1$ is revolved about the y-axis. Compute the surface area.

45. Suppose that the triangle with vertices $(-1, -1)$, $(0, 1)$ and $(1, -1)$ is revolved about the y-axis. Compute the surface area.

46. Sketch the square, circle and triangle of exercises 43-45 on the same axes. Show that the relative surface areas of the solids of revolution (cylinder, sphere and cone, respectively) are $3:2:\tau$, where τ is the **golden mean** defined by $\tau = \dfrac{1 + \sqrt{5}}{2}$.

47. Sketch and compute the length of the **astroid** defined by $x^{2/3} + y^{2/3} = 1$.

48. (exploratory) In exercises 39 and 40, you explored the length of $y = x^n$ on the interval $[0, 1]$. In this exercise, you will look at more general polynomials on the same interval. First, guess what the largest possible arc length for a polynomial would be on a given interval. Now, check the arc length for some selected parabolas. Explain why $f_c(x) = cx(1 - x)$ is a downward-opening parabola with x-intercepts 0 and 1 and vertex at $x = 1/2$. What will happen to the arc length as c gets larger? Suppose we limit ourselves to polynomials that have function values lying between -1 and 1. Verify that on the interval $[0, 1]$ the functions $g_2(x) = 4x(1 - x)$, $g_3(x) = 20x(1/2 - x)(1 - x)$ and $g_4(x) = 80x(1/3 - x)(2/3 - x)(1 - x)$ all have this property. Compare the arc lengths of $g_2(x)$, $g_3(x)$ and $g_4(x)$. What will be the limiting value of such arc lengths? Construct a function $g_5(x)$ which continues this pattern.

49. (exploratory) In this exercise, you will explore a famous paradox. Suppose that the curve $y = 1/x$, for $1 \leq x \leq R$, is revolved about the x-axis (where R is a large positive constant). Compute the enclosed volume and the surface area of the resulting surface. (In both cases, antiderivatives can be found, although you may need help from your CAS to get the surface area.) Determine the limit of the volume and surface area as $R \to \infty$. Now for the paradox. Based on your answers, you should have a solid with finite volume, but infinite surface area. Thus, the three-dimensional solid could be completely filled with a finite amount of paint but the outside surface could never be completely painted.

5.5   Projectile Motion

In several earlier sections, we discussed the motion of an object moving back and forth along a straight line (i.e., rectilinear motion). You should recall that we determined that if we knew the position of an object at any time t, then we could determine its velocity and acceleration, by differentiation. Specifically, the derivative of a position function is velocity and the derivative of velocity is acceleration. A much more important problem is to go backward, that is, to find the position and velocity of an object, given its acceleration. Mathematically, this means that, starting with the derivative of a function, we must find the original function. Now that we have integration at our disposal, we can accomplish this with ease. In this section, we also consider the case of an object moving along a *curve* in two dimensions, rather than the simpler case of rectilinear motion considered earlier.

You may already be familiar with **Newton's second law of motion,** which says that

$$\boxed{F = m\,a,}$$

where F is the sum of the forces acting on an object, m is the mass of the object and a is the acceleration of the object. This says that if you know all of the forces acting on an object, you can determine its acceleration. More importantly, once you know some calculus, you can determine an object's velocity and position from its acceleration.

Start by imagining that you take a dive from a high diving platform. The primary force acting on you throughout the dive is gravity. The force due to gravity is your own weight, which is related to mass by $W = mg$, where g is the gravitational constant. (Common approximations of g, accurate near sea level are 32 ft/s^2 and 9.8 m/s^2.) To keep the problem simple mathematically, we will ignore any other forces, such as air resistance, acting on you during your dive.

Let $h(t)$ be your height above the water t seconds after jumping. Then the force due to gravity is $F = -mg$, where the minus sign indicates that the force is acting downward, in the negative direction. From our earlier work, the acceleration is $a(t) = h''(t)$. Newton's second law gives us $-mg = mh''(t)$ or

$$h''(t) = -g.$$

Notice that the derivation of this equation depended only on the identification of gravity as the sole force of interest. The position function of *any* object (regardless of its mass) subject to gravity and no other forces will satisfy the same equation. The only differences from situation to situation are the initial conditions (the initial velocity and initial position) and the questions being asked.

Example 5.1   Finding the Velocity of a Diver at Impact

    If a diving board is 15 feet above the surface of the water and a diver starts with initial velocity 8 ft/s (in the upward direction), what is the diver's velocity at impact (assuming no air resistance)?

Solution   If the height above the water at time t is given by $h(t)$, Newton's second law gives us $h''(t) = -32$ (since the height is given in feet). Further, we have the initial conditions $h(0) = 15$ (the diver starts 15 feet above the water) and $h'(0) = 8$. As you'll see, finding $h(t)$ takes little more than elementary integration. We have

$$\int h''(t)\, dt = \int -32\, dt$$

or

$$h'(t) = -32t + c.$$

Notice that from the initial velocity, we have

$$8 = h'(0) = -32(0) + c = c,$$

so that the velocity at any time t is given by

$$h'(t) = -32t + 8.$$

To find the velocity at impact, you first need to find the *time* of impact. Notice that the diver will impact the water when $h(t) = 0$ (i.e., when the height above the water is 0), so you need to perform one more integration to find the height function. Integrating the velocity function gives us

$$\int h'(t)\, dt = \int (-32t + 8)\, dt$$

or

$$h(t) = -16t^2 + 8t + c.$$

From the initial height, we have

$$15 = h(0) = -16(0)^2 + 8(0) + c = c,$$

so that

$$h(t) = -16t^2 + 8t + 15.$$

This gives the height above the water at any time t. Impact then occurs when

$$0 = h(t) = -16t^2 + 8t + 15$$
$$= -(4t + 3)(4t - 5),$$

so that $t = \frac{5}{4}$ is the time of impact. (You may ignore the extraneous solution $t = -\frac{3}{4}$. Why?). When $t = \frac{5}{4}$, the velocity is $h'(\frac{5}{4}) = -32\left(\frac{5}{4}\right) + 8 = -32$ ft/s (impact velocity).

To put this in more familiar units of velocity, multiply by 3600/5280 to convert to miles per hour. In this case, the impact velocity is about 22 mph. (You probably don't want to come down in the wrong position at that speed!)

■

In example 5.1, the negative sign of the velocity indicated that the diver was coming down. In many situations, both upward and downward motions are important.

Example 5.2   An Equation for the Vertical Motion of a Ball
A ball is propelled from the ground straight upward with initial velocity 64 ft/s. Ignoring air resistance, find an equation for the height of the ball at any time t. Also, determine the maximum height and the amount of time the ball spends in the air.

Solution   With gravity as the only force, the height $h(t)$ satisfies $h''(t) = -32$. The initial conditions are $h'(0) = 64$ and $h(0) = 0$. We then have

$$\int h''(t)\,dt = \int -32\,dt$$

or

$$h'(t) = -32t + c.$$

From the initial velocity, we have

$$64 = h'(0) = -32(0) + c = c$$

and so,

$$h'(t) = 64 - 32t.$$

Integrating one more time gives us

$$\int h'(t)\,dt = \int (64 - 32t)\,dt$$

or

$$h(t) = 64t - 16t^2 + c.$$

From the initial altitude we have

$$0 = h(0) = 64(0) - 16(0)^2 + c = c,$$

and so,

$$h(t) = 64t - 16t^2.$$

Since the height function is quadratic, the maximum occurs at the one time when $h'(t) = 0$. (You should also consider the physics of the situation: what happens physically when $h'(t) = 0$?) Solving $64 - 32t = 0$ gives $t = 2$ (the time at the maximum altitude) and the corresponding height is $h(2) = 64(2) - 16(2)^2 = 64$ feet. Again, the ball lands when $h(t) = 0$. Solving

$$0 = h(t) = 64t - 16t^2 = 16t(4 - t)$$

gives $t = 0$ (launch time) and $t = 4$ (landing time). The time of flight is thus 4 seconds.

■

FIGURE 5.43: Height of the ball at time t.

You can observe an interesting property of projectile motion by graphing the height function from example 5.2 along with the line $y = 48$ (see Figure 5.43). Notice that the graphs intersect at $t = 1$ and $t = 3$. Further, the time interval $[1, 3]$ corresponds to exactly half the total time in the air. Notice that this says that the ball is in the top one-fourth of its height for half of its time in the air. You may have marveled at how some athletes jump so high that they seem to "hang in the air." As this calculation suggests, all objects tend to hang in the air.

Example 5.3   Finding the Initial Velocity Required to Reach a Certain Height

It has been reported that basketball star Michael Jordan has a vertical leap of $54''$. Ignoring air resistance, what is the initial velocity required to jump this high?

Solution   Once again, Newton's second law leads us to the equation $h''(t) = -32$ for the height $h(t)$. Further, the initial velocity and initial position are given by $h'(0) = v_0$ and $h(0) = 0$, respectively. Our task is to determine the value of v_0 that will give a maximum altitude of $54''$. As before, we integrate to get

$$h'(t) = -32t + c.$$

Using the initial velocity, we get

$$v_0 = h'(0) = -32\,(0) + c = c.$$

This gives us the velocity function

$$h'(t) = v_0 - 32t.$$

Integrating once again and using the initial position $h(0) = 0$, we get

$$h(t) = v_0 t - 16t^2.$$

The maximum height occurs when $h'(t) = 0$. (Why?) Setting

$$0 = h'(t) = v_0 - 32t,$$

gives us $t = \dfrac{v_0}{32}$. The height at this time (i.e., the maximum altitude) is then

$$h\left(\frac{v_0}{32}\right) = v_0\left(\frac{v_0}{32}\right) - 16\left(\frac{v_0}{32}\right)^2 = \frac{v_0^2}{32} - \frac{v_0^2}{64} = \frac{v_0^2}{64}.$$

A jump of $54'' = 4.5'$ requires $\dfrac{v_0^2}{64} = 4.5$ or $v_0^2 = 288$, so that $v_0 = \sqrt{288} \approx 17$ ft/s. Note that this initial velocity is equivalent to roughly 11.6 mph.

∎

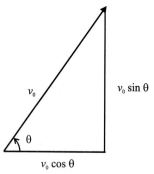

FIGURE 5.44a: Path of projectile.

FIGURE 5.44b: Vertical and horizontal components of velocity.

So far, our projectiles have only traveled in the vertical direction. Most applications of projectile motion must also consider movement in the horizontal direction. If we ignore air resistance, these calculations are also relatively straightforward. The basic principle is that we can apply Newton's second law separately for the horizontal and vertical components of the motion. If $y(t)$ represents the vertical position, then we have $y''(t) = -g$, as before. Ignoring air resistance, there are no forces acting horizontally on the projectile. So, if $x(t)$ represents the horizontal position, Newton's second law gives us $x''(t) = 0$.

The initial conditions are slightly more complicated here. In general, we want to consider projectiles that are launched with an initial speed v_0 at an angle θ from the horizontal. Figure 5.44a shows a projectile fired with $\theta > 0$. Notice that an initial angle of $\theta < 0$ would mean a downward initial velocity.

As shown in Figure 5.44b, the initial velocity can be separated into horizontal and vertical components. From elementary trigonometry, the horizontal component of the initial velocity is $v_x = v_0 \cos\theta$ and the vertical component is $v_y = v_0 \sin\theta$.

Example 5.4   The Motion of a Projectile in Two Dimensions

An object is launched at angle $\theta = \pi/6$ from the horizontal with initial speed $v_0 = 98$ m/s. Determine the time of flight and the (horizontal) range of the projectile.

Solution   Starting with the vertical component of the motion, we have $y''(t) = -9.8$ (since the initial speed is given in terms of meters per second). Referring to Figure 5.44b, notice that the vertical component of the initial velocity is $y'(0) = 98 \sin \pi/6 = 49$ and the initial altitude is $y(0) = 0$. A pair of simple integrations give us the velocity function $y'(t) = -9.8t + 49$ and the position function $y(t) = -4.9t^2 + 49t$. The ball hits the ground when $y(t) = 0$ (i.e., when its height above the ground is 0). Solving

$$0 = y(t) = -4.9t^2 + 49t = 49t(1 - 0.1t)$$

gives $t = 0$ (launch time) and $t = 10$ (landing time). The time of flight is then 10 seconds. The horizontal component of motion is determined from the equation $x''(t) = 0$ with initial velocity $x'(0) = 98 \cos \pi/6 = 49\sqrt{3}$ and initial position $x(0) = 0$. Integration gives us $x'(t) = 49\sqrt{3}$ and $x(t) = (49\sqrt{3})t$. In Figure 5.45, we give a plot of the path of the ball. (You can easily do the same by using the parametric plot mode on your graphing calculator or CAS. Simply enter the separate equations for $x(t)$ and $y(t)$ and set the range of t-values to be $0 \le t \le 10$. Alternatively, you could easily solve for t in terms of x and replace t by $\frac{1}{49\sqrt{3}}x$, to see that the curve is simply a parabola.) Notice that the horizontal location of the ball when it lands is the value of $x(t)$ at $t = 10$ (the landing time). The range is then

$$x(10) = (49\sqrt{3})(10) = 490\sqrt{3} \approx 849 \text{ meters.}$$

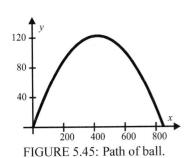

FIGURE 5.45: Path of ball.

Remark 5.1   We urge you to resist the temptation to reduce this section to a few memorized formulas. It is true that if you ignore air resistance, the vertical component of position will always turn out to be $y(t) = -\frac{1}{2}gt^2 + (v_0 \sin \theta)t + y(0)$. (We'll pursue this further in the exercises.) However, your understanding of the process and your chances of finding the correct answer will improve dramatically if you start each problem with Newton's second law and work through the integrations (which are not difficult).

One reason to start with the Newton's second law is to force yourself to consider the forces that are (and are not) being considered. For example, we have thus far ignored air resistance. It is important to realize that this is a simplification of reality. Some calculations using such simplified equations are reasonably accurate. Others, such as in the following example, are not.

Example 5.5   An Example Where Air Resistance Can't be Ignored

Suppose a rain drop falls from a cloud 3000 feet above the ground. Ignoring air resistance, how fast would the raindrop be falling when it hits the ground?

Solution   If the height of the raindrop at time t is given by $y(t)$, Newton's second law of motion tells us that $y''(t) = -32$. Further, we have the initial velocity $y'(0) = 0$ (since the drop falls as opposed to being thrown down) and the initial altitude $y(0) = 3000$. Integrating and using the initial conditions gives us $y'(t) = -32t$ and $y(t) = 3000 - 16t^2$. The raindrop hits the ground when $y(t) = 0$. Setting

$$0 = y(t) = 3000 - 16t^2$$

gives us $t = \sqrt{3000/16} \approx 13.693$ seconds. The velocity at this time is then

$$y'(\sqrt{3000/16}) = -32\sqrt{3000/16} \approx -438.18 \text{ ft/s.}$$

This corresponds to nearly 300 mph! Fortunately, air resistance does play a significant role in the fall of a raindrop, which has an actual landing speed of about 10 mph.

■

FIGURE 5.46:
Cross section of a wing.

The obvious lesson from example 5.5 is that it is not always reasonable to ignore air resistance. Unfortunately, we will have to wait two more chapters to develop the mathematical tools necessary to analyze projectile motion with air resistance.

The air resistance (more precisely, *air drag*) that slows the raindrop down is only one of the ways in which air can affect the motion of an object. The **Magnus force**, produced by the spinning of an object or asymmetries (i.e., lack of symmetry) in the shape of an object, can cause the object to change directions and curve. Perhaps the most common example of a Magnus force occurs on an airplane. One side of an airplane wing is curved and the other side is comparatively flat (see Figure 5.46). The lack of symmetry causes the air to move over the top of the wing faster than it moves over the bottom. This produces a Magnus force in the upward direction (lift), lifting the airplane into the air.

A more down-to-earth example of a Magnus force occurs in an unusual baseball pitch called the knuckleball. To throw this pitch, the pitcher grips the ball with all five fingers and tries to throw the ball with as little spin as possible (despite the name, the knuckles are not used). Baseball players claim that the knuckleball "dances" around unpredictably and is exceptionally hard to hit or catch. There still is no complete agreement on why the knuckleball moves so much, but we will present one current theory due to physicists Robert Watts and Terry Bahill.

The cover of the baseball is sewed on with stitches that are raised up slightly from the rest of the ball. These curved stitches act much like an airplane wing, creating a Magnus force that affects the ball. The direction of the Magnus force depends on the exact orientation of the ball's stitches. Measurements by Watts and Bahill indicate that the lateral force (left/right from the pitcher's perspective) is approximately $F_m = -0.1 \sin(4\theta)$ lb, where θ is the angle in radians of the ball's position rotated from a particular starting position.

Regulation baseball, showing stitching.

Since gravity does not affect the lateral motion of the ball, the only force acting on the ball laterally is the Magnus force. Newton's second law applied to the lateral motion of the knuckleball gives $mx''(t) = -0.1 \sin(4\theta)$. The mass of a baseball is about 0.01 (slugs) and so, we have

$$x''(t) = -10 \sin(4\theta).$$

If the ball is spinning at the rate of ω radians per second, then $4\theta = 4\omega t + \theta_0$, where the initial angle θ_0 depends on where the pitcher grips the ball. Our model is then

$$x''(t) = -10 \sin(4\omega t + \theta_0), \tag{5.1}$$

with initial conditions $x'(0) = 0$ and $x(0) = 0$. For a typical knuckleball speed of 60 mph, it takes about 0.68 second for the pitch to reach home plate.

Example 5.6   An Equation for the Motion of a Knuckleball

For a spin rate of $\omega = 2$ radians per second and $\theta_0 = 0$, find an equation for the motion of the knuckleball and graph it for $0 \leq t \leq 0.68$. Repeat this for $\theta_0 = \pi/2$.

Solution   For $\theta_0 = 0$, Newton's second law gives us $x''(t) = -10 \sin 8t$, from (5.1). Integrating this and using the initial condition $x'(0) = 0$ gives us

$$x'(t) = -\frac{10}{8} \left[-\cos 8t - (-\cos 0) \right] = 1.25(\cos 8t - 1).$$

Integrating once again and using the second initial condition $x(0) = 0$, we get

$$x(t) = 1.25 \left(\frac{1}{8} \right) (\sin 8t - 0) - 1.25t = 0.15625 \sin 8t - 1.25t.$$

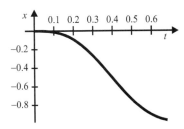

FIGURE 5.47a: Lateral Motion of a knuckleball for $\theta_0 = 0$.

A graph of this function shows the lateral motion of the ball (see Figure 5.47a). The graph shows the path of the pitch as it would look viewed from above. Notice that after starting out straight, this pitch breaks nearly a foot away from the center of home plate!

For the case where $\theta_0 = \pi/2$, we have from (5.1) that

$$x''(t) = -10 \sin(8t + \pi/2).$$

Integrating this and using the first initial condition gives us

$$x'(t) = -\frac{10}{8} \left\{ -\cos(8t + \pi/2) - [-\cos(0 + \pi/2)] \right\} = 1.25 \cos(8t + \pi/2).$$

Integrating a second time yields

$$x(t) = 1.25 \left(\frac{1}{8} \right) [\sin(8t + \pi/2) - \sin(\pi/2)] = 0.15625[\sin(8t + \pi/2) - 1].$$

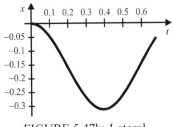

FIGURE 5.47b: Lateral motion of a knuckleball for $\theta_0 = \pi/2$.

A graph of the lateral motion in this case is shown in Figure 5.47b. Notice that this pitch breaks nearly 4 inches to the pitcher's right and then curves back over the plate for a strike! You can see that in theory, the knuckleball is very sensitive to spin and initial position, and why it is very difficult to hit when thrown properly.

■

Exercises 5.5

1. (writing) In example 5.4, the assumption that air resistance can be ignored is obviously invalid. Discuss the validity of this assumption in examples 5.1 and 5.3.

2. (writing) In the discussion preceding example 5.3, we showed that Michael Jordan (and any other human) spends half of his air-time in the top one-fourth of the height. Compare his velocities at various points in the jump to explain why relatively more time is spent at the top than at the bottom.

3. (writing) In example 5.4, we derived separate equations for the horizontal and vertical components of position. To discover one consequence of this separation, consider the following situation. Two people are standing next to each other with arms raised to the same height. One person fires a bullet horizontally from a gun. At the same time, the other person drops a bullet. Explain why the bullets will hit the ground at the same time.

4. (writing) For the falling raindrop in example 5.5, a more accurate model would be $y''(t) = -32 + f(t)$, where $f(t)$ represents the force due to air resistance (divided by the mass). If $v(t)$ is the downward velocity of the raindrop, explain why this equation is equivalent to $v'(t) = 32 - f(t)$. Explain in physical terms why the larger $v(t)$ is, the larger $f(t)$ is. Thus, a model such as $f(t) = v(t)$ or $f(t) = [v(t)]^2$ would be reasonable. (In most situations, it turns out that $[v(t)]^2$ matches the experimental data better.)

In exercises 5–8, identify the initial conditions $y(0)$ and $y'(0)$.

5. An object is dropped from a height of 80 feet.

6. An object is dropped from a height of 100 feet.

7. An object is released from a height of 60 feet with an upward velocity of 10 ft/s.

8. An object is released from a height of 20 feet with a downward velocity of 4 ft/s.

In exercises 9–49, ignore air resistance.

9. A diver starts 30 feet above the water (about the height of an Olympic platform diver). What is the diver's velocity at impact?

10. A diver starts 120 feet above the water (about the height of divers at the Acapulco Cliff Diving competition). What is the diver's velocity at impact?

11. Compare the impact velocities of objects falling from 30 feet (exercise 9), 120 feet (exercise 10) and 3000 feet (example 5.5). If height is increased by a factor of h, by what factor does the impact velocity increase?

12. The Washington monument is 555 feet, $5\frac{1}{8}$ inches high. In a famous experiment, a baseball was dropped from the top of the monument to see if a player could catch it. How fast would the ball be going?

13. A certain not-so-wily coyote discovers that he just stepped off the edge of a cliff. Four seconds later, he hits the ground in a puff of dust. How high was the cliff?

14. A large boulder dislodged by the falling coyote in exercise 13 falls for 3 seconds before landing on the coyote. How far did the boulder fall? What was its velocity when it flattened the coyote?

15. The coyote's next scheme involves launching himself into the air with an Acme catapult. If the coyote is propelled vertically from the ground with initial velocity 64 ft/s, find an equation for the height of the coyote at any time t. Find his maximum height, the amount of time spent in the air and his velocity when he smacks back into the catapult.

16. On the rebound, the coyote in exercise 15 is propelled to a height of 256 feet. What is the initial velocity required to reach this height?

17. One of the authors has a vertical "jump" of 20 inches. What is the initial velocity required to jump this high? How does this compare to Michael Jordan's velocity, found in example 5.3?

18. If the author underwent an exercise program and increased his initial velocity by 10%, by what percentage would he increase his vertical jump?

19. A typical flea jumps vertically with an initial velocity of about 5 ft/s. How high does the flea jump?

20. Compute the ratio of the vertical jump of the flea in exercise 19 to its height (about $\frac{1}{12}$-inch). If the (6-foot-tall) author in exercise 17 could maintain the same ratio, how high would he jump?

21. Show that an object dropped from a height of H feet will hit the ground at time $T = \frac{1}{4}\sqrt{H}$ seconds with impact velocity $V = -8\sqrt{H}$ ft/s.

22. Show that an object propelled from the ground with initial velocity v_0 ft/s will reach a maximum height of $v_0^2/64$ ft.

In exercises 23–36, sketch the parametric graphs to indicate the flight path.

23. An object is launched at angle $\theta = \pi/3$ radians from the horizontal with an initial speed of 98 m/s. Determine the time of flight and the horizontal range.

24. Compare the launch angles, initial speeds and ranges of the objects in exercise 23 and example 5.4.

25. Find the time of flight and horizontal range of an object launched at angle 30° with initial speed 120 ft/s.

26. Find the time of flight and horizontal range of an object launched at angle 60° with initial speed 120 ft/s. Compare the ranges in exercises 25 and 26.

27. A baseball pitcher releases the ball horizontally from a height of 6 ft with an initial speed of 130 ft/s. Find the height of the ball when it reaches home plate 60 feet away. (Hint: Determine the time of flight from the x-equation, then use the y-equation to determine the height.)

28. Repeat exercise 27 with an initial speed of 80 ft/s. (Hint: Carefully interpret the negative answer.)

29. A baseball player throws a ball toward first base 120 feet away. The ball is released from a height of 5 feet with an initial speed of 120 ft/s at an angle of 5° above the horizontal. Find the height of the ball when it reaches first base.

30. Repeat exercise 29 with an initial angle of 10° above the horizontal.

31. Find by trial and error the angle at which the ball in exercises 29 and 30 will reach first base at the catchable height of 5 feet. At this angle, how far above the first baseman's head would the thrower be aiming?

32. In exercise 29, if the ball is released horizontally, how far from first base would the ball first bounce?

33. A daredevil plans to jump over 25 cars. If the cars are all compact cars with a width of 5 feet and the ramp angle is 30°, determine the initial velocity required to complete the jump successfully.

34. Repeat exercise 33 with a takeoff angle of 45°. In spite of the reduced initial velocity requirement, why might the daredevil prefer an angle of 30° to 45°?

35. A plane at an altitude of 256 feet wants to drop supplies to a specific location on the ground. If the plane has a horizontal velocity of 100 ft/s, how far away from the target should the plane release the supplies in order to hit the target location? (Hint: Use the y-equation to determine the time of flight, then use the x-equation to determine how far the supplies will drift.)

36. Repeat exercise 35 for an altitude of 144 feet.

37. Consider a knuckleball (see example 5.6) with lateral motion satisfying the initial value problem $x''(t) = -25\sin(4\omega t + \theta_0)$, $x'(0) = x(0) = 0$. With $\theta_0 = 0$ and $\omega = 1$, find an equation for $x(t)$ and graph the solution for $0 \le t \le 0.68$.

38. Repeat exercise 37 for $\theta_0 = \pi/2$ and $\omega = 1$.

39. Repeat exercise 37 for $\theta_0 = \pi/4$ and $\omega = 2$.

40. Repeat exercise 37 for $\theta_0 = \pi/4$ and $\omega = 1$.

41. For the Olympic diver in exercise 9, what would be the average angular velocity (measured in radians per second) necessary to complete $2\frac{1}{2}$ somersaults?

42. In the Flying Zucchini circus' human cannonball act, a performer is shot out of a cannon from a height of 10 feet at an angle of $45°$ with an initial speed of 160 ft/s. If the safety net stands 5 feet above the ground, how far should the safety net be placed from the cannon? If the safety net can only withstand an impact velocity of 160 ft/s, will the Flying Zucchini land safely or come down squash?

43. In a basketball free throw, a ball is shot from a height of h feet toward a basket 10 feet above the ground at a horizontal distance of 15 feet. If $h = 6$, $\theta = 52°$ and $v_0 = 25$ ft/s, show that the free throw is good. Since the basket is larger than the ball, a free throw has a margin of error of several inches. If any shot that passes through height 10 ft with $14.65 \le x \le 15.35$ is good, show that, for the given speed v_0, the margin of error is $48° \le \theta \le 57°$. Sketch parametric graphs to show several of these free throws.

44. For the basketball shot in exercise 43, fix $h = 6$ and $\theta = 52°$. Find the range of speeds v_0 for which the free throw will be good. Given that it requires a force of $F = 0.01v_0^2$ pounds to launch a free throw with speed v_0, how much margin of error is there in the applied force?

45. To train astronauts to operate in a weightless environment, NASA sends them up in a special plane (nicknamed the *vomit comet*). To allow the passengers to experience weightlessness, the vertical acceleration of the plane must exactly match the acceleration due to gravity. If $y''(t)$ is the vertical acceleration of the plane, then $y''(t) = -g$. Show that, for a constant horizontal velocity, the plane follows a parabolic path. NASA's plane flies parabolic paths of approximately 2500 feet in height (2500 feet up and 2500 down). The time to complete such a path is the amount of weightless time for the passengers. Compute this time.

46. In a typical golf tee shot, the ball is launched at an angle of $9.3°$ at an initial speed of 220 ft/s. In the absence of air resistance, how far (horizontally) would this shot carry? The actual carry on such a shot is about 240 yards (720 feet)! In

this case, a backspin of 4000 rpm gives the ball a huge upward Magnus force, which offsets most of the air resistance and gravity.

47. In the 1992 Summer Olympics in Barcelona, Spain, an archer lit the Olympic flame by shooting an arrow towards a cauldron at a distance of about 70 meters horizontally and 30 meters vertically. If the arrow reached the cauldron at the peak of its trajectory, determine the initial speed and angle of the arrow. (Hint: Show that $y'(t) = 0$ if $t = (v_0\sin\theta)/9.8$. For this t, show that $\dfrac{x(t)}{y(t)} = 2\cot\theta = \dfrac{7}{3}$ and solve for θ. Then solve for v_0.)

48. Professional jugglers generally agree that 10 is the maximum number of balls that a human being can successfully maintain. To get an idea why, suppose that it takes $\frac{1}{2}$-second to catch and toss a ball up in the air (in other words, the juggler can process 2 balls per second). To juggle 10 balls, then, each ball would have to be in the air for 5 seconds. How high would the ball have to be tossed to stay in the air this long? Given this height, does $\frac{1}{2}$-second per ball sound like much time? How much higher would the balls have to be tossed to juggle 11 balls?

49. (exploratory) In the text and exercises 37–40, we discussed the differential equation $x''(t) = -25\sin(4\omega t + \theta_0)$ for the lateral motion of a knuckleball. Integrate and apply the initial conditions $x'(0) = 0$ and $x(0) = 0$ to derive the general equation $x(t) = \dfrac{25}{16\omega^2}\sin(4\omega t + \theta_0) - \left(\dfrac{25}{4\omega}\cos\theta_0\right)t - \dfrac{25}{16\omega^2}\sin\theta_0$. If you have access to three-dimensional graphics, graph $x(t,\omega)$ for $\theta_0 = 0$ with $0 \le t \le 0.68$ and $0.01 \le \omega \le 10$. (Note: Some plotters will have trouble with $\omega = 0$.) Repeat with $\theta_0 = \pi/4$, $\theta_0 = \pi/2$ and two choices of your own for θ_0. A pitcher wants the ball to move as much as possible back and forth but end up near home plate ($x = 0$). Based on these criteria, pick the combinations of θ_0 and ω that produce the four best pitches. Graph these pitches in two dimensions with $x = x(t)$ as in Figures 5.47a and 5.47b.

50. (exploratory) Although we have commented on some inadequacies of the gravity-only model of projectile motion, we have not presented any alternatives. Such models tend to be somewhat more mathematically complex. The model explored in this exercise takes into account air resistance in a way that is mathematically tractable but still not very realistic. Assume that the force of air resistance is proportional to the speed and acts in the opposite direction of velocity. For a horizontal motion (no gravity), we have $a(t) = F(t)/m = -cv(t)$ for some constant c. Explain what the minus sign indicates. Since $a(t) = v'(t)$, the model is $v'(t) = -cv(t)$. Show that the function $v(t) = v_0 e^{-ct}$ satisfies the equation $v'(t) = -cv(t)$ and the initial condition $v(0) = v_0$. If an object starts at $x(0) = a$, integrate $v(t) = v_0 e^{-ct}$ to find its position at any time t. Show that the amount of time needed to reach $x = b$

(for $a < b$) is given by $T = -\dfrac{1}{c}\ln\left(1 - c\dfrac{b-a}{v_0}\right)$. For a baseball ($c = 0.15$) thrown at 125 ft/s from $a = 0$, determine how long it takes to reach $b = 60$ and compute its velocity at that point. By what percentage has the velocity decreased? In baseball, two different types of radar guns are used to mea-

sure the velocity of a pitch. One measures the speed of the ball right as it leaves the pitcher's hand. The other measures the speed of the ball part way to home plate. If the first gun registers 94 mph and the second registers 89 mph, how far from the plate does the second gun makes it measurement?

5.6   Work, Moments and Hydrostatic Force

In this section, we explore several applications of integration in physics. In each case, we will define a basic concept to help solve a specific problem. We'll then use the definite integral to generalize the concept to solve a much wider range of problems. This use of integration is an excellent example of how the mathematical theory helps you find solutions of practical problems.

Imagine that you are at the bottom of a snow-covered hill with a sled. To get a good ride, you want to push the sled as far up the hill as you can. A physicist would say that the higher up you are, the more **potential energy** you have. Sliding down the hill converts the potential energy into **kinetic energy.** (This is the fun part!) But pushing the sled up the hill requires you to do some work: you must exert a force over a long distance.

Physicists have studied how things like force and speed are related. While it's certainly clear that increasing the force in the direction of motion will increase the speed, there is much more to it than that. One way in which physicists simplify the process is by identifying a small number of key quantities to keep track of. In this section, we will discuss three of these: work, impulse and moments.

Our first task is to quantify work. If you push a sled up a hill, you're doing work, but can you give a measure of how much? Certainly, if you push twice the weight (i.e., exert twice the force), you're doing twice the work. So, it should seem reasonable that work is directly proportional to the force you're exerting. Further, if you push the sled twice as far, you've done twice the work. That is, work is directly proportional to the distance over which you exert the force. In view of these observations, for any constant force F applied over a distance d, we define the **work** W done as

$$W = Fd.$$

Unfortunately, forces are generally not constant. For instance, think of a rocket that burns its fuel in flight. Since its mass is not constant, neither is the force exerted on the rocket by gravity. We can extend this notion of work to the case of a nonconstant force $F(x)$ applied on the interval $[a, b]$ as follows. First, we divide the interval $[a, b]$ into n equal subintervals, each of width $\Delta x = \dfrac{b-a}{n}$ and consider the work done on each subinterval. Notice that if Δx is small, then the force $F(x)$ applied on the interval $[x_{i-1}, x_i]$ can be approximated by the constant force $F(c_i)$ for some point $c_i \in [x_{i-1}, x_i]$. The work done moving the object along the subinterval is then approximately $F(c_i)\Delta x$. The total work done W is then approximately

$$W \approx \sum_{i=1}^{n} F(c_i)\,\Delta x.$$

You should recognize that this is a Riemann sum and that as n gets larger, the Riemann sums should approach the actual work. Taking the limit as $n \to \infty$ gives us

$$\boxed{W = \lim_{n\to\infty} \sum_{i=1}^{n} F(c_i)\,\Delta x = \int_a^b F(x)\,dx.}\tag{6.1}$$

We take (6.1) as our definition of **work**.

You've probably noticed that the further a spring is compressed (or stretched) from its natural resting position, the more force is required to further compress (or stretch) the spring. According to **Hooke's Law**, the force required to maintain a spring in a given position is proportional to the distance it's compressed (or stretched). That is, if x is the distance a spring is compressed (or stretched) from its natural length, the force $F(x)$ exerted by the spring is given by

$$F(x) = kx, \tag{6.2}$$

for some constant k (the **spring constant**).

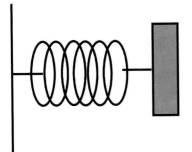

FIGURE 5.48:
Compressed spring.

Example 6.1   Computing the Work Done Stretching a Spring

A force of 3 pounds is required to hold a spring that has been stretched $\frac{1}{4}$ foot from its natural length (see Figure 5.48). Find the work done in stretching the spring 6 inches beyond its natural length.

Solution   First, we have from Hooke's Law (6.2) that

$$3 = F(1/4) = k\left(\frac{1}{4}\right),$$

so that $k = 12$ and $F(x) = 12x$. From (6.1), the work done in stretching the spring 6 inches (1/2 foot) is then

$$W = \int_0^{1/2} F(x)\, dx = \int_0^{1/2} 12x\, dx = \frac{3}{2} \text{ foot-pounds.}$$

In this case, notice that stretching the spring transfers potential energy to the spring (if the spring is later released, it springs back to its resting position and converts the potential energy to kinetic energy).

∎

Example 6.2   Computing the Work Done by a Weightlifter

A weightlifter lifts a 200-pound barbell a distance of 3 feet. How much work was done? Also, determine the work done by the weightlifter if the weight is raised 4 feet above the ground and then lowered back into place.

Solution   Since the force (the weight) is constant here, we simply have

$$W = Fd = 200 \times 3 = 600 \text{ foot-pounds.}$$

On the other hand, find the amount of work done if the weightlifter lifts the same weight 4 feet from the ground and then lowers it back into place. It may seem strange, but since the barbell ends up in the same place as it started, the net distance covered is zero and the work done is zero. Of course, it would feel like work to the weightlifter, but this is where the mathematical notion of work differs from the usual use of the word. As we have defined it, work accounts for the energy change in the object having work done on it. Since the barbell has the same kinetic and potential energy that it started with, the total work done on it is zero.

∎

In the following example, both the force and the distance are nonconstant. This presents some unique challenges and we'll need to first approximate the work and then recognize the definite integral this approximation process generates.

FIGURE 5.49a: Sperical tank.

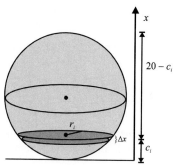

FIGURE 5.49b: The ith slice of water.

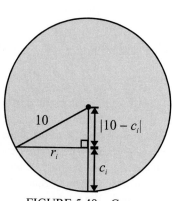

FIGURE 5.49c: Cross section of tank.

Example 6.3   Computing the Work Required to Pump the Water Out of a Tank

A spherical tank of radius 10 feet is filled with water. Find the work done in pumping all of the water out through the top of the tank.

Solution   You should observe that the basic formula $W = Fd$ does not directly apply here, for several reasons. The most obvious reason is that the distance traveled by the water in each part of the tank is different, as the water toward the bottom of the tank must be pumped all the way to the top, while the water near the top of the tank must only be pumped a short distance. To envision this, we must partition the water tank into n thin layers, each of depth $\Delta x = \frac{20}{n}$. Let x represent distance as measured from the bottom of the tank, as indicated in Figure 5.49a. Then, the entire tank corresponds to the interval $0 \le x \le 20$, which we partition into

$$0 = x_0 < x_1 < \cdots < x_n = 20,$$

where $x_i - x_{i-1} = \Delta x$, for each $i = 1, 2, \ldots, n$. Notice that this partitions the tank into n layers, each corresponding to an interval $[x_{i-1}, x_i]$ (see Figure 5.49b). You can think of the water in the layer corresponding to $[x_{i-1}, x_i]$ as being approximately cylindrical, of height Δx. This layer must be pumped a distance of approximately $20 - c_i$, for some $c_i \in [x_{i-1}, x_i]$. (You might think of taking c_i as the midpoint of the interval, although the choice of c_i doesn't really matter.) Notice from Figure 5.49b that the radius of the ith layer depends on the value of x. From Figure 5.49c, (where we show a cross section of the tank) the radius r_i corresponding to a depth of $x = c_i$ is the base of a right triangle with hypotenuse 10 and height $|10 - c_i|$. From the Pythagorean Theorem, we now have

$$(10 - c_i)^2 + r_i^2 = 10^2.$$

Solving this for r_i^2, we have

$$r_i^2 = 10^2 - (10 - c_i)^2 = 100 - \left(100 - 20c_i + c_i^2\right)$$
$$= 20c_i - c_i^2.$$

The force F_i required to move the ith layer is then simply the force exerted on the water by gravity (i.e., its weight). For this we will need to know the weight density of water, 62.4 lb/ft³. We have

$$F_i \approx (\text{Volume of cylindrical slice}) \, (\text{weight of water per unit volume})$$
$$= \left(\pi r_i^2 h\right) \left(62.4 \text{ lb/ft}^3\right)$$
$$= 62.4 \, \pi \left(20c_i - c_i^2\right) \Delta x.$$

The work required to pump out the ith slice is then given approximately by

$$W_i \approx (\text{Force}) \, (\text{distance})$$
$$= 62.4 \, \pi \left(20c_i - c_i^2\right) \Delta x \, (20 - c_i)$$
$$= 62.4 \, \pi \, c_i \, (20 - c_i)^2 \, \Delta x.$$

The work required to pump out all of the water is then the sum of the work required for each of the n slices:

$$W \approx \sum_{i=1}^{n} 62.4 \, \pi \, c_i \, (20 - c_i)^2 \, \Delta x.$$

Finally, taking the limit as $n \to \infty$ gives the exact work, which you should recognize as a definite integral:

$$W = \lim_{n \to \infty} \sum_{i=1}^{n} 62.4\,\pi\, c_i\, (20 - c_i)^2\, \Delta x = \int_0^{20} 62.4\pi x\, (20 - x)^2\, dx$$

$$= 62.4\,\pi \int_0^{20} \left(400x - 40x^2 + x^3\right)\, dx$$

$$= 62.4\,\pi \left[400\,\frac{x^2}{2} - 40\,\frac{x^3}{3} + \frac{x^4}{4} \right]_0^{20}$$

$$= 62.4\,\pi \left(\frac{40,\,000}{3} \right) \approx 2.\,613\,805 \times 10^6 \;\; \text{foot-pounds.}$$

$\blacksquare$

Impulse is a physical quantity closely related to work. Instead of relating force and distance to account for changes in energy, impulse relates force and time to account for changes in velocity. First, suppose that a constant force F is applied to an object from time $t = 0$ to time $t = T$. If the position of the object at time t is given by $x(t)$, then Newton's second law says that $F = ma = mx''(t)$. Integrating this equation once with respect to t gives us

$$\int_0^T F\, dt = m \int_0^T x''\,(t)\, dt,$$

or

$$F\,(T - 0) = m[x'(T) - x'(0)].$$

Recall that $x'(t)$ is the velocity $v(t)$, so that

$$F\,T = m[v(T) - v(0)]$$

or $F\,T = m\Delta v$, where $\Delta v = v(T) - v(0)$ is the change in velocity. The quantity $F\,T$ is called the **impulse**, $mv(t)$ is the **momentum** at time t and the equation relating the impulse to the change in velocity is called the **impulse-momentum equation.**

Since we defined the impulse for a constant force, we must now generalize the notion to the case of a nonconstant force. Think about this some and try to guess what the definition should be.

We define the **impulse** J of a force $F(t)$ applied over the time interval $[a, b]$ to be

$$J = \int_a^b F(t)\, dt.$$

Note the similarities and differences between work and impulse. We leave the derivation of the impulse integral for the case of a nonconstant force as an exercise. The impulse-momentum equation also generalizes to the case of a nonconstant force:

$$J = m[v(b) - v(a)].$$

Example 6.4   Estimating the Impulse for a Baseball

Suppose that a baseball traveling at 130 ft/s (about 90 mph) collides with a bat. The following data (adapted from *The Physics of Baseball* by Robert Adair) shows the force exerted by the bat on the ball at 0.0001 second intervals.

t (sec):	0	0.0001	0.0002	0.0003	0.0004	0.0005	0.0006	0.0007
$F(t)$ (lbs)	0	1250	4250	7500	9000	5500	1250	0

Estimate the impulse of the bat on the ball and (using $m = 0.01$ slugs) the speed of the ball after impact.

Solution   In this case, the impulse J is given by $\int_0^{0.0007} F(t)\,dt$. However, you can't evaluate the integral, since we don't know the force function $F(t)$. Since we're given only a fixed number of measurements of $F(t)$, the best we can do is approximate the integral numerically (e.g., using Simpson's Rule). Recall that Simpson's Rule requires an odd value for $n+1$, the number of points in the partition, which means that you need an even number n of subintervals. (Think about this some!) Using $n = 8$ and adding a 0 function value at $t = 0.0008$ (why is it fair to do this?), Simpson's rule gives us

$$J \approx [0 + 4(1250) + 2(4250) + 4(7500) + 2(9000) + 4(5500)$$
$$+ 2(1250) + 4(0) + 0]\frac{0.0001}{3}$$
$$\approx 2.866.$$

In this case, the impulse-momentum equation $J = m\Delta v$ becomes $2.866 = 0.01\Delta v$ or $\Delta v = 286.6$ ft/s. Since the ball started out with a speed of 130 ft/s in one direction and it ended up traveling in the opposite direction, it ended up traveling at a speed of 156.6 ft/s.

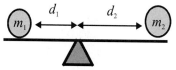

FIGURE 5.50: Balancing
two masses.

The concept of the **first moment**, like work, involves force and distance. Moments are used to solve problems of balance and rotation. We start with a simple problem involving two children on a playground seesaw (or teeter-totter). Suppose that the child on the left in Figure 5.50 is heavier (i.e., has larger mass) than the child on the right. If the children sit an equal distance from the pivot point, you know what will happen: the left side will be pulled down and the lighter child will be raised into the air. You may also know that the children can balance each other if the heavier child moves closer to the pivot point. That is, the balance is determined both by weight (force) and distance. In turns out that balance is easy to describe mathematically. If the children have masses m_1 and m_2 and are sitting at distances d_1 and d_2, respectively, from the pivot point, then they balance each other if and only if

$$m_1 d_1 = m_2 d_2. \tag{6.3}$$

FIGURE 5.51: Two
point-masses.

Let's turn the problem around slightly. Suppose there are two objects, of mass m_1 and m_2, located at different places, say x_1 and x_2, respectively, with $x_1 < x_2$. We consider the objects to be **point-masses**. That is, each is a single point, with all of the mass concentrated at that point (see Figure 5.51).

Suppose that you want to find the **center of mass** $\bar{x}$, that is, the location at which you could place the pivot of a seesaw and have the objects balance. From the balance equation (6.3), you'll need $m_1(\bar{x} - x_1) = m_2(x_2 - \bar{x})$. Solving this equation for $\bar{x}$ gives us

$$\bar{x} = \frac{m_1 x_1 + m_2 x_2}{m_1 + m_2}.$$

Notice that the denominator in this equation for the center of mass is the total mass of the "system" (in this case, the mass of the two objects). The numerator of this expression is called the **first moment** of the system.

More generally, for a system of n masses $m_1, m_2, \ldots, m_n$, located at $x = x_1, x_2, \ldots, x_n$, respectively, the center of mass $\bar{x}$ is given by the first moment divided by the total mass, that is,

$$\bar{x} = \frac{m_1 x_1 + m_2 x_2 + \cdots + m_n x_n}{m_1 + m_2 + \cdots + m_n}.$$

Rather than compute the mass of a system of discrete objects, suppose instead that we wish to find the mass and center of mass of an object of variable density that extends from $x = a$ to $x = b$. Here, we assume that the density function $\rho(x)$ (measured in units of mass per unit length) is known. In the simplest possible case, note that if the density is a constant ρ, the mass of the object is simply given by $m = \rho L$, where L is the length of the object ($L = b - a$). On the other hand, if the density varies throughout the object, its mass is less clear. Since we don't know how to find this exactly, we first settle for finding it approximately. As we have done many times now, we begin by dividing the interval $[a, b]$ into n pieces of equal width $\Delta x = \dfrac{b-a}{n}$. On each subinterval $[x_{i-1}, x_i]$, notice that the mass is approximately $\rho(c_i)\Delta x$, where c_i is a point in the subinterval and Δx is the length of the subinterval. The total mass is then approximately

$$m \approx \sum_{i=1}^{n} \rho(c_i)\,\Delta x.$$

You should observe that this is a Riemann sum that approaches the total mass as $n \to \infty$. Taking the limit as $n \to \infty$, we get that

$$m = \lim_{n \to \infty} \sum_{i=1}^{n} \rho(c_i)\,\Delta x = \int_{a}^{b} \rho(x)\,dx. \tag{6.4}$$

Example 6.5   **Computing the Mass of a Baseball Bat**

A 30-inch baseball bat can be represented approximately by an object extending from $x = 0$ to $x = 30$ inches, with density $\rho(x) = \left(\dfrac{1}{46} + \dfrac{x}{690}\right)^2$ slugs per inch. (If you are unfamiliar with slugs, these are the units of measurement of *mass* in the English system. To get the more familiar *weight* in pounds, simply multiply the mass by $g = 32$.) The density takes into account the fact that a baseball bat is similar to an elongated cone. Find the mass of the object.

Solution   From (6.4), the mass is given by

$$
\begin{aligned}
m &= \int_{0}^{30} \left(\frac{1}{46} + \frac{x}{690}\right)^2 dx \\[2mm]
&= \frac{690}{3} \left(\frac{1}{46} + \frac{x}{690}\right)^3 \bigg|_{0}^{30} = \frac{690}{3}\left[\left(\frac{1}{46} + \frac{30}{690}\right)^3 - \left(\frac{1}{46}\right)^3\right] \\[2mm]
&= 6.\,14366\,7296 \times 10^{-2} \text{ slugs.}
\end{aligned}
$$

To make the result more meaningful, you might compute the weight (in ounces). To do this, multiply the mass by $32 \cdot 16$. You should find that the bat weighs roughly 31.5 ounces.

■

To compute the first moment for an object of nonconstant density $\rho(x)$ extending from $x = a$ to $x = b$, we again divide the interval into n equal pieces. From our earlier argument, for each $i = 1, 2, \ldots, n$, the mass of the ith slice of the object is approximately $\rho(c_i)\Delta x$, for any choice of $c_i \in [x_{i-1}, x_i]$. We then represent that portion of the object with a particle of mass $\rho(c_i)\Delta x$ located at $x = c_i$. We can now think of the original object as having been approximated by n distinct point-masses, as indicated in Figure 5.52. Notice that the first moment M_n of this approximate system is

m_1 m_2 m_3 m_4 m_5 m_6

c_1 c_2 c_3 c_4 c_5 c_6

FIGURE 5.52: Six point-masses.

$$
\begin{aligned}
M_n &= [\rho(c_1)\Delta x]c_1 + [\rho(c_2)\Delta x]c_2 + \cdots + [\rho(c_n)\Delta x]c_n \\[2mm]
&= [c_1\rho(c_1) + c_2\rho(c_2) + \cdots + c_n\rho(c_n)]\,\Delta x = \sum_{i=1}^{n} c_i\,\rho(c_i)\,\Delta x
\end{aligned}
$$

Taking the limit as $n \to \infty$, the sum approaches the first moment

$$M = \lim_{n \to \infty} \sum_{i=1}^{n} c_i\,\rho(c_i)\,\Delta x = \int_{a}^{b} x\rho(x)\,dx. \tag{6.5}$$

The center of mass of the object is then given by

$$\bar{x} = \frac{M}{m} = \frac{\displaystyle\int_a^b x\rho(x)\,dx}{\displaystyle\int_a^b \rho(x)\,dx}. \tag{6.6}$$

Example 6.6   Finding the Center of Mass (Sweet Spot) of a Baseball Bat
   Find the center of mass of the baseball bat from example 6.5.

Solution   From (6.5), the first moment is given by

$$M = \int_0^{30} x\left(\frac{1}{46} + \frac{x}{690}\right)^2 dx = \left[\frac{x^2}{4232} + \frac{x^3}{47\,610} + \frac{x^4}{1\,904\,400}\right]_0^{30} = 1.20510397.$$

Recall that we had already found the mass to be $m = 6.\,143\,667\,296 \times 10^{-2}$ slugs and so, from (6.6), the center of mass of the bat is

$$\bar{x} = \frac{M}{m} = \frac{1.20510397}{6.143\,667\,296 \times 10^{-2}} \approx 19.6 \text{ inches.}$$

Note that for a baseball bat, the center of mass is one candidate for the so-called "sweet spot" of the bat, the best place to hit the ball.

■

For our final application of integration in this section, we consider hydrostatic force. Imagine a dam holding back a lake full of water. What force must the dam be built to withstand?

Hoover Dam

As usual, we solve a simpler problem first. If you have a flat rectangular plate oriented horizontally underwater, notice that the force exerted on the plate by the water (the **hydrostatic force**) is simply the weight of the water lying above the plate. To find this, we must only find the volume of the water lying above the plate and multiply this by the weight density of water (62.4 lb/ft^3). If the area of the plate is A ft^2 and it lies d ft below the surface (see Figure 5.53), then the force on the plate is

$$F = 62.4\,A\,d.$$

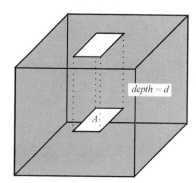

FIGURE 5.53: A plate of area A submerged to depth d.

According to Pascal's Principle, the pressure at a given depth d in a fluid is the same in all directions. This says that if a flat plate is submerged in a fluid, then the pressure on one side of the plate at any given point is $\rho \cdot d$, where ρ is the weight density of the fluid and

d is the depth. In particular, this says that it's irrelevant whether the plate is submerged vertically, horizontally or otherwise (see Figure 5.54).

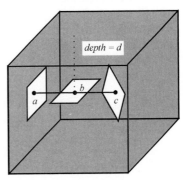

FIGURE 5.54: Pressure at a given depth is the same, regardless of the orientation.

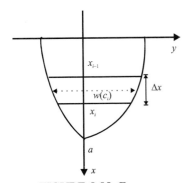

FIGURE 5.55: Force acting on a dam.

Consider now a vertically oriented wall (a dam) holding back a lake. It is convenient to orient the x-axis vertically with $x = 0$ located at the surface of the water and the bottom of the wall at $x = a > 0$ (see Figure 5.55). In this way, notice that x measures the depth of a section of the dam. Suppose $w(x)$ is the width of the wall at depth x (where all distances are measured in feet).

Partition the interval $[0, a]$ into n subintervals of equal width $\Delta x = \dfrac{a}{n}$. Observe that this has the effect of slicing the dam into n slices, each of width Δx. We now consider the force acting on each slice of the dam. For each $i = 1, 2, \ldots, n$, observe that the area of the ith slice is approximately $w(c_i)\Delta x$, where c_i is some point in the subinterval $[x_{i-1}, x_i]$. Further, the depth at every point on this slice is approximately c_i. We can then approximate the force F_i acting on this slice of the dam by the weight of the water lying above a plate the size of this portion but which is oriented horizontally:

$$F_i \approx \underbrace{62.4}_{\text{weight density}} \underbrace{w(c_i)}_{\text{length}} \underbrace{\Delta x}_{\text{width}} \underbrace{c_i}_{\text{depth}} = 62.4\, c_i\, w(c_i)\, \Delta x.$$

Adding together the forces acting on each slice, notice that the total force F on the dam is approximately

$$F \approx \sum_{i=1}^{n} 62.4\, c_i\, w(c_i)\, \Delta x.$$

Taking the limit as $n \to \infty$, the Riemann sums approach the total hydrostatic force on the dam,

$$F = \lim_{n \to \infty} \sum_{i=1}^{n} 62.4\, c_i\, w(c_i)\, \Delta x = \int_0^a 62.4\, x\; w(x)\, dx, \qquad (6.7)$$

where you should recognize the integral as the limit of a Riemann sum.

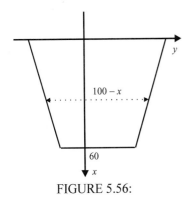

FIGURE 5.56: Trapezoidal dam.

Example 6.7   Finding the Hydrostatic Force on a Dam

A dam is shaped like a trapezoid with height 60 ft. The width at the top is 100 ft and the width at the bottom is 40 ft. (see Figure 5.56). Find the maximum hydrostatic force the wall will need to withstand. Find the hydrostatic force if a drought lowers the water level by 10 ft.

Solution   Notice that the width function is a linear function of depth with $w(0) = 100$ and $w(60) = 40$. The slope is then $\frac{60}{-60} = -1$ and so, $w(x) = 100 - x$. From (6.7), the hydrostatic force is then

$$F = \int_0^{60} \underbrace{62.4}_{weight\ density} \underbrace{x}_{depth} \underbrace{(100 - x)}_{width} dx$$

$$= 3120x^2 - 62.4\frac{x^3}{3}\Big|_0^{60} = 6,739,200 \text{ lb.}$$

If the water level dropped 10 ft, the width of the dam at the water level would be 90 ft. Lowering the origin by 10 ft, the new width function satisfies $w(0) = 90$ and $w(50) = 40$. The slope is still -1 and so, the width is given by $w(x) = 90 - x$. From (6.7), the hydrostatic force is now

$$F = \int_0^{50} \underbrace{62.4}_{weight\ density} \underbrace{x}_{depth} \underbrace{(90 - x)}_{width} dx$$

$$= 2808x^2 - 62.4\frac{x^3}{3}\Big|_0^{50} = 4,420,000 \text{ lb.}$$

Notice that this represents a reduction in force of over 34%.

■

Exercises 5.6

1. (writing) For each of work, impulse and the first moment: identify the quantities in the definition (e.g., force and distance) and the calculations for which it is used (e.g., change in velocity).

2. (writing) The center of mass is not always the location at which half the mass is on one side and half the mass is on the other side. Give an example where more than half the mass is on one side (see examples 6.5 and 6.6) and explain why the object balances at the center of mass.

3. (writing) People who play catch have a seemingly instinctive method of pulling their hand back as they catch the ball. To catch a ball, you must apply an impulse equal to the mass times velocity of the ball. By pulling your hand back, you increase the amount of time in which you decelerate the ball. Use the impulse-momentum equation to explain why this reduces the average force on your hand.

4. (writing) A tennis ball comes toward you at 100 mph. After you hit the ball, it is moving away from you at 100 mph. Work measures changes in energy. Explain why work has been done by the tennis racket on the ball even though the ball has the same speed before and after the hit.

5. A force of 5 pounds stretches a spring 4 inches. Find the work done in stretching this spring 6 inches beyond its natural length.

6. A force of 10 pounds stretches a spring 2 inches. Find the work done in stretching this spring 3 inches beyond its natural length.

7. A force of 20 pounds stretches a spring $\frac{1}{2}$-foot. Find the work done in stretching this spring 1 foot beyond its natural length.

8. A force of 12 pounds stretches a spring $\frac{1}{3}$-foot. Find the work done in stretching this spring 4 inches beyond its natural length.

9. A weightlifter lifts 250 pounds a distance of 20 inches. Find the work done (as measured in foot-pounds).

10. A wrestler lifts his 300-pound opponent overhead, a height of 6 feet. Find the work done (as measured in foot-pounds).

11. A person lifts a 100-pound couch up 3 feet onto a truck. Find the work done.

12. A forklift raises 2000 pounds of bricks up 8 feet. Find the work done.

13. A rocket full of fuel weighs 10,000 pounds at launch. After launch, the rocket gains altitude and loses weight as the fuel burns. Assume that the rocket loses 1 pound of fuel for every 15 feet of altitude gained. Explain why the work done raising the rocket to an altitude of 30,000 feet is $\int_0^{30,000} (10,000 - x/15)\, dx$ and compute the integral.

14. Referring to exercise 13, suppose that a rocket weighs 8000 pounds at launch and loses 1 pound of fuel for every

10 feet of altitude gained. Find the work needed to raise the rocket to a height of 10,000 feet.

15. Suppose that a car engine exerts a force of $800x(1-x)$ pounds when the car is at position x miles, $0 \le x \le 1$. Compute the work done.

16. Horsepower measures the rate of work done as a function of time. For the situation in exercise 15, explain why this is not equal to $800x(1-x)$. If the car in exercise 15 takes 80 seconds to travel the mile, compute the average horsepower (1 hp= 550 ft-lb/s).

17. A water tower is spherical in shape with radius 50 feet, extending from 200 feet to 300 feet above ground. Compute the work done in filling the tank.

18. Compute the work done in filling the tank of exercise 17 half way.

19. Compute the work done in pumping half of the water out of the top of the tank in example 6.3.

20. A water tank is in the shape of a right circular cone of altitude 10 feet and base radius 5 feet, with its vertex at the ground. (Think of an ice cream cone with its point facing down.) If the tank is full, find the work done in pumping all of the water out the top of the tank.

21. In example 6.4, suppose that the baseball was traveling at 100 ft/s. The force exerted by the bat on the ball would change to the values in the table. Estimate the impulse and the speed of the ball after impact.

t (sec)	0	0.0001	0.0002	0.0003	0.0004
F (lb)	0	1000	2100	4000	5000
t (sec)	0.0005	0.0006	0.0007	0.0008	
F (lb)	5200	2500	1000	0	

22. In exercise 21, suppose that the baseball was travelling at 85 ft/s. The force exerted by the bat on the ball would change to the values in the table. Estimate the impulse and the speed of the ball after impact.

t (sec)	0	0.0001	0.0002	0.0003	0.0004
F (lb)	0	600	1200	2000	2500
t (sec)	0.0005	0.0006	0.0007	0.0008	
F (lb)	3000	2500	1100	300	

23. Suppose that a crash test is performed on a vehicle. The force of the wall on the front bumper is shown below. Estimate the impulse and the speed of the vehicle (assume $m = 200$).

t (sec)	0	0.1	0.2	0.3	0.4	0.5	0.6
F (lb)	0	8000	16,000	24,000	15,000	9000	0

24. Two football players collide. The force of the defensive player on the offensive player is given below. Estimate the impulse. If the offensive player has mass $m = 7$ and velocity of 29 ft/s before the collision, does the defensive player stop the offensive player?

t (sec)	0	0.1	0.2	0.3	0.4	0.5
F (lb)	0	300	500	400	250	150
t (sec)	0.6	0.7	0.8	0.9	1.0	
F (lb)	100	100	80	40	0	

25. Compute the mass and center of mass of an object with density $\rho(x) = \dfrac{x}{6} + 2$ kg/m, $0 \le x \le 6$. Briefly explain in terms of the density function why the center of mass is not at $x = 3$.

26. Compute the mass and center of mass of an object with density $\rho(x) = 3 - \dfrac{x}{6}$ kg/m, $0 \le x \le 6$. Briefly explain in terms of the density function why the center of mass is not at $x = 3$.

27. Compute the mass and center of mass of an object with density $\rho(x) = 4 + \dfrac{x^2}{4}$ kg/m, $-2 \le x \le 2$.

28. Compute the mass and center of mass of an object with density $\rho(x) = 4 - \sin\dfrac{\pi x}{4}$ kg/m, $0 \le x \le 4$.

29. Compute the weight in ounces of an object extending from $x = -3$ to $x = 27$ with density $\rho(x) = \left(\dfrac{1}{46} + \dfrac{x+3}{690}\right)^2$ slugs/in.^{2}.

30. Compute the weight in ounces of an object extending from $x = 0$ to $x = 32$ with density $\rho(x) = \left(\dfrac{1}{46} + \dfrac{x+3}{690}\right)^2$ slugs/in.^{2}.

31. Compute the center of mass of the object in exercise 29. This object models the baseball bat of example 6.5 "choked up" (held 3 inches up the handle). Compare the masses and centers of mass of the two bats.

32. Compute the center of mass of the object in exercise 30. This object models a baseball bat that is 2 inches longer than the bat of example 6.5. Compare the masses and centers of mass of the two bats.

33. Compute the mass and weight in ounces and center of mass of an object extending from $x = 0$ to $x = 30$ with density $\rho(x) = 0.00468 \left(\dfrac{3}{16} + \dfrac{x}{60}\right)$ slugs/in.

34. The object in exercise 33 models an aluminum baseball bat (hollow and $\frac{1}{4}$-inch thick). Compare the mass and center of mass to the wooden bat of example 6.5. Baseball experts claim that it is easier to hit an inside pitch (small x value) with an aluminum bat. Explain why your calculations indicate that this is true.

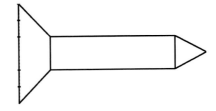

35. The figure above shows the outline of a model rocket. Assume that the vertical scale is 3 units high and the horizontal scale is 6 units wide. Use basic geometry to compute the area of each of the three regions of the rocket. Assuming a constant density ρ, locate the x-coordinate of the center of mass of each region. (Hint: The first region can be thought of as extending from $x = 0$ to $x = 1$ with density $\rho(3 - 2x)$. The third region extends from $x = 5$ to $x = 6$ with density $\rho(6 - x)$.)

36. For the model rocket in exercise 35, replace the rocket with 3 particles, one for each region. Assume that the mass of each particle equals the area of the region, and the location of the particle on the x-axis equals the center of mass of the region. Find the center of mass of the 3-particle system. [Rockets are designed with bottom fins large enough that the center of mass is shifted near the bottom (or, in the figure above, left) of the rocket. This improves the flight stability of the rocket.]

37. A dam is in the shape of a trapezoid with height 60 feet. The width at the top is 40 feet and the width at the bottom is 100 feet. Find the maximum hydrostatic force the wall would have to withstand. Explain why the force is so much greater than the force in example 6.7.

38. Find the maximum hydrostatic force in exercise 37 if a drought lowers the water level by 10 feet.

39. An underwater viewing window is installed at an aquarium. The window is circular with radius 5 feet. The center of the window is 40 feet below the surface of the water. Find the hydrostatic force on the window.

40. An underwater viewing window is rectangular with width 40 feet. The window extends from the surface of the water to a depth of 10 feet. Find the hydrostatic force on the window.

41. The camera's window on a robotic submarine is circular with radius 3 inches. How much hydrostatic force would the window have to withstand to descend to a depth of 1000 feet?

42. A diver wears a watch to a depth of 60 feet. The face of the watch is circular with a radius of 1 inch. How much hydrostatic force will the face have to withstand if the watch is to keep on ticking?

43. In the text, we mentioned that work calculates changes in energy. For example, a 200-pound pole vaulter is propelled by a pole to a height of 20 feet. The work done by the pole, equal to 4000 ft-lb, gives the vaulter a large potential energy. To see what this means, compute the speed v of the vaulter when he reaches the ground. Show that the kinetic energy at impact, given by $\frac{1}{2}mv^2$ ($m = 200/32$), also equals 4000 ft-lb. This illustrates the concept of conservation of energy.

44. Compute the speed of the falling vaulter in exercise 43 at the 10-foot mark. Show that the sum of the potential energy ($32mh$) and the kinetic energy ($\frac{1}{2}mv^2$) equals 4000 ft-lb.

45. Compute the horsepower needed to move a 100-ton blue whale at 20 mph (1 hp = 550 ft-lb/s). Such whales swim so efficiently that they can maintain this speed with an output of 60–70 hp.

46. For a constant force F exerted over a length of time t, impulse is defined by $F \cdot t$. For a variable force $F(t)$, derive the impulse formula $J = \int_a^b F(t)\, dt$.

47. (exploratory) As equipment has improved, heights cleared in the pole vault have increased. A crude estimate of the maximum pole vault possible can be derived from conservation of energy principles. Assume that the maximum speed a pole-vaulter could run carrying a long pole is 25 mph. Convert this speed to ft/s. The kinetic energy of this vaulter would be $\frac{1}{2}mv^2$ (leave m as an unknown for the time being). This initial kinetic energy would equal the potential energy at the top of the vault minus whatever energy is absorbed by the pole (which we will ignore). Set the potential energy, $32mh$, equal to the kinetic energy and solve for h. This represents the maximum amount the vaulter's center of mass could be raised. Add 3 feet for the height of the vaulter's center of mass and you have an estimate of the maximum vault possible. Compare this to Sergei Bubka's 1994 world record vault of $20'1\frac{3}{4}''$.

48. (exploratory) The first moment of a solid of density $\rho(x)$ is $\int_a^b x\rho(x)\, dx$. The second moment, defined by $\int_a^b x^2\rho(x)\, dx$, is also important in applications. The larger this number is, the more difficult it is to rotate the solid about the y-axis. Compute the second moments of the baseball bats in example 6.5 and exercise 29. Choking up on a bat makes it easier to swing (and control). Compute the percentage by which the second moment is reduced by choking up 3 inches.

49. (exploratory) Occasionally, baseball players illegally "cork" their bats by drilling out a portion of wood from the end of the bats and filling the hole with a light substance such as cork. The advantage of this procedure is that the second moment is significantly reduced. To model this, take the bat of example 6.5 and change the density to

$$\rho(x) = \begin{cases} \left(\frac{1}{46} + \frac{x}{690}\right)^2 & \text{if } 0 \le x \le 28 \\ \left(\frac{1}{92} + \frac{x}{690}\right)^2 & \text{if } 28 < x \le 30 \end{cases},$$

representing a hole of radius $\frac{1}{4}''$ and length $2''$. Compute the mass and second moment of the corked bat, and compare to the original bat.

5.7   Probability (Optional)

You may have suspected this for some time, but in real-world applications, you are rarely simply handed a function to analyze. More than likely, you will need to go out and take numerous measurements and/or develop a detailed theory about the process you're studying. It is often quite challenging to identify an appropriate function(s) to study, especially if the measurements are subject to random or unpredictable errors. The analysis of random processes is the focus of the mathematical fields of probability and statistics. In this section, we give a brief introduction to the use of calculus in probability theory. It may surprise you to learn that calculus provides insight into random processes, but this is in fact, a very important application of integrals. We urge you to continue your studies by taking a course(s) in probability and statistics.

We begin with a simple example involving coin-tossing. Suppose that you toss two coins, each of which has a 50% chance of coming up heads. Because of the randomness involved, you cannot calculate exactly how many heads you will get on a given number of tosses. But you *can* calculate the *likelihood* of each of the possible outcomes. If we denote heads by H and tails by T, then the four possible outcomes from tossing two coins are HH, HT, TH and TT. Each of these four outcomes is equally likely, so we can say that each has probability $\frac{1}{4}$. This means that, *on average*, each of these events will occur in one-fourth of your tries. Said a different way, the **relative frequency** with which each event occurs in a large number of trials will be approximately $\frac{1}{4}$.

Suppose that we are primarily interested in recording the number of heads. Based on our calculations above, the probability of getting two heads is $\frac{1}{4}$, the probability of getting one head is $\frac{2}{4}$ (since there are two ways for this to happen: HT and TH) and the probability of getting zero heads is $\frac{1}{4}$. We often summarize such information by displaying it in a **histogram** or bar graph (see Figure 5.57).

Suppose that we instead toss eight coins. The probabilities for getting a given number of heads are given in the accompanying table and the corresponding histogram is shown in Figure 5.58. You should notice that the sum of all the probabilities is 1 (or 100%, since it's certain that **one** of the possible outcomes will occur on a given try). This is one of the defining properties of probability theory. Another basic property is called the **addition principle**: to compute the probability of getting 6, 7 or 8 heads (or any other mutually exclusive outcomes), simply add together the individual probabilities:

$$P(6, \ 7 \text{ or } 8 \text{ heads}) = \frac{28}{256} + \frac{8}{256} + \frac{1}{256} = \frac{37}{256} \approx 0.144.$$

A graphical interpretation of this calculation is very revealing. In the histogram in Figure 5.58, notice that each bar is a rectangle of width 1. Then the probability associated with each bar equals the area of the rectangle. In graphical terms,

- The total area in such a histogram is 1.
- The probability of getting between 6 and 8 heads (inclusive) equals the sum of the areas of the rectangles located between 6 and 8 (inclusive).

Not all probability events have the nice theoretical structure of coin-tossing. For instance, the situation is somewhat different if we want to find the probability that a randomly chosen person will have a height of $5'9''$ or $5'10''$. There is no easy theory we can use here to compute the probabilities (since not all heights are equally likely). In this case, we would need to use the correspondence between probability and relative frequency. If we collected information about the heights of a large number of adults (for instance, from driver's license data), we might find the following.

FIGURE 5.57:
Histogram for two-coin toss.

# of heads	probability
0	1/256
1	8/256
2	28/256
3	56/256
4	70/256
5	56/256
6	28/256
7	8/256
8	1/256

FIGURE 5.58:
Histogram for eight-coin toss.

height	$< 64''$	$64''$	$65''$	$66''$	$67''$	$68''$	$69''$	$70''$	$71''$	$72''$	$73''$	$> 73''$
# people	23	32	61	94	133	153	155	134	96	62	31	26

Since the total number of people in the survey is 1000, the relative frequency of the height $5'9''$ ($69''$) is $\frac{155}{1000} = 0.155$ and the relative frequency of the height $5'10''$ ($70''$) is $\frac{134}{1000} =$

FIGURE 5.59:
Histogram for relative
frequency of heights.

0.134. An estimate of the probability of being $5'9''$ or $5'10''$ is then $0.155 + 0.134 = 0.289$. A histogram is shown in Figure 5.59.

Suppose that we want to be more specific: for example, what is the probability that a randomly chosen person is $5'8\frac{1}{2}''$ or $5'9''$? To answer this question, we would need to have our data broken down further, as in the following partial table.

$66\frac{1}{2}''$	$67''$	$67\frac{1}{2}''$	$68''$	$68\frac{1}{2}''$	$69''$	$69\frac{1}{2}''$	$70''$	$70\frac{1}{2}''$	$71''$
52	61	72	71	82	81	74	69	65	58

The probability that a person is $5'9''$ can be estimated by the relative frequency of $5'9''$ people in our survey, which is $\frac{81}{1000} = 0.081$. Similarly, the probability that a person is $5'8\frac{1}{2}''$ is approximately $\frac{82}{1000} = 0.082$. The probability of being $5'8\frac{1}{2}''$ or $5'9''$ is then approximately $0.081 + 0.082 = 0.163$. A histogram for this portion of the data is shown in Figure 5.60a.

FIGURE 5.60a: Histogram for relative
frequency of heights.

FIGURE 5.60b: Histogram showing
double the relative frequency.

Notice that since each bar of the histogram now represents a half-inch range of height, we can no longer interpret area in the histogram as the probability (we had assumed each bar has width 1). We will modify the histogram to make the area connection clearer. In Figure 5.60b, we have labeled the horizontal axis with the height in inches, and the vertical axis shows twice the relative frequency. The bar at $69''$ inches has height 0.162 and width $\frac{1}{2}$. Its area, $\frac{1}{2}(0.162) = 0.081$, corresponds to the relative frequency (or probability) of the height $5'9''$.

Of course, we could continue subdividing the height intervals into smaller and smaller pieces. Think of doing this while modifying the scale on the vertical axis so that the area of each rectangle (length times width of interval) always gives the relative frequency (probability) of that height interval. For example, suppose that there are n height intervals between $5'8''$ and $5'9''$. Let x represent height in inches and $f(x)$ equal the height of the histogram bar for the interval containing x. Let $x_1 = 68 + \frac{1}{n}$, $x_2 = 68 + \frac{2}{n}$ and so on, so that $x_i = 68 + \frac{i}{n}$, for $1 \le i \le n$ and let $\Delta x = \frac{1}{n}$. For a randomly selected person, the probability that their height is between $5'8''$ and $5'9''$ is estimated by the sum of the areas of the corresponding histogram rectangles, given by

FIGURE 5.61:
Histogram for heights.

$$P(68 \le x \le 69) \approx f(x_1)\,\Delta x + f(x_2)\,\Delta x + \cdots + f(x_n)\,\Delta x = \sum_{i=1}^{n} f(x_i)\,\Delta x. \quad (7.1)$$

Observe that as n increases, the histogram of Figure 5.61 will "smooth out", approaching a curve like the one shown in Figure 5.62. We call this limiting function $f(x)$, the **probability density function (pdf)** for heights. Notice that for any given $i = 1, 2, \ldots, n$, $f(x_i)$ does not give the probability that a person's height equals x_i. Instead, for small values of Δx, the quantity $f(x_i)\,\Delta x$ is an approximation of the probability that the height is in the range $[x_{i-1}, x_i]$.

FIGURE 5.62: Probability
density function and
histogram for heights.

Look carefully at (7.1) and think about what will happen as n increases. The Riemann sum on the right should approach an integral $\int_a^b f(x)\,dx$. In this particular case, the limits of integration are 68 (5′8″) and 69 (5′9″). We have

$$\lim_{n\to\infty} \sum_{i=1}^{n} f(x_i)\, \Delta x = \int_{68}^{69} f(x)\, dx.$$

Notice that by adjusting the function values so that probability corresponds to area (given by an integral), we have found a familiar and direct technique for computing probabilities. We now summarize our discussion with some definitions. The preceding examples are of **discrete probability distributions** (the term *discrete* indicates that the quantity being measured can only assume values from a certain finite set or from a fixed sequence of values). For instance, in coin-tossing, the number of heads must be an integer. By contrast, many distributions are **continuous**. That is, the quantity of interest (the random variable) assumes values from a continuous range of numbers (an interval). For instance, suppose you are measuring the height of people in a certain group. Although height is normally rounded off to the nearest integer number of inches, a person's actual height can be any number. (For instance, one of the authors lists his height as 6′0″, while his actual height is closer to 5′11.8″, depending on how good his posture is).

For continuous distributions, the graph corresponding to a histogram is the graph of a **probability density function** (pdf). We now give a precise definition of a pdf.

Definition 7.1   Suppose that X is a random variable that may assume any value x with $a \le x \le b$. A probability density function for X is a function $f(x)$ satisfying

(i) $f(x) \ge 0$ for $a \le x \le b$,

(ii) $\displaystyle\int_{a}^{b} f(x)\, dx = 1$ and

(iii)   The probability that the (observed) value of X falls between c and d is given by the area under the graph of the pdf on that interval. That is,

$$P(c \le X \le d) = \int_{c}^{d} f(x)\, dx.$$

The properties in Definition 7.1 state the basic rules for probability density functions: they are never negative, the total probability is 1 (100%), and probability corresponds to area under the curve.

Example 7.1   **Verifying That a Function Is a pdf on an Interval**

Show that $f(x) = 3x^2$ defines a pdf on the interval $[0, 1]$ by verifying properties (i) and (ii) of Definition 7.1.

Solution   Clearly, $f(x) \ge 0$. For property (ii), we integrate the pdf over its domain. We have

$$\int_{0}^{1} 3x^2\, dx = x^3 \Big|_{0}^{1} = 1.$$

■

Example 7.2   **Using a pdf to Estimate Probabilities**

Suppose that $f(x) = \dfrac{0.4}{\sqrt{2\pi}}\, e^{-0.08(x-68)^2}$ is the probability density function for the heights in inches of adult American males. Find the probability that a randomly selected adult American male will be between 5'8'' and 5'9''. Also, find the probability that a randomly selected adult American male will be between 6'2'' and 6'4''.

Solution   To compute the probabilities, you first need to convert the specified heights into inches. The probability of being between 68 and 69 inches tall is

$$P(68 \leq X \leq 69) = \int_{68}^{69} \frac{0.4}{\sqrt{2\pi}} e^{-0.08(x-68)^2} \, dx \approx 0.15542.$$

Here, we approximated the value of the integral numerically, since we don't know an antiderivative for the integrand. (You can use Simpson's Rule or the numerical integration method built into your calculator or CAS.) Similarly, the probability of being between 74 and 76 inches is

$$P(74 \leq X \leq 76) = \int_{74}^{76} \frac{0.4}{\sqrt{2\pi}} e^{-0.08(x-68)^2} \, dx \approx 0.00751,$$

where we have again approximated the value of the integral numerically.

FIGURE 5.63:
Heights of adult males.

According to data in Gyles Brandreth's *Your Vital Statistics*, the pdf for the heights of adult males in the United States looks like the graph of $f(x) = \frac{0.4}{\sqrt{2\pi}} e^{-0.08(x-68)^2}$ shown in Figure 5.63 and used in example 7.2. You probably have seen *bell-shaped curves* like this before. This distribution is referred to as a **normal distribution**. Besides the normal distribution, there are many other probability distributions that are important in applications.

Example 7.3   **Computing Probability with an Exponential pdf**

Suppose that the lifetime in years of a certain brand of lightbulb is exponentially distributed with pdf $f(x) = 4e^{-4x}$. Find the probability that a given light bulb lasts 3 months or less.

Solution   First, since the random variable measures lifetime in years, convert 3 months to $\frac{1}{4}$ year. The probability is then

$$P\left(0 \leq X \leq \frac{1}{4}\right) = \int_0^{1/4} 4e^{-4x} \, dx = 4\left(-\frac{1}{4}\right) e^{-4x}\Big|_0^{1/4}$$
$$= -e^{-1} + e^0 = 1 - e^{-1} \approx 0.63212.$$

In some cases, there may be theoretical reasons for assuming that a pdf has a certain form. In this event, the first task is to determine the values of any constants to achieve the properties of a pdf.

Example 7.4   **Determining the Coefficient of a pdf**

Suppose that the pdf for a random variable has the form $f(x) = ce^{-3x}$ for some constant c, with $0 \leq x \leq 1$. Find the value of c which makes this a pdf.

Solution   To be a pdf, we first need that $f(x) = ce^{-3x} \geq 0$, for all $x \in [0,1]$. (This will be the case as long as $c \geq 0$.) Also, the integral over the domain must equal 1.

So, we set

$$1 = \int_0^1 ce^{-3x}\, dx = c\left(-\frac{1}{3}\right) e^{-3x}\Big|_0^1 = -\frac{c}{3} e^{-3} + \frac{c}{3} = \frac{c}{3}(1 - e^{-3}).$$

It now follows that $c = \dfrac{3}{1 - e^{-3}} \approx 3.1572$.

■

Given a pdf, it is possible to compute various statistics to summarize the properties of the random variable. The most common statistic is the **mean,** the best-known measure of average value. If you wanted to average test scores of 85, 89, 93 and 93, you would probably compute the mean, given by $\frac{85 + 89 + 93 + 93}{4} = 90$.

Notice here that there were three different test scores recorded: 85, which has a relative frequency of $\frac{1}{4}$, 89, also with a relative frequency of $\frac{1}{4}$, and 93, with a relative frequency of $\frac{2}{4}$. We can also compute the mean by multiplying each value by its relative frequency and then summing: $(85)\frac{1}{4} + (89)\frac{1}{4} + (93)\frac{2}{4} = 90$.

Now, suppose we wanted to compute the mean height of the people in the following table.

height	63″	64″	65″	66″	67″	68″	69″	70″	71″	72″	73″	74″
number	23	32	61	94	133	153	155	134	96	62	31	26

It would be silly to write out the heights of all 1000 people, add and divide by 1000. It is much easier to multiply each value by its relative frequency and add the results. Following this route, the mean m is given by

$$m = (63)\tfrac{23}{1000} + (64)\tfrac{32}{1000} + (65)\tfrac{61}{1000} + (66)\tfrac{94}{1000} + (67)\tfrac{133}{1000} + \cdots + (74)\tfrac{26}{1000} = 68.523.$$

If we denote the heights by x_1, x_2, and let $f(x_i)$ be the corresponding relative frequencies or probabilities, the mean then has the form

$$m = x_1 f(x_1) + x_2 f(x_2) + x_3 f(x_3) + \cdots + x_{12} f(x_{12}).$$

If the heights in our data set were given for every half-inch or tenth-of-an-inch, we would compute the mean by multiplying x_i by the probabilities $f(x_i)\Delta x$, where Δx is the fraction of an inch between data points. The mean now has the form

$$m = [x_1 f(x_1) + x_2 f(x_2) + x_3 f(x_3) + \cdots + x_n f(x_n)]\, \Delta x = \sum_{i=1}^n x_i\, f(x_i)\, \Delta x,$$

where n is the number of data points. Notice that, as n increases and Δx approaches 0, the Riemann sum approaches the integral $\int_a^b x f(x)\, dx$. This gives us the following definition.

Definition 7.2   The **mean** μ of a random variable with pdf $f(x)$ on the interval $[a, b]$ is given by

$$\mu = \int_a^b x f(x)\, dx. \tag{7.2}$$

Although the mean is commonly used to report the average value of a random variable, it is important to realize that it is not the only measure of *average* used by statisticians. An alternative measurement of average is the **median,** the x-value that divides the probability in half. (That is, half of all values of the random variable lie at or below the median and half lie at or above the median.) In the following example and in the exercises, you will explore situations in which each measure provides a better or worse indication about the average of a random variable.

Example 7.5   Finding the Mean Age and Median Age of a Group of Cells

Suppose that the age in days of a type of single-celled organism has pdf $f(x) = (\ln 2)e^{-kx}$ where $k = \frac{1}{2}\ln 2$. The domain is $0 \le x \le 2$ (the assumption here is that upon reaching an age of 2 days, each cell divides into two daughter cells). Find (a) the mean age of the cells, (b) the proportion of cells that are younger than the mean and (c) the median age of the cells.

Solution   For part (a), we have from (7.2) that the mean is given by

$$\mu = \int_0^2 x \ln 2\, e^{-(\ln 2)x/2}\, dx \approx 0.88539 \text{ days,}$$

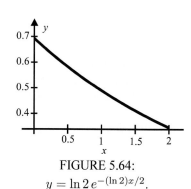

FIGURE 5.64:
$y = \ln 2\, e^{-(\ln 2)x/2}$.

where we have approximated the value of the integral numerically, since at present we do not know how to find an antiderivative for this integrand. Notice that even though the cells range in age from 0 to 2 days, the mean is not 1. Look at the graph of the pdf in Figure 5.64 to see why the mean should be less than 1. The graph shows that younger ages are more likely than older ages and this causes the mean to be less than 1.

For part (b), notice that the proportion of cells younger than the mean is the same as the probability that a randomly selected cell is younger than the mean. This probability is given by

$$P\left(0 \le X \le \mu\right) = \int_0^{0.88539} \ln 2\, e^{-(\ln 2)x/2}\, dx \approx 0.52848,$$

where we have again approximated the value of the integral numerically. Therefore, the proportion of cells younger than the mean is about 53%. Notice that in this case the mean does not represent the 50% mark for probabilities. In other words, the mean is not the same as the median.

To find the median in part (c), we must solve for the constant c such that

$$0.5 = \int_0^c \ln 2\, e^{-(\ln 2)x/2}\, dx.$$

Since an antiderivative of $e^{-(\ln 2)x/2}$ is $-\frac{2}{\ln 2}e^{-(\ln 2)x/2}$, we have

$$0.5 = \int_0^c \ln 2\, e^{-(\ln 2)x/2}\, dx$$
$$= \ln 2\left[-\frac{2}{\ln 2}e^{-(\ln 2)x/2}\right]_0^c$$
$$= -2e^{-(\ln 2)c/2} + 2.$$

Subtracting 2 from both sides, we have

$$-1.5 = -2e^{-(\ln 2)c/2},$$

so that dividing by -2 yields

$$0.75 = e^{-(\ln 2)c/2}.$$

Taking the natural log of both sides, gives us

$$\ln 0.75 = -(\ln 2)c/2.$$

Finally, solving for c gives us

$$c = \frac{-2\ln 0.75}{\ln 2},$$

so that the median is $-2\ln 0.75/\ln 2 \approx 0.83$. We can now conclude that half of the cells are younger than 0.83 days and half the cells are older than 0.83 days.

∎

Exercises 5.7

1. (writing) In the text, we stated that the probability of tossing two fair coins and getting two heads is $\frac{1}{4}$. If you try this experiment four times, explain why you will not always get two heads exactly one out of four times. If probability doesn't give precise predictions, what is its usefulness? To answer this question, discuss the information conveyed by knowing that in the above experiment the probability of getting one head and one tail is $\frac{1}{2}$ (twice as big as $\frac{1}{4}$).

2. (writing) Suppose you toss two coins numerous times (or simulate this on your calculator or computer). Theoretically, the probability of getting two heads is $\frac{1}{4}$. In the long run (as the coins are tossed more and more often), what proportion of the time should two heads occur? Try this and discuss how your results compare to the theoretical calculation.

3. (writing) Based on Figures 5.57 and 5.58, describe what you expect the histogram to look like for larger numbers of coins. Compare to Figure 5.63.

4. (writing) The height of a person is determined by numerous factors, both hereditary and environmental (e.g., diet). Explain why this might produce a histogram similar to that produced by tossing a large number of coins.

In exercises 5–12, show that the given function is a pdf on the indicated interval.

5. $f(x) = 4x^3$, $[0, 1]$

6. $f(x) = \frac{3}{8}x^2$, $[0, 2]$

7. $f(x) = x + 2x^3$, $[0, 1]$

8. $f(x) = \frac{3}{2}x^2$, $[-1, 1]$

9. $f(x) = \frac{1}{2}\sin x$, $[0, \pi]$

10. $f(x) = \cos x$, $[0, \pi/2]$

11. $f(x) = e^{-x/2}$, $[0, \ln 4]$

12. $f(x) = \dfrac{1}{4 - 8/e}xe^{-x/2}$, $[0, 2]$

In exercises 13–18, find a value of c for which $f(x)$ is a pdf on the indicated interval.

13. $f(x) = cx^3$, $[0, 1]$

14. $f(x) = cx + x^2$, $[0, 1]$

15. $f(x) = ce^{-4x}$, $[0, 1]$

16. $f(x) = ce^{-x/2}$, $[0, 2]$

17. $f(x) = 2ce^{-cx}$, $[0, 2]$

18. $f(x) = 2ce^{-cx}$, $[0, 4]$

In exercises 19–22, use the pdf in example 7.2 to find the probability that a randomly selected American male has height in the indicated range.

19. Between $5'10''$ and $6'$.

20. Between $6'6''$ and $6'10''$.

21. Between $7'$ and $10'$.

22. Between $2'$ and $5'$.

In exercises 23–26, find the indicated probabilities, given that the lifetime of a lightbulb is exponentially distributed with pdf $f(x) = 6e^{-6x}$ (with x measured in years).

23. The lightbulb lasts less than 3 months.

24. The lightbulb lasts less than 6 months.

25. The lightbulb lasts between 1 and 2 years.

26. The lightbulb lasts between 3 and 10 years.

In exercises 27–30, find the indicated probabilities, given that the lifetime of a lightbulb is exponentially distributed with pdf $f(x) = 8e^{-8x}$ (with x measured in years).

27. The lightbulb lasts between 1 and 2 months.

28. The lightbulb lasts between 5 and 6 months.

29. The lightbulb lasts less than 6 months.

30. The lightbulb lasts more than 6 months.

In exercises 31–34, suppose the lifetime of an organism has pdf $f(x) = 4xe^{-2x}$ (with x measured in years).

31. Find the probability that the organism lives less than 1 year.

32. Find the probability that the organism lives between 1 and 2 years.

33. Find the mean lifetime ($0 \le x \le 10$).

34. Graph the pdf and compare the maximum value of the pdf to the mean.

In exercises 35–42, find (a) the mean and (b) the median of the random variable with the given pdf.

35. $f(x) = 3x^2$, $0 \le x \le 1$

36. $f(x) = 4x^3, 0 \le x \le 1$

37. $f(x) = \frac{1}{2}\sin x, 0 \le x \le \pi$

38. $f(x) = \cos x, 0 \le x \le \pi/2$

39. $f(x) = \frac{1}{2}(\ln 3)e^{-kx}, k = \frac{1}{3}\ln 3, 0 \le x \le 3$

40. $f(x) = \frac{1}{3}(\ln 4)e^{-kx}, k = \frac{1}{4}\ln 4, 0 \le x \le 4$

41. $f(x) = \frac{4}{1 - e^{-4}}e^{-4x}, 0 \le x \le 1$

42. $f(x) = \frac{6}{1 - e^{-6}}e^{-6x}, 0 \le x \le 1$

43. For $f(x) = ce^{-4x}$, find c so that $f(x)$ is a pdf on the interval $[0, b]$ for $b > 0$. What happens to c as $b \to \infty$?

44. For the pdf of exercise 43, find the mean exactly (use a CAS for the antiderivative). As b increases, what happens to the mean?

45. Repeat exercises 43 and 44 for $f(x) = ce^{-6x}$.

46. Based on the results of exercises 43–45, conjecture the values for c and the mean as $b \to \infty$, for $f(x) = ce^{-ax}$, $a > 0$.

47. For eight coins being tossed, the probabilities of getting a given numbers of heads is shown in the table. Use the addition principle to find the probability of each event indicated below.

# Heads	0	1	2	3	4
Probability	$\frac{1}{256}$	$\frac{8}{256}$	$\frac{28}{256}$	$\frac{56}{256}$	$\frac{70}{256}$
# Heads	5	6	7	8	
Probability	$\frac{56}{256}$	$\frac{28}{256}$	$\frac{8}{256}$	$\frac{1}{256}$	

(a) 3 or fewer heads   (b) more heads than tails
(c) all heads or all tails   (d) an odd number of heads

48. In one version of the game of keno, you choose 10 numbers between 1 and 80. A random drawing selects 20 numbers between 1 and 80. Your payoff depends on how many of your numbers are selected. Use the given probabilities (rounded to 4 digits) to find the probability of each event indicated below. (To win, at least 5 of your numbers must be selected. On a $2 bet, you win $40 or more if 6 or more of your numbers are selected.)

# selected	0	1	2	3	4
Probability	0.0458	0.1796	0.2953	0.2674	0.1473
# selected	5	6	7	8	9	10
Probability	0.0514	0.0115	0.0016	0.0001	0.0	0.0

(a) winning (at least 5 selected)
(b) losing (4 or fewer selected)
(c) winning big (6 or more)
(d) 3 or 4 numbers selected

49. In the baseball World Series, two teams play games until one team or the other wins four times. Suppose team A should win each game with probability 0.6. The probabilities for team A's record (given as wins/losses) in the World Series are shown. Find the probability of each event indicated below.

Wins/Losses	0/4	1/4	2/4	3/4
Probability	0.0256	0.0615	0.0922	0.1106
Wins/Losses	4/3	4/2	4/1	4/0
Probability	0.1659	0.2073	0.2073	0.1296

(a) Team A wins the World Series
(b) Team B wins the World Series
(c) One team wins all 4 games
(d) The teams play 6 or 7 games

50. Suppose a basketball player makes 70% of her free throws. If she shoots three free throws and the probability of making each one is 0.7, the probabilities for the total number made are as shown. Find the probability of each event indicated below.

# made	0	1	2	3
Probability	0.027	0.189	0.441	0.343

(a) She makes 2 or 3   (b) She makes at least 1

51. In any given time period, some species become extinct. Mass extinctions (such as that of the dinosaurs) are relatively rare. Fossil evidence indicates that the probability that p percent $(1 < p < 100)$ of the species become extinct in a 4-million-year period is approximately $e(p) = cp^{-2}$ for some constant c. Find c to make $e(p)$ a pdf and estimate the probability that in a 4-million-year period 60–70% of the species will become extinct.

52. In example 7.5, we noted that the probability that a given cell is of average (mean) age or older is 0.47152, not 0.5 as most people would expect. By trial and error, find the median, the age such that the probability of being older (or younger) is exactly 0.5.

53. The pdf in example 7.2 is the pdf for a normally distributed random variable. The mean is easily read off of $f(x)$; in example 7.2, the mean is 68. The mean and a number called the **standard deviation** characterize normal distributions. As Figure 5.63 indicates, the graph of the pdf has a maximum at the mean and has two inflection points located on opposite sides of the mean. The standard deviation equals the distance from the mean to an inflection point. Find the standard deviation in example 7.2.

54. In exercise 53, you found the standard deviation for the pdf in example 7.2. Denoting the mean as μ and the standard deviation as σ, find the probability that a given height is between $\mu - \sigma$ and $\mu + \sigma$ (that is, within one standard deviation of the mean). Find the probability that a given height is within two standard deviations of the mean ($\mu - 2\sigma$ to $\mu + 2\sigma$), and within three standard deviations of the mean. These probabilities are the same for any normal distribution. If you know the mean and standard deviation of a normally distributed random variable, you automatically know these probabilities.

55. When a set of numbers is "curved" to achieve a predetermined mean, it can be difficult to evaluate the quality of the process producing those numbers. For example, suppose a math class consists of students whose scores on the final exam are 400, 480, 550, 620 and 690. Suppose the middle score (median) is always given a C. Argue that if 550 is a C, then 690 should be an A. Now, suppose the students score 680, 680, 685, 690 and 690. If 685 is a C, can 690 be an A? Explain why, under these rules, the absence of A's doesn't necessarily mean that the class performed poorly.

56. Renowned biologist Stephen Jay Gould has used the argument in exercise 55 to explain a baseball fact (see his book *Full House*). A baseball player's batting average records the proportion of times he makes a hit. Traditionally written with three decimal places, a batting average of .300 (hits 30% of the time) is an indicator of a good batter. In the early 1900s, several batters had averages over .400, but this mark has not been achieved since 1941. The rules of baseball are constantly being adjusted, with one of the goals being to maintain an overall batting average of about .270. If all players today are almost as good as the best players of the early 1900s, explain why it would be more difficult to average .400 today than it was in the past.

57. (exploratory) The mathematical theory of chaos indicates that numbers generated by very simple algorithms can look random. Chaos researchers look at a variety of graphs to try to distinguish randomness from deterministic chaos. For example, iterate the function $f(x) = 4x(1 - x)$ starting at $x = 0.1$. That is, compute $f(0.1) = 0.36$, $f(0.36) = 0.9216$, $f(0.9216) = 0.289$ and so on. Iterate 50 times and record how many times each first digit occurs (so far, we've got a 1, a 3, a 9 and a 2). If the process were truly random, the digits would occur about the same number of times. Does this seem to be happening? To unmask this process as nonrandom, you can draw a **phase portrait**. To do this, take consecutive iterates as coordinates of a point (x, y) and plot the points. The first three points are $(0.1, 0.36)$, $(0.36, 0.9216)$ and $(0.9216, 0.289)$. Describe the (nonrandom) pattern that appears, identifying it as precisely as possible.

58. (exploratory) Suppose that a spring is oscillating up and down with vertical position given by $u(t) = \sin t$. If you pick a random time and look at the position of the spring, would you be more likely to find the spring near an extreme ($u = 1$ or $u = -1$) or near the middle ($u = 0$)? The pdf is inversely proportional to speed (why is this reasonable?). Show that speed is given by $|\cos t| = \sqrt{1 - u^2}$, so the pdf is $f(u) = c/\sqrt{1 - u^2}$, $-1 \le u \le 1$, for some constant c. Show that $c = 1/\pi$, then graph $f(x)$ and describe what positions the spring is likely to be found in.

Chapter Review Exercises

In exercises 1–8, find the indicated area exactly if possible (estimate if necessary).

1. The area between $y = x^2 + 2$ and $y = \sin x$ for $0 \le x \le \pi$

2. The area between $y = e^x$ and $y = e^{-x}$ for $0 \le x \le 1$

3. The area between $y = x^3$ and $y = 2x^2 - x$

4. The area between $y = x^2 - 3$ and $y = -x^2 + 5$

5. The area between $y = e^{-x}$ and $y = 2 - x^2$

6. The area between $x = y^2$ and $y = 1 - x$

7. The area of the region bounded by $y = x^2$, $y = 2 - x$ and $y = 0$

8. The area of the region bounded by $y = x^2$, $y = 0$ and $x = 2$

9. A town has a population of 10,000 with a birth rate of $10 + 2t$ people per year and a death rate of $4 + t$ people per year. Compute the town's population after 6 years.

10. From the given data, estimate the area between the curves for $0 \le x \le 2$.

x	0.0	0.2	0.4	0.6	0.8	1.0
$f(x)$	3.2	3.6	3.8	3.7	3.2	3.4
$g(x)$	1.2	1.5	1.6	2.2	2.0	2.4

x	1.2	1.4	1.6	1.8	2.0
$f(x)$	3.0	2.8	2.4	2.9	3.4
$g(x)$	2.2	2.1	2.3	2.8	2.4

11. Find the volume of the solid with cross-sectional area $A(x) = \pi(3 + x)^2$ for $0 \le x \le 2$.

12. A swimming pool viewed from above has an outline given by $y = \pm(5 + x)$ for $0 \le x \le 2$. The depth is given by $4 + x$ (all measurements in feet). Compute the volume.

13. The cross-sectional areas of an underwater object are given. Estimate the volume.

x :	0	0.4	0.8	1.2	1.6	2.0	2.4	2.8
$A(x)$:	0.4	1.4	1.8	2.0	2.1	1.8	1.1	0.4

In exercises 14–18, find the volume of the indicated solid of revolution.

14. The region bounded by $y = x^2$, $y = 0$ and $x = 1$ revolved about (a) the x-axis; (b) the y-axis; (c) $x = 2$; (d) $y = -2$

15. The region bounded by $y = x^2$ and $y = 4$ revolved about (a) the x-axis; (b) the y-axis; (c) $x = 2$; (d) $y = -2$

16. The region bounded by $y = x$, $y = 2x$ and $x = 2$ revolved about (a) the x-axis; (b) the y-axis; (c) $x = -1$; (d) $y = 4$

17. The region bounded by $y = x$, $y = 2 - x$ and $y = 0$ revolved about (a) the x-axis; (b) the y-axis; (c) $x = -1$; (d) $y = 4$

18. The region bounded by $x = 4 - y^2$ and $x = y^2 - 4$ revolved about (a) the x-axis; (b) the y-axis; (c) $x = 4$; (d) $y = 4$

In exercises 19–22, set up an integral for the arc length and numerically estimate the integral.

19. The portion of $y = x^4$ for $-1 \le x \le 1$

20. The portion of $y = x^2 + x$ for $-1 \le x \le 0$

21. The portion of $y = e^{x/2}$ for $-2 \le x \le 2$

22. The portion of $y = \sin 2x$ for $0 \le x \le \pi$

In exercises 23 and 24, set up an integral for the surface area and numerically estimate the integral.

23. The surface generated by revolving $y = 1 - x^2$, $0 \le x \le 1$, about the x-axis

24. The surface generated by revolving $y = x^3$, $0 \le x \le 1$, about the x-axis

In exercises 25–32, ignore air resistance.

25. A diver drops from a height of 64 feet. What is the velocity at impact?

26. If the diver in exercise 25 has an initial upward velocity of 4 ft/s, what will be the impact velocity?

27. An object is launched from the ground at an angle of $20°$ with an initial speed of 48 ft/s. Find the time of flight and the horizontal range.

28. Repeat exercise 27 for an object launched from a height of 6 feet.

29. A football is thrown from a height of 6 feet with initial speed 80 ft/s at an angle of $8°$. A person stands 40 yards downfield in the direction of the throw. Is it possible to catch the ball?

30. Repeat exercise 29 with a launch angle of $24°$. By trial and error, find the range of angles (rounded to the nearest degree) which produce a catchable throw.

31. Find the initial velocity needed to propel an object to a height of 128 feet. Find the object's velocity at impact.

32. A plane at an altitude of 120 ft drops supplies to a location on the ground. If the plane has a horizontal velocity of 100 ft/s, how far from the target should the supplies be released.

33. A force of 60 pounds stretches a spring 1 foot. Find the work done to stretch the spring 8 inches beyond its natural length.

34. A car engine exerts a force of $800 + 2x$ pounds when the car is at position x miles. Find the work done as the car moves from $x = 0$ to $x = 8$.

35. Compute the mass and center of mass of an object with density $\rho(x) = x^2 - 2x + 8$ for $0 \le x \le 4$. Explain why the center of mass is not at $x = 2$.

36. Compute the mass and center of mass of an object with density $\rho(x) = x^2 - 2x + 8$ for $0 \le x \le 2$. Explain why the center of mass is at $x = 1$.

37. A dam has the shape of a trapezoid with height 80 feet. The width at the top of the dam is 60 feet and the width at the bottom of the dam is 140 feet. Find the maximum hydrostatic force that the dam will need to withstand.

38. An underwater viewing window is a rectangle with width 20 feet extending from 5 feet below the surface to 10 feet below the surface. Find the maximum hydrostatic force which the window will need to withstand.

39. The force exerted by a bat on a ball over time is shown in the table. Use the data to estimate the impulse. If the ball (mass $m = 0.01$ slugs) had speed 120 ft/s before the collision, estimate its speed after the collision.

t (sec)	0	0.0001	0.0002	0.0003	0.0004
$F(t)$ (lb)	0	800	1600	2400	3000

t (sec)	0.0005	0.0006	0.0007	0.0008
$F(t)$ (lb)	3600	2200	1200	0

40. If a wall applies a force of $f(t) = 3000t(2 - t)$ pounds to a car for $0 \le t \le 2$, find the impulse. If the car (mass $m = 100$ slugs) is motionless after the collision, compute its speed before the collision.

41. Show that $f(x) = x + 2x^3$ is a pdf on the interval $[0, 1]$.

42. Show that $f(x) = \frac{8}{3}e^{-2x}$ is a pdf on the interval $[0, \ln 2]$.

43. Find the value of c such that $f(x) = \dfrac{c}{x^2}$ is a pdf on the interval $[1, 2]$.

44. Find the value of c such that $f(x) = ce^{-2x}$ is a pdf on the interval $[0, 4]$.

45. The lifetime of a lightbulb has pdf $f(x) = 4e^{-4x}$ (x in years). Find the probability that the lightbulb lasts (a) less than 6 months; (b) between 6 months and 1 year.

46. The lifetime of an organism has pdf $f(x) = 9xe^{-3x}$ (x in years). Find the probability that the organism lasts (a) less than 2 months; (b) between 3 months and 1 year.

47. Find the (a) mean and (b) median of a random variable with pdf $f(x) = x + 2x^3$ on the interval $[0, 1]$.

48. Find the (a) mean and (b) median of a random variable with pdf $f(x) = \frac{8}{3}e^{-2x}$ on the interval $[0, \ln 2]$.

49. (exploratory) As indicated in section 5.5, general formulas can be derived for many important quantities in projectile motion. For an object launched from the ground at angle θ_0 with initial speed v_0 ft/s, find the horizontal range R ft and use the trig identity $\sin(2\theta_0) = 2\sin\theta_0\cos\theta_0$ to show that $R = \dfrac{v_0^2\sin(2\theta_0)}{32}$. Conclude that the maximum range is achieved with angle $\theta_0 = 45°$.

50. (exploratory) To follow up on exercise 49, suppose that the ground makes an angle of $A°$ with the horizontal. If $A > 0$ (i.e., the projectile is being launched uphill), explain why the maximum range would be achieved with an angle larger than $45°$. If $A < 0$ (launching downhill), explain why the maximum range would be achieved with an angle less than $45°$. To determine the exact value of the optimal angle, first argue that the ground can be represented by the line $y = (\tan A)x$. Show that the projectile reaches the ground at time $t = v_0\dfrac{\sin\theta_0 - \tan A\cos\theta_0}{16}$. Compute $x(t)$ for this value of t and use a trig identity to replace the quantity $\sin\theta_0\cos A - \sin A\cos\theta_0$ with $\sin(\theta_0 - A)$. Then use another trig identity to replace $\cos\theta_0\sin(\theta_0 - A)$ with $\sin(2\theta_0 - A) - \sin A$. At this stage, the only term involving θ_0 will be $\sin(2\theta_0 - A)$. To maximize the range, maximize this term by taking $\theta_0 = \dfrac{\pi}{4} + \dfrac{1}{2}A$.

Chapter 6
Exponentials, Logarithms and Other Transcendental Functions

The International Space Station is one of the most ambitious engineering projects ever attempted. Despite the obvious challenges, a construction project in space offers a few advantages over construction on earth. For one, the weightlessness and lack of atmospheric conditions reduces the need for structural strength, which is such a fundamental constraint on earth. In fact, the International Space Station is a structure that could not support its own weight if it were constructed on earth! Nevertheless, the design of the space station has provided numerous special challenges. For instance, the beams holding the station's solar panels are long, yet relatively thin and flexible. One problem with this is that the great length and flexibility can magnify a minor tremor into a dangerous vibration, due to the phenomenon of resonance. Consequently, this design requires a system to maintain the stability of the structure. This is one area where calculus plays a critical role in the design of the space station.

International Space Station

One way to think of such a stability problem is to imagine yourself operating a joy stick, where moving the joy stick applies a force at one of the beam's joints. The goal is to apply the appropriate forces to keep the beam from vibrating. For instance, if the beam starts moving to the left, you might move your joystick to the right, applying an opposing force. If we think of this process as a mathematical function, you are supplying the input (the force) which determines the output (the motion of the beam). Your task is then to solve an **inverse problem**. That is, you know the desired output (stability) and must determine the correct input (force) that produces it. We discuss inverse functions in section 6.2.

Engineers analyzing vibration in beams begin with a **differential equation**, an equation involving an unknown function and one or more of its derivatives. A solution of a differential equation is a function; in our case, a function describing the motion of the beam. Very often, solutions of differential equations involve exponential functions of the form e^{kt}. In fact, simple cases of our vibrating beam problem can be solved in terms of such functions. If e^{kt} represents the displacement of the beam from its resting position (at a given point), notice that for $k < 0$, e^{kt} approaches 0 as t tends to ∞. This would represent a stable beam. On the other hand, if $k > 0$, e^{kt} tends to ∞ as t tends to ∞. This would be bad news for our space station. In view of this, we would then need to make certain that k is negative, in order to ensure the stability of the structure. In many situations, the value of k is very close to 0, so that a small change in k can turn a stable system into an unstable one.

In the first four sections of this chapter, we carefully develop and explore the properties and some basic applications of the natural exponential function. Of course, we have already studied exponentials in earlier chapters, but the prominent role that these functions play in engineering and science makes the extra time and effort worthwhile. Many important applications of exponential functions involve differential equations. Accordingly, in section 6.5 we present a particular type of differential equation for which we can find solutions. In section 6.6, we introduce a simple method for numerically approximating solutions of more general differential equations. In the final three sections of the chapter, we introduce several new functions that are important in calculus. The inverse trigonometric functions developed in sections 6.7 and 6.8 apply the notion of inverse functions (from section 6.2) to the trigonometric functions. The hyperbolic functions introduced in section 6.9 are related to both the exponential function and the trigonometric functions.

6.1   The Natural Logarithm

Recall that we first defined the natural logarithm in Chapter 0. There, we said that the natural logarithm is the common logarithm with base e. That is,

$$\ln x = \log_e x,$$

where e is the (so far) mysterious transcendental number, $e \approx 2.718\ldots$. So, why would this be called *natural*? A logarithm with a transcendental base would seem to be anything but natural. Further, it is as yet unclear why anyone would ever be interested in such a seemingly unusual function. We will resolve both of these issues in this section.

First, recall the power rule for integrals,

$$\int x^n \, dx = \frac{x^{n+1}}{n+1} + c, \text{ for } n \neq -1.$$

Of course, this rule doesn't hold for $n = -1$, since this would result in division by zero, which is undefined. Assume, for the moment, that we have not yet defined $\ln x$. (We are about to do this very carefully, for the first time.) Then, what can we say about

$$\int \frac{1}{x} \, dx?$$

(Remember: although we found this integral in section 4.1, our observation hinged on the (as yet) unproven conjecture that $\frac{d}{dx} \ln x = \frac{1}{x}$.) From Theorem 4.1 in Chapter 4, we know that since $f(x) = \frac{1}{x}$ is continuous for $x \neq 0$, it must be integrable on any interval not including $x = 0$. We simply don't know how to find an antiderivative. Notice, though, that by Part II of the Fundamental Theorem of Calculus,

$$\int_1^x \frac{1}{t} \, dt$$

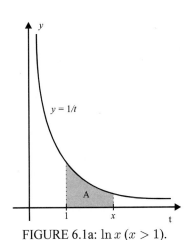

FIGURE 6.1a: $\ln x \ (x > 1)$.

is an antiderivative of $\frac{1}{x}$ for $x > 0$. We give this new (and naturally arising) function the name **natural logarithm**, written

$$\ln x = \int_1^x \frac{1}{t} \, dt, \text{ for } x > 0. \tag{1.1}$$

We'll see later in this chapter that this definition is, in fact, consistent with how we defined the function in Chapter 0. First, let's interpret this function graphically. Notice that for $x > 1$, this definite integral corresponds to the area A under the curve $y = \frac{1}{t}$ from 1 to x, as indicated in Figure 6.1a. That is,

$$\ln x = \int_1^x \frac{1}{t} \, dt = A.$$

Similarly, for $0 < x < 1$, notice from Figure 6.1b that for the area A under the curve $y = \frac{1}{t}$ from x to 1, we have

$$\ln x = \int_1^x \frac{1}{t} \, dt = -\int_x^1 \frac{1}{t} \, dt = -A.$$

Using this new definition (1.1), we get by Part II of the Fundamental Theorem of Calculus that

$$\frac{d}{dx} \ln x = \frac{d}{dx} \int_1^x \frac{1}{t} \, dt = \frac{1}{x}, \text{ for } x > 0. \tag{1.2}$$

FIGURE 6.1b: $\ln x \ (0 < x < 1)$.

Notice that (1.2) is the same derivative formula as we had conjectured in section 2.6.

Example 1.1   Differentiating a Logarithm
Find the derivative of $\ln(x^3 + 7x^2)$.

Solution   Using the chain rule, we have

$$\frac{d}{dx} \ln(x^3 + 7x^2) = \left(\frac{1}{x^3 + 7x^2} \right) \frac{d}{dx}(x^3 + 7x^2)$$

$$= \left(\frac{1}{x^3 + 7x^2} \right) (3x^2 + 14x).$$

∎

Recall that we had shown in section 4.1 that for any $x \neq 0$, we can extend (1.2) to obtain $\frac{d}{dx} \ln |x| = \frac{1}{x}$. This, in turn gives us the integration rule

$$\int \frac{1}{x} \, dx = \ln |x| + c.$$

We now remind you of some of the ways in which this integration rule arises.

Example 1.2   Several Integrals Involving Logarithms
Evaluate the integrals (a) $\displaystyle\int \frac{x^2}{x^3 + 7} \, dx$ and (b) $\displaystyle\int \tan x \, dx$.

Solution   For (a), notice that the numerator is nearly the derivative of the denominator:

$$\frac{d}{dx} \left(x^3 + 7 \right) = 3x^2.$$

This suggests the substitution $u = x^3 + 7$, so that $du = 3x^2 dx$:

$$\int \frac{x^2}{x^3 + 7} \, dx = \frac{1}{3} \int \frac{3x^2}{x^3 + 7} \, dx$$

$$= \frac{1}{3} \int \frac{1}{u} \, du = \frac{1}{3} \ln |u| + c$$

$$= \frac{1}{3} \ln \left| x^3 + 7 \right| + c.$$

For (b), we must first rewrite the integrand in terms of $\sin x$ and $\cos x$, to reveal

$$\int \tan x \, dx = \int \frac{\sin x}{\cos x} \, dx.$$

Here, you should quickly observe that setting $u = \cos x$ gives us $du = -\sin x \, dx$, so that

$$\int \tan x \, dx = - \int \frac{-\sin x}{\cos x} \, dx = - \int \frac{1}{u} \, du$$

$$= - \ln |u| + c = - \ln |\cos x| + c.$$

∎

Notice that since $\ln x$ is defined by a definite integral, we can use Simpson's rule (or any other convenient numerical integration method) to compute approximate values of the function. For instance,

$$\ln 2 = \int_1^2 \frac{1}{t} \, dt \approx 0.693147$$

and

$$\ln 3 = \int_1^3 \frac{1}{t}\, dt \approx 1.09861.$$

We leave these approximations as exercises. (You should also check the values with the ln key on your calculator.)

We now briefly sketch a graph of $y = \ln x$. As we've already observed, the domain of $f(x) = \ln x$ is $(0, \infty)$. Further, recall that

$$\ln x \begin{cases} < 0 & \text{for } 0 < x < 1 \\ = 0 & \text{for } x = 1 \\ > 0 & \text{for } x > 1 \end{cases}$$

and that

$$f'(x) = \frac{1}{x} > 0, \text{ for } x > 0,$$

so that f is increasing throughout its domain. Next,

$$f''(x) = -\frac{1}{x^2} < 0, \text{ for } x > 0,$$

and hence, the graph is concave down everywhere. You can easily use Simpson's rule or the Trapezoid rule (this is left as an exercise) to make the conjectures

$$\lim_{x \to \infty} \ln x = \infty \qquad (1.3)$$

and

$$\lim_{x \to 0^+} \ln x = -\infty. \qquad (1.4)$$

We postpone the proof of (1.3) until after Theorem 1.1. The proof of (1.4) is left as an exercise. We now obtain the graph pictured in Figure 6.2.

Now, it remains for us to explain why this function should be called a *logarithm*. The answer is simple: it satisfies all of the properties satisfied by common logarithms. Since $\ln x$ behaves like any other logarithm, we call it (what else?) a logarithm. We summarize this in the following theorem.

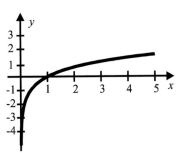

FIGURE 6.2: $y = \ln x$.

Theorem 1.1　For any real numbers $a, b > 0$ and any rational number r,

(i)　$\ln 1 = 0$

(ii)　$\ln(ab) = \ln a + \ln b$

(iii)　$\ln\left(\frac{a}{b}\right) = \ln a - \ln b$ and

(iv)　$\ln(a^r) = r \ln a$.

Proof. (i) By definition,

$$\ln 1 = \int_1^1 \frac{1}{t}\, dt = 0.$$

(ii) Also from the definition, we have

$$\ln(ab) = \int_1^{ab} \frac{1}{t}\, dt = \int_1^a \frac{1}{t}\, dt + \int_a^{ab} \frac{1}{t}\, dt,$$

from part (iv) of Theorem 4.2 in section 4.4. Make the substitution $u = \frac{t}{a}$ in the second integral only. This gives us $du = \frac{1}{a}\, dt$. Finally, the limits of integration must be changed to reflect the new variable (when $t = a$, we have $u = \frac{a}{a} = 1$ and when $t = ab$, we have

$u = \frac{ab}{a} = b$) to yield

$$\ln(ab) = \int_1^a \frac{1}{t}\,dt + \int_a^{ab} \underbrace{\frac{a}{t}}_{\frac{1}{u}}\underbrace{\frac{1}{a}\,dt}_{du}$$

From the definition (1.1).

$$= \int_1^a \frac{1}{t}\,dt + \int_1^b \frac{1}{u}\,du = \ln a + \ln b.$$

(iv) Note that

From (1.2) and the chain rule.

$$\frac{d}{dx}\ln(x^r) = \frac{1}{x^r}\frac{d}{dx}x^r$$

From the power rule.

$$= \frac{1}{x^r}\,r\,x^{r-1} = \frac{r}{x}.$$

Likewise,

$$\frac{d}{dx}[r\,\ln x] = r\,\frac{d}{dx}\,(\ln x) = \frac{r}{x}.$$

Now, since $\ln(x^r)$ and $r\ln x$ have the same derivative, it follows from Corollary 9.1 in section 2.9 that for all $x > 0$,

$$\ln(x^r) = r\,\ln x + k,$$

for some constant, k. In particular, taking $x = 1$, we find that

$$\ln(1^r) = r\,\ln 1 + k,$$

where since $1^r = 1$ and $\ln 1 = 0$, we have

$$0 = r\,(0) + k.$$

So, $k = 0$ and $\ln(x^r) = r\,\ln x$, for all $x > 0$.
(iii) follows from (ii) and (iv) and is left as an exercise.

∎

Example 1.3   Rewriting a Logarithmic Expression
Simplify $\ln\frac{1}{\sqrt[n]{a}}$, where n is a positive integer.

Solution   For any integer $n > 0$,

$$\ln\frac{1}{\sqrt[n]{a}} = \ln\left(a^{-\frac{1}{n}}\right) = -\left(\frac{1}{n}\right)\ln a.$$

∎

CAUTION:

$$\ln\frac{1}{\sqrt[n]{a}} \neq (\ln a)^{-\frac{1}{n}}.$$

(What is wrong with this, anyway?)

Using the properties of logarithms will often simplify the calculation of certain derivatives. This is an example of what is more generally referred to as **logarithmic differentiation**. We illustrate this in the following example.

Example 1.4   Using Logarithmic Differentiation
Find the derivative of $\ln\sqrt{\dfrac{(x-2)^3}{x^2+5}}$.

Solution   Rather than directly differentiating this expression, by applying the chain rule and the quotient rule and the like, notice that we can considerably simplify our work by using the properties of logarithms. We have

$$\frac{d}{dx} \ln \sqrt{\frac{(x-2)^3}{x^2+5}} = \frac{d}{dx} \ln \left[\frac{(x-2)^3}{x^2+5} \right]^{\frac{1}{2}}$$

From Theorem 1.1 (iv).

$$= \frac{1}{2} \frac{d}{dx} \ln \left[\frac{(x-2)^3}{x^2+5} \right]$$

From Theorem 1.1 (iii).

$$= \frac{1}{2} \frac{d}{dx} \left[\ln(x-2)^3 - \ln(x^2+5) \right]$$

From Theorem 1.1 (iv).

$$= \frac{1}{2} \frac{d}{dx} \left[3 \ln(x-2) - \ln(x^2+5) \right]$$

From (1.1) and the chain rule.

$$= \frac{1}{2} \left[3 \left(\frac{1}{x-2} \right) \frac{d}{dx}(x-2) - \left(\frac{1}{x^2+5} \right) \frac{d}{dx}(x^2+5) \right]$$

$$= \frac{1}{2} \left(\frac{3}{x-2} - \frac{2x}{x^2+5} \right).$$

Of course, you could simply compute the derivative directly using the original expression. Try this for yourself. It's not pretty (there are multiple chain rules and a quotient rule; don't say we didn't warn you), but it's entirely equivalent. The bottom line is that the rules of logarithms can save you untold complication, so use them where appropriate.

■

Example 1.5   **Examining the Limiting Behavior of** $\ln x$

Use the properties of logarithms in Theorem 1.1 to prove that

$$\lim_{x \to \infty} \ln x = \infty.$$

Solution   We can verify this as follows. First, recall that $\ln 3 \approx 1.0986 > 1$. Further, by the rules of logarithms, we have that for any integer n

$$\ln 3^n = n \ln 3.$$

So,

$$\lim_{n \to \infty} \ln 3^n = \lim_{n \to \infty} (n \ln 3) = +\infty.$$

Finally, since $\ln x$ is a strictly increasing function, the conjecture follows.

■

Exercises 6.1

1. (writing) Explain why it is (mathematically) legal to define $\ln x$ as $\int_1^x \frac{1}{t} dt$. For some, this type of definition is not very satisfying. Even though this is probably the first function you have seen defined this way, try the following comparison. Clearly, it is easier to compute function values for x^2 than for $\ln x$, and therefore x^2 is easier to understand. However, compare how you would compute (without a calculator) function values for $\sin x$ versus function values for $\ln x$. Describe which is more "natural" and easy to understand.

2. (writing) In this section, we gave two different "definitions" of $\ln x$. Explain why it is logically invalid to give different definitions unless you can show that they define the same thing. If they define the same object, either definition is equally valid and you should use whichever definition is clearer for the task at hand. Explain why, in this section, the integral definition is more convenient than the base e logarithm.

3. (writing) Use the integral definition of $\ln x$ (interpreted as area) to explain why it is reasonable that $\lim\limits_{x \to 0^+} \ln x = -\infty$ and $\lim\limits_{x \to \infty} \ln x = \infty$.

4. (writing) Use the integral definition of $\ln x$ (interpreted as area) to explain why the graph of $\ln x$ is concave down.

In exercises 5–8, approximate the indicated value using Simpson's Rule or some other numerical method.

5. $\ln 4$ **6.** $\ln 5$ **7.** $\ln 8.2$ **8.** $\ln 24$

In exercises 9–16, find the derivative of the function.

9. $\ln 4x$

10. $\ln(x^2)$

11. $\ln(\cos x)$

12. $\ln(\sec x)$

13. $x \ln x$

14. $\frac{\ln x}{x}$

15. $\ln(x^2 + 2)$

16. $\ln(\sec x + \tan x)$

In exercises 17–20, use the properties of logarithms to rewrite the expression as a single term.

17. $\ln \sqrt{2} + 3\ln 2$

18. $\ln 8 - 2\ln 2$

19. $2\ln 3 - \ln 9 + \ln \sqrt{3}$

20. $2\ln(\frac{1}{3}) - \ln 3 + \ln(\frac{1}{9})$

In exercises 21–34, evaluate the integral.

21. $\displaystyle\int \frac{2x}{x^2 + 1}\,dx$

22. $\displaystyle\int \frac{3x^3}{x^4 + 5}\,dx$

23. $\displaystyle\int \frac{\sin x}{\cos x}\,dx$

24. $\displaystyle\int \frac{\cos x}{\sin x}\,dx$

25. $\displaystyle\int \frac{x + 1}{x^2 + 2x - 1}\,dx$

26. $\displaystyle\int \frac{x^2 - 1}{x^3 - 3x + 5}\,dx$

27. $\displaystyle\int \frac{1}{x \ln x}\,dx$

28. $\displaystyle\int \frac{1}{\sqrt{x}(\sqrt{x} + 1)}\,dx$

29. $\displaystyle\int \frac{(\ln x + 1)^2}{x}\,dx$

30. $\displaystyle\int \frac{\cos(\ln x)}{x}\,dx$

31. $\displaystyle\int_0^2 \frac{1}{x + 1}\,dx$

32. $\displaystyle\int_0^1 \frac{x^2}{x^3 + 1}\,dx$

33. $\displaystyle\int_0^1 \frac{x^2}{x^3 - 4}\,dx$

34. $\displaystyle\int_1^2 \frac{\ln x}{x}\,dx$

In exercises 35–40, evaluate the derivative or integral using properties of logarithms where needed.

35. $\frac{d}{dx}\left(\ln \sqrt{x^2 + 1}\right)$

36. $\frac{d}{dx}[\ln(x^5 \sin x \cos x)]$

37. $\frac{d}{dx}\left(\ln \frac{x^4}{x^5 + 1}\right)$

38. $\frac{d}{dx}\left(\ln \sqrt{\frac{x^3}{x^5 + 1}}\right)$

39. $\displaystyle\int \frac{1}{x \ln x^2}\,dx$

40. $\displaystyle\int \frac{\sin(\ln x^3)}{x}\,dx$

In exercises 41–46, graph the function.

41. $y = \ln(x - 2)$

42. $y = \ln(3x + 5)$

43. $y = \ln(x^2 + 1)$

44. $y = \ln(x^3 + 1)$

45. $y = x \ln x$

46. $y = x^2 \ln x$

47. Use property (ii) of Theorem 1.1 to prove property (iii) that $\ln\left(\frac{a}{b}\right) = \ln a - \ln b$.

48. Starting with $e^x = \lim\limits_{n \to \infty} \left(1 + \frac{x}{n}\right)^n$, show that $\ln x = \lim\limits_{n \to \infty} \left[n(x^{1/n} - 1)\right]$.

49. A telegraph cable is made of an outer winding around an inner core. If x is defined as the core radius divided by the outer radius, the transmission speed is proportional to $s(x) = x^2 \ln(1/x)$. Find an x that maximizes the transmission speed.

50. (exploratory) A special function used in many applications is the **error function**, defined by $\text{erf}(x) = \frac{2}{\sqrt{\pi}} f(x)$, where $f(x) = \displaystyle\int_0^x e^{-u^2}\,du$. Explore the function $f(x)$. For which x's is $f(x)$ positive? negative? increasing? decreasing? concave up? concave down? Estimate some function values for large x. Conjecture $\lim\limits_{x \to \infty} f(x)$. Sketch a graph of $f(x)$.

51. (exploratory) Verify that

$$\int \sec x\,dx = \ln|\sec x + \tan x| + c.$$

(Hint: Differentiate the suspected antiderivative and show that you get the integrand.) This integral appears in the construction of a special type of map called a Mercator map. On this map, the latitude lines are not equally spaced. Instead, they are placed so that straight lines on a Mercator map correspond to paths of constant heading. (If you travel due northeast, your path on a map with equally spaced latitude will appear to curve due to the curvature of the earth.) Let R be the (average) radius of the earth. Assuming the earth is a sphere, the actual distance from the equator to a place at latitude $b°$ is $\frac{\pi}{180} Rb$. On a Mercator map, this distance is scaled to $\frac{\pi}{180} R \displaystyle\int_0^b \sec x\,dx$. Tampa, Florida, has latitude $28°$ north. Moscow, Russia, is twice as far from the equator at $56°$ north. What is the spacing for Tampa and Moscow on a Mercator map?

6.2   Inverse Functions

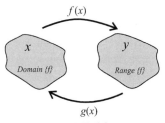

FIGURE 6.3:
$g(x) = f^{-1}(x)$.

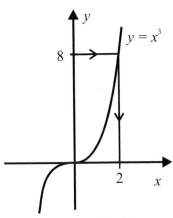

FIGURE 6.4: Finding
the x-value corresponding
to $y = 8$.

The notion of an *inverse* relationship is basic to many areas of science, although the term is only infrequently used. The number of common inverse problems is immense. As only one example, take the case of the electrocardiogram (EKG). In an EKG, technicians connect a series of electrodes to a patient's chest and use measurements of electrical activity on the surface of the body to infer something about the electrical activity on the surface of the heart. This is referred to as an *inverse* problem, since physicians are attempting to determine what *inputs* (i.e., the electrical activity on the surface of the heart) cause an observed *output* (the measured electrical activity on the surface of the chest).

The mathematical notion of inverse is much the same as that described above. Given an output (i.e., a value in the range of a given function), we wish to find the input (the value in the domain) that produced the observed output. That is, given a $y \in Range\{f\}$, find the $x \in Domain\{f\}$ for which $y = f(x)$. (See the illustration of the inverse function $g(x)$ shown in Figure 6.3.)

For instance, suppose that $f(x) = x^3$ and $y = 8$. Can you find an x such that $x^3 = 8$? That is, can you find the x-value corresponding to $y = 8$ (see Figure 6.4). Of course, you know the solution of this particular equation: $x = \sqrt[3]{8} = 2$. In fact, in general, if $x^3 = y$, then $x = \sqrt[3]{y}$. In light of this, we say that the cube root function is the **inverse** of $f(x) = x^3$.

Example 2.1   Two Functions That Reverse the Action of Each Other
If $f(x) = x^3$ and $g(x) = x^{1/3}$, show that

$$f(g(x)) = x \text{ and } g(f(x)) = x,$$

for all x.

Solution   For all real numbers x, we have

$$f(g(x)) = f(x^{1/3}) = (x^{1/3})^3 = x$$

and

$$g(f(x)) = g(x^3) = (x^3)^{1/3} = x.$$

$\blacksquare$

Notice in example 2.1 that the action of f undoes the action of g and vice versa. We may take this as the definition of an inverse function. More precisely, we have the following definition.

Definition 2.1   Assume that f and g have domains A and B, respectively, and that $f(g(x))$ is defined for all $x \in B$ and $g(f(x))$ is defined for all $x \in A$. If

$$f(g(x)) = x, \text{ for all } x \in B \text{ and}$$

$$g(f(x)) = x, \text{ for all } x \in A,$$

we say that g is the **inverse** of f, written $g = f^{-1}$. Equivalently, f is the inverse of g, $f = g^{-1}$.

CAUTION: Pay close attention to the notation. Notice that $f^{-1}(x)$ does **not** mean $\dfrac{1}{f(x)}$.

We can, however write

$$[f(x)]^{-1} = \frac{1}{f(x)}.$$

Observe that many familiar functions have no inverse.

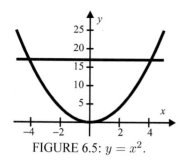

FIGURE 6.5: $y = x^2$.

Example 2.2   A Function with No Inverse

Show that $f(x) = x^2$ has no inverse on the interval $(-\infty, \infty)$.

Solution   Notice that $f(2) = 4$ and $f(-2) = 4$. That is, there are **two** x values that produce the same y value. So, if we were to try to define an inverse of f, how would we define $f^{-1}(4)$? Look at the graph of $y = x^2$ (see Figure 6.5) to see what the problem is. For each $y > 0$, there are **two** x-values for which $y = x^2$. Such functions do not have an inverse.

■

Remark 2.1   For $f(x) = x^2$, it is tempting to jump to the conclusion that $g(x) = \sqrt{x}$ is the inverse of $f(x)$. Notice that although $f(g(x)) = (\sqrt{x})^2 = x$ for all $x \geq 0$ (i.e., for all x in the domain of $g(x)$), it is **not** generally true that $g(f(x)) = \sqrt{x^2} = x$. In fact, this last equality holds **only** for $x \geq 0$. However, for $f(x) = x^2$ restricted to the domain $x \geq 0$, we do have that $f^{-1}(x) = \sqrt{x}$.

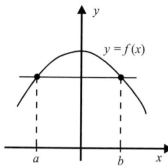

FIGURE 6.6a: $f(a) = f(b)$, for $a \neq b$. So, f does not pass the horizontal line test and is not one-to-one.

Definition 2.2   A function f is called **one-to-one** when for every $y \in Range\{f\}$, there is **exactly one** $x \in Domain\{f\}$ for which $y = f(x)$.

Remark 2.2   Observe that an equivalent definition of one-to-one is the following. A function $f(x)$ is one-to-one if and only if the equality $f(a) = f(b)$ implies $a = b$. This version of the definition is often useful for proofs involving one-to-one functions.

It is also most helpful to think of the concept of one-to-one in graphical terms. Notice that a function f is one-to-one if and only if every horizontal line intersects the graph in at most one point. This is usually referred to as the **horizontal line test**. We illustrate this in Figures 6.6a and 6.6b. One can then prove the following result.

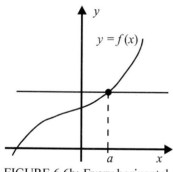

FIGURE 6.6b: Every horizontal line intersects the curve in at most one point. So, f passes the horizontal line test and is one-to-one.

Theorem 2.1   A function f has an inverse if and only if it is one-to-one.

This simply says that one-to-one functions have an inverse, but says nothing about how to find them. For very simple functions, we can find inverses by solving equations.

Example 2.3   Finding an Inverse Function

Find the inverse of $f(x) = x^3 - 5$.

Solution   You will show in the exercises that f is one-to-one and therefore has an inverse. Note that it is fairly unclear from the graph (see Figure 6.7) whether or not f passes the horizontal line test. To find the inverse function, write $y = f(x)$ and solve for x (i.e., solve for the input x which produced the observed output y). We have

$$y = x^3 - 5.$$

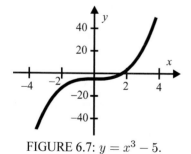

FIGURE 6.7: $y = x^3 - 5$.

Adding 5 to both sides and taking the cube root gives us

$$(y + 5)^{1/3} = (x^3)^{1/3} = x.$$

So, we have that $x = f^{-1}(y) = (y+5)^{1/3}$. Reversing the variables x and y (think about why this makes sense), we have

$$f^{-1}(x) = (x+5)^{1/3}.$$

∎

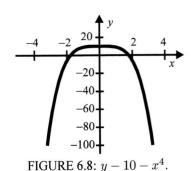

FIGURE 6.8: $y - 10 - x^4$.

Example 2.4   A Function That Is Not One-to-One
Show that $f(x) = 10 - x^4$ does not have an inverse.

Solution   You can see from a graph (see Figure 6.8) that f is not one-to-one and consequently, does not have an inverse.

∎

Remark 2.3   Most often, we cannot find a formula for an inverse function and must be satisfied with simply knowing that the inverse function exists. The following example is typical of this situation.

Example 2.5   Showing That a Function Has an Inverse
Show that $f(x) = x^5 + 7x^3 + 4x + 1$ has an inverse. Also, find $f^{-1}(1)$ and $f^{-1}(13)$.

Solution   From the graph shown in Figure 6.9, the function certainly looks one-to-one, but how can we be certain of this? (Remember that graphs can be deceptive!) Observe that if a function is strictly increasing or strictly decreasing on its entire domain, then it is one-to-one and hence, has an inverse. In this case, we have that

$$f'(x) = 5x^4 + 21x^2 + 4 > 0, \text{ for all } x.$$

So, f is strictly increasing and consequently, has an inverse. It would certainly be ideal to find the inverse function algebraically. In this case, that means solving the equation

$$y = x^5 + 7x^3 + 4x + 1$$

for x. Consider this for a few moments; you should quickly realize that there is no way of solving for x in terms of y here. That is, there is no way to find a formula for the inverse function. Still, since we have shown that f is one-to-one, we know that it has an inverse.

Turning to the problem of finding $f^{-1}(1)$ and $f^{-1}(13)$, you might wonder if this is possible, since we were unable to find a formula for $f^{-1}(x)$. While it's certainly true that we have no such formula, you might observe that $f(0) = 1$, so that $f^{-1}(1) = 0$. By trial and error, you might also discover that $f(1) = 13$ and so, $f^{-1}(13) = 1$.

∎

FIGURE 6.9:
$y = x^5 + 7x^3 + 4x + 1$.

In example 2.5, we saw a function that we knew had an inverse, although we could not find that inverse algebraically. Even when we can't find an inverse function explicitly, we can say something graphically. Notice that if (a, b) is a point on the graph of $y = f(x)$ and f has an inverse, f^{-1}, then since

$$b = f(a),$$

we have

$$f^{-1}(b) = f^{-1}(f(a)) = a.$$

That is, (b, a) is a point on the graph of $y = f^{-1}(x)$. This tells us a great deal about the inverse function. In particular, we can immediately obtain any number of points on the graph of $y = f^{-1}(x)$, simply by inspection. Further, notice that the point (b, a) is the reflection of the point (a, b) through the line $y = x$ (see Figure 6.10). It now follows that

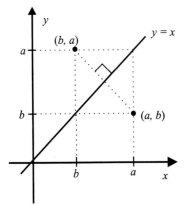

FIGURE 6.10:
Reflection through $y = x$.

given the graph of any one-to-one function, you can draw the graph of its inverse simply by reflecting the entire graph through the line $y = x$. One consequence of this symmetry is the following result.

Theorem 2.2   Suppose that f is a one-to-one continuous function. Then, f^{-1} is also continuous.

In the following example, we illustrate the symmetry of a function and its inverse.

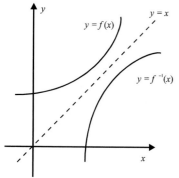

FIGURE 6.11: $y = x^3$
and $y = x^{1/3}$.

Example 2.6   **The Graph of a Function and its Inverse**
Draw a graph of $f(x) = x^3$ and its inverse.

Solution   From example 2.1, the inverse of $f(x) = x^3$ is $f^{-1}(x) = x^{1/3}$. Notice the symmetry of their graphs shown in Figure 6.11.   ∎

Observe that we can use this symmetry principle to draw the graph of an inverse function, even when we don't have a formula for that function (see Figure 6.12).

Example 2.7   **Drawing the Graph of an Unknown Inverse Function**
Draw a graph of $f(x) = x^5 + 7x^3 + 4x + 1$ and its inverse.

Solution   In example 2.5, we showed that f was strictly increasing and hence, was one-to-one, but we were unable to find a formula for the inverse function. Despite this, we can draw a graph of f^{-1} with ease. One way to do this would be to plot a few points on the graph of $y = f^{-1}(x)$ by hand, but we suggest that you use the parametric plotting feature of your graphing utility. To write down parametric equations for the curve $y = f(x)$, we introduce the parameter t and observe that

$$x = t \quad \text{and} \quad y = f(t) \tag{2.1}$$

FIGURE 6.12: Graph of f
and f^{-1}.

are equivalent parametric equations. Notice that parametric equations for $y = f^{-1}(x)$ are then simply

$$x = f(t) \quad \text{and} \quad y = t. \tag{2.2}$$

We used the two pairs of parametric equations (2.1) and (2.2) to produce the graphs of $y = f(x)$ and $y = f^{-1}(x)$ shown in Figure 6.13. We also added a dashed line for the line $y = x$, entered parametrically as

$$x = t \quad \text{and} \quad y = t.$$

∎

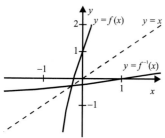

FIGURE 6.13: Graph of f
and f^{-1}.

We make one final observation regarding inverse functions. Suppose that f is a one-to-one and differentiable function. Then, as a consequence of the symmetry of the function and its inverse, notice that, as long as $f'\left(f^{-1}(a)\right) \neq 0$ (i.e., the tangent line is not horizontal), then $f^{-1}(x)$ should be differentiable at $x = a$. We develop this notion carefully in the following theorem.

Theorem 2.3   Suppose that f is a one-to-one and differentiable function. Then, as long as $f'(f^{-1}(x)) \neq 0$,

$$\frac{d}{dx} f^{-1}(x) = \frac{1}{f'(f^{-1}(x))}.$$

Proof. Let $g(x) = f^{-1}(x)$. Then for any fixed $x = a$, we have from the alternative definition of derivative that

$$g'(a) = \lim_{x \to a} \frac{g(x) - g(a)}{x - a}. \tag{2.3}$$

Since $g = f^{-1}$, we have that $y = g(x)$ if and only if $f(y) = x$ and $b = g(a)$ if and only if $f(b) = a$. Further, since f is differentiable, it is continuous and so, from Theorem 2.2, g must be continuous, also. In particular, this says that as $x \to a$, we must also have $g(x) \to g(a)$, so that $y \to b$. From (2.3), we now have

$$g'(a) = \lim_{x \to a} \frac{g(x) - g(a)}{x - a} = \lim_{y \to b} \frac{y - b}{f(y) - f(b)}$$

$$= \lim_{y \to b} \frac{1}{\frac{f(y)-f(b)}{y-b}} = \frac{1}{\lim_{y \to b} \frac{f(y)-f(b)}{y-b}} = \frac{1}{f'(b)}$$

$$= \frac{1}{f'(f^{-1}(a))},$$

as desired, since the limit in the denominator is nonzero. ■

As an alternative to the proof given above, if we know that an inverse function is differentiable, we can find its derivative using implicit differentiation, as follows. If $y = f^{-1}(x)$, then

$$f(y) = x. \tag{2.4}$$

Differentiating both sides of (2.4) with respect to x, we find that

$$\frac{d}{dx} f(y) = \frac{d}{dx} x,$$

so that

$$f'(y) \frac{dy}{dx} = 1,$$

from the chain rule. (Notice that in this last step, we needed to assume that $\frac{dy}{dx}$ exists.) Thus, so long as $f'(y) \neq 0$, we have

$$\frac{dy}{dx} = \frac{1}{f'(y)}$$

or

$$\frac{d}{dx} f^{-1}(x) = \frac{1}{f'(y)},$$

as we had determined above.

Example 2.8   Finding a Tangent Line to the Graph of an Inverse Function

Find the equation of the tangent line to the graph of $y = f^{-1}(x)$ at $x = 3$, where $f(x) = x^3 - 5$.

Solution   First, note that $f'(x) = 3x^2$ and $f(2) = 3$, so that $f^{-1}(3) = 2$. From Theorem 2.3, we have

$$\frac{d}{dx} f^{-1}(x) \Big|_{x=3} = \frac{1}{f'(f^{-1}(3))}$$

$$= \frac{1}{f'(2)} = \frac{1}{12}.$$

So, the tangent line has slope $\frac{1}{12}$ and passes through the point with coordinates $x = 3$ and $y = f^{-1}(3) = 2$. An equation of the tangent line is then

$$y = \frac{1}{12}(x - 3) + 2.$$

Recall that for the function in example 2.8, we had found (in example 2.3) a formula for the inverse function,

$$f^{-1}(x) = (x + 5)^{1/3}.$$

You should differentiate this directly and verify that the value of the derivative at $x = 3$ is the same either way we compute it. Of course, the primary value of Theorem 2.3 is for the most common case where we cannot find a formula for the inverse function.

In the following example, we apply our theoretical knowledge of inverse functions in a medical setting.

FIGURE 6.14a: Hormone data.

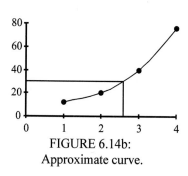

FIGURE 6.14b:
Approximate curve.

Example 2.9   Determining the Proper Dosage of a Drug

Suppose that the injection of a certain drug raises the level of a key hormone in the body. Physicians want to determine the dosage that produces a healthy hormone level. Dosages of 1, 2, 3 and 4 mg produce hormone levels of 12, 20, 40 and 76, respectively. If the desired hormone level is 30, what is the proper dosage?

Solution   A plot of the points $(1, 12)$, $(2, 20)$, $(3, 40)$ and $(4, 76)$ summarizes the data (see Figure 6.14a). The problem is an inverse problem: given $y = 30$, what is x? It is tempting to argue the following: since 30 is halfway between 20 and 40, the x-value should be halfway between 2 and 3: $x = 2.5$. This method of solution is called **linear interpolation**, since the point $x = 2.5$, $y = 30$ lies on the line through the points $(2, 20)$ and $(3, 40)$. While this provides a primitive estimate of the needed dosage, we should be able to do better. After all, this estimate does not take into account all of the information we have. (Think about what's missing here.) The points in Figure 6.14a suggest a curve that is concave up. If this is the case, $x = 2.6$ or $x = 2.7$ may be a better estimate of the required dosage. In Figure 6.14b, we have sketched a smooth curve through the data points and indicated a graphical solution of the problem. There are numerous more advanced techniques (e.g., polynomial interpolation) developed by mathematicians to make the estimate of such quantities as accurate as possible.

Exercises 6.2

1. (writing) Explain in words (and a picture) why the following is true: if $f(x)$ is differentiable and $f'(x) > 0$ for all x, then f has an inverse.

2. (writing) Suppose the graph of a function passes the horizontal line test. Explain why you know that the function has an inverse.

3. (writing) Radar works by bouncing a high-frequency electromagnetic pulse off of a moving object, then measuring the disturbance in the pulse as it is bounced back. Explain why this is an inverse problem by identifying the input and output.

4. (writing) Each human disease has a set of symptoms associated with it. Physicians attempt to solve an inverse problem: given the symptoms, they try to identify the disease causing the symptoms. Explain why this is not a well-defined inverse problem (i.e., logically it is not always possible to correctly identify diseases from symptoms alone).

In exercises 5–16, determine whether or not the function is one-to-one.

5. $f(x) = x^3 - 2$

6. $f(x) = x^2 - 3$

7. $f(x) = x^3 - 2x - 1$

8. $f(x) = x^3 + 2x - 1$

9. $f(x) = \sin x$

10. $f(x) = \cos x$

11. $f(x) = \ln x$

12. $f(x) = \ln x^2$

13. $f(x) = x^5 + 2x^3 - 2$

14. $f(x) = x^5 - 2x^3 - 2$

15. $f(x) = \sqrt{x^3 + 2x}$ **16.** $f(x) = \sqrt{x^2 + 4}$

In exercises 17–24, determine whether or not the function is one-to-one. If it is, find the inverse and graph both the function and its inverse.

17. $f(x) = x^3 - 2$ **18.** $f(x) = x^3 + 4$

19. $f(x) = x^5 - 1$ **20.** $f(x) = x^5 + 4$

21. $f(x) = x^4 + 2$ **22.** $f(x) = x^4 - 2x - 1$

23. $f(x) = \sqrt{x^3 + 1}$ **24.** $f(x) = \sqrt{x^2 + 1}$

In exercises 25–30, the function has an inverse. Without solving for the inverse, find the derivative of the inverse function at $x = a$ and graph the inverse function.

25. $f(x) = x^3 + 4x - 1, a = -1$

26. $f(x) = x^3 + 2x + 1, a = 1$

27. $f(x) = x^5 + 3x^3 + x, a = 5$

28. $f(x) = x^5 + 4x - 2, a = 3$

29. $f(x) = \sqrt{x^3 + 2x + 1}, a = 2$

30. $f(x) = \sqrt{x^5 + 4x^3 + 3x + 1}, a = 3$

31. Find an equation of the tangent line for exercise 25.

32. Find an equation of the tangent line for exercise 26.

In exercises 33–36, use linear interpolation as in example 2.9 to estimate $f^{-1}(b)$. Use the concavity to conjecture whether the estimate is too high or too low.

33. $(1, 12), (2, 20), (3, 26), (4, 30), b = 23$

34. $(1, 12), (2, 10), (3, 6), (4, 0), b = 8$

35. $(1, 12), (2, 6), (3, 2), (4, 0), b = 5$

36. $(1, 12), (2, 20), (3, 36), (4, 50), b = 32$

In exercises 37–42, discuss whether or not the function described has an inverse.

37. The income of a company varies with time.

38. The height of a person varies with time.

39. For a dropped ball, its height varies with time.

40. For a ball thrown upward, its height varies with time.

41. The shadow made by an object depends on its three-dimensional shape.

42. The number of calories burned depends on how fast a person runs.

43. Find all values of k such that $f(x) = x^3 + kx + 1$ is one-to-one.

44. Find all values of k such that $f(x) = x^3 + 2x^2 + kx - 1$ is one-to-one.

45. Suppose that your boss informs you that you have been awarded a 10% raise. The next week, your boss announces that due to circumstances beyond her control, all employees will have their salaries cut by 10%. Show that adding 10% and subtracting 10% are not inverses. Find the inverse for adding ten percent. (Hint: To add 10% to a quantity you can multiply by 1.10.)

46. Suppose that the graph of profit versus time of a rival company is increasing and concave up. Explain why you would prefer to look at the graph of the inverse function.

47. (exploratory) The idea of an inverse can be extended from functions to operators. That is, suppose there is an operator D such that if f is a function of a certain type, Df is a another function of the same type. The operator D would have an inverse D^{-1} if $D^{-1}Df = DD^{-1}f = f$ for all functions f of the right type. Define $Df = f'(x)$ for all functions for which derivatives of all orders exist. Show that the operator I defined by $If = \int_0^x f(t)\, dt$ is **not** the inverse operator of D. However, if you only consider functions for which $f(0) = 0$, the two operators **are** inverses of each other.

6.3   The Exponential Function

We now take the time to revisit the natural exponential function, e^x. As we did with the natural logarithm in section 6.1, we will now carefully define this function and develop its properties. First, recall that in Chapter 0, we gave the (usual) mysterious description of e as an irrational number $e = 2.71828\ldots$, without attempting to explain why this number is significant. We then proceeded to define $\ln x$ as $\log_e x$, the logarithm with base e. Now that we have carefully defined $\ln x$ (independent of the definition of e), we can clearly define e, as well as calculate its approximate value. We define e to be that number for which

$$\ln e = 1.$$

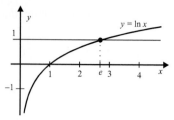

FIGURE 6.15: Definition of e.

That is, e is the x-coordinate of the point of intersection of the graphs of $y = \ln x$ and $y = 1$ (see Figure 6.15). In other words, e is the solution of the equation

$$\ln x - 1 = 0.$$

You can solve this approximately (e.g., using Newton's method) to obtain

$$e \approx 2.71828182846.$$

So, having defined the irrational number e, you might wonder what the big deal is with defining e^x? Of course, there's no problem at all, when x is rational. For instance, we have

$$e^2 = e \cdot e$$
$$e^3 = e \cdot e \cdot e$$
$$e^{\frac{1}{2}} = \sqrt{e}$$
$$e^{\frac{5}{7}} = \sqrt[7]{e^5},$$

and so on. In fact, for any rational power, $x = \frac{p}{q}$ (where p and q are integers), we have

$$e^x = e^{\frac{p}{q}} = \sqrt[q]{e^p}.$$

But, if x is irrational, what do we mean by e^x? What could it mean to raise a number to an **irrational** power? For instance, could you give any meaning to e^π? Make no mistake about it, this is a serious issue. You might recall that in Chapter 0, we gave a necessarily vague and incomplete answer to this important question. It's now time to give a complete and meaningful answer.

First, observe that for $f(x) = \ln x$ $(x > 0)$, $f'(x) = \frac{1}{x} > 0$. So, f is a strictly increasing function and consequently must be one-to-one and therefore, has an inverse, f^{-1}. As is often the case, there is no algebraic method of solving for the inverse function. However, from Theorem 1.1 (iv), we have that for any rational power x,

$$\ln(e^x) = x \ln e = x,$$

since we had defined e so that $\ln e = 1$. Observe that this says that

$$f^{-1}(x) = e^x, \text{ for } x \text{ rational.}$$

That is, the (otherwise unknown) inverse function, $f^{-1}(x)$ agrees with e^x at every rational number x. Since e^x so far has no meaning when x is irrational, we now **define** it to be the value of $f^{-1}(x)$. That is, for x irrational, we define

$$y = e^x \text{ if and only if } \ln y = \ln(e^x) = x. \tag{3.1}$$

In other words, for any irrational exponent x, we define e^x to be that real number for which $\ln(e^x) = x$. According to this definition, notice that for any $x > 0$, $e^{\ln x}$ is that real number for which

$$\ln(e^{\ln x}) = \ln x. \tag{3.2}$$

Since $\ln x$ is a one-to-one function, (3.2) says that

$$e^{\ln x} = x. \tag{3.3}$$

Notice that (3.3) says that

$$\ln x = \log_e x.$$

That is, the integral definition of $\ln x$, given in section 6.1 is consistent with our earlier definition of $\ln x$ as $\log_e x$, given in Chapter 0. Observe also that with this definition of the exponential, we have

$$\ln(e^x) = x, \text{ for all } x \in (-\infty, \infty)$$

and

$$e^{\ln x} = x, \text{ for all } x > 0.$$

Thus, we have that e^x and $\ln x$ are inverse functions. Keep in mind that for x irrational, e^x is defined **only** through the inverse function relationship given in (3.1). We now state some familiar laws of exponents and prove that they hold even for the case of irrational exponents.

Theorem 3.1   For r, s any real numbers and t any rational number,

    (i) $e^r e^s = e^{r+s}$

    (ii) $\dfrac{e^r}{e^s} = e^{r-s}$ and

    (iii) $\left(e^r\right)^t = e^{rt}$.

Proof. These laws are all obvious when the exponents are rational. If the exponent is irrational though, we only know the value of these exponentials indirectly, through the inverse function relationship with $\ln x$, given in (3.1)

(i) Note that using the rules of logarithms,

$$\ln(e^r e^s) = \ln(e^r) + \ln(e^s) = r + s = \ln(e^{r+s}).$$

Since $\ln x$ is one-to-one, it must follow that

$$e^r e^s = e^{r+s}.$$

The proofs of (ii) and (iii) are similar and are left as exercises.

∎

Derivative of the Exponential

In Chapter 2, we found the derivative of e^x using the limit definition of derivative. You may recall that the derivation was complete, except for the evaluation of the limit

$$\lim_{h \to 0} \frac{e^h - 1}{h}.$$

At the time, we conjectured that the value of this limit was 1, but we were unable to prove it, with the tools at our disposal. We revisit this limit as an exercise, at the end of this section. We now present an alternative derivation, based on our new, refined definition of the exponential function. Again, from (3.1), we have that

$$y = e^x \text{ if and only if } \ln y = x.$$

Differentiating this last equation with respect to x gives us

$$\frac{d}{dx}\ln y = \frac{d}{dx}x = 1.$$

From the chain rule, we then have

$$1 = \frac{d}{dx}\ln y = \frac{1}{y}\frac{dy}{dx}. \tag{3.4}$$

Multiplying both sides of (3.4) by y, we have

$$\frac{dy}{dx} = y = e^x$$

or

$$\frac{d}{dx}\left(e^x\right) = e^x. \tag{3.5}$$

Note that (3.5) is the same derivative formula as that conjectured in Chapter 2, but we have now carefully verified it. Of course, this also verifies the corresponding integration rule

$$\int e^x \, dx = e^x + c.$$

Although we have been computing derivatives (and antiderivatives) involving exponentials for some time now, we give several short examples as a reminder.

Example 3.1   Computing the Derivative of Some Exponentials
Compute the derivatives of the functions (a) e^{3x} and (b) $e^{\cos x}$.

Solution   For (a), we have

$$\frac{d}{dx}e^{3x} = e^{3x}\underbrace{\frac{d}{dx}(3x)}_{\text{chain rule}} = e^{3x}\,(3) = 3e^{3x}.$$

Similarly, for (b), we get

$$\frac{d}{dx}e^{\cos x} = e^{\cos x}\underbrace{\frac{d}{dx}(\cos x)}_{\text{chain rule}} = e^{\cos x}\,(-\sin x) = -\sin x\, e^{\cos x}.$$

∎

Example 3.2   Evaluating the Integral of Some Exponentials
Evaluate the integrals (a) $\int e^{-5x}\,dx$ and $\int x^3 e^{x^4}\,dx$.

Solution   We can resolve these with simple u-substitutions. For (a), we have

$$\int e^{-5x}\,dx = -\frac{1}{5}\int \underbrace{e^{-5x}}_{e^u}\underbrace{(-5)\,dx}_{du} = -\frac{1}{5}e^{-5x} + c.$$

For (b), we get

$$\int x^3 e^{x^4}\,dx = \frac{1}{4}\int \underbrace{e^{x^4}}_{e^u}\underbrace{(4x^3)\,dx}_{du} = \frac{1}{4}e^{x^4} + c.$$

∎

We now have the tools to revisit the graph of $f(x) = e^x$. Since $e = 2.718\ldots > 1$, we have

$$\lim_{x\to\infty} e^x = \infty \quad\text{and}\quad \lim_{x\to-\infty} e^x = 0.$$

We also have that

$$f'(x) = e^x > 0,$$

so that f is increasing for all x and

$$f''(x) = e^x > 0,$$

so that the graph is concave up everywhere. You should now readily obtain the graph in Figure 6.16.

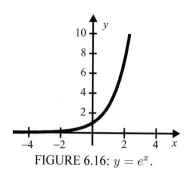

FIGURE 6.16: $y = e^x$.

Similarly, for $f(x) = e^{-x}$, we have

$$\lim_{x \to \infty} e^{-x} = 0 \text{ and } \lim_{x \to -\infty} e^{-x} = \infty.$$

Further, from the chain rule,

$$f'(x) = -e^{-x} < 0,$$

so that f is decreasing for all x. We also have

$$f''(x) = e^{-x} > 0,$$

so that the graph is concave up everywhere. You should easily obtain the graph in Figure 6.17.

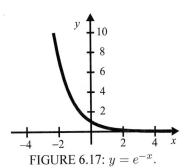

FIGURE 6.17: $y = e^{-x}$.

Notation: We will occasionally write $e^x = \exp(x)$. This is particularly helpful when the exponent is a complex expression. For example,

$$\exp(x^3 - 5x^2 + 2x + 7) = e^{x^3 - 5x^2 + 2x + 7},$$

where the former is more easily read than the latter.

You might be wondering about derivatives of more general exponential functions, such as $f(x) = b^x$, for any base $b > 0$. These functions are easy to express in terms of the natural exponential function, as follows. Notice that for any constant $b > 0$, we have by the usual rules of logs and exponentials that

$$e^{x \ln b} = e^{\ln(b^x)} = b^x.$$

In particular, observe that this says that

$$\frac{d}{dx} b^x = \frac{d}{dx} e^{x \ln b} = e^{x \ln b} \frac{d}{dx} (x \ln b)$$
$$= e^{x \ln b} (\ln b) = b^x (\ln b),$$

as we had conjectured in section 2.6. Similarly, we can use the properties of the natural exponential to evaluate integrals involving more general exponentials, as follows. For $b > 0$ $(b \neq 1)$, we have

$$\int b^x \, dx = \int e^{x \ln b} \, dx = \frac{1}{\ln b} \int e^{\overbrace{x \ln b}^{u}} \underbrace{(\ln b) \, dx}_{du}$$
$$= \frac{1}{\ln b} e^{x \ln b} + c = \frac{1}{\ln b} b^x + c.$$

Although of less significance than the natural exponential function, you can now see that the general exponential functions are easily dealt with in terms of the natural exponential. In fact, you should not bother to memorize the formulas for the derivatives and integrals of general exponentials. Rather, each time you are confronted with the exponential function $f(x) = b^x$, simply rewrite it as $f(x) = e^{x \ln b}$ and then use the familiar rules for the derivative and integral of the natural exponential and the chain rule.

In a similar way, we can use our knowledge of the natural logarithm to discuss more general logarithms. First, recall that for any base $a > 0$ $(a \neq 1)$ and any $x > 0$, $y = \log_a x$ if and only if $x = a^y$. Taking the natural logarithm of both sides of this equation, we have

$$\ln x = \ln (a^y) = y \ln a.$$

Solving for y gives us

$$y = \frac{\ln x}{\ln a}.$$

This proves the following theorem.

Theorem 3.2   For any base $a > 0$ $(a \neq 1)$ and any $x > 0$, $\log_a x = \dfrac{\ln x}{\ln a}$.

Among other things, Theorem 3.2 allows us to use a calculator to evaluate logarithms with any base. Calculators typically do not have built-in functions for evaluation of general logarithms, opting instead for only $\ln x$ and $\log_{10} x$ keys. Notice that evaluating general logarithms is now easy. For instance, we have

$$\log_7 3 = \frac{\ln 3}{\ln 7} \approx 0.564575.$$

More importantly, observe that we can use Theorem 3.2 to express derivatives of general logarithms in terms of the familiar derivative of the natural logarithm. In particular, for any base $a > 0$ $(a \neq 1)$, we have

$$\frac{d}{dx} \log_a x = \frac{d}{dx} \left(\frac{\ln x}{\ln a} \right) = \frac{1}{\ln a} \frac{d}{dx} (\ln x)$$

$$= \frac{1}{\ln a} \left(\frac{1}{x} \right) = \frac{1}{x \ln a}.$$

As with the derivative formula for general exponentials, there is little point in learning this as a new differentiation rule. Rather, when you need to differentiate the general logarithmic function $f(x) = \log_a x$, simply rewrite it first as $f(x) = \dfrac{\ln x}{\ln a}$ and use the familiar derivative of the natural logarithm.

Exercises 6.3

1. (writing) Thinking of e as a number (larger than 1), explain why $\lim\limits_{x \to \infty} e^x = \infty$ and $\lim\limits_{x \to -\infty} e^x = 0$.

2. (writing) Explain why the graph of $y = e^{-x}$ in Figure 6.17 is the mirror image (about the y-axis) of the graph of $y = e^x$ in Figure 6.16.

3. (writing) In the text, we gave a precise but technical definition of e^π. In this exercise, we ask you to think about this further. Carefully describe what is meant by e^3, $e^{3.1}$ and $e^{3.14}$. Then, explain how e^π should relate to these numbers.

4. (writing) There are a variety of equivalent ways of defining the number e. Discuss the definition given in this section versus the definition given in section 0.4. Different definitions were given in different sections for convenience, based on the mathematics we wanted to develop in each section. Explain why the limit definition was more convenient in a precalculus section, whereas the inverse function definition is more convenient now.

In exercises 5–12, graph the indicated function.

5. $3e^{2x}$

6. $2e^{3x}$

7. $2e^{-x}$

8. $3e^{-2x}$

9. $3xe^{-2x}$

10. $2xe^{-3x}$

11. $2e^{-x^2}$

12. e^{-2x^2}

In exercises 13–26, find the derivative.

13. $4e^{3x}$

14. $2e^{3x}$

15. $3xe^{-2x}$

16. $2xe^{-3x}$

17. $2e^{-x^2}$

18. e^{-2x^2}

19. $x^2 e^{-x}$

20. $2x^2 e^{-3x}$

21. $2e^{x^3 - 3x}$

22. $e^{2x - x^2}$

23. $\sqrt{e^{2x}}$

24. $(e^{3x} + 1)^2$

25. $e^{\cos x}$

26. $e^{\sqrt{x}}$

In exercises 27–44, evaluate the integral.

27. $\displaystyle\int e^{3x} \, dx$

28. $\displaystyle\int e^{2-3x} \, dx$

29. $\displaystyle\int xe^{x^2} \, dx$

30. $\displaystyle\int 2xe^{-x^2} \, dx$

31. $\displaystyle\int \sin x e^{\cos x} \, dx$

32. $\displaystyle\int e^x \cos(e^x) \, dx$

33. $\displaystyle\int \frac{e^{1/x}}{x^2} \, dx$

34. $\displaystyle\int \frac{e^x}{1 + e^x} \, dx$

35. $\displaystyle\int (1+e^x)^2\, dx$

36. $\displaystyle\int e^x(1+e^x)^2\, dx$

37. $\displaystyle\int \frac{4}{e^{2x}}\, dx$

38. $\displaystyle\int \frac{e^{2x}}{4}\, dx$

39. $\displaystyle\int \ln e^{x^2}\, dx$

40. $\displaystyle\int e^{-\ln x}\, dx$

41. $\displaystyle\int_0^1 e^{3x}\, dx$

42. $\displaystyle\int_0^2 e^{-2x}\, dx$

43. $\displaystyle\int_{-2}^2 xe^{-x^2}\, dx$

44. $\displaystyle\int_0^1 \frac{e^x-1}{e^{2x}}\, dx$

In exercises 45–48, find all extrema and inflection points.

45. xe^{-2x}

46. xe^{-4x}

47. x^2e^{-2x}

48. $\dfrac{e^x}{x}$

49. Prove parts (ii) and (iii) of Theorem 3.1.

50. The atmospheric pressure at height h feet above sea level is approximately $p = 2116e^{-0.0000318h}$. If a balloon is at height 1000 feet and rising at the rate of 160 ft/s, at what rate is the atmospheric pressure changing?

In exercises 51–58, use your graphing calculator or CAS to graph each function.

51. 3^x

52. 2^x

53. $\left(\dfrac{1}{2}\right)^x$

54. $\left(\dfrac{3}{4}\right)^x$

55. 3^{-x}

56. 2^{-x}

57. $\left(\dfrac{1}{2}\right)^{-x}$

58. $\left(\dfrac{3}{4}\right)^{-x}$

59. Based on exercises 51–58, describe the graph of $y = b^x$ for $b > 1; 0 < b < 1$.

60. Based on exercises 51–58, describe the graph of $y = b^{-x}$ for $b > 1; 0 < b < 1$.

In exercises 61–70, find the derivative.

61. 2^x

62. 3^x

63. 3^{2x}

64. 5^{-2x}

65. 3^{x^2}

66. 4^{-x}

67. $\log_3 x$

68. $\log_8 x$

69. $\log_4 x^2$

70. $\log_6(x^2 + 5)$

In exercises 71–76, evaluate each integral.

71. $\displaystyle\int 2^x\, dx$

72. $\displaystyle\int 3^x\, dx$

73. $\displaystyle\int x2^{x^2}\, dx$

74. $\displaystyle\int 2x3^{-x^2}\, dx$

75. $\displaystyle\int 4^{3x}\, dx$

76. $\displaystyle\int 4^{1-x}\, dx$

77. The **sigmoid function** $f(x) = \dfrac{1}{1+e^{-x}}$ is used to model situations with a threshold. For example, in the brain each neuron receives inputs from numerous other neurons and fires only after its total input crosses some threshold. Graph $y = f(x)$ and find $\lim_{x\to\infty} f(x)$ and $\lim_{x\to-\infty} f(x)$. Define the function $g(x)$ to be the value of $f(x)$ rounded off to the nearest integer. What value of x is the threshold for this function to switch from "off" (0) to "on" (1)? How could you modify the function to move the threshold to $x = 4$ instead?

78. A human being starts with a single fertilized egg cell, which divides into 2 cells, which then divide into 4 cells and so on, until the newborn infant has about 1 quadrillion (10^{15}) cells. Without doing any calculations, guess how many divisions are required to reach 10^{15}. Then, determine the value of n such that $2^n \approx 10^{15}$. Are you surprised?

79. Suppose a certain type of cell grows for 3 days and then divides into two cells. The distribution of ages of cells will have a probability distribution function (pdf) of the form $f(x) = 2ke^{-kx}$ for $0 \le x \le 3$. Find the value of k such that $f(x)$ is a pdf; that is, $\displaystyle\int_0^3 f(x)\, dx = 1$. Then find the probability that a given cell is between 1 and 2 days old.

80. Suppose you have a 1-in-10 chance of winning a prize with some purchase (like a lottery). If you make 10 purchases (i.e., you get 10 tries), the probability of winning at least one prize is $1 - (9/10)^{10}$. If the prize had probability 1-in-20 and you tried 20 times, would the probability of winning at least once be higher or lower? Compare $1 - (9/10)^{10}$ and $1 - (19/20)^{20}$ to find out. To see what happens for larger and larger odds, compute $\lim_{n\to\infty} [1 - ((n-1)/n)^n]$.

81. In the text, we referred the proof of $\lim_{h\to 0} \dfrac{e^h - 1}{h} = 1$ to the exercises. In this exercise, we guide you through one possible proof. (Another proof is given in the next exercise.) Starting with $h > 0$, write $h = \ln e^h = \displaystyle\int_1^{e^h} \frac{1}{x}\, dx$. Use the Integral Mean Value Theorem to write $\displaystyle\int_1^{e^h} \frac{1}{x}\, dx = \frac{(e^h - 1)}{\bar{x}}$ for some number $\bar{x}$ between 1 and e^h. This gives you $\dfrac{e^h - 1}{h} = \bar{x}$. Now, take the limit as $h \to 0^+$. For $h < 0$, repeat this argument, with h replaced with $-h$.

82. In this exercise, we guide you through a different proof of $\lim_{h\to 0} \dfrac{e^h - 1}{h} = 1$. Start with $f(x) = \ln x$ and the fact that $f'(1) = 1$. Using the alternative definition of derivative, we write this as $f'(1) = \lim_{x\to 1} \dfrac{\ln x - \ln 1}{x - 1} = 1$. Explain why this implies that $\lim_{x\to 1} \dfrac{x - 1}{\ln x} = 1$. Finally, substitute $x = e^h$.

83. (exploratory) Find the number of intersections of $y = x^a$ and $y = a^x$ for $a = 3$, $a = e$ and $a = 2$. The number of intersections changes between $a = 2$ and $a = 3$. Repeat this for $a = 2.1$, $a = 2.2$ and so on, to explore whether other a-values have different numbers of intersections. Try to prove it. You will want to use the following fact: if $f(c) = g(c)$, $f'(x) > g'(x)$ for $x > c$ and $f'(x) < g'(x)$ for $x < c$, then $x = c$ is the only intersection of f and g.

84. (exploratory) Determine the general properties (positive/negative, increasing/decreasing, concave up/down) of the functions xe^{-bx} and $x^2 e^{-bx}$ for a constant $b > 0$. In a number of applications, mathematicians need to find a function for which $f(0) = 0$ and (for $x > 0$) $f(x)$ rises to a single maximum at $x = c$ and gradually drops with $\lim\limits_{x \to \infty} f(x) = 0$. Show that both families of functions meet these criteria and find the appropriate values for b. Taking $c = 2$, graph xe^{-bx} and $x^2 e^{-bx}$. Compare and contrast the graphs. Explain what an applied mathematician might use to decide which one is a more realistic function for a given application.

6.4   Growth and Decay Problems

In this age, we are all keenly aware of how infection by food-borne microorganisms causes disease. Certain strains of the bacterium *Escherichia coli* (*E. coli*) have become particularly notorious. Most people are unaware of the mechanisms by which bacteria make us sick. Many organisms (like *E. coli*) produce a toxin, which in sufficient quantity can cause sickness or even death. Some bacteria can reproduce in our bodies at a surprisingly fast rate. If they reproduce quickly enough, they can overwhelm our bodies' natural defenses, growing to such large numbers that the sheer volume of toxin they are producing causes serious illness. The table shown in the margin indicates the number of *E. coli* bacteria (in millions of bacteria per ml) in a laboratory culture measured at 30 minute (1/2 hour) intervals during the course of an experiment. We have plotted the number of bacteria per milliliter versus time in Figure 6.18. What would you say the graph most resembles? If you said, "an exponential," you guessed right. Careful analysis of experimental data has shown that certain populations grow at a rate proportional to their current level. This is quite easily observed in bacterial cultures, where the bacteria reproduce by binary fission (i.e., each cell reproduces by dividing into two cells). In this case, the rate at which the bacterial culture grows is directly proportional to the current population (until such time as resources become scarce or overcrowding becomes a limiting factor). If we let $y(t)$ represent the number of bacteria in a culture at time t, then the rate of change of the population with respect to time is $y'(t)$. Thus, since $y'(t)$ is proportional to $y(t)$, we have

$$y'(t) = k\, y(t), \tag{4.1}$$

for some constant of proportionality k (the **growth constant**). Since equation (4.1) involves the derivative of an unknown function, we call it a **differential equation**. Our aim is to **solve** the differential equation, that is, find the *function* $y(t)$. Assuming that $y(t) > 0$ (this is a reasonable assumption, since $y(t)$ represents a population), we have

$$\frac{y'(t)}{y(t)} = k. \tag{4.2}$$

Integrating both sides of equation 4.2 with respect to t, we obtain

$$\int \frac{y'(t)}{y(t)}\, dt = \int k\, dt. \tag{4.3}$$

You should notice that we can substitute $u = y(t)$ in the integral on the left-hand side. For this choice of u, we have $du = y'(t)\, dt$ and so, (4.3) becomes

$$\int \frac{1}{u}\, du = \int k\, dt.$$

Evaluating these integrals, we obtain

$$\ln|u| = k\, t + c,$$

Time (hours)	Number of bacteria (millions per ml)
0	1.2
0.5	2.5
1	5.1
1.5	11.0
2	23.0
2.5	45.0
3	91.0
3.5	180.0
4	350.0

FIGURE 6.18: Growth of bacteria.

where c is a constant of integration. Since $u = y(t) > 0$, we have

$$\ln y(t) = k\,t + c$$

and taking exponentials of both sides, we get

$$y(t) = e^{\ln y(t)} = e^{kt+c} = e^{kt}\,e^{c}.$$

Since c is an arbitrary constant, we write $A = e^{c}$ and hence, we have

$$y(t) = A\,e^{kt}. \tag{4.4}$$

For $k > 0$, equation (4.4) is called an **exponential growth law** and for $k < 0$, it is an **exponential decay law** (think about the distinction).

In the following example, we examine how an exponential growth law predicts the number of cells in a bacterial culture.

___

Example 4.1   Exponential Growth of a Bacterial Colony
    A freshly inoculated bacterial culture of *Streptococcus A* (a common group of microorganisms that cause strep throat) contains 100 cells. When the culture is checked 60 minutes later, it is determined that there are 450 cells present. Determine the number of cells present at any time t (measured in minutes), assuming exponential growth.

Solution   Exponential growth means that

$$y'(t) = k\,y(t)$$

and hence, from (4.4),

$$y(t) = A\,e^{kt}, \tag{4.5}$$

where A and k are constants to be determined. Notice that

$$y(0) = 100, \tag{4.6}$$

where we have set the starting time to be $t = 0$. Equation (4.6) is called an **initial condition**. Setting $t = 0$ in (4.5), we now have

$$100 = y(0) = A\,e^{0} = A$$

and hence,

$$y(t) = 100\,e^{kt}.$$

We can use the second observation to determine the value of the growth constant k. We have

$$450 = y(60) = 100\,e^{60k}.$$

Dividing both sides by 100, we have

$$4.5 = e^{60k}$$

and taking the natural logarithm of both sides,

$$\ln 4.5 = \ln e^{60k} = 60\,k.$$

Thus,

$$k = \frac{\ln 4.5}{60} \approx 0.02507.$$

FIGURE 6.19:
$y = 100\, e^{\left(\frac{\ln 4.5}{60} t\right)}$.

We now have a formula for the number of cells present at any time t:

$$y(t) = 100\, e^{kt} = 100\, \exp\left(\frac{\ln 4.5}{60}\, t\right).$$

See Figure 6.19 for a graph of the projected bacterial growth over the first 120 minutes. One further question of interest to microbiologists is the **doubling time**, that is, the time it takes for the number of cells to double. We can find this by solving for the time t for which $y(t) = 200$. We have

$$200 = y(t) = 100\, \exp\left(\frac{\ln 4.5}{60}\, t\right).$$

Dividing both sides by 100 and taking logs, we obtain

$$\ln 2 = \frac{\ln 4.5}{60}\, t,$$

so that

$$t = \frac{60 \ln 2}{\ln 4.5} \approx 27.65.$$

So, the doubling time for this culture of *Streptococcus A* is about 28 minutes. Note that the doubling time for a bacterium is a function of the specific strain of bacteria as well as the quality and quantity of the food supply, the temperature and other environmental factors. However, it is not dependent on the initial population. In this example, to double from 80 to 160 would require

$$160 = 80 \exp\left(\frac{\ln 4.5}{60}\, t\right).$$

Dividing by 80 and taking the natural logarithm of both sides would produce the same equation

$$\ln 2 = \frac{\ln 4.5}{60}\, t,$$

solved above. Thus, if an initial population of 100 doubles to 200 in approximately 28 minutes, then it will double again (to 400) in the next 28 minutes, and so on.

■

There are any number of physical phenomena that satisfy a growth or decay law. Another of these is found in the decay of radioactive elements. Experiments have shown that the rate at which a radioactive element decays is directly proportional to the amount present. (Recall that radioactive elements are chemically unstable elements that gradually decay into other, more stable elements.) Let $y(t)$ be the amount (mass) of a radioactive element present at time t. Then, we have that the rate of change (rate of decay) of $y(t)$ satisfies

$$y'(t) = k\, y(t). \tag{4.7}$$

Note that (4.7) is precisely the same differential equation as (4.1), encountered in example 4.1 for the growth of bacteria and hence, from (4.4), we have that

$$y(t) = A\, e^{kt},$$

for some constants A and k (the **decay constant**) to be determined.

It is common to discuss the decay rate of a radioactive element in terms of its half-life. The **half-life** of an element is the time required for half of the initial quantity to decay into other elements. For instance, scientists have calculated that the half life of carbon-14 (^{14}C) is approximately 5730 years. That is, if you have 2 grams of ^{14}C today and you come back in 5730 years, you will have approximately 1 gram of ^{14}C remaining. It is this long half-life and the fact that every living creature continually takes in ^{14}C (until it dies), which makes ^{14}C measurements useful for radiocarbon dating (see the exercise set for more on this important application).

Example 4.2   Radioactive Decay

If you have 50 grams of ^{14}C today, how much will be left in 100 years?

Solution   Let $y(t)$ be the number of grams of ^{14}C present at time t. Then, we have

$$y'(t) = k\, y(t)$$

and as we have already seen,

$$y(t) = A\, e^{kt}.$$

Our initial condition is $y(0) = 50$ and so,

$$50 = y(0) = A\, e^0 = A.$$

Thus,

$$y(t) = 50\, e^{kt}.$$

In order to determine the value of the decay constant k, we use the half-life:

$$25 = y(5730) = 50\, e^{5730k}.$$

Dividing both sides by 50 and taking logs gives us

$$\ln \frac{1}{2} = \ln e^{5730k} = 5730\, k$$

and so,

$$k = \frac{\ln \frac{1}{2}}{5730} \approx -1.20968 \times 10^{-4}.$$

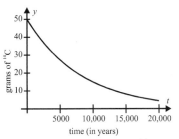

FIGURE 6.20: Decay of ^{14}C.

A graph of the mass of ^{14}C as a function of time is seen in Figure 6.20. Notice the extremely large time scale shown. This should give you an idea of the incredibly slow rate of decay of ^{14}C. Notice that if we start with 50 grams, then the amount left after 100 years is

$$y(100) = 50\, e^{100\, k} \approx 49.3988 \text{ grams}.$$

∎

Another easily observable and mathematically similar physical principle is **Newton's Law of Cooling**. You might notice that if you introduce a hot object into cool surroundings (or equivalently, a cold object into warm surroundings) the rate at which the object cools (warms) is not proportional to its temperature, but rather, is proportional to the difference in temperature between the object and its surroundings. Symbolically, if we let $y(t)$ be the temperature of the object at time t and let T_a be the temperature of the surroundings (the **ambient temperature**, which we assume to be constant), we have the differential equation

$$y'(t) = k\,[y(t) - T_a]. \tag{4.8}$$

Notice that (4.8) is not the same as the differential equation describing exponential growth or decay. (Compare these; what's the difference?) Even so, we can approach finding a solution in the same way. In the case of cooling, we assume that

$$T_a < y(t).$$

(Why is it fair to assume this?) If we divide both sides of equation (4.8) by $y(t) - T_a$ and then integrate both sides, we obtain

$$\int \frac{y'(t)}{y(t) - T_a} \, dt = \int k \, dt = k\,t + c. \tag{4.9}$$

Notice that we can evaluate the integral on the left-hand side by making the substitution $u = y(t) - T_a$, so that $du = y'(t)dt$. Thus, we have

$$\int \frac{y'(t)}{y(t) - T_a} \, dt = \int \frac{1}{u} \, du$$
$$= \ln|u| = \ln|y(t) - T_a|$$
$$= \ln[y(t) - T_a],$$

since $y(t) - T_a > 0$. From (4.9), we now have

$$\ln[y(t) - T_a] = kt + c.$$

You'll notice that we can solve for $y(t)$ by taking exponentials of both sides, to obtain

$$y(t) - T_a = e^{kt+c} = e^{kt}\,e^c.$$

Finally, for convenience, we write $A = e^c$, and obtain

$$y(t) = A\,e^{kt} + T_a,$$

where A and k are constants to be determined.

We illustrate Newton's Law of Cooling in the following example.

Example 4.3   Newton's Law of Cooling for a Cup of Coffee

A cup of fast food coffee is 180°F when freshly poured. After 2 minutes in a room at 70°F, the coffee has cooled to 165°F. Find the temperature at any time t.

Solution   Here, letting $y(t)$ be the temperature of the coffee at time t, we have the differential equation

$$y'(t) = k\,[y(t) - 70].$$

Proceeding as above, we obtain

$$y(t) = A\,e^{kt} + 70.$$

Observe that the initial condition here is the initial temperature, $y(0) = 180$. This gives us

$$180 = y(0) = A\,e^0 + 70 = A + 70,$$

so that $A = 110$. Thus,

$$y(t) = 110\,e^{kt} + 70.$$

We can now use the second measured temperature to solve for the constant k. We have

$$165 = y(2) = 110\,e^{2k} + 70.$$

Subtracting 70 from both sides and dividing by 110, we have

$$e^{2k} = \frac{165 - 70}{110} = \frac{95}{110}.$$

Taking logs of both sides yields

$$2\,k = \ln\left(\frac{95}{110}\right)$$

FIGURE 6.21: Temperature of coffee.

and hence,

$$k = \frac{1}{2} \ln\left(\frac{95}{110}\right) \approx -0.0733017.$$

A graph of the projected temperature against time is shown in Figure 6.21. A further, reasonable question might be to ask at what time the coffee's temperature has dropped to a tolerable 120°F. From Figure 6.21, you might observe that the temperature appears to have fallen to 120°F after about 10 minutes. We can solve this symbolically by finding the time t for which

$$120 = y(t) = 110\,e^{kt} + 70.$$

It is not hard to solve this to obtain

$$t \approx 10.76 \text{ minutes.}$$

This is left as an exercise.

■

Compound Interest

Suppose that a bank agrees to pay you 8% (annual) interest on your investment of $10,000. Then, at the end of a year, you will have

$$\$10,000 + (0.08)\,\$10,000 = \$10,000\,(1 + 0.08) = \$10,800.$$

On the other hand, if the bank agrees to pay you interest twice a year at the same 8% annual rate, at the end of the year, you will have

$$\$10,000\left(1 + \frac{0.08}{2}\right)\left(1 + \frac{0.08}{2}\right) = \$10,000\left(1 + \frac{0.08}{2}\right)^2$$
$$= \$10,816.$$

Continuing in this fashion, notice that paying (compounding) interest monthly would pay $\frac{8}{12}\%$ each month (period), resulting in a balance of

$$\$10,000\left(1 + \frac{0.08}{12}\right)^{12} = \$10,830.00.$$

Further, if interest is compounded daily, you would end up with

$$\$10,000\left(1 + \frac{0.08}{365}\right)^{365} = \$10,832.78.$$

It should be evident that the more often interest is compounded, the greater the interest will be. A reasonable question to ask is if there is a limit to how much interest can accrue on a given investment at a given interest rate. If n is the number of times per year that interest is compounded, we wish to calculate the annual percentage rate (APR) under **continuous compounding**,

$$\text{APR} = \lim_{n \to \infty}\left(1 + \frac{.08}{n}\right)^n - 1.$$

To determine this limit, you must recall (see Chapter 0) that

$$e = \lim_{m \to \infty}\left(1 + \frac{1}{m}\right)^m.$$

Notice that if we make the change of variable $m = (0.08)\, n$, then we have

$$\text{APR} = \lim_{m \to \infty} \left(1 + \frac{0.08}{0.08m}\right)^{0.08m} - 1$$

$$= \left[\lim_{m \to \infty}\left(1 + \frac{1}{m}\right)^m\right]^{0.08} - 1$$

$$= e^{0.08} - 1 \approx 0.083287.$$

Under continuous compounding, you would thus earn

$$\$10,000\left(e^{0.08} - 1\right) = \$832.87$$

in interest, leaving your investment with a total value of \$10,832.87. More generally, suppose that you invest \$$P$ at an annual interest rate r, compounded n times per year. Then the value of your investment after t years is

$$\$P\left(1 + \frac{r}{n}\right)^{nt}.$$

Under continuous compounding (i.e., taking the limit as $n \to \infty$), this becomes

$$\$Pe^{rt}.$$

Example 4.4   Comparing Forms of Compounding Interest

If you invest \$7000 at an annual interest rate of 5.75%, compare the value of your investment after 5 years under various forms of compounding.

Solution   With annual compounding, the value is

$$\$7000\left(1 + \frac{0.0575}{1}\right)^5 = \$9257.63.$$

With monthly compounding, this becomes

$$\$7000\left(1 + \frac{0.0575}{12}\right)^{12 \cdot 5} = \$9325.23.$$

With daily compounding, this yields

$$\$7000\left(1 + \frac{0.0575}{365}\right)^{365 \cdot 5} = \$9331.42.$$

Finally, with continuous compounding, the value is

$$\$7000\, e^{0.0575 \cdot 5} = \$9331.63.$$

■

Exercises 6.4

1. (writing) A linear function is defined by constant slope. If a population showed a constant numerical increase year by year, explain why the population could be represented by a linear function. If the population showed a constant *percentage* increase instead, explain why the population could be represented by an exponential function.

2. (writing) If a population has a constant birth rate and a constant death rate (smaller than the birth rate), describe what the population would look like over time. In the United States, is the death rate increasing, decreasing or staying the same? Given, this, why is there concern about reducing the birth rate?

3. (writing) Explain, in monetary terms, why for a given interest rate the more times the interest is compounded the more money is in the account at the end of a year.

4. (writing) In the growth-and-decay examples, the constant A turned out to be equal to the initial value. In the cooling examples, the constant A did not equal the initial value. Explain why the cooling example worked differently.

In exercises 5–12, solve the initial value problem.

5. $y' = 4y$, $y(0) = 2$ **6.** $y' = 3y$, $y(0) = -2$

7. $y' = -3y$, $y(0) = 5$ **8.** $y' = -2y$, $y(0) = -6$

9. $y' = 2y$, $y(1) = 2$ **10.** $y' = -y$, $y(1) = 2$

11. $y' = -3$, $y(0) = 3$ **12.** $y' = -2$, $y(0) = -8$

13. Suppose a bacterial culture doubles in population every 4 hours. If the population is initially 100, find an equation for the population at any time. Determine when the population will reach 6000.

14. Suppose a bacterial culture triples in population every 5 hours. If the population is initially 200, find an equation for the population at any time. Determine when the population will reach 20,000.

15. Suppose a bacterial culture initially has 400 cells. After 1 hour, the population has increased to 800. Find an equation for the population at any time. What will the population be after 10 hours?

16. Suppose a bacterial culture initially has 100 cells. After 2 hours, the population has increased to 400. Find an equation for the population at any time. What will the population be after 8 hours?

17. A bacterial culture grows exponentially with growth constant 0.44 hour^{-1}. Find its doubling time.

18. A bacterial culture grows exponentially with growth constant 0.12 hour^{-1}. Find its doubling time.

19. The radioactive element iodine-131 has a decay constant of -1.3863 day^{-1}. Find its half-life.

20. The radioactive element cesium-137 has a decay constant of -0.023 year^{-1}. Find its half-life.

21. Suppose some quantity is increasing exponentially (e.g., the number of cells in a bacterial culture) with growth rate r. Show that the doubling time is $\dfrac{\ln 2}{r}$.

22. Suppose some quantity is decaying exponentially with decay constant r. Show that the half-life is $-\dfrac{\ln 2}{r}$. What is the difference between the half-life here and the doubling time in exercise 21?

23. The half-life of morphine in the human bloodstream is 3 hours. If initially there is 0.4 mg of morphine in the bloodstream, find an equation for the amount in the bloodstream at any time. When does the amount drop below 0.01 mg?

24. Given a half-life of 3 hours, by what percentage will the amount of morphine in the bloodstream have decreased in 1 day?

25. Suppose that a population of *E. coli* doubles every 20 minutes. A treatment of the infection removes 90% of the *E. coli* present, and is timed to accomplish the following. The population starts at size 10^8, grows for T minutes, the treatment is applied and the population returns to size 10^8. Find the time T.

26. Research by Meadows, Meadows, Randers and Behrens indicates that the earth has 3.2×10^9 acres of arable land available. The world population of 1950 required 10^9 acres to sustain it, and the population of 1980 required 2×10^9 acres. If the population grows at a constant percentage rate, in what year will it reach the maximum sustainable size?

27. Strontium-90 is a dangerous radioactive isotope. Because of its similarity to calcium, it is easily absorbed into human bones. The half-life of strontium-90 is 28 years. If a certain amount is absorbed into the bones due to exposure to a nuclear explosion, what percentage will remain after 50 years?

28. The half-life of uranium U-235 is approximately 0.7×10^9 years. If 50 grams are buried at a nuclear waste site, how much will remain after 100 years?

29. Scientists dating a fossil estimate that 20% of the original amount of carbon-14 is present. Recalling that the half-life is 5730 years, approximately how old is the fossil?

30. If a fossil is 1 million years old, what percentage of its original carbon-14 should remain?

31. A bowl of porridge at $200°$ (too hot) is placed in a $70°$ room. One minute later the porridge has cooled to $180°$. When will the temperature be $120°$ (just right)?

32. A smaller bowl of porridge served at $200°$ cools to $160°$ in 1 minute. What temperature (too cold) will this porridge be when the bowl of exercise 31 has reached $120°$ (just right)?

33. A cold drink is poured out at $50°$. After 2 minutes of sitting in a $70°$ room, the temperature has risen to $56°$. Find the drink's temperature at any time t.

34. For the cold drink in exercise 33, what will the temperature be after 10 minutes? When will the drink have warmed to $66°$?

35. At 10:07 P.M. you find a secret agent murdered. Next to him is a martini that got shaken before the secret agent could stir it. Room temperature is $70°$. The martini warms from $60°$ to $61°$ in the 2 minutes from 10:07 to 10:09. If the secret agent's martinis are always served at $40°$, what was the time of death?

36. Twenty minutes after being served a cup of fast food coffee it is still too hot to drink at $160°$. Two minutes later, the temperature has dropped to $158°$. Your friend, whose coffee is also too hot to drink, speculates that since the temperature is dropping an average of 1 degree per minute, it was served at $180°$. Explain what is wrong with this logic. Was the actual serving temperature hotter or cooler than $180°$?

37. Find the actual serving temperature in exercise 36 if room temperature is $68°$.

38. For the cup of coffee in example 4.3, suppose that the goal is to have the coffee cool to $120°$ in 5 minutes. At what temperature should the coffee be served?

39. If you invest $1000 at an annual interest rate of 8%, compare the value of the investment after 1 year under the following forms of compounding: annual, monthly, daily, continuous.

40. Repeat exercise 39 for the value of the investment after 5 years.

41. Person A invests $10,000 in 1990 and person B invests $20,000 in 2000. If both receive 12% interest (compounded continuously), what are the values of the investments in 2010?

42. Repeat exercise 41 for an interest rate of 4%. Then determine the interest rate such that person A ends up exactly even with person B. (Hint: You want person A to have $20,000 in 2000.)

43. One of the authors bought a set of basketball trading cards in 1985 for $34. In 1995, the "book price" for this set was $9800. Assuming a constant percentage return on this investment, find an equation for the worth of the set at time t (where $t = 0$ corresponds to 1985). At this rate of return, what will the set be worth in 2005?

44. The author also bought a set of baseball cards in 1985, costing $22. In 1995, this set was worth $32. How much will this set be worth in 2005?

45. Financial analyses are often confusing because of the difficulty of making valid comparisons. Due to inflation, a dollar today is not equivalent to a dollar next year. As an example, we try to make a simple income tax comparison here. In 1975, income between $16,000 and $20,000 was taxed at 28%. In 1988, income between $16,000 and $20,000 was taxed at 15%. This makes it seem like taxes went down considerably between 1975 and 1988. Briefly explain why this is not a valid comparison.

46. To make the comparison in exercise 45 a little more fair, note that income above $30,000 was taxed at 28% in 1988 and assume that inflation averaged 5.5% between 1975 and 1988. Adjust $16,000 for inflation by computing its equivalent after increasing continuously at 5.5% for 13 years. Based on this calculation, how do the tax rates compare?

47. Suppose the income tax structure is as follows: the first $30,000 is taxed at 15%, the remainder is taxed at 28%. Compute the tax T_1 on an income of $40,000. Now, suppose that

inflation is 5% and you receive a cost of living (5%) raise to $42,000. Compute the tax T_2 on this income. To compare the taxes, you should adjust the tax T_1 for inflation (add 5%).

48. In exercise 47, the tax code stayed the same, but (adjusted for inflation) the tax owed did not stay the same. Briefly explain why this happened. What could be done to make the tax owed remain constant?

49. One of the mysteries in population biology is how populations regulate themselves. The most famous myth involves lemmings diving off of cliffs at times of overpopulation. It **is** true that lemming populations rise and fall dramatically, for whatever reason (not including suicide). Animal ecologists draw graphs to visualize the rise and falls of animal populations. Instead of graphing population versus time, ecologists graph the logarithm of population versus time. To understand why, note that a population drop from 1000 to 500 would represent the same percentage decrease as a drop from 10 to 5. Show that the slopes of the drops are different, so that these drops would appear to be different on a population/time graph. However, show that the slopes of the drops in the logarithms (e.g., $\ln 1000$ to $\ln 500$) are the same. In general, if a population were changing at a constant percentage rate, what would the graph of population versus time look like? What would the graph of the logarithm of population versus time look like?

50. (exploratory) In the text, we briefly discussed the use of the radioactive isotope carbon-14 to date fossils. We elaborate on that here. The amount of carbon-14 in the atmosphere is largely determined by cosmic ray bombardment, with nuclear testing also playing a role. Living organisms maintain a constant level of carbon-14 through exchanges with the environment. At death, the organism no longer takes in carbon-14, so the carbon-14 level decreases with the 5730-year half-life. Scientists can measure the rate of disintegration of carbon-14 (you might visualize a Geiger counter). If $y(t)$ is the amount of carbon-14 remaining, the rate of disintegration is $y'(t) = ky(t)$. The main assumption is that the rate of disintegration at the time of death is the same as it is now for living organisms. Mathematically, this rate is $ky(0)$. The ratio of disintegration rates is $y(t)/y(0)$. Given this ratio, describe how to determine the time t. In particular, suppose that $ky(t) = 2.4$ disintegrations per minute and $ky(0) = 6.7$ disintegrations per minute. Solve for t. Different inaccuracies can creep into this process. First, suppose the assumption of constant carbon-14 levels is wrong. If $ky(0)$ is decreased by 5%, by what percentage does the estimate of the time t change? There may also be inaccuracies in the measurement of the current disintegration rate. If $ky(t)$ is decreased by 5%, by what percentage does the estimate of the time change? Roughly, how do errors in the measurements translate to errors in the estimate of the time?

51. (exploratory) Differential equations other than those studied in this section are also important. In particular, **second order equations** (those with a second derivative) play a central role in physics and engineering. The equation

$y''(x) + 4y'(x) + 3y(x) = 0$ could represent the motion of a mechanical spring, an electrical circuit or the motion of a projectile. This is also an equation you can solve. First, read the equation: you are looking for a function whose second derivative plus four times the first derivative exactly cancels three times the function. Of the functions you have studied, explain why the exponential is the most likely candidate for a solution. Now, look for solutions of the form $y = e^{rx}$ by computing two derivatives and plugging into the equation. You should conclude that you need $r^2 + 4r + 3 = 0$, so that $r = -3$ or $r = -1$. Show that e^{-3x} and e^{-x} are indeed solutions of the differential equation. Then show that for any constants c_1 and c_2, the function $c_1 e^{-3x} + c_2 e^{-x}$ is also a solution. This is called the **general solution,** because all solutions can be written in this form. Find the general solution of the second order equations $y''(x) - 6y'(x) + 8y(x) = 0$ and $y''(x) - 2y'(x) - 3y(x) = 0$.

6.5   Separable Differential Equations

In section 6.4, we solved the differential equation

$$y'(t) = k\, y(t).$$

This is an example of a **separable** differential equation. We will examine this type of equation at some length in this section. First, we consider the more general **first order ordinary differential equation**

$$y' = f(x, y). \tag{5.1}$$

Here, the derivative y' of some unknown function $y(x)$ is given as a function f of both x and y. Our objective is to find some function $y(x)$ (a **solution**) that satisfies equation (5.1). The equation is **first order** since it involves only the first derivative of the unknown function. We will consider the case where the x's and y's can be separated. We call equation (5.1) **separable** if we can separate the variables, i.e., if we can rewrite it in the form

$$g(y)\, y' = h(x),$$

where all of the x's are on one side of the equation and all of the y's are on the other side.

Example 5.1   A Separable Differential Equation
        Determine if the differential equation

$$y' = xy^2 - 2xy$$

is separable.

Solution   Notice that this equation is separable, since we may rewrite it as

$$y' = x(y^2 - 2y)$$

and then divide by $(y^2 - 2y)$ (assuming this is not zero), to obtain

$$\frac{1}{y^2 - 2y}\, y' = x.$$

∎

Example 5.2   An Equation That is Not Separable
        The equation

$$y' = xy^2 - 2x^2 y$$

is not separable, as there is no way to separate the x's and the y's. (Try this for yourself!)

∎

Essentially, the x's and y's must be separated by multiplication or division in order for a differential equation to be separable. Notice that in example 5.2, you can factor to get $y' = xy(y - 2x)$. It is the subtraction $y - 2x$ that keeps this equation from being separable.

Separable differential equations are of interest because there is a very simple means of solving them. Notice that if we integrate both sides of

$$g(y) \, y'(x) = h(x)$$

with respect to x, we get

$$\int g(y) \, y'(x) \, dx = \int h(x) \, dx. \tag{5.2}$$

Since $dy = y'(x) \, dx$, the integral on the left-hand side of (5.2) becomes

$$\int g(y) \, \underbrace{y'(x) \, dx}_{dy} = \int g(y) \, dy.$$

Consequently, from (5.2), we have

$$\int g(y) \, dy = \int h(x) \, dx.$$

So, provided we can evaluate both of the integrals above, we have an equation relating x and y, which no longer involves y'.

___

Example 5.3   Solving a Separable Equation

Solve the differential equation

$$y' = \frac{x^2 + 7x + 3}{y^2}.$$

Solution   Separating variables, observe that we have

$$y^2 \, y' = x^2 + 7x + 3.$$

Integrating both sides with respect to x, we obtain

$$\int y^2 \, y'(x) \, dx = \int (x^2 + 7x + 3) \, dx$$

or

$$\int y^2 \, dy = \int (x^2 + 7x + 3) \, dx.$$

Performing the indicated integrations yields

$$\frac{y^3}{3} + c_1 = \frac{x^3}{3} + 7\frac{x^2}{2} + 3x + c_2.$$

Subtracting c_1 from both sides, we get

$$\frac{y^3}{3} = \frac{x^3}{3} + 7\frac{x^2}{2} + 3x + (c_2 - c_1)$$

$$= \frac{x^3}{3} + 7\frac{x^2}{2} + 3x + c,$$

where we have combined the two constants of integration into one on the right-hand side.

FIGURE 6.22: A family
of solutions.

We can now solve for y:

$$y = \sqrt[3]{x^3 + \frac{21}{2}x^2 + 9x + 3c}.$$

Notice that for each value of c, we get a different solution of the differential equation. This is called a one-parameter **family of solutions** (the **general solution**) of the differential equation. In Figure 6.22, we have plotted a number of the members of this family of solutions.

■

You should notice that in the general case, the solution of a first order separable equation will depend on an arbitrary constant (the constant of integration). To select just one of these solution curves, we must simply specify a single point that the solution curve must pass through, say (x_0, y_0). That is, we require that

$$y(x_0) = y_0.$$

This is called an **initial condition** (since this condition often specifies the initial state of a physical system). Such a differential equation together with an initial condition is referred to as an **initial value problem** (IVP).

Example 5.4   **Solving an Initial Value Problem**

Solve the IVP

$$y' = \frac{x^2 + 7x + 3}{y^2}, \; y(0) = 3.$$

Solution   In example 5.3, we found that the general solution of the differential equation was

$$y = \sqrt[3]{x^3 + \frac{21}{2}x^2 + 9x + 3c}.$$

From the initial condition, we now have

$$3 = y(0) = \sqrt[3]{0 + 3c} = \sqrt[3]{3c}$$

and hence, $c = 9$. The solution of the IVP is then

$$y = \sqrt[3]{x^3 + \frac{21}{2}x^2 + 9x + 27}.$$

■

We are not always as fortunate as we were in example 5.4. There, we were able to obtain an explicit representation of the solution (i.e., we found a formula for y in terms of x). Most often, we must settle for an **implicit** representation of the solution, that is, an equation relating x and y, which cannot be solved for y in terms of x alone.

Example 5.5   **An Initial Value Problem That Has Only an Implicit Solution**

Find the solution of the IVP

$$y' = \frac{9x^2 - \sin x}{\cos y + 5e^y}, \; y(0) = \pi. \tag{5.3}$$

Solution   First, note that the differential equation in (5.3) is separable, since we can rewrite it as

$$(\cos y + 5e^y)\, y'(x) = 9x^2 - \sin x. \tag{5.4}$$

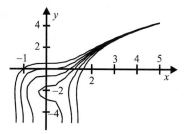

FIGURE 6.23a: A family
of solutions.

Integrating both sides of (5.4) with respect to x, we find

$$\int (\cos y + 5e^y)\, dy = \int (9x^2 - \sin x)\, dx.$$

Evaluating the integrals, we obtain

$$\sin y + 5e^y = 3x^3 + \cos x + c. \qquad (5.5)$$

Notice that there is no way to solve this equation explicitly for y in terms of x. However, you can still picture the graphs of some members of this family of solutions by using the implicit plot mode on your graphing utility. Several of these are plotted in Figure 6.23a. Even though we have not solved for y explicitly in terms of x, we can still use the initial condition. We simply substitute $x = 0$ and $y = \pi$ into equation (5.5). We have

$$\sin \pi + 5e^\pi = 0 + \cos 0 + c$$

or

$$5e^\pi - 1 = c.$$

This leaves us with

$$\sin y + 5e^y = 3x^3 + \cos x + 5e^\pi - 1$$

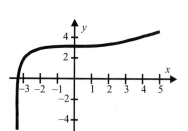

FIGURE 6.23b: The solution
of the IVP.

as an implicit representation of the solution of the IVP. We still cannot solve for y in terms of x alone. However, given any particular value for x, we can use Newton's method (or some other numerical method) to solve for an approximate value for the corresponding y. This is essentially what your CAS is doing (with many, many points) when you use it to plot a graph in implicit mode. We plot the solution of the IVP in Figure 6.23b.

■

Logistic Growth

In section 6.4, we introduced the differential equation

$$y' = k\,y$$

as a model of population growth. At that time, we made the point that this model was valid for populations growing with unlimited resources and with unlimited room for growth. Of course, **all** populations have factors that limit their growth. Thus, this particular model generally provides useful information only for relatively short periods of time.

As an alternative model of population growth, consider the following. It is reasonable to assume that populations grow at a rate proportional to their present size, but that there is a maximum sustainable population, M (called the **carrying capacity**) determined by the available resources. Further, as the population size approaches M (and available resources become more scarce), the population growth will slow. Thus, we might assume that the rate of growth of a population is jointly proportional to the present population level and the difference between the current level and the maximum, M. That is, if $y(t)$ is the population at time, t, we assume that

$$y'(t) = k\,y\,(M - y).$$

Two special solutions of this differential equation are apparent. The constant functions $y = 0$ and $y = M$ are both solutions of this differential equation. These are called **equilibrium solutions**, since under the assumption of logistic growth, once a population hits one of these levels, it remains there for all time. If $y \neq 0$ and $y \neq M$, we can solve the differential equation, since it is separable, as

$$\frac{1}{y\,(M - y)}\, y'(t) = k. \qquad (5.6)$$

Integrating both sides of (5.6) with respect to t, we obtain

$$\int \frac{1}{y\,(M-y)}\, y'(t)\, dt = \int k\, dt$$

or

$$\int \frac{1}{y\,(M-y)}\, dy = \int k\, dt. \tag{5.7}$$

Observe that we can add the following two fractions to yield:

$$\frac{1}{M\,y} + \frac{1}{M\,(M-y)} = \frac{M-y}{My\,(M-y)} + \frac{y}{My\,(M-y)} = \frac{1}{y\,(M-y)}. \tag{5.8}$$

Although you probably do not see how we came up with the two fractions on the left side of (5.8), notice that the expression on the left is simple to integrate. This rewriting of a rational expression as a sum of two simpler rational expressions is an example of a **partial fraction decomposition**. This is a useful technique that we present in some detail in section 7.4. For the moment, simply observe from (5.7) and (5.8) that we now have

$$\int \left[\frac{1}{M\,y} + \frac{1}{M\,(M-y)} \right] dy = \int k\, dt.$$

Carrying out the integrations gives us

$$\frac{1}{M} \ln|y| - \frac{1}{M} \ln|M-y| = kt + c.$$

Multiplying both sides by M, and using the fact that $0 < y < M$, we have

$$\ln y - \ln(M-y) = kMt + Mc.$$

Taking exponentials of both sides, we find

$$\exp\left[\ln y - \ln(M-y)\right] = e^{kMt+Mc} = e^{kMt}\, e^{Mc}.$$

Next, using rules of exponentials and logs and replacing the constant term e^{Mc} by a new constant A, we obtain

$$\frac{y}{M-y} = A\, e^{kMt}. \tag{5.9}$$

If we multiply both sides of equation (5.9) by $(M-y)$, we obtain

$$y = A\, e^{kMt}\, (M-y)$$
$$= A\, M\, e^{kMt} - A\, e^{kMt}\, y.$$

Combining the two y terms, we find

$$y\left(1 + A\, e^{kMt}\right) = A\, M\, e^{kMt}.$$

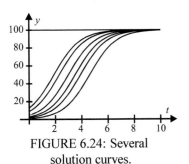

FIGURE 6.24: Several
solution curves.

Finally, we can isolate y, to obtain an explicit representation of the general solution of the logistic equation,

$$y = \frac{A\, M\, e^{kMt}}{1 + A\, e^{kMt}}. \tag{5.10}$$

In Figure 6.24, we plot a number of these solution curves for various values of A (for the case where $M = 100$ and $k = 0.01$), along with the equilibrium solutions $y = 0$ and $y = 100$. You can see from Figure 6.24 that logistic growth consists of nearly exponential growth initially, followed by the graph becoming concave down and then asymptotically approaching the maximum, M.

Example 5.6   **Solving a Logistic Growth Problem**

Given a maximum sustainable population of $M = 1000$ (this could be measured in millions or tons, etc.) and growth rate $r = 0.007$, find an expression for the population at any time, t, given an initial population of $y(0) = 350$.

Solution   From the solution (5.10) of the logistic equation, we have

$$y = \frac{1000\, A\, e^{7t}}{1 + A\, e^{7t}}.$$

Next, the initial condition gives us

$$350 = y(0) = \frac{1000\, A}{1 + A}.$$

Solving for A, we obtain $A = \frac{35}{65}$, which gives us the solution of the IVP

$$y = \frac{35,000 e^{7t}}{65 + 35 e^{7t}}.$$

This solution is plotted in Figure 6.25.

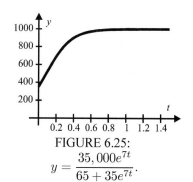

FIGURE 6.25:
$$y = \frac{35,000 e^{7t}}{65 + 35 e^{7t}}.$$

Note that in practice, the values of M and r are not known and must be estimated from a careful study of the particular population. We explore these issues further in the exercises.   ∎

Exercises 6.5

1. (writing) Discuss the differences between solving algebraic equations (e.g., $x^2 - 1 = 0$) and solving differential equations. Especially note what type of mathematical object you are solving for. ·

2. (writing) A differential equation is not separable if it can't be written in the form $g(y)\, y' = h(x)$. If you have an equation that you can't write in this form, how do you know whether it's really impossible or you just haven't figured it out yet? Discuss some general forms (e.g., $x + y$ and xy) that give you clues as to whether the equation is likely to separate or not.

3. (writing) The solution curves in Figures 6.22, 6.23a and 6.24 do not appear to cross. This is correct. If solution curves crossed at the point (x_1, y_1), then there would be two solutions satisfying the initial condition $y(x_1) = y_1$. Explain why this does not happen. In terms of the logistic equation as a model of population growth, explain why it is important to know that this does not happen.

4. (writing) The logistic equation includes a term in the differential equation that slows population growth as the population increases. Discuss some of the reasons why this occurs in real populations (human, animal and plant).

In exercises 5–12, determine whether or not the differential equation is separable.

5. $y' = (3x + 1)\cos y$

6. $y' = 2x(y - 1)$

7. $y' = (3x + y)\cos y$

8. $y' = 2x(y - x)$

9. $y' = x^2 y + y \cos x$

10. $y' = 2x \cos y - xy^3$

11. $y' = x^2 y - x \cos y$

12. $y' = x^3 - 2x + 1$

In exercises 13–26, the differential equation is separable. Find the general solution, in an explicit form if possible. Sketch several members of the family of solutions.

13. $y' = (x^2 + 1)y$

14. $y' = 2x(y - 1)$

15. $y' = 2x^2 y^2$

16. $y' = 2(y - 1)$

17. $y' = \dfrac{3}{y(1 + x^2)}$

18. $y' = \dfrac{3x}{y + 1}$

19. $y' = \dfrac{4x^2}{y^2}$

20. $y' = \dfrac{x^2 + 1}{y}$

21. $y' = y^2 - y$

22. $y' = e^{-y}$

23. $y' = \dfrac{xy}{1 + x^2}$

24. $y' = \dfrac{1 + y^2}{x + 1}$

25. $y' = \dfrac{4x \sin x}{y^3}$

26. $y' = \dfrac{x(y^2 + 1)}{2y}$

In exercises 27–34, solve the IVP, explicitly if possible.

27. $y' = 3(x + 1)^2 y,\ y(0) = 1$

28. $y' = \dfrac{x - 1}{y^2},\ y(0) = 2$

29. $y' = \dfrac{4x^2}{y}, y(0) = 2$ **30.** $y' = \dfrac{x-1}{y}, y(0) = -2$

31. $y' = \dfrac{4y}{x+3}, y(-2) = 1$ **32.** $y' = \dfrac{3x}{4y+1}, y(1) = 4$

33. $y' = \dfrac{4x}{\cos y}, y(0) = 0$ **34.** $y' = \dfrac{\tan y}{x}, y(1) = \dfrac{\pi}{2}$

In exercises 35–40, use equation (5.8) to help solve the IVP.

35. $y' = 3y(2-y), y(0) = 1$

36. $y' = y(3-y), y(0) = 2$

37. $y' = 2y(5-y), y(0) = 4$

38. $y' = 3y(2-y), y(0) = 1$

39. $y' = y(1-y), y(0) = \frac{3}{4}$

40. $y' = y(3-y), y(0) = 0$

41. The logistic equation is sometimes written in the form $y'(t) = r\,y(t)\,(1 - y(t)/M)$. Show that this is equivalent to equation (5.6) with $r/M = k$. Biologists have measured the values of the carrying capacity M and growth rate r for a variety of fish. Just for the halibut, approximate values are $r = 0.71$ year^{-1} and $M = 8 \times 10^7$ kg. If the initial biomass of halibut is $y(0) = 2 \times 10^7$ kg, find an equation for the biomass of halibut at any time. Sketch a graph of the biomass as a function of time.

42. Estimate how long it will take for the biomass of the halibut in exercise 41 to get within 10% of the carrying capacity.

43. Find the solution of equation (5.6) if $y(t) > M$.

44. Use the solution found in exercise 43 and the parameters for the halibut in exercise 41 to answer the following question. If the halibut biomass explodes to 3×10^8 kg, how long will it take for the population to drop back to within 10% of the carrying capacity?

In exercises 45–50, find each equilibrium solution. An equilibrium solution is a constant solution, which therefore is a solution of the differential equation $y' = 0$. Find all values of y such that $y' = 0$.

45. $y' = 3y(2-y)$ **46.** $y' = y(3-y)$

47. $y' = 3y(4-y^2)$ **48.** $y' = 3y(1-y^2)$

49. $y' = e^y(1-y)$ **50.** $y' = \cos y - y^2$

51. To emphasize what the equilibrium solutions in exercises 45–50 represent, find the solution of the logistic equation in exercise 45 subject to the following initial conditions: (a) $y(0) = 0$, (b) $y(0) = 2$.

52. Find the solution of the logistic equation in exercise 46 with the initial conditions (a) $y(0) = 0$, (b) $y(0) = 3$

53. The downward velocity of a falling object is modeled by the differential equation $\dfrac{dv}{dt} = 32 - 0.4v^2$. If $v(0) = 0$, the velocity will increase to a **terminal velocity**. The terminal velocity is an equilibrium solution where the upward air drag exactly cancels the downward gravitational force. Find the terminal velocity.

54. An object that is more aerodynamically sound than the object of exercise 53 would have a reduced drag coefficient, producing an equation such as $\dfrac{dv}{dt} = 32 - 0.2v^2$. Find the terminal velocity of this object.

55. In some chemical reactions, one molecule of substance A reacts with one molecule of substance B to form one molecule of a third substance X. If there is initially 10 g of substance A available and 12 g of substance B available, then the amount $x(t)$ of substance X formed by time t satisfies the IVP $x'(t) = r(10-x)(12-x), x(0) = 0$. Explain why, physically, it makes sense that $0 \le x \le 10$. Solve the IVP and indicate where you need this assumption.

56. The differential equation $y' = -ay\ln(y/b)$ (for positive constants a and b) arises in the study of the growth of some animal tumors. Solve the differential equation and sketch several members of the family of solutions. What adjective (e.g., rapid, moderate, slow) would you use to characterize this type of growth?

57. It is an interesting fact that the inflection point in the solution of a logistic equation (see Figure 6.24) occurs at $y = \frac{1}{2}M$. To verify this, you do not want to compute two derivatives of equation (5.10) and solve $y'' = 0$. This would be quite ugly and would give you the solution in terms of t instead of y. Instead, a more abstract approach works well. Start with the differential equation $y' = ky(M - y)$ and take derivatives of both sides. (Hint: use the product and chain rules on the right-hand side.) You should find that $y'' = ky'(M - 2y)$. Then, $y'' = 0$ if and only if $y' = 0$ or $y = \frac{1}{2}M$. Rule out $y' = 0$ by describing what the solution behaves like at the equilibrium values.

58. Use the technique of exercise 57 to show that solutions of $y' = -ay\ln(y/b)$ for positive constants a and b have at most one inflection point, which occurs at $y = b/e$.

59. (exploratory) Look up the census figures for the U.S. population starting in 1790. (You can find this information in any library, in virtually any almanac or encyclopedia.) Plot the data on a graph of population versus time. Does this look like the solution of a logistic equation? Briefly explain. If you wanted to model this data with a logistic function, you would need to estimate values for k and M. As shown in exercise 57, M equals twice the height of the inflection point. Explain why (for the logistic curve) the inflection point represents the point of maximum slope. Estimate this for the population data. To estimate r, note that for small populations, the logistic equation $y' = ry(1 - y/k) \approx ry$. Then r equals the rate of exponential increase. Show that for the first 50 years the U.S. population growth was approximately exponential and find the percentage increase as an estimate of r. With these values of r and M and the initial population in 1790, find a function describing the population. Test this model by comparing actual populations to predicted populations for several years.

60. (exploratory) An object traveling through the air is acted on by gravity (acting vertically), air resistance (acting in the direction opposite velocity) and other forces (such as a motor). An equation for the horizontal motion of a jet plane is $v' = c - f(v)/m$, where c is the thrust of the motor and $f(v)$ is the air resistance force. For some ranges of velocity, the air resistance actually **drops** substantially for higher velocities as the air around the object becomes turbu-

lent. For example, suppose that $v' = 32,000 - f(v)$ where

$$f(v) = \begin{cases} 0.8v^2 & \text{if } 0 \le v \le 100 \\ 0.2v^2 & \text{if } 100 < v \end{cases}.$$ To solve the initial value

problem $v' = 32,000 - f(v)$, $v(0) = 0$, start with the initial value problem $v' = 32,000 - 0.8v^2$, $v(0) = 0$. Solve this IVP

$$\left(\text{Hint: } \frac{1}{40,000 - v^2} = \frac{1}{400}\left[\frac{1}{200 + v} + \frac{1}{200 - v} \right] \right)$$

and determine the time t such that $v(t) = 100$. From this time forward, the equation becomes $v' = 32,000 - 0.2v^2$. Solve the IVP $v' = 32,000 - 0.2v^2$, $v(0) = 100$. Put this solution together with the previous solution to piece together a solution valid for all time.

6.6   Euler's Method

In section 6.5, we saw how to solve some simple first order differential equations, namely those that are separable. While there are numerous other special cases of differential equations whose solutions are known (you will encounter many of these in any beginning course in differential equations), the vast majority cannot be solved exactly. For instance, the equation

$$y' = x^2 + y^2 + 1$$

is not separable and cannot be solved using our current techniques. Nevertheless, **some** information about the solution(s) can be determined. In particular, since $y' = x^2 + y^2 + 1 > 0$, we can conclude that every solution is an increasing function. This type of information is called **qualitative**, since it tells us about some quality of the solution without providing any specific quantitative information.

In this section, we examine first order differential equations in a more general setting. We consider the more general first order equation of the form

$$y' = f(x, y). \tag{6.1}$$

While we cannot solve all such equations, it turns out that there are many numerical methods available for approximating the solution of such problems. We will study one simple method here, called Euler's method.

We begin by observing that any solution of equation (6.1) is a function $y = y(x)$ whose slope at any particular point (x, y) is given by $f(x, y)$. In order to get a handle on what a solution curve looks like, we draw a short line segment through each of a sequence of points (x, y), with slope $f(x, y)$, respectively. This collection of line segments is called the **direction field** or **slope field** of the differential equation. Notice that if a particular solution curve passes through a given point (x, y), then its slope at that point is $f(x, y)$. Thus, the direction field gives an indication of the behavior of the family of solutions of a differential equation.

Example 6.1   **Constructing a Direction Field**
Construct the direction field for

$$y' = \frac{1}{2}y. \tag{6.2}$$

Solution   All that needs to be done is to plot a number of points and through each point (x, y) a short line segment with slope $f(x, y)$ is then drawn. For example, at the point $(0, 1)$, draw a short line segment with slope

$$y'(0) = f(0, 1) = \frac{1}{2}(1) = \frac{1}{2}.$$

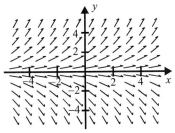

FIGURE 6.26a: Direction field
for $y' = \frac{1}{2}y$.

Draw similar segments at 25 to 30 points. This is a bit tedious to do by hand, but a good graphing utility can do this for you with minimal effort. See Figure 6.26a for the direction field for equation (6.2). Notice that equation (6.2) is separable. We leave it as an exercise to produce the general solution

$$y = A\,e^{\frac{1}{2}x}.$$

We plot a number of the curves in this family of solutions in Figure 6.26b using the same window we used for Figure 6.26a. Notice that if you connected some of the line segments in Figure 6.26a, you would obtain a close approximation to the exponential curves depicted in Figure 6.26b. What is significant about this is that Figure 6.26a was constructed using elementary algebra, **without** first solving the differential equation. That is, by constructing the direction field, we can obtain a reasonably good picture of how the solution curves behave. Since most differential equations are not solvable exactly, this is an extremely useful observation. This is what we mean by qualitative information about the solution: we get a graphical idea of how solutions behave, but no details, such as the value of a solution at specific points. We'll see later in this section that we can obtain approximate values of the solution of an IVP numerically.

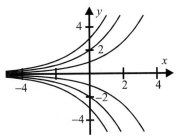

FIGURE 6.26b: Several solutions
of $y' = \frac{1}{2}y$.

As we have already seen, differential equations are used to describe a wide variety of phenomena in science and engineering. Among many other applications, differential equations are used to find flow lines or equipotential lines for electromagnetic fields. In such cases, it is very helpful to visualize solutions graphically, so as to gain an intuitive understanding of the behavior of such solutions and the physical phenomenon they are modeling.

Example 6.2   Using a Direction Field to Visualize the Behavior of Solutions
Construct the direction field for

$$y' = x + e^{-y}.$$

FIGURE 6.27a: Direction field
for $y' = x + e^{-y}$.

Solution   There's really no trick to this; just draw a number of line segments with the correct slope. Again, we let our CAS do this for us and obtained the direction field in Figure 6.27a. Unlike example 6.1, you do not know how to solve this differential equation exactly. Even so, you should be able to clearly see from the direction field how solutions behave. For example, solutions that start out in the second quadrant initially decrease very rapidly, may dip into the third quadrant and then get pulled into the first quadrant and increase quite rapidly toward infinity. This is quite a bit of information to determine using little more than elementary algebra. In Figure 6.27b, we have plotted the solution of the differential equation that also satisfies the initial condition $y(-4) = 2$. We'll see how to generate such an approximate solution later in this section. Note how well this corresponds with what you get by connecting a few of the line segments in Figure 6.27a.

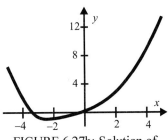

FIGURE 6.27b: Solution of
$y' = x + e^{-y}$ passing through
$(-4, 2)$.

You have already seen (in sections 6.4 and 6.5) how differential equation models can provide important information about how populations change over time. A model that includes a **critical threshold** is

$$P'(t) = -2\left[1 - P(t)\right]\left[2 - P(t)\right]P(t),$$

where $P(t)$ represents the size of a population at time t.

A simple context in which to understand a critical threshold is with the problem of the sudden infestations of pests. For instance, suppose that you have some method for removing ants from your home. As long as the reproductive rate of the ants is lower than your removal rate, you will keep the ant population under control. However, as soon as the ant reproductive rate becomes larger than your removal rate (i.e., crosses the critical threshold), you won't be able to keep up with the extra ants and you will suddenly be faced with a big ant problem. We see this type of behavior in the following example.

FIGURE 6.28:
Direction field for $P'(t) = -2\left[1 - P(t)\right]\left[2 - P(t)\right]P(t)$.

Example 6.3   Population Growth with a Critical Threshold
Draw the direction field for

$$P'(t) = -2\left[1 - P(t)\right]\left[2 - P(t)\right]P(t)$$

and discuss the eventual size of the population.

Solution   The direction field is particularly easy to sketch here since the right-hand side depends on P, but not on t. If $P(t) = 0$, then $P'(t) = 0$, also, so that the direction field is horizontal. The same is true for $P(t) = 1$ and $P(t) = 2$. If $0 < P(t) < 1$, then $P'(t) < 0$ and the solution decreases. If $1 < P(t) < 2$, then $P'(t) > 0$ and the solution increases. Finally, if $P(t) > 2$, then $P'(t) < 0$ and the solution decreases. Putting all of these pieces together, we get the direction field seen in Figure 6.28. The constant solutions $P(t) = 0$, $P(t) = 1$ and $P(t) = 2$ are called **equilibrium solutions**. $P(t) = 1$ is called an **unstable equilibrium**, since populations that start near 1 don't remain close to 1. $P(t) = 0$ and $P(t) = 2$ are called **stable equilibria**, since populations either rise to 2 or drop to 0 (extinction), depending on which side of the critical threshold $P(t) = 1$ they are on. (Look again at Figure 6.28.)

In cases where you are interested in finding a particular solution, the numerous arrows of a direction field can be distracting. Euler's method, developed below, allows you to approximate a single solution curve. The method is quite simple, based almost entirely on the idea of a direction field. However, Euler's method does not provide particularly accurate approximations. More accurate, but more complicated methods will be explored in the exercises.

Consider the IVP

$$y' = f(x, y), \quad y(x_0) = y_0.$$

We must emphasize once again that, assuming there is a solution $y = y(x)$, the differential equation tells us that the **slope** of the tangent line to the solution curve at any point (x, y) is given by $f(x, y)$. Remember that the tangent line to a curve stays close to that curve near the point of tangency. This idea was the key to both Newton's method and differentials. Notice that we already know one point on the graph of $y = y(x)$, namely, the initial point (x_0, y_0). Referring to Figure 6.29, if we would like to approximate the value of the solution at $x = x_1$ [i.e., $y(x_1)$] and if x_1 is not too far from x_0, then we could follow the tangent line at (x_0, y_0) to the point corresponding to $x = x_1$ and use the $y-$value at that point (call it y_1) as an approximation to $y(x_1)$. Again, this is virtually the same thinking we employed when we devised Newton's method and differential (tangent line) approximations. It's a simple matter to obtain the equation of the tangent line at $x = x_0$:

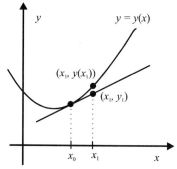

FIGURE 6.29: Tangent line approximation.

$$m_{\text{tan}} = y'(x_0) = f(x_0, y_0) = \frac{y - y_0}{x - x_0}$$

or

$$y = y_0 + y'(x_0)(x - x_0).$$

Thus, an approximation to the value of the solution at $x = x_1$ is the y-coordinate of the point on the tangent line corresponding to $x = x_1$, that is,

$$y(x_1) \approx y_1 = y_0 + y'(x_0)\,(x_1 - x_0). \tag{6.3}$$

You have only to glance at Figure 6.29 to realize that this approximation is valid only when x_1 is close to x_0. In solving an IVP, though, we are usually interested in finding the value of the solution on an interval $[a, b]$ of the x-axis. Since this can only be done when we can find an exact solution, we settle for finding an approximate solution at a sequence of points in the interval $[a, b]$. Still, this tangent line approximation is valid only for points very close to the initial point. We can approximate the solution at other points of interest, as follows. First, we partition the interval $[a, b]$ into n equal-sized pieces (a **regular partition**; where did you see this notion before?):

$$a = x_0 < x_1 < x_2 < \cdots < x_n = b,$$

where

$$x_{i+1} - x_i = h,$$

for all $i = 0, 1, \ldots, n - 1$. We call h the **step size**. From the tangent line approximation (6.3), we already have

$$\begin{aligned} y(x_1) \approx y_1 &= y_0 + y'(x_0)\,(x_1 - x_0) \\ &= y_0 + h\,f(x_0, y_0), \end{aligned}$$

where we have replaced $(x_1 - x_0)$ by the step size, h and used the differential equation to write $y'(x_0) = f(x_0, y_0)$. Next, we would like to approximate the value of $y(x_1)$. Certainly, we could use the tangent line at the point $(x_1, y(x_1))$ to produce a tangent line approximation, but we don't even know the y-coordinate of the point of tangency, $y(x_1)$. We do, however, have an approximation for this, produced in the preceding step. So, we make the further approximation

$$\begin{aligned} y(x_2) \approx y(x_1) + y'(x_1)\,(x_2 - x_1) \\ = y(x_1) + h\,f(x_1, y(x_1)), \end{aligned}$$

where we have used the differential equation to replace $y'(x_1)$ by $f(x_1, y(x_1))$ and used the fact that $x_2 - x_1 = h$. Finally, we approximate $y(x_1)$ by the approximation obtained in the previous step, y_1, to obtain

$$\begin{aligned} y(x_2) &\approx y(x_1) + h\,f(x_1, y(x_1)) \\ &\approx y_1 + h\,f(x_1, y_1) = y_2. \end{aligned}$$

Continuing in this way, we obtain the sequence of approximate values

$$\boxed{y(x_{i+1}) \approx y_{i+1} = y_i + h\,f(x_i, y_i),\ \text{for}\ i = 0, 1, 2, \ldots} \tag{6.4}$$

This tangent line method of approximation is called **Euler's method**.

Example 6.4   Using Euler's Method

Use Euler's method to approximate the solution of the IVP

$$y' = y,\ \ y(0) = 1.$$

Solution   Of course, this is just about the simplest differential equation imaginable. You can probably solve it by inspection, but if not, notice that it's separable and that the solution of the IVP is $y = y(x) = e^x$. We will use this exact solution as a basis for comparison for the performance of Euler's method. From (6.4) with $f(x,y) = y$ and taking $h = 1$, we have

$$y(x_1) \approx y_1 = y_0 + h \, f(x_0, y_0)$$
$$= y_0 + h \, y_0 = 1 + 1 \, (1) = 2.$$

Likewise, for further approximations, we have

$$y(x_2) \approx y_2 = y_1 + h \, f(x_1, y_1)$$
$$= y_1 + h \, y_1 = 2 + 1 \, (2) = 4,$$

$$y(x_3) \approx y_3 = y_2 + h \, f(x_2, y_2)$$
$$= y_2 + h \, y_2$$
$$= 4 + 1 \, (4) = 8$$

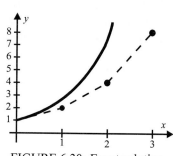

FIGURE 6.30: Exact solution versus the approximate solution (dashed line).

and so on. In this way, we construct a sequence of approximate values of the solution function. Remember, we are applying this approximate method to a problem with a known exact solution, so that we might compare the approximate and the exact values. In Figure 6.30, we have plotted the exact solution (solid line) against the approximate solution obtained from Euler's method (dashed line). Notice how the error grows, as x gets farther and farther from the initial point. This is characteristic of Euler's method (and other similar methods). This growth in error becomes even more apparent if we look at a table of values of the approximate and exact solutions together. We display these in the table that follows where we have used $h = 0.1$ (values are displayed to seven digits).

x	$Euler$	$Exact$	$Error = exact - Euler$
0.1	1.1	1.1051709	0.0051709
0.2	1.21	1.2214028	0.0114028
0.3	1.331	1.3498588	0.0188588
0.4	1.4641	1.4918247	0.0277247
0.5	1.61051	1.6487213	0.0382113
0.6	1.71561	1.8221188	0.1065088
0.7	1.9487171	2.0137527	0.0650356
0.8	2.1435888	2.2255409	0.0819521
0.9	2.3579477	2.4596031	0.1016554
1.0	2.5937425	2.7182818	0.1245393

As you might expect from our development of Euler's method, the smaller we make h, the more accurate the approximation at a given point tends to be. As well, the smaller the value of h, the more steps it takes to reach a given value of x. In the following table, we display the Euler's method approximation, the error and the number of steps needed to reach $x = 1.0$. Here, the exact value of the solution is $y = e^1 \approx 2.718281828459$.

h	$Euler$	$Error = Exact - Euler$	$Number\ of\ steps$
1.0	2	0.7182818	1
0.5	2.25	0.4682818	2
0.25	2.4414063	0.2768756	4
0.125	2.5657845	0.1524973	8
0.0625	2.6379285	0.0803533	16
0.03125	2.6769901	0.0412917	32
0.015625	2.697345	0.0209369	64
0.0078125	2.707739	0.0105428	128
0.00390625	2.7129916	0.0052902	256

From the table, you should observe that each time the step size h is cut in half, the error is also cut roughly in half. This increased accuracy though, comes at the cost of the additional steps of Euler's method required to reach a given point (doubled when h is halved).

■

The point of having a numerical method, of course, is to find meaningful approximations to the solution of problems that we do not know how to solve exactly. Our final example is of this type.

FIGURE 6.31: Direction field for $y' = x^2 + y^2$.

Example 6.5   Finding an Approximate Solution

Find an approximate solution for the IVP

$$y' = x^2 + y^2, \quad y(-1) = -\frac{1}{2}.$$

Solution   First, let's take a look at the direction field, so that we can see how solutions to this differential equation should behave (see Figure 6.31). Using Euler's method, with $h = 0.1$, we get

$$y(x_1) \approx y_1 = y_0 + h\,f(x_0, y_0)$$
$$= y_0 + h\,(x_0^2 + y_0^2)$$
$$= -\frac{1}{2} + 0.1 \left[(-1)^2 + \left(-\frac{1}{2}\right)^2\right] = -0.375$$

and

$$y(x_2) \approx y_2 = y_1 + h\,f(x_1, y_1)$$
$$= y_1 + h\,(x_1^2 + y_1^2)$$
$$= -0.375 + 0.1 \left[(-0.9)^2 + (-0.375)^2\right] = -0.2799375$$

and so on. Continuing in this way, we generate the table of values that follows.

x	Euler	x	Euler	x	Euler
−0.9	−0.375	0.1	−0.0575822	1.1	0.3369751
−0.8	−0.2799375	0.2	−0.0562506	1.2	0.4693303
−0.7	−0.208101	0.3	−0.0519342	1.3	0.6353574
−0.6	−0.1547704	0.4	−0.0426645	1.4	0.8447253
−0.5	−0.116375	0.5	−0.0264825	1.5	1.1120813
−0.4	−0.0900207	0.6	−0.0014123	1.6	1.4607538
−0.3	−0.0732103	0.7	0.0345879	1.7	1.9301340
−0.2	−0.0636743	0.8	0.0837075	1.8	2.5916757
−0.1	−0.0592689	0.9	0.1484082	1.9	3.587354
0	−0.0579176	1.0	0.2316107	2.0	5.235265

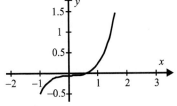

FIGURE 6.32: Approximate solution of $y' = x^2 + y^2$, passing through $\left(-1, -\frac{1}{2}\right)$.

In Figure 6.32, we have displayed a graph of the data in the preceding table. Take particular note of how well this corresponds with the direction field in Figure 6.31. Since this corresponds so well with the behavior you expect from the direction field, you should expect that there are no gross errors in this approximate solution. (Certainly, there is always some level of round-off and other numerical errors.)

■

Exercises 6.6

1. (writing) For Euler's method, explain why using a smaller step size should produce a better approximation.

2. (writing) Look back at the direction field in Figure 6.31 and the Euler's method solution in Figure 6.32. Describe how the direction field gives you a more accurate sense of the exact solution. Given this, explain why Euler's method is important. (Hint: How would you get the table of numbers below Figure 6.32 from a direction field?)

3. (writing) In the situation of example 6.3, if you only needed to know the stability of an equilibrium solution, explain why a qualitative method is preferred over trying to solve the differential equation. Describe one situation in which you would need to solve the equation.

4. (writing) Imagine superimposing solution curves over Figure 6.27a. Explain why the Euler's method approximation takes you from one solution curve to a nearby one. Use one of the examples in this section to describe how such a small error could lead to very large errors in approximations for large values of x.

In exercises 5–16, construct four of the direction field arrows by hand and use your CAS or calculator to do the rest. Describe the general pattern of solutions.

5. $y' = x + 4y$ **6.** $y' = \sqrt{x^2 + y^2}$

7. $y' = 2y - y^2$ **8.** $y' = y^3 - 1$

9. $y' = 2xy - y^2$ **10.** $y' = y^3 - x$

11. $y' = \cos y + 1/x$ **12.** $y' = \sin y - x^2$

13. $y' = 1 - y^2 - e^{-x}$ **14.** $y' = 1 - x^2 + e^{-y}$

15. $y' = 1 - y^2/x$ **16.** $y' = y^3/x - 1$

In exercises 17–24, use Euler's method with $h = 0.1$ and $h = 0.05$ to approximate $y(1)$ and $y(2)$. Show the first two steps by hand.

17. $y' = 2xy,\ y(0) = 1$ **18.** $y' = y^2,\ y(0) = 2$

19. $y' = 4y - y^2,\ y(0) = 1$ **20.** $y' = x/y^2,\ y(0) = 2$

21. $y' = 1 - y + e^{-x},\ y(0) = 3$

22. $y' = \sin y - x^2,\ y(0) = 1$ **23.** $y' = \sqrt{x + y},\ y(0) = 1$

24. $y' = \sqrt{x^2 + y^2},\ y(0) = 4$

25. Find the exact solutions in exercises 17 and 18 and compare $y(1)$ and $y(2)$ to the Euler's method approximations.

26. Find the exact solutions in exercises 19 and 20 and compare $y(1)$ and $y(2)$ to the Euler's method approximations.

27. Sketch the direction fields for exercises 21 and 22, highlight the curve corresponding to the given initial condition and compare the Euler's method approximations to the location of the curve at $x = 1$ and $x = 2$.

28. Sketch the direction fields for exercises 23 and 24, highlight the curve corresponding to the given initial condition and compare the Euler's method approximations to the location of the curve at $x = 1$ and $x = 2$.

In exercises 29–34, find the equilibrium solutions and determine which are stable and which are unstable.

29. $y' = 2y - y^2$ **30.** $y' = y^3 - 1$

31. $y' = y^2 - y^4$ **32.** $y' = e^{-y} - 1$

33. $y' = (1 - y)\sqrt{1 + y^2}$ **34.** $y' = \sqrt{1 - y^2}$

35. Zebra stripes and patterns on butterfly wings are thought to be the result of gene-activated chemical processes. Suppose $g(t)$ is the amount of gene that is activated at time t. The differential equation $g' = -g + \dfrac{3g^2}{1 + g^2}$ has been used to model the process. Show that there are three equilibrium solutions, 0 and two positive solutions a and b, with $a < b$. Show that $g' > 0$ if $a < g < b$ and $g' < 0$ if $0 < g < a$ or $g > b$. Explain why $\lim_{t \to \infty} g(t)$ could be 0 or b, depending on the initial amount of activated gene. Suppose that a patch of zebra skin extends from $x = 0$ to $x = 4\pi$ with an initial activated-gene distribution $g(0) = \frac{3}{2} + \frac{3}{2}\sin x$ at location x. If black corresponds to an eventual activated-gene level of 0 and white corresponds to an eventual activated-gene level of b, show what the zebra stripes will look like.

36. Many species of trees are plagued by sudden infestations of worms. Let $x(t)$ be the population of a species of worm on a particular tree. For some species, a model for population change is $x' = 0.1x(1 - x/k) - x^2/(1 + x^2)$ for some positive constant k. If $k = 10$, show that there is only one positive equilibrium solution. If $k = 50$, show that there are three positive equilibrium solutions. Sketch the direction field for $k = 50$. Explain why the middle equilibrium value is called a *threshold*. An outbreak of worms corresponds to crossing the threshold for a large value of k (k is determined by the resources available to the worms).

37. (exploratory) There are several ways of deriving the Euler's method formula. One benefit of having an alternative derivation is that it may inspire an improvement of the method. Here, we use an alternative form of Euler's method to derive a method known as the **Improved Euler's method**. Start with the differential equation $y'(x) = f(x, y(x))$ and integrate both sides from $x = x_n$ to $x = x_{n+1}$. Show that

$$y(x_{n+1}) = y(x_n) + \int_{x_n}^{x_{n+1}} f(x, y(x))\, dx.$$ Given $y(x_n)$, then, you can estimate $y(x_{n+1})$ by estimating the integral $\int_{x_n}^{x_{n+1}} f(x, y(x))\, dx$. One such estimate is a Riemann sum using left-endpoint evaluation, given by $f(x_n, y(x_n))\Delta x$. Show that with this estimate you get Euler's method. There are numerous ways of getting better estimates of the integral. One is to use the Trapezoid rule,

$$\int_{x_n}^{x_{n+1}} f(x, y(x))\, dx \approx \frac{f(x_n, y(x_n)) + f(x_{n+1}, y(x_{n+1}))}{2}\Delta x.$$

The drawback with this estimate is that you know $y(x_n)$ but you do **not** know $y(x_{n+1})$. Briefly explain why this statement is correct. The way out is to use Euler's method; you do not know $y(x_{n+1})$ but you can approximate it by $y(x_{n+1}) \approx y(x_n) + h\, f(x_n, y(x_n))$. Put all of this together to get the Improved Euler's method

$$y_{n+1} = y_n + \frac{h}{2}[f(x_n, y_n) + f(x_n + h, y_n + h f(x_n, y_n))].$$

Use the Improved Euler's method for the IVP $y' = y$, $y(0) = 1$ with $h = 0.1$ to compute y_1, y_2 and y_3. Compare to the exact values and the Euler's method approximations given in example 6.4.

38. (exploratory). In sections 6.4 and 6.5, you explored some differential equation models of population growth. An obvious flaw in those models was the consideration of a single

species in isolation. In this exercise, you will investigate a **predator-prey** model. In this case, there are two species, X and Y, with populations $x(t)$ and $y(t)$, respectively. The general form of the model is

$$x'(t) = b\, x(t)\, y(t) - a\, x(t)$$
$$y'(t) = -b\, x(t)\, y(t) + c\, y(t)$$

for positive constants a, b and c. First, look carefully at the equations. The term $b\, x(t)\, y(t)$ is included to represent the effects of encounters between the species. This effect is negative on species Y and positive on species X. If $b = 0$, the species don't interact at all. In this case, show that species X dies out (with death rate a) and species Y thrives (with growth rate c). Given all of this, explain why X must be the predator and Y the prey. Next, you should find the equilibrium point for coexistence. That is, find positive values $\bar{x}$ and $\bar{y}$ such that both $x'(t) = 0$ and $y'(t) = 0$. For this problem, think of Y as an insect that damages farmers' crops and X as a natural predator (e.g., a bat). A farmer might decide to use a pesticide to reduce the damage caused by the Y's. Briefly explain why the effects of the pesticide might be to decrease the value of c and increase the value of a. Now, determine how these changes affect the equilibrium values. Show that the pest population Y actually increases and the predator population X decreases. Explain, in terms of the interaction between predator and prey, how this could happen. The moral is that the long-range effects of pesticides can be the exact opposite of the short-range (and desired) effects.

6.7   The Inverse Trigonometric Functions

In this section, we enlarge the set of functions available to you by defining inverses to the trigonometric functions. Before we get started, though, let's again look at a graph of $y = \sin x$ (see Figure 6.33). Notice that we cannot define an inverse function, since $\sin x$ is not one-to-one. We can remedy this by looking at only a portion of the domain. If we restrict the domain to the interval $\left[-\frac{\pi}{2}, \frac{\pi}{2}\right]$, then $y = \sin x$ is one-to-one there (see Figure 6.34) and hence, has an inverse. We thus define the inverse sine function by

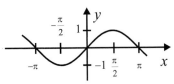

FIGURE 6.33: $y = \sin x$.

$$\boxed{y = \sin^{-1} x \text{ if and only if } \sin y = x \textbf{ and } -\tfrac{\pi}{2} \leq y \leq \tfrac{\pi}{2}.} \tag{7.1}$$

It is convenient to think of this definition as follows. If $y = \sin^{-1} x$, then y is the angle (between $-\frac{\pi}{2}$ and $\frac{\pi}{2}$) for which $\sin y = x$. You should note that we could have selected any interval on which $\sin x$ is one-to-one, but $\left[-\frac{\pi}{2}, \frac{\pi}{2}\right]$ is the most convenient. To see that these are indeed inverse functions, you should note that

$$\sin\left(\sin^{-1} x\right) = x, \text{ for all } x \in [-1, 1]$$

and

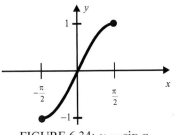

FIGURE 6.34: $y = \sin x$ on $\left[-\frac{\pi}{2}, \frac{\pi}{2}\right]$.

$$\sin^{-1}\left(\sin x\right) = x, \text{ for all } x \in \left[-\frac{\pi}{2}, \frac{\pi}{2}\right]. \tag{7.2}$$

Read equation (7.2) very carefully. It **does not** say that $\sin^{-1}(\sin x) = x$ for **all** x, but rather, **only** for those in the restricted domain, $\left[-\frac{\pi}{2}, \frac{\pi}{2}\right]$.

Remark 7.1   Mathematicians often use the notation $\arcsin x$ in place of $\sin^{-1} x$. People will read $\sin^{-1} x$ interchangeably as "inverse sine of x" or "arcsine of x."

Example 7.1   Evaluating an Inverse Sine

Evaluate $\sin^{-1}\left(\frac{\sqrt{3}}{2}\right)$.

Solution   We look for the angle θ in the interval $\left[-\frac{\pi}{2}, \frac{\pi}{2}\right]$ for which $\sin \theta = \frac{\sqrt{3}}{2}$. Note that since $\sin\left(\frac{\pi}{3}\right) = \frac{\sqrt{3}}{2}$ and $\frac{\pi}{3} \in \left[-\frac{\pi}{2}, \frac{\pi}{2}\right]$, we have that $\sin^{-1}\left(\frac{\sqrt{3}}{2}\right) = \frac{\pi}{3}$.

■

Example 7.2   Evaluating an Inverse Sine with a Negative Argument

Evaluate $\sin^{-1}\left(-\frac{1}{2}\right)$.

Solution   Here, note that $\sin\left(-\frac{\pi}{6}\right) = -\frac{1}{2}$ and $-\frac{\pi}{6} \in \left[-\frac{\pi}{2}, \frac{\pi}{2}\right]$. Thus,

$$\sin^{-1}\left(-\frac{1}{2}\right) = -\frac{\pi}{6}.$$

■

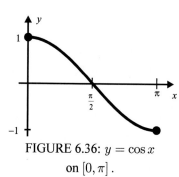

FIGURE 6.35: $y = \sin^{-1} x$.

Judging by the preceding two examples, you might think that (7.1) is a roundabout way of defining a function. If so, you've got the idea exactly. In fact, we want to emphasize that what we know about the inverse sine function is principally through reference to the sine function. We will not have an independent definition of arcsine, nor are there any algebraic formulas for this function. (These things are true of most inverse functions.) Further, you should recall from our discussion in section 6.2 that we can draw a graph of $y = \sin^{-1} x$ simply by reflecting the graph of $y = \sin x$ on the interval $\left[-\frac{\pi}{2}, \frac{\pi}{2}\right]$ (from Figure 6.34) through the line $y = x$ (see Figure 6.35).

Turning to $y = \cos x$, can you think of how to restrict the domain to make the function one-to-one? Notice that restricting the domain to the interval $\left[-\frac{\pi}{2}, \frac{\pi}{2}\right]$, as we did for the inverse sine function will not work here. (Why not?) The simplest way to do this is to restrict its domain to the interval $[0, \pi]$ (see Figure 6.36). Consequently, we define

FIGURE 6.36: $y = \cos x$ on $[0, \pi]$.

$$\boxed{y = \cos^{-1} x \text{ if and only if } \cos y = x \textbf{ and } 0 \leq y \leq \pi.} \tag{7.3}$$

Note that here, we have

$$\cos\left(\cos^{-1} x\right) = x, \text{ for all } x \in [-1, 1]$$

and

$$\cos^{-1}\left(\cos x\right) = x, \text{ for all } x \in [0, \pi].$$

As with the definition of arcsine, it is helpful to think of $\cos^{-1} x$ as that angle θ in $[0, \pi]$ for which $\cos \theta = x$. As with $\sin^{-1} x$, it is common to use $\cos^{-1} x$ and $\arccos x$ interchangeably.

Example 7.3   Evaluating an Inverse Cosine

Evaluate $\cos^{-1}(0)$.

Solution   You will need to find that angle θ in $[0, \pi]$ for which $\cos \theta = 0$. It's not hard to see that $\cos^{-1}(0) = \frac{\pi}{2}$. If you calculate this on your calculator and get 90, your calculator is in degrees mode. In this case, you should immediately change it to radians mode.

■

FIGURE 6.37: $y = \cos^{-1} x$.

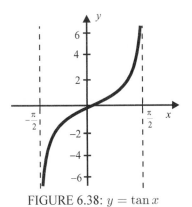

FIGURE 6.38: $y = \tan x$
on $\left[-\frac{\pi}{2}, \frac{\pi}{2}\right]$.

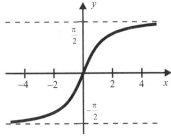

FIGURE 6.39: $y = \tan^{-1} x$.

FIGURE 6.40:
$y = \sec x$ on $[0, \pi]$.

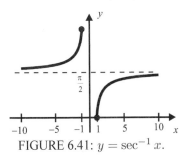

FIGURE 6.41: $y = \sec^{-1} x$.

Example 7.4   Evaluating an Inverse Cosine with a Negative Argument

Evaluate $\cos^{-1}\left(-\frac{\sqrt{2}}{2}\right)$.

Solution   Here, look for the angle $\theta \in [0, \pi]$ for which $\cos \theta = -\frac{\sqrt{2}}{2}$. Notice that $\cos\left(\frac{3\pi}{4}\right) = -\frac{\sqrt{2}}{2}$ and $\frac{3\pi}{4} \in [0, \pi]$. Consequently,

$$\cos^{-1}\left(-\frac{\sqrt{2}}{2}\right) = \frac{3\pi}{4}.$$

∎

Once again, we obtain the graph of this inverse function by reflecting the graph of $y = \cos x$ on the interval $[0, \pi]$ (seen in Figure 6.36) through the line $y = x$ (see Figure 6.37).

We can define inverses for each of the four remaining trig functions in similar ways. For $y = \tan x$, we restrict the domain to the interval $\left(-\frac{\pi}{2}, \frac{\pi}{2}\right)$. Think about why the endpoints of this interval are not included (see Figure 6.38). Having done this, you should readily see that we define

$$\boxed{y = \tan^{-1} x \text{ if and only if } \tan y = x \textbf{ and } -\tfrac{\pi}{2} < y < \tfrac{\pi}{2}.} \tag{7.4}$$

The graph of $y = \tan^{-1} x$ is then as seen in Figure 6.39, found by reflecting the graph in Figure 6.38 through the line $y = x$.

Example 7.5   Evaluating an Inverse Tangent

Evaluate $\tan^{-1}(1)$.

Solution   You must look for the angle, θ on the interval $\left(-\frac{\pi}{2}, \frac{\pi}{2}\right)$ for which $\tan \theta = 1$. This is easy enough. Since $\tan\left(\frac{\pi}{4}\right) = 1$ and $\frac{\pi}{4} \in \left(-\frac{\pi}{2}, \frac{\pi}{2}\right)$, we have that $\tan^{-1}(1) = \frac{\pi}{4}$.

∎

We now turn to defining an inverse for $\sec x$. First, we must issue a disclaimer. As we have indicated, there are any number of ways to suitably restrict the domains of the trigonometric functions in order to make them one-to-one. With the first three we've seen, there has been an obvious choice of how to do this and there is general agreement among mathematicians on the choice of these intervals. In the case of $\sec x$, this is **not** true. There are several reasonable ways in which to suitably restrict the domain and different authors restrict these differently. We have (arbitrarily) chosen to restrict the domain to be $\left[0, \frac{\pi}{2}\right) \cup \left(\frac{\pi}{2}, \pi\right]$. You might initially think that this looks strange. Why not use all of $[0, \pi]$? You need only think about the definition of $\sec x$ to see why we had to exclude the point $x = \frac{\pi}{2}$. See Figure 6.40 for a graph of the $\sec x$ on this domain. (Note the vertical asymptote at $x = \frac{\pi}{2}$.) Consequently, we define

$$\boxed{y = \sec^{-1} x \text{ if and only if } \sec y = x \textbf{ and } y \in \left[0, \tfrac{\pi}{2}\right) \cup \left(\tfrac{\pi}{2}, \pi\right].} \tag{7.5}$$

A graph of $\sec^{-1} x$ is shown in Figure 6.41.

Example 7.6   Evaluating an Inverse Secant

Evaluate $\sec^{-1}(-\sqrt{2})$.

Solution   You must look for an angle θ with $\theta \in \left[0, \frac{\pi}{2}\right) \cup \left(\frac{\pi}{2}, \pi\right]$, for which $\sec \theta = -\sqrt{2}$. Notice that if $\sec \theta = -\sqrt{2}$, then $\cos \theta = -\frac{1}{\sqrt{2}} = -\frac{\sqrt{2}}{2}$. Recall from example 7.4

that $\cos \frac{3\pi}{4} = -\frac{\sqrt{2}}{2}$. Further, the angle $\frac{3\pi}{4}$ is in the interval $\left(\frac{\pi}{2}, \pi\right]$ and so, $\sec^{-1}(-\sqrt{2}) = \frac{3\pi}{4}$.

∎

Calculators do not usually have built-in functions for $\sec x$ or $\sec^{-1} x$. In this case, you must convert the desired secant value to a cosine value and use the inverse cosine function.

Remark 7.2   We can likewise define inverses to $\cot x$ and $\csc x$. As these functions are used only infrequently, we will omit them here and examine them in the exercises.

Often, as a part of a larger problem (for example, evaluating an integral by some of the methods discussed in Chapter 7) you will need to recognize some relationship between the trigonometric functions and their inverses. We present several clues here. When you are faced with these problems, our best advice is to keep in mind the definitions of the inverse functions and then draw a picture.

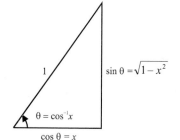

FIGURE 6.42: $\theta = \cos^{-1} x$.

Example 7.7   Simplifying Expressions Involving Inverse Trigonometric Functions
Simplify $\sin\left(\cos^{-1} x\right)$ and $\tan\left(\cos^{-1} x\right)$.

Solution   Do not look for some arcane formula to help you out. **Think** first: $\cos^{-1} x$ is the angle (call it θ) for which $x = \cos \theta$. First, consider the case where $x > 0$. Looking at Figure 6.42, we have drawn a right triangle, with hypotenuse 1 and adjacent angle θ. From the definition of the sine and cosine, then, we have that the base of the triangle is $\cos \theta = x$ and the altitude is $\sin \theta$, which by the Pythagorean Theorem is

$$\sin \theta = \sqrt{1 - x^2}.$$

There are two subtle points that we should illuminate here. First, what if anything in Figure 6.42 changes if $x < 0$? (Think about this one.) Second, since $\theta = \cos^{-1} x$, we have that $\theta \in [0, \pi]$ and hence, $\sin \theta \geq 0$. Finally, you can also read from Figure 6.42 that

$$\tan\left(\cos^{-1} x\right) = \tan \theta = \frac{\sin \theta}{\cos \theta} = \frac{\sqrt{1 - x^2}}{x}.$$

Note that this last identity is valid, regardless of whether $x = \cos \theta$ is positive or negative.

∎

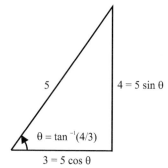

FIGURE 6.43: $\theta = \tan^{-1}\left(\frac{4}{3}\right)$.

Example 7.8   Simplifying an Expression Involving an Inverse Tangent
Simplify $\sin\left(\tan^{-1} \frac{4}{3}\right)$.

Solution   Once again, keep in mind that $\tan^{-1} \frac{4}{3}$ is the angle θ, in the interval $\left(-\frac{\pi}{2}, \frac{\pi}{2}\right)$ for which $\tan \theta = \frac{4}{3}$. As an aid, we visualize the triangle shown in Figure 6.43. Note that we have made the side opposite the angle, θ to be of length 4 and the adjacent side of length 3, so that the tangent of the angle is $\frac{4}{3}$, as desired. Of course, this makes the hypotenuse 5, by the Pythagorean Theorem. Using the picture as a guide, we can read off the desired value.

$$\sin \theta = \sin\left(\tan^{-1} \frac{4}{3}\right) = \frac{4}{5}.$$

While we're at it, we should also observe that we have found that

$$\cos \theta = \cos\left(\tan^{-1} \frac{4}{3}\right) = \frac{3}{5}.$$

∎

Exercises 6.7

1. (writing) Discuss how to compute $\sec^{-1} x$, $\csc^{-1} x$ and $\cot^{-1} x$ on a calculator that only has built-in functions for $\sin^{-1} x$, $\cos^{-1} x$ and $\tan^{-1} x$.

2. (writing) Give a different range for $\sec^{-1} x$ than that given in the text. For which x's would the value of $\sec^{-1} x$ change? Using the calculator discussion in exercise 1, give one reason why we might have chosen the range that we did.

3. (writing) Inverse functions are necessary for solving equations. The restricted range we had to use to define inverses of the trigonometric functions also restricts their usefulness in equation-solving. Explain how to use $\sin^{-1} x$ to find all solutions of the equation $\sin u = x$.

4. (writing) The idea of restricting ranges can be used to define inverses for a variety of functions. Explain how to define an inverse for $f(x) = x^2$.

In exercises 5–18, evaluate the inverse function by sketching a unit circle and locating the correct angle on the circle.

5. $\sin^{-1} 0$ **6.** $\cos^{-1} 0$

7. $\tan^{-1} 1$ **8.** $\tan^{-1} 0$

9. $\sin^{-1}(-1)$ **10.** $\cos^{-1}(1)$

11. $\sec^{-1} 1$ **12.** $\csc^{-1} 0$

13. $\tan^{-1}(-1)$ **14.** $\cot^{-1} 1$

15. $\sec^{-1} 2$ **16.** $\csc^{-1} 2$

17. $\sin^{-1}(-1/2)$ **18.** $\cos^{-1}(1/2)$

In exercises 19–32, use a triangle to simplify each expression. State the range of x's for which the simplification holds.

19. $\cos(\sin^{-1} x)$ **20.** $\cos(\tan^{-1} x)$

21. $\sec^{-1}(\cos x)$ **22.** $\csc^{-1}(\sin x)$

23. $\tan(\sec^{-1} x)$ **24.** $\cot(\cos^{-1} x)$

25. $\sec(\tan^{-1} x)$ **26.** $\csc(\tan^{-1} x)$

27. $\sin(\cos^{-1} 1/2)$ **28.** $\cos(\sin^{-1} 1/2)$

29. $\cos(\tan^{-1} 1)$ **30.** $\sec(\sin^{-1} 3/5)$

31. $\tan(\cos^{-1} 3/5)$ **32.** $\csc(\sin^{-1} 2/3)$

In exercises 33–38, sketch a graph of the function.

33. $\cos^{-1}(2x)$ **34.** $\cos^{-1}(x^3)$

35. $\sin^{-1}(3x)$ **36.** $\sin^{-1}(x/4)$

37. $\tan^{-1}(5x)$ **38.** $\tan^{-1}(x^2)$

In exercises 39–48, find all solutions.

39. $\cos(2x) = 0$ **40.** $\cos(3x) = 1$

41. $\sin(3x) = 0$ **42.** $\sin(x/4) = 1$

43. $\tan(5x) = 1$ **44.** $\tan(2x) = -1$

45. $\sin(3x) = 1/2$ **46.** $\sin(x/4) = -1/2$

47. $\sec(2x) = 2$ **48.** $\csc(2x) = -2$

49. Give precise definitions of $\csc^{-1} x$ and $\cot^{-1} x$.

50. In baseball, outfielders are able to easily track down and catch fly balls that have very long and high trajectories. Physicists have argued for years about how this is done. A recent explanation involves the following geometry.

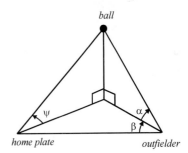

The player can catch the ball by running to keep the angle ψ constant (this makes it appear that the ball is moving in a straight line). Assuming that all triangles shown are right triangles, show that $\tan \psi = \dfrac{\tan \alpha}{\tan \beta}$ and then solve for ψ.

51. A picture hanging in an art gallery has a frame $20''$ high and the bottom of the frame is 6 feet above the floor. A person who is 6 feet tall stands x feet from the wall. Let A be the angle formed by the ray from the person to the bottom of the frame and the ray from the person to the top of the frame. Write A as a function of x.

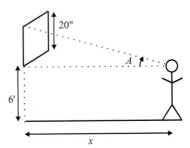

52. In golf, the goal is to hit a ball into a hole of diameter 4.5 inches. Suppose a golfer stands x feet from the hole trying to putt the ball into the hole. A first approximation of the margin of error in a putt is to measure the angle A formed by the ray from the ball to the right edge of the hole and the ray from the ball to the left edge of the hole. Find A as a function of x.

53. (exploratory) An oil tank with circular cross sections lies on its side. A stick is placed in a hole at the top and used to measure the depth d of oil in the tank. Based on this measurement, the goal is to compute the percentage of oil left in the tank.

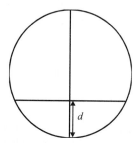

To simplify calculations, suppose the circle is a unit circle with center at $(0,0)$. Sketch radii extending from the origin to the top of the oil. The area of oil at the bottom equals the area of the portion of the circle bounded by the radii minus the

area of the triangle formed above.

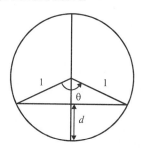

Start with the triangle, which has area one-half base times height. Explain why the height is $1 - d$. Find a right triangle in the figure (there are two of them) with hypotenuse 1 (the radius of the circle) and one vertical side of length $1 - d$. The horizontal side has length equal to one-half the base of the larger triangle. Show that this equals $\sqrt{1 - (1 - d)^2}$. The area of the portion of the circle equals $\pi\theta/2\pi = \theta/2$ where θ is the angle at the top of the triangle. Find this angle as a function of d. (Hint: Go back to the right triangle used above with upper angle $\theta/2$.) Then find the area filled with oil and divide by π to get the portion of the tank filled with oil.

6.8   The Calculus of the Inverse Trigonometric Functions

Now that we have defined the inverse trigonometric functions in section 6.7, we will examine the calculus of these functions briefly in the present section. Our first job is to find the derivative of $\sin^{-1} x$. You might think of the definition of derivative,

$$\frac{d}{dx} \sin^{-1} x = \lim_{h \to 0} \frac{\sin^{-1}(x + h) - \sin^{-1} x}{h},$$

but we don't know enough about $\sin^{-1} x$ to make this limit resolvable. So, what **do** you know about this function? Hopefully, you'll realize that, beyond the definition and a few isolated values, not much is known. Thus, while you might compute this limit approximately, for fixed values of x, you have little hope of resolving this limit symbolically. But remember, this is an inverse function and if you focus on that, you'll see the following analysis. Recall the definition of $\sin^{-1} x$ given in (7.1):

$$y = \sin^{-1} x \text{ if and only if } \sin y = x \text{ and } y \in \left[-\frac{\pi}{2}, \frac{\pi}{2} \right].$$

Differentiating implicitly, we have

$$\frac{d}{dx} \sin y = \frac{d}{dx} x$$

and so,

$$\cos y \frac{dy}{dx} = 1.$$

Solving this for $\frac{dy}{dx}$, we find (for $\cos y \neq 0$) that

$$\frac{dy}{dx} = \frac{1}{\cos y}.$$

This is not entirely satisfactory, though, since this gives us the derivative in terms of y.

Notice, though, that for $y \in \left[-\frac{\pi}{2}, \frac{\pi}{2}\right]$, $\cos y \geq 0$ and hence,

$$\cos y = \sqrt{1 - \sin^2 y} = \sqrt{1 - x^2}.$$

This leaves us with

$$\frac{dy}{dx} = \frac{1}{\cos y} = \frac{1}{\sqrt{1 - x^2}},$$

for $-1 < x < 1$. That is,

$$\boxed{\frac{d}{dx} \sin^{-1} x = \frac{1}{\sqrt{1 - x^2}},}$$

for $-1 < x < 1$. Similarly, we can show that

$$\boxed{\frac{d}{dx} \cos^{-1} x = \frac{-1}{\sqrt{1 - x^2}},}$$

also for $-1 < x < 1$. We leave the derivation of this as an exercise.

Likewise, to find $\frac{d}{dx} \tan^{-1} x$, we rely on its definition in (7.4). Recall that we have

$$y = \tan^{-1} x \text{ if and only if } \tan y = x \text{ and } y \in \left(-\frac{\pi}{2}, \frac{\pi}{2}\right).$$

Using implicit differentiation, we then have

$$\frac{d}{dx} \tan y = \frac{d}{dx} x$$

and so,

$$\left(\sec^2 y\right) \frac{dy}{dx} = 1.$$

We solve this for $\frac{dy}{dx}$, to obtain

$$\frac{dy}{dx} = \frac{1}{\sec^2 y}$$
$$= \frac{1}{1 + \tan^2 y}$$
$$= \frac{1}{1 + x^2}.$$

That is,

$$\boxed{\frac{d}{dx} \tan^{-1} x = \frac{1}{1 + x^2}.}$$

You can likewise show that

$$\boxed{\frac{d}{dx} \sec^{-1} x = \frac{1}{|x| \sqrt{x^2 - 1}},}$$

for $|x| > 1$. This is left as an exercise. The derivatives of the two remaining inverse trigonometric functions are not important and are not discussed here.

Example 8.1   Finding the Derivative of an Inverse Trigonometric Function

Compute the derivative of (a) $\cos^{-1}(3x^2)$ and (b) $\left(\sec^{-1} x\right)^2$.

Solution   (a) From the chain rule, we have

$$\frac{d}{dx}\cos^{-1}(3x^2) = \frac{-1}{\sqrt{1-(3x^2)^2}}\frac{d}{dx}(3x^2)$$

$$= \frac{-6x}{\sqrt{1-9x^4}}.$$

(b) Also from the chain rule, we have

$$\frac{d}{dx}\left(\sec^{-1}x\right)^2 = 2\left(\sec^{-1}x\right)\frac{d}{dx}\left(\sec^{-1}x\right)$$

$$= 2\left(\sec^{-1}x\right)\frac{1}{|x|\sqrt{x^2-1}}.$$

■

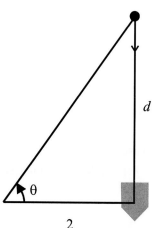

FIGURE 6.44:
A ballplayer's gaze.

Example 8.2   Modeling the Rate of Change of a Ballplayer's Gaze
    One of the guiding principles of most sports is to "keep your eye on the ball." In baseball, a batter stands 2 feet from home plate as a pitch is thrown with a velocity of 130 ft/s (about 90 mph). At what rate does the batter's angle of gaze change when the ball crosses home plate?

Solution   First, look at the triangle shown in Figure 6.44. We denote the distance from the ball to home plate by d and the angle of gaze by θ. Since the distance is changing with time, we write $d = d(t)$. The velocity of 130 ft/s means that $d'(t) = -130$. (Why would $d'(t)$ be negative?) From Figure 6.44, notice that

$$\theta(t) = \tan^{-1}\left[\frac{d(t)}{2}\right].$$

The rate of change of the angle is then

$$\theta'(t) = \frac{1}{1+\left[\frac{d(t)}{2}\right]^2}\frac{d'(t)}{2}$$

$$= \frac{2d'(t)}{4+[d(t)]^2} \text{ radians/second.}$$

When $d(t) = 0$ (i.e., when the ball is crossing home plate), the rate of change is then

$$\theta'(t) = \frac{2(-130)}{4} = -65 \text{ radians/second.}$$

One problem with this is that most humans can accurately track objects only at the rate of 3 radians/second. Keeping your eye on the ball in this case is thus physically impossible.

■

Integrals Involving the Inverse Trigonometric Functions

You may have already guessed the next step. Each of our new differentiation formulas gives rise to a new integration formula. First, since

$$\frac{d}{dx}\sin^{-1}x = \frac{1}{\sqrt{1-x^2}},$$

we also have that

$$\boxed{\int\frac{1}{\sqrt{1-x^2}}\,dx = \sin^{-1}x + c.} \tag{8.1}$$

Likewise, since

$$\frac{d}{dx}\cos^{-1}x = \frac{-1}{\sqrt{1-x^2}},$$

we also have that

$$\int \frac{1}{\sqrt{1-x^2}}\,dx = -\cos^{-1}x + c. \qquad (8.2)$$

However, since the integral in (8.2) is the same integral as in (8.1), we will ignore (8.2). In the same way, we obtain

$$\int \frac{1}{1+x^2}\,dx = \tan^{-1}x + c$$

and

$$\int \frac{1}{|x|\sqrt{x^2-1}}\,dx = \sec^{-1}x + c.$$

Example 8.3   **An Integral Related to $\tan^{-1}x$**

Evaluate

$$\int \frac{1}{9+x^2}\,dx.$$

Solution   Notice that the integrand is nearly the derivative of $\tan^{-1}x$. With this in mind, we rewrite the integral as

$$\int \frac{1}{9+x^2}\,dx = \frac{1}{9}\int \frac{1}{1+\left(\frac{x}{3}\right)^2}\,dx.$$

If we let $u = \frac{x}{3}$, then $du = \frac{1}{3}dx$ and hence,

$$\int \frac{1}{9+x^2}\,dx = \frac{1}{9}\int \frac{1}{1+\left(\frac{x}{3}\right)^2}\,dx$$

$$= \frac{1}{3}\int \frac{1}{\underbrace{1+\left(\frac{x}{3}\right)^2}_{1+u^2}}\underbrace{\frac{1}{3}dx}_{du}$$

$$= \frac{1}{3}\int \frac{1}{1+u^2}\,du$$

$$= \frac{1}{3}\tan^{-1}u + c$$

$$= \frac{1}{3}\tan^{-1}\left(\frac{x}{3}\right) + c.$$

∎

We leave it as an exercise to prove the more general formula:

$$\int \frac{1}{a^2+x^2}\,dx = \frac{1}{a}\tan^{-1}\left(\frac{x}{a}\right) + c.$$

Example 8.4   **An Integral Requiring a Simple Substitution**

Evaluate $\displaystyle\int \frac{e^x}{1+e^{2x}}\,dx.$

470 Chapter 6 Exponentials, Logarithms and Other Transcendental Functions

Solution   Think about how you might approach this. You probably won't recognize an antiderivative immediately. Remember that it often helps to look for terms that are derivatives of other terms. You should also recognize that $e^{2x} = (e^x)^2$. With this in mind, we let $u = e^x$, so that $du = e^x dx$. We then have

$$\int \frac{e^x}{1 + e^{2x}} dx = \int \frac{1}{\underbrace{1 + (e^x)^2}_{1+u^2}} \underbrace{e^x \, dx}_{du}$$

$$= \int \frac{1}{1 + u^2} du$$

$$= \tan^{-1} u + c$$

$$= \tan^{-1} (e^x) + c.$$

∎

Example 8.5   **Another Integral Requiring a Simple Substitution**

Evaluate $\int \frac{x}{\sqrt{1 - x^4}} dx$.

Solution   Look carefully at the integrand and recognize that you don't know an antiderivative. However, you might observe that

$$\int \frac{x}{\sqrt{1 - x^4}} dx = \int \frac{x}{\sqrt{1 - (x^2)^2}} dx.$$

Once you've recognized this one small step, you can complete the problem with a simple substitution. Letting $u = x^2$, we have $du = 2x \, dx$ and

$$\int \frac{x}{\sqrt{1 - x^4}} dx = \frac{1}{2} \int \frac{2x}{\sqrt{1 - (x^2)^2}} dx$$

$$= \frac{1}{2} \int \frac{1}{\sqrt{1 - u^2}} du$$

$$= \frac{1}{2} \sin^{-1} u + c$$

$$= \frac{1}{2} \sin^{-1} (x^2) + c.$$

∎

You will explore integrals involving the inverse trigonometric functions further in the exercises. Whenever dealing with these functions, it is easiest if you keep the basic definitions in mind (including the domains and ranges). You will only need to know the derivative formulas for $\sin^{-1} x$, $\tan^{-1} x$ and $\sec^{-1} x$. With these basic formulas, you can quickly develop anything else that you need, using elementary techniques (such as u-substitution).

Exercises 6.8

1. (writing) Explain why, as is stated in the text, antiderivatives of three of the six inverse trigonometric functions "can be ignored."

2. (writing) From equations (8.1) and (8.2), explain why it follows that $\sin^{-1} x = -\cos^{-1} x + c$. From the graphs of $y = \sin x$ and $y = \cos x$, explain why this is plausible and identify the constant c for $0 < x < \pi/2$.

In exercises 3–16, find the derivative of the function.

3. $\sin^{-1}(3x^2)$ **4.** $\sin^{-1}(x^2 + 1)$

5. $\cos^{-1}(x^3)$

6. $\cos^{-1}(x^2+1)$

7. $\tan^{-1}(x^2)$

8. $\cot^{-1}(\sqrt{x})$

9. $\sec^{-1}(x^2)$

10. $\csc^{-1}(\sqrt{x})$

11. $x\cos^{-1}(2x)$

12. $\sin x \sin^{-1}(2x)$

13. $\cos^{-1}(\sin x)$

14. $\tan^{-1}(\cos x)$

15. $\tan^{-1}(\sec x)$

16. $\cot^{-1}(\sin x)$

In exercises 17–32, evaluate the given integral.

17. $\displaystyle\int \frac{6}{1+x^2}\,dx$

18. $\displaystyle\int \frac{2}{9+x^2}\,dx$

19. $\displaystyle\int \frac{4}{\sqrt{1-x^2}}\,dx$

20. $\displaystyle\int \frac{2}{|x|\sqrt{x^2-1}}\,dx$

21. $\displaystyle\int \frac{2x}{1+x^4}\,dx$

22. $\displaystyle\int \frac{3x^2}{1+x^6}\,dx$

23. $\displaystyle\int \frac{4x}{\sqrt{1-x^4}}\,dx$

24. $\displaystyle\int \frac{2x^2}{\sqrt{1-x^6}}\,dx$

25. $\displaystyle\int \frac{2x}{x^2\sqrt{x^4-1}}\,dx$

26. $\displaystyle\int \frac{3}{|x|\sqrt{x^6-1}}\,dx$

27. $\displaystyle\int \frac{2}{4+x^2}\,dx$

28. $\displaystyle\int \frac{2x}{4+x^2}\,dx$

29. $\displaystyle\int \frac{e^x}{\sqrt{1-e^{2x}}}\,dx$

30. $\displaystyle\int \frac{\cos x}{\sqrt{1-\sin^2 x}}\,dx$

31. $\displaystyle\int \frac{\cos x}{4+\sin^2 x}\,dx$

32. $\displaystyle\int \frac{2x}{9+9x^4}\,dx$

33. Derive the formula $\dfrac{d}{dx}\cos^{-1}x = \dfrac{-1}{\sqrt{1-x^2}}$.

34. Derive the formula $\dfrac{d}{dx}\sec^{-1}x = \dfrac{1}{|x|\sqrt{x^2-1}}$.

35. Derive the formula $\displaystyle\int \frac{1}{a^2+x^2}\,dx = \frac{1}{a}\tan^{-1}\left(\frac{x}{a}\right)+c$ for any positive constant a.

36. In this exercise, we give a different derivation from the text for the derivative of $\sin^{-1}x$. Use Theorem 2.3 to show that $\dfrac{d}{dx}\sin^{-1}x = \dfrac{1}{\cos(\sin^{-1}x)}$. Then, evaluate $\cos(\sin^{-1}x)$.

37. In example 8.2, it was shown that by the time the baseball reached home plate, the rate of rotation of the player's gaze (θ') was too fast for humans to track. Given a maximum rotational rate of $\theta' = -3$ radians per second, find d such that $\theta' = -3$. That is, find how close to the plate a player *can* track the ball. In a major league setting, the player must start swinging by the time the pitch is halfway ($30'$) to home plate. How does this correspond to the distance at which the player loses track of the ball?

38. Suppose the pitching speed x' in example 8.2 is different. Then θ' will be different and the value of x for which $\theta' = -3$ will change. Find x as a function of x' for x' ranging from 30 ft/s (slowpitch softball) to 140 ft/s (major league fastball) and sketch the graph.

39. Suppose a painting hangs on a wall. The frame extends from 6 feet to 8 feet above the floor. A 5-foot person stands x feet from the wall and views the painting, with a viewing angle A formed by the ray from the person to the top of the frame and the ray from the person to the bottom of the frame. Find the value of x that maximizes the viewing angle A.

40. What changes in exercise 39 if the person is 6 feet tall?

41. (exploratory) When the astronauts blasted off from the moon, the lunar excursion module (LEM) burned its rockets briefly. Suppose the acceleration of the LEM was $a(t) = \begin{cases} 40 \text{ if } 0 \le t \le 2 \\ 0 \;\; \text{if } 2 < t \end{cases}$ ft/s^2. A camera was set up on the moon to allow viewers at home to watch the takeoff. Obviously, there was no camera operator, but the motion of the camera had to be preprogrammed. Suppose that the camera was 200 feet from the LEM. Find the camera angle between the ground and LEM as a function of the height h of the LEM. Integrate the equation $h''(t) = a(t)$ to find the height and hence the camera angle as a function of time. To rotate the camera to the correct angle, a torque would have to be applied at the pivot point of the camera. Torque is proportional to the second derivative of the angle. Find this derivative.

42. (exploratory) You have already seen that one of the important features of calculus is that one technique (e.g., differentiation) can be used to solve a wide variety of problems. Mathematicians take this laziness principle (don't rework a problem that has already been solved) to an extreme. The idea of a **homeomorphism** helps to identify when superficially different sets are essentially the same. For example, the intervals $(0,1)$, $(0,\pi)$ and $(-\pi/2, \pi/2)$ are all finite open intervals. They are *homeomorphic*. By definition, intervals I_1 and I_2 are homeomorphic if there is a function f from I_1 onto I_2 that has an inverse with f and f^{-1} both being continuous. A homeomorphism from $(0,1)$ to $(0,\pi)$ is $f(x) = \pi x$. Show that this is a homeomorphism by finding its inverse and verifying that both are continuous. Find a homeomorphism from $(0,\pi)$ to $(-\pi/2, \pi/2)$. (Hint: sketch a picture, and decide how you could move the interval $(0,\pi)$ to produce the interval $(-\pi/2, \pi/2)$.) This will take some thinking, but try to find a homeomorphism for any two finite open intervals (a,b) and (c,d). It remains to decide whether the interval $(-\infty, \infty)$ is different because it is infinite or the same because it is open. In fact, $(-\infty, \infty)$ is homeomorphic to $(-\pi/2, \pi/2)$ and hence to all other open intervals. Show that $\tan^{-1}x$ is a homeomorphism from $(-\infty, \infty)$ to $(-\pi/2, \pi/2)$.

6.9   The Hyperbolic Functions

The Gateway Arch in Saint Louis is one of the most distinctive and recognizable architectural structures in the United States. There are several interesting features of its shape. For instance, is the arch taller or wider? Most people think that it is taller, but this is the result of a common optical illusion. In fact, the arch has the same width as height. A slightly less mysterious illusion of the arch's shape is that it is **not** a parabola. Its shape corresponds to the graph of the hyperbolic cosine function. This function and the other five hyperbolic functions are introduced in this section.

   You may be wondering why we need more functions. Well, these functions are not entirely new. They are simply common combinations of exponentials. We study them because of their usefulness in applications (e.g., the Gateway Arch) and their convenience in solving equations (in particular, differential equations).

   The **hyperbolic sine** function is defined by

$$\sinh x = \frac{e^x - e^{-x}}{2}.$$

The **hyperbolic cosine** function is defined by

$$\cosh x = \frac{e^x + e^{-x}}{2}.$$

You can easily use the preceding definitions to verify the important identity

$$\cosh^2 u - \sinh^2 u = 1, \tag{9.1}$$

for any value of u. (We leave this as an exercise.) In light of this identity, notice that if $x = \cosh u$ and $y = \sinh u$, then

$$x^2 - y^2 = \cosh^2 u - \sinh^2 u = 1,$$

which you should recognize as the equation of a hyperbola. This identity is the source of the name "hyperbolic" for these functions. You should also notice some parallel with the trigonometric functions $\cos x$ and $\sin x$. This will become even more apparent below.

   The remaining four hyperbolic functions are defined in terms of the hyperbolic sine and hyperbolic cosine functions, in a manner analogous to their trigonometric counterparts. That is, we define the hyperbolic tangent function $\tanh x$, the hyperbolic cotangent function $\coth x$, the hyperbolic secant function $\operatorname{sech} x$ and the hyperbolic cosecant function csch x as follows:

$$\tanh x = \frac{\sinh x}{\cosh x}, \quad \coth x = \frac{\cosh x}{\sinh x}$$
$$\operatorname{sech} x = \frac{1}{\cosh x}, \quad \operatorname{csch} x = \frac{1}{\sinh x}.$$

These functions are remarkably easy to deal with, and we can readily determine their behavior, using what we have already learned about exponentials. First, note that

$$\frac{d}{dx} \sinh x = \frac{d}{dx} \left(\frac{e^x - e^{-x}}{2} \right) = \frac{e^x + e^{-x}}{2} = \cosh x.$$

Similarly, we can derive the remaining derivative formulas:

$$\frac{d}{dx} \cosh x = \sinh x, \qquad \frac{d}{dx} \tanh x = \operatorname{sech}^2 x$$
$$\frac{d}{dx} \coth x = -\operatorname{csch}^2 x, \qquad \frac{d}{dx} \operatorname{sech} x = -\operatorname{sech} x \tanh x$$

$$\text{and} \quad \frac{d}{dx} \operatorname{csch} x = -\operatorname{csch} x \coth x.$$

The Gateway Arch, St. Louis, MO

These are all elementary applications of earlier derivative rules and are left as exercises. As it turns out, only the first three of these are of much significance.

Example 9.1   Computing the Derivative of a Hyperbolic Function

Compute the derivative of $f(x) = \sinh^2(3x)$.

Solution   From the chain rule, we have

$$f'(x) = \frac{d}{dx}\sinh^2(3x) = \frac{d}{dx}\left[\sinh(3x)\right]^2$$

$$= 2\sinh(3x)\frac{d}{dx}\left[\sinh(3x)\right]$$

$$= 2\sinh(3x)\cosh(3x)\frac{d}{dx}(3x)$$

$$= 2\sinh(3x)\cosh(3x)(3)$$

$$= 6\sinh(3x)\cosh(3x).$$

$\blacksquare$

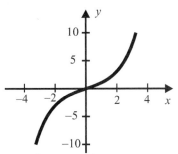

FIGURE 6.45: $y = \sinh x$.

Example 9.2   An Integral Involving a Hyperbolic Function

Evaluate $\int x\cosh(x^2)\,dx$.

Solution   Notice that you can evaluate this integral using a substitution. If we let $u = x^2$, we get $du = 2x\,dx$ and so,

$$\int x\cosh(x^2)\,dx = \frac{1}{2}\int \underbrace{\cosh(x^2)}_{\cosh u}\underbrace{(2x)\,dx}_{du}$$

$$= \frac{1}{2}\int \cosh u\,du = \frac{1}{2}\sinh u + c$$

$$= \frac{1}{2}\sinh(x^2) + c.$$

$\blacksquare$

For $f(x) = \sinh x$, note that

$$f(x) = \sinh x = \frac{e^x - e^{-x}}{2} \quad \begin{cases} > 0 \text{ if } x > 0 \\ < 0 \text{ if } x < 0. \end{cases}$$

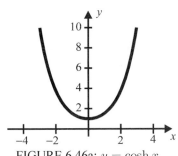

FIGURE 6.46a: $y = \cosh x$.

This is left as an exercise. Further, since $f'(x) = \cosh x > 0$, $\sinh x$ is increasing for all x. Next, note that $f''(x) = \sinh x$. Thus, the graph is concave down for $x < 0$ and concave up for $x > 0$. Finally, you can easily verify that

$$\lim_{x\to\infty}\sinh x = \infty \quad \text{and} \quad \lim_{x\to-\infty}\sinh x = -\infty.$$

It is now a simple matter to produce the graph seen in Figure 6.45. Similarly, you should be able to produce the graphs of $\cosh x$ and $\tanh x$ seen in Figure 6.46a and 6.46b, respectively. We leave the graphs of the remaining three hyperbolic functions to the exercises.

If a flexible cable or wire (such as a power line or telephone line) hangs between two towers, it will assume the shape of a **catenary** curve (derived from the Latin word *catena* meaning "chain"). As we will show at the end of this section, this naturally occurring curve corresponds to the graph of the hyperbolic cosine function $f(x) = a\cosh\left(\frac{x}{a}\right)$.

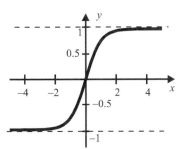

FIGURE 6.46b: $y = \tanh x$.

Example 9.3   Finding the Amount of Sag in a Hanging Cable

For the catenary $f(x) = 20\cosh\left(\frac{x}{20}\right)$, for $-20 \leq x \leq 20$, find the amount of sag in the cable and the arc length.

Solution   From the graph of the function in Figure 6.47, it appears that the minimum value of the function is at the midpoint $x = 0$, with the maximum at $x = -20$ and $x = 20$. To verify this observation, note that

$$f'(x) = \sinh\left(\frac{x}{20}\right)$$

and hence, $f'(0) = 0$, while $f'(x) < 0$ for $x < 0$ and $f'(x) > 0$, for $x > 0$. Thus, f decreases to a minimum at $x = 0$. Further, $f(-20) = f(20) \approx 30.86$ is the maximum for $-20 \leq x \leq 20$ and $f(0) = 20$, so that the cable sags approximately 10.86 feet. From the usual formula for arc length, developed in section 5.4, the length of the cable is

$$L = \int_{-20}^{20} \sqrt{1 + [f'(x)]^2}\, dx = \int_{-20}^{20} \sqrt{1 + \sinh^2\left(\frac{x}{20}\right)}\, dx.$$

Notice that from (9.1), we have

$$1 + \sinh^2 x = \cosh^2 x.$$

Using this identity, the arc length integral simplifies to

$$L = \int_{-20}^{20} \sqrt{1 + \sinh^2\left(\frac{x}{20}\right)}\, dx = \int_{-20}^{20} \cosh\left(\frac{x}{20}\right)\, dx$$

$$= 20\sinh\left(\frac{x}{20}\right)\Big|_{-20}^{20} = 20\left[\sinh(1) - \sinh(-1)\right]$$

$$= 40\sinh(1) \approx 47 \text{ feet.}$$

∎

The Inverse Hyperbolic Functions

You should note from the graphs of $\sinh x$ and $\tanh x$ that these functions are one-to-one. Also, $\cosh x$ is one-to-one for $x \geq 0$. Thus, we can define inverses for these functions, as follows. For any $x \in \mathbb{R}$, we define the **inverse hyperbolic sine** by

$$y = \sinh^{-1} x \text{ if and only if } \sinh y = x.$$

For any $x \geq 1$, we define the **inverse hyperbolic cosine** by

$$y = \cosh^{-1} x \text{ if and only if } \cosh y = x, \text{ and } y \geq 0.$$

Finally, for any $x \in (-1, 1)$, we define the **inverse hyperbolic tangent** by

$$y = \tanh^{-1} x \text{ if and only if } \tanh y = x.$$

Inverses for the remaining three hyperbolic functions can be defined similarly and are left to the exercises. We show the graphs of $y = \sinh^{-1} x$, $y = \cosh^{-1} x$ and $y = \tanh^{-1} x$ in Figures 6.48a, 6.48b and 6.48c, respectively. (As usual, you can obtain these by reflecting the graph of the original function through the line $y = x$.)

We can find derivatives for the inverse hyperbolic functions using implicit differentiation, just as we have for previous inverse functions. We have that

$$y = \sinh^{-1} x \text{ if and only if } \sinh y = x. \tag{9.2}$$

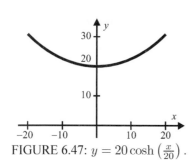
FIGURE 6.47: $y = 20\cosh\left(\frac{x}{20}\right)$.

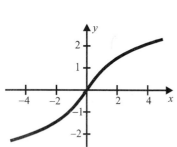
FIGURE 6.48a: $y = \sinh^{-1} x$.

FIGURE 6.48b: $y = \cosh^{-1} x$.

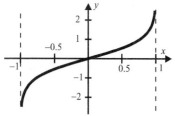
FIGURE 6.48c: $y = \tanh^{-1} x$.

Differentiating both sides of this last equation with respect to x yields

$$\frac{d}{dx}\sinh y = \frac{d}{dx}x$$

or

$$(\cosh y)\,\frac{dy}{dx} = 1.$$

Solving for the derivative, we find

$$\frac{dy}{dx} = \frac{1}{\cosh y} = \frac{1}{\sqrt{1+\sinh^2 y}} = \frac{1}{\sqrt{1+x^2}},$$

since we know that

$$\cosh^2 y - \sinh^2 y = 1,$$

from (9.1). That is, we have shown that

$$\boxed{\frac{d}{dx}\sinh^{-1}x = \frac{1}{\sqrt{1+x^2}}.}$$

Note the similarity with the derivative formula for $\sin^{-1}x$. We can likewise derive the derivatives of the other five inverse hyperbolic functions. We list these below for the sake of completeness.

$$\frac{d}{dx}\sinh^{-1}x = \frac{1}{\sqrt{1+x^2}} \qquad \frac{d}{dx}\cosh^{-1}x = \frac{1}{\sqrt{x^2-1}}$$

$$\frac{d}{dx}\tanh^{-1}x = \frac{1}{1-x^2} \qquad \frac{d}{dx}\coth^{-1}x = \frac{1}{1-x^2}$$

$$\frac{d}{dx}\operatorname{sech}^{-1}x = \frac{-1}{x\sqrt{1-x^2}} \qquad \frac{d}{dx}\operatorname{csch}^{-1}x = \frac{-1}{|x|\sqrt{1+x^2}}$$

Before closing this section, we wish to point out that the inverse hyperbolic functions have a significant advantage over earlier inverse functions we have discussed. It turns out that we can solve for the inverse functions explicitly in terms of more elementary functions.

Example 9.4   Finding a Formula for an Inverse Hyperbolic Function

Find an explicit formula for $\sinh^{-1}x$.

Solution   Recall from (9.2) that

$$y = \sinh^{-1}x \text{ if and only if } \sinh y = x.$$

Using this definition, we have

$$x = \sinh y = \frac{e^y - e^{-y}}{2}. \qquad\qquad (9.3)$$

We can solve this equation for y, as follows. First, recall also that

$$\cosh y = \frac{e^y + e^{-y}}{2}.$$

Now, notice that adding these last two equations and using the identity (9.1), we have

$$e^y = \sinh y + \cosh y = \sinh y + \sqrt{\cosh^2 y}$$

$$= \sinh y + \sqrt{1 + \sinh^2 y}$$

$$= x + \sqrt{1 + x^2},$$

from (9.3). Finally, taking the natural logarithm of both sides, we get

$$y = \ln\left(e^y\right) = \ln\left(x + \sqrt{x^2 + 1}\right).$$

That is, we have found a formula for the inverse sine function:

$$\sinh^{-1} x = \ln\left(x + \sqrt{x^2 + 1}\right).$$

■

Similarly, we can show that for $x \geq 1$,

$$\cosh^{-1} x = \ln\left(x + \sqrt{x^2 - 1}\right)$$

and for $-1 < x < 1$,

$$\tanh^{-1} x = \frac{1}{2} \ln\left(\frac{1 + x}{1 - x}\right).$$

We leave it to the exercises to derive these formulas and corresponding formulas for the remaining inverse hyperbolic functions. There is little point in memorizing any of these formulas. You need only realize that these are always available by performing some elementary algebra.

Derivation of the Catenary

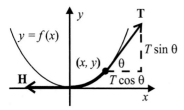

FIGURE 6.49: Forces acting on a section of hanging cable.

We close this section by deriving a formula for the catenary. As you follow the steps, pay special attention to the variety of calculus results that we use.

In Figure 6.49, we assume that the lowest point of the catenary curve is located at the origin. We further assume that the cable has constant linear density ρ (measured in units of weight per unit length) and that the function $y = f(x)$ is twice continuously differentiable. We focus on the portion of the cable from the origin to the general point (x, y) indicated in the figure. Since this section is not moving, the horizontal and vertical forces must be balanced. Horizontally, this section of cable is pulled to the left by the tension H at the origin and is pulled to the right by the horizontal component $T \cos \theta$ of the tension T at the point (x, y). Notice that these forces are balanced if

$$H = T \cos \theta \tag{9.4}$$

Vertically, the section of cable is pulled up by the vertical component $T \sin \theta$ of the tension. The section of cable is pulled down by the weight of the section. Notice that the weight of the section is given by the product of the density ρ (weight per unit length) and the length of the section. Recall from your study of arc length in Chapter 5 that the arc length of this section of cable is given by $\int_0^x \sqrt{1 + [f'(t)]^2}\, dt$. So, the vertical forces will balance if

$$T \sin \theta = \rho \int_0^x \sqrt{1 + [f'(t)]^2}\, dt. \tag{9.5}$$

We can combine equations (9.4) and (9.5) by multiplying (9.4) by $\tan \theta$ to get $H \tan \theta = T \sin \theta$, then using (9.5) to conclude that

$$H \tan \theta = \rho \int_0^x \sqrt{1 + [f'(t)]^2}\, dt.$$

Notice from Figure 6.49 that $\tan \theta = f'(x)$, so that we have

$$H f'(x) = \rho \int_0^x \sqrt{1 + [f'(t)]^2}\, dt.$$

Differentiating both sides of this equation, the Fundamental Theorem of Calculus gives us

$$H\,f''(x) = \rho\sqrt{1 + [f'(x)]^2}.\qquad (9.6)$$

Now, divide both sides of the equation by H and name $b = \frac{\rho}{H}$. Further, substitute $u(x) = f'(x)$. Equation (9.6) then becomes

$$u'(x) = b\sqrt{1 + [u(x)]^2},$$

which you should recognize as a separable differential equation. Putting together all of the u terms and integrating with respect to x gives us

$$\int \frac{1}{\sqrt{1 + [u(x)]^2}}\, u'(x)\, dx = \int b\, dx.$$

You should recognize the integral on the left-hand side as $\sinh^{-1}(u(x))$, so that we now have

$$\sinh^{-1}(u(x)) = bx + c.$$

Notice that since $f(x)$ has a minimum at $x = 0$, we must have that $u(0) = f'(0) = 0$. So, taking $x = 0$, we get $c = \sinh^{-1}(0) = 0$. From $\sinh^{-1}(u(x)) = bx$, we obtain $u(x) = \sinh(bx)$. Now, recall that $u(x) = f'(x)$, so that $f'(x) = \sinh(bx)$. Integrating this gives us

$$f(x) = \int \sinh(bx)\, dx$$
$$= \tfrac{1}{b}\cosh(bx) + c.$$

Finally, recall that $f(0) = 0$ and so, we must have $c = 0$. This leaves us with $f(x) = \tfrac{1}{b}\cosh(bx)$. Finally, writing $a = \tfrac{1}{b}$, we obtain the catenary equation $f(x) = a\cosh\left(\frac{x}{a}\right)$.

Exercises 6.9

1. (writing) Compare the derivatives and integrals of the trigonometric functions to the derivatives and integrals of the hyperbolic functions. Also note that the trigonometric identity $\cos^2 x + \sin^2 x = 1$ differs only by a minus sign from the corresponding hyperbolic identity $\cosh^2 x - \sinh^2 x = 1$.

2. (writing) As noted in the text, the hyperbolic functions are not really new functions. They provide names for useful combinations of exponential functions. Explain why it is advantageous to assign special names to these functions instead of leaving them as exponentials.

3. (writing) Briefly describe the graphs of $\sinh x$, $\cosh x$ and $\tanh x$. Which simple polynomials do the graphs of $\sinh x$ and $\cosh x$ resemble?

4. (writing) The catenary (hyperbolic cosine) is the shape assumed by a hanging cable because this distributes the weight of the cable most evenly throughout the cable. Knowing this, why was it smart to build the Gateway Arch in this shape? Why would you suspect that the profile of an egg has this same shape?

In exercises 5–12, sketch the graph of each function.

5. $\cosh 2x$

6. $\sinh 3x$

7. $\tanh 4x$

8. $\tanh x^2$

9. $\cosh 2x \sinh 2x$

10. $e^{-x}\sinh x$

11. $x^2 \sinh 2x$

12. $x^3 \sinh x$

In exercises 13–24, find the derivative of each function.

13. $\cosh 4x$

14. $\cosh x^2$

15. $\sinh 2x$

16. $\sinh \sqrt{x}$

17. $\tanh 4x$

18. $\tanh x^2$

19. $\cosh^{-1} 2x$

20. $\sinh^{-1} 3x$

21. $x^2 \sinh 2x$

22. $x^3 \sinh x$

23. $\tanh^{-1} 4x$

24. $\sinh^{-1} x^2$

In exercises 25–36, evaluate each integral.

25. $\displaystyle\int \cosh 6x\, dx$

26. $\displaystyle\int \sinh 2x\, dx$

27. $\displaystyle\int \tanh 3x\, dx$

28. $\displaystyle\int \operatorname{sech}^2 dx$

29. $\displaystyle\int_0^1 \frac{e^{4x} - e^{-4x}}{2}\, dx$

30. $\displaystyle\int_0^1 \frac{e^{2x} - e^{-2x}}{e^{2x} + e^{-2x}}\, dx$

31. $\displaystyle\int \frac{2}{\sqrt{1+x^2}}\,dx$

32. $\displaystyle\int \frac{2x}{\sqrt{1+x^4}}\,dx$

33. $\displaystyle\int \cos x \sinh(\sin x)\,dx$

34. $\displaystyle\int x\cosh(x^2)\,dx$

35. $\displaystyle\int_0^1 \cosh x\, e^{\sinh x}\,dx$

36. $\displaystyle\int_0^1 \frac{\cosh 2x}{3+\sinh 2x}\,dx$

37. Derive the formulas $\dfrac{d}{dx}\cosh x = \sinh x$ and $\dfrac{d}{dx}\tanh x = \operatorname{sech}^2 x$.

38. Derive the formulas for the derivatives of $\coth x$, $\operatorname{sech} x$ and $\operatorname{csch} x$.

39. Using the properties of exponential functions, prove that $\sinh x > 0$ if $x > 0$ and $\sinh x < 0$ if $x < 0$.

40. Use the first and second derivatives to explain the properties of the graph of $\tanh x$.

41. Use the first and second derivatives to explain the properties of the graph of $\cosh x$.

42. Prove that $\cosh^2 x - \sinh^2 x = 1$.

43. Find an explicit formula, as in example 9.4, for $\cosh^{-1} x$.

44. Find an explicit formula, as in example 9.4, for $\tanh^{-1} x$.

45. Suppose that a hanging cable has the shape $10\cosh(x/10)$ for $-20 \le x \le 20$. Find the amount of sag in the cable.

46. Find the length of the cable in exercise 45.

47. Suppose that a hanging cable has the shape $15\cosh(x/15)$ for $-25 \le x \le 25$. Find the amount of sag in the cable.

48. Find the length of the cable in exercise 47.

49. Suppose that a hanging cable has the shape $a\cosh(x/a)$ for $-b \le x \le b$. Show that the amount of sag is given by $a\cosh(b/a) - a$ and the length of the cable is $2a\sinh(b/a)$.

50. (exploratory) The Saint Louis Gateway Arch is both 630 feet wide and 630 feet tall. Its shape looks very much like a parabola, but is actually a hyperbolic cosine. You will explore the difference between the two functions in this exercise. First, consider the model $y = 757.7 - 127.7\cosh(x/127.7)$ for $y \ge 0$. Find the x- and y-intercepts and show that this model (approximately) matches the arch's measurements of 630 feet wide and 630 feet tall. What would the 127.7 in the model have to be to match the measurements exactly? Now, consider a parabolic model. To have x-intercepts $x = -315$ and $x = 315$, explain why the model must have the form $y = -c(x+315)(x-315)$ for some positive constant c. Then find c to match the desired y-intercept of 630. Graph the parabola and hyperbolic cosine on the same axes for $-315 \le x \le 315$. How much difference is there between the graphs? Find the maximum distance between the curves. The authors have seen mathematics books where the arch is modeled by a parabola. How wrong is it to do this?

51. (exploratory) Suppose a person jumps out of an airplane from a great height. There are two primary forces acting on the skydiver: gravity and air resistance. In this situation, the air resistance would be proportional to the square of the velocity. Then an equation for the (downward) velocity would be $v' = g - cv^2$ where g is the gravitational constant and c is a constant determined by the orientation of the jumper's body. Replace c with g/v_T^2 and explain why the initial condition $v(0) = 0$ is reasonable. Then show that the solution of the IVP can be written in the form $v = v_T \tanh(gt/v_T)$. Show that v, the downward velocity, is an increasing function and find the limiting velocity, usually called the **terminal velocity**, as $t \to \infty$. As mentioned above, the constant c depends on the position of the jumper's body. If spread-eagle represents a c-value four times as large as a head-first dive, compare the corresponding terminal velocities. You may have seen video of skydivers jumping out of a plane at different times but catching up to each other and forming a circle. Explain how one diver could catch up to someone who jumped out of the plane earlier. Now, integrate the velocity function to obtain the distance function. Finally, answer the following two-part question. How much time and height does it take for a skydiver to reach 90% of terminal velocity?

Chapter Review Exercises

In exercises 1–16, find the derivative of the function.

1. $\ln(x^3 + 5)$

2. $\ln(1 - \sin x)$

3. $\ln\sqrt{x^4 + x}$

4. $\ln\frac{x^2-1}{x^3+2x+1}$

5. e^{-x^2}

6. $e^{\tan x}$

7. 4^{x^3}

8. 3^{1-2x}

9. $\sin^{-1} 2x$

10. $\cos^{-1} x^2$

11. $\tan^{-1}(\cos 2x)$

12. $\sec^{-1}(3x^2)$

13. $\cosh\sqrt{x}$

14. $\sinh(e^{2x})$

15. $\sinh^{-1} 3x$

16. $\tanh^{-1}(3x+1)$

In exercises 17–40, evaluate the integral.

17. $\displaystyle\int \frac{x^2}{x^3+4}\,dx$

18. $\displaystyle\int \frac{e^{2x}}{e^{2x}+4}\,dx$

19. $\int_0^1 \dfrac{x}{x^2+1}\,dx$

20. $\int_{\pi/12}^{\pi/4} \dfrac{\cos 2x}{\sin 2x}\,dx$

21. $\int \dfrac{\sin(\ln x)}{x}\,dx$

22. $\int \dfrac{\ln x + 1}{x}\,dx$

23. $\int e^{-4x}\,dx$

24. $\int e^{2x}\cos(e^{2x})\,dx$

25. $\int \dfrac{e^{\sqrt{x}}}{\sqrt{x}}\,dx$

26. $\int xe^{-x^2}\,dx$

27. $\int_0^2 e^{3x}\,dx$

28. $\int_{-2}^0 e^{-3x}\,dx$

29. $\int 3^{4x}\,dx$

30. $\int 2^{-5x}\,dx$

31. $\int \dfrac{3}{x^2+4}\,dx$

32. $\int \dfrac{6}{\sqrt{4-x^2}}\,dx$

33. $\int \dfrac{x^2}{\sqrt{1-x^6}}\,dx$

34. $\int \dfrac{e^{-x}}{1+e^{-2x}}\,dx$

35. $\int \dfrac{9x}{x^2\sqrt{x^4-1}}\,dx$

36. $\int \dfrac{9x^3}{\sqrt{x^4-1}}\,dx$

37. $\int \dfrac{4}{\sqrt{1+x^2}}\,dx$

38. $\int \dfrac{4}{x^2-1}\,dx$

39. $\int \cosh 4x\,dx$

40. $\int \tanh 3x\,dx$

In exercises 41–44, determine if the function is one-to-one. If so, find its inverse.

41. x^3-1

42. e^{-4x}

43. e^{2x^2}

44. x^3-2x+1

In exercises 45–48, do both parts without solving for the inverse: (a) find the derivative of the inverse at $x=a$ and (b) graph the inverse.

45. $x^5+2x^3-1,\ a=2$

46. $x^3+5x+2,\ a=2$

47. $\sqrt{x^3+4x},\ a=4$

48. $e^{x^3+2x},\ a=1$

In exercises 49–54, solve the IVP.

49. $y'=2y,\ y(0)=3$

50. $y'=-3y,\ y(0)=2$

51. $y'=\dfrac{2x}{y},\ y(0)=2$

52. $y'=-3xy^2,\ y(0)=4$

53. $y'=\sqrt{xy},\ y(1)=4$

54. $y'=x+y^2x,\ y(0)=1$

55. A bacterial culture has an initial population 10^4 and doubles every 2 hours. Find an equation for the population at any time t and determine when the population reaches 10^6.

56. An organism has population 100 at time $t=0$ and population 140 at time $t=2$. Find an equation for the population at any time and determine the population at time $t=6$.

57. The half-life of nicotine in the human bloodstream is 2 hours. If there is initially 2 mg of nicotine present, find an equation for the amount at any time t and determine when the nicotine level reaches 0.1 mg.

58. If the half-life of a radioactive material is 3 hours, what percentage of the material will be left after 9 hours? 11 hours?

59. If you invest \$2000 at 8% compounded continuously, how long will it take the investment to double?

60. If you invest \$4000 at 6% compounded continuously, how much will the investment be worth in 10 years?

61. A cup of coffee is served at $180°$ in a room with temperature $68°$. After 1 minute, the temperature has dropped to $176°$. Find an equation for the temperature at any time and determine when the temperature will reach $120°$.

62. A cold drink is served at $46°$ in a room with temperature $70°$. After 4 minutes, the temperature has increased to $48°$. Find an equation for the temperature at any time and determine when the temperature will reach $58°$.

In exercises 63–66, solve each separable equation, explicitly if possible.

63. $y'=2x^3y$

64. $y'=\dfrac{y}{\sqrt{1-x^2}}$

65. $y'=\dfrac{4}{(y^2+y)(1+x^2)}$

66. $y'=e^{x+y}$

In exercises 67–70, find all equilibrium solutions and determine which are stable and which are unstable.

67. $y'=3y(2-y)$

68. $y'=y(1-y^2)$

69. $y'=-y\sqrt{1+y^2}$

70. $y'=y+\dfrac{2y}{1-y}$

In exercises 71–74, sketch the direction field.

71. $y'=-x(4-y)$

72. $y'=4x-y^2$

73. $y'=2xy-y^2$

74. $y'=4x-y$

In exercises 75 and 76, use Euler's method with (a) $h=0.1$ and (b) $h=0.05$ to approximate $y(1)$ and $y(2)$. Show the first two steps by hand.

75. $y'=2x-y,\ y(0)=3$

76. $y'=3xy,\ y(0)=1$

In exercises 77–80, evaluate the quantity using the unit circle.

77. $\sin^{-1}1$

78. $\cos^{-1}\left(-\tfrac{1}{2}\right)$

79. $\tan^{-1}(-1)$

80. $\csc^{-1}(-2)$

In exercises 81–84, simplify the expression using a triangle.

81. $\sin(\sec^{-1} 2)$

82. $\tan(\cos^{-1}(4/5))$

83. $\sin^{-1}(\sin(3\pi/4))$

84. $\cos^{-1}(\sin(-\pi/4))$

In exercises 85 and 86, find all solutions of the equation.

85. $\sin 2x = 1$

86. $\cos 3x = \frac{1}{2}$

In exercises 87–92, sketch a graph.

87. $y = \cosh 2x$

88. $y = \tanh^{-1} 3x$

89. $y = \sin^{-1} 2x$

90. $y = \tan^{-1} 3x$

91. $y = e^{-x^2}$

92. $y = \ln 2x$

93. A hanging cable assumes the shape of $y = 20 \cosh \dfrac{x}{20}$ for $-25 \le x \le 25$. Find the amount of sag in the cable.

94. Find the arc length of the cable in exercise 93.

95. (exploratory) In tennis, a serve must clear the net and then land inside of a box drawn on the other side of the net. In this exercise, you will explore the margin of error for successfully serving. First, consider a straight serve (this essentially means a serve hit infinitely hard) struck 9 feet above the ground. Call the starting point $(0, 9)$. The back of the service box is 60 feet away, at $(60, 0)$. The top of the net is 3 feet above the ground and 39 feet from the server, at $(39, 3)$. Find the service angle θ (i.e., the angle as measured from the horizontal) for the triangle formed by these three points. Of course, most serves curve down due to gravity. Ignoring air resistance, the path of the ball struck at angle θ and initial speed v ft/s is

$y = -\dfrac{16}{(v \cos \theta)^2} x^2 - (\tan \theta)x + 9$. To hit the back of the service line, you need $y = 0$ when $x = 60$. Substitute in these values along with $v = 120$. Multiply by $\cos^2 \theta$ and replace $\sin \theta$ with $\sqrt{1 - \cos^2 \theta}$. Replacing $\cos \theta$ with z gives you an algebraic equation in z. Numerically estimate z. Similarly, substitute $x = 40$ and $y = 3$ and find an equation for $w = \cos \theta$. Numerically estimate w. The margin of error for the serve is given by $\cos^{-1} z < \theta < \cos^{-1} w$.

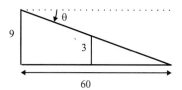

96. (exploratory) Baseball players often say that an unusually fast pitch rises or even hops up as it reaches the plate. One explanation of this illusion involves the players' inability to track the ball all the way to the plate. The player must compensate by predicting where the ball will be when it reaches the plate. Suppose the height of a pitch when it reaches home plate is $h = -(240/v)^2 + 6$ feet for a pitch with velocity v ft/s (this equation takes into consideration gravity but not air resistance). Halfway to the plate the height would be $h = -(120/v)^2 + 6$ feet. Compare the halfway heights for pitches with $v = 132$ and $v = 139$ (about 90 and 95 mph, respectively). Would a batter be able to tell much difference between them? Now compare the heights at the plate. Why might the batter think that the faster pitch hopped up right at the plate? How many inches did the faster pitch hop?

Chapter 7
Integration Techniques

If you have ever pondered the spin cycle of a washing machine, you have probably recognized that it is designed to squeeze the excess water out of your clothes before drying. The tub of the washer spins rapidly and the clothes are pressed against the wall of the tub by the resulting centripetal force. When the load is unusually large or a bulky item is in the washer, the distribution of weight in the tub may become unequal and cause the washing machine to visibly vibrate or even go totally out of balance. An important concern in the engineering of a washing machine (and countless other devices, as well) is the strength and placement of springs and other shock-absorbing devices that dampen vibration. A good design must balance the competing goals of minimizing vibrations and minimizing the force transmitted to the floor. A simplified schematic for a machine like the washer is shown below.

Machine schematic

The figure shows a vertical force F pushing downward against a spring with spring constant k (describing the stiffness of the spring) and a shock absorbing device with damping constant b (describing how much the shock absorber resists the motion). The mathematical analysis of such a system typically begins with Newton's second law of motion, which leads us to a differential equation describing the motion of the machine. We then solve the differential equation or (if the equation is too difficult to be solved), use the equation to determine important properties of the solution. Engineers often use a technique based on something called the **Laplace transform** to analyze such differential equations (see, for example, Wan's *Mathematical Models and Their Analysis* or an elementary differential equations text).

The mathematics that we develop in this chapter is essential for solving engineering problems like the analysis of a vibrating machine. You should recall from our brief introduction to differential equations in Chapter 6 that integration is an indispensable skill for solving differential equations. The first five sections of this chapter are devoted to reviewing integration and developing more advanced techniques for finding antiderivatives. In the last two sections of the chapter, we develop the ideas needed to define Laplace transforms and other important mathematical objects. In particular, we investigate how to compute definite integrals over an infinite domain. We define such **improper integrals** in section 7.7 and use a new limit technique (called l'Hôpital's Rule) developed in section 7.6 to calculate some of the limits arising from the evaluation of these improper integrals.

Although we finish the chapter in a position to define and use Laplace transforms, we must leave an in-depth study of these to future courses. We simply offer this as one example of the direction that mathematics classes take after calculus.

7.1   Review of Formulas and Techniques

In this brief section we draw together all of the integration formulas and the one integration technique (u-substitution) that we have developed so far. We use these to develop some more general formulas, as well as to solve more complicated integration problems. First, look over the following table of basic integration formulas.

$$\int x^n \, dx = \frac{x^{n+1}}{n+1} + c, \text{ for } n \neq -1 \text{ (power rule)} \qquad \int \frac{1}{x} \, dx = \ln |x| + c, \text{ for } x \neq 0$$

$$\int \sin x \, dx = -\cos x + c \qquad \int \cos x \, dx = \sin x + c$$

$$\int \sec^2 x \, dx = \tan x + c \qquad \int \sec x \tan x \, dx = \sec x + c$$

$$\int \csc^2 x \, dx = -\cot x + c \qquad \int \csc x \cot x \, dx = -\csc x + c$$

$$\int e^x \, dx = e^x + c \qquad \int e^{-x} \, dx = -e^{-x} + c$$

$$\int \tan x \, dx = -\ln |\cos x| + c \qquad \int \frac{1}{\sqrt{1 - x^2}} \, dx = \sin^{-1} x + c$$

$$\int \frac{1}{1 + x^2} \, dx = \tan^{-1} x + c \qquad \int \frac{1}{|x| \sqrt{x^2 - 1}} \, dx = \sec^{-1} x + c$$

Recall that each of these follows from a corresponding differentiation rule. So far, we have expanded this list slightly by using the method of u-substitution, as in the following example.

Example 1.1   A Simple Substitution

Evaluate $\int \sin(ax) \, dx$, for $a \neq 0$.

Solution   You may already know the answer here, but take a moment to make certain that you see why that answer is correct. The obvious choice is to let $u = ax$, so that $du = a \, dx$. This gives us

$$\int \sin(ax) \, dx = \frac{1}{a} \int \underbrace{\sin(ax)}_{\sin u} \underbrace{a \, dx}_{du} = \frac{1}{a} \int \sin u \, du$$

$$= -\frac{1}{a} \cos u + c = -\frac{1}{a} \cos(ax) + c.$$

∎

Observe that there is no need to memorize general rules like the one given in example 1.1, although it is often convenient to do so. Realize that you can reproduce such general rules any time you need them using u-substitution. Another general rule is found in the next example.

Example 1.2   Generalizing a Basic Integration Rule

Evaluate $\int \frac{1}{a^2 + x^2} \, dx$, for $a \neq 0$.

Solution   Notice that this is nearly the same as $\int \frac{1}{1 + x^2} \, dx$ and we can write

$$\int \frac{1}{a^2 + x^2} \, dx = \frac{1}{a^2} \int \frac{1}{1 + \left(\frac{x}{a}\right)^2} \, dx.$$

Now, letting $u = \frac{x}{a}$, we have $du = \frac{1}{a} dx$ and so,

$$\int \frac{1}{a^2 + x^2} dx = \frac{1}{a^2} \int \frac{1}{1 + \left(\frac{x}{a}\right)^2} dx = \frac{1}{a} \int \frac{1}{\underbrace{1 + \left(\frac{x}{a}\right)^2}_{1+u^2}} \underbrace{\left(\frac{1}{a}\right) dx}_{du}$$

$$= \frac{1}{a} \int \frac{1}{1 + u^2} du = \frac{1}{a} \tan^{-1} u + c = \frac{1}{a} \tan^{-1}\left(\frac{x}{a}\right) + c.$$

■

The basic integral formulas given above together with u-substitution and some ingenuity will allow you to evaluate a large number of integrals. However, u-substitution will not resolve all of your integration difficulties, as we see in the following example.

Example 1.3   An Integrand That Must be Expanded

Evaluate $\int \left(x^2 - 5\right)^2 dx$.

Solution   Note that you can't evaluate this as it stands. Your first impulse might be to substitute $u = x^2 - 5$. However, this fails, as we don't have $du = 2x\, dx$ in the integral. (We can force the constant 2 into the integral, but we can't get the x in there.) On the other hand, you can always multiply out the binomial to obtain

$$\int \left(x^2 - 5\right)^2 dx = \int \left(x^4 - 10x^2 + 25\right) dx = \frac{x^5}{5} - 10\frac{x^3}{3} + 25x + c.$$

■

The moral of example 1.3 is to make certain you don't overlook more primitive methods. The most general rule in integration is to keep trying. Sometimes, you will need to do some algebra before you can recognize the form of the integrand.

Example 1.4   An Integral Where We Must Complete the Square

Evaluate $\int \frac{1}{\sqrt{-5 + 6x - x^2}} dx$.

Solution   Probably very little comes to mind here. Substitution for either the entire denominator or the quantity under the square root does not work. (Why not?) So, what's left? Recall that there are essentially only two things you can do to a quadratic polynomial: either factor it or complete the square. Here, doing the latter sheds some light on the integral. We have

$$\int \frac{1}{\sqrt{-5 + 6x - x^2}} dx = \int \frac{1}{\sqrt{-5 - (x^2 - 6x + 9) + 9}} dx = \int \frac{1}{\sqrt{4 - (x - 3)^2}} dx.$$

Notice how much this looks like $\int \frac{1}{\sqrt{1 - x^2}} dx = \sin^{-1} x + c$. If we factor out the 4 in the square root, we get

$$\int \frac{1}{\sqrt{-5 + 6x - x^2}} dx = \int \frac{1}{\sqrt{4 - (x - 3)^2}} dx = \int \frac{1}{\sqrt{1 - \left(\frac{x-3}{2}\right)^2}} \frac{1}{2} dx.$$

Now, let $u = \frac{x-3}{2}$, so that $du = \frac{1}{2}\,dx$. This gives us

$$\int \frac{1}{\sqrt{-5+6x-x^2}}\,dx = \int \frac{1}{\underbrace{\sqrt{1-\left(\frac{x-3}{2}\right)^2}}_{\sqrt{1-u^2}}} \underbrace{\frac{1}{2}\,dx}_{du} = \int \frac{1}{\sqrt{1-u^2}}\,du$$

$$= \sin^{-1}u + c = \sin^{-1}\left(\frac{x-3}{2}\right) + c.$$

■

The following example illustrates the value of perseverance.

Example 1.5   An Integral Requiring Some Imagination

Evaluate $\displaystyle\int \frac{4x+1}{2x^2+4x+10}\,dx$.

Solution   As with most integrals, you cannot evaluate this as it stands. Notice that the numerator is very nearly the derivative of the denominator (but not quite). So ask yourself what else you can do with an algebraic expression such as this. In the denominator, there's a quadratic polynomial, but it doesn't factor. The only remaining algebraic technique, then, is to complete the square. Notice that

$$\int \frac{4x+1}{2x^2+4x+10}\,dx = \int \frac{4x+1}{2(x^2+2x+1)-2+10}\,dx = \int \frac{4x+1}{2(x+1)^2+8}\,dx.$$

So, you've completed the square. Now what can you do? When the next step is not obvious, we recommend that you look for a pattern. Here, you should focus on the denominator. It nearly looks like the denominator in $\displaystyle\int \frac{1}{1+x^2}\,dx = \tan^{-1}x + c$. If we factor out an 8, it will look even more like this, as follows.

$$\int \frac{4x+1}{2x^2+4x+10}\,dx = \int \frac{4x+1}{2(x+1)^2+8}\,dx$$

$$= \frac{1}{8}\int \frac{4x+1}{\frac{1}{4}(x+1)^2+1}\,dx$$

$$= \frac{1}{8}\int \frac{4x+1}{\left(\frac{x+1}{2}\right)^2+1}\,dx.$$

Now, make the substitution $u = \frac{x+1}{2}$, so that $du = \frac{1}{2}\,dx$ and $x = 2u-1$. We then have

$$\int \frac{4x+1}{2x^2+4x+10}\,dx = \frac{1}{8}\int \frac{4x+1}{\left(\frac{x+1}{2}\right)^2+1}\,dx = \frac{1}{4}\int \frac{\overbrace{4x+1}^{4(2u-1)+1}}{\underbrace{\left(\frac{x+1}{2}\right)^2+1}_{u^2+1}}\underbrace{\frac{1}{2}\,dx}_{du}$$

$$= \frac{1}{4}\int \frac{4(2u-1)+1}{u^2+1}\,du = \frac{1}{4}\int \frac{8u-3}{u^2+1}\,du$$

$$= \frac{4}{4}\int \frac{2u}{u^2+1}\,du - \frac{3}{4}\int \frac{1}{u^2+1}\,du$$

$$= \ln(u^2+1) - \frac{3}{4}\tan^{-1}u + c$$

$$= \ln\left[\left(\frac{x+1}{2}\right)^2+1\right] - \frac{3}{4}\tan^{-1}\left(\frac{x+1}{2}\right) + c.$$

■

This last example was tedious, but reasonably straightforward. The real issue in integration is to recognize what pieces are present in a given integral and to see how you might rewrite the integral in a more familiar form. In the next section, we develop a powerful tool for evaluating integrals.

Exercises 7.1

1. (writing). In example 1.2, explain how you should know to write the denominator as $a^2\left[1+\left(\frac{x}{a}\right)^2\right]$. Would this still be a good first step if the numerator was x instead of 1? What would you do if the denominator was $\sqrt{a^2-x^2}$?

2. (writing) In both examples 1.4 and 1.5, we completed the square and found antiderivatives involving $\sin^{-1}x$, $\tan^{-1}x$ and $\ln(x^2+1)$. Briefly describe how the presence of an x in the numerator or a square root in the denominator affects which of these functions will be in the antiderivative.

In exercises 3–40, find the antiderivative.

3. $\displaystyle\int \sin 6x\,dx$

4. $\displaystyle\int 3\cos 4x\,dx$

5. $\displaystyle\int \sec 2x \tan 2x\,dx$

6. $\displaystyle\int x\sec x^2 \tan x^2\,dx$

7. $\displaystyle\int e^{3-2x}\,dx$

8. $\displaystyle\int \frac{3}{e^{6x}}\,dx$

9. $\displaystyle\int \frac{\sin\sqrt{x}}{\sqrt{x}}\,dx$

10. $\displaystyle\int \frac{\cos 1/x}{x^2}\,dx$

11. $\displaystyle\int \cos x e^{\sin x}\,dx$

12. $\displaystyle\int \sec^2 x\, e^{\tan x}\,dx$

13. $\displaystyle\int \frac{\sin x}{\cos^2 x}\,dx$

14. $\displaystyle\int \frac{1}{\sin^2 x}\,dx$

15. $\displaystyle\int \frac{3}{16+x^2}\,dx$

16. $\displaystyle\int \frac{2}{4+4x^2}\,dx$

17. $\displaystyle\int \frac{x^2}{1+x^6}\,dx$

18. $\displaystyle\int \frac{x^5}{1+x^6}\,dx$

19. $\displaystyle\int \frac{1}{\sqrt{4-x^2}}\,dx$

20. $\displaystyle\int \frac{e^x}{\sqrt{1-e^{2x}}}\,dx$

21. $\displaystyle\int \frac{x}{\sqrt{1-x^4}}\,dx$

22. $\displaystyle\int \frac{2x^3}{\sqrt{1-x^4}}\,dx$

23. $\displaystyle\int \frac{4}{5+2x+x^2}\,dx$

24. $\displaystyle\int \frac{4x+4}{5+2x+x^2}\,dx$

25. $\displaystyle\int \frac{4x}{5+2x+x^2}\,dx$

26. $\displaystyle\int \frac{x+1}{x^2+2x+4}\,dx$

27. $\displaystyle\int (x^2+4)^2\,dx$

28. $\displaystyle\int x(x^2+4)^2\,dx$

29. $\displaystyle\int \frac{1}{\sqrt{3-2x-x^2}}\,dx$

30. $\displaystyle\int \frac{x+1}{\sqrt{3-2x-x^2}}\,dx$

31. $\displaystyle\int \frac{2}{\sqrt{4x-x^2}}\,dx$

32. $\displaystyle\int \frac{3}{\sqrt{5-4x-x^2}}\,dx$

33. $\displaystyle\int e^{\ln(x^2+1)}\,dx$

34. $\displaystyle\int e^{2\ln x}\,dx$

35. $\displaystyle\int x\sqrt{x-3}\,dx$

36. $\displaystyle\int x(x-3)^2\,dx$

37. $\displaystyle\int \frac{e^x}{1+e^{2x}}\,dx$

38. $\displaystyle\int e^x \cot(e^x)\csc(e^x)\,dx$

39. $\displaystyle\int \frac{x^2+1}{\sqrt{x}}\,dx$

40. $\displaystyle\int xe^{-x^2}\,dx$

In exercises 41–48, find the antiderivatives (using formulas not reviewed in this section, but covered elsewhere in the text).

41. $\displaystyle\int \sinh 2x\,dx$

42. $\displaystyle\int \cosh 4x\,dx$

43. $\displaystyle\int \tanh 3x\,dx$

44. $\displaystyle\int e^x \cosh e^x\,dx$

45. $\displaystyle\int x\cosh x^2\,dx$

46. $\displaystyle\int \sec^2 x \cosh(\tan x)\,dx$

47. $\displaystyle\int \frac{3}{\sqrt{x^2-1}}\,dx$

48. $\displaystyle\int \frac{3}{\sqrt{x^2+1}}\,dx$

49. $\displaystyle\int \frac{3x}{x^2\sqrt{x^4-1}}\,dx$

50. $\displaystyle\int \frac{\cos x}{\sqrt{\sin^2 x+1}}\,dx$

In exercises 51–56, you are given a pair of antiderivatives. Evaluate the integral that can be worked using the techniques covered so far (the other cannot).

51. $\displaystyle\int \frac{5}{3+x^2}\,dx$ and $\displaystyle\int \frac{5}{3+x^3}\,dx$

52. $\displaystyle\int \sin 2x\,dx$ and $\displaystyle\int \sin^2 x\,dx$

53. $\displaystyle\int \ln x\,dx$ and $\displaystyle\int \frac{\ln x}{2x}\,dx$

54. $\displaystyle\int \frac{x^3}{1+x^8}\,dx$ and $\displaystyle\int \frac{x^4}{1+x^8}\,dx$

55. $\displaystyle\int e^{-x^2}\,dx$ and $\displaystyle\int xe^{-x^2}\,dx$

56. $\displaystyle\int \sec x\,dx$ and $\displaystyle\int \sec^2 x\,dx$

57. (exploratory) Evaluate both integrals in exercise 51 with your CAS. Also try $\int \frac{5}{3+x^4}\,dx$ and $\int \frac{5}{3+x^5}\,dx$. Describe any patterns that you see. Try to decide whether or not an antiderivative can be found for $\int \frac{5}{3+x^n}\,dx$ any power n.

58. (exploratory) Evaluate both integrals in exercise 52 with your CAS. Also try $\int \sin^3 x\,dx$ and $\int \sin^4 x\,dx$. Describe any patterns that you see. Try to decide whether or not antiderivatives can be found for any power of $\sin x$.

7.2   Integration by Parts

At this point, you will have recognized that differentiation and integration, although intimately connected, are vastly different processes. You have at your disposal rules for computing the derivative of nearly any function you can imagine. By contrast, you can compute antiderivatives for a relatively small number of functions. Thus far, we have developed a collection of basic integration formulas, most of which followed directly from the corresponding differentiation formulas. Beyond the basic formulas, you have one technique, u-substitution, that you have used to rewrite certain integrals in a simpler form. Unfortunately, this leaves us with many integrals that cannot be evaluated given your current knowledge. For instance,

$$\int x \sin x\,dx$$

cannot be evaluated with what you already know. We improve this situation in the current section by introducing a powerful tool called **integration by parts**.

We have said any number of times that every differentiation rule gives rise to a corresponding integration rule. So, recall the product rule:

$$\frac{d}{dx}\left[f(x)\,g(x)\right] = f'(x)\,g(x) + f(x)\,g'(x).$$

If we integrate both sides of this equation, we get

$$\int \frac{d}{dx}\left[f(x)\,g(x)\right]\,dx = \int f'(x)\,g(x)\,dx + \int f(x)\,g'(x)\,dx.$$

Of course, the integral on the left-hand side is simply $f(x)\,g(x)$. Solving for the second integral on the right-hand side then yields

$$\int f(x)\,g'(x)\,dx = f(x)\,g(x) - \int f'(x)\,g(x)\,dx.$$

This rule is called **integration by parts**. (It may be beneficial for you to differentiate both sides to verify the equality directly.) Alright, you're probably asking what this is all about. Admittedly, the significance of this new rule is hard to see at first. We'll let the examples convince you of the power of this technique. First, it's usually convenient to write this using the notation $u = f(x)$ and $v = g(x)$. Notice that this gives us

$$du = f'(x)\,dx \text{ and } dv = g'(x)\,dx,$$

so that

$$\boxed{\int u\,dv = u\,v - \int v\,du.} \tag{2.1}$$

To apply integration by parts, you need to make a judicious choice of u and dv so that the integral on the right-hand side of (2.1) is one that you know how to evaluate.

Example 2.1   Integration by Parts

Evaluate $\displaystyle\int x \sin x\,dx$.

Solution   First, observe that you cannot evaluate this as it stands. That is, it is not one of our basic integrals and no u-substitution will help. To use integration by parts, you will need to choose u (something to differentiate) and dv (something to integrate). Notice that if we let

$$u = x \quad \text{and } dv = \sin x \, dx,$$

then $du = dx$ and integrating dv, we have

$$v = \int \sin x \, dx = -\cos x + k.$$

In performing integration by parts, we drop this constant of integration (think about why it makes sense to do this). Also, we usually write this information as the block:

$$u = x \qquad dv = \sin x \, dx$$
$$du = dx \qquad v = -\cos x$$

This gives us

$$\int \underbrace{x}_{u} \underbrace{\sin x \, dx}_{dv} = \int u \, dv = u \, v - \int v \, du$$

$$= -x \cos x - \int (-\cos x) \, dx$$

$$= -x \cos x + \sin x + c. \tag{2.2}$$

It's a simple matter to differentiate the expression on the right-hand side of (2.2) and verify that you have indeed found an antiderivative of $x \sin x$.

∎

You should quickly realize that the choice of u and dv is critical. Observe what happens if we switch the choice of u and dv made in example 2.1.

Example 2.2   **A Poor Choice of u and dv**

Consider $\int x \sin x \, dx$ as in example 2.1, but this time, reverse the choice of u and dv.

Solution   Here, we let

$$u = \sin x \qquad dv = x \, dx$$
$$du = \cos x \, dx \qquad v = \tfrac{1}{2}x^2$$

This gives us

$$\int \underbrace{\sin x}_{u} \underbrace{x \, dx}_{dv} = u \, v - \int v \, du = \tfrac{1}{2}x^2 \cos x - \tfrac{1}{2} \int x^2 \cos x \, dx.$$

Notice that the last integral is one that we do **not** know how to calculate any better than the original one. In fact, we have made the situation worse in that the power of x in the new integral is higher than in the original integral.

∎

Remark 2.1   When using integration by parts, keep in mind that you are splitting up the integrand into two pieces. One of these pieces, u will be differentiated and the other, dv will be integrated. Since you can differentiate virtually every function you run across, you should think in terms of choosing a dv that you know how to integrate, as well as a choice of both that will result in an easier integral. Unfortunately, it's not always so easy to see the problem through from beginning to end. You will learn what works best by working through lots of problems. Even if you don't see how the problem is going to end up, **try something**! (At least you'll learn what doesn't work.)

Example 2.3   An Integrand with a Single Term

Evaluate $\int \ln x \, dx$.

Solution   This may look like it should be simple, but it's not one of our basic integrals and there's no substitution that will simplify it. That leaves us with integration by parts. Remember that you must pick u (to be differentiated) and dv (to be integrated). You obviously can't pick $dv = \ln x \, dx$, since the problem here is to find a way to integrate this very term. So, try

$$u = \ln x \qquad dv = dx$$
$$du = \tfrac{1}{x} \, dx \qquad v = x$$

From the integration by parts algorithm, we have

$$\int \underbrace{\ln x}_{u} \underbrace{dx}_{dv} = uv - \int v \, du = x \ln x - \int x \left(\frac{1}{x}\right) dx$$

$$= x \ln x - \int 1 \, dx = x \ln x - x + c.$$

$\blacksquare$

Frequently, an integration by parts results in an integral that we cannot evaluate directly, but instead, one that we can evaluate only by repeating integration by parts one or more times.

Example 2.4   Repeated Integration by Parts

Evaluate $\int x^2 \sin x \, dx$.

Solution   Certainly, you cannot evaluate this as it stands and there is no simplification or substitution that will help. We choose

$$u = x^2 \qquad dv = \sin x \, dx$$
$$du = 2x \, dx \qquad v = -\cos x$$

With this choice, integration by parts yields

$$\int \underbrace{x^2}_{u} \underbrace{\sin x \, dx}_{dv} = -x^2 \cos x + 2 \int x \cos x \, dx.$$

Of course, this last integral cannot be evaluated as it stands, but we could do it using a further integration by parts. We now choose

$$u = x \qquad dv = \cos x \, dx$$
$$du = dx \qquad v = \sin x$$

Applying integration by parts to the last integral, we now have

$$\int x^2 \sin x \, dx = -x^2 \cos x + 2 \int \underbrace{x}_{u} \underbrace{\cos x \, dx}_{dv}$$

$$= -x^2 \cos x + 2 \left(x \sin x - \int \sin x \, dx \right)$$

$$= -x^2 \cos x + 2x \sin x + 2 \cos x + c.$$

$\blacksquare$

Based on our work in example 2.4, try to figure out how many integrations by parts would be required to evaluate $\int x^n \sin x\, dx$, for a positive integer n. (There will be more on this in the exercises.)

Remark 2.2   In the second integration by parts in example 2.4, if you choose $u = \cos x$ and $dv = x\, dx$, then integration by parts will fail and leave you with the less than astounding conclusion that the integral equals itself. (Try this as an exercise.)

Repeated integration by parts sometimes takes you back to the integral you started with. This can be bad news (see Remark 2.2), or this can give us a clever way of evaluating an integral, as in the following example.

Example 2.5   **Repeated Integration by Parts with a Twist**

Evaluate $\int e^{2x} \sin x\, dx$.

Solution   Again, you should verify that none of our elementary methods works on this integral. This leaves us with integration by parts. In this case, there are two viable choices for u and dv. We take

$$u = e^{2x} \qquad\qquad dv = \sin x\, dx$$
$$du = 2e^{2x}\, dx \qquad\qquad v = -\cos x$$

(The opposite choice also works. Try this as an exercise.) Applying integration by parts yields

$$\int \underbrace{e^{2x}}_{u}\ \underbrace{\sin x\, dx}_{dv} = -e^{2x}\cos x + 2\int e^{2x}\cos x\, dx.$$

The remaining integral again requires integration by parts. We choose

$$u = e^{2x} \qquad\qquad dv = \cos x\, dx$$
$$du = 2e^{2x}\, dx \qquad\qquad v = \sin x$$

It now follows that

$$\int e^{2x}\sin x\, dx = -e^{2x}\cos x + 2\int \underbrace{e^{2x}}_{u}\ \underbrace{\cos x\, dx}_{dv}$$

$$= -e^{2x}\cos x + 2\left(e^{2x}\sin x - 2\int e^{2x}\sin x\, dx\right)$$

$$= -e^{2x}\cos x + 2e^{2x}\sin x - 4\int e^{2x}\sin x\, dx. \qquad (2.3)$$

Look carefully at the last line and you'll see the integral that we started with! If you treat the integral $\int e^{2x}\sin x\, dx$ as the unknown you're trying to solve for, you can add $4\int e^{2x}\sin x\, dx$ to both sides of equation (2.3). This leaves us with

$$5\int e^{2x}\sin x\, dx = -e^{2x}\cos x + 2e^{2x}\sin x + K,$$

where we have added the constant of integration K on the right side. Dividing both sides by 5 then gives us

$$\int e^{2x}\sin x\, dx = -\frac{1}{5}e^{2x}\cos x + \frac{2}{5}e^{2x}\sin x + c,$$

where we have replaced the arbitrary constant of integration $\frac{K}{5}$ by c.

■

Remark 2.3   For integrals like $\int e^{2x} \cos x \, dx$ (or related integrals like $\int e^{-3x} \sin x \, dx$), repeated integration by parts as in example 2.5 will produce an antiderivative. The first choice of u and dv is up to you (either choice will work) but your choice of u and dv in the second integration by parts must be consistent with your first choice. For instance, in example 2.5, our initial choice of $u = e^{2x}$ commits us to using $u = e^{2x}$ for the second integration by parts. If you want to know why, rework the second integral taking $u = \cos x$ and see what happens!

Note that to evaluate a definite integral, it is always possible to apply integration by parts to the corresponding indefinite integral and then simply evaluate the resulting antiderivative between the limits of integration. Whenever possible, however (i.e., when the integration is not too involved), you should apply integration by parts directly to the definite integral. Observe that the integration by parts algorithm for definite integrals is simply

$$\int_{x=a}^{x=b} u \, dv = u \, v \Big|_{x=a}^{x=b} - \int_{x=a}^{x=b} v \, du,$$

where we have written the limits of integration as we have to remind you that these refer to the values of x. (Recall that we derived the integration by parts formula by taking u and v both to be functions of x.)

Example 2.6   **Integration by Parts for a Definite Integral**

Evaluate $\displaystyle\int_{1}^{2} x^2 \ln x \, dx$.

Solution   Again, since more elementary methods are fruitless, we try integration by parts. Since we do not know how to integrate $\ln x$ (except via integration by parts), we choose

$$u = \ln x \qquad dv = x^2 \, dx$$
$$du = \tfrac{1}{x} \, dx \qquad v = \tfrac{1}{3} x^3$$

and hence, we have

$$\int_{1}^{2} \underbrace{\ln x}_{u} \; \underbrace{x^2 \, dx}_{dv} = u \, v \Big|_{1}^{2} - \int_{1}^{2} v \, du = \frac{1}{3} x^3 \ln x \Big|_{1}^{2} - \frac{1}{3} \int_{1}^{2} x^3 \left(\frac{1}{x} \right) dx$$

$$= \frac{1}{3} \left(2^3 \ln 2 - 1^3 \ln 1 \right) - \frac{1}{3} \int_{1}^{2} x^2 \, dx$$

$$= \frac{8 \ln 2}{3} - 0 - \frac{1}{9} x^3 \Big|_{1}^{2} = \frac{8 \ln 2}{3} - \frac{1}{9} \left(2^3 - 1^3 \right)$$

$$= \frac{8 \ln 2}{3} - \frac{1}{9} (8 - 1) = \frac{8 \ln 2}{3} - \frac{7}{9}.$$

∎

Integration by parts is the most powerful tool in our integration arsenal. In order to master its use, you will need to work through many problems. We provide a wide assortment of these in the exercise set that follows.

Exercises 7.2

1. (writing) Discuss your best strategy for determining which part of the integrand should be u and which part should be dv.

2. (writing) Integration by parts comes from the product rule for derivatives. Which integration technique comes from the chain rule? Briefly discuss why there is no commonly used integration technique derived from the quotient rule.

In exercises 3–32, evaluate the integrals.

3. $\displaystyle\int x \cos x \, dx$

4. $\displaystyle\int x \sin 4x \, dx$

5. $\displaystyle\int x e^{2x} \, dx$

6. $\displaystyle\int x \ln x \, dx$

7. $\displaystyle\int x^2 \ln x \, dx$

8. $\displaystyle\int \frac{\ln x}{x} \, dx$

9. $\displaystyle\int x^2 e^{-3x} \, dx$

10. $\displaystyle\int x^2 e^{x^3} \, dx$

11. $\displaystyle\int x^2 \cos 2x \, dx$

12. $\displaystyle\int x^2 \sin 3x \, dx$

13. $\displaystyle\int e^x \sin 4x \, dx$

14. $\displaystyle\int e^{2x} \cos x \, dx$

15. $\displaystyle\int \cos x \cos 2x \, dx$

16. $\displaystyle\int \sin x \sin 2x \, dx$

17. $\displaystyle\int x \sec^2 x \, dx$

18. $\displaystyle\int x^3 e^{x^2} \, dx$

19. $\displaystyle\int (\ln x)^2 \, dx$

20. $\displaystyle\int x^2 e^{3x} \, dx$

21. $\displaystyle\int \cos x \ln(\sin x) \, dx$

22. $\displaystyle\int x \sin x^2 \, dx$

23. $\displaystyle\int x \cosh 2x \, dx$

24. $\displaystyle\int 2x \sinh 3x \, dx$

25. $\displaystyle\int_0^1 x \sin 2x \, dx$

26. $\displaystyle\int_0^\pi 2x \cos x \, dx$

27. $\displaystyle\int_0^1 x \cos \pi x \, dx$

28. $\displaystyle\int_0^1 x e^{3x} \, dx$

29. $\displaystyle\int_0^1 x \sin \pi x \, dx$

30. $\displaystyle\int_0^1 x \cos 2\pi x \, dx$

31. $\displaystyle\int_1^{10} \ln x \, dx$

32. $\displaystyle\int_1^2 x \ln x \, dx$

In exercises 33–36, evaluate the integral using integration by parts (as we recommended in the text, "Try something!").

33. $\displaystyle\int \cos^{-1} x \, dx$

34. $\displaystyle\int \tan^{-1} x \, dx$

35. $\displaystyle\int \sin \sqrt{x} \, dx$

36. $\displaystyle\int e^{\sqrt{x}} \, dx$

37. How many times would integration by parts need to be performed to evaluate $\displaystyle\int x^n \sin x \, dx$ (where n is a positive integer)?

38. How many times would integration by parts need to be performed to evaluate $\displaystyle\int x^n \ln x \, dx$ (where n is a positive integer)?

39. Several useful integration formulas called **reduction formulas** are used to automate the process of performing multiple integrations by parts. Prove that for any positive integer n,

$$\int \cos^n x \, dx = \frac{1}{n} \cos^{n-1} x \sin x + \frac{n-1}{n} \int \cos^{n-2} x \, dx.$$

(Use integration by parts with $u = \cos^{n-1} x$ and $dv = \cos x \, dx$.)

40. Use integration by parts to prove that for any positive integer n,

$$\int \sin^n x \, dx = -\frac{1}{n} \sin^{n-1} x \cos x + \frac{n-1}{n} \int \sin^{n-2} x \, dx.$$

In exercises 41–48, evaluate the integral using the reduction formulas from exercises 39 and 40. Example: With $n = 2$, exercise 39 gives us

$$\int \cos^2 x \, dx = \frac{1}{2} \cos x \sin x + \frac{1}{2} \int \cos^0 x \, dx$$
$$= \frac{1}{2} \cos x \sin x + \frac{1}{2} x + c.$$

41. $\displaystyle\int \cos^3 x \, dx$

42. $\displaystyle\int \cos^5 x \, dx$

43. $\displaystyle\int \sin^3 x \, dx$

44. $\displaystyle\int \sin^4 x \, dx$

45. $\displaystyle\int_0^{\pi/2} \cos^2 x \, dx$

46. $\displaystyle\int_0^{\pi/2} \sin^4 x \, dx$

47. $\displaystyle\int_0^{\pi/2} \sin^5 x \, dx$

48. $\displaystyle\int_0^{\pi/2} \sin^6 x \, dx$

49. Based on exercises 46–48, conjecture a formula for $\int_0^{\pi/2} \sin^m x\, dx$. (Note: You will need different formulas for m odd and for m even).

50. Conjecture a formula for $\int_0^{\pi/2} \cos^m x\, dx$.

51. The excellent movie *Stand and Deliver* tells the story of mathematics teacher Jaime Escalante, who developed a remarkable AP calculus program in inner-city Los Angeles. In one scene, Escalante shows a student how to evaluate the integral $\int x^2 \sin x\, dx$. He forms a chart like the following:

$$
\begin{array}{lll}
 & \sin x & \\
x^2 & -\cos x & + \\
2x & -\sin x & - \\
2 & \cos x & +
\end{array}
$$

Multiplying across each full row, the antiderivative is $-x^2 \cos x + 2x \sin x + 2 \cos x + c$. Explain where each column comes from and why the method works on this problem.

In exercises 52–57, use the method of exercise 51 to evaluate the integral.

52. $\int x^4 \sin x\, dx$ **53.** $\int x^4 \cos x\, dx$

54. $\int x^4 e^x\, dx$ **55.** $\int x^4 e^{2x}\, dx$

56. $\int x^5 \cos 2x\, dx$ **57.** $\int x^3 e^{-3x}\, dx$

58. You should be aware that the method of exercise 51 doesn't always work, especially if both the derivative and antiderivative columns have powers of x. Show that the method doesn't work on $\int x^2 \ln x\, dx$.

59. Show that $\int_{-\pi}^{\pi} \cos(mx) \cos(nx)\, dx = 0$ for positive integers $m \neq n$.

60. Show that $\int_{-\pi}^{\pi} \sin(mx) \sin(nx)\, dx = 0$ for positive integers $m \neq n$.

61. Show that $\int_{-\pi}^{\pi} \cos(mx) \sin(nx)\, dx = 0$ for positive integers m and n.

62. Show that $\int_{-\pi}^{\pi} \cos^2 nx\, dx = \int_{-\pi}^{\pi} \sin^2 nx\, dx = \pi$ for any positive integer n.

63. (exploratory) Integration by parts can be used to compute coefficients for important functions called **Fourier series**. We cover Fourier series in detail in Chapter 8. Here, you will discover what some of the fuss is about. Start by computing $a_n = \frac{2}{\pi} \int_{-\pi}^{\pi} x \sin nx\, dx$ for an unspecified positive integer n. Write out the specific values for a_1, a_2, a_3 and a_4 and then form the function

$$f(x) = a_1 \sin x + a_2 \sin 2x + a_3 \sin 3x + a_4 \sin 4x.$$

Compare the graphs of $y = x$ and $y = f(x)$ on the interval $[-\pi, \pi]$. From writing out a_1 through a_4, you should notice a nice pattern. Use it to form the function

$$g(x) = f(x) + a_5 \sin 5x + a_6 \sin 6x + a_7 \sin 7x + a_8 \sin 8x.$$

Compare the graphs of $y = x$ and $y = g(x)$ on the interval $[-\pi, \pi]$. Is it surprising that you can add sine functions together and get something close to a straight line? It turns out that Fourier series can be used to find cosine and sine approximations to any continuous function on a closed interval.

64. (exploratory) Along with giving us a technique to compute antiderivatives, integration by parts is very important theoretically. In this context, it can be thought of as a technique for moving derivatives off of one function and onto another. To see what we mean, suppose that $f(x)$ and $g(x)$ are functions with $f(0) = g(0) = 0$, $f(1) = g(1) = 0$ and with continuous second derivatives $f''(x)$ and $g''(x)$. Use integration by parts twice to show that

$$\int_0^1 f''(x) g(x)\, dx = \int_0^1 f(x) g''(x)\, dx.$$

7.3   Trigonometric Techniques of Integration

Integrals Involving Powers of Trigonometric Functions

Evaluating an integral whose integrand contains powers of one or more trigonometric functions often involves making a clever substitution. These integrals are sufficiently common that we present them here as a group.

Our first aim is to evaluate integrals of the form

$$\int \sin^m x \cos^n x\, dx,$$

where m and n are positive integers.

Case 1: m or n is an odd positive integer.

If m is odd, first isolate one factor of $\sin x$. (You'll need this for du.) Then, replace any factors of $\sin^2 x$ with $1 - \cos^2 x$ and make the substitution $u = \cos x$. If n is odd, first isolate one factor of $\cos x$. (You'll need this for du.) Then, replace any factors of $\cos^2 x$ with $1 - \sin^2 x$ and make the substitution $u = \sin x$.

We illustrate this for the case where m is odd in the following example.

Example 3.1   A Typical Substitution

Evaluate $\displaystyle\int \cos^4 x \sin x \, dx$.

Solution   There is nothing particularly new about this integral. Since you cannot evaluate it as it stands, you might think of any available substitutions. (Hint: Look for terms that are derivatives of other terms.) You should quickly settle on letting $u = \cos x$, so that $du = -\sin x \, dx$. This gives us

$$\int \cos^4 x \sin x \, dx = -\int \underbrace{\cos^4 x}_{u^4} \, \underbrace{(-\sin x) \, dx}_{du} = -\int u^4 \, du$$

$$= -\frac{u^5}{5} + c = -\frac{\cos^5 x}{5} + c.$$

■

While this first example was not a particularly challenging one, it should help to give you an idea of what to do with the next example.

Example 3.2   An Integrand with an Odd Power of Sine

Evaluate $\displaystyle\int \cos^4 x \sin^3 x \, dx$.

Solution   If you're looking for terms that are derivatives of other terms, you should see both sine and cosine terms, but which do you substitute for? Consider $u = \cos x$, so that $du = -\sin x \, dx$. Notice that we can now rewrite the integral as

$$\int \cos^4 x \sin^3 x \, dx = \int \cos^4 x \sin^2 x \sin x \, dx = -\int \cos^4 x \sin^2 x \, (-\sin x) \, dx$$

$$= -\int \underbrace{\cos^4 x \left(1 - \cos^2 x\right)}_{u^4(1-u^2)} \underbrace{(-\sin x) \, dx}_{du} = -\int u^4 \left(1 - u^2\right) \, du$$

$$= -\int \left(u^4 - u^6\right) \, du = -\left(\frac{u^5}{5} - \frac{u^7}{7}\right) + c$$

$$= -\frac{\cos^5 x}{5} + \frac{\cos^7 x}{7} + c.$$

Pay close attention to how we did this. We took the odd power (in this case, $\sin^3 x$) and factored out one power of $\sin x$ (to use for du). The remaining (even) powers of $\sin x$ were rewritten in terms of $\cos x$ using the Pythagorean identity

$$\sin^2 x + \cos^2 x = 1.$$

■

The ideas used in example 3.2 can be applied to any integral of the specified form.

Example 3.3   An Integrand with an Odd Power of Cosine

Evaluate $\int \sqrt{\sin x} \, \cos^5 x \, dx$.

Solution   We can rewrite this as

$$\int \sqrt{\sin x} \, \cos^5 x \, dx = \int \sqrt{\sin x} \, \cos^4 x \cos x \, dx = \int \sqrt{\sin x} \left(1 - \sin^2 x\right)^2 \cos x \, dx.$$

Substitute $u = \sin x$, so that $du = \cos x \, dx$ and we have

$$\int \sqrt{\sin x} \cos^5 x \, dx = \int \underbrace{\sqrt{\sin x} \left(1 - \sin^2 x\right)^2}_{\sqrt{u} \, (1-u^2)^2} \underbrace{\cos x \, dx}_{du}$$

$$= \int \sqrt{u} \left(1 - u^2\right)^2 du = \int u^{1/2} \left(1 - 2u^2 + u^4\right) du$$

$$= \int \left(u^{1/2} - 2u^{5/2} + u^{9/2}\right) du = \frac{2}{3} u^{3/2} - 2 \left(\frac{2}{7}\right) u^{7/2} + \frac{2}{11} u^{11/2} + c$$

$$= \frac{2}{3} \sin^{3/2} x - \frac{4}{7} \sin^{7/2} x + \frac{2}{11} \sin^{11/2} x + c.$$

Although the details of calculation here got a bit messy, you should see through the details to the main point: all integrals of this form are calculated in essentially the same way.

$\blacksquare$

Case 2: m and n are both even positive integers.

In this case, we can use the double angle formulas for sine and cosine, $\sin^2 x = \frac{1}{2}(1 - \cos 2x)$ and $\cos^2 x = \frac{1}{2}(1 + \cos 2x)$ to reduce the powers in the integrand.
    We illustrate this case in the following example.

Example 3.4   An Integrand with an Even Power of Sine

Evaluate $\int \sin^2 x \, dx$.

Solution   Using the double angle formula, we can rewrite the integral as

$$\int \sin^2 x \, dx = \frac{1}{2} \int (1 - \cos 2x) \, dx.$$

We can evaluate this last integral by using the u-substitution $u = 2x$, so that $du = 2 \, dx$. This gives us

$$\int \sin^2 x \, dx = \frac{1}{2} \left(\frac{1}{2}\right) \int \underbrace{(1 - \cos 2x)}_{1 - \cos u} \underbrace{2 \, dx}_{du} = \frac{1}{4} \int (1 - \cos u) \, du$$

$$= \frac{1}{4} (u - \sin u) + c = \frac{1}{4} (2x - \sin 2x) + c.$$

$\blacksquare$

With some integrals, you may need to apply the double angle formulas several times, as in the following example.

Example 3.5   An Integrand with an Even Power of Cosine

Evaluate $\int \cos^4 x \, dx$.

Solution   Using the double angle formula for cosine, we have

$$\int \cos^4 x \, dx = \int \left(\cos^2 x\right)^2 \, dx = \frac{1}{4} \int \left(1 + \cos 2x\right)^2 \, dx$$

$$= \frac{1}{4} \int \left(1 + 2\cos 2x + \cos^2 2x\right) \, dx.$$

Using the double angle formula again, on the last term in the integrand, we get

$$\int \cos^4 x \, dx = \frac{1}{4} \int \left[1 + 2\cos 2x + \frac{1}{2}\left(1 + \cos 4x\right)\right] \, dx$$

$$= \frac{3}{8} x + \frac{1}{4} \sin 2x + \frac{1}{32} \cos 4x + c.$$

We leave it as an (easy) exercise to obtain this last expression.

∎

Our next aim is to devise a strategy for evaluating integrals of the form

$$\int \tan^m x \sec^n x \, dx,$$

where m and n are integers.

Case 1: m is an odd positive integer.
First, isolate one factor of $\sec x \tan x$. (You'll need this for du.) Then, replace any factors of $\tan^2 x$ with $\sec^2 x - 1$ and make the substitution $u = \sec x$.
    We illustrate this in the following example.

Example 3.6   An Integrand with an Odd Power of Tangent
     Evaluate $\int \tan^3 x \sec^3 x \, dx$.

Solution   Looking for terms that are derivatives of other terms, we rewrite the integral
as

$$\int \tan^3 x \sec^3 x \, dx = \int \tan^2 x \sec^2 x \, (\sec x \tan x) \, dx$$

$$= \int \left(\sec^2 x - 1\right) \sec^2 x \, (\sec x \tan x) \, dx,$$

where we have used the Pythagorean identity

$$\tan^2 x = \sec^2 x - 1.$$

You should see the substitution now. We let $u = \sec x$, so that $du = \sec x \tan x \, dx$ and
hence,

$$\int \tan^3 x \sec^3 x \, dx = \int \underbrace{\left(\sec^2 x - 1\right) \sec^2 x}_{(u^2-1)\,u^2} \underbrace{(\sec x \tan x) \, dx}_{du}$$

$$= \int \left(u^2 - 1\right) u^2 \, du = \int \left(u^4 - u^2\right) \, du$$

$$= \frac{1}{5} u^5 - \frac{1}{3} u^3 + c = \frac{1}{5} \sec^5 x - \frac{1}{3} \sec^3 x + c.$$

∎

Case 2: n is an even positive integer.
First, isolate one factor of $\sec^2 x$. (You'll need this for du.) Then, replace any remaining factors of $\sec^2 x$ with $1 + \tan^2 x$ and make the substitution $u = \tan x$.
    We illustrate this in the following example.

Example 3.7   An Integrand with an Even Power of Secant

Evaluate $\int \tan^2 x \sec^4 x \, dx$.

Solution   Since $\frac{d}{dx} \tan x = \sec^2 x$, we rewrite the integral as

$$\int \tan^2 x \sec^4 x \, dx = \int \tan^2 x \sec^2 x \sec^2 x \, dx = \int \tan^2 x \left(1 + \tan^2 x\right) \sec^2 x \, dx.$$

Now, we let $u = \tan x$, so that $du = \sec^2 x \, dx$ and

$$\int \tan^2 x \sec^4 x \, dx = \int \underbrace{\tan^2 x \left(1 + \tan^2 x\right)}_{u^2(1+u^2)} \underbrace{\sec^2 x \, dx}_{du}$$

$$= \int u^2 \left(1 + u^2\right) du = \int \left(u^2 + u^4\right) du$$

$$= \frac{1}{3} u^3 + \frac{1}{5} u^5 + c = \frac{1}{3} \tan^3 x + \frac{1}{5} \tan^5 x + c.$$

∎

Case 3: m is an even positive integer and n is an odd positive integer.
Replace any factors of $\tan^2 x$ with $\sec^2 x - 1$ and then use a special **reduction formula** (given in the exercises) to evaluate integrals of the form $\int \sec^n x \, dx$. This complicated case will be covered briefly in the exercises. Much of this depends on the following example.

Example 3.8   An Unusual Integral

Evaluate the integral $\int \sec x \, dx$.

Solution   Think about this integral some and you'll probably realize that you don't have a clue as to how to evaluate it. Finding an antiderivative depends on an unusual observation. Notice that if we multiply the integrand by the fraction $\frac{\sec x + \tan x}{\sec x + \tan x}$ (which is of course equal to 1), we get

$$\int \sec x \, dx = \int \sec x \left(\frac{\sec x + \tan x}{\sec x + \tan x}\right) dx$$

$$= \int \frac{\sec^2 x + \sec x \tan x}{\sec x + \tan x} \, dx.$$

Now, observe that the numerator is exactly the derivative of the denominator. That is,

$$\frac{d}{dx} \left(\sec x + \tan x\right) = \sec x \tan x + \sec^2 x,$$

so that taking $u = \sec x + \tan x$ gives us

$$\int \sec x \, dx = \int \frac{\sec^2 x + \sec x \tan x}{\sec x + \tan x} \, dx$$

$$= \int \frac{1}{u} \, du = \ln |u| + c$$

$$= \ln |\sec x + \tan x| + c.$$

∎

Trigonometric Substitution

If an integral contains a term of the form $\sqrt{a^2 - x^2}$, $\sqrt{a^2 + x^2}$ or $\sqrt{x^2 - a^2}$, for some $a > 0$, you can often evaluate the integral by making a substitution involving a trig function (hence, the name *trigonometric substitution*).

First, suppose that an integrand contains a term of the form $\sqrt{a^2 - x^2}$, for some $a > 0$. If we let $x = a \sin \theta$, where $-\frac{\pi}{2} \leq \theta \leq \frac{\pi}{2}$, then we can eliminate the square root, as follows. Notice that we now have

$$\sqrt{a^2 - x^2} = \sqrt{a^2 - (a \sin \theta)^2} = \sqrt{a^2 - a^2 \sin^2 \theta}$$
$$= a \sqrt{1 - \sin^2 \theta} = a \sqrt{\cos^2 \theta} = a \cos \theta,$$

since for $-\frac{\pi}{2} \leq \theta \leq \frac{\pi}{2}$, $\cos \theta \geq 0$. The following example is typical of how these substitutions are used.

Example 3.9   An Integral Involving $\sqrt{a^2 - x^2}$

Evaluate $\displaystyle\int \frac{1}{x^2 \sqrt{4 - x^2}}\, dx$.

Solution   You should always first consider whether an integral can be done directly, with a simple substitution or by parts. Since none of these methods help here, we consider trigonometric substitution. Keep in mind that the objective here is to eliminate the square root term. So, make a substitution which will accomplish that. We let

$$x = 2 \sin \theta, \quad -\frac{\pi}{2} < \theta < \frac{\pi}{2}.$$

(Why do we need **strict** inequalities here?) This gives us

$$dx = 2 \cos \theta\, d\theta$$

and hence,

$$\int \frac{1}{x^2 \sqrt{4 - x^2}}\, dx = \int \frac{2 \cos \theta}{(2 \sin \theta)^2 \sqrt{4 - (2 \sin \theta)^2}}\, d\theta$$

$$= \int \frac{2 \cos \theta}{4 \sin^2 \theta \sqrt{4 - 4 \sin^2 \theta}}\, d\theta$$

$$= \int \frac{\cos \theta}{(2 \sin^2 \theta)\, 2 \sqrt{1 - \sin^2 \theta}}\, d\theta$$

$$= \int \frac{\cos \theta}{4 \sin^2 \theta \cos \theta}\, d\theta = \frac{1}{4} \int \csc^2 \theta\, d\theta$$

$$= -\frac{1}{4} \cot \theta + c.$$

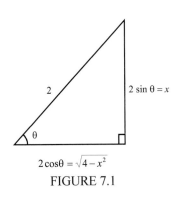

$2 \sin \theta = x$

$2 \cos \theta = \sqrt{4 - x^2}$

FIGURE 7.1

The only remaining problem is that the antiderivative is written in terms of the variable θ. When converting back to the original variable $x = 2 \sin \theta$, we urge you to draw a diagram, as in Figure 7.1. Since the substitution was $x = 2 \sin \theta$, we labelled the hypotenuse as 2. The side opposite the angle θ is then $2 \sin \theta$. By the Pythagorean Theorem, we get that the adjacent side is $\sqrt{4 - x^2}$, as indicated. So, we have

$$\cot \theta = \frac{\cos \theta}{\sin \theta} = \frac{\sqrt{4 - x^2}}{x}.$$

It now follows that

$$\int \frac{1}{x^2\sqrt{4-x^2}}\,dx = -\frac{1}{4}\cot\theta + c = -\frac{1}{4}\frac{\sqrt{4-x^2}}{x} + c.$$

∎

Next, suppose that an integrand contains a term of the form $\sqrt{a^2+x^2}$, for some $a > 0$. If we let $x = a\tan\theta$, where $-\frac{\pi}{2} < \theta < \frac{\pi}{2}$, then we can eliminate the square root, as follows. Notice that in this case, we have

$$\sqrt{a^2+x^2} = \sqrt{a^2 + (a\tan\theta)^2} = \sqrt{a^2 + a^2\tan^2\theta}$$

$$= a\sqrt{1+\tan^2\theta} = a\sqrt{\sec^2\theta} = a\sec\theta,$$

since for $-\frac{\pi}{2} < \theta < \frac{\pi}{2}$, $\sec\theta \geq 0$. The following example is typical of how these substitutions are used.

Example 3.10   An Integral Involving $\sqrt{a^2+x^2}$

Evaluate the integral $\displaystyle\int \frac{1}{\sqrt{9+x^2}}\,dx$.

Solution   First, you should recognize that you do not know how to evaluate this integral directly. Of course, the big problem here is the square root of the sum. You can eliminate this by letting $x = 3\tan\theta$, for $-\frac{\pi}{2} < \theta < \frac{\pi}{2}$. This gives us $dx = 3\sec^2\theta\,d\theta$, so that

$$\int \frac{1}{\sqrt{9+x^2}}\,dx = \int \frac{1}{\sqrt{9+(3\tan\theta)^2}}\,3\sec^2\theta\,d\theta$$

$$= \int \frac{3\sec^2\theta}{\sqrt{9+9\tan^2\theta}}\,d\theta$$

$$= \int \frac{3\sec^2\theta}{3\sqrt{1+\tan^2\theta}}\,d\theta$$

$$= \int \frac{\sec^2\theta}{\sec\theta}\,d\theta$$

$$= \int \sec\theta\,d\theta$$

$$= \ln|\sec\theta + \tan\theta| + c,$$

from example 3.8. We're not done here, since we first need to express the integral in terms of the original variable x. Observe that we had $x = 3\tan\theta$, so that $\tan\theta = \frac{x}{3}$. It remains only to solve for $\sec\theta$. The simplest way to do this is to recognize that for $-\frac{\pi}{2} < \theta < \frac{\pi}{2}$,

$$\sec\theta = \sqrt{1+\tan^2\theta} = \sqrt{1 + \left(\frac{x}{3}\right)^2}.$$

This leaves us with

$$\int \frac{1}{\sqrt{9+x^2}}\,dx = \ln|\sec\theta + \tan\theta| + c$$

$$= \ln\left|\sqrt{1 + \left(\frac{x}{3}\right)^2} + \frac{x}{3}\right| + c.$$

∎

Finally, suppose that an integrand contains a term of the form $\sqrt{x^2-a^2}$, for some $a > 0$. If we let $x = a\sec\theta$, where $\theta \in \left[0, \frac{\pi}{2}\right) \cup \left(\frac{\pi}{2}, \pi\right]$, then we can eliminate the square root, as follows. Notice that in this case, we have

$$\sqrt{x^2 - a^2} = \sqrt{(a\sec\theta)^2 - a^2} = \sqrt{a^2\sec^2\theta - a^2}$$
$$= a\sqrt{\sec^2\theta - 1} = a\sqrt{\tan^2\theta} = a\,|\tan\theta|\,.$$

Notice that the absolute values are needed, as $\tan\theta$ can be both positive and negative on $\left[0, \frac{\pi}{2}\right) \cup \left(\frac{\pi}{2}, \pi\right]$. The following example is typical of how these substitutions are used.

Example 3.11   **An Integral Involving $\sqrt{x^2 - a^2}$**

Evaluate the integral $\displaystyle\int \frac{\sqrt{x^2 - 25}}{x}\,dx$, for $x > 5$.

Solution   Here, we make the substitution $x = 5\sec\theta$, for $\theta \in \left[0, \frac{\pi}{2}\right)$. Notice that we chose the first half of the domain $\left[0, \frac{\pi}{2}\right) \cup \left(\frac{\pi}{2}, \pi\right]$, so that we would have $x = 5\sec\theta > 5$. (If we had $x < -5$, we would have chosen $\theta \in \left(\frac{\pi}{2}, \pi\right]$.) Note that this gives us $dx = 5\sec\theta\tan\theta\,d\theta$ and the integral then becomes:

$$\int \frac{\sqrt{x^2 - 25}}{x}\,dx = \int \frac{\sqrt{(5\sec\theta)^2 - 25}}{5\sec\theta}\,(5\sec\theta\tan\theta)\,d\theta$$

$$= \int \sqrt{25\sec^2\theta - 25}\,\tan\theta\,d\theta$$

$$= \int 5\sqrt{\sec^2\theta - 1}\,\tan\theta\,d\theta$$

$$= 5\int \tan^2\theta\,d\theta$$

$$= 5\int \left(\sec^2\theta - 1\right)d\theta$$

$$= 5\left(\tan\theta - \theta\right) + c.$$

Finally, observe that since $x = 5\sec\theta$, for $\theta \in \left[0, \frac{\pi}{2}\right)$, we have that

$$\tan\theta = \sqrt{\sec^2\theta - 1} = \sqrt{\left(\frac{x}{5}\right)^2 - 1} = \frac{1}{5}\sqrt{x^2 - 25}$$

and $\theta = \sec^{-1}\left(\frac{x}{5}\right)$. We now have

$$\int \frac{\sqrt{x^2 - 25}}{x}\,dx = 5\left(\tan\theta - \theta\right) + c$$

$$= \sqrt{x^2 - 25} - 5\sec^{-1}\left(\frac{x}{5}\right) + c.$$

∎

You will find a number of additional integrals requiring trigonometric substitution in the exercises. The principal idea here is to see that you can eliminate certain square root terms in an integrand by making use of a carefully chosen trigonometric substitution.

Exercises 7.3

1. (writing) Suppose a friend in your calculus class tells you that this section just has too many rules to memorize. (By the way, the authors would agree.) Help your friend out by making it clear that each rule indicates when certain substitutions will work. In turn, a substitution $u(x)$ works if the expression $u'(x)$ appears in the integrand and the resulting integral is easier to integrate. In each of the rules covered in the text, identify $u'(x)$ and point out why n has to be odd (or whatever the rule says) for the remaining integrand to be workable. Without memorizing rules, you can remember a small number of potential substitutions and see which one works for a given problem.

2. (writing) In the text, we suggested that when the integrand contains a term of the form $\sqrt{4 - x^2}$, you might try the trigonometric substitution $x = 2\sin\theta$. We should admit now that this does not always work. How can you tell whether or not this substitution will work?

In exercises 3–40, evaluate the integrals.

3. $\displaystyle\int \cos x \sin^4 x \, dx$ **4.** $\displaystyle\int \cos^3 x \sin^4 x \, dx$

5. $\displaystyle\int \cos x \sin^3 x \, dx$ **6.** $\displaystyle\int \cos^3 x \sin^3 x \, dx$

7. $\displaystyle\int \cos^2 x \sin x \, dx$ **8.** $\displaystyle\int \cos x \sin^2 x \, dx$

9. $\displaystyle\int \cos x \sin x \, dx$ **10.** $\displaystyle\int \cos^3 x \sin x \, dx$

11. $\displaystyle\int \cos^2 x \, dx$ **12.** $\displaystyle\int \sin^4 x \, dx$

13. $\displaystyle\int \tan x \sec^3 x \, dx$ **14.** $\displaystyle\int \tan^3 x \sec x \, dx$

15. $\displaystyle\int \tan^4 x \sec^4 x \, dx$ **16.** $\displaystyle\int \tan^4 x \sec^2 x \, dx$

17. $\displaystyle\int \cos^2 x \sin^2 x \, dx$ **18.** $\displaystyle\int (\cos^2 x + \sin^2 x) \, dx$

19. $\displaystyle\int \cot^3 x \csc^3 x \, dx$ **20.** $\displaystyle\int \cot x \csc^4 x \, dx$

21. $\displaystyle\int \sqrt{\cos x} \sin^3 x \, dx$ **22.** $\displaystyle\int \cos x / \sqrt{\sin x} \, dx$

23. $\displaystyle\int \cot^2 x \csc^4 x \, dx$ **24.** $\displaystyle\int \cot^4 x \csc^4 x \, dx$

25. $\displaystyle\int \tan^3 x \sec^4 x \, dx$ **26.** $\displaystyle\int \tan^3 2x \sec^3 2x \, dx$

27. $\displaystyle\int \frac{1}{x^2\sqrt{9 - x^2}} \, dx$ **28.** $\displaystyle\int \frac{1}{x^2\sqrt{16 - x^2}} \, dx$

29. $\displaystyle\int \sqrt{4 - x^2} \, dx$ **30.** $\displaystyle\int \frac{x}{\sqrt{4 - x^2}} \, dx$

31. $\displaystyle\int \frac{x^3}{\sqrt{x^2 - 4}} \, dx$ **32.** $\displaystyle\int x^3\sqrt{x^2 - 4} \, dx$

33. $\displaystyle\int \frac{2}{\sqrt{x^2 - 4}} \, dx$ **34.** $\displaystyle\int \frac{x}{\sqrt{x^2 - 4}} \, dx$

35. $\displaystyle\int \frac{x^3}{\sqrt{x^2 + 4}} \, dx$

36. $\displaystyle\int x^3\sqrt{x^2 + 4} \, dx$

37. $\displaystyle\int \sqrt{x^2 + 4} \, dx$

38. $\displaystyle\int \frac{x}{\sqrt{x^2 + 4}} \, dx$

39. $\displaystyle\int x\sqrt{x^2 + 4} \, dx$

40. $\displaystyle\int x^2\sqrt{x^2 + 4} \, dx$

In exercises 41 and 42, evaluate the integral using both substitutions $u = \tan x$ and $u = \sec x$ and compare the results.

41. $\displaystyle\int \tan x \sec^4 x \, dx$

42. $\displaystyle\int \tan^3 x \sec^4 x \, dx$

43. Show that for any integer $n > 1$,

$$\int \sec^n x \, dx = \frac{1}{n-1} \sec^{n-2} x \tan x + \frac{n-2}{n-1} \int \sec^{n-2} x \, dx.$$

44. Evaluate (a) $\displaystyle\int \sec^3 x \, dx$, (b) $\displaystyle\int \sec^4 x \, dx$ and (c) $\displaystyle\int \sec^5 x \, dx$.

45. In an AC circuit, the current has the form $i(t) = I\cos(\omega t)$ for constants I and ω. The power is defined as Ri^2 for a constant R. Find the average value of the power by integrating over the interval $[0, 2\pi/\omega]$.

46. The area of the ellipse $\dfrac{x^2}{a^2} + \dfrac{y^2}{b^2} = 1$ is given by $\dfrac{4b}{a}\displaystyle\int_0^a \sqrt{a^2 - x^2} \, dx$. Compute this integral.

47. (exploratory) In section 7.2, you were asked to show that for positive integers m and n with $m \neq n$, each of the following integrals equals 0. $\displaystyle\int_{-\pi}^{\pi} \cos mx \cos nx \, dx = 0$, $\displaystyle\int_{-\pi}^{\pi} \sin mx \sin nx \, dx = 0$. Also, $\displaystyle\int_{-\pi}^{\pi} \cos^2 nx \, dx = \displaystyle\int_{-\pi}^{\pi} \sin^2 nx \, dx = \pi$. Finally, $\displaystyle\int_{-\pi}^{\pi} \cos mx \sin nx \, dx = 0$ for any positive integers m and n. We will use these formulas to explain how a radio can tune in an AM station.

Amplitude modulation (or AM) radio sends a signal (e.g., music) that modulates the carrier frequency. For example, if the signal is $2\sin t$ and the carrier frequency is 16, then the radio sends out the modulated signal $2\sin t \sin 16t$. The graphs of $y = 2\sin t$, $y = -2\sin t$ and $y = 2\sin t \sin 16t$ are shown below.

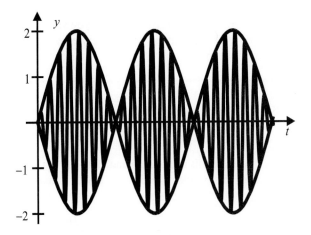

The graph of $y = 2\sin t \sin 16t$ has the same period as the carrier $\sin 16t$, but the amplitude varies between $2\sin t$ and $-2\sin t$ (hence the term amplitude modulation). The radio's problem is to tune in the frequency 16 and recover the signal $2\sin t$. The difficulty is that other radio stations are broadcasting simultaneously. A radio receives all the signals mixed together. To see how this works, suppose a second station broadcasts the signal $3\sin t$ at frequency 32. The combined

signal that the radio receives is $2\sin t \sin 16t + 3\sin t \sin 32t$. We will decompose this signal. The first step is to rewrite the signal using the identity

$$\sin A \sin B = \frac{1}{2}\cos(B - A) - \frac{1}{2}\cos(B + A).$$

The signal then equals

$$f(t) = \cos 15t - \cos 17t + \frac{3}{2}\cos 31t - \frac{3}{2}\cos 33t.$$

If the radio "knows" that the signal has the form $c\sin t$ for some constant c, it can determine the constant c at frequency 16 by computing the integral $\int_{-\pi}^{\pi} f(t)\cos 15t\, dt$ and multiplying by $2/\pi$. Show that $\int_{-\pi}^{\pi} f(t)\cos 15t\, dt = \pi$, so that the correct constant is $c = \pi\,(2/\pi) = 2$. The signal is then $2\sin t$. To recover the signal sent out by the second station, compute $\int_{-\pi}^{\pi} f(t)\cos 31t\, dt$ and multiply by $2/\pi$. Show that you correctly recover the signal $3\sin t$.

7.4   Integration of Rational Functions Using Partial Fractions

The notion we introduce in this section is less a technique of integration than it is an algebraic method for rewriting certain rational functions. As we will see, it is very useful in integration as well as in other applications. We begin with a simple observation. Note that

$$\frac{3}{x+2} - \frac{2}{x-5} = \frac{3(x-5) - 2(x+2)}{(x+2)(x-5)} = \frac{x-19}{x^2 - 3x - 10}. \tag{4.1}$$

Observe further that if we want to evaluate the integral of the expression on the right-hand side of (4.1), this provides us with a simple means of evaluating it. We have

$$\int \frac{x-19}{x^2 - 3x - 10}\, dx = \int \left(\frac{3}{x+2} - \frac{2}{x-5}\right) dx = 3\ln|x+2| - 2\ln|x-5| + c.$$

On the other hand, without knowing that we could write the integral on the left as the integral on the right, we would have a tough time evaluating the first integral. The second integrand,

$$\frac{3}{x+2} - \frac{2}{x-5}$$

is called a **partial fractions decomposition** of the first integrand. You should be wondering if this is a fluke or if this kind of expansion can be done more generally. It turns out that we can decompose **any** rational function (i.e., any quotient of two polynomials) into a sum of simpler rational functions. We develop this method in this section. Equation (4.1) is an example of a special case of partial fractions. More generally, if the three factors $a_1x + b_1$, $a_2x + b_2$ and $a_3x + b_3$ are all distinct (i.e., no one is a constant multiple of another), then we can write

$$\frac{a_1x + b_1}{(a_2x + b_2)(a_3x + b_3)} = \frac{A}{a_2x + b_2} + \frac{B}{a_3x + b_3},$$

for some choice of constants A and B to be determined. Notice that if you wanted to integrate this expression, the partial fractions on the right-hand side are very easy to integrate, just as they were in the introductory example above.

Example 4.1   Partial Fractions: Distinct Linear Factors

Evaluate $\int \dfrac{1}{x^2 + x - 2}\, dx.$

Solution   First, note that you can't evaluate this as it stands and all of our earlier methods fail to help (consider each of these for this problem). However, we can make a partial fractions decomposition, as follows.

$$\frac{1}{x^2 + x - 2} = \frac{1}{(x-1)(x+2)} = \frac{A}{x-1} + \frac{B}{x+2} = \frac{A(x+2) + B(x-1)}{(x-1)(x+2)}.$$

Looking at the first and last of the fractions above, notice that their denominators are the same. So, to be equal, their numerators must be the same, too. Thus,

$$1 = A(x+2) + B(x-1). \tag{4.2}$$

We would like to solve this equation for A and B. The key is to realize that this equation must be true for all x, including $x = 1$ and $x = -2$. (We single out these two values because they will make one or the other of the terms in (4.2) zero and thereby allow us to easily solve for the unknowns A and B.) In particular, for $x = 1$, notice that from (4.2), we have

$$1 = A(1+2) + B(1-1) = 3A,$$

so that $A = \frac{1}{3}$. Likewise, taking $x = -2$, we have

$$1 = A(-2+2) + B(-2-1) = -3B,$$

so that $B = -\frac{1}{3}$. Thus, we have

$$\int \frac{1}{x^2 + x - 2}\, dx = \int \left[\frac{1}{3} \left(\frac{1}{x-1} \right) - \frac{1}{3} \left(\frac{1}{x+2} \right) \right] dx$$

$$= \frac{1}{3} \ln |x - 1| - \frac{1}{3} \ln |x + 2| + c.$$

■

We can do the same as we did above whenever a rational expression has a denominator that factors into n distinct linear factors, as follows. If the degree of $P(x) < n$ and the factors $(a_i x + b_i)$, for $i = 1, 2, \ldots, n$ are all distinct, then we can write

$$\frac{P(x)}{(a_1 x + b_1)(a_2 x + b_2) \cdots (a_n x + b_n)} = \frac{c_1}{a_1 x + b_1} + \frac{c_2}{a_2 x + b_2} + \cdots + \frac{c_n}{a_n x + b_n},$$

for some constants $c_1, c_2, \ldots, c_n$.

Example 4.2   Partial Fractions: Three Distinct Linear Factors

Evaluate $\int \dfrac{3x^2 - 7x - 2}{x^3 - x}\, dx.$

Solution   Once again, our earlier methods fail us, but we can rewrite the integrand using partial fractions. We have

$$\frac{3x^2 - 7x - 2}{x^3 - x} = \frac{3x^2 - 7x - 2}{x(x - 1)(x + 1)} = \frac{A}{x} + \frac{B}{x - 1} + \frac{C}{x + 1}$$

$$= \frac{A(x - 1)(x + 1) + Bx(x + 1) + Cx(x - 1)}{x(x - 1)(x + 1)}.$$

Again, we equate the numerators of the fractions on the far left and on the far right, to get

$$3x^2 - 7x - 2 = A(x - 1)(x + 1) + Bx(x + 1) + Cx(x - 1). \qquad (4.3)$$

In this case, notice that taking $x = 0$, $x = 1$ or $x = -1$ will make two of the three terms on the right side of (4.3) zero. Specifically, for $x = 0$, we get

$$-2 = A(-1)(1) = -A,$$

so that $A = 2$. Likewise, taking $x = 1$, we find $B = -3$ and taking $x = -1$, we find $C = 4$. Thus, we have

$$\int \frac{3x^2 - 7x - 2}{x^3 - x} \, dx = \int \left(\frac{2}{x} - \frac{3}{x - 1} + \frac{4}{x + 1} \right) dx$$

$$= 2 \ln |x| - 3 \ln |x - 1| + 4 \ln |x + 1| + c.$$

■

NOTE: If the numerator of a rational expression has the same or higher degree than the denominator, you must first perform a long division and follow this with a partial fractions decomposition of the remaining (proper) fraction.

Example 4.3   Partial Fractions Where Long Division is Required

Find the indefinite integral of $f(x) = \dfrac{2x^3 - 4x^2 - 15x + 5}{x^2 - 2x - 8}$ using a partial fractions decomposition.

Solution   Since the degree of the numerator exceeds that of the denominator, first divide. We indicate the long division below. (You should perform the division however you are most comfortable.)

$$
\begin{array}{r}
2x \\
x^2 - 2x - 8 \overline{)\, 2x^3 - 4x^2 - 15x + 5} \\
\underline{2x^3 - 4x^2 - 16x} \\
x + 5
\end{array}
$$

Thus, we have

$$f(x) = \frac{2x^3 - 4x^2 - 15x + 5}{x^2 - 2x - 8} = 2x + \frac{x + 5}{x^2 - 2x - 8}.$$

The remaining proper fraction can be expanded as

$$\frac{x + 5}{x^2 - 2x - 8} = \frac{x + 5}{(x - 4)(x + 2)} = \frac{A}{x - 4} + \frac{B}{x + 2}.$$

It is a simple matter to solve for the constants: $A = \frac{3}{2}$ and $B = -\frac{1}{2}$. (This is left as an exercise). We now have

$$\int \frac{2x^3 - 4x^2 - 15x + 5}{x^2 - 2x - 8}\, dx = \int \left[2x + \frac{3}{2}\left(\frac{1}{x-4} \right) - \frac{1}{2}\left(\frac{1}{x+2} \right) \right] dx$$

$$= x^2 + \frac{3}{2}\ln|x-4| - \frac{1}{2}\ln|x+2| + c.$$

∎

You may already have begun to wonder what happens when the denominator of a rational expression contains repeated linear factors. For example, is there a partial fractions decomposition of an expression like

$$\frac{2x+3}{(x-1)^2}?$$

The answer is yes and the decomposition looks like the following. If the degree of $P(x) < n$, then we can write

$$\frac{P(x)}{(ax+b)^n} = \frac{c_1}{(ax+b)} + \frac{c_2}{(ax+b)^2} + \cdots + \frac{c_n}{(ax+b)^n},$$

for constants $c_1, c_2, \ldots, c_n$ to be determined. The following example is typical.

Example 4.4   Partial Fractions with a Repeated Linear Factor

Use a partial fractions decomposition to find an antiderivative of $f(x) = \dfrac{5x^2 + 20x + 6}{x^3 + 2x^2 + x}$.

Solution   First, note that there is a repeated linear factor in the denominator. We have

$$\frac{5x^2 + 20x + 6}{x^3 + 2x^2 + x} = \frac{5x^2 + 20x + 6}{x(x+1)^2} = \frac{A}{x} + \frac{B}{x+1} + \frac{C}{(x+1)^2}$$

$$= \frac{A(x+1)^2 + Bx(x+1) + Cx}{x(x+1)^2}.$$

Comparing numerators, we have

$$5x^2 + 20x + 6 = A(x+1)^2 + Bx(x+1) + Cx.$$

Taking $x = 0$, we find $A = 6$. Likewise, taking $x = -1$, we find that $C = 9$. To determine B, substitute any convenient value for x, say $x = 1$. (Unfortunately, notice that there is no choice of x that will make the two terms containing A and C both zero, without also making the term containing B zero.) You should find that $B = -1$. So, we have

$$\int \frac{5x^2 + 20x + 6}{x^3 + 2x^2 + x}\, dx = \int \left[\frac{6}{x} - \frac{1}{x+1} + \frac{9}{(x+1)^2} \right] dx$$

$$= 6\ln|x| - \ln|x+1| - 9(x+1)^{-1} + c.$$

∎

We can extend the notion of partial fraction decomposition to rational expressions with denominators containing irreducible quadratic factors (i.e., quadratic factors that have no real factorization). If the degree of $P(x) < 2n$ (the degree of the denominator), we can write

$$\frac{P(x)}{(a_1x^2 + b_1x + c_1)(a_2x^2 + b_2x + c_2)\cdots(a_nx^2 + b_nx + c_n)}$$

$$= \frac{A_1x + B_1}{(a_1x^2 + b_1x + c_1)} + \frac{A_2x + B_2}{(a_2x^2 + b_2x + c_2)} + \cdots + \frac{A_nx + B_n}{(a_nx^2 + b_nx + c_n)}. \quad (4.4)$$

You can think of this in terms of irreducible quadratic denominators getting linear numerators, while linear denominators get constant numerators. If you think this looks messy, you're right; it is. But only the algebra is messy (and you can always use a CAS to do the algebra for you). You should note that the partial fractions on the right hand side of (4.4) are integrated comparatively easily using u-substitution together with possibly completing the square.

Example 4.5   Partial Fractions with a Quadratic Factor

Use a partial fractions decomposition to find an antiderivative of $f(x) = \dfrac{2x^2 - 5x + 2}{x^3 + x}$.

Solution   First, note that

$$\frac{2x^2 - 5x + 2}{x^3 + x} = \frac{2x^2 - 5x + 2}{x(x^2 + 1)} = \frac{A}{x} + \frac{Bx + C}{x^2 + 1} = \frac{A(x^2 + 1) + (Bx + C)x}{x(x^2 + 1)}.$$

Equating the numerators gives us

$$2x^2 - 5x + 2 = A(x^2 + 1) + (Bx + C)x$$
$$= (A + B)x^2 + Cx + A.$$

Rather than substitute numbers for x (notice that there are no convenient values to plug in, except for $x = 0$) we instead match up the coefficients of like powers of x:

$$2 = A + B$$
$$-5 = C$$
$$\text{and } 2 = A.$$

This leaves us with $B = 0$ and so,

$$\int \frac{2x^2 - 5x + 2}{x^3 + x}\, dx = \int \left(\frac{2}{x} - \frac{5}{x^2 + 1} \right) dx = 2\ln|x| - 5\tan^{-1} x + c.$$

Often, partial fractions decompositions involving irreducible quadratic terms lead to expressions that require further massaging (such as completing the square) before we can find an antiderivative. We illustrate this in the following example.

Example 4.6   Partial Fractions with a Quadratic Factor

Use a partial fractions decomposition to find an antiderivative for

$$f(x) = \frac{5x^2 + 6x + 2}{(x + 2)(x^2 + 2x + 5)}.$$

Solution   First, notice that the quadratic factor in the denominator does not factor, so that the correct decomposition is

$$\frac{5x^2 + 6x + 2}{(x + 2)(x^2 + 2x + 5)} = \frac{A}{x + 2} + \frac{Bx + C}{x^2 + 2x + 5}$$
$$= \frac{A(x^2 + 2x + 5) + (Bx + C)(x + 2)}{(x + 2)(x^2 + 2x + 5)}.$$

If we equate the numerators, we get

$$5x^2 + 6x + 2 = A(x^2 + 2x + 5) + (Bx + C)(x + 2).$$

Matching up the coefficients of like powers of x, we get

$$5 = A + B$$
$$6 = 2A + 2B + C$$
$$\text{and } 2 = 5A + 2C.$$

You'll need to solve this by elimination. We leave it as an exercise to show that $A = 2$, $B = 3$ and $C = -4$. Integrating, we have

$$\int \frac{5x^2 + 6x + 2}{(x + 2)(x^2 + 2x + 5)}\, dx = \int \left(\frac{2}{x + 2} + \frac{3x - 4}{x^2 + 2x + 5} \right) dx. \qquad (4.5)$$

The integral of the first term is easy, but what about the second term? Since the denominator doesn't factor, you have very few choices. Try substituting for the denominator: let $u = x^2 + 2x + 5$, so that $du = (2x + 2)\, dx$. Notice that we can now write the integral of the second term as

$$\int \frac{3x - 4}{x^2 + 2x + 5}\, dx = \int \frac{3(x + 1) - 7}{x^2 + 2x + 5}\, dx = \int \left[\left(\frac{3}{2} \right) \frac{2(x + 1)}{x^2 + 2x + 5} - \frac{7}{x^2 + 2x + 5} \right] dx$$

$$= \frac{3}{2} \int \frac{2(x + 1)}{x^2 + 2x + 5}\, dx - \int \frac{7}{x^2 + 2x + 5}\, dx$$

$$= \frac{3}{2} \ln \left(x^2 + 2x + 5 \right) - \int \frac{7}{x^2 + 2x + 5}\, dx. \qquad (4.6)$$

To see how to evaluate the remaining integral in (4.6), you'll need to complete the square in the denominator. Notice that we can write

$$\int \frac{7}{x^2 + 2x + 5}\, dx = \int \frac{7}{(x + 1)^2 + 4}\, dx = \frac{7}{2} \tan^{-1} \left(\frac{x + 1}{2} \right) + c.$$

(We leave the details of this last integration as an exercise.) Putting this together with (4.5) and (4.6), we now have

$$\int \frac{5x^2 + 6x + 2}{(x + 2)(x^2 + 2x + 5)}\, dx = 2 \ln |x + 2| + \frac{3}{2} \ln \left(x^2 + 2x + 5 \right) - \frac{7}{2} \tan^{-1} \left(\frac{x + 1}{2} \right) + c.$$

∎

Rational expressions with repeated irreducible quadratic factors in the denominator are explored in the exercises. The idea of these is the same as the preceding decompositions, but the algebra (without a CAS) is even messier.

Remark 4.1   After mastering decompositions involving repeated irreducible quadratic factors, you will be able to find the partial fractions decomposition of **any** rational function. The denominator of such a function (a polynomial) can always be factored into linear and quadratic factors, some of which may be repeated. Then, use the techniques covered in this section! You should recognize that while this observation is certainly true, in practice you may require a CAS to accurately complete the calculations.

Remark 4.2   Most CASs include commands for performing partial fraction decomposition. Even so, we urge you to work through the exercises in this section by hand. Once you have the idea of how these decompositions work, by all means, use your CAS to do the drudge work for you. Until that time, be patient and work carefully by hand.

Exercises 7.4

1. (writing) There is a shortcut for determining the constants for linear terms in a partial fractions decomposition. For example, take

$$\frac{x-1}{(x+1)(x-2)} = \frac{A}{x+1} + \frac{B}{x-2}.$$

To compute A, take the original fraction on the left, cover up the $x+1$ in the denominator and replace x with -1: $A = \frac{-1-1}{-1-2} = \frac{2}{3}$. Similarly, to solve for B, cover up the $x-2$ and replace x with 2: $B = \frac{2-1}{2+1} = \frac{1}{3}$. Explain why this works and practice it for yourself.

2. (writing) For partial fractions, there is a big distinction between quadratic functions that factor into linear terms and quadratic functions that are irreducible. Recall that a quadratic function factors as $(x-a)(x-b)$ if and only if a and b are zeros of the function. Explain how you can use the quadratic formula to determine whether or not if a given quadratic function is irreducible.

In exercises 3–34, find the partial fractions decomposition and an antiderivative. If you have a CAS available, use it to check your answer.

3. $\dfrac{x-5}{x^2-1}$

4. $\dfrac{5x-2}{x^2-4}$

5. $\dfrac{6x}{x^2-x-2}$

6. $\dfrac{4}{x^2-1}$

7. $\dfrac{x+1}{x^2-x-6}$

8. $\dfrac{3x}{x^2-3x-4}$

9. $\dfrac{-x+5}{x^3-x^2-2x}$

10. $\dfrac{3x+8}{x^3+5x^2+6x}$

11. $\dfrac{x^3+x+2}{x^2+2x-8}$

12. $\dfrac{x^2+1}{x^2-5x-6}$

13. $\dfrac{-3x-1}{x^3-x}$

14. $\dfrac{x^2+3}{x^3-x}$

15. $\dfrac{2x+3}{(x+2)^2}$

16. $\dfrac{3x-5}{(x-1)^2}$

17. $\dfrac{x-1}{x^3+4x^2+4x}$

18. $\dfrac{4x-5}{x^3-3x^2}$

19. $\dfrac{x+4}{x^3+3x^2+2x}$

20. $\dfrac{-2x^2+4}{x^3+3x^2+2x}$

21. $\dfrac{x+2}{x^3+x}$

22. $\dfrac{1}{x^3+4x}$

23. $\dfrac{4x-2}{x^4-1}$

24. $\dfrac{3x+7}{x^4-16}$

25. $\dfrac{3x^2-6}{x^2-x-2}$

26. $\dfrac{x^3+x}{x^2-1}$

27. $\dfrac{2x+3}{x^2+2x+1}$

28. $\dfrac{2x}{x^2-6x+9}$

29. $\dfrac{x^2+2x+1}{x^3+x}$

30. $\dfrac{2x^2-3x+2}{x^3+x}$

31. $\dfrac{4x^2+3}{x^3+x^2+x}$

32. $\dfrac{4x+4}{x^4+x^3+2x^2}$

33. $\dfrac{3x^3+1}{x^3-x^2+x-1}$

34. $\dfrac{2x^4+9x^2+x-4}{x^3+4x}$

35. In this exercise, we find the partial fractions decomposition of $\dfrac{4x^2+2}{(x^2+1)^2}$. Consistent with the form for repeated linear factors, the form for the decomposition is $\dfrac{Ax+B}{x^2+1} + \dfrac{Cx+D}{(x^2+1)^2}$. We set

$$\frac{4x^2+2}{(x^2+1)^2} = \frac{Ax+B}{x^2+1} + \frac{Cx+D}{(x^2+1)^2}$$

Multiplying through by $(x^2+1)^2$, we get

$$4x^2+2 = (Ax+B)(x^2+1) + Cx+D$$
$$= Ax^3 + Bx^2 + Ax + B + Cx + D$$

As in example 4.5, we match up coefficients of like powers of x. For x^3, we have $0 = A$. For x^2, we have $4 = B$. Match the coefficients of x and the constants to finish the decomposition.

In exercises 36–40, find the partial fractions decomposition. (Refer to exercise 35.)

36. $\dfrac{x^3+2}{(x^2+1)^2}$

37. $\dfrac{2x^2+4}{(x^2+4)^2}$

38. $\dfrac{2x^3-x^2}{(x^2+1)^2}$

39. $\dfrac{4x^2+3}{(x^2+x+1)^2}$

40. $\dfrac{x^4+x^3}{(x^2+4)^2}$

41. To solve the logistic differential equation $y' = ay(1-y)$ in section 6.5, we wrote it in the separable form $\dfrac{1}{y(1-y)}y' = a$ and integrated. Perform the partial fractions decomposition of the integrand, integrate $\displaystyle\int \frac{1}{y(1-y)}\,dy$ and then solve the differential equation.

42. In certain chemical reactions, two molecules of substance A react with one molecule of substance B to form one molecule of a third substance X. This type of reaction is found in the breakdown of the ozone layer. If there are a total of a molecules of A available and a total of b molecules of B available, the amount $x(t)$ of X formed obeys the equation

$$x'(t) = k[a - 2x(t)]^2[b - x(t)]$$

for some rate constant k. Write this as a separable equation and use partial fractions to find an implicit representation of the solution.

43. (exploratory) In this exercise, you explore a more general form of the logistic equation than that analyzed in section 6.5. Suppose that $y' = ay - by^2$. Note that if $a = b$, this is the same as the logistic equation previously studied. Assuming that $0 < y < a/b$, show that $y(t) = \dfrac{a}{Ce^{-at} + b}$, where the constant C equals $\dfrac{a}{y(0)} - b$. What happens to $y(t)$ as $t \to \infty$? How does this compare to the logistic solutions found previously? Solve the equation if $y > a/b$.

44. (exploratory) In developing the definite integral, we looked at sums such as $\displaystyle\sum_{i=1}^{n} \frac{2}{i^2 + i}$. As with Riemann sums, we are especially interested in the limit as $n \to \infty$. Write out several terms of the sum and try to guess what the limit is. It turns out that this is one of the few sums for which a precise formula exists, because this is a **telescoping sum**. To find out what this means, write out the partial fractions decomposition for $\dfrac{2}{i^2 + i}$. Using the partial fractions form, write out several terms of the sum and notice how much cancellation there is. Briefly describe why the term *telescoping* is appropriate, and determine $\displaystyle\sum_{i=1}^{n} \frac{2}{i^2 + i}$. Then find the limit as $n \to \infty$. Repeat this process for the telescoping sum $\displaystyle\sum_{i=1}^{n} \frac{4i}{i^2 - 1}$.

7.5   Integration Tables and Computer Algebra Systems

Ask anyone who has ever needed to evaluate a large number of integrals as part of their work (this includes innumerable engineers, mathematicians, physicists and others) and they will tell you that they have made extensive use of integral tables and/or a computer algebra system. These are extremely powerful tools for the professional user of mathematics. However, they do **not** take the place of learning all the basic techniques of integration that we've been discussing. To use a table, you often must first rewrite the integral in the form of one of the integrals in the table. This may require you to perform some algebraic manipulation or to make a substitution. While a CAS will report an antiderivative, it will occasionally report it in an inconvenient form. More significantly, a CAS will from time to time report an answer that is (at least technically) incorrect. We will point out some of these shortcomings in the examples which follow.

Using Tables of Integrals

We include a small table of indefinite integrals in Appendix B. A larger table can be found in the *CRC Standard Mathematical Tables*. An amazingly extensive table is found in the book *Table of Integrals, Series and Products*, compiled by Gradshteyn and Ryzhik. If you can't find an integral there, you're not likely to find it anywhere.

Example 5.1   Using an Integral Table

Use a table to evaluate $\displaystyle\int \frac{\sqrt{3 + 4x^2}}{x} \, dx$.

Solution   Certainly, you could evaluate this integral using trigonometric substitution. However, if you look in our integral table, you will find

$$\int \frac{\sqrt{a^2 + u^2}}{u} \, du = \sqrt{a^2 + u^2} - a \ln \left| \frac{a + \sqrt{a^2 + u^2}}{u} \right| + c. \tag{5.1}$$

Unfortunately, the integral in question is not quite in the form of (5.1). However, we can fix this with the substitution $u = 2x$, so that $du = 2 dx$. This gives us

$$\int \frac{\sqrt{3 + 4x^2}}{x} \, dx = \int \frac{\sqrt{3 + (2x)^2}}{2x} \, (2) \, dx = \int \frac{\sqrt{3 + u^2}}{u} \, du$$

$$= \sqrt{3 + u^2} - \sqrt{3} \ln \left| \frac{\sqrt{3} + \sqrt{3 + u^2}}{u} \right| + c$$

$$= \sqrt{3 + 4x^2} - \sqrt{3} \ln \left| \frac{\sqrt{3} + \sqrt{3 + 4x^2}}{2x} \right| + c.$$

A number of the formulas in the table are called **reduction formulas**. These are of the form

$$\int f(u) \, du = g(u) + \int h(u) \, du,$$

where the second integral is simpler than the first. These are often applied repeatedly, as in the following example.

Example 5.2   Using a Reduction Formula

Use a reduction formula to evaluate $\int \sin^6 x \, dx$.

Solution   You should recognize that this integral can be evaluated using techniques you already know. (How?) However, for any integer $n \geq 1$, we have the reduction formula

$$\int \sin^n u \, du = -\frac{1}{n} \sin^{n-1} u \cos u + \frac{n-1}{n} \int \sin^{n-2} u \, du. \qquad (5.2)$$

If we apply (5.2) with $n = 6$, we get

$$\int \sin^6 x \, dx = -\frac{1}{6} \sin^5 x \cos x + \frac{5}{6} \int \sin^4 x \, dx.$$

We can apply the same reduction formula (this time for $n = 4$) to evaluate $\int \sin^4 x \, dx$. We have

$$\int \sin^6 x \, dx = -\frac{1}{6} \sin^5 x \cos x + \frac{5}{6} \int \sin^4 x \, dx$$

$$= -\frac{1}{6} \sin^5 x \cos x + \frac{5}{6} \left(-\frac{1}{4} \sin^3 x \cos x + \frac{3}{4} \int \sin^2 x \, dx \right).$$

Finally, for $\int \sin^2 x \, dx$, we can use (5.2) once again (for $n = 2$), or evaluate the integral using a double angle formula. We choose the former here and obtain

$$\int \sin^6 x \, dx = -\frac{1}{6} \sin^5 x \cos x + \frac{5}{6} \left(-\frac{1}{4} \sin^3 x \cos x + \frac{3}{4} \int \sin^2 x \, dx \right)$$

$$= -\frac{1}{6} \sin^5 x \cos x - \frac{5}{24} \sin^3 x \cos x + \frac{15}{24} \left(-\frac{1}{2} \sin x \cos x + \frac{1}{2} \int dx \right)$$

$$= -\frac{1}{6} \sin^5 x \cos x - \frac{5}{24} \sin^3 x \cos x - \frac{15}{48} \sin x \cos x + \frac{15}{48} x + c.$$

We should remind you at this point that there are many different ways to find an antiderivative. Antiderivatives found through different means may look quite different, even though they are equivalent. For instance, notice that if an antiderivative has the form $\sin^2 x + c$, then an equivalent antiderivative is $-\cos^2 x + c$, since we can write

$$\sin^2 x + c = 1 - \cos^2 x + c = -\cos^2 x + (1 + c).$$

Finally, since c is an arbitrary constant, so is $1 + c$. In example 5.2, observe that the first three terms all have factors of $\sin x \cos x$, which equals $\frac{1}{2}\sin 2x$. Using this and other identities, you can show that our solution in example 5.2 is equivalent to the following solution obtained from a popular CAS:

$$\int \sin^6 x\, dx = \frac{5}{16}x - \frac{15}{64}\sin 2x + \frac{3}{64}\sin 4x - \frac{1}{192}\sin 6x + c.$$

So, do not panic if your answer differs from a friend's or from one in the back of the book. Both answers may be correct. If you're unsure which one is right, try computing the derivative of your answer. If you get the integrand, you're right (regardless of what your friend or the back of the book says).

You will sometimes want to apply several different reduction formulas at different points in a given problem.

Example 5.3   Making a Substitution Before Using a Reduction Formula

Evaluate $\int x^3 \sin 2x\, dx$.

Solution   From our table, we have the reduction formula

$$\int u^n \sin u\, du = -u^n \cos u + n \int u^{n-1} \cos u\, du. \qquad (5.3)$$

In order to use (5.3), we must first make the substitution $u = 2x$, so that $du = 2\, dx$, which gives us

$$\int x^3 \sin 2x\, dx = \frac{1}{2}\int \frac{(2x)^3}{2^3}\sin 2x\,(2)\,dx = \frac{1}{16}\int u^3 \sin u\, du$$

$$= \frac{1}{16}\left(-u^3 \cos u + 3\int u^2 \cos u\, du\right),$$

where we have used the reduction formula with $n = 3$. Now, to evaluate this last integral, we use the reduction formula

$$\int u^n \cos u\, du = u^n \sin u - n \int u^{n-1} \sin u\, du,$$

with $n = 2$ to get

$$\int x^3 \sin 2x\, dx = -\frac{1}{16}u^3 \cos u + \frac{3}{16}\int u^2 \cos u\, du$$

$$= -\frac{1}{16}u^3 \cos u + \frac{3}{16}\left(u^2 \sin u - 2\int u \sin u\, du\right).$$

Applying the first reduction formula (5.3) one more time (this time, with $n = 1$), we get

$$\int x^3 \sin 2x\, dx = -\frac{1}{16}u^3 \cos u + \frac{3}{16}u^2 \sin u - \frac{3}{8}\int u \sin u\, du$$

$$= -\frac{1}{16}u^3 \cos u + \frac{3}{16}u^2 \sin u - \frac{3}{8}\left(-u \cos u + \int u^0 \cos u\, du\right)$$

$$= -\frac{1}{16}u^3 \cos u + \frac{3}{16}u^2 \sin u + \frac{3}{8}u \cos u - \frac{3}{8}\sin u + c$$

$$= -\frac{1}{16}(2x)^3 \cos 2x + \frac{3}{16}(2x)^2 \sin 2x + \frac{3}{8}(2x)\cos 2x - \frac{3}{8}\sin 2x + c$$

$$= -\frac{1}{2}x^3 \cos 2x + \frac{3}{4}x^2 \sin 2x + \frac{3}{4}x \cos 2x - \frac{3}{8}\sin 2x + c.$$

∎

As we'll see in the following example, some integrals require some insight before using an integral table.

Example 5.4   Making a Substitution Before Using an Integral Table

Evaluate $\displaystyle\int \frac{\sin 2x}{\sqrt{4\cos x - 1}}\, dx.$

Solution   You won't find this integral or any one particularly close to it in our integral table. However, with a little fiddling, we can rewrite this in a simpler form. First, use the double angle formula to rewrite the numerator of the integrand. We get

$$\int \frac{\sin 2x}{\sqrt{4\cos x - 1}}\, dx = 2\int \frac{\sin x \cos x}{\sqrt{4\cos x - 1}}\, dx.$$

Remember to always be on the lookout for terms that are derivatives of other terms. (We did not intend for you to abandon this advice after completing the section on u-substitution.) Here, let $u = \cos x$, so that $du = -\sin x\, dx$. This gives us

$$\int \frac{\sin 2x}{\sqrt{4\cos x - 1}}\, dx = 2\int \frac{\sin x \cos x}{\sqrt{4\cos x - 1}}\, dx = -2\int \frac{u}{\sqrt{4u - 1}}\, du.$$

From our table, notice that

$$\int \frac{u}{\sqrt{a + bu}}\, du = \frac{2}{3b^2}(bu - 2a)\sqrt{a + bu} + c. \tag{5.4}$$

Taking $a = -1$ and $b = 4$ in (5.4), we have

$$\int \frac{\sin 2x}{\sqrt{4\cos x - 1}}\, dx = -2\int \frac{u}{\sqrt{4u - 1}}\, du = (-2)\frac{2}{3(4^2)}(4u + 2)\sqrt{4u - 1} + c$$

$$= -\frac{1}{12}(4\cos x + 2)\sqrt{4\cos x - 1} + c.$$

∎

Integration Using a Computer Algebra System

Computer algebra systems are some of the most powerful new tools to arrive on the mathematical scene in the last 20 years. They run the gamut from handheld calculators (e.g., the TI-89) to powerful software systems (e.g., Maple), which will run on nearly any personal computer.

The examples that follow focus on some of the rare problems you may encounter using a CAS. We admit that we intentionally searched for CAS mistakes. The bad news is that we found several problems. The good news is that the mistakes were very uncommon, and the CAS you're using won't necessarily make any of them. The bottom line here is that a CAS is "taught" many rules by its programmers. If it applies the wrong rule to the wrong problem, you get an incorrect answer. Be aware that these are software bugs and the next version of your CAS may be given a more complete set of rules. As an intelligent user of technology, you need to be aware of common errors and have the calculus skills to catch mistakes when they occur.

The first thing you notice when using a CAS to evaluate an indefinite integral is that they supply *an* antiderivative, instead of the most general one (the indefinite integral) by leaving off the constant of integration. This is not a big deal; it's just something to keep in mind.

Example 5.5   A Shortcoming of Some Computer Algebra Systems

Use a computer algebra system to evaluate $\displaystyle\int \frac{1}{x}\, dx$.

Solution   Many CASs evaluate

$$\int \frac{1}{x}\, dx = \ln x.$$

(Actually, one CAS reports the integral as $\log x$, where it is using the notation $\log x$ to denote the natural logarithm.) Not only is this missing the constant of integration, but notice that this antiderivative is only valid for $x > 0$. The TI-92 calculator returns the more general antiderivative

$$\int \frac{1}{x}\, dx = \ln |x| ,$$

which, while still missing the constant of integration, at least is valid for all $x \neq 0$. On the other hand, all of the CASs referred to above correctly evaluate

$$\int_{-2}^{-1} \frac{1}{x}\, dx = -\ln 2,$$

even though the reported antiderivative for all but the TI-92 is not defined at the limits of integration.

■

Sometimes the antiderivative reported by a CAS is not valid, as written for **any** values of x, as in the following example.

Example 5.6   An Incorrect Antiderivative

Use a computer algebra system to evaluate $\displaystyle\int \frac{\cos x}{\sin x - 2}\, dx$.

Solution   One CAS reports the incorrect antiderivative

$$\int \frac{\cos x}{\sin x - 2}\, dx = \ln (\sin x - 2) .$$

At first glance, this may not appear to be wrong, especially since the chain rule seems to indicate that it's correct:

$$\frac{d}{dx} \ln(\sin x - 2) = \frac{\cos x}{\sin x - 2}.$$

The error is more fundamental (and subtle) than a misuse of the chain rule. Notice that the expression $\ln (\sin x - 2)$ is undefined for **all** x, as $\sin x - 2 < 0$, for all x. You should also recall that a general antiderivative rule that applies here is

$$\int \frac{f'(x)}{f(x)}\, dx = \ln |f(x)| + c$$

and realize that the absolute value is important.   The correct antiderivative is $\ln |\sin x - 2| + c$, which can also be written as $\ln(2 - \sin x) + c$ since $2 - \sin x > 0$ for all x.

■

Probably the most common errors you will run into are actually your own. If you give your CAS a problem in the wrong form, it will solve a different problem than you intended. One simple, but common, mistake is shown in the following example.

Example 5.7   A Problem Where the CAS Misinterprets What You Enter

Use a computer algebra system to evaluate $\int 4x8x\,dx$.

Solution   After entering the integrand as $4x8x$, one CAS returned the odd answer

$$\int 4x8x\,dx = 4x8xx.$$

You can easily evaluate the integral (first, rewrite the integrand as $32x^2$) and show this is incorrect, but what was the error? The answer lies in an impressive ability of the CAS to evaluate integrals such as $\int cx\,dx$ for unspecified constants c. This type of problem is why the CAS requires you to tell it the variable of integration. In our problem above, the CAS interpreted the integrand as four times a variable named $x8x$, which is unrelated to the variable of integration, x. Its answer is of the form $\int 4c\,dx = 4cx$.

∎

The form of the antiderivative reported by a CAS will not always be the most convenient.

Example 5.8   An Inconvenient Form of an Antiderivative

Use a computer algebra system to evaluate $\int x(x^2+3)^5\,dx$.

Solution   Several CASs will evaluate

$$\int x(x^2+3)^5\,dx = \frac{1}{12}x^{12} + \frac{3}{2}x^{10} + \frac{45}{4}x^8 + 45x^6 + \frac{405}{4}x^4 + \frac{243}{2}x^2,$$

while others return the much simpler (and equivalent) expression

$$\int x(x^2+3)^5\,dx = \frac{\left(x^2+3\right)^6}{12}.$$

∎

Notably, a CAS will perform even lengthy integrations with ease.

Example 5.9   Some Good Integrals for Using a Computer Algebra System

Use a computer algebra system to evaluate $\int x^3 \sin 2x\,dx$ and $\int x^{10} \sin 2x\,dx$.

Solution   Using a CAS, you can get in one step

$$\int x^3 \sin 2x\,dx = -\frac{1}{2}x^3 \cos 2x + \frac{3}{4}x^2 \sin 2x - \frac{3}{8}\sin 2x + \frac{3}{4}x\cos 2x + c.$$

With the same effort, you can obtain

$$\int x^{10} \sin 2x\,dx = -\frac{1}{2}x^{10} \cos 2x + \frac{5}{2}x^9 \sin 2x + \frac{45}{4}x^8 \cos 2x - 45x^7 \sin 2x$$
$$-\frac{315}{2}x^6 \cos 2x + \frac{945}{2}x^5 \sin 2x + \frac{4725}{4}x^4 \cos 2x$$
$$-\frac{4725}{2}x^3 \sin 2x - \frac{14175}{4}x^2 \cos 2x + \frac{14175}{8}\cos 2x$$
$$+\frac{14175}{4}x\sin 2x + c.$$

If you wanted to, you could even evaluate

$$\int x^{100} \sin 2x \, dx,$$

although the large number of terms makes displaying the result prohibitive. Think about doing this by hand, using a staggering 100 integrations by parts or by applying a reduction formula 100 times.

■

You should get the idea by now: a CAS can perform repetitive calculations (numerical or symbolic) that you could never dream of doing by hand. It is difficult to find a function that has an elementary antiderivative that your CAS cannot find. Consider the following example of a hard integral.

Example 5.10   A Very Hard Integral

Evaluate $\int x^7 e^x \sin x \, dx$.

Solution   Consider what you would need to do to evaluate this integral by hand and then use a computer algebra system. For instance, one CAS reports the antiderivative

$$\int x^7 e^x \sin x \, dx = \left(-\frac{1}{2}x^7 + \frac{7}{2}x^6 - \frac{21}{2}x^5 + 105x^3 - 315x^2 + 315x \right) e^x \cos x$$

$$+ \left(\frac{1}{2}x^7 - \frac{21}{2}x^5 + \frac{105}{2}x^4 - 105x^3 + 315x - 315 \right) e^x \sin x.$$

Don't try this by hand unless you have plenty of time and patience. However, based on your experience, observe that this antiderivative is plausible.

■

A CAS can save you much time and effort when you need to evaluate an integral. Just be aware of the few shortcomings. You may ask why we've spent so much time on integration techniques when you can always let a CAS do the work for you. No, we're not going to launch into that tired old cliche "but what if you were ship-wrecked on a desert island and you didn't have access to a CAS?" In truth, your CAS can solve virtually all of the computational problems that will arise in this text. On rare occasions, however, a CAS-generated answer will be incorrect and you need to be prepared to catch these errors. More importantly, many of the insights at the heart of science and engineering are arrived at through a precise use of several integration techniques. To share these insights and make progress in your field, you need to understand *how* the integration techniques transform one set of symbols into another. Computers are faster and more accurate at symbol manipulation than humans will ever be. However, our special ability as humans is to understand the **intent** as well as the logic of the techniques and apply the right technique at the right time to make a surprising connection or important discovery. So, what do **you** need to know? You need to have a good understanding of what an integral (definite or indefinite) is. Further, you need to have a basic facility for computing integrals. This includes u-substitution and integration by parts, which have important uses beyond the calculation of specific integrals. As well, you need at least a passing knowledge of other techniques (e.g., partial fractions).

Exercises 7.5

1. (writing) Suppose that you are hired by a company to develop a new CAS. Outline a strategy for symbolic integration. Include provisions for formulas in the Table of Integrals in Appendix B and the various techniques you have studied.

2. (writing) In the text, we discussed the importance of knowing general rules for integration. Consider the integral in example 5.4, $\int \dfrac{\sin 2x}{\sqrt{4\cos x - 1}}\, dx$. Can your CAS evaluate this integral? For many integrals like this that **do** show up in applications (there are harder ones in exercises 39 and 40 below), you have to do some work before the technology can finish the task. For this purpose, discuss the importance of recognizing basic forms and understanding how substitution works.

In exercises 3–30, use the Table of Integrals in Appendix B to find an antiderivative. Note: when checking the back of the book or a CAS for answers, beware of functions that look very different but that are equivalent (through a trig identity, for instance).

3. $\displaystyle\int \frac{x}{(2 + 4x)^2}\, dx$

4. $\displaystyle\int \frac{x^2}{(2 + 4x)^2}\, dx$

5. $\displaystyle\int e^{2x}\sqrt{1 + e^x}\, dx$

6. $\displaystyle\int e^{3x}\sqrt{1 + e^{2x}}\, dx$

7. $\displaystyle\int \frac{x^2}{\sqrt{1 + 4x^2}}\, dx$

8. $\displaystyle\int \frac{\cos x}{\sin^2 x(3 + 2\sin x)}\, dx$

9. $\displaystyle\int x^8\sqrt{4 - x^6}\, dx$

10. $\displaystyle\int \sqrt{16 - e^{2x}}\, dx$

11. $\displaystyle\int \frac{e^x}{\sqrt{e^{2x} + 4}}\, dx$

12. $\displaystyle\int \frac{x\sqrt{x^4 - 9}}{x^2}\, dx$

13. $\displaystyle\int \frac{\sqrt{6x - x^2}}{(x - 3)^2}\, dx$

14. $\displaystyle\int \frac{\sec^2 x}{\tan x\sqrt{8\tan x - \tan^2 x}}\, dx$

15. $\displaystyle\int \tan^6 x\, dx$

16. $\displaystyle\int \csc^4 x\, dx$

17. $\displaystyle\int \frac{\cos x}{\sin x\sqrt{4 + \sin x}}\, dx$

18. $\displaystyle\int \frac{x^5}{\sqrt{4 + x^2}}\, dx$

19. $\displaystyle\int x^3\cos x^2\, dx$

20. $\displaystyle\int x\sin 3x^2\cos 4x^2\, dx$

21. $\displaystyle\int \frac{\sin x\cos x}{\sqrt{1 + \cos x}}\, dx$

22. $\displaystyle\int \frac{x\sqrt{1 + 4x^2}}{x^4}\, dx$

23. $\displaystyle\int \frac{\sin^2 x\cos x}{\sqrt{\sin^2 x + 4}}\, dx$

24. $\displaystyle\int \frac{\ln\sqrt{x}}{\sqrt{x}}\, dx$

25. $\displaystyle\int \frac{e^{-2/x^2}}{x^3}\, dx$

26. $\displaystyle\int x^3 e^{2x^2}\, dx$

27. $\displaystyle\int \frac{x}{\sqrt{4x - x^2}}\, dx$

28. $\displaystyle\int e^{5x}\cos 3x\, dx$

29. $\displaystyle\int e^x\tan^{-1}(e^x)\, dx$

30. $\displaystyle\int (\ln 4x)^3\, dx$

31. Check your CAS against all examples in this section. Discuss which errors, if any, your CAS makes.

32. Find out how your CAS evaluates $\int x\sin x\, dx$ if you fail to leave a space between x and $\sin x$.

33. Have your CAS evaluate $\int (\sqrt{1 - x} + \sqrt{x - 1})\, dx$. If you get an answer, explain why it's wrong.

34. To find out if your CAS "knows" integration by parts, try $\int x^3\cos 3x\, dx$ and $\int x^3 e^{5x}\cos 3x\, dx$. To see if it "knows" reduction formulas, try $\int \sec^5 x\, dx$.

35. To find out how many trigonometric techniques your CAS "knows," try $\int \sin^6 x\, dx$, $\int \sin^4 x\cos^3 x\, dx$ and $\int \tan^4 x\sec^3 x\, dx$.

36. Find out if your CAS has a special command (e.g., APART in Mathematica) to do partial fractions decompositions. Also, try $\int \dfrac{x^2 + 2x - 1}{(x - 1)^2(x^2 + 4)}\, dx$ and $\int \dfrac{3x}{(x^2 + x + 2)^2}\, dx$.

37. To find out if your CAS "knows" how to do substitution, try $\int \dfrac{1}{x^2(3 + 2x)}\, dx$ and $\int \dfrac{\cos x}{\sin^2 x(3 + 2\sin x)}\, dx$. Try to find one that your CAS can't do: start with a basic formula like $\int \dfrac{1}{|x|\sqrt{x^2 - 1}}\, dx = \sec^{-1} x + c$ and substitute your favorite function. With $x = e^u$, the preceding integral becomes $\int \dfrac{e^u}{e^u\sqrt{e^{2u} - 1}}\, du$, which you can use to test your CAS.

38. To compute the area of the ellipse $\dfrac{x^2}{a^2} + \dfrac{y^2}{b^2} = 1$, note that the upper-right quarter of the ellipse is given by

$$y = b\sqrt{1 - x^2/a^2}$$

for $0 \le x \le a$. Thus, the area of the ellipse is $4b\displaystyle\int_0^a \sqrt{1 - x^2/a^2}\, dx$. Try this integral on your CAS. The (implicit) assumption we usually make is that $a > 0$, but your CAS should not make this assumption for you. Does your CAS give you πab or $\pi b|a|$?

39. (exploratory) This exercise explores two aspects of a very famous problem (solved in the 1600s; when you finish the problem, think about solving this problem before calculators, computers or even much of calculus was invented). The idea is to imagine a bead sliding down a thin wire that extends in some shape from the point $(0,0)$ to the point $(\pi, -2)$. Assume that gravity pulls the bead down but that there is no friction or other force acting on the bead. This situation is easiest to analyze using **parametric equations** where we have functions $x(u)$ and $y(u)$ giving the horizontal and vertical position of the bead in terms of the variable u. Examples of paths the bead might follow are $\begin{cases} x(u) = \pi u \\ y(u) = -2u \end{cases}$ and $\begin{cases} x(u) = \pi u \\ y(u) = 2(u-1)^2 - 2 \end{cases}$ and $\begin{cases} x(u) = \pi u - \sin \pi u \\ y(u) = \cos \pi u - 1 \end{cases}$. In each case, the bead starts at $(0,0)$ for $u = 0$ and finishes at $(\pi, -2)$ for $u = 1$. You should graph each path on your graphing calculator. The first path is a line, the second is a parabola and the third is a special curve called a **brachistochrone**. For a given path, the time it takes the bead to travel the path is given by

$$T = (1/\sqrt{g}) \int_0^1 \sqrt{\frac{[x'(u)]^2 + [y'(u)]^2}{-2y(u)}}\, du,$$

where g is the gravitational constant. Compute this quantity for the line and the parabola. Explain why the parabola would be a faster path for the bead to slide down, even though the line is shorter in distance (think of which would be a faster hill to ski down). It can be shown that the brachistochrone is the fastest path possible. Try to get your CAS to compute the optimal time. Comparing the graphs of the parabola and brachistochrone, what important advantage does the brachistochrone have at the start of the path?

40. (exploratory) It turns out that the brachistochrone in the exercise 39 has an amazing property, along with providing

the fastest time (which is essentially what the term *brachistochrone* means). The path is shown below.

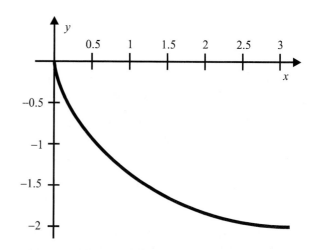

Suppose that instead of starting the bead at the point $(0,0)$, you start the bead partway down the path at $x = c$. How would the time to reach the bottom from $x = c$ compare to the total time from $x = 0$? Note that the answer is **not** obvious, since the farther down you start, the less speed the bead will gain. If $x = c$ corresponds to $u = a$, the time to reach the bottom is given by $(\pi/\sqrt{g}) \int_a^1 \sqrt{\frac{1 - \cos \pi u}{\cos a\pi - \cos \pi u}}\, du$. If $a = 0$ (that is, the bead starts at the top), the time is $\pi/\sqrt{g}$ (the integral equals 1). If you have a very good CAS, try to evaluate the integral for various values of a between 0 and 1. If your CAS can't handle it, approximate the integral numerically. You should discover the amazing fact that *the integral always equals 1*. The brachistochrone is also the **tautochrone**, a curve for which the time to reach the bottom is the same regardless of where you start.

7.6   Indeterminate Forms and L'Hôpital's Rule

In this section, we change the pace somewhat by reconsidering the problem of computing limits. This section might seem a bit out of place. After all, this is a chapter on techniques of integration. In fact, there is no connection between this section and the rest of the chapter, except that we need the tools developed here to deal with some of the limits we'll encounter in the following section on improper integrals.

You have frequently seen limits of the form

$$\lim_{x \to a} \frac{f(x)}{g(x)},$$

where $\lim_{x \to a} f(x) = \lim_{x \to a} g(x) = 0$ or where $\lim_{x \to a} f(x) = \lim_{x \to a} g(x) = \infty$ (or $-\infty$). Recall that from either of these forms ($\frac{0}{0}$ or $\frac{\infty}{\infty}$, called **indeterminate forms**), we cannot determine the value of the limit, or even whether or not the limit exists. You already know how to evaluate a small number of limits of this type. For instance, note that

$$\lim_{x \to 1} \frac{x^2 - 1}{x - 1} = \lim_{x \to 1} \frac{(x-1)(x+1)}{x-1} = \lim_{x \to 1} \frac{x+1}{1} = \frac{2}{1} = 2,$$

even though the limit initially looks like $\frac{0}{0}$. Similarly, we have

$$\lim_{x \to 1} \frac{x-1}{x^2 - 1} = \lim_{x \to 1} \frac{x-1}{(x-1)(x+1)} = \lim_{x \to 1} \frac{1}{x+1} = \frac{1}{2}.$$

Of course, the expression $\frac{0}{0}$ is mathematically meaningless. It only indicates that both numerator and denominator tend to zero and that we'll need to dig deeper to find the value of the limit.

CAUTION: We will frequently write $\left(\frac{0}{0}\right)$ or $\left(\frac{\infty}{\infty}\right)$ next to an expression, for instance,

$$\lim_{x \to 1} \frac{x-1}{x^2 - 1} \qquad \left(\frac{0}{0}\right).$$

We use this shorthand to indicate that the limit has the indicated indeterminate form. This notation **does not** mean that the value of the limit is $\frac{0}{0}$. You should take care to avoid writing $\lim_{x \to a} f(x) = \frac{0}{0}$, as this is a **meaningless** expression.

Sometimes, a limit with an indeterminate form does not exist, as in the following. Observe that

$$\lim_{x \to 1} \frac{x-1}{x^2 - 2x + 1} = \lim_{x \to 1} \frac{x-1}{(x-1)^2} = \lim_{x \to 1} \frac{1}{x-1} \text{ which does not exist.}$$

Very often, limits with an indeterminate form are much harder to resolve than those seen above. For example, in section 2.5, we proved that

$$\lim_{x \to 0} \frac{\sin x}{x} = 1$$

using an intricate geometric argument. If faced with a limit like

$$\lim_{x \to 0} \frac{1 - \cos x}{\sin x} \qquad \left(\frac{0}{0}\right),$$

are we forced to make a similarly sophisticated geometric argument? Fortunately, the answer is no.

Before we state the main result, recall that you have seen another type of indeterminate form, namely $\frac{\infty}{\infty}$. Each of the following limits have the indeterminate form $\frac{\infty}{\infty}$. First, we have

$$\lim_{x \to \infty} \frac{x^2 + 1}{x^3 + 5} = \lim_{x \to \infty} \frac{(x^2 + 1)\left(\frac{1}{x^3}\right)}{(x^3 + 5)\left(\frac{1}{x^3}\right)} = \lim_{x \to \infty} \frac{\frac{1}{x} + \frac{1}{x^3}}{1 + \frac{5}{x^3}} = \frac{0}{1} = 0.$$

The rule of thumb we have followed in cases like this was to divide top and bottom by the highest power of x in the denominator. Likewise, we have

$$\lim_{x \to \infty} \frac{x^3 + 5}{x^2 + 1} = \lim_{x \to \infty} \frac{(x^3 + 5)\left(\frac{1}{x^2}\right)}{(x^2 + 1)\left(\frac{1}{x^2}\right)} = \lim_{x \to \infty} \frac{x + \frac{5}{x^2}}{1 + \frac{1}{x^2}} = \infty.$$

Finally, you saw limits like

$$\lim_{x \to \infty} \frac{2x^2 + 3x - 5}{x^2 + 4x - 11} = \lim_{x \to \infty} \frac{(2x^2 + 3x - 5)\left(\frac{1}{x^2}\right)}{(x^2 + 4x - 11)\left(\frac{1}{x^2}\right)} = \lim_{x \to \infty} \frac{2 + \frac{3}{x} - \frac{5}{x^2}}{1 + \frac{4}{x} - \frac{11}{x^2}} = \frac{2}{1} = 2.$$

So, once again, if a limit has the form $\frac{\infty}{\infty}$, we must dig deeper to determine the value of the limit or whether or not the limit even exists. As with the indeterminate form $\frac{0}{0}$, while we know how to resolve such limits in a small number of cases, we do not know how to deal with the vast majority of such cases, such as

$$\lim_{x \to \infty} \frac{e^x}{x} \quad \left(\frac{\infty}{\infty}\right).$$

The following powerful result will help us to resolve a wide range of limit problems.

Theorem 6.1   (l'Hôpital's Rule) Suppose that f and g are differentiable on the interval (a, b), except possibly at some fixed point $c \in (a, b)$ and that $g'(x) \neq 0$, on (a, b), except possibly at c. Suppose further that $\lim\limits_{x \to c} \dfrac{f(x)}{g(x)}$ has the indeterminate form $\dfrac{0}{0}$ or $\dfrac{\infty}{\infty}$ and that $\lim\limits_{x \to c} \dfrac{f'(x)}{g'(x)} = L$ (or $\pm\infty$). Then,

$$\lim_{x \to c} \frac{f(x)}{g(x)} = \lim_{x \to c} \frac{f'(x)}{g'(x)}.$$

Proof. (Special Case) The proof of this theorem in the general case is somewhat intricate. We leave the general case for the form $\frac{0}{0}$ to Appendix A, while leaving the $\frac{\infty}{\infty}$ case to more advanced texts. Here, we prove only the special case where f, f', g and g' are all continuous on all of (a, b) and $g'(c) \neq 0$. First, recall the alternative form of the definition of derivative (found in section 2.2):

$$f'(c) = \lim_{x \to c} \frac{f(x) - f(c)}{x - c}.$$

Working backward, we have by continuity that

$$\lim_{x \to c} \frac{f'(x)}{g'(x)} = \frac{f'(c)}{g'(c)} = \frac{\lim\limits_{x \to c} \dfrac{f(x) - f(c)}{x - c}}{\lim\limits_{x \to c} \dfrac{g(x) - g(c)}{x - c}} = \lim_{x \to c} \frac{\dfrac{f(x) - f(c)}{x - c}}{\dfrac{g(x) - g(c)}{x - c}} = \lim_{x \to c} \frac{f(x) - f(c)}{g(x) - g(c)}.$$

Further, since f and g are continuous, we get that

$$f(c) = \lim_{x \to c} f(x) = 0 \quad \text{and} \quad g(c) = \lim_{x \to c} g(x) = 0,$$

so that

$$\lim_{x \to c} \frac{f'(x)}{g'(x)} = \lim_{x \to c} \frac{f(x) - f(c)}{g(x) - g(c)} = \lim_{x \to c} \frac{f(x)}{g(x)},$$

which is what we wanted.

∎

Remark 6.1   The conclusion of the theorem also holds if $\lim\limits_{x \to c} \dfrac{f(x)}{g(x)}$ is replaced with any of the limits $\lim\limits_{x \to c^+} \dfrac{f(x)}{g(x)}, \lim\limits_{x \to c^-} \dfrac{f(x)}{g(x)}, \lim\limits_{x \to \infty} \dfrac{f(x)}{g(x)}$ or $\lim\limits_{x \to -\infty} \dfrac{f(x)}{g(x)}$. (In each case, we must make appropriate adjustments to the hypotheses.)

Example 6.1   **The Indeterminate Form $\dfrac{0}{0}$**

Evaluate $\lim\limits_{x \to 0} \dfrac{1 - \cos x}{\sin x}$.

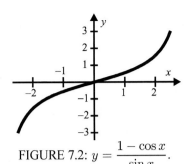

FIGURE 7.2: $y = \dfrac{1 - \cos x}{\sin x}$.

Solution   First, note that this has the indeterminate form $\frac{0}{0}$ and that both $(1 - \cos x)$ and $\sin x$ are continuous and differentiable everywhere and $\frac{d}{dx} \sin x = \cos x \neq 0$ in some interval containing $x = 0$. (Can you determine what one such interval is?) From the graph of $f(x) = \dfrac{1 - \cos x}{\sin x}$ seen in Figure 7.2, it appears that $f(x) \to 0$, as $x \to 0$. We can confirm this with l'Hôpital's Rule, as

$$\lim_{x \to 0} \frac{1 - \cos x}{\sin x} = \lim_{x \to 0} \frac{\frac{d}{dx}(1 - \cos x)}{\frac{d}{dx}(\sin x)} = \lim_{x \to 0} \frac{\sin x}{\cos x} = \frac{0}{1} = 0.$$

∎

Remark 6.2   A very common error in applying l'Hôpital's Rule is to apply the quotient rule (i.e., you compute the derivative of the quotient), rather than computing the quotient of the two derivatives. Another common error is to apply l'Hôpital's Rule indiscriminately, without first checking that the limit has the form $\frac{0}{0}$ or $\frac{\infty}{\infty}$. Be very careful here.

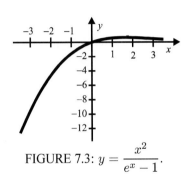

FIGURE 7.3: $y = \dfrac{x^2}{e^x - 1}$.

Example 6.2   **The Indeterminate Form $\dfrac{0}{0}$**

Evaluate $\lim\limits_{x \to 0} \dfrac{x^2}{e^x - 1}$.

Solution   This limit looks like $\frac{0}{0}$ and it's easy to verify that the hypotheses of the theorem hold here. From the graph of $f(x) = \dfrac{x^2}{e^x - 1}$ in Figure 7.3, it appears as if $f(x) \to 0$, as $x \to 0$. From l'Hôpital's Rule, we get

$$\lim_{x \to 0} \frac{x^2}{e^x - 1} = \lim_{x \to 0} \frac{\frac{d}{dx}(x^2)}{\frac{d}{dx}(e^x - 1)} = \lim_{x \to 0} \frac{2x}{e^x} = \frac{0}{1} = 0,$$

as expected.

∎

Often, you must apply l'Hôpital's Rule more than once in order to resolve a limit, as in the following example.

FIGURE 7.4: $y = \dfrac{e^x - 1 - x}{x^2}$.

Example 6.3   **A Limit Requiring Two Applications of L'Hôpital's Rule**

Evaluate $\lim\limits_{x \to 0} \dfrac{e^x - 1 - x}{x^2}$.

Solution   Again, the limit has the indeterminate form $\frac{0}{0}$ and you can easily verify the hypotheses of the theorem. From the graph in Figure 7.4, it appears that the limit is somewhere around 0.5. We have

$$\lim_{x \to 0} \frac{e^x - 1 - x}{x^2} = \lim_{x \to 0} \frac{\frac{d}{dx}(e^x - 1 - x)}{\frac{d}{dx}(x^2)} = \lim_{x \to 0} \frac{e^x - 1}{2x},$$

which still has the form $\frac{0}{0}$. If we apply l'Hôpital's Rule again (being careful to again check that the hypotheses are satisfied), we find

$$\lim_{x \to 0} \frac{e^x - 1 - x}{x^2} = \lim_{x \to 0} \frac{e^x - 1}{2x} = \lim_{x \to 0} \frac{\frac{d}{dx}(e^x - 1)}{\frac{d}{dx}(2x)} = \lim_{x \to 0} \frac{e^x}{2} = \frac{1}{2},$$

which is consistent with what we expected from Figure 7.4.

∎

L'Hôpital's Rule is equally easy to apply with limits of the form $\frac{\infty}{\infty}$.

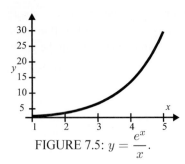

FIGURE 7.5: $y = \dfrac{e^x}{x}$.

Example 6.4   **The Indeterminate Form $\dfrac{\infty}{\infty}$**

Evaluate $\displaystyle\lim_{x \to \infty} \frac{e^x}{x}$.

Solution   This has the form $\frac{\infty}{\infty}$ and from the graph in Figure 7.5, it appears that the function grows larger and larger, without bound, as $x \to \infty$. Applying l'Hôpital's Rule confirms our suspicions, as

$$\lim_{x \to \infty} \frac{e^x}{x} = \lim_{x \to \infty} \frac{\frac{d}{dx}\left(e^x\right)}{\frac{d}{dx}\left(x\right)} = \lim_{x \to \infty} \frac{e^x}{1} = \infty.$$

You may need to apply l'Hôpital's Rule any number of times. You should always be careful to verify the hypotheses at each step.

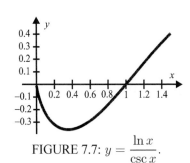

FIGURE 7.6: $y = \dfrac{x^2}{e^x}$.

Example 6.5   **A Limit Requiring Two Applications of L'Hôpital's Rule**

Evaluate $\displaystyle\lim_{x \to \infty} \frac{x^2}{e^x}$.

Solution   First, note that this limit has the form $\frac{\infty}{\infty}$. From the graph in Figure 7.6, it seems that the function tends to 0, as $x \to \infty$ (and does so very rapidly, at that). We have

$$\lim_{x \to \infty} \frac{x^2}{e^x} = \lim_{x \to \infty} \frac{\frac{d}{dx}\left(x^2\right)}{\frac{d}{dx}\left(e^x\right)} = \lim_{x \to \infty} \frac{2x}{e^x} \quad \left(\frac{\infty}{\infty}\right)$$

$$= \lim_{x \to \infty} \frac{\frac{d}{dx}\left(2x\right)}{\frac{d}{dx}\left(e^x\right)} = \lim_{x \to \infty} \frac{2}{e^x} = 0,$$

as expected.

Sometimes an application of l'Hôpital's Rule must be followed by some simplification, as we see in the following example.

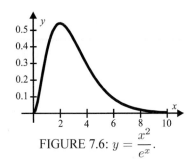

FIGURE 7.7: $y = \dfrac{\ln x}{\csc x}$.

Example 6.6   **Simplification of the Indeterminate Form $\dfrac{\infty}{\infty}$**

Evaluate $\displaystyle\lim_{x \to 0^+} \frac{\ln x}{\csc x}$.

Solution   First, notice that this limit has the form $\frac{\infty}{\infty}$. From the graph in Figure 7.7, it appears that the function tends to 0, as $x \to 0^+$. Applying l'Hôpital's Rule, we have

$$\lim_{x \to 0^+} \frac{\ln x}{\csc x} = \lim_{x \to 0^+} \frac{\frac{d}{dx}\left(\ln x\right)}{\frac{d}{dx}\left(\csc x\right)} = \lim_{x \to 0^+} \frac{\frac{1}{x}}{-\csc x \cot x} \quad \left(\frac{\infty}{\infty}\right).$$

This last limit still has the indeterminate form $\frac{\infty}{\infty}$, but rather than apply l'Hôpital's Rule again, observe that we can rewrite the expression. (Do this wherever possible when a limit expression gets too complicated.) We have

$$\lim_{x \to 0^+} \frac{\ln x}{\csc x} = \lim_{x \to 0^+} \frac{\frac{1}{x}}{-\csc x \cot x} = \lim_{x \to 0^+} \left(-\frac{\sin x}{x} \tan x\right) = (-1)(0) = 0,$$

as expected, where we have used the fact (established in section 2.5) that

$$\lim_{x \to 0} \frac{\sin x}{x} = 1.$$

(You can also establish this by using l'Hôpital's Rule.) Notice that if we had simply blasted away with further applications of l'Hôpital's Rule to $\lim\limits_{x \to 0^+} \dfrac{\frac{1}{x}}{-\csc x \cot x}$, we would **never** have resolved the limit. (Why not?)

■

Other Indeterminate Forms

There are five additional indeterminate forms to consider: $\infty - \infty$, 0^0, $0 \cdot \infty$, 1^∞ and ∞^0. Look closely at each of these to see why they are indeterminate. When evaluating a limit of this type, the objective is to somehow reduce it to one of the indeterminate forms $\frac{0}{0}$ or $\frac{\infty}{\infty}$, at which point we can apply l'Hôpital's Rule.

FIGURE 7.8: $y = \dfrac{1}{x^2} - \dfrac{1}{x^4}$.

Example 6.7   **The Indeterminate Form $\infty - \infty$**

Evaluate $\lim\limits_{x \to 0} \left(\dfrac{1}{x^2} - \dfrac{1}{x^4} \right)$.

Solution   First, notice that the limit has the form $(\infty - \infty)$. From the graph of the function in Figure 7.8, it appears that the function values tend to $-\infty$, as $x \to 0$. If we add the fractions, we get

$$\lim_{x \to 0} \left(\frac{1}{x^2} - \frac{1}{x^4} \right) = \lim_{x \to 0} \left(\frac{x^2 - 1}{x^4} \right) = -\infty,$$

as conjectured, where we have resolved the limit without using l'Hôpital's Rule, which does not apply here. (Why not?)

■

Example 6.8   **The Indeterminate Form $\infty - \infty$**

Evaluate $\lim\limits_{x \to 0} \left[\dfrac{1}{\ln(x+1)} - \dfrac{1}{x} \right]$.

Solution   In this case, the limit has the form $(\infty - \infty)$. From the graph in Figure 7.9, it appears that the limit is somewhere around 0.5. If we add the fractions, we get a form to which we can apply l'Hôpital's Rule. We have

$$\lim_{x \to 0} \left[\frac{1}{\ln(x+1)} - \frac{1}{x} \right] = \lim_{x \to 0} \frac{x - \ln(x+1)}{\ln(x+1)\,x} \qquad \left(\frac{0}{0} \right)$$

$$= \lim_{x \to 0} \frac{\frac{d}{dx}\left[x - \ln(x+1) \right]}{\frac{d}{dx}\left[\ln(x+1)\,x \right]}$$

$$= \lim_{x \to 0} \frac{1 - \frac{1}{x+1}}{\left(\frac{1}{x+1} \right) x + \ln(x+1)\,(1)} \qquad \left(\frac{0}{0} \right).$$

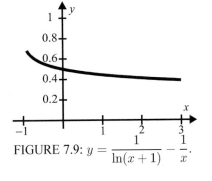

FIGURE 7.9: $y = \dfrac{1}{\ln(x+1)} - \dfrac{1}{x}$.

Rather than apply l'Hôpital's Rule to this last expression, we first simplify to get rid of the fractions. Multiplying top and bottom by $(x + 1)$, we have

$$
\lim_{x \to 0} \left[\frac{1}{\ln(x+1)} - \frac{1}{x} \right] = \lim_{x \to 0} \frac{1 - \frac{1}{x+1}}{\frac{1}{x+1}(x) + \ln(x+1)(1)} \left(\frac{x+1}{x+1} \right)
$$

$$
= \lim_{x \to 0} \frac{(x+1) - 1}{x + (x+1)\ln(x+1)} \qquad \left(\frac{0}{0} \right)
$$

$$
= \lim_{x \to 0} \frac{\frac{d}{dx}(x)}{\frac{d}{dx}[x + (x+1)\ln(x+1)]}
$$

$$
= \lim_{x \to 0} \frac{1}{1 + (1)\ln(x+1) + (x+1)\frac{1}{(x+1)}} = \frac{1}{2},
$$

which is consistent with Figure 7.9.

■

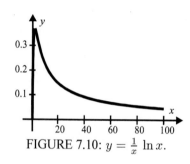

FIGURE 7.10: $y = \frac{1}{x}\ln x$.

Example 6.9   The Indeterminate Form $0 \cdot \infty$

Evaluate $\lim\limits_{x \to \infty} \left(\frac{1}{x} \ln x \right)$.

Solution   This limit has the indeterminate form $(0 \cdot \infty)$. From the graph in Figure 7.10, it appears that the function is decreasing very slowly toward 0, as $x \to \infty$. (However, its rate of decrease is so slow that it's difficult to see whether the function is approaching 0, or some value close to 0.) It's easy to rewrite this in the form $\frac{\infty}{\infty}$, from which we can use l'Hôpital's Rule. Note that

$$
\lim_{x \to \infty} \left(\frac{1}{x} \ln x \right) = \lim_{x \to \infty} \frac{\ln x}{x} \qquad \left(\frac{\infty}{\infty} \right)
$$

$$
= \lim_{x \to \infty} \frac{\frac{d}{dx}\ln x}{\frac{d}{dx}x} = \lim_{x \to \infty} \frac{\frac{1}{x}}{1} = \frac{0}{1} = 0.
$$

■

NOTE: If $\lim\limits_{x \to c}[f(x)]^{g(x)}$ has one of the indeterminate forms $0^0, \infty^0$ or 1^∞, then, letting $y = [f(x)]^{g(x)}$, we have for $f(x) > 0$ that

$$
\ln y = \ln[f(x)]^{g(x)} = g(x)\ln[f(x)],
$$

so that

$$
\lim_{x \to a} \ln y = \lim_{x \to a} \{g(x)\ln[f(x)]\}
$$

will have the indeterminate form $0 \cdot \infty$, which we can deal with as above.

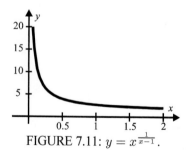

FIGURE 7.11: $y = x^{\frac{1}{x-1}}$.

Example 6.10   The Indeterminate Form 1^∞

Evaluate $\lim\limits_{x \to 1^+} x^{\frac{1}{x-1}}$.

Solution   First, note that this limit has the indeterminate form (1^∞). From the graph in Figure 7.11, it appears that the limit is somewhere around 3. We define $y = x^{\frac{1}{x-1}}$, so that

$$
\ln y = \ln x^{\frac{1}{x-1}} = \frac{1}{x-1}\ln x.
$$

We now consider the limit

$$\lim_{x \to 1^+} \ln y = \lim_{x \to 1^+} \frac{1}{x-1} \ln x \qquad (0 \cdot \infty)$$

$$= \lim_{x \to 1^+} \frac{\ln x}{x-1} \qquad \left(\frac{0}{0}\right)$$

$$= \lim_{x \to 1^+} \frac{\frac{d}{dx}(\ln x)}{\frac{d}{dx}(x-1)} = \lim_{x \to 1^+} \frac{x^{-1}}{1} = 1.$$

Be careful; we have found that $\lim\limits_{x \to 1^+} \ln y = 1$, but this is not the original limit. We want

$$\lim_{x \to 1^+} y = \lim_{x \to 1^+} e^{\ln y} = e^1,$$

which is consistent with Figure 7.11.

∎

Often, the computation of limits in this form requires several applications of l'Hôpital's Rule. Just be careful (in particular, verify the hypotheses at every step) and do not lose sight of the original problem.

Example 6.11   The Indeterminate Form 0^0
Evaluate $\lim\limits_{x \to 0^+} (\sin x)^x$.

Solution   Notice that this limit has the indeterminate form (0^0). Looking at the graph in Figure 7.12, it appears that the limit is approximately 1. We let $y = (\sin x)^x$, so that

$$\ln y = \ln(\sin x)^x = x \ln(\sin x).$$

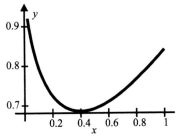

FIGURE 7.12: $y = (\sin x)^x$.

Now consider the limit

$$\lim_{x \to 0^+} \ln y = \lim_{x \to 0^+} \ln(\sin x)^x = \lim_{x \to 0^+} [x \ln(\sin x)] \qquad (0 \cdot \infty)$$

$$= \lim_{x \to 0^+} \frac{\ln(\sin x)}{\left(\frac{1}{x}\right)} \qquad \left(\frac{\infty}{\infty}\right)$$

$$= \lim_{x \to 0^+} \frac{\frac{d}{dx}[\ln(\sin x)]}{\frac{d}{dx}(x^{-1})}$$

$$= \lim_{x \to 0^+} \frac{(\sin x)^{-1} \cos x}{-x^{-2}} \qquad \left(\frac{\infty}{\infty}\right).$$

As we have seen several times earlier, we should rid the expression of the negative exponents before proceeding with any additional applications of l'Hôpital's Rule. Here, we multiply top and bottom by $x^2 \sin x$ to get

$$\lim_{x \to 0^+} \ln y = \lim_{x \to 0^+} \frac{(\sin x)^{-1} \cos x}{-x^{-2}} \left(\frac{x^2 \sin x}{x^2 \sin x}\right)$$

$$= \lim_{x \to 0^+} \frac{-x^2 \cos x}{\sin x} \qquad \left(\frac{0}{0}\right)$$

$$= \lim_{x \to 0^+} \frac{\frac{d}{dx}(-x^2 \cos x)}{\frac{d}{dx}(\sin x)}$$

$$= \lim_{x \to 0^+} \frac{-2x \cos x + x^2 \sin x}{\cos x} = \frac{0}{1} = 0.$$

Again, we have not yet found the original limit. However,

$$\lim_{x\to 0^+} y = \lim_{x\to 0^+} e^{\ln y} = e^0 = 1,$$

which is consistent with Figure 7.12.

■

Example 6.12   **The Indeterminate Form ∞^0**

Evaluate $\displaystyle\lim_{x\to\infty} (x+1)^{2/x}$.

Solution   Notice that this limit has the indeterminate form (∞^0). From the graph in Figure 7.13, it appears that the function tends to a limit around 1, as $x \to \infty$. We let $y = (x+1)^{2/x}$ and consider

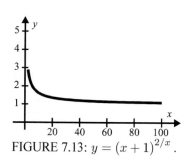

FIGURE 7.13: $y = (x+1)^{2/x}$.

$$\lim_{x\to\infty} \ln y = \lim_{x\to\infty} \ln (x+1)^{2/x} = \lim_{x\to\infty} \left[\frac{2}{x} \ln (x+1)\right] \quad (0\cdot\infty)$$

$$= \lim_{x\to\infty} \frac{2\ln (x+1)}{x} \quad \left(\frac{\infty}{\infty}\right)$$

$$= \lim_{x\to\infty} \frac{\frac{d}{dx}[2\ln (x+1)]}{\frac{d}{dx}x} = \lim_{x\to\infty} \frac{2(x+1)^{-1}}{1}$$

$$= \lim_{x\to\infty} \frac{2}{x+1} = 0.$$

We now have that

$$\lim_{x\to\infty} y = \lim_{x\to\infty} e^{\ln y} = e^0 = 1,$$

as expected.

■

Exercises 7.6

1. (writing) L'Hôpital's Rule states that, in certain situations, the ratios of function values approaches the same limit as the ratios of corresponding derivatives (rates of change). Graphically, this may be hard to understand. To get a handle on this, consider $\dfrac{f(x)}{g(x)}$ where both $f(x) = ax + b$ and $g(x) = cx + d$ are linear functions. Explain why the value of $\displaystyle\lim_{x\to\infty} \frac{f(x)}{g(x)}$ should depend on the relative sizes of the slopes of the lines; that is, it should be equal to $\displaystyle\lim_{x\to\infty} \frac{f'(x)}{g'(x)}$.

2. (writing) Think of a limit of 0 as actually meaning "getting very small" and a limit of ∞ as meaning "getting very large." Discuss whether the following limit forms are indeterminate or not, and explain your answer: $\infty - \infty$, $\frac{1}{0}$, $0 \cdot \infty$, $\infty \cdot \infty$, ∞^0, 0^∞ and 0^0.

In exercises 3–40, find the indicated limits.

3. $\displaystyle\lim_{x\to -2} \frac{x+2}{x^2-4}$

4. $\displaystyle\lim_{x\to 2} \frac{x^2-4}{x^2-3x+2}$

5. $\displaystyle\lim_{x\to -2} \frac{x+1}{x^2+4x+3}$

6. $\displaystyle\lim_{x\to 2} \frac{x-2}{x^2+x-4}$

7. $\displaystyle\lim_{x\to\infty} \frac{3x^2+2}{x^2-4}$

8. $\displaystyle\lim_{x\to -\infty} \frac{2x^3-4x+1}{x^3-3x+2}$

9. $\displaystyle\lim_{x\to -\infty} \frac{x+1}{x^2+4x+3}$

10. $\displaystyle\lim_{x\to\infty} \frac{x^2-2}{3x^2+x-6}$

11. $\displaystyle\lim_{x\to 0} \frac{\sin x}{x}$

12. $\displaystyle\lim_{x\to 0} \frac{1-\cos x}{x^2}$

13. $\displaystyle\lim_{x\to 0} \frac{\sin x - x}{x^3}$

14. $\displaystyle\lim_{x\to 0} \frac{\tan x - x}{x^3}$

15. $\displaystyle\lim_{x\to 1} \frac{\sqrt{x}-1}{x-1}$

16. $\displaystyle\lim_{x\to 1} \frac{\ln x}{x-1}$

17. $\displaystyle\lim_{x\to\infty} \frac{x^3}{e^x}$

18. $\displaystyle\lim_{x\to\infty} \frac{e^x}{x^4}$

19. $\displaystyle\lim_{x\to 0} \frac{e^x-1}{x}$

20. $\displaystyle\lim_{x\to 0} \frac{\sin x}{e^x-1}$

21. $\displaystyle\lim_{x\to 1} \frac{\sin \pi x}{x-1}$

22. $\displaystyle\lim_{x\to 1} \frac{e^{x-1}-1}{x^2-1}$

23. $\lim\limits_{x\to\infty} \dfrac{\ln x}{x^2}$

24. $\lim\limits_{x\to\infty} \dfrac{\ln x}{\sqrt{x}}$

25. $\lim\limits_{x\to\infty} xe^{-x}$

26. $\lim\limits_{x\to\infty} x\sin(1/x)$

27. $\lim\limits_{x\to 0} \dfrac{x\sin x}{\cos x - 1}$

28. $\lim\limits_{x\to 0} \dfrac{\sec x - 1}{x^2}$

29. $\lim\limits_{x\to 0^+} x\ln x$

30. $\lim\limits_{x\to 0^+} \dfrac{\sin x}{\sqrt{x}}$

31. $\lim\limits_{x\to 0^+} \dfrac{\ln x}{\cot x}$

32. $\lim\limits_{x\to 0^+} \dfrac{\sqrt{x}}{\ln x}$

33. $\lim\limits_{x\to\infty} (\sqrt{x^2+1} - x)$

34. $\lim\limits_{x\to\infty} (\ln x - x)$

35. $\lim\limits_{x\to\infty} (1+\tfrac{1}{x})^x$

36. $\lim\limits_{x\to\infty} (1/x)^{1/x}$

37. $\lim\limits_{x\to 0^+} \left(\ln x + \tfrac{1}{x}\right)$

38. $\lim\limits_{x\to 0^+} (1/x - 1/\sin x)$

39. $\lim\limits_{x\to 0^+} (1/x)^x$

40. $\lim\limits_{x\to 0^+} (\cos x)^{1/x}$

In exercises 41 and 42, determine which function "dominates," where we say that the function $f(x)$ dominates the function $g(x)$ as $x\to\infty$ if $\lim\limits_{x\to\infty} f(x) = \lim\limits_{x\to\infty} g(x) = \infty$ and either $\lim\limits_{x\to\infty}\dfrac{f(x)}{g(x)} = \infty$ or $\lim\limits_{x\to\infty}\dfrac{g(x)}{f(x)} = 0$.

41. e^x or x^n (n = any positive integer)

42. $\ln x$ or x

In exercises 43 and 44, find the error(s) in the incorrect calculations

43. $\lim\limits_{x\to 0}\dfrac{\cos x}{x^2} = \lim\limits_{x\to 0}\dfrac{-\sin x}{2x} = \lim\limits_{x\to 0}\dfrac{-\cos x}{2} = -\dfrac{1}{2}.$

44. $\lim\limits_{x\to 0}\dfrac{e^x - 1}{x^2} = \lim\limits_{x\to 0}\dfrac{e^x}{2x} = \lim\limits_{x\to 0}\dfrac{e^x}{2} = \dfrac{1}{2}.$

45. (exploratory) In this exercise, you take a quick look at what we call **Taylor series** in Chapter 8. Start with the limit $\lim\limits_{x\to 0}\dfrac{\sin x}{x} = 1$. Briefly explain why this means that for x close to 0, $\sin x \approx x$. Graph $y = \sin x$ and $y = x$ to see why this is true. If you look far enough away from $x = 0$, the graph of $y = \sin x$ eventually curves noticeably. Let's see if we can find polynomials of higher order to match this curving. Show that $\lim\limits_{x\to 0}\dfrac{\sin x - x}{x^2} = 0$. This means that $\sin x - x \approx 0$ or (again) $\sin x \approx x$. Show that $\lim\limits_{x\to 0}\dfrac{\sin x - x}{x^3} = -\dfrac{1}{6}$. This says that if x is close to 0, then $\sin x - x \approx -\dfrac{1}{6}x^3$ or $\sin x \approx x - \dfrac{1}{6}x^3$. Graph these two functions to see how well they match up. To continue, compute $\lim\limits_{x\to 0}\dfrac{\sin x - (x - x^3/6)}{x^4}$ and $\lim\limits_{x\to 0}\dfrac{\sin x - f(x)}{x^5}$ for the appropriate approximation $f(x)$. At this point, look at the pattern of terms you have (Hint: $6 = 3!$ and $120 = 5!$). Using this pattern, approximate $\sin x$ with an 11th-degree polynomial and graph the two functions.

46. (exploratory) A **zero** of a function $f(x)$ is a solution of the equation $f(x) = 0$. Clearly, not all zeros are created equal. For example, $x = 1$ is a zero of $f(x) = x - 1$ but in some ways it seems that $x = 2$ should count as two zeros of $f(x) = (x-2)^2$. To quantify this, we say that $x = 2$ is a **zero of multiplicity 2** of $f(x) = (x-2)^2$. The precise definition is $x = c$ is a **zero of multiplicity** n of $f(x)$ if $f(c) = 0$ and $\lim\limits_{x\to c}\dfrac{f(x)}{(x-c)^n}$ exists and is nonzero. Thus, $x = 0$ is a zero of multiplicity 2 of $x\sin x$ since $\lim\limits_{x\to 0}\dfrac{x\sin x}{x^2} = \lim\limits_{x\to 0}\dfrac{\sin x}{x} = 1$. Find the multiplicity of each zero of the following functions: $x^2\sin x$, $x\sin x^2$, $x^4\sin x^3$, $(x-1)\ln x$, $\ln(x-1)^2$, $e^x - 1$ and $\cos x - 1$.

7.7   Improper Integrals

Improper Integrals with a Discontinuous Integrand

We're willing to bet that you have heard the saying "familiarity breeds contempt" more than once. This phrase has particular relevance for us in this section. You have been using the Fundamental Theorem of Calculus for quite some time now. Do you always check to see that the hypotheses of the theorem are met before applying it? What hypotheses, you ask? We won't make you look back, but before we give you the answer, try to see what is wrong with the following **erroneous** calculation.

$$\int_{-1}^{2} \frac{1}{x^2}\,dx = \left.\frac{x^{-1}}{-1}\right|_{-1}^{2} = -\frac{3}{2}. \quad \textbf{(This is incorrect!)}$$

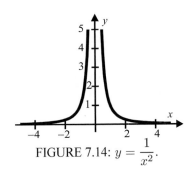

FIGURE 7.14: $y = \dfrac{1}{x^2}$.

There is something fundamentally wrong with the preceding "calculation." The problem is that the integrand is not continuous over the interval of integration. (See Figure 7.14 for a graph of $y = \frac{1}{x^2}$.) Since the Fundamental Theorem assumes a continuous integrand, our attempt to apply the theorem is invalid and our answer is **incorrect**. Observe that the integrand $\frac{1}{x^2}$ blows up to ∞ at $x = 0$, which is in the interval over which we're trying to integrate. This says that the integral is not even defined (in the conventional sense, at least). Further, note that an answer of $-\frac{3}{2}$ is especially suspicious given that the integrand $\frac{1}{x^2}$ is always positive.

Recall that in Chapter 4, we defined the Riemann integral by

$$\int_a^b f(x)\,dx = \lim_{n\to\infty} \sum_{i=1}^{n} f(c_i)\,\Delta x,$$

where c_i was taken to be any point in the subinterval $[x_{i-1}, x_i]$, for $i = 1, 2, \ldots, n$ and where the limit had to be the same for any choice of these c_i's. So, if $f(x) \to \infty$ [or $f(x) \to -\infty$] at some point in $[a, b]$, then the limit defining $\int_a^b f(x)\,dx$ is meaningless. (How would we add $f(c_i)$ to the sum, if c_i were the point where f blows up?) In this case, we call this integral an **improper integral** and we will need to carefully define what we mean by such an integral. First, we examine a somewhat simpler case.

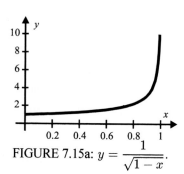

FIGURE 7.15a: $y = \dfrac{1}{\sqrt{1-x}}$.

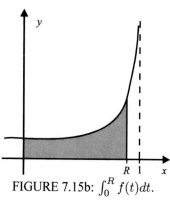

FIGURE 7.15b: $\int_0^R f(t)\,dt$.

Consider $\int_0^1 \dfrac{1}{\sqrt{1-x}}\,dx$. Observe that this is **not** a proper Riemann integral, as the integrand is discontinuous at $x = 1$. (In fact, the integrand blows up to ∞ as $x \to 1^-$.) Look at the graph of $f(x) = \dfrac{1}{\sqrt{1-x}}$ on the interval $[0, 1)$ given in Figure 7.15a. How much area is there under the curve on the interval $[0, 1]$? The area could be infinite, but suppose for the moment that it is finite. How could we find that area? This is a hard question to answer directly. One general rule in mathematics is that if you can't see how to solve the problem you want to solve, then try to find an approximate solution. Notice from Figure 7.15b that for $0 < R < 1$, we can approximate the area by $\int_0^R \dfrac{1}{\sqrt{1-x}}\,dx$. This is a proper Riemann integral, since for $0 < x < R < 1$, $f(x)$ is continuous. Further, the closer R is to 1, the better the approximation should be. In the accompanying table, we compute some approximate values of $\int_0^R \dfrac{1}{\sqrt{1-x}}\,dx$, for a sequence of values of R approaching 1. Although you could compute each of these integrals exactly, we computed these values using the numerical integration program built into our CAS package. (You can do the same using Simpson's Rule, but be aware that as R gets closer to 1, you will need to increase the number of points used in the approximation.)

From the table, the sequence of integrals seems to be approaching 2, as $R \to 1^-$. Notice that since we know how to compute $\int_0^R \dfrac{1}{\sqrt{1-x}}\,dx$, for any $0 < R < 1$, we can compute this limiting value exactly. We have

R	$\displaystyle\int_0^R \frac{1}{\sqrt{1-x}}\,dx$
0.9	1.367544
0.99	1.8
0.999	1.936754
0.9999	1.98
0.99999	1.993675
0.999999	1.998
0.9999999	1.999368
0.99999999	1.9998

$$\lim_{R\to1^-} \int_0^R \frac{1}{\sqrt{1-x}}\,dx = \lim_{R\to1^-} \left[-2\,(1-x)^{1/2}\right]_0^R$$

$$= \lim_{R\to1^-} \left[-2\,(1-R)^{1/2} + 2\,(1-0)^{1/2}\right] = 2.$$

From this computation, we can see that the area under the curve is the limiting value, 2.

In general, suppose that f is continuous on the interval $[a, b)$ and $|f(x)| \to \infty$, as $x \to b^-$ (i.e., as x approaches b from the left). Then we can approximate $\int_a^b f(x)\,dx$ by $\int_a^R f(x)\,dx$, for some $R < b$, but close to b. (Recall that since f is continuous on $[a, R]$, for any $a < R < b$, $\int_a^R f(x)\,dx$ is defined.) Further, as in our introductory example above, the closer R is to b, the better the approximation should be. See Figure 7.16 for a graphical representation of this approximation.

Finally, let $R \to b^-$; if $\int_a^R f(x)\,dx$ approaches some value, L, then we define the **improper integral** $\int_a^b f(x)\,dx$ to be this limiting value. That is, if f is continuous on the

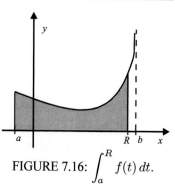

FIGURE 7.16: $\displaystyle\int_a^R f(t)\,dt$.

interval $[a, b)$ and $|f(x)| \to \infty$, as $x \to b^-$, we define

$$\int_a^b f(x)\, dx = \lim_{R \to b^-} \int_a^R f(x)\, dx. \tag{7.1}$$

If the limit exists, we say that the improper integral **converges** (to L). If the limit does not exist, we say that the improper integral **diverges**.

Example 7.1   An Integrand That Blows Up at the Right Endpoint

Determine whether $\displaystyle\int_0^1 \frac{1}{\sqrt{1-x}}\, dx$ converges or diverges.

Solution   Based on the work we did above,

$$\int_0^1 \frac{1}{\sqrt{1-x}}\, dx = \lim_{R \to 1^-} \int_0^R \frac{1}{\sqrt{1-x}}\, dx = 2$$

and so, the improper integral converges to 2.   ∎

The following example illustrates a divergent improper integral closely related to this section's introductory example.

Example 7.2   An Integrand That Blows Up at the Right Endpoint

Determine whether the improper integral $\displaystyle\int_{-1}^0 \frac{1}{x^2}\, dx$ converges or diverges.

Solution   From the definition (7.1), we have

$$\int_{-1}^0 \frac{1}{x^2}\, dx = \lim_{R \to 0^-} \int_{-1}^R \frac{1}{x^2}\, dx = \lim_{R \to 0^-} \left(\frac{x^{-1}}{-1} \right)\Big|_{-1}^R$$

$$= \lim_{R \to 0^-} \left(-\frac{1}{R} - \frac{1}{1} \right) = \infty.$$

Since the defining limit does not exist, the improper integral diverges.   ∎

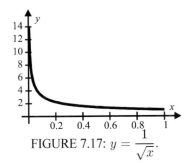

FIGURE 7.17: $y = \dfrac{1}{\sqrt{x}}$.

R	$\displaystyle\int_R^1 \frac{1}{\sqrt{x}}\, dx$
0.1	1.367544
0.01	1.8
0.001	1.936754
0.0001	1.98
0.00001	1.993675
0.000001	1.998
0.0000001	1.999368
0.00000001	1.9998

Similarly, we would like to interpret $\displaystyle\int_0^1 \frac{1}{\sqrt{x}}\, dx$. Once again, this is not a proper Riemann integral, as $f(x) = \dfrac{1}{\sqrt{x}} \to \infty$, as $x \to 0^+$. The question is essentially the same as earlier: is there a finite area under the curve on the interval $(0, 1]$ or is the area infinite? (See Figure 7.17 for a graph of the integrand on the interval in question.) Assuming the area is finite, notice that we could approximate it by $\displaystyle\int_R^1 \frac{1}{\sqrt{x}}\, dx$, for some $0 < R < 1$. (Recall that since $f(x) = \dfrac{1}{\sqrt{x}}$ is continuous on $[R, 1]$, for any $0 < R < 1$, this is a proper Riemann integral.) In the table given in the margin, we compute some approximate values of this integral for a sequence of values of $R \to 0^+$. (Again, we have computed these using the numerical integration package built into our CAS. You can compute these values as well using Simpson's Rule, if you also take care to increase the number of points used in the approximation, as $R \to 0^+$.)

From the computed values, it appears that the integrals are approaching 2, as $R \to 0^+$. Since we know an antiderivative for the integrand, we can compute these integrals exactly, for any fixed $0 < R < 1$. We have

$$\lim_{R \to 0^+} \int_R^1 \frac{1}{\sqrt{x}}\,dx = \lim_{R \to 0^+} \left. \frac{x^{1/2}}{\left(\frac{1}{2}\right)} \right|_R^1 = \lim_{R \to 0^+} 2 \left[1^{1/2} - R^{1/2} \right] = 2.$$

Again, these calculations indicate that the area under the curve is 2.

In general, suppose that f is continuous on the interval $(a, b]$ and $|f(x)| \to \infty$, as $x \to a^+$ (i.e., as x approaches a from the right). Notice that for any number $R \in (a, b)$, $\int_R^b f(x)\,dx$ is defined, since f is continuous on $[R, b]$. Let $R \to a^+$ and if $\int_R^b f(x)\,dx$ approaches some value, L, then we define the **improper integral** $\int_a^b f(x)\,dx$ to be this limiting value. That is, if f is continuous on the interval $(a, b]$ and $|f(x)| \to \infty$, as $x \to a^+$, we define

$$\int_a^b f(x)\,dx = \lim_{R \to a^+} \int_R^b f(x)\,dx. \tag{7.2}$$

As before, if the limit exists, we say that the improper integral **converges** (to L); if the limit does not exist, we say that the improper integral **diverges**.

Example 7.3   A Convergent Improper Integral

Determine whether the improper integral $\int_0^1 \frac{1}{\sqrt{x}}\,dx$ converges or diverges.

Solution   Based on the above computations,

$$\int_0^1 \frac{1}{\sqrt{x}}\,dx = \lim_{R \to 0^+} \int_R^1 \frac{1}{\sqrt{x}}\,dx = 2$$

and so, the improper integral converges to 2.

∎

Example 7.4   A Divergent Improper Integral

Determine whether the improper integral $\int_1^2 \frac{1}{x - 1}\,dx$ converges or diverges.

Solution   From the definition (7.2), we have

$$\int_1^2 \frac{1}{x - 1}\,dx = \lim_{R \to 1^+} \int_R^2 \frac{1}{x - 1}\,dx = \lim_{R \to 1^+} \left. \ln|x - 1| \right|_R^2$$

$$= \lim_{R \to 1^+} \left[\ln|2 - 1| - \ln|R - 1| \right] = \infty,$$

so that the improper integral diverges.

∎

The introductory example in this section represents a third type of improper integral, one where the integrand blows up at a point in the interior of the interval (a, b). We can define such an integral as follows. Suppose that f is continuous on the interval $[a, b]$, except at some $c \in (a, b)$ and $|f(x)| \to \infty$, as $x \to c$. Again, the integral is improper and we write

$$\int_a^b f(x)\,dx = \int_a^c f(x)\,dx + \int_c^b f(x)\,dx. \tag{7.3}$$

If **both** $\int_a^c f(x)\,dx$ and $\int_c^b f(x)\,dx$ converge (to L_1 and L_2, respectively), we say that the improper integral $\int_a^b f(x)\,dx$ **converges**, also (to $L_1 + L_2$). If **either** of the improper integrals $\int_a^c f(x)\,dx$ or $\int_c^b f(x)\,dx$ diverges, then we say that the improper integral $\int_a^b f(x)\,dx$ **diverges**, also.

We can now return to our introductory example.

Example 7.5   An Integrand That Blows Up in the Middle of an Interval

Determine whether the improper integral $\displaystyle\int_{-1}^{2} \frac{1}{x^2}\,dx$ converges or diverges.

Solution   From the definition (7.3), we have

$$\int_{-1}^{2} \frac{1}{x^2}\,dx = \int_{-1}^{0} \frac{1}{x^2}\,dx + \int_{0}^{2} \frac{1}{x^2}\,dx.$$

In example 7.2, we determined that $\int_{-1}^{0} \frac{1}{x^2}\,dx$ diverges. Thus, $\int_{-1}^{2} \frac{1}{x^2}\,dx$ also diverges. Note that you do **not** need to consider $\int_{0}^{2} \frac{1}{x^2}\,dx$ (although it's an easy exercise to show that this, too diverges). Keep in mind that if either of the two improper integrals defining this type of improper integral diverges, then the original integral diverges, too.

∎

Improper Integrals with an Infinite Limit of Integration

Another type of improper integral that is frequently encountered in applications is one where one of the limits of integration is infinite. For instance, $\int_0^\infty e^{-x^2}\,dx$ is of **fundamental** importance in probability and statistics.

So, given a continuous function f defined on $[a, \infty)$, what could we mean by $\int_a^\infty f(x)\,dx$? Once again, recall that we had defined the Riemann integral by

$$\int_a^b f(x)\,dx = \lim_{n \to \infty} \sum_{i=1}^{n} f(c_i)\,\Delta x,$$

where c_i was taken to be any point in the subinterval $[x_{i-1}, x_i]$, for $i = 1, 2, \ldots, n$ and $\Delta x = \frac{b-a}{n}$. If we replace b by ∞, this definition makes no sense. (What does it mean to multiply by $\frac{\infty - a}{n}$?) As with improper integrals with a discontinuous integrand, since the definition makes no sense, the Fundamental Theorem and any other results about integrals do not apply. We should define $\int_a^\infty f(x)\,dx$ in some way consistent with what we already know about integrals.

First, consider what meaning we could give to $\int_1^\infty \frac{1}{x^2}\,dx$. Since $f(x) = \frac{1}{x^2}$ is positive and continuous on the interval $[1, \infty)$, this integral should correspond to area under the curve (assuming this area is in fact, finite.) From the graph of $y = \frac{1}{x^2}$ shown in Figure 7.18, it should appear at least plausible that the area under this curve is finite.

Assuming the area is finite, how could you approximate it? A reasonable approximation should be $\int_1^R \frac{1}{x^2}\,dx$, for some large value R. (Notice that this is a proper Riemann integral, as long as R is finite.) A sequence of values of this integral for increasingly large values of R is displayed in the table shown in the margin.

The sequence of approximating proper integrals seems to be approaching 1, as $R \to \infty$. In this case, we can compute this limit explicitly. We have

$$\lim_{R \to \infty} \int_1^R x^{-2}\,dx = \lim_{R \to \infty} \left. \frac{x^{-1}}{-1} \right|_1^R = \lim_{R \to \infty} \left(-\frac{1}{R} + 1 \right) = 1.$$

Thus, the area under the curve on the interval $[1, \infty)$ is seen to be 1, even though the interval is infinite.

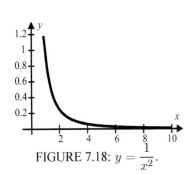

FIGURE 7.18: $y = \dfrac{1}{x^2}$.

R	$\displaystyle\int_1^R \frac{1}{x^2}\,dx$
10	0.9
100	0.99
1000	0.999
10000	0.9999
100000	0.99999
1000000	0.999999

More generally, if f is continuous on the interval $[a, \infty)$, we define the **improper integral** $\int_a^\infty f(x)\,dx$ to be

$$\boxed{\int_a^\infty f(x)\,dx = \lim_{R\to\infty}\int_a^R f(x)\,dx.}$$ (7.4)

If the limit exists (and equals some value L), we say that the improper integral **converges** (to L). If the limit does not exist, we say that the improper integral **diverges**.

Example 7.6   An Integral with an Infinite Limit of Integration

Determine whether the improper integral $\displaystyle\int_1^\infty \frac{1}{x^2}\,dx$ converges or diverges.

Solution   From our work above, observe that the improper integral

$$\int_1^\infty \frac{1}{x^2}\,dx = \lim_{R\to\infty}\int_1^R x^{-2}\,dx = 1,$$

so that the improper integral converges to 1.

∎

You may have already observed that in order for $\int_a^\infty f(x)\,dx$ to converge, it must be the case that $f(x) \to 0$, as $x \to \infty$. (Think about this in terms of area.) However, the reverse need **not** be true. That is, even though $f(x) \to 0$, as $x \to \infty$, the improper integral may diverge, as we see in the next example.

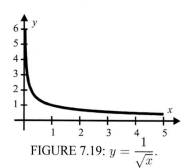

FIGURE 7.19: $y = \dfrac{1}{\sqrt{x}}$.

Example 7.7   A Divergent Improper Integral

Determine whether $\displaystyle\int_1^\infty \frac{1}{\sqrt{x}}\,dx$ converges or diverges.

Solution   Note that $\frac{1}{\sqrt{x}} \to 0$ as $x \to \infty$. Further, from the graph in Figure 7.19, it should seem at least plausible that the area under the curve is finite. However, from the definition (7.4), we have that

$$\int_1^\infty \frac{1}{\sqrt{x}}\,dx = \lim_{R\to\infty}\int_1^R x^{-1/2}\,dx = \lim_{R\to\infty}\frac{x^{1/2}}{\frac{1}{2}}\Big|_1^R = \lim_{R\to\infty}\left(2R^{1/2} - 2\right) = \infty.$$

This says that the improper integral diverges.

∎

You may need to utilize l'Hôpital's Rule to evaluate the defining limit, as in the following example.

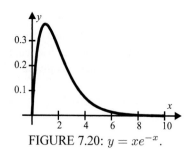

FIGURE 7.20: $y = xe^{-x}$.

Example 7.8   A Convergent Improper Integral

Determine whether $\displaystyle\int_0^\infty xe^{-x}\,dx$ converges or diverges.

Solution   Judging by the graph of $y = xe^{-x}$ in Figure 7.20, it appears plausible that there might be a finite area under the graph. From the definition (7.4), we have

$$\int_0^\infty xe^{-x}\,dx = \lim_{R\to\infty}\int_0^R xe^{-x}\,dx.$$

To evaluate the last integral, you will need integration by parts. Let

$$u = x \qquad dv = e^{-x}\,dx$$
$$du = dx \qquad v = -e^{-x}$$

We then have

$$\int_0^\infty xe^{-x}\,dx = \lim_{R\to\infty}\int_0^R xe^{-x}\,dx = \lim_{R\to\infty}\left(-xe^{-x}\Big|_0^R + \int_0^R e^{-x}\,dx\right)$$

$$= \lim_{R\to\infty}\left[(-Re^{-R}+0) - e^{-x}\Big|_0^R\right] = \lim_{R\to\infty}\left(-Re^{-R} - e^{-R} + e^0\right).$$

Note that the limit $\displaystyle\lim_{R\to\infty} Re^{-R}$ has the indeterminate form $\infty \cdot 0$. We resolve this with l'Hôpital's Rule, as follows:

$$\lim_{R\to\infty} Re^{-R} = \lim_{R\to\infty}\frac{R}{e^R} \quad \left(\frac{\infty}{\infty}\right)$$

$$= \lim_{R\to\infty}\frac{\frac{d}{dR}R}{\frac{d}{dR}e^R} = \lim_{R\to\infty}\frac{1}{e^R} = 0.$$

Returning to the improper integral, we now have

$$\int_0^\infty xe^{-x}\,dx = \lim_{R\to\infty}\left(-Re^{-R} - e^{-R} + e^0\right) = 0 - 0 + 1 = 1.$$

■

Similarly, if f is continuous on $(-\infty, a]$, we define the improper integral

$$\boxed{\int_{-\infty}^a f(x)\,dx = \lim_{R\to-\infty}\int_R^a f(x)\,dx.} \qquad (7.5)$$

Once again, if the limit exists, we say that the integral **converges** and if not, it **diverges**.

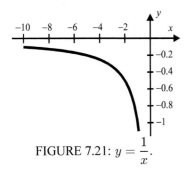

FIGURE 7.21: $y = \dfrac{1}{x}$.

Example 7.9   An Integral with an Infinite Lower Limit of Integration

Determine whether $\displaystyle\int_{-\infty}^{-1}\frac{1}{x}\,dx$ converges or diverges.

Solution   In Figure 7.21, it appears plausible that there might be a finite area bounded between the graph of $y = \frac{1}{x}$ and the x-axis, on the interval $(-\infty, -1]$. However, from the definition (7.5), we have

$$\int_{-\infty}^{-1}\frac{1}{x}\,dx = \lim_{R\to-\infty}\int_R^{-1}\frac{1}{x}\,dx = \lim_{R\to-\infty}\ln|x|\Big|_R^{-1} = \lim_{R\to-\infty}\left[\ln|-1| - \ln|R|\right] = -\infty$$

and hence, the improper integral diverges.

■

Example 7.10   A Convergent Improper Integral

Determine whether $\displaystyle\int_{-\infty}^0\frac{1}{(x-1)^2}\,dx$ converges or diverges.

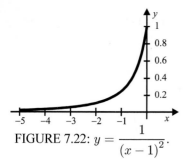

FIGURE 7.22: $y = \dfrac{1}{(x-1)^2}$.

Solution   The graph in Figure 7.22 gives us hope to believe that the improper integral might converge. From the definition (7.5), we have

$$\int_{-\infty}^{0} \frac{1}{(x-1)^2}\,dx = \lim_{R\to-\infty}\int_{R}^{0}(x-1)^{-2}\,dx = \lim_{R\to-\infty}\frac{(x-1)^{-1}}{-1}\bigg|_{R}^{0}$$

$$= \lim_{R\to-\infty}\left[1 + \frac{1}{(R-1)}\right] = 1$$

and hence, the improper integral converges.

■

A final type of improper integral is $\int_{-\infty}^{\infty} f(x)\,dx$. If f is continuous on $(-\infty, \infty)$, we write

$$\boxed{\int_{-\infty}^{\infty} f(x)\,dx = \int_{-\infty}^{a} f(x)\,dx + \int_{a}^{\infty} f(x)\,dx,\ \text{for any } a,} \qquad (7.6)$$

where $\int_{-\infty}^{\infty} f(x)\,dx$ converges if and only if **both** $\int_{-\infty}^{a} f(x)\,dx$ and $\int_{a}^{\infty} f(x)\,dx$ converge. If either one diverges, the original improper integral also diverges. You can choose a to be any real number. So, choose it to be something convenient (usually 0).

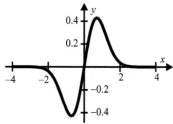

FIGURE 7.23: $y = xe^{-x^2}$.

Example 7.11   An Integral with Two Infinite Limits of Integration

Determine whether $\displaystyle\int_{-\infty}^{\infty} xe^{-x^2}\,dx$ converges or diverges.

Solution   Notice from the graph of the integrand in Figure 7.23, that since the function tends to 0 relatively quickly (both as $x \to \infty$ and as $x \to -\infty$) it appears plausible that there is a finite area bounded by the graph of the function and the x-axis. From the definition (7.6), we have

$$\int_{-\infty}^{\infty} xe^{-x^2}\,dx = \int_{-\infty}^{0} xe^{-x^2}\,dx + \int_{0}^{\infty} xe^{-x^2}\,dx. \qquad (7.7)$$

You must evaluate each of the improper integrals on the right side of (7.7) separately. First, we have

$$\int_{-\infty}^{0} xe^{-x^2}\,dx = \lim_{R\to-\infty}\int_{R}^{0} xe^{-x^2}\,dx.$$

Letting $u = -x^2$, we have $du = -2x\,dx$, and so,

$$\int_{-\infty}^{0} xe^{-x^2}\,dx = -\frac{1}{2}\lim_{R\to-\infty}\int_{R}^{0} e^{-x^2}(-2x)\,dx = -\frac{1}{2}\lim_{R\to-\infty}\int_{-R^2}^{0} e^{u}\,du$$

$$= -\frac{1}{2}\lim_{R\to-\infty} e^{u}\bigg|_{-R^2}^{0} = -\frac{1}{2}\lim_{R\to-\infty}\left(e^0 - e^{-R^2}\right) = -\frac{1}{2}.$$

Similarly, we get (you should fill in the details)

$$\int_{0}^{\infty} xe^{-x^2}\,dx = \lim_{R\to\infty}\int_{0}^{R} xe^{-x^2}\,dx = -\frac{1}{2}\lim_{R\to\infty} e^{u}\bigg|_{0}^{R^2} = -\frac{1}{2}\lim_{R\to\infty}\left(e^{-R^2} - e^0\right) = \frac{1}{2}.$$

Since both of the preceding improper integrals converge, we get that the original integral also converges, to

$$\int_{-\infty}^{\infty} xe^{-x^2}\,dx = \int_{-\infty}^{0} xe^{-x^2}\,dx + \int_{0}^{\infty} xe^{-x^2}\,dx = -\frac{1}{2} + \frac{1}{2} = 0.$$

■

Example 7.12   An Integral with Two Infinite Limits of Integration

Determine whether $\displaystyle\int_{-\infty}^{\infty} e^{-x}\,dx$ converges or diverges.

Solution   From the definition (7.6), we write

$$\int_{-\infty}^{\infty} e^{-x}\,dx = \int_{-\infty}^{0} e^{-x}\,dx + \int_{0}^{\infty} e^{-x}\,dx.$$

It's easy to show that $\int_{0}^{\infty} e^{-x}\,dx$ converges (this is left as an exercise). However,

$$\int_{-\infty}^{0} e^{-x}\,dx = \lim_{R\to-\infty} \int_{R}^{0} e^{-x}\,dx = \lim_{R\to-\infty} -e^{-x}\Big|_{R}^{0} = \lim_{R\to-\infty} \left(-e^{0} + e^{-R}\right) = \infty.$$

This says that $\int_{-\infty}^{0} e^{-x}\,dx$ diverges and hence, $\int_{-\infty}^{\infty} e^{-x}\,dx$ diverges, also, even though $\int_{0}^{\infty} e^{-x}\,dx$ converges.

■

WARNING: Do **not** write

$$\int_{-\infty}^{\infty} f(x)\,dx = \lim_{R\to\infty} \int_{-R}^{R} f(x)\,dx.$$

It's certainly tempting to write this, especially since this will often give a correct answer, with about half of the work. Unfortunately, this will often give incorrect answers, too, frequently suggesting that a divergent integral is convergent. We explore this issue further in the exercises.

We can't emphasize enough that you should verify the continuity of the integrand for every single integral you evaluate. In the next example, we see another reminder of why you must do this.

Example 7.13   An Integral That is Improper for Two Reasons

Determine the convergence or divergence of the improper integral $\displaystyle\int_{0}^{\infty} \frac{1}{(x-1)^2}\,dx$.

Solution   Try to see what is wrong with the following erroneous calculation:

$$\int_{0}^{\infty} \frac{1}{(x-1)^2}\,dx = \lim_{R\to\infty} \int_{0}^{R} \frac{1}{(x-1)^2}\,dx. \quad \textbf{(This is incorrect!)}$$

Look carefully at the integrand and observe that it is **not** continuous on $[0,\infty)$. In fact, the integrand blows up at $x = 1$, which is in the interval over which you're trying to integrate. Thus, this integral is improper for several different reasons. In order to deal with the discontinuity at $x = 1$, we must break up the integral into several pieces, as in (7.3). We write

$$\int_{0}^{\infty} \frac{1}{(x-1)^2}\,dx = \int_{0}^{1} \frac{1}{(x-1)^2}\,dx + \int_{1}^{\infty} \frac{1}{(x-1)^2}\,dx.$$

The second integral above must be further broken into two pieces, since it is improper, both at the left endpoint and by virtue of having an infinite limit of integration. You can pick any point on $(1,\infty)$ at which to break up the interval. We'll simply choose $x = 2$. We have

$$\int_{0}^{\infty} \frac{1}{(x-1)^2}\,dx = \int_{0}^{1} \frac{1}{(x-1)^2}\,dx + \int_{1}^{2} \frac{1}{(x-1)^2}\,dx + \int_{2}^{\infty} \frac{1}{(x-1)^2}\,dx.$$

Each of these three improper integrals must be evaluated separately, using the corresponding limit definitions. We leave it as an exercise to show that the first two integrals diverge, while the third one converges. This says that the original improper integral diverges (a conclusion you would miss if you did not notice that the integrand blows-up at $x = 1$).

■

A Comparison Test

We have now defined several different types of improper integrals, each as a limit of a proper Riemann integral. In order to compute such a limit, we first need to find an antiderivative. But, what if no antiderivative is available? For instance, how would you determine the convergence or divergence of $\int_0^\infty e^{-x^2}\,dx$, when there is no (elementary) antiderivative for e^{-x^2}? (Note that this is a significant question, as this integral is of fundamental importance in the study of probability and statistics.) An answer lies in the following result.

Given two functions f and g which are continuous on the interval $[a, \infty)$, suppose that

$$0 \le f(x) \le g(x), \quad \text{for all } x \ge a.$$

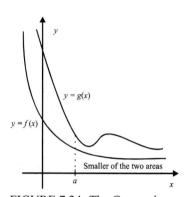

FIGURE 7.24: The Comparison Test.

We see this situation displayed in Figure 7.24. In this case, $\int_a^\infty f(x)\,dx$ and $\int_a^\infty g(x)\,dx$ correspond to the areas under the respective curves. Notice that if $\int_a^\infty g(x)\,dx$ (corresponding to the larger area) converges, then this says that there is a finite area under the curve $y = g(x)$ on the interval $[a, \infty)$. Since $y = f(x)$ lies **below** $y = g(x)$, there can only be a finite area under the curve $y = f(x)$, as well. (How could the area be infinite when $0 \le f(x) \le g(x)$ and $\int_a^\infty g(x)\,dx$ is finite?) Thus, $\int_a^\infty f(x)\,dx$ converges also.

On the other hand, if $\int_a^\infty f(x)\,dx$ (corresponding to the smaller area) diverges, the area under the curve $y = f(x)$ is infinite. Since $y = g(x)$ lies **above** $y = f(x)$, there must be an infinite area under the curve $y = g(x)$, also, so that $\int_a^\infty g(x)\,dx$ diverges, as well. This comparison of improper integrals based on the relative size of their integrands is called a **comparison test** (one of several) and is spelled out in the following theorem.

Theorem 7.1 **(Comparison Test)** Suppose that f and g are continuous on $[a, \infty)$ and $0 \le f(x) \le g(x)$, for all $x \in [a, \infty)$.
  (i) If $\int_a^\infty g(x)\,dx$ converges, then $\int_a^\infty f(x)\,dx$ converges, also.
  (ii) If $\int_a^\infty f(x)\,dx$ diverges, then $\int_a^\infty g(x)\,dx$ diverges, also.

We omit the proof of the theorem, leaving it to stand on the intuitive argument made above.

Remark 7.1 There are corresponding comparison tests that we can state for improper integrals of the form $\int_{-\infty}^a f(x)dx$, where f is continuous on $(-\infty, a]$, as well as for integrals which are improper owing to a discontinuity in the integrand.

The idea of a comparison test is to compare a given improper integral to an improper integral whose convergence or divergence is already known (or can be more easily determined). If you use a comparison to establish that an improper integral converges, then you can always approximate its value numerically. If you use a comparison to establish that an improper integral diverges, then, there's nothing more to do.

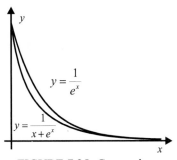

FIGURE 7.25: Comparing $y = \dfrac{1}{e^x}$ and $y = \dfrac{1}{x + e^x}$.

R	$\displaystyle\int_0^R \frac{1}{x + e^x}\, dx$
10	0. 806 350 2
20	0. 806 395 6
30	0. 806 395 6
40	0. 806 395 6

Example 7.14   Using the Comparison Test for an Improper Integral

Determine the convergence or divergence of $\displaystyle\int_0^\infty \frac{1}{x + e^x}\, dx$.

Solution   First, note that you do not know an antiderivative for $\dfrac{1}{x + e^x}$ and so, there is no way to compute the improper integral directly. However, notice that for $x \geq 0$,

$$0 \leq \frac{1}{x + e^x} \leq \frac{1}{e^x}$$

(see Figure 7.25 for a graph of both functions). It's an easy exercise to show that $\displaystyle\int_0^\infty \frac{1}{e^x}\, dx$ converges (to 1). From Theorem 7.1, it now follows that $\displaystyle\int_0^\infty \frac{1}{x + e^x}\, dx$ converges, also. So, we know that the integral is convergent, but to what value does it converge? The comparison test only helps us to determine whether or not the integral converges. Notice that it does **not** help to find the value of the integral. We can, however use numerical integration (e.g., Simpson's Rule) to approximate $\displaystyle\int_0^R \frac{1}{x + e^x}\, dx$, for a sequence of values of R. The accompanying table illustrates some approximate values of $\displaystyle\int_0^R \frac{1}{x + e^x}\, dx$, produced using the numerical integration package built into our CAS. [If you use Simpson's Rule for this, note that you will need to increase the value of n (the number of subintervals in the partition) as R increases.] Notice that as R gets larger and larger, the approximate values for the corresponding integrals seem to be approaching 0.8063956, so we take this as an approximate value for the improper integral,

$$\int_0^\infty \frac{1}{x + e^x}\, dx \approx 0.8063956.$$

You should calculate approximate values for even larger values of R to show that this estimate is accurate.

$\blacksquare$

In the following example, we examine an integral that has important applications in probability and statistics.

Example 7.15   Using the Comparison Test for an Improper Integral

Determine the convergence or divergence of $\displaystyle\int_0^\infty e^{-x^2}\, dx$.

Solution   Once again, notice that you do not know an antiderivative for the integrand e^{-x^2}. For $x > 1$, notice that $e^{-x^2} < e^{-x}$. (Draw a graph of the two functions and do some calculus to see why this is true.) As we observed in example 7.14, $\int_0^\infty e^{-x}\, dx$ converges. By the Comparison Test, it then follows that $\int_0^\infty e^{-x^2}\, dx$ also converges. We leave it as an exercise to show that

$$\int_0^\infty e^{-x^2}\, dx \approx 0.8862269.$$

Using advanced techniques of integration, it is possible to prove the surprising result that $\int_0^\infty e^{-x^2}\, dx = \frac{\sqrt{\pi}}{2}$.

$\blacksquare$

Our comparison test can be used with equal ease to show that an improper integral is divergent.

Example 7.16   Using the Comparison Test: A Divergent Integral

Determine the convergence or divergence of $\int_1^\infty \frac{2 + \sin x}{\sqrt{x}}\,dx$.

Solution   As in example 7.14, you do not know an antiderivative for the integrand and so, your only hope for determining whether or not the integral converges is to use a comparison. First, recall that

$$-1 \le \sin x \le 1, \text{ for all } x.$$

We then have that

$$\frac{1}{\sqrt{x}} = \frac{2 - 1}{\sqrt{x}} \le \frac{2 + \sin x}{\sqrt{x}}, \text{ for } 1 \le x < \infty$$

FIGURE 7.26: Comparing $y = \dfrac{1}{\sqrt{x}}$ and $y = \dfrac{2 + \sin x}{\sqrt{x}}$.

(see Figure 7.26 for a graph of the two functions). Recall that we showed in example 7.7 that $\int_1^\infty \frac{1}{\sqrt{x}}\,dx$ diverges. The Comparison Test now tells us that $\int_1^\infty \frac{2 + \sin x}{\sqrt{x}}\,dx$ must diverge, also.

∎

The big question, of course, is how to find an improper integral to compare to a given integral. Look carefully at the integrand to see if it resembles any functions whose antiderivative you might know (or at least have a hope of finding using our various techniques of integration). Beyond this, our best answer is that this comes with experience. Comparisons are typically done by the seat of your pants. We provide ample exercises on this topic to give you some experience with finding appropriate comparisons. Look hard for comparisons and don't give up too easily.

Exercises 7.7

1. (writing) For many students, our emphasis on working through the limit process for an improper integral may seem unnecessarily careful. Explain, using examples from this section, why it is important to have and use precise definitions.

2. (writing) Identify the following statement as true or false (meaning not always true) and explain why: If the integrand $f(x) \to \infty$ as $x \to a^+$ or $x \to b^-$ then the area $\int_a^b f(x)\,dx$ is infinite; that is, $\int_a^b f(x)\,dx$ diverges.

In exercises 3–10, determine whether or not the integral is improper.

3. $\int_0^2 x^{-2/5}\,dx$

4. $\int_1^2 x^{-2/5}\,dx$

5. $\int_0^2 x^{2/5}\,dx$

6. $\int_0^\infty x^{2/5}\,dx$

7. $\int_{-2}^2 \frac{3}{x}\,dx$

8. $\int_{-2}^{-1} \frac{3}{x}\,dx$

9. $\int_2^\infty \frac{3}{x}\,dx$

10. $\int_{-\infty}^{-1} \frac{3}{x}\,dx$

In exercises 11–46, determine whether the integral converges or diverges. Find the value of the integral if it converges.

11. $\int_0^1 x^{-1/3}\,dx$

12. $\int_0^1 x^{-4/5}\,dx$

13. $\int_0^1 x^{-4/3}\,dx$

14. $\int_0^1 x^{-6/5}\,dx$

15. $\int_1^\infty x^{-1/3}\,dx$

16. $\int_1^\infty x^{-4/5}\,dx$

17. $\int_1^\infty x^{-4/3}\,dx$

18. $\int_1^\infty x^{-6/5}\,dx$

19. $\int_1^3 1/\sqrt{x-1}\,dx$

20. $\int_5^{10} 2/\sqrt{x-5}\,dx$

21. $\int_0^1 \ln x\,dx$

22. $\int_0^{\pi/2} \cot x\,dx$

23. $\int_{-1}^1 1/x\,dx$

24. $\int_{-4}^4 1/\sqrt[3]{x}\,dx$

25. $\int_0^\infty xe^x\,dx$

26. $\int_0^\infty xe^{-2x}\,dx$

27. $\displaystyle\int_{-\infty}^{0} x e^{x}\,dx$

28. $\displaystyle\int_{-\infty}^{0} x e^{-2x}\,dx$

29. $\displaystyle\int_{-\infty}^{\infty} 1/x^{2}\,dx$

30. $\displaystyle\int_{-\infty}^{\infty} 1/\sqrt[3]{x}\,dx$

31. $\displaystyle\int_{-\infty}^{\infty} e^{-2x}\,dx$

32. $\displaystyle\int_{-\infty}^{\infty} e^{3x}\,dx$

33. $\displaystyle\int_{-\infty}^{\infty} \frac{1}{1+x^{2}}\,dx$

34. $\displaystyle\int_{-\infty}^{\infty} \frac{1}{x^{2}-1}\,dx$

35. $\displaystyle\int_{0}^{\pi} \tan x\,dx$

36. $\displaystyle\int_{0}^{\pi} \sec^{2} x\,dx$

37. $\displaystyle\int_{0}^{2} \frac{x}{x^{2}-1}\,dx$

38. $\displaystyle\int_{0}^{2} \frac{2}{x^{2}+1}\,dx$

39. $\displaystyle\int_{0}^{1} \frac{2}{\sqrt{1-x^{2}}}\,dx$

40. $\displaystyle\int_{0}^{1} \frac{2}{x\sqrt{1-x^{2}}}\,dx$

41. $\displaystyle\int_{0}^{\infty} \frac{1}{(x-2)^{2}}\,dx$

42. $\displaystyle\int_{-\infty}^{0} \frac{1}{(x-2)^{2}}\,dx$

43. $\displaystyle\int_{0}^{\infty} 1/\sqrt{x}\,dx$

44. $\displaystyle\int_{0}^{\infty} \tan x\,dx$

45. $\displaystyle\int_{0}^{\infty} \cos x\,dx$

46. $\displaystyle\int_{0}^{\infty} \cos x e^{-\sin x}\,dx$

47. Based on exercises 11–14, conjecture a value of r for which $\displaystyle\int_{0}^{1} x^{-n}\,dx$ converges if and only if $n < r$.

48. Based on exercises 15–18, conjecture a value of r for which $\displaystyle\int_{1}^{\infty} x^{-n}\,dx$ converges if and only if $n > r$.

49. Based on exercises 25–28, conclude that the exponential term controls the convergence or divergence of $\displaystyle\int_{0}^{\infty} x e^{cx}\,dx$ and $\displaystyle\int_{-\infty}^{0} x e^{cx}\,dx$. For which values of c do these integrals converge?

50. At the beginning of this section, we indicated that the calculation $\displaystyle\int_{-1}^{2} 1/x^{2}\,dx = -3/2$ was incorrect. Without any calculations, explain how you should immediately recognize that **negative** 3/2 is not a correct value.

In exercises 51–60, use a comparison to determine whether the integral converges or diverges.

51. $\displaystyle\int_{1}^{\infty} \frac{x}{1+x^{3}}\,dx$

52. $\displaystyle\int_{1}^{\infty} \frac{x^{2}-2}{x^{4}+3}\,dx$

53. $\displaystyle\int_{2}^{\infty} \frac{x}{x^{3/2}-1}\,dx$

54. $\displaystyle\int_{1}^{\infty} \frac{2+\sec^{2} x}{x}\,dx$

55. $\displaystyle\int_{0}^{\infty} \frac{3}{x+e^{x}}\,dx$

56. $\displaystyle\int_{1}^{\infty} e^{-x^{3}}\,dx$

57. $\displaystyle\int_{0}^{\infty} \frac{\sin^{2} x}{1+e^{x}}\,dx$

58. $\displaystyle\int_{2}^{\infty} \frac{\ln x}{e^{x}+1}\,dx$

59. $\displaystyle\int_{2}^{\infty} \frac{x^{2}e^{x}}{\ln x}\,dx$

60. $\displaystyle\int_{1}^{\infty} e^{x^{2}+x+1}\,dx$

In exercises 61 and 62, use integration by parts and l'Hôpital's Rule.

61. $\displaystyle\int_{0}^{1} x \ln 4x\,dx$

62. $\displaystyle\int_{0}^{\infty} x e^{-2x}\,dx$

63. In this exercise, you will look at an interesting pair of calculations known as the "horn paradox." The horn is formed by taking the curve $y = 1/x$ for $x \ge 1$ and revolving it about the x-axis. Show that the volume is finite (i.e., the integral converges), but that the surface area is infinite (i.e., the integral diverges). The paradox is that this would seem to indicate that the horn could be filled with a finite amount of paint but that the outside of the horn could not be covered with any finite amount of paint.

64. Show that $\displaystyle\int_{-\infty}^{\infty} x^{3}\,dx$ diverges but $\displaystyle\lim_{R\to\infty} \int_{-R}^{R} x^{3}\,dx = 0$.

65. A function $f(x) \ge 0$ is a probability density function (pdf) on the interval $[0, \infty)$ if $\displaystyle\int_{0}^{\infty} f(x)\,dx = 1$. Find the value of the constant k to make each of the following pdf's on the interval $[0, \infty)$.
(a) $k e^{-2x}$ (b) $k e^{-4x}$ (c) $k e^{-rx}$

66. Find the value of the constant k to make each of the following pdf's on the interval $[0, \infty)$.
(a) $k x e^{-2x}$ (b) $k x e^{-4x}$ (c) $k x e^{-rx}$

67. The **mean** μ (one measure of average) of a random variable with pdf $f(x)$ on the interval $[0, \infty)$ is $\mu = \displaystyle\int_{0}^{\infty} x f(x)\,dx$. Find the mean if $f(x) = r e^{-rx}$.

68. Find the mean of a random variable with pdf $f(x) = r^{2} x e^{-rx}$.

69. For the mean μ found in exercise 67, compute the probability that the random variable is greater than μ. This probability is given by $\displaystyle\int_{\mu}^{\infty} r e^{-rx}\,dx$. Do you think that it is odd that the probability is *not* equal to 1/2?

70. Find the **median** (another measure of average) for a random variable with pdf $r e^{-rx}$ on $x \ge 0$. The median is the 50% mark on probability, that is, the value m such that

$$\int_{m}^{\infty} r e^{-rx}\,dx = 1/2.$$

71. (exploratory) The so-called **Boltzmann integral**

$$I(p) = \int_{0}^{1} p(x) \ln p(x)\,dx$$

is important in the mathematical field of **information theory**. Here, $p(x)$ is a pdf on the interval $[0, 1]$. Graph the pdf's $p_1(x) = 1$ and

$$p_2(x) = \begin{cases} 4x & \text{if } 0 \le x \le 1/2 \\ 4 - 4x & \text{if } 1/2 \le x \le 1 \end{cases}$$

and compute the integrals $\int_0^1 p_1(x)\, dx$ and $\int_0^1 p_2(x)\, dx$ to verify that they are pdf's. Then compute the Boltzmann integrals $I(p_1)$ and $I(p_2)$. Suppose that you are trying to determine the value of a quantity which you know is between 0 and 1. If the pdf for this quantity is $p_1(x)$, then all values are equally likely. What would a pdf of $p_2(x)$ indicate? Noting that $I(p_2) > I(p_1)$, explain why it is fair to say that the Boltzmann integral measures the amount of *information* available. Given this interpretation, sketch a pdf $p_3(x)$ which would have a larger Boltzmann integral than $p_2(x)$.

72. (exploratory) The **Laplace transform** is an invaluable tool in many engineering disciplines. As the name indicates, the transform turns a function $f(t)$ into a different function $F(s)$. By definition, the Laplace transform of the function $f(t)$ is

$$F(s) = \int_0^\infty f(t) e^{-st}\, dt.$$

To find the Laplace transform of $f(t) = 1$, compute

$$\int_0^\infty (1) e^{-st}\, dt = \int_0^\infty e^{-st}\, dt.$$

Show that the integral equals $1/s$, for $s > 0$. We write $L\{1\} = 1/s$. Show that

$$L\{t\} = \int_0^\infty t e^{-st}\, dt = 1/s^2,$$

for $s > 0$. Compute $L\{t^2\}$ and $L\{t^3\}$ and conjecture the general formula for $L\{t^n\}$.

Chapter Review Exercises

In exercises 1–44, evaluate the integral.

1. $\displaystyle\int \frac{e^{\sqrt{x}}}{\sqrt{x}}\, dx$

2. $\displaystyle\int \frac{\sin(1/x)}{x^2}\, dx$

3. $\displaystyle\int \frac{x^2}{\sqrt{1-x^2}}\, dx$

4. $\displaystyle\int \frac{2}{\sqrt{9-x^2}}\, dx$

5. $\displaystyle\int x^2 e^{-3x}\, dx$

6. $\displaystyle\int x^2 e^{-x^3}\, dx$

7. $\displaystyle\int \frac{x}{1+x^4}\, dx$

8. $\displaystyle\int \frac{x^3}{1+x^4}\, dx$

9. $\displaystyle\int \frac{x^3}{4+x^4}\, dx$

10. $\displaystyle\int \frac{x}{4+x^4}\, dx$

11. $\displaystyle\int e^{2\ln x}\, dx$

12. $\displaystyle\int \cos 4x\, dx$

13. $\displaystyle\int_0^1 x \sin 3x\, dx$

14. $\displaystyle\int_0^1 x \sin 4x^2\, dx$

15. $\displaystyle\int_0^{\pi/2} \sin^4 x\, dx$

16. $\displaystyle\int_0^{\pi/2} \cos^3 x\, dx$

17. $\displaystyle\int_{-1}^1 x \sin \pi x\, dx$

18. $\displaystyle\int_0^1 x^2 \cos \pi x\, dx$

19. $\displaystyle\int_1^2 x^3 \ln x\, dx$

20. $\displaystyle\int_0^{\pi/4} \sin x \cos x\, dx$

21. $\displaystyle\int \cos x \sin^2 x\, dx$

22. $\displaystyle\int \cos x \sin^3 x\, dx$

23. $\displaystyle\int \cos^3 x \sin^3 x\, dx$

24. $\displaystyle\int \cos^4 x \sin^3 x\, dx$

25. $\displaystyle\int \tan^2 x \sec^4 x\, dx$

26. $\displaystyle\int \tan^3 x \sec^2 x\, dx$

27. $\displaystyle\int \sqrt{\sin x} \cos^3 x\, dx$

28. $\displaystyle\int \tan^3 x \sec^3 x\, dx$

29. $\displaystyle\int \frac{2}{8+4x+x^2}\, dx$

30. $\displaystyle\int \frac{3}{\sqrt{-2x-x^2}}\, dx$

31. $\displaystyle\int \frac{2}{x^2\sqrt{4-x^2}}\, dx$

32. $\displaystyle\int \frac{x}{\sqrt{9-x^2}}\, dx$

33. $\displaystyle\int \frac{x^3}{\sqrt{9-x^2}}\, dx$

34. $\displaystyle\int \frac{x^3}{\sqrt{x^2-9}}\, dx$

35. $\displaystyle\int \frac{x^3}{\sqrt{x^2+9}}\, dx$

36. $\displaystyle\int \frac{4}{\sqrt{x+9}}\, dx$

37. $\displaystyle\int \frac{x+4}{x^2+3x+2}\, dx$

38. $\displaystyle\int \frac{5x+6}{x^2+x-12}\, dx$

39. $\displaystyle\int \frac{4x^2+6x-12}{x^3-4x}\, dx$

40. $\displaystyle\int \frac{5x^2+2}{x^3+x}\, dx$

41. $\displaystyle\int e^x \cos 2x\, dx$

42. $\displaystyle\int x^3 \sin x^2\, dx$

43. $\displaystyle\int x\sqrt{x^2+1}\, dx$

44. $\displaystyle\int \sqrt{1-x^2}\, dx$

In exercises 45–50, find the partial fractions decomposition.

45. $\dfrac{4}{x^2 - 3x - 4}$

46. $\dfrac{2x}{x^2 + x - 6}$

47. $\dfrac{-6}{x^3 + x^2 - 2x}$

48. $\dfrac{x^2 - 2x - +2}{x^3 + x}$

49. $\dfrac{x - 2}{x^2 + 4x + 4}$

50. $\dfrac{x^2 - 2}{(x^2 + 1)^2}$

In exercises 51–60, use the Table of Integrals to find the integral.

51. $\displaystyle\int e^{3x}\sqrt{4 + e^{2x}}\,dx$

52. $\displaystyle\int x\sqrt{x^4 - 4}\,dx$

53. $\displaystyle\int \sec^4 x\,dx$

54. $\displaystyle\int \tan^5 x\,dx$

55. $\displaystyle\int \dfrac{4}{x(3 - x)^2}\,dx$

56. $\displaystyle\int \dfrac{\cos x}{\sin^2 x(3 + 4\sin x)}\,dx$

57. $\displaystyle\int \dfrac{\sqrt{9 + 4x^2}}{x^2}\,dx$

58. $\displaystyle\int \dfrac{x^2}{\sqrt{4 - 9x^2}}\,dx$

59. $\displaystyle\int \dfrac{\sqrt{4 - x^2}}{x}\,dx$

60. $\displaystyle\int \dfrac{x^2}{(x^6 - 4)^{3/2}}\,dx$

In exercises 61–68, find the limit.

61. $\displaystyle\lim_{x\to 1} \dfrac{x^3 - 1}{x^2 - 1}$

62. $\displaystyle\lim_{x\to 0} \dfrac{\sin x}{x^2 + 3x}$

63. $\displaystyle\lim_{x\to\infty} \dfrac{e^{2x}}{x^4 + 2}$

64. $\displaystyle\lim_{x\to\infty} \left(x^2 e^{-3x}\right)$

65. $\displaystyle\lim_{x\to\infty} \left(1 + \tfrac{3}{x}\right)^x$

66. $\displaystyle\lim_{x\to 0} \dfrac{\ln x^2}{x}$

67. $\displaystyle\lim_{x\to 0} \dfrac{\tan x}{x}$

68. $\displaystyle\lim_{x\to 0} \dfrac{\tan x - x}{x^3}$

In exercises 69–76, determine whether the integral converges or diverges. If it converges, find the limit.

69. $\displaystyle\int_0^1 \dfrac{x}{x^2 - 1}\,dx$

70. $\displaystyle\int_4^{10} \dfrac{2}{\sqrt{x - 4}}\,dx$

71. $\displaystyle\int_1^\infty \dfrac{3}{x^2}\,dx$

72. $\displaystyle\int_1^\infty xe^{-3x}\,dx$

73. $\displaystyle\int_0^\infty \dfrac{4}{4 + x^2}\,dx$

74. $\displaystyle\int_{-\infty}^\infty xe^{-x^2}\,dx$

75. $\displaystyle\int_{-2}^2 \dfrac{3}{x^2}\,dx$

76. $\displaystyle\int_{-2}^2 \dfrac{x}{1 - x^2}\,dx$

77. (exploratory) In this exercise, you will try to determine whether or not $\displaystyle\int_0^1 \sin(1/x)\,dx$ converges. Since $|\sin(1/x)| \le 1$, the integral does not diverge to ∞, but that does not necessarily mean it converges. Explain why the integral $\displaystyle\int_0^\infty \sin x\,dx$ diverges (not to ∞, but by oscillating indefinitely). You need to determine whether a similar oscillation occurs for $\displaystyle\int_0^1 \sin(1/x)\,dx$. First, estimate $\displaystyle\int_R^1 \sin(1/x)\,dx$ numerically for $R = 1/\pi$, $1/(2\pi)$, $1/(3\pi)$ and so on. Note that once you have $\displaystyle\int_{1/\pi}^1 \sin(1/x)\,dx$, you can get $\displaystyle\int_{1/(2\pi)}^1 \sin(1/x)\,dx$ by "adding" $\displaystyle\int_{1/(2\pi)}^{1/\pi} \sin(1/x)\,dx$. We put this in quotes because this new integral is negative. Verify that the integrals $\displaystyle\int_{1/(2\pi)}^{1/\pi} \sin(1/x)\,dx$, $\displaystyle\int_{1/(3\pi)}^{1/(2\pi)} \sin(1/x)\,dx$ and so on, are alternately negative and positive, so that the sum $\displaystyle\int_R^1 \sin(1/x)\,dx$ does seem to converge as $R \to \infty$. It turns out that the limit does converge if the additional integrals $\displaystyle\int_{1/((n+1)\pi)}^{1/(n\pi)} \sin(1/x)\,dx$ tend to 0 as $n \to \infty$. Show that this is true.

Chapter 8
Infinite Series

Thus far in our study of calculus, we have developed three major mathematical concepts: the limit, the derivative and the definite integral. As you recall, the derivative and the definite integral are each defined in terms of a limit. In this chapter, we explore a new concept, called a **series** (or **infinite series**), defined in terms of (what else?) a limit. Unlike the derivative and the definite integral, which were motivated by trying to calculate the physical quantities of velocity and area, respectively, we have no clear physical motivation for the development of series. However, infinite series are essential to countless important applications.

You have probably had the experience of putting on a favorite music CD and adjusting the bass and treble controls until you got the sound just right. While your primary concern here is with getting a solid bass feeling or a brighter sound, you may be interested to learn that there is a great deal of physics and engineering that goes into the design of musical equipment. What we hear as a stinging guitar or soulful saxophone is actually a precise combination of sounds of many different frequencies. Modern stereos and music synthesizers are essentially special purpose computers designed to reproduce complex musical waveforms with maximum fidelity. Your stereo adjustments alter the combination of high-frequency and low-frequency vibrations with which your speakers produce sound.

The mathematical problem we explore in this chapter is closely connected to the musical problem described above. We want to find a combination of basic elements (in this case, certain simple functions) that produce a desired result. Our need is a practical one: we need to compute with an ever-expanding list of functions. Unfortunately, our ability to do so is extremely limited for transcendental functions (e.g., $\sin x$, $\cos x$, $\ln x$, e^x, etc.). In fact, let's be completely honest: we just don't know how to compute with these functions very well. For instance, how would you compute $\sin(1.234567)$? Yes, we're aware that any scientific calculator can give you an answer, but that's not really the question. For a polynomial, say

$$f(x) = 6x^4 - 15x^3 + 3x^2 + \frac{12}{7}x - 142,$$

you can compute the value of $f(x)$, for any x, simply by performing the indicated arithmetic. But, there is **no** arithmetic for computing $\sin(1.234567)$. So, how is it done, then? While we won't discuss the sophisticated algorithms calculators use to compute approximate values of transcendental functions, we will, over the course of this chapter, develop the mathematical machinery you need to calculate highly accurate approximations to any of these functions. This is no small matter, as you will see, but the benefits of having infinite series available are substantial. Series are used widely in applications, for instance, in the solution of differential equations. Constructing series (specifically, power series) will yield an explosion of new transcendental functions. In fact, many functions of significance in physics and engineering (e.g., Bessel functions) are **defined** as series.

Music synthesis and the evaluation of transcendental functions are only two of countless applications of infinite series. In our increasingly digital world, you can find these ideas everywhere: digital television, digital photography and so on, as well as more traditional scientific research areas such as holography, tomography and spectroscopy all rely heavily on the concept of infinite series. Although it takes us six sections in this chapter to develop the mathematical tools necessary to understand infinite series, stick with us. The applications in the last two sections (and in future courses) are well worth your effort!

8.1   Sequences of Real Numbers

The mathematical notion of **sequence** is not much different from the common English usage of the word. For example, if you were asked to describe the sequence of events that led up to a traffic accident, you'd not only need to list the events, but you'd need to do so in a specific **order** (hopefully, the order in which they actually occurred). In mathematics, we use the term **sequence** to mean an infinite collection of real numbers, written in a specific order.

We have already seen sequences several times now (although we have not formally introduced the notion). For instance, you have found approximate solutions to nonlinear equations like $\tan x - x = 0$, by first making an initial guess, x_0 and then using Newton's method to compute a **sequence** of successively improved approximations, $x_1, x_2,...,x_n,...$

Definition of Sequence

By **sequence** we mean any function whose domain is the set of integers starting with some integer n_0 (often 0 or 1). For instance, the function $a(n) = \frac{1}{n}$, for $n = 1, 2, 3, \ldots$ defines the sequence

$$\frac{1}{1}, \frac{1}{2}, \frac{1}{3}, \frac{1}{4}, \ldots$$

Here, $\frac{1}{1}$ is called the first **term**, $\frac{1}{2}$ is the second **term** and so on. We call $a(n) = \frac{1}{n}$ the **general term**, since it gives a (general) formula for computing all the terms of the sequence. Further, we usually use subscript notation, in lieu of function notation and write a_n instead of $a(n)$.

Example 1.1   The Terms of a Sequence

Write out the terms of the sequence whose general term is given by $a_n = \dfrac{n+1}{n}$, for $n = 1, 2, 3, \ldots$

Solution   We have the sequence

$$\frac{2}{1}, \frac{3}{2}, \frac{4}{3}, \frac{5}{4}, \ldots$$

∎

We often use set notation to denote a sequence. For instance, for the sequence with general term $a_n = \dfrac{1}{n^2}$, for $n = 1, 2, 3, \ldots$, we denote the sequence by

$$\{a_n\}_{n=1}^{\infty} = \left\{\frac{1}{n^2}\right\}_{n=1}^{\infty}$$

or equivalently, by listing the terms of the sequence:

$$\left\{\frac{1}{1}, \frac{1}{2^2}, \frac{1}{3^2}, \ldots, \frac{1}{n^2}, \ldots\right\}.$$

FIGURE 8.1: $a_n = \dfrac{1}{n^2}$.

To graph this sequence, we plot a number of discrete points, since a sequence is a function defined only on the integers (see Figure 8.1). You have likely already noticed something about the sequence $\left\{\dfrac{1}{n^2}\right\}_{n=1}^{\infty}$. As n gets larger and larger, the terms of the sequence, $a_n = \dfrac{1}{n^2}$ get closer and closer to zero. In this case, we say that the sequence **converges** to 0 and write

$$\lim_{n \to \infty} a_n = \lim_{n \to \infty} \frac{1}{n^2} = 0.$$

In general, we say that the sequence $\{a_n\}_{n=1}^{\infty}$ **converges** to L $\left(\text{i.e., } \lim\limits_{n \to \infty} a_n = L\right)$ if we can make a_n as close to L as desired, simply by making n sufficiently large. You may notice that this language parallels that used in the definition of the limit

$$\lim_{x \to \infty} f(x) = L$$

for a function of a real variable x given in section 1.5. The only difference is that n can take on only integer values, while x can take on any real value (integer, rational or irrational).

When we say that we can make a_n as close to L as desired (i.e., arbitrarily close), just how close must we be able to make a_n to L? Well, if you pick any (small) real number, $\varepsilon > 0$, you must be able to make a_n within a distance ε of L, simply by making n sufficiently large. Think about this carefully; a_n is within distance ε of L if

$$L - \varepsilon < a_n < L + \varepsilon.$$

This is equivalent to

$$-\varepsilon < a_n - L < \varepsilon \quad \text{or} \quad |a_n - L| < \varepsilon.$$

We summarize this in the following definition.

Definition 1.1   The sequence $\{a_n\}_{n=n_0}^{\infty}$ **converges** to L if and only if given any number $\varepsilon > 0$, there is a number $N > n_0$ for which

$$|a_n - L| < \varepsilon, \text{ for every } n > N.$$

If there is no such number L, we say that the sequence **diverges**.

We illustrate Definition 1.1 in Figure 8.2. Notice that the definition says that the sequence $\{a_n\}_{n=1}^{\infty}$ converges to L if given any number $\varepsilon > 0$, we can find an integer $N > n_0$, so that the terms of the sequence stay between $L - \varepsilon$ and $L + \varepsilon$ for all values of $n > N$.

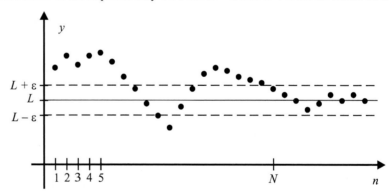

FIGURE 8.2: Convergence of a sequence.

In the following example, we show how to use Definition 1.1 to prove that a sequence converges.

Example 1.2   **Proving That a Sequence Converges**

Use definition 1.1 to show that the sequence $\left\{ \dfrac{1}{n^2} \right\}_{n=1}^{\infty}$ converges to 0.

Solution   In this case, we must show that we can make $\dfrac{1}{n^2}$ as close to 0 as desired, just by making n sufficiently large. So, given any $\varepsilon > 0$, we must find $N > 0$ sufficiently large so that for every $n > N$,

$$\left| \frac{1}{n^2} - 0 \right| < \varepsilon.$$

Notice that this is equivalent to

$$\frac{1}{n^2} < \varepsilon. \tag{1.1}$$

Observe that since n^2 and ε are positive, we can divide both sides of (1.1) by ε and multiply by n^2, to obtain

$$\frac{1}{\varepsilon} < n^2.$$

Taking square roots gives us

$$\sqrt{\frac{1}{\varepsilon}} < n.$$

So, if we choose $N \geq \sqrt{\frac{1}{\varepsilon}}$, then $n > N$ implies that $\frac{1}{n^2} < \varepsilon$, as desired.

■

Most of the usual rules for computing limits of functions of a real variable also apply to computing the limit of a sequence, as we see in the following result.

Theorem 1.1   Suppose that $\{a_n\}_{n=n_0}^{\infty}$ and $\{b_n\}_{n=n_0}^{\infty}$ both converge (to L_1 and L_2, respectively). Then

   (i) $\displaystyle\lim_{n\to\infty}(a_n + b_n) = \lim_{n\to\infty} a_n + \lim_{n\to\infty} b_n = L_1 + L_2$,

   (ii) $\displaystyle\lim_{n\to\infty}(a_n - b_n) = \lim_{n\to\infty} a_n - \lim_{n\to\infty} b_n = L_1 - L_2$,

   (iii) $\displaystyle\lim_{n\to\infty}(a_n\, b_n) = \left(\lim_{n\to\infty} a_n\right)\left(\lim_{n\to\infty} b_n\right) = L_1 L_2$ and

   (iv) $\displaystyle\lim_{n\to\infty} \frac{a_n}{b_n} = \frac{\lim_{n\to\infty} a_n}{\lim_{n\to\infty} b_n} = \frac{L_1}{L_2}$ (assuming $L_2 \neq 0$).

The proof of Theorem 1.1 is virtually identical to the proof of the corresponding theorem about limits of a function of a real variable (see Theorem 2.3 in section 1.2 and Appendix A) and is omitted.

Remark 1.1   To find the limit of a sequence, you should work largely the same as when computing the limit of a function of a real variable, but keep in mind that sequences are defined **only** for integer values of the variable.

FIGURE 8.3: $a_n = \dfrac{5n + 7}{3n - 5}$.

Example 1.3   **Finding the Limit of a Sequence**

Evaluate $\displaystyle\lim_{n\to\infty} \frac{5n + 7}{3n - 5}$.

Solution   Of course, this has the indeterminate form $\dfrac{\infty}{\infty}$. From the graph in Figure 8.3, it looks like the sequence tends to some limit around 2. Note that we cannot apply l'Hôpital's Rule here, since the functions in the numerator and the denominator are not continuous. (They are only defined for integer values of n, even though you *could* define these expressions for any real values of n.) You can of course, use the more primitive method of dividing numerator and denominator by the highest power of n in the denominator. We have

$$\lim_{n\to\infty} \frac{5n + 7}{3n - 5} = \lim_{n\to\infty} \frac{(5n + 7)\left(\frac{1}{n}\right)}{(3n - 5)\left(\frac{1}{n}\right)} = \lim_{n\to\infty} \frac{5 + \frac{7}{n}}{3 - \frac{5}{n}} = \frac{5}{3}.$$

■

NOTE: If you (incorrectly) apply l'Hôpital's Rule in example 1.3, you get the right answer. (Go ahead and try it; nobody's looking.) Unfortunately, you will not always be so lucky. It's a lot like trying to cross a busy highway: while there are times when you can safely cross with your eyes closed, it's not generally recommended. Theorem 1.2 (below) describes how you can safely use l'Hôpital's Rule.

In the following example, we see a sequence that diverges by virtue of its terms tending to $+\infty$.

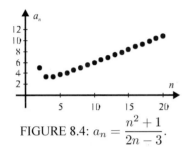

FIGURE 8.4: $a_n = \dfrac{n^2+1}{2n-3}$.

Example 1.4   A Divergent Sequence

Evaluate $\displaystyle\lim_{n\to\infty}\frac{n^2+1}{2n-3}$.

Solution   Again, this has the indeterminate form $\dfrac{\infty}{\infty}$, but from the graph in Figure 8.4, the sequence appears to be increasing without bound. Dividing top and bottom by n (the highest power of n in the denominator) we have

$$\lim_{n\to\infty}\frac{n^2+1}{2n-3}=\lim_{n\to\infty}\frac{\left(n^2+1\right)\left(\frac{1}{n}\right)}{(2n-3)\left(\frac{1}{n}\right)}=\lim_{n\to\infty}\frac{n+\frac{1}{n}}{2-\frac{3}{n}}=\infty$$

and so, the sequence $\left\{\dfrac{n^2+1}{2n-3}\right\}_{n=1}^{\infty}$ diverges.

∎

In the following example, we see that a sequence doesn't need to tend to $\pm\infty$ in order to diverge.

FIGURE 8.5: $a_n = (-1)^n$.

Example 1.5   A Divergent Sequence Whose Terms Do Not Blow Up

Determine the convergence or divergence of the sequence $\{(-1)^n\}_{n=1}^{\infty}$.

Solution   If we write out the terms of the sequence, we have

$$\{-1, 1, -1, 1, -1, 1, \ldots\}.$$

That is, the terms of the sequence alternate back and forth between -1 and 1 and so, the sequence diverges. To see this graphically, we plot the first few terms of the sequence in Figure 8.5. Notice that the points do not approach any limit (a horizontal line).

∎

You can use an advanced tool like l'Hôpital's Rule to find the limit of a sequence, but you must be careful. The following theorem says that if $f(x)\to L$, as $x\to\infty$ through all real values, then $f(n)$ must approach L, too as $n\to\infty$ through integer values. (See Figure 8.6 for a graphical representation of this.)

FIGURE 8.6: $a_n = f(n)$, where $f(x)\to 2$, as $x\to\infty$.

Theorem 1.2   Suppose that $\displaystyle\lim_{x\to\infty}f(x)=L$. Then, $\displaystyle\lim_{n\to\infty}f(n)=L$, also.

FIGURE 8.7a: $a_n = \cos(2\pi n)$.

Remark 1.2   The converse of Theorem 1.2 is false. That is, if $\displaystyle\lim_{n\to\infty}f(n)=L$, it need **not** be true that $\displaystyle\lim_{x\to\infty}f(x)=L$. This is clear from the following observation. Note that

$$\lim_{n\to\infty}\cos(2\pi n)=1,$$

since $\cos(2\pi n)=1$ for every **integer** n (see Figure 8.7a).

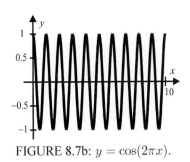

FIGURE 8.7b: $y = \cos(2\pi x)$.

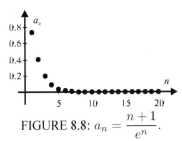

FIGURE 8.8: $a_n = \dfrac{n+1}{e^n}$.

However,

$$\lim_{x \to \infty} \cos(2\pi x) \text{ does not exist,}$$

since as $x \to \infty$, $\cos(2\pi x)$ oscillates between -1 and 1 (see Figure 8.7b).

Example 1.6   Applying L'Hôpital's Rule to a Related Function

Evaluate $\displaystyle \lim_{n \to \infty} \frac{n+1}{e^n}$.

Solution   This has the indeterminate form $\dfrac{\infty}{\infty}$, but from the graph in Figure 8.8, it appears that the sequence converges to 0. However, there appears to be no way to resolve this, except by l'Hôpital's Rule (which does **not** apply to limits of sequences). So, we instead consider the limit of the corresponding function of a real variable to which we may apply l'Hôpital's Rule. (Be sure you check the hypotheses.) We have

$$\lim_{x \to \infty} \frac{x+1}{e^x} = \lim_{x \to \infty} \frac{\frac{d}{dx}(x+1)}{\frac{d}{dx}(e^x)} = \lim_{x \to \infty} \frac{1}{e^x} = 0.$$

From Theorem 1.2, we get that

$$\lim_{n \to \infty} \frac{n+1}{e^n} = 0, \text{ also.}$$

∎

Although we now have a few tools for computing the limit of a sequence, most interesting sequences resist our attempts to find their limit. In many cases (including infinite series, which we study in the remainder of this chapter), we don't even have an explicit formula for the general term. In such circumstances, we must test the sequence for convergence in some indirect way. The first indirect tool we present corresponds to the result (of the same name) for limits of functions of a real variable presented in section 1.2.

Theorem 1.3 (Squeeze Theorem) Suppose $\{a_n\}_{n=n_0}^{\infty}$ and $\{b_n\}_{n=n_0}^{\infty}$ are convergent sequences, both converging to the limit, L. If there is an integer $n_1 \geq n_0$ such that for all $n \geq n_1$, $a_n \leq c_n \leq b_n$, then $\{c_n\}_{n=n_0}^{\infty}$ converges to L, too.

In the following example, we demonstrate how to apply the Squeeze Theorem to a sequence. Observe that the trick here is to find two sequences, one on either side of the given sequence (i.e., one larger and one smaller), which have the same limit.

Example 1.7   Applying the Squeeze Theorem to a Sequence

Determine the convergence or divergence of $\left\{ \dfrac{\sin n}{n^2} \right\}_{n=1}^{\infty}$.

FIGURE 8.9: $a_n = \dfrac{\sin n}{n^2}$.

Solution   From the graph in Figure 8.9, the sequence appears to converge to 0, despite the oscillation. Further, note that you cannot compute this limit using the rules we have established so far. (Try it!) However, recall that

$$-1 \leq \sin n \leq 1, \text{ for all } n.$$

Dividing by n^2, this gives us

$$\frac{-1}{n^2} \leq \frac{\sin n}{n^2} \leq \frac{1}{n^2}, \text{ for all } n \geq 1.$$

Finally, observe that

$$\lim_{n\to\infty} \frac{-1}{n^2} = 0 = \lim_{n\to\infty} \frac{1}{n^2}.$$

From the Squeeze Theorem, we now have that

$$\lim_{n\to\infty} \frac{\sin n}{n^2} = 0,$$

also.

■

The following useful result follows immediately from Theorem 1.3.

Corollary 1.1   If $\lim\limits_{n\to\infty} |a_n| = 0$, then $\lim\limits_{n\to\infty} a_n = 0$, also.

Proof. Notice that for all n,

$$-|a_n| \le a_n \le |a_n|$$

and

$$\lim_{n\to\infty} |a_n| = 0 \text{ and } \lim_{n\to\infty} (-|a_n|) = -\lim_{n\to\infty} |a_n| = 0.$$

So, from the Squeeze Theorem, $\lim\limits_{n\to\infty} a_n = 0$, too.

■

Corollary 1.1 is particularly useful for sequences with both positive and negative terms, as in the following example.

FIGURE 8.10: $a_n = \dfrac{(-1)^n}{n}$.

Example 1.8   **A Sequence with Terms of Alternating Signs**

Determine the convergence or divergence of $\left\{ \dfrac{(-1)^n}{n} \right\}_{n=1}^{\infty}$.

Solution   From the graph of the sequence in Figure 8.10, it seems that the sequence oscillates but still may be converging to 0. Since $(-1)^n$ oscillates back and forth between -1 and 1, we cannot compute $\lim\limits_{n\to\infty} \dfrac{(-1)^n}{n}$ directly. However, notice that

$$\left| \frac{(-1)^n}{n} \right| = \frac{1}{n}$$

and

$$\lim_{n\to\infty} \frac{1}{n} = 0.$$

From Corollary 1.1, we get that $\lim\limits_{n\to\infty} \dfrac{(-1)^n}{n} = 0$, too.

■

We remind you of the following definition, which we use throughout the chapter.

Definition 1.2   For any integer $n \ge 1$, the **factorial**, $n!$ is defined as the product of the first n positive integers,

$$n! = 1 \cdot 2 \cdot 3 \cdots \cdot n.$$

We define $0! = 1$.

The following example shows a sequence whose limit would be extremely difficult to find without the Squeeze Theorem.

Example 1.9   An Indirect Proof of Convergence

Investigate the convergence of $\left\{ \dfrac{n!}{n^n} \right\}_{n=1}^{\infty}$.

Solution   First, notice that we have no means of computing $\displaystyle\lim_{n\to\infty} \dfrac{n!}{n^n}$ directly. (Try this!) From the graph of the sequence in Figure 8.11, it appears that the sequence is converging to 0. Notice that the general term of the sequence satisfies

$$0 < \frac{n!}{n^n} = \underbrace{\frac{1 \cdot 2 \cdot 3 \cdots\cdots n}{n \cdot n \cdot n \cdots\cdots n}}_{n \text{ factors}}$$

$$= \left(\frac{1}{n}\right) \underbrace{\frac{2 \cdot 3 \cdots\cdots n}{n \cdot n \cdots\cdots n}}_{n-1 \text{ factors}} \leq \left(\frac{1}{n}\right)(1) = \frac{1}{n}. \qquad (1.2)$$

From the Squeeze Theorem and (1.2), we have that since

$$\lim_{n\to\infty} \frac{1}{n} = 0 \text{ and } \lim_{n\to\infty} 0 = 0,$$

then

$$\lim_{n\to\infty} \frac{n!}{n^n} = 0, \text{ also.}$$

■

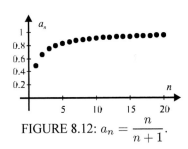

FIGURE 8.11: $a_n = \dfrac{n!}{n^n}$.

Just as we did with functions of a real variable, we need to distinguish between sequences that are increasing and decreasing. The definitions are quite straight-forward.

Definition 1.3   (i) The sequence $\{a_n\}_{n=1}^{\infty}$ is **increasing** if

$$a_1 \leq a_2 \leq \cdots \leq a_n \leq a_{n+1} \leq \cdots.$$

(ii) The sequence $\{a_n\}_{n=1}^{\infty}$ is **decreasing** if

$$a_1 \geq a_2 \geq \cdots \geq a_n \geq a_{n+1} \geq \cdots.$$

If a sequence is either increasing or decreasing, it is called **monotonic**.

There are any number of ways to show that a sequence is monotonic. Regardless of which method you use, you will need to show that either $a_n \leq a_{n+1}$ for all n (increasing) or $a_{n+1} \leq a_n$ for all n (decreasing). One very useful method is to look at the ratio of the two successive terms a_n and a_{n+1}. We illustrate this in the following example.

FIGURE 8.12: $a_n = \dfrac{n}{n+1}$.

Example 1.10   An Increasing Sequence

Investigate whether the sequence $\left\{ \dfrac{n}{n+1} \right\}_{n=1}^{\infty}$ is increasing, decreasing or neither.

Solution   From the graph in Figure 8.12, it appears as if the sequence is increasing. However, you should not be deceived by looking at the first few terms of a sequence.

Instead, we look at the ratio of two successive terms. So, if we define $a_n = \dfrac{n}{n+1}$, we have $a_{n+1} = \dfrac{n+1}{n+2}$ and

$$
\begin{aligned}
\frac{a_{n+1}}{a_n} &= \frac{\left(\dfrac{n+1}{n+2}\right)}{\left(\dfrac{n}{n+1}\right)} = \frac{n+1}{n+2}\,\frac{n+1}{n} \\[2mm]
&= \frac{n^2+2n+1}{n^2+2n} = 1 + \frac{1}{n^2+2n} > 1.
\end{aligned}
\tag{1.3}
$$

Since $a_n > 0$, notice that we can multiply both sides of (1.3) by a_n, to obtain

$$a_{n+1} > a_n,$$

for all n and so, the sequence is increasing.

■

FIGURE 8.13: $a_n = \dfrac{n!}{e^n}$.

Since $(n+1)! = (n+1) \cdot n!$ and $e^{n+1} = e \cdot e^n$.

Example 1.11   A Sequence That is Increasing for $n \geq 2$

Investigate whether the sequence $\left\{\dfrac{n!}{e^n}\right\}_{n=1}^{\infty}$ is increasing, decreasing or neither.

Solution   From the graph of the sequence in Figure 8.13, it appears that the sequence is increasing (and rather rapidly, at that). Here, for $a_n = \dfrac{n!}{e^n}$, we have $a_{n+1} = \dfrac{(n+1)!}{e^{n+1}}$, so that

$$
\begin{aligned}
\frac{a_{n+1}}{a_n} &= \frac{\left[\dfrac{(n+1)!}{e^{n+1}}\right]}{\left(\dfrac{n!}{e^n}\right)} = \frac{(n+1)!}{e^{n+1}}\,\frac{e^n}{n!} \\[2mm]
&= \frac{(n+1)\,n!\,e^n}{e\,(e^n)\,n!} = \frac{n+1}{e} > 1, \text{ for } n \geq 2.
\end{aligned}
\tag{1.4}
$$

Multiplying both sides of (1.4) by $a_n > 0$, we get

$$a_{n+1} > a_n, \text{ for } n \geq 2.$$

Notice that in this case, although the sequence is not increasing for all n, it is increasing for $n \geq 2$. Keep in mind that it doesn't really matter what the first few terms do, anyway. We are only concerned with the behavior of a sequence as $n \to \infty$.

■

We need to define one additional property of sequences.

Definition 1.4   We say that the sequence $\{a_n\}_{n=n_0}^{\infty}$ is **bounded** if there is a number $M > 0$ (called a **bound**) for which $|a_n| \leq M$, for all n.

There is often some slight confusion here. A bound is **not** the same as a limit, although the two may coincide. The limit of a convergent sequence is the value that the terms of the sequence are approaching as $n \to \infty$. On the other hand, a bound is any number that is greater than or equal to the absolute value of every term. This says that a given sequence may have any number of bounds (e.g., if $|a_n| \leq 10$ for all n, then $|a_n| \leq 20$, for all n, too). However, a sequence may have only **one** limit (or no limit, in the case of a divergent sequence).

Example 1.12   A Bounded Sequence

Show that the sequence $\left\{ \dfrac{3 - 4n^2}{n^2 + 1} \right\}_{n=1}^{\infty}$ is bounded.

Solution   We use the fact that $4n^2 - 3 > 0$, for all $n \geq 1$, to get

$$|a_n| = \left| \frac{3 - 4n^2}{n^2 + 1} \right| = \frac{4n^2 - 3}{n^2 + 1} < \frac{4n^2}{n^2 + 1} < \frac{4n^2}{n^2} = 4.$$

So, this sequence is bounded by 4. (We might also say in this case that the sequence is bounded between -4 and 4.) Further, note that we could also use any number greater than 4 as a bound.

■

We should emphasize that the reason we are looking at whether a sequence is monotonic or bounded is that very often we cannot compute the limit of a given sequence directly and must rely on indirect means to determine whether or not the sequence is convergent. The following theorem provides a powerful tool for the investigation of sequences.

Theorem 1.4   Every bounded, monotonic sequence converges.

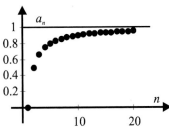

FIGURE 8.14a: A bounded and increasing sequence.

FIGURE 8.14b: A bounded and decreasing sequence.

A typical bounded and increasing sequence is illustrated in Figure 8.14a, while a bounded and decreasing sequence is illustrated in Figure 8.14b. In both figures, notice that a bounded and monotonic sequence has nowhere to go and consequently, must converge. The proof of Theorem 1.4 is rather involved and we leave it to the end of the section.

In the very common case where we do not know how to compute the limit of a sequence, this theorem says that if we can show that a sequence is bounded and monotonic, then it must also be convergent, although to what, we often won't know. Once we establish that a sequence converges, we can approximate its limit by computing a sufficient number of terms, as in the following example.

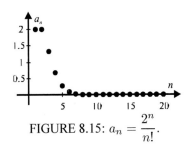

FIGURE 8.15: $a_n = \dfrac{2^n}{n!}$.

Since $2^{n+1} = 2 \cdot 2^n$
and $(n + 1)! = (n + 1) \cdot n!$

Example 1.13   An Indirect Proof of Convergence

Investigate the convergence of the sequence $\left\{ \dfrac{2^n}{n!} \right\}_{n=1}^{\infty}$.

Solution   First, note that we do not know how to compute $\displaystyle\lim_{n \to \infty} \frac{2^n}{n!}$. This has the indeterminate form $\dfrac{\infty}{\infty}$, but we cannot apply l'Hôpital's Rule to it. (Why not?) The graph in Figure 8.15 suggests that the sequence converges to some number close to 0. To confirm this suspicion, we first show that the sequence is monotonic. We have

$$\frac{a_{n+1}}{a_n} = \frac{\left[\dfrac{2^{n+1}}{(n+1)!} \right]}{\left(\dfrac{2^n}{n!} \right)} = \frac{2^{n+1}}{(n+1)!} \frac{n!}{2^n}$$

$$= \frac{2 \, (2^n) \, n!}{(n+1) \, n! \, 2^n} = \frac{2}{n+1} \leq 1, \text{ for } n \geq 1. \qquad (1.5)$$

Multiplying both sides of (1.5) by a_n gives us $a_{n+1} \leq a_n$, for all n and so, the sequence is decreasing. Next, since we have already shown that the sequence is decreasing, observe that

$$0 < \frac{2^n}{n!} \leq \frac{2^1}{1!} = 2,$$

for $n \geq 1$ (i.e., the sequence is bounded by 2). Since the sequence is both bounded and monotonic, it must be convergent, by Theorem 1.4. To get an approximation of the limit of the sequence, we display a number of terms of the sequence in the table that follows.

n	$a_n = \dfrac{2^n}{n!}$		n	$a_n = \dfrac{2^n}{n!}$
1	2		11	0.0000513
2	2		12	0.0000086
3	1.333333		13	0.0000013
4	0.666667		14	1.88×10^{-7}
5	0.266667		15	2.51×10^{-8}
6	0.088889		16	3.13×10^{-9}
7	0.025397		17	3.69×10^{-10}
8	0.006349		18	4.09×10^{-11}
9	0.001411		19	4.31×10^{-12}
10	0.000282		20	4.31×10^{-13}

From the table, it appears that the sequence is converging to approximately 0. We can make a slightly stronger statement, though. Since we have established that the sequence is *decreasing* and convergent, we have from our computations that

$$0 \leq a_n \leq a_{20} \approx 4.31 \times 10^{-13}, \text{ for } n \geq 20.$$

Further, the limit, L must also satisfy the inequality

$$0 \leq L \leq 4.31 \times 10^{-13}.$$

So, even if we can't conclude that the sequence converges to 0, we can conclude that it converges to some number extremely close to zero. For most purposes, such an estimate is entirely adequate. Of course, if you need greater precision, you can always compute a few more terms.

■

Remark 1.3   Do not underestimate the importance of Theorem 1.4. This indirect way of testing a sequence for convergence takes on additional significance when we study infinite series (a special type of sequence which is the topic of the remainder of this chapter).

Proof of Theorem 1.4 (Optional)

Before we can prove Theorem 1.4, we need to state one of the properties of the real number system.

The Completeness Axiom

If a nonempty set S of real numbers has a lower bound, then it has a *greatest lower bound*. Equivalently, if it has an upper bound, it has a *least upper bound*.

This axiom says that if a nonempty set S has an upper bound, that is, a number M such that

$$x \leq M, \text{ for all } x \in S,$$

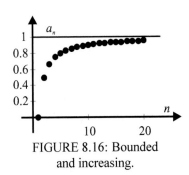

FIGURE 8.16: Bounded and increasing.

then there is an upper bound L, for which

$$L \le M \text{ for \textbf{all} upper bounds, } M,$$

with a corresponding statement holding for lower bounds.

The Completeness Axiom allows us to prove Theorem 1.4.

Proof. (Increasing sequence) Suppose that $\{a_n\}_{n=n_0}^{\infty}$ is increasing and bounded. This is illustrated in Figure 8.16, where you can see an increasing sequence bounded by 1. We have

$$a_1 \le a_2 \le a_3 \le \cdots \le a_n \le a_{n+1} \le \cdots$$

and for some number $M > 0$, $|a_n| \le M$ for all n. This is the same as saying that

$$-M \le a_n \le M, \text{ for all } n.$$

Now, let S be the set containing all of the terms of the sequence, $S = \{a_1, a_2, \ldots, a_n, \ldots\}$. Notice that M is an upper bound for the set S. From the Completeness Axiom, S must have a least upper bound, L. That is, L is the **smallest** number for which

$$a_n \le L, \text{ for all } n. \tag{1.6}$$

Notice that for any $\varepsilon > 0$, $L - \varepsilon < L$ and so, $L - \varepsilon$ is **not** an upper bound, since L is the **least** upper bound. Since $L - \varepsilon$ is not an upper bound for S, there is some element, a_N of S for which

$$L - \varepsilon < a_N.$$

Since $\{a_n\}$ is increasing, we have that for $n \ge N$, $a_N \le a_n$. Finally, from (1.6) and the fact that L is an upper bound for S and since $\varepsilon > 0$, we have

$$L - \varepsilon < a_N \le a_n \le L < L + \varepsilon,$$

or more simply

$$L - \varepsilon < a_n < L + \varepsilon,$$

for $n \ge N$. This is equivalent to

$$|a_n - L| < \varepsilon, \text{ for } n \ge N.$$

This says that $\{a_n\}$ converges to L. The proof for the case of a decreasing sequence is similar and is left as an exercise.   ■

Exercises 8.1

1. (writing) Compare and contrast $\lim\limits_{x \to \infty} \sin \pi x$ and $\lim\limits_{n \to \infty} \sin \pi n$. Indicate the domains of the two functions and how this affects the limits.

2. (writing) Explain why Theorem 1.2 should be true, taking into account the respective domains and their effect on the limits.

3. (writing) In words, explain why a decreasing bounded sequence must converge.

4. (writing) A sequence is said to diverge if it does not converge. The word "diverge" is well chosen for sequences that diverge to ∞, but is less descriptive of sequences such as $\{1, 2, 1, 2, 1, 2, \ldots\}$ and $\{1, 2, 3, 1, 2, 3, \ldots\}$. Briefly describe the limiting behavior of these sequences and discuss other possible limiting behaviors of divergent sequences.

In exercises 5–8, write out the terms $a_1, a_2, \ldots, a_6$ of the given sequence.

5. $a_n = \dfrac{2n - 1}{n^2}$

6. $a_n = \dfrac{3}{n + 4}$

7. $a_n = \dfrac{4}{n!}$

8. $a_n = (-1)^n \dfrac{n}{n + 1}$

In exercises 9–14, (a) find the limit of each sequence and (b) use the definition to show that the sequence converges.

9. $a_n = \dfrac{1}{n^3}$

10. $a_n = \dfrac{2}{n^2}$

11. $a_n = \dfrac{n}{n+1}$ **12.** $a_n = \dfrac{2n+1}{n}$

13. $a_n = \dfrac{2}{\sqrt{n}}$ **14.** $a_n = \dfrac{4}{\sqrt{n+1}}$

In exercises 15–34, determine whether the sequence converges or diverges.

15. $a_n = \dfrac{3n^2+1}{2n^2-1}$ **16.** $a_n = \dfrac{5n^3-1}{2n^3+1}$

17. $a_n = \dfrac{n^2+1}{n+1}$ **18.** $a_n = \dfrac{n^2+1}{n^3+1}$

19. $a_n = \dfrac{n+2}{3n-1}$ **20.** $a_n = \dfrac{n+4}{n+1}$

21. $a_n = (-1)^n \dfrac{n+2}{3n-1}$ **22.** $a_n = (-1)^n \dfrac{n+4}{n+1}$

23. $a_n = (-1)^n \dfrac{n+2}{n^2+4}$ **24.** $a_n = (-1)^n \dfrac{4}{n+1}$

25. $a_n = \cos \pi n$ **26.** $a_n = \sin \pi n$

27. $a_n = n e^{-n}$ **28.** $a_n = \dfrac{\cos n}{e^n}$

29. $a_n = \dfrac{e^n+2}{e^{2n}-1}$ **30.** $a_n = \dfrac{e^{2n}}{e^n+1}$

31. $a_n = \dfrac{3^n}{e^n+1}$ **32.** $a_n = \dfrac{n2^n}{3^n}$

33. $a_n = \dfrac{\cos n}{n!}$ **34.** $a_n = \dfrac{n!}{2^n}$

In exercises 35–38, use the Squeeze Theorem and Corollary 1.1 to prove that the sequence converges to 0 (given that $\lim\limits_{n\to\infty} \frac{1}{n} = \lim\limits_{n\to\infty} \frac{1}{n^2} = 0$).

35. $a_n = \dfrac{\cos n}{n^2}$ **36.** $a_n = \dfrac{\cos n\pi}{n^2}$

37. $a_n = (-1)^n \dfrac{e^{-n}}{n}$ **38.** $a_n = (-1)^n \dfrac{\ln n}{n^2}$

In exercises 39–46, determine whether the sequence is increasing, decreasing or neither.

39. $a_n = \dfrac{n+3}{n+2}$ **40.** $a_n = \dfrac{n-1}{n+1}$

41. $a_n = \dfrac{e^n}{n}$ **42.** $a_n = \dfrac{n}{2^n}$

43. $a_n = \dfrac{2^n}{(n+1)!}$ **44.** $a_n = \dfrac{3^n}{(n+2)!}$

45. $a_n = \dfrac{10^n}{n!}$ **46.** $a_n = \dfrac{n!}{5^n}$

In exercises 47–50, show that the sequence is bounded.

47. $a_n = \dfrac{3n^2-2}{n^2+1}$ **48.** $a_n = \dfrac{6n-1}{n+3}$

49. $a_n = \dfrac{\sin(n^2)}{n+1}$ **50.** $a_n = e^{1/n}$

51. A packing company works with $12''$ square boxes. Show that for $n = 1, 2, 3, \ldots$, a total of n^2 disks of diameter $\frac{12''}{n}$ fit into a box. Let a_n be the wasted area in a box with n^2 disks. Compute a_n.

52. The pattern of a sequence can't always be determined from the first few terms. Start with a circle, pick two points on the circle and connect them with a line segment. The circle is divided into $a_1 = 2$ regions. Add a third point, connect all points and show that there are now $a_2 = 4$ regions. Add a fourth point, connect all points and show that there are $a_3 = 8$ regions. Is the pattern clear? Show that $a_4 = 16$ and then compute a_5 for a surprise!

53. You have heard about the "population explosion." The following dramatic warning is adapted from the article "Doomsday: Friday 13 November 2026" by Foerster, Mora and Amiot in *Science* (Nov. 1960). Start with $a_0 = 3.049$ to indicate that the world population in 1960 was approximately 3.049 billion. Then compute $a_1 = a_0 + 0.005a_0^{2.01}$ to estimate the population in 1961. Compute $a_2 = a_1 + 0.005a_1^{2.01}$ to estimate the population in 1962, then $a_3 = a_2 + 0.005a_2^{2.01}$ for 1963, and so on. Continue iterating and compare your calculations to the actual populations in 1970 (3.721 billion), 1980 (4.473 billion) and 1990 (5.333 billion). Then project ahead to the year 2035. Frightening, isn't it?

54. The so-called **hailstone sequence** is defined by

$$x_k = \begin{cases} 3x_{k-1} + 1 & \text{if } x_{k-1} \text{ is odd} \\ \frac{1}{2}x_{k-1} & \text{if } x_{k-1} \text{ is even} \end{cases} .$$

If you start with $x_1 = 2^n$ for some positive integer n, show that $x_{n+1} = 1$. The question (an unsolved research problem) is whether you eventually reach 1 from **any** starting value. Try several odd values for x_1 and show that you always reach 1.

55. A different population model was studied by Fibonacci, an Italian mathematician of the thirteenth century. He imagined a population of rabbits starting with a pair of newborns. For one month, they grow and mature. The second month, they have a pair of newborn baby rabbits. We count the number of pairs of rabbits. Thus far, $a_0 = 1$, $a_1 = 1$ and $a_2 = 2$. The rules are: adult rabbit pairs give birth to a pair of newborns every month, newborns take one month to mature and no rabbits die. Show that $a_3 = 3$, $a_4 = 5$ and in general $a_n = a_{n-1} + a_{n-2}$. This sequence of numbers, known as the **Fibonacci sequence**, occurs in an amazing number of applications.

56. In this exercise, we visualize the Fibonacci sequence. Start with two squares of side 1 placed next to each other (see Figure A). Place a square on the long side of the resulting rectangle (see Figure B); this square has side 2. Continue placing squares on the long sides of the rectangles: a square of side 3 is added in Figure C, then a square of side 5 is added to the bottom of Figure C, and so on.

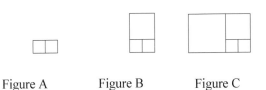

Figure A   Figure B   Figure C

Argue that the sides of the squares are determined by the Fibonacci sequence of exercise 55.

57. Determine whether the sequence $a_n = \sqrt[n]{n}$ converges or diverges. (Hint: Use $n^{1/n} = e^{(1/n)\ln n}$.)

58. Suppose that $a_1 = 1$ and $a_{n+1} = \frac{1}{2}\left(a_n + \frac{4}{a_n}\right)$. Show numerically that the sequence converges to 2. To find this limit analytically, let $L = \lim_{n\to\infty} a_{n+1} = \lim_{n\to\infty} a_n$ and solve the equation $L = \frac{1}{2}\left(L + \frac{4}{L}\right)$.

59. As in exercise 58, determine the limit of the sequence defined by $a_1 = 1$, $a_{n+1} = \frac{1}{2}\left(a_n + \frac{c}{a_n}\right)$ for $c > 0$ and $a_n > 0$.

60. (exploratory) Suppose that a ball is launched from the ground with initial velocity v. Ignoring air resistance, it will rise to a height of $v^2/(2g)$ and fall back to the ground at time $t = 2v/g$. Depending on how "lively" the ball is, the next bounce will only rise to a fraction of the previous height. The **coefficient of restitution** r, defined as the ratio of landing velocity to rebound velocity, measures the liveliness of the ball. The second bounce has launch velocity rv, the third bounce

has launch velocity r^2v and so on. It might seem that the ball will bounce forever. To see that it does not, argue that the time to complete two bounces is $a_2 = \frac{2v}{g}(1 + r)$, the time to complete three bounces is $a_3 = \frac{2v}{g}(1 + r + r^2)$, etc. Take $r = .5$ and numerically determine the limit of this sequence (we study this type of sequence in detail in section 8.2). In particular, show that $(1 + 0.5) = \frac{3}{2}$, $(1 + 0.5 + 0.5^2) = \frac{7}{4}$ and $(1 + 0.5 + 0.5^2 + 0.5^3) = \frac{15}{8}$, find a general expression for a_n and determine the limit of the sequence. Argue that at the end of this amount of time, the ball has stopped bouncing.

61. (exploratory) A surprising follow-up to the bouncing ball problem of exercise 60 is found in *An Experimental Approach to Nonlinear Dynamics and Chaos* by Tufillaro, Abbott and Reilly. Suppose the ball is bouncing on a moving table that oscillates up and down according to the equation $A\cos\omega t$ for some amplitude A and frequency ω. Without the motion of the table, the ball will quickly reach a height of 0 as in exercise 60. For different values of A and ω, however, the ball can settle into an amazing variety of patterns. To understand this, explain why the collision between table and ball could subtract or add velocity to the ball (what happens if the table is going up? down?). A simplified model of the velocity of the ball at successive collisions with the table is $v_{n+1} = 0.8v_n - 10\cos(v_0 + v_1 + \cdots + v_n)$. Starting with $v_0 = 5$, compute $v_1, v_2, \ldots, v_{15}$. In this case, the ball never settles into a pattern; its motion is chaotic.

8.2   Infinite Series

Recall that we write the decimal expansion of $\frac{1}{3}$ as the repeating decimal

$$\frac{1}{3} = 0.33333333\ldots,$$

where we understand that the 3's in this expansion go on for ever and ever. An alternative way to think of this is as

$$\frac{1}{3} = 0.3 + 0.03 + 0.003 + 0.0003 + 0.00003 + \cdots$$
$$= 3\,(0.1) + 3\,(0.1)^2 + 3\,(0.1)^3 + 3\,(0.1)^4 + \cdots + 3\,(0.1)^k + \cdots. \qquad (2.1)$$

For convenience, we write (2.1) using summation notation as

$$\frac{1}{3} = \sum_{k=1}^{\infty} \left[3\,(0.1)^k\right] \qquad (2.2)$$

But, what exactly could we mean by the *infinite sum* indicated in (2.2)? Of course, you can't add infinitely many things together (you can only add two things at a time), but by this expression, we mean that as you add together more and more terms, the sum gets closer and closer to $\frac{1}{3}$.

In general, for any sequence $\{a_k\}_{k=1}^{\infty}$, suppose we start adding the terms together. We define the individual sums by

$$S_1 = a_1,$$
$$S_2 = a_1 + a_2 = S_1 + a_2,$$
$$S_3 = \underbrace{a_1 + a_2}_{S_2} + a_3 = S_2 + a_3,$$
$$S_4 = \underbrace{a_1 + a_2 + a_3}_{S_3} + a_4 = S_3 + a_4, \qquad\qquad (2.3)$$

$$\vdots$$

$$S_n = \underbrace{a_1 + a_2 + \cdots + a_{n-1}}_{S_{n-1}} + a_n = S_{n-1} + a_n \qquad\qquad (2.4)$$

and so on. We refer to S_n as the nth **partial sum**. Note that we can compute any one of these as the sum of two numbers: the nth term, a_n and the previous partial sum, S_{n-1}, as indicated in (2.4).

For instance, for the sequence $\left\{\dfrac{1}{2^k}\right\}_{k=1}^{\infty}$, consider the partial sums

$$S_1 = \frac{1}{2},$$
$$S_2 = \frac{1}{2} + \frac{1}{2^2} = \frac{3}{4},$$
$$S_3 = \frac{3}{4} + \frac{1}{2^3} = \frac{7}{8},$$
$$S_4 = \frac{7}{8} + \frac{1}{2^4} = \frac{15}{16},$$
$$S_5 = \frac{15}{16} + \frac{1}{2^5} = \frac{31}{32}$$

and so on. Look at these carefully and you might notice that $S_2 = \dfrac{3}{4} = 1 - \dfrac{1}{2^2}$, $S_3 = \dfrac{7}{8} = 1 - \dfrac{1}{2^3}$, $S_4 = \dfrac{15}{16} = 1 - \dfrac{1}{2^4}$ and so on, so that $S_n = 1 - \dfrac{1}{2^n}$, for each $n = 1, 2, \ldots$. If we were to consider the convergence or divergence of the sequence $\{S_n\}_{n=1}^{\infty}$ of partial sums, observe that we now have

$$\lim_{n \to \infty} S_n = \lim_{n \to \infty} \left(1 - \frac{1}{2^n}\right) = 1.$$

Think about what this says: as we add together more and more terms of the sequence $\left\{\dfrac{1}{2^k}\right\}_{k=1}^{\infty}$, the partial sums are drawing closer and closer to 1. In this instance, we write

$$\sum_{k=1}^{\infty} \frac{1}{2^k} = 1. \qquad\qquad (2.5)$$

It's very important to understand what's going on here. This new mathematical object, $\sum_{k=1}^{\infty} \dfrac{1}{2^k}$ is called a **series** (or **infinite series**). It is **not a sum** in the usual sense of the word, but rather, the **limit** of the sequence of partial sums. Equation (2.5) says that as we add together more and more terms, the sums are approaching the limit of 1.

In general, for any sequence, $\{a_k\}_{k=1}^{\infty}$, we can write down the series

$$a_1 + a_2 + \cdots + a_k + \cdots = \sum_{k=1}^{\infty} a_k.$$

If the sequence of partial sums, $S_n = \sum_{k=1}^{n} a_k$ converges (to some number S), then we say that the series $\sum_{k=1}^{\infty} a_k$ **converges** (to S). We write

Definition of Infinite Series

$$\boxed{\sum_{k=1}^{\infty} a_k = \lim_{n \to \infty} \sum_{k=1}^{n} a_k = \lim_{n \to \infty} S_n = S.} \qquad (2.6)$$

In this case, we call S the **sum** of the series. On the other hand, if the sequence of partial sums, $\{S_n\}_{n=1}^{\infty}$ diverges (i.e., $\lim_{n \to \infty} S_n$ does not exist), then we say that the series **diverges**.

Example 2.1   A Convergent Series

Determine if the series $\sum_{k=1}^{\infty} \dfrac{1}{2^k}$ converges or diverges.

Solution   From our work on the introductory example, observe that

$$\sum_{k=1}^{\infty} \frac{1}{2^k} = \lim_{n \to \infty} \sum_{k=1}^{n} a_k = \lim_{n \to \infty} \left(1 - \frac{1}{2^n} \right) = 1.$$

In this case, we say that the series converges to 1.

$\blacksquare$

In the following example, we examine a simple divergent series.

Example 2.2   A Divergent Series

Investigate the convergence or divergence of the series $\sum_{k=1}^{\infty} k^2$.

Solution   Here, we have the nth partial sum

$$S_n = \sum_{k=1}^{n} k^2 = 1^2 + 2^2 + \cdots + n^2$$

and

$$\lim_{n \to \infty} S_n = \lim_{n \to \infty} \left(1^2 + 2^2 + \cdots + n^2 \right) = \infty.$$

Since the sequence of partial sums diverges, the series diverges also.

$\blacksquare$

Determining the convergence or divergence of a series is only rarely as simple as it was in examples 2.1 and 2.2.

FIGURE 8.17:
$$S_n = \sum_{k=1}^{n} \frac{1}{k(k+1)}.$$

Example 2.3   A Series with a Simple Expression for the Partial Sums

Investigate the convergence or divergence of the series $\sum_{k=1}^{\infty} \dfrac{1}{k(k+1)}$.

Solution   In Figure 8.17, we have plotted the first 20 partial sums. In the accompanying table, we list a number of partial sums of the series.

n	$S_n = \sum\limits_{k=1}^{n} \frac{1}{k(k+1)}$
10	0.90909091
100	0.99009901
1000	0.999001
10000	0.99990001
100000	0.99999
1×10^6	0.999999
1×10^7	0.9999999

From both the graph and the table, it appears that the partial sums are approaching 1, as $n \to \infty$. However, we must urge caution. It is extremely difficult to look at a graph or a table of any finite number of the partial sums and decide whether a given series is converging or diverging. In the present case, we are fortunate that we can find a simple expression for the partial sums. We leave it as an exercise to find the partial fraction decomposition of the general term of the series:

$$\frac{1}{k(k+1)} = \frac{1}{k} - \frac{1}{k+1}. \tag{2.7}$$

Now, consider the nth partial sum. From (2.7), we have

$$S_n = \sum_{k=1}^{n} \frac{1}{k(k+1)} = \sum_{k=1}^{n}\left(\frac{1}{k} - \frac{1}{k+1}\right)$$
$$= \left(\frac{1}{1} - \frac{1}{2}\right) + \left(\frac{1}{2} - \frac{1}{3}\right) + \left(\frac{1}{3} - \frac{1}{4}\right) + \cdots + \left(\frac{1}{n-1} - \frac{1}{n}\right) + \left(\frac{1}{n} - \frac{1}{n+1}\right).$$

Notice how nearly every term in the partial sum is canceled by another term in the sum (the next term). For this reason, such a sum is referred to as **telescoping** (or **collapsing**). We now have

$$S_n = 1 - \frac{1}{n+1}$$

and so,

$$\lim_{n\to\infty} S_n = \lim_{n\to\infty}\left(1 - \frac{1}{n+1}\right) = 1.$$

This says that the series converges to 1, as conjectured from the graph and the table.

■

It is relatively rare that we know the sum of a convergent series exactly. Usually, we must test a series for convergence using some indirect method and then approximate the sum by calculating some partial sums. We now consider one additional type of series whose sum is known exactly. The series we considered in example 2.1, $\sum\limits_{k=1}^{\infty} \frac{1}{2^k}$ is an example of a type of series called **geometric series**. We have the following result.

Theorem 2.1 For $a \neq 0$, the **geometric series**, $\sum\limits_{k=0}^{\infty} ar^k$ converges to $\frac{a}{1-r}$, if $|r| < 1$ and diverges if $|r| \geq 1$. (Here, r is referred to as the **ratio**.)

Proof. The proof relies on a clever observation. Note that the first term of the series corresponds to $k=0$ and so, the nth partial sum (the sum of the first n terms) is

$$S_n = a + ar^1 + ar^2 + \cdots + ar^{n-1}. \tag{2.8}$$

Multiplying (2.8) by r, we get

$$r S_n = ar^1 + ar^2 + ar^3 + \cdots + ar^n. \tag{2.9}$$

Subtracting (2.9) from (2.8), we get

$$(1-r) S_n = \left(a + ar^1 + ar^2 + \cdots + ar^{n-1}\right) - \left(ar^1 + ar^2 + ar^3 + \cdots + ar^n\right)$$
$$= a - ar^n = a\left(1 - r^n\right).$$

Dividing both sides by $(1 - r)$ gives us

$$S_n = \frac{a\,(1 - r^n)}{1 - r}.$$

If $|r| < 1$, we have that $r^n \to 0$, as $n \to \infty$ and so,

$$\lim_{n \to \infty} S_n = \lim_{n \to \infty} \frac{a\,(1 - r^n)}{1 - r} = \frac{a}{1 - r}.$$

We leave it as an exercise to show that if $|r| \geq 1$, $\lim_{n \to \infty} S_n$ does not exist.

■

FIGURE 8.18:
$$S_n = \sum_{k=2}^{n+1} 5 \cdot \left(\frac{1}{3}\right)^k.$$

Example 2.4   **A Convergent Geometric Series**

Investigate the convergence or divergence of the series $\displaystyle\sum_{k=2}^{\infty} 5 \left(\frac{1}{3}\right)^k$.

Solution   The first 20 partial sums are plotted in Figure 8.18. It appears from the graph that the sequence of partial sums is converging to some number around 0.8. Further evidence is found in the following table, showing the first 20 partial sums.

n	$S_n = \sum_{k=2}^{n+1} 5 \left(\frac{1}{3}\right)^k$	n	$S_n = \sum_{k=2}^{n+1} 5 \left(\frac{1}{3}\right)^k$
1	0.55555556	11	0.83332863
2	0.74074074	12	0.83333177
3	0.80246914	13	0.83333281
4	0.82304527	14	0.83333316
5	0.82990398	15	0.83333328
6	0.83219021	16	0.83333331
7	0.83295229	17	0.83333331
8	0.83320632	18	0.83333331
9	0.833291	19	0.83333331
10	0.83331922	20	0.83333331

The table suggests that the series converges to approximately 0.83333331. Again, we must urge caution. Graphical and numerical evidence can be very unreliable when examining series. Some sequences and series converge (or diverge) far too slowly to observe graphically or numerically. You must **always** confirm your suspicions with careful mathematical analysis. In the present case, note that while the series is not quite written in the usual form, it is a geometric series, as follows:

$$\sum_{k=2}^{\infty} 5 \left(\frac{1}{3}\right)^k = 5 \left(\frac{1}{3}\right)^2 + 5 \left(\frac{1}{3}\right)^3 + 5 \left(\frac{1}{3}\right)^4 + \cdots + 5 \left(\frac{1}{3}\right)^n + \cdots$$

$$= 5 \left(\frac{1}{3}\right)^2 \left[1 + \frac{1}{3} + \left(\frac{1}{3}\right)^2 + \cdots \right]$$

$$= \sum_{k=0}^{\infty} \left\{ 5 \left(\frac{1}{3}\right)^2 \left(\frac{1}{3}\right)^k \right\}.$$

You can now see that this is a geometric series with ratio $r = \frac{1}{3}$ and $a = 5 \left(\frac{1}{3}\right)^2$. Further, since

$$|r| = \frac{1}{3} < 1,$$

we have from Theorem 2.1 that the series converges to

$$\frac{a}{1-r} = \frac{5\left(\frac{1}{3}\right)^2}{1-\left(\frac{1}{3}\right)} = \frac{\left(\frac{5}{9}\right)}{\left(\frac{2}{3}\right)} = \frac{5}{6} = 0.8333333\overline{3},$$

which is consistent with the graph and the table of partial sums.

■

FIGURE 8.19:

$$S_n = \sum_{k=0}^{n-1} 6 \cdot \left(-\frac{7}{2}\right)^k.$$

Example 2.5   A Divergent Geometric Series

Investigate the convergence or divergence of the series $\sum\limits_{k=0}^{\infty} 6\left(-\frac{7}{2}\right)^k$.

Solution   A graph showing the first 20 partial sums (see Figure 8.19) is not particularly helpful, until you look at the vertical scale. The following table showing the values of the first 20 partial sums is more revealing.

n	$S_n = \sum\limits_{k=0}^{n-1} 6\left(-\frac{7}{2}\right)^k$	n	$S_n = \sum\limits_{k=0}^{n-1} 6\left(-\frac{7}{2}\right)^k$
1	6	11	1.29×10^6
2	-15	12	-4.5×10^6
3	58.5	13	1.6×10^7
4	-198.75	14	-5.5×10^7
5	701.63	15	1.9×10^8
6	-2449.7	16	-6.8×10^8
7	8579.9	17	2.4×10^9
8	-30024	18	-8.3×10^9
9	1.05×10^5	19	2.9×10^{10}
10	-3.68×10^5	20	-1×10^{11}

Note that while the partial sums are oscillating back and forth between positive and negative values, they are growing larger and larger in absolute value. We can confirm our suspicions by observing that this is a geometric series with ratio $r = -\frac{7}{2}$. Since

$$|r| = \left|-\frac{7}{2}\right| = \frac{7}{2} > 1,$$

the series is divergent as we suspected.

■

You will find that determining whether a series is convergent or divergent usually involves a lot of hard work. The following simple observation provides us with a very useful test.

Theorem 2.2   If $\sum\limits_{k=1}^{\infty} a_k$ converges, then $\lim\limits_{k\to\infty} a_k = 0$.

Proof. Suppose that $\sum\limits_{k=1}^{\infty} a_k$ converges to some number L. This means that the sequence of partial sums defined by $S_n = \sum\limits_{k=1}^{n} a_k$ also converges to L. Notice that

$$S_n = \sum_{k=1}^{n} a_k = \sum_{k=1}^{n-1} a_k + a_n = S_{n-1} + a_n.$$

Subtracting S_{n-1} from both sides, we have

$$a_n = S_n - S_{n-1}.$$

This gives us

$$\lim_{n \to \infty} a_n = \lim_{n \to \infty} (S_n - S_{n-1}) = \lim_{n \to \infty} S_n - \lim_{n \to \infty} S_{n-1} = L - L = 0,$$

as desired.

■

The following very useful test follows directly from Theorem 2.2.

kth-**Term Test for Divergence:** If $\lim_{k \to \infty} a_k \neq 0$, then the series $\sum_{k=1}^{\infty} a_k$ diverges.

The kth-term test is so simple, you should use it to test every series you run into. If the terms don't tend to zero, the series is divergent and there's nothing more to do. However, as we'll soon see, if the terms **do** tend to zero, there is no guarantee that the series converges and additional testing is needed.

FIGURE 8.20:
$$S_n = \sum_{k=1}^{n} \frac{k}{k+1}.$$

Example 2.6   A Series Whose Terms Do Not Tend to Zero

Investigate the convergence or divergence of the series $\sum_{k=1}^{\infty} \frac{k}{k+1}$.

Solution   A graph showing the first 20 partial sums is shown in Figure 8.20. The partial sums appear to be increasing without bound as n increases. A table of values would indicate the same sort of growth. (Try this!) We can resolve the question of convergence quickly by observing that

$$\lim_{k \to \infty} \frac{k}{k+1} = 1 \neq 0.$$

From the kth-term test for divergence, the series must diverge.

■

WARNING: The converse of Theorem 2.2 is **false**. That is, having $\lim_{k \to \infty} a_k = 0$ does **not** guarantee that the series $\sum_{k=1}^{\infty} a_k$ converges. **Be very clear about this point. This is a very common misconception.**

This next example shows an important series whose terms tend to 0 as $k \to \infty$, but which diverges, nonetheless.

FIGURE 8.21: $S_n = \sum_{k=1}^{n} \frac{1}{k}$.

Example 2.7   The Harmonic Series

Investigate the convergence or divergence of the **harmonic series**: $\sum_{k=1}^{\infty} \frac{1}{k}$.

Solution   In Figure 8.21, we see the first 20 partial sums of the series. In the following table, we display the first 20 partial sums. (You could do the same with the first 200 or

2000 partial sums and there would be little difference in your conclusion.)

n	$S_n = \sum_{k=1}^{n} \dfrac{1}{k}$	n	$S_n = \sum_{k=1}^{n} \dfrac{1}{k}$
1	1	11	3.01988
2	1.5	12	3.10321
3	1.83333	13	3.18013
4	2.08333	14	3.25156
5	2.28333	15	3.31823
6	2.45	16	3.38073
7	2.59286	17	3.43955
8	2.71786	18	3.49511
9	2.82897	19	3.54774
10	2.92897	20	3.59774

The table and the graph suggest that the series might converge to a number around 3.6. As always with sequences and series, we need to confirm this suspicion. From our test for divergence, we have

$$\lim_{k \to \infty} a_k = \lim_{k \to \infty} \frac{1}{k} = 0.$$

Be careful: once again, this does **not** say that the series converges. If the limit had been nonzero, we would have stopped and concluded that the series diverges. In the present case, where the limit is 0, we can only conclude that further study is needed. (That is, the series *may* converge, but we will need to investigate further.)

The following clever proof provides a preview of things to come. Consider the nth partial sum

$$S_n = \sum_{k=1}^{n} \frac{1}{k} = \frac{1}{1} + \frac{1}{2} + \frac{1}{3} + \cdots + \frac{1}{n}.$$

Note that S_n corresponds to the sum of the areas of the n rectangles superimposed on the graph of $y = \frac{1}{x}$ shown in Figure 8.22 for the case where $n = 7$.

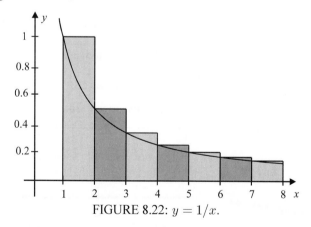

FIGURE 8.22: $y = 1/x$.

Notice that since each of the indicated rectangles lies partly above the curve, we have

$$S_n = \text{Sum of areas of } n \text{ rectangles}$$

$$\geq \text{Area under the curve} = \int_{1}^{n+1} \frac{1}{x}\, dx$$

$$= \ln|x| \Big|_{1}^{n+1} = \ln(n+1). \tag{2.10}$$

However, the sequence $\{\ln(n+1)\}_{n=1}^{\infty}$ diverges, as

$$\lim_{n\to\infty} \ln(n+1) = \infty.$$

Since $S_n \geq \ln(n+1)$, for all n [from (2.10)], we must also have that $\lim_{n\to\infty} S_n = \infty$.

From the definition of convergence of a series, we now have that $\sum_{k=1}^{\infty} \frac{1}{k}$ **diverges**, too, even

though $\lim_{k\to\infty} \frac{1}{k} = 0$.

■

We complete this section with several unsurprising results.

Theorem 2.3   (i) If $\sum_{k=1}^{\infty} a_k$ converges to A and $\sum_{k=1}^{\infty} b_k$ converges to B, then the series

$\sum_{k=1}^{\infty} (a_k \pm b_k)$ converges to $A \pm B$ and $\sum_{k=1}^{\infty} (c\,a_k)$ converges to cA, for any constant, c.

(ii) If $\sum_{k=1}^{\infty} a_k$ converges and $\sum_{k=1}^{\infty} b_k$ diverges, then $\sum_{k=1}^{\infty} (a_k \pm b_k)$ diverges.

The proof of the theorem is left as an exercise.

Exercises 8.2

1. (writing) Suppose that your friend is confused about the difference between the convergence of a sequence and the convergence of a series. Carefully explain the difference between convergence or divergence of the sequence $a_k = \dfrac{k}{k+1}$ and

the series $\sum_{k=1}^{\infty} \dfrac{k}{k+1}$.

2. (writing) Explain in words why the kth term test for divergence is true. Explain why it is **not** true that if $\lim_{k\to\infty} a_k = 0$

then $\sum_{k=1}^{\infty} a_k$ necessarily converges. In your explanation, include an important example that proves that this is not true and comment on the fact that the convergence of a_k to 0 can be slow or fast.

3. (writing) In Theorems 2.2 and 2.3, the series starts at $k = 1$,

as in $\sum_{k=1}^{\infty} a_k$. Explain why the conclusions of the theorems

hold if the series starts at $k = 2$, or $k = 3$ or at any positive integer.

4. (writing) We emphasized in the text that numerical and graphical evidence for the convergence of a series can be misleading. Suppose your calculator carries 14 digits in its calculations. Explain why for large enough values of n, the term $\dfrac{1}{n}$

will be too small to change the partial sum $\sum_{k=1}^{n} \dfrac{1}{k}$. Thus, the

calculator would incorrectly indicate that the harmonic series converges.

In exercises 5–28, determine if the series converges or diverges. For convergent series, find the sum of the series.

5. $\sum_{k=0}^{\infty} 3\left(\frac{1}{5}\right)^k$

6. $\sum_{k=0}^{\infty} \frac{1}{3}\left(5\right)^k$

7. $\sum_{k=0}^{\infty} \frac{1}{2}\left(-\frac{1}{3}\right)^k$

8. $\sum_{k=0}^{\infty} 4\left(\frac{1}{2}\right)^k$

9. $\sum_{k=0}^{\infty} \frac{1}{2}\left(3\right)^k$

10. $\sum_{k=0}^{\infty} 5\left(-\frac{1}{3}\right)^k$

11. $\sum_{k=1}^{\infty} \frac{4}{k(k+2)}$

12. $\sum_{k=1}^{\infty} \frac{4k}{k+2}$

13. $\sum_{k=1}^{\infty} \frac{3k}{k+4}$

14. $\sum_{k=1}^{\infty} \frac{9}{k(k+3)}$

15. $\sum_{k=1}^{\infty} \frac{2}{k}$

16. $\sum_{k=0}^{\infty} \frac{4}{k+1}$

17. $\sum_{k=1}^{\infty} \frac{2k+1}{k^2(k+1)^2}$

18. $\sum_{k=0}^{\infty} \left[e^{-k} - e^{-(k+1)}\right]$

19. $\displaystyle\sum_{k=0}^{\infty} 3^{-k}$

20. $\displaystyle\sum_{k=0}^{\infty} 2e^{-k}$

21. $\displaystyle\sum_{k=0}^{\infty} \left(\frac{1}{2^k} - \frac{1}{k}\right)$

22. $\displaystyle\sum_{k=0}^{\infty} \left(\frac{1}{2^k} - \frac{1}{3^k}\right)$

23. $\displaystyle\sum_{k=0}^{\infty} \left(\frac{2}{3^k} + \frac{1}{2^k}\right)$

24. $\displaystyle\sum_{k=0}^{\infty} \left(\frac{1}{k} - \frac{1}{4^k}\right)$

25. $\displaystyle\sum_{k=0}^{\infty} (-1)^k \frac{3}{2^k}$

26. $\displaystyle\sum_{k=0}^{\infty} (-1)^{k+1} \frac{4}{3^k}$

27. $\displaystyle\sum_{k=0}^{\infty} (-1)^{k+1} \frac{3k}{k+1}$

28. $\displaystyle\sum_{k=0}^{\infty} (-1)^k \frac{k^3}{k^2+1}$

In exercises 29–34, use graphical and numerical evidence to conjecture the convergence or divergence of the series.

29. $\displaystyle\sum_{k=1}^{\infty} \frac{1}{k^2}$

30. $\displaystyle\sum_{k=1}^{\infty} \frac{1}{\sqrt{k}}$

31. $\displaystyle\sum_{k=1}^{\infty} \frac{3}{k!}$

32. $\displaystyle\sum_{k=1}^{\infty} \frac{2^k}{k}$

33. $\displaystyle\sum_{k=1}^{\infty} \frac{4^k}{k^2}$

34. $\displaystyle\sum_{k=1}^{\infty} \frac{2^k}{k!}$

35. Prove that if $\displaystyle\sum_{k=1}^{\infty} a_k$ converges, then $\displaystyle\sum_{k=m}^{\infty} a_k$ converges for any positive integer m. In particular, if $\displaystyle\sum_{k=1}^{\infty} a_k$ converges to L, what does $\displaystyle\sum_{k=m}^{\infty} a_k$ converge to?

36. Prove that if $\displaystyle\sum_{k=1}^{\infty} a_k$ diverges, then $\displaystyle\sum_{k=m}^{\infty} a_k$ diverges for any positive integer m.

37. Prove Theorem 2.3 (i).

38. Prove Theorem 2.3 (ii).

39. The harmonic series is probably the single most important series to understand. In this exercise, we guide you through another proof of the divergence of this series. Let $S_n = \displaystyle\sum_{k=1}^{n} \frac{1}{k}$. Show that $S_1 = 1$ and $S_2 = \frac{3}{2}$. Since $\frac{1}{3} > \frac{1}{4}$, we have $\frac{1}{3} + \frac{1}{4} > \frac{1}{4} + \frac{1}{4} = \frac{1}{2}$. Therefore, $S_4 > \frac{3}{2} + \frac{1}{2} = 2$. Similarly, $\frac{1}{5} + \frac{1}{6} + \frac{1}{7} + \frac{1}{8} > \frac{1}{8} + \frac{1}{8} + \frac{1}{8} + \frac{1}{8} = \frac{1}{2}$, so $S_8 > \frac{5}{2}$. Show that $S_{16} > 3$ and $S_{32} > \frac{7}{2}$. For which n can you guarantee that $S_n > 4$? $S_n > 5$? For any positive integer m, determine n such that $S_n > m$. Conclude that the harmonic series diverges.

40. Compute several partial sums of the series $1 + 1 - 1 + 1 - 1 + 1 - 1 + \cdots$. Argue that the limit of the sequence of partial sums does not exist, so that the series diverges. Also, write this series as a geometric series and use Theorem 2.1 to conclude that the series diverges. Finally, use the kth-term test for divergence to conclude that the series diverges.

41. Write $0.9999\overline{9} = 0.9 + 0.09 + 0.009 + \ldots$ and sum the geometric series to prove that $0.9999\overline{9} = 1$.

42. As in exercise 41, prove that $0.19999\overline{9} = 0.2$.

43. Suppose you have n boards of length L. Place the first board with length $\frac{L}{2n}$ hanging over the edge of the table. Place the next board with length $\frac{L}{2(n-1)}$ hanging over the edge of the first board. The next board should hang $\frac{L}{2(n-2)}$ over the edge of the second board. Continuing on until the last board hangs $\frac{L}{2}$ over the edge of the $(n-1)$st board. Theoretically, this stack will balance (in practice, don't use quite as much overhang). With $n = 8$, compute the total overhang of the stack. Determine the number of boards n such that the total overhang is greater than L. This means that the last board is entirely beyond the edge of the table. What is the limit of the total overhang as $n \to \infty$?

44. Have you ever felt that the line you're standing in moves more slowly than the other lines? In *An Introduction to Probability Theory and Its Applications*, William Feller proved just how bad your luck is. Let N be the number of people who get in line until someone waits longer than you do (you're the first, so $N \geq 2$). The probability that $N = k$ is given by $p(k) = \dfrac{1}{k(k-1)}$. Prove that the total probability equals 1; that is, $\displaystyle\sum_{k=2}^{\infty} \frac{1}{k(k-1)} = 1$. From probability theory, the average (mean) number of people who must get in line before someone has waited longer than you is given by $\displaystyle\sum_{k=2}^{\infty} k \frac{1}{k(k-1)}$. Prove that this diverges to ∞. Talk about bad luck!

45. If $0 < r < y_2$, show that $1 + 2r + 4r^2 + \cdots + (2r)^n + \cdots = \dfrac{1}{1 - 2r}$. Replace r with $\frac{1}{1000}$ and discuss what's interesting about the decimal representation of $\frac{500}{499}$.

46. In exercise 60 of section 8.1, you showed that a particular bouncing ball takes 2 seconds to complete its infinite number of bounces. In general, the total time it takes for a ball to complete its bounces is $\dfrac{2v}{g} \displaystyle\sum_{k=0}^{\infty} r^k$ and the total distance the ball moves is $\dfrac{v^2}{g} \displaystyle\sum_{k=0}^{\infty} r^{2k}$. Assuming $0 < r < 1$, find the sums of these geometric series.

47. To win a deuce tennis game, one player or the other must win the next two points. If the deuce starts over. If you win each point with probability p, the probability that you win the next two points is p^2. The probability that you win one of the next two points is

$2p(1 - p)$. The probability that you win a deuce game is then $p^2 + 2p(1 - p)p^2 + [2p(1 - p)]^2 p^2 + [2p(1 - p)]^3 p^2 + \cdots$. Explain what each term represents, explain why the geometric series converges and find the sum of the series. If $p = 0.6$, you're a better player than your opponent. Show that you are more likely to win a deuce game than you are a single point. The slightly strange scoring rules in tennis make it more likely that the better player wins.

48. On an analog clock, at 1:00 the minute hand points to 12 and the hour hand points to 1. When the minute hand reaches 1, the hour hand has progressed to $1 + \frac{1}{12}$. When the minute hand reaches $1 + \frac{1}{12}$, the hour hand has moved to $1 + \frac{1}{12} + \frac{1}{12^2}$. Find the sum of a geometric series to determine the time at which the minute hand and hour hand are in the same location.

49. Two bicyclists are 40 miles apart, riding toward each other at 20 mph (each). A fly starts at one bicyclist and flies toward the other bicyclist at 60 mph. When it reaches the bike, it turns around and flies back to the first bike. It continues flying back and forth until the bikes meet. Determine the distance flown on each leg of the fly's journey and find the sum of the geometric series to get the total distance flown. Verify that this is the right answer by solving the problem the easy way.

50. Give an example where $\displaystyle\sum_{k=1}^{\infty} a_k$ and $\displaystyle\sum_{k=1}^{\infty} b_k$ both diverge but $\displaystyle\sum_{k=1}^{\infty} (a_k + b_k)$ converges.

51. Suppose $100,000 of counterfeit money is introduced into the economy. Each time the money is used, 25% of the remaining money is identified as counterfeit and removed from circulation. Determine the total amount of counterfeit money

successfully used in transactions. This is an example of the **multiplier effect** in economics.

52. In exercise 51, suppose that a new marking scheme on dollar bills helps raise the detection rate to 40%. Determine the reduction in the total amount of counterfeit money successfully spent.

53. A dosage d of a drug is given at times $t = 0, 1, 2, \ldots$. The drug decays exponentially with rate r in the bloodstream. The amount in the bloodstream after $n + 1$ doses is $d + de^{-r} + de^{-2r} + \cdots + de^{-nr}$. Show that the eventual level of the drug (after an "infinite" number of doses) is $\dfrac{d}{1 - e^{-r}}$. If $r = .1$, find the dosage needed to maintain a drug level of 2.

54. (exploratory) **Infinite products** are also of great interest to mathematicians. Numerically explore the convergence or divergence of the infinite product $\left(1 - \frac{1}{4}\right)\left(1 - \frac{1}{9}\right) \times \left(1 - \frac{1}{25}\right)\left(1 - \frac{1}{49}\right)\cdots = \displaystyle\prod_{p=\text{prime}} \left(1 - \frac{1}{p^2}\right)$. Note that the product is taken over the prime numbers, not all integers. Compare your results to the number $\dfrac{6}{\pi^2}$.

55. (exploratory) In example 2.7, we showed that $1 + \frac{1}{2} + \frac{1}{3} + \cdots + \frac{1}{n} > \ln(n + 1)$. Superimpose the graph of $f(x) = \frac{1}{x-1}$ onto Figure 8.22 and show that $\frac{1}{2} + \frac{1}{3} + \cdots + \frac{1}{n} < \ln(n)$. Conclude that $\ln(n + 1) < 1 + \frac{1}{2} + \frac{1}{3} + \cdots + \frac{1}{n} < 1 + \ln(n)$. **Euler's number** is defined by

$$\gamma = \lim_{n\to\infty}\left[1 + \frac{1}{2} + \frac{1}{3} + \cdots + \frac{1}{n} - \ln(n)\right].$$

Look up the value of γ. Use γ to estimate $\displaystyle\sum_{i=1}^{n} \frac{1}{i}$ for $n = 10,000$ and $n = 100,000$.

8.3 The Integral Test and Comparison Tests

As we have observed a number of times now, we are usually unable to determine the sum of a convergent series. In fact, for most series we cannot determine whether they converge or diverge by simply looking at the sequence of partial sums. Most of the time, we will need to test a series for convergence in some indirect way. If we find that the series is convergent, we can then approximate its sum by numerically computing some partial sums. In this section, we will develop additional tests for convergence of series. The first of these is a generalization of the method we used to show that the harmonic series was divergent in section 8.2.

For a given series $\displaystyle\sum_{k=1}^{\infty} a_k$, suppose that there is a function f for which

$$f(k) = a_k, \text{ for } k = 1, 2, \ldots,$$

where f is continuous, decreasing and $f(x) \geq 0$ for all $x \geq 1$. We consider the nth partial sum

$$S_n = \sum_{k=1}^{n} a_k = a_1 + a_2 + \cdots + a_n.$$

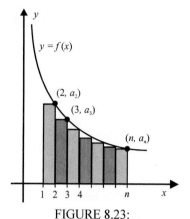

FIGURE 8.23:
$(n-1)$ rectangles, lying
beneath the curve.

Look carefully at Figure 8.23. We have constructed $(n-1)$ rectangles on the interval $[1, n]$, each of width 1 and with height equal to the value of the function at the right hand endpoint of the subinterval on which the rectangle is constructed. Notice that since each rectangle lies completely beneath the curve, the sum of the areas of the $(n-1)$ rectangles shown is less than the area under the curve from $x = 1$ to $x = n$. That is,

$$0 \leq \textit{Sum of areas of } (n-1) \textit{ rectangles} \leq \textit{area under the curve} = \int_1^n f(x)\,dx. \quad (3.1)$$

Note that the area of the first rectangle is $length \times width = (1)\,(a_2)$, the area of the second rectangle is $(1)\,(a_3)$ and so on. We get that the sum of the areas of the $(n-1)$ rectangles indicated in Figure 8.23 is

$$a_2 + a_3 + a_4 + \cdots + a_n = S_n - a_1,$$

since

$$S_n = a_1 + a_2 + \cdots + a_n.$$

Together with (3.1), this gives us

$$0 \leq \textit{Sum of areas of } (n-1) \textit{ rectangles}$$
$$= S_n - a_1 \leq \textit{area under the curve} = \int_1^n f(x)\,dx. \quad (3.2)$$

Now, suppose that the improper integral $\int_1^\infty f(x)\,dx$ converges. Then, from (3.2), we have

$$0 \leq S_n - a_1 \leq \int_1^n f(x)\,dx \leq \int_1^\infty f(x)\,dx.$$

Adding a_1 to all the terms gives us

$$a_1 \leq S_n \leq a_1 + \int_1^\infty f(x)\,dx.$$

This says that the sequence of partial sums $\{S_n\}_{n=1}^\infty$ is bounded. Since $\{S_n\}_{n=1}^\infty$ is also monotonic (why is that?), $\{S_n\}_{n=1}^\infty$ is convergent by Theorem 1.4 and so, the series $\sum_{k=1}^\infty a_k$ is also convergent.

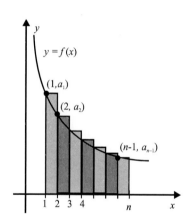

FIGURE 8.24:
$(n-1)$ rectangles, partially
above the curve.

In Figure 8.24, we have constructed $(n-1)$ rectangles on the interval $[1, n]$, each of width 1 but with height equal to the value of the function at the left-hand endpoint of the subinterval on which the rectangle is constructed. In this case, the sum of the areas of the $(n-1)$ rectangles shown is greater than the area under the curve. That is,

$$0 \leq \textit{area under the curve} = \int_1^n f(x)\,dx$$
$$\leq \textit{sum of areas of } (n-1) \textit{ rectangles}. \quad (3.3)$$

Further, note that the area of the first rectangle is $length \times width = (1)\,(a_1)$, the area of the second rectangle is $(1)\,(a_2)$ and so on. We get that the sum of the areas of the $(n-1)$ rectangles indicated in Figure 8.24 is

$$a_1 + a_2 + \cdots + a_{n-1} = S_{n-1}.$$

Together with (3.3), this gives us

$$0 \leq \textit{area under the curve} = \int_1^n f(x)\,dx$$
$$\leq \textit{sum of areas of } (n-1) \textit{ rectangles} = S_{n-1}. \quad (3.4)$$

Now, suppose that the improper integral $\int_1^\infty f(x)\,dx$ diverges. Since $f(x) \geq 0$, this says that $\lim_{n\to\infty} \int_1^n f(x)\,dx = \infty$. From (3.4), we have that

$$\int_1^n f(x)\,dx \leq S_{n-1}.$$

This says that

$$\lim_{n\to\infty} S_{n-1} = \infty,$$

also. So, the sequence of partial sums $\{S_n\}_{n=1}^\infty$ diverges and hence, the series $\sum_{k=1}^\infty a_k$ diverges, too.

We summarize the results of this analysis in the following theorem.

Theorem 3.1   (Integral Test) If $f(k) = a_k$ for all $k = 1, 2, \ldots$ and f is continuous, decreasing and $f(x) \geq 0$, for $x \geq 1$, then $\int_1^\infty f(x)\,dx$ and $\sum_{k=1}^\infty a_k$ either **both** converge or **both** diverge.

It is important to recognize that while the Integral Test might say that a given series and improper integral either both converge or both diverge, it does **not** say that they will converge to the same thing. In fact, this is generally not true, as we see in the following example.

Example 3.1   Using the Integral Test

Investigate the convergence or divergence of the series $\sum_{k=0}^\infty \dfrac{1}{k^2 + 1}$.

Solution   The graph of the first 20 partial sums shown in Figure 8.25 suggests that the series converges to some value around 2. In the following table, you can find some selected partial sums. We have shown so many partial sums due to the very slow rate of convergence of this series.

FIGURE 8.25:

$$S_n = \sum_{k=0}^{n-1} \frac{1}{k^2 + 1}.$$

n	$S_n = \sum_{k=0}^{n-1} \dfrac{1}{k^2 + 1}$	n	$S_n = \sum_{k=0}^{n-1} \dfrac{1}{k^2 + 1}$
10	1.97189	200	2.07166
20	2.02545	300	2.07334
30	2.04279	400	2.07417
40	2.05136	500	2.07467
50	2.05648	600	2.07501
60	2.05987	700	2.07524
70	2.06229	800	2.07542
80	2.0641	900	2.07556
90	2.0655	1000	2.07567
100	2.06662	2000	2.07617

Based on our computations, we cannot say whether the series is converging very slowly to a limit around 2.076 or whether the series is instead diverging very slowly, as we had seen for the harmonic series. To determine which is the case, we must test the series further. Define $f(x) = \dfrac{1}{x^2 + 1}$. Note that f is continuous and positive everywhere and $f(k) = \dfrac{1}{k^2 + 1} = a_k$, for all $k \geq 1$. Further,

$$f'(x) = (-1)\,(x^2 + 1)^{-2}\,(2x) < 0,$$

for $x \in (0, \infty)$, and so, f is decreasing. This says that the Integral Test applies to this series. So, we consider the improper integral

$$\int_0^\infty \frac{1}{x^2 + 1}\,dx = \lim_{R \to \infty} \int_0^R \frac{1}{x^2 + 1}\,dx = \lim_{R \to \infty} \tan^{-1} x \Big|_0^R$$

$$= \lim_{R \to \infty} \left(\tan^{-1} R - \tan^{-1} 0 \right) = \frac{\pi}{2} - 0 = \frac{\pi}{2}.$$

By the Integral Test, we have that since the improper integral converges, the series must converge, also. Since we have now established that the series is convergent, we can use our earlier calculations to arrive at the estimated sum 2.076. Notice that this is **not** the same as the value of the corresponding improper integral, which is $\dfrac{\pi}{2} \approx 1.5708$. ∎

In the following example, we discuss an important type of series.

Example 3.2   **The p-Series**

Determine for which values of p the series $\displaystyle\sum_{k=1}^\infty \frac{1}{k^p}$ (a p-**series**) converges.

Solution   First, notice that for $p = 1$, this is the divergent harmonic series. For $p > 1$, define $f(x) = \dfrac{1}{x^p} = x^{-p}$. Notice that for $x \geq 1$, f is continuous and positive. Further,

$$f'(x) = -p\,x^{-p-1} < 0,$$

so that f is decreasing for $x \geq 1$. This says that the Integral Test applies. We now consider

Since $p > 1$ implies that $-p + 1 < 0$.

$$\int_1^\infty x^{-p}\,dx = \lim_{R \to \infty} \int_1^R x^{-p}\,dx = \lim_{R \to \infty} \frac{x^{-p+1}}{-p+1} \Big|_1^R$$

$$= \lim_{R \to \infty} \left(\frac{R^{-p+1}}{-p+1} - \frac{1}{-p+1} \right) = \frac{1}{-p+1}.$$

In this case, the improper integral converges and so too, must the series. In the case where $p < 1$, we leave it as an exercise to show that the series diverges. ∎

Comparison Tests

We next present two results which will allow us to compare a given series with one which is already known to be convergent or divergent, much as we did with improper integrals in section 7.7.

Theorem 3.2   (Comparison Test) Suppose that $0 \leq a_k \leq b_k$, for all k.

(i) If $\displaystyle\sum_{k=1}^\infty b_k$ converges, then $\displaystyle\sum_{k=1}^\infty a_k$ converges, too.

(ii) If $\displaystyle\sum_{k=1}^\infty a_k$ diverges, then $\displaystyle\sum_{k=1}^\infty b_k$ diverges, too.

Intuitively, this theorem should make abundant sense: if the "larger" series converges, then the "smaller" one must also converge. Likewise, if the "smaller" series diverges, the "larger" one must diverge, too.

Proof. Given that $0 \leq a_k \leq b_k$ for all k, observe that the nth partial sums of the two series satisfy

$$0 \leq S_n = a_1 + a_2 + \cdots + a_n \leq b_1 + b_2 + \cdots + b_n.$$

(i) If $\sum\limits_{k=1}^{\infty} b_k$ converges (say to B), this says that

$$0 \leq a_1 + a_2 + \cdots + a_n \leq b_1 + b_2 + \cdots + b_n \leq \sum_{k=1}^{\infty} b_k = B, \qquad (3.5)$$

for all $n \geq 1$. From (3.5), the sequence $\{S_n\}_{n=1}^{\infty}$ of partial sums of $\sum\limits_{k=1}^{\infty} a_k$ is bounded. $\{S_n\}_{n=1}^{\infty}$ is also increasing. (Why?) Since every bounded, monotonic sequence is convergent, we get that $\sum\limits_{k=1}^{\infty} a_k$ is convergent, too.

(ii) If $\sum\limits_{k=1}^{\infty} a_k$ is divergent, we have (since all of the terms of the series are nonnegative) that

$$\lim_{n \to \infty} (b_1 + b_2 + \cdots + b_n) \geq \lim_{n \to \infty} (a_1 + a_2 + \cdots + a_n) = \infty.$$

Thus, $\sum\limits_{k=1}^{\infty} b_k$ must be divergent, also. ∎

You can use the Comparison Test to test the convergence of series that look similar to series that you already know are convergent or divergent (notably, geometric series or p-series).

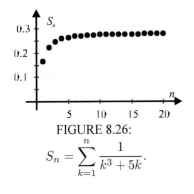

FIGURE 8.26:
$$S_n = \sum_{k=1}^{n} \frac{1}{k^3 + 5k}.$$

Example 3.3   Using the Comparison Test for a Convergent Series

Investigate the convergence or divergence of $\sum\limits_{k=1}^{\infty} \dfrac{1}{k^3 + 5k}$.

Solution   From the graph of the first 20 partial sums shown in Figure 8.26, it appears that the series converges to some value near 0.3. To confirm such a conjecture, we must carefully test the series. Note that for large values of k, the general term of the series looks like $\dfrac{1}{k^3}$, since when k is large, k^3 is much larger than $5k$. This is significant in that you should recognize that $\sum\limits_{k=1}^{\infty} \dfrac{1}{k^3}$ is a convergent p-series ($p = 3 > 1$). Further, observe that

$$\frac{1}{k^3 + 5k} \leq \frac{1}{k^3},$$

for all $k \geq 1$. Since $\sum\limits_{k=1}^{\infty} \dfrac{1}{k^3}$ converges, the Comparison Test says that $\sum\limits_{k=1}^{\infty} \dfrac{1}{k^3 + 5k}$ converges, too. As with the Integral Test, although the Comparison Test tells us that both series converge, the two series **need not** converge to the same sum. A quick calculation of a few partial sums should convince you that $\sum\limits_{k=1}^{\infty} \dfrac{1}{k^3}$ converges to approximately 1.202, while $\sum\limits_{k=1}^{\infty} \dfrac{1}{k^3 + 5k}$ converges to approximately 0.2798. (Note that this is consistent with what we saw in Figure 8.26.) ∎

FIGURE 8.27:
$$S_n = \sum_{k=1}^{n} \frac{5^k + 1}{2^k - 1}.$$

Example 3.4   Using the Comparison Test for a Divergent Series

Investigate the convergence or divergence of $\displaystyle\sum_{k=1}^{\infty} \frac{5^k + 1}{2^k - 1}$.

Solution   From the graph of the first 20 partial sums seen in Figure 8.27, it appears that the partial sums are growing very rapidly. On this basis, we would conjecture that the series diverges. Of course, we can only determine this with further testing. Notice that for k large, the general term looks like $\dfrac{5^k}{2^k} = \left(\dfrac{5}{2}\right)^k$ and we know that $\displaystyle\sum_{k=1}^{\infty} \left(\dfrac{5}{2}\right)^k$ is a divergent geometric series ($|r| = \dfrac{5}{2} > 1$.) Further,

$$\frac{5^k + 1}{2^k - 1} \geq \frac{5^k}{2^k - 1} \geq \frac{5^k}{2^k} = \left(\frac{5}{2}\right)^k.$$

By the Comparison Test, $\displaystyle\sum_{k=1}^{\infty} \frac{5^k + 1}{2^k - 1}$ diverges, too.

There are plenty of series whose general term looks like the general term of a familiar series, but for which it is unclear how to get the inequality required for the Comparison Test to go in the right direction.

FIGURE 8.28:
$$S_n = \sum_{k=3}^{n} \frac{1}{k^3 - 5k}.$$

Example 3.5   A Comparison That Does Not Work

Investigate the convergence or divergence of the series $\displaystyle\sum_{k=3}^{\infty} \frac{1}{k^3 - 5k}$.

Solution   Note that this is nearly identical to example 3.3, except that there is a '$-$' sign in the denominator instead of a '$+$' sign. The graph of the first 20 partial sums seen in Figure 8.28 looks somewhat similar to the graph in Figure 8.26, except that the series appears to be converging to about 12.05. In this case, however, we have the inequality

$$\frac{1}{k^3 - 5k} \geq \frac{1}{k^3}, \text{ for all } k \geq 3.$$

Unfortunately, this inequality goes the **wrong way**: we know that $\displaystyle\sum_{k=3}^{\infty} \frac{1}{k^3}$ is a convergent p-series, but since $\displaystyle\sum_{k=3}^{\infty} \frac{1}{k^3 - 5k}$ is "larger" than this convergent series, the Comparison Test says nothing. Think about it this way: while you might observe that

$$k^2 \geq \frac{1}{k^3}, \text{ for all } k \geq 1,$$

and you know that $\displaystyle\sum_{k=1}^{\infty} \frac{1}{k^3}$ is convergent, the Comparison Test says nothing about the "larger" series $\displaystyle\sum_{k=1}^{\infty} k^2$. In fact, we know that this last series is divergent (by the kth-term test for divergence, since $\displaystyle\lim_{k\to\infty} k^2 = \infty \neq 0$). To resolve this difficulty for the present problem, we will need to either make a more appropriate comparison or use the Limit Comparison Test that follows.

Theorem 3.3   (Limit Comparison Test) Suppose that $a_k, b_k > 0$ and that for some (finite) value, L, $\lim_{k \to \infty} \dfrac{a_k}{b_k} = L > 0$. Then, either $\sum\limits_{k=1}^{\infty} a_k$ and $\sum\limits_{k=1}^{\infty} b_k$ **both** converge or **both** diverge.

NOTE: When we say $\lim\limits_{k \to \infty} \dfrac{a_k}{b_k} = L > 0$, we mean that the limit exists and is positive. In particular, we mean that $\lim\limits_{k \to \infty} \dfrac{a_k}{b_k} \neq \pm\infty$.

Proof. If $\lim\limits_{k \to \infty} \dfrac{a_k}{b_k} = L > 0$, this says that we can make $\dfrac{a_k}{b_k}$ as close to L as desired. So, in particular, we can make $\dfrac{a_k}{b_k}$ within distance $\dfrac{L}{2}$ of L. That is, for some number $N > 0$,

$$L - \frac{L}{2} < \frac{a_k}{b_k} < L + \frac{L}{2}, \text{ for } k > N$$

or

$$\frac{L}{2} < \frac{a_k}{b_k} < \frac{3L}{2}. \tag{3.6}$$

Multiplying inequality (3.6) through by b_k (recall that $b_k > 0$), we get

$$\frac{L}{2} b_k < a_k < \frac{3L}{2} b_k, \text{ for } k \geq N.$$

Note that this says that if $\sum\limits_{k=1}^{\infty} a_k$ converges, then the "smaller" series $\sum\limits_{k=1}^{\infty} \left(\frac{L}{2} b_k\right) = \frac{L}{2} \sum\limits_{k=1}^{\infty} b_k$ must also converge, by the Comparison Test. Likewise, if $\sum\limits_{k=1}^{\infty} a_k$ diverges, the "larger" series $\sum\limits_{k=1}^{\infty} \left(\frac{3L}{2} b_k\right) = \frac{3L}{2} \sum\limits_{k=1}^{\infty} b_k$ must also diverge. In the same way, if $\sum\limits_{k=1}^{\infty} b_k$ converges, then $\sum\limits_{k=1}^{\infty} \left(\frac{3L}{2} b_k\right)$ converges and so, too must the "smaller" series $\sum\limits_{k=1}^{\infty} a_k$. Finally, if $\sum\limits_{k=1}^{\infty} b_k$ diverges, then $\sum\limits_{k=1}^{\infty} \left(\frac{L}{2} b_k\right)$ diverges and hence, the "larger" series $\sum\limits_{k=1}^{\infty} a_k$ must diverge, also. ∎

We can now use the Limit Comparison Test to test the series from example 3.5 whose convergence we have so far been unable to confirm.

Example 3.6   Using the Limit Comparison Test

Investigate the convergence or divergence of the series $\sum\limits_{k=3}^{\infty} \dfrac{1}{k^3 - 5k}$.

Solution   Recall that we had already observed in example 3.5 that the general term $a_k = \dfrac{1}{k^3 - 5k}$ "looks like" $b_k = \dfrac{1}{k^3}$, for k large. We then consider the limit

$$\lim_{k \to \infty} \frac{a_k}{b_k} = \lim_{k \to \infty} \frac{1}{(k^3 - 5k)} \frac{1}{\left(\frac{1}{k^3}\right)} = \lim_{k \to \infty} \frac{1}{1 - \frac{5}{k^2}} = 1 > 0.$$

Since $\displaystyle\sum_{k=1}^{\infty}\frac{1}{k^3}$ is a convergent p-series ($p = 3 > 1$), the Limit Comparison Test says that

$\displaystyle\sum_{k=1}^{\infty}\frac{1}{k^3 - 5k}$ is also convergent, as we had originally suspected.

∎

The Limit Comparison Test can be used to resolve convergence questions for a great many series. The first step in using this (like the Comparison Test) is to find another series (whose convergence or divergence is known) that "looks like" the series in question.

FIGURE 8.29:
$S_n = \displaystyle\sum_{k=1}^{n}\frac{k^2-2k+7}{k^5+5k^4-3k^3+2k-1}.$

Example 3.7   Using the Limit Comparison Test

Investigate the convergence or divergence of the series

$$\sum_{k=1}^{\infty}\frac{k^2 - 2k + 7}{k^5 + 5k^4 - 3k^3 + 2k - 1}.$$

Solution   The graph of the first 20 partial sums in Figure 8.29 suggests that the series converges to a limit of about 1.61. The following table of partial sums supports this conjecture.

n	$S_n = \displaystyle\sum_{k=1}^{n}\frac{k^2 - 2k + 7}{k^5 + 5k^4 - 3k^3 + 2k - 1}$	n	$S_n = \displaystyle\sum_{k=1}^{n}\frac{k^2 - 2k + 7}{k^5 + 5k^4 - 3k^3 + 2k - 1}$
5	1.60522	55	1.61447
10	1.61145	60	1.61449
15	1.61301	65	1.61451
20	1.61365	70	1.61452
25	1.61397	75	1.61453
30	1.61415	80	1.61454
35	1.61427	85	1.61455
40	1.61434	90	1.61456
45	1.6144	95	1.61457
50	1.61444	100	1.61457

Notice that for k large, the general term looks like $\dfrac{k^2}{k^5} = \dfrac{1}{k^3}$ (since the terms with the largest exponents tend to dominate the expression, for large values of k.) From the Limit Comparison Test, we have

$$\lim_{k\to\infty}\frac{a_k}{b_k} = \lim_{k\to\infty}\frac{k^2 - 2k + 7}{k^5 + 5k^4 - 3k^3 + 2k - 1}\frac{1}{\left(\frac{1}{k^3}\right)}$$

$$= \lim_{k\to\infty}\frac{\left(k^2 - 2k + 7\right)}{\left(k^5 + 5k^4 - 3k^3 + 2k - 1\right)}\frac{k^3}{1}$$

$$= \lim_{k\to\infty}\frac{\left(k^5 - 2k^4 + 7k^3\right)}{\left(k^5 + 5k^4 - 3k^3 + 2k - 1\right)}\frac{\left(\frac{1}{k^5}\right)}{\left(\frac{1}{k^5}\right)}$$

$$= \lim_{k\to\infty}\frac{1 - \frac{2}{k} + \frac{7}{k^2}}{1 + \frac{5}{k} - \frac{3}{k^2} + \frac{2}{k^4} - \frac{1}{k^5}} = 1 > 0.$$

Since $\displaystyle\sum_{k=1}^{\infty}\frac{1}{k^3}$ is a convergent p-series ($p = 3 > 1$), the Limit Comparison Test says

that $\displaystyle\sum_{k=1}^{\infty}\frac{k^2 - 2k + 7}{k^5 + 5k^4 - 3k^3 + 2k - 1}$ converges, also. Finally, now that we have established that the series is in fact, convergent, we can use our table of computed partial sums to approximate the sum of the series as 1.61457.

■

Exercises 8.3

1. (writing) Notice that the Comparison Test doesn't always give us information about convergence or divergence. If $a_k \leq b_k$ for each k and $\displaystyle\sum_{k=1}^{\infty} b_k$ diverges, explain why you can't tell whether or not $\displaystyle\sum_{k=1}^{\infty} a_k$ diverges.

2. (writing) Explain why the Limit Comparison Test works. In particular, if $\displaystyle\lim_{k\to\infty} \frac{a_k}{b_k} = 1$, explain how a_k and b_k compare and conclude that $\displaystyle\sum_{k=1}^{\infty} a_k$ and $\displaystyle\sum_{k=1}^{\infty} b_k$ either both converge or both diverge.

3. (writing) In the Limit Comparison Test, if $\displaystyle\lim_{k\to\infty} \frac{a_k}{b_k} = 0$ and $\displaystyle\sum_{k=1}^{\infty} a_k$ converges, explain why you can't tell whether or not $\displaystyle\sum_{k=1}^{\infty} b_k$ converges.

4. (writing) The p-series converges if $p > 1$ and diverges if $p < 1$. What happens for $p = 1$? If your friend knows that the harmonic series diverges, explain an easy way to remember the rest of the conclusion of the p-series test.

In exercises 5–40, determine convergence or divergence of the series.

5. $\displaystyle\sum_{k=1}^{\infty} \frac{4}{\sqrt[3]{k}}$

6. $\displaystyle\sum_{k=1}^{\infty} k^{-9/10}$

7. $\displaystyle\sum_{k=1}^{\infty} k^{-11/10}$

8. $\displaystyle\sum_{k=1}^{\infty} \frac{4}{\sqrt{k}}$

9. $\displaystyle\sum_{k=0}^{\infty} \frac{k+1}{k^2 + 2k + 3}$

10. $\displaystyle\sum_{k=0}^{\infty} \frac{k^2 + 1}{k^3 + 3k + 2}$

11. $\displaystyle\sum_{k=1}^{\infty} \frac{4}{2 + 4k}$

12. $\displaystyle\sum_{k=1}^{\infty} \frac{4}{(2 + 4k)^2}$

13. $\displaystyle\sum_{k=2}^{\infty} \frac{2}{k \ln k}$

14. $\displaystyle\sum_{k=2}^{\infty} \frac{3}{k(\ln k)^2}$

15. $\displaystyle\sum_{k=1}^{\infty} \frac{2k}{k^3 + 1}$

16. $\displaystyle\sum_{k=0}^{\infty} \frac{\sqrt{k}}{k^2 + 1}$

17. $\displaystyle\sum_{k=1}^{\infty} \frac{e^{1/k}}{k^2}$

18. $\displaystyle\sum_{k=1}^{\infty} \frac{\sqrt{1 + 1/k}}{k^2}$

19. $\displaystyle\sum_{k=1}^{\infty} \frac{e^{-\sqrt{k}}}{\sqrt{k}}$

20. $\displaystyle\sum_{k=1}^{\infty} \frac{ke^{-k^2}}{4 + e^{-k}}$

21. $\displaystyle\sum_{k=1}^{\infty} \frac{3k}{k^{3/2} + 2}$

22. $\displaystyle\sum_{k=1}^{\infty} \frac{2k^2}{k^{5/2} + 2}$

23. $\displaystyle\sum_{k=0}^{\infty} \frac{2}{\sqrt{k^2 + 4}}$

24. $\displaystyle\sum_{k=0}^{\infty} \frac{4}{\sqrt{k^3 + 1}}$

25. $\displaystyle\sum_{k=0}^{\infty} \frac{k^2 + 1}{\sqrt{k^5 + 1}}$

26. $\displaystyle\sum_{k=0}^{\infty} \frac{k + 2}{\sqrt[3]{k^5 + 4}}$

27. $\displaystyle\sum_{k=1}^{\infty} \frac{\tan^{-1} k}{1 + k^2}$

28. $\displaystyle\sum_{k=1}^{\infty} \frac{\sin^{-1}(1/k)}{k^2}$

29. $\displaystyle\sum_{k=1}^{\infty} \frac{1}{\cos^2 k}$

30. $\displaystyle\sum_{k=1}^{\infty} \frac{1}{\sin^2 k}$

31. $\displaystyle\sum_{k=1}^{\infty} \frac{\sin k + 2}{k^2}$

32. $\displaystyle\sum_{k=1}^{\infty} \frac{e^{1/k} + 1}{k^3}$

33. $\displaystyle\sum_{k=2}^{\infty} \frac{\ln k}{k}$

34. $\displaystyle\sum_{k=1}^{\infty} \frac{2 + \cos k}{k}$

35. $\displaystyle\sum_{k=1}^{\infty} \frac{k^4 + 2k - 1}{k^5 + 3k^2 + 1}$

36. $\displaystyle\sum_{k=0}^{\infty} \frac{k^3 + 2k + 3}{k^4 + 2k^2 + 4}$

37. $\displaystyle\sum_{k=1}^{\infty} \frac{k + 1}{k + 2}$

38. $\displaystyle\sum_{k=1}^{\infty} \frac{k + 1}{k^2 + 2}$

39. $\displaystyle\sum_{k=1}^{\infty} \frac{k + 1}{k^3 + 2}$

40. $\displaystyle\sum_{k=1}^{\infty} \frac{k + 1}{k^4 + 2}$

41. In our statement of the Comparison Test, we required that $a_k \leq b_k$ for all k. Explain why the conclusion would remain true if $a_k \leq b_k$ for $k \geq 100$.

42. If $a_k > 0$ and $\displaystyle\sum_{k=1}^{\infty} a_k$ converges, prove that $\displaystyle\sum_{k=1}^{\infty} a_k^2$ converges.

43. Prove the following extension of the Limit Comparison Test: if $\displaystyle\lim_{k\to\infty} \frac{a_k}{b_k} = 0$ and $\displaystyle\sum_{k=1}^{\infty} b_k$ converges, then $\displaystyle\sum_{k=1}^{\infty} a_k$ converges.

44. Prove the following extension of the Limit Comparison Test: if $\lim\limits_{k\to\infty}\dfrac{a_k}{b_k}=\infty$ and $\sum\limits_{k=1}^{\infty}b_k$ diverges, then $\sum\limits_{k=1}^{\infty}a_k$ diverges.

In exercises 45–48, determine all values of p for which the series converges.

45. $\sum\limits_{k=2}^{\infty}\dfrac{1}{k(\ln k)^p}$

46. $\sum\limits_{k=0}^{\infty}\dfrac{1}{(a+bk)^p}$, $a>0, b>0$

47. $\sum\limits_{k=2}^{\infty}\dfrac{\ln k}{k^p}$

48. $\sum\limits_{k=1}^{\infty}k^{p-1}e^{k^p}$

49. Prove that the every-other-term harmonic series $1+\frac{1}{3}+\frac{1}{5}+\frac{1}{7}+\cdots$ diverges. (Hint: Write the series as $\sum\limits_{k=0}^{\infty}\dfrac{1}{2k+1}$ and use the Limit Comparison Test.)

50. Would the every-third term harmonic series $1+\frac{1}{4}+\frac{1}{7}+\frac{1}{10}+\cdots$ diverge? How about the every-fourth-term harmonic series $1+\frac{1}{5}+\frac{1}{9}+\frac{1}{13}+\cdots$? Make as general a statement as possible about such series.

51. The **Riemann-zeta function** is defined by $\zeta(x)=\sum\limits_{k=1}^{\infty}\dfrac{1}{k^x}$ for $x>1$. Explain why the restriction $x>1$ is necessary. (Leonhard Euler, considered to be one of the greatest mathematicians ever, proved the remarkable result that $\zeta(x)=\prod\limits_{p\text{ prime}}\dfrac{1}{\left(1-\frac{1}{p^x}\right)}$.)

52. Estimate $\zeta(2)$ numerically. Compare your result with that of exercise 54 of section 8.2.

In exercises 53–56, numerically estimate the sum of the convergent p-series and identify x such that the sum equals $\zeta(x)$ for the Riemann-zeta function of exercise 51.

53. $\sum\limits_{k=1}^{\infty}\dfrac{1}{k^4}$

54. $\sum\limits_{k=1}^{\infty}\dfrac{1}{k^6}$

55. $\sum\limits_{k=1}^{\infty}\dfrac{1}{k^8}$

56. $\sum\limits_{k=1}^{\infty}\dfrac{1}{k^{10}}$

57. (exploratory) Numerically investigate the p-series $\sum\limits_{k=1}^{\infty}\dfrac{1}{k^{0.9}}$ and $\sum\limits_{k=1}^{\infty}\dfrac{1}{k^{1.1}}$ and for other values of p close to 1. Can you distinguish convergent from divergent series numerically?

58. (exploratory) You know that $\sum\limits_{k=1}^{\infty}\dfrac{1}{k}$ diverges. This is the "smallest" p-series that diverges, in the sense that $\frac{1}{k}<\frac{1}{k^p}$ for $p<1$. Show that $\sum\limits_{k=1}^{\infty}\dfrac{1}{k\ln k}$ diverges and $\frac{1}{k\ln k}<\frac{1}{k}$. Show that $\sum\limits_{k=1}^{\infty}\dfrac{1}{k\ln k\ln(\ln k)}$ diverges and $\frac{1}{k\ln k\ln(\ln k)}<\frac{1}{k\ln k}$. Find a series such that $\sum\limits_{k=1}^{\infty}a_k$ diverges and $a_k<\dfrac{1}{k\ln k\ln(\ln k)}$. Is there a smallest divergent series?

8.4   Alternating Series

So far, we have focused our attention on positive term series, that is, series all of whose terms are positive. Before we consider the general case, we spend some time in this section examining **alternating series**, that is, series whose terms alternate back and forth from positive to negative. There are several reasons for doing this. First, alternating series appear frequently in applications. Second, alternating series are surprisingly simple to deal with and studying them should yield significant insight into how series work.

An **alternating series** is any series of the form
$$\sum_{k=1}^{\infty}(-1)^{k+1}a_k=a_1-a_2+a_3-a_4+a_5-a_6+\cdots,$$
where $a_k>0$, for all k.

Example 4.1   The Alternating Harmonic Series
Investigate the convergence or divergence of the **alternating harmonic series**
$$\sum_{k=1}^{\infty}\dfrac{(-1)^{k+1}}{k}=1-\dfrac{1}{2}+\dfrac{1}{3}-\dfrac{1}{4}+\dfrac{1}{5}-\dfrac{1}{6}+\cdots.$$

FIGURE 8.30:

$$S_n = \sum_{k=1}^{n} \frac{(-1)^{k+1}}{k}.$$

Solution   The graph of the first 20 partial sums seen in Figure 8.30 suggests (although the evidence is less than compelling) that the series might converge to about 0.7. We now calculate the first few partial sums by hand. Note that

$$S_1 = 1,$$

$$S_2 = 1 - \frac{1}{2} = \frac{1}{2},$$

$$S_3 = \frac{1}{2} + \frac{1}{3} = \frac{5}{6},$$

$$S_4 = \frac{5}{6} - \frac{1}{4} = \frac{7}{12},$$

$$S_5 = \frac{7}{12} + \frac{1}{5} = \frac{47}{60},$$

$$S_6 = \frac{47}{60} - \frac{1}{6} = \frac{37}{60},$$

$$S_7 = \frac{37}{60} + \frac{1}{7} = \frac{319}{420},$$

$$S_8 = \frac{319}{420} - \frac{1}{8} = \frac{533}{840}$$

and so on. We have plotted these first 8 partial sums on the number line shown in Figure 8.31.

FIGURE 8.31: Partial sums of $\displaystyle\sum_{k=1}^{\infty} \frac{(-1)^{k+1}}{k}$.

You should notice that the partial sums are bouncing back and forth, but seem to be zeroing-in on some value. (Could it be the sum of the series?) This should not be surprising, since as each new term is added or subtracted, we are adding or subtracting less than we *subtracted or added* (why did we reverse the order here?) to get the previous partial sum. You should notice this same zeroing-in in the following table displaying the first 20 partial sums of the series.

n	$S_n = \sum_{k=1}^{n} \frac{(-1)^{k+1}}{k}$	n	$S_n = \sum_{k=1}^{n} \frac{(-1)^{k+1}}{k}$
1	1	11	0.73654
2	0.5	12	0.65321
3	0.83333	13	0.73013
4	0.58333	14	0.65871
5	0.78333	15	0.72537
6	0.61667	16	0.66287
7	0.75952	17	0.7217
8	0.63452	18	0.66614
9	0.74563	19	0.71877
10	0.64563	20	0.66877

Based on the behavior of the partial sums, it is reasonable to conjecture that the series converges to some value between 0.66877 and 0.71877. We can resolve the question of convergence definitively with the theorem which follows.

∎

Theorem 4.1   (Alternating Series Test) Suppose that $\lim\limits_{k\to\infty} a_k = 0$ and $0 < a_{k+1} \le a_k$ for all $k \ge 1$. Then, the alternating series $\sum\limits_{k=1}^{\infty}(-1)^{k+1}a_k$ converges.

Before considering the proof of the theorem, make sure that you have a clear idea of what it is saying. In the case of an alternating series satisfying the hypotheses of the theorem, we start with 0 and add $a_1 > 0$ to get the first partial sum S_1. To get the next partial sum, S_2, we subtract a_2 from S_1, where $a_2 < a_1$. This says that S_2 will be between 0 and S_1. We illustrate this situation in Figure 8.32.

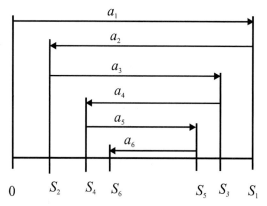

FIGURE 8.32: Convergence of the partial sums of an alternating series.

Continuing in this fashion, we add a_3 to S_2 to get S_3. Since $a_3 < a_2$, we must have that $S_2 < S_3 < S_1$. Referring to Figure 8.32, notice that

$$S_2 < S_4 < S_6 < \cdots < S_5 < S_3 < S_1.$$

In particular, this says that **all** of the odd-indexed partial sums (i.e., S_{2n+1}, for $n = 0, 1, 2, \ldots$) are larger than **all** of the even-indexed partial sums (i.e., S_{2n}, for $n = 1, 2, \ldots$). As the partial sums oscillate back and forth, they should be drawing closer and closer to some limit S, somewhere between all of the even-indexed partial sums and the odd-indexed partial sums,

$$S_2 < S_4 < S_6 < \cdots < S < \cdots < S_5 < S_3 < S_1. \tag{4.1}$$

Proof. Notice from Figure 8.32 that the even- and odd-indexed partial sums seem to behave somewhat differently. First, we consider the even-indexed partial sums. We have

$$S_2 = a_1 - a_2 > 0$$

and

$$S_4 = S_2 + (a_3 - a_4) \ge S_2,$$

since $(a_3 - a_4) \ge 0$. Likewise, for any n, we can write

$$S_{2n} = S_{2n-2} + (a_{2n-1} - a_{2n}) \ge S_{2n-2},$$

since $(a_{2n-1} - a_{2n}) \ge 0$. This says that the sequence of even-indexed partial sums $\{S_{2n}\}_{n=1}^{\infty}$ is increasing (as we saw in Figure 8.32). Further, observe that

$$0 < S_{2n} = a_1 + (-a_2 + a_3) + (-a_4 + a_5) + \cdots + (-a_{2n-2} + a_{2n-1}) - a_{2n} \le a_1,$$

for all n, since every term in parentheses is negative. Thus, $\{S_{2n}\}_{n=1}^{\infty}$ is both bounded (by a_1) and monotonic (increasing). By Theorem 1.4, $\{S_{2n}\}_{n=1}^{\infty}$ must be convergent to some number, say L.

Turning to the sequence of odd-indexed partial sums, notice that we have

$$S_{2n+1} = S_{2n} + a_{2n+1}.$$

From this, we have

$$\lim_{n \to \infty} S_{2n+1} = \lim_{n \to \infty} (S_{2n} + a_{2n+1}) = \lim_{n \to \infty} S_{2n} + \lim_{n \to \infty} a_{2n+1} = L + 0 = L,$$

since $\lim_{n \to \infty} a_n = 0$. Since both the sequence of odd-indexed partial sums $\{S_{2n+1}\}_{n=0}^{\infty}$ and the sequence of even-indexed partial sums $\{S_{2n}\}_{n=1}^{\infty}$ converge to the same limit, L, we have that

$$\lim_{n \to \infty} S_n = L,$$

also.

■

Example 4.2   Using the Alternating Series Test

Reconsider the convergence of the alternating harmonic series $\displaystyle\sum_{k=1}^{\infty} \frac{(-1)^{k+1}}{k}$.

Solution   Notice that

$$\lim_{k \to \infty} a_k = \lim_{k \to \infty} \frac{1}{k} = 0.$$

Further,

$$0 < a_{k+1} = \frac{1}{k+1} \le \frac{1}{k} = a_k, \text{ for all } k \ge 1.$$

By the Alternating Series Test, the series converges. (You can use the calculations from example 4.1 to arrive at an approximate sum.)

■

The Alternating Series Test is certainly the easiest test we've discussed so far for determining the convergence of a series. It's straightforward, but you will sometimes need to work a bit to verify the conditions.

FIGURE 8.33:
$$S_n = \sum_{k=1}^{n} \frac{(-1)^k (k+3)}{k(k+1)}.$$

Example 4.3   Using the Alternating Series Test

Investigate the convergence or divergence of the alternating series

$$\sum_{k=1}^{\infty} \frac{(-1)^k (k+3)}{k(k+1)}.$$

Solution   The graph of the first 20 partial sums seen in Figure 8.33 suggests that the series converges to some value around -1.5. A table of values of the partial sums suggests the same conclusion.

n	$S_n = \sum_{k=1}^{n} \frac{(-1)^k(k+3)}{k(k+1)}$	n	$S_n = \sum_{k=1}^{n} \frac{(-1)^k(k+3)}{k(k+1)}$
50	-1.45545	51	-1.47581
100	-1.46066	101	-1.47076
150	-1.46237	151	-1.46908
200	-1.46322	201	-1.46824
250	-1.46372	251	-1.46774
300	-1.46406	301	-1.46741
350	-1.4643	351	-1.46717
400	-1.46448	401	-1.46699
450	-1.46462	451	-1.46685
500	-1.46473	501	-1.46674

We can verify that the series converges by first checking that

$$\lim_{k \to \infty} a_k = \lim_{k \to \infty} \frac{(k+3)}{k(k+1)} \frac{\frac{1}{k^2}}{\frac{1}{k^2}} = \lim_{k \to \infty} \frac{\frac{1}{k} + \frac{3}{k^2}}{1 + \frac{1}{k}} = 0.$$

Next, consider the ratio of two consecutive terms:

$$\frac{a_{k+1}}{a_k} = \frac{(k+4)}{(k+1)(k+2)} \frac{k(k+1)}{(k+3)} = \frac{k^2 + 4k}{k^2 + 5k + 6} < 1,$$

for all $k \geq 1$. From this, it follows that $a_{k+1} < a_k$, for all $k \geq 1$ and so, by the Alternating Series Test, the series converges. Finally, from the preceding table, we can see that the series converges to a sum between -1.46473 and -1.46674. (How can you be sure that the sum is in this interval?)

■

Example 4.4　A Divergent Alternating Series

Determine whether the alternating series $\sum_{k=1}^{\infty} \frac{(-1)^k \, k}{k+2}$ converges or diverges.

Solution　First, notice that

$$\lim_{k \to \infty} a_k = \lim_{k \to \infty} \frac{k}{k+2} = 1 \neq 0.$$

So, this alternating series is divergent, since by the kth term test for divergence, the terms must tend to zero, in order for the series to be convergent.

■

Estimating the Sum of an Alternating Series

We have repeatedly remarked that once you know that a series converges, you can always approximate the sum of the series by computing some partial sums. However, in finding an approximate sum of a convergent series, how close is close enough? Realize that answers to such questions of accuracy are not "one size fits all", but rather are highly context-sensitive. For instance, if the sum of the series is to be used to find the angle from the ground at which you throw a ball from right field to third base, you might accept one answer. On the other hand, if the sum of that same series is to be used to find the angle at which to aim your spacecraft to ensure a safe reentry into the earth's atmosphere, you would likely insist on greater precision (at least, we would).

So far, we have calculated approximate sums of series by observing that a number of successive partial sums of the series are within a given distance of one another. The underlying assumption here is that when this happens, the partial sums are also within that same distance of the sum of the series. Unfortunately, this is simply not true, in general (although it is true for some series). What's a mathematician to do? For the case of alternating series, we are quite fortunate to have available a simple way to get a handle on the accuracy. Note that the error in approximating the sum S by the nth partial sum S_n is $S - S_n$.

Take a look back at Figure 8.32. Recall that we had observed from the figure that all of the even-indexed partial sums, S_n of the convergent alternating series, $\sum_{k=1}^{\infty} (-1)^{k+1} a_k$ lie below the sum, S, while all of the odd-indexed partial sums all lie above S, that is, [as in (4.1)],

$$S_2 < S_4 < S_6 < \cdots < S < \cdots < S_5 < S_3 < S_1.$$

This says that for n even,

$$S_n \leq S \leq S_{n+1}.$$

Subtracting S_n from all terms, we get

$$0 \leq S - S_n \leq S_{n+1} - S_n = a_{n+1}.$$

Since $a_{n+1} > 0$, we have

$$-a_{n+1} \leq 0 \leq S - S_n \leq a_{n+1},$$

or

$$|S - S_n| \leq a_{n+1}, \text{ for } n \text{ even.} \tag{4.2}$$

Similarly, for n odd, we have that

$$S_{n+1} \leq S \leq S_n.$$

Again subtracting S_n, we get

$$-a_{n+1} = S_{n+1} - S_n \leq S - S_n \leq 0 \leq a_{n+1}$$

or

$$|S - S_n| \leq a_{n+1}, \text{ for } n \text{ odd.} \tag{4.3}$$

Since (4.2) and (4.3) (these are called **error bounds**) are the same, we have the same error bound whether n is even or odd. This establishes the following result.

Theorem 4.2   Suppose that $\lim_{k \to \infty} a_k = 0$ and $0 < a_{k+1} \leq a_k$ for all $k \geq 1$. Then, the alternating series $\sum_{k=1}^{\infty} (-1)^{k+1} a_k$ converges to some number S and the error in approximating S by the nth partial sum S_n satisfies

$$|S - S_n| \leq a_{n+1}. \tag{4.4}$$

Notice that this says that the absolute value of the error in approximating S by S_n does not exceed a_{n+1} (the absolute value of the first neglected term).

Example 4.5   Estimating the Sum of an Alternating Series

Approximate the sum of the alternating series $\sum_{k=1}^{\infty} \frac{(-1)^{k+1}}{k^4}$ by the 40th partial sum and estimate the error in this approximation.

Solution   We leave it as an exercise to show that this series is convergent. We then approximate the sum by

$$S \approx S_{40} = 0.9470326439.$$

From our error estimate (4.4), we have

$$|S - S_{40}| \leq a_{41} = \frac{1}{41^4} = 3.54 \times 10^{-7}.$$

This says that our approximation $S \approx 0.9470326439$ is off by no more than $\pm 3.54 \times 10^{-7}$.

∎

A much more interesting question than the one asked in example 4.5 is the following. For a given convergent alternating series, how many terms must we take in order to guarantee that our approximation is accurate to a given level? We use the same estimate of error from (4.4) to answer this question, as in the following example.

Example 4.6   Finding the Number of Terms Needed for a Given Accuracy

For the convergent alternating series $\sum_{k=1}^{\infty} \frac{(-1)^{k+1}}{k^4}$, how many terms are needed to guarantee that S_n is within 1×10^{-10} of the actual sum S?

Solution   In this case, we want to find the number of terms, n for which

$$|S - S_n| \leq 1 \times 10^{-10}.$$

From (4.4), we have that

$$|S - S_n| \leq a_{n+1} = \frac{1}{(n+1)^4}.$$

So, we look for n so that

$$\frac{1}{(n+1)^4} \leq 1 \times 10^{-10}.$$

Solving for n, we get

$$10^{10} \leq (n+1)^4,$$

so that

$$\sqrt[4]{10^{10}} \leq n + 1$$

or

$$n \geq \sqrt[4]{10^{10}} - 1 \approx 315.2.$$

So, if we take $n \geq 316$, we will guarantee an error of no more than 1×10^{-10}. Using the suggested number of terms, we get the approximate sum

$$S \approx S_{316} = 0.947032829447,$$

which we now know to be correct to within 1×10^{-10}.

∎

Exercises 8.4

1. (writing) If $a_k \geq 0$, explain in terms of partial sums why $\displaystyle\sum_{k=1}^{\infty}(-1)^{k+1}a_k$ is more likely to converge than $\displaystyle\sum_{k=1}^{\infty}a_k$.

2. (writing) Explain why in Theorem 4.2 we need the assumption that $a_{k+1} \leq a_k$. That is, what would go wrong with the proof if $a_{k+1} > a_k$?

3. (writing) The Alternating Series Test was stated for the series $\displaystyle\sum_{k=1}^{\infty}(-1)^{k+1}a_k$. Explain the difference between $\displaystyle\sum_{k=1}^{\infty}(-1)^{k}a_k$ and $\displaystyle\sum_{k=1}^{\infty}(-1)^{k+1}a_k$ and explain why we could have stated the theorem for $\displaystyle\sum_{k=1}^{\infty}(-1)^{k}a_k$.

4. (writing) A common mistake is to think that if $\displaystyle\lim_{k\to\infty} a_k = 0$, then $\displaystyle\sum_{k=1}^{\infty}a_k$ converges. Explain why this is not true for positive-term series. This is also not true for alternating series **unless** you add one more hypothesis. State the extra hypothesis and explain why it's needed.

In exercises 5–30, determine if the series is convergent or divergent.

5. $\displaystyle\sum_{k=1}^{\infty}(-1)^{k+1}\frac{3}{k}$

6. $\displaystyle\sum_{k=1}^{\infty}(-1)^{k+1}\frac{4}{k+1}$

7. $\displaystyle\sum_{k=1}^{\infty}(-1)^{k}\frac{2}{k^2}$

8. $\displaystyle\sum_{k=1}^{\infty}(-1)^{k}\frac{4}{\sqrt{k}}$

9. $\displaystyle\sum_{k=1}^{\infty}(-1)^{k+1}\frac{k^2}{k+1}$

10. $\displaystyle\sum_{k=1}^{\infty}(-1)^{k+1}\frac{2k^2-1}{k}$

11. $\displaystyle\sum_{k=1}^{\infty}(-1)^{k}\frac{k}{k^2+2}$

12. $\displaystyle\sum_{k=1}^{\infty}(-1)^{k}\frac{2k-1}{k^3}$

13. $\displaystyle\sum_{k=1}^{\infty}(-1)^{k+1}\frac{k}{2^k}$

14. $\displaystyle\sum_{k=1}^{\infty}(-1)^{k+1}\frac{3^k}{k}$

15. $\displaystyle\sum_{k=1}^{\infty}(-1)^{k}\frac{4^k}{k^2}$

16. $\displaystyle\sum_{k=1}^{\infty}(-1)^{k}\frac{k+2}{4^k}$

17. $\displaystyle\sum_{k=1}^{\infty}\frac{2k}{k+1}$

18. $\displaystyle\sum_{k=1}^{\infty}\frac{4k^2}{k^2+2k+2}$

19. $\displaystyle\sum_{k=1}^{\infty}(-1)^{k}\frac{3}{\sqrt{k+1}}$

20. $\displaystyle\sum_{k=1}^{\infty}(-1)^{k}\frac{k+1}{k^3}$

21. $\displaystyle\sum_{k=1}^{\infty}(-1)^{k+1}\frac{2}{k!}$

22. $\displaystyle\sum_{k=1}^{\infty}(-1)^{k+1}\frac{k!}{3^k}$

23. $\displaystyle\sum_{k=1}^{\infty}(-1)^{k}\frac{k!}{2^k}$

24. $\displaystyle\sum_{k=1}^{\infty}(-1)^{k}\frac{4^k}{k!}$

25. $\displaystyle\sum_{k=0}^{\infty}(-1)^{k+1}2e^{-k}$

26. $\displaystyle\sum_{k=0}^{\infty}(-1)^{k+1}3e^{1/k}$

27. $\displaystyle\sum_{k=2}^{\infty}(-1)^{k}\ln k$

28. $\displaystyle\sum_{k=2}^{\infty}(-1)^{k}\frac{1}{\ln k}$

29. $\displaystyle\sum_{k=0}^{\infty}(-1)^{k+1}\frac{1}{2^k}$

30. $\displaystyle\sum_{k=0}^{\infty}(-1)^{k+1}2^k$

In exercises 31–38, estimate the sum of each convergent series to within 0.01.

31. $\displaystyle\sum_{k=1}^{\infty}(-1)^{k+1}\frac{4}{k^3}$

32. $\displaystyle\sum_{k=1}^{\infty}(-1)^{k+1}\frac{2}{k^3}$

33. $\displaystyle\sum_{k=1}^{\infty}(-1)^{k}\frac{k}{2^k}$

34. $\displaystyle\sum_{k=1}^{\infty}(-1)^{k}\frac{k^2}{10^k}$

35. $\displaystyle\sum_{k=0}^{\infty}(-1)^{k}\frac{3}{k!}$

36. $\displaystyle\sum_{k=0}^{\infty}(-1)^{k+1}\frac{2}{k!}$

37. $\displaystyle\sum_{k=1}^{\infty}(-1)^{k+1}\frac{4}{k^4}$

38. $\displaystyle\sum_{k=1}^{\infty}(-1)^{k+1}\frac{3}{k^5}$

In exercises 39–44, determine how many terms are needed to estimate the sum of the series to within 0.0001.

39. $\displaystyle\sum_{k=1}^{\infty}(-1)^{k+1}\frac{2}{k}$

40. $\displaystyle\sum_{k=1}^{\infty}(-1)^{k+1}\frac{4}{\sqrt{k}}$

41. $\displaystyle\sum_{k=0}^{\infty}(-1)^{k}\frac{2^k}{k!}$

42. $\displaystyle\sum_{k=0}^{\infty}(-1)^{k}\frac{10^k}{k!}$

43. $\displaystyle\sum_{k=1}^{\infty}(-1)^{k+1}\frac{k!}{k^k}$

44. $\displaystyle\sum_{k=1}^{\infty}(-1)^{k+1}\frac{4^k}{k^k}$

45. In the text, we showed you one way to verify that a sequence is decreasing. As an alternative, explain why if $a_k = f(k)$ and $f'(k) < 0$, then the sequence a_k is decreasing. Use this method to prove that $a_k = \dfrac{k}{k^2+2}$ is decreasing.

46. Use the method of exercise 45 to prove that $a_k = \dfrac{k}{2^k}$ is decreasing.

47. In this exercise, you will discover why the Alternating Series Test requires that $a_{k+1} \leq a_k$. If $a_k = \begin{cases} 1/k & \text{if } k \text{ is odd} \\ 1/k^2 & \text{if } k \text{ is even} \end{cases}$, argue that $\sum_{k=1}^{\infty} (-1)^{k+1} a_k$ diverges to ∞. Thus, an alternating series can diverge even if $\lim_{k \to \infty} a_k = 0$.

48. Verify that the series $\sum_{k=0}^{\infty} (-1)^k \frac{1}{2k+1} = 1 - \frac{1}{3} + \frac{1}{5} - \frac{1}{7} + \cdots$ converges. It can be shown that the sum of this series is $\frac{\pi}{4}$. Given this result, we could use this series to obtain an approximation of π. How many terms would be necessary to get eight digits of π correct?

49. A person starts walking from home (at $x = 0$) toward a friend's house (at $x = 1$). Three-fourths of the way there, he changes his mind and starts walking back home. Three-fourths of the way home, he changes his mind again and starts walking back to his friend's house. If he continues this pattern of indecision, always turning around at the three-fourths mark, what will be the eventual outcome? A similar problem appeared in a national magazine and created a minor controversy due to the ambiguous wording of the problem. It is clear that the first turnaround is at $x = \frac{3}{4}$ and the second turnaround is at $\frac{3}{4} - \frac{3}{4} \left(\frac{3}{4} \right) = \frac{3}{16}$. But is the third turnaround three-fourths of the way to $x = 1$ or $x = \frac{3}{4}$? The magazine writer assumed the latter. Show that with this assumption, the person's location forms a geometric series. Find the sum of the series to find where the person ends up.

50. If the problem of exercise 49 is interpreted differently, a more interesting answer results. As before, let $x_1 = \frac{3}{4}$ and

$x_2 = \frac{3}{16}$. If the next turnaround is three-fourths of the way from x_2 to 1, then $x_3 = \frac{3}{16} + \frac{3}{4} \left(1 - \frac{3}{16} \right) = \frac{3}{4} + \frac{1}{4} x_2 = \frac{51}{64}$. Three-fourths of the way back to $x = 0$ would put us at $x_4 = x_3 - \frac{3}{4} x_3 = \frac{1}{4} x_3 = \frac{51}{264}$. Show that if n is even, then $x_{n+1} = \frac{3}{4} + \frac{1}{4} x_n$ and $x_{n+2} = \frac{1}{4} x_{n+1}$. Show that the person ends up walking back and forth between two specific locations.

51. (exploratory) In this exercise, you will determine whether or not the improper integral $\int_0^1 \sin(1/x)\, dx$ converges. Argue that $\int_{1/\pi}^1 \sin(1/x)\, dx$, $\int_{1/(2\pi)}^{1/\pi} \sin(1/x)\, dx$, $\int_{1/(3\pi)}^{1/(2\pi)} \sin(1/x)\, dx$, ... exist and that (if it exists),

$$\int_0^1 \sin(1/x)\, dx = \int_{1/\pi}^1 \sin(1/x)\, dx + \int_{1/(2\pi)}^{1/\pi} \sin(1/x)\, dx$$
$$+ \int_{1/(3\pi)}^{1/(2\pi)} \sin(1/x)\, dx + \cdots$$

Verify that the series is an alternating series and show that the hypotheses of the Alternating Series Test are met. Thus, the series and the improper integral both converge.

52. (exploratory) Consider the series $\sum_{k=1}^{\infty} (-1)^k \frac{x^k}{k}$ where x is a constant. Show that the series converges for $x = 1/2$; $x = -1/2$; any x such that $-1 < x \leq 1$. Show that the series diverges if $x = -1$, $x < -1$ or $x > 1$. We see in section 8.6 that when the series converges, it converges to $\ln(1+x)$. Verify this numerically for $x = 1/2$ and $x = -1/2$.

8.5   Absolute Convergence and the Ratio Test

You should note that, outside of the Alternating Series Test presented in section 8.4, our other tests for convergence of series (i.e., the Integral Test and the two comparison tests) apply only to series all of whose terms are **positive**. So, what do we do if we're faced with a series that has both positive and negative terms, but that is not an alternating series? For instance, look at the series

$$\sum_{k=1}^{\infty} \frac{\sin k}{k^3} = \sin 1 + \frac{1}{8} \sin 2 + \frac{1}{27} \sin 3 + \frac{1}{64} \sin 4 + \cdots$$

This has both positive and negative terms, but the terms do not alternate signs. (Calculate the first five or six terms of the series to see this for yourself.) For any such series $\sum_{k=1}^{\infty} a_k$, we can get around this problem by checking if the series of absolute values, $\sum_{k=1}^{\infty} |a_k|$ is convergent. When this happens, we say that the original series $\sum_{k=1}^{\infty} a_k$ is **absolutely convergent** (or **converges absolutely**). You should note that to test the convergence of the series of absolute values, $\sum_{k=1}^{\infty} |a_k|$ (all of whose terms are positive), we have all of our earlier tests for positive term series available to us. How does this help us with the original series? In Theorem 5.1, we see that every absolutely convergent series is also convergent.

So, if we can establish that a given series converges absolutely, then we also know that it converges. Absolute convergence also has other significant implications, as we see later.

Example 5.1   Testing for Absolute Convergence

Determine if $\displaystyle\sum_{k=1}^{\infty} \frac{(-1)^{k+1}}{2^k}$ is absolutely convergent.

Solution   It is easy to show that this alternating series is convergent. (Try it!) From the graph of the first 20 partial sums in Figure 8.34, it appears that the series converges to approximately 0.35. To determine absolute convergence, we need to determine whether or not the series of absolute values, $\displaystyle\sum_{k=1}^{\infty} \left| \frac{(-1)^{k+1}}{2^k} \right|$ is convergent. We have

FIGURE 8.34:
$$S_n = \sum_{k=1}^{n} \frac{(-1)^{k+1}}{2^k}.$$

$$\sum_{k=1}^{\infty} \left| \frac{(-1)^{k+1}}{2^k} \right| = \sum_{k=1}^{\infty} \frac{1}{2^k} = \sum_{k=1}^{\infty} \left(\frac{1}{2} \right)^k,$$

which you should recognize as a convergent geometric series ($|r| = \frac{1}{2} < 1$). This says that the original series $\displaystyle\sum_{k=1}^{\infty} \left| \frac{(-1)^{k+1}}{2^k} \right|$ converges absolutely.

■

You might be wondering what the relationship is between convergence and absolute convergence. We'll prove shortly that every absolutely convergent series is also convergent (as in example 5.1). However, the reverse is not true; there are many series that are convergent, but not absolutely convergent. These are called **conditionally convergent** series. Can you think of an example of such a series? If so, it's probably the example that follows.

Example 5.2   A Conditionally Convergent Series

Determine if the alternating harmonic series $\displaystyle\sum_{k=1}^{\infty} \frac{(-1)^{k+1}}{k}$ is absolutely convergent.

Solution   In example 4.2, we showed that this series is convergent. To test this for absolute convergence, we consider the series of absolute values,

$$\sum_{k=1}^{\infty} \left| \frac{(-1)^{k+1}}{k} \right| = \sum_{k=1}^{\infty} \frac{1}{k},$$

which is the harmonic series. We showed in section 8.2 (example 2.7) that the harmonic series diverges. This says that $\displaystyle\sum_{k=1}^{\infty} \frac{(-1)^{k+1}}{k}$ converges, but does not converge absolutely (i.e., it converges conditionally).

■

Theorem 5.1   If $\displaystyle\sum_{k=1}^{\infty} |a_k|$ converges, then $\displaystyle\sum_{k=1}^{\infty} a_k$ converges.

This result says that if a series converges absolutely, then it must also converge. Because of this, when we test series, we first test for absolute convergence. If the series converges absolutely, then we need not test any further to establish convergence.

Proof. Notice that for any real number, x, we can say that $-|x| \leq x \leq |x|$. So, for any k, we have

$$-|a_k| \leq a_k \leq |a_k|.$$

Adding $|a_k|$ to all the terms, we get

$$0 \leq a_k + |a_k| \leq 2 |a_k|. \tag{5.1}$$

Since $\sum\limits_{k=1}^{\infty} a_k$ is absolutely convergent, we have that $\sum\limits_{k=1}^{\infty} |a_k|$ and hence, also $\sum\limits_{k=1}^{\infty} 2 |a_k| = 2 \sum\limits_{k=1}^{\infty} |a_k|$ is convergent. Define $b_k = a_k + |a_k|$. From (5.1),

$$0 \leq b_k \leq 2 |a_k|,$$

and so, by the Comparison Test, $\sum\limits_{k=1}^{\infty} b_k$ is convergent. Observe that we may write

$$\sum_{k=1}^{\infty} a_k = \sum_{k=1}^{\infty} (a_k + |a_k| - |a_k|) = \sum_{k=1}^{\infty} \underbrace{(a_k + |a_k|)}_{b_k} - \sum_{k=1}^{\infty} |a_k|$$

$$= \sum_{k=1}^{\infty} b_k - \sum_{k=1}^{\infty} |a_k|.$$

Since the two series on the right-hand side are convergent, it follows that $\sum\limits_{k=1}^{\infty} a_k$ must also be convergent.

∎

FIGURE 8.35:
$$S_n = \sum_{k=1}^{n} \frac{\sin k}{k^3}.$$

Example 5.3   Testing for Absolute Convergence

Determine whether $\sum\limits_{k=1}^{\infty} \dfrac{\sin k}{k^3}$ is convergent or divergent.

Solution   Notice that while this is not a positive term series, it is also not an alternating series. Because of this, our only choice (given what we know) is to test the series for absolute convergence. From the graph of the first 20 partial sums seen in Figure 8.35, it appears that the series is converging to some value around 0.94. To test for absolute convergence, we consider the series of absolute values, $\sum\limits_{k=1}^{\infty} \left| \dfrac{\sin k}{k^3} \right|$. Notice that

$$\left| \frac{\sin k}{k^3} \right| = \frac{|\sin k|}{k^3} \leq \frac{1}{k^3}, \tag{5.2}$$

since $|\sin k| \leq 1$, for all k. Of course, $\sum\limits_{k=1}^{\infty} \dfrac{1}{k^3}$ is a convergent p-series ($p = 3 > 1$). By the Comparison Test and (5.2), $\sum\limits_{k=1}^{\infty} \left| \dfrac{\sin k}{k^3} \right|$ converges too and so, the original series $\sum\limits_{k=1}^{\infty} \dfrac{\sin k}{k^3}$ converges absolutely (and hence, converges also).

∎

The Ratio Test

We now introduce a very powerful tool for testing a series for absolute convergence. This test can be applied to a wide range of series, including the extremely important case of power series that we discuss in section 8.6. As you'll see, this test is also remarkably easy to use.

Theorem 5.2   **(Ratio Test)** Given $\sum\limits_{k=1}^{\infty} a_k$ ($a_k \neq 0$, for all k), suppose that

$$\lim_{k \to \infty} \left| \frac{a_{k+1}}{a_k} \right| = L.$$

Then,

(i)   if $L < 1$, the series converges absolutely,

(ii) if $L > 1$ (or $L = \infty$), the series diverges and

(iii) if $L = 1$, there is no conclusion.

Proof. (i) For $L < 1$, pick any number r with $L < r < 1$. Then, we have

$$\lim_{k \to \infty} \left| \frac{a_{k+1}}{a_k} \right| = L < r.$$

For this to occur, there must be some number $N > 0$, such that for $k \geq N$,

$$\left| \frac{a_{k+1}}{a_k} \right| < r. \tag{5.3}$$

Multiplying both sides of (5.3) by $|a_k|$, gives us

$$|a_{k+1}| < r \, |a_k|.$$

In particular, taking $k = N$ gives us

$$|a_{N+1}| < r \, |a_N|$$

and taking $k = N + 1$ gives us

$$|a_{N+2}| < r \, |a_{N+1}| < r^2 \, |a_N|.$$

Likewise,

$$|a_{N+3}| < r \, |a_{N+2}| < r^3 \, |a_N|$$

and so on. We have

$$|a_{N+k}| < r^k \, |a_N|, \text{ for } k = 1, 2, 3, \ldots.$$

Notice that $\sum\limits_{k=1}^{\infty} |a_N| \, r^k = |a_N| \sum\limits_{k=1}^{\infty} r^k$ is a convergent geometric series, since $0 < r < 1$.

By the Comparison Test, it follows that $\sum\limits_{k=1}^{\infty} |a_{N+k}| = \sum\limits_{n=N+1}^{\infty} |a_n|$ converges, too. This

says that $\sum\limits_{n=N+1}^{\infty} a_n$ converges absolutely. Finally, since

$$\sum_{n=1}^{\infty} a_n = \sum_{n=1}^{N} a_n + \sum_{n=N+1}^{\infty} a_n,$$

we also get that $\sum\limits_{n=1}^{\infty} a_n$ converges absolutely.

(ii) For $L > 1$, we have

$$\lim_{k \to \infty} \left| \frac{a_{k+1}}{a_k} \right| = L > 1.$$

This says that there must be some number $N > 0$, such that for $k \geq N$,

$$\left| \frac{a_{k+1}}{a_k} \right| > 1. \tag{5.4}$$

Multiplying both sides of (5.4) by $|a_k|$, we get

$$|a_{k+1}| > |a_k| > 0, \text{ for all } k \geq N.$$

Notice that if this is the case, then

$$\lim_{k \to \infty} a_k \neq 0.$$

By the kth-term test for divergence, we have that $\sum\limits_{k=1}^{\infty} a_k$ diverges.

◼

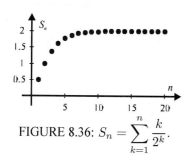

FIGURE 8.36: $S_n = \sum\limits_{k=1}^{n} \dfrac{k}{2^k}$.

Since $2^{k+1} = 2^k \cdot 2^1$.

Example 5.4   Using the Ratio Test

Test $\sum\limits_{k=1}^{\infty} \dfrac{(-1)^k k}{2^k}$ for convergence.

Solution   From the graph of the first 20 partial sums of the series of absolute values, $\sum\limits_{k=1}^{\infty} \dfrac{k}{2^k}$ seen in Figure 8.36, it appears that the series of absolute values converges to about 2. From the Ratio Test, we have

$$\lim_{k \to \infty} \left| \frac{a_{k+1}}{a_k} \right| = \lim_{k \to \infty} \frac{\frac{k+1}{2^{k+1}}}{\frac{k}{2^k}} = \lim_{k \to \infty} \frac{k+1}{2^{k+1}} \frac{2^k}{k} = \frac{1}{2} \lim_{k \to \infty} \frac{k+1}{k} = \frac{1}{2} < 1$$

and so, the series converges absolutely, as expected from Figure 8.36.

◼

The Ratio Test is particularly useful when the general term of a series contains an exponential term, as in example 5.4 or a factorial, as in the following example.

FIGURE 8.37:
$$S_n = \sum\limits_{k=0}^{n-1} \frac{(-1)^k k!}{e^k}.$$

Since $(k+1)! = (k+1) \cdot k!$ and $e^{k+1} = e^k \cdot e^1$.

Example 5.5   Using the Ratio Test

Test $\sum\limits_{k=0}^{\infty} \dfrac{(-1)^k k!}{e^k}$ for convergence.

Solution   From the graph of the first 20 partial sums of the series seen in Figure 8.37, it appears that the series is diverging. (Look closely at the scale on the y-axis and compute a table of values for yourself.) We can confirm this suspicion with the Ratio Test. We have

$$\lim_{k \to \infty} \left| \frac{a_{k+1}}{a_k} \right| = \lim_{k \to \infty} \frac{\frac{(k+1)!}{e^{k+1}}}{\frac{k!}{e^k}} = \lim_{k \to \infty} \frac{(k+1)!}{e^{k+1}} \frac{e^k}{k!}$$

$$= \lim_{k \to \infty} \frac{(k+1) k!}{e \, k!} = \frac{1}{e} \lim_{k \to \infty} \frac{k+1}{1} = \infty.$$

By the Ratio Test, the series diverges, as we suspected.

◼

Recall that in the statement of the Ratio Test (Theorem 5.2), we said that if

$$\lim_{k \to \infty} \left| \frac{a_{k+1}}{a_k} \right| = 1,$$

then the Ratio Test yields no conclusion. By this, we mean that in such cases, the series may or may not converge and further testing is required.

Example 5.6   **A Divergent Series for Which the Ratio Test Fails**

Use the Ratio Test for the harmonic series $\displaystyle\sum_{k=0}^{\infty} \frac{1}{k}$.

Solution   We have

$$\lim_{k \to \infty} \left| \frac{a_{k+1}}{a_k} \right| = \lim_{k \to \infty} \frac{\dfrac{1}{k+1}}{\dfrac{1}{k}} = \lim_{k \to \infty} \frac{k}{k+1} = 1.$$

In this case, the Ratio Test yields no conclusion, although we already know that the harmonic series diverges.

∎

Example 5.7   **A Convergent Series for Which the Ratio Test Fails**

Use the Ratio Test to test the series $\displaystyle\sum_{k=0}^{\infty} \frac{1}{k^2}$.

Solution   Here, we have

$$\lim_{k \to \infty} \left| \frac{a_{k+1}}{a_k} \right| = \lim_{k \to \infty} \frac{1}{(k+1)^2} \frac{k^2}{1}$$

$$= \lim_{k \to \infty} \frac{k^2}{k^2 + 2k + 1} = 1.$$

So again, the Ratio Test yields no conclusion, although we already know that this is a convergent p-series.

∎

By this point in your study of series, it may seem as if we have thrown at you a dizzying array of different series and tests for convergence or divergence. Just how are you to keep all of these straight? The only suggestion we have is that you work through **many** problems. We provide a good assortment in the exercise set which follows this section. Some of these require the methods of this section, while others are drawn from the preceding sections (just to keep you thinking about the big picture). For the sake of convenience, we summarize our convergence tests in the table that follows.

Test	When to use	Conclusions	Section
Geometric Series	$$\sum_{k=0}^{\infty} a\,r^k$$	Converges to $\dfrac{a}{1-r}$ if $\lvert r \rvert < 1$; diverges if $\lvert r \rvert \geq 1$	8.2
*k*th **Term Test**	all series	If $\lim\limits_{k\to\infty} a_k \neq 0$, the series diverges	8.2
Integral Test	$$\sum_{k=1}^{\infty} a_k \text{ where } f(k) = a_k \text{ and }$$ f is continuous, decreasing and $f(x) \geq 0$	$$\sum_{k=1}^{\infty} a_k \text{ and } \int_1^{\infty} f(x)dx$$ both converge or both diverge	8.3
p-**series**	$$\sum_{k=1}^{\infty} \frac{1}{k^p}$$	Converges for $p > 1$; diverges for $p \leq 1$	8.3
Comparison Test	$$\sum_{k=1}^{\infty} a_k \text{ and } \sum_{k=1}^{\infty} b_k,$$ where $0 \leq a_k \leq b_k$	If $\sum\limits_{k=1}^{\infty} b_k$ converges, then $\sum\limits_{k=1}^{\infty} a_k$ converges. If $\sum\limits_{k=1}^{\infty} a_k$ diverges, then $\sum\limits_{k=1}^{\infty} b_k$ diverges.	8.3
Limit Comparison Test	$$\sum_{k=1}^{\infty} a_k \text{ and } \sum_{k=1}^{\infty} b_k, \text{ where}$$ $a_k, b_k > 0$ and $\lim\limits_{k\to\infty} \dfrac{a_k}{b_k} = L > 0$	$$\sum_{k=1}^{\infty} a_k \text{ and } \sum_{k=1}^{\infty} b_k$$ **both** converge or **both** diverge.	8.3
Alternating Series Test	$$\sum_{k=1}^{\infty} (-1)^k a_k$$ where $a_k > 0$ for all k	If $\lim\limits_{k\to\infty} a_k = 0$ and $a_{k+1} \leq a_k$ for all k, then the series converges.	8.4
Absolute Convergence	Series with some positive and some negative terms (including alternating series)	If $\sum\limits_{k=1}^{\infty} \lvert a_k \rvert$ converges, then $\sum\limits_{k=1}^{\infty} a_k$ converges (absolutely).	8.5
Ratio Test	Any series (especially those involving exponentials and/or factorials)	For $\lim\limits_{k\to\infty} \dfrac{a_{k+1}}{a_k} = L$, if $L < 1$, $\sum\limits_{k=1}^{\infty} a_k$ converges absolutely if $L > 1$, $\sum\limits_{k=1}^{\infty} a_k$ diverges, if $L = 1$, no conclusion.	8.5

Exercises 8.5

1. (writing) Suppose that two series have identical terms except that in series A all terms are positive and in series B some terms are positive and some terms are negative. Explain why series B is more likely to converge. In light of this, explain why Theorem 5.1 is true.

2. (writing) In the Ratio Test, if $\lim\limits_{k\to\infty} \left\lvert \dfrac{a_{k+1}}{a_k} \right\rvert > 1$, which is bigger, $\lvert a_{k+1} \rvert$ or $\lvert a_k \rvert$? Explain why this implies that the series $\sum\limits_{k=1}^{\infty} a_k$ diverges.

3. (writing) In the Ratio Test, if $\lim\limits_{k\to\infty} \left\lvert \dfrac{a_{k+1}}{a_k} \right\rvert = L < 1$, which is bigger, $\lvert a_{k+1} \rvert$ or $\lvert a_k \rvert$? This inequality could also hold if $L = 1$. Compare the relative sizes of $\lvert a_{k+1} \rvert$ and $\lvert a_k \rvert$ if $L = 0.8$ versus $L = 1$. Explain why $L = 0.8$ would be more likely to correspond to a convergent series than $L = 1$.

4. (writing) In many series of interest, the terms of the series involve powers of k (e.g., k^2), exponentials (e.g., 2^k) or factorials (e.g., $k!$). For which type(s) of terms is the Ratio Test likely to produce a result (i.e., a limit different than 1)? Briefly explain.

In exercises 5–36, determine if the series is absolutely convergent, conditionally convergent or divergent.

5. $\displaystyle\sum_{k=0}^{\infty} (-1)^k \frac{3}{k!}$

6. $\displaystyle\sum_{k=0}^{\infty} (-1)^k \frac{6}{k!}$

7. $\displaystyle\sum_{k=0}^{\infty} (-1)^k 2^k$

8. $\displaystyle\sum_{k=0}^{\infty} (-1)^k \frac{2}{3^k}$

9. $\displaystyle\sum_{k=1}^{\infty} (-1)^{k+1} \frac{k}{k^2+1}$

10. $\displaystyle\sum_{k=1}^{\infty} (-1)^{k+1} \frac{k^2+1}{k}$

11. $\displaystyle\sum_{k=0}^{\infty} (-1)^k \frac{3^k}{k!}$

12. $\displaystyle\sum_{k=0}^{\infty} (-1)^k \frac{10^k}{k!}$

13. $\displaystyle\sum_{k=1}^{\infty} (-1)^{k+1} \frac{k}{2k+1}$

14. $\displaystyle\sum_{k=1}^{\infty} (-1)^{k+1} \frac{4}{2k+1}$

15. $\displaystyle\sum_{k=1}^{\infty} (-1)^k \frac{k2^k}{3^k}$

16. $\displaystyle\sum_{k=1}^{\infty} (-1)^k \frac{k^2 3^k}{2^k}$

17. $\displaystyle\sum_{k=1}^{\infty} \frac{-2}{k}$

18. $\displaystyle\sum_{k=1}^{\infty} \frac{4}{k}$

19. $\displaystyle\sum_{k=1}^{\infty} (-1)^{k+1} \frac{\sqrt{k}}{k+1}$

20. $\displaystyle\sum_{k=1}^{\infty} (-1)^{k+1} \frac{k}{k^3+1}$

21. $\displaystyle\sum_{k=1}^{\infty} \frac{k^2}{e^k}$

22. $\displaystyle\sum_{k=1}^{\infty} k^3 e^{-k}$

23. $\displaystyle\sum_{k=1}^{\infty} \frac{\sin k}{k^2}$

24. $\displaystyle\sum_{k=1}^{\infty} \frac{\cos k}{k^3}$

25. $\displaystyle\sum_{k=1}^{\infty} \frac{\cos k\pi}{k}$

26. $\displaystyle\sum_{k=1}^{\infty} \frac{\sin k\pi}{k}$

27. $\displaystyle\sum_{k=2}^{\infty} \frac{(-1)^k}{\ln k}$

28. $\displaystyle\sum_{k=2}^{\infty} \frac{(-1)^k}{k \ln k}$

29. $\displaystyle\sum_{k=1}^{\infty} \frac{(-1)^k}{k\sqrt{k}}$

30. $\displaystyle\sum_{k=1}^{\infty} \frac{(-1)^{k+1}}{\sqrt{k}}$

31. $\displaystyle\sum_{k=1}^{\infty} (-1)^{k+1} \frac{k!}{4^k}$

32. $\displaystyle\sum_{k=1}^{\infty} (-1)^{k+1} \frac{k^2 4^k}{k!}$

33. $\displaystyle\sum_{k=1}^{\infty} (-1)^{k+1} \frac{k^{10}}{(2k)!}$

34. $\displaystyle\sum_{k=0}^{\infty} (-1)^k \frac{4^k}{(2k+1)!}$

35. $\displaystyle\sum_{k=0}^{\infty} \frac{(-2)^k(k+1)}{5^k}$

36. $\displaystyle\sum_{k=1}^{\infty} \frac{(-3)^k}{k^2 4^k}$

In exercises 37–42, use the Root Test to determine whether the series converges or diverges.

The Root Test states that if $\displaystyle\lim_{k\to\infty} \sqrt[k]{a_k} = L$, then conclusions (i)–(iii) of the Ratio Test hold. For example, to test $\displaystyle\sum_{k=1}^{\infty} \left(\frac{2}{3k+1}\right)^k$, you would compute

$$\lim_{k\to\infty} \sqrt[k]{\left(\frac{2}{3k+1}\right)^k} = \lim_{k\to\infty} \left(\frac{2}{3k+1}\right) = 0.$$

Since $0 < 1$, we conclude that the series converges absolutely.

37. $\displaystyle\sum_{k=1}^{\infty} \left(\frac{4k}{5k+1}\right)^k$

38. $\displaystyle\sum_{k=1}^{\infty} \left(\frac{1-3k}{4k}\right)^k$

39. $\displaystyle\sum_{k=2}^{\infty} \frac{e^{3k}}{k^{3k}}$

40. $\displaystyle\sum_{k=1}^{\infty} \left(\frac{e^k}{k^2}\right)^k$

41. $\displaystyle\sum_{k=1}^{\infty} \frac{3}{k^k}$

42. $\displaystyle\sum_{k=0}^{\infty} \frac{2k}{3^k}$

43. In the 1910s, the Indian mathematician Srinivasa Ramanujan discovered the formula

$$\frac{1}{\pi} = \frac{\sqrt{8}}{9801} \sum_{k=0}^{\infty} \frac{(4k)!(1103 + 26390k)}{(k!)^4 396^{4k}}.$$

Approximate the series with only the $k = 0$ term and show that you get 6 digits of π correct. Approximate the series using the $k = 0$ and $k = 1$ terms and show that you get 14 digits of π correct. In general, each term of this remarkable series increases the accuracy by 8 digits.

44. Prove that Ramanujan's series in exercise 43 converges.

45. To show that $\displaystyle\sum_{k=1}^{\infty} \frac{k!}{k^k}$ converges, use the Ratio Test and the fact that

$$\lim_{k\to\infty} \left(\frac{k+1}{k}\right)^k = \lim_{k\to\infty} \left(1 + \frac{1}{k}\right)^k = e.$$

46. Determine whether $\displaystyle\sum_{k=1}^{\infty} \frac{k!}{1 \cdot 3 \cdot 5 \cdots (2k-1)}$ converges or diverges.

47. (exploratory) One reason that it is important to distinguish absolutely from conditionally convergent series is the rearrangement of series, to be explored in this exercise. Show that the series $\displaystyle\sum_{k=0}^{\infty} \frac{(-1)^k}{2^k}$ is absolutely convergent and find its sum S. Find the sum S_+ of the positive terms of the series. Find the sum S_- of the negative terms of the series. Verify that $S = S_+ + S_-$. This may seem obvious, since for the finite sums you are most familiar with, the order of addition never matters. However, you cannot separate the positive and

negative terms for conditionally convergent series. For example, show that $\sum_{k=0}^{\infty} \frac{(-1)^k}{k+1}$ converges (conditionally) but that the series of positive terms and the series of negative terms both diverge. Explain in words why this will always happen for conditionally convergent series. Thus, the order of terms matters for conditionally convergent series. By exploring further, we can uncover a truly remarkable fact: for conditionally convergent series, you can reorder the terms so that the partial sums converge to **any** real number. To illustrate this, suppose we want to reorder the series $\sum_{k=0}^{\infty} \frac{(-1)^k}{k+1}$ so that the partial sums converge to $\frac{\pi}{2}$. Start by pulling out positive terms

$(1 + \frac{1}{3} + \frac{1}{5} + \cdots)$ such that the partial sum is within 0.1 of $\frac{\pi}{2}$. Next, take the first negative term $(-\frac{1}{2})$ and positive terms such that the partial sum is within 0.05 of $\frac{\pi}{2}$. Then take the next negative term $(-\frac{1}{4})$ and positive terms such that the partial sum is within 0.01 of $\frac{\pi}{2}$. Argue that you could continue in this fashion to reorder the terms so that the partial sums converge to $\frac{\pi}{2}$. (Especially explain why you will never "run out of" positive terms.) Then explain why you cannot do the same with the absolutely convergent series $\sum_{k=0}^{\infty} \frac{(-1)^k}{2^k}$.

8.6   Power Series

We now want to expand our discussion of series to the case where the terms of the series are functions of the variable x. Pay close attention to what we are about to introduce, for this is the culmination of all your hard work in the preceding five sections. The primary reason for studying series is that we can use them to represent functions. This opens up all kinds of possibilities for us, from approximating the values of transcendental functions to calculating derivatives and integrals of such functions, to studying differential equations. As well, **defining** functions as convergent series produces an explosion of new functions available to us. In fact, many functions of great significance in applications (for instance, Bessel functions) are defined as a series. We take the first few steps in this section.

As a start, consider the series

$$\sum_{k=0}^{\infty} (x-2)^k = 1 + (x-2) + (x-2)^2 + (x-2)^3 + \cdots.$$

Notice that for each fixed x, this is a geometric series with $r = (x-2)$. Recall that this says that the series will converge whenever $|r| = |x-2| < 1$ and will diverge whenever $|r| = |x-2| \geq 1$. Further, for each x with $|x-2| < 1$ (i.e., $1 < x < 3$), the series converges to

$$\frac{a}{1-r} = \frac{1}{1-(x-2)} = \frac{1}{3-x}.$$

That is, for each x in the interval $(1,3)$, we have

$$\sum_{k=0}^{\infty} (x-2)^k = \frac{1}{3-x}.$$

For all other values of x, the series diverges. In Figure 8.38, we show a graph of $f(x) = \frac{1}{3-x}$, along with the first three partial sums P_n, of this series, where

$$P_n(x) = \sum_{k=0}^{n} (x-2)^k = 1 + (x-2) + (x-2)^2 + \cdots + (x-2)^n,$$

on the interval $[1,3]$. Notice that as n gets larger, $P_n(x)$ tends to stay closer to $f(x)$ for a larger range of x-values.

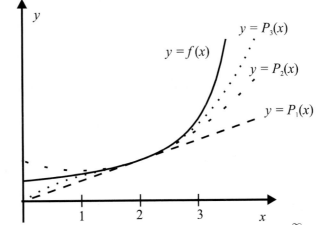

FIGURE 8.38: $y = \dfrac{1}{3-x}$ and the first three partial sums of $\displaystyle\sum_{k=0}^{\infty}(x-2)^k$.

Make certain that you understand what we've observed here: we have taken a series and noticed that it is equivalent to (i.e., it converges to) a **known** function on a certain interval. You might ask why anyone would care if you could do that. Certainly, $f(x) = \dfrac{1}{3-x}$ is a perfectly good function and anything you'd want to do with it will most definitely be easier using the algebraic expression $\dfrac{1}{3-x}$ than using the equivalent series representation, $\displaystyle\sum_{k=0}^{\infty}(x-2)^k$. However, imagine what benefits you might find if you could take a given function (say, one that you don't know a whole lot about) and find an equivalent series representation. This is precisely what we are going to do in section 8.7. For instance, we will be able to show that for all x,

$$e^x = \sum_{k=0}^{\infty}\frac{x^k}{k!} = 1 + x + \frac{x^2}{2!} + \frac{x^3}{3!} + \frac{x^4}{4!} + \cdots. \tag{6.1}$$

So, who cares? Well, suppose you wanted to calculate $e^{1.234567}$. How would you do that? Of course, you'd use a calculator. But, haven't you ever wondered how your calculator does it? The problem is that e^x is not an algebraic function. That is, we can't compute its values by using algebraic methods (i.e., addition, subtraction, multiplication, division and nth roots). Over the next few sections, we begin to explore this question. For the moment, let us say this: if we have the series representation (6.1) for e^x, then for any given x, we can compute an approximation to e^x, simply by summing the first few terms of the equivalent power series. This is easy to do, since the partial sums of the series are simply polynomials.

In general, any series of the form

$$\sum_{k=0}^{\infty}b_k(x-c)^k = b_0 + b_1(x-c) + b_2(x-c)^2 + b_3(x-c)^3 + \cdots$$

is called a **power series** in powers of $(x-c)$. We refer to the constants b_k, $k = 0, 1, 2, \ldots$ as the **coefficients** of the series. The first question is: for what values of x does the series converge? Saying this another way, the power series $\displaystyle\sum_{k=0}^{\infty}b_k(x-c)^k$ defines a function of x. Its domain is the set of all x for which the series converges. The primary tool for investigating the convergence or divergence of a power series is the Ratio Test. Notice, again that the partial sums of a power series are all polynomials (the simplest functions around).

Example 6.1   Determining Where a Power Series Converges

Determine the values of x for which the power series $\sum_{k=0}^{\infty} \frac{k}{3^{k+1}} x^k$ converges.

Solution   Using the Ratio Test, we have

Since $x^{k+1} = x^k \cdot x^1$
and $3^{k+2} = 3^{k+1} \cdot 3^1$.

$$\lim_{k \to \infty} \left| \frac{a_{k+1}}{a_k} \right| = \lim_{k \to \infty} \left| \frac{(k+1)\, x^{k+1}}{3^{k+2}} \frac{3^{k+1}}{k\, x^k} \right|$$

$$= \lim_{k \to \infty} \frac{(k+1)\, |x|}{3\, k} = \frac{|x|}{3} \lim_{k \to \infty} \frac{k+1}{k}$$

$$= \frac{|x|}{3} < 1,$$

for $|x| < 3$ or $-3 < x < 3$. So, the series converges absolutely for $-3 < x < 3$ and diverges for $|x| > 3$ (i.e., for $x > 3$ or $x < -3$). Since the Ratio Test gives no conclusion for the endpoints $x = \pm 3$, we must test these separately.

For $x = 3$, we have the series

$$\sum_{k=0}^{\infty} \frac{k}{3^{k+1}} x^k = \sum_{k=0}^{\infty} \frac{k}{3^{k+1}} 3^k = \sum_{k=0}^{\infty} \frac{k}{3}.$$

Since

$$\lim_{k \to \infty} \frac{k}{3} = \infty \neq 0,$$

the series diverges by the kth-term test for divergence. The series diverges when $x = -3$, for the same reason. Thus, the power series converges for all x in the interval $(-3, 3)$ and diverges for all x outside this interval.

■

Observe that example 6.1 has something in common with the introductory example. In both cases, the series have the form $\sum_{k=0}^{\infty} b_k (x - c)^k$ and there is an interval of the form $(c-r, c+r)$ on which the series converges and outside of which the series diverges. (In the case of example 6.1, notice that $c = 0$.) This interval on which a power series converges is called the **interval of convergence**. The constant r is called the **radius of convergence** (i.e., r is half the length of the interval of convergence). It turns out that there is such an interval for every power series. We have the following result.

Theorem 6.1   Given any power series, $\sum_{k=0}^{\infty} b_k (x-c)^k$, there are exactly three possibilities:

(i) The series converges for **all** $x \in (-\infty, \infty)$ and the radius of convergence is $r = \infty$;

(ii) The series converges **only** for $x = c$ (and diverges for all other values of x) and the radius of convergence is $r = 0$; or

(iii) The series converges for $x \in (c-r, c+r)$ and diverges for $x < c-r$ and $x > c+r$, for some number r with $0 < r < \infty$.

The proof of the theorem can be found in Appendix A.

Example 6.2   Finding the Interval and Radius of Convergence

Determine the interval and radius of convergence for the power series

$$\sum_{k=0}^{\infty} \frac{10^k}{k!} (x - 1)^k.$$

Solution   From the Ratio Test, we have

$$\lim_{k \to \infty} \left| \frac{a_{k+1}}{a_k} \right| = \lim_{k \to \infty} \left| \frac{10^{k+1} (x-1)^{k+1}}{(k+1)!} \frac{k!}{10^k (x-1)^k} \right|$$

$$= 10 \, |x-1| \lim_{k \to \infty} \frac{k!}{(k+1) \, k!}$$

$$= 10 \, |x-1| \lim_{k \to \infty} \frac{1}{k+1} = 0 < 1,$$

Since
$(x-1)^{k+1} = (x-1)^k \cdot (x-1)^1$
and $(k+1)! = (k+1) \cdot k!$

for **all** x. This says that the series converges absolutely for all x. Thus, the interval of convergence for this series is $(-\infty, \infty)$ and the radius of convergence is $r = \infty$.

∎

The interval of convergence for a power series can be a closed interval, an open interval or a half-open interval, as in the following example.

Example 6.3   **A Half-Open Interval of Convergence**

Determine the interval and radius of convergence for the power series $\displaystyle\sum_{k=1}^{\infty} \frac{x^k}{k \, 4^k}$.

Solution   From the Ratio Test, we have

$$\lim_{k \to \infty} \left| \frac{a_{k+1}}{a_k} \right| = \lim_{k \to \infty} \left| \frac{x^{k+1}}{(k+1) \, 4^{k+1}} \frac{k \, 4^k}{x^k} \right|$$

$$= \frac{|x|}{4} \lim_{k \to \infty} \frac{k}{k+1} = \frac{|x|}{4} < 1.$$

So, we are guaranteed absolute convergence for $|x| < 4$ and divergence for $|x| > 4$. It remains only to test the endpoints of the interval: $x = \pm 4$. For $x = 4$, we have

$$\sum_{k=1}^{\infty} \frac{x^k}{k \, 4^k} = \sum_{k=1}^{\infty} \frac{4^k}{k \, 4^k} = \sum_{k=1}^{\infty} \frac{1}{k},$$

which you will recognize as the harmonic series, which diverges. For $x = -4$, we have

$$\sum_{k=1}^{\infty} \frac{x^k}{k \, 4^k} = \sum_{k=1}^{\infty} \frac{(-4)^k}{k \, 4^k} = \sum_{k=1}^{\infty} \frac{(-1)^k}{k},$$

which is the alternating harmonic series, which we know converges (see example 4.2). So, in this case, the interval of convergence is the half-open interval $[-4, 4)$ and the radius of convergence is $r = 4$.

∎

Notice that (as stated in Theorem 6.1) every power series, $\displaystyle\sum_{k=0}^{\infty} a_k(x-c)^k$ converges at least for $x = c$ since for $x = c$, we have the trivial case

$$\sum_{k=0}^{\infty} a_k(x-c)^k = \sum_{k=0}^{\infty} a_k(c-c)^k = a_0 + \sum_{k=1}^{\infty} a_k \, 0^k = a_0 + 0 = a_0.$$

Example 6.4   **A Power Series That Converges at Only One Point**

Determine the radius of convergence for the power series $\displaystyle\sum_{k=0}^{\infty} k!(x-5)^k$.

Solution   From the Ratio Test, we have

$$\lim_{k \to \infty} \left| \frac{a_{k+1}}{a_k} \right| = \lim_{k \to \infty} \left| \frac{(k+1)!(x-5)^{k+1}}{k!(x-5)^k} \right|$$

$$= \lim_{k \to \infty} \frac{(k+1)\, k!\, |x-5|}{k!}$$

$$= \lim_{k \to \infty} \left[(k+1)\, |x-5| \right]$$

$$= \begin{cases} 0, & \text{if } x = 5 \\ \infty, & \text{if } x \neq 5 \end{cases}.$$

Thus, this power series converges only for $x = 5$ and so, its radius of convergence is $r = 0$.

■

Suppose that the power series $\sum_{k=0}^{\infty} b_k (x-c)^k$ has radius of convergence $r > 0$. Then the series converges absolutely for all x in the interval $(c - r, c + r)$ and possibly at the endpoints, $x = c - r$ and $x = c + r$. Notice that since the series converges for each $x \in (c - r, c + r)$, it defines a function f on the interval $(c - r, c + r)$,

$$f(x) = \sum_{k=0}^{\infty} b_k (x-c)^k = b_0 + b_1(x-c) + b_2(x-c)^2 + b_3(x-c)^3 + \cdots.$$

It turns out that such a function is continuous and differentiable, although the proof is beyond the level of this course. In fact, we differentiate exactly the way you might expect,

Differentiating a Power Series

$$f'(x) = \frac{d}{dx} f(x) = \frac{d}{dx} \left[b_0 + b_1(x-c) + b_2(x-c)^2 + b_3(x-c)^3 + \cdots \right]$$

$$= b_1 + 2b_2(x-c) + 3b_3(x-c)^2 + \cdots = \sum_{k=0}^{\infty} b_k\, k\, (x-c)^{k-1},$$

where the radius of convergence of the resulting series is also r. Since we find the derivative by differentiating each term in the series, we call this **term by term** differentiation. Likewise, we can integrate a convergent power series term by term,

Integrating a Power Series

$$\int f(x)\, dx = \int \sum_{k=0}^{\infty} b_k (x-c)^k\, dx = \sum_{k=0}^{\infty} b_k \int (x-c)^k\, dx$$

$$= \sum_{k=0}^{\infty} b_k \frac{(x-c)^{k+1}}{k+1} + K,$$

where the radius of convergence of the resulting series is again r and where K is a constant of integration. The proof of these two results can be found in a text on advanced calculus. It's important to recognize that these two results are **not** obvious. They are not simply an application of the rule that a derivative or integral of a sum is simply the sum of the derivatives or integrals, respectively, since a series is not a sum, but rather, a limit of a sum. (What's the difference, anyway?) Further, these results are true for power series, but are **not** true for series in general.

Example 6.5   A Convergent Series Whose Series of Derivatives Diverges

Find the interval of convergence of the series $\sum_{k=1}^{\infty} \frac{\sin(k^3 x)}{k^2}$ and show that the series of derivatives does not converge for any x.

Solution   Notice that

$$\left|\frac{\sin(k^3 x)}{k^2}\right| \le \frac{1}{k^2}, \text{ for all } x,$$

since $\left|\sin(k^3 x)\right| \le 1$. Further, $\displaystyle\sum_{k=1}^{\infty} \frac{1}{k^2}$ is a convergent p-series ($p = 2 > 1$) and so, it

follows from the Comparison Test that $\displaystyle\sum_{k=0}^{\infty} \frac{\sin(k^3 x)}{k^2}$ converges absolutely, for all x. On

the other hand, the series of derivatives (found by differentiating the series term by term) is

$$\sum_{k=1}^{\infty} \frac{d}{dx}\left[\frac{\sin(k^3 x)}{k^2}\right] = \sum_{k=1}^{\infty} \frac{k^3 \cos(k^3 x)}{k^2} = \sum_{k=1}^{\infty} \left[k \, \cos(k^3 x)\right],$$

which **diverges** for all x, by the kth-term test for divergence, since the terms do not tend to zero as $k \to \infty$, for any x. ∎

We want to emphasize that $\displaystyle\sum_{k=1}^{\infty} \frac{\sin(k^3 x)}{k^2}$ is not a power series. (Why not?) The result

of example 6.5 (a convergent series whose series of derivatives diverges) **cannot** occur with any power series with radius of convergence $r > 0$.

In the following example, we find that once we have a convergent power series representation for a given function, we can use this to obtain power series representations for any number of other functions, by differentiating and integrating the series term by term.

Example 6.6   **Differentiating and Integrating a Power Series**

Notice that $\displaystyle\sum_{k=0}^{\infty} (-1)^k x^k = \sum_{k=0}^{\infty} (-x)^k$ is a geometric series with ratio $r = -x$. This

series converges, then, whenever $|r| = |-x| = |x| < 1$, to

$$\frac{a}{1-r} = \frac{1}{1-(-x)} = \frac{1}{1+x}.$$

That is, for $-1 < x < 1$,

$$\frac{1}{1+x} = \sum_{k=0}^{\infty} (-1)^k x^k. \tag{6.2}$$

Differentiating both sides of (6.2), we get

$$\frac{-1}{(1+x)^2} = \sum_{k=0}^{\infty} (-1)^k k x^{k-1}, \text{ for } -1 < x < 1.$$

Multiplying both sides by -1 gives us a new power series representation:

$$\frac{1}{(1+x)^2} = \sum_{k=0}^{\infty} (-1)^{k+1} k x^{k-1},$$

valid for $-1 < x < 1$. Notice that we can also obtain a new power series from (6.2) by substitution. For instance, if we replace x with x^2, we get

$$\frac{1}{1+x^2} = \sum_{k=0}^{\infty} (-1)^k (x^2)^k = \sum_{k=0}^{\infty} (-1)^k x^{2k}, \tag{6.3}$$

valid for $-1 < x^2 < 1$ (which is equivalent to having $x^2 < 1$ or $-1 < x < 1$.)

Integrating both sides of (6.3) gives us

$$\int \frac{1}{1+x^2}\, dx = \sum_{k=0}^{\infty} (-1)^k \int x^{2k}\, dx = \sum_{k=0}^{\infty} \frac{(-1)^k x^{2k+1}}{2k+1} + c. \tag{6.4}$$

You should recognize the integral on the left-hand side of (6.4) as $\tan^{-1} x$. That is,

$$\tan^{-1} x = \sum_{k=0}^{\infty} \frac{(-1)^k x^{2k+1}}{2k+1} + c, \text{ for } -1 < x < 1. \tag{6.5}$$

Taking $x = 0$ gives us

$$\tan^{-1} 0 = \sum_{k=0}^{\infty} \frac{(-1)^k 0^{2k+1}}{2k+1} + c = c,$$

so that $c = \tan^{-1} 0 = 0$. Equation (6.5) now gives us a power series representation for $\tan^{-1} x$, namely:

$$\tan^{-1} x = \sum_{k=0}^{\infty} \frac{(-1)^k x^{2k+1}}{2k+1} = x - \frac{1}{3}x^3 + \frac{1}{5}x^5 - \frac{1}{7}x^7 + \cdots, \text{ for } -1 < x < 1.$$

$\blacksquare$

Notice that working as in example 6.6, we can produce power series representations of any number of functions. In section 8.7, we present a systematic method for producing power series representations for a wide range of functions.

Exercises 8.6

1. (writing) Power series have the form $\sum_{k=0}^{\infty} a_k(x-c)^k$. Explain why the further x is from c, the larger the terms of the series are and the less likely the series is to converge. Describe how this general trend relates to the radius of convergence.

2. (writing) Applying the Ratio Test to $\sum_{k=0}^{\infty} a_k(x-c)^k$ requires you to evaluate $\lim\limits_{k\to\infty} \left| \frac{a_{k+1}}{a_k}(x-c) \right|$. For $x = c$, this limit equals 0 and the series converges. As x increases or decreases, $|x-c|$ increases. If the series has a finite radius of convergence $r > 0$, what is the value of the limit when $|x-c| = r$? Explain how the limit changes when $|x-c| < r$ and $|x-c| > r$ and how this determines the convergence or divergence of the series.

3. (writing) As shown in example 6.2, $\sum_{k=0}^{\infty} \frac{10^k}{k!}(x-1)^k$ converges for all x. If $x = 1001$, the value of $(x-1)^k = 1000^k$ gets very large very fast as k increases. Explain why, for the series to converge, the value of $k!$ must get large faster than 1000^k. To illustrate how fast the factorial grows, compute 50!, 100! and 200! (if you can).

4. (writing) In a power series representation of $\sqrt{x+1}$ about $c = 0$, explain why the radius of convergence cannot be greater than 1 (think about the domain of $\sqrt{x+1}$).

In exercises 5–14, find a power series representation of $f(x)$ about $c = 0$ (refer to example 6.6). Also, determine the radius and interval of convergence and graph $f(x)$ together with the partial sums $\sum\limits_{k=0}^{3} a_k x^k$ and $\sum\limits_{k=0}^{6} a_k x^k$.

5. $f(x) = \dfrac{2}{1-x}$

6. $f(x) = \dfrac{3}{x-1}$

7. $f(x) = \dfrac{3}{1+x^2}$

8. $f(x) = \dfrac{2}{1-x^2}$

9. $f(x) = \dfrac{2x}{1-x^3}$

10. $f(x) = \dfrac{3x}{1+x^2}$

11. $f(x) = \dfrac{4}{1+4x}$

12. $f(x) = \dfrac{3}{1-4x}$

13. $f(x) = \dfrac{2}{4+x}$

14. $f(x) = \dfrac{3}{6-x}$

In exercises 15–20, determine the interval of convergence and the function to which the given power series converges.

15. $\sum\limits_{k=0}^{\infty} (x+2)^k$

16. $\sum\limits_{k=0}^{\infty} (x-3)^k$

17. $\sum\limits_{k=0}^{\infty} (2x-1)^k$

18. $\sum\limits_{k=0}^{\infty} (3x+1)^k$

19. $\displaystyle\sum_{k=0}^{\infty}(-1)^k\left(\frac{x}{2}\right)^k$

20. $\displaystyle\sum_{k=0}^{\infty}3\left(\frac{x}{4}\right)^k$

In exercises 21–38, determine the radius and interval of convergence.

21. $\displaystyle\sum_{k=0}^{\infty}\frac{2^k}{k!}(x-2)^k$

22. $\displaystyle\sum_{k=0}^{\infty}\frac{3^k}{k!}x^k$

23. $\displaystyle\sum_{k=0}^{\infty}\frac{k}{4^k}x^k$

24. $\displaystyle\sum_{k=0}^{\infty}\frac{k}{2^k}x^k$

25. $\displaystyle\sum_{k=1}^{\infty}\frac{(-1)^k}{k3^k}(x-1)^k$

26. $\displaystyle\sum_{k=1}^{\infty}\frac{(-1)^{k+1}}{k4^k}(x+2)^k$

27. $\displaystyle\sum_{k=0}^{\infty}k!(x+1)^k$

28. $\displaystyle\sum_{k=0}^{\infty}k!(x-2)^k$

29. $\displaystyle\sum_{k=1}^{\infty}\frac{1}{k}(x-1)^k$

30. $\displaystyle\sum_{k=0}^{\infty}k(x-2)^k$

31. $\displaystyle\sum_{k=0}^{\infty}k^2(x-3)^k$

32. $\displaystyle\sum_{k=1}^{\infty}\frac{1}{k^2}(x+2)^k$

33. $\displaystyle\sum_{k=0}^{\infty}\frac{k!}{(2k)!}x^k$

34. $\displaystyle\sum_{k=0}^{\infty}\frac{(k!)^2}{(2k)!}x^k$

35. $\displaystyle\sum_{k=1}^{\infty}\frac{2^k}{k^2}(x+2)^k$

36. $\displaystyle\sum_{k=0}^{\infty}\frac{k^2}{k!}(x+1)^k$

37. $\displaystyle\sum_{k=1}^{\infty}\frac{4^k}{\sqrt{k}}x^k$

38. $\displaystyle\sum_{k=1}^{\infty}\frac{(-1)^k x^k}{\sqrt{k}}$

In exercises 39–46, find a power series representation and radius of convergence by integrating or differentiating one of the series from exercises 5-14.

39. $f(x)=3\tan^{-1}x$

40. $f(x)=2\ln(1-x)$

41. $f(x)=\dfrac{2x}{(1-x^2)^2}$

42. $f(x)=\dfrac{3}{(x-1)^2}$

43. $f(x)=\ln(1+x^2)$

44. $f(x)=\ln(1+4x)$

45. $f(x)=\dfrac{1}{(1+4x)^2}$

46. $f(x)=\dfrac{2}{(4+x)^2}$

In exercises 47–50, find the interval of convergence of the (non-power) series and the corresponding series of derivatives.

47. $\displaystyle\sum_{k=1}^{\infty}\frac{\cos(k^3x)}{k^2}$

48. $\displaystyle\sum_{k=1}^{\infty}\frac{\cos(x/k)}{k}$

49. $\displaystyle\sum_{k=0}^{\infty}e^{kx}$

50. $\displaystyle\sum_{k=0}^{\infty}e^{-2kx}$

51. For any constants a and $b>0$, determine the interval and radius of convergence of $\displaystyle\sum_{k=0}^{\infty}\frac{(x-a)^k}{b^k}$.

52. Prove that if $\displaystyle\sum_{k=0}^{\infty}a_kx^k$ has radius of convergence r, with $0<r<\infty$, then $\displaystyle\sum_{k=0}^{\infty}a_kx^{2k}$ has radius of convergence $\sqrt{r}$.

53. If $\displaystyle\sum_{k=0}^{\infty}a_kx^k$ has radius of convergence r, with $0<r<\infty$, determine the radius of convergence of $\displaystyle\sum_{k=0}^{\infty}a_k(x-c)^k$ for any constant c.

54. If $\displaystyle\sum_{k=0}^{\infty}a_kx^k$ has radius of convergence r, with $0<r<\infty$, determine the radius of convergence of $\displaystyle\sum_{k=0}^{\infty}a_k\left(\frac{x}{b}\right)^k$ for any constant $b\neq0$.

55. Show that $f(x)=\dfrac{x+1}{(1-x)^2}=\dfrac{\frac{2x}{1-x}+1}{1-x}$ has the power series representation $f(x)=1+3x+5x^2+7x^3+9x^4+\cdots$. Find the radius of convergence. Set $x=\frac{1}{1000}$ and discuss the interesting decimal representation of $\frac{1,001,000}{998,001}$.

56. Use long division to show that $\frac{1}{1-x}=1+x+x^2+x^3+\cdots$.

57. Even great mathematicians can make mistakes. Leonhard Euler started with the equation $\frac{x}{x-1}+\frac{x}{1-x}=0$, rewrote it as $\frac{1}{1-1/x}+\frac{x}{1-x}=0$, found power series representations for each function and concluded that $\cdots+\frac{1}{x^2}+\frac{1}{x}+1+x+x^2+\cdots=0$. Substitute $x=1$ to show that the conclusion is false, then find the mistake in Euler's derivation.

58. (exploratory) For each series $f(x)$, compare the intervals of convergence of $f(x)$ and $\int f(x)\,dx$, where the antiderivative is taken term by term. (a) $f(x)=\displaystyle\sum_{k=0}^{\infty}(-1)^kx^k$; (b) $f(x)=\displaystyle\sum_{k=0}^{\infty}\sqrt{k}x^k$; (c) $f(x)=\displaystyle\sum_{k=0}^{\infty}\frac{1}{k}x^k$. As stated in the text, the radius of convergence remains the same after integration (or differentiation). Based on the examples in this exercise, does integration make it more or less likely that the series will converge at the endpoints? Conversely, will differentiation make it more or less likely that the series will converge at the endpoints?

8.7   Taylor Series

You may still be wondering what series are used for (at least, we hope you're wondering this). Each time we have developed a new concept, we have worked hard to build a case for why we want to do what we're doing. For example, in developing the derivative, we set out to find the slope of a tangent line and to find instantaneous velocity, only to find that they were essentially the same thing. When we developed the definite integral, we did so in the course of trying to find area under the curve. But, we have not yet said why we're pursuing series, even though we've been developing them for more than five sections now. Well, the punchline is close at hand. In this section, we develop a compelling reason for considering series. They are not merely another mathematical curiosity, but rather, are an essential means for exploring and computing with transcendental functions (e.g., $\sin x$, $\cos x$, $\ln x$, e^x, etc.).

Suppose that the power series $\sum_{k=0}^{\infty} b_k (x - c)^k$ has radius of convergence $r > 0$. As we've observed, this means that the series converges absolutely to some function f on the interval $(c - r, c + r)$. We have

$$f(x) = \sum_{k=0}^{\infty} b_k (x - c)^k = b_0 + b_1(x - c) + b_2(x - c)^2 + b_3(x - c)^3 + b_4(x - c)^4 + \cdots,$$

for each $x \in (c - r, c + r)$. Differentiating term by term, we get that

$$f'(x) = \sum_{k=0}^{\infty} b_k \, k \, (x - c)^{k-1} = b_1 + 2b_2(x - c) + 3b_3(x - c)^2 + 4b_4(x - c)^3 + \cdots,$$

again, for each $x \in (c - r, c + r)$. Likewise, we get

$$f''(x) = \sum_{k=0}^{\infty} b_k \, k \, (k - 1) \, (x - c)^{k-2} = 2 \, b_2 + 3 \cdot 2 \, b_3 (x - c) + 4 \cdot 3 \, b_4 (x - c)^2 + \cdots$$

and

$$f'''(x) = \sum_{k=0}^{\infty} b_k \, k \, (k - 1) \, (k - 2) \, (x - c)^{k-3} = 3 \cdot 2 \, b_3 + 4 \cdot 3 \cdot 2 \, b_4 \, (x - c) + \cdots$$

and so on (all valid for $c - r < x < c + r$). Notice that if we plug in $x = c$, then in each of the above derivatives, all the terms of the series drop out, except one. We get

$$f(c) = b_0,$$
$$f'(c) = b_1,$$
$$f''(c) = 2 \, b_2,$$
$$f'''(c) = 3! \, b_3$$

and so on. Observe that in general, we have

$$f^{(k)}(c) = k! \, b_k. \tag{7.1}$$

Solving (7.1) for b_k, we have

$$b_k = \frac{f^{(k)}(c)}{k!}, \text{ for } k = 0, 1, 2, \ldots.$$

To summarize, we found that if $\sum_{k=0}^{\infty} b_k (x - c)^k$ is a convergent power series with radius of convergence $r > 0$, then the series converges to some function f which we can write as

$$f(x) = \sum_{k=0}^{\infty} b_k (x - c)^k = \sum_{k=0}^{\infty} \frac{f^{(k)}(c)}{k!} \, (x - c)^k, \text{ for } x \in (c - r, c + r).$$

Now, think about this problem from another angle. Instead of starting with a series, suppose that you start with an infinitely differentiable function, f (i.e., f can be differentiated infinitely often). Then, we can construct the series

$$\sum_{k=0}^{\infty} \frac{f^{(k)}(c)}{k!} (x-c)^k,$$

called a **Taylor series** expansion for f. There are several important questions we need to answer.

- Does a series constructed in this way converge? If so, what is its radius of convergence?
- If the series converges, it converges to a function. What is that function (e.g., could it be f)?

We can answer the first of these questions on a case-by-case basis, usually by applying the Ratio Test. The second question will require more insight.

Example 7.1   **Constructing a Taylor Series Expansion**
Construct the Taylor series expansion for $f(x) = e^x$, about $x = 0$ (i.e., take $c = 0$).

Solution   Here, we have the extremely simple case where

$$f'(x) = e^x, \; f''(x) = e^x \text{ and so on, } f^{(k)}(x) = e^x, \text{ for } k = 0, 1, 2, \ldots.$$

This gives us the Taylor series

$$\sum_{k=0}^{\infty} \frac{f^{(k)}(0)}{k!} (x-0)^k = \sum_{k=0}^{\infty} \frac{e^0}{k!} x^k = \sum_{k=0}^{\infty} \frac{1}{k!} x^k.$$

From the Ratio Test, we have

$$\lim_{k \to \infty} \left| \frac{a_{k+1}}{a_k} \right| = \lim_{k \to \infty} \frac{|x|^{k+1}}{(k+1)!} \frac{k!}{|x|^k} = |x| \lim_{k \to \infty} \frac{k!}{(k+1)\,k!}$$

$$= |x| \lim_{k \to \infty} \frac{1}{k+1} = |x|\,(0) = 0 < 1, \text{ for all } x.$$

So, the Taylor series $\displaystyle\sum_{k=0}^{\infty} \frac{1}{k!} x^k$ converges for all real numbers x. At this point, though, we do not know to what function the series converges. (Could it be e^x?) ∎

Remark 7.1   The special case of a Taylor series expansion about $x = 0$ is often called a **Maclaurin series**. That is, the series $\displaystyle\sum_{k=0}^{\infty} \frac{f^{(k)}(0)}{k!} x^k$ is the Maclaurin series expansion for f.

Before we present any further examples of Taylor series, let's see if we can determine the function to which a given Taylor series converges. First, notice that the partial sums of a Taylor series (like any power series) are simply polynomials. We define

$$P_n(x) = \sum_{k=0}^{n} \frac{f^{(k)}(c)}{k!} (x-c)^k = f(c) + f'(c)(x-c) + \frac{f''(c)}{2!} (x-c)^2 + \cdots + \frac{f^{(n)}(c)}{n!} (x-c)^n.$$

Observe that $P_n(x)$ is a polynomial of degree n, as $\dfrac{f^{(k)}(c)}{k!}$ is a constant for each k. We refer to P_n as the **Taylor polynomial of degree** n for f expanded about $x = c$.

Example 7.2   Constructing and Graphing Taylor Polynomials

For $f(x) = e^x$, find the Taylor polynomial of degree n expanded about $x = 0$.

Solution   As in example 7.1, we have that $f^{(k)}(x) = e^x$, for all k. So, we have the nth degree Taylor polynomial is

$$P_n(x) = \sum_{k=0}^{n} \frac{f^{(k)}(0)}{k!}(x-0)^k = \sum_{k=0}^{n} \frac{e^0}{k!}x^k$$

$$= \sum_{k=0}^{n} \frac{1}{k!}x^k = 1 + x + \frac{x^2}{2!} + \frac{x^3}{3!} + \cdots + \frac{x^n}{n!}.$$

Since we established in example 7.1 that the Taylor series for $f(x) = e^x$ about $x = 0$ converges for all x, this says that the sequence of partial sums (i.e., the sequence of Taylor polynomials) converges. In an effort to determine the function to which the Taylor polynomials are converging, we have plotted $P_1(x)$, $P_2(x)$, $P_3(x)$ and $P_4(x)$, together with the graph of $f(x) = e^x$ in Figures 8.39a–d, respectively.

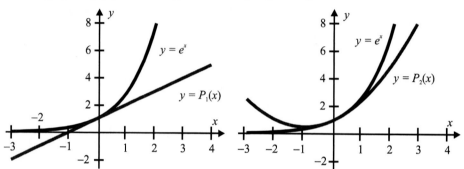

FIGURE 8.39a: $y = e^x$ and $y = P_1(x)$.   FIGURE 8.39b: $y = e^x$ and $y = P_2(x)$.

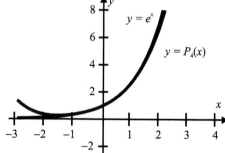

FIGURE 8.39c: $y = e^x$ and $y = P_3(x)$.   FIGURE 8.39d: $y = e^x$ and $y = P_4(x)$.

Notice that as n gets larger, the graphs of $P_n(x)$ appear (at least on the interval displayed) to be approaching the graph of $f(x) = e^x$. Since we know that the Taylor series converges and the graphical evidence suggests that the partial sums of the series are approaching $f(x) = e^x$, it is reasonable to conjecture that the series converges to e^x. This is, in fact, exactly what is happening, as we can prove using the following two theorems.

■

Theorem 7.1   (Taylor's Theorem) Suppose that f has $(n + 1)$ derivatives on the interval $(c - r, c + r)$, for some $r > 0$. Then, for $x \in (c - r, c + r)$, $f(x) \approx P_n(x)$ and the error in using $P_n(x)$ to approximate $f(x)$ is

$$R_n(x) = f(x) - P_n(x) = \frac{f^{(n+1)}(z)}{(n + 1)!} (x - c)^{n+1}, \tag{7.2}$$

for some number z between x and c.

The error term $R_n(x)$ in (7.2) is often called the **remainder term**. Note that this term looks very much like the first neglected term of the Taylor series, except that $f^{(n+1)}$ is evaluated at some (unknown) number z between x and c, instead of at c. This remainder term serves two purposes: it allows us to get a handle on the error in using a Taylor polynomial to approximate a given function and as we'll see in the next theorem, it gives us the means to prove that a Taylor series for a given function f, converges to f.

The proof of Taylor's Theorem is somewhat technical and so we leave it for the end of the section.

Remark 7.2   Observe that for $n = 0$, Taylor's Theorem simplifies to a very familiar result. We have

$$R_0(x) = f(x) - P_0(x) = \frac{f'(z)}{(0 + 1)!} (x - c)^{0+1}.$$

Since $P_0(x) = f(c)$, we have simply

$$f(x) - f(c) = f'(z) (x - c).$$

Dividing by $(x - c)$, gives us

$$\frac{f(x) - f(c)}{x - c} = f'(z),$$

which is the conclusion of the Mean Value Theorem. In this way, notice that Taylor's Theorem is a generalization of the Mean Value Theorem.

NOTE: If we could show that

$$\lim_{n \to \infty} R_n(x) = 0, \text{ for } \textbf{all } x \text{ in } (c - r, c + r),$$

then we would have that

$$0 = \lim_{n \to \infty} R_n(x) = \lim_{n \to \infty} [f(x) - P_n(x)] = f(x) - \lim_{n \to \infty} P_n(x)$$

or

$$\lim_{n \to \infty} P_n(x) = f(x), \text{ for } \textbf{all } x \in (c - r, c + r).$$

That is, the sequence of partial sums of the Taylor series (i.e., the sequence of Taylor polynomials) converges to $f(x)$ for each $x \in (c - r, c + r)$. We summarize this in the following theorem.

Theorem 7.2   Suppose f has derivatives of all orders in the interval $(c - r, c + r)$, for some $r > 0$ and $\lim_{n \to \infty} R_n(x) = 0$, for all x in $(c - r, c + r)$. Then, the Taylor series for f

expanded about $x = c$ converges to $f(x)$, that is,

$$f(x) = \sum_{k=0}^{\infty} \frac{f^{(k)}(c)}{k!} (x - c)^k,$$

for all x in $(c - r, c + r)$.

We now return to the Taylor series expansion of $f(x) = e^x$ about $x = 0$, constructed in example 7.1 and investigated further in example 7.2 and prove that it converges to e^x, as we had suspected.

Example 7.3   Proving That a Taylor Series Converges to the Desired Function

Show that the Taylor series for $f(x) = e^x$ expanded about $x = 0$ converges to e^x.

Solution   We already found the indicated Taylor series, $\displaystyle\sum_{k=0}^{\infty} \frac{1}{k!} x^k$ in example 7.1. Here, we have $f^{(k)}(x) = e^x$, for all $k = 0, 1, 2, \ldots$. This gives us the remainder term

$$R_n(x) = \frac{f^{(n+1)}(z)}{(n+1)!}(x - 0)^{n+1} = \frac{e^z}{(n+1)!} x^{n+1}, \tag{7.3}$$

where z is somewhere between x and 0 (and depends also on the value of n). We first find a bound on the size of e^z. Notice that if $x > 0$, then $0 < z < x$ and so,

$$e^z < e^x.$$

If $x \leq 0$, then $x \leq z \leq 0$, so that

$$e^z \leq e^0 = 1.$$

We define M to be the larger of these two bounds on e^z. That is, we let

$$M = \max\{e^x, 1\}.$$

Then, for any x and any n, we have

$$e^z \leq M.$$

Together with (7.3), this gives us the error estimate

$$|R_n(x)| = \frac{e^z}{(n+1)!} |x|^{n+1} \leq M \frac{|x|^{n+1}}{(n+1)!}. \tag{7.4}$$

To prove that the Taylor series converges to e^x, we want to use (7.4) to show that $\displaystyle\lim_{n\to\infty} R_n(x) = 0$, for all x. However, for any given x, how can we compute $\displaystyle\lim_{n\to\infty} \frac{|x|^{n+1}}{(n+1)!}$? While we cannot do so directly, we can use the following indirect approach. We consider the **series** $\displaystyle\sum_{n=0}^{\infty} \frac{|x|^{n+1}}{(n+1)!}$. Applying the Ratio Test, we have

$$\lim_{n\to\infty} \left| \frac{a_{n+1}}{a_n} \right| = \lim_{n\to\infty} \frac{|x|^{n+2}}{(n+2)!} \frac{(n+1)!}{|x|^{n+1}} = |x| \lim_{n\to\infty} \frac{1}{n+2} = 0,$$

for all x. This then says that the series $\displaystyle\sum_{n=0}^{\infty} \frac{|x|^{n+1}}{(n+1)!}$ converges absolutely for all x. By the kth-term test for divergence, since this last series converges, its general term must tend to

0 as $n \to \infty$. That is,

$$\lim_{n \to \infty} \frac{|x|^{n+1}}{(n+1)!} = 0$$

and so, from (7.4), $\lim_{n \to \infty} R_n(x) = 0$, for all x. From Theorem 7.2, we now have that the Taylor series converges to e^x, that is,

$$e^x = \sum_{k=0}^{\infty} \frac{1}{k!} x^k = 1 + x + \frac{x^2}{2!} + \frac{x^3}{3!} + \frac{x^4}{4!} + \cdots. \qquad (7.5)$$

■

The trick, if there is one, in finding a Taylor series expansion is in *accurately* calculating enough derivatives for you to recognize the general form of the nth derivative. So, take your time and **BE CAREFUL**! Once this is done, you need only show that $R_n(x) \to 0$, as $n \to \infty$, for all x to ensure that the series converges to the function you are expanding.

One of the reasons for calculating Taylor series is that we can use their partial sums to compute approximate values of a function.

Example 7.4   Using a Taylor Series to Obtain an Approximation of e

Use the Taylor series for e^x in (7.5) to obtain an approximation to the number e.

Solution   We have

$$e = e^1 = \sum_{k=0}^{\infty} \frac{1}{k!} 1^k = \sum_{k=0}^{\infty} \frac{1}{k!}.$$

We list some partial sums of this series in the accompanying table. From this we get the very accurate approximation

$$e \approx 2.718281828.$$

■

M	$\sum_{k=0}^{M} \frac{1}{k!}$
5	2. 71666 6667
10	2. 71828 1801
15	2. 71828 1828
20	2. 71828 1828

Example 7.5   A Taylor Series Expansion of $\sin x$

Find the Taylor series for $f(x) = \sin x$, expanded about $x = \frac{\pi}{2}$ and prove that the series converges to $\sin x$ for all x.

Solution   In this case, the Taylor series is

$$\sum_{k=0}^{\infty} \frac{f^{(k)}\left(\frac{\pi}{2}\right)}{k!} \left(x - \frac{\pi}{2}\right)^k.$$

First, we compute some derivatives.

$$\begin{aligned}
f(x) &= \sin x & f\left(\tfrac{\pi}{2}\right) &= 1, \\
f'(x) &= \cos x & f'\left(\tfrac{\pi}{2}\right) &= 0, \\
f''(x) &= -\sin x & f''\left(\tfrac{\pi}{2}\right) &= -1, \\
f'''(x) &= -\cos x & f'''\left(\tfrac{\pi}{2}\right) &= 0, \\
f^{(4)}(x) &= \sin x & f^{(4)}\left(\tfrac{\pi}{2}\right) &= 1
\end{aligned}$$

and so on. Recognizing that every other term is zero and every other term is ± 1, we see that the Taylor series is

$$\sum_{k=0}^{\infty} \frac{f^{(k)}\left(\frac{\pi}{2}\right)}{k!} \left(x - \frac{\pi}{2}\right)^k = 1 - \frac{1}{2}\left(x - \frac{\pi}{2}\right)^2 + \frac{1}{4!}\left(x - \frac{\pi}{2}\right)^4 - \frac{1}{6!}\left(x - \frac{\pi}{2}\right)^6 + \cdots$$

$$= \sum_{k=0}^{\infty} \frac{(-1)^k}{(2k)!} \left(x - \frac{\pi}{2}\right)^{2k}.$$

In order to test the series for convergence, we consider the remainder term

$$|R_n(x)| = \left| \frac{f^{(n+1)}(z)}{(n+1)!} \left(x - \frac{\pi}{2}\right)^{n+1} \right|, \tag{7.6}$$

for some z between x and $\frac{\pi}{2}$. From our derivative calculations, note that

$$f^{(n+1)}(z) = \begin{cases} \pm \cos z, & \text{if } n \text{ is even} \\ \pm \sin z, & \text{if } n \text{ is odd} \end{cases}.$$

From this, it follows that

$$\left| f^{(n+1)}(z) \right| \le 1,$$

for every n. (Notice that this is true whether n is even or odd.) From (7.6), we now have

$$|R_n(x)| = \left| \frac{f^{(n+1)}(z)}{(n+1)!} \right| \left|x - \frac{\pi}{2}\right|^{n+1} \le \frac{1}{(n+1)!} \left|x - \frac{\pi}{2}\right|^{n+1} \to 0,$$

as $n \to \infty$, for every x, as in example 7.3. This says that

$$\sin x = \sum_{k=0}^{\infty} \frac{(-1)^k}{(2k)!} \left(x - \frac{\pi}{2}\right)^{2k} = 1 - \frac{1}{2}\left(x - \frac{\pi}{2}\right)^2 + \frac{1}{4!}\left(x - \frac{\pi}{2}\right)^4 - \cdots,$$

for all x. In Figures 8.40a–d, we show graphs of $f(x) = \sin x$ together with the Taylor polynomials $P_2(x)$, $P_4(x)$, $P_6(x)$ and $P_8(x)$ (the first few partial sums of the series).

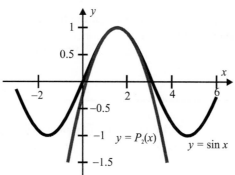

FIGURE 8.40a: $y = \sin x$ and $y = P_2(x)$.

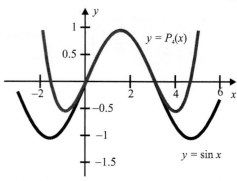

FIGURE 8.40b: $y = \sin x$ and $y = P_4(x)$.

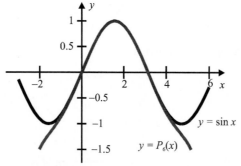

FIGURE 8.40c: $y = \sin x$ and $y = P_6(x)$.

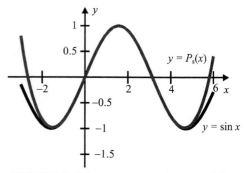

FIGURE 8.40d: $y = \sin x$ and $y = P_8(x)$.

In the following example, we illustrate how to use Taylor's Theorem to estimate the error in using a Taylor polynomial to approximate the value of a function.

Example 7.6   Estimating the Error in a Taylor Polynomial Approximation

Use a Taylor polynomial to approximate the value of $\ln(1.1)$ and estimate the error in this approximation.

Solution   First, note that since $\ln 1$ is known exactly and 1 is close to 1.1 (why would this matter?), we expand $f(x) = \ln x$ in a Taylor series about $x = 1$. We compute an adequate number of derivatives so that the pattern becomes clear. We have

$$
\begin{array}{ll}
f(x) = \ln x & f(1) = 0 \\
f'(x) = x^{-1} & f'(1) = 1 \\
f''(x) = -x^{-2} & f''(1) = -1 \\
f'''(x) = 2x^{-3} & f'''(1) = 2 \\
f^{(4)}(x) = -3 \cdot 2x^{-4} & f^{(4)}(1) = -3! \\
f^{(5)}(x) = 4!\, x^{-5} & f^{(5)}(1) = 4! \\
\quad\vdots & \quad\vdots \\
f^{(k)}(x) = (-1)^{k+1}(k-1)!\, x^{-k} & f^{(k)}(1) = (-1)^{k+1}(k-1)!
\end{array}
$$

We get the Taylor series

$$
\sum_{k=0}^{\infty} \frac{f^{(k)}(1)}{k!} (x-1)^k
$$

$$
= (x-1) - \frac{1}{2}(x-1)^2 + \frac{2}{3!}(x-1)^3 + \cdots + (-1)^{k+1}\frac{(k-1)!}{k!}(x-1)^k + \cdots
$$

$$
= \sum_{k=1}^{\infty} \frac{(-1)^{k+1}}{k} (x-1)^k .
$$

We could use the remainder term to show that the series converges to $f(x) = \ln x$, for $0 < x < 2$, but this is not the original question here. (This is left as an exercise.) As an illustration, we construct the Taylor polynomial, $P_4(x)$,

$$
P_4(x) = \sum_{k=0}^{4} \frac{f^{(k)}(1)}{k!} (x-1)^k
$$

$$
= (x-1) - \frac{1}{2}(x-1)^2 + \frac{2}{3!}(x-1)^3 - \frac{3!}{4!}(x-1)^4.
$$

We show a graph of $y = \ln x$ and $y = P_4(x)$ in Figure 8.41. Taking $x = 1.1$ gives us the approximation

$$
\ln(1.1) \approx P_4(1.1) = 0.1 - \frac{1}{2}(0.1)^2 + \frac{1}{3}(0.1)^3 - \frac{1}{4}(0.1)^4 \approx 0.095308333.
$$

We can use the remainder term to estimate the error in the above approximation. We have

$$
|Error| = |\ln(1.1) - P_4(1.1)| = |R_4(1.1)|
$$

$$
= \left| \frac{f^{(4+1)}(z)}{(4+1)!} (1.1-1)^{4+1} \right| = \frac{4!\, |z|^{-5}}{5!}(0.1)^5,
$$

where z is between 1 and 1.1. This gives us the following bound on the error:

$$
|Error| = \frac{(0.1)^5}{5\, z^5} < \frac{(0.1)^5}{5\,(1^5)} = 0.000002,
$$

since $1 < z < 1.1$ implies that $\dfrac{1}{z} < \dfrac{1}{1} = 1$. Another way to think of this is that our approximation of $\ln(1.1) \approx 0.095308333$ is off by no more than ± 0.000002.

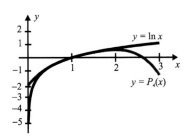

FIGURE 8.41: $y = \ln x$ and $y = P_4(x)$.

A more interesting problem related to example 7.6 is to determine how many terms of the Taylor series are needed in order to guarantee a given accuracy. We use the remainder term to accomplish this in the following example.

Example 7.7   Finding the Number of Terms Needed for a Given Accuracy

Find the number of terms in the Taylor series for $f(x) = \ln x$ expanded about $x = 1$ which will guarantee an accuracy of at least 1×10^{-10} in the approximation of $\ln(1.1)$.

Solution   From our calculations in example 7.6 and (7.2), we have that for some number z between 1 and 1.1,

$$|R_n(1.1)| = \left| \frac{f^{(n+1)}(z)}{(n+1)!} (1.1-1)^{n+1} \right|$$

$$= \frac{n! \, |z|^{-n-1}}{(n+1)!} (0.1)^{n+1} = \frac{(0.1)^{n+1}}{(n+1) \, z^{n+1}} < \frac{(0.1)^{n+1}}{(n+1)},$$

since $1 < z < 1.1$ implies that $\dfrac{1}{z} < \dfrac{1}{1} = 1$. Further, since we want the error to be less than 1×10^{-10}, we require that

$$|R_n(1.1)| < \frac{(0.1)^{n+1}}{n+1} < 1 \times 10^{-10}.$$

You can solve this inequality for n by trial and error to find that $n = 9$ will guarantee the required accuracy. Notice that larger values of n will also guarantee this accuracy, since $\frac{(0.1)^{n+1}}{n+1}$ is a decreasing function of n. We then have the approximation

$$\ln(1.1) \approx P_9(1.1) = \sum_{k=0}^{9} \frac{f^{(k)}(1)}{k!} (1.1-1)^k$$

$$= (0.1) - \frac{1}{2}(0.1)^2 + \frac{2}{3!}(0.1)^3 - \frac{3!}{4!}(0.1)^4 + \frac{4!}{5!}(0.1)^5$$

$$- \frac{5!}{6!}(0.1)^6 + \frac{6!}{7!}(0.1)^7 - \frac{7!}{8!}(0.1)^8 + \frac{8!}{9!}(0.1)^9$$

$$\approx 0.095310179813,$$

which from our error estimate we know is correct to within 1×10^{-10}. We show a graph of $y = \ln x$ and $y = P_9(x)$ in Figure 8.42.

■

FIGURE 8.42: $y = \ln x$ and $y = P_9(x)$.

For your convenience, we have compiled a list of common Taylor series in the following table.

Taylor Series		Interval of Convergence	Where to find
$e^x \quad = \sum\limits_{k=0}^{\infty} \dfrac{1}{k!}\, x^k$	$= 1 + x + \frac{1}{2}x^2 + \frac{1}{3!}x^3 + \frac{1}{4!}x^4 + \cdots$	$(-\infty, \infty)$	examples 7.1 and 7.3
$\sin x \ = \sum\limits_{k=0}^{\infty} \dfrac{(-1)^k}{(2k+1)!}\, x^{2k+1}$	$= x - \frac{1}{3!}x^3 + \frac{1}{5!}x^5 - \frac{1}{7!}x^7 + \cdots$	$(-\infty, \infty)$	exercise 9
$\cos x \ = \sum\limits_{k=0}^{\infty} \dfrac{(-1)^k}{(2k)!}\, x^{2k}$	$= 1 - \frac{1}{2}x^2 + \frac{1}{4!}x^4 - \frac{1}{6!}x^6 + \cdots$	$(-\infty, \infty)$	exercise 7
$\sin x \ = \sum\limits_{k=0}^{\infty} \dfrac{(-1)^k}{(2k)!}\, (x - \frac{\pi}{2})^{2k}$	$= 1 - \frac{1}{2}(x - \frac{\pi}{2})^2 + \frac{1}{4!}(x - \frac{\pi}{2})^4 - \cdots$	$(-\infty, \infty)$	example 7.5
$\ln x \ = \sum\limits_{k=1}^{\infty} \dfrac{(-1)^{k+1}}{k}\, (x - 1)^k$	$= (x - 1) - \frac{1}{2}(x - 1)^2 + \frac{1}{3}(x - 1)^3 - \cdots$	$(0, 2)$	examples 7.6, 7.7
$\tan^{-1} x = \sum\limits_{k=0}^{\infty} \dfrac{(-1)^k}{2k+1}\, x^{2k+1}$	$= x - \frac{1}{3}x^3 + \frac{1}{5}x^5 - \frac{1}{7}x^7 + \cdots$	$(-1, 1)$	example 6.6

Notice that once you have found a Taylor series expansion for a given function, you can find any number of other Taylor series simply by making a change of variable.

Example 7.8   **Finding New Taylor Series from Old Ones**

Find Taylor series in powers of x for e^{2x}, e^{x^2} and e^{-x}.

Solution   Rather than compute the Taylor series for these functions from scratch, recall that we had established in example 7.3 that

$$e^t = \sum_{k=0}^{\infty} \frac{1}{k!}\, t^k = 1 + t + \frac{1}{2!}\, t^2 + \frac{1}{3!}\, t^3 + \frac{1}{4!}\, t^4 + \cdots, \tag{7.7}$$

for all $t \in (-\infty, \infty)$. We use the variable t here instead of x, so that we can more easily make substitutions. Taking $t = 2x$ in (7.7), we get the new Taylor series:

$$e^{2x} = \sum_{k=0}^{\infty} \frac{1}{k!}\, (2x)^k = \sum_{k=0}^{\infty} \frac{2^k}{k!}\, x^k = 1 + 2\, x + \frac{2^2}{2!}\, x^2 + \frac{2^3}{3!}\, x^3 + \cdots.$$

Similarly, letting $t = x^2$ in (7.7), we get the Taylor series

$$e^{x^2} = \sum_{k=0}^{\infty} \frac{1}{k!}\, (x^2)^k = \sum_{k=0}^{\infty} \frac{1}{k!}\, x^{2k} = 1 + x^2 + \frac{1}{2!}\, x^4 + \frac{1}{3!}\, x^6 + \cdots.$$

Finally, taking $t = -2x$ in (7.7), we get

$$e^{-2x} = \sum_{k=0}^{\infty} \frac{1}{k!}\, (-2x)^k = \sum_{k=0}^{\infty} \frac{(-1)^k}{k!}\, 2^k\, x^k = 1 - 2\, x + \frac{2^2}{2!}\, x^2 - \frac{2^3}{3!}\, x^3 + \cdots.$$

Notice that all of these last three series converge for all $x \in (-\infty, \infty)$. (Why is that?)

■

Proof of Taylor's Theorem

Recall that we had observed that the Mean Value Theorem was a special case of Taylor's Theorem. As it turns out, the proof of Taylor's Theorem parallels that of the Mean Value Theorem. Both make use of Rolle's Theorem: If g is continuous on the interval $[a, b]$, differentiable on (a, b) and $g(a) = g(b)$, then there is a number $c \in (a, b)$ for which $g'(c) = 0$. As with the proof of the Mean Value Theorem, for a **fixed** $x \in (c - r, c + r)$, we define the function

$$g(t) = f(x) - f(t) - f'(t)(x - t) - \frac{1}{2!}f''(t)(x - t)^2 - \frac{1}{3!}f'''(t)(x - t)^3$$

$$- \cdots - \frac{1}{n!}f^{(n)}(t)(x - t)^n - R_n(x)\frac{(x - t)^{n+1}}{(x - c)^{n+1}},$$

where $R_n(x)$ is the remainder term, $R_n(x) = f(x) - P_n(x)$. If we take $t = x$, notice that

$$g(x) = f(x) - f(x) - 0 - 0 - \cdots - 0 = 0$$

and if we take $t = c$, we get

$$g(c) = f(x) - f(c) - f'(c)(x - c) - \frac{1}{2!}f''(c)(x - c)^2 - \frac{1}{3!}f'''(c)(x - c)^3$$

$$- \cdots - \frac{1}{n!}f^{(n)}(c)(x - c)^n - R_n(x)\frac{(x - c)^{n+1}}{(x - c)^{n+1}}$$

$$= f(x) - P_n(x) - R_n(x) = R_n(x) - R_n(x) = 0.$$

By Rolle's Theorem, there must be some number z between x and c for which $g'(z) = 0$. Differentiating our expression for $g(t)$ (with respect to t!), we get (beware of all the product rules!)

$$g'(t) = 0 - f'(t) - f'(t)(-1) - f''(t)(x - t) - \frac{1}{2}f''(t)(2)(x - t)(-1)$$

$$- \frac{1}{2}f'''(t)(x - t)^2 - \cdots - \frac{1}{n!}f^{(n)}(t)(n)(x - t)^{n-1}(-1)$$

$$- \frac{1}{n!}f^{(n+1)}(t)(x - t)^n - R_n(x)\frac{(n + 1)(x - t)^n(-1)}{(x - c)^{n+1}}$$

$$= -\frac{1}{n!}f^{(n+1)}(t)(x - t)^n + R_n(x)\frac{(n + 1)(x - t)^n}{(x - c)^{n+1}},$$

after most of the terms cancel. So, taking $t = z$, we have that

$$0 = g'(z) = -\frac{1}{n!}f^{(n+1)}(z)(x - z)^n + R_n(x)\frac{(n + 1)(x - z)^n}{(x - c)^{n+1}}.$$

Solving this for the remainder term, $R_n(x)$, we get

$$R_n(x)\frac{(n + 1)(x - z)^n}{(x - c)^{n+1}} = \frac{1}{n!}f^{(n+1)}(z)(x - z)^n$$

and finally,

$$R_n(x) = \frac{1}{n!}f^{(n+1)}(z)(x - z)^n\frac{(x - c)^{n+1}}{(n + 1)(x - z)^n}$$

$$= \frac{f^{(n+1)}(z)}{(n + 1)\,n!}(x - c)^{n+1}$$

$$= \frac{f^{(n+1)}(z)}{(n + 1)!}(x - c)^{n+1},$$

as we had claimed.   ∎

Exercises 8.7

1. (writing) Describe how the Taylor polynomial with $n = 1$ compares to the linear approximation (see section 3.1). Give an analogous interpretation of the Taylor polynomial with $n = 2$. That is, how do various graphical properties (position, slope, concavity) of the Taylor polynomial compare with those of the function $f(x)$ at $x = c$?

2. (writing) Briefly discuss how a computer might use Taylor polynomials to compute $\sin(1.2)$. In particular, how would the computer know how many terms to compute? How would the number of terms necessary to compute $\sin(1.2)$ compare to the number needed to compute $\sin(100)$? Describe a trick that would make it much easier for the computer to compute $\sin(100)$. (Hint: the sine function is periodic.)

3. (writing) Taylor polynomials are built-up from a knowledge of $f(c)$, $f'(c)$, $f''(c)$ and so on. Explain in graphical terms why information at one point (e.g., position, slope, concavity, etc.) can be used to construct the graph of the function on the entire interval of convergence.

4. (writing) If $f(c)$ is the position of an object at time $t = c$, then $f'(c)$ is the object's velocity and $f''(c)$ is the object's acceleration at time c. Explain in physical terms how knowledge of these values at one time (plus $f'''(c)$, etc.) can be used to predict the position of the object on the interval of convergence.

5. (writing) Our table of common Taylor series lists two different series for $\sin x$. Explain how the same function could have two different Taylor series representations. For a given problem (e.g., approximate $\sin 2$), explain how you would choose which Taylor series to use.

6. (writing) Explain why the Taylor series with center $c = 0$ of $f(x) = x^2 - 1$ is simply $x^2 - 1$.

In exercises 7–14, find the Maclaurin series (i.e., Taylor series about $c = 0$) and its interval of convergence.

7. $f(x) = \cos x$ **8.** $f(x) = e^{2x}$

9. $f(x) = \sin x$ **10.** $f(x) = \cos 2x$

11. $f(x) = \ln(1 + x)$ **12.** $f(x) = \sqrt{1 + x}$

13. $f(x) = 1/(1 + x)^2$ **14.** $f(x) = 1/(1 - x)$

In exercises 15–20, find the Taylor series about the indicated center and determine the interval of convergence.

15. $f(x) = e^{x-1}$, $c = 1$ **16.** $f(x) = \cos x$, $c = -\pi/2$

17. $f(x) = \ln x$, $c = e$ **18.** $f(x) = e^x$, $c = 2$

19. $f(x) = 1/x$, $c = 1$ **20.** $f(x) = 1/x$, $c = -1$

In exercises 21–28, graph $f(x)$ and the Taylor polynomials for the indicated center c and degree n.

21. $f(x) = \cos x$, $c = 0$, $n = 5$; $n = 9$

22. $f(x) = \ln x$, $c = 1$, $n = 4$; $n = 8$

23. $f(x) = \sqrt{x}$, $c = 1$, $n = 3$; $n = 6$

24. $f(x) = \dfrac{1}{1 + x}$, $c = 0$, $n = 4$; $n = 8$

25. $f(x) = e^x$, $c = 2$, $n = 3$; $n = 6$

26. $f(x) = \cos x$, $c = \pi/2$, $n = 4$; $n = 8$

27. $f(x) = \dfrac{1}{\sqrt{x}}$, $c = 4$, $n = 2$; $n = 4$

28. $f(x) = \sqrt{1 + x^2}$, $c = 0$, $n = 2$; $n = 4$

In exercises 29–32, prove that the Taylor series converges to $f(x)$ by showing that $R_n(x) \to 0$ as $n \to \infty$.

29. $\sin x = \displaystyle\sum_{k=0}^{\infty} (-1)^k \frac{x^{2k+1}}{(2k+1)!}$

30. $\cos x = \displaystyle\sum_{k=0}^{\infty} (-1)^k \frac{x^{2k}}{(2k)!}$

31. $\ln x = \displaystyle\sum_{k=1}^{\infty} (-1)^{k+1} \frac{(x-1)^k}{k}$, $1 < x < 2$

32. $e^{-x} = \displaystyle\sum_{k=0}^{\infty} (-1)^k \frac{x^k}{k!}$

In exercises 33–38, (a) use a Taylor polynomial of degree 4 to approximate the given number, (b) estimate the error in the approximation, (c) estimate the number of terms needed in a Taylor polynomial to guarantee an accuracy of 10^{-10}.

33. $\ln(1.05)$ **34.** $\ln(0.9)$

35. $\sqrt{1.1}$ **36.** $\sqrt{1.2}$

37. $e^{0.1}$ **38.** $e^{-0.1}$

In exercises 39–44, use a Taylor series to verify the given formula.

39. $\displaystyle\sum_{k=0}^{\infty} \frac{2^k}{k!} = e^2$ **40.** $\displaystyle\sum_{k=0}^{\infty} \frac{(-1)^k}{k!} = e^{-1}$

41. $\displaystyle\sum_{k=0}^{\infty} \frac{(-1)^k \pi^{2k+1}}{(2k+1)!} = 0$　　**42.** $\displaystyle\sum_{k=0}^{\infty} \frac{(-1)^k (\pi/2)^{2k+1}}{(2k+1)!} = 1$

43. $\displaystyle\sum_{k=0}^{\infty} \frac{(-1)^k}{2k+1} = \frac{\pi}{4}$　　**44.** $\displaystyle\sum_{k=1}^{\infty} \frac{(-1)^{k+1} 2^k}{k} = \ln 3$

In exercises 45–52, use a known Taylor series to find the Taylor series about $c = 0$ for the given function and find its radius of convergence.

45. $f(x) = e^{-2x}$　　　　　　**46.** $f(x) = e^{3x}$

47. $f(x) = xe^{-x^2}$　　　　　**48.** $f(x) = \dfrac{e^x - 1}{x}$

49. $f(x) = \sin x^2$　　　　　**50.** $f(x) = x \sin 2x$

51. $f(x) = \cos 3x$　　　　　**52.** $f(x) = \cos x^3$

In exercises 53–58, use a known Taylor series to conjecture the value of the limit.

53. $\displaystyle\lim_{x \to 0} \frac{\cos x^2 - 1}{x^4}$　　**54.** $\displaystyle\lim_{x \to 0} \frac{\sin x^2 - x^2}{x^6}$

55. $\displaystyle\lim_{x \to 1} \frac{\ln x - (x-1)}{(x-1)^2}$　　**56.** $\displaystyle\lim_{x \to 0} \frac{\tan^{-1} x - x}{x^3}$

57. $\displaystyle\lim_{x \to 0} \frac{e^x - 1}{x}$　　**58.** $\displaystyle\lim_{x \to 0} \frac{e^{-2x} - 1}{x}$

In exercises 59–64, use a known Taylor polynomial of the indicated degree n to estimate the value of the integral.

59. $\displaystyle\int_{-1}^{1} \frac{\sin x}{x}\, dx, n = 3$　　**60.** $\displaystyle\int_{-\sqrt{\pi}}^{\sqrt{\pi}} \cos x^2\, dx, n = 4$

61. $\displaystyle\int_{-1}^{1} e^{-x^2}\, dx, n = 4$　　**62.** $\displaystyle\int_{0}^{1} \tan^{-1} x\, dx, n = 5$

63. $\displaystyle\int_{1}^{2} \ln x\, dx, n = 4$　　**64.** $\displaystyle\int_{0}^{1} e^{\sqrt{x}}\, dx, n = 4$

65. You may have wondered why it is necessary to show that $\displaystyle\lim_{n \to \infty} R_n(x) = 0$ to conclude that a Taylor series converges to $f(x)$. Consider $f(x) = \begin{cases} e^{-1/x^2}, & \text{if } x \neq 0 \\ 0, & \text{if } x = 0 \end{cases}$. Show that $f'(0) = f''(0) = 0$. It turns out that $f^{(n)}(0) = 0$ for all n. Thus, the Taylor series of $f(x)$ about $c = 0$ equals 0, a convergent "series" which does not converge to $f(x)$.

66. In many applications the **error function** $\text{erf}(x) = \dfrac{2}{\sqrt{\pi}} \displaystyle\int_{0}^{x} e^{-u^2}\, du$ is important. Compute and graph the fourth-order Taylor polynomial for $\text{erf}(x)$ about $c = 0$.

67. Suppose that a plane is at location $f(0) = 10$ miles with velocity $f'(10) = 10$ miles/min, acceleration $f''(0) = 2$ miles/min^2 and $f'''(0) = -1$ miles/min^3. Predict the location of the plane at time $t = 2$ min.

68. Suppose that an astronaut is at $(0,0)$ and the moon is represented by a circle of radius 1 centered at $(10, 5)$. The astronaut's capsule follows a path $y = f(x)$ with current position $f(0) = 0$, slope $f'(0) = 1/5$, concavity $f''(0) = -1/10$, $f'''(0) = 1/25$, $f^{(4)}(0) = 1/25$ and $f^{(5)}(0) = -1/50$. Graph a Taylor polynomial approximation of $f(x)$. Based on your current information, do you advise the astronaut to change paths? How confident are you in the accuracy of your approximation?

69. Einstein's theory of relativity states that the mass of an object traveling at velocity v is $m(v) = m_0/\sqrt{1 - v^2/c^2}$, where m_0 is the rest mass of the object and c is the speed of light. Show that $m \approx m_0 + \left(\dfrac{m_0}{2c^2}\right) v^2$. Use this approximation to estimate how large v would have to be to increase the mass by 10%.

70. The weight (force due to gravity) of an object of mass m and altitude x miles above the surface of the earth is $w(x) = \dfrac{mgR^2}{(R+x)^2}$ where R is the radius of the earth and g is the acceleration due to gravity. Show that $w(x) \approx mg(1 - 2x/R)$. Estimate how large x would have to be to reduce the weight by 10%.

71. Find the Taylor series of e^x about a general center c.

72. (exploratory) In many applications, including the study of planetary orbits, **Bessel functions** play an important role. There are a variety of Bessel functions, including

$$J_n(x) = \sum_{k=0}^{\infty} \frac{(-1)^k}{k!(k+n)!} \left(\frac{x}{2}\right)^{2k+n},$$

where n is a nonnegative integer. For $n = 0$ and $n = 1$, verify that the series converges. Compute and graph four-term approximations to $J_0(x)$ and $J_1(x)$. Illustrate graphically the fact that between two consecutive positive zeros of $J_1(x)$ is a zero of $J_0(x)$.

73. (exploratory) The Bessel functions of exercise 72 and **Legendre polynomials** are examples of the so-called **special functions**. For nonnegative integers n, the Legendre polynomials are defined by

$$P_n(x) = 2^{-n} \sum_{k=0}^{[n/2]} \frac{(-1)^k (2n - 2k)!}{(n-k)! k! (n-2k)!} x^{n-2k}.$$

Here, $[n/2]$ is the greatest integer less than or equal to $n/2$ (for example, $[1/2] = 0$ and $[2/2] = 1$). Show that $P_0(x) = 1$, $P_1(x) = x$ and $P_2(x) = \frac{3}{2}x^2 - \frac{1}{2}$. Show that for these three functions,

$$\int_{-1}^{1} P_m(x) P_n(x)\, dx = 0, \text{ for } m \neq n.$$

This fact, which is true for all Legendre polynomials, is called the **orthogonality condition**. Orthogonal functions are commonly used to provide simple representations of complicated functions.

8.8   Fourier Series

Over the last several sections, we saw how we could represent a function by a power series (i.e., a Taylor series). This was an extraordinary development, in particular because we can use the partial sums of such a series expansion (i.e., the Taylor polynomials) to compute approximate values of the function for values of x close to the point c about which you expanded. For the first time, this gave us the means of calculating approximate values of transcendental functions like e^x, $\ln x$ and $\sin x$, which we otherwise could not compute. Although this is a good reason for studying series expansions of functions, it is not the only reason we need them.

Observe that many phenomena we encounter in the world around us are periodic in nature. That is, they repeat themselves over and over again. For instance, light, sound, radio waves and x-rays (to mention only a few) are all periodic. For such phenomena, Taylor polynomial approximations have some obvious shortcomings. Look back at any of the graphs you (or we) constructed of the partial sums of a convergent Taylor series expansion of a periodic function and you'll notice that as x gets farther away from c (the point you expanded about), the greater the difference between the function and a given Taylor polynomial grows. Such behavior, as we illustrate in Figure 8.43 for the case of $f(x) = \sin x$ expanded about $x = \frac{\pi}{2}$, is typical of the convergence of Taylor series.

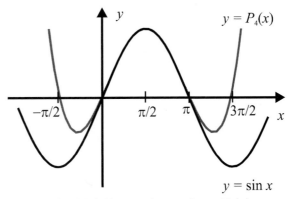

FIGURE 8.43: $y = \sin x$ and $y = P_4(x)$

Because the Taylor polynomials provide an accurate approximation only in the vicinity of c, we say that they are accurate *locally*. In general, no matter how large you make n, the approximation is still only valid locally. In many situations, notably in communications, we have the need to find an approximation to a given periodic function that is valid *globally* (i.e., for all x). Consider the Taylor polynomial graphed in Figure 8.44 and convince yourself that Taylor polynomials will not satisfy this need. For this reason, we construct a different type of series expansion for periodic functions, one where each of the terms in the expansion is periodic.

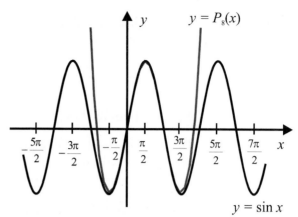

FIGURE 8.44: $y = \sin x$ and $y = P_8(x)$

We say that a function f is **periodic** of **period** T, if $f(x + T) = f(x)$, for all x in the domain of f. Can you think of any periodic functions? Surely, $\sin x$ and $\cos x$ come to mind. These are both periodic of period 2π. Further, $\sin(2x)$, $\cos(2x)$, $\sin(3x)$, $\cos(3x)$ and so on are all periodic of period 2π. In fact, the simplest periodic functions you can think of (aside from constant functions) are the functions

$$\sin(kx) \text{ and } \cos(kx), \text{ for } k = 1, 2, 3, \ldots.$$

Note that each of these is periodic of period 2π, as follows. For any integer k, let $f(x) = \sin(kx)$. We then have

$$f(x + 2\pi) = \sin\left[k(x + 2\pi)\right] = \sin(kx + 2k\pi) = \sin(kx) = f(x).$$

Likewise, you can show that $\cos(kx)$ has period 2π.

So, if you wanted to expand a periodic function in a series, the simplest periodic functions to use in the terms of the series are just these functions. We consider a series of the form

$$\frac{a_0}{2} + \sum_{k=1}^{\infty} \left[a_k \cos(kx) + b_k \sin(kx)\right].$$

Such a series is called a **Fourier series**. Notice that if the series converges, it will converge to a periodic function whose period is 2π, since every term in the series has period 2π. The coefficients of the series, $a_0, a_1, a_2, \ldots$ and $b_1, b_2, \ldots$ are called the **Fourier coefficients**. You may have noticed the unusual way in which we wrote the leading term of the series ($\frac{a_0}{2}$). We did this in order to simplify the formulas for computing these coefficients, as we'll see later.

There are a number of important questions we must address.

• What functions can be expanded in a Fourier series?
• How do we compute the Fourier coefficients?
• Does the Fourier series converge? If so, to what function does the series converge?

We begin our investigation much as we did with power series. Suppose that a given Fourier series converges on the interval $[-\pi, \pi]$. It then represents a function f on that interval,

$$f(x) = \frac{a_0}{2} + \sum_{k=1}^{\infty} \left[a_k \cos(kx) + b_k \sin(kx)\right], \tag{8.1}$$

where f must be periodic outside of $[-\pi, \pi]$. Although some of the details of the proof are beyond the level of this course, we want to give you some idea of how the Fourier coefficients are computed. If we integrate both sides of equation (8.1) with respect to x on the interval $[-\pi, \pi]$, we get

$$\int_{-\pi}^{\pi} f(x)\,dx = \int_{-\pi}^{\pi} \frac{a_0}{2}\,dx + \int_{-\pi}^{\pi} \sum_{k=1}^{\infty} [a_k \cos(kx) + b_k \sin(kx)]\,dx$$

$$= \int_{-\pi}^{\pi} \frac{a_0}{2}\,dx + \sum_{k=1}^{\infty} \left[a_k \int_{-\pi}^{\pi} \cos(kx)\,dx + b_k \int_{-\pi}^{\pi} \sin(kx)\,dx \right], \quad (8.2)$$

assuming we can interchange the order of integration and summation. In general, it's **not** alright to do so (this is beyond the level of this course), but for many Fourier series, doing so is permissible. Observe that for every $k = 1, 2, 3, \dots$, we have

$$\int_{-\pi}^{\pi} \cos(kx)\,dx = \left. \frac{1}{k} \sin(kx) \right|_{-\pi}^{\pi} = \frac{1}{k} [\sin(k\pi) - \sin(-k\pi)] = 0$$

and

$$\int_{-\pi}^{\pi} \sin(kx)\,dx = \left. -\frac{1}{k} \cos(kx) \right|_{-\pi}^{\pi} = -\frac{1}{k} [\cos(k\pi) - \cos(-k\pi)] = 0.$$

This reduces equation (8.2) to simply

$$\int_{-\pi}^{\pi} f(x)\,dx = \int_{-\pi}^{\pi} \frac{a_0}{2}\,dx = a_0\,\pi.$$

Solving this for a_0, we have

$$a_0 = \frac{1}{\pi} \int_{-\pi}^{\pi} f(x)\,dx. \quad (8.3)$$

If we multiply both sides of equation (8.1) by $\cos(nx)$, and then integrate with respect to x on the interval $[-\pi, \pi]$, observe that we get

$$\int_{-\pi}^{\pi} f(x) \cos(nx)\,dx$$

$$= \int_{-\pi}^{\pi} \frac{a_0}{2} \cos(nx)\,dx$$

$$+ \int_{-\pi}^{\pi} \sum_{k=1}^{\infty} [a_k \cos(kx) \cos(nx) + b_k \sin(kx) \cos(nx)]\,dx$$

$$= \frac{a_0}{2} \int_{-\pi}^{\pi} \cos(nx)\,dx$$

$$+ \sum_{k=1}^{\infty} \left[a_k \int_{-\pi}^{\pi} \cos(kx) \cos(nx)\,dx + b_k \int_{-\pi}^{\pi} \sin(kx) \cos(nx)\,dx \right], \quad (8.4)$$

again assuming we can interchange the order of integration and summation. Next, recall that

$$\int_{-\pi}^{\pi} \cos(nx)\,dx = 0, \text{ for all } n = 1, 2, \dots.$$

It's an easy exercise to show that

$$\int_{-\pi}^{\pi} \sin(kx) \cos(nx)\,dx = 0, \text{ for all } n = 1, 2, \dots \text{ and for all } k = 1, 2, \dots$$

and that

$$\int_{-\pi}^{\pi} \cos(kx) \cos(nx)\,dx = \begin{cases} 0, & \text{if } n \neq k \\ \pi, & \text{if } n = k. \end{cases}$$

Notice that this says that every term in the series in equation (8.4) except one (the term corresponding to $k = n$) is zero and equation (8.4) reduces to simply

$$\int_{-\pi}^{\pi} f(x) \cos(nx)\, dx = a_n\, \pi.$$

This gives us (after substituting k for n)

$$a_k = \frac{1}{\pi} \int_{-\pi}^{\pi} f(x) \cos(kx)\, dx, \ \text{for } k = 1, 2, 3, \ldots. \tag{8.5}$$

Likewise, multiplying both sides of equation (8.1) by $\sin(nx)$ and integrating from $-\pi$ to π gives us

$$b_k = \frac{1}{\pi} \int_{-\pi}^{\pi} f(x) \sin(kx)\, dx, \ \text{for } k = 1, 2, 3, \ldots. \tag{8.6}$$

Equations (8.3), (8.5) and (8.6) are called the **Euler-Fourier formulas**. Notice that equation (8.3) is the same as (8.5) with $k = 0$. (This was the reason we chose the leading term of the series to be $\dfrac{a_0}{2}$, instead of simply a_0.)

Let's summarize what we've done so far. We observed that if a Fourier series converges on some interval, then it converges to a function f where the Fourier coefficients satisfy the Euler-Fourier formulas (8.3), (8.5) and (8.6).

Just as we did with power series, given any integrable function f, we can compute the coefficients in (8.3), (8.5) and (8.6) and write down a Fourier series. But, will the series converge and if it does, to what function will it converge? We'll answer these questions shortly. For the moment, let's try to compute the terms of a Fourier series and see what we can observe.

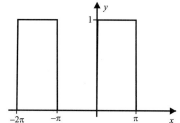

FIGURE 8.45: Square-wave function.

Example 8.1   Finding a Fourier Series Expansion
Find the Fourier series corresponding to the **square-wave** function

$$f(x) = \begin{cases} 0, \text{ if } -\pi < x \le 0 \\ 1, \text{ if } \ \ 0 < x \le \pi, \end{cases}$$

where f is assumed to be periodic outside of the interval $[-\pi, \pi]$ (see the graph in Figure 8.45).

Solution   Notice that even though a_0 satisfies the same formula as a_k, for $k \ge 1$, we must always compute a_0 separately from the rest of the a_k's. From equation (8.5), we get

$$a_0 = \frac{1}{\pi} \int_{-\pi}^{\pi} f(x)\, dx = \frac{1}{\pi} \int_{-\pi}^{0} 0\, dx + \frac{1}{\pi} \int_{0}^{\pi} 1\, dx = 0 + \frac{\pi}{\pi} = 1.$$

We also have for $k \ge 1$,

$$a_k = \frac{1}{\pi} \int_{-\pi}^{\pi} f(x) \cos(kx)\, dx = \frac{1}{\pi} \int_{-\pi}^{0} (0) \cos(kx)\, dx + \frac{1}{\pi} \int_{0}^{\pi} (1) \cos(kx)\, dx$$

$$= \frac{1}{\pi k} \sin(kx) \Big|_{0}^{\pi} = \frac{1}{\pi k} \left[\sin(k\pi) - \sin(0) \right] = 0.$$

Finally, from (8.6), we have

$$b_k = \frac{1}{\pi} \int_{-\pi}^{\pi} f(x) \sin(kx)\, dx = \frac{1}{\pi} \int_{-\pi}^{0} (0) \sin(kx)\, dx + \frac{1}{\pi} \int_{0}^{\pi} (1) \sin(kx)\, dx$$

$$= -\left. \frac{1}{\pi k} \cos(kx) \right|_{0}^{\pi} = -\frac{1}{\pi k} \left[\cos(k\pi) - \cos(0) \right] = -\frac{1}{\pi k} \left[(-1)^k - 1 \right]$$

$$= \begin{cases} 0, & \text{if } k \text{ is even} \\ \dfrac{2}{\pi k}, & \text{if } k \text{ is odd.} \end{cases}$$

We then have the Fourier series

Since $b_{2k} = 0$ and $b_{2k-1} = \dfrac{2}{(2k-1)\pi}$.

$$\frac{a_0}{2} + \sum_{k=1}^{\infty} [a_k \cos(kx) + b_k \sin(kx)] = \frac{1}{2} + \sum_{k=1}^{\infty} b_k \sin(kx) = \frac{1}{2} + \sum_{k=1}^{\infty} b_{2k-1} \sin(2k-1)x$$

$$= \frac{1}{2} + \sum_{k=1}^{\infty} \frac{2}{(2k-1)\pi} \sin\left[(2k-1)x\right]$$

$$= \frac{1}{2} + \frac{2}{\pi} \sin x + \frac{2}{3\pi} \sin(3x) + \frac{2}{5\pi} \sin(5x) + \cdots.$$

Notice that none of our existing convergence tests are appropriate for Fourier series. Since we can't test this, we consider the graphs of the first few partial sums of the series defined by

$$F_n(x) = \frac{1}{2} + \sum_{k=1}^{n} \frac{2}{(2k-1)\pi} \sin\left[(2k-1)x\right].$$

In Figures 8.46a–d, we graph a number of these partial sums.

FIGURE 8.46a: $y = F_4(x)$.

FIGURE 8.46b: $y = F_8(x)$.

FIGURE 8.46c: $y = F_{20}(x)$.

FIGURE 8.46d: $y = F_{50}(x)$.

Notice that as n gets larger and larger, the graph of $F_n(x)$ appears to be approaching the graph of the square-wave function $f(x)$ seen in Figure 8.45. Based on this, we might conjecture that the Fourier series converges to the function $f(x)$. As it turns out, this is not quite correct. We'll see soon that the series converges to $f(x)$, everywhere, *except* at points of discontinuity.

■

Next, we give an example of constructing a Fourier series for another common waveform.

FIGURE 8.47: Triangular wave.

Example 8.2   A Fourier Series Expansion for the Triangular Wave Function
Find the Fourier series expansion of

$$f(x) = |x|, \text{ for } -\pi \le x \le \pi,$$

where f is assumed to be periodic outside of the interval $[-\pi, \pi]$, of period 2π.

Solution   In this case, f is the **triangular wave** function graphed in Figure 8.47. From the Euler-Fourier formulas, we have

$$a_0 = \frac{1}{\pi} \int_{-\pi}^{\pi} |x| \, dx = \frac{1}{\pi} \int_{-\pi}^{0} -x \, dx + \frac{1}{\pi} \int_{0}^{\pi} x \, dx$$

$$= -\frac{1}{\pi} \frac{x^2}{2} \Big|_{-\pi}^{0} + \frac{1}{\pi} \frac{x^2}{2} \Big|_{0}^{\pi} = \frac{\pi}{2} + \frac{\pi}{2} = \pi.$$

Similarly, for each $k \ge 1$, we get

$$a_k = \frac{1}{\pi} \int_{-\pi}^{\pi} |x| \cos(kx) \, dx = \frac{1}{\pi} \int_{-\pi}^{0} (-x) \cos(kx) \, dx + \frac{1}{\pi} \int_{0}^{\pi} x \cos(kx) \, dx.$$

Both integrals require the same integration by parts. We let

$$u = x \quad dv = \cos(kx) \, dx$$
$$du = dx \quad v = \tfrac{1}{k} \sin(kx)$$

so that

$$a_k = -\frac{1}{\pi} \int_{-\pi}^{0} x \cos(kx) \, dx + \frac{1}{\pi} \int_{0}^{\pi} x \cos(kx) \, dx$$

$$= -\frac{1}{\pi} \left[\frac{x}{k} \sin(kx) \right]_{-\pi}^{0} + \frac{1}{\pi k} \int_{-\pi}^{0} \sin(kx) \, dx + \frac{1}{\pi} \left[\frac{x}{k} \sin(kx) \right]_{0}^{\pi} - \frac{1}{\pi k} \int_{0}^{\pi} \sin(kx) \, dx$$

$$= -\frac{1}{\pi} \left[0 + \frac{\pi}{k} \sin(-\pi k) \right] - \frac{1}{\pi k^2} \cos(kx) \Big|_{-\pi}^{0}$$

$$+ \frac{1}{\pi} \left[\frac{\pi}{k} \sin(\pi k) - 0 \right] + \frac{1}{\pi k^2} \cos(kx) \Big|_{0}^{\pi}$$

$$= 0 - \frac{1}{\pi k^2} \left[\cos 0 - \cos(-k\pi) \right] + 0 + \frac{1}{\pi k^2} \left[\cos(k\pi) - \cos 0 \right]$$

Since $\cos(k\pi) = 1$, when k is even and $\cos(k\pi) = -1$, when k is odd.

$$= \frac{2}{\pi k^2} \left[\cos(k\pi) - 1 \right] = \begin{cases} 0, & \text{if } k \text{ is even} \\ \dfrac{-4}{\pi k^2}, & \text{if } k \text{ is odd.} \end{cases}$$

We leave it as an exercise to show that

$$b_k = 0, \text{ for all } k.$$

This gives us the Fourier series

Since $b_k = 0$ for all k.

$$\frac{a_0}{2} + \sum_{k=1}^{\infty} [a_k \cos(kx) + b_k \sin(kx)] = \frac{\pi}{2} + \sum_{k=1}^{\infty} a_k \cos(kx) = \frac{\pi}{2} + \sum_{k=1}^{\infty} a_{2k-1} \cos(2k-1)x$$

Since $a_{2k} = 0$ and
$a_{2k-1} = \dfrac{4}{(2k-1)^2 \pi}$.

$$= \frac{\pi}{2} - \sum_{k=1}^{\infty} \frac{4}{(2k-1)^2 \pi} \cos(2k-1)x$$

$$= \frac{\pi}{2} - \frac{4}{\pi} \cos x - \frac{4}{9\pi} \cos(3x) - \frac{4}{25\pi} \cos(5x) - \cdots.$$

You can show that this series converges absolutely for all x, by using the Comparison Test, since

$$|a_k| = \left| \frac{4}{\pi(2k-1)^2} \cos(2k-1)x \right| \le \frac{4}{\pi(2k-1)^2}$$

and the series $\displaystyle\sum_{k=1}^{\infty} \frac{4}{\pi(2k-1)^2}$ converges. (Hint: Compare this last series to the convergent p-series $\displaystyle\sum_{k=1}^{\infty} \frac{1}{k^2}$, using the Limit Comparison Test). To get an idea of the function to which the series is converging, we plot several of the partial sums of the series,

$$F_n(x) = \frac{\pi}{2} - \sum_{k=1}^{n} \frac{4}{\pi(2k-1)^2} \cos(2k-1)x.$$

See if you can conjecture the sum of the series, by looking at Figures 8.48a–d.

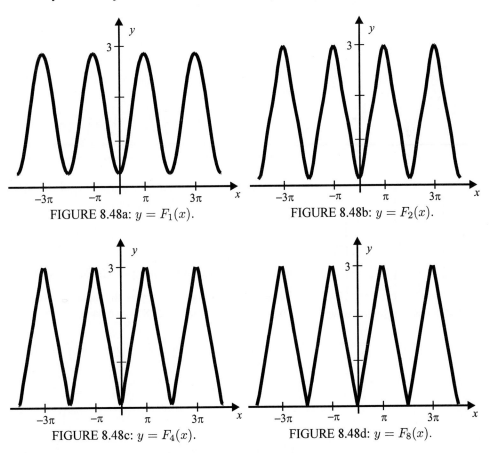

FIGURE 8.48a: $y = F_1(x)$.

FIGURE 8.48b: $y = F_2(x)$.

FIGURE 8.48c: $y = F_4(x)$.

FIGURE 8.48d: $y = F_8(x)$.

Notice how quickly the partial sums of the series appear to converge to the triangular wave function f (see Figure 8.47). As it turns out, the graph of the partial sums will not change appreciably if you plot $F_n(x)$ for much larger values of n. (Try this!) We'll see later how to be sure that the Fourier series converges to $f(x)$ for all x. There's something further to note here: the accuracy of the approximation is fairly uniform. That is, the difference between a given partial sum and f is roughly the same for each x. Take care to distinguish this behavior from that of Taylor polynomial approximations, where the farther you get away from the point about which you've expanded, the worse the approximation tends to get.

∎

Functions of Period Other than 2π

Now, suppose you have a function f that is periodic of period T, but $T \neq 2\pi$. In this case, we want to expand f in a series of simple functions of period T. First, define $l = \frac{T}{2}$ and notice that

$$\cos\left(\frac{k\pi x}{l}\right) \text{ and } \sin\left(\frac{k\pi x}{l}\right)$$

are periodic of period $T = 2l$, for each $k = 1, 2, \ldots$. (Hint: To prove this, let $f(x) = \cos\left(\frac{k\pi x}{l}\right)$ and show that $f(x + 2l) = f(x)$, for all x and for each $k = 1, 2, \ldots$. Likewise, for $g(x) = \sin\left(\frac{k\pi x}{l}\right)$, show that $g(x + 2l) = g(x)$ for all x and for $k = 1, 2, \ldots$.) The Fourier series expansion of f of period $2l$ is then

$$\frac{a_0}{2} + \sum_{k=1}^{\infty} \left[a_k \cos\left(\frac{k\pi x}{l}\right) + b_k \sin\left(\frac{k\pi x}{l}\right)\right].$$

We leave it as an exercise to show that the Fourier coefficients in this case are given by the Euler-Fourier formulas:

$$a_k = \frac{1}{l} \int_{-l}^{l} f(x) \cos\left(\frac{k\pi x}{l}\right) dx, \text{ for } k = 0, 1, 2, \ldots \qquad (8.7)$$

and

$$b_k = \frac{1}{l} \int_{-l}^{l} f(x) \sin\left(\frac{k\pi x}{l}\right) dx, \text{ for } k = 1, 2, 3, \ldots. \qquad (8.8)$$

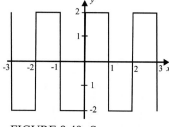

FIGURE 8.49: Square wave.

Example 8.3   A Fourier Series Expansion for a Square Wave Function
Find a Fourier series expansion for the function

$$f(x) = \begin{cases} -2, \text{ if } -1 < x \leq 0 \\ 2, \quad \text{if } 0 < x \leq 1, \end{cases}$$

where f is defined so that it is periodic of period 2 outside of the interval $[-1, 1]$.

Solution   The graph of f is the square-wave seen in Figure 8.49. From the Euler-Fourier formulas (8.7) and (8.8), we have

$$a_0 = \frac{1}{1} \int_{-1}^{1} f(x)\, dx = \int_{-1}^{0} (-2)\, dx + \int_{0}^{1} 2\, dx = 0.$$

Likewise, we get

$$a_k = \frac{1}{1} \int_{-1}^{1} f(x) \cos\left(\frac{k\pi x}{1}\right) dx = 0, \text{ for } k = 1, 2, 3, \ldots.$$

(This is left as an exercise.)

Finally, we have

$$b_k = \frac{1}{1} \int_{-1}^{1} f(x) \sin\left(\frac{k\pi x}{1}\right) \, dx = \int_{-1}^{0} (-2)\sin(k\pi x)\, dx + \int_{0}^{1} 2\sin(k\pi x)\, dx$$

$$= \frac{2}{k\pi}\cos(k\pi x)\Big|_{-1}^{0} - \frac{2}{k\pi}\cos(k\pi x)\Big|_{0}^{1} = \frac{4}{k\pi}\left[\cos 0 - \cos(-k\pi)\right]$$

Since $\cos(k\pi) = 1$, when k is even and $\cos(k\pi) = -1$, when k is odd.

$$= \frac{4}{k\pi}\left[1 - \cos(k\pi)\right] = \begin{cases} 0, & \text{if } k \text{ is even} \\ \dfrac{8}{k\pi}, & \text{if } k \text{ is odd.} \end{cases}$$

This gives us the Fourier series

Since $b_{2k} = 0$ and $b_{2k-1} = \dfrac{8}{(2k-1)\pi}$.

$$\frac{a_0}{2} + \sum_{k=1}^{\infty}\left[a_k \cos(k\pi x) + b_k \sin(k\pi x)\right] = \sum_{k=1}^{\infty} b_k \sin(k\pi x) = \sum_{k=1}^{\infty} b_{2k-1}\sin\left[(2k-1)\pi x\right]$$

$$= \sum_{k=1}^{\infty} \frac{8}{(2k-1)\pi}\sin\left[(2k-1)\pi x\right].$$

Although we as yet have no tools for determining the convergence or divergence of this series, we graph a few of the partial sums of the series,

$$F_n(x) = \sum_{k=1}^{n} \frac{8}{(2k-1)\pi}\sin\left[(2k-1)\pi x\right]$$

in Figures 8.50a–d.

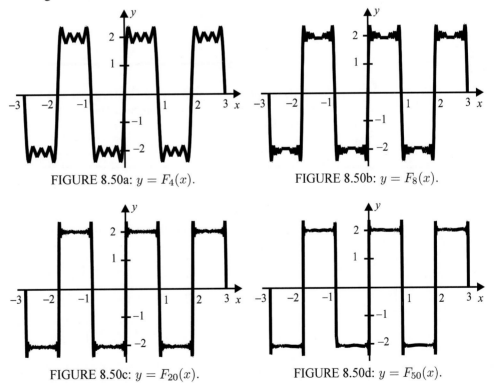

FIGURE 8.50a: $y = F_4(x)$.

FIGURE 8.50b: $y = F_8(x)$.

FIGURE 8.50c: $y = F_{20}(x)$.

FIGURE 8.50d: $y = F_{50}(x)$.

From the graphs, it appears that the series is converging to the square wave function f, except at the points of discontinuity, $x = 0, \pm 1, \pm 2, \pm 3, \ldots$. At those, points, the series appears to converge to 0. You can easily verify this by observing that

$$\frac{8}{(2k-1)\pi}\sin\left[(2k-1)\pi x\right] = 0, \text{ for integer values of } x.$$

Since each term in the series is zero, the series converges to 0 at all integer values of x. You might think of this as follows: at the points of discontinuity, the series is converging to the average of the two function values on either side of the discontinuity. As we will see, this is typical of the convergence of Fourier series.

■

We now state the following major result on the convergence of Fourier series without proof.

Theorem 8.1 **(Fourier Convergence Theorem)** Suppose that f is periodic of period $2l$ and that f and f' are continuous on the interval $[-l, l]$, except for at most a finite number of jump discontinuities. Then, f has a convergent Fourier series expansion. Further, the series converges to $f(x)$, when f is continuous at x and to

$$\frac{1}{2} \left[\lim_{t \to x^+} f(t) + \lim_{t \to x^-} f(t) \right]$$

at any points x where f is discontinuous.

Proof. The proof of the theorem is beyond the level of this text, and can be found in texts on advanced calculus or Fourier analysis.

Remark 8.1   The Fourier Convergence Theorem says that a Fourier series may converge to a discontinuous function, even though every term in the series is continuous (and differentiable) for all x.

Example 8.4   Proving Convergence of a Fourier Series
     Use the Fourier Convergence Theorem to prove that the Fourier series expansion of period 2π,

$$\frac{\pi}{2} - \sum_{k=1}^{\infty} \frac{4}{(2k-1)^2 \pi} \cos(2k-1)x,$$

derived in example 8.2, for $f(x) = |x|$, for $-\pi \leq x \leq \pi$ and periodic outside of $[-\pi, \pi]$, converges to $f(x)$ everywhere.

Solution   First, note that f is continuous everywhere (see Figure 8.47). We also have that since

$$f(x) = |x| = \begin{cases} -x, \text{ if } -\pi \leq x < 0 \\ x, \quad \text{ if } 0 \leq x < \pi \end{cases}$$

and is periodic outside $[-\pi, \pi]$, then

$$f'(x) = \begin{cases} -1, -\pi < x < 0 \\ 1, \quad 0 < x < \pi. \end{cases}$$

So, f' is also continuous on $[-\pi, \pi]$, except for jump discontinuities at $x = 0$ and $x = \pm\pi$. From the Fourier Convergence Theorem, we now have that the Fourier series converges to f everywhere (since f is continuous everywhere). Because of this, we can write

$$f(x) = \frac{\pi}{2} - \sum_{k=1}^{\infty} \frac{4}{(2k-1)^2 \pi} \cos(2k-1)x,$$

for all x.

■

As you can see from the Fourier Convergence Theorem, Fourier series do not always converge to the function you are expanding.

Example 8.5   Proving Convergence of a Fourier Series

Use the Fourier Convergence Theorem to investigate the convergence of the Fourier series

$$\sum_{k=1}^{\infty} \frac{8}{(2k-1)\pi} \sin\left[(2k-1)\pi x\right],$$

derived as an expansion of the square-wave function

$$f(x) = \begin{cases} -2, & \text{if } -1 < x \le 0 \\ 2, & \text{if } \;\; 0 < x \le 1 \end{cases},$$

where f is taken to be periodic outside of $[-1, 1]$ (see example 8.3).

Solution   First, note that f is continuous, except for jump discontinuities at $x = 0, \pm1, \pm2, \dots$. Further,

$$f'(x) = \begin{cases} 0, & \text{if } -1 < x < 0 \\ 0, & \text{if } \;\; 0 < x < 1 \end{cases}$$

and is periodic outside of $[-1, 1]$. Thus, f' is also continuous everywhere, except for jump discontinuities at integer values of x. From the Fourier Convergence Theorem, the Fourier series will converge to $f(x)$ everywhere, except at the discontinuities, $x = 0, \pm1, \pm2, \dots$, where the series converges to the average of the one-sided limits, that is, 0. (Why 0?) Since the series does not converge to f everywhere, we cannot say that the function and the series are *equal*. In this case, we usually write

$$f(x) \sim \sum_{k=1}^{\infty} \frac{8}{(2k-1)\pi} \sin\left[(2k-1)\pi x\right]$$

to indicate that the series *corresponds* to f (but is not necessarily equal to f). In the case of Fourier series, this says that the series converges to $f(x)$ at every x where f is continuous and to the average of the one-sided limits at any jump discontinuities. Notice that this is the behavior seen in the graphs of the partial sums of the series seen in Figures 8.50a–d. ■

Fourier Series and Music Synthesizers

You may be wondering why we have taken the trouble to investigate Fourier series. There must be some significant payoff for all of this work. Most people find it surprising that series of sines and cosines can converge to form line segments, parabolas and other shapes on finite intervals. As interesting as this is, this is not why we have pursued this study. As it turns out, Fourier series are widely used in engineering, physics, chemistry and so on. We give you a sense of how Fourier series are used with the following brief discussion of music synthesizers and a variety of exercises.

Suppose that you had a music machine that could generate pure tones at various pitches and volumes. What types of sounds could you synthesize by combining several pure tones together? To answer this question, we first translate the problem into mathematics. A pure tone can be modeled by $A \sin \omega t$, where the amplitude A determines the volume and the frequency ω determines the pitch. For example, to mimic a saxophone, you must match the characteristic waveform of a saxophone (see Figure 8.51). The shape of the waveform affects the **timbre** of the tone, a quality most humans readily discern (a saxophone *sounds* different than a trumpet, doesn't it?).

FIGURE 8.51: Saxophone waveform.

Consider the following music synthesizer problem. Given a waveform such as the one shown in Figure 8.51, can you add together several pure tones of the form $A \sin \omega t$ to approximate the waveform? Note that if the pure tones are of the form $b_1 \sin t$, $b_2 \sin 2t$, $b_3 \sin 3t$ and so on, this is essentially a Fourier series problem. That is, we want to approximate a given wave function $f(t)$ by a sum of these pure tones, as follows:

$$f(t) \approx b_1 \sin t + b_2 \sin 2t + b_3 \sin 3t + \cdots + b_n \sin nt.$$

Although the cosine terms are all missing, notice that this is the partial sum of a Fourier series. (Such series are called **Fourier sine series** and are explored in the exercises.) For music synthesizers, the Fourier coefficients are simply the amplitudes of the various harmonics in a given waveform. In this context, you can think of the bass and treble knobs on a stereo as manipulating the amplitudes of different terms in a Fourier series. Cranking up the bass emphasizes low frequency terms (i.e., increases the coefficients of the first few terms of the Fourier series), while turning up the treble emphasizes the high frequency terms. An equalizer (see Figure 8.52) gives you more direct control of individual frequencies.

FIGURE 8.52: A graphic equalizer.

In general, the idea of analyzing a wave phenomenon by breaking the wave down into its component frequencies is essential to much of modern science and engineering. This type of **spectral analysis** is used in numerous scientific disciplines.

Exercises 8.8

1. (writing) Explain why the Fourier series of $f(x) = 1 + 3\cos x - \sin 2x$ on the interval $[-\pi, \pi]$ is simply $1 + 3\cos x - \sin 2x$. (Hint: Explain what the goal of a Fourier series representation is and note that in this case no work needs to be done.) Would this change if the interval were $[-1, 1]$ instead?

2. (writing) Polynomials are built up from the basic operations of arithmetic. We often use Taylor series to rewrite an awkward function (e.g., $\sin x$) into arithmetic form. Many natural phenomena are waves, which are well-modeled by sines and cosines. Discuss the extent to which the following statement is true: Fourier series allow us to rewrite algebraic functions (e.g., x^2) into a natural (wave) form.

3. (writing) Theorem 8.1 states that a Fourier series may converge to a function with jump discontinuities. In examples 8.1 and 8.3, identify the locations of the jump discontinuities and the values to which the Fourier series converges at these points. In what way are these values reasonable compromises?

4. (writing) Carefully examine Figures 8.46 and 8.50. For which x's does the Fourier series seem to converge rapidly? slowly? Note that for every n the partial sum $F_n(x)$ passes *exactly* through the limiting point for jump discontinuities. Describe the behavior of the partial sums *near* the jump discontinuities. This overshoot/undershoot behavior is referred to as the **Gibbs phenomenon** (see exercise 49 below).

In exercises 5–12, find the Fourier series of the function on the interval $[-\pi, \pi]$. **Graph the function and the partial sums $F_4(x)$ and $F_8(x)$ on the interval $[-2\pi, 2\pi]$.**

5. $f(x) = x$

6. $f(x) = x^2$

7. $f(x) = 2|x|$

8. $f(x) = 3x$

9. $f(x) = \begin{cases} 1, & \text{if } -\pi < x < 0 \\ -1, & \text{if } 0 < x < \pi \end{cases}$

10. $f(x) = \begin{cases} 1, & \text{if } -\pi < x < 0 \\ 0, & \text{if } 0 < x < \pi \end{cases}$

11. $f(x) = 3\sin 2x$

12. $f(x) = 2\sin 3x$

In exercises 13–18, find the Fourier series of the function on the given interval.

13. $f(x) = -x$, $[-1, 1]$

14. $f(x) = |x|$, $[-1, 1]$

15. $f(x) = x^2$, $[-1, 1]$

16. $f(x) = 3x$, $[-2, 2]$

17. $f(x) = \begin{cases} 0, & \text{if } -1 < x < 0 \\ x, & \text{if } 0 < x < 1 \end{cases}$

18. $f(x) = \begin{cases} 0, & \text{if } -1 < x < 0 \\ 1 - x, & \text{if } 0 < x < 1 \end{cases}$

In exercises 19–24, do not compute the Fourier series but graph the function to which the Fourier series converges, showing at least three full periods of the limit function.

19. $f(x) = x$, $[-2, 2]$

20. $f(x) = x^2$, $[-3, 3]$

21. $f(x) = \begin{cases} -x, & \text{if } -1 < x < 0 \\ 0, & \text{if } 0 < x < 1 \end{cases}$

22. $f(x) = \begin{cases} 1, & \text{if } -2 < x < 0 \\ 3, & \text{if } 0 < x < 2 \end{cases}$

23. $f(x) = \begin{cases} -1, & \text{if } -2 < x < -1 \\ 0, & \text{if } -1 < x < 1 \\ 1, & \text{if } 1 < x < 2 \end{cases}$

24. $f(x) = \begin{cases} 2, & \text{if } -2 < x < -1 \\ -2, & \text{if } -1 < x < 1 \\ 0, & \text{if } 1 < x < 2 \end{cases}$

25. Substitute $x = 1$ into the Fourier series formula of exercise 15 to prove that $\sum_{k=0}^{\infty} \frac{1}{k^2} = \frac{\pi^2}{6}$.

26. Use the Fourier series of example 8.1 to prove that $\sum_{k=1}^{\infty} \frac{\sin(2k-1)}{(2k-1)} = \frac{\pi}{4}$.

27. Use the Fourier series of example 8.2 to prove that $\sum_{k=1}^{\infty} \frac{1}{(2k-1)^2} = \frac{\pi^2}{8}$.

28. Combine the results of exercises 25 and 27 to find $\sum_{k=1}^{\infty} \frac{1}{(2k)^2}$.

29. You have undoubtedly noticed that many Fourier series consist of only cosine or only sine terms. This can be easily understood in terms of even and odd functions. A function $f(x)$ is **even** if $f(-x) = f(x)$ for all x. A function is **odd** if $f(-x) = -f(x)$ for all x. Show that $\cos x$ is even, $\sin x$ is odd and $\cos x + \sin x$ is neither.

30. If $f(x)$ is even, show that $g(x) = f(x) \cos x$ is even and $h(x) = f(x) \sin x$ is odd.

31. If $f(x)$ is odd, show that $g(x) = f(x) \cos x$ is odd and $h(x) = f(x) \sin x$ is even.

32. If f and g are even, what can you say about $f\,g$?

33. If f is even and g is odd, what can you say about $f\,g$?

34. If f and g are odd, what can you say about $f\,g$?

35. Prove the general Euler-Fourier formulas (8.7) and (8.8).

36. If $g(x)$ is an odd function (see exercise 29), show that $\int_{-l}^{l} g(x)\,dx = 0$ for any (positive) constant l. (Hint: Compare $\int_{-l}^{0} g(x)\,dx$ and $\int_{0}^{l} g(x)\,dx$. You will need to make the change of variable: $t = -x$ in one of the integrals.) Using the results of exercise 30, show that if $f(x)$ is even then $b_k = 0$ for all k and the Fourier series of $f(x)$ consists only of a constant and cosine terms. If $f(x)$ is odd, show that $a_k = 0$ for all k and the Fourier series of $f(x)$ consists only of sine terms.

In exercises 37–42, use the even/odd properties of $f(x)$ to predict (don't compute) whether the Fourier series will contain only cosine, only sine terms or both.

37. $f(x) = x^3$

38. $f(x) = x^4$

39. $f(x) = e^x$

40. $f(x) = |x|$

41. $f(x) = \begin{cases} 0, & \text{if } -1 < x < 0 \\ x, & \text{if } 0 < x < 1 \end{cases}$

42. $f(x) = \begin{cases} -1, & \text{if } -2 < x < 0 \\ 3, & \text{if } 0 < x < 2 \end{cases}$

43. The function $f(x) = \begin{cases} -1, & \text{if } -2 < x < 0 \\ 3, & \text{if } 0 < x < 2 \end{cases}$ is neither even nor odd, but can be written as $f(x) = g(x) + 1$ where $g(x) = \begin{cases} -2, & \text{if } -2 < x < 0 \\ 2, & \text{if } 0 < x < 2 \end{cases}$. Explain why the Fourier series of $f(x)$ will contain sine terms and the constant 1, but no cosine terms.

44. Suppose that you want to find the Fourier series of $f(x) = x + x^2$. Explain why to find b_k you would only need to integrate $x \sin\left(\frac{k\pi x}{l}\right)$ and to find a_k you would only need to integrate $x^2 \cos\left(\frac{k\pi x}{l}\right)$.

Exercises 45–48 are adapted from the owner's manual of a high-end music synthesizer.

45. A fundamental choice to be made when generating a new tone on a music synthesizer is the waveform. The options are sawtooth, square and pulse. You worked with the sawtooth wave in exercise 13. Graph the limiting function for the function in exercise 13 on the interval $[-4, 4]$. Explain why "sawtooth" is a good name. A square wave is shown in Figure 8.49. A pulse wave with width $1/n$ is generated by $f(x) = \begin{cases} -2, & \text{if } 1/n < |x| < 1 \\ 2, & \text{if } |x| \le 1/n \end{cases}$. Graph pulse waves of width $1/3$ and $1/4$ on the interval $[-4, 4]$.

46. The **harmonic content** of a wave equals the ratio of integral harmonic waves to the fundamental wave. To understand what this means, write the Fourier series of exercise 13 as $\frac{2}{\pi}(\sin \pi x + \frac{1}{2} \sin 2\pi x + \frac{1}{3} \sin 3\pi x + \frac{1}{4} \sin 4\pi x + \cdots)$. The harmonic content of the sawtooth wave is $\frac{1}{n}$. Explain how this relates to the relative sizes of the Fourier coefficients. The harmonic content of the square wave is $\frac{1}{n}$ with even-numbered harmonics missing. Compare this description to the Fourier series of example 8.3. The harmonic content of the pulse wave of width $\frac{1}{3}$ is $\frac{1}{n}$ with every third harmonic missing. Without computing the Fourier coefficients, write out the general form of the Fourier series of $f(x) = \begin{cases} -2, & \text{if } 1/3 < |x| < 1 \\ 2, & \text{if } |x| \le 1/3 \end{cases}$.

47. The cutoff frequency setting on a music synthesizer has a dramatic effect on the timbre of the tone produced. In terms of harmonic content (see exercise 46), when the cutoff frequency is set at $n > 0$, all harmonics beyond the nth harmonic are set equal to 0. In Fourier series terms, explain how this corresponds to the partial sum $F_n(x)$. For the sawtooth and square waves, graph the waveforms with the cutoff frequency set at 4. Compare these to the waveforms with the cutoff frequency set at 2. As the setting is lowered, you hear more of a "pure" tone. Briefly explain why.

48. The resonance setting on a music synthesizer also changes timbre significantly. Set at 1, you get the basic waveform (e.g., sawtooth or square). Set at 2, the harmonic content of the first four harmonics are divided by 2, the fifth harmonic is multiplied by $\frac{3}{4}$, the sixth harmonic is left the same, the seventh harmonic is divided by 2 and the remaining harmonics are set to 0. Graph the sawtooth and square waves with resonance set to 2. Which one is starting to resemble the saxophone waveform of Figure 8.51?

49. The Gibbs phenomenon is the tendency of approximating partial sums of Fourier series to badly undershoot and overshoot the limit function near jump discontinuities (places where the limit function changes rapidly). A black-and-white photograph can be digitized by partitioning the photograph into small rectangles and assigning each rectangle a number. For example, a pure white rectangle might be a 1 and a pure black rectangle a 10 with grey rectangles assigned values between 1 and 10. The digitized photograph can then be approximated by a Fourier series. A sharp edge in a photograph would have a rapid change from black to white, similar to a jump discontinuity. The rapid change in values from black to white requires high frequency components in the Fourier series (i.e., the terms of the series for large values of k.) The photograph on the left is out of focus, to the point that we can't read the markings on the plane (you can imagine this would be important in many tense situations). By increasing the coefficients in the high frequencies in the Fourier series representation of the photograph, you can get the sharper photograph on the right. This photograph also has a distinct "halo effect" around the plane. Explain how the halo could be related to the Gibbs phenomenon. (photos reprinted by permission from *Visualization* by Friedhoff and Benzon).

(Images courtesy of Vicom Systems.)

50. Piano tuning is relatively simple due to the phenomenon studied in this exercise. Compare the graphs of $\sin 8t + \sin 8.2t$ and $2\sin 8t$. Note especially that the amplitude of $\sin 8t + \sin 8.2t$ appears to slowly rise and fall. In the trig identity $\sin 8t + \sin 8.2t = [2\cos(0.2t)]\sin(8.1t)$, think of $2\cos(0.2t)$ as the amplitude of $\sin(8.1t)$ and explain why the amplitude varies slowly. Piano tuners often start by striking a tuning fork of a certain pitch (e.g., $\sin 8t$) and then striking the corresponding piano note. If the piano is slightly out-of-tune (e.g., $\sin 8.2t$), the tuning fork plus piano produces a combined tone that noticeably increases and decreases in volume. Use your graph to explain why this occurs.

51. The function $\sin 8\pi t$ represents a 4 Hz signal (1 Hz equals 1 cycle per second) if t is measured in seconds. If you received this signal, your task might be to take your measurements of the signal and try to reconstruct the function. For example, if you measured three samples per second, you would have the data $f(0) = 0$, $f(1/3) = \sqrt{3}/2$, $f(2/3) = -\sqrt{3}/2$ and $f(1) = 0$. Knowing the signal is of the form $A\sin Bt$, you would use the data to try to solve for A and B. In this case, you don't have enough information to guarantee getting the right values for A and B. Prove this by finding several values of A and B with $B \neq 8\pi$ that match the data. A famous result of H. Nyquist from 1928 states that to reconstruct a signal of frequency f you need at least $2f$ samples.

52. (exploratory) Suppose you wanted to approximate a waveform with sine functions (no cosines), as in the music synthesizer problem. Such a **Fourier sine series** will be derived in this exercise. You essentially use Fourier series with a trick to guarantee sine terms only. Start with your waveform as a function defined on the interval $[0, l]$, for some length l. Then define a function $g(x)$ that equals $f(x)$ on $[0, l]$ **and** that is an odd function. Show that $g(x) = \begin{cases} f(x) & \text{if } 0 \leq x \leq l \\ -f(-x) & \text{if } -l < x < 0 \end{cases}$ works. Explain why the Fourier series expansion of $g(x)$ on $[-l, l]$ would contain sine terms only. This series is the sine series expansion of $f(x)$. Show the following helpful shortcut: the sine series coefficients are

$$b_k = \frac{1}{l}\int_{-l}^{l} g(x)\sin\left(\frac{k\pi}{l}\right)dx = \frac{2}{l}\int_0^l f(x)\sin\left(\frac{k\pi}{l}\right)dx.$$

Then compute the sine series expansion of $f(x) = x^2$ on $[0, 1]$ and graph the limit function on $[-3, 3]$.

53. (exploratory) Analogous to exercise 52, develop a **Fourier cosine series** and find the cosine series expansion of $f(x) = x$ on $[0, 1]$.

54. (exploratory) Fourier series is a part of the field of **Fourier analysis,** which is central to many engineering applications. Fourier analysis includes the Fourier transforms (and the FFT or Fast Fourier Transform) and inverse Fourier transforms, which you will get a brief introduction to in this exercise. Given measurements of a signal (waveform), the goal is to construct the Fourier series of a function. To start with a simple version of the problem, suppose the signal has the form $f(x) = \frac{a_0}{2} + a_1\cos\pi x + a_2\cos 2\pi x + b_1\sin\pi x + b_2\sin 2\pi x$ and you have the measurements $f(-1) = 0$, $f(-\frac{1}{2}) = 1$, $f(0) = 2$, $f(\frac{1}{2}) = 1$ and $f(1) = 0$. Substituting into the general equation for $f(x)$, show that

$$f(-1) = \frac{a_0}{2} - a_1 + a_2 = 0$$

$$f\left(-\frac{1}{2}\right) = \frac{a_0}{2} - a_2 - b_1 = 1$$

$$f(0) = \frac{a_0}{2} + a_1 + a_2 = 2$$

$$f\left(\frac{1}{2}\right) = \frac{a_0}{2} - a_2 + b_1 = 1$$

$$\text{and}\quad f(1) = \frac{a_0}{2} - a_1 + a_2 = 0.$$

Note that the first and last equations are identical and that b_2 never appears in an equation. Thus, you have four equations and four unknowns. Solve the equations. (Hint: Start by comparing the second and fourth equations, then the third and fifth equations.) You should conclude that $f(x) = 1 + \cos \pi x$, with no information about b_2. To determine b_2, we would need another function value. In general, the number of measurements determines how many terms you can find in the Fourier series (see exercise 51). Fortunately, there is an easier way of determining the Fourier coefficients. Recall that

$a_k = \int_{-1}^{1} f(x) \cos n\pi x \, dx$ and $b_k = \int_{-1}^{1} f(x) \sin n\pi x \, dx$. You can estimate the integral using function values at $x = -1/2$, $x = 0$, $x = 1/2$ and $x = 1$. Find a version of a Riemann sum approximation that gives $a_0 = 2$, $a_1 = 1$, $a_2 = 0$ and $b_1 = 0$. What value is given for b_2? Use this Riemann sum rule to find the appropriate coefficients for the data $f(-\frac{3}{4}) = \frac{3}{4}$, $f(-\frac{1}{2}) = \frac{1}{2}$, $f(-\frac{1}{4}) = \frac{1}{4}$, $f(0) = 0$, $f(\frac{1}{4}) = -\frac{1}{4}$, $f(\frac{1}{2}) = -\frac{1}{2}$, $f(\frac{3}{4}) = -\frac{3}{4}$ and $f(1) = -1$. Compare to the Fourier series of exercise 13.

Chapter Review Exercises

In exercises 1–8, determine whether the sequence converges or diverges. If it converges, give the limit.

1. $a_n = \dfrac{4}{3+n}$

2. $a_n = \dfrac{3n}{1+n}$

3. $a_n = (-1)^n \dfrac{n}{n^2 + 4}$

4. $a_n = (-1)^n \dfrac{n}{n+4}$

5. $a_n = \dfrac{4^n}{n!}$

6. $a_n = \dfrac{n!}{n^n}$

7. $a_n = \cos \pi n$

8. $a_n = \dfrac{\cos n\pi}{n}$

In exercises 9–12, find the sum of the convergent series.

9. $\displaystyle\sum_{k=0}^{\infty} 4\left(\tfrac{1}{2}\right)^k$

10. $\displaystyle\sum_{k=1}^{\infty} \dfrac{4}{k(k+2)}$

11. $\displaystyle\sum_{k=0}^{\infty} 4^{-k}$

12. $\displaystyle\sum_{k=0}^{\infty} (-1)^k \dfrac{3}{4^k}$

In exercises 13 and 14, estimate the sum of the series to within 0.01.

13. $\displaystyle\sum_{k=0}^{\infty} (-1)^k \dfrac{k}{k^4 + 1}$

14. $\displaystyle\sum_{k=0}^{\infty} (-1)^{k+1} \dfrac{3}{k!}$

In exercises 15–34, determine if the series converges or diverges.

15. $\displaystyle\sum_{k=0}^{\infty} \dfrac{2k}{k+3}$

16. $\displaystyle\sum_{k=0}^{\infty} (-1)^k \dfrac{2k}{k+3}$

17. $\displaystyle\sum_{k=0}^{\infty} (-1)^k \dfrac{4}{\sqrt{k+1}}$

18. $\displaystyle\sum_{k=0}^{\infty} \dfrac{4}{\sqrt{k+1}}$

19. $\displaystyle\sum_{k=1}^{\infty} 3k^{-7/8}$

20. $\displaystyle\sum_{k=1}^{\infty} 2k^{-8/7}$

21. $\displaystyle\sum_{k=1}^{\infty} \dfrac{\sqrt{k}}{k^3 + 1}$

22. $\displaystyle\sum_{k=1}^{\infty} \dfrac{k}{\sqrt{k^3 + 1}}$

23. $\displaystyle\sum_{k=1}^{\infty} (-1)^k \dfrac{4^k}{k!}$

24. $\displaystyle\sum_{k=1}^{\infty} (-1)^k \dfrac{2^k}{k}$

25. $\displaystyle\sum_{k=1}^{\infty} (-1)^k \ln(1 + \tfrac{1}{k})$

26. $\displaystyle\sum_{k=1}^{\infty} \dfrac{\cos k\pi}{\sqrt{k}}$

27. $\displaystyle\sum_{k=1}^{\infty} \dfrac{2}{(k+3)^2}$

28. $\displaystyle\sum_{k=2}^{\infty} \dfrac{4}{k \ln k}$

29. $\displaystyle\sum_{k=1}^{\infty} \dfrac{k!}{3^k}$

30. $\displaystyle\sum_{k=1}^{\infty} \dfrac{k}{3^k}$

31. $\displaystyle\sum_{k=1}^{\infty} \dfrac{e^{1/k}}{k^2}$

32. $\displaystyle\sum_{k=1}^{\infty} \dfrac{1}{k\sqrt{\ln k + 1}}$

33. $\displaystyle\sum_{k=1}^{\infty} \dfrac{4^k}{(k!)^2}$

34. $\displaystyle\sum_{k=1}^{\infty} \dfrac{k^2 + 4}{k^3 + 3k + 1}$

In exercises 35–38, determine if the series converges absolutely, converges conditionally or diverges.

35. $\displaystyle\sum_{k=1}^{\infty} (-1)^k \dfrac{k}{k^2 + 1}$

36. $\displaystyle\sum_{k=1}^{\infty} (-1)^k \dfrac{3}{k+1}$

37. $\displaystyle\sum_{k=1}^{\infty} \dfrac{\sin k}{k^{3/2}}$

38. $\displaystyle\sum_{k=1}^{\infty} (-1)^{k+1} \dfrac{3}{\ln k + 1}$

In exercises 39 and 40, find all values of p for which the series converges.

39. $\displaystyle\sum_{k=1}^{\infty} \dfrac{2}{(3+k)^p}$

40. $\displaystyle\sum_{k=1}^{\infty} e^{kp}$

In exercises 41 and 42, determine the number of terms necessary to estimate the sum of the series to within 10^{-6}.

41. $\displaystyle\sum_{k=1}^{\infty} (-1)^k \dfrac{3}{k^2}$

42. $\displaystyle\sum_{k=1}^{\infty} (-1)^k \dfrac{2^k}{k!}$

In exercises 43–46, find a power series representation for the function. Find the radius of convergence.

43. $\dfrac{1}{4+x}$ **44.** $\dfrac{2}{6-x}$

45. $\dfrac{3}{3+x^2}$ **46.** $\dfrac{2}{1+4x^2}$

In exercises 47 and 48, use the series from exercises 43 and 44 to find a power series and its radius of convergence.

47. $\ln(4+x)$ **48.** $\ln(6-x)$

In exercises 49–56, find the interval of convergence.

49. $\displaystyle\sum_{k=0}^{\infty}(-1)^k 2x^k$ **50.** $\displaystyle\sum_{k=0}^{\infty}(-1)^k(2x)^k$

51. $\displaystyle\sum_{k=0}^{\infty}(-1)^k\frac{2}{k}x^k$ **52.** $\displaystyle\sum_{k=0}^{\infty}\frac{-3}{\sqrt{k}}\left(\frac{x}{2}\right)^k$

53. $\displaystyle\sum_{k=0}^{\infty}\frac{4}{k!}(x-2)^k$ **54.** $\displaystyle\sum_{k=0}^{\infty}k^2(x+3)^k$

55. $\displaystyle\sum_{k=0}^{\infty}3^k(x-2)^k$ **56.** $\displaystyle\sum_{k=0}^{\infty}\frac{k}{4^k}(x+1)^k$

In exercises 57 and 58, derive the Taylor series of $f(x)$ about the center $x = c$.

57. $f(x) = \sin x,\ c = 0$ **58.** $f(x) = \dfrac{1}{x},\ c = 1$

In exercises 59 and 60, find the Taylor polynomial $P_4(x)$. Graph $f(x)$ and $P_4(x)$.

59. $f(x) = \ln x,\ c = 1$ **60.** $f(x) = \dfrac{1}{\sqrt{x}},\ c = 1$

In exercises 61 and 62, use the Taylor polynomials from exercises 59 and 60 to estimate the given values. Determine the order of the Taylor polynomial needed to estimate the value to within 10^{-8}.

61. $\ln 1.2$ **62.** $\dfrac{1}{\sqrt{1.1}}$

In exercises 63 and 64, use a known Taylor series to find a Taylor series of the function and find its radius of convergence.

63. e^{-3x^2} **64.** $\sin 4x$

In exercises 65 and 66, use a fourth-order polynomial from a known Taylor series to estimate the value of the integral.

65. $\displaystyle\int_0^1 \tan^{-1} x\, dx$ **66.** $\displaystyle\int_0^2 e^{-3x^2}\, dx$

In exercises 67 and 68, derive the Fourier series of the function.

67. $f(x) = x,\ -2 \le x \le 2$

68. $f(x) = \begin{cases} 0 & \text{if } -\pi < x \le 0 \\ 1 & \text{if } 0 < x \le \pi \end{cases}$

In exercises 69–72, graph at least three periods of the function to which the Fourier series converges.

69. $f(x) = x^2,\ -1 \le x \le 1$

70. $f(x) = 2x,\ -2 \le x \le 2$

71. $f(x) = \begin{cases} -1 & \text{if } -1 < x \le 0 \\ 1 & \text{if } 0 < x \le 1 \end{cases}$

72. $f(x) = \begin{cases} 0 & \text{if } -2 < x \le 0 \\ x & \text{if } 0 < x \le 2 \end{cases}$

73. Suppose you and your friend take turns tossing a coin. The first one to get a head wins. Obviously, the person who goes first has an advantage, but how much of an advantage is it? If you go first, the probability that you win on your first toss is $\frac{1}{2}$, the probability that you win on your second toss is $\frac{1}{8}$, the probability that you win on your third toss is $\frac{1}{32}$ and so on. Sum a geometric series to find the probability that you win.

74. In a game similar to that of exercise 73, the first one to roll a 4 on a six-sided die wins. Is this game fairer than the previous game. The probabilities of winning on the first, second and third roll are $\frac{1}{6}$, $\frac{25}{216}$ and $\frac{625}{7776}$, respectively. Sum a geometric series to find the probability that you win.

75. Recall the Fibonacci sequence defined by $a_0 = 1$, $a_1 = 1$, $a_2 = 2$ and $a_{n+1} = a_n + a_{n-1}$. Prove the following fact: $\displaystyle\lim_{n\to\infty}\frac{a_{n+1}}{a_n} = \frac{1+\sqrt{5}}{2}$ (this number, known to the ancient Greeks, is called the **golden ratio**). (Hint: Start with $a_{n+1} = a_n + a_{n-1}$ and divide by a_n. If $r = \displaystyle\lim_{n\to\infty}\frac{a_{n+1}}{a_n}$, argue that $\displaystyle\lim_{n\to\infty}\frac{a_{n-1}}{a_n} = 1/r$ and then solve the equation $r = 1 + 1/r$.)

76. The Fibonacci sequence can be visualized with the following construction. Start with two side-by-side squares of side 1 (Figure A). Above them, draw a square (Figure B), which will have side 2. To the left of that, draw a square (Figure C), which will have side 3. Continue to spiral around, drawing squares which have sides given by the Fibonacci sequence. For each bounding rectangle in Figures A–C, compute the ratio of the sides of the rectangle. (Hint: Start with $\frac{2}{1}$ and then $\frac{3}{2}$.) Find the limit of the ratios as the construction process continues. The Greeks proclaimed this to be the most "pleasing" of all rectangles, building the Parthenon and other important buildings with these proportions (see *The Divine Proportion* by H. E. Huntley).

Figure A   Figure B   Figure C

77. Another type of sequence studied by mathematicians is the **continued fraction**. Numerically explore the sequence

$1 + \dfrac{1}{1}, 1 + \dfrac{1}{1 + \frac{1}{1}}, 1 + \dfrac{1}{1 + \frac{1}{1 + \frac{1}{1}}}$ and so on. This is yet an-

other occurrence of the golden ratio. Viscount Brouncker, a seventeenth-century English mathematician, showed that the

sequence $1 + \dfrac{1^2}{2}, 1 + \dfrac{1^2}{2 + \frac{3^2}{2}}, 1 + \dfrac{1^2}{2 + \frac{3^2}{2 + \frac{5^2}{2}}}$ and so on, con-

verges to $\dfrac{4}{\pi}$ (see *A History of Pi* by Petr Beckmann). Explore this sequence numerically.

78. For the power series $\dfrac{1}{1 - x - x^2} = c_1 + c_2 x + c_3 x^2 + \cdots$, show that the constants c_i are the Fibonacci numbers. Substitute $x = \dfrac{1}{1000}$ to find the interesting decimal representation for $\dfrac{1,000,000}{998,999}$.

Chapter 9
Parametric Equations and Polar Coordinates

For hundreds of years, people believed that the earth was the center of the universe, with the sun and other heavenly bodies revolving around the earth. In 1543, the astronomer Copernicus shocked the world with the (posthumous) publication of his theory that the earth and the other known planets revolve in circular orbits about the sun. By the end of the sixteenth century, the astronomer Tycho Brahe had spent 20 years painstakingly recording his observations of the planets and hundreds of stars (made **before** the invention of the telescope). It was left to his student, Johannes Kepler to analyze Brahe's abundant data. Kepler's analysis led him to conclude that in fact, each planet moved in an elliptical orbit, with the sun located at one focus. About one hundred years later, Isaac Newton used his newly created calculus to show that Kepler's conclusions follow directly from Newton's **universal law of gravitation**.

An elliptical orbit.

As an experienced user of calculus, you might expect to be prepared to precisely analyze some properties of the path of a planet. For instance, you might want to find a tangent line (this would indicate the direction of the planet's motion at any given point in time) or you might want to find a function describing its velocity (this would indicate both its direction and its speed). You also might want to compute the length of an orbit or the area enclosed by the orbital path. (In fact, this last calculation is essential to Kepler's second law of planetary motion.)

There are several roadblocks standing in the way of your analysis of something as complex as planetary motion. First, a planet's path is **not** the graph of any function. (Why is that?) Virtually all of the concepts and techniques of calculus that we have developed so far apply only to functions. While you might think of piecing together the graphs of several functions to form the ellipse, using this to analyze the motion of a planet is sloppy, at best. Second, the position (x, y) of a planet whose orbit lies in the xy-plane is a function of yet another variable, **time**. In this chapter, we extend the concepts of calculus to include curves defined by **parametric equations** and **polar coordinates**. For the case of planetary motion, it is quite useful to describe a planet's position (x, y) as a function of the parameter t (time), that is, to write it in the form $(x(t), y(t))$, where $x(t)$ and $y(t)$ are **functions** to which our existing techniques of calculus can be applied. Alternatively, we'll show how to use polar coordinates to represent such a path, not as a set of points (x, y), corresponding to horizontal and vertical distances from a fixed reference point (the origin), but rather, by specifying the distance from a fixed point and an angle representing the direction from the fixed point to the planet.

These alternative descriptions of curves bring us a great deal of needed flexibility in attacking many problems. Often, even very complicated looking curves have a simple description in terms of parametric equations or polar coordinates. We explore a variety of interesting curves in this chapter and see how to extend the methods of calculus to such curves.

9.1   Plane Curves and Parametric Equations

We often find it convenient to describe the location of a point (x, y) in the plane in terms of a parameter. For instance, in tracking the movement of a satellite, we would naturally want to give its location in terms of time. In this way, we not only know the path it follows, but we also know **when** it passes through each point.

Given any pair of functions $g(t)$ and $h(t)$ defined on the same domain D, the equations

$$x = g(t), \quad y = h(t)$$

are called **parametric equations**. Notice that for each choice of t, the parametric equations specify a point $(x, y) = (g(t), h(t))$ in the xy-plane. The collection of all such points is called the **graph** of the parametric equations. In the case where g and h are continuous functions and D is an interval of the real line, the graph is a curve in the xy-plane, referred to as a **plane curve**.

The choice of the letter t to denote the independent variable (called the **parameter**) should make you think of *time*. In many situations, this is exactly what the parameter represents. For instance, we might represent the position $(g(t), h(t))$ of an object as a function of the time t. In fact, you might recall that in section 5.5, we used a pair of equations of this type to describe two-dimensional projectile motion. In many applications, the parameter has an interpretation other than time; in others, it has no physical meaning at all. In general, the parameter can be **any** quantity that is convenient for describing the relationship between x and y. In our first example, we can simplify our discussion by eliminating the parameter.

Example 1.1   Graphing a Plane Curve

Sketch the plane curve defined by the parametric equations

$$x = 6 - t^2, \ y = t/2, \text{ for } -2 \leq t \leq 4.$$

Solution   In the table that follows, we list a number of values of the parameter t and the corresponding values of x and y.

t	-2	-1	0	1	2	3	4
x	2	5	6	5	2	-3	-10
y	-1	$-1/2$	0	$1/2$	1	$3/2$	2

We have plotted these points and connected them with a smooth curve in Figure 9.1. You might also notice that we can easily eliminate the parameter here, by solving for t in terms of y. We have $t = 2y$, so that $x = 6 - 4y^2$. The graph of this last equation is a parabola opening to the left. The plane curve we're looking for is the portion of this parabola corresponding to $-2 \leq t \leq 4$. From the table, notice that this corresponds to $-1 \leq y \leq 2$, so that the plane curve is the portion of the parabola indicated in Figure 9.1.

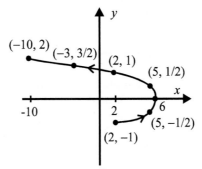

FIGURE 9.1: $x = 6 - t^2, \ y = t/2, \ -2 \leq t \leq 4.$

■

You probably noticed the small arrows drawn on top of the plane curve in Figure 9.1. These indicate the **orientation** of the curve (i.e., the direction of increasing t). In the case where t represents time and the curve represents the path of an object, notice that the orientation indicates the direction followed by the object as it traverses the path, as in the following example.

Example 1.2   The Path of a Projectile

Find the path of a projectile thrown horizontally with initial speed of 20 ft/s from a height of 64 feet.

Solution   Following our discussion in section 5.5, the path is defined by the parametric equations

$$x = 20t, \ y = 64 - 16t^2, \ \text{for } 0 \le t \le 2,$$

where t represents time (in seconds). This describes the plane curve shown in Figure 9.2. Note that in this case, the orientation indicated in the graph gives the direction of motion. Although we could eliminate the parameter, as in example 1.1, the parametric equations provide us with more information. Observe that while the corresponding x-y equation $y = 64 - 16 \left(\frac{x^2}{20^2} \right)$ describes the path followed by the projectile, the parametric equations also tell us **when** the object is located at a given point and indicates the **direction** of motion.

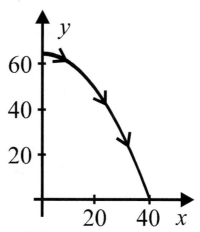

FIGURE 9.2: Path of projectile.

∎

Graphing calculators and computer algebra systems sketch a plane curve by plotting points corresponding to a large number of values of t and then connecting the plotted points with a smooth curve. The appearance of the resulting graph depends greatly on the graphing window used **and also** on the particular choice of t-values. This can be seen in the following example.

Example 1.3   Parametric Equations Involving Sines and Cosines

Sketch the plane curve defined by the parametric equations

$$x = 2 \cos t, \ y = 2 \sin t, \ \text{for } 0 \le t \le 2\pi. \tag{1.1}$$

Solution   The default graph produced by most graphing calculators looks something like the curve shown in Figure 9.3a (where we have added arrows indicating the orientation). With some thought, we can improve this sketch. First, notice that since $x = 2 \cos t$, x ranges between -2 and 2. Similarly, y ranges between -2 and 2. Changing the graphing window to $-2.1 \le x \le 2.1$ and $-2.1 \le y \le 2.1$ produces the curve shown in Figure 9.3b. This is certainly a big improvement over Figure 9.3a. The curve still looks like an ellipse, but with some more thought we can identify it as a circle. Rather than eliminate the parameter by solving for t in terms of either x or y, instead notice from (1.1) that

$$x^2 + y^2 = 4 \cos^2 t + 4 \sin^2 t = 4 \left(\cos^2 t + \sin^2 t \right) = 4.$$

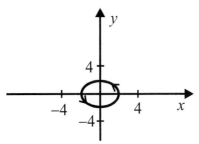

FIGURE 9.3a: $x = 2\cos t$, $y = 2\sin t$.

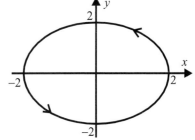

FIGURE 9.3b: $x = 2\cos t$, $y = 2\sin t$.

FIGURE 9.3c: A circle.

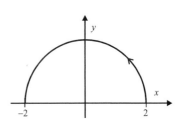

FIGURE 9.3d: Top semicircle.

So, the plane curve lies on the circle of radius 2 centered at the origin. In fact, it's the whole circle, as we can see by recognizing what the parameter represents in this case. Recall from the definition of sine and cosine that if (x, y) is a point on the unit circle and θ is the angle from the positive x-axis to the line segment joining (x, y) and the origin, then we define $\cos\theta = x$ and $\sin\theta = y$. Since we have $x = 2\cos t$ and $y = 2\sin t$, the parameter t corresponds to the angle θ. Further, the curve is the entire circle of radius 2, traced out as the angle t ranges from 0 to 2π. A "square" graphing window is one with the same scale on the x and y axes (not necessarily the same x and y ranges, but the same scales!). Some programs generate such a window when you specify an **aspect ratio** of 1. Such a square window gives us the circle seen in Figure 9.3c. Finally, what would change if the domain were limited to $0 \le t \le \pi$? Since we've identified t as the angle as measured from the positive x-axis, it should be clear that you will now get the top half of the unit circle, as shown in Figure 9.3d.

■

Remark 1.1   To sketch a parametric graph on a CAS, you may need to write the equations in **vector format**. For instance, in the case of example 1.3, instead of entering $x = 2\cos t$ and $y = 2\sin t$, you would enter the ordered pair of functions $(2\cos t, 2\sin t)$.

Simple modifications to the parametric equations in example 1.3 will produce a variety of circles and ellipses. We explore this in the following example and the exercises.

Example 1.4   **More Circles and Ellipses Defined by Parametric Equations**
Identify the plane curves

$$\text{(a)} \quad x = 2\cos t, \quad y = 3\sin t, \tag{1.2}$$

$$\text{(b)} \quad x = 2 + 4\cos t, \quad y = 3 + 4\sin t$$

and

$$\text{(c)} \quad x = 3\cos 2t, \quad y = 3\sin 2t,$$

all for $0 \le t \le 2\pi$.

Solution   A computer-generated sketch of (a) is shown in Figure 9.4a. It's difficult to determine from the sketch whether the curve is an ellipse or simply a distorted graph of a circle. You should rule out a circle, since the parametric equations produce x-values between -2 and 2 and y-values between -3 and 3. To verify that this is an ellipse, notice from (1.2) that

$$\frac{x^2}{4} + \frac{y^2}{9} = \frac{4\cos^2 t}{4} + \frac{9\sin^2 t}{9} = \cos^2 t + \sin^2 t = 1.$$

A computer-generated sketch of (b) is shown in Figure 9.4b. You should verify that this is the circle $(x-2)^2 + (y-3)^2 = 16$. Finally, a computer sketch of (c) is shown in Figure 9.4c. You should verify that this is the circle $x^2 + y^2 = 9$. So, what is the role of the 2 in the argument of cosine and sine? If you sketched this on a calculator, you may have noticed that the circle was completed long before the calculator finished graphing. Because of the 2, a complete circle corresponds to $0 \le 2t \le 2\pi$ or $0 \le t \le \pi$. With the domain $0 \le t \le 2\pi$, the circle is traced out twice. You might say that the factor of 2 in the argument doubles the speed of the trace.

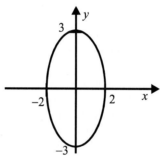

FIGURE 9.4a: $x = 2\cos t, y = 3\sin t$.

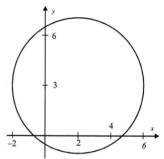

FIGURE 9.4b: $x = 2 + 4\cos t$, $y = 3 + 4\sin t$.

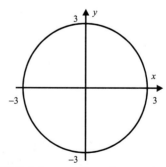

FIGURE 9.4c: $x = 3\cos 2t, y = 3\sin 2t$.

■

Remark 1.2   Look carefully at the plane curves in examples 1.3 and 1.4 until you can identify the roles of each of the constants in the equations $x = a + b\cos ct, \ y(t) = d + e\sin ct$. These interpretations are important in applications.

In the following example, we see how to find parametric equations for a line segment.

Example 1.5   Parametric Equations for a Line Segment

Find parametric equations for the line segment joining the points $(1, 2)$ and $(4, 7)$.

Solution   For a line segment, notice that the parametric equations can be linear functions, that is,

$$x = a + bt, \ \ y = c + dt,$$

for some constants a, b, c and d. (Eliminate the parameter t to see why this generates a line.) The simplest way to choose these constants is to have $t = 0$ correspond to the starting point $(1, 2)$. Note that if $t = 0$, the equations reduce to $x = a$ and $y = c$. To start our segment at $x = 1$ and $y = 2$, we set $a = 1$ and $c = 2$. Now note that with $t = 1$, the equations are $x = a + b$ and $y = c + d$. To produce the endpoint $(4, 7)$, we must have $a + b = 4$ and $c + d = 7$. With $a = 1$ and $c = 2$, solve to get $b = 3$ and $d = 5$. We now have that

$$x = 1 + 3t, \ \ y = 2 + 5t, \ \text{for } 0 \le t \le 1$$

is a pair of parametric equations describing the line segment.

■

In general, for parametric equations of the form $x = a + bt$, $y = c + dt$, notice that you can always choose a and c to be the x- and y-coordinates, respectively, of the starting point (since $x = a$, $y = b$ corresponds to $t = 0$). Then b is the difference in x-coordinates (endpoint minus starting point) and d is the difference in y-coordinates. With these choices, the line segment is always sketched out for $0 \le t \le 1$.

Remark 1.3   Note that in example 1.5, as for all parametric equations, there are an infinite number of choices of parameters that produce a given curve. For instance, you can easily verify that the parametric equations

$$x = -2 + 3t, \ \ y = -3 + 5t, \ \text{for } 1 \le t \le 2$$

and

$$x = t, \ \ y = (1 + 5t)/3, \ \text{for } 1 \le t \le 4$$

also produce the line segment of example 1.5. We say that each of these is a different **parameterization** of the curve.

As we illustrate with the following example, every equation of the form $y = f(x)$ can be simply written in parametric equations.

Example 1.6   Parametric Equations from an x-y Equation

Find parametric equations for the portion of the parabola $y = x^2$ from $(-1, 1)$ to $(3, 9)$.

Solution   Any equation of the form $y = f(x)$ can be converted to parametric form by simply letting x equal t. Here, this gives us $y = x^2 = t^2$. Then

$$x = t, \ \ y = t^2, \ \text{for } -1 \le t \le 3,$$

is a parametric representation of the curve. (Of course, you can use the letter x instead of the letter t if you prefer.)

■

Besides indicating an orientation, parametric representations of curves often also carry with them a built-in restriction on the portion of the curve included, as we see in the following example.

Example 1.7   Parametric Representations of a Curve with a Subtle Difference
Sketch the plane curves

$$\text{(a) } x = t - 1, \quad y = t^2 - 2$$

and

$$\text{(b) } x = t^2 - 1, \quad y = t^4 - 2.$$

Solution   Since there is no restriction placed on t, we can assume that t can be any real number. Eliminating the parameter in (a), we get $t = x + 1$, so that the parametric equations in (a) correspond to the parabola $y = (x + 1)^2 - 2$, shown in Figure 9.5a. You should verify that the graph includes the entire parabola by noting that if t can be any real number, $x = t - 1$ can be any real number. (If your calculator sketch doesn't show both sides of the parabola, adjust the range of t-values in the plot.) The importance of this check is shown by (b). When we eliminate the parameter, we get $t^2 = x + 1$ and so, $y = (x + 1)^2 - 2$. This gives the same parabola as in (a). However, the initial computer sketch of the parametric equations shown in Figure 9.5b appears to show only the right half of the parabola. To verify that this is correct, note that since $x = t^2 - 1$, we have that $x \geq -1$ for every real number t. Therefore, the curve is only the right half of the parabola $y = (x + 1)^2 - 2$, as indicated.

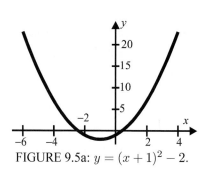

FIGURE 9.5a: $y = (x + 1)^2 - 2$.

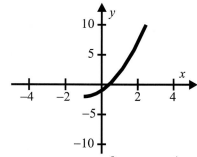

FIGURE 9.5b: $x = t^2 - 1$, $y = t^4 - 2$.

We can describe many plane curves parametrically that are unlike anything you've seen so far in your study of calculus. Many of these are difficult to draw by hand, but can be easily plotted with a graphing calculator or CAS.

Example 1.8   Some Unusual Plane Curves
Sketch the plane curves

$$\text{(a) } x = t^2 - 2, \quad y = t^3 - t$$

and

$$\text{(b) } x = t^3 - t, \quad y = t^4 - 5t^2 + 4.$$

Solution   A sketch of (a) is shown in Figure 9.6a. This is clearly not the graph of any function (consider the vertical line test to see why), so converting to an x-y equation would not be particularly helpful. You should, however, examine the parametric equations to see if important portions of the graph have been left out (e.g., is there supposed to be anything to the left of $x = -2$?). In this case, $x = t^2 - 2 \geq -2$ for all t and $y = t^3 - t$ has no maximum or minimum (think about why). It seems that most of the graph is indeed shown in Figure 9.6a. A computer sketch of (b) is shown in Figure 9.6b. Again, this is not the graph of a function. To get an idea of the scope of the graph, note that $x = t^3 - t$ has no maximum or minimum. To find the minimum of $y = t^4 - 5t^2 + 4$, note that critical numbers are at $t = 0$ and $t = \pm\sqrt{5/2}$ with corresponding function values 4 and $-9/4$, respectively. You should conclude that $y \geq -9/4$ as shown in Figure 9.6b.

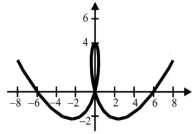

FIGURE 9.6a: $x = t^2 - 2$, $y = t^3 - t$.   FIGURE 9.6b: $x = t^3 - t$, $y = t^4 - 5t^2 + 4$.

■

You should now have some idea of the flexibility of parametric equations. This flexibility alone increases our ability to solve problems and makes the study of parametric equations worthwhile. But more importantly, a large number of applications translate simply into parametric equations.

Bear in mind that parametric equations communicate more information than do the corresponding x-y equations. We illustrate this with the following example.

Example 1.9   Intercepting a Missile in Flight

Suppose that a missile is fired toward your location from 500 miles away and follows a flight path given by the parametric equations

$$x = 100t, \quad y = 80t - 16t^2, \text{ for } 0 \leq t \leq 5.$$

Two minutes later, you fire an interceptor missile from your location following the flight path

$$x = 500 - 200(t - 2), \quad y = 80(t - 2) - 16(t - 2)^2, \text{ for } 2 \leq t \leq 7.$$

Determine if the interceptor missile hits its target.

Solution   In Figure 9.7a, we have plotted the flight paths for both missiles simultaneously. The two paths clearly intersect, but this does **not** necessarily mean that the two missiles collide. For that to happen, they need to be at the same point **at the same time**.

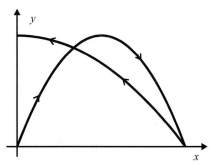

FIGURE 9.7a: Missile flight paths.

To determine whether there are any values of t for which both paths are simultaneously passing through the same point, we set the two x-values equal:

$$100t = 500 - 200\,(t-2)$$

and obtain one solution: $t = 3$. Note that this simply says that the two missiles have the same x-coordinate when $t = 3$. Unfortunately, the y-coordinates are not the same here, since when $t = 3$, we have

$$80t - 16t^2 = 96 \text{ but } 80\,(t-2) - 16\,(t-2)^2 = 64.$$

You can see this graphically by plotting the two paths simultaneously for $0 \le t \le 3$ only, as we have done in Figure 9.7b. From the graph, you can clearly see that the two missiles pass one another without colliding. By the time the interceptor missile intersects the flight path of the incoming missile, it is long gone! Another very nice way to observe this behavior is to plot the two sets of parametric equations on your graphing calculator in "simultaneous plot" mode. With this, you can animate the flight paths and watch the missiles pass by one another.

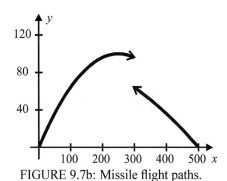
FIGURE 9.7b: Missile flight paths.

Exercises 9.1

1. (writing) Interpret in words the roles of each of the constants in the parametric equations $\begin{cases} x = a_1 + b_1 \cos(ct) \\ y = a_2 + b_2 \cos(ct) \end{cases}$.

2. (writing) An algorithm was given in example 1.5 for finding parametric equations of a line segment. Discuss the advantages that this method has over the other methods presented in Remark 1.3.

3. (writing) As indicated in remark 1.3, a given curve can be described by numerous sets of parametric equations. Explain why several different equations can all be correct. (Hint: Emphasize the fact that t is a dummy variable.)

4. (writing) In example 1.9, you saw that missiles don't collide even though their paths intersect. If you wanted to determine the intersection point of the graphs, explain why you would need to solve for values s and t (possibly different) such that $100t = 500 - 200(s-2)$ and $80t - 16t^2 = 80(s-2) - 16(s-2)^2$.

In exercises 5–16, sketch the plane curve defined by the given parametric equations and find a corresponding x-y equation for the curve.

5. $\begin{cases} x = 3\cos t \\ y = 3\sin t \end{cases}$ **6.** $\begin{cases} x = 2\cos t \\ y = 3\sin t \end{cases}$

7. $\begin{cases} x = 1 + 2\cos t \\ y = -2 + 2\sin t \end{cases}$ **8.** $\begin{cases} x = 2\sin t \\ y = 3\cos t \end{cases}$

9. $\begin{cases} x = -1 + 2t \\ y = 3t \end{cases}$ **10.** $\begin{cases} x = 4 + 3t \\ y = 2 - 4t \end{cases}$

11. $\begin{cases} x = 1 + t \\ y = t^2 + 2 \end{cases}$ **12.** $\begin{cases} x = 2 - t \\ y = t^2 + 1 \end{cases}$

13. $\begin{cases} x = t^2 - 1 \\ y = 2t \end{cases}$ **14.** $\begin{cases} x = t^2 + 1 \\ y = t - 1 \end{cases}$

15. $\begin{cases} x = t^2 - 1 \\ y = t^2 + 1 \end{cases}$ **16.** $\begin{cases} x = \cos t \\ y = 3\cos t - 1 \end{cases}$

In exercises 17–34, sketch the plane curves defined by the given parametric equations.

17. $\begin{cases} x = t^3 - 2t \\ y = t^2 - 3 \end{cases}$
18. $\begin{cases} x = t^3 - 2t \\ y = t^2 - 3t \end{cases}$

19. $\begin{cases} x = t^2 - 1 \\ y = t^4 - 4t \end{cases}$
20. $\begin{cases} x = t^2 - 1 \\ y = t^4 - 4t^2 \end{cases}$

21. $\begin{cases} x = \cos t - \frac{1}{2}\cos 3t \\ y = \sin t - \frac{1}{2}\sin 3t \end{cases}$
22. $\begin{cases} x = \cos t - \frac{1}{2}\cos 4t \\ y = \sin t - \frac{1}{2}\sin 4t \end{cases}$

23. $\begin{cases} x = \cos 2t \\ y = \sin 7t \end{cases}$
24. $\begin{cases} x = \cos 2t \\ y = \sin \pi t \end{cases}$

25. $\begin{cases} x = t\cos t \\ y = t\sin t \end{cases}$
26. $\begin{cases} x = t\cos 2t \\ y = t\sin t \end{cases}$

27. $\begin{cases} x = t^2 \cos 2t \\ y = t^2 \sin 2t \end{cases}$
28. $\begin{cases} x = t^2 \cos 2t \\ y = t^2 \sin 3t \end{cases}$

29. $\begin{cases} x = 3\cos 2t + \sin 5t \\ y = 3\sin 2t + \cos 5t \end{cases}$
30. $\begin{cases} x = 3\cos 2t + \sin 6t \\ y = 3\sin 2t + \cos 6t \end{cases}$

31. $\begin{cases} x = \cos 3t \\ y = \sin t \end{cases}$
32. $\begin{cases} x = \cos 3t \\ y = \sin 2t \end{cases}$

33. $\begin{cases} x = e^t \\ y = e^{-2t} \end{cases}$
34. $\begin{cases} x = e^t \\ y = e^{2t} \end{cases}$

In exercises 35–40, match the parametric equations with the corresponding plane curve displayed below. Give reasons for your choices.

35. $\begin{cases} x = t^2 - 1 \\ y = t^4 \end{cases}$
36. $\begin{cases} x = t - 1 \\ y = t^3 \end{cases}$

37. $\begin{cases} x = t^2 - 1 \\ y = \sin t \end{cases}$
38. $\begin{cases} x = t^2 - 1 \\ y = \sin 2t \end{cases}$

39. $\begin{cases} x = \cos 3t \\ y = \sin 2t \end{cases}$
40. $\begin{cases} x = 3\cos t \\ y = 2\sin t \end{cases}$

Figure A   Figure B

Figure C   Figure D

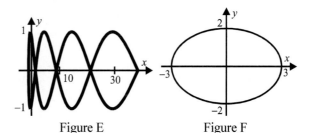

Figure E   Figure F

In exercises 41–50, find parametric equations describing the given curve.

41. The line segment from $(0, 1)$ to $(3, 4)$

42. The line segment from $(3, 1)$ to $(1, 3)$

43. The line segment from $(-2, 4)$ to $(6, 1)$

44. The line segment from $(4, -2)$ to $(2, -1)$

45. The portion of the parabola $y = x^2 + 1$ from $(1, 2)$ to $(2, 5)$

46. The portion of the parabola $y = 2x^2 - 1$ from $(0, -1)$ to $(2, 7)$

47. The portion of the parabola $y = 2 - x^2$ from $(2, -2)$ to $(0, 2)$

48. The portion of the parabola $y = x^2 + 1$ from $(1, 2)$ to $(-1, 2)$

49. The circle of radius 3 centered at $(2, 1)$

50. The circle of radius 5 centered at $(-1, 3)$

In exercises 51–54, find all points of intersection of the two curves.

51. $\begin{cases} x = t \\ y = t^2 - 1 \end{cases}$ and $\begin{cases} x = 1 + s \\ y = 4 - s \end{cases}$

52. $\begin{cases} x = t^2 \\ y = t + 1 \end{cases}$ and $\begin{cases} x = 2 + s \\ y = 1 - s \end{cases}$

53. $\begin{cases} x = t + 3 \\ y = t^2 \end{cases}$ and $\begin{cases} x = 1 + s \\ y = 2 - s \end{cases}$

54. $\begin{cases} x = t^2 + 3 \\ y = t^3 + t \end{cases}$ and $\begin{cases} x = 2 + s \\ y = 1 - s \end{cases}$

55. Rework example 1.9 with the interceptor missile following the flight path $x = 500 - 500(t - 2)$ and $y = 208(t - 2) - 16(t - 2)^2$.

56. Rework example 1.9 with the interceptor missile following the flight path $x = 500 - 100t$ and $y = 80t - 16t^2$.

57. In example 1.9 and exercise 55, explain why the 2 in the term $t - 2$ represents the time delay between the launches of the two missiles. For the equations in example 1.9, find a value of the time delay such that the two missiles do collide.

58. Explain why the missile path in exercise 56 must produce a collision (compare the y-equations) but is unrealistic.

59. Compare the graphs of $\begin{cases} x = \cos 2t \\ y = \sin t \end{cases}$ and $\begin{cases} x = \cos t \\ y = \sin 2t \end{cases}$. Use the identities $\cos 2t = \cos^2 t - \sin^2 t$ and $\sin 2t = 2\cos t \sin t$ to find x-y equations for each graph.

60. Sketch the graph of $\begin{cases} x = \cosh t \\ y = \sinh t \end{cases}$. Use the identity $\cosh^2 t - \sinh^2 t = 1$ to find an x-y equation for the graph. Explain where the "hyperbolic" in hyperbolic sine and hyperbolic cosine might come from.

61. Sketch the graph of $\begin{cases} x = \frac{1}{2}\cos t - \frac{1}{4}\cos 2t \\ y = \frac{1}{2}\sin t - \frac{1}{4}\sin 2t \end{cases}$. This heart-shaped region is the largest feature of the **Mandelbrot set**, one of the most famous mathematical sets. Portions of the Mandelbrot set have been turned into colorful T-shirts and posters that you may have seen.

Mandelbrot set.   Mandelbrot zoom

To progress further on a sketch of the Mandelbrot set, add the circle $\begin{cases} x = -1 + \frac{1}{4}\cos t \\ y = \frac{1}{4}\sin t \end{cases}$ to your initial sketch.

62. Determine parametric equations for the curves defined by $x^{2n} + y^{2n} = r^{2n}$ for integers n. (Hint: Start with $n = 1$, $x^2 + y^2 = r^2$, then think of the general equation as $(x^n)^2 + (y^n)^2 = r^{2n}$.) Sketch the graphs for $n = 1$, $n = 2$ and $n = 3$ and predict what the curve will look like for large values of n.

63. (exploratory) Many carnivals have a version of the double Ferris wheel. A large central arm rotates clockwise. At each end of the central arm is a Ferris wheel that rotates clockwise around the arm.

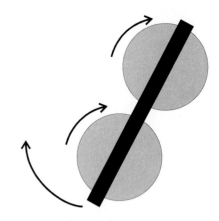

Assume that the central arm has length 200 feet and rotates about its center. Also assume that the wheels have radius 40 feet and rotate at the same speed as the central arm. Find parametric equations for the position of a rider and graph the rider's path. Adjust the speed of rotation of the wheels to improve the ride.

64. (exploratory) The Flying Zucchini Circus Troupe has a human cannonball act, shooting a performer from a cannon into a specially padded seat of a turning Ferris wheel. The Ferris wheel has a radius of 40 feet and rotates counterclockwise at one revolution per minute. The special seat starts at ground level. Carefully explain why parametric equations for the seat are $\begin{cases} x = 40\cos\left(\frac{\pi}{30}t - \frac{\pi}{2}\right) \\ y = 40 + 40\sin\left(\frac{\pi}{30}t - \frac{\pi}{2}\right) \end{cases}$. The cannon is located 200 feet left of the Ferris wheel with the muzzle 10 feet above ground. The performer is launched 35 seconds after the wheel starts turning with an initial velocity of 100 ft/s at an angle of $\frac{\pi}{5}$ above the horizontal. Carefully explain why parametric equations for the human cannonball are $\begin{cases} x = \left(100\cos\frac{\pi}{5}\right)(t - 35) - 200 \\ y = -16(t-35)^2 + \left(100\sin\frac{\pi}{5}\right)(t - 35) + 10 \end{cases}$ $(t \geq 35)$. Determine whether the ride is safe or the Flying Zucchini comes down squash.

65. (exploratory) Rework exercise 64 with initial velocity 135 ft/s, launch angle 30° and a 27-second delay. How close does the Flying Zucchini get to the special seat? Given that a Ferris wheel seat actually has height, width and depth, do you think that this is close enough? Repeat with (a) initial velocity 75 ft/s, launch angle 47° and 47.25-second delay; (b) initial velocity 118 ft/s, launch angle 35° and 28-second delay. Develop criteria for a safe and exciting human cannonball act. Consider each of the following: Should the launch velocity be large or small? Should the seat be high or low when the cannonball lands? Should the human have a positive or negative vertical velocity at landing? How close (vertically and horizontally) should the human need to get to the center of the seat? Based on your criteria, which of the launches in this exercise is the best? Find an initial velocity, launch angle and launch delay that is better.

9.2   Calculus and Parametric Equations

FIGURE 9.8: The Scrambler.

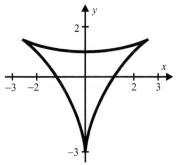

FIGURE 9.9:
Path of a Scrambler rider.

The Scrambler is a popular carnival ride consisting of two sets of rotating arms (see Figure 9.8). Suppose that the inner arms have length 2 and rotate counterclockwise. In this case, we can describe the location (x_i, y_i) of the end of one of the inner arms by the parametric equations $x_i = 2\cos t$, $y_i = 2\sin t$. At the end of each inner arm, a set of outer arms rotate clockwise at roughly twice the speed. If the outer arms have length 1, parametric equations describing the outer arm rotation are $x_o = \sin 2t$, $y_o = \cos 2t$. Here, the reversal of sine and cosine terms indicates that the rotation is clockwise and the factor of 2 inside the sine and cosine terms indicates that the speed of the rotation is double that of the inner arms. The position of a person riding the Scrambler is the sum of the two component motions; that is,

$$x = 2\cos t + \sin 2t, \quad y = 2\sin t + \cos 2t.$$

The graph of these parametric equations is shown in Figure 9.9. Passengers on the Scrambler feel like they rapidly accelerate to the outside of the ride, momentarily stop, then change direction and accelerate to a different point on the outside of the ride. Figure 9.9 suggests that this is an accurate description of the ride, but we need to develop the calculus of parametric equations to determine whether the riders actually come to a complete stop.

Our initial aim is to find a way to determine the slopes of tangent lines to curves which are defined parametrically. First, recall that for a function $y = f(x)$, the slope of the tangent line at the point $x = a$ is given by $f'(a)$. Written in Leibniz notation, the slope is $\dfrac{dy}{dx}(a)$. In the case of the Scrambler ride, both x and y are functions of the parameter t. Notice that if $x = g(t)$ and $y = h(t)$ both have derivatives which are continuous at $t = c$, the chain rule gives us

$$\frac{dy}{dt} = \frac{dy}{dx}\frac{dx}{dt}.$$

In the case where $\dfrac{dx}{dt}(c) \neq 0$, we then have

$$\boxed{\frac{dy}{dx}(a) = \frac{\dfrac{dy}{dt}(c)}{\dfrac{dx}{dt}(c)} = \frac{h'(c)}{g'(c)},} \qquad (2.1)$$

where $a = f(c)$. In the case where $g'(c) = h'(c) = 0$, we define

$$\frac{dy}{dx}(a) = \lim_{t\to c}\frac{\dfrac{dy}{dt}}{\dfrac{dx}{dt}}(t) = \lim_{t\to c}\frac{h'(t)}{g'(t)},$$

provided the limit exists.

Remark 2.1   Be careful with how you interpret this formula. The primes on the right side of equation (2.1) refer to derivatives with respect to the parameter t. We recommend that you (at least initially) use the Leibniz notation, which also gives you a simple way to accurately remember the chain rule.

We can use (2.1) to calculate second (as well as higher order) derivatives. Notice that if we replace y by $\dfrac{dy}{dx}$, we get

$$\frac{d^2y}{dx^2} = \frac{d}{dx}\left(\frac{dy}{dx}\right) = \frac{\dfrac{d}{dt}\left(\dfrac{dy}{dx}\right)}{\dfrac{dx}{dt}}. \tag{2.2}$$

CAUTION: Look carefully at (2.2) and convince yourself that

$$\frac{d^2y}{dx^2} \neq \frac{\dfrac{d^2y}{dt^2}}{\dfrac{d^2x}{dt^2}}.$$

Equating these two expressions is a common error. You should be careful to avoid this trap.

Example 2.1   Slopes of Tangent Lines to the Path of the Scrambler

Find the slope of the tangent line to the path of the Scrambler

$$x = 2\cos t + \sin 2t, \quad y = 2\sin t + \cos 2t$$

at (a) $t = 0$; (b) $t = \pi$; (c) the point $(0, -3)$.

Solution   (a) First, note that

$$\frac{dx}{dt} = -2\sin t + 2\cos 2t \text{ and } \frac{dy}{dt} = 2\cos t - 2\sin 2t.$$

From (2.1), the slope of the tangent line at $t = 0$ is then

$$\left.\frac{dy}{dx}\right|_{t=0} = \frac{\dfrac{dy}{dt}(0)}{\dfrac{dx}{dt}(0)} = \frac{2\cos 0 - 2\sin 0}{-2\sin 0 + 2\cos 0} = 1.$$

(b) The slope of the tangent line at $t = \dfrac{\pi}{4}$ is

$$\left.\frac{dy}{dx}\right|_{t=\pi/4} = \frac{\dfrac{dy}{dt}\left(\frac{\pi}{4}\right)}{\dfrac{dx}{dt}\left(\frac{\pi}{4}\right)} = \frac{2\cos \frac{\pi}{4} - 2\sin \frac{\pi}{2}}{-2\sin \frac{\pi}{4} + 2\cos \frac{\pi}{2}} = \frac{\sqrt{2} - 2}{-\sqrt{2}}.$$

(c) To determine the slope at the point $(0, -3)$, we must first determine a value of t which corresponds to that point. In this case, notice that $t = 3\pi/2$ gives $x = 0$ and $y = -3$. Here, we have

$$\frac{dx}{dt}\left(\frac{3\pi}{2}\right) = \frac{dy}{dt}\left(\frac{3\pi}{2}\right) = 0.$$

Using l'Hôpital's Rule, we have

$$\frac{dy}{dx}\left(\frac{3\pi}{2}\right) = \lim_{t\to 3\pi/2} \frac{2\cos t - 2\sin 2t}{-2\sin t + 2\cos 2t} = \lim_{t\to 3\pi/2} \frac{-2\sin t - 4\cos 2t}{-2\cos t - 4\sin 2t},$$

which does not exist, since the limit in the numerator is 6 and the limit in the denominator is 0. This says that the slope of the tangent line at $t = \dfrac{3\pi}{2}$ is undefined. Examine Figure 9.9 to see that the tangent line at the point $(0, -3)$ appears to be vertical.

■

For the passenger on the Scrambler of example 2.1, notice that the slope of the tangent line indicates the *direction* of motion and does not correspond to speed, which we discuss below.

Finding slopes of tangent lines can help us identify many points of interest.

Example 2.2   Finding Vertical and Horizontal Tangent Lines

Identify all points at which the plane curve

$$x = \cos 2t, \quad y = \sin 3t$$

has a horizontal or vertical tangent line.

Solution   A sketch of the curve is shown in Figure 9.10. There appear to be two locations (the top and bottom of the bow) with horizontal tangent lines and one point (the far right edge of the bow) with a vertical tangent line. Recall that horizontal tangent lines occur where $\dfrac{dy}{dx} = 0$. From (2.1), we then have $\dfrac{dy}{dx} = \dfrac{y'(t)}{x'(t)} = 0$. Of course, this can only occur when

$$0 = y'(t) = 3\cos 3t,$$

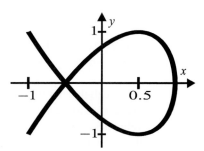

FIGURE 9.10: $x = \cos 2t, y = \sin 3t$.

provided that $x'(t) = -2\sin 2t \neq 0$ for the same value of t. Since $\cos\theta = 0$ only when θ is an odd multiple of $\frac{\pi}{2}$, we have that $y'(t) = 3\cos 3t = 0$, only when $3t = \frac{\pi}{2}, \frac{3\pi}{2}, \frac{5\pi}{2}, \dots$ and so, $t = \frac{\pi}{6}, \frac{3\pi}{6}, \frac{5\pi}{6}, \dots$. The corresponding points on the curve are then

$$\left(x\left(\frac{\pi}{6}\right), y\left(\frac{\pi}{6}\right)\right) = \left(\cos\frac{\pi}{3}, \sin\frac{\pi}{2}\right) = \left(\frac{1}{2}, 1\right),$$

$$\left(x\left(\frac{3\pi}{6}\right), y\left(\frac{3\pi}{6}\right)\right) = \left(\cos\pi, \sin\frac{3\pi}{2}\right) = (-1, -1),$$

$$\left(x\left(\frac{7\pi}{6}\right), y\left(\frac{7\pi}{6}\right)\right) = \left(\cos\frac{7\pi}{3}, \sin\frac{7\pi}{2}\right) = \left(\frac{1}{2}, -1\right)$$

and

$$\left(x\left(\frac{9\pi}{6}\right), y\left(\frac{9\pi}{6}\right)\right) = \left(\cos 3\pi, \sin\frac{9\pi}{2}\right) = (-1, 1).$$

Note that $t = \frac{5\pi}{6}$ and $t = \frac{11\pi}{6}$ reproduce the first and third points, respectively and so on. The points $\left(\frac{1}{2}, 1\right)$ and $\left(\frac{1}{2}, -1\right)$ are on the top and bottom of the bow, respectively, where there clearly are horizontal tangents. The points $(-1, -1)$ and $(-1, 1)$ should not seem quite right, though. These points are on the extreme ends of the bow and certainly don't look like they have vertical or horizontal tangents. In fact, they don't. Notice that at both $t = \frac{\pi}{2}$ and $t = \frac{3\pi}{2}$, we have $x'(t) = y'(t) = 0$ and so, the slope must be computed as a limit. We leave it as an exercise to show that the slopes at $t = \frac{\pi}{2}$ and $t = \frac{3\pi}{2}$ are $\frac{9}{4}$ and $-\frac{9}{4}$, respectively.

To find points where there is a vertical tangent, we need to see where $x'(t) = 0$, but $y'(t) \neq 0$. Setting $0 = x'(t) = -2\sin 2t$, we get $\sin 2t = 0$ which occurs if $2t = 0, \pi, 2\pi, \ldots$ or $t = 0, \frac{\pi}{2}, \pi, \ldots$. The corresponding points are

$$(x(0), y(0)) = (\cos 0, \sin 0) = (1, 0),$$

$$(x(\pi), y(\pi)) = (\cos 2\pi, \sin 3\pi) = (1, 0)$$

and the points corresponding to $t = \frac{\pi}{2}$ and $t = \frac{3\pi}{2}$, which we have already discussed (where $y'(t) = 0$, also). So, there is a vertical tangent line only at the point $(1, 0)$.

∎

The following result generalizes what we observed in example 2.2.

Theorem 2.1   Suppose that $g'(t)$ and $h'(t)$ are continuous. Then for the curve defined by the parametric equations $x = g(t)$ and $y = h(t)$,
 (i) if $h'(c) = 0$ and $g'(c) \neq 0$, there is a horizontal tangent line at the point $(g(c), h(c))$;
 (ii) if $g'(c) = 0$ and $h'(c) \neq 0$, there is a vertical tangent line at the point $(g(c), h(c))$.

Proof. The proof depends on the calculation of derivatives for parametric curves and is left as an exercise.

∎

Recall that our introductory question about the Scrambler was whether or not the rider ever comes to a complete stop. To answer this question, we will need to be able to compute velocities. Recall that if the position of an object moving along a straight line is given by the differentiable function $f(t)$, the object's velocity is given by $f'(t)$. The situation with parametric equations is completely analogous. If the position is given by $(g(t), h(t))$ for differentiable functions $g(t)$ and $h(t)$, then the **horizontal component of velocity** is given by $g'(t)$ and the **vertical component of velocity** is given by $h'(t)$. We define the **speed** to be $\sqrt{[g'(t)]^2 + [h'(t)]^2}$.

Example 2.3   Velocity of the Scrambler
For the path of the Scrambler

$$x = 2\cos t + \sin 2t, \quad y = 2\sin t + \cos 2t,$$

find the horizontal and vertical components of velocity and speed at times $t = 0$ and $t = \pi/2$ and indicate the direction of motion. Also determine all times at which the speed is zero.

Solution   Here, the horizontal component of velocity is $\dfrac{dx}{dt} = -2\sin t + 2\cos 2t$ and the vertical component is $\dfrac{dy}{dt} = 2\cos t - 2\sin 2t$. At $t = 0$, the horizontal and vertical components of velocity both equal 2 and the speed is $\sqrt{4 + 4} = \sqrt{8}$. The rider is located at the point $(x(0), y(0)) = (2, 1)$ and is moving to the right and up. At $t = \pi/2$, the velocity has components -4 (horizontal) and 0 (vertical) and the speed is $\sqrt{16 + 0} = 4$. At this time, the rider is located at the point $(0, -1)$ and is moving to the left.

In general, the speed of the rider at time t is given by

$$s(t) = \sqrt{\left(\frac{dx}{dt}\right)^2 + \left(\frac{dy}{dt}\right)^2} = \sqrt{(-2\sin t + 2\cos 2t)^2 + (2\cos t - 2\sin 2t)^2}$$

$$= \sqrt{4\sin^2 t - 8\sin t \cos 2t + 4\cos^2 2t + 4\cos^2 t - 8\cos t \sin 2t + 4\sin^2 2t}$$

$$= \sqrt{8 - 8\sin t \cos 2t - 8\cos t \sin 2t}$$

$$= \sqrt{8 - 8\sin 3t},$$

since $\sin^2 t + \cos^2 t = 1$, $\cos^2 2t + \sin^2 2t = 1$ and $\sin t \cos 2t + \sin 2t \cos t = \sin 3t$. Notice that this says that the speed equals 0 whenever $\sin 3t = 1$. This occurs when $3t = \frac{\pi}{2}, \frac{5\pi}{2}, \frac{9\pi}{2}, \ldots$, or $t = \frac{\pi}{6}, \frac{5\pi}{6}, \frac{9\pi}{6}, \ldots$. The corresponding points on the curve are $\left(x\left(\frac{\pi}{6}\right), y\left(\frac{\pi}{6}\right)\right) = \left(\frac{3}{2}\sqrt{3}, \frac{3}{2}\right)$, $\left(x\left(\frac{5\pi}{6}\right), y\left(\frac{5\pi}{6}\right)\right) = \left(-\frac{3}{2}\sqrt{3}, \frac{3}{2}\right)$ and $\left(x\left(\frac{9\pi}{6}\right), y\left(\frac{9\pi}{6}\right)\right) = (0, -3)$. You can easily verify that these points are the three tips of the path seen in Figure 9.9.

∎

We just showed that riders in the Scrambler of Figure 9.9 actually come to a brief stop at the outside of each loop. As you will explore in the exercises, for similar Scrambler paths, the riders slow down but have a positive speed at the outside of each loop. This is true of the Scrambler at most carnivals, for which a more complicated path makes up for the lack of stopping.

Notice that the Scrambler path shown in Figure 9.9 begins and ends at the same point and so, encloses an area. An interesting question is to determine the area enclosed by such a curve. Computing areas in parametric equations is a straightforward extension of our original development of integration. Recall that for a continuous function f defined on $[a, b]$, where $f(x) \geq 0$ on $[a, b]$, the area under the curve $y = f(x)$ for $a \leq x \leq b$ is given by

$$A = \int_a^b f(x)\, dx = \int_a^b y\, dx.$$

Now, suppose that this same curve is described parametrically by $x = g(t)$ and $y = h(t)$, where the curve is traversed exactly once for $c \leq t \leq d$. We can then compute the area by making the substitution $x = g(t)$. It then follows that $dx = g'(t)\, dt$ and so, the area is given by

$$A = \int_a^b \underbrace{y}_{h(t)} \underbrace{dx}_{g'(t)\, dt} = \int_c^d h(t)\, g'(t)\, dt,$$

where you should notice that we have also changed the limits of integration to match the new variable of integration. We generalize this result in the following theorem.

Theorem 2.2   Suppose that the parametric equations $x = g(t)$ and $y = h(t)$ with $c \leq t \leq d$, describe a curve that is traced out **clockwise** exactly once, as t increases from c to d and where the curve does not intersect itself, except that the initial and terminal points are the same (i.e., $g(c) = g(d)$ and $h(c) = h(d)$). Then the enclosed area is given by

$$A = \int_c^d y(t)\, x'(t)\, dt = -\int_c^d x(t) y'(t)\, dt. \qquad (2.3)$$

If the curve is traced out **counterclockwise**, then the enclosed area is given by

$$A = -\int_c^d y(t)\, x'(t)\, dt = \int_c^d x(t)\, y'(t)\, dt. \qquad (2.4)$$

Proof. This result is a special case of Green's Theorem, which we will develop in section 14.4.

∎

The new area formulas given in Theorem 2.2 turn out to be quite useful. As we see in the following example, we can use these to find the area enclosed by a parametric curve, without the need to eliminate the parameter.

Example 2.4   Finding the Area Enclosed by a Curve
Find the area enclosed by the path of the Scrambler

$$x = 2\cos t + \sin 2t, \quad y = 2\sin t + \cos 2t.$$

Solution   Notice that the curve is traced out counterclockwise once for $0 \le t \le 2\pi$. From (2.4), the area is then

$$A = \int_0^{2\pi} x(t)\, y'(t)\, dt = \int_0^{2\pi} (2\cos t + \sin 2t)(2\cos t - 2\sin 2t)\, dt$$

$$= \int_0^{2\pi} (4\cos^2 t - 2\cos t \sin 2t - 2\sin^2 2t)\, dt = 2\pi,$$

where we evaluated the integral using a CAS.

∎

In the following example, we use Theorem 2.2 to derive a formula for the area enclosed by an ellipse. Pay particular attention to how much easier this is to do with parametric equations than it is to do with the original x-y equation.

Example 2.5   Finding the Area Enclosed by an Ellipse
Find the area enclosed by the ellipse $\dfrac{x^2}{a^2} + \dfrac{y^2}{b^2} = 1$ (for constants $a, b > 0$).

Solution   One way to compute the area is to solve the equation for y to obtain $y = \pm b\sqrt{1 - \dfrac{x^2}{a^2}}$ and then integrate:

$$A = \int_{-a}^{a} \left[b\sqrt{1 - \frac{x^2}{a^2}} - \left(-b\sqrt{1 - \frac{x^2}{a^2}} \right) \right] dx.$$

You can evaluate this integral by trigonometric substitution or by using a CAS, but a simpler, more elegant way to compute the area is to use parametric equations. Notice that the ellipse is described parametrically by $x = a\cos t,\, y = b\sin t$, for $0 \le t \le 2\pi$. The ellipse is then traced out counterclockwise exactly once for $0 \le t \le 2\pi$, so that the area is given by (2.4) to be

$$A = -\int_0^{2\pi} y(t)\, x'(t)\, dt = -\int_0^{2\pi} (b\sin t)(-a\sin t)\, dt = ab\int_0^{2\pi} \sin^2 t\, dt = ab\,\pi,$$

where this last integral can be evaluated by using the half-angle formula:

$$\sin^2 t = \frac{1}{2}(1 - \cos 2t).$$

We leave the details of this calculation as an exercise.

∎

Exercises 9.2

1. (writing) In the derivation of parametric equations for the Scrambler, we used the fact that reversing the sine and cosine functions to $\begin{cases} x = \sin t \\ y = \cos t \end{cases}$ causes the circle to be traced out clockwise. Explain why this is so by starting at $t = 0$ and following the graph as t increases to 2π.

2. (writing) Explain why Theorem 2.1 makes sense. (Hint: if $y'(c) = 0$, what does that say about the change in y-coordinates on the graph? Why do you also need $x'(c) \neq 0$ to guarantee a horizontal tangent?)

3. (writing) Imagine an object with position given by $x(t)$ and $y(t)$. If a right triangle has a horizontal leg of length $x'(t)$ and a vertical leg of length $y'(t)$, what would the length of the hypotenuse represent? Explain why this makes sense.

4. (writing) Explain why the sign ($\pm$) of $\int_c^d y(t)x'(t)\, dt$ in Theorem 2.2 is different for curves traced out clockwise and counterclockwise.

In exercises 5–10, find the slopes of the tangent lines to the given curves at the indicated points.

5. $\begin{cases} x = t^2 - 2 \\ y = t^3 - t \end{cases}$ (a) $t = -1$, (b) $t = 1$, (c) $(-2, 0)$

6. $\begin{cases} x = t^3 - t \\ y = t^4 - 5t^2 + 4 \end{cases}$ (a) $t = -1$, (b) $t = 1$, (c) $(0, 4)$

7. $\begin{cases} x = 2\cos t \\ y = 3\sin t \end{cases}$ (a) $t = \frac{\pi}{4}$, (b) $t = \frac{\pi}{2}$, (c) $(0, 3)$

8. $\begin{cases} x = 2\cos 2t \\ y = 3\sin 2t \end{cases}$ (a) $t = \frac{\pi}{4}$, (b) $t = \frac{\pi}{2}$, (c) $(-2, 0)$

9. $\begin{cases} x = \cos 2t \\ y = \sin 4t \end{cases}$ (a) $t = \frac{\pi}{4}$, (b) $t = \frac{\pi}{2}$, (c) $\left(\frac{\sqrt{2}}{2}, 1\right)$

10. $\begin{cases} x = \cos 3t \\ y = \sin 5t \end{cases}$ (a) $t = 0$, (b) $t = \frac{\pi}{2}$, (c) $(0, 1)$

In exercises 11 and 12, sketch the graph and find all slopes of the curves at the given point.

11. $\begin{cases} x = t^2 - 2 \\ y = t^3 - t \end{cases}$ at $(-1, 0)$

12. $\begin{cases} x = t^3 - t \\ y = t^4 - 5t^2 + 4 \end{cases}$ at $(0, 0)$

In exercises 13–20, identify all points at which the curve has (a) a horizontal tangent and (b) a vertical tangent.

13. $\begin{cases} x = 2\cos t \\ y = 4\sin t \end{cases}$

14. $\begin{cases} x = 2\cos t \\ y = 7\sin t \end{cases}$

15. $\begin{cases} x = \cos 2t \\ y = \sin 4t \end{cases}$

16. $\begin{cases} x = \cos 2t \\ y = \sin 7t \end{cases}$

17. $\begin{cases} x = t^2 - 1 \\ y = t^4 - 4t \end{cases}$

18. $\begin{cases} x = t^2 - 1 \\ y = t^4 - 4t^2 \end{cases}$

19. $\begin{cases} x = 2\cos t + \sin 2t \\ y = 2\sin t + \cos 2t \end{cases}$

20. $\begin{cases} x = 2\cos 2t + \sin t \\ y = 2\sin 2t + \cos t \end{cases}$

In exercises 21–26, parametric equations for the position of an object are given. Find the object's velocity and speed at the given times, and describe its motion.

21. $\begin{cases} x = 2\cos t \\ y = 2\sin t \end{cases}$ (a) $t = 0$, (b) $t = \frac{\pi}{2}$

22. $\begin{cases} x = 2\cos 2t \\ y = 2\sin 2t \end{cases}$ (a) $t = 0$, (b) $t = \frac{\pi}{2}$

23. $\begin{cases} x = 20t \\ y = 30 - 2t - 16t^2 \end{cases}$ (a) $t = 0$, (b) $t = 2$

24. $\begin{cases} x = 40t + 5 \\ y = 20 + 3t - 16t^2 \end{cases}$ (a) $t = 0$, (b) $t = 2$

25. $\begin{cases} x = 2\cos 2t + \sin 5t \\ y = 2\sin 2t + \cos 5t \end{cases}$ (a) $t = 0$, (b) $t = \frac{\pi}{2}$

26. $\begin{cases} x = 3\cos t + \sin 3t \\ y = 3\sin t + \cos 3t \end{cases}$ (a) $t = 0$, (b) $t = \frac{\pi}{2}$

In exercises 27–34, find the area enclosed by the given curve.

27. $\begin{cases} x = 3\cos t \\ y = 2\sin t \end{cases}$

28. $\begin{cases} x = 6\cos t \\ y = 2\sin t \end{cases}$

29. $\begin{cases} x = \frac{1}{2}\cos t - \frac{1}{4}\cos 2t \\ y = \frac{1}{2}\sin t - \frac{1}{4}\sin 2t \end{cases}$

30. $\begin{cases} x = 2\cos 2t + \cos 4t \\ y = 2\sin 2t + \sin 4t \end{cases}$

31. $\begin{cases} x = \cos t \\ y = \sin 2t \end{cases}, \ \frac{\pi}{2} \leq t \leq \frac{3\pi}{2}$

32. $\begin{cases} x = t\sin t \\ y = t\cos t \end{cases}, \ -\frac{\pi}{2} \leq t \leq \frac{\pi}{2}$

33. $\begin{cases} x = t^3 - 4t \\ y = t^2 - 3 \end{cases}$, $-2 \le t \le 2$

34. $\begin{cases} x = t^3 - 4t \\ y = t^4 - 1 \end{cases}$, $-2 \le t \le 2$

In exercises 35 and 36, find the speed of the object each time it crosses the x-axis.

35. $\begin{cases} x = 2\cos^2 t + 2\cos t - 1 \\ y = 2(1 - \cos t)\sin t \end{cases}$

36. $\begin{cases} x = 6\cos t + 5\cos 3t \\ y = 6\sin t - 5\sin 3t \end{cases}$

37. A modification of the Scrambler in example 2.1 is $\begin{cases} x = 2\cos 3t + \sin 5t \\ y = 2\sin 3t + \cos 5t \end{cases}$. In example 2.1, the ratio of the speed of the outer arms to the speed of the inner arms is 2-to-1. What is the ratio in this version of the Scrambler? Sketch a graph showing the motion of this new Scrambler.

38. Compute the speed of the Scrambler in exercise 37. Using trigonometric identities as in example 2.3, show that the speed is at a minimum when $\sin 8t = 1$ but that the speed is never zero. Show that the minimum speed is reached at the outer points of the path.

39. Find parametric equations for a Scrambler that is the same as in example 2.1 except that the outer arms rotate three times as fast as the inner arms. Sketch a graph of its motion and determine its minimum and maximum speeds.

40. Find parametric equations for a Scrambler that is the same as in example 2.1 except that the inner arms have length 3. Sketch a graph of its motion and determine its minimum and maximum speeds.

41. Suppose an object follows the path $\begin{cases} x = \sin 4t \\ y = -\cos 4t \end{cases}$. Show that its speed is constant. Show that, at any time t, the tangent line is perpendicular to a line connecting the origin and the object.

42. A Ferris wheel has height 100 feet and completes one revolution in 3 minutes at a constant speed. Compute the speed of a rider in the Ferris wheel.

43. Suppose you are standing at the origin watching an object that has position $(x(t), y(t))$ at time t. Show that, from your perspective, the object is moving clockwise if $\left(\frac{y(t)}{x(t)}\right)' < 0$ and is moving counterclockwise if $\left(\frac{y(t)}{x(t)}\right)' > 0$.

44. In the Ptolemaic model of planetary motion, the earth was at the center of the solar system and the sun and planets orbited the earth. Circular orbits, which were preferred for aesthetic reasons, could not account for the actual motion of the planets as viewed from the earth. Ptolemy modified the circles into epicycloids, which are circles on circles similar to the Scrambler of example 2.1. Suppose that a planet's motion is given by $\begin{cases} x = 10\cos 16\pi t + 20\cos 4\pi t \\ y = 10\sin 16\pi t + 20\sin 4\pi t \end{cases}$. Using the result of exercise 43, find the intervals in which the planet rotates clockwise and the intervals in which the planet rotates counterclockwise.

45. (exploratory) By varying the speed of the outer arms, the Scrambler of example 2.1 can be generalized to $\begin{cases} x = 2\cos t + \sin kt \\ y = 2\sin t + \cos kt \end{cases}$ for some positive constant k. Show that the minimum speed for any such Scrambler is reached at the outside of a loop. Show that the only value of k that actually produces a speed of 0 is $k = 2$. By varying the lengths of the arms, you can further generalize the Scrambler to $\begin{cases} x = r\cos t + \sin kt \\ y = r\sin t + \cos kt \end{cases}$ for positive constants r and k. Sketch the paths for several such Scramblers and determine the relationship between r and k needed to produce a speed of 0.

46. (exploratory) **Bézier curves** are essential in almost all areas of modern engineering design. (For instance, Bézier curves were used for sketching many of the figures for this book). One version of a Bézier curve starts with **control points** at (a, y_a), (b, y_b), (c, y_c) and (d, y_d). The Bézier curve passes through the points (a, y_a) and (d, y_d). The tangent line at $x = a$ passes through (b, y_b) and the tangent line at $x = d$ passes through (c, y_c). Show that these criteria are met, for $0 \le t \le 1$, with

$$\begin{cases} x = (a + b - c - d)t^3 + (2d - 2b + c - a)t^2 \\ \quad + (b - a)t + a \\ y = (y_a + y_b - y_c - y_d)t^3 + (2y_d - 2y_b + y_c - y_a)t^2 \\ \quad + (y_b - y_a)t + y_a \end{cases}$$

Use this formula to find and graph the Bézier curve with control points $(0, 0)$, $(1, 2)$, $(2, 3)$ and $(3, 0)$. Explore the effect of moving the middle control points, for example moving them up to $(1, 3)$ and $(2, 4)$, respectively.

9.3　Arc Length and Surface Area in Parametric Equations

In section 9.2, you learned how to find slopes of tangent lines and areas for the case of curves defined parametrically. In this section, we investigate arc length and surface area for such curves. Along the way, we explore one of the most famous and interesting curves in mathematics.

Let C be the curve defined by the parametric equations $x = g(t)$ and $y = h(t)$, for $a \leq t \leq b$ (see Figure 9.11), where g, g', h and h' are continuous on the interval $[a, b]$. We further assume that the curve does not intersect itself, except possibly at a finite number of points. That is, there are at most a finite number of points on the curve that are produced by more than one value of t. As we did in section 5.4 for the case of the curve $y = f(x)$, we want to find a way of computing the length of the curve (the **arc length**). As we have done countless times now, we begin by constructing an approximation.

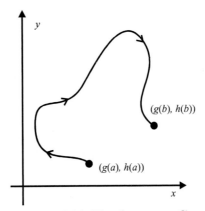

FIGURE 9.11: The plane curve C.

First, we divide the t-interval $[a, b]$ into n equal pieces,

$$a = t_0 < t_1 < t_2 < \cdots < t_n = b,$$

where $t_i - t_{i-1} = \Delta t = \dfrac{b - a}{n}$, for each $i = 1, 2, 3, \ldots$. (That is, each of the subintervals $[t_{i-1}, t_i]$ has the **same** length Δt.) For each subinterval $[t_{i-1}, t_i]$, we approximate the arc length s_i of the portion of the curve joining the point $(g(t_{i-1}), h(t_{i-1}))$ to the point $(g(t_i), h(t_i))$ with the length of the line segment joining these points. This approximation is shown in Figure 9.12 for the case where $n = 4$. We have

$$s_i \approx d\{(g(t_{i-1}), h(t_{i-1})), (g(t_i), h(t_i))\}$$
$$= \sqrt{[g(t_i) - g(t_{i-1})]^2 + [h(t_i) - h(t_{i-1})]^2}.$$

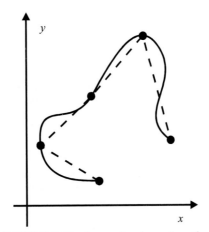

FIGURE 9.12: Approximate arc length.

Recall that from the Mean Value Theorem (see section 2.9 and make sure you know why we can apply this here), we have that

$$g(t_i) - g(t_{i-1}) = g'(c_i)(t_i - t_{i-1}) = g'(c_i)\Delta t$$

and

$$h(t_i) - h(t_{i-1}) = h'(d_i)\,(t_i - t_{i-1}) = h'(d_i)\,\Delta t,$$

where c_i and d_i are some points in the interval (t_{i-1}, t_i). This gives us

$$s_i \approx \sqrt{[g(t_i) - g(t_{i-1})]^2 + [h(t_i) - h(t_{i-1})]^2}$$

$$= \sqrt{[g'(c_i)\,\Delta t]^2 + [h'(d_i)\,\Delta t]^2}$$

$$= \sqrt{[g'(c_i)]^2 + [h'(d_i)]^2}\,\Delta t.$$

Notice that if Δt is small, then c_i and d_i are close together. So, we can make the further approximation

$$s_i \approx \sqrt{[g'(c_i)]^2 + [h'(c_i)]^2}\,\Delta t,$$

for each $i = 1, 2, \ldots, n$. The total arc length is then approximately

$$s \approx \sum_{i=1}^{n} \sqrt{[g'(c_i)]^2 + [h'(c_i)]^2}\,\Delta t.$$

Taking the limit as $n \to \infty$ should then give us the exact arc length, which you should recognize as an integral:

$$s = \lim_{n \to \infty} \sum_{i=1}^{n} \sqrt{[g'(c_i)]^2 + [h'(c_i)]^2}\,\Delta t = \int_a^b \sqrt{[g'(t)]^2 + [h'(t)]^2}\,dt.$$

We summarize this discussion in the following result.

Theorem 3.1   For the curve defined parametrically by $x = g(t)$, $y = h(t)$, $a \le t \le b$, if g, g', h and h' are continuous on $[a, b]$ and the curve does not intersect itself (except possibly at a finite number of points), then the arc length s of the curve is given by

$$s = \int_a^b \sqrt{[g'(t)]^2 + [h'(t)]^2}\,dt = \int_a^b \sqrt{\left(\frac{dx}{dt}\right)^2 + \left(\frac{dy}{dt}\right)^2}\,dt. \qquad (3.1)$$

In the following example, we illustrate the use of (3.1) to find the arc length of the Scrambler curve from example 2.1.

Example 3.1   Finding the Arc Length of a Plane Curve

Find the arc length of the Scrambler curve $x = g(t) = 2\cos t + \sin 2t$, $y = h(t) = 2\sin t + \cos 2t$, for $0 \le t \le 2\pi$. Also, find the average speed of the Scrambler over this interval.

Solution   The curve is shown in Figure 9.13. First, note that g, g', h and h' are all continuous on the interval $[0, 2\pi]$. From (3.1), we then have

$$s = \int_a^b \sqrt{\left(\frac{dx}{dt}\right)^2 + \left(\frac{dy}{dt}\right)^2}\, dt = \int_0^{2\pi} \sqrt{(-2\sin t + 2\cos 2t)^2 + (2\cos t - 2\sin 2t)^2}\, dt$$

$$= \int_0^{2\pi} \sqrt{4\sin^2 t - 8\sin t \cos 2t + 4\cos^2 2t + 4\cos^2 t - 8\cos t \sin 2t + 4\sin^2 2t}\, dt$$

$$= \int_0^{2\pi} \sqrt{8 - 8\sin t \cos 2t - 8\cos t \sin 2t}\, dt = \int_0^{2\pi} \sqrt{8 - 8\sin 3t}\, dt \approx 16,$$

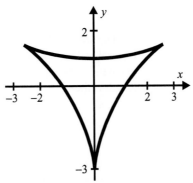

FIGURE 9.13: $x = 2\cos t + \sin 2t$, $y = 2\sin t + \cos 2t$, $0 \le t \le 2\pi$.

since $\sin^2 t + \cos^2 t = 1$, $\cos^2 2t + \sin^2 2t = 1$ and

$$\sin t \cos 2t + \sin 2t \cos t = \sin(t + 2t) = \sin 3t,$$

and where we have approximated the last integral numerically. To find the average speed over the given interval, we simply divide the arc length (i.e., the distance traveled), by the total time, 2π, to obtain

$$s_{\text{ave}} \approx \frac{16}{2\pi} \approx 2.546.$$

∎

We want to emphasize that Theorem 3.1 allows the curve to intersect itself at a **finite** number of points. We cannot, however, allow a curve to intersect itself over an entire interval of values of the parameter t. To see why this requirement is needed, notice that the parametric equations $x = \cos t$, $y = \sin t$, for $0 \le t \le 4\pi$ describes the circle of radius 1 centered at the origin. However, the circle is traversed **twice** as t ranges from 0 to 4π. If you were to apply (3.1) to this curve, you'd obtain

$$\int_0^{4\pi} \sqrt{\left(\frac{dx}{dt}\right)^2 + \left(\frac{dy}{dt}\right)^2}\, dt = \int_0^{4\pi} \sqrt{(-\sin t)^2 + \cos^2 t}\, dt = 4\pi,$$

which corresponds to twice the arc length (circumference) of the circle. As you can see, if a curve intersects itself over an entire interval of values of t, the arc length of such a portion of the curve is counted twice by (3.1).

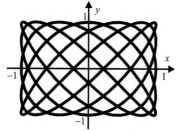

FIGURE 9.14: A Lissajous curve.

Example 3.2   Finding the Arc Length of a Complicated Plane Curve

Find the length of the plane curve $x = \cos 5t$, $y = \sin 7t$, for $0 \leq t \leq 2\pi$.

Solution   This unusual curve (an example of a **Lissajous curve**) is sketched in Figure 9.14. We leave it as an exercise to verify that the hypotheses of Theorem 3.1 are met. From (3.1), we then have that

$$s = \int_0^{2\pi} \sqrt{\left[\frac{dx}{dt}\right]^2 + \left[\frac{dy}{dt}\right]^2}\, dt = \int_0^{2\pi} \sqrt{(-5\sin 5t)^2 + (7\cos 7t)^2}\, dt \approx 36.5,$$

where the integral has been approximated numerically. This is a long curve to be confined within the rectangle $-1 \leq x \leq 1$, $-1 \leq y \leq 1$!

■

The arc length formula (3.1) should seem familiar to you. First, compare the parametric integral in Theorem 3.1 to the formula derived in section 5.4 for a function $y = f(x)$. Note that parametric equations for such a curve are $x = t$, $y = f(t)$ and that from (3.1), the arc length is then

$$s = \int_a^b \sqrt{\left(\frac{dx}{dt}\right)^2 + \left(\frac{dy}{dt}\right)^2}\, dt = \int_a^b \sqrt{1 + [f'(t)]^2}\, dt,$$

which is the arc length integral derived in section 5.4. In other words, the formula developed in section 5.4 is a special case of (3.1).

You might notice that the speed of the Scrambler calculated in example 2.3 and the length of the Scrambler curve found in example 3.1 both depend on the same quantity: $\sqrt{\left(\frac{dx}{dt}\right)^2 + \left(\frac{dy}{dt}\right)^2}$. Observe that if the parameter t represents time, then $\sqrt{\left(\frac{dx}{dt}\right)^2 + \left(\frac{dy}{dt}\right)^2}$ represents speed and from Theorem 3.1, the arc length (i.e., the distance traveled) is the integral of the speed with respect to time.

We can use our notion of arc length to address a famous problem called the **brachistochrone problem**. We state this problem in the context of a downhill skier. Consider a ski slope consisting of a tilted plane, where a skier wishes to get from a point A at the top of the slope to a point B down the slope (but **not** directly beneath A) in the least time possible (see Figure 9.15). Suppose the path taken by the skier is given by the parametric equations $x = g(u)$ and $y = h(u)$, $0 \leq u \leq 1$, where x and y determine the position of the skier in the plane of the ski slope. (For simplicity, we orient the positive y-axis so that it points down. Also, we chose the parameter u since u will, in general, **not** represent time.)

FIGURE 9.15: Downhill Skier.

To derive a formula for the time required to get from point A to point B, start with the simple formula $d = r \cdot t$ relating the distance to the time and the rate. As seen in the derivation of the arc length formula (3.1), for a small section of the curve, the distance is approximately $\sqrt{[g'(u)]^2 + [h'(u)]^2}$. The rate is harder to identify since we aren't given position as a function of time. For simplicity, we assume that the only effect of friction is to keep the skier on the path and that $y(t) \geq 0$. In this case, using the principle of conservation of energy from physics, it can be shown that the skier's speed is given by $\dfrac{\sqrt{h(u)}}{k}$ for some constant $k \geq 0$. Putting the pieces together, the total time from point A to point B is given by

$$Time = \int_0^1 k\sqrt{\frac{[g'(u)]^2 + [h'(u)]^2}{h(u)}}\, du. \qquad (3.2)$$

Your first thought is probably that the shortest path from point A to point B is along a straight line. If you're thinking of short in terms of *distance*, you're right, of course. However, if you think of short in terms of *time* (This is how most skiers would think of it!), this is not true. In the following example, we show that the fastest path from point A to point B is in fact, **not** along a straight line, by exhibiting a faster path.

Example 3.3   A Faster Path than a Straight Line Segment

If point A in the preceding skiing example is $(0,0)$ and point B is $(\pi, 2)$, show that the **cycloid** defined by

$$x = \pi u - \sin \pi u, y = 1 - \cos \pi u$$

is faster than the line segment connecting the points. Explain the result in physical terms.

Solution   First, note that the line segment connecting the points is given by $x = \pi u$, $y = 2u$, for $0 \leq u \leq 1$. Further, both curves meet the endpoint requirements that $(x(0), y(0)) = (0,0)$ and $(x(1), y(1)) = (\pi, 2)$. For the cycloid, we have from (3.2) that

$$Time = \int_0^1 k\sqrt{\frac{[x'(u)]^2 + [y'(u)]^2}{y(u)}}\, du$$

$$= k\int_0^1 \sqrt{\frac{(\pi - \pi \cos \pi u)^2 + (\pi \sin \pi u)^2}{1 - \cos \pi u}}\, du$$

$$= k\sqrt{2\pi}\int_0^1 \sqrt{\frac{1 - \cos \pi u}{1 - \cos \pi u}}\, du$$

$$= k\sqrt{2}\pi.$$

Similarly, for the line segment, we have that

$$Time = \int_0^1 k\sqrt{\frac{[x'(u)]^2 + [y'(u)]^2}{y(u)}}\, du$$

$$= k\int_0^1 \sqrt{\frac{\pi^2 + 2^2}{2u}}\, du$$

$$= k\sqrt{2}\sqrt{\pi^2 + 4}.$$

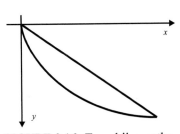

FIGURE 9.16: Two skiing paths.

Notice that both times are multiples of the constant $k\sqrt{2}$. The cycloid route is faster since $\pi < \sqrt{\pi^2 + 4}$. The two paths are shown in Figure 9.16. You should observe that the cycloid is very steep at the beginning. This would allow a skier to go faster following the cycloid than following the straight line. As it turns out, the greater speed of the cycloid more than compensates for the shorter distance of the straight line path.

■

We will ask you to construct some skiing paths of your own in the exercises. However, it has been proven that the cycloid is the plane curve with the shortest time (which is what the Greek root words for **brachistochrone** mean). This problem has an interesting history focused on brothers Jakob and Johann Bernoulli, who solved the problem in 1697 (along with Newton, Leibniz and L'Hôpital) and argued incessantly about who deserved credit. In addition, we will give you an opportunity to discover another remarkable property of the cycloid, relating to another famous problem, the **tautochrone problem**.

Much as we did in section 5.4, we can use our arc length formula to find a formula for the surface area of a surface of revolution. Recall that if the curve $y = f(x)$ for $c \leq x \leq d$ is revolved about the x-axis (see Figure 9.17), the surface area is given by

$$Surface\ Area = \int_c^d 2\pi \underbrace{|f(x)|}_{radius}\ \underbrace{\sqrt{1 + [f'(x)]^2}}_{arc\ length}\ dx.$$

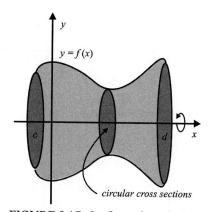

FIGURE 9.17: Surface of revolution.

Let C be the curve defined by the parametric equations $x = g(t)$ and $y = h(t)$ with $a \leq t \leq b$, where g, g', h and h' are continuous and where the curve does not intersect itself for $a \leq t \leq b$. We leave it as an exercise to derive the corresponding formula for parametric equations:

$$Surface\ Area = \int_a^b 2\pi \underbrace{|h(t)|}_{radius}\ \underbrace{\sqrt{[g'(t)]^2 + [h'(t)]^2}}_{arc\ length}\ dt.$$

More generally, we have that if the curve is revolved about the line $y = c$, the surface area is given by

$$Surface\ Area = \int_a^b 2\pi \underbrace{|h(t) - c|}_{radius}\ \underbrace{\sqrt{[g'(t)]^2 + [h'(t)]^2}}_{arc\ length}\ dt. \tag{3.3}$$

Likewise, if we revolve the curve about the line $x = d$, the surface area is given by

$$Surface\ Area = \int_a^b 2\pi \underbrace{|g(t) - d|}_{radius}\ \underbrace{\sqrt{[g'(t)]^2 + [h'(t)]^2}}_{arc\ length}\ dt. \tag{3.4}$$

Look carefully at what all of the surface area formulas have in common. That is, in each case, the surface area is given by

$$\boxed{Surface\ Area = \int_a^b 2\pi\ (radius)\ (arc\ length)\ dt.} \tag{3.5}$$

All you need to do is to look carefully at the graph of the curve and the axis you are revolving about to see how to fill in the blanks in (3.5). As we observed in section 5.4, it is very important that you draw a picture here.

Example 3.4   **Finding Surface Area with Parametric Equations**

Find the surface area of the surface formed by revolving the half-ellipse $\dfrac{x^2}{9} + \dfrac{y^2}{4} = 1$, $y \geq 0$, about the x-axis (see Figure 9.18).

FIGURE 9.18: $y = 2\sqrt{1 - x^2/9}$.

Solution   It would truly be a mess to set up the integral for $y = f(x) = 2\sqrt{1 - x^2/9}$. (Think about this!) Instead, notice that you can represent the curve by the parametric equations $x = 3\cos t$, $y = 2\sin t$, for $0 \leq t \leq \pi$. From (3.3), the surface area is then given by

$$Surface\ Area = \int_0^{\pi} 2\pi \underbrace{(2\sin t)}_{radius} \underbrace{\sqrt{(-3\sin t)^2 + (2\cos t)^2}}_{arc\,length}\ dt$$

$$= 4\pi \int_0^{\pi} \sin t \sqrt{9\sin^2 t + 4\cos^2 t}\ dt$$

$$= 4\pi\ \frac{9\sqrt{5}\sin^{-1}(\sqrt{5}/3) + 10}{5} \approx 67.7,$$

where we leave the details of the integration to you.

■

Example 3.5   **Revolving about a Line Other than a Coordinate Axis**

Find the surface area of the surface formed by revolving the curve $x = \sin 2t$, $y = \cos 3t$, for $0 \leq t \leq \pi/3$, about the line $x = 2$.

Solution   A sketch of the curve is shown in Figure 9.19. Since the x-values on the curve are all less than 2, the radius of the solid of revolution is $(2 - \sin t)$ and so, from (3.4), the surface area is given by

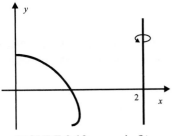

FIGURE 9.19: $x = \sin 2t$, $y = \cos 3t$.

$$Surface\ Area = \int_0^{\pi/3} 2\pi \underbrace{(2 - \sin 2t)}_{radius} \underbrace{\sqrt{[2\cos 2t]^2 + [-3\sin 3t]^2}}_{arc\,length}\ dt \approx 20.1,$$

where we have approximated the value of the integral numerically.

■

Exercises 9.3

1. (writing) In the derivation preceding Theorem 3.1, we justified the equation

$$g(t_i) - g(t_{i-1}) = g'(c_i) \, \Delta t.$$

Thinking of $g(t)$ as position and $g'(t)$ as velocity, explain why this makes sense.

2. (writing) The curve in example 3.2 was a long curve contained within a small rectangle. What would you guess would be the maximum length for a curve contained in such a rectangle? Briefly explain.

3. (writing) In example 3.3, we noted that the steeper initial slope of the cycloid would allow the skier to build up more speed than the straight line path. The cycloid takes this idea to the limit by having a vertical tangent line at the origin. Explain why despite the vertical tangent line, it is still physically possible for the skier to stay on this slope. (Hint: How do the two dimensions of the path relate to the three dimensions of the ski slope?)

4. (writing) The tautochrone problem discussed below involves starting on the same curve at two different places and comparing the times required to reach the end. For the cycloid, compare the speed of the skier starting at the origin versus one starting halfway to the bottom. Explain why it is not clear whether starting halfway down would get you to the bottom faster.

In exercises 5–18, find the arc length of the curve; approximate numerically, if needed.

5. $\begin{cases} x = 2\cos t \\ y = 4\sin t \end{cases}$ **6.** $\begin{cases} x = 6\cos t \\ y = 3\sin t \end{cases}$

7. $\begin{cases} x = 1 - 2\cos t \\ y = 2 + 3\sin t \end{cases}$ **8.** $\begin{cases} x = \cos 3t \\ y = \sin 3t \end{cases}$

9. $\begin{cases} x = t^3 - 4t \\ y = t^2 - 3 \end{cases}, -2 \le t \le 2$

10. $\begin{cases} x = t^3 - 4t \\ y = t^2 - 3t \end{cases}, -2 \le t \le 2$

11. $\begin{cases} x = \cos 2t \\ y = \sin 4t \end{cases}$

12. $\begin{cases} x = \cos 7t \\ y = \sin 11t \end{cases}$

13. $\begin{cases} x = t\cos t \\ y = t\sin t \end{cases}, -1 \le t \le 1$

14. $\begin{cases} x = t^2\cos t \\ y = t^2\sin t \end{cases}, -1 \le t \le 1$

15. $\begin{cases} x = \sin 2t \cos t \\ y = \sin 2t \sin t \end{cases}, 0 \le t \le \pi/2$

16. $\begin{cases} x = \sin 4t \cos t \\ y = \sin 4t \sin t \end{cases}, 0 \le t \le \pi/2$

17. $\begin{cases} x = \sin t \\ y = \sin \pi t \end{cases}, 0 \le t \le \pi$

18. $\begin{cases} x = \sin t \\ y = \sin \sqrt{2}t \end{cases}, 0 \le t \le \pi$

In exercises 19–22, show that the curve starts at the origin at $t = 0$ and reaches the point $(\pi, 2)$ at $t = 1$. Then use the time formula from example 3.3 to determine how long it would take a skier to take the given path.

19. $\begin{cases} x = \pi t \\ y = 2\sqrt{t} \end{cases}$

20. $\begin{cases} x = \pi t \\ y = 2\sqrt[4]{t} \end{cases}$

21. $\begin{cases} x = -\frac{1}{2}\pi(\cos \pi t - 1) \\ y = 2t + \frac{7}{10}\sin \pi t \end{cases}$

22. $\begin{cases} x = \pi t - 0.6\sin \pi t \\ y = 2t + 0.4\sin \pi t \end{cases}$

In exercises 23–26, find the slope at the origin and the arc length for the curve in the indicated exercise. Compare to the cycloid.

23. exercise 19   **24.** exercise 20

25. exercise 21   **26.** exercise 22

In exercises 27–32, compute the surface area of the surface obtained by revolving the given curve about the indicated axis.

27. $\begin{cases} x = t^2 - 1 \\ y = t^3 - 4t \end{cases}, -2 \le t \le 0$, about the x-axis

28. $\begin{cases} x = t^2 - 1 \\ y = t^3 - 4t \end{cases}, 0 \le t \le 2$, about the x-axis

29. $\begin{cases} x = t^2 - 1 \\ y = t^3 - 4t \end{cases}, -1 \le t \le 1$, about the y-axis

30. $\begin{cases} x = t^2 - 1 \\ y = t^3 - 4t \end{cases}, -2 \le t \le 0$, about $x = -1$

31. $\begin{cases} x = t^3 - 4t \\ y = t^2 - 3 \end{cases}, 0 \le t \le 2$, about the y-axis

32. $\begin{cases} x = t^3 - 4t \\ y = t^2 - 3 \end{cases}, 0 \le t \le 2$, about $y = 2$

33. (exploratory) For the brachistochrone problem, two criteria for the fastest curve are: (1) steep slope at the origin and (2) concave down. Explain why these criteria make sense, and identify other criteria. Then find parametric equations for a curve (different from the cycloid or those of exercises 19–22) which meet all the criteria. Use the formula of example 3.3 to find out how fast your curve is. You can't beat the cycloid, but get as close as you can!

34. (exploratory) The **tautochrone** problem is another surprising problem, which was studied and solved by the same seventeenth century mathematicians as the brachistochrone problem. (See *Journey Through Genius* by William Dunham for a description of this interesting piece of history, featuring the brilliant yet combative Bernoulli brothers.) Recall that the cycloid of example 3.3 runs from $(0,0)$ to $(\pi, 2)$. It takes the skier $k\sqrt{2\pi} = \pi/g$ seconds to ski the path. How long would it take the skier starting part way down the path, for instance, at $(\pi/2 - 1, 1)$? Find the slope of the cycloid at this point and compare it to the slope at $(0,0)$. Explain why the skier would build up less speed starting at this new point. Graph the speed function for the cycloid with $0 \leq u \leq 1$ and explain why the farther down the slope you start, the less speed you'll have. To see how speed and distance balance, use the time formula

$$T = \frac{\pi}{g} \int_a^1 \frac{\sqrt{1 - \cos \pi u}}{\sqrt{\cos a\pi - \cos \pi u}} \, du$$

for the time it takes to ski the cycloid starting at the point $(\pi a - \sin \pi a, 1 - \cos \pi a)$, $0 < a < 1$. What is the remarkable property that the cycloid has?

9.4 Polar Coordinates

You've probably heard the cliche about how difficult it is to try to fit a round peg into a square hole. In some respects, we have faced this problem on several occasions so far in our study of calculus. For instance, if we were to use an integral to calculate the area of the circle $x^2 + y^2 = 9$, we would have

$$A = \int_{-3}^{3} \left[\sqrt{9 - x^2} - \left(-\sqrt{9 - x^2} \right) \right] dx = 2 \int_{-3}^{3} \sqrt{9 - x^2} \, dx. \tag{4.1}$$

Note that you can evaluate this integral by making the trigonometric substitution $x = 3 \sin t$. However, the authors would prefer to use a CAS to evaluate this integral for us. (It's a good thing that we already know a simple formula for finding the area of a circle!) A better plan might be to use parametric equations, such as $x = 3 \cos t$, $y = 3 \sin t$, for $0 \leq t \leq 2\pi$ to describe the circle. From what we learned in section 9.2, the area is given by

$$\int_0^{2\pi} x(t) \, y'(t) \, dt = \int_0^{2\pi} (3 \cos t) \, (3 \cos t) \, dt$$

$$= 9 \int_0^{2\pi} \cos^2 t \, dt.$$

This is certainly better than the integral in (4.1), but it still requires some effort to evaluate this. The basic problem is that circles do not translate well into the usual x-y coordinate system. We often refer to this system as a system of **rectangular coordinates,** because a point is described in terms of the horizontal and vertical distances from the origin (see Figure 9.20).

An alternative description of a point in the xy-plane consists of specifying the distance r from the point to the origin and the angle θ (in radians) measured from the positive x-axis counterclockwise to the ray connecting the point and the origin (see Figure 9.21). We describe the point by the ordered pair (r, θ) and refer to r and θ as **polar coordinates** for the point.

FIGURE 9.20:
Rectangular coordinates.

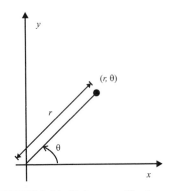

FIGURE 9.21: Polar coordinates.

Example 4.1 Converting from Polar to Rectangular Coordinates

Plot the points with the indicated polar coordinates and determine the corresponding rectangular coordinates (x, y), for: (a) $(2, 0)$, (b) $\left(3, \frac{\pi}{2}\right)$, (c) $\left(-3, \frac{\pi}{2}\right)$ and (d) $(2, \pi)$.

Solution   (a) Notice that the angle $\theta = 0$ locates the point on the positive x-axis. At a distance of $r = 2$ from the origin, this corresponds to the point $(2, 0)$ in rectangular coordinates (see Figure 9.22a).

FIGURE 9.22a: The point $(2, 0)$
in polar coordinates.

FIGURE 9.22b: The points $(3, \pi/2)$
and $(-3, \pi/2)$ in polar coordinates.

(b) The angle $\theta = \frac{\pi}{2}$ locates points on the positive y-axis. At a distance of $r = 3$ from the origin, this corresponds to the point $(0, 3)$ in rectangular coordinates (see Figure 9.22b).

(c) The angle is the same as in (b), but a negative value of r indicates that the point is located 3 units in the opposite direction, at the point $(0, -3)$ in rectangular coordinates (see Figure 9.22b).

(d) The angle $\theta = \pi$ corresponds to the negative x-axis. The distance of $r = 2$ from the origin gives us the point $(-2, 0)$ in rectangular coordinates (see Figure 9.22c).

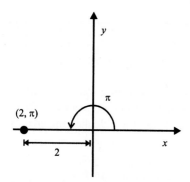

FIGURE 9.22c: The point $(2, \pi)$
in polar coordinates.

Example 4.2   Converting from Rectangular to Polar Coordinates
Find a polar coordinate representation of the rectangular point $(1, 1)$.

Solution   From Figure 9.23a, notice that the point lies on the line $y = x$, which makes an angle of $\frac{\pi}{4}$ with the positive x-axis. From the distance formula, we get that $r = \sqrt{1^2 + 1^2} = \sqrt{2}$. This says that we can write the point as $\left(\sqrt{2}, \frac{\pi}{4}\right)$ in polar coordinates. Referring to Figure 9.23b, notice that we can specify the same point by using a negative value of r, $r = -\sqrt{2}$ with the angle $\frac{5\pi}{4}$ (think about this some).

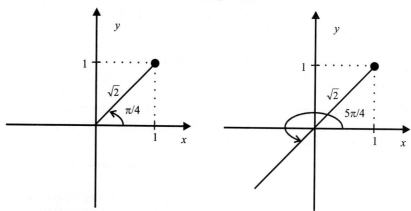

FIGURE 9.23a: Polar coordinates
for the point $(1, 1)$.

FIGURE 9.23b: An alternative
polar representation of $(1, 1)$.

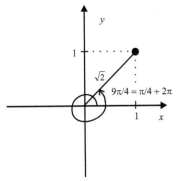

FIGURE 9.23c: Another polar representation of the point $(1, 1)$.

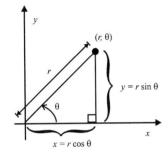

FIGURE 9.24: Converting from polar to rectangular coordinates.

Notice further, that the angle $\frac{9\pi}{4} = \frac{\pi}{4} + 2\pi$ corresponds to the same ray shown in Figure 9.23a (see Figure 9.23c). In fact, all of the polar points $\left(\sqrt{2}, \frac{\pi}{4} + 2n\pi\right)$ and $\left(-\sqrt{2}, \frac{5\pi}{4} + 2n\pi\right)$ for any integer n correspond to the same point in the xy-plane. ∎

Remark 4.1   Given a distance r and angle θ, there is exactly one point in the xy-plane with the polar coordinates (r, θ). However, as we saw in example 4.2, for a given point (x, y) in the plane, there are an infinite number of possible polar coordinate representations. In particular, you can use both positive and negative values of r. Also, for a given angle θ, the angles $\theta \pm 2\pi$, $\theta \pm 4\pi$ and so on, all correspond to the same ray and can also be used. For convenience, we use the notation $\theta + 2n\pi$ (for any integer n) to represent all of these possible angles.

Referring to Figure 9.24, notice that it is a simple matter to find the rectangular coordinates (x, y) of a point specified in polar coordinates as (r, θ). From the usual definitions for $\sin\theta$ and $\cos\theta$, we get

$$\boxed{x = r\cos\theta \text{ and } y = r\sin\theta.} \tag{4.2}$$

As we've already observed, a given point (x, y) in the plane will have infinitely many polar coordinate representations. From equations (4.2), notice that

$$x^2 + y^2 = r^2\cos^2\theta + r^2\sin^2\theta = r^2\left(\cos^2\theta + \sin^2\theta\right) = r^2$$

and

$$\frac{y}{x} = \frac{r\sin\theta}{r\cos\theta} = \frac{\sin\theta}{\cos\theta} = \tan\theta.$$

That is, every polar coordinate representation (r, θ) of the point (x, y) must satisfy

$$\boxed{r^2 = x^2 + y^2 \text{ and } \tan\theta = \frac{y}{x}.} \tag{4.3}$$

Notice that since there's more than one choice of r and θ, we cannot actually solve equations (4.3) to produce formulas for r and θ. In particular, while you might be tempted to write $\theta = \tan^{-1}\left(\frac{y}{x}\right)$, this is not the only possible choice. Remember that for (r, θ) to be a polar representation of the point (x, y), θ can be **any** angle for which $\tan\theta = \frac{y}{x}$, while $\tan^{-1}\left(\frac{y}{x}\right)$ gives you only an angle θ in the interval $\left(-\frac{\pi}{2}, \frac{\pi}{2}\right)$. Finding polar coordinates for a given point is typically a process involving some graphing and some thought.

Example 4.3   Converting from Rectangular to Polar Coordinates

Find all polar coordinate representations for the rectangular points (a) $(2, 3)$ and (b) $(-3, 1)$.

Solution   (a) With $x = 2$ and $y = 3$, we have from (4.3) that

$$r^2 = x^2 + y^2 = 2^2 + 3^2 = 13,$$

so that $r = \pm\sqrt{13}$. Also,

$$\tan\theta = \frac{y}{x} = \frac{3}{2}.$$

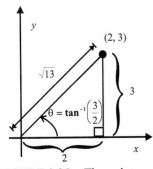

FIGURE 9.25a: The point $(2, 3)$.

One solution of this (the most obvious solution) is $\theta = \tan^{-1}\left(\frac{3}{2}\right) \approx 0.98$ radians. To determine which choice of r corresponds to this angle, note that $(2, 3)$ is located in the first quadrant (see Figure 9.25a). Since 0.98 radians also puts you in the first quadrant, this angle corresponds to the positive value of r, so that $\left(\sqrt{13}, \tan^{-1}\left(\frac{3}{2}\right)\right)$ is one polar representation of the point. The negative choice of r corresponds to an angle one half-circle (i.e., π

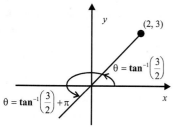

FIGURE 9.25b: Negative value of r.

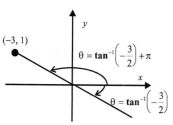

FIGURE 9.26: The point $(-3, 1)$.

radians), away (see Figure 9.25b), so that another representation is $\left(-\sqrt{13}, \tan^{-1}\left(\frac{3}{2}\right) + \pi\right)$. Every other polar representation is found by adding multiples of 2π to the two angles used above. That is, every polar representation of the point $(2, 3)$ must have the form $\left(\sqrt{13}, \tan^{-1}\left(\frac{3}{2}\right) + 2n\pi\right)$ or $\left(-\sqrt{13}, \tan^{-1}\left(\frac{3}{2}\right) + \pi + 2n\pi\right)$, for some integer choice of n.

(b) For the point $(-3, 1)$, we have $x = -3$ and $y = 1$. From (4.3), we have

$$r^2 = x^2 + y^2 = (-3)^2 + 1^2 = 10,$$

so that $r = \pm\sqrt{10}$. Further,

$$\tan\theta = \frac{y}{x} = \frac{1}{-3},$$

so that the most obvious choice for the polar angle is $\theta = \tan^{-1}\left(-\frac{1}{3}\right) \approx -0.32$, which lies in the fourth quadrant. Since the point $(-3, 1)$ is in the second quadrant, this choice of the angle corresponds to the negative value of r (see Figure 9.26). The positive value of r then corresponds to the angle $\theta = \tan^{-1}\left(-\frac{1}{3}\right) + \pi$. Observe that all polar coordinate representations must then be of the form $\left(-\sqrt{10}, \tan^{-1}\left(-\frac{1}{3}\right) + 2n\pi\right)$ or $\left(\sqrt{10}, \tan^{-1}\left(-\frac{1}{3}\right) + \pi + 2n\pi\right)$, for some integer choice of n. ∎

Remark 4.2   Notice that for any point (x, y) specified in rectangular coordinates ($x \neq 0$), we can always write the point in polar coordinates using either of the polar angles $\tan^{-1}\left(\frac{y}{x}\right)$ or $\tan^{-1}\left(\frac{y}{x}\right) + \pi$. You can determine which angle corresponds to $r = \sqrt{x^2 + y^2}$ and which corresponds to $r = -\sqrt{x^2 + y^2}$ by looking at which quadrant the point lies in.

Observe that the conversion from polar coordinates to rectangular coordinates is completely straightforward, as in the following example.

Example 4.4   Converting from Polar to Rectangular Coordinates

Find the rectangular coordinates for the polar points (a) $\left(3, \frac{\pi}{6}\right)$ and (b) $(-2, 3)$.

Solution   For (a), we have from (4.2) that

$$x = r\cos\theta = 3\cos\frac{\pi}{6} = \frac{3\sqrt{3}}{2}$$

and

$$y = r\sin\theta = 3\sin\frac{\pi}{6} = \frac{3}{2}.$$

The rectangular point is then $\left(\frac{3\sqrt{3}}{2}, \frac{3}{2}\right)$. For (b), we have

$$x = r\cos\theta = -2\cos 3 \approx 1.98$$

and

$$y = r\sin\theta = -2\sin 3 \approx -0.28.$$

The rectangular point is $(-2\cos 3, -2\sin 3)$, which is located at approximately $(1.98, -0.28)$. ∎

The **graph** of a polar equation $r = f(\theta)$ is the set of all points (x, y) for which $x = r\cos\theta$, $y = r\sin\theta$ and $r = f(\theta)$. In other words, the graph of a polar equation is a graph in the xy-plane of all those points whose polar coordinates satisfy the given equation. We begin by sketching two very simple (and familiar) graphs. The key to drawing the graph of a polar equation is to always keep in mind what the polar coordinates represent.

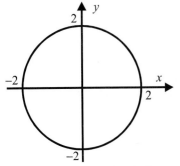

FIGURE 9.27a: The circle
$r = 2$.

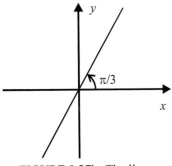

FIGURE 9.27b: The line
$\theta = \pi/3$.

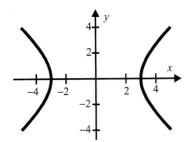

FIGURE 9.28: $x^2 - y^2 = 9$.

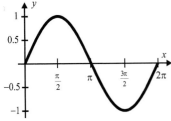

FIGURE 9.29a: $y = \sin x$
plotted in **rectangular**
coordinates.

Example 4.5   Some Simple Graphs in Polar Coordinates

Sketch the graphs of (a) $r = 2$ and (b) $\theta = \pi/3$.

Solution   For (a), notice that $2 = r = \sqrt{x^2 + y^2}$, and so, we want all points whose distance from the origin is 2 (with **any** polar angle θ). Of course, this is the definition of a circle of radius 2 with center at the origin (see Figure 9.27a). For (b), notice that $\theta = \pi/3$ specifies all points with a polar angle of $\pi/3$ from the positive x-axis (at **any** distance r from the origin). Including negative values for r, this defines a line with slope $\tan \pi/3 = \sqrt{3}$ (see Figure 9.27b).

■

Remark 4.3   Calculators sketch polar graphs of $r = f(\theta)$ by computing the value of $f(\theta)$ for numerous values of θ at regularly spaced intervals and then plotting the resulting points (x, y). You should be aware that the appearance of a calculator plot depends on the x-y graphing window specified, and also on the range of displayed values of θ. When drawing polar graphs, you should identify any values of θ corresponding to $r = 0$ or to where r reaches a maximum or minimum. In addition, you should identify the range of values of θ which produce one copy of the polar curve, when this is appropriate.

It turns out that many familiar curves have simple polar equations.

Example 4.6   Converting an Equation from Rectangular to Polar Coordinates

Find the polar equation(s) corresponding to the hyperbola $x^2 - y^2 = 9$ (see Figure 9.28).

Solution   From (4.2), we have

$$9 = x^2 - y^2 = r^2 \cos^2 \theta - r^2 \sin^2 \theta$$
$$= r^2 \left(\cos^2 \theta - \sin^2 \theta\right) = r^2 \cos 2\theta.$$

Solving for r, we get

$$r^2 = \frac{9}{\cos 2\theta} = 9 \sec 2\theta,$$

so that

$$r = \pm 3\sqrt{\sec 2\theta}.$$

Notice that in order to keep $\sec 2\theta > 0$, we can restrict 2θ to lie in the interval $-\frac{\pi}{2} < 2\theta < \frac{\pi}{2}$, so that $-\frac{\pi}{4} < \theta < \frac{\pi}{4}$. Observe that with this range of values of θ, the hyperbola is drawn exactly once, where $r = 3\sqrt{\sec 2\theta}$ corresponds to the right branch of the hyperbola and $r = -3\sqrt{\sec 2\theta}$ corresponds to the left branch.

■

Example 4.7   A Surprisingly Simple Polar Graph

Sketch the graph of the polar equation $r = \sin \theta$.

Solution   For reference, we first sketch a graph of the sine function in **rectangular** coordinates on the interval $[0, 2\pi]$ (see Figure 9.29a). Notice that on the interval $0 \leq \theta \leq \frac{\pi}{2}$, $\sin \theta$ increases from 0 to its maximum value of 1. Then, on the interval $\frac{\pi}{2} \leq \theta \leq \pi$, $\sin \theta$ decreases from 1 to 0. When plotting the polar graph, keep in mind that $r = 0$ corresponds to the origin. Next, on the interval $\pi \leq \theta \leq \frac{3\pi}{2}$, $\sin \theta$ decreases from 0 to its minimum value of -1. Since the values of r are negative, remember that this means

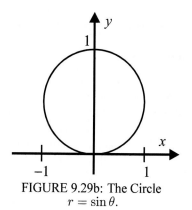

FIGURE 9.29b: The Circle
$r = \sin\theta$.

that the points plotted are in the **opposite** quadrant (i.e., the first quadrant). Notice that this traces out the same curve in the first quadrant as we've already drawn for $0 \leq \theta \leq \frac{\pi}{2}$. Likewise, taking θ in the interval $\frac{3\pi}{2} \leq \theta \leq 2\pi$ retraces the portion of the curve in the second quadrant. Since $\sin\theta$ is periodic of period 2π, taking further values of θ simply retraces portions of the curve that we have already traced. A sketch of the polar graph is shown in Figure 9.29b. We now verify that this curve is actually a circle. Notice that if we multiply the equation $r = \sin\theta$ through by r, we get

$$r^2 = r\sin\theta.$$

You should immediately recognize from (4.2) and (4.3) that $y = r\sin\theta$ and $r^2 = x^2 + y^2$. This gives us the rectangular equation

$$x^2 + y^2 = y$$

or

$$0 = x^2 + y^2 - y.$$

Completing the square, we get

$$0 = x^2 + \left(y^2 - y + \frac{1}{4}\right) - \frac{1}{4}$$

or, adding $\frac{1}{4}$ to both sides,

$$\left(\frac{1}{2}\right)^2 = x^2 + \left(y - \frac{1}{2}\right)^2.$$

This is the rectangular equation for the circle of radius $\frac{1}{2}$ centered at the point $\left(0, \frac{1}{2}\right)$, which is what we see in Figure 9.29b.

■

The graphs of many polar equations are not the graphs of **any** functions, as in the following example.

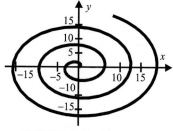

FIGURE 9.30: The spiral
$r = \theta, \theta \geq 0$.

Example 4.8   An Archimedian Spiral
    Sketch the graph of the polar equation $r = \theta$, for $\theta \geq 0$.

Solution   Notice that here, as θ increases, so too does r. That is, as the polar angle increases, the distance from the origin also increases accordingly. This produces the spiral (an example of an **Archimedian spiral**) seen in Figure 9.30.

■

The graphs in the following three examples are all in the general class known as **limaçons**. This class of graphs is defined by $r = a \pm b\sin\theta$ or $r = a \pm b\cos\theta$ for positive constants a and b. If $a = b$, the graphs are called **cardioids**.

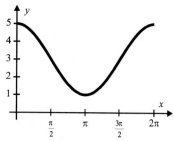

FIGURE 9.31: $y = 3 + 2\cos x$
in rectangular coordinates.

Example 4.9   A Limaçon
    Sketch the graph of the polar equation $r = 3 + 2\cos\theta$.

Solution   We begin by sketching the graph of $y = 3 + 2\cos x$ in rectangular coordinates on the interval $[0, 2\pi]$ to use as a reference (see Figure 9.31). Notice that in this case, we have $r = 3 + 2\cos\theta > 0$ for all values of θ. Further, the maximum value of r is 5 (corresponding to when $\cos\theta = 1$ at $\theta = 0, 2\pi$, etc.) and the minimum value of r is 1 (corresponding to when $\cos\theta = -1$ at $\theta = \pi, 3\pi$, etc.). In this case, the graph is traced out with $0 \le \theta \le 2\pi$. We summarize the intervals of increase and decrease for r in the following table.

Interval	$\cos\theta$	$r = 3 + 2\cos\theta$
$\left[0, \frac{\pi}{2}\right]$	decreases from 1 to 0	decreases from 5 to 3
$\left[\frac{\pi}{2}, \pi\right]$	decreases from 0 to -1	decreases from 3 to 1
$\left[\pi, \frac{3\pi}{2}\right]$	increases from -1 to 0	increases from 1 to 3
$\left[\frac{3\pi}{2}, 2\pi\right]$	increases from 0 to 1	increases from 3 to 5

In Figures 9.32a–9.32d, we show how the sketch progresses through each interval indicated in the table, with the completed figure (called a **limaçon**) shown in Figure 9.32d.

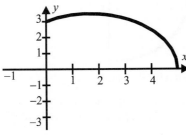

FIGURE 9.32a: $0 \le \theta \le \pi/2$.

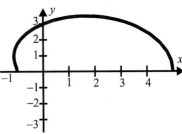

FIGURE 9.32b: $0 \le \theta \le \pi$.

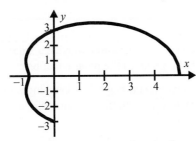

FIGURE 9.32c: $0 \le \theta \le 3\pi/2$.

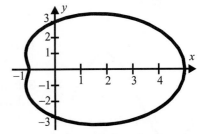

FIGURE 9.32d: $0 \le \theta \le 2\pi$.

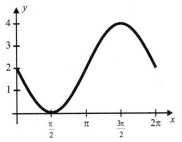

FIGURE 9.33: $y = 2 - 2\sin x$
in rectangular coordinates.

Example 4.10   The Graph of a Cardioid
Sketch the graph of the polar equation $r = 2 - 2\sin\theta$.

Solution   As we have done several times now, we first sketch a graph of $y = 2 - 2\sin x$ in rectangular coordinates, on the interval $[0, 2\pi]$, as in Figure 9.33. We summarize the intervals of increase and decrease in the following table.

Interval	$\sin\theta$	$r = 2 - 2\sin\theta$
$\left[0, \frac{\pi}{2}\right]$	increases from 0 to1	decreases from 2 to 0
$\left[\frac{\pi}{2}, \pi\right]$	decreases from 1 to 0	increases from 0 to 2
$\left[\pi, \frac{3\pi}{2}\right]$	decreases from 0 to -1	increases from 2 to 4
$\left[\frac{3\pi}{2}, 2\pi\right]$	increases from -1 to 0	decreases from 4 to 2

Again, we sketch the graph in stages, corresponding to each of the intervals indicated in the table, as seen in Figures 9.34a – 9.34d.

FIGURE 9.34a: $0 \leq \theta \leq \pi/2$.   FIGURE 9.34b: $0 \leq \theta \leq \pi$.

FIGURE 9.34c: $0 \leq \theta \leq 3\pi/2$.   FIGURE 9.34d: $0 \leq \theta \leq 2\pi$.

The completed graph appears in Figure 9.34d and is sketched-out for $0 \leq \theta \leq 2\pi$. You can see why this figure is called a **cardioid** ("heart-like").

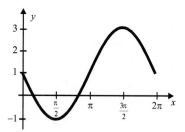

FIGURE 9.35: $y = 1 - 2\sin x$ in rectangular coordinates.

Example 4.11   A Limaçon with a Loop
Sketch the graph of the polar equation $r = 1 - 2\sin\theta$.

Solution   We again begin by sketching a graph of $y = 1 - 2\sin x$ in rectangular coordinates, as in Figure 9.35. We summarize the intervals of increase and decrease in the following table.

Interval	$\sin\theta$	$r = 1 - 2\sin\theta$
$\left[0, \frac{\pi}{2}\right]$	increases from 0 to1	decreases from 1 to -1
$\left[\frac{\pi}{2}, \pi\right]$	decreases from 1 to 0	increases from -1 to 1
$\left[\pi, \frac{3\pi}{2}\right]$	decreases from 0 to -1	increases from 1 to 3
$\left[\frac{3\pi}{2}, 2\pi\right]$	increases from -1 to 0	decreases from 3 to 1

Notice that since r assumes both positive and negative values in this case, we need to exercise a bit more caution, as negative values for r cause us to draw that portion of the graph in the **opposite** quadrant. Observe that $r = 0$ when $1 - 2\sin\theta = 0$, that is, when $\sin\theta = \frac{1}{2}$. This will occur when $\theta = \frac{\pi}{6}$ and when $\theta = \frac{5\pi}{6}$. For this reason, we expand the above table, to include more intervals and where we also indicate the quadrant where the graph is to be drawn, as follows:

Interval	$\sin\theta$	$r = 1 - 2\sin\theta$	Quadrant
$\left[0, \frac{\pi}{6}\right]$	increases from 0 to $\frac{1}{2}$	decreases from 1 to 0	1st
$\left[\frac{\pi}{6}, \frac{\pi}{2}\right]$	increases from $\frac{1}{2}$ to 1	decreases from 0 to -1	3rd
$\left[\frac{\pi}{2}, \frac{5\pi}{6}\right]$	decreases from 1 to $\frac{1}{2}$	increases from -1 to 0	4th
$\left[\frac{5\pi}{6}, \pi\right]$	decreases from $\frac{1}{2}$ to 0	increases from 0 to 1	2nd
$\left[\pi, \frac{3\pi}{2}\right]$	decreases from 0 to -1	increases from 1 to 3	3rd
$\left[\frac{3\pi}{2}, 2\pi\right]$	increases from -1 to 0	decreases from 3 to 1	4th

We sketch the graph in stages in Figures 9.36a – 9.36f, corresponding to each of the intervals indicated in the table.

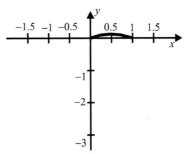

FIGURE 9.36a: $0 \le \theta \le \frac{\pi}{6}$.

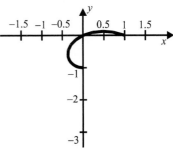

FIGURE 9.36b: $0 \le \theta \le \frac{\pi}{2}$.

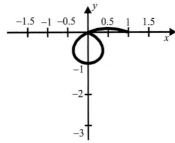

FIGURE 9.36d: $0 \le \theta \le \pi$.

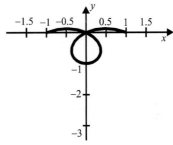

FIGURE 9.36d: $0 \le \theta \le \pi$.

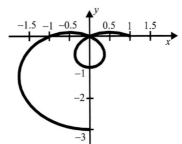

FIGURE 9.36e: $0 \le \theta \le \frac{3\pi}{2}$.

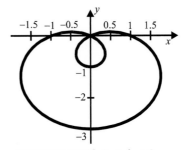

FIGURE 9.36f: $0 \le \theta \le 2\pi$.

The completed graph appears in Figure 9.36f and is sketched out for $0 \le \theta \le 2\pi$. You should observe from this the importance of determining where $r = 0$, as well as where r is increasing and decreasing.

■

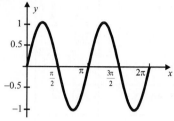

FIGURE 9.37: $y = \sin 2x$ in rectangular coordinates.

Example 4.12   A Four-Leaf Rose

Sketch the graph of the polar equation $r = \sin 2\theta$.

Solution   As usual, we will first draw a graph of $y = \sin 2x$ in rectangular coordinates on the interval $[0, 2\pi]$, as seen in Figure 9.37. Notice that the period of $\sin 2\theta$ is only π. We summarize the intervals on which the function is increasing and decreasing in the following table.

Interval	$r = \sin 2\theta$	Quadrant
$\left[0, \frac{\pi}{4}\right]$	increases from 0 to 1	1st
$\left[\frac{\pi}{4}, \frac{\pi}{2}\right]$	decreases from 1 to 0	1st
$\left[\frac{\pi}{2}, \frac{3\pi}{4}\right]$	decreases from 0 to -1	4th
$\left[\frac{3\pi}{4}, \pi\right]$	increases from -1 to 0	4th
$\left[\pi, \frac{5\pi}{4}\right]$	increases from 0 to 1	3rd
$\left[\frac{5\pi}{4}, \frac{3\pi}{2}\right]$	decreases from 1 to 0	3rd
$\left[\frac{3\pi}{2}, \frac{7\pi}{4}\right]$	decreses from 0 to -1	2nd
$\left[\frac{7\pi}{4}, 2\pi\right]$	increases from -1 to 0	2nd

We sketch the graph in stages in Figures 9.38a – 9.38h, each one corresponding to the intervals indicated in the table, where we have also indicated the lines $y = \pm x$, as a guide.

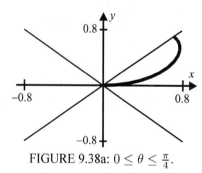

FIGURE 9.38a: $0 \leq \theta \leq \frac{\pi}{4}$.

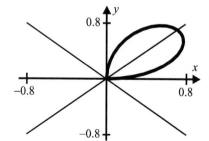

FIGURE 9.38b: $0 \leq \theta \leq \frac{\pi}{2}$.

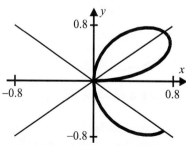

FIGURE 9.38c: $0 \leq \theta \leq \frac{3\pi}{4}$.

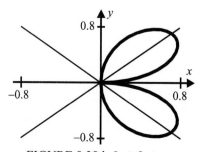

FIGURE 9.38d: $0 \leq \theta \leq \pi$.

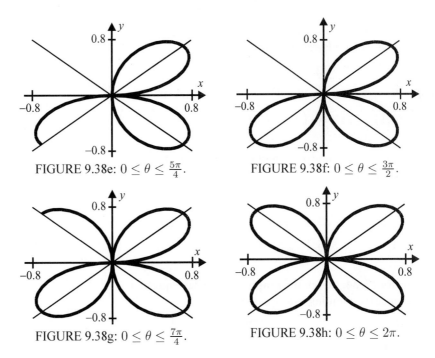

FIGURE 9.38e: $0 \le \theta \le \frac{5\pi}{4}$.

FIGURE 9.38f: $0 \le \theta \le \frac{3\pi}{2}$.

FIGURE 9.38g: $0 \le \theta \le \frac{7\pi}{4}$.

FIGURE 9.38h: $0 \le \theta \le 2\pi$.

This is an interesting curve known as a **four-leaf rose**. Notice again the significance of the points corresponding to $r = 0$, or $\sin 2\theta = 0$. Also, notice that r reaches a maximum of 1 when $2\theta = \frac{\pi}{2}, \frac{5\pi}{2}, \ldots$ or $\theta = \frac{\pi}{4}, \frac{5\pi}{4}, \ldots$ and r reaches a minimum of -1 when $2\theta = \frac{3\pi}{2}, \frac{7\pi}{2}, \ldots$ or $\theta = \frac{3\pi}{4}, \frac{7\pi}{4}, \ldots$ Further, you must keep in mind that when the value of r is negative, this causes us to draw the graph in the opposite quadrant.

Remark 4.4   Note that in example 4.12, even though the period of the function $\sin 2\theta$ is π, it took θ's ranging from 0 to 2π to sketch the entire curve $r = \sin 2\theta$. By contrast, the period of the function $\sin \theta$ is 2π, but the circle $r = \sin \theta$ was completed with $0 \le \theta \le \pi$. To determine the range of θ's producing a graph, you need to carefully identify important points as we did in example 4.12. The Trace feature found on graphing calculators can be very helpful for getting an idea of the θ-range, but remember that Trace values are only approximate.

You will explore a variety of other interesting graphs in the exercises.

Exercises 9.4

1. (writing) Suppose a point has polar representation (r, θ). Explain why another polar representation of the same point is $(-r, \theta + \pi)$.

2. (writing) After working with rectangular coordinates for so long, the idea of polar representations may seem slightly awkward. However, polar representations are entirely natural in many settings. For instance, if you were on a ship at sea and another ship was approaching you, explain whether you would use a polar representation (distance and bearing) or a rectangular representation (distance east-west and distance north-south).

3. (writing) In example 4.7, the graph (a circle) of $r = \sin \theta$ is completely traced out with $0 \le \theta \le \pi$. Explain why

graphing $r = \sin \theta$ with $\pi \le \theta \le 2\pi$ would produce the same full circle.

4. (writing) Two possible advantages of introducing a new coordinate system are making previous problems easier to solve and allowing new problems to be solved. Give two examples of graphs for which the polar equation is simpler than the rectangular equation. Give two examples of polar graphs for which you have not seen a rectangular equation.

In exercises 5–10, plot the given polar points (r, θ) and find their rectangular representation.

5. $(2, 0)$ **6.** $(2, \pi)$ **7.** $(-2, \pi)$

8. $\left(-3, \frac{3\pi}{2}\right)$ **9.** $(3, -\pi)$ **10.** $\left(5, -\frac{\pi}{2}\right)$

In exercises 11–18, find all polar coordinates representations of the given rectangular point.

11. $(2, -2)$ **12.** $(-1, 1)$ **13.** $(0, 3)$

14. $(0, -4)$ **15.** $(2, -1)$ **16.** $(3, 4)$

17. $\left(-\sqrt{3}, 1\right)$ **18.** $\left(-2, -\sqrt{5}\right)$

In exercises 19–26, find rectangular coordinates for the given polar point.

19. $\left(2, -\frac{\pi}{3}\right)$ **20.** $\left(-1, \frac{\pi}{3}\right)$ **21.** $(0, 3)$

22. $\left(0, -\frac{\pi}{6}\right)$ **23.** $\left(3, \frac{\pi}{8}\right)$ **24.** $\left(4, \frac{\pi}{10}\right)$

25. $(-3, 1)$ **26.** $(-2, 5)$

In exercises 27–34, sketch the graph of the polar equation and find a corresponding x-y equation.

27. $r = 4$ **28.** $r = \sqrt{3}$

29. $\theta = \pi/6$ **30.** $\theta = 3\pi/4$

31. $r = \cos\theta$ **32.** $r = 2\cos\theta$

33. $r = 3\sin\theta$ **34.** $r = 2\sin\theta$

In exercises 35–60, sketch the graph and identify all values of θ where $r = 0$ and a range of values of θ that produces one copy of the graph.

35. $r = \cos 2\theta$ **36.** $r = \cos 3\theta$

37. $r = \sin 3\theta$ **38.** $r = \sin 2\theta$

39. $r = 3 + 2\sin\theta$ **40.** $r = 2 - 2\cos\theta$

41. $r = 2 - 4\sin\theta$ **42.** $r = 2 + 4\cos\theta$

43. $r = 2 + 2\sin\theta$ **44.** $r = 3 - 6\cos\theta$

45. $r = \frac{1}{4}\theta$ **46.** $r = e^{\theta/4}$

47. $r = 2\cos(\theta - \pi/4)$ **48.** $r = 2\sin(3\theta - \pi)$

49. $r = \cos\theta + \sin\theta$ **50.** $r = \cos\theta + \sin 2\theta$

51. $r = \tan^{-1} 2\theta$ **52.** $r = \theta/\sqrt{\theta^2 + 1}$

53. $r = 2 + 4\cos 3\theta$ **54.** $r = 2 - 4\sin 4\theta$

55. $r = \dfrac{2}{1 + \sin\theta}$ **56.** $\dfrac{3}{1 - \sin\theta}$

57. $r = \dfrac{2}{1 + \cos\theta}$ **58.** $\dfrac{3}{1 - \cos\theta}$

59. $r = \theta\cos\theta, \ -2\pi \le \theta \le 2\pi$

60. $r = \theta^2\cos\theta, \ -2\pi \le \theta \le 2\pi$

61. Based on your graphs in exercises 31 and 32, conjecture the graph of $r = a\cos\theta$ for any positive constant a.

62. Based on your graphs in exercises 33 and 34, conjecture the graph of $r = a\sin\theta$ for any positive constant a.

63. Based on the graphs in exercises 35 and 36 and others (try $r = \cos 4\theta$ and $r = \cos 5\theta$), conjecture the graph of $r = \cos n\theta$ for any positive integer n.

64. Based on the graphs in exercises 37 and 38 and others (try $r = \sin 4\theta$ and $r = \sin 5\theta$), conjecture the graph of $r = \sin n\theta$ for any positive integer n.

In exercises 65–70, find a polar equation corresponding to the given rectangular equation.

65. $y^2 - x^2 = 4$ **66.** $x^2 + y^2 = 9$

67. $x^2 + y^2 = 16$ **68.** $x^2 + y^2 = x$

69. $y = 3$ **70.** $x = 2$

71. Sketch the graph of $r = \cos\frac{11}{12}\theta$ first for $0 \le \theta \le \pi$, then for $0 \le \theta \le 2\pi$, then for $0 \le \theta \le 3\pi$, …, and finally for $0 \le \theta \le 24\pi$. Discuss any patterns that you find and predict what will happen for larger domains.

72. Sketch the graph of $r = \cos\pi\theta$ first for $0 \le \theta \le 1$, then for $0 \le \theta \le 2$, then for $0 \le \theta \le 3$, …, and finally for $0 \le \theta \le 20$. Discuss any patterns that you find and predict what will happen for larger domains.

73. One situation where polar coordinates apply directly to sports is in making a golf putt. The two factors that the golfer tries to control are distance (determined by speed) and direction (usually called the "line"). Suppose a putter is d feet from the hole, which has radius $h = \frac{1}{6}'$. Show that the path of the ball will intersect the hole if the angle A in the figure satisfies $-\sin^{-1}(h/d) < A < \sin^{-1}(h/d)$.

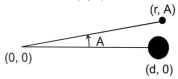

74. The distance r that the golf ball in exercise 73 travels also needs to be controlled. The ball must reach the front of the hole. In rectangular coordinates, the hole has equation $(x - d)^2 + y^2 = h^2$, so the left side of the hole is $x = d - \sqrt{h^2 - y^2}$. Show that this converts in polar coordinates to $r = d\cos\theta - \sqrt{d^2\cos^2\theta - (d^2 - h^2)}$. (Hint: Substitute for x and y, isolate the square root term, square both sides, combine r^2 terms and use the quadratic formula.)

75. The golf putt in exercises 73 and 74 will not go in the hole if it is hit too hard. Suppose that the putt would go $r = d + c$ feet if it did not go in the hole ($c > 0$). For a putt hit toward the center of the hole, define b to be the largest value of c such that the putt goes in (i.e., if the ball is hit more than b feet past the hole, it is hit too hard). Experimental evidence (see Dave Pelz's *Putt Like the Pros*) shows that at other angles A, the distance r must be less than $d + b\left(1 - \left[\dfrac{A}{\sin^{-1}(h/d)}\right]^2\right)$.

The results of exercises 73 and 74 define limits for the angle A and distance r of a successful putt. Identify the functions

$r_1(A)$ and $r_2(A)$ such that $r_1(A) < r < r_2(A)$ and constants A_1 and A_2 such that $A_1 < A < A_2$.

76. Take the general result of exercise 75 and apply it to a putt of $d = 15$ feet with a value of $b = 4$ feet. Visualize this by graphing the region

$$15 \cos \theta - \sqrt{225 \cos^2 \theta - (225 - 1/36)}$$
$$< r < 15 + 4 \left(1 - \left[\frac{\theta}{\sin^{-1}(1/90)} \right]^2 \right)$$

with $-\sin^{-1}(1/90) < \theta < \sin^{-1}(1/90)$. A good choice of graphing windows is $13.8 \le x \le 19$ and $-0.5 \le y \le 0.5$.

77. (exploratory) In this exercise, you will explore the role of the constants a, b and c in the graph of $r = af(b\theta + c)$. To start with, sketch $r = \sin \theta$ followed by $r = 2\sin \theta$ and $r = 3 \sin \theta$. What does the constant a affect? Then sketch $r = \sin (\theta + \pi/2)$ and $r = \sin(\theta - \pi/4)$. What does the constant c affect? Now for the tough one. Sketch $r = \sin 2\theta$ and $r = \sin 3\theta$. What does the constant b seem to affect? Test all of your hypotheses on the base function $r = 1 + 2 \cos \theta$ and several functions of your choice.

78. (exploratory) The polar curve $r = ae^{b\theta}$ is sometimes called an **equiangular curve**. To see why, sketch the curve and then show that $\dfrac{dr}{d\theta} = br$. A somewhat complicated geometric argument shows that $\dfrac{dr}{d\theta} = r \cot \alpha$, where α is the angle between the tangent line and the line connecting the point on the curve to the origin. Comparing equations, conclude that the angle α is constant (hence "equiangular"). To illustrate this property, compute α for the points at $\theta = 0$ and $\theta = \pi$ for $r = e^{\theta}$. This type of spiral shows up often in nature, possibly because the equal-angle property can be easily achieved. Spirals can be found among shellfish (the picture below is of an ammonite fossil from about 350 million years ago) and the florets of the common daisy. Other examples, including the connection to sunflowers, the Fibonacci sequence and the musical scale, can be found in H. E. Huntley's *The Divine Proportion*.

(Photo courtesy of Roland Minton.)

9.5   Calculus and Polar Coordinates

Now that we have introduced you to polar coordinates and looked at a variety of polar graphs, our next step is to extend the techniques of calculus to the case of polar coordinates. In this section, we focus on tangent lines, area and arc length. Surface area and other applications will be examined in the exercises.

Notice that you can think of the graph of the polar equation $r = f(\theta)$ as the graph of the parametric equations $x = f(t) \cos t$, $y(t) = f(t) \sin t$ (where we have used the parameter $t = \theta$), since from (4.2)

$$x = r \cos \theta = f(\theta) \cos \theta \tag{5.1}$$

and

$$y = r \sin \theta = f(\theta) \sin \theta. \tag{5.2}$$

In view of this, we can now take any results already derived for parametric equations and extend these to the special case of polar coordinates.

We start by computing the slope of a tangent line to the polar curve $r = f(\theta)$. From our study of parametric equations in section 9.2, we know that the slope of the tangent line at the point corresponding to $\theta = a$ is given [from (2.1)] to be

$$\left. \frac{dy}{dx} \right|_{\theta=a} = \frac{\dfrac{dy}{d\theta}(a)}{\dfrac{dx}{d\theta}(a)}. \tag{5.3}$$

From the product rule and (5.1) and (5.2), we have

$$\frac{dy}{d\theta} = f'(\theta)\sin\theta + f(\theta)\cos\theta$$

and

$$\frac{dx}{d\theta} = f'(\theta)\cos\theta - f(\theta)\sin\theta.$$

Putting these together with (5.3), we get

$$\frac{dy}{dx}\bigg|_{\theta=a} = \frac{f'(a)\sin a + f(a)\cos a}{f'(a)\cos a - f(a)\sin a}. \qquad (5.4)$$

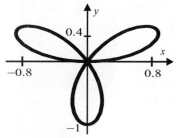

FIGURE 9.39a: Three-leaf rose.

Example 5.1   Finding the Slope of the Tangent Line to a Three-Leaf Rose

Find the slope of the tangent line to the three-leaf rose $r = \sin 3\theta$ at $\theta = 0$ and $\theta = \frac{\pi}{4}$.

Solution   A sketch of the curve is shown in Figure 9.39a. From (4.2), we have

$$y = r\sin\theta = \sin 3\theta \, \sin\theta$$

and

$$x = r\cos\theta = \sin 3\theta \, \cos\theta.$$

Using (5.3), we have

$$\frac{dy}{dx}(\theta) = \frac{\dfrac{dy}{d\theta}}{\dfrac{dx}{d\theta}} = \frac{(3\cos 3\theta)\sin\theta + \sin 3\theta(\cos\theta)}{(3\cos 3\theta)\cos\theta - \sin 3\theta(\sin\theta)}. \qquad (5.5)$$

At $\theta = 0$, this gives us

$$\frac{dy}{dx}(0) = \frac{(3\cos 0)\sin 0 + \sin 0(\cos 0)}{(3\cos 0)\cos 0 - \sin 0(\sin 0)} = \frac{0}{3} = 0.$$

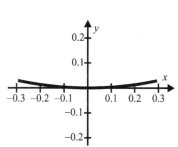

FIGURE 9.39b: $-0.1 \le \theta \le 0.1$.

In Figure 9.39b, we sketch $r = \sin 3\theta$ for $-0.1 \le \theta \le 0.1$, in order to isolate the portion of the curve around $\theta = 0$. Notice that from this figure, a slope of 0 seems reasonable.

Similarly, at $\theta = \frac{\pi}{4}$, we have from (5.5) that

$$\frac{dy}{dx}\left(\frac{\pi}{4}\right) = \frac{(3\cos\frac{3\pi}{4})\sin\frac{\pi}{4} + \sin\frac{3\pi}{4}(\cos\frac{\pi}{4})}{(3\cos\frac{3\pi}{4})\cos\frac{\pi}{4} - \sin\frac{3\pi}{4}(\sin\frac{\pi}{4})} = \frac{-\frac{3}{2}+\frac{1}{2}}{-\frac{3}{2}-\frac{1}{2}} = \frac{1}{2}.$$

In Figure 9.39c, we show the section of $r = \sin 3\theta$ for $0 \le \theta \le \frac{\pi}{3}$, along with the tangent line at $\theta = \frac{\pi}{4}$.

FIGURE 9.39c: The tangent line at $\theta = \frac{\pi}{4}$.

Recall that for functions $y = f(x)$, horizontal tangents were especially significant for locating maximum and minimum points. For polar graphs, the significant points are often places where r has reached a maximum or minimum, which may or may not correspond to a horizontal tangent. We explore this idea further in the following example.

Example 5.2   Polar Graphs and Horizontal Tangent Lines

For the three-leaf rose $r = \sin 3\theta$, find the locations of all horizontal tangent lines and interpret the significance of these points. Further, at the three points where $|r|$ is a maximum, show that the tangent line is perpendicular to the line segment connecting the point to the origin.

Solution   From (5.3) and (5.4), we have

$$\frac{dy}{dx} = \frac{\dfrac{dy}{d\theta}}{\dfrac{dx}{d\theta}} = \frac{f'(\theta)\,\sin\theta + f(\theta)\,\cos\theta}{f'(\theta)\,\cos\theta - f(\theta)\,\sin\theta}.$$

Here, $f(\theta) = \sin 3\theta$ and so, to have $\dfrac{dy}{dx} = 0$, we must have

$$0 = \frac{dy}{d\theta} = 3\cos 3\theta\,\sin\theta + \sin 3\theta\,\cos\theta.$$

FIGURE 9.40a:
$y = 3\cos 3x \sin x + \sin 3x \cos x$.

Solving this equation is not an easy matter. As a start, we graph $f(x) = 3\cos 3x \sin x + \sin 3x \cos x$ with $0 \le x \le \pi$ (see Figure 9.40a). You should observe that there appear to be 5 solutions. Three of the solutions can be found exactly: $\theta = 0$, $\theta = \frac{\pi}{2}$ and $\theta = \pi$. You can find the remaining two numerically: $\theta \approx 0.659$ and $\theta \approx 2.48$. The corresponding points on the curve $r = \sin 3\theta$ (specified in rectangular coordinates) are $(0,0)$, $(0.73, 0.56)$, $(0, -1)$, $(-0.73, 0.56)$, and $(0,0)$. The point $(0, -1)$ lies at the bottom of a leaf. This is the familiar situation of a horizontal tangent line at a local (and in fact, absolute) minimum. The point $(0,0)$ is a little more tricky to interpret. As seen in Figure 9.39b, if we graph a small piece of the curve with θ near 0 (or π), the point $(0,0)$ is a minimum point. However, this is not true for other values of θ (e.g., $\pi/3$) where the curve passes through the point $(0,0)$. The tangent lines at the points $(\pm 0.73, 0.56)$ are shown in Figure 9.40b. Note that these points correspond to points where the y-coordinate is a maximum. However, referring to the graph, these points do not appear to be of particular interest. Rather, the tips of the leaves represent the extreme points of most interest. Notice that the tips are where $|r|$ is a maximum. For $r = \sin 3\theta$, this occurs when $\sin 3\theta = \pm 1$, that is, where $3\theta = \frac{\pi}{2}, \frac{3\pi}{2}, \frac{5\pi}{2}, \ldots$, or $\theta = \frac{\pi}{6}, \frac{\pi}{2}, \frac{5\pi}{6}, \ldots$. From (5.4), the slope of the tangent line to the curve at $\theta = \frac{\pi}{6}$ is given by

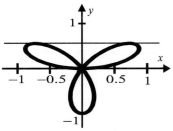

FIGURE 9.40b: Horizontal tangent lines.

$$\frac{dy}{dx}\bigg|_{\theta = \pi/6} = \frac{\left(3\cos\frac{3\pi}{6}\right)\sin\frac{\pi}{6} + \sin\frac{3\pi}{6}\left(\cos\frac{\pi}{6}\right)}{\left(3\cos\frac{3\pi}{6}\right)\cos\frac{\pi}{6} - \sin\frac{3\pi}{6}\left(\sin\frac{\pi}{6}\right)} = \frac{0 + \frac{\sqrt{3}}{2}}{0 - \frac{1}{2}} = -\sqrt{3}.$$

The rectangular point corresponding to $\theta = \frac{\pi}{6}$ is given by

$$\left(1\cos\frac{\pi}{6}, 1\sin\frac{\pi}{6}\right) = \left(\frac{\sqrt{3}}{2}, \frac{1}{2}\right).$$

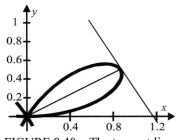

FIGURE 9.40c: The tangent line at the tip of a leaf.

A quick calculation shows that the slope of the line segment joining this point to the origin is $\frac{1}{\sqrt{3}}$. Observe that the line segment from the origin to the point is perpendicular to the tangent line since the product of the slopes ($-\sqrt{3}$ and $\frac{1}{\sqrt{3}}$) is -1. This is illustrated in Figure 9.40c. Similarly, the slope of the tangent line at $\theta = \frac{5\pi}{6}$ is $\sqrt{3}$, which again makes the tangent line at that point perpendicular to the line segment from the origin to the point $\left(-\frac{\sqrt{3}}{2}, \frac{1}{2}\right)$. Finally, we have already shown that the slope of the tangent line at $\theta = \frac{\pi}{2}$ is 0 and a horizontal tangent line is perpendicular to the vertical line from the origin to the point $(0, -1)$.

∎

For polar curves like the three-leaf rose seen in Figure 9.39a, it is reasonable to expect that we could compute the area enclosed by the curve. Since such a graph is **not** the graph of a function, we cannot use the usual area formulas developed in Chapter 5. However, we can convert our area formulas for parametric equations (from Theorem 2.2) into polar coordinates. A simpler approach to get the same result uses the following geometric argument.

FIGURE 9.41: Circular sector.

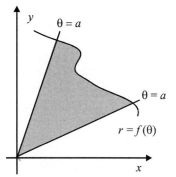

FIGURE 9.42a: Area of
a polar region.

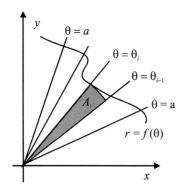

FIGURE 9.42b: Approximating
the area of a polar region.

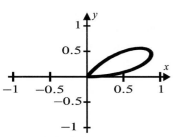

FIGURE 9.43: One leaf of
$r = \sin 3\theta$.

First, recall that the area of a circle of radius r is πr^2. Next, consider a sector of a circle of radius r and central angle θ (see Figure 9.41). Notice that the sector contains a fraction $\left(\dfrac{\theta}{2\pi}\right)$ of the area of the entire circle and so, the area of the sector is given by

$$A = \pi r^2 \frac{\theta}{2\pi} = \frac{1}{2} r^2 \theta.$$

Now, consider the area enclosed by the polar curve defined by the equation $r = f(\theta)$ and the rays $\theta = a$ and $\theta = b$ (see Figure 9.42a), where f is continuous and positive on the interval $a \leq \theta \leq b$. As we did when we defined the definite integral, we begin by partitioning the θ-interval into n equal pieces:

$$a = \theta_0 < \theta_1 < \theta_2 < \cdots < \theta_n = b.$$

The width of each of these subintervals is then $\Delta\theta = \theta_i - \theta_{i-1} = \dfrac{b-a}{n}$. (Does this look familiar?) On each subinterval $[\theta_{i-1}, \theta_i]$ $(i = 1, 2, \ldots, n)$, we approximate the curve with the circular arc $r = f(\theta_i)$ (see Figure 9.42b). The area A_i enclosed by the curve on this subinterval is then approximately the same as the area of the circular sector of radius $f(\theta_i)$ and central angle $\Delta\theta$:

$$A_i \approx \frac{1}{2} r^2 \Delta\theta = \frac{1}{2} \left[f(\theta_i) \right]^2 \Delta\theta.$$

The total area A enclosed by the curve is then approximately the same as the sum of the areas of all such circular sectors:

$$A \approx \sum_{i=1}^{n} A_i = \sum_{i=1}^{n} \frac{1}{2} \left[f(\theta_i) \right]^2 \Delta\theta.$$

As we have seen numerous times now, we can improve the approximation by making n larger. Taking the limit as $n \to \infty$ gives us a definite integral:

$$A = \lim_{n\to\infty} \sum_{i=1}^{n} \tfrac{1}{2} \left[f(\theta_i) \right]^2 \Delta\theta = \int_a^b \tfrac{1}{2} \left[f(\theta) \right]^2 \, d\theta. \qquad (5.6)$$

Example 5.3   The Area of One Leaf of a Three-Leaf Rose

Find the area of one leaf of the rose $r = \sin 3\theta$.

Solution   Notice that one leaf of the rose is traced out with $0 \leq \theta \leq \frac{\pi}{3}$ (see Figure 9.43). Notice that from (5.6), the area is given by

$$A = \int_0^{\pi/3} \frac{1}{2} \left(\sin 3\theta \right)^2 \, d\theta = \frac{1}{2} \int_0^{\pi/3} \sin^2 3\theta \, d\theta$$

$$= \frac{1}{4} \int_0^{\pi/3} (1 - \cos 6\theta) \, d\theta = \frac{1}{4} \left(\theta - \frac{1}{6}\sin 6\theta \right) \Big|_0^{\pi/3} = \frac{\pi}{12},$$

where here, we have used the half-angle formula $\sin^2 \alpha = \frac{1}{2}(1 - \cos 2\alpha)$ to simplify the integrand.

∎

Often, the most challenging part of finding the area of a polar region is determining the limits of integration.

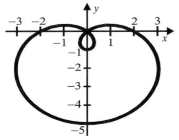

FIGURE 9.44: $r = 2 - 3\sin\theta$.

Example 5.4   The Area of the Inner Loop of a Limaçon

Find the area of the inner loop of the limaçon $r = 2 - 3\sin\theta$.

Solution   A sketch of the limaçon is shown in Figure 9.44. Starting at $\theta = 0$, the curve starts at the point $(1, 0)$, passes through the origin, traces out the inner loop, passes back through the origin and finally traces out the outer loop. Thus, the inner loop is formed by θ-values between the first and second occurrences of $r = 0$. Solving $r = 0$, we get $\sin\theta = \frac{2}{3}$. The smallest two positive solutions are $\theta = \sin^{-1}\left(\frac{2}{3}\right)$ and $\theta = \pi - \sin^{-1}\left(\frac{2}{3}\right)$. Numerically, these are approximately equal to $\theta = 0.73$ and $\theta = 2.41$. From (5.6), the area is approximately (here the area is only approximate, owing to the approximate limits of integration)

$$A \approx \int_{0.73}^{2.41} \frac{1}{2}\left[2 - 3\sin\theta\right]^2 \, d\theta = \frac{1}{2}\int_{0.73}^{2.41}\left(4 - 12\sin\theta + 9\sin^2\theta\right) \, d\theta$$

$$= \frac{1}{2}\int_{0.73}^{2.41}\left[4 - 12\sin\theta + \frac{9}{2}\left(1 - \cos 2\theta\right)\right] \, d\theta \approx 0.44,$$

where we have used the half-angle formula $\sin^2\theta = \frac{1}{2}(1 - \cos 2\theta)$ to simplify the integrand.

∎

When finding the area lying between two polar graphs, we use the familiar device of subtracting one area from another. Although the calculations in the following example aren't too messy, finding the points of intersection of two polar curves often provides the greatest challenge.

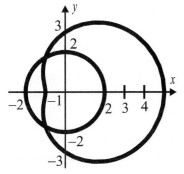

FIGURE 9.45a: $r = 3 + 2\cos\theta$
and $r = 2$.

FIGURE 9.45b: $\frac{2\pi}{3} \le \theta \le \frac{4\pi}{3}$.

Example 5.5   Finding the Area between Two Polar Graphs

Find the area inside the limaçon $r = 3 + 2\cos\theta$ and outside the circle $r = 2$.

Solution   We show a sketch of the two curves in Figure 9.45a. Notice that the limits of integration correspond to the values of θ where the two curves intersect. So, we must first solve the equation $3 + 2\cos\theta = 2$. Notice that since $\cos\theta$ is periodic, there are infinitely many solutions of this equation. Consequently, it is absolutely necessary to consult the graph to determine which solutions you are interested in! In this case, we want the least negative and the smallest positive solutions. (Look carefully at Figure 9.45b, where we have plotted only those portions of the graphs corresponding to θ between $\frac{2\pi}{3}$ and $\frac{4\pi}{3}$, the first two positive solutions. This portion of the graphs corresponds to the area **outside** the limaçon and **inside** the circle!) With $3 + 2\cos\theta = 2$, we have $\cos\theta = -\frac{1}{2}$, which occurs at $\theta = -\frac{2\pi}{3}$ and $\theta = \frac{2\pi}{3}$. From (5.6), the area enclosed by the portion of the limaçon on this interval is given by

$$\int_{-2\pi/3}^{2\pi/3} \frac{1}{2}\left(3 + 2\cos\theta\right)^2 \, d\theta = \frac{33\sqrt{3} + 44\pi}{6}.$$

Similarly, the area enclosed by the circle on this interval is given by

$$\int_{-2\pi/3}^{2\pi/3} \frac{1}{2}\left(2\right)^2 \, d\theta = \frac{8\pi}{3}.$$

The area **inside** the limaçon and **outside** the circle is then given by

$$A = \int_{-2\pi/3}^{2\pi/3} \frac{1}{2}\,(3 + 2\cos\theta)^2\,d\theta - \int_{-2\pi/3}^{2\pi/3} \frac{1}{2}\,(2)^2\,d\theta$$

$$= \frac{33\sqrt{3} + 44\pi}{6} - \frac{8\pi}{3} = \frac{33\sqrt{3} + 28\pi}{6} \approx 24.2.$$

Here, we have left the (routine) details of the integrations to you.

■

In cases where r takes on both positive and negative values, finding the intersection points of two curves is more complicated.

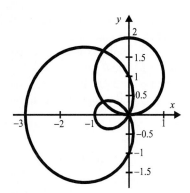

FIGURE 9.46a: $r = 1 - 2\cos\theta$
and $r = 2\sin\theta$.

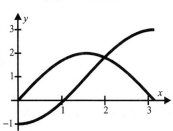

FIGURE 9.46b: Rectangular plot:
$y = 1 - 2\cos x,\ y = 2\sin x,$
$0 \le x \le \pi.$

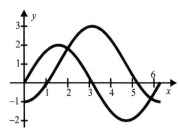

FIGURE 9.46c: Rectangular plot:
$y = 1 - 2\cos x,\ y = 2\sin x,$
$0 \le x \le 2\pi.$

Example 5.6   Finding Intersections of Polar Curves Where r Can be Negative
Find all intersections of the limaçon $r = 1 - 2\cos\theta$ and the circle $r = 2\sin\theta$.

Solution   We show a sketch of the two curves Figure 9.46a. Notice from the sketch that there are three intersections of the two curves. Since $r = 2\sin\theta$ is completely traced with $0 \le \theta \le \pi$, you might reasonably expect to find three solutions of the equation $1 - 2\cos\theta = 2\sin\theta$ on the interval $0 \le \theta \le \pi$. However, if we draw a rectangular plot of the two curves: $y = 1 - 2\cos x$ and $y = 2\sin x$, on the interval $0 \le x \le \pi$ (see Figure 9.46b), we can clearly see that there is only one solution in this range, at approximately $\theta \approx 1.99$. (Recall that you can use Newton's method or your calculator's solver to obtain an accurate approximation.) The corresponding rectangular point is $(r\cos\theta, r\sin\theta) \approx (-0.74, 1.67)$. Looking at Figure 9.46a, observe that there is another intersection located below this point. One way to find this point is to look at a rectangular plot of the two curves corresponding to an expanded range of values of θ (see Figure 9.46c). Notice that there is a second solution of the equation $1 - 2\cos\theta = 2\sin\theta$ near $\theta = 5.86$, which corresponds to the point $(-0.74, .34)$. Note that this point is on the inner loop of $r = 1 - 2\cos\theta$ and corresponds to a negative value of r. Finally, there appears to be a third intersection at or near the origin. Notice that this does not arise from any solution of the equation $1 - 2\cos\theta = 2\sin\theta$. This is because, while both curves pass through the origin (You should verify this!), they each do so for **different** values of θ. You need to realize that the origin corresponds to the point $(0, \theta)$, in polar coordinates, for **any** angle θ. Notice that $1 - 2\cos\theta = 0$ for $\theta = \pi/3$ and $2\sin\theta = 0$ for $\theta = 0$. So, although the curves intersect at the origin, they each pass through the origin for different values of θ.

■

Remark 5.1   To find points of intersection of two polar curves $r = f(\theta)$ and $r = g(\theta)$, you must keep in mind that points have more than one representation in polar coordinates. In particular, this says that points of intersection need not correspond to solutions of $f(\theta) = g(\theta)$.

In the following example, we see an application that is far simpler to set up in polar coordinates than in rectangular coordinates.

FIGURE 9.47a: A cylindrical
oil tank.

Example 5.7   Finding the Volume of a Partially Filled Cylinder
A cylindrical oil tank with a radius of 2 feet is lying on its side. A measuring stick shows that the oil is 1.8 feet deep (see Figure 9.47a). What percentage of a full tank is left?

Solution   Notice that since we wish to find only the percentage of oil remaining in the tank, the length of the tank has no bearing on this problem. (Think about this some.) We need only consider a cross section of the tank, which we represent as a circle of radius 2 centered at the origin. The proportion of oil remaining is given by the area of that portion of the circle lying beneath the line $y = -0.2$, divided by the total area of the circle. The

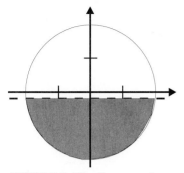

FIGURE 9.47b: Cross-section
of tank.

area of the circle is 4π, so we need only find the area of the shaded region in Figure 9.47b. Computing this area in rectangular coordinates is a mess (try it!), but it is straightforward in polar coordinates. First, notice that the line $y = -0.2$ corresponds to $r\sin\theta = -0.2$ or $r = -0.2\csc\theta$. The area beneath the line and inside the circle is then given by (5.6) as

$$Area = \int_{\theta_1}^{\theta_2} \frac{1}{2}\,(2)^2\,d\theta - \int_{\theta_1}^{\theta_2} \frac{1}{2}\,(-0.2\csc\theta)^2\,d\theta,$$

where θ_1 and θ_2 are the appropriate intersections of $r = 2$ and $r = -0.2\csc\theta$. Using Newton's method, the first two positive solutions of $2 = -0.2\csc\theta$ are $\theta_1 \approx 3.242$ and $\theta_2 \approx 6.183$. The area is then

$$Area = \int_{\theta_1}^{\theta_2} \frac{1}{2}\,(2)^2\,d\theta - \int_{\theta_1}^{\theta_2} \frac{1}{2}\,(-0.2\csc\theta)^2\,d\theta$$

$$= (2\theta + 0.02\cot\theta)\Big|_{\theta_1}^{\theta_2} \approx 5.485.$$

The fraction of oil remaining in the tank is then approximately $5.485/4\pi \approx 0.43648$ or about 43.6% of the total capacity of the tank.

■

We close this section with a brief discussion of arc length for polar curves. Recall that from (3.1), the arc length of a curve defined parametrically by $x = x(t)$, $y = y(t)$, for $a \le t \le b$, is given by

$$s = \int_a^b \sqrt{\left(\frac{dx}{dt}\right)^2 + \left(\frac{dy}{dt}\right)^2}\,dt. \tag{5.7}$$

Once again thinking of a polar curve as a parametric representation (where the parameter is θ), we have that for the polar curve $r = f(\theta)$,

$$x = r\cos\theta = f(\theta)\cos\theta \quad\text{and}\quad y = r\sin\theta = f(\theta)\sin\theta.$$

This give us

$$\left(\frac{dx}{d\theta}\right)^2 + \left(\frac{dy}{d\theta}\right)^2 = [f'(\theta)\cos\theta - f(\theta)\sin\theta]^2 + [f'(\theta)\sin\theta + f(\theta)\cos\theta]^2$$

$$= [f'(\theta)]^2\,(\cos^2\theta + \sin^2\theta) + f'(\theta)\,f(\theta)\,(-2\cos\theta\,\sin\theta + 2\sin\theta\,\cos\theta)$$
$$+ [f(\theta)]^2\,(\cos^2\theta + \sin^2\theta)$$
$$= [f'(\theta)]^2 + [f(\theta)]^2.$$

From (5.7), we get that the arc length is given by

$$\boxed{s = \int_a^b \sqrt{[f'(\theta)]^2 + [f(\theta)]^2}\,d\theta.} \tag{5.8}$$

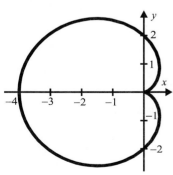

FIGURE 9.48: $r = 2 - 2\cos\theta$.

Example 5.8   Arc Length of a Polar Curve
Find the arc length of the cardioid $r = 2 - 2\cos\theta$.

Solution   A sketch of the cardioid is shown in Figure 9.48. First, notice that the curve is traced out with $0 \le \theta \le 2\pi$. From (5.8), the arc length is given by

$$s = \int_a^b \sqrt{[f'(\theta)]^2 + [f(\theta)]^2}\,d\theta = \int_0^{2\pi} \sqrt{(2\sin\theta)^2 + (2 - 2\cos\theta)^2}\,d\theta$$

$$= \int_0^{2\pi} \sqrt{4\sin^2\theta + 4 - 8\cos\theta + 4\cos^2\theta}\,d\theta = \int_0^{2\pi} \sqrt{8 - 8\cos\theta}\,d\theta = 16,$$

where we leave the details of the integration as an exercise. (Hint: Use the half-angle formula: $\sin^2 x = \frac{1}{2}(1 - \cos 2x)$ to simplify the integrand. Be careful: remember that $\sqrt{x^2} = |x|$!)

∎

Exercises 9.5

1. (writing) Explain why the tangent line is perpendicular to the radius line at any point at which r is a local maximum (see example 5.2).

2. (writing) In example 5.5, explain why integrating from $\frac{2\pi}{3}$ to $\frac{4\pi}{3}$ would give the area shown in Figure 9.45b, not the desired area.

3. (writing) Referring to example 5.6, explain why intersections can occur in each of the cases $f(\theta) = g(\theta)$, $f(\theta) = -g(\theta + \pi)$ and $f(\theta_1) = g(\theta_2) = 0$.

4. (writing) In example 5.7, explain why the length of the tank doesn't matter. If the problem were to compute the **amount** of oil left, would the length matter?

In exercises 5–14, find the slope of the tangent line to the polar curve at the given point.

5. $r = \sin 3\theta$ at $\theta = \frac{\pi}{3}$

6. $r = \sin 3\theta$ at $\theta = \frac{\pi}{2}$

7. $r = \cos 2\theta$ at $\theta = 0$

8. $r = \cos 2\theta$ at $\theta = \frac{\pi}{4}$

9. $r = 3 \sin \theta$ at $\theta = 0$

10. $r = 3 \sin \theta$ at $\theta = \frac{\pi}{2}$

11. $r = \sin 4\theta$ at $\theta = \frac{\pi}{4}$

12. $r = \sin 4\theta$ at $\theta = \frac{\pi}{16}$

13. $r = \cos 3\theta$ at $\theta = \frac{\pi}{6}$

14. $r = \cos 3\theta$ at $\theta = \frac{\pi}{3}$

In exercises 15–18, find all points at which $|r|$ is a maximum and show that the tangent line is perpendicular to the radius connecting the point to the origin.

15. $r = \sin 3\theta$

16. $r = \cos 4\theta$

17. $r = 2 - 4 \sin 2\theta$

18. $r = 2 + 4 \sin 2\theta$

In exercises 19–34, find the area of the indicated region.

19. One leaf of $r = \cos 3\theta$

20. One leaf of $r = \sin 4\theta$

21. Inner loop of $r = 3 - 4 \sin \theta$

22. Inner loop of $r = 1 - 2 \cos \theta$

23. Bounded by $r = 2 \cos \theta$

24. Bounded by $r = 2 - 2 \cos \theta$

25. Small loop of $r = 1 + 2 \sin 2\theta$

26. Large loop of $r = 1 + 2 \sin 2\theta$

27. Inner loop of $r = 2 + 3 \sin 3\theta$

28. Outer loop of $r = 2 + 3 \sin 3\theta$

29. Inside of $r = 3 + 2 \sin \theta$ and outside of $r = 2$

30. Inside of $r = 2$ and outside of $r = 2 - 2 \sin \theta$

31. Inside of $r = 2$ and outside of $r = 1 + 2 \sin \theta$

32. Inside of $r = 2 \sin 2\theta$ and outside $r = 1$

33. Inside of both $r = 1 + \cos \theta$ and $r = 1$

34. Inside of both $r = 1 + \sin \theta$ and $r = 1 + \cos \theta$

In exercises 35–38, find all points at which the two curves intersect.

35. $r = 1 - 2 \sin \theta$ and $r = 2 \cos \theta$

36. $r = 1 + 3 \cos \theta$ and $r = -2 + 5 \sin \theta$

37. $r = 1 + \sin \theta$ and $r = 1 + \cos \theta$

38. $r = 1 + \sqrt{3} \sin \theta$ and $r = 1 + \cos \theta$

In exercises 39–44, find the arc length of the given curve.

39. $r = 2 - 2 \sin \theta$

40. $r = 3 - 3 \cos \theta$

41. $r = \sin 3\theta$

42. $r = 2 \cos 3\theta$

43. $r = 1 + 2 \sin 2\theta$

44. $r = 2 + 3 \sin 3\theta$

45. Repeat example 5.7 for the case where the oil stick shows a depth of 1.4.

46. Repeat example 5.7 for the case where the oil stick shows a depth of 1.0.

47. Repeat example 5.7 for the case where the oil stick shows a depth of 2.4.

48. Repeat example 5.7 for the case where the oil stick shows a depth of 2.6.

49. The problem of finding the slope of $r = \sin 3\theta$ at the point $(0, 0)$ is not a well-defined problem. To see what we mean, show that the curve passes through the origin at $\theta = 0$, $\theta = \frac{\pi}{3}$ and $\theta = \frac{2\pi}{3}$ and find the slopes at these angles. Briefly explain why they are different even though the point is the same.

50. For each of the three slopes found in exercise 49, illustrate with a sketch of $r = \sin 3\theta$ for θ-values near the given values (e.g., $-\frac{\pi}{6} \leq \theta \leq \frac{\pi}{6}$ to see the slope at $\theta = 0$).

51. If the polar curve $r = f(\theta)$, $a \leq \theta \leq b$, has length L, show that $r = cf(\theta)$, $a \leq \theta \leq b$, has length $|c|\, L$ for any constant c.

52. If the polar curve $r = f(\theta)$, $a \leq \theta \leq b$, encloses area A, show that for any constant c, $r = cf(\theta)$, $a \leq \theta \leq b$, encloses area $c^2 A$.

53. (exploratory) In this exercise, you will discover a remarkable property about the area underneath the graph of $y = \frac{1}{x}$. First, show that a polar representation of this curve is $r^2 = \dfrac{1}{\sin\theta \cos\theta}$. We will find the area bounded by $y = \frac{1}{x}$, $y = mx$ and $y = 2mx$ for $x > 0$, where m is a positive constant. Sketch graphs for $m = 1$ (the area bounded by $y = \frac{1}{x}$, $y = x$ and $y = 2x$) and $m = 2$ (the area bounded by $y = \frac{1}{x}$, $y = 2x$ and $y = 4x$). Which area looks larger? To find out, you should integrate. Explain why this would be a very difficult integration in rectangular coordinates. Then convert all curves to polar coordinates and compute the polar area. You should discover that the area equals $\frac{1}{2}\ln 2$ for any value of m. (Are you surprised?)

9.6   Conic Sections

So far in this chapter, we have introduced a variety of interesting curves, many of which are not graphs of a function $y = f(x)$ in rectangular coordinates. Among the most important curves are the **conic sections**, which we explore here. The conic sections include parabolas, ellipses and hyperbolas, which are undoubtedly already familiar to you. In this section, we focus on geometric properties that are not always easily determined in rectangular coordinates.

We visualize the conic sections as the intersection of a plane with a right circular cone (see Figures 9.49a – 9.49c).

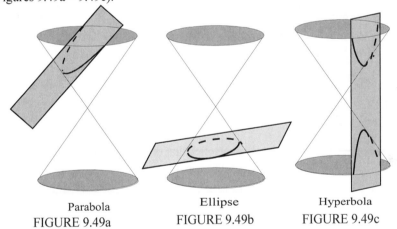

Parabola
FIGURE 9.49a

Ellipse
FIGURE 9.49b

Hyperbola
FIGURE 9.49c

Depending on the orientation of the plane, the resulting curve can be a parabola, ellipse or hyperbola.

Parabolas

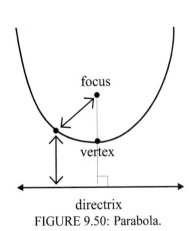

FIGURE 9.50: Parabola.

We define a **parabola** (see Figure 9.50) to be the set of all points that are equidistant from a fixed point (called the **focus**) and a line (called the **directrix**). A special point on the parabola is the **vertex**, the midpoint of the perpendicular segment from the focus to the directrix.

If the focus of the parabola lies on the y-axis and the directrix is a horizontal line, it is possible to find a simple rectangular equation of the parabola.

Example 6.1   Finding the Equation of a Parabola

Find an equation of the parabola with focus at the point $(0, 2)$ and whose directrix is the line $y = -2$.

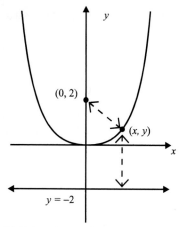

FIGURE 9.51: The parabola with vertex at $(0, 2)$ and directrix $y = -2$.

Solution   By definition, any point (x, y) on the parabola must be equidistant from the focus and the directrix (see Figure 9.51). From the distance formula, the distance from (x, y) to the focus is given by $\sqrt{x^2 + (y - 2)^2}$ and the distance to the directrix is $|y - (-2)|$. Since these distances must be equal, the parabola is defined by the equation

$$\sqrt{x^2 + (y - 2)^2} = |y + 2|.$$

Squaring both sides, we get

$$x^2 + (y - 2)^2 = (y + 2)^2.$$

Expanding this out and simplifying, we get

$$x^2 + y^2 - 4y + 4 = y^2 + 4y + 4$$

or

$$x^2 = 8y.$$

Solving for y gives us $y = \frac{1}{8}x^2$.

In general, the following relationship holds.

Theorem 6.1   The graph of the equation $y = a(x - b)^2 + c$ is a parabola, with vertex at the point (b, c), focus at $\left(b, c + \frac{1}{4a}\right)$ and directrix given by the line $y = c - \frac{1}{4a}$.

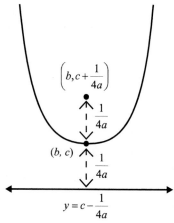

FIGURE 9.52: Parabola.

Proof. Given the focus $\left(b, c + \frac{1}{4a}\right)$ and directrix $y = c - \frac{1}{4a}$, the vertex is the midpoint (b, c) (see Figure 9.52). For any point (x, y) on the parabola, its distance to the focus is given by $\sqrt{(x - b)^2 + \left(y - c - \frac{1}{4a}\right)^2}$, while its distance to the directrix is given by $\left|y - c + \frac{1}{4a}\right|$. Setting these equal and squaring as in example 6.1, we have

$$(x - b)^2 + \left(y - c - \frac{1}{4a}\right)^2 = \left(y - c + \frac{1}{4a}\right)^2.$$

Expanding this out and simplifying, we get the more familiar form of the equation: $y = a(x - b)^2 + c$, as desired.   ∎

Notice that the roles of x and y can be reversed. We leave the proof of the following result as an exercise.

Theorem 6.2   The graph of the equation $x = a(y - b)^2 + c$ is a parabola, with vertex at the point (c, b), focus at $\left(c + \frac{1}{4a}, b\right)$ and directrix given by the line $x = c - \frac{1}{4a}$.

We illustrate Theorem 6.2 in the following example.

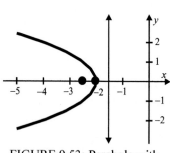

FIGURE 9.53: Parabola with focus at $\left(-\frac{5}{2}, 0\right)$ and directrix $x = -\frac{3}{2}$.

Example 6.2   A Parabola Opening to the Left

For the parabola $4x + 2y^2 + 8 = 0$, find the vertex, focus and directrix.

Solution   To put this into the form of the equation given in Theorem 6.2, we must first solve for x. We have $x = -\frac{1}{2}y^2 - 2$. The vertex is then at $(-2, 0)$. The focus and directrix are shifted left and right, respectively from the vertex by $\frac{1}{4a} = -\frac{1}{2}$. This puts the focus at $\left(-2 - \frac{1}{2}, 0\right) = \left(-\frac{5}{2}, 0\right)$ and the directrix at $x = -2 - \left(-\frac{1}{2}\right) = -\frac{3}{2}$. We show a sketch of the parabola in Figure 9.53.   ∎

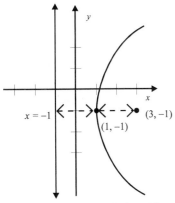

FIGURE 9.55: Parabola with focus at $(3, -1)$ and directrix at $x = -1$.

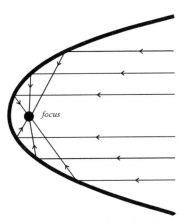

FIGURE 9.56a: Reflection of rays.

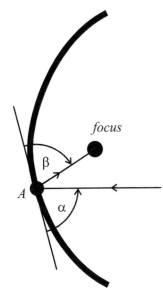

FIGURE 9.56b: The reflective property.

Example 6.3   Finding the Equation of a Parabola

Find an equation relating all points that are equidistant from the point $(3, 2)$ and the line $y = 6$.

Solution   Referring to Figure 9.54, notice that the vertex must be at the point $(3, 4)$ (i.e., the midpoint of the perpendicular line segment connecting the focus to the directrix) and the parabola opens down. From the vertex, the focus is shifted vertically by $\frac{1}{4a} = -2$ units, so $a = \frac{1}{(-2)4} = -\frac{1}{8}$. An equation is then

$$y = -\frac{1}{8}(x - 3)^2 + 4.$$

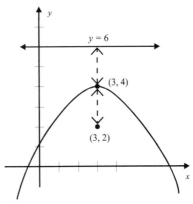

FIGURE 9.54: Parabola with focus at $(3, 2)$ and directrix at $y = 6$. ∎

Example 6.4   A Parabola Opening to the Right

Find an equation relating all points that are equidistant from the point $(3, -1)$ and the line $x = -1$.

Solution   Referring to Figure 9.55, notice that the vertex must be halfway between the focus $(3, -1)$ and the directrix $x = -1$, that is, at the point $(1, -1)$ and the parabola opens to the right. From the vertex, the focus is shifted horizontally by $\frac{1}{4a} = 2$ units, so that $a = \frac{1}{8}$. An equation is then

$$x = \frac{1}{8}(y + 1)^2 + 1.$$ ∎

You see parabolas nearly every day. As we discussed in section 5.5, the motion of many projectiles is approximately parabolic. In addition, parabolas have a **reflective property** that is extremely useful in many important applications. This can be seen as follows. For the parabola $x = ay^2$ indicated in Figure 9.56a, draw a horizontal line that intersects the parabola at the point A. Then, one can show that the acute angle α between the horizontal line and the tangent line at A is the same as the acute angle β between the tangent line and the line segment joining A to the focus. You may already have recognized that light rays are reflected from a surface in exactly the same fashion. In Figure 9.56b, we indicate a number of rays (you can think of them as light rays, although they could represent other forms of energy) traveling horizontally until they strike the parabola. As indicated, **all** rays striking the parabola are reflected **through the focus** of the parabola.

Due to this reflective property, satellite dishes are usually built with a parabolic shape and have a microphone located at the focus to receive all signals. This reflective property works in both directions. That is, energy emitted from the focus will reflect off the parabola and travel in parallel rays. For this reason, flashlights utilize parabolic reflectors to direct their light in a beam of parallel rays.

Example 6.5   Design of a Flashlight

A parabolic reflector for a flashlight has the shape $x = 2y^2$. Where should the light-bulb be located?

Solution   Based on the reflective property of parabolas, the lightbulb should be located at the focus of the parabola. The vertex is at $(0, 0)$ and the focus is shifted to the right from the vertex $\frac{1}{4a} = \frac{1}{8}$ units, so the lightbulb should be located at the point $(\frac{1}{8}, 0)$.

■

Ellipses

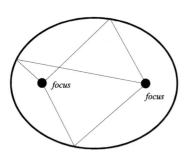

FIGURE 9.57a: Definition of ellipse.

The second conic section we study is the **ellipse**. We define an ellipse to be the set of all points for which the sum of the distances to two fixed points (called **foci**, the plural of focus) is constant. This definition is illustrated in Figure 9.57a. We define the **center** of an ellipse to be the midpoint of the line segment connecting the foci.

The familiar equation of an ellipse can be derived from this definition. For convenience, we assume that the foci lie at the points $(c, 0)$ and $(-c, 0)$ for some positive constant c (i.e., they lie on the x-axis, at the same distance from the origin). For any point (x, y) on the ellipse, the distance from (x, y) to the focus $(c, 0)$ is $\sqrt{(x - c)^2 + y^2}$ and the distance to the focus $(-c, 0)$ is $\sqrt{(x + c)^2 + y^2}$. The sum of these distances must equal a constant that we'll call k. We then have

$$\sqrt{(x - c)^2 + y^2} + \sqrt{(x + c)^2 + y^2} = k.$$

Subtracting the first square root from both sides and then squaring, we get

$$\left(\sqrt{(x + c)^2 + y^2}\right)^2 = \left(k - \sqrt{(x - c)^2 + y^2}\right)^2.$$

Expanding this out gives us

$$x^2 + 2cx + c^2 + y^2 = k^2 - 2k\sqrt{(x - c)^2 + y^2} + x^2 - 2cx + c^2 + y^2.$$

Now, solving for the remaining term with the radical gives us

$$2k\sqrt{(x - c)^2 + y^2} = k^2 - 4cx.$$

In order to get rid of the radical, we square both sides and expand, to get

$$4k^2 x^2 - 8k^2 cx + 4k^2 c^2 + 4k^2 y^2 = k^4 - 8k^2 cx + 16c^2 x^2$$

or

$$(4k^2 - 16c^2)x^2 + 4k^2 y^2 = k^4 - 4k^2 c^2.$$

To simplify this expression, we set $k = 2a$, to obtain

$$(16a^2 - 16c^2)x^2 + 16a^2 y^2 = 16a^4 - 16a^2 c^2.$$

Dividing both sides of the equation by 16 and defining $b^2 = a^2 - c^2$, we get

$$b^2 x^2 + a^2 y^2 = a^2 b^2.$$

Finally, dividing by $a^2 b^2$ leaves us with the familiar equation

$$\frac{x^2}{a^2} + \frac{y^2}{b^2} = 1.$$

In this equation, notice that x can assume values from $-a$ to a and y can assume values from $-b$ to b. The points $(a, 0)$ and $(-a, 0)$ are called the **vertices** of the ellipse (see

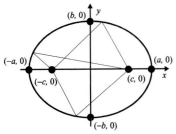

FIGURE 9.57b: Ellipse with foci at $(c, 0)$ and $(-c, 0)$.

Figure 9.57b). Since $a > b$, we call the line segment joining the vertices the **major axis** and we call the line segment joining the points $(0, b)$ and $(0, -b)$ the **minor axis**. Notice that the length of the major axis is $2a$ and the length of the minor axis is $2b$.

More generally, we have the following.

Theorem 6.3   The equation

$$\frac{(x - x_0)^2}{a^2} + \frac{(y - y_0)^2}{b^2} = 1 \qquad (6.1)$$

with $a > b > 0$ describes an ellipse with foci at $(x_0 - c, y_0)$ and $(x_0 + c, y_0)$, where $c = \sqrt{a^2 - b^2}$. The **center** of the ellipse is at the point (x_0, y_0) and the **vertices** are located at $(x_0 \pm a, y_0)$ on the major axis. The endpoints of the minor axis are located at $(x_0, y_0 \pm b)$.

The equation

$$\frac{(x - x_0)^2}{b^2} + \frac{(y - y_0)^2}{a^2} = 1 \qquad (6.2)$$

with $a > b > 0$ describes an ellipse with foci at $(x_0, y_0 - c)$ and $(x_0, y_0 + c)$ where $c = \sqrt{a^2 - b^2}$. The **center** of the ellipse is at the point (x_0, y_0) and the **vertices** are located at $(x_0, y_0 \pm a)$ on the major axis. The endpoints of the minor axis are located at $(x_0 \pm b, y_0)$.

In the following example, we use Theorem 6.3 to identify the features of an ellipse whose major axis runs along the x-axis.

Example 6.6   **Identifying the Features of an Ellipse**

Identify the center, foci and vertices of the ellipse $\dfrac{x^2}{16} + \dfrac{y^2}{9} = 1$.

Solution   From (6.1), the equation describes an ellipse with center at the origin $(0, 0)$. The values of a^2 and b^2 are 16 and 9, respectively, and so, $c = \sqrt{a^2 - b^2} = \sqrt{7}$. Since the major axis is parallel to the x-axis, the foci are shifted c units to the left and right of the center. That is, the foci are located at $(-\sqrt{7}, 0)$ and $(\sqrt{7}, 0)$. The vertices here are the x-intercepts (i.e., the intersections of the ellipse with the major axis). With $y = 0$, we have $x^2 = 16$ and so, the vertices are at $(\pm 4, 0)$. Taking $x = 0$, we get $y^2 = 9$ so that $y = \pm 3$. The y-intercepts are then $(0, -3)$ and $(0, 3)$. The ellipse is sketched in Figure 9.58.

∎

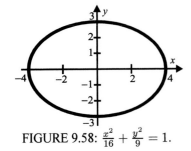

FIGURE 9.58: $\frac{x^2}{16} + \frac{y^2}{9} = 1$.

Example 6.7   **An Ellipse with Major Axis Parallel to the y-axis**

For the ellipse $\dfrac{(x - 2)^2}{4} + \dfrac{(y + 1)^2}{25} = 1$, identify the center, foci and vertices.

Solution   From (6.2), the center is at $(2, -1)$. The values of a^2 and b^2 are 25 and 4, respectively, so that $c = \sqrt{21}$. Since the major axis is parallel to the y-axis, the foci are shifted c units above and below the center, at $(2, -1 - \sqrt{21})$ and $(2, -1 + \sqrt{21})$. Notice that in this case, the vertices are the intersections of the ellipse with the line $x = 2$. With $x = 2$, we have $(y + 1)^2 = 25$, so that $y = -1 \pm 5$ and the vertices are $(2, -6)$ and $(2, 4)$. Finally, the endpoints of the minor axis are found by setting $y = -1$. We have

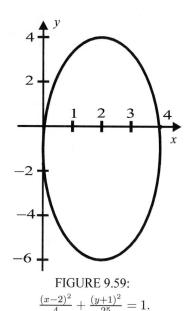

FIGURE 9.59:
$\frac{(x-2)^2}{4} + \frac{(y+1)^2}{25} = 1.$

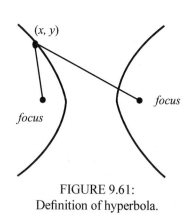

FIGURE 9.60: The reflective
property of ellipses.

$(x - 2)^2 = 4$, so that $x = 2 \pm 2$ and these endpoints are $(0, -1)$ and $(4, -1)$. The ellipse is sketched in Figure 9.59.

∎

Example 6.8   Finding the Equation of an Ellipse

Find an equation of the ellipse with foci at $(2, 3)$ and $(2, 5)$ and vertices $(2, 2)$ and $(2, 6)$.

Solution   Recall that the center is the midpoint of the foci, in this case $(2, 4)$. You can now see that the foci are shifted $c = 1$ unit from the center. The vertices are shifted $a = 2$ units from the center. From $c^2 = a^2 - b^2$, we get $b^2 = 4 - 1 = 3$. Notice that the major axis is parallel to the y-axis, so that $a^2 = 4$ is the divisor of the y-term. From (6.2), the ellipse has the equation

$$\frac{(x - 2)^2}{3} + \frac{(y - 4)^2}{4} = 1.$$

∎

Much like parabolas, ellipses have some useful reflective properties. As illustrated in Figure 9.60, a line segment joining one focus to a point A on the ellipse makes the same acute angle with the tangent line at A as does the line segment joining the other focus to A. Again, this is the same way in which light and sound reflect off a surface, so that a ray originating at one focus will always reflect off the ellipse toward the other focus. A surprising application of this principle is found in the so-called "whispering room" of the U. S. Senate. The ceiling of this room is elliptical, so that by standing at one focus you can hear everything said on the other side of the room at the other focus. You probably never imagined how much of a role mathematics could play in political intrigue.

Example 6.9   A Medical Application of the Reflective Property of Ellipses

A recently developed medical procedure called **shockwave lithotripsy** is used to break up kidney stones that are too large or irregular to be passed. In this procedure, shockwaves emanating from a transducer located at one focus are bounced off of an elliptical reflector to the kidney stone located at the other focus. Suppose that the reflector is described by the equation $\frac{x^2}{112} + \frac{y^2}{48} = 1$ (in units of inches). Where should the transducer be placed?

Solution   In this case,

$$c = \sqrt{a^2 - b^2} = \sqrt{112 - 48} = 8,$$

so that the foci are 16 inches apart. Since the transducer must be located at one focus, it should be placed 16 inches away from the kidney stone and aligned so that the line segment from the kidney stone to the transducer lies along the major axis of the elliptical reflector.

∎

Hyperbolas

The third type of conic section is the **hyperbola**. We define a hyperbola to be the set of all points such that the **difference** of the distances between two fixed points (called the **foci**) is a constant. This definition is illustrated in Figure 9.61. Notice that it is nearly identical to the definition of the ellipse, except that we subtract the distances instead of add them.

The familiar equation of the hyperbola can be derived from the definition. The derivation is almost identical to that of the ellipse, except that the quantity $a^2 - c^2$ is now negative.

FIGURE 9.61:
Definition of hyperbola.

We leave the details of the derivation of this as an exercise. An equation of the hyperbola with foci at $(\pm c, 0)$ and parameter $2a$ (equal to the difference of the distances) is

$$\frac{x^2}{a^2} - \frac{y^2}{b^2} = 1,$$

where $b^2 = c^2 - a^2$. An important feature of hyperbolas that is not shared by ellipses is the presence of asymptotes. For the hyperbola $\frac{x^2}{a^2} - \frac{y^2}{b^2} = 1$, we have $\frac{y^2}{b^2} = \frac{x^2}{a^2} - 1$ or $y^2 = \frac{b^2}{a^2}x^2 - b^2$. Notice that

$$\lim_{x \to \pm\infty} \frac{y^2}{x^2} = \lim_{x \to \pm\infty} \left(\frac{b^2}{a^2} - \frac{b^2}{x^2} \right) = \frac{b^2}{a^2}.$$

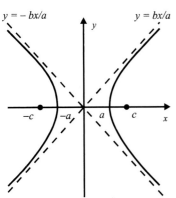

$y = -bx/a$ $y = bx/a$

That is, as $x \to \pm\infty$, $\frac{y^2}{x^2} \to \frac{b^2}{a^2}$, so that $\frac{y}{x} \to \pm\frac{b}{a}$ and so, $y = \pm\frac{b}{a}x$ are the (slant) asymptotes as shown in Figure 9.62.

More generally, we have the following.

FIGURE 9.62: Hyperbola, shown with its asymptotes.

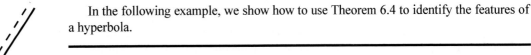

Theorem 6.4   The equation

$$\boxed{\frac{(x - x_0)^2}{a^2} - \frac{(y - y_0)^2}{b^2} = 1}$$ (6.3)

describes a hyperbola with foci at the points $(x_0 - c, y_0)$ and $(x_0 + c, y_0)$, where $c = \sqrt{a^2 + b^2}$. The **center** of the hyperbola is at the point (x_0, y_0) and the **vertices** are located at $(x_0 \pm a, y_0)$. The asymptotes are $y = \pm\frac{b}{a}(x - x_0) + y_0$.

The equation

$$\boxed{\frac{(y - y_0)^2}{a^2} - \frac{(x - x_0)^2}{b^2} = 1}$$ (6.4)

describes a hyperbola with foci at the points $(x_0, y_0 - c)$ and $(x_0, y_0 + c)$ where $c = \sqrt{a^2 + b^2}$. The **center** of the hyperbola is at the point (x_0, y_0) and the **vertices** are located at $(x_0, y_0 \pm a)$. The asymptotes are $y = \pm\frac{a}{b}(x - x_0) + y_0$.

In the following example, we show how to use Theorem 6.4 to identify the features of a hyperbola.

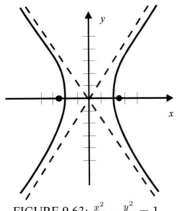

FIGURE 9.63: $\frac{x^2}{4} - \frac{y^2}{9} = 1$.

Example 6.10   **Identifying the Features of a Hyperbola**

For the hyperbola $\frac{x^2}{4} - \frac{y^2}{9} = 1$, find the center, vertices, foci and asymptotes.

Solution   From (6.3), we can see that the center is at $(0, 0)$. Further, the vertices lie on the x-axis, where $\frac{x^2}{4} = 1$ (set $y = 0$), so that $x = \pm 2$. The vertices are then located at $(2, 0)$ and $(-2, 0)$. The foci are shifted by $c = \sqrt{a^2 + b^2} = \sqrt{4 + 9} = \sqrt{13}$ units from the center, to $(\pm\sqrt{13}, 0)$. Finally, the asymptotes are $y = \pm\frac{3}{2}x$. A sketch of the hyperbola is shown in Figure 9.63.   ∎

Example 6.11   **Identifying the Features of a Hyperbola**

For the hyperbola $\frac{(y - 1)^2}{9} - \frac{(x + 1)^2}{16} = 1$, find the center, vertices, foci and asymptotes.

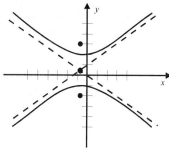

FIGURE 9.64:

$$\frac{(y-1)^2}{9} - \frac{(x+1)^2}{16} = 1.$$

Solution   Notice that from (6.4), the center is at $(-1, 1)$. Setting $x = -1$, we find that the vertices are shifted vertically by $a = 3$ units from the center, to $(-1, -2)$ and $(-1, 4)$. The foci are shifted vertically by $c = \sqrt{a^2 + b^2} = \sqrt{25} = 5$ units from the center, to $(-1, -4)$ and $(-1, 6)$. The asymptotes are $y = \pm\frac{3}{4}(x+1) + 1$. A sketch of the hyperbola is shown in Figure 9.64.

■

Example 6.12   Finding the Equation of a Hyperbola

Find an equation of the hyperbola with center at $(-2, 0)$, vertices at $(-4, 0)$ and $(0, 0)$ and foci at $(-5, 0)$ and $(1, 0)$.

Solution   Notice that since the center, vertices and foci all lie on the x-axis, the hyperbola must have an equation of the form of (6.3). Here, the vertices are shifted $a = 2$ units from the center, and the foci are shifted $c = 3$ units from the center. Then, we have $b^2 = c^2 - a^2 = 5$. Following (6.3), we have the equation

$$\frac{(x+2)^2}{4} - \frac{y^2}{5} = 1.$$

■

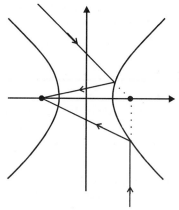

FIGURE 9.65: The reflective property of hyperbolas.

Much like parabolas and ellipses, hyperbolas have a reflective property that is useful in applications. It can be shown that a ray directed toward one focus will reflect off the hyperbola toward the other focus. We illustrate this in Figure 9.65.

Example 6.13   An Application to Hyperbolic Mirrors

A hyperbolic mirror is constructed in the shape of the top half of the hyperbola $(y + 2)^2 - \dfrac{x^2}{3} = 1$. Toward what point will light rays following the paths $y = kx$ reflect (where k is a constant)?

Solution   For the given hyperbola, we have $c = \sqrt{a^2 + b^2} = \sqrt{1+3} = 2$. Notice that the center is at $(0, -2)$, and the foci are at $(0, 0)$ and $(0, -4)$. Since rays of the form $y = kx$ will pass through the focus at $(0, 0)$, they will be reflected toward the focus at $(0, -4)$.

■

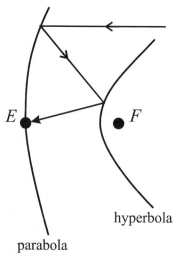

FIGURE 9.66: A combination of parabolic and hyperbolic mirrors.

As a final note on the reflective properties of the conic sections, we briefly discuss a clever use of parabolic and hyperbolic mirrors in telescope design. In Figure 9.66, a parabolic mirror to the left and a hyperbolic mirror to the right are arranged so that they have a common focus at the point F. The vertex of the parabola is located at the other focus of the hyperbola, at the point E, where there is an opening for the eye or a camera. Notice that light entering the telescope from the right (and passing around the hyperbolic mirror) will reflect off the parabola directly toward its focus at F. Since F is also a focus of the hyperbola, the light will reflect off the hyperbola toward its other focus at E. In combination, the mirrors cause all incoming light to be focused at the point E.

Exercises 9.6

1. (writing) Each fixed point referred to in the definitions of the conic sections is called a **focus**. Use the reflective properties of the conic sections to explain why this is an appropriate name.

2. (writing) A hyperbola looks somewhat like a pair of parabolas facing opposite directions. Discuss the differences between a parabola and one half of a hyperbola (recall that hyperbolas have asymptotes).

3. (writing) Carefully explain why in example 6.8 (or for any other ellipse) the sum of the distances from a point on the ellipse to the two foci equals $2a$.

4. (writing) Imagine playing a game of pool on an elliptical pool table with a single hole located at one focus. If a ball rests near the other focus, which is clearly marked, describe an easy way to hit the ball into the hole.

In exercises 5–16, find an equation for the indicated conic section.

5. Parabola with focus $(0, -1)$ and directrix $y = 1$

6. Parabola with focus $(1, 2)$ and directrix $y = -2$

7. Parabola with focus $(3, 0)$ and directrix $x = 1$

8. Parabola with focus $(2, 0)$ and directrix $x = -2$

9. Ellipse with foci $(0, 1)$ and $(0, 5)$ and vertices $(0, -1)$ and $(0, 7)$

10. Ellipse with foci $(1, 2)$ and $(1, 4)$ and vertices $(1, 1)$ and $(1, 5)$

11. Ellipse with foci $(2, 1)$ and $(6, 1)$ and vertices $(0, 1)$ and $(8, 1)$

12. Ellipse with foci $(3, 2)$ and $(5, 2)$ and vertices $(2, 2)$ and $(6, 2)$

13. Hyperbola with foci $(0, 0)$ and $(4, 0)$ and vertices $(1, 0)$ and $(3, 0)$

14. Hyperbola with foci $(-2, 2)$ and $(6, 2)$ and vertices $(0, 2)$ and $(4, 2)$

15. Hyperbola with foci $(2, 2)$ and $(2, 6)$ and vertices $(2, 3)$ and $(2, 5)$

16. Hyperbola with foci $(0, -2)$ and $(0, 4)$ and vertices $(0, 0)$ and $(0, 2)$

In exercises 17–28, identify the conic section and find each vertex, focus and directrix.

17. $y = 2(x + 1)^2 - 1$

18. $y = -2(x + 2)^2 - 1$

19. $\dfrac{(x - 1)^2}{4} + \dfrac{(y - 2)^2}{9} = 1$

20. $\dfrac{(x + 2)^2}{16} + \dfrac{y^2}{4} = 1$

21. $\dfrac{(x - 1)^2}{9} - \dfrac{y^2}{4} = 1$

22. $\dfrac{(x + 1)^2}{4} - \dfrac{(y - 3)^2}{4} = 1$

23. $\dfrac{(y + 1)^2}{16} - \dfrac{(x + 2)^2}{4} = 1$

24. $\dfrac{y^2}{4} - \dfrac{(x + 2)^2}{9} = 1$

25. $(x - 2)^2 + 9y^2 = 9$

26. $4x^2 + (y + 1)^2 = 16$

27. $(x + 1)^2 - 4(y - 2) = 16$

28. $4(x + 2) - (y - 1)^2 = -4$

In exercises 29–34, graph the conic section and find an equation.

29. All points equidistant from the point $(2, 1)$ and the line $y = -3$

30. All points equidistant from the point $(-1, 0)$ and the line $y = 4$

31. All points such that the sum of the distances to the points $(0, 2)$ and $(4, 2)$ equals 8

32. All points such that the sum of the distances to the points $(3, 1)$ and $(-1, 1)$ equals 6

33. All points such that the difference of the distances to the points $(0, 4)$ and $(0, -2)$ equals 4

34. All points such that the difference of the distances to the points $(2, 2)$ and $(6, 2)$ equals 2

35. A parabolic flashlight reflector has the shape $x = 4y^2$. Where should the lightbulb be placed?

36. A parabolic flashlight reflector has the shape $x = \frac{1}{2}y^2$. Where should the lightbulb be placed?

37. A parabolic satellite dish has the shape $y = 2x^2$. Where should the microphone be placed?

38. A parabolic satellite dish has the shape $y = 4x^2$. Where should the microphone be placed?

39. In example 6.9, if the shape of the reflector is $\dfrac{x^2}{124} + \dfrac{y^2}{24} = 1$, how far from the kidney stone should the transducer be placed?

40. In example 6.9, if the shape of the reflector is $\dfrac{x^2}{44} + \dfrac{y^2}{125} = 1$, how far from the kidney stone should the transducer be placed?

41. If a hyperbolic mirror is in the shape of the top half of $(y + 4)^2 - \dfrac{x^2}{15} = 1$, to which point will light rays following the path $y = cx$ $(y < 0)$ reflect?

42. If a hyperbolic mirror is in the shape of the bottom half of $(y - 3)^2 - \dfrac{x^2}{8} = 1$, to which point will light rays following the path $y = cx$ $(y > 0)$ reflect?

43. If a hyperbolic mirror is in the shape of the right half of $\dfrac{x^2}{3} - y^2 = 1$, to which point will light rays following the path $y = c(x - 2)$ reflect?

44. If a hyperbolic mirror is in the shape of the left half of $\dfrac{x^2}{8} - y^2 = 1$, to which point will light rays following the path $y = c(x + 3)$ reflect?

45. If the ceiling of a room has the shape $\dfrac{x^2}{400} + \dfrac{y^2}{100} = 1$, where should you place the desks so that you can sit at one desk and hear everything said at the other desk?

46. If the ceiling of a room has the shape $\dfrac{x^2}{900} + \dfrac{y^2}{100} = 1$, where should you place the desks so that you can sit at one desk and hear everything said at the other desk?

47. A spectator at the 2000 Summer Olympics Games throws an object. After 2 seconds, the object is 28 meters from the spectator. After 4 seconds, the object is 48 meters from the spectator. If the object's distance from the spectator is a quadratic function of time, find an equation for the position of the object. Sketch a graph of the path. What is the object?

48. Halley's comet follows an elliptical path with $a = 17.79$ (Au) and $b = 4.53$ (Au). Compute the distance the comet travels in one orbit. Given that Halley's comet completes an orbit in approximately 76 years, what is the average speed of the comet?

49. (exploratory) All of the equations of conic sections that we have seen so far have been of the form $Ax^2 + Cy^2 + Dx +$ $Ey + F = 0$. In this exercise, you will classify the conic sections for different values of the constants. First, assume that $A > 0$ and $C > 0$. Which conic section will you get? Next, try $A > 0$ and $C < 0$. Which conic section is it this time? How about $A < 0$ and $C > 0$? $A < 0$ and $C < 0$? Finally, suppose that either A or C (not both) equals 0; which conic section is it? In all cases, the values of the constants D, E and F do not affect the classification. Explain what effect these constants have.

50. (exploratory) In this exercise, you will generalize the results of exercise 49 by exploring the equation $Ax^2 + Bxy + Cy^2 + Dx + Ey + F = 0$. (In exercise 49, the coefficient of xy was 0.) You will need to have software that will graph such equations. Make up several examples with $B^2 - 4AC = 0$ (e.g., $B = 2$, $A = 1$ and $C = 1$). Which conic section results? Now, make up several examples with $B^2 - 4AC < 0$ (e.g., $B = 1$, $A = 1$ and $C = 1$). Which conic section do you get? Finally, make up several examples with $B^2 - 4AC > 0$ (e.g., $B = 4$, $A = 1$ and $C = 1$). Which conic section is this?

9.7   Conic Sections in Polar Coordinates

The conic sections are among the most studied curves in mathematics. As we saw in section 9.6, their unique properties allow us to make use of them in a variety of important applications. In addition to the definitions given in section 9.6, there are a variety of alternative definitions of the conic sections. One such alternative, utilizing an important quantity called **eccentricity**, is especially convenient for studying conic sections in polar coordinates. We introduce this definition in this section, and review some options for parametric representations of conic sections.

For a fixed point P (the **focus**) and a fixed line l not containing P (the **directrix**), consider the set of all points whose distance to the focus is a constant multiple of their distance to the directrix. The constant multiple $e > 0$ is called the **eccentricity**. Note that if $e = 1$, this is the usual definition of a parabola. As shown below, for other values of e we get the other conic sections.

Theorem 7.1   The set of all points whose distance to the focus is the product of the eccentricity e and the distance to the directrix is
  (i)   an ellipse if $0 < e < 1$,
  (ii) a parabola if $e = 1$ or
  (iii) a hyperbola if $e > 1$.

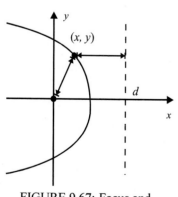

FIGURE 9.67: Focus and directrix.

Proof. We can simplify the algebra greatly by assuming that the focus is located at the origin and the directrix is the line $x = d > 0$. (We illustrate this in Figure 9.67 for the case of a parabola.) For any point (x, y) on the curve, observe that the distance to the focus is given by $\sqrt{x^2 + y^2}$ and the distance to the directrix is $d - x$. We then have

$$\sqrt{x^2 + y^2} = e\,(d - x). \tag{7.1}$$

Squaring both sides gives us

$$x^2 + y^2 = e^2\,(d^2 - 2dx + x^2).$$

Finally, if we gather together the like terms, we get

$$x^2 \left(1 - e^2\right) + 2\, d\, e^2\, x + y^2 = e^2\, d^2. \tag{7.2}$$

Note that (7.2) has the form of the equation of a conic section. In particular, if $e = 1$, (7.2) becomes

$$2\, d\, x + y^2 = d^2,$$

which is the equation of a parabola. If $0 < e < 1$, notice that $(1 - e^2) > 0$ and so, (7.2) is the equation of an ellipse (with center shifted to the left by the x-term). Finally, if $e > 1$, then $(1 - e^2) < 0$ and so, (7.2) is the equation of a hyperbola.   ∎

Notice that the original form of the defining equation (7.1) of these conic sections includes the term $\sqrt{x^2 + y^2}$. You should think of polar coordinates anytime you see this expression. Recall that in polar coordinates, $r = \sqrt{x^2 + y^2}$ and $x = r\cos\theta$. Equation (7.1) now becomes

$$r = e(d - r\cos\theta)$$

or

$$r(e\cos\theta + 1) = ed.$$

Finally, solving for r, we have

$$\boxed{r = \frac{ed}{e\cos\theta + 1}.}$$

This is the polar form of an equation for the conic sections with focus and directrix oriented as in Figure 9.67. As you will show in the exercises, different orientations of the focus and directrix can produce different forms of the polar equation. The possibilities are summarized in the following result.

Theorem 7.2   The conic section with eccentricity $e > 0$, focus at $(0, 0)$ and the indicated directrix has the polar equation

(i) $r = \dfrac{ed}{e\cos\theta + 1}$, if the directrix is the line $x = d > 0$,

(ii) $r = \dfrac{ed}{e\cos\theta - 1}$, if the directrix is the line $x = d < 0$,

(iii) $r = \dfrac{ed}{e\sin\theta + 1}$, if the directrix is the line $y = d > 0$ or

(iv) $r = \dfrac{ed}{e\sin\theta - 1}$, if the directrix is the line $y = d < 0$.

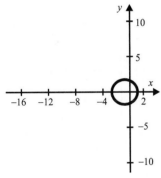

FIGURE 9.68a: $e = 0.4$.

Notice that we proved part (i) above. The remaining parts are derived in similar fashion and are left as exercises. In the following example, we illustrate how the eccentricity effects the graph of a conic section.

Example 7.1   The Effect of Various Eccentricities
Find polar equations of the conic sections with focus $(0, 0)$, directrix $x = 4$ and eccentricities (a) $e = 0.4$, (b) $e = 0.8$, (c) $e = 1$, (d) $e = 1.2$ and (e) $e = 2$.

Solution   By Theorem 7.1, (a) and (b) are ellipses, (c) is a parabola and (d) and (e) are hyperbolas. By Theorem 7.2, all have polar equations of the form $r = \dfrac{4e}{e\cos\theta + 1}$. The graphs of $r = \dfrac{1.6}{0.4\cos\theta + 1}$ and $r = \dfrac{3.2}{0.8\cos\theta + 1}$ are shown in Figures 9.68a and 9.68b,

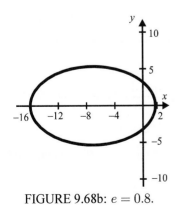

FIGURE 9.68b: $e = 0.8$.

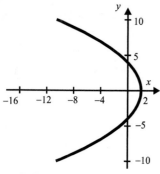

FIGURE 9.68c: $e = 1$.

FIGURE 9.68d: $e = 1.2$

FIGURE 9.68e: $e = 2$.

respectively. Note that the ellipse with the smaller eccentricity is much more nearly circular than the ellipse with the larger eccentricity. Further, the ellipse with $e = 0.8$ opens up much farther to the left. In fact, as the value of e approaches 1, the ellipse will open up farther to the left, approaching the parabola with $e = 1$, $r = \dfrac{4}{\cos\theta + 1}$, as shown in Figure 9.68c. For values of $e > 1$, the graph is a hyperbola, opening up to the right and left. For instance, with $e = 1.2$, we have the hyperbola shown in Figure 9.68d, $r = \dfrac{4.8}{1.2\cos\theta + 1}$.

With $e = 2$, we have the hyperbola shown in Figure 9.68e, $r = \dfrac{8}{2\cos\theta + 1}$. Notice how this hyperbola approaches its asymptotes much more rapidly than the hyperbola in Figure 9.68d.

■

Example 7.2   The Effect of Various Directrixes

Find polar equations of the conic sections with focus $(0,0)$, eccentricity $e = 0.5$ and directrix given by (a) $y = 2$, (b) $y = -3$ and (c) $x = -2$.

Solution   First, note that with an eccentricity of $e = 0.5$, each of these conic sections is an ellipse. From Theorem 7.2, we know that (a) has the form $r = \dfrac{1}{0.5\sin\theta + 1}$. A sketch is shown in Figure 9.69a.

For (b), we have $r = \dfrac{-1.5}{0.5\sin\theta - 1}$ and show a sketch in Figure 9.69b. For (c), the directrix is parallel to the x-axis and so, from Theorem 7.2, we have $r = \dfrac{-1}{0.5\cos\theta - 1}$. A sketch is shown in Figure 9.69c.

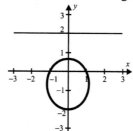

FIGURE 9.69a: Directrix:
$y = 2$.

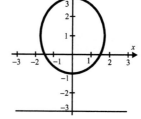

FIGURE 9.69b: Directrix:
$y = -3$.

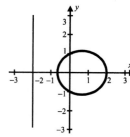

FIGURE 9.69c: Directrix:
$x = -2$.

■

The results of Theorem 7.2 apply only to conic sections with a focus at the origin. Recall that in rectangular coordinates, it's easy to translate the center of a conic section. Unfortunately, this is not true in polar coordinates.

In the following example, we see how to write some conic sections parametrically.

Example 7.3   Parametric Equations for Some Conic Sections

Find parametric equations of the conic sections (a) $\dfrac{(x-1)^2}{4} + \dfrac{(y+2)^2}{9} = 1$ and (b) $\dfrac{(x+2)^2}{9} - \dfrac{(y-3)^2}{16} = 1$.

Solution   Notice that the curve in (a) is an ellipse with center at $(1, -2)$ and major axis parallel to the y-axis. Parametric equations for the ellipse are

$$\begin{cases} x = 2\cos t + 1 \\ y = 3\sin t - 2 \end{cases} \quad \text{with } 0 \le t \le 2\pi.$$

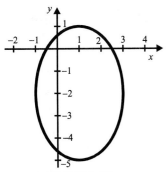

FIGURE 9.70a:

$$\frac{(x-1)^2}{4} + \frac{(y+2)^2}{9} = 1$$

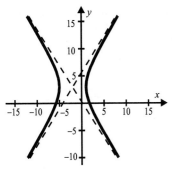

FIGURE 9.70b:

$$\frac{(x+2)^2}{9} - \frac{(y-3)^2}{16} = 1.$$

We show a sketch in Figure 9.70a. You should recognize that the curve in (b) is a hyperbola. It is convenient to use hyperbolic functions in its parametric representation. The parameters are $a^2 = 9$ ($a = 3$) and $b^2 = 16$ ($b = 4$) and the center is $(-2, 3)$. Parametric equations are

$$\begin{cases} x = 3\cosh t - 2 \\ y = 4\sinh t + 3, \end{cases}$$

for the right half of the hyperbola and

$$\begin{cases} x = -3\cosh t - 2 \\ y = 4\sinh t + 3, \end{cases}$$

for the left half. We leave it as an exercise to verify that this is a correct parametrization. We sketch the hyperbola in Figure 9.70b.

∎

We opened this chapter with a discussion of Kepler's laws of planetary motion. Although we must delay a more complete presentation of these laws until Chapter 11, we are now in a position to illustrate one of Kepler's laws. Kepler's second law states that, measuring from the sun to the planet, equal areas are swept out in equal times. As we see in the following example, this implies that planets speed up and slow down as they orbit the sun.

Example 7.4   Kepler's Second Law of Planetary Motion

Suppose that a planet's orbit follows the elliptical path $r = \dfrac{2}{\sin\theta + 2}$ with the sun located at the origin (one of the foci), as illustrated in Figure 9.71a. Show that roughly equal areas are swept out from $\theta = 0$ to $\theta = \pi$ and from $\theta = 3\pi/2$ to $\theta = 5.224895$. Then, find the corresponding arc lengths and compare the average speeds of the planet on these arcs.

Solution   First, note that the area swept out by the planet from $\theta = 0$ to $\theta = \pi$ is the area bounded by the polar graphs $r = f(\theta) = \frac{2}{\sin\theta+2}$, $\theta = 0$ and $\theta = \pi$ (see Figure 9.71b). From (5.6), this is given by

$$A = \frac{1}{2}\int_0^\pi [f(\theta)]^2 \, d\theta = \frac{1}{2}\int_0^\pi \left(\frac{2}{\sin\theta + 2}\right)^2 d\theta \approx 0.9455994.$$

Similarly, the area swept out from $\theta = 3\pi/2$ to $\theta = 5.224895$ (see Figure 9.71c) is given by

$$A = \frac{1}{2}\int_{3\pi/2}^{5.224895} \left(\frac{2}{\sin\theta + 2}\right)^2 d\theta \approx 0.9455995.$$

FIGURE 9.71a: Ellptical orbit.

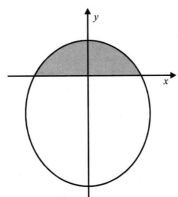

FIGURE 9.71b: Area swept out by the orbit from $\theta = 0$ to $\theta = \pi$.

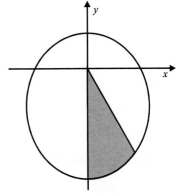

FIGURE 9.71c: Area swept out by the orbit from $\theta = 3\pi/2$ to $\theta = 5.224895$.

From (5.8), the arc length of the portion of the curve on the interval from $\theta = 0$ to $\theta = \pi$ is given by

$$s_1 = \int_0^\pi \sqrt{[f'(\theta)]^2 + [f(\theta)]^2}\, d\theta$$

$$= \int_0^\pi \sqrt{\frac{4\cos^2\theta}{(\sin\theta + 2)^4} + \frac{4}{(\sin\theta + 2)^2}}\, d\theta \approx 2.53,$$

while the arc length of the portion of the curve on the interval from $\theta = 3\pi/2$ to $\theta = 5.224895$ is given by

$$s_2 = \int_{3\pi/2}^{5.224895} \sqrt{\frac{4\cos^2\theta}{(\sin\theta + 2)^4} + \frac{4}{(\sin\theta + 2)^2}}\, d\theta \approx 1.02.$$

Notice that this says that the average speed on the portion of the orbit from $\theta = 0$ to $\theta = \pi$ is roughly two-and-a-half times the average speed on the portion of the orbit from $\theta = 3\pi/2$ to $\theta = 5.224895$.

■

Exercises 9.7

1. (writing) Based on Theorem 7.1, we might say that parabolas are the rarest of the conic sections, since they occur only for $e = 1$ exactly. Referring to Figure 9.49, explain why it takes a fairly precise cut of the cone to produce a parabola.

2. (writing) Describe how the ellipses in Figure 9.68 "open up" into a parabola as e increases to $e = 1$. What happens as e decreases to $e = 0$?

In exercises 3–18, find polar equations for, and graph the conic section with focus $(0,0)$ and the given directrix and eccentricity.

3. directrix $x = 2$, $e = 0.6$ **4.** directrix $x = 2$, $e = 1.2$

5. directrix $x = 2$, $e = 1$ **6.** directrix $x = 2$, $e = 2$

7. directrix $y = 2$, $e = 0.6$ **8.** directrix $y = 2$, $e = 1.2$

9. directrix $y = 2$, $e = 1$ **10.** directrix $y = 2$, $e = 2$

11. directrix $x = -2$, $e = 0.4$ **12.** directrix $x = -2$, $e = 1$

13. directrix $x = -2$, $e = 2$ **14.** directrix $x = -2$, $e = 4$

15. directrix $y = -2$, $e = 0.4$

16. directrix $y = -2$, $e = 0.9$

17. directrix $y = -2$, $e = 1$

18. directrix $y = -2$, $e = 1.1$

In exercises 19–24, graph and interpret the conic section.

19. $r = \dfrac{4}{2\cos(\theta - \pi/6) + 1}$

20. $r = \dfrac{4}{4\sin(\theta - \pi/6) + 1}$

21. $r = \dfrac{-6}{\sin(\theta - \pi/4) - 2}$

22. $r = \dfrac{-4}{\cos(\theta - \pi/4) - 4}$

23. $r = \dfrac{-3}{2\cos(\theta + \pi/4) - 2}$

24. $r = \dfrac{3}{2\cos(\theta + \pi/4) + 2}$

In exercises 25–30, find parametric equations of the conic sections.

25. $\dfrac{(x+1)^2}{9} + \dfrac{(y-1)^2}{4} = 1$

26. $\dfrac{(x-2)^2}{9} - \dfrac{(y+1)^2}{16} = 1$

27. $\dfrac{(x+1)^2}{16} - \dfrac{y^2}{9} = 1$

28. $\dfrac{x^2}{4} + y^2 = 1$

29. $\dfrac{x^2}{4} + y = 1$

30. $x - \dfrac{y^2}{4} = 1$

31. Repeat example 7.4 with $0 \le \theta \le \frac{\pi}{2}$ and $\frac{3\pi}{2} \le \theta \le 4.953$.

32. Repeat example 7.4 with $\frac{\pi}{2} \le \theta \le \pi$ and $4.471 \le \theta \le \frac{3\pi}{2}$.

33. Prove Theorem 7.2 (ii).

34. Prove Theorem 7.2 (iii).

35. Prove Theorem 7.2 (iv).

36. (exploratory) Earth's orbit is approximately elliptical with the sun at one focus, a minor axis of length 93 million miles and eccentricity $e = 0.017$. Find a polar equation for earth's orbit.

37. (exploratory) If Neptune's orbit is given by

$$r = \frac{1.82 \times 10^{14}}{343 \cos(\theta - 0.77) + 40,000}$$

and Pluto's orbit is given by

$$r = \frac{5.52 \times 10^{13}}{2481 \cos(\theta - 3.91) + 10,000}$$

show that Pluto is sometimes closer and sometimes further from the sun than Neptune. Based on these equations, will the planets ever collide?

Chapter Review Exercises

In exercises 1–4, sketch the plane curve defined by the parametric equations and find a corresponding x-y equation for the curve.

1. $\begin{cases} x = -1 + 3\cos t \\ y = 2 + 3\sin t \end{cases}$ **2.** $\begin{cases} x = 2 - t \\ y = 1 + 3t \end{cases}$

3. $\begin{cases} x = t^2 + 1 \\ y = t^4 \end{cases}$ **4.** $\begin{cases} x = \cos t \\ y = \cos^2 t - 1 \end{cases}$

In exercises 5–8, sketch the plane curves defined by the parametric equations.

5. $\begin{cases} x = \cos 2t \\ y = \sin 6t \end{cases}$ **6.** $\begin{cases} x = \cos 6t \\ y = \sin 2t \end{cases}$

7. $\begin{cases} x = \cos 2t \cos t \\ y = \cos 2t \sin t \end{cases}$ **8.** $\begin{cases} x = \cos 2t \cos 3t \\ y = \cos 2t \sin 3t \end{cases}$

In exercises 9–12, match the parametric equations with the corresponding plane curve.

9. $\begin{cases} x = t^2 - 1 \\ y = t^3 \end{cases}$ **10.** $\begin{cases} x = t^3 \\ y = t^2 - 1 \end{cases}$

11. $\begin{cases} x = \cos 2t \cos t \\ y = \cos 2t \sin t \end{cases}$ **12.** $\begin{cases} x = \cos(t + \cos t) \\ y = \cos(t + \sin t) \end{cases}$

Figure A

Figure B

Figure C

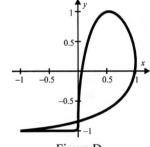

Figure D

In exercises 13 and 14, find parametric equations for the given curve.

13. The line segment from $(2, 1)$ to $(4, 7)$.

14. The portion of the parabola $y = x^2 + 1$ from $(1, 2)$ to $(3, 10)$.

In exercises 15 and 16, find the slopes of the curves at the points (a) $t = 0$, (b) $t = 1$ and (c) $(2, 3)$.

15. $\begin{cases} x = t^3 - 3t \\ y = t^2 - t + 1 \end{cases}$ **16.** $\begin{cases} x = t^2 - 2 \\ y = t + 2 \end{cases}$

In exercises 17 and 18, parametric equations for the position of an object are given. Find the object's velocity and speed at time $t = 0$ and describe its motion.

17. $\begin{cases} x = t^3 - 3t \\ y = t^2 + 2t \end{cases}$ **18.** $\begin{cases} x = t^3 - 3t \\ y = t^2 + 2 \end{cases}$

In exercises 19–22, find the area enclosed by the curve.

19. $\begin{cases} x = 3\sin t \\ y = 2\cos t \end{cases}$

20. $\begin{cases} x = 4\sin 3t \\ y = 3\cos 3t \end{cases}$

21. $\begin{cases} x = \cos 2t \\ y = \sin \pi t \end{cases}, \; -1 \le t \le 1$

22. $\begin{cases} x = t^2 - 1 \\ y = t^3 - t \end{cases}, -1 \le t \le 1$

In exercises 23–26, find the arc length of the curve (approximate numerically, if needed).

23. $\begin{cases} x = \cos 2t \\ y = \sin \pi t \end{cases}, -1 \le t \le 1$

24. $\begin{cases} x = t^2 - 1 \\ y = t^3 - 4t \end{cases}, -1 \le t \le 1$

25. $\begin{cases} x = \cos 4t \\ y = \sin 5t \end{cases}$ **26.** $\begin{cases} x = \sin 10t \\ y = t^2 - 1 \end{cases}, -\pi \le t \le \pi$

In exercises 27 and 28, compute the surface area of the surface obtained by revolving the curve about the indicated axis.

27. $\begin{cases} x = t^3 - 4t \\ y = t^4 - 4t \end{cases}, -1 \le t \le 1$, about the x-axis

28. $\begin{cases} x = t^3 - 4t \\ y = t^4 - 4t \end{cases}, -1 \le t \le 1$, about $y = 2$

In exercises 29 and 30, sketch the graph of the polar equation and find a corresponding x-y equation.

29. $r = 3 \cos \theta$ **30.** $r = 2 \sec \theta$

In exercises 31–38, sketch the graph and identify all values of θ where $r = 0$ and a range of values of θ which produces one copy of the graph.

31. $r = 2 \sin \theta$ **32.** $r = 2 - 2 \sin \theta$

33. $r = 2 - 3 \sin \theta$ **34.** $r = \cos 3\theta + \sin 2\theta$

35. $r^2 = 4 \sin 2\theta$ **36.** $r = e^{\cos \theta} - 2 \cos 4\theta$

37. $r = \dfrac{2}{1 + 2 \sin \theta}$ **38.** $\dfrac{2}{1 + 2 \cos \theta}$

In exercises 39 and 40, find a polar equation corresponding to the rectangular equation.

39. $x^2 + y^2 = 9$ **40.** $(x - 3)^2 + y^2 = 9$

In exercises 41 and 42, find the slope of the tangent line to the polar curve at the given point.

41. $r = \cos 3\theta$ at $\theta = \frac{\pi}{6}$ **42.** $r = 1 - \sin \theta$ at $\theta = 0$

In exercises 43–48, find the area of the indicated region.

43. One leaf of $r = \sin 5\theta$

44. One leaf of $r = \cos 2\theta$

45. Inner loop of $r = 1 - 2 \sin \theta$

46. Bounded by $r = 3 \sin \theta$

47. Inside of $r = 1 + \sin \theta$ and outside of $r = 1 + \cos \theta$

48. Inside of $r = 1 + \cos \theta$ and outside of $r = 1 + \sin \theta$

In exercises 49 and 50, find the arc length of the curve.

49. $r = 3 - 4 \sin \theta$ **50.** $r = \sin 4\theta$

In exercises 51–53, find an equation for the conic section.

51. Parabola with focus $(1, 2)$ and directrix $y = 0$

52. Ellipse with foci $(2, 1)$ and $(2, 3)$ and vertices $(2, 0)$ and $(2, 4)$

53. Hyperbola with foci $(2, 0)$ and $(2, 4)$ and vertices $(2, 1)$ and $(2, 3)$

In exercises 54–58, identify the conic section and find each vertex, focus and directrix.

54. $y = 3(x - 2)^2 + 1$

55. $\dfrac{(x + 1)^2}{9} + \dfrac{(y - 3)^2}{25} = 1$

56. $\dfrac{x^2}{9} - \dfrac{(y + 2)^2}{4} = 1$

57. $(x - 1)^2 + y = 4$

58. $(x - 1)^2 + 4y^2 = 4$

59. A parabolic satellite dish has the shape $y = \frac{1}{2}x^2$. Where should the microphone be placed?

60. If a hyperbolic mirror is in the shape of the top half of $(y + 2)^2 - \dfrac{x^2}{3} = 1$, to which point will light rays following the path $y = cx$ $(y < 0)$ reflect?

In exercises 61–64, find a polar equation and graph the conic section with focus $(0, 0)$ and the given directrix and eccentricity.

61. directrix $x = 3$, $e = 0.8$ **62.** directrix $y = 3$, $e = 1$

63. directrix $y = 2$, $e = 1.4$ **64.** directrix $x = 1$, $e = 2$

In exercises 65 and 66, find parametric equations of the conic sections.

65. $\dfrac{(x + 1)^2}{9} + \dfrac{(y - 3)^2}{25} = 1$ **66.** $\dfrac{x^2}{9} - \dfrac{(y + 2)^2}{4} = 1$

67. (exploratory) Sketch several polar graphs of the form $r = 1 + a \cos \theta$ and $r = 1 + a \sin \theta$ using some constants a that are positive and some that are negative, greater than 1, equal to 1 and less than 1 (for example, $a = -2$, $a = -1$, $a = -1/2$, $a = 1/2$, $a = 1$ and $a = 2$). Discuss all patterns you find.

Appendix   A
Proofs of Selected Theorems

In this appendix, we provide the proofs of selected theorems from the body of the text. These are results that were not proven in the body of the text for one reason or another.

The first several results require the formal (ε-δ) definition of limit, which was not discussed until the optional section 1.5. Given what we've done in section 1.5, we are now in a position to prove these results. The first of these results concerns our routine rules for calculating limits and appeared as Theorem 2.3 in section 1.2.

Theorem 0.1 Suppose that $\lim_{x \to a} f(x)$ and $\lim_{x \to a} g(x)$ both exist and equal L_1 and L_2, respectively and let c be any constant. The following then apply:

(i) $\lim_{x \to a} [cf(x)] = c \lim_{x \to a} f(x) = cL_1,$

(ii) $\lim_{x \to a} [f(x) \pm g(x)] = \lim_{x \to a} f(x) \pm \lim_{x \to a} g(x) = L_1 \pm L_2,$

(iii) $\lim_{x \to a} [f(x)\, g(x)] = \left[\lim_{x \to a} f(x)\right]\left[\lim_{x \to a} g(x)\right] = L_1 L_2$ and

(iv) $\lim_{x \to a} \dfrac{f(x)}{g(x)} = \dfrac{\lim_{x \to a} f(x)}{\lim_{x \to a} g(x)} = \dfrac{L_1}{L_2}$ (if $L_2 \neq 0$).

Proof. (i) Since $\lim_{x \to a} f(x) = L_1$, we know by the precise definition of limit that given any number $\varepsilon_1 > 0$, there is a number $\delta_1 > 0$ for which

$$|f(x) - L_1| < \varepsilon_1, \text{ whenever } 0 < |x - a| < \delta_1. \tag{A.1}$$

In order to show that $\lim_{x \to a} [cf(x)] = cL_1$, we need to be able to make $cf(x)$ as close to cL_1 as needed. We have

$$|cf(x) - cL_1| = |c|\,|f(x) - L_1|.$$

We already know that we can make $|f(x) - L_1|$ as small as needed. Specifically, given any number $\varepsilon_1 > 0$, there is a number $\delta_1 > 0$ for which

$$|cf(x) - cL_1| = |c|\,|f(x) - L_1| < |c|\,\varepsilon_1, \text{ whenever } 0 < |x - a| < \delta_1.$$

Taking $\varepsilon_1 = \frac{\varepsilon}{|c|}$ and $\delta = \delta_1$, we get

$$|cf(x) - cL_1| < |c|\,\varepsilon_1 = |c|\,\frac{\varepsilon}{|c|} = \varepsilon, \text{whenever } 0 < |x - a| < \delta.$$

This says that $\lim_{x \to a} [cf(x)] = c\,L_1$, as desired.

(ii) Likewise, since $\lim_{x \to a} g(x) = L_2$, we know that given any number $\varepsilon_2 > 0$, there is a number $\delta_2 > 0$ for which

$$|g(x) - L_2| < \varepsilon_2 \text{ whenever } 0 < |x - a| < \delta_2. \tag{A.2}$$

Now, in order to verify that

$$\lim_{x \to a} [f(x) + g(x)] = (L_1 + L_2),$$

we must show that, given any number $\varepsilon > 0$, there is a number $\delta > 0$ such that

$$|[f(x) + g(x)] - (L_1 + L_2)| < \varepsilon, \text{whenever } 0 < |x - a| < \delta.$$

Notice that

$$|[f(x) + g(x)] - (L_1 + L_2)| = |[f(x) - L_1] + [g(x) - L_2]|$$
$$\leq |f(x) - L_1| + |g(x) - L_2|, \qquad (A.3)$$

by the triangle inequality. Of course, both terms on the right-hand side of (A.3) can be made arbitrarily small, from (A.1) and (A.2). In particular, if we take $\varepsilon_1 = \varepsilon_2 = \frac{\varepsilon}{2}$, then as long as

$$0 < |x - a| < \delta_1 \text{ and } 0 < |x - a| < \delta_2,$$

we get from (A.1), (A.2) and (A.3) that

$$|[f(x) + g(x)] - (L_1 + L_2)| \leq |f(x) - L_1| + |g(x) - L_2|$$
$$< \frac{\varepsilon}{2} + \frac{\varepsilon}{2} = \varepsilon,$$

as desired. Of course, this will happen if we take

$$0 < |x - a| < \delta = \min\{\delta_1, \delta_2\},$$

where taking $\delta = \min\{\delta_1, \delta_2\}$ simply means to pick δ to be the smaller of δ_1 and δ_2. (Recognize that if $0 < |x - a| < \delta = \min\{\delta_1, \delta_2\}$, then $0 < |x - a| < \delta_1$ **and** $0 < |x - a| < \delta_2$.)

(iii) In this case, we need to show that for any given $\varepsilon > 0$, we can find a $\delta > 0$, such that

$$|f(x)\, g(x) - L_1 L_2| < \varepsilon, \text{ whenever } 0 < |x - a| < \delta.$$

The object, then, is to make $|f(x)\, g(x) - L_1 L_2|$ as small as needed. Notice that we have

$$|f(x)\, g(x) - L_1 L_2| = |f(x)\, g(x) - g(x)\, L_1 + g(x)\, L_1 - L_1 L_2|$$
$$= |[f(x) - L_1]\, g(x) + L_1\, [g(x) - L_2]|$$
$$\leq |f(x) - L_1|\, |g(x)| + |L_1|\, |g(x) - L_2|, \qquad (A.4)$$

by the triangle inequality. Now, notice that we can make $|f(x) - L_1|$ and $|g(x) - L_2|$ as small as we like. If we make both of the terms in (A.4) less than $\frac{\varepsilon}{2}$, then the sum will be less that ε, as desired. In particular, we know that there is a number $\delta_2 > 0$, such that

$$|g(x) - L_2| < \frac{\varepsilon}{2\, |L_1|}, \text{ whenever } 0 < |x - a| < \delta_2,$$

assuming $L_1 \neq 0$, so that

$$|L_1|\, |g(x) - L_2| < |L_1|\, \frac{\varepsilon}{2\, |L_1|} = \frac{\varepsilon}{2}.$$

If $L_1 = 0$, then,

$$|L_1|\, |g(x) - L_2| = 0 < \frac{\varepsilon}{2}.$$

So, no matter the value of L_1, we have that

$$|L_1|\, |g(x) - L_2| < \frac{\varepsilon}{2}, \text{ whenever } 0 < |x - a| < \delta_2. \qquad (A.5)$$

Notice that for the first term in (A.4), we have a slightly more complicated job. Here, we must also estimate the size of $|g(x)|$. Notice that

$$|g(x)| = |g(x) - L_2 + L_2| \leq |g(x) - L_2| + |L_2|. \qquad (A.6)$$

Since $\lim_{x \to a} g(x) = L_2$, there is a number $\delta_3 > 0$, such that

$$|g(x) - L_2| < 1, \text{ whenever } 0 < |x - a| < \delta_3.$$

From (A.6), we now have that

$$|g(x)| \leq |g(x) - L_2| + |L_2| < 1 + |L_2|.$$

Returning to the first term in (A.4), we have that for $0 < |x - a| < \delta_3$,

$$|f(x) - L_1|\, |g(x)| < |f(x) - L_1|\, (1 + |L_2|). \tag{A.7}$$

Now, since $\lim_{x \to a} f(x) = L_1$, given any $\varepsilon > 0$, there is a number $\delta_1 > 0$ such that for $0 < |x - a| < \delta_1$,

$$|f(x) - L_1| < \frac{\varepsilon}{2(1 + |L_2|)}.$$

From (A.7), we then have

$$\begin{aligned}
|f(x) - L_1|\, |g(x)| &< |f(x) - L_1|\, (1 + |L_2|) \\
&< \frac{\varepsilon}{2(1 + |L_2|)}\, (1 + |L_2|) \\
&= \frac{\varepsilon}{2},
\end{aligned}$$

whenever $0 < |x - a| < \delta_1$ **and** $0 < |x - a| < \delta_3$. Together with (A.4) and (A.5), this tells us that for $\delta = \min\{\delta_1, \delta_2, \delta_3\}$, if $0 < |x - a| < \delta$, then

$$\begin{aligned}
|f(x)\, g(x) - L_1 L_2| &\leq |f(x) - L_1|\, |g(x)| + |L_1|\, |g(x) - L_2| \\
&< \frac{\varepsilon}{2} + \frac{\varepsilon}{2} = \varepsilon,
\end{aligned}$$

which proves (iii).

(iv) We first show that

$$\lim_{x \to a} \frac{1}{g(x)} = \frac{1}{L_2},$$

for $L_2 \neq 0$. Notice that in this case, we need to show that we can make $\left|\frac{1}{g(x)} - \frac{1}{L_2}\right|$ as small as possible. We have

$$\left|\frac{1}{g(x)} - \frac{1}{L_2}\right| = \left|\frac{L_2 - g(x)}{L_2\, g(x)}\right|. \tag{A.8}$$

Of course since $\lim_{x \to a} g(x) = L_2$, we can make the numerator of the fraction on the right-hand side as small as needed. We must also consider what happens to the denominator, though. Recall that given any $\varepsilon_2 > 0$, there is a $\delta_2 > 0$ such that

$$|g(x) - L_2| < \varepsilon_2, \quad \text{whenever} \ \ 0 < |x - a| < \delta_2.$$

In particular, for $\varepsilon_2 = \frac{|L_2|}{2}$, this says that

$$|g(x) - L_2| < \frac{|L_2|}{2}.$$

Notice that we can now say that

$$|L_2| = |L_2 - g(x) + g(x)| \leq |L_2 - g(x)| + |g(x)| < \frac{|L_2|}{2} + |g(x)|.$$

Subtracting $\frac{|L_2|}{2}$ from both sides now gives us

$$\frac{|L_2|}{2} < |g(x)|,$$

so that

$$\frac{2}{|L_2|} > \frac{1}{|g(x)|}.$$

From (A.8), we now have

$$\left| \frac{1}{g(x)} - \frac{1}{L_2} \right| = \left| \frac{L_2 - g(x)}{L_2\, g(x)} \right| < \frac{2\, |L_2 - g(x)|}{L_2^2}. \tag{A.9}$$

Further, given any $\varepsilon > 0$, there is a $\delta_3 > 0$ so that

$$|L_2 - g(x)| < \frac{\varepsilon\, L_2^2}{2}, \quad \text{whenever } 0 < |x - a| < \delta_3.$$

From (A.9), we now have that for $\delta = \min\{\delta_2, \delta_3\}$,

$$\left| \frac{1}{g(x)} - \frac{1}{L_2} \right| < \frac{2\, |L_2 - g(x)|}{L_2^2} < \varepsilon,$$

whenever $0 < |x - a| < \delta$, as desired. We have now established that

$$\lim_{x \to a} \frac{1}{g(x)} = \frac{1}{L_2}.$$

From (iii), we now have that

$$\lim_{x \to a} \frac{f(x)}{g(x)} = \lim_{x \to a} \left[f(x)\, \frac{1}{g(x)} \right] = \left[\lim_{x \to a} f(x) \right] \left[\lim_{x \to a} \frac{1}{g(x)} \right]$$
$$= L_1 \left(\frac{1}{L_2} \right) = \frac{L_1}{L_2},$$

which proves the last part of the theorem.

■

The following result appeared as Theorem 2.5 in section 1.2.

Theorem 0.2   (i) For any integer $n > 0$, and any real number a,

$$\lim_{x \to a} \sqrt[n]{x} = \sqrt[n]{a},$$

where for n even, we must assume that $a > 0$. (ii) Further, suppose that $\lim_{x \to a} f(x) = L$. Then,

$$\lim_{x \to a} \sqrt[n]{f(x)} = \sqrt[n]{\lim_{x \to a} f(x)} = \sqrt[n]{L},$$

where for n even, we assume that $L > 0$.

Proof. (i) Given any number $\varepsilon > 0$, we must find a number $\delta > 0$, such that

$$\left| \sqrt[n]{x} - \sqrt[n]{a} \right| < \varepsilon, \quad \text{whenever } 0 < |x - a| < \delta.$$

Notice that this is equivalent to

$$\sqrt[n]{a} - \varepsilon < \sqrt[n]{x} < \sqrt[n]{a} + \varepsilon.$$

Raising all terms to the nth power, we now have

$$\left(\sqrt[n]{a} - \varepsilon \right)^n < x < \left(\sqrt[n]{a} + \varepsilon \right)^n$$

and subtracting a from all terms gives us

$$\left(\sqrt[n]{a} - \varepsilon\right)^n - a < x - a < \left(\sqrt[n]{a} + \varepsilon\right)^n - a. \tag{A.10}$$

Remember that $\varepsilon > 0$ is taken to be a small positive number, so we further assume that $\varepsilon < \sqrt[n]{a}$. Observe that this says that $0 < \sqrt[n]{a} - \varepsilon < \sqrt[n]{a}$. Now, take $\delta = \min\left\{a - \left(\sqrt[n]{a} - \varepsilon\right)^n, \left(\sqrt[n]{a} + \varepsilon\right)^n - a\right\}$. Then $0 < |x - a| < \delta$ is equivalent to $-\delta < x - a < \delta$, from which it follows that

$$-\left[a - \left(\sqrt[n]{a} - \varepsilon\right)^n\right] \le -\delta < x - a < \delta \le \left(\sqrt[n]{a} + \varepsilon\right)^n - a,$$

which is equivalent to (A.10).

(ii) Since $\lim_{x \to a} f(x) = L$, we know that given any number $\varepsilon_1 > 0$, there is a number $\delta_1 > 0$, so that

$$|f(x) - L| < \varepsilon_1, \quad \text{whenever } 0 < |x - a| < \delta_1.$$

To show that $\lim_{x \to a} \sqrt[n]{f(x)} = \sqrt[n]{L}$, we need to show that given any $\varepsilon > 0$, there is a $\delta > 0$ such that

$$\left|\sqrt[n]{f(x)} - \sqrt[n]{L}\right| < \varepsilon, \quad \text{whenever } 0 < |x - a| < \delta.$$

Notice that this is equivalent to having

$$\sqrt[n]{L} - \varepsilon < \sqrt[n]{f(x)} < \sqrt[n]{L} + \varepsilon.$$

Now, raising all sides to the nth power, we have

$$\left(\sqrt[n]{L} - \varepsilon\right)^n < f(x) < \left(\sqrt[n]{L} + \varepsilon\right)^n.$$

Subtracting L from all terms gives us

$$\left(\sqrt[n]{L} - \varepsilon\right)^n - L < f(x) - L < \left(\sqrt[n]{L} + \varepsilon\right)^n - L.$$

Since ε is taken to be small, we now assume that $\varepsilon < \sqrt[n]{L}$. Observe that in this case, $0 < \sqrt[n]{L} - \varepsilon < \sqrt[n]{L}$. Let $\varepsilon_1 = \min\left\{\left(\sqrt[n]{L} + \varepsilon\right)^n - L, L - \left(\sqrt[n]{L} - \varepsilon\right)^n\right\}$. Then, since $\lim_{x \to a} f(x) = L$, we know that there is a number $\delta > 0$, so that

$$-\varepsilon_1 < f(x) - L < \varepsilon_1, \quad \text{whenever } 0 < |x - a| < \delta.$$

It then follows that

$$\left(\sqrt[n]{L} - \varepsilon\right)^n - L \le -\varepsilon_1 < f(x) - L < \varepsilon_1 \le \left(\sqrt[n]{L} + \varepsilon\right)^n - L,$$

whenever $0 < |x - a| < \delta$. Reversing the above sequence of steps gives us $\lim_{x \to a} \sqrt[n]{f(x)} = \sqrt[n]{L}$, as desired.   ∎

The following result appeared in section 1.2, as Theorem 2.7.

Theorem 0.3   (Squeeze Theorem) Suppose that

$$f(x) \le g(x) \le h(x) \tag{A.11}$$

for all x in some interval (c, d), except possibly at the point $a \in (c, d)$ and that

$$\lim_{x \to a} f(x) = \lim_{x \to a} h(x) = L,$$

for some number L. Then, it follows that

$$\lim_{x \to a} g(x) = L, \text{also.}$$

Proof. To show that $\lim_{x \to a} g(x) = L$, we must prove that given any $\varepsilon > 0$, there is a $\delta > 0$, such that

$$|g(x) - L| < \varepsilon, \quad \text{whenever } 0 < |x - a| < \delta.$$

Since $\lim_{x \to a} f(x) = L$, we have that given any $\varepsilon > 0$, there is a $\delta_1 > 0$, such that

$$|f(x) - L| < \varepsilon, \quad \text{whenever } 0 < |x - a| < \delta_1.$$

Likewise, since $\lim_{x \to a} h(x) = L$, we have that given any $\varepsilon > 0$, there is a $\delta_2 > 0$, such that

$$|h(x) - L| < \varepsilon, \quad \text{whenever } 0 < |x - a| < \delta_2.$$

Now, choose $\delta = \min\{\delta_1, \delta_2\}$. Then, if $0 < |x - a| < \delta$, it follows that $0 < |x - a| < \delta_1$ **and** $0 < |x - a| < \delta_2$, so that

$$|f(x) - L| < \varepsilon \text{ and } |h(x) - L| < \varepsilon.$$

Equivalently, we can say that

$$L - \varepsilon < f(x) < L + \varepsilon \text{ and } L - \varepsilon < h(x) < L + \varepsilon. \tag{A.12}$$

It now follows from (A.11) and (A.12) that if $0 < |x - a| < \delta$, then

$$L - \varepsilon < f(x) \le g(x) \le h(x) < L + \varepsilon,$$

which gives us

$$L - \varepsilon < g(x) < L + \varepsilon$$

or $|g(x) - L| < \varepsilon$, and it follows that $\lim_{x \to a} g(x) = L$, as desired. ∎

In Chapter 7, we proved l'Hôpital's rule only for a special case. Here, we present a general proof for the $\frac{0}{0}$ case. First, we need the following generalization of the Mean Value Theorem.

Theorem 0.4   (Generalized Mean Value Theorem) Suppose that f and g are continuous on the interval $[a, b]$ and differentiable on the interval (a, b) and that $g'(x) \ne 0$, for all x on (a, b). Then, there is a number $z \in (a, b)$, such that

$$\frac{f(b) - f(a)}{g(b) - g(a)} = \frac{f'(z)}{g'(z)}.$$

Notice that the Mean Value Theorem is simply the special case where $g(x) = x$.

Proof. First, observe that since $g'(x) \ne 0$, for all x on (a, b), we must have that $g(b) - g(a) \ne 0$. This follows from Rolle's Theorem, since if $g(a) = g(b)$, there would be some number $c \in (a, b)$ for which $g'(c) = 0$. Now, define

$$h(x) = [f(b) - f(a)] \, g(x) - [g(b) - g(a)] \, f(x).$$

Notice that h is continuous on $[a, b]$ and differentiable on (a, b), since both f and g are continuous on $[a, b]$ and differentiable on (a, b). Further, we have

$$
\begin{aligned}
h(a) &= [f(b) - f(a)]\, g(a) - [g(b) - g(a)]\, f(a) \\
&= f(b)\, g(a) - g(b)\, f(a)
\end{aligned}
$$

and

$$
\begin{aligned}
h(b) &= [f(b) - f(a)]\, g(b) - [g(b) - g(a)]\, f(b) \\
&= g(a)\, f(b) - f(a)\, g(b),
\end{aligned}
$$

so that $h(a) = h(b)$. In view of this, Rolle's Theorem says that there must be a number $z \in (a, b)$ for which

$$
0 = h'(z) = [f(b) - f(a)]\, g'(z) - [g(b) - g(a)]\, f'(z)
$$

or

$$
\frac{f(b) - f(a)}{g(b) - g(a)} = \frac{f'(z)}{g'(z)},
$$

as desired.

∎

We can now give a general proof of l'Hôpital's rule for the $\frac{0}{0}$ case. The proof of the $\frac{\infty}{\infty}$ case can be found in a more advanced text.

Theorem 0.5   (l'Hôpital's Rule) Suppose that f and g are differentiable on the interval (a, b), except possibly at some fixed point $c \in (a, b)$ and that $g'(x) \neq 0$, on (a, b), except possibly at c. Suppose further that $\displaystyle \lim_{x \to c} \frac{f(x)}{g(x)}$ has the indeterminate form $\frac{0}{0}$ or $\frac{\infty}{\infty}$ and that $\displaystyle \lim_{x \to c} \frac{f'(x)}{g'(x)} = L$ (or $\pm\infty$). Then,

$$
\lim_{x \to c} \frac{f(x)}{g(x)} = \lim_{x \to c} \frac{f'(x)}{g'(x)}.
$$

Proof. ($\frac{0}{0}$ case) In this case, we have that $\displaystyle \lim_{x \to c} f(x) = \lim_{x \to c} g(x) = 0$. Define

$$
F(x) = \begin{cases} f(x) & \text{if } x \neq c \\ 0 & \text{if } x = c \end{cases} \quad \text{and} \quad G(x) = \begin{cases} g(x) & \text{if } x \neq c \\ 0 & \text{if } x = c \end{cases}.
$$

Notice that

$$
\lim_{x \to c} F(x) = \lim_{x \to c} f(x) = 0 = F(c)
$$

and

$$
\lim_{x \to c} G(x) = \lim_{x \to c} g(x) = 0 = G(c),
$$

so that both F and G are continuous on all of (a, b). Further, observe that for $x \neq c$, $F'(x) = f'(x)$ and $G'(x) = g'(x)$ and so, both F and G are differentiable on each of the intervals (a, c) and (c, b). We first consider the interval (c, b). Notice that F and G are continuous on $[c, b]$ and differentiable on (c, b) and so, by the Generalized Mean Value

Theorem, for any $x \in (c, b)$, we have that there is some number z, with $c < z < x$, for which

$$\frac{F'(z)}{G'(z)} = \frac{F(x) - F(c)}{G(x) - G(c)} = \frac{F(x)}{G(x)} = \frac{f(x)}{g(x)},$$

where we have used the fact that $F(c) = G(c) = 0$. Notice that as $x \to c^+$, $z \to c^+$, also, since $c < z < x$. Taking the limit as $x \to c^+$, we now have

$$\lim_{x \to c^+} \frac{f(x)}{g(x)} = \lim_{z \to c^+} \frac{F'(z)}{G'(z)} = \lim_{z \to c^+} \frac{f'(z)}{g'(z)} = L.$$

Similarly, by focusing on the interval (a, c), we can show that $\lim\limits_{x \to c^-} \frac{f(x)}{g(x)} = L$, which proves that $\lim\limits_{x \to c} \frac{f(x)}{g(x)} = L$ (since both one-sided limits agree). ∎

The following theorem corresponds to Theorem 6.1 in section 8.6.

Theorem 0.6   Given any power series, $\sum\limits_{k=0}^{\infty} b_k (x - c)^k$, there are exactly three possibilities:

(i) the series converges for **all** $x \in (-\infty, \infty)$ and the radius of convergence is $r = \infty$;

(ii) the series converges **only** for $x = c$ (and diverges for all other values of x) and the radius of convergence is $r = 0$ or

(iii) the series converges for $x \in (c - r, c + r)$ and diverges for $x < c - r$ and $x > c + r$, for some number r with $0 < r < \infty$.

In order to prove the theorem, we first introduce and prove two simpler results.

Theorem 0.7   (i) If the power series $\sum\limits_{k=0}^{\infty} b_k x^k$ converges for $x = a \neq 0$, then it also converges for all x with $|x| < |a|$.

(ii) If the power series $\sum\limits_{k=0}^{\infty} b_k x^k$ diverges for $x = d$, then it also diverges for all x with $|x| > |d|$.

Proof.   (i) Suppose that $\sum\limits_{k=0}^{\infty} b_k a^k$ converges. Then, by Theorem 2.2 in section 8.2, $\lim\limits_{k \to \infty} b_k a^k = 0$. For this to occur, we must be able to make $\left| b_k a^k \right|$ as small as desired, just by making k sufficiently large. In particular, there must be a number $N > 0$, such that $\left| b_k a^k \right| < 1$, for all $k > N$. So, for $k > N$, we must have

$$\left| b_k x^k \right| = \left| b_k a^k \left(\frac{x^k}{a^k} \right) \right| = \left| b_k a^k \right| \left| \frac{x}{a} \right|^k < \left| \frac{x}{a} \right|^k.$$

If $|x| < |a|$, then $\left| \frac{x}{a} \right| < 1$ and so, $\sum\limits_{k=0}^{\infty} \left| \frac{x}{a} \right|^k$ is a convergent geometric series. It then follows from the Comparison Test that $\sum\limits_{k=0}^{\infty} \left| b_k x^k \right|$ converges and hence, $\sum\limits_{k=0}^{\infty} b_k x^k$ converges absolutely.

(ii) Suppose that $\sum_{k=0}^{\infty} b_k \, d^k$ diverges. Notice that if x is any number with $|x| > |d|$, then $\sum_{k=0}^{\infty} b_k \, x^k$ must diverge, since if it converged, we would have by part (i) that $\sum_{k=0}^{\infty} b_k \, d^k$ would also converge, which contradicts our assumption. ∎

Next, we state and prove a slightly simpler version of Theorem 0.6.

Theorem 0.8   Given any power series, $\sum_{k=0}^{\infty} b_k \, x^k$, there are exactly three possibilities:
  (i) the series converges for **all** $x \in (-\infty, \infty)$ and the radius of convergence is $r = \infty$;
  (ii) the series converges **only** for $x = 0$ (and diverges for all other values of x) and the radius of convergence is $r = 0$ or
  (iii) the series converges for $x \in (-r, r)$ and diverges for $x < -r$ and $x > r$, for some number r with $0 < r < \infty$.

Proof. If neither (i) nor (ii) is true, then there must be nonzero numbers a and d such that the series converges for $x = a$ and diverges for $x = d$. From Theorem 0.7, observe that $\sum_{k=0}^{\infty} b_k \, x^k$ diverges for all values of x with $|x| > |d|$. Define the set S to be the set of all values of x for which the series converges. Since the series converges for $x = a$, S is nonempty. Further, $|d|$ is an upper bound on S, since the series diverges for all values of x with $|x| > |d|$. By the Completeness Axiom (see section 8.1), S must have a least upper bound r. So, if $|x| > r$, then $\sum_{k=0}^{\infty} b_k \, x^k$ diverges. Further, if $|x| < r$, then $|x|$ is not an upper bound for S and there must be a number t in S with $|x| < t$. Since $t \in S$, $\sum_{k=0}^{\infty} b_k \, t^k$ converges and by Theorem 0.7, $\sum_{k=0}^{\infty} b_k \, x^k$ converges for $|x| < t \leq |t|$. This proves the result. ∎

We can now prove the original result (Theorem 0.6).

Proof. Let $t = x - c$ and the power series $\sum_{k=0}^{\infty} b_k \, (x - c)^k$ becomes simply $\sum_{k=0}^{\infty} b_k \, t^k$. By Theorem 0.8, we know that either the series converges for all t (i.e., for all x) or only for $t = 0$ (i.e., only for $x = c$) or there is a number $r > 0$ such that the series converges for $|t| < r$ (i.e., for $|x - c| < r$) and diverges for $|t| > r$ (i.e., for $|x - c| > r$). This proves the original result. ∎

Appendix   B
Table of Integrals

Forms involving $a + bu$

1. $\displaystyle \int \frac{1}{a + bu}\, du = \frac{1}{b} \ln|a + bu| + c$

2. $\displaystyle \int \frac{u}{a + bu}\, du = \frac{1}{b^2}\left(a + bu - a \ln|a + bu|\right) + c$

3. $\displaystyle \int \frac{u^2}{a + bu}\, du = \frac{1}{2b^3}\left[(a + bu)^2 - 4a(a + bu) + 2a^2 \ln|a + bu|\right] + c$

4. $\displaystyle \int \frac{1}{u(a + bu)}\, du = \frac{1}{a} \ln\left|\frac{u}{a + bu}\right| + c$

5. $\displaystyle \int \frac{1}{u^2(a + bu)}\, du = \frac{b}{a^2} \ln\left|\frac{a + bu}{u}\right| - \frac{1}{au} + c$

Forms involving $(a + bu)^2$

6. $\displaystyle \int \frac{1}{(a + bu)^2}\, du = \frac{-1}{b(a + bu)} + c$

7. $\displaystyle \int \frac{u}{(a + bu)^2}\, du = \frac{1}{b^2}\left(\frac{a}{a + bu} + \ln|a + bu|\right) + c$

8. $\displaystyle \int \frac{u^2}{(a + bu)^2}\, du = \frac{1}{b^3}\left(a + bu - \frac{a^2}{a + bu} - 2a \ln|a + bu|\right) + c$

9. $\displaystyle \int \frac{1}{u(a + bu)^2}\, du = \frac{1}{a(a + bu)} + \frac{1}{a^2} \ln\left|\frac{u}{a + bu}\right| + c$

10. $\displaystyle \int \frac{1}{u^2(a + bu)^2}\, du = \frac{2b}{a^3} \ln\left|\frac{a + bu}{u}\right| - \frac{a + 2bu}{a^2u(a + bu)} + c$

Forms involving $\sqrt{a + bu}$

11. $\displaystyle \int u\sqrt{a + bu}\, du = \frac{2}{15b^2}(3bu - 2a)(a + bu)^{3/2} + c$

12. $\displaystyle \int u^2\sqrt{a + bu}\, du = \frac{2}{105b^3}(15b^2u^2 - 12abu + 8a^2)(a + bu)^{3/2} + c$

13. $\displaystyle \int u^n\sqrt{a + bu}\, du = \frac{2}{b(2n + 3)}u^n(a + bu)^{3/2} - \frac{2na}{b(2n + 3)}\int u^{n-1}\sqrt{a + bu}\, du$

14. $\displaystyle \int \frac{\sqrt{a + bu}}{u}\, du = 2\sqrt{a + bu} + a\int \frac{1}{u\sqrt{a + bu}}\, du$

15. $\displaystyle\int \frac{\sqrt{a+bu}}{u^n}\,du = \frac{-1}{a(n-1)}\frac{(a+bu)^{3/2}}{u^{n-1}} - \frac{(2n-5)b}{2a(n-1)}\int \frac{\sqrt{a+bu}}{u^{n-1}}\,du, n \neq 1$

16a. $\displaystyle\int \frac{1}{u\sqrt{a+bu}}\,du = \frac{1}{\sqrt{a}}\ln\left|\frac{\sqrt{a+bu}-\sqrt{a}}{\sqrt{a+bu}+\sqrt{a}}\right| + c, a > 0$

16b. $\displaystyle\int \frac{1}{u\sqrt{a+bu}}\,du = \frac{2}{\sqrt{-a}}\tan^{-1}\sqrt{\frac{a+bu}{-a}} + c, a < 0$

17. $\displaystyle\int \frac{1}{u^n\sqrt{a+bu}}\,du = \frac{-1}{a(n-1)}\frac{\sqrt{a+bu}}{u^{n-1}} - \frac{(2n-3)b}{2a(n-1)}\int \frac{1}{u^{n-1}\sqrt{a+bu}}\,du, n \neq 1$

18. $\displaystyle\int \frac{u}{\sqrt{a+bu}}\,du = \frac{2}{3b^2}(bu-2a)\sqrt{a+bu} + c$

19. $\displaystyle\int \frac{u^2}{\sqrt{a+bu}}\,du = \frac{2}{15b^3}(3b^2u^2 - 4abu + 8a^2)\sqrt{a+bu} + c$

20. $\displaystyle\int \frac{u^n}{\sqrt{a+bu}}\,du = \frac{2}{(2n+1)b}u^n\sqrt{a+bu} - \frac{2na}{(2n+1)b}\int \frac{u^{n-1}}{\sqrt{a+bu}}\,du$

Forms involving $\sqrt{a^2+u^2}, a > 0$

21. $\displaystyle\int \sqrt{a^2+u^2}\,du = \tfrac{1}{2}u\sqrt{a^2+u^2} + \tfrac{1}{2}a^2\ln\left|u+\sqrt{a^2+u^2}\right| + c$

22. $\displaystyle\int u^2\sqrt{a^2+u^2}\,du = \tfrac{1}{8}u(a^2+2u^2)\sqrt{a^2+u^2} - \tfrac{1}{8}a^4\ln\left|u+\sqrt{a^2+u^2}\right| + c$

23. $\displaystyle\int \frac{\sqrt{a^2+u^2}}{u}\,du = \sqrt{a^2+u^2} - a\ln\left|\frac{a+\sqrt{a^2+u^2}}{u}\right| + c$

24. $\displaystyle\int \frac{\sqrt{a^2+u^2}}{u^2}\,du = \ln\left|u+\sqrt{a^2+u^2}\right| - \frac{\sqrt{a^2+u^2}}{u} + c$

25. $\displaystyle\int \frac{1}{\sqrt{a^2+u^2}}\,du = \ln\left|u+\sqrt{a^2+u^2}\right| + c$

26. $\displaystyle\int \frac{u^2}{\sqrt{a^2+u^2}}\,du = \frac{1}{2}u\sqrt{a^2+u^2} - \frac{1}{2}a^2\ln\left|u+\sqrt{a^2+u^2}\right| + c$

27. $\displaystyle\int \frac{1}{u\sqrt{a^2+u^2}}\,du = \frac{1}{a}\ln\left|\frac{u}{a+\sqrt{a^2+u^2}}\right| + c$

28. $\displaystyle\int \frac{1}{u^2\sqrt{a^2+u^2}}\,du = -\frac{\sqrt{a^2+u^2}}{a^2u} + c$

Forms involving $\sqrt{a^2-u^2}, a > 0$

29. $\displaystyle\int \sqrt{a^2-u^2}\,du = \tfrac{1}{2}u\sqrt{a^2-u^2} + \tfrac{1}{2}a^2\sin^{-1}\tfrac{u}{a} + c$

30. $\displaystyle\int u^2\sqrt{a^2-u^2}\,du = \tfrac{1}{8}u(2u^2-a^2)\sqrt{a^2-u^2} + \tfrac{1}{8}a^4\sin^{-1}\tfrac{u}{a} + c$

31. $\displaystyle\int \frac{\sqrt{a^2 - u^2}}{u}\, du = \sqrt{a^2 - u^2} - a\ln\left|\frac{a + \sqrt{a^2 - u^2}}{u}\right| + c$

32. $\displaystyle\int \frac{\sqrt{a^2 - u^2}}{u^2}\, du = -\frac{\sqrt{a^2 - u^2}}{u} - \sin^{-1}\frac{u}{a} + c$

33. $\displaystyle\int \frac{1}{\sqrt{a^2 - u^2}}\, du = \sin^{-1}\frac{u}{a} + c$

34. $\displaystyle\int \frac{1}{u\sqrt{a^2 - u^2}}\, du = -\frac{1}{a}\ln\left|\frac{a + \sqrt{a^2 - u^2}}{u}\right| + c$

35. $\displaystyle\int \frac{u^2}{\sqrt{a^2 - u^2}}\, du = -\frac{1}{2}u\sqrt{a^2 - u^2} + \frac{1}{2}a^2 \sin^{-1}\frac{u}{a} + c$

36. $\displaystyle\int \frac{1}{u^2\sqrt{a^2 - u^2}}\, du = -\frac{\sqrt{a^2 - u^2}}{a^2 u} + c$

Forms involving $\sqrt{u^2 - a^2}, a > 0$

---

37. $\displaystyle\int \sqrt{u^2 - a^2}\, du = \tfrac{1}{2}u\sqrt{u^2 - a^2} - \tfrac{1}{2}a^2 \ln\left|u + \sqrt{u^2 - a^2}\right| + c$

38. $\displaystyle\int u^2\sqrt{u^2 - a^2}\, du = \tfrac{1}{8}u(2u^2 - a^2)\sqrt{u^2 - a^2} - \tfrac{1}{8}a^4 \ln\left|u + \sqrt{u^2 - a^2}\right| + c$

39. $\displaystyle\int \frac{\sqrt{u^2 - a^2}}{u}\, du = \sqrt{u^2 - a^2} - a\sec^{-1}\frac{|u|}{a} + c$

40. $\displaystyle\int \frac{\sqrt{u^2 - a^2}}{u^2}\, du = \ln\left|u + \sqrt{u^2 - a^2}\right| - \frac{\sqrt{u^2 - a^2}}{u} + c$

41. $\displaystyle\int \frac{1}{\sqrt{u^2 - a^2}}\, du = \ln\left|u + \sqrt{u^2 - a^2}\right| + c$

42. $\displaystyle\int \frac{u^2}{\sqrt{u^2 - a^2}}\, du = \frac{1}{2}u\sqrt{u^2 - a^2} + \frac{1}{2}a^2 \ln\left|u + \sqrt{u^2 - a^2}\right| + c$

43. $\displaystyle\int \frac{1}{u\sqrt{u^2 - a^2}}\, du = \frac{1}{a}\sec^{-1}\frac{|u|}{a} + c$

44. $\displaystyle\int \frac{1}{u^2\sqrt{u^2 - a^2}}\, du = \frac{\sqrt{u^2 - a^2}}{a^2 u} + c$

Forms involving $\sqrt{2au - u^2}$

---

45. $\displaystyle\int \sqrt{2au - u^2}\, du = \frac{1}{2}(u - a)\sqrt{2au - u^2} + \frac{1}{2}a^2 \cos^{-1}\left(\frac{a - u}{a}\right) + c$

46. $\displaystyle\int u\sqrt{2au - u^2}\, du = \frac{1}{6}(2u^2 - au - 3a^2)\sqrt{2au - u^2} + \frac{1}{2}a^3 \cos^{-1}\left(\frac{a - u}{a}\right) + c$

47. $\displaystyle\int \frac{\sqrt{2au - u^2}}{u}\, du = \sqrt{2au - u^2} + a\cos^{-1}\left(\frac{a - u}{a}\right) + c$

48. $\displaystyle\int \frac{\sqrt{2au - u^2}}{u^2}\, du = -\frac{2\sqrt{2au - u^2}}{u} - \cos^{-1}\left(\frac{a - u}{a}\right) + c$

49. $\displaystyle\int \frac{1}{\sqrt{2au - u^2}}\, du = \cos^{-1}\left(\frac{a - u}{a}\right) + c$

50. $\displaystyle\int \frac{u}{\sqrt{2au - u^2}}\, du = -\sqrt{2au - u^2} + a\cos^{-1}\left(\frac{a - u}{a}\right) + c$

51. $\displaystyle\int \frac{u^2}{\sqrt{2au - u^2}}\, du = -\frac{1}{2}(u + 3a)\sqrt{2au - u^2} + \frac{3}{2}a^2 \cos^{-1}\left(\frac{a - u}{a}\right) + c$

52. $\displaystyle\int \frac{1}{u\sqrt{2au - u^2}}\, du = -\frac{\sqrt{2au - u^2}}{au} + c$

Forms involving $\sin u$ or $\cos u$

53. $\displaystyle\int \sin u\, du = -\cos u + c$

54. $\displaystyle\int \cos u\, du = \sin u + c$

55. $\displaystyle\int \sin^2 u\, du = \tfrac{1}{2}u - \tfrac{1}{2}\sin u \cos u + c$

56. $\displaystyle\int \cos^2 u\, du = \tfrac{1}{2}u + \tfrac{1}{2}\sin u \cos u + c$

57. $\displaystyle\int \sin^3 u\, du = -\tfrac{2}{3}\cos u - \tfrac{1}{3}\sin^2 u \cos u + c$

58. $\displaystyle\int \cos^3 u\, du = \tfrac{2}{3}\sin u + \tfrac{1}{3}\sin u \cos^2 u + c$

59. $\displaystyle\int \sin^n u\, du = -\tfrac{1}{n}\sin^{n-1} u \cos u + \tfrac{n-1}{n}\int \sin^{n-2} u\, du$

60. $\displaystyle\int \cos^n u\, du = \tfrac{1}{n}\cos^{n-1} u \sin u + \tfrac{n-1}{n}\int \cos^{n-2} u\, du$

61. $\displaystyle\int u \sin u\, du = \sin u - u \cos u + c$

62. $\displaystyle\int u \cos u\, du = \cos u + u \sin u + c$

63. $\displaystyle\int u^n \sin u\, du = -u^n \cos u + n\int u^{n-1} \cos u\, du + c$

64. $\displaystyle\int u^n \cos u\, du = u^n \sin u - n\int u^{n-1} \sin u\, du + c$

65. $\displaystyle\int \frac{1}{1 + \sin u}\, du = \tan u - \sec u + c$

66. $\displaystyle\int \frac{1}{1 - \sin u}\, du = \tan u + \sec u + c$

67. $\displaystyle\int \frac{1}{1+\cos u}\,du = -\cot u + \csc u + c$

68. $\displaystyle\int \frac{1}{1-\cos u}\,du = -\cot u - \csc u + c$

69. $\displaystyle\int \sin(mu)\sin(nu)\,du = \frac{\sin(m-n)u}{2(m-n)} - \frac{\sin(m+n)u}{2(m+n)} + c$

70. $\displaystyle\int \cos(mu)\cos(nu)\,du = \frac{\sin(m-n)u}{2(m-n)} + \frac{\sin(m+n)u}{2(m+n)} + c$

71. $\displaystyle\int \sin(mu)\cos(nu)\,du = \frac{\cos(n-m)u}{2(n-m)} - \frac{\cos(m+n)u}{2(m+n)} + c$

72. $\displaystyle\int \sin^m u \cos^n u\,du = -\frac{\sin^{m-1} u \cos^{n+1} u}{m+n} + \frac{m-1}{m+n}\int \sin^{m-2} u \cos^n u\,du$

Forms involving other trigonometric functions

73. $\displaystyle\int \tan u\,du = -\ln|\cos u| + c = \ln|\sec u| + c$

74. $\displaystyle\int \cot u\,du = \ln|\sin u| + c$

75. $\displaystyle\int \sec u\,du = \ln|\sec u + \tan u| + c$

76. $\displaystyle\int \csc u\,du = \ln|\csc u - \cot u| + c$

77. $\displaystyle\int \tan^2 u\,du = \tan u - u + c$

78. $\displaystyle\int \cot^2 u\,du = -\cot u - u + c$

79. $\displaystyle\int \sec^2 u\,du = \tan u + c$

80. $\displaystyle\int \csc^2 u\,du = -\cot u + c$

81. $\displaystyle\int \tan^3 u\,du = \tfrac{1}{2}\tan^2 u + \ln|\cos u| + c$

82. $\displaystyle\int \cot^3 u\,du = -\tfrac{1}{2}\cot^2 u - \ln|\sin u| + c$

83. $\displaystyle\int \sec^3 u\,du = \tfrac{1}{2}\sec u \tan u + \tfrac{1}{2}\ln|\sec u + \tan u| + c$

84. $\displaystyle\int \csc^3 u\,du = -\tfrac{1}{2}\csc u \cot u + \tfrac{1}{2}\ln|\csc u - \cot u| + c$

85. $\displaystyle\int \tan^n u\,du = \frac{1}{n-1}\tan^{n-1} u - \int \tan^{n-2} u\,du,\ n \neq 1$

86. $\int \cot^n u \, du = -\frac{1}{n-1} \cot^{n-1} u - \int \cot^{n-2} u \, du, n \neq 1$

87. $\int \sec^n u \, du = \frac{1}{n-1} \sec^{n-2} u \tan u + \frac{n-2}{n-1} \int \sec^{n-2} u \, du, n \neq 1$

88. $\int \csc^n u \, du = -\frac{1}{n-1} \csc^{n-2} u \cot u + \frac{n-2}{n-1} \int \csc^{n-2} u \, du, n \neq 1$

89. $\int \frac{1}{1 \pm \tan u} \, du = \frac{1}{2} u \pm \ln |\cos u \pm \sin u| + c$

90. $\int \frac{1}{1 \pm \cot u} \, du = \frac{1}{2} u \mp \ln |\sin u \pm \cos u| + c$

91. $\int \frac{1}{1 \pm \sec u} \, du = u + \cot u \mp \csc u + c$

92. $\int \frac{1}{1 \pm \csc u} \, du = u - \tan u \pm \sec u + c$

Forms involving inverse trigonometric functions

93. $\int \sin^{-1} u \, du = u \sin^{-1} u + \sqrt{1-u^2} + c$

94. $\int \cos^{-1} u \, du = u \cos^{-1} u - \sqrt{1-u^2} + c$

95. $\int \tan^{-1} u \, du = u \tan^{-1} u - \ln \sqrt{1+u^2} + c$

96. $\int \cot^{-1} u \, du = u \cot^{-1} u + \ln \sqrt{1+u^2} + c$

97. $\int \sec^{-1} u \, du = u \sec^{-1} u - \ln |u + \sqrt{u^2-1}| + c$

98. $\int \csc^{-1} u \, du = u \csc^{-1} u + \ln |u + \sqrt{u^2-1}| + c$

99. $\int u \sin^{-1} u \, du = \frac{1}{4}(2u^2-1) \sin^{-1} u + \frac{1}{4} u \sqrt{1-u^2} + c$

100. $\int u \cos^{-1} u \, du = \frac{1}{4}(2u^2-1) \cos^{-1} u - \frac{1}{4} u \sqrt{1-u^2} + c$

Forms involving e^u

101. $\int e^{au} \, du = \frac{1}{a} e^{au} + c$

102. $\int u e^{au} \, du = \left(\frac{1}{a} u - \frac{1}{a^2}\right) e^{au} + c$

103. $\int u^2 e^{au} \, du = \left(\frac{1}{a} u^2 - \frac{2}{a^2} u + \frac{2}{a^3}\right) e^{au} + c$

104. $\displaystyle\int u^n e^{au}\, du = \frac{1}{a} u^n e^{au} - \frac{n}{a}\int u^{n-1} e^{au}\, du$

105. $\displaystyle\int e^{au} \sin bu\, du = \frac{1}{a^2 + b^2}(a \sin bu - b \cos bu)e^{au} + c$

106. $\displaystyle\int e^{au} \cos bu\, du = \frac{1}{a^2 + b^2}(a \cos bu + b \sin bu)e^{au} + c$

Forms involving $\ln u$

107. $\displaystyle\int \ln u\, du = u \ln u - u + c$

108. $\displaystyle\int u \ln u\, du = \tfrac{1}{2} u^2 \ln u - \tfrac{1}{4} u^2 + c$

109. $\displaystyle\int u^n \ln u\, du = \frac{1}{n+1} u^{n+1} \ln u - \frac{1}{(n+1)^2} u^{n+1} + c$

110. $\displaystyle\int \frac{1}{u \ln u}\, du = \ln|\ln u| + c$

111. $\displaystyle\int (\ln u)^2\, du = u(\ln u)^2 - 2u \ln u + 2u + c$

112. $\displaystyle\int (\ln u)^n\, du = u(\ln u)^n - n\int (\ln u)^{n-1}\, du$

Forms involving hyperbolic functions

113. $\displaystyle\int \sinh u\, du = \cosh u + c$

114. $\displaystyle\int \cosh u\, du = \sinh u + c$

115. $\displaystyle\int \tanh u\, du = \ln|\cosh u| + c$

116. $\displaystyle\int \coth u\, du = \ln|\sinh u| + c$

117. $\displaystyle\int \operatorname{sech} u\, du = \tan^{-1}|\sinh u| + c$

118. $\displaystyle\int \operatorname{csch} u\, du = \ln\left|\tanh \tfrac{1}{2}u\right| + c$

119. $\displaystyle\int \operatorname{sech}^2 u\, du = \tanh u + c$

120. $\displaystyle\int \operatorname{csch}^2 u\, du = -\coth u + c$

121. $\displaystyle\int \operatorname{sech} u \tanh u\, du = -\operatorname{sech} u + c$

122. $\displaystyle\int \operatorname{csch} u \coth u \, du = -\operatorname{csch} u + c$

123. $\displaystyle\int \frac{1}{\sqrt{x^2+1}} \, dx = \sinh^{-1} x + c$

124. $\displaystyle\int \frac{1}{\sqrt{x^2-1}} \, dx = \cosh^{-1} x + c$

125. $\displaystyle\int \frac{1}{1-x^2} \, dx = \tanh^{-1} x + c$

126. $\displaystyle\int \frac{1}{|x|\sqrt{x^2+1}} \, dx = -\operatorname{csch}^{-1} x + c$

127. $\displaystyle\int \frac{1}{x\sqrt{1-x^2}} \, dx = -\operatorname{sech}^{-1} x + c$

Appendix C
Answers to Select Exercises

CHAPTER 0

Exercises 0.1 ■ page 8

5. $x < 3$ **7.** $x < \frac{11}{2}$ **9.** $x > -\frac{2}{3}$ **11.** $3 \le x < 6$
13. $-\frac{1}{2} < x < 2$ **15.** $x < -4$ or $x > 1$
17. $-2 < x < 3$ **19.** all reals **21.** $-1 < x < 7$
23. $2 < x < 4$ **25.** $x < -\frac{3}{2}$ or $x > \frac{1}{2}$
27. $x < -2$ or $x > 2$
29. $x < -4$ or $-4 < x < -1$ or $x > 2$
31. $x < -1$ or $x > 0$ **33.** $\sqrt{13}$ **35.** 4 **37.** $\sqrt{20}$
39. yes **41.** yes
43. increases by $550, 650, 750$
predict $3200 + 850 = 4050$
45. decreases by $10, 30, 50$
predict $3910 - 70 = 3840$
47. finite number of digits terminate
49. $\dfrac{\left(\frac{1}{2}\right)^7 - \left(\frac{2}{3}\right)^{12}}{\left(\frac{1}{2}\right)^7} = 0.013$
51. P: $0.551, 0.587, 0.404, 0.538, 0.605$
win%: $0.568, 0.593, 0.414, 0.556, 0.615$

Exercises 0.2 ■ page 19

5. yes **7.** yes **9.** no **11.** no **13.** 2 **15.** $\frac{1}{2}$
17. $-\frac{5}{2}$ **19.** $-\frac{5}{7}$ **21.** $(2, 5)$, $y = 2(x - 1) + 3$
23. $(2, 2)$, $y = -2(x - 1) + 4$ **25.** $(0, 1)$, $y = 1$
27. $(4, 2)$, $y = \frac{1}{2}(x - 2) + 1$
29. $(3.3, 2.3)$, $y = 1.2(x - 2.3) + 1.1$
31. parallel **33.** perpendicular **35.** neither
37. neither **39.** parallel
41. (a) $y = 2(x - 2) + 1$ (b) $y = -\frac{1}{2}(x - 2) + 1$
43. (a) $y = -1$ (b) $x = 0$
45. (a) $y = 2(x - 3) + 1$ (b) $y = -\frac{1}{2}(x - 3) + 1$
47. $y = 2(x - 1) + 1$; 7 **49.** $y = -2(x - 1) + 3$; -3
51. $y = \frac{1}{2}(x + 2) + 1$; 4 **53.** yes **55.** no **57.** yes
59. both **61.** rational **63.** neither
65. both **67.** $x \ge -2$ **69.** all reals
71. $x \ne \pm 1$ **73.** all reals **75.** $-1, 1, 11, -\frac{5}{4}$
77. $1, 2, 0, \sqrt{\frac{3}{2}}$ **79.** $50 < x$
81. $0 \le x \le$ number made **83.** no: many y's for one x
85. no: many y's for one x **87.** days 1, 9
89. constant, increasing, decreasing;
graph going down; graph going up
95. 63,000 feet **97.** 9550 rpm

Exercises 0.3 ■ page 29

5.

7.

9.

11.

13.

15.

17.

19.

21.

23.

25.

27.

29.

31.

33.

35.

37.

39.

41.   **43.**   **21.**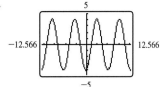

45. $x = -2, x = 2, y = -4$ **47.** $y = 4$
49. $x = -1, x = 4, y = -4$
51. $x = -1, x = 1, x = 2, y = 2$ **53.** $x = -1, y = -\frac{1}{4}$
55. $x = -2, x = 2$ **57.** $x = -5, x = 2$ **59.** none
61. $x = -2, x = -1, x = 0$ **63.** $x = -3, x = 3$

23.   **25.**

65.   **67.**

27.   **29.**

69. $y = 0$ **71.** $y = 3$ **73.** $y = 1$
75. possible answer: $-9 \le x \le 11, -17 \le x \le 23$
77. parabola $y = \frac{1}{4}x^2 + 1$ **79.** lead $v_T = \sqrt{2}$ (wood v_T)

31. $A = 3$, period $= \pi$, frequency $= \dfrac{1}{\pi}$

Exercises 0.4 ■ page 35

5. $1, 3$ **7.** $-3, 5$ **9.** $2 - \sqrt{2}, 2 + \sqrt{2}$

11. $-1 + \dfrac{\sqrt{6}}{2}, -1 - \dfrac{\sqrt{6}}{2}$ **13.** $0, 1, 2$ **15.** $-2, -1, 2$

17. $-1, 1$ **19.** $-\sqrt[3]{2}, 1$ **21.** $-1.879, 0.347, 1.532$
23. -1.325 **25.** $-2.831, -1$ **27.** 0.125
29. $-5.248, 10.006$ **31.** $2; x = 1, x \approx 1.206$
33. $1; x = 1$ **35.** $2; x = 0, x \approx 9.534$ **37.** $1; x = 0$
39. $2; x \approx -1.17, x \approx 1.17$ **45.** 3 (true); -1 (extra)
47. $-1 - 2i, -1 + 2i$ **49.** $-1 - \sqrt{6}, -1 + \sqrt{6}$

51. $1, -\dfrac{1}{2} - \dfrac{\sqrt{3}}{2}i, -\dfrac{1}{2} + \dfrac{\sqrt{3}}{2}i$ **53.** $\dfrac{1}{2} + \dfrac{\sqrt{5}}{2}, \dfrac{1}{2} - \dfrac{\sqrt{5}}{2}$

55. $-1, 0, 1$ **59.** 86.2 ft/s ≈ 58.8 mph   **61.** 81 feet

33. $A = 5$, period $= \dfrac{2\pi}{3}, f = \dfrac{3}{2\pi}$

35. $A = 3$, period $= \pi, f = \dfrac{1}{\pi}$

37. $A = 4$, period $= 2\pi, f = \dfrac{1}{2\pi}$ **43.** $\beta \approx 0.6435$

45. no   **47.** no   **49.** yes, 2π
51. $2 \tan 20° \approx 0.73$ miles
53. $100 \tan 50° \approx 119$ feet
55. $\dfrac{1}{2}$ **57.** $\dfrac{\sqrt{8}}{3}$ **59.** $-\dfrac{\sqrt{3}}{2}$

61. $3; x = 0, x \approx 1.109, x \approx 3.698$
63. $2; x \approx -1.455, \approx 1.455$ **65.** $1; x \approx 1.249$
67. $2; x = 0, x \approx 0.8777$
69. $f = \dfrac{30}{\pi}, \dfrac{170}{\sqrt{2}} \approx 120.2$ volts   **71.** \$24,000 per year

Exercises 0.5 ■ page 43

5. (a) $45°$ (b) $60°$ (c) $30°$ (d) $240°$

7. (a) π (b) $\dfrac{3\pi}{2}$ (c) $\dfrac{2\pi}{3}$ (d) $\dfrac{\pi}{6}$

Exercises 0.6 ■ page 55

5. $\dfrac{1}{8}$ **7.** $\sqrt{3}$ **9.** $\sqrt[3]{25}$ **11.** $\dfrac{1}{\sqrt[3]{16}}$ **13.** x^{-2}

15. $x^{1/2}$ **17.** $2x^{-3}$ **19.** $\frac{1}{2}x^{-1/2}$ **21.** 8 **23.** 2
25. 7.389 **27.** 1.213 **29.** 4.415 **31.** 9.900

9.   **11.**

33.

33.   **35.**

13.   **15.**

37.   **39.**

17.   **19.**

41.

43.

45.

47.

49.

51. $\frac{1}{2}\ln 2$ **53.** $-\frac{1}{2}\ln\frac{1}{2}$ **55.** $\frac{1}{2}e^4$ **57.** $\frac{1}{4}e^{7/2}$
59. $\frac{1}{2}-\frac{1}{2}\ln\frac{3}{2}$ **61.** 2   **63.** (a) 2   (b) 3   (c) -3
65. $\ln\frac{3}{4}$ **67.** 0   **69.** $\ln 12$ **71.** $2e^{(1/2\ln 3)x}$
73. $4e^{(1/2\ln 1/2)x}$ **75.** $-2e^{(\ln 3/2)x}$ **77.** 0.651
79. $1 - e \approx 0.632$

81.

$m = 2,\ b \approx 1.0986,\ a = c^b = 3$

83.

no, exponential

85. $10^{-7},10^{-8},10^{-9},\frac{1}{10}$ **87.** $10^{10.4},10^{11.9},10^{13.4},10^{1.5}\approx 31.6$
89. $10^{-4},10^{-3},10^{-2},10$ **91.** $x = 1,\frac{1}{2},\frac{1}{3},\frac{1}{k}$

Exercises 0.7 ■ page 63

5. $(f\circ g)(x) = \sqrt{x-3}+1,\ x\geq 3$
   $(g\circ f)(x) = \sqrt{x}-2,\ x\geq 2$
7. $(f\circ g)(x) = x,\ x>0$
   $(g\circ f)(x) = x,$ all reals
9. $(f\circ g)(x) = \sin^2 x + 1,$ all reals
   $(g\circ f)(x) = \sin(x^2+1),$ all reals
11. possible answer: $f(x) = \sqrt{x},\ g(x) = x^4+1$
13. possible answer: $f(x) = \dfrac{1}{x},\ g(x) = x^2+1$
15. possible answer: $f(x) = x^2+3,\ g(x) = 4x+1$
17. possible answer: $f(x) = x^3,\ g(x) = \sin x$
19. possible answer: $f(x) = \cos x,\ g(x) = 4x$
21. possible answer: $f(x) = e^x,\ g(x) = x^2+1$

23. possible answer: $f(x) = \sqrt{x},\ g(x) = e^x+1$
25. possible answer: $f(x) = \ln x,\ g(x) = 3x-5$
27.

29.

31.

33.

35.

37.

39.

41.

43. $y = (x+1)^2$, shift left one
45. $y = (x+1)^2 + 3$, shift left one, up three
47. $y = 2[(x+1)^2 + 1]$, shift left one, up one, double vertical scale
49. upside down, vertical scale doubled
51. upside down, vertical scale tripled, shifted up two
53. mirror image across y-axis
55. mirror image, shifted up one
57. mirror image across y-axis, vertical scale times $|c|$
59. oscillations dying out
63. go to 0
65. 0.739085

Exercises 0.8 ■ page 69

3. 2   **5.** 0   **7.** 0   **9.** $\frac{1}{2}$ **11.** 1   **13.** 1
15. 4.6267, 4.6417, 4.64678   **17.** 1.90626, 1.90913, 1.91010
19. 3.16732, 3.16770, 3.16784   **21.** 9.15298, 9.25344, 9.29357
23. $1,\frac{4}{3},\frac{7}{5},\frac{10}{7};\frac{3}{2}$ **25.** $0,\frac{1}{2},\frac{2}{3},\frac{3}{4};1$ **27.** $2,\frac{3}{2},\frac{4}{5},\frac{5}{10};0$
29. $0,\frac{2}{5},\frac{8}{8},\frac{18}{13};2$ **31.** $2,1,\frac{2}{3},\frac{1}{2};0$ **33.** $0,\frac{1}{2},\frac{2}{3},\frac{3}{4};1$
35. $e^{-1},e^{-2},e^{-3},e^{-4};0$
37. $\cos(1),\cos(\frac{1}{2}),\cos(\frac{1}{3}),\cos(\frac{1}{4});1$
39. $0,\dfrac{1}{\sqrt{2}},\dfrac{2}{\sqrt{5}},\dfrac{3}{\sqrt{10}};1$ **41.** $3,4,4\frac{1}{9},4\frac{1}{8};4$
43. $\displaystyle\lim_{n\to\infty} a_n = 0$

Chapter 0 Review Exercises ■ **page 71**

1. $x > 1$
3. $x < -2$ or $x > 4$
5. $-1 < x < 2$ **7.** $\sqrt{20}$ **9.** -2 **11.** parallel
13. no   **15.** $y = \frac{1}{2}(x-1)+1, y = \frac{5}{2}$
17. $y = -\frac{1}{3}(x+1)-1$ **19.** yes   **21.** $-2 \le x \le 2$
23. **25.**

27. **29.**

31. **33.**

35.

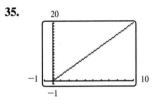

37. $x = -4, x = 2, y = -8$
39. vertical $x = -2$, horizontal $y = 4$ **41.** $-2, 5$
43. $1, 1 + \sqrt{3}, 1 - \sqrt{3}$ **45.** 3   **47.** yes
49. $50 \tan 34° = 33.7$ feet

51. (a) $\dfrac{1}{\sqrt{5}}$ (b) $\frac{1}{9}$ **53.** $\ln 2$ **55.** $\frac{1}{2}\ln\frac{8}{3}$
57. $(f \circ g)(x) = x - 1, x \ge 1$
    $(g \circ f)(x) = \sqrt{x^2 - 1}, x \le -1$ or $x \ge 1$
59. $f(x) = e^x, g(x) = 3x^2 + 2$
61. $(x-2)^2 - 3$, shift two right and three down   **63.** 2
65. (a) 1.05799   (b) 1.05806   **67.** $3, \frac{1}{2}, \frac{3}{13}, \frac{3}{22}; 0$

CHAPTER 1

Exercises 1.1 ■ **page 81**

5. (a) -2 (b) 2   (c) does not exist   (d) 1
    (e) $\frac{1}{2}$ (f) -1 (g) 3   (h) does not exist
    (i) 2   (j) 2
7.
    (a) 4   (b) 4   (c) 4   (d) 2

9.
    (a) 2   (b) -2
    (c) does not exist   (d) 1

11. 2.2247, 2.0488, 2.0049, 2.0004 → 2;
    1.7071, 1.9486, 1.9949, 1.9994 → 2
13. 0   **15.** 1   **17.** 1   **19.** no—jump   **21.** yes
23. no—vertical asymptote   **25.** no—vertical asymptote
27. does not exist; can't tell
29. **31.**

33. $m = \frac{1}{2}$ **35.** 2.7182818   **37.** 0

Exercises 1.2 ■ **page 91**

5. 1   **7.** 2   **9.** $-\frac{5}{4}$ **11.** 5   **13.** $\frac{3}{4}$ **15.** $\frac{3}{4}$
17. 1   **19.** 0   **21.** 2   **23.** 2   **25.** 9
27. $0, f(x) = -x^2, h(x) = x^2$
29. $f(x) = 0, h(x) = \sqrt{x}$
31. does not exist—undefined for $x > 4$
33. does not exist—see #31   **35.** 0   **37.** $\sqrt{\frac{1}{2}}$
39. $h(a)$ **43.** $m = \frac{1}{2}$ **47.** $f(x) = \dfrac{1}{x}, g(x) = -\dfrac{1}{x}$
49. yes   **51.** 0, does not exist
53. for $2 \le x < 3, [x] = 2$ and for $3 \le x < 4, [x] = 3$,
    so $\lim\limits_{x\to 3^-}[x] \ne \lim\limits_{x\to 3^+}[x]$.

Exercises 1.3 ■ **page 102**

5. $x = -2, x = 2$ **7.** $x = -2, x = 1, x = 4$
9. $x = -2, x = 2, x = 4$ **11.** $f(1)$ undefined
13. $f(1)$ undefined   **15.** $\lim\limits_{x\to 2} f(x) \ne f(2)$
17. $x = 1$ (removable), $x = -1$ **19.** none
21. $\dfrac{n\pi}{2}$ for odd integers n
23. $x = n\pi$ for nonzero integers n, $x = 0$ is removable
25. $x = 1$ **27.** $x = 1$ **29.** $[-3, \infty)$ **31.** $(-\infty, \infty)$
33. $(-\infty, \infty)$ **35.** $(-1, \infty)$ **37.** -700
39. $34{,}573.5 + 0.36(x - 128{,}100)$
41. $f(2) < 0, f(3) > 0, [\,2\frac{20}{32}, 2\frac{21}{32}\,]$
43. $f(-1) > 0, f(0) < 0, [\,-\frac{18}{32}, -\frac{17}{32}\,]$
45. $f(0) > 0, f(1) < 0, [\,\frac{23}{32}, \frac{24}{32}\,]$ **47.** yes   **49.** no
51. #49 is   **53.** $g(T) = 100 - 25(T - 30)$
55.
    $x = 100$,
    force that moves box

57. $M(t) =$ distance from home at time t on Monday
$T(t) =$ distance from home at time t on Tuesday
$M(1{:}59) - T(1{:}59) > 0$ and $M(7{:}13) - T(7{:}13) < 0$
59. raises at 3-month marks, $f(t) = s(t)$ when t is a multiple of 3

61. If $f(d) < 0$, then $f(x) = 0$ for some x between c and d.

Exercises 1.4 ■ page 112

5. ∞ **7.** does not exist   **9.** $-\infty$ **11.** ∞ **13.** ∞
15. $-\infty$ **17.** ∞ **19.** 1   **21.** $\frac{1}{3}$ **23.** ∞ **25.** 1
27. $\frac{3}{4}$ **29.** 0   **31.** ∞ **33.** 0   **35.** 0   **37.** ∞
39. 0   **41.** $y = -1, y = 1$ **43.** $x = -2, x = 2, y = 0$
45. $x = -2, x = 2, y = -x$ **47.** $y = x$
49. $y = 2x + 2$ **51.** $y = 0$ as $x \to \infty$, ∞ crosses
55. 30 mm, 300 mm   **57.** 40 mm, 6 mm
59. 66 rad/s $\gg \dfrac{\pi}{2}$ rad/s, no   **61.** ∞, c

63. $t = 4.6$ for A, $t = 17.8$ for B

65. no— $\dfrac{\sec x}{x}$ at $x = \dfrac{\pi}{2}$

Exercises 1.5 ■ page 125

5. $\sqrt{0.1} \approx 0.32$ **7.** 0.45   **9.** 0.39   **11.** 0.02   **13.** $\dfrac{\varepsilon}{3}$

15. $\dfrac{\varepsilon}{3}$ **17.** $\dfrac{\varepsilon}{4}$ **19.** 2ε **21.** $\min\left(1, \dfrac{\varepsilon}{3}\right)$

23. $\min\left(1, \dfrac{\varepsilon}{5}\right)$ **25.** $\dfrac{\varepsilon}{|m|}$, no   **29.** 0.02   **31.** 0.01

33. 0.0001   **35.** 12   **37.** -10 **39.** 1.15

41. $m = \sqrt[3]{\dfrac{1}{\varepsilon}}$ **43.** $m = \sqrt[k]{\dfrac{1}{\varepsilon}}$ **45.** $N = -\sqrt{\dfrac{1}{\varepsilon} - 2}$

47. $\delta = \sqrt[4]{-\dfrac{1}{N}}$ **49.** $\delta = \sqrt{\dfrac{1}{M}}$ **51.** 1   **53.** 1

55. $\min\left(1, \dfrac{\varepsilon}{10}\right)$

Exercises 1.6 ■ page 133

5. $\frac{1}{4}$; 600,000; $\dfrac{x}{\sqrt{4x^2 + 1} + 2x}$

7. $1; 10^{13}$; $\dfrac{2\sqrt{x}}{\sqrt{x + 4} + \sqrt{x + 2}}$

9. $1; 6{,}000{,}000$; $\dfrac{2x}{\sqrt{x^2 + 4} + \sqrt{x^2 + 2}}$

11. $\frac{1}{6}; 9 \times 10^{-7}$; $\dfrac{\sin^2 2x}{12x^2(1 + \cos 2x)}$ **13.** $3, \pm\infty$

15. $f(x) = 0$, $g(x) = 0.0016, -0.0159, -0.1586, -0.9997$
17. 20, 0

Chapter 1 Review Exercises ■ page 134

1. does not exist
3. (a) 1   (b) -2 (c) does not exist   (d) 0
5. $x = -1, x = 1$ **7.** $\frac{1}{4}$ **9.** does not exist
11. 0   **13.** does not exist   **15.** $\sqrt{2}$ **17.** ∞
19. $\frac{1}{3}$
21. 0   **23.** ∞ **25.** 0   **29.** $x = -3, x = 1$ (removable)
31. $x = 2$ **33.** $(-\infty, -2) \cup (-2, 3) \cup (3, \infty)$
35. $(-\infty, \infty)$ **37.** $x = 1, x = 2, y = 0$
39. $x = -1, x = 1, y = 1$ **41.** 0.4   **43.** $\dfrac{\varepsilon}{2}$ **45.** 3

CHAPTER 2

Exercises 2.1 ■ page 146

5.   **7.**

9. -1 **11.** C, B, A, D
13. (a) 6   (b) 18   (c) 8.25   (d) 14.25   (e) 10.41
     (f) 11.61   (g) 11
15. (a) -1.19 (b) -0.26 (c) -0.05 (d) 3.3
     (e) 2.4   (f) 3.5   (g) 3
17. 0   **19.** 9   **21.** $y = 2(x - 1) - 1$
23. $y = -7(x + 2) + 10$ **25.** $y = -\frac{1}{2}(x - 1) + 1$
27. $y = \frac{1}{2}(x + 2) + 1$ **29.** no: corner point   **31.** yes: 0
33. no: jump discontinuity
35. (a) 32   (b) 48   (c) 62.4   (d) 63.84   (e) 64
37. (a) 2.236   (b) 1.472   (c) 1.350   (d) 1.342
     (e) 1.341
39. -32 **41.** $\frac{1}{8}$
43. about 1.75 hours; 1.5 hours; 4 hours; rest   **47.** 15 rad/s
51. 0.359 million people per year

Exercises 2.2 ■ page 157

5. 3   **7.** 6   **9.** $\frac{3}{4}$ **11.** 3   **13.** $6x$
15. $-\dfrac{3}{(x + 1)^2}$ **17.** $\dfrac{3}{2\sqrt{3x + 1}}$ **19.** $3x^2 + 2$
21. c   **23.** a   **25.** b
27.   **29.**

31.

33. $x = -3, x = -1, x = 3$ **35.** 5
37. 10   **41.** $D_+f(0) = 3, D_-f(0) = 2$
43. $D_-f(0) \neq g'(0)$ **49.** $f(x) = 2x - 3$ for $x \leq 3$
51. $0.016s; 0.008s$

Exercises 2.3   ■   page 165

5. $3x^2 - 2$ **7.** $6x$ **9.** 0 **11.** $9t^2 - \dfrac{1}{\sqrt{t}}$

13. $-\dfrac{3}{x^2} - 8$ **15.** $-5x^{-3/2} - 2$ **17.** $3s^{1/2} + s^{-4/3}$

19. $\frac{2}{3}x^{-2/3}$ **21.** $9x^2 - \frac{3}{2}x^{1/2}$ **23.** $\frac{3}{2} - \frac{1}{2}x^{-2}$

25. $12x^2 + 6$ **27.** $30x^4 + \frac{1}{4}x^{-3/2}$ **29.** $-96t^{-5}$

31. 24 **33.** $-\frac{3}{8}x^{-3/2} - \frac{3}{8}x^{-5/2} - \frac{15}{8}x^{-7/2}$

35. $v(t) = -32t + 40$, $a(t) = -32$
37. $v(t) = \frac{1}{2}t^{-1/2} + 2$, $a(t) = -\frac{1}{4}t^{-3/2}$
39. $v(1) = 8$ (going up); $a(1) = -32$ (slowing down)
41. $v(2) = 16$ (going up); $a(2) = 20$ (speeding up)
43. $y = 2$ **45.** $y = -2$
47. $x = -1$ (peak) $x = 1$ (trough)
49. $x = 0$ (minimum)
51. (a) $f(x)$ (b) $f''(x)$ (c) $f'(x)$

53. $(-1)^{(n-1)}\dfrac{1 \cdot 3 \ldots (2n-3)}{2^n}x^{-(2n-1)/2}$ **55.** $\frac{3}{2}x^2 + 2x - 2$

57. $1 = x^0$ **59.** second; not much

61. $b > \dfrac{4}{9c^2}, \dfrac{\Delta w}{\Delta b}$ **63.** x^4 **65.** $\frac{1}{5}x^5$ **67.** $\frac{3}{2}x^2 - x$

69. $\frac{2}{3}x^{3/2}$

Exercises 2.4   ■   page 174

5. $2x(x^3 - 3x + 1) + (x^2 + 3)(3x^2 - 3)$
7. $3(x^3 - 2x^2 + x) + (3x + 4)(3x^2 - 4x + 1)$

9. $\left(\dfrac{1}{2}x^{-\frac{1}{2}} + 3\right)\left(5x^2 - \dfrac{3}{x}\right) + (\sqrt{x} + 3x)(10x + 3x^{-2})$

11. $\dfrac{3(5x + 1) - (3x - 2)5}{(5x + 1)^2} = \dfrac{13}{(5x + 1)^2}$

13. $\dfrac{1(x^2 + x + 1) - (x - 2)(2x + 1)}{(x^2 + x + 1)^2} = \dfrac{-x^2 + 4x + 3}{(x^2 + x + 1)^2}$

15. $\dfrac{(3 - 3x^{-\frac{1}{2}})(5x^2 - 2) - (3x - 6\sqrt{x})10x}{(5x - 2)^2}$

17. $\dfrac{(2x - 1)(x^2 - 5x + 1) - (x^2 - x - 2)(2x - 5)}{(x^2 - 5x + 1)^2}$

$= \dfrac{-4x^2 + 6x - 11}{(x^2 - 5x + 1)^2}$

19. $\frac{3}{2}x^{1/2} + \frac{3}{2}x^{-1/2} + x^{-3/2}$ **21.** $\frac{4}{3}x^{1/3} + 3$

23. $2x\dfrac{x^3 + 3x^2}{x^2 + 2}$

$+ (x^2 - 1)\dfrac{(3x^2 + 6x)(x^2 + 2) - (x^3 + 3x^2)(2x)}{(x^2 + 2)^2}$

25. $f'(x)g(x)h(x) + f(x)g'(x)h(x) + f(x)g(x)h'(x)$
27. $\frac{2}{3}x^{-1/3}(x^2 - 2)(x^3 - x + 1) + x^{2/3}(2x)(x^3 - x + 1)$

$+ x^{2/3}(x^2 - 2)(3x^2 - 1)$

29. $(x^3 + 4x)(x^5 - 3x^2 + 1)$

$+ (x + 1)(3x^2 + 4)(x^5 - 3x^2 + 1)$

$+ (x + 1)(x^3 + 4x)(5x^4 - 6x)$

31. $P'(t) = 0.03P(t)$; $3 - 4 = -1$
33. \$65,000 per year.

35. rate of change of $\tan \alpha$ as you move backward
37. $F'''(x) = f'''(x)g(x) + 3f''(x)g'(x)$
$+ 3f'(x)g''(x) + f(x)g'''(x)$

41. $T_c = \dfrac{8a}{27Rb}, V_c = 3nb, P_c = \dfrac{a}{27b^2}$; 647 K

Exercises 2.5   ■   page 182

5. $4\cos x - 1$ **7.** $\sec^2 x + \csc x \cot x$

9. $\cos x - x\sin x$ **11.** $2x^{-1/2} - 2\cos x$

13. $\dfrac{x\cos x - \sin x}{x^2}$ **15.** $-3\csc t \cot t + 8\csc^2 t$

17. $\dfrac{-x^2\sin x - (\cos x - 1)2x}{x^4}$ **19.** $2\cos^2 x - 2\sin^2 x$

21. $8x\tan x + 4x^2\sec^2 x$ **23.** $8\sin x \cos x$ **25.** $y = 1$

27. $y = -x + \dfrac{\pi}{2}$ **29.** -1 **31.** $\dfrac{1}{\pi^2}$

33. $y(t) = \sin t$; $x(t) = 3\cos t$, $y(t) = 3\sin t$;
$x(t) = \cos 2t$, $y(t) = \sin 2t$; $x(t) = \cos \pi t$, $y(t) = \sin \pi t$;
$x(t) = \cos t$, $y = -\sin t$;

$x(t) = \cos\left(t + \dfrac{\pi}{2}\right), y(t) = \sin\left(t + \dfrac{\pi}{2}\right)$

35. $4\cos t$ **37.** 4 amps, 2.62 amps   **41.** $-\cos x$; $-\sin x$
45. (a) 3   (b) $\frac{1}{4}$ (c) 0   (d) 1   (e) 2   (f) 0

Exercises 2.6   ■   page 189

5. $4e^x - 1$ **7.** $e^x + xe^x$ **9.** $1 + (\ln 2)2^x$
11. $2e^{x+1}$ **13.** $(\ln\frac{1}{3})(\frac{1}{3})^x$ **15.** $-(\ln 4)4^{-x+1}$

17. $\dfrac{xe^x - e^x}{x^2}$ **19.** $\dfrac{1}{x}$ **21.** $\dfrac{2}{x}$ **23.** $2e^x$

25. $2xe^{-x} - x^2e^{-x}$ **27.** $\dfrac{1 - \ln x}{x^2}$

29. $y = 3e(x - 1) + 3e$ **31.** $y = 3\ln 3(x - 1) + 3$
33. $y = 2e(x - 1) + e$ **35.** $y = x - 1$
37. $100\ln 3 = 109.86\%$ **39.** 100%
41. (a) $e^{.05} - 1 \to 5.1\%$ (b) 10.5%   (c) 22.1%
(d) 100%   (e) 171.8%

43. $-e^{-t}\cos t - e^{-t}\sin t$; vel. is 0 at $t = \dfrac{3\pi}{4}, \dfrac{7\pi}{4}, \ldots$;

max vel. at $t = \pi, 3\pi, \ldots$; min vel. at $t = 0, 2\pi, \ldots$
45. $a = 1, b = \frac{1}{2}, c = -\frac{1}{2}$

Exercises 2.7   ■   page 195

5. $6x^2(x^3 - 1)$ **7.** $-2e^{-2x}$ **9.** $x(x^2 + 4)^{-1/2}$
11. $3(3x^2 + 1)(x^3 + x - 1)^2$ **13.** $4x\cos(2x^2 + 3)$
15. $-3\sin x \cos^2 x$ **17.** $2\tan x \sec^2 x$ **19.** $2xe^{x^2}$

21. $-\dfrac{1}{x^2}e^{1/x}$ **23.** $\dfrac{2x}{x^2 + 1}$ **25.** $2x\sin 4x + 4x^2\cos 4x$

27. $12\sec^3 4x \tan 4x$ **29.** $\dfrac{2x^3\cos x^2 - 2x\sin x^2}{x^4}$

31. $\dfrac{3x^2 + 3}{x^3 + 3x}$ **33.** $x\cos x^2(\sin x^2)^{-1/2}$ **35.** $\dfrac{8}{x}(\ln x)^7$

37. $-2x\sin x^2$ **39.** $-3e^{-3x}\sin 2x + 2e^{-3x}\cos 2x$
41. $y = \frac{3}{5}(x - 3) + 5$
43. $e^{-8}(3\cos 6 - 4\cos 6) \approx 0.00134$

45. $-(2x + 1)^{-3/2}, 3(2x + 1)^{-5/2}, -15(2x + 1)^{-7/2}$,
$(-1)^{n+1}3 \cdot 5 \ldots (2n - 3)(2x + 1)^{-(2n-1)/2}$

47. $4e^{2x}, 8e^{2x}, 16e^{2x}, 2^n e^{2x}$ **49.** $\frac{1}{3}(x^2 + 3)^3$

51. $-\frac{3}{2}\cos 2x$

53. max at $x = 0$

Exercises 2.8 ■ page 202

5. $-\frac{1}{2}$ **7.** 0 **9.** $\dfrac{4 - 2xy^2}{3 + 2x^2y}$ **11.** $\dfrac{y}{16y\sqrt{xy} - x}$

13. $\dfrac{y - 4y^2}{x + 3 + 2y^3}$ **15.** $\dfrac{1 - 2xye^{x^2y}}{x^2 e^{x^2y} - e^y}$

17. $\dfrac{16x\sqrt{x + y} - 1}{1 - 2\sqrt{x + y}}$ **19.** $\dfrac{2y}{4ye^{4y} - 1}$

21. $y = \frac{1}{2}(x - 2) + 1$ **23.** $y = \frac{1}{3}(x - 2) + 1$

25. $y = -(x - 2) + 1$ **27.** $y = -\frac{9}{2}(x - 1) + 3$

29. -65 rad/s   **31.** 0.03 rad/s   **33.** 6π mm²/hr

35. $24\sqrt{101} \approx 241$ mph

37. $-\dfrac{100}{\sqrt{5}}$, 44.7 mph is more accurate

41. -2 dollars per year

43. $x' \approx 2, s'(20) \approx 0.00182$ thousand $ / year   **45.** 1 ft/s

49. horizontal: $(0, 2.2)$ vertical: $(\pm\sqrt{6}, 1), (\pm\sqrt{2}, -1)$

51. $\dfrac{-y^2 - x^2}{y^3}$

53. $x = 1.9, y \approx 1 - 4(-0.1) = 1.4$
$x = 2.1, y \approx 1 - 4(0.1) = 0.6$

55. $y' = \dfrac{2xy}{2 - x^2}$, vertical asymptotes at $x = \pm\sqrt{2}$,
horizontal asymptote at $y = 0$

Exercises 2.9 ■ page 210

5. discontinuous at $x = 0$ **7.** discontinuous at $x = \dfrac{\pi}{2}$

9. $c = 0$ **11.** $c = \dfrac{\sqrt{7} - 1}{3}$ **13.** $c = \cos^{-1}\left(\dfrac{2}{\pi}\right)$

17. increasing   **19.** decreasing   **21.** increasing
23. increasing   **25.** $3x^2 + 5 > 0$
27. $f'(x) = 4x^3 + 6x$ has one zero   **29.** $3x^2 + a > 0$
31. $5x^4 + 3ax^2 + b > 0$ **35.** $\frac{1}{3}x^3 + c$ **37.** $\frac{1}{5}x^5 + c$

39. $-\dfrac{1}{x} + c$ **41.** $-\cos x + c$

43. $f(x) > 0$ in an interval $(b, 0)$ for some $b > 0$
45. discontinuous at $x = 0$

Chapter 2 Review Exercises ■ page 211

1. 0.8 **3.** 2 **5.** $\frac{1}{2}$ **7.** $3x^2 + 1$ **9.** $y = 2x - 2$

11. $y = 6x + 3$ **13.** $y = -3(x - 1) + 1$
15. $v(t) = -32t + 40; a(t) = -32$
17. $v(t) = 10e^{-2t}(4\cos 4t - 2\sin 4t)$;
$a(t) = -40e^{-2t}(4\cos 4t + 3\sin 4t)$
19. 8 ft/s going up, -24 ft/s coming down

21. (a) 0.3178 (b) 0.3338 (c) 0.3492 (d) 0.35
23. $4x^3 - 9x^2 + 2$ **25.** $-\frac{3}{2}x^{-3/2} - 10x^{-3}$

27. $2t(t + 2)^3 + 3t^2(t + 2)^2$ **29.** $\dfrac{(3x^2 - 1) - x(6x)}{(3x^2 - 1)^2}$

31. $2x\sin x + x^2\cos x$ **33.** $\frac{1}{2}x^{-1/2}\sec^2\sqrt{x}$

35. $\csc t - t\csc t \cot t$ **37.** $-4xe^{-x^2}$ **39.** $\ln x^2 + 2$

41. $2\cos 4x(\sin 4x)^{-1/2}$ **43.** $2\left(\dfrac{x + 1}{x - 1}\right)\dfrac{-2}{(x - 1)^2}$

45. $e^{4t} + 4te^{4t}$ **49.** $12x^2 - 18x + 4$ **51.** $(12 + 8x)e^{2x}$
53. $2\sec^2x \tan x$ **55.** $-3^{26}\sin 3x$ **57.** $-\$2400$ per year
59. (a) $f(t) = \pm 4$ (b) $f(t) = 0$ (c) $f(t) = 0$

61. $\dfrac{2x - 2xy}{x^2 - 9y^2}$ **63.** $\dfrac{\sec^2x + \dfrac{y}{(x + 1)^2}}{\dfrac{1}{(x + 1)} - 3}$

65. $\dfrac{170}{\sqrt{5}} \approx 76$ mph

67. (a) $(0, 1)$ and $(4, -31)$ (b) none
69. (a) $(0, 0)$ (b) none   **71.** $3x^2 + 7 > 0$
73. $c = 1$ **75.** $x^3 - \sin x + c$

CHAPTER 3

Exercises 3.1 ■ page 223

5. $1 + \frac{1}{2}(x - 1)$ **7.** $3 + \frac{1}{3}x$ **9.** $3x$ **11.** $1 + 2x$
13. 0.01 and 0.100003; 0.1 and 0.1003; 1 and 1.557
15. 2.0025 and 2.00249; 2.025 and 2.0248; 2.25 and 2.236

17. $\dfrac{\sqrt{3}}{2} + \dfrac{1}{2}\left(1 - \dfrac{\pi}{3}\right)$ **19.** $2 + \frac{1}{32}(0.04) = 2.00125$

21. $2 + \frac{1}{32}(0.16) = 2.005$
23. $0.00000117; 0.00000467; 0.0000186$
25. $x = 1$ **27.** $\frac{2}{3}, \frac{79}{144}; 0.53208$ **29.** $\frac{1}{2}, \frac{5}{8}; 0.61803$
31. -4.685779 **33.** 0.525260 **35.** 0.739085
37. -0.636732 or 1.409624 **39.** -0.567143
41. $f'(0) = 0$, roots $x = -0.3454, 0.4362, 1.6591$
43. $f'(0) = 0$, no root
45. $f'(-1)$ does not exist, roots $x = 0.1339, 1.8660$

47. if $x_0 < \sqrt{a}$, then $\dfrac{a}{x_0} > \sqrt{a}$

49. $P - \dfrac{2P}{R}x; x = 0.005R = 104{,}500$ feet

51. $L \approx 20.19$,
$y = 4.493 - 4.493x - 4.493\cos 4.493x + \sin 4.493x$

Exercises 3.2 ■ page 234

5. $-\frac{5}{2}$, minimum   **7.** -1, local maximum; 1, local minimum
9. 1, neither   **11.** 0, neither; $\frac{9}{4}$, minimum
13. -1, minimum; 0, local maximum; 1, minimum
15. $-\frac{2}{3}$, local maximum; 2, local minimum

17. $\dfrac{\pi}{4}$, maximum; $\dfrac{3\pi}{4}$, minimum; $\dfrac{5\pi}{4}$, maximum; $\dfrac{7\pi}{4}$, minimum

19. none   **21.** -1, minimum; 1, maximum
23. 0, minimum   **25.** 0, neither
27. $-\frac{2}{3}$, minimum; -1, maximum   **29.** 0, maximum

31. 0, local minimum; $\sqrt{\dfrac{\pi}{2}}$, maximum;

$\sqrt{\dfrac{3\pi}{2}}$, minimum; $\sqrt{\dfrac{5\pi}{2}}$, maximum

33. max = 3 at $x = 2$, min = -1 at $x = 1$
35. max = 11 at $x = -3$, min = -14 at $x = -2$
37. max = $\sqrt[3]{16}$ at $x = -4$, min = $\sqrt[3]{4}$ at $x = -2$
39. no extrema
41. max = $\sqrt{2}$ at $x = \dfrac{\pi}{4}$, min = $-\sqrt{2}$ at $x = \dfrac{5\pi}{4}$
43. max = 4.8197 at $x = 2.0287$, min = -1.8144 at $x = 4.9131$
45. (ex. 33) no max, min = -1 at $x = 1$
(ex. 35) no max, min = -14 at $x = -2$
(ex. 37) no extrema
51. $x = \sqrt{\frac{1}{3}}$ **53.** $x = \sqrt{\frac{1}{2}}$ **55.** $f(x) = \sec^2 x$
57. $x = -1.575$, top; $x = 1.575$, bottom; steepest at $x = \pm\sqrt{\frac{6}{5}}$
59. $W'(t) = abe^{-be^{-t}-t}$; max $W' = ae^{-1}$ at $t = \ln b$

Exercises 3.3 ■ page 242

5. increasing: $x < -1$, $x > 1$; decreasing: $-1 < x < 1$
7. increasing: $x < 1$, $x > 3$; decreasing: $1 < x < 3$
9. increasing: $-2 < x < 0$, $x > 2$; decreasing: $x < -2$, $0 < x < 2$
11. increasing: $-\dfrac{\pi}{6} < x < \dfrac{\pi}{6}, \dfrac{3\pi}{6} < x < \dfrac{5\pi}{6}$, etc;
decreasing: $\dfrac{\pi}{6} < x < \dfrac{3\pi}{6}, \dfrac{5\pi}{6} < x < \dfrac{7\pi}{6}$, etc
13. increasing: $x > 0$; decreasing: $x < 0$
15. local max at $x = -\dfrac{2}{3} - \dfrac{\sqrt{7}}{3}$; local min at $x = -\dfrac{2}{3} + \dfrac{\sqrt{7}}{3}$

17. min at $x = 0.2367$

19. no extrema

21. max at $x = \frac{1}{2}$

23. local max at $x = -0.3689$; local min at $x = 9.0355$

23. local max at $x = -0.3689$; local min at $x = 9.0355$

25. no extrema

27. no extrema

29. local max at $x = -\sqrt{3}$, local min at $x = \sqrt{3}$

31. max at $x = \dfrac{\pi}{4} + 2\pi n$, min at $x = \dfrac{5\pi}{4} + 2\pi n$

33. local max at $x = -2$, min at $x = 0$

35. local min at $x = -1$

45. If $f(x) = 2\sqrt{x}$ and $g(x) = 3 - \dfrac{1}{x}$, $f(1) = g(1) = 2$ and for $x > 1$ $f'(x) = \dfrac{1}{\sqrt{x}} > \dfrac{1}{x^2} = g'(x)$
47. If $f(x) = e^x$ and $g(x) = x + 1$, $f(0) = g(0) = 1$ and for $x > 1$ $f'(x) = e^x > 1 = g'(x)$
49. $f(x) = 3 + e^{-x}$ has no zeros
51. $s'(t) = \dfrac{1}{2\sqrt{t+4}}$ = rate of increase of sales function

Exercises 3.4 ■ page 250

5. up: $x < -\frac{1}{2}$, $x > \frac{1}{2}$; down: $-\frac{1}{2} < x < \frac{1}{2}$
7. up: $x > 1$; down: $x < 1$
9. local max at $x = 0$, inflection point at $x = 1$, local min at $x = 2$

11. local min at $x = -1$ and $x = 1$, local max at $x = 0$, inflection points at $x = -\sqrt{\frac{1}{3}}, \sqrt{\frac{1}{3}}$

13. local max at $x = -1$,
local min at $x = 1$,
no inflection points

15. local max at $x = -\sqrt{2}$,
inflection point at $x = 0$,
local min at $x = \sqrt{2}$

33. inflection point at $x = -\frac{1}{2}$

35. local min at $x = -0.1129$ and $x = 19.4993$,
local max at $x = 0.1135$, inflection points at $x = 0$,
$x = 13$

17. min at $x = -3$,
inflection points at
$x = -2$, $x = 0$

19. max at $x = 1$,
inflection point at $x = 2$

37. local max at $x = -2$, local min at $x = 2$

21. inflection points when
$x^2 = \frac{1}{4}(27 + \sqrt{297})$;
$x = \pm 3.325$

23. min at $x = 0$,
no inflection points

39. local min at $x = 0.8952$ and $x = 9.9987$,
local max at $x = 1.106$, inflection points at $x = 1$, $x = 7$

25. local max at $x = 0$

47. min at $x = 0$
49. local max at $x = -1$, inflection point at $x = 0$,
local min at $x = 1$
51. need to know   **53.** $x = 30$ **55.** $x = 600$
57. min at $x = -\frac{3}{4}c$, second inflection point at $x = -\frac{c}{2}$
59. $f(x) = -1 - x^2$ **61.** tangent line points above sun

27. max at $x = \dfrac{\pi}{4}$, inflection point at $x = \dfrac{3\pi}{4}$
min at $x = \dfrac{5\pi}{4}$, inflection point at $x = \dfrac{7\pi}{4}$, etc.

Exercises 3.5 ■ page 261

5. inflection point at $x = 1$

29. min at $x = 0$

31. inflection point at $x = 0$

7. local min at $x = -\dfrac{1}{2} - \dfrac{\sqrt{3}}{2}$ and $x = 1$
local max at $x = -\dfrac{1}{2} + \dfrac{\sqrt{3}}{2}$, inflection points at
$x = \pm\sqrt{\frac{1}{2}}$

9. local max at $x = -\sqrt{\frac{6}{5}}$, local min at $x = \sqrt{\frac{6}{5}}$,
inflection points at $x = -\sqrt{\frac{3}{5}}$, $x = 0$, $x = \sqrt{\frac{3}{5}}$

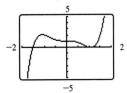

11. inflection points at $x = -\sqrt{6}$, $x = \sqrt{6}$

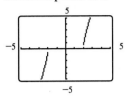

13. local max at $x = -2$, local min at $x = 2$

15. inflection point at $x = \dfrac{\pi}{4}$, max at $x = \dfrac{3\pi}{4}$

inflection point at $x = \dfrac{5\pi}{4}$, min at $x = \dfrac{7\pi}{4}$

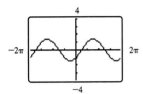

17. max at $x = 0$, inflection points at $x = -\sqrt{2}$ and $x = \sqrt{2}$

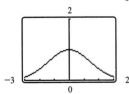

19. min at $x = e^{-1}$

21. min at $x = 0$

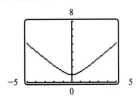

23. min at $x = 0$, inflection points at $x = -\sqrt{\frac{2}{3}}$, $x = \sqrt{\frac{2}{3}}$
horizontal asymptote at $y = 2$

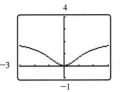

25. local min at $x = 0$, local max at $x = \sqrt[3]{2}$,
inflection points at $x = 0.5264$, 1.8995,
vertical asymptote at $x = -1$,
horizontal asymptote at $y = 0$

27. local min at $x = -1$, local max at $x = 1$,
inflection points at $x = -1.879$, $x = 0.347$, $x = 1.532$
horizontal asymptote at $y = 0$

29. inflection points at $x = 0$, $x = 1$

31. local max at $x = -0.2318 + \pi n$; local min at
$x = -0.2318 + \dfrac{n\pi}{2}$, n odd; inflection points at
$x = 0.3217 + \dfrac{n\pi}{2}$ for all n; horizontal asymptote at
$y = 0$.

33. local max at $x = -1$, inflection point at $x = 0$,
local min at $x = 1$

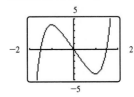

35. inflection point at $x = 0$, horizontal asymptotes at $y = -2, y = 2$

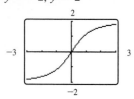

37. min at $x = -\frac{1}{6}$, inflection points at $x = 0$ and $x = \frac{2}{3}$

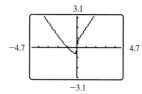

39. max at $x = \dfrac{2\pi}{3} + 2\pi n$, min at $x = \dfrac{4\pi}{3} + 2\pi n$, inflections points at $x = n\pi$, $x = 1.318 + 2\pi n$, and $x = -1.318 + 2\pi n$

41. inflection point at $x = 1$, vertical asymptote at $x = 0$, horizontal asymptote at $y = 1$

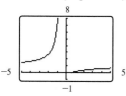

43. min at $x = -1$, max at $x = 1$, inflection points at $x = -\sqrt{3}$, $x = \sqrt{3}$, $x = 0$, horizontal asymptote at $y = 0$

45. $c < 0$: 3 extrema, 2 inflection points; $c \geq 0$: 1 extremum, 0 inflection points; as $c \to -\infty$, the graph widens and lowers; as $c \to +\infty$, the graph narrows

47. inflection points at $x = \pm\dfrac{c}{\sqrt{3}}$; graph widens as $c^2 \to \infty$; $y = 1$ for $c = 0$

49. $|c| =$ frequency of oscillation

51. min at $x = \dfrac{3\pi}{2} + 2\pi n$, max at $x = \dfrac{\pi}{2} + 2\pi n$

53. max at $x = \dfrac{1}{b}$; $\dfrac{1}{b} =$ most common lifespan

55. $y = 3x$ **57.** $y = x - 2$ **59.** $y = x$

61. $f(x) = \dfrac{3x^2}{(x-1)(x-2)}$

63. $f(x) = \dfrac{2x}{\sqrt{(x-1)(x+1)}}$

Exercises 3.6 ■ page 271

5. $30' \times 60'$ **7.** $20' \times 30'$ **11.** $(\sqrt{\frac{1}{2}}, \frac{1}{2})$ or $(-\sqrt{\frac{1}{2}}, \frac{1}{2})$

13. $(0, 1)$ **17.** $1.2137 = \dfrac{8}{3} - \dfrac{\sqrt{19}}{3}$

19. $\frac{15}{7} \approx 2.143$ miles east of first development

21. 1.2529 miles east of bridge; \$1.963 million

23. 1.894 miles east of bridge; \$234,800

27. $r = 1.1989''$, $h = 4.7957''$ **29.** $r = \frac{2}{3}r_0$, contracts

31. $x = R$ **33.** $2' \times 2'$ **35.** $\sqrt{46}'' \times 2\sqrt{46}''$

37. plant 2 acres   **39.** (a) $50°$ (b) $45°$ (c) $40°$

41. 2.744 sec; $x = 150$; relay; relay becomes longer

43. $x = 25$; times equal with 0.007 sec delay

Exercises 3.7 ■ page 281

5. $e^{-2t}(-8\cos 3t + \sin 3t)$ amps

7. $Q(t) \to 4\sin 3t$; transient $e^{-3t}(-3\cos 2t - 2\sin 2t)$ steady-state $12\cos 3t$

9. $x(t) = 2$

11. If $0 < x(t) < 4$, then $x'(t) > 0$ and if $x(t) > 4$, then $x'(t) < 0$; $x(t) = 5$ is max.

13. $r = ck$ **15.** 0; same   **17.** 0; a; increases from 0 to a

19. $4 - \cos x$; less dense at ends   **21.** 4; homogenous

23. healthy   **25.** in danger

27. $3x^2 + 40x + 90$; 9590 vs. 9421

29. $3x^2 + 42x + 110$; 34,310 vs. 33,990

31. $x = 10$; costs rise more sharply

33. $f'(t) = \left(\dfrac{96e^{-0.4t}}{(1 + 3e^{-0.4t})^2} \right) > 0$

35. 43; 4.49, about 4.5 points; 3.017, about 3 points

37. 23; 3.105 = percent to hear rumor in 3rd hour; 70 = percent to eventually hear rumor

39. $f'(x) = \dfrac{-816x^{-1.4}}{(4x^{-0.4} + 15)^2} < 0$

41. $C'(100) = 42$, $\overline{C}(100) = 77$; $\overline{C}(101) = 76.65 < \overline{C}(100)$

43. min at $x = 600$; $C'(600) = \overline{C}(600) = 52$

45. $P'(x) = 0$ if $R'(x) = C'(x)$ **47.** $f = \frac{1}{2}$, pH$\to\infty$

Chapter 3 Review Exercises ■ page 284

1. $1 + 3x$ **3.** $2 + \frac{1}{12}(7.96 - 8) \approx 1.99666$

5. 0.198437   **7.** $f'(1) = 0$

9. (a) $x = -3, 1$
(b) increase: $x < -3$, $x > 1$; decrease: $-3 < x < 1$
(c) local max at $x = -3$, local min at $x = 1$
(d) up: $x > -1$, down: $x < -1$
(e) $x = -1$

11. (a) $x = 0, 3$
(b) increase: $x > 3$; decrease: $x < 3$
(c) min at $x = 3$
(d) up: $x < 0$, $x > 2$; down: $0 < x < 2$
(e) $x = 0$, $x = 2$

13. (a) $x = \frac{1}{4}$
(b) increase: $x < \frac{1}{4}$; decrease: $x > \frac{1}{4}$
(c) local max at $x = \frac{1}{4}$
(d) up: $x > \frac{1}{2}$; down: $x < \frac{1}{2}$
(e) $x = \frac{1}{2}$

15. (a) none
  (b) increase: $x < -2, x > 2$
  (c) none
  (d) up: $-\sqrt{6} < x < -2, x > \sqrt{6}$,
    down: $x < -\sqrt{6}, 2 < x < \sqrt{6}$
  (e) $x = -\sqrt{6}, x = \sqrt{6}$
17. (a) $x = -2, 2$
  (b) increase: $-2 < x < 2$, decrease: $x < -2, x > 2$
  (c) min at $x = -2$, max at $x = 2$
  (d) up: $-\sqrt{12} < x < 0, x > \sqrt{12}$;
    down: $x < -\sqrt{12}, 0 < x < \sqrt{12}$
  (e) $x = -\sqrt{12}, x = 0, x = \sqrt{12}$
19. min $= -5$ at $x = 1$, max $= 76$ at $x = 4$
21. min $= 0$ at $x = 0$, max $= 3^{4/5}$ at $x = 3$
23. local max at $x = -\dfrac{4}{3} - \dfrac{\sqrt{10}}{3}$,

  local min at $x = -\dfrac{4}{3} + \dfrac{\sqrt{10}}{3}$
25. local max at $x \approx 0.2553$, local min at $x \approx 0.8227$
29. min at $x = -3$, inflection points at $x = -2, x = 0$

31. min at $x = -1$

33. min at $x = -1$, max at $x = 1$,
  inflection points at $x = -\sqrt{3}, x = 0$, and $x = \sqrt{3}$,
  horizontal asymptote at $y = 0$

35. min at $x = 0$, inflection points at $x = -\sqrt{\frac{1}{3}}, x = \sqrt{\frac{1}{3}}$
  horizontal asymptote at $y = 1$

37. local max at $x = -\sqrt{3}$ local min at $x = \sqrt{3}$,
  inflection point at $x = 0$,
  vertical asymptotes at $x = -1$ and $x = 1$

39. $(0.8237, 1.3570)$ **41.** 1.136 miles east of point A
43. $r = 1.663, h = 3.325$
45. $e^{-3t}(2 \cos 2t - 3 \sin 2t)$ amps
47. $2x$, denser to the right
49. $0.04x + 20, 20.8$ versus 20.78

CHAPTER 4

Exercises 4.1 ■ page 295
5. $\frac{3}{5}x^5 + c$ **7.** $\frac{3}{5}x^5 - \frac{3}{2}x^2 + c$ **9.** $2x^{3/2} + c$
11. $3x + \frac{1}{3}x^{-3} + c$ **13.** $-x^{-1} + \frac{3}{2}x^{-2} + c$
15. $-2 \cos x + \sin x + c$ **17.** $2 \sec x + c$
19. $5 \tan x + c$ **21.** $3e^x - 2x + c$
23. $3 \sin x - \ln|x| + c$ **25.** $2 \ln|x^2 + 4| + c$
27. $\frac{5}{2}x^2 + 3e^{-x} + c$ **29.** $-\frac{5}{2} \cos 2x + c$
31. $\frac{1}{3}e^{3x} - \frac{1}{2}x^2 + c$ **33.** $\frac{3}{2} \sec 2x + c$
35. $\ln|e^x + 3| + c$ **37.** $x - 3e^{-x} + c$
39. $\frac{1}{3}x^3 - \frac{3}{2}x^2 - 4x + c$ **41.** N/A
43. $3x + 4x^{-1} + c$ **45.** N/A   **47.** $-\frac{1}{2} \cos 4x + c$
49. N/A   **51.** $-x^{-1} - x + c$ **53.** $\frac{4}{3}x^3 - x + 2$
55. $3e^x + \frac{1}{2}x^2 + 1$ **57.** $6x^2 + 2x + 3$
59. $-3 \sin x + \frac{1}{3}x^4 + c_1 x + c_2$

61. $\frac{2}{3}x^3 - \ln|x| + \dfrac{c_1}{2}x^2 + c_2 x + c_3$

63. $3t - 6t^2 + 3$ **65.** $-3 \sin t + \frac{1}{2}t^2 + 3t + 4$
67. $a = \frac{1}{720}, s = \frac{2}{45}$ miles
69.   **71.** $t \approx 76$ sec, $v \approx -2532$ ft/s

Exercises 4.2 ■ page 303
3. $\displaystyle\sum_{i=1}^{50} i^2 = 42{,}925$ **5.** $\displaystyle\sum_{i=1}^{10} \sqrt{i} \approx 22.47$
7. $3 + 12 + 27 + 48 + 75 + 108 = 273$
9. $12 + 20 + 30 + 42 + 56 = 160$
11. $-1 + 2 + 5 + 8 + 11 + 14 + 17 + 20 = 76$
13. $\sqrt{2} + 2 + \sqrt{3} + 3 + \sqrt{4} + 4 + \sqrt{5} + 5$
  $= 16 + \sqrt{2} + \sqrt{3} + \sqrt{5} \approx 21.38$
15. 7385 **17.** $-21{,}980$ **19.** $323{,}400$ **21.** $-2{,}746{,}200$
23. $\dfrac{n(n + 1)(2n + 1)}{6} - 3n$

25. $\dfrac{4n(n + 1)(2n + 1)}{6} - \dfrac{n(n + 1)}{2}$

27. $\dfrac{(n + 1)(2n + 1)}{6n^2} + \dfrac{n + 1}{n} \to \dfrac{4}{3}$

29. $\dfrac{8}{3}\dfrac{(n+1)(2n+1)}{n^2} - \dfrac{n+1}{n} \to \dfrac{13}{3}$

31. 375 miles   **33.** $\frac{74}{3}$ miles   **35.** 2870   **37.** 24,825,800

Exercises 4.3 ■ page 311

3. 0.125, 0.375, 0.625, 0.875; 1.328125

5. 1.125, 1.375, 1.625, 1.875; 2.7265625

7. $\dfrac{\pi}{8}, \dfrac{3\pi}{8}, \dfrac{5\pi}{8}, \dfrac{7\pi}{8}$; 2.05234

9. −0.75, −0.25, 0.25, 0.75; 7.375

11. 0.3320   **13.** 0.6875   **15.** 6.3343   **17.** 4.0991

19. 1.00004   **21.** 16.5   **23.** 17.9996   **25.** −2.02

31. left: 1.81, right: 1.67   **33.** left: 3.8, right: 4.32

35. left: 1.18, right: 1.26   **41.** $a - \frac{1}{2}\Delta x + i\Delta x$ $i = 1, \dots, n$

Exercises 4.4 ■ page 322

5. 24.75   **7.** 0.25   **9.** 0.77   **11.** 1   **13.** $\frac{8}{3}$

15. 6   **17.** $\displaystyle\int_{-2}^{2}(4 - x^2)\,dx$ **19.** $-\displaystyle\int_{-2}^{2}(x^2 - 4)\,dx$

21. $\displaystyle\int_{0}^{2} x^2\,dx$ **23.** $\displaystyle\int_{0}^{\pi}\sin x\,dx$

25. $\displaystyle\int_{0}^{1}(x^3 - 3x^2 + 2x)\,dx - \displaystyle\int_{1}^{2}(x^3 - 3x^2 + 2x)\,dx$

27. 88   **29.** 140   **31.** $\displaystyle\int_{0}^{3} f(x)\,dx$ **33.** $\displaystyle\int_{0}^{1} f(x)\,dx$

37. positive   **39.** negative

41. $x = 1$ is not an evaluation point.

43. 28   **45.** 22.5   **47.** $t < 40; t < 40; t > 40; t = 40$

51. 6.93   **53.** 5   **55.** 3.33   **57.** 0.33

59. 0.64   **61.** 9000 lb; 360 ft/s

Exercises 4.5 ■ page 331

5. −2   **7.** 0   **9.** $\frac{88}{3}$ **11.** $\frac{12}{5}$ **13.** 2   **15.** $\sqrt{2} - 1$

17. 3   **19.** $e + e^{-1} - 2$ **21.** $\frac{3}{2}e^6 - \frac{21}{2}$

23. $e^2 + 4 - e^{-2}$ **25.** $3 - 3\ln 4$ **27.** $\frac{16}{3}$

29. $\frac{77}{4} + \cos 3$ **31.** 2.96   **33.** $\frac{1}{2}e^2 + 2e - \frac{3}{2}$

35. 1.71   **37.** 0.772   **39.** $\sqrt{2} - 1$ **41.** $x^2 - 3x + 2$

43. $(e^{-x^4} + 1)2x$ **45.** $-\ln(x^2 + 1)$ **47.** $y = 0$

49. $y = x - 2$

51. relative max at $x = 1$, relative min at $x = 2$

53. (a) $\dfrac{1}{x^2} \geq 0$ (b) $\dfrac{1}{x^2}$ is discontinuous at $x = 0$

55. $40t + \cos t + 1$ **57.** $25(t + \frac{1}{2}e^{-2t} - \frac{1}{2})$

59. $2t^2 - \frac{1}{6}t^3 + 8t$ **61.** $12t^2 + e^{-t} + t - 1$

63. 8 rad/s; 24 ft/s; 3.2 rad   **65.** 14.4 billion barrels

67. 236.36 hp   **69.** $\frac{10}{3}$ **71.** $\frac{1}{3}$ **73.** $\dfrac{2}{\pi}$

75. $\displaystyle\int_{0}^{3} f(x)\,dx, \displaystyle\int_{0}^{2} f(x)\,dx, \displaystyle\int_{0}^{1} f(x)\,dx$

Exercises 4.6 ■ page 340

5. $\frac{2}{9}(x^3 + 2)^{3/2} + c$ **7.** $\frac{1}{2}(\sqrt{x} + 2)^4 + c$

9. $\frac{1}{10}(x^2 - 3)^5 + c$ **11.** $\frac{1}{4}(x^2 + x)^4 + c$

13. $\frac{2}{3}(\sin x + 1)^{3/2} + c$ **15.** $-2\sqrt{\cos x} + c$

17. $\frac{1}{3}\sin x^3 + c$ **19.** $-\frac{4}{7}(\cos x + 3)^{7/4} + c$

21. $\frac{1}{2}e^{x+1} + c$ **23.** $2e^{\sqrt{x}} + c$

25. $\frac{2}{3}\sqrt{x^3 - 2} + c$ **27.** $\frac{1}{3}(\ln x + 2)^3 + c$

29. $\ln|x^2 + x - 1| + c$ **31.** $-2(\sqrt{x} + 1)^{-1} + c$

33. $e^{\sin x} + c$ **35.** $-\frac{1}{4}(\cos x - 1)^4 + c$

37. $-4(\ln x + 1)^{-1} + c$ **39.** $\ln(e^x + e^{-x}) + c$

41. $2\ln|\ln x| + c$

43. $2(x + 7) - 11\ln|x + 7| + c$

45. $\frac{3}{8}(x + 3)^{8/3} - \frac{18}{5}(x + 3)^{5/3} + \frac{27}{2}(x + 2)^{2/3} + c$

47. $\frac{5}{3}\sqrt{5} - \frac{1}{3}$ **49.** 0   **51.** −2   **53.** $-\ln\dfrac{\sqrt{2}}{2} = \dfrac{1}{2}\ln 2$

55. $\frac{8}{3}$ **57.** 0.77   **59.** 0   **61.** 3.0   **63.** 1.414

65. 0.881

67. $\cos(x) = \cos(-x), (-x)\sin(-x) = x\sin x$
$\sin(-x) = -\sin x, (-x)\cos(-x) = -x\cos x$

69. odd, 0   **71.** even, $\frac{16}{15}$ **73.** even, 0.48   **75.** neither, $\frac{2}{3}$

77. $x\sqrt{4 - x^2}$ is odd, $4 - x^2$ is even, $\bar{y} = \dfrac{8}{3\pi}$

Exercises 4.7 ■ page 349

5. midpoint 1.328, trapezoid 1.343, Simpson 1.333

7. midpoint 1.089, trapezoid 1.116, Simpson 1.1

9. (a) 0.75   (b) 0.8   (c) 0.75

11.

n	midpoint	trapezoid	Simpson
10	0.5537	0.5888	0.5660
20	0.5628	0.5713	0.5654
50	0.5652	0.5665	0.5656

13.

n	midpoint	trapezoid	Simpson
10	0.88220	0.88183	0.88207
20	0.88211	0.88202	0.88208
50	0.88208	0.88207	0.88208

15.

n	midpoint	trapezoid	Simpson
10	3.9774	3.9774	3.9774
20	3.9774	3.9774	3.9774
50	3.9774	3.9774	3.9774

17.

n	midpoint error	trapezoid error	Simpson error
10	0.00831	−0.01665	−0.00006
20	0.00208	−0.00416	-4.1×10^{-6}
40	0.00052	−0.00104	-2.6×10^{-7}
80	0.00013	−0.00026	-1.6×10^{-8}

19.

n	midpoint error	trapezoid error	Simpson error
10	-5.5×10^{-17}	0	0
20	-2.7×10^{-17}	1.6×10^{-16}	1.1×10^{-16}
40	-2.9×10^{-16}	-6.9×10^{-17}	-1.3×10^{-16}
80	-1.7×10^{-16}	-3.1×10^{-16}	1.5×10^{-16}

21. 4, 4, 16   **25.** 1.66, 128,841   **27.** (a) 9.1   (b) 9.033

29. (a) 6.96   (b) 6.96   **31.** 6193 ft^2 **33.** 529 ft

35. 2.6 liters   **37.** (a) under   (b) over   (c) can't tell

39. (a) over   (b) under   (c) can't tell

41. (a) under   (b) over   (c) exact

45. parabolas curve

47. if $T_n - I \approx -2(M_n - I)$ then $\frac{1}{3}T_n + \frac{2}{3}M_n \approx I$

Chapter 4 Review Exercises ■ page 352

1. $\frac{4}{3}x^3 - 3x + c$ **3.** $4\ln|x| + c$ **5.** $-\frac{1}{2}\cos 4x + c$
7. $\frac{1}{2}x^2 - \frac{1}{4}e^{4x} + c$ **9.** $\frac{1}{2}x^2 + 4\ln|x| + c$
11. $e^x - x + c$ **13.** $\frac{1}{3}(x^2 + 4)^{3/2} + c$
15. $2\sin x^3 + c$ **17.** $-e^{1/x} + c$ **19.** $-\ln|\cos x| + c$
21. $x^3 + x + 2$ **23.** $-16t^2 + 10t + 2$
25. $4 + 10 + 18 + 28 + 40 + 54 = 154$ **27.** 338,250
29. $\dfrac{n(n+1)(2n-1)}{6n^3} - \dfrac{n(n+2)}{2n^3} \to \dfrac{1}{3}$ **31.** 2.65625
33. 4.668
35. (a) 2.84   (b) 2.92   (c) 2.88   (d) 2.906
37. Simpson   **39.** $\frac{2}{3}$ **41.** $\displaystyle\int_0^3 (3x - x^2)dx = \frac{9}{2}$
43. 25   **45.** $\frac{1}{2}(e^2 - 1) \approx 3.19$ **47.** $-\frac{4}{3}$ **49.** 1
51. $6 + 4e^{-5/2}$ **53.** $\frac{1}{2}\ln 5$ **55.** $\frac{1}{3}(8^{3/2} - 8)$
57. $\frac{15}{2} + \frac{1}{2}e^2 - 4e$ **59.** $\sin x^2 - 2$
61. (a) 2.079   (b) 2.082   (c) 2.080
63. (a) 2.0841, 2.0845   (b) 2.08055, 2.08048
     (c) 2.0846, 2.08046

CHAPTER 5

Exercises 5.1 ■ page 359

5. $\frac{40}{3}$ **7.** $5 - e^{-2}$ **9.** 3   **11.** $\frac{29}{2}$ **13.** $\frac{64}{3}$
15. $\frac{1}{6}$ **17.** $\frac{27}{4}$ **19.** $\frac{1}{12}$ **21.** 0.08235   **23.** 0.135697
25. 4.01449   **27.** $\displaystyle\int_0^1 (2 - 2y)\,dy = 1$
29. $\displaystyle\int_1^2 (3y - 2 - y^2)\,dy = \frac{1}{6}$ **31.** $\displaystyle\int_0^1 2x\,dx = 1$
33. $\displaystyle\int_0^2 (6 - 2y)\,dy = 8$
35. 3; $\displaystyle\int_0^{\sqrt{3}} (3 - x^2)\,dx = \int_{\sqrt{3}}^3 (x^2 - 3)\,dx = 2\sqrt{3}$
37. $14.4 \times 365 = 5257$ million barrels
39. 2.45 million people   **41.** 35.08%   **43.** 93.02%

Exercises 5.2 ■ page 372

5. 12   **7.** $\frac{56\pi}{3}$ **9.** 123.8 ft^3 **11.** 93,750,000 ft^3
13. $\frac{215}{2}$ ft^3 **15.** $33\pi^2 + 32\pi$ in^3 **17.** 0.2467 cm^3
19. 2.5 ft^3 **21.** $\frac{8\pi}{3}$ **23.** $\frac{28\pi}{3}$ **25.** $\frac{32\pi}{5}$
27. $\frac{224\pi}{15}$ **29.** $2\pi e^2 + 2\pi$ **31.** $2\pi - \frac{\pi^2}{4}$ **33.** $\frac{2\pi}{5}$
35. $\frac{\pi}{10}$
37. (a) 9π (b) 9π (c) 18π (d) 36π (e) 18π
     (f) 36π
39. (a) $\frac{\pi}{2}$ (b) $\frac{\pi}{5}$ (c) $\frac{\pi}{6}$ (d) $\frac{7\pi}{15}$ (e) $\frac{7\pi}{6}$
     (f) $\frac{13\pi}{15}$

41. $\dfrac{\pi h^2}{2a} = \frac{1}{2}\pi h\left(\sqrt{\dfrac{h}{a}}\right)^2$ **43.** $\displaystyle\int_{-1}^1 \pi(1)^2\,dy = 2\pi$
45. $\displaystyle\int_{-1}^1 \pi\left(\dfrac{1-y}{2}\right)^2 dy = \dfrac{2\pi}{3}$

Exercises 5.3 ■ page 379

5. $r = 2 - x, h = x^2, V = \dfrac{8\pi}{3}$
7. $r = x, h = 2x, V = \dfrac{4\pi}{3}$
9. $r = 2 - y, h = 2y, V = \dfrac{8\pi}{3}$
11. $r = y, h = \sqrt{1 - (y-1)^2}, V = \pi^2 \approx 9.8696$
13. $\dfrac{32\pi}{3}$ **15.** $\dfrac{16\pi}{3}$ **17.** $\dfrac{27\pi}{2}$ **19.** $\dfrac{8\pi}{3}$
21. (a) $\dfrac{80\pi}{3}$ (b) 16π (c) 16π (d) $\dfrac{16\pi}{3}$
23. (a) $\dfrac{625\pi}{6}$ (b) $\dfrac{625\pi}{3}$ (c) $\dfrac{875\pi}{6}$ (d) $\dfrac{500\pi}{3}$
25. (a) 16.723   (b) 12.635   (c) 4.088   (d) 1.496
27. (a) $\dfrac{32\pi}{15}$ (b) $\dfrac{5\pi}{6}$ (c) $\dfrac{3\pi}{2}$ (d) $\dfrac{38\pi}{15}$
29. (a) $\dfrac{5\pi}{6}$ (b) $\dfrac{64\pi}{15}$ **31.** $\dfrac{\sqrt{3}\pi}{2}$
33. $\dfrac{\sqrt{100 - 30\sqrt{10}}}{10} = 0.2265$

Exercises 5.4 ■ page 386

5. 1.4604; 1.4743   **7.** 3.7242; 3.7900
9. $\displaystyle\int_{-1}^1 \sqrt{1 + 9x^4}\,dx = 3.0957$
11. $\displaystyle\int_0^2 \sqrt{1 + (2 - 2x)^2}\,dx = 2.9578$
13. $\displaystyle\int_0^3 \sqrt{1 + (3x^2 + 1)^2}\,dx = 30.3665$
15. $\displaystyle\int_0^\pi \sqrt{1 + \sin^2 x}\,dx = 3.8201$
17. $\displaystyle\int_0^\pi \sqrt{1 + (\sin x + x\cos x)^2}\,dx = 5.0404$
19. 27.512 feet   **21.** 5.43 feet
23. 60 yards; 60 yards; 139.4 yards; 104.55 ft/s
25. $\displaystyle\int_0^1 2\pi x^2\sqrt{1 + 4x^2}\,dx = 3.8097$
27. $\displaystyle\int_0^2 2\pi(2x - x^2)\sqrt{1 + (2 - 2x)^2}\,dx = 10.9654$
29. $\displaystyle\int_0^1 2\pi e^x\sqrt{1 + e^{2x}}\,dx = 22.9430$
31. $\displaystyle\int_0^{\pi/2} 2\pi \cos x\sqrt{1 + \sin^2 x}\,dx = 7.2117$
33. 0.9997   **35.** 0.9784   **37.** 0.99995
39. $1.672, 1.720, 1.754 \to 2$ **41.** $4x^3 > 2x$ if $x > \sqrt{\frac{1}{2}}$
43. 6π **45.** $\pi\sqrt{5} + \pi$ **47.** 6

Exercises 5.5 ■ page 393

5. $y(0) = 80, y'(0) = 0$ **7.** $y(0) = 60, y'(0) = 10$
9. $-8\sqrt{30}$ ft/s   **11.** $\sqrt{h}$ **13.** 256 feet

15. $-16t^2 + 64t$; 64 feet; 4 sec; -64 ft/s   **17.** $8\sqrt{\frac{5}{3}}$

19. $\frac{25}{64}$ ft ≈ 4.7 in.   **23.** $10\sqrt{3} \approx 17$ sec, $490\sqrt{3} \approx 849$ m

25. $\frac{15}{4}$ sec, $225\sqrt{3} \approx 390$ ft   **27.** 2.59 feet

29. ball bounces ($h \approx -0.62$)

31. $7.7°, 120 \tan(7.7°) \approx 16.2$ ft   **33.** $40\dfrac{5}{\sqrt{3}} \approx 68$ ft/s

35. 400 ft   **37.** $\frac{25}{16} \sin 4t - \frac{25}{4}t$

39. $\frac{25}{64} \sin\left(8t + \dfrac{\pi}{4}\right) - \dfrac{25\sqrt{2}}{16}t - \dfrac{25\sqrt{2}}{128}$

41. $\dfrac{20\pi}{\sqrt{30}} \approx 11.5$ rad/s   **45.** 25 sec   **47.** $40.6°, 37.26$ m/s

Exercises 5.6 ■ page 404

5. $\frac{15}{8}$ ft-lb   **7.** 20 ft-lb   **9.** $\frac{1250}{3}$ ft-lb   **11.** 300 ft-lb
13. 270,000,000 ft-lb   **15.** 704,000 ft-lb
17. 8,168,140,899 ft-lb   **19.** 816,814 ft-lb
21. $J \approx 2.133$; 113 ft/s   **23.** $J \approx 7533.3$; 37.7 ft/s
25. $m = 15$ kg, $\overline{x} = \frac{16}{5}$ m ; heavier to right of center
27. $m = \frac{52}{3}$ kg, $\overline{x} = 0$ m   **29.** 0.0614 slugs, 31.5 oz
31. 16.6 in.; same mass, $\overline{x}$ differs by 3
33. 0.0614 slugs, 31.4 oz; $\overline{x} = 17.8$ in.
35. $2, 4, \frac{1}{2}$; 0.4167, 3, 5.33   **37.** 8,985,600 lb
39. 196,035 lb   **41.** 12,252 lb   **43.** $-16\sqrt{5}$ ft/s
45. 10,667 hp

Exercises 5.7 ■ page 413

13. 4   **15.** $\dfrac{4}{1 - e^{-4}}$ **17.** $\dfrac{\ln 2}{2} \approx 0.346$ **19.** 0.157

21. 7.76×10^{-11} **23.** $1 - e^{-3/2} \approx 0.77687$
25. $e^{-6} - e^{-12} \approx 0.00247$ **27.** $e^{-2/3} - e^{-4/3} \approx 0.2498$
29. $1 - e^{-4} \approx 0.9817$ **31.** 0.594   **33.** 0.9999995

35. (a) $\frac{3}{4}$ (b) $\sqrt[3]{\frac{1}{2}}$ **37.** (a) 1.57   (b) $\dfrac{\pi}{2} \approx 1.57$

39. (a) 1.23   (b) $-\dfrac{3 \ln \frac{2}{3}}{\ln 3} \approx 1.11$

41. (a) 0.2313   (b) $-\frac{1}{4} \ln\left(\dfrac{e^{-4} + 1}{2}\right) \approx 0.1687$

43. $c = \dfrac{4}{1 - e^{-4b}} \to 4$ **45.** $\dfrac{6}{1 - e^{-6b}} \to 6$

47. (a) $\frac{93}{256}$ (b) $\frac{93}{256}$ (c) $\frac{2}{256}$ (d) $\frac{1}{2}$
49. (a) 0.7101   (b) 0.2899   (c) 0.1552   (d) 0.576
51. $\frac{100}{99}$, 0.0024   **53.** $\frac{5}{2}$

Chapter 5 Review Exercises ■ page 415

1. $\dfrac{\pi^3}{3} + 2\pi - 2$ **3.** $\frac{1}{12}$ **5.** 1.452   **7.** $\frac{5}{6}$

9. 10,054   **11.** $\dfrac{98\pi}{3}$ **13.** 4.37

15. (a) $\dfrac{256\pi}{5}$ (b) 8π (c) $\dfrac{128\pi}{3}$ (d) $\dfrac{1408\pi}{15}$

17. (a) $\dfrac{2\pi}{3}$ (b) 2π (c) 4π (d) $\dfrac{22\pi}{3}$

19. $\displaystyle\int_{-1}^{1} \sqrt{1 + 16x^6}\, dx \approx 3.2$

21. $\displaystyle\int_{-2}^{2} \sqrt{1 + \frac{1}{4}e^x}\, dx \approx 4.767$

23. $\displaystyle\int_{0}^{1} 2\pi(1 - x^2)\sqrt{1 + 4x^2}\, dx \approx 5.48$

25. -64 ft/s   **27.** $3 \sin 20° \approx 1.026$ sec, 46.3 ft
29. no, ball bounces   **31.** $64\sqrt{2}$ ft/s, $-64\sqrt{2}$ ft/s
33. $\frac{40}{3}$ ft-lb
35. $m = \frac{112}{3}, \overline{x} = \frac{16}{7}$, heavier to right of $x = 1$
37. 22,630,400 lb   **39.** 32 ft/s   **43.** 2
45. (a) $1 - e^{-2} \approx 0.864$ (b) $e^{-2} - e^{-4} \approx 0.117$
47. (a) $\frac{11}{15}$ (b) 0.786

CHAPTER 6

Exercises 6.1 ■ page 423

5. 1.386   **7.** 2.104   **9.** $\dfrac{1}{x}$ **11.** $-\tan x$

13. $\ln x + 1$ **15.** $\dfrac{2x}{x^2 + 2}$ **17.** $\frac{7}{2} \ln 2$ **19.** $\frac{1}{2} \ln 3$

21. $\ln(x^2 + 1) + c$ **23.** $-\ln|\cos x| + c$
25. $\frac{1}{2} \ln|x^2 + 2x - 1| + c$ **27.** $\ln|\ln x| + c$
29. $\frac{1}{3}(\ln x + 1)^3 + c$ **31.** $\ln 3$ **33.** $\frac{1}{3} \ln \frac{3}{4}$ **35.** $\dfrac{x}{x^2 + 1}$

37. $\dfrac{4}{x} - \dfrac{5x^4}{x^5 + 1}$ **39.** $\frac{1}{2} \ln\big|\ln|x|\big| + c$

41.

43.

45.

49. $e^{-1/2}$

Exercises 6.2 ■ page 430

5. yes   **7.** no   **9.** no   **11.** yes   **13.** yes   **15.** yes
17. $\sqrt[3]{x + 2}$

19. $\sqrt[5]{x + 1}$

21. not one-to-one

23. $\sqrt[3]{x^2 - 1}$, $x \geq 0$;

25. $\dfrac{1}{f'(0)} = \dfrac{1}{4}$ **27.** $\dfrac{1}{f'(1)} = \dfrac{1}{15}$

29. $\dfrac{1}{f'(1)} = \dfrac{4}{5}$ **31.** $y = \frac{1}{4}(x + 1)$ **33.** 2.5, high

35. 2.25, high   **37.** probably not one-to-one

39. inverse exists (before bounce)   **41.** not one-to-one

43. $k \geq 0$ **45.** $0.9(1.1)x = 0.99x; \frac{1}{1.1}x$

Exercises 6.3 ■ page 436

5.

7.

9.

11.

13. $12e^{3x}$ **15.** $3e^{-2x} - 6xe^{-2x}$

17. $-4xe^{-x^2}$ **19.** $2xe^{-x} - x^2e^{-x}$ **21.** $2(3x^2 - 3)e^{x^3 - 3x}$

23. e^x **25.** $-\sin x \, e^{\cos x}$ **27.** $\frac{1}{3}e^{3x} + c$ **29.** $\frac{1}{2}e^{x^2} + c$

31. $-e^{\cos x} + c$ **33.** $-e^{1/x} + c$ **35.** $x + 2e^x + \frac{1}{2}e^{2x} + c$

37. $-2e^{-2x} + c$ **39.** $\frac{1}{3}x^3 + c$ **41.** $\frac{1}{3}e^3 - \frac{1}{3}$ **43.** 0

45. max at $x = \frac{1}{2}$, inflection point at $x = 1$

47. min at $x = 0$, local max at $x = 1$,

inflection points at $x = 1 \pm \dfrac{\sqrt{2}}{2}$

51.

53.

55.

57.

59. increasing, concave up; decreasing, concave up

61. $(\ln 2)2^x$ **63.** $2(\ln 3)3^{2x}$ **65.** $2x(\ln 3)3^{x^2}$

67. $\dfrac{1}{x \ln 3}$ **69.** $\dfrac{2}{x \ln 4}$ **71.** $\dfrac{1}{\ln 2}2^x + c$

73. $\dfrac{1}{2 \ln 2}2^{x^2} + c$ **75.** $\dfrac{1}{3 \ln 4}4^{3x} + c$

77.

$1; 0; x = 0; \dfrac{1}{1 + e^{-(x-4)}}$

79. $k = \frac{1}{3}\ln 2; 2^{2/3} - 2^{1/3} \approx 0.327$

Exercises 6.4 ■ page 444

5. $2e^{4x}$ **7.** $5e^{-3x}$ **9.** $2e^{2(x-1)}$ **11.** $-3x + 3$

13. $100e^{(\ln 2/4)t}$, 23.6 hours   **15.** $400e^{(\ln 2)t}$, 409,600

17. $\dfrac{\ln 2}{0.44} \approx 1.575$ hours   **19.** $\dfrac{\ln 2}{1.3863} \approx 0.5$ day

23. $0.4e^{-(\ln 2/3)t}$, 15.96 hours

25. $20\,\dfrac{\ln 10}{\ln 2} \approx 66.4$ minutes   **27.** 29%   **29.** 13,305 years

31. $\dfrac{\ln\left(\frac{5}{13}\right)}{\ln\left(\frac{11}{13}\right)} \approx 5.72$ minutes   **33.** $70 - 20e^{kt}, k = \dfrac{\ln 0.7}{2}$

35. 9:46   **37.** 182.6°

39. $1080, $1083, $1083.28, $1083.29

41. $A = \$110{,}231.76, B = \$66{,}402.34$

43. $34e^{kt}, k = \frac{1}{10}\ln\left(\frac{9800}{34}\right), \$2{,}824{,}705.88$

47. $7300; $7860 vs. $7665   **49.** exponential, linear

Exercises 6.5 ■ page 452

5. yes   **7.** no   **9.** yes   **11.** no   **13.** $ce^{x+x^3/3}$

15. $\dfrac{1}{\frac{2x^3}{3} + c}$ **17.** $\pm\sqrt{6\tan^{-1}x + c}$ **19.** $\sqrt[3]{4x^3 + c}$

21. $\dfrac{1}{1 + ce^x}$ **23.** $c\sqrt{1 + x^2}$

25. $\pm\sqrt[4]{16\sin x - 16x\cos x + c}$

27. $e^{-1}e^{(x+1)^3}$ **29.** $\sqrt{\frac{8}{3}x^3 + 4}$ **31.** $(x + 3)^4$

33. $\sin y = 2x^2$ **35.** $\dfrac{2e^{6t}}{1 + e^{6t}}$ **37.** $\dfrac{20e^{10t}}{1 + 4e^{10t}}$

39. $\dfrac{3e^t}{1 + 3e^t}$ **41.** $\dfrac{(8 \times 10^7)e^{0.71t}}{3 + e^{0.71t}}$ **43.** $\dfrac{AMe^{kMt}}{Ae^{kMt} - 1}$

45. 0, 2   **47.** 0, −2, 2   **49.** 1

51. (a) $y = 0$ (b) $y = 2$ **53.** $\sqrt{80}$ ft/s

55. $\dfrac{10e^{2rt} - 10}{e^{2rt} - \frac{5}{6}}$

Exercises 6.6 ■ page 460

5.

7.

9.

11.

13.

15.

17. $h = 0.1$: $y_1 = 1$, $y_2 = 1.02$,
$y(1) \approx 2.3346$, $y(2) \approx 29.4986$
$h = 0.05$: $y_1 = 1$, $y_2 = 1.005$,
$y(1) \approx 2.5106$, $y(2) \approx 39.0930$

19. $h = 0.1$: $y_1 = 1.3$, $y_2 = 1.651$,
$y(1) \approx 3.8478$, $y(2) \approx 3.9990$
$h = 0.05$: $y_1 = 1.15$, $y_2 \approx 1.3139$
$y(1) \approx 3.8188$, $y(2) \approx 3.9978$

21. $h = 0.1$: $y_1 = 2.9$, $y_2 \approx 2.8005$,
$y(1) \approx 2.0943$, $y(2) \approx 1.5275$
$h = 0.05$: $y_1 = 2.95$, $y_2 \approx 2.9001$,
$y(1) \approx 2.0990$, $y(2) \approx 1.5346$

23. $h = 0.1$: $y_1 = 1.1$, $y_2 \approx 1.2095$,
$y(1) \approx 2.3960$, $y(2) \approx 4.5688$
$h = 0.05$: $y_1 = 1.05$, $y_2 \approx 1.1024$,
$y(1) \approx 2.4210$, $y(2) \approx 4.6203$

25. $y(1) = e \approx 2.7183$, $y(2) = e^4 \approx 54.5982$;
$y(1) = -2$, $y(2) = -\frac{2}{3}$

27.

29. $y = 0$ (unstable), $y = 2$ (stable)
31. $y = 0$ (unstable), $y = -1$ (unstable), $y = 1$ (stable)
33. $y = 1$ (stable)   **35.** $y = 0$ and $y = b$ are stable

Exercises 6.7 ■ **page 465**

5. 0 **7.** $\dfrac{\pi}{4}$ **9.** $-\dfrac{\pi}{2}$ **11.** 0 **13.** $-\dfrac{\pi}{4}$ **15.** $\dfrac{\pi}{3}$

17. $-\dfrac{\pi}{6}$ **19.** $\sqrt{1 - x^2}$, $-1 \le x \le 1$

21. $\begin{cases} 0 \text{ if } x = n\pi \ (n \text{ even}) \\ \pi \text{ if } x = n\pi \ (n \text{ odd}) \end{cases}$

23. $\sqrt{x^2 - 1}$, $x \ge 1$ or $-\sqrt{x^2 - 1}$, $x \le -1$

25. $\sqrt{x^2 + 1}$, $x \in (-\infty, \infty)$

27. $\dfrac{\sqrt{3}}{2}$ **29.** $\dfrac{\sqrt{2}}{2}$ **31.** $\frac{4}{3}$

33.

35.

37.

39. $\dfrac{n\pi}{4}$, odd n

41. $\dfrac{n\pi}{3}$ **43.** $\dfrac{\pi}{20} + \dfrac{n\pi}{5}$ **45.** $\dfrac{\pi}{18} + \dfrac{2\pi n}{3}, \dfrac{5\pi}{18} + \dfrac{2\pi n}{3}$

47. $\pm\dfrac{\pi}{6} + \pi n$

51. $\tan^{-1}\left(\dfrac{5}{3x}\right)$

Exercises 6.8 ■ **page 470**

3. $\dfrac{6x}{\sqrt{1 - 9x^4}}$ **5.** $\dfrac{-3x^2}{\sqrt{1 - x^6}}$ **7.** $\dfrac{2x}{1 + x^4}$

9. $\dfrac{2}{x\sqrt{x^4 - 1}}$ **11.** $\cos^{-1}2x - \dfrac{2x}{\sqrt{1 - 4x^2}}$

13. $\dfrac{-\cos x}{\sqrt{1 - \sin^2 x}} = \pm 1$ **15.** $\dfrac{\sec x \tan x}{1 + \sec^2 x}$

17. $6 \tan^{-1}x + c$ **19.** $4 \sin^{-1}x + c$ **21.** $\tan^{-1}x^2 + c$

23. $2 \sin^{-1}x^2 + c$ **25.** $\sec^{-1}x^2 + c$ **27.** $\tan^{-1}\left(\dfrac{x}{2}\right) + c$

29. $\sin^{-1}(e^x) + c$ **31.** $\dfrac{1}{2} \tan^{-1}\left(\dfrac{\sin x}{2}\right) + c$

37. $\sqrt{\dfrac{248}{3}} \approx 9.09$ feet   **39.** $\sqrt{3} \approx 1.73$ feet

Exercises 6.9 ■ **page 477**

5.

7.

9.

11.

13. $4 \sinh 4x$ **15.** $2 \cosh 2x$

17. $4 \operatorname{sech}^2 4x$ **19.** $\dfrac{2}{\sqrt{4x^2 - 1}}$

21. $2x \sinh 2x + 2x^2 \cosh 2x$ **23.** $\dfrac{4}{1 - 16x^2}$

25. $\frac{1}{6} \sinh 6x + c$ **27.** $\frac{1}{3} \ln (\cosh 3x) + c$

29. $\frac{1}{4} \cosh 4 - \frac{1}{4}$ **31.** $2 \sinh^{-1} x + c$

33. $\cosh(\sin x) + c$ **35.** $e^{\sinh 1} - 1$

41. $f'(x) < 0$ for $x < 0$, $f'(x) > 0$ for $x > 0$,
$f''(x) > 0$ for all x

43. $\ln (x + \sqrt{x^2 - 1})$ **45.** 27.62 **47.** 26.13

Chapter 6 Review Exercises ■ page 478

1. $\dfrac{3x^2}{x^3 + 5}$ **3.** $\dfrac{4x^3 + 1}{2(x^4 + x)}$ **5.** $-2xe^{-x^2}$

7. $(3x^2 \ln 4)4^{x^3}$ **9.** $\dfrac{2}{\sqrt{1 - 4x^2}}$ **11.** $\dfrac{-2 \sin 2x}{1 + \cos^2 2x}$

13. $\frac{1}{2} x^{-1/2} \sinh \sqrt{x}$ **15.** $\dfrac{3}{\sqrt{9x^2 + 1}}$

17. $\frac{1}{3} \ln |x^3 + 4| + c$ **19.** $\frac{1}{2} \ln 2$ **21.** $-\cos(\ln x) + c$

23. $-\frac{1}{4} e^{-4x} + c$ **25.** $2e^{\sqrt{x}} + c$ **27.** $\frac{1}{3} e^6 - \frac{1}{3}$

29. $\dfrac{1}{4 \ln 3} 3^{4x} + c$ **31.** $\dfrac{3}{2} \tan^{-1}\!\left(\dfrac{x}{2}\right) + c$

33. $\frac{1}{3} \sin^{-1}(x^3) + c$ **35.** $\frac{9}{2} \sec^{-1} x^2 + c$

37. $4 \sinh^{-1} x + c$ **39.** $\frac{1}{4} \sinh 4x + c$ **41.** yes, $\sqrt[3]{x + 1}$

43. no   **45.** $\dfrac{1}{f'(1)} = \dfrac{1}{11}$ **47.** $\dfrac{1}{f'(2)} = \dfrac{1}{2}$ **49.** $3e^{2x}$

51. $\sqrt{2x^2 + 4}$ **53.** $(\frac{1}{3} x^{3/2} + \frac{5}{3})^2$

55. $10^4 e^{(\ln 2/2)t}; \dfrac{2 \ln 100}{\ln 2} \approx 13.29$ hours

57. $2e^{-(\ln 2/2)t}; \dfrac{-2 \ln .05}{\ln 2} \approx 8.64$ hours   **59.** $\dfrac{\ln 2}{0.08} \approx 8.66$ years

61. $112 e^{(\ln 108/112)t} + 68; 21.10$ minutes   **63.** $ce^{1/2x^4}$

65. $\frac{1}{3} y^3 + \frac{1}{2} y^2 = 4 \tan^{-1} x + c$

67. $y = 0$ (unstable), $y = 2$ (stable)   **69.** $y = 0$ (stable)

71.

73.

75. $h = 0.1$: $y_1 = 2.7$, $y_2 = 2.45$;
$y(1) \approx 1.7434$, $y(2) \approx 2.6079$
$h = 0.05$: $y_1 = 2.85$, $y_2 = 2.7125$;
$y(1) \approx 1.7924$, $y(2) \approx 2.6426$

77. $\dfrac{\pi}{2}$ **79.** $-\dfrac{\pi}{4}$ **81.** $\dfrac{\sqrt{3}}{2}$ **83.** $\dfrac{\pi}{4}$ **85.** $\dfrac{\pi}{4} + \pi n$

87.

89.

91.

y = e^{-x^2}

93. 17.77

CHAPTER 7

Exercises 7.1 ■ page 485

3. $-\frac{1}{6} \cos 6x + c$ **5.** $\frac{1}{2} \sec 2x + c$ **7.** $-\frac{1}{2} e^{3-2x} + c$

9. $-2 \cos \sqrt{x} + c$ **11.** $e^{\sin x} + c$ **13.** $\sec x + c$

15. $\frac{3}{4} \tan^{-1}\!\left(\dfrac{x}{4}\right) + c$ **17.** $\frac{1}{3} \tan^{-1} x^3 + c$ **19.** $\sin^{-1} \dfrac{x}{2} + c$

21. $\frac{1}{2} \sin^{-1} x^2 + c$ **23.** $2 \tan^{-1}\!\left(\dfrac{x + 1}{2}\right) + c$

25. $2 \ln |x^2 + 2x + 5| - 2 \tan^{-1}\!\left(\dfrac{x + 1}{2}\right) + c$

27. $\frac{1}{5} x^5 + \frac{8}{3} x^3 + 16x + c$ **29.** $\sin^{-1}\!\left(\dfrac{x + 1}{2}\right) + c$

31. $2 \sin^{-1}\!\left(\dfrac{x - 2}{2}\right) + c$ **33.** $\frac{1}{3} x^3 + x + c$

35. $\frac{2}{5} (x - 3)^{5/2} + 2(x - 3)^{3/2} + c$ **37.** $\tan^{-1}(e^x) + c$

39. $\frac{2}{5} x^{5/2} + 2x^{1/2} + c$ **41.** $\dfrac{1}{2} \cosh 2x + c$

43. $\frac{1}{3} \ln (\cosh 3x) + c$ **45.** $\frac{1}{2} \sinh x^2 + c$

47. $3 \cosh^{-1} x + c$ **49.** $\frac{2}{3} \sec^{-1} x^2 + c$

51. $\dfrac{5}{\sqrt{3}} \tan^{-1}\!\left(\dfrac{x}{\sqrt{3}}\right) + c$ **53.** $\dfrac{1}{4} (\ln x)^2 + c$

55. $-\frac{1}{2} e^{-x^2} + c$

Exercises 7.2 ■ page 491

3. $x \sin x + \cos x + c$ **5.** $\frac{1}{2} x e^{2x} - \frac{1}{4} e^{2x} + c$

7. $\frac{1}{3} x^3 \ln x - \frac{1}{9} x^3 + c$ **9.** $-\frac{1}{3} x^2 e^{-3x} - \frac{2}{9} x e^{-3x} - \frac{2}{27} e^{-3x} + c$

11. $\frac{1}{2} x^2 \sin 2x + \frac{1}{2} x \cos 2x - \frac{1}{4} \sin 2x + c$

13. $\frac{1}{17} e^x \sin 4x - \frac{4}{17} e^x \cos 4x + c$

15. $\frac{2}{3} \sin 2x \cos x - \frac{1}{3} \cos 2x \sin x + c$

17. $x \tan x + \ln|\cos x| + c$

19. $x(\ln x)^2 - 2x \ln x + 2x + c$

21. $\sin x \ln(\sin x) - \sin x + c$

23. $\frac{1}{2} x \sinh 2x - \frac{1}{4} \cosh 2x + c$ **25.** $\frac{1}{4} \sin 2 - \frac{1}{2} \cos 2$

27. $-\dfrac{2}{\pi^2}$ **29.** $\dfrac{1}{\pi}$ **31.** $10 \ln 10 - 9$

33. $x \cos^{-1} x - \sqrt{1 - x^2} + c$

35. $-2\sqrt{x} \cos \sqrt{x} + 2 \sin \sqrt{x} + c$ **37.** n times

41. $\frac{1}{3} \cos^2 x \sin x + \frac{2}{3} \sin x + c$

43. $-\frac{1}{3} \sin^2 x \cos x - \frac{2}{3} \cos x + c$ **45.** $\dfrac{\pi}{4}$ **47.** $\dfrac{8}{15}$

49. m even: $\dfrac{(m-1)(m-3)\dots 1}{m(m-2)\dots 2}\cdot\dfrac{\pi}{2}$

m odd: $\dfrac{(m-1)(m-3)\dots 2}{m(m-2)\dots 3}$

51. first column: derivatives; second column: antiderivatives

53. $x^4\sin x + 4x^3\cos x - 12x^2\sin x - 24x\cos x + 24\sin x + c$

55. $(\frac{1}{2}x^4 - x^3 + \frac{3}{2}x^2 - \frac{3}{2}x + \frac{3}{4})e^{2x} + c$

57. $-\frac{1}{3}x^3e^{-3x} - \frac{1}{3}x^2e^{-3x} - \frac{2}{9}xe^{-3x} - \frac{2}{27}e^{-3x} + c$

Exercises 7.3 ■ page 499

3. $\frac{1}{5}\sin^5 x + c$ **5.** $\frac{1}{4}\sin^4 x + c$ **7.** $-\frac{1}{3}\cos^3 x + c$

9. $\frac{1}{2}\sin^2 x + c$ **11.** $\frac{1}{2}x + \frac{1}{4}\sin 2x + c$ **13.** $\frac{1}{3}\sec^3 x + c$

15. $\frac{1}{7}\tan^7 x + \frac{1}{5}\tan^5 x + c$ **17.** $\frac{1}{8}x - \frac{1}{32}\sin 4x + c$

19. $-\frac{1}{5}\csc^5 x + \frac{1}{3}\csc^3 x + c$ **21.** $\frac{2}{7}\cos^{7/2}x - \frac{2}{3}\cos^{3/2}x + c$

23. $-\frac{1}{5}\cot^5 x - \frac{1}{3}\cot^3 x + c$ **25.** $\frac{1}{6}\tan^6 x + \frac{1}{4}\tan^4 x + c$

27. $-\frac{1}{9}\dfrac{\sqrt{9-x^2}}{x} + c$ **29.** $2\sin^{-1}\left(\dfrac{x}{2}\right) + \dfrac{x\sqrt{4-x^2}}{2} + c$

31. $\frac{1}{3}(x^2-4)^{3/2} + 4\sqrt{x^2-4} + c$

33. $2\cosh^{-1}\left(\dfrac{x}{2}\right) + c$

35. $\frac{1}{3}(x^2+4)^{3/2} - 4\sqrt{x^2+4} + c$

37. $\frac{1}{2}x\sqrt{x^2+4} + 2\ln\left|\dfrac{x}{2} + \dfrac{\sqrt{x^2+4}}{2}\right| + c$

39. $\frac{1}{3}(x^2+4)^{3/2} + c$

41. $\frac{1}{4}\tan 4x + \frac{1}{2}\tan^2 x + c; \frac{1}{4}\sec^4 x + c$ **45.** $\frac{1}{2}RI^2$

Exercises 7.4 ■ page 507

3. $\dfrac{3}{x+1} - \dfrac{2}{x-1}; 3\ln|x+1| - 2\ln|x-1| + c$

5. $\dfrac{2}{x+1} + \dfrac{4}{x-2}; 2\ln|x+1| + 4\ln|x-2| + c$

7. $\dfrac{\frac{1}{5}}{x+2} + \dfrac{\frac{4}{5}}{x-3}; \frac{1}{5}\ln|x+2| + \frac{4}{5}\ln|x-3| + c$

9. $\dfrac{2}{x+1} + \dfrac{\frac{1}{2}}{x-2} - \dfrac{\frac{5}{2}}{x};$
$2\ln|x+1| + \frac{1}{2}\ln|x-2| - \frac{5}{2}\ln|x| + c$

11. $\dfrac{11}{x+4} + \dfrac{2}{x-2} + x - 2;$
$11\ln|x+4| + 2\ln|x-2| + \frac{1}{2}x^2 - 2x + c$

13. $\dfrac{1}{x+1} - \dfrac{2}{x-1} + \dfrac{1}{x};$
$\ln|x+1| - 2\ln|x-1| + \ln|x| + c$

15. $\dfrac{2}{x+2} - \dfrac{1}{(x+2)^2}; 2\ln|x+2| + (x+2)^{-1} + c$

17. $\dfrac{\frac{1}{4}}{x+2} + \dfrac{\frac{3}{2}}{(x+2)^2} - \dfrac{\frac{1}{4}}{x};$
$\frac{1}{4}\ln|x+2| - \frac{3}{2}(x+2)^{-1} - \frac{1}{4}\ln|x| + c$

19. $\dfrac{1}{x+2} - \dfrac{3}{x+1} + \dfrac{2}{x};$
$\ln|x+2| - 3\ln|x+1| + 2\ln|x| + c$

21. $\dfrac{-2x+1}{x^2+1} + \dfrac{2}{x}; -\ln(x^2+1) + \tan^{-1}x + 2\ln|x| + c$

23. $\dfrac{\frac{3}{2}}{x+1} + \dfrac{\frac{1}{2}}{x-1} + \dfrac{-2x+1}{x^2+1};$
$\frac{3}{2}\ln|x+1| + \frac{1}{2}\ln|x-1| - \ln(x^2+1) + \tan^{-1}x + c$

25. $3 + \dfrac{1}{x+1} + \dfrac{2}{x-2}; 3x + \ln|x+1| + 2\ln|x-2| + c$

27. $\dfrac{2}{x+1} + \dfrac{1}{(x+1)^2}; 2\ln|x+1| - (x+1)^{-1} + c$

29. $\dfrac{2}{x^2+1} + \dfrac{1}{x}; 2\tan^{-1}x + \ln|x| + c$

31. $\dfrac{3}{x} + \dfrac{x-3}{x^2+x+1};$
$3\ln|x| + \frac{1}{2}\ln|x^2+x+1| - \dfrac{7}{\sqrt 3}\tan^{-1}\left(\dfrac{2x+1}{\sqrt 3}\right) + c$

33. $3 + \dfrac{2}{x-1} + \dfrac{x-2}{x^2+1};$
$3x + 2\ln|x-1| + \frac{1}{2}\ln(x^2+1) - 2\tan^{-1}x + c$

35. $\dfrac{4}{x^2+1} - \dfrac{2}{(x^2+1)^2}$ **37.** $\dfrac{2}{x^2+4} - \dfrac{4}{(x^2+4)^2}$

39. $\dfrac{4}{x^2+x+1} - \dfrac{4x+1}{(x^2+x+1)^2}$

41. $\dfrac{1}{y} + \dfrac{1}{1-y}; \dfrac{ce^{at}}{1+ce^{at}}$

Exercises 7.5 ■ page 515

3. $\dfrac{1}{8(2+4x)} + \dfrac{1}{16}\ln|2+4x| + c$

5. $\frac{2}{15}(3e^x - 2)(1 + e^x)^{3/2} + c$

7. $\frac{1}{8}x\sqrt{1+4x^2} - \frac{1}{16}\ln(2x + \sqrt{1+4x^2}) + c$

9. $\dfrac{x^3}{24}(2x^6-4)\sqrt{4-x^6} + \frac{2}{3}\sin^{-1}\left(\dfrac{x^3}{2}\right) + c$

11. $\ln(e^x + \sqrt{4+e^{2x}}) + c$

13. $\dfrac{-1}{x-3}\sqrt{9-(x-3)^2} - \sin^{-1}\left(\dfrac{x-3}{3}\right) + c$

15. $\frac{1}{5}\tan^5 x - \frac{1}{3}\tan^3 x + \tan x - x + c$

17. $\dfrac{1}{2}\ln\left|\dfrac{\sqrt{4+\sin x}-2}{\sqrt{4+\sin x}+2}\right| + c$

19. $\frac{1}{2}\cos x^2 + \frac{1}{2}x^2\sin x^2 + c$

21. $-\frac{2}{3}(\cos x - 2)\sqrt{1+\cos x} + c$

23. $\frac{1}{2}\sin x\sqrt{4+\sin^2 x} - 2\ln(\sin x + \sqrt{4+\sin^2 x}) + c$

25. $\frac{1}{4}e^{-2/x^2} + c$ **27.** $-\sqrt{4x-x^2} + 2\cos^{-1}\left(\dfrac{2-x}{2}\right) + c$

29. $e^x\tan^{-1}(e^x) - \frac{1}{2}\ln(1+e^{2x}) + c$

Exercises 7.6 ■ page 524

3. $-\frac{1}{4}$ **5.** 1   **7.** 3   **9.** 0   **11.** 1   **13.** $-\frac{1}{6}$
15. $\frac{1}{2}$ **17.** 0   **19.** 1   **21.** $-\pi$ **23.** 0   **25.** 0
27. –2   **29.** 0   **31.** 0   **33.** 0   **35.** e **37.** ∞
39. 1   **41.** e^x **43.** not indeterminate

Exercises 7.7 ■ page 536

3. improper   **5.** not   **7.** improper   **9.** improper
11. $\frac{3}{2}$ **13.** diverges   **15.** diverges   **17.** 3
19. $2\sqrt 2$ **21.** –1   **23.** diverges   **25.** diverges
27. –1   **29.** diverges   **31.** diverges   **33.** π
35. diverges   **37.** diverges   **39.** π **41.** diverges
43. diverges   **45.** diverges   **47.** $r = 1$

49. $c < 0; c > 0$ **51.** $\dfrac{x}{1+x^3} < \dfrac{1}{x^2}$, converges

53. $\dfrac{x}{x^{3/2}-1} > \dfrac{1}{\sqrt x}$, diverges

55. $\dfrac{3}{x+e^x} < \dfrac{3}{e^x}$, converges

57. $\dfrac{\sin x}{1+e^x} < \dfrac{1}{e^x}$, converges   **59.** $\dfrac{x^2 e^x}{\ln x} > e^x$, diverges

61. $\frac{1}{2}\ln 4 - \frac{1}{4}$ **65.** (a) 2   (b) 4   (c) r **67.** $\dfrac{1}{r}$

69. $e^{-1} \approx .37$

Chapter 7 Review Exercises   ■   page 538

1. $2e^{\sqrt{x}} + c$ **3.** $\frac{1}{2}\sin^{-1}x - \frac{1}{2}x\sqrt{1-x^2} + c$

5. $-\frac{1}{3}x^2 e^{-3x} - \frac{2}{9}xe^{-3x} - \frac{2}{27}e^{-3x} + c$ **7.** $\frac{1}{2}\tan^{-1}x^2 + c$

9. $\frac{1}{4}\ln(4+x^4) + c$ **11.** $\frac{1}{3}x^3 + c$

13. $\frac{1}{9}\sin 3 - \frac{1}{3}\cos 3$ **15.** $\dfrac{3\pi}{16}$ **17.** $\dfrac{2}{\pi}$ **19.** $4\ln 2 - \frac{15}{16}$

21. $\frac{1}{3}\sin^3 x + c$ **23.** $\frac{1}{4}\sin^4 x - \frac{1}{6}\sin^6 x + c$

25. $\frac{1}{5}\tan^5 x + \frac{1}{3}\tan^3 x + c$ **27.** $\frac{2}{3}(\sin x)^{3/2} - \frac{2}{7}(\sin x)^{7/2} + c$

29. $\tan^{-1}\left(\dfrac{x+2}{2}\right) + c$ **31.** $-\dfrac{\sqrt{4-x^2}}{2x} + c$

33. $-\dfrac{x^2}{3}\sqrt{9-x^2} - 6\sqrt{9-x^2} + c$

35. $\dfrac{x^2}{3}\sqrt{x^2+9} - 6\sqrt{x^2+9} + c$

37. $3\ln|x+1| - 2\ln|x+2| + c$

39. $3\ln|x| + 2\ln|x-2| - \ln|x+2| + c$

41. $\frac{1}{5}e^x \cos 2x + \frac{2}{5}e^x \sin 2x + c$ **43.** $\frac{1}{3}(x^2+1)^{3/2} + c$

45. $\dfrac{\frac{4}{5}}{x-4} - \dfrac{\frac{4}{5}}{x+1}$ **47.** $\dfrac{3}{x} - \dfrac{1}{x+2} - \dfrac{2}{x-1}$

49. $\dfrac{1}{x+2} - \dfrac{4}{(x+2)^2}$

51. $\frac{1}{8}e^x(4+2e^{2x})\sqrt{4+e^{2x}} - 2\ln(e^x + \sqrt{4+e^{2x}}) + c$

53. $\frac{1}{3}\tan x \sec^2 x + \frac{2}{3}\tan x + c$

55. $\dfrac{4}{3(3-x)} - \dfrac{4}{9}\ln\left|\dfrac{3-x}{x}\right| + c$

57. $-\dfrac{\sqrt{9+4x^2}}{x} + 2\ln\left(x + \sqrt{\frac{9}{4}+x^2}\right) + c$

59. $\sqrt{4-x^2} - 2\ln\left|\dfrac{2+\sqrt{4-x^2}}{x}\right| + c$ **61.** $\frac{3}{2}$

63. ∞ **65.** e^3 **67.** 1   **69.** diverges   **71.** 3

73. π **75.** diverges

CHAPTER 8

Exercises 8.1   ■   page 551

5. $1, \frac{3}{4}, \frac{5}{9}, \frac{7}{16}, \frac{9}{25}, \frac{11}{36}$ **7.** $4, 2, \frac{2}{3}, \frac{1}{6}, \frac{1}{30}, \frac{1}{180}$ **9.** converges to 0

11. converges to 1   **13.** converges to 0

15. converges to $\frac{3}{2}$ **17.** diverges

19. converges to $\frac{1}{3}$ **21.** diverges   **23.** converges to 0

25. diverges   **27.** converges to 0   **29.** converges to 0

31. diverges   **33.** converges to 0   **39.** decreasing

41. increasing   **43.** decreasing   **45.** neither

47. $|a_n| < 3$ **49.** $|a_n| < \frac{1}{2}$

51. $a_n = 144 - 36\pi \approx 30.9$

53. 3.096, 3.144, 3.194; 3.594 < 3.721; 4.376 < 4.473;
5.589 > 5.33; 4.131 × 10^{53} billion

57. converges to 1   **59.** $\sqrt{c}$

Exercises 8.2   ■   page 561

5. converges to $\frac{15}{4}$ **7.** converges to $\frac{3}{8}$ **9.** diverges

11. converges to 3   **13.** diverges   **15.** diverges

17. converges to 1   **19.** converges to $\frac{3}{2}$ **21.** diverges

23. converges to 5   **25.** converges to 2   **27.** diverges

29. converges   **31.** converges   **33.** diverges

39. 64; 256; 4^{m-1} **41.** $\dfrac{0.9}{1-0.1} = 1$ **43.** 1.3589L

45. 1.002004008...   **47.** $\dfrac{p^2}{1-2p(1-p)} > p$ if $p > \frac{1}{2}$; 0.692

49. $30 + 15 + \frac{15}{2} + \cdots = 60$ miles   **51.** \$400,000

53. $2(1 - e^{-0.1}) \approx 0.19$

Exercises 8.3   ■   page 571

5. diverges   **7.** converges   **9.** diverges   **11.** diverges

13. diverges   **15.** converges   **17.** converges

19. converges   **21.** diverges   **23.** diverges

25. diverges   **27.** converges   **29.** diverges

31. converges   **33.** diverges   **35.** diverges

37. diverges   **39.** converges   **45.** $p > 1$ **47.** $p > 1$

51. series converges for $x > 1$ **53.** 1.0823, $x = 4$

55. 1.0041, $x = 8$

Exercises 8.4   ■   page 579

5. convergent   **7.** convergent   **9.** divergent

11. convergent   **13.** convergent   **15.** divergent

17. divergent   **19.** convergent   **21.** convergent

23. divergent   **25.** convergent   **27.** divergent

29. convergent   **31.** 3.60   **33.** −0.23   **35.** 1.10

37. 3.78   **39.** 20,000   **41.** 11   **43.** 11

45. $f'(k) < 0$ for $k \geq 2$

47. positives diverge, negatives converge

49. $\dfrac{\frac{3}{4}}{1+\frac{3}{4}} = \dfrac{3}{7}$

Exercises 8.5   ■   page 586

5. absolutely convergent   **7.** divergent

9. conditionally convergent   **11.** absolutely convergent

13. divergent   **15.** absolutely convergent   **17.** divergent

19. conditionally convergent   **21.** absolutely convergent

23. absolutely convergent   **25.** conditionally convergent

27. conditionally convergent   **29.** absolutely convergent

31. divergent   **33.** absolutely convergent

35. absolutely convergent   **37.** absolutely convergent

39. absolutely convergent   **41.** absolutely convergent

Exercises 8.6   ■   page 594

5. $\displaystyle\sum_{k=0}^{\infty} 2x^k, r = 1, (-1, 1)$

7. $\displaystyle\sum_{k=0}^{\infty} (-1)^k 3x^{2k}, r = 1, (-1, 1)$

9. $\displaystyle\sum_{k=0}^{\infty} 2x^{3k+1}, r = 1, (-1, 1)$

11. $\displaystyle\sum_{k=0}^{\infty} (-1)^k 4^{k+1} x^k, r = \frac{1}{4}, \left(-\frac{1}{4}, \frac{1}{4}\right)$

13. $\displaystyle\sum_{k=0}^{\infty} (-1)^k \frac{1}{2^{2k+1}} x^k, r = 4, (-4, 4)$ **15.** $(-3, -1), \dfrac{-1}{1+x}$

17. $(0, 1), \dfrac{1}{2-2x}$ **19.** $(-2, 2), \dfrac{2}{2+x}$ **21.** $\infty, (-\infty, \infty)$

23. 4, $(-4, 4)$ **25.** 3, $(-2, 4]$ **27.** 0, $\{x = -1\}$

29. 1, $[0, 2)$ **31.** 1, $(2, 4)$ **33.** ∞, $(-\infty, \infty)$

35. $\frac{1}{2}$, $[-\frac{5}{2}, -\frac{3}{2}]$ **37.** $\frac{1}{4}$, $[-\frac{1}{4}, \frac{1}{4})$

39. $\displaystyle\sum_{k=0}^{\infty} (-1)^k \frac{3}{2k+1} x^{2k+1}, r = 1$

41. $\displaystyle\sum_{k=1}^{\infty} 2kx^{2k-1}, r = 1$ **43.** $\displaystyle\sum_{k=0}^{\infty} (-1)^k \frac{1}{k+1} x^{2k+2}, r = 1$

45. $\displaystyle\sum_{k=1}^{\infty} (-1)^{k+1} 4^{k-1} k x^{k-1}, r = \frac{1}{4}$ **47.** $(-\infty, \infty)$, $\{x = 0\}$

49. $(-\infty, 0), (-\infty, 0)$ **51.** $(a - b, a + b), r = b$

53. r **55.** $r = 1, 1.003005007...$

57. no x's for which both series converge

Exercises 8.7   ■   page 607

7. $\displaystyle\sum_{k=0}^{\infty} (-1)^k \frac{x^{2k}}{(2k)!}, (-\infty, \infty)$

9. $\displaystyle\sum_{k=0}^{\infty} (-1)^k \frac{x^{2k+1}}{(2k+1)!}, (-\infty, \infty)$

11. $\displaystyle\sum_{k=0}^{\infty} (-1)^k \frac{x^{k+1}}{k+1}, (-1, 1]$

13. $\displaystyle\sum_{k=0}^{\infty} (-1)^k (k+1) x^k, (-1, 1)$

15. $\displaystyle\sum_{k=0}^{\infty} \frac{(x-1)^k}{k!}, (-\infty, \infty)$

17. $1 + \displaystyle\sum_{k=1}^{\infty} (-1)^{k+1} \frac{e^{-k}}{k} (x - e)^k, (0, 2e]$

19. $\displaystyle\sum_{k=0}^{\infty} (-1)^k (x - 1)^k, (0, 2)$

21. $1 - \frac{1}{2}x^2 + \frac{1}{24}x^4 - \frac{1}{720}x^6 + \frac{1}{40320}x^8$

23. $1 + \frac{1}{2}(x - 1) - \frac{1}{8}(x - 1)^2 + \frac{1}{16}(x - 1)^3 - \frac{5}{128}(x - 1)^4$
$+ \frac{7}{256}(x - 1)^5 - \frac{21}{1024}(x - 1)^6$

25. $e^2 [1 + (x - 2) + \frac{1}{2}(x - 2)^2 + \frac{1}{6}(x - 2)^3 + \frac{1}{24}(x - 2)^4$
$+ \frac{1}{120}(x - 2)^5 + \frac{1}{720}(x - 2)^6]$

27. $\frac{1}{2} - \frac{1}{16}(x - 4) + \frac{3}{256}(x - 4)^2 - \frac{5}{2048}(x - 4)^3$
$+ \frac{35}{65536}(x - 4)^4$

29. $|R_n(x)| \le \dfrac{|x|^{n+1}}{(n+1)!} \to 0$

31. $|R_n(x)| = \dfrac{1}{n+1} \left| \dfrac{x-1}{z} \right|^{n+1} \to 0$

33. (a) 0.04879   (b) $\dfrac{0.05^5}{5}$ (c) 7

35. (a) 1.0488   (b) $\dfrac{7(0.1)^5}{256}$ (c) 8

37. (a) 1.10517   (b) $e^{0.1} \dfrac{(0.1)^5}{5!}$ (c) 6

39. e^x with $x = 2$

41. $\sin x$ with $x = \pi$ **43.** $\tan^{-1} x$ with $x = 1$

45. $\displaystyle\sum_{k=0}^{\infty} (-1)^k \frac{(2x)^k}{k!}; r = \infty$ **47.** $\displaystyle\sum_{k=0}^{\infty} (-1)^k \frac{x^{2k+1}}{k!}; r = \infty$

49. $\displaystyle\sum_{k=0}^{\infty} (-1)^k \frac{x^{4k+2}}{(2k+1)!}; r = \infty$

51. $\displaystyle\sum_{k=0}^{\infty} (-1)^k \frac{9^k x^{2k}}{(2k)!}; r = \infty$

53. $-\frac{1}{2}$ **55.** $-\frac{1}{2}$ **57.** 1   **59.** $\frac{17}{9}$ **61.** $\frac{23}{15}$ **63.** $\frac{11}{30}$

67. $\frac{98}{3}$ miles   **69.** $\sqrt{.2}c \approx 83{,}000$ miles per second

71. $e^c \displaystyle\sum_{k=0}^{\infty} \frac{(x - c)^k}{k!}$

Exercises 8.8   ■   page 620

5. $\displaystyle\sum_{k=1}^{\infty} (-1)^{k+1} \frac{2}{k} \sin kx$

7. $\pi - \displaystyle\sum_{k=1}^{\infty} \frac{8}{(2k-1)^2 \pi} \cos(2k - 1)x$

9. $\displaystyle\sum_{k=1}^{\infty} \frac{-4}{\pi(2k-1)} \sin(2k-1)x$ **11.** $3 \sin 2x$

13. $\displaystyle\sum_{k=1}^{\infty} (-1)^k \frac{2}{k\pi} \sin k\pi x$ **15.** $\frac{1}{3} + \displaystyle\sum_{k=1}^{\infty} (-1)^k \frac{4}{k^2 \pi^2} \cos k\pi x$

17. $\frac{1}{4} + \displaystyle\sum_{k=1}^{\infty} \frac{-2}{(2k-1)^2 \pi^2} \cos(2k - 1)\pi x$
$+ \displaystyle\sum_{k=1}^{\infty} (-1)^{k+1} \frac{1}{k\pi} \sin k\pi x$

19.   **21.**

23.

33. odd   **37.** sine   **39.** both   **41.** both

45.

47.

49. overshoot at edges

51. $A = -1, B = 4\pi; A = 1, B = 2\pi$

Chapter 8 Review Exercises ■ **page 623**

1. converges to 0 **3.** converges to 0 **5.** converges to 0
7. diverges   **9.** 8 **11.** $\frac{4}{3}$ **13.** -0.40 **15.** diverges
17. converges   **19.** diverges   **21.** converges
23. converges   **25.** converges   **27.** converges
29. diverges   **31.** converges   **33.** converges
35. converges conditionally   **37.** converges absolutely

39. $p > 1$ **41.** 1732 **43.** $\displaystyle\sum_{k=0}^{\infty} (-1)^k \frac{x^k}{4^{k+1}}, r = 4$

45. $\displaystyle\sum_{k=0}^{\infty} (-1)^k \frac{x^{2k}}{3^k}, r = \sqrt{3}$

47. $\displaystyle\sum_{k=0}^{\infty} (-1)^k \frac{x^{k+1}}{(k+1)4^{k+1}}, r = 4$ **49.** $(-1, 1)$

51. $(-1, 1]$ **53.** $(-\infty, \infty)$ **55.** $\left(\frac{5}{3}, \frac{7}{3}\right)$

57. $\displaystyle\sum_{k=0}^{\infty} (-1)^k \frac{x^{2k+1}}{(2k+1)!}$

59. $(x - 1) - \frac{1}{2}(x - 1)^2 + \frac{1}{3}(x - 1)^3 - \frac{1}{4}(x - 1)^4$

61. $0.1822666, 10$ **63.** $\displaystyle\sum_{k=0}^{\infty} (-1)^k \frac{3^k x^{2k}}{k!}, r = \infty$ **65.** $\frac{5}{12}$

67. $\displaystyle\sum_{k=1}^{\infty} (-1)^{k+1} \frac{4}{k\pi} \sin\left(\frac{k\pi}{2}x\right)$

69.   **71.**

73. $\frac{2}{3}$ **75.** $r = \dfrac{1 + \sqrt{2}}{2}$

77. $2, 1.5, 1.67, 1.6, 1.625 \rightarrow 1.618\ldots$

CHAPTER 9

Exercises 9.1 ■ **page 634**

5. 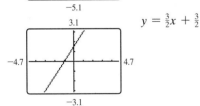 $x^2 + y^2 = 9$

7. $(x - 1)^2 + (y + 2)^2 = 4$

9. $y = \frac{3}{2}x + \frac{3}{2}$

11. $y = x^2 - 2x + 3$

13. 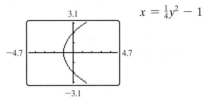 $x = \frac{1}{4}y^2 - 1$

15. $y = x + 2, x \geq -1$

17.

19.

21.

23.

25.

27.

29.

31.

33.

35. C   **37.** B   **39.** A

41. $x = 3t, y = 1 + 3t, 0 \le t \le 1$
43. $x = -2 + 8t, y = 4 - 3t, 0 \le t \le 1$
45. $x = t, y = t^2 + 1, 1 \le t \le 2$
47. $x = 2 - t, y = -t^2 + 4t - 2, 0 \le t \le 2$
49. $x = 2 + 3 \cos t, y = 1 + 3 \sin t, 0 \le t \le 2\pi$
51. $(2, 3)$ and $(-3, 8)$ **53.** $(2, 1)$ and $(3, 0)$
55. yes, at $(250, 100)$
57. $y = 0$ at $t = 2, d = 0$ or $d = 5$
59. $x = 1 - 2y^2, y = \pm 2x\sqrt{1 - x^2}$
61.

Exercises 9.2 ■ page 643

5. (a) -1 (b) 1 (c) undefined
7. (a) $-\frac{3}{2}$ (b) 0 (c) 0
9. (a) 2 (b) undefined   (c) 0
11. $t = 1$ or -1: 1 and -1
13. (a) $(0, 4), (0, -4)$ (b) $(-2, 0), (2, 0)$
15. (a) $\left(\frac{\sqrt{2}}{2}, 1\right), \left(\frac{\sqrt{2}}{2}, -1\right), \left(-\frac{\sqrt{2}}{2}, 1\right)\left(-\frac{\sqrt{2}}{2}, -1\right)$
    (b) $(1, 0), (-1, 0)$
17. (a) $(0, -3)$ (b) $(-1, 0)$
19. (a) $(0, 1)$ (b) $(0, -3)$
21. (a) $x'(0) = 0, y'(0) = 2$, speed $= 2$, up
    (b) $x'\left(\frac{\pi}{2}\right) = -2, y'\left(\frac{\pi}{2}\right) = 0$, speed $= 2$, left
23. (a) $x'(0) = 20, y'(0) = -2$, speed $= 2\sqrt{101}$, right/down
    (b) $x'(2) = 20, y'(2) = -66$, speed $= \sqrt{4756}$, right/down
25. (a) $x'(0) = 5, y'(0) = 4$, speed $= \sqrt{41}$, right/up
    (b) $x'\left(\frac{\pi}{2}\right) = 0, y'\left(\frac{\pi}{2}\right) = -9$, speed $= 9$, down
27. 6π **29.** $\frac{3\pi}{8}$ **31.** $\frac{4}{3}$ **33.** $\frac{256}{15}$
35. at $(-1, 0)$, speed is 4; at $(3, 0)$, speed is 0
37. $5:3$,

39. $x = 2 \cos t + \sin 3t, y = 2 \sin t + \cos 3t$;

Min/max speeds: $1, 5$

41. speed $= 4, (\tan 4t)(-\cot 4t) = -1$

Exercises 9.3 ■ page 652

5. 19.37 **7.** 15.86 **9.** 15.69 **11.** 9.43 **13.** 2.29
15. 2.42 **17.** 6.91 **19.** $4.4864k$ **21.** $4.4569k$
23. undefined; 3.889 **25.** undefined; 4.066 **27.** 85.8
29. 29.7 **31.** 85.8

Exercises 9.4 ■ page 663

5. $(2, 0)$ **7.** $(2, 0)$ **9.** $(-3, 0)$
11. $\left(2\sqrt{2}, -\frac{\pi}{4} + 2\pi n\right), \left(-2\sqrt{2}, \frac{3\pi}{4} + 2\pi n\right)$
13. $\left(3, \frac{\pi}{2} + 2\pi n\right), \left(-3, \frac{3\pi}{2} + 2\pi n\right)$
15. $\left(\sqrt{5}, -\tan^{-1}\left(\frac{1}{2}\right) + 2\pi n\right),$
    $\left(-\sqrt{5}, -\tan^{-1}\left(\frac{1}{2}\right) + (2n + 1)\pi\right)$
17. $\left(2, \frac{5\pi}{6} + 2\pi n\right), \left(-2, -\frac{\pi}{6} + 2\pi n\right)$ **19.** $(1, -\sqrt{3})$
21. $(0, 0)$ **23.** $\left(3 \cos \frac{\pi}{8}, 3 \sin \frac{\pi}{8}\right) \approx (2.77, 1.15)$
25. $(-3 \cos 1, -3 \sin 1) \approx (-1.62, -2.52)$
27.

$x^2 + y^2 = 16$

29.

$y = \frac{1}{\sqrt{3}} x$

31.

$x^2 + y^2 = x$

33.

$x^2 + y^2 = 3y$

35. $r = 0$ at $\theta = \dfrac{k\pi}{4}$ (k odd),

$0 \leq \theta \leq 2\pi$

37. $r = 0$ at $\theta = \dfrac{n\pi}{3}, 0 \leq \theta \leq \pi$

39. $r > 0, 0 \leq \theta \leq 2\pi$

41. $r = 0$ at $\theta = \dfrac{\pi}{6} + 2\pi n$,

$\dfrac{5\pi}{6} + 2\pi n; 0 \leq \theta \leq 2\pi$

43. $r = 0$ at $\theta = \dfrac{3\pi}{2} + 2\pi n$,

$0 \leq \theta \leq 2\pi$

45. $r = 0$ at $\theta = 0, -\infty < \theta < \infty$

47. $r = 0$ at $\theta = \dfrac{3\pi}{4} + \pi n$

$0 \leq \theta \leq \pi$

49. $r = 0$ at $\theta = \dfrac{3\pi}{4} + \pi n$

$0 \leq \theta \leq \pi$

51. $r = 0$ at $\theta = 0, -\infty < \theta < \infty$

53. $r = 0$ at

$\theta = \dfrac{2\pi}{9} + \dfrac{2\pi n}{3}, \dfrac{4\pi}{9} + \dfrac{2\pi n}{3},$

$0 \leq \theta \leq 2\pi$

55. $r \neq 0, -\dfrac{\pi}{2} < \theta < \dfrac{3\pi}{2}$

57. 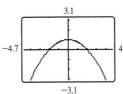 $r \neq 0, -\pi < \theta < \pi$

59. $r = 0$ at $\theta = 0, \pm\dfrac{\pi}{2}, \pm\dfrac{3\pi}{2}$

61. circle, radius $\dfrac{a}{2}$, center $\left(\dfrac{a}{2}, 0\right)$

63. rose with n leaves (n odd) or $2n$ leaves (n even)

65. $r = \pm2\sqrt{-\sec 2\theta}$ **67.** $r = 4$ **69.** $r = 3 \csc \theta$

71. n wide overlapping petals on a flower when $0 \leq \theta \leq n\pi$, up to $n = 24$; graph repeats for larger domains.

75. $r_1(A) = d \cos A - \sqrt{d^2 \cos^2 A - (d^2 - h^2)}$

$r_2(A) = d + b\left(1 - \left(\dfrac{A}{\sin^{-1}\left(\frac{h}{d}\right)}\right)^2\right)$

$A_1 = -\sin^{-1}\left(\dfrac{h}{d}\right), A_2 = \sin^{-1}\left(\dfrac{h}{d}\right)$

Exercises 9.5 ■ page 672

5. $\sqrt{3}$ **7.** undefined   **9.** 0   **11.** 1   **13.** $\dfrac{1}{\sqrt{3}}$

15. $\left(\dfrac{\sqrt{3}}{2}, \dfrac{1}{2}\right), \left(-\dfrac{\sqrt{3}}{2}, \dfrac{1}{2}\right), (0, -1)$

17. $(3\sqrt{2}, -3\sqrt{2}), (-3\sqrt{2}, 3\sqrt{2})$

19. $\dfrac{\pi}{12}$ **21.** 0.3806   **23.** π **25.** 0.2718   **27.** 0.147

29. 24.187   **31.** 6.968   **33.** 1.927

35. $(0, 0), (0.3386, -0.75), (1.6614, -0.75)$

37. $(0, 0)(1.2071, 1.2071), (-0.2071, -0.2071)$ **39.** 16

41. 6.6824   **43.** 20.0158   **45.** 31.2%   **47.** 62.6%

49. $\theta = 0: 0, \theta = \dfrac{\pi}{3}: \sqrt{3}, \theta = \dfrac{2\pi}{3}: -\sqrt{3}$

Exercises 9.6 ■ page 680

5. $y = -\frac{1}{4}x^2$ **7.** $x = \frac{1}{4}y^2 + 2$ **9.** $\dfrac{x^2}{12} + \dfrac{(y-3)^2}{16} = 1$

11. $\dfrac{(x-4)^2}{16} + \dfrac{(y-1)^2}{12} = 1$ **13.** $\dfrac{(x-2)^2}{1} - \dfrac{y^2}{3} = 1$

15. $\dfrac{(y-4)^2}{1} - \dfrac{(x-2)^2}{3} = 1$

17. parabola, $(-1, -1)$, $(-1, -\frac{7}{8})$, $y = -\frac{9}{8}$

19. ellipse, $(1, -1)$ and $(1, 5)$, $(1, 2 - \sqrt{5})$ and $(1, 2 + \sqrt{5})$

21. hyperbola, $(-2, 0)$ and $(4, 0)$, $(1 - \sqrt{13}, 0)$ and $(1 + \sqrt{13}, 0)$

23. hyperbola, $(-2, -5)$ and $(-2, 3)$, $(-2, -1 - \sqrt{20})$ and $(-2, -1 + \sqrt{20})$

25. ellipse, $(-1, 0)$ and $(5, 0)$, $(2 - \sqrt{8}, 0)$ and $(2 + \sqrt{8}, 0)$

27. parabola, $(-1, -2)$, $(-1, -1)$, $y = -3$

29. $y = \frac{1}{8}(x - 2)^2 - 1$

31. $\dfrac{(x-2)^2}{16} + \dfrac{(y-2)^2}{12} = 1$

33. $\dfrac{(y-1)^2}{4} - \dfrac{x^2}{5} = 1$

35. $(\frac{1}{16}, 0)$ **37.** $(0, \frac{1}{8})$ **39.** 20 inches   **41.** $(0, -8)$

43. $(-2, 0)$ **45.** $(\sqrt{300}, 0)$ and $(-\sqrt{300}, 0)$

47. $-t^2 + 16t$, boomerang

Exercises 9.7   ■   page 686

3. $r = \dfrac{1.2}{0.6 \cos \theta + 1}$

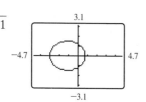

5. $r = \dfrac{2}{\cos \theta + 1}$

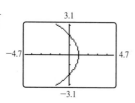

7. $r = \dfrac{1.2}{0.6 \sin \theta + 1}$

9. $r = \dfrac{2}{\sin \theta + 1}$

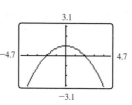

11. $r = \dfrac{-0.8}{0.4 \cos \theta - 1}$

13. $r = \dfrac{-4}{2 \cos \theta - 1}$

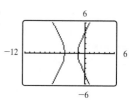

15. $r = \dfrac{-0.8}{0.4 \sin \theta - 1}$

17. $r = \dfrac{-2}{\sin \theta - 1}$

19.

          hyperbola rotated

21.

          ellipse rotated

23.   parabola rotated

25. $x = -1 + 3 \cos t$, $y = 1 + 2 \sin t$, $0 \le t \le 2\pi$

27. $x = -1 + 4 \cosh t$, $y = 3 \sinh t$ for the right half;
$x = -1 - 4 \cosh t$, $y = 3 \sinh t$ for the left half.

29. $x = t$, $y = -\frac{1}{4}t^2 + 1$ **31.** 2.6 times as fast

Chapter 9 Review Exercises ■ **page 687**

1. $(x + 1)^2 + (y - 2)^2 = 9$

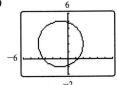

3. $y = x^2 - 2x + 1$

5.

7.

9. C   **11.** B   **13.** $x = 2 + 2t, y = 1 + 6t, 0 \leq t \leq 1$
15. (a) $\frac{1}{3}$
    (b) undefined
    (c) undefined at $t = -1$; $\frac{1}{3}$ at $t = 2$
17. $x'(0) = -3, y'(0) = 2$, speed $= \sqrt{13}$, left/up
19. 6π **21.** 1.946   **23.** 5.249   **25.** 13.593
27. 128.075
29. $x^2 + y^2 = 3x$

31.

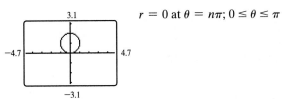

$r = 0$ at $\theta = n\pi$; $0 \leq \theta \leq \pi$

33.

$r = 0$ at
$\theta = \sin^{-1}\frac{2}{3} + 2\pi n,$
$\quad \pi - \sin^{-1}\frac{2}{3} + 2\pi n;$
$\quad 0 \leq \theta \leq 2\pi$

35.

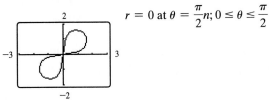

$r = 0$ at $\theta = \dfrac{\pi}{2}n$; $0 \leq \theta \leq \dfrac{\pi}{2}$

37.

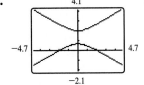

$r \neq 0$; $0 \leq \theta \leq 2\pi$
$\left(\theta \neq \dfrac{7\pi}{6}, \dfrac{11\pi}{6}\right)$

39. $r = 3$ **41.** $\dfrac{1}{\sqrt{3}}$ **43.** 0.157   **45.** 0.543
47. 2.828   **49.** 28.814   **51.** $y = \frac{1}{4}(x - 1)^2 + 1$
53. $\dfrac{(y - 2)^2}{1} - \dfrac{(x - 2)^2}{3} = 1$
55. ellipse, $(-1, -2)$ and $(-1, 8)$, $(-1, -1)$ and $(-1, 7)$
57. parabola, $(1, 4)$, $(1, \frac{15}{4})$, $y = \frac{17}{4}$ **59.** $(0, \frac{1}{2})$
61. $r = \dfrac{2.4}{0.8 \cos \theta + 1}$ **63.** $r = \dfrac{2.8}{1.4 \sin \theta + 1}$
65. $x = -1 + 3 \cos t, y = 3 + 5 \sin t, 0 \leq t \leq 2\pi$

Bibliography

This section is part bibliography and part acknowledgments. We don't intend to give a comprehensive list of related books that are worth reading. Instead, we wish to give credit to many of the sources from which we have derived inspiration. Even this list is nowhere near complete, as we could not possibly list all of the books or thank all of the professors and students from whom we've learned mathematics. We have organized the list by chapter and topic.

CHAPTER 0

Fibonacci sequence: a search on the Internet will turn up several good sites. Two of the many books with interesting discussions of the nautilus shell spiral are:

[1] Huntley, H.E., *The Divine Proportion*, Dover Publications, 1970. (Subtitled *A Study in Mathematical Beauty*, with several interesting connections to art, architecture and nature.)

[2] Stewart, Ian, *Nature's Numbers*, BasicBooks, 1995. (A collection of easy-to-read essays on current topics in mathematics. Stewart is one of the best popular mathematics authors.)

Music and mathematics: a very strong connection between mathematics and music dates back to Pythagoras. The material we develop is adapted from [3] and [4].

[3] Pierce, John R., *The Science of Musical Sound*, W. H. Freeman and Company, 1992. (This is a strong introduction to the science of music. It is fairly technical, but has good explanations and many graphs and diagrams.)

[4] Levenson, Thomas, *Measure to Measure*, Touchstone Books, 1994. (The subtitle, *A Musical History of Science*, describes the author's goal. It is readable and well written.)

[5] Hofstadter, Douglas, *Gödel, Escher, Bach*, Vintage Books, 1980. (A Pulitzer Prize-winning classic of puzzles and explorations of mathematics, art and music.)

Baseball and statistics: sadly, there has not been a *Baseball Abstract* since 1988. Each *Abstract* was filled with insights into baseball and the mathematical analysis of sports.

[6] James, Bill, *The Bill James Baseball Abstract 1988*, Ballantine Books, 1988.

Physiology: Clark's book is an interesting investigation into how (and why) the human body works the way it does.

[7] Clark, William, *Sex and the Origins of Death*, Oxford University Press, 1996.

Golf: most of what we discuss at various places in the book comes from one of the following books.

[8] Cochran, A.J. and M.R. Farrally, eds., *Science and Golf II*, E & FN Spon, 1994. (A collection of research papers on a variety of golf-related topics and at varying degrees of complexity. This is the proceedings of the 1994 World Scientific Congress of Golf.)

[9] Jorgensen, Theodore, *The Physics of Golf*, AIP Press, 1994. (A technical but practical investigation into all aspects of the golf shot.)

Graphing calculators: the manuals that come with most graphing calculators are very helpful. They include interesting examples to work out as well as basic keystroke instructions. Keep your manual and use it!

Terminal velocity: this and many other physics and sports topics are presented in Brancazio's excellent book. Most college-level physics book also treat this subject.

[10] Brancazio, Peter, *Sport Science*, Simon and Schuster, 1984. (Brancazio enthusiastically uses sports to illustrate physics concepts and physics to explain sports techniques. Thoroughly enjoyable.)

Random dot pictures: the excitement over these pictures depends largely on whether or not you can see them.

[11] Kinsman, Andrew, *Random Dot Stereograms*, Kinsman Physics, 1992. (Explains how the pictures work, with computer code for several different styles of pictures.)

[12] *Stare-EO Workshop*, N.E. Thing Enterprises. (Easy-to-use software used to create our picture.)

[13] Terrell, Maria and Robert Terrell, "Behind the Scenes of a Random Dot Stereogram," *The American Mathematics Monthly*, Mathematical Association of America, October 1994. (Extends the usage of random dot pictures to graphs of functions.)

Chaos: one of the fastest-growing research areas in applied mathematics in the 1980s and 1990s. The Gleick book is the best-known general interest chaos reader. The other books cited here are mathematics textbooks.

[14] Gleick, James, *Chaos: The Making of a New Science*, Penguin Books, 1987. (Very popular book, giving a sense of the broad scope of early chaos research.)

[15] Devaney, Robert, *Chaos, Fractals and Dynamics*, Addison-Wesley, 1990. (A nice introduction to the mathematics of chaos. The subtitle *Computer Experiments in Mathematics* is accurate.)

[16] Gulick, Denny, *Encounters with Chaos*, McGraw-Hill, 1992. (A good textbook on the mathematics of chaos theory. Some of the mathematics is at an advanced calculus level.)

[17] Barnsley, Michael, *Fractals Everywhere* , Academic Press, 1988. (Not an easy book to read, but it includes several great images related to the chaos game. Barnsley founded a successful company applying the mathematics of the chaos game to image processing.)

Circumference of the earth: the video series *The Ring of Truth* is excellent, highlighting several demonstrations of how knowledge is obtained. The book is a very good companion to the video series.

[18] Morrison, Philip and Phylis, *The Ring of Truth*, Vintage Books, 1987.

CHAPTER 1

Size of pupils: a technical reference for the equation we use is:

[19] Crawford, B.H., "The Dependence of Pupil Size upon the External Light Stimulus under Static and Variable Condition ," *Proceedings of the Royal Society Series B*, **121** (1937).

Knuckleball, keep your eye on the ball: these are two topics we return to throughout the book. The original research is presented in a highly readable manner in Watts and Bahill's book, which is the best of several books available on baseball and physics.

[20] Watts, Robert and A. Terry Bahill, *Keep Your Eye on the Ball*, W. H. Freeman, 1990.

Mississippi alligators: one of several examples in Cohen and Stewart's book illustrating the complicated interaction between hereditary and environment.

[21] Cohen, Jack and Ian Stewart, *The Collapse of Chaos*, Viking Penguin, 1994.

History of mathematics: many of our historical notes are adapted from workshop material from Professor V. Frederick Rickey. There are several good history of mathematics books available.

[22] Simmons, George F., *Calculus Gems*, McGraw-Hill, 1992.

[23] Kramer, Edna E., *The Nature and Growth of Modern Mathematics*, Princeton University Press, 1970.

[24] Smith, David E., *A Source Book in Mathematics*, Dover Publications, 1959.

Butterfly effect: the motto of chaos theory (see references [14]–[17]), the name comes from an analogy used by meteorologist and chaos pioneer Edward Lorenz.

[25] Lorenz, Edward, *The Essence of Chaos*, University of Washington Press, 1993.

CHAPTER 2

Speed of runner: an excellent discussion is in Brancazio [10]. Also see
[26] Griffing, David F., *The Dynamics of Sports*, The Dalog Company, 1982.

Hammer throw, collisions: see [10] and [26].

Desert mirage: this optical illusion is beautifully demonstrated in the video
[27] Greenler, Robert, "The Mirage, the Discovery of Greenland and the Green Flash," Blue Sky Associates, 1990.

Mental rotation: there is good evidence that some mental processes and language skills are innate. Very readable and interesting.
[28] Pinker, Steven, *The Language Instinct*, Harper Collins, 1994.

Extinction probability: one of the topics in the genetics section of a book that also explores artificial life and ecology.
[29] Sigmund, Karl, *Games of Life*, Oxford University Press, 1993.

Bishop Berkeley, Leibniz: see [22]–[24].

Plane descent: adapted from a problem in this interesting calculus text.
[30] Strang, Gilbert, *Calculus*, Wellesley-Cambridge Press, 1991.

Richard Feynman: one of the most influential scientists of the twentieth century.
[31] Feynman, Richard, *Surely You're Joking, Mr. Feynman*, Bantam Books, 1986.

Baseball, catching a fly ball: this paper critiques a previous theory of how players judge a fly ball, and introduces a new theory.
[32] McBeath, Shaffer and Kaiser, "How Baseball Outfielders Determine Where to Run to Catch Fly Balls," *Science* , April 28, 1995.

Mississippi alligators: see [21].

Chaos theory: see [14]–[16] and [25].

CHAPTER 3

Medical imaging: good, brief explanations of medical imaging techniques and other forms of digital processing can be found in this general interest book.
[33] Friedhoff, Richard and William Benzon, *Visualization*, Abrams Publishers, 1989.

Basin of attraction: more details and some nice color images can be found in
[34] Devaney, Robert and Linda Keen, ed., *Chaos and Fractals*, American Mathematical Society, 1988.

Kepler's wine cask: adapted from
[35] Balk, M.B., "The Secret of the Venerable Wine Cooper," *Quantum*, May 1990.

AIDS epidemiology: a number of preliminary studies have been published about this frightening disease, including
[36] Brookmeyer, Ron and Mitchell H. Gail, *AIDS Epidemiology: A Quantitative Approach*, Oxford University Press, 1994.
[37] Kirschner, Denise, "Using Mathematics to Understand HIV Immune Dynamics," *Notices of AMS*, February 1996.

Extinction probability: see [29].

Tree infestation: this problem is discussed in both books:
[38] Murray, J.D., *Mathematical Biology*, Springer-Verlag, 1989. (A wealth of examples in population dynamics, diffusion patterns, neural patterns and epidemiology are presented. Many models use calculus, some are more advanced.)
[39] Strogatz, Steven, *Nonlinear Dynamics and Chaos*, Addison-Wesley, 1994. (Excellent introduction to chaos with great examples throughout.)

Flying wing aircraft:
[40] Renz, Peter, "Thoughts on *Innumeracy*: Mathematics Versus the World?" *The American Mathematics Monthly* , Mathematical Association of America, October 1993.
[41] Biddle, Wayne, "Skeleton Alleged in the Stealth Bomber's Closet," *Science*, May 12, 1989.

Heart rates: chaos theory has had a profound influence on human physiology in general and cardiology in particular. The interplay between order and disorder is fascinating.
[42] Glass, Leon and Michael Mackey, *From Clocks to Chaos*, Princeton University Press, 1988. (Subtitled *Rhythms of Life*, there are innumerable graphs and some calculus and differential equations used to explain physiological processes.)

[43] Kaplan, Daniel and Leon Glass, *Understanding Nonlinear Dynamics*, Springer-Verlag, 1995. (Similar examples to [42] , but more focused on the mathematics.)

Autocatalytic reactions: the following is a collection of brief mathematical models of a variety of biological and chemical processes.

[44] Thrall, Robert M., ed., *Some Mathematical Models in Biology*, University of Michigan, 1967.

Chemical titrations: covered in this quantitative analysis chemistry book.

[45] Harris, Daniel, *Quantitative Chemical Analysis*, 3rd edition, W. H. Freeman, 1991.

Laser field: see [39].

CHAPTER 4

HIV epidemiology: see [36]–[38]

Space shuttle landing: general projectile motion with some specific equations for the space shuttle can be found in this article.

[46] Long, Lyle and Howard Weiss, "The Velocity Dependence of Aerodynamic Drag: A Primer for Mathematicians," *The American Mathematical Monthly*, Mathematical Association of America, February 1999.

Predator-prey dynamics: a common topic in differential equations texts, the pesticide paradox is in this whimsically named book.

[47] Harte, John, *Consider a Spherical Cow*, University Science Books, 1988.

Synchrony and fire flies: see [39] and this video series.

[48] *The Trials of Life: Talking to Strangers*, Ambrose Video Publishing.

CHAPTER 5

Tennis racket design: brief discussions can be found in [10] and [49], more thorough treatments in [50] and [51].

[49] Schrier, Eric and William Allman, ed., *Newton at the Bat*, Charles Scribner's Sons, 1987. (An interesting collection of essays on a variety of sports.)

[50] Torrey, Lee, *Stretching the Limits*, Dodd, Mead and Company, 1985. (Another interesting collection of essays on a variety of sports.)

[51] Brody, Howard, *Tennis Science for Tennis Players*, University of Pennsylvania Press, 1987. (A thorough and informative look at all aspects of tennis. Numerous graphs.)

Baseball forces: Adair's book is a comprehensive discussion of the physics of baseball. See [20], also.

[52] Adair, Robert, *The Physics of Baseball* , Harper and Row, 1990.

Wallaby tendons and human feet: two examples from an excellent discussion of animal movements from crawling and walking to swimming and flying.

[53] Alexander, R. McNeill, *Exploring Biomechanics*, Scientific American Library, 1992.

Projectile motion: many of our best examples come from Brancazio [10] and Alexander [53]. A general but mathematically advanced discussion is in

[54] deMestre, Neville, *The Mathematics of Projectiles in Sports*, Cambridge University Press, 1990.

Air resistance: see [46] and the following well-written module.

[55] Donley, H. Edward, "The Drag Force on a Sphere," UMAP Module 712, COMAP, Inc., 1991.

Knuckleball: see [20].

Weightless training: an entertaining description of the "vomit comet" and the making of the movie *Apollo 13* is

[56] Kluger, Jeffrey, *The Apollo Adventure*, Pocket Books, 1995.

Sweet spot: see [10], [20], [50] and [51].

Height statistics: thousands of facts about our world.

[57] Brandreth, Gyles, *Your Vital Statistics* , Citadel Press, 1986.

Batting .400: one of several examples of how the variance of a data set can be more meaningful than its mean.

[58] Gould, Stephen Jay, *Full House*, Harmony Books, 1996.

Model rockets: sophisticated scientific principles go into the design of rockets.

[59] Stine, G. Harry, *Handbook of Model Rocketry*, 6th edition, John Wiley and Sons, 1994.

Pole vault: one of several sports discussed in this overview of Summer Olympics sports.

[60] Mallette, Vincent, *The Science of the Summer Games*, Charles River Media, 1996.

CHAPTER 6

International space station: the NASA web site is excellent.

Mercator map: a variety of map-making techniques and their uses and abuses are in this unique book.

[61] Monmonier, Mark, *How to Lie with Maps*, University of Chicago Press, 1991.

The number e: the long and interesting history of e and the exponential function.

[62] Maor, Eli, *e: The Story of a Number*, Princeton University Press, 1994.

Sigmoid functions: the important role this class of functions plays in modeling human physiology is detailed in [42].

Lemmings: an exploration of various attempts to explain changes in lemming populations is the focus of this honest portrayal of population ecology. Subtitled *Beautiful Hypotheses and Ugly Facts*.

[63] Chitty, Dennis, *Do Lemmings Commit Suicide?* Oxford University Press, 1996.

Carbon-14 dating: a common example in calculus and differential equations texts, but Braun explains how it's actually done.

[64] Braun, Martin, *Differential Equations and Applications*, Springer-Verlag, 1975.

Critical threshold, spruce budworms: see [38] and [39].

Air resistance: see [46] and [55].

Zebra stripes: see [39].

Baseball, catching a fly ball: see [32].

Gateway arch: the Jefferson National Expansion Memorial has an excellent web site. Some interesting calculations are done in the module.

[65] Thayer, William, "The St. Louis Arch Problem ," UMAP Module 638, COMAP, 1984.

CHAPTER 7

Laplace transforms: most differential equations texts (see [64]) contain a chapter.

Mathematical models: several textbooks apply differential equation models to a variety of physical problems.

[66] Wan, Frederic, *Mathematical Models and Their Analysis*, Harper and Row, 1989.

[67] Giordano, Frank and Maurice Weir, *A First Course in Mathematical Modeling*, Brooks/Cole, 1985.

Jaime Escalante: dramatized in the movie *Stand and Deliver*, the life of this outstanding mathematics teacher is well told in this book.

[68] Mathews, Jay, *Escalante*, Henry Holt and Company, 1988.

Ozone layer: one of several chemistry examples from Dr. Ben Huddle of Roanoke College.

Integral tables: a CAS such as Mathematica may have better tables than these classic references.

[69] Beyer, W.H., ed., *Standard Mathematical Tables and Formulas,* 29th edition, CRC, 1991.

[70] Jeffrey, Alan, ed., *Gradshteyn and Ryzhik's Tables of Integrals, Series and Products,* 4th edition, Academic Press, 1995.

Brachistochrone and tautochrone: two classic mathematics problems. One of the best descriptions of the mathematics and personalities is in

[71] Dunham, William, *Journey Through Genius* , John Wiley and Sons, 1990.

CHAPTER 8

Music synthesizers: see [3] and [4]. We used the owner's manual for a Roland Synth-Plus 60 synthesizer.

Population explosion: this is one of the most extreme population models.

[72] Forster, Mora and Amiot, "Doomsday, Friday, 13 Nov A.D. 2026," *Science*, November 1960.

Fibonacci sequence: see [1] and [2].

Bouncing ball: a collection of well-developed and interesting chaos models with software.

[73] Tufillaro, Nicholas, Tyler Abbott and Jeremiah Reilly, *An Experimental Approach to Nonlinear Dynamics and Chaos*, Addison-Wesley, 1992.

Waiting in line: this is the second of the classic series on probability theory.

[74] Feller, William, *An Introduction to Probability Theory and Its Applications, volume 2*, John Wiley, 1966.

Ramanujan: perhaps the most interesting mathematician in history, the life of this self-trained mathematical genius is told in this excellent book.

[75] Kanigel, Robert, *The Man Who Knew Infinity*, Washington Square Press, 1991.

Unusual fractions: several interesting decimal representations and hints on how to construct your own.

[76] Kreminski, Richard, "Fun Fractions," *Mathematics Teacher*, National Council of Teachers of Mathematics, July 1998.

Gibbs phenomenon and image processing: see [33].

Continued fractions: a very lively, often irreverent, look at several aspects of the mathematics involving π.

[77] Beckmann, Petr, *A History of π*, St. Martin's Press, 1971.

CHAPTER 9

Kepler's laws: a good brief description of the history of planetary analysis, updated to include new discoveries of chaotic orbits.

[78] Peterson, Ivars, *Newton's Clock*, W. H. Freeman, 1993.

Mandelbrot set: good descriptions can be found in [14], [15] and [34]. We used the share-ware software Fractint to generate our images.

Bézier curves: computer code for many types of curves is provided in this book.

[79] Glassner, Andrew, ed., *Graphics Gems*, Academic Press, 1990.

Skiing: physics and chemistry applied to all aspects of skiing.

[80] Lind, David and Scott P. Sanders, *The Physics of Skiing*, AIP Press, 1997.

The Bernoulli brothers: see [71].

Golf putting: The paper "How to Lower Your Putting Score without Improving" by B. Hoadley in [8] focuses on the specific model we use in our book. Pelz provides a helpful, mostly nonmathematical tutorial covering all aspects of putting.

[81] Pelz, Dave, *Putt Like the Pros*, Harper and Row, 1989.

Nautilus shell: see [1].

Index

TRIGONOMETRY

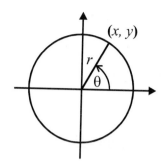

$$\sin\theta = \frac{y}{r} \qquad \sin\theta = \frac{\text{opp}}{\text{hyp}}$$

$$\cos\theta = \frac{x}{r} \qquad \cos\theta = \frac{\text{adj}}{\text{hyp}}$$

$$\tan\theta = \frac{y}{x} \qquad \tan\theta = \frac{\text{opp}}{\text{adj}}$$

hyp   **opp**

θ

adj

Reciprocals	Definitions	Pythagorean	Cofunction
$\cot\theta = \dfrac{1}{\tan\theta}$	$\cot\theta = \dfrac{\cos\theta}{\sin\theta}$	$\sin^2\theta + \cos^2\theta = 1$	$\sin\left(\frac{\pi}{2} - \theta\right) = \cos\theta$
$\sec\theta = \dfrac{1}{\cos\theta}$	$\sec\theta = \dfrac{1}{\cos\theta}$	$\tan^2\theta + 1 = \sec^2\theta$	$\cos\left(\frac{\pi}{2} - \theta\right) = \sin\theta$
$\csc\theta = \dfrac{1}{\sin\theta}$	$\csc\theta = \dfrac{1}{\sin\theta}$	$1 + \cot^2\theta = \csc^2\theta$	$\tan\left(\frac{\pi}{2} - \theta\right) = \cot\theta$

Even/Odd	Double-angle	Half-angle
$\sin(-\theta) = -\sin\theta$	$\sin 2\theta = 2\sin\theta\cos\theta$	$\sin^2\theta = \dfrac{1 - \cos 2\theta}{2}$
$\cos(-\theta) = \cos\theta$	$\cos 2\theta = \cos^2\theta - \sin^2\theta$	$\cos^2\theta = \dfrac{1 + \cos 2\theta}{2}$
$\tan(-\theta) = -\tan\theta$	$\cos 2\theta = 1 - 2\sin^2\theta$	

Addition $\sin(a + b) = \sin a \cos b + \cos a \sin b$ $\cos(a + b) = \cos a \cos b - \sin a \sin b$

Subtraction $\sin(a - b) = \sin a \cos b - \cos a \sin b$ $\cos(a - b) = \cos a \cos b + \sin a \sin b$

Sum $\sin u + \sin v = 2\sin\frac{u+v}{2}\cos\frac{u-v}{2}$ $\cos u + \cos v = 2\cos\frac{u+v}{2}\cos\frac{u-v}{2}$

Product $\sin u \sin v = \frac{1}{2}\left[\cos(u - v) - \cos(u + v)\right]$ $\cos u \cos v = \frac{1}{2}\left[\cos(u - v) + \cos(u + v)\right]$

$\sin u \cos v = \frac{1}{2}\left[\sin(u + v) + \sin(u - v)\right]$ $\cos u \sin v = \frac{1}{2}\left[\sin(u + v) - \sin(u - v)\right]$

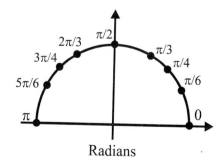

Radians

$\sin(0) = 0$	$\cos(0) = 0$	$\sin\left(\frac{\pi}{6}\right) = \frac{1}{2}$	$\cos\left(\frac{\pi}{6}\right) = \frac{\sqrt{3}}{2}$
$\sin\left(\frac{\pi}{4}\right) = \frac{\sqrt{2}}{2}$	$\cos\left(\frac{\pi}{4}\right) = \frac{\sqrt{2}}{2}$	$\sin\left(\frac{\pi}{3}\right) = \frac{\sqrt{3}}{2}$	$\cos\left(\frac{\pi}{3}\right) = \frac{1}{2}$
$\sin\left(\frac{\pi}{2}\right) = 1$	$\cos\left(\frac{\pi}{2}\right) = 0$	$\sin\left(\frac{2\pi}{3}\right) = \frac{\sqrt{3}}{2}$	$\cos\left(\frac{2\pi}{3}\right) = -\frac{1}{2}$
$\sin\left(\frac{3\pi}{4}\right) = \frac{\sqrt{2}}{2}$	$\cos\left(\frac{3\pi}{4}\right) = -\frac{\sqrt{2}}{2}$	$\sin\left(\frac{5\pi}{6}\right) = \frac{1}{2}$	$\cos\left(\frac{5\pi}{6}\right) = -\frac{\sqrt{3}}{2}$
$\sin(\pi) = 0$	$\cos(\pi) = -1$	$\sin(2\pi) = 0$	$\cos(2\pi) = 1$

DERIVATIVE FORMULAS

General Rules

$\frac{d}{dx}[f(x) + g(x)] = f'(x) + g'(x)$

$\frac{d}{dx}[f(x) - g(x)] = f'(x) - g'(x)$

$\frac{d}{dx}[cf(x)] = c\,f'(x)$

$\frac{d}{dx}[f(g(x)] = f'(g(x))\,g'(x)$

$\frac{d}{dx}[f(x)\,g(x)] = f'(x)\,g(x) + f(x)\,g'(x)$

$\frac{d}{dx}\left[\frac{f(x)}{g(x)}\right] = \dfrac{f'(x)\,g(x) - f(x)g'(x)}{[g(x)]^2}$

Power Rules

$\frac{d}{dx}(x^n) = nx^{n-1}$

$\frac{d}{dx}(c) = 0$

$\frac{d}{dx}(cx) = c$

$\frac{d}{dx}(\sqrt{x}) = \dfrac{1}{2\sqrt{x}}$

Exponential

$\frac{d}{dx}[e^x] = e^x$

$\frac{d}{dx}[a^x] = a^x \ln a$

$\frac{d}{dx}[e^{u(x)}] = e^{u(x)}\,u'(x)$

$\frac{d}{dx}[e^{rx}] = r\,e^{rx}$

Trigonometric

$\frac{d}{dx}(\sin x) = \cos x$

$\frac{d}{dx}(\cos x) = -\sin x$

$\frac{d}{dx}(\tan x) = \sec^2 x$

$\frac{d}{dx}(\cot x) = -\csc^2 x$

$\frac{d}{dx}(\sec x) = \sec x \tan x$

$\frac{d}{dx}(\csc x) = -\csc x \cot x$

Inverse Trigonometric

$\frac{d}{dx}(\sin^{-1} x) = \dfrac{1}{\sqrt{1 - x^2}}$

$\frac{d}{dx}(\cos^{-1} x) = -\dfrac{1}{\sqrt{1 - x^2}}$

$\frac{d}{dx}(\tan^{-1} x) = \dfrac{1}{1 + x^2}$

$\frac{d}{dx}(\cot^{-1} x) = -\dfrac{1}{1 + x^2}$

$\frac{d}{dx}(\sec^{-1} x) = \dfrac{1}{|x|\sqrt{x^2 - 1}}$

$\frac{d}{dx}(\csc^{-1} x) = -\dfrac{1}{|x|\sqrt{x^2 - 1}}$

Hyperbolic

$\frac{d}{dx}(\sinh x) = \cosh x$

$\frac{d}{dx}(\cosh x) = \sinh x$

$\frac{d}{dx}(\tanh x) = \operatorname{sech}^2 x$

$\frac{d}{dx}(\coth x) = -\operatorname{csch}^2 x$

$\frac{d}{dx}(\operatorname{sech} x) = -\operatorname{sech} x \tanh x$

$\frac{d}{dx}(\operatorname{csch} x) = -\operatorname{csch} x \coth x$

Inverse Hyperbolic

$\frac{d}{dx}(\sinh^{-1} x) = \dfrac{1}{\sqrt{1 + x^2}}$

$\frac{d}{dx}(\cosh^{-1} x) = \dfrac{1}{\sqrt{x^2 - 1}}$

$\frac{d}{dx}(\tanh^{-1} x) = \dfrac{1}{1 - x^2}$

$\frac{d}{dx}(\coth^{-1} x) = \dfrac{1}{1 - x^2}$

$\frac{d}{dx}(\operatorname{sech}^{-1} x) = -\dfrac{1}{x\sqrt{1 - x^2}}$

$\frac{d}{dx}(\operatorname{csch}^{-1} x) = -\dfrac{1}{|x|\sqrt{x^2 + 1}}$